Geology of Illinois

Geology of Illinois

Dennis R. Kolata and Cheryl K. Nimz, Editors

University of Illinois at Urbana-Champaign
Institute of Natural Resource Sustainability
Illinois State Geological Survey

University of Illinois at Urbana-Champaign
Institute of Natural Resource Sustainability
William W. Shilts, Executive Director
Illinois State Geological Survey
E. Donald McKay III, Director
615 East Peabody Drive
Champaign, IL 61820-6964
217-333-4747
www.isgs.illinois.edu

©2010 University of Illinois Board of Trustees. All rights reserved.
For permissions information, contact the Illinois State Geological Survey.

Printed in the United States of America.

1 2 3

ISBN: 978-0-615-41739-4

Library of Congress Control Number: 2010916656

This book is dedicated to the people of Illinois

Our entire society rests upon—and is dependent upon—our water, our land, our forests, and our minerals. How we use these resources influences our health, security, economy, and well-being.

– John F. Kennedy

Contents

PREFACE xi

ACKNOWLEDGMENTS xiii

PERSPECTIVES ON ILLINOIS GEOLOGY 1
William W. Shilts

    1 **History of Investigations on Illinois Geology** 3
        Morris W. Leighton

    2 **Overview of Illinois Geology** 59
        Dennis R. Kolata

TECTONICS AND STRUCTURAL GEOLOGY OF ILLINOIS 75
Dennis R. Kolata

    3 **Tectonic History** 77
        Dennis R. Kolata and W. John Nelson

    4 **Structural Features** 90
        W. John Nelson

    5 **Neotectonics** 105
        W. John Nelson and John H. McBride

GEOLOGICAL HISTORY AND THE STRATIGRAPHIC RECORD OF ILLINOIS 121
Dennis R. Kolata

    6 **The Precambrian Crust** 123
        John H. McBride, Dennis R. Kolata, Michael L. Sargent,
        and Thomas G. Hildenbrand

    7 **Cambrian and Ordovician Systems** (Sauk Sequence and Tippecanoe I Subsequence) 136
        Dennis R. Kolata

    8 **Silurian System and Lower Devonian Series** (Tippecanoe II Subsequence) 158
        Donald G. Mikulic, Joanne Kluessendorf, and Rodney D. Norby

    9 **Middle Devonian Series through Mississippian System** (Kaskaskia Sequence) 167
        Joseph A. Devera, W. John Nelson, and Rodney D. Norby

    10 **Pennsylvanian Subsystem and Permian System** (Absaroka Sequence) 187
        W. John Nelson and Russell J. Jacobson

    11 **Mesozoic and Tertiary Eras** 206
        W. John Nelson

    12 **Quaternary Period** 216
        Ardith K. Hansel and E. Donald McKay III

    13 **Quaternary Paleoclimate** 248
        B. Brandon Curry, Hong Wang, Samuel V. Panno, and Keith C. Hackley

MINERAL RESOURCES OF ILLINOIS   261
Subhash B. Bhagwat

- **14   Coal   263**
  W. John Nelson, Russell J. Jacobson, Scott D. Elrick, Gary B. Dreher, and William R. Roy

- **15   Oil and Gas Geology   283**
  Bryan G. Huff and Beverly Seyler

- **16   Lead, Zinc, and Fluorite Mining   299**
  Zakaria Lasemi

- **17   Industrial Minerals   309**
  Zakaria Lasemi, Donald G. Mikulic, Randall E. Hughes, Timothy J. Kemmis, Subhash B. Bhagwat, and Karan S. Keith

GROUNDWATER RESOURCES OF ILLINOIS   323
Beverly L. Herzog

- **18   Aquifers   325**
  David R. Larson and Beverly L. Herzog

- **19   Geological Influences on Groundwater Quality   337**
  Samuel V. Panno and Keith C. Hackley

- **20   Protecting Groundwater Resources from Contamination   351**
  Richard C. Berg

- **21   Wetlands Geology   361**
  James J. Miner and Michael V. Miller

GEOLOGICAL APPLICATIONS: LAND USE AND ENVIRONMENTAL HAZARDS IN ILLINOIS   371
Beverly L. Herzog

- **22   Soils   373**
  Michael L. Barnhardt

- **23   Geological Factors in Siting and Design of Facilities and Infrastructure   385**
  Robert A. Bauer, Wen-June Su, and Nelson Kawamura

- **24   Geological Perspectives on Flooding   395**
  Michael J. Chrzastowski and Richard C. Berg

- **25   The Illinois Coast of Lake Michigan   404**
  Michael J. Chrzastowski

- **26   Natural Radiation   418**
  Richard A. Cahill

- **27   Earthquakes   424**
  Timothy H. Larson and Robert A. Bauer

- **28   Karst Terrane   432**
  Samuel V. Panno and C. Pius Weibel

- **29   Pollution of Groundwater and Surface Water   443**
  Samuel V. Panno, Richard C. Berg, and Walton R. Kelly

- **30   Surface Mine Reclamation   458**
  Timothy J. Kemmis, Robert A. Bauer, and Zakaria Lasemi

EPILOGUE   465

APPENDIX I: EXPLANATION OF LITHOLOGIC SYMBOLS   467

APPENDIX II: ILLINOIS COUNTIES AND SELECTED CITIES   468

LIST OF CONTRIBUTORS   469

INDEX   475

Preface

The state of Illinois, if it were a separate country, would rank among the 15 largest national economies in the world. Illinois is home to one of the world's largest metropolitan areas and the nation's third largest city. Its agricultural industry is arguably the most productive in the world. Illinois is nearly surrounded by waterways, which have given it access to continental and international markets for over 150 years.

Illinois' prominent position among the world's countries and states did not come about by chance. Its society was shaped and nurtured on and by her geological foundation, which itself is a product of the state's unique geological history within the North American continent. The geological base on which Illinois society has evolved and prospered has been the subject of scientific investigations for almost 200 years. During that time, many of the insights of Illinois geologists have led to the increased understanding of geological phenomena around the globe. In particular, ground-breaking publications on Illinois' glacial history, clay mineralogy, and coal-bearing strata have provided scientific models that are still the basis for geological reasoning in settings far from the state. Illinois' universities and surveys have provided the homes for such internationally celebrated geologists as T. C. Chamberlin, Gilbert Cady, M. King Hubbert, Ralph Grim, Paul Potter, and many other scientists whose names are legendary in the history of geology. The disciplines of environmental geology and clay mineralogy originated and were nurtured in Illinois, and engineering geology, hydrogeology, geophysics, and glacial stratigraphy all had roots in the geological institutions of this state.

Until now, no agency or institution has been able to synthesize the information about the state's geology and the contributions that Illinois geologists have given to the world. Two new major publications were suggested during discussions leading up to 2005 celebrations commemorating the 100th anniversary of the rebirth of the Illinois State Geological Survey (ISGS) in 1905 (the first Survey operated from 1851-1875). As the director of the ISGS at that time, I authorized the production of a new map, the *Bedrock Geology of Illinois*. The map was to be a special centennial publication printed in cooperation with the Illinois Association of Aggregate Producers. This map proved to be a major undertaking and was coordinated by Dr. Dennis Kolata. As the map was being compiled and revised from several earlier versions, dating back to the original Survey in the nineteenth century, I urged ISGS scientists and editors to prepare an accompanying major publication summarizing the geology of Illinois, as it is known in the early twenty-first century. This book is the realization of that charge, and it will be obvious to the reader that compiling its chapters has been a long and difficult task. Authors were asked to write their chapters to be understood by an informed lay audience, and they have been successful in accomplishing that goal.

A glance at the *Bedrock Geology of Illinois* map shows a deceptively simple picture—a gently warped basin comprising beds of marine and terrestrial sedimentary rocks, gently inclined toward the basin's center. There are faults and disturbances on the brinks of the basin, but the geology is "simple," at least compared with the contorted, intruded, metamorphosed rocks of the North American mountain ranges and the Canadian Shield. But this apparent simplicity belies the strategic difficulty of discovering a bedrock landscape that is 75% hidden under thick, unconsolidated glacial sediments. In about half of Illinois' 101 counties, there is not a single outcrop of bedrock, and geological understanding is based primarily on borehole records and artificial exposures.

The story of Illinois' hidden geology, glimpsed on the bedrock map, unfolds in more detail in the *Geology of Illinois* volume. Among those details are descriptions of

- the global tectonic forces that shaped Illinois' part of the bedrock landscape that emerged after the first billions of years of the Earth's history;
- the sediments that were laid down on that rugged landscape in episodic shallow seas;
- the ancient swamps that eventually fringed those seas and in which were deposited the organic detritus that ultimately hardened into the coal that provided fuel for Illinois' growth and sustenance;
- the geological structures and materials that contain, protect, and make available the state's abundant supplies of water, mineral, and fuel resources; and
- the skyscraper-high glaciers that ground down, catastrophically flooded, and deposited the muds and sands and silts that form the rich soils of most of our modern landscape.

An understanding of the complexities of the geological framework and its intimate interconnection to the humans who impact and are affected by it helps us meet the challenges related to land use, water location and safety, mineral resource extraction, energy, and natural hazards. That understanding historically has been—and still is—critical to ensuring economic development, environmental security, and quality of life for Illinois' citizens.

In this book, these events and issues are expanded on and illuminated by a talented and articulate group of geologist authors. I hope that their stories will captivate you as much as they captivated me.

–William W. Shilts, Ph.D.
Executive Director,
Institute of Natural Resource Sustainability

Acknowledgments

The valuable contributions of Illinois State Geological Survey (ISGS) staff in production of this book are greatly appreciated. Enormous thanks are given to Pamella K. Carrillo for layout and production of this volume. Special thanks are given to Jennifer K. Hines and Joel A. Steinfeldt for assistance in technical editing; to librarians Anne M. Huber and Mary Krick for assistance in obtaining materials; to the librarians and technical editors Mark Zulauf and Thomas Rice for the arduous task of reference verification; to photographer Joel M. Dexter who provided many images; to graphic artists Cynthia A. Briedis and Pamella K. Carrillo for book and cover design; to graphic artists Cynthia A. Briedis, Daniel L. Byers, Pamella K. Carrillo, Jacquelyn L. Hannah, and Michael W. Knapp for drafting of technical illustrations; and to geologists Curtis C. Abert, Adrianne Knight, Donald E. Luman, Bobbie Robinson, and Barbara J. Stiff for production of maps and LIDAR images.

Technical and scientific reviews were kindly provided by the following individuals, whose inclusion here does not necessarily indicate full agreement with all material in this book, but to whom we are grateful. External peer reviewers included Daniel Barkley, Andrew R. Benziger, Colin J. Booth, Rick Cobb, G. Czapar, Robert Darmody, Garland Dever, James A. Drahovzal, Cortland F. Eble, David Fullerton, Steve Greb, Richard Harrison, Linda M. Hiltabrand, Daniel Injerd, P. Kremmel, Brian D. Keith, Ralph Langenheim Jr., David Mickelson, John T. Popp, Ira D. Sasowsky, Charles Shabica, John M. Shafer, Nelson R. Shafer, Carol Thompson, Terry R. West, Randi T. Wille, and Brian J. Witzke.

Peer reviewers from the ISGS included Marie-France Dufour, Robert J. Finley, Leon R. Follmer, Jonathan H. Goodwin, Eric C. Grimm, Edward Mehnert, Thomas R. Moore, David G. Morse, Rodney D. Norby, André Pugin, Michael L. Sargent, Dean Spindler, and Jack A. Simon. Notably, detailed reviews of several chapters were contributed by coauthors, including Richard C. Berg, Gary B. Dreher, Ardith K. Hansel, Beverly L. Herzog, Walton R. Kelly, David R. Larson, John H. McBride, E. Donald McKay III, W. John Nelson, Samuel V. Panno, Beverly Seyler, William W. Shilts, and C. Pius Weibel.

Thanks are also expressed to approximately 50 ISGS staff members and former staff members who were interviewed for insights on the development of milestones and concepts of Illinois' rich history of geological investigations and to respondents of a questionnaire mailed to Illinois educational institutions.

Permissions

Acknowledged also are the many contributors who generously granted their permission to use the photographs and illustrations credited to them. Individuals seeking further use of images in this volume must contact the copyright holders directly. Special thanks are given to the American Association of Petroleum Geologists, American Geological Institute, Chicago Historical Society, Geological Society of America, and U.S. Geological Survey for their generous use of copyrighted images referenced to them. Requests for permissions for further use of these images should be directed to those organizations.

Research Support

The research reported in Chapter 6 (Precambrian Crust) was supported in part by a grant from the National Science Foundation under Award Number EAR-0307539. Also gratefully acknowledged is the support of this research by Landmark Graphics via the Landmark University Grant Program at Brigham Young University and the University of Illinois at Urbana-Champaign. Data processing for this study was performed using Landmark's ProMAX2D™. The authors express appreciation to Seismic Micro-technology (Kingdom Suite™), which kindly provided a University Grant of its visualization and mapping software. This project was made possible in part by the kind release of seismic reflection data to the Illinois State Geological Survey by Seismic Exchange Inc.

The Illinois Department of Transportation and the Illinois Nature Preserves Commission have been long-term sponsors of research presented in Chapter 21 (Wetlands Geology).

Centennial Donors

The Illinois State Geological Survey gratefully acknowledges the generosity of the following individuals and corporations who made contributions or provided in-kind services to support the activities of the ISGS Centennial Celebration, which included the printing of this volume. (List as of September 2010.)

Diamond Level Donors ($5,000 or more)

BP America
Exxon Mobil Foundation
Fox Development Corporation
Illinois Association of Aggregate Producers
  (John Henriksen)
Isotech Laboratories (Dennis D. Coleman)

Morris W. Leighton
William A. Newton
Arthur F. Preston
Jack A. Simon
Waste Management of Illinois
  (William Schubert)

Gold Level ($1,000 or more)

Ameren Energy Fuels & Services
  (Michael G. Mueller, Vice President)
Bi-Petro, Inc. (John F. Homeier)
Bradford Supply Co. (W. Jack Chamblin)
Thomas C. Buschbach Caterpillar Inc.
Ceja Corporation (Donald L. Carpenter)
Christopher B. Burke Engineering, Ltd.
Coal Network, Inc. (Ramesh Malhotra, President)
Charles W. Collinson
DAKFAM, Inc. (Peter L. Dakuras)
Paul B. and Dollie DuMontelle
Jonathan H. Goodwin

David L. Gross
Roy Helfinstine
Illinois American Water Co. (Barry Suits)
Illinois Basin Section,
  Society of Petroleum Engineers
Illinious Coal Association (Phillip M. Gonet)
Illinois Geological Society
Illinois Oil and Gas Association
Material Service Corporation
Linda and E. Donald McKay III
Midwest Arc Users Group
Oelze Production Co., LLC

Podolsky Oil Co., LLC (Bernard Podolsky and
  Michael D. Podolsky)
Paul Edwin Potter
Philip C. and Rita Reed
Michael and Maralyn Reilly
William W. Shilts
Edmund B. Thornton
U. S. Silica Company
University of Illinois Department of Geology
  (Stephen Marshak)
Paul A. Witherspoon
Wood Energy, Inc. (J. Nelson Wood, Vice President)

Silver Level ($500 or more)

Richard C. Anderson
Anna Quarries, Inc.
Robert A. Bauer
Margaret J. Bergstrom and children
James A. Bier
Brent Burgess
Casper Stolle Quarry and Contracting, Inc.
  (John E. Cramer)
Mark and Rhonda Cloos
Lester W. and Virginia K. Clutter
James C. Cobb
Dennis D. and Eileen Coleman
Continental Resources of Illinois, Inc.
Countrymark Cooperative
Heinz H. Damberger
Donald R. Dickerson
Kari E. Downey family
Dynamic Separation Inc.
Robert J. and Sandra Finley
Foundation Coal Corporation
Fox River Stone Company
Scott M. Frailey

Harold J. Gluskoter
Donald L. Graf
Claudia L. Gross
Ardith K. Hansel
Walter E. Hanson
Richard D. and Jenny Harvey
Beverly L. Herzog and Craig W. Cutbirth
Illinois Corn Marketing Board
Illinois GIS Association
Illinois State Water Survey
Thomas Johnson
James S. and Barbara C. Kahn
John P. and Betty Kempton
Inez Kettles
Knight Hawk Coal, LLC
Dennis R. Kolata
Lamamco Drilling Company
Layne Christensen (Gregg Buffington)
F. Beach, Morris W., and Richard T. Leighton
Jean B. Leighton
Lincoln Orbit Earth Science Society
Martin Marietta Materials (Al Witty)

John M. Masters
Haydn Murray
W. John Nelson
Oelze Production Co., LLC
Paar Instrument Co. (Michael Steffenson)
Russell A. Peppers
Petco Petroleum Corporation (J. D. Bergman)
Royal Drilling and Producing, Inc.
  (James R. Cantrell)
Shabica and Associates, Inc. (Charles Shabica)
Shulman Brothers, Inc.
Stewart Producers, Inc. (Robert G. Stewart)
Team Energy, LLC (Dennis Swager)
U. S. Geological Survey,
  Illinois Water Science Center
Vulcan Materials Company (Charles W. King)
Waste Management, Inc. (William Schubert)
Harriet Weller
Stuart Weller
L. E. Workman Family (Rollin, James, Nathan, and
  Miriam Alberg)

Bronze Level and Friends (up to $500)

Allen F. Agnew
Walter Anderson
Margaret H. Bargh
Lawrence E. Bengal
Richard C. Berg
Craig Bethke
Booth Oil Co., Inc.
Brehm Oil, Inc.
Ross D. Brower
Chen-Lin Chou
Colonial Brick Co. (Daniel A. Swartz)
Columbia Quarry Co.
Joan E. Crockett
Marshall E. and Patti Daniel
Dee Drilling Company
Ilham Demir
Sally L. Denhart
Joseph A. Devera
The Discovery Group (Robert M. Cluff)
Joe B. Dixon
William G. Dixon, Jr.
James and Alicia Eidel
Anne L. Erdmann
Farnsworth Group, Inc.
Feltes Sand and Gravel Co., Inc. (Timothy
  J. Feltes)
Gary Fleeger
Edward and Marjorie Foley

Franklin Well Services, Inc.
David A. Grimley
John Grube
Steven R. Gustison
Keith Hackley
Judith Weller Harvey
Jack Healy
Henigman Oil Co., Inc.
M E Hopkins
Bryan G. Huff
Donald A. Keefer
Karan S. Keith
Myrna M. Killey
James E. King
James Kirkpatrick
Christopher Korose
Ivan G. Krapac
Robert Krumm (Illinois GISA)
Phillip E. LaMoureaux
Ralph L. Langenheim Jr.
David R. Larson
Zak Lasemi
Julian H. and Virginia Lauchner
Richard L. Leary
Alison B. Lecouris
Jim and Kathy Lee
Hannes E. Leetaru
Jon C. and Judith S. Liebman

Richard E. Lounsbury
Marino Engineering Associates, Inc.
  (Jerry Marino)
Maria Matalerz
Edward Mehnert
Kristi Mercer
Daniel Merriam
John E. Moore
David G. Morse
National Ground Water Association
  (Kevin McCray)
Nature of Illinois Foundation
Daniel O. Nelson
Cheryl K. Nimz
Donald F. and Theresa Z. Oltz
James E. Palmer
Walter E. Parham
Richard and Estelle Parizek
Peoples Energy
Pioneer Oil Company, Inc.
Charles Porterfield
Doug Pottorff
Kevin W. Reimer, Consulting
David and Nancy Reinertsen
Republic Oil Company, Inc.
Larry and Karen Ritchie
Gary A. Roberts
Robert S. Roth

William R. Roy
Rodney R. Ruch
Schwartz Oilfeld Services, Inc.
Thomas K. Searight
Beverly Seyler
John and Denise Sieving
Paul K. Sims
Sloan's Water Well Service, Inc.
Edward C. Smith
John Steinmetz
Streator Brick Company
Wen-June Su
Daniel Thurston
Colin Treworgy
Janis D. Treworgy
Michael D. and Tracey T. Tringali
John E. Utgaard
Charles Weaver
John Marvin Weller
William E. Wilson III
Paul Witherspoon
Suzanne Wyness
Timothy Young
Darwin and Alberta Zachay
Arthur J. Zeisel

Perspectives on Illinois Geology

William W. Shilts

The mostly unseen geology of Illinois is quite literally the foundation for all life—and the type of life—that can survive on the state's land surface. Since the time of the paleoindians and early explorers, the area's residents have struggled to understand the environment around them to ensure their survival and prosperity. Even today, understanding the complex layers of rock that lie beneath the state's glacial blanket is essential to the use, conservation, and protection of Illinois' rich legacy of natural resources: fertile agricultural soils, plentiful water supplies, and vast mineral resources.

The book begins with a brief history of some of the outstanding individuals who led the way, with major scientific contributions, to important discoveries and advancements in knowledge about Illinois geology. These individual contributions fueled the state's industrial expansion, strengthened its economy, addressed societal issues, and supported a healthy environment. Throughout the state's history, innovative methods and new, powerful technologies continued to be developed to improve geological interpretations and manage and share ever-increasing amounts of data. Research directions, methods, and technologies responded to a changing society and shifting demands for resources, energy, and environmental information.

The introduction concludes with an overview of the state's geological record, the forces that affect it, and the methods geologists use to study, classify, and date rock and earth materials. Subsequent sections of the book focus on Illinois' tectonic setting and structural geology, geological history and stratigraphic record, mineral and groundwater resources, and land use and environmental hazards. Illinois geology is the story of grand and dynamic forces: tectonic movements, transgressions and withdrawals of shallow seas, and glacial episodes. The organization of the book is designed to help the reader understand the approach taken in classifying and describing those forces and their impacts throughout time and continuing to the present.

Photograph by Joel M. Dexter

Image on previous page: Bluffs of St. Peter Sandstone, Starved Rock State Park, La Salle County, Illinois.

History of Investigations on Illinois Geology

Morris W. Leighton

INTRODUCTION

This brief history does not attempt to cover all of the significant geological concepts, breakthroughs, and milestones that have occurred over more than a century of work or mention all of the contributors to the many investigations on Illinois geology. Despite the selectivity, it is hoped that the reader will obtain a sense of the direction of history and understand some of the driving forces behind the geological investigations in Illinois, driving forces that began with the needs of our earliest ancestors, the paleoindians, who needed to understand their surroundings simply to stay alive and defend themselves.

Economics and the concerns for economic development have driven many of the investigations that occurred since the 1800s. Wars and disasters had major impacts. The Great Depression in the late 1920s and early 1930s, the socioeconomics of the 1960s, and decades of environmental legislation took geology in distinctly new directions. Changing technologies and computerization influenced research studies by expanding the kind of information that could be gathered, enhancing its quality, and increasing the value and impact of the geological information obtained and shared with the Illinois public.

Peer motivation has been a strong incentive throughout time. Multidisciplinary studies began to be emphasized during the 1930s and team dynamics during the 1970s. Multiple authors and interdisciplinary involvement increased throughout the 1980s, 1990s, and into the twenty-first century. Despite this major trend, individual creativity and individual contributions have continued to play a dominant role throughout history.

Illinois institutions and their leaders have also exercised a profound influence on the nature and direction of geological investigations (Figure 1-1). During the last 50 years, scientific protocols, legislation, and regulations have had major impacts on the nature and direction of investigations of Illinois geology. Finally, public funding sources have always controlled in part the amount and type of work done on Illinois geology, but the importance of state and federal government grants and contracts has grown substantially since the 1970s.

EARLIEST BEGINNINGS: PREHISTORIC TIMES TO 1805

Native Americans and Their Ancestors

The paleoindians and Native Americans were keen observers of their geological surroundings and natural resources, both of which were necessary for their survival. These peoples located sources of flint for making fire and sources of chert and stones for fashioning a variety of tools and weapons, from scrapers and axes to knives and spears (Berkson and Wiant 2001, Illinois State Museum 2002). The earliest paleoindians, from 10,000 B.C. or so, left behind distinctive spears and stone tools at camp sites. Much later, Native Americans made long, pointed knives using a particular stone from Union County called Mill Creek chert. Bows and arrows were used extensively for hunting or fighting just prior to and following 1,000 A.D., and arrowheads were chipped out of chert.

Native Americans are known to have made pottery using Illinois clays from around 600 to 200 B.C. Carved pipes, dating from around 2,000 years ago, were made from flint clays called pipestone (Figure 1-2). Coal, galena, fluorite (the latter found as ornamental or religious artifacts), and salt springs were also known and utilized by Native Americans in Illinois.

Early Explorers and Cartographers (1670 to 1805)

Early explorers and cartographers made a number of observations about Illinois resources, physiography, soils, and rocks, stimulating interest and development of the Illinois country. These were the beginnings of documented, geologically related observations in Illinois. Among the earliest were the French explorers, Pere Marquette and Louis Jolliet in 1673 (Gluskoter 1982), and a Jesuit priest, Father Louis Hennepin, during 1679 to 1682 (Hennepin 1698). Those early explorers noted *"charbon de terre"*— coal— along the Illinois River. During the last few decades of the seventeenth century, several French explorers, including Pierre-Charles Le Sueur and Nicolas Perrot, visited northwestern Illinois and southwestern Wisconsin where

Figure 1-1 Some of the distinguished scientists of the past in geological investigations of Illinois. Photographs used with permission: Bretz and Hubbert (Pettijohn 1984), University of Chicago Press; Owen (Shaver 1987), Indiana Geological Survey; Cady (Simon 1974), Risser (Simon 1977b), Horberg (Fryxell 1962), and Schopf (Kosanke 1979), Geological Society of America; Hall,

HISTORY OF INVESTIGATIONS

Rensselaer Alumni Hall of Fame; Libby, The Nobel Foundation (© The Nobel Foundation); and Sloss, Stuart-Rodgers Photography. All other photographs are from the Illinois State Geological Survey collection.

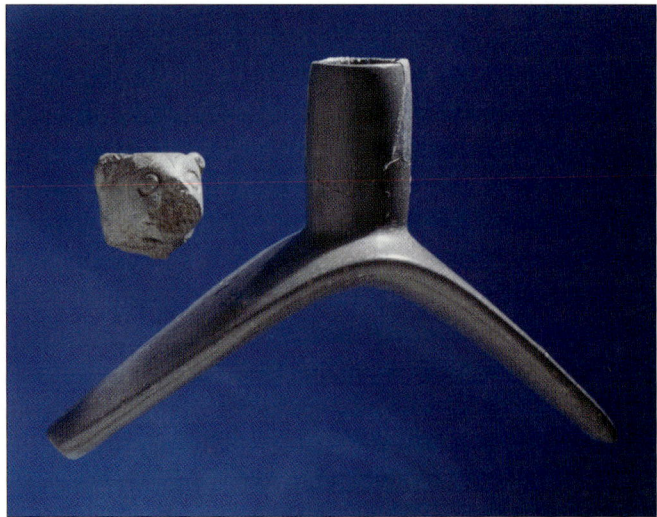

Figure 1-2 Native American artifacts made of Illinois pipestone. (Photograph by Joel M. Dexter.)

they traded for lead from the Native Americans (Risser and Major 1968). Their reports indicate that small amounts of lead were being mined by Native Americans prior to 1700 near Galena.

Thomas Hutchins, a military engineer and surveyor who traveled between 1764 and 1775, published maps and notes on an area that included the Illinois country (Hutchins 1778). Hutchins (p. 43) wrote that "the Illinois country is in general of a superior soil to any other part of North America that I have seen." One of his maps is annotated and refers to a "Coal Mine half a Mile long in the Bank of the [Illinois] River" on its northwest side just north of the tributary of the "Varmullion River."

By 1787, the Northwest Ordinance had placed the Illinois country in the Northwest Territory (Petterchak 1999). Thomas Jefferson (1788) published notes referring to limestone on the Mississippi and Ohio Rivers and to lead, salt, and coal deposits in Ohio, Kentucky, and the Illinois country. In 1800, Congress created the Indiana Territory, which included the Illinois country. It was not until 1809, though, that Congress established the Illinois Territory (Petterchak 1999).

Salt was produced from the Illinois area long before it became a state (Risser and Major 1968). First the Native Americans and later the explorers and settlers obtained salt by evaporating the salty brine that came from springs. The main operations were near the present town of Equality in Gallatin County. In 1803, as a result of a treaty with seven tribes, the U.S. Government obtained the land containing the salt springs at Equality, promising to furnish the Native Americans with as much as 150 bushels of salt each year (Risser and Major 1968).

Birth of Geology in Illinois: 1805 to 1850

Geological Investigations before Illinois Statehood (1805 to 1818)

James Mease (1807) included the Mississippi Valley in his *Geological Account of the United States*. He noted that the rocks in this area were essentially horizontal and included a great deal of limestone. Although geology was already being practiced in Europe, Mease's account was the first book with "geological" in the title to be published in the United States (White and Slanker 1962). The science of geology was just taking hold in the United States. In 1809, William Maclure, sometimes called the Father of American Geology, presented a paper, "Observations on the Geology of the United States, Explanatory of a Geologic Map," before the American Philosophical Society (Maclure 1809). Maclure's map, the first geological map of the United States, extended into the Illinois Territory and emphasized the distribution of rock types.

Explorers continued to travel into and across the Illinois Territory. One of those explorers was John Bradbury, who traversed the region during 1809 to 1811. Bradbury's (1817) references to oolitic limestone, galena, and fluorite near the Ohio River are among the earliest on record. He also recognized that the horizontal strata in Illinois had been deposited under marine conditions, that the valleys were stream-eroded, and that a long period of time would have been necessary for the deposition and later erosion of the rocks he saw (White 1967).

Awakening to Illinois' Resource Potential (1818 to 1850)

At the time considered to be one of the "western" states, Illinois officially became the twenty-first state in the nation in 1818. During the same year, Benjamin Silliman, professor of geology at Yale College, founded the *American Journal of Science*, which became the oldest continuously published journal in America and an important influence in the development of the scientific community in Illinois and the United States. In the first series of this new journal, Silliman (1819) published a communication from Joseph Baldwin of Shawnee Town on a new locality of fluorspar and galena in Illinois. This publication helped mark the beginning of the awakening to Illinois' resource potential. In 1823, Colonel James Johnson brought with him a large number of workers from southern Illinois and Kentucky to open a new mine in the Galena district in northwestern Illinois. This mine was the start of the large-scale, systematic mining of lead (Risser and Major 1968) and the beginning of a significant industry in Illinois.

Many firsts accompanied a paper by Shepard (1838), a mineralogist and chemist with an interest in paleontology. Shepard was the first to publish a description of the geology of Chicago (White and Slanker 1962). He described the raised beaches of Lake Michigan and discussed the engineering geology and construction of the Illinois and Michigan Canal. He made illustrations of fossils that were characteristic of the "Magnesium Limestone" near Chicago, and he identified the northern margin of coal-bearing strata and described some of its fossils. Shepard also recognized the superior clays found near Peru and the "extensive beds of pure white sand," the St. Peter Sandstone. For both resources, he predicted a promising economic future, a prediction that has been "brilliantly fulfilled" (White and Slanker 1962). That paper also contained a map and a section showing the disturbed strata of the La Salle Anticlinorium near Peru and La Salle.

Prior to 1830, there were very few field geologists in a nation just awakening to the potential of its natural resources (Johnson 1977). The new School of Industry at New Harmony, Indiana, under David Dale Owen, established an outdoor laboratory for the on-the-job training of field geologists. In 1839, the federal government, which then owned all mineral rights in the United States, engaged Owen to survey parts of Illinois, Iowa, and Wisconsin, focusing on the Galena area, to determine the value of the land. Owen's instructions were to explore each quarter section in an area of about 11,000 square miles (about 28,490 km^2) and to complete the work before the coming winter, a seemingly impossible task. Assisted by John Locke, Owen recruited and trained a 139-member field party for the tri-state survey, including 20 volunteers from New Harmony who had attended Owen's lectures. The men were divided into 24 five- and six-man groups, each of which was to survey seven to eight sections a day. By November 24, 1839, the entire area had been surveyed, and the group had been disbanded. Lane (1966) observed that this group was undoubtedly the largest geological field group ever assembled in the United States, and its successful completion, according to Merrill (1906), was "a feat of generalship which has never been equaled in American geological history."

Locke (1840) and Owen (1844) submitted written reports to Congress; these were the first geological reports on the Galena Lead-Zinc District. Together, these men accurately mapped, described, and correlated the rocks; worked out the relationships of ore deposits to stratigraphy and structure; and determined which lands were mineral-bearing. Owen's 1844 report included a map of the Illinois coal field. This map was the first to show a number of the major geological features of the state with any accuracy (Weller 1939). As the results of Owen's study became widely known and as new mines continued to open, including the first fluorspar mine at Rosiclare in 1842 (Risser and Major 1968), the awakening to Illinois' mineral resources was in full force.

From 1843 to 1846, the first true geological maps of Illinois made their appearance, one by James Hall (1843a) and one by David Dale Owen (1844, 1846). Hall (1843b) had another very important paper, describing a cross section from Cleveland, Ohio, to the Mississippi River; the paper included remarks on the formations and their fossils. Hall's correlations were based in part on lithologic similarities, but mostly on the similarity of fossils (White and Slanker 1962). Reflecting on Hall's (1843b) paper, White and Slanker (1962) noted, "The pioneer days were coming to an end, and professional geologists were taking over."

In 1848, a group of geologists, led in part by Louis Agassiz, the famed Swiss geologist who introduced the "Ice Age" concept in 1840, formed the American Association for the Advancement of Science (AAAS) with the goal of increasing the professionalism of American science. The following year, a committee of 15 scientists working on behalf of the AAAS drafted a memorandum supporting the establishment of state geological surveys. The memorandum was signed and sent to all states that did not have active surveys. It arrived on Illinois Governor French's desk a few months later in 1850.

A Period of Rapid Growth: 1850 to 1905

The First Illinois Geological Survey (1851 to 1875)

Governor Augustus C. French incorporated a recommendation for establishing a geological survey of Illinois in his message to the Illinois legislature on January 7, 1851. The message stated, "We have unmistakable evidence that this state is scarcely exceeded in the extent of her mineral riches, and all that seems wanting to render them richly productive is to point attention to them" (M. M. Leighton 1966). As cited by McLure (1962), the *Illinois State Journal*, January 31, 1851, (p. 2) reported that the responsible Illinois legislative committee urged the formation of a Survey "for the purposes of practical utility . . . and more especially to . . . develop the resources of the state. . . ." The law authorizing the Illinois Geological Survey was passed in February 1851. The law also authorized the hiring of a geologist of known integrity and practical skill to make a geological and mineralogical survey of the entire territory of the state.

Joseph G. Norwood Years (1851 to 1858)

Governor French appointed Joseph G. Norwood as the Survey's first director on July 29, 1851. Work began that fall in southern Illinois as directed by the legislature. Norwood collected an extensive number of mineral and fossil specimens but published little, except for minor paleontological works and a little known, small-scale geological map of Illinois in 1858 that showed a "fair degree of accuracy" (J. M. Weller 1939).

A. H. Worthen Years (1858 to 1875)

Amos Henry Worthen and J. H. McChesney, former assistants to Norwood, stood in rivalry to succeed him. Governor William H. Bissell commissioned Worthen as state geologist on March 24, 1858 (McLure 1962). McChesney (1860), having been appointed as a professor of chemistry, geology, mineralogy, and agriculture at the first University of Chicago, went on to publish descriptions of fossils in Illinois, which, according to R. L. Langenheim (December 6, 2004, personal communication), were the first substantial paleontological works in the state.

Worthen, acutely aware of the need for tangible Survey results, published eight volumes on the geology and paleontology of Illinois (Figure 1-3). His first volume, held up by lack of appropriations, was published in 1866. Worthen recognized the legislature's lack of understanding of the practical value of paleontology, and so his first volume (Worthen et al. 1866) described the physical features of the state, including the topography and principal streams, general geological principles, surface geology, and the succession of stratigraphic units (geological systems). The volume also included special reports by others on the lead district, coal fields, and prairies. Worthen noted several "principal axes of disturbance," some reflected in a striking cross section along the Mississippi River from one end of the state to the other. Worthen and other individuals he selected provided a number of geological descriptions of individual counties in the state.

Having established solidly the principles of geology in his first volume, Worthen issued a second volume devoted entirely to paleontology (Newberry et al. 1866). This publication provided descriptions and plates illustrating species of vertebrates, invertebrates, and plants; the emphasis was on the "Carboniferous Formations" in Illinois. The remaining volumes focused on both geology and paleontology. Volumes, 3, 4, 5, and 6 followed (Worthen et al. 1868, 1870, 1873, 1875). Volume 6, which contained Worthen's 1875 geological map of Illinois, was the final volume completed during the existence of the first Illinois Geological Survey. Volume 7 was published (Worthen et al. 1883) after the demise of the Survey in 1875, and Volume 8 (Worthen et al. 1890) after Worthen's death in 1888. By 1890, 101 of 102 counties—all except De Witt County—had been de-

Figure 1-3 Historic 1873 engraving of Lower Carboniferous limestone, Jersey Landing on the Mississippi River. (Plate H from Worthen et al. 1875; engraving by Paulus Roetter, Delaware, Western Engraving Co., Chicago.)

scribed geologically. Worthen's efforts and works, including his 1875 map, were monumental achievements and formed the basis for much of the ensuing work on Illinois geology. They still are cited as references.

State Institutions Fill the Survey Gap (1875 to 1905)

A huge gap was created in the investigations of Illinois geology by the demise of the first Illinois Geological Survey in 1875 due to the lack of funds. Fortunately for Illinois, other institutions and individuals helped to fill the gap.

In 1877, the collections of the Survey and the Illinois Natural History Museum were combined in the State Historical Library and Natural History Museum (Thompson 1988). A. H. Worthen was appointed as the first curator of the new state museum and was also designated as the state geologist. Worthen et al. (1883) were able to find funds from the state museum to publish some of the investigations begun with the first geological survey. After Worthen's death in 1888, his successor, Joshua Lindahl, a professor from Augustana College, edited and published Worthen's eighth volume. According to M. D. Thompson (1988, p. 19–20), files from the late 1890s and early 1900s indicated

> an extensive interest by companies, county commissioners, and private land owners in the location of coal, lead, zinc, clay, and marble, and on building stones and gravel. There were many requests for geological maps, bulletins, and reports of the Illinois Geological Survey from educators, libraries, other geological surveys, and institutions of higher learning.

However, Fannie Fisher, the assistant curator, wrote, "from this time on no scientific work was done here. . ." and "The financial wherewithal is absolutely wanting" (Thompson 1988, p. 21). Obviously, State efforts to pursue focused investigations on Illinois geology began withering in 1875 and had essentially died by the early 1890s.

Scientific curiosity remained alive and well in academic institutions and scientific societies, especially concerning glacial geology and paleontology. The chartering of Augustana Seminary in 1860, later to become Augustana College, the founding of the University of Illinois in 1867 (following the 1862 Morrill Act, which initiated the land-grant university system), the establishment of the U. S. Geological Survey in 1879, and the formation of the "new" University of Chicago in 1891 after the first was burned in the Great Chicago Fire of 1871—all of these contributed greatly to stimulating interest in furthering geological investigations in the state during this time.

Glacial Geology Investigations (1850 to 1905)

Investigations of surficial materials blossomed. As the 1850s opened, the origin of the deposits lying on the bedrock was not understood, although these unconsolidated deposits had become known as "drift." Investigators expounded two fundamentally different theories to explain drift's origins. Contrary to the views of Agassiz in Europe and others in the eastern United States, most geologists in the western states thought early on that the drift was a water-laid deposit, laid down as ice-rafted deposits in an inland sea or lake or by drifting icebergs, leaving large boulders and coarse debris as they melted (Worthen et al. 1866; Andrews 1867, 1869; Bannister in Worthen et al. 1868 [p. 241]; Bradley in Worthen et al. 1868 [p. 192]; Newberry 1870, 1874). The opposing theory, advanced initially by Whittlesey (1867) for Illinois, attributed drift to the moving forces of glaciation. Shaw supported this view (Worthen et al. 1873), as did T. C. Chamberlin (1877, 1878, 1882).

The Extent and Number of Glaciations

A paper by Chamberlin (1878) contained a map showing the approximate southern limit of the drift area, including moraines. This paper illustrated two early and important concepts: the existence of multiple glaciations and their extent. White (1973) wrote that this was "the first map to show explicitly drift of more than one ice age." Thwaites (1927), in a historical review of multiple glaciation research studies, observed that Chamberlin's was the first study to apply topographic forms and weathering to distinguish relative ages, methods that stood in strong contrast to attempts to treat drift deposits as if they were stratified rock. Twelve criteria for determining the different ages of glacial drift were published by R. D. Salisbury (1893).

Processes in Glacial Geology

The Driftless Area is an area untouched by glaciation in northwestern Illinois and the adjacent area in southwestern Wisconsin. Glaciers moved around this area, and glacial drift was not deposited there. The concepts of the origin of the Driftless Area range from the one proposed by Chamberlin and Salisbury (1885) to the one proposed by Hobbs (1999). None is more poetic than the description by Chamberlin and Salisbury (1885, p. 322):

> Diverted by highlands, led away by valleys, consumed by wastage where weak, self-perpetuating where strong, the fingers of the mer de glace closed around the ancient Jardin of the Upper Mississippi Valley, but failed to close upon it.

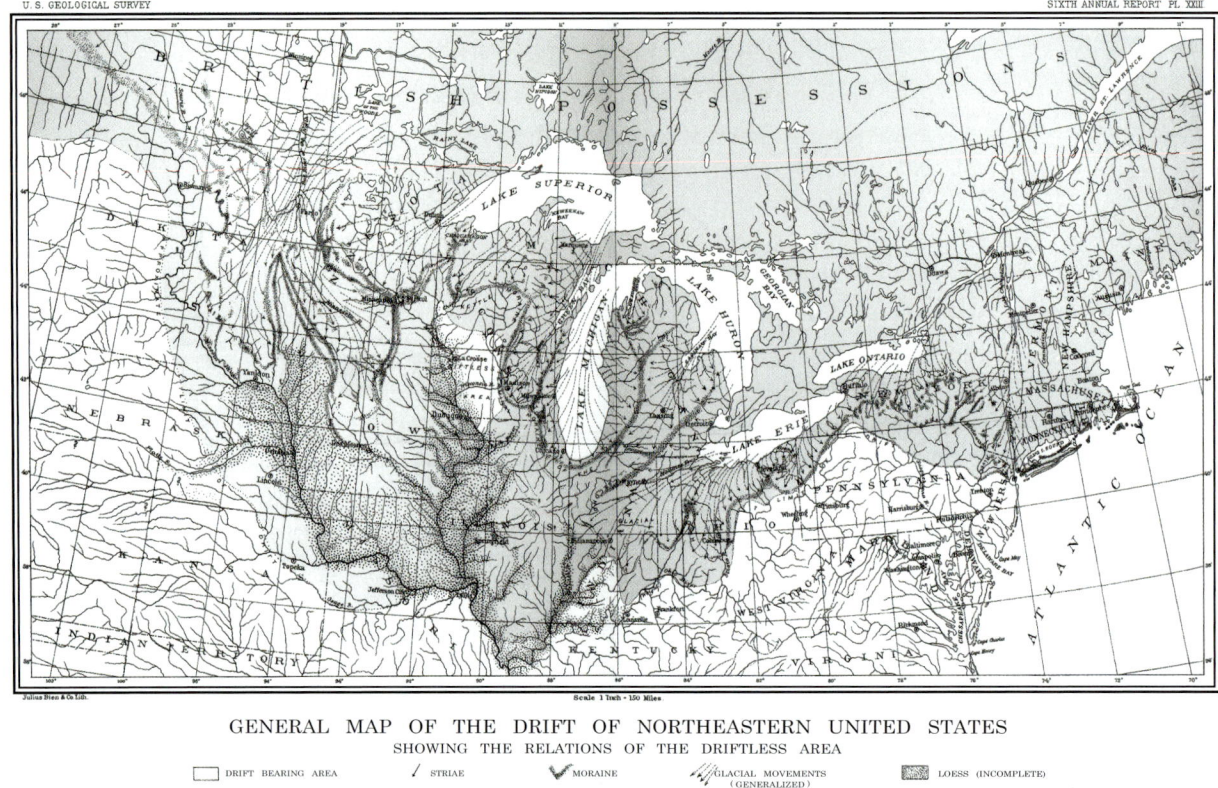

Figure 1-4 Early map of the glacial drift of the northeastern United States, illustrating four major advances in geological concepts: the recognition and delineation of (1) moraines and multiple glaciations, (2) the extent of glaciation in Illinois and the Midwest, (3) the widespread distribution of loess, and (4) the Driftless Area in northwestern Illinois and southwestern Wisconsin. (Plate XXIII in Chamberlin and Salisbury 1885; courtesy of the U.S. Geological Survey.)

The power of glaciers to divert, block, and alter the courses of major rivers and lesser streams was described by Leverett (1899) and others, using as examples the Mississippi River as it once flowed through Illinois and the Ohio River, which was diverted from the Cache River valley in southern Illinois.

The Classification and Mapping of Glacial Deposits

The precursors to a classification of glacial deposits were established in another classic paper (Chamberlin 1882) and follow-up article (Chamberlin and Salisbury 1885) that recognized two main glacial epochs separated in time by an interglacial epoch. Then Chamberlin (1894) introduced the use of geographic names for the drifts of different ages and added geographic names to the interglacial stages (Chamberlin 1895). Using stage and substage instead of epoch, he introduced the beginnings of a classification scheme for the Pleistocene. Frank Leverett, a protégé of T. C. Chamberlin, put forth prodigious efforts during the 1890s to map the glacial deposits in Illinois, resulting in a classic U.S. Geological Survey (USGS) monograph (Leverett 1899). In it, Leverett developed a glacial map of the Illinois ice lobe,

which distinctly showed the glacial moraines of the Wisconsin Episode and the Kansan and Illinoian drift sheets. This remarkable map first demonstrated the full extent of glacial deposits in Illinois, separating them into 15 different

Figure 1-5 Geologists visiting exposures east of Peoria, comparing the Peoria and Sangamon weathered zones in 1898. Left to right: S. W. Beyer, Iowa State College, Ames; Johan A. Udden, Augustana College, Rock Island; Thomas C. Chamberlin, University of Chicago; Samuel Calvin, University of Iowa; and Frank Leverett, U.S. Geological Survey, May 1898 (Leverett 1899). (Photograph was taken by H. Foster Bain, who was the first state geologist of Illinois. Photograph from the Illinois State Geological Survey collection.)

stages and substages. The unique position of Illinois at the apex of glaciation in North America (Figure 1-4) attracted geologists to study the area, thus helping to fill the void left by the demise of the first Illinois Geological Survey in 1875 (Figure 1-5).

Paleontological Investigations

During the early part of the period from 1875 to 1905, in parallel with and partly as a result of the first Illinois Geological Survey, paleontological work grew explosively. Between 1850 and 1875, over 45% of the articles written on Illinois geology (Willman et al. 1968) dealt largely or wholly with paleontology, mainly descriptions of new species. Initially, mastodons in Illinois were hot topics (Lathrop 1851, Le Conte 1854, Foster 1857, Wilber 1861). F. B. Meek, mainly with A. H. Worthen, was a prolific contributor, publishing 25 articles between 1861 and 1874 describing Paleozoic fossils. Numerous authors described hundreds of new species of crinoids, brachiopods, bryozoans, blastoids, scorpions, fish, reptiles, amphibians, and the flora of the "Coal Measures."

Toward the end of the century, synopses and compilations of paleontological works began to appear (Keyes 1895, Ulrich and Schuchert 1902, Klem 1904). The change in emphasis from descriptive paleontology to stratigraphic paleontology was well under way by 1904.

Mineral Resource Investigations

Illinois realized amazing population growth from 1818 to 1904, increasing from just over 40,000 in 1818 to over 850,000 by 1850 and to almost 5,000,000 by 1900 (U.S. Census Bureau 2005). Chicago grew spectacularly from fewer than 4,000 at the time of statehood to almost 30,000 by 1850 and to nearly 1,700,000 by 1900 (Chicago Census Records 2002). After the Civil War, industrialization, which had started much earlier in England, was in full force in the United States, including Illinois. To meet the needs for raw materials to fuel industries, demands on Illinois mineral resources grew rapidly.

Coal, Oil, and Gas

Coal mining expanded in response to the growing demand for coal for steam locomotives and for residential and industrial purposes. Worthen and others devoted considerable efforts to describing the "Coal Measures" and gathering data from coal mine shafts.

Following the discovery of oil and gas in Pennsylvania in 1859 by Edwin Drake, oil excitement spread west, and, in 1865, wildcat drilling was undertaken in Clark County. A small amount of oil was found there, but the well was

Figure 1-6 Oil well at Casey, Illinois, 1905. (Photograph from the Illinois State Geological Survey collection.)

not commercially viable and was abandoned. Interest in oil and gas in Illinois faded. Nevertheless, small commercial discoveries were found near Litchfield and Sparta in the 1880s (Huff and Goodwin 1999, Crockett and Helpingstine 1996). In 1904, as interest was renewed in Crawford County, the Westfield pool was discovered at the shallow depth of about 280 feet (about 85 m) in Pennsylvanian sands near the original wildcat well drilled in the 1860s (Figure 1-6).

Fluorspar, Lead, and Zinc

After 1888, a new and important use was found for fluorspar: as flux in open hearth steel furnaces. Fluorspar demand rose rapidly (Risser and Major 1968), leading to serious field investigations for fluorspar deposits in the 1890s (Emmons 1893). Fluorspar studies continued into the 1900s, mainly by the USGS (Bain 1905) in the Kentucky-Illinois Fluorspar District. While working for the USGS,

Figure 1-7 Supplies delivered by horse and wagon arrive at Old Timers Lead Mine, northern Illinois, 1927. (Photograph contributed by H. B. Willman to the Illinois State Geological Survey collection.)

Bain (1904) also focused on the zinc and lead deposits in the Galena District (Figure 1-7).

The Illinois State Geological Survey: Initiated 1905

For several years, geologists and engineers urged the formation of a second Illinois Geological Survey to assemble and provide information on resource development. C. W. Rolfe, the third professor of geology at the University of Illinois, was one of the major advocates (Langenheim 2001). In 1905, through the efforts of T. C. Chamberlin, University of Chicago; A. Bement, a member of the Society of Western Engineers; the Society itself; and the University of Illinois, Governor Deneen (a friend and neighbor of Chamberlin) was convinced of the need for a geological survey. The governor requested a bill to initiate a Survey to focus on investigations helpful to the state's economy and development (Rolfe 1931). The General Assembly promptly passed the necessary legislation to form the modern Illinois State Geological Survey (ISGS).

The scientific advances of the ISGS can be divided into four 25-year periods:

1. 1905–1930, conducting new investigations on Illinois geology;
2. 1930–1955, recognizing the economic value of geology;
3. 1955–1980, addressing societal issues through geology; and
4. 1980–2004, balancing a healthy environment with a secure economy.

New Investigations on Illinois Geology: 1905 to 1930

During the period from 1905 to 1930, new studies on Illinois geology began at the newly established ISGS. Also during this period, the nation moved from the optimism of continued industrial expansion and invention, including the introduction of the automobile, to the trauma of World War I and the stock market crash of 1929.

Fundamental Geological Investigations

Topographic Mapping

The topographic mapping of Illinois became a priority of the new Survey. A petition from the Society of Western Engineers to the General Assembly on February 11, 1905, pressed for a topographical map of the state as "of the first economic importance. . . ." H. Foster Bain, upon his appointment in 1905 as the first director of the new Survey, immediately initiated a cooperative federal-state mapping agreement with the USGS (Herron 1909, Bain 1931). By 1927, detailed topographic maps of the Chicago region at a scale of 1:24,000 (1 inch [2.5 cm] on the map represents 24,000 inches, or 2,000 feet [about 610 m] on the ground) were completed. By 1930, about 60% of the state had been topographically mapped (Leighton 1928, 1931a) at map scales ranging from 1:48,000 to 1:24,000.

Geological Mapping

Detailed geological mapping followed on the heels of the completed topographic maps. W. C. Alden, USGS, had published a map of the Chicago region in 1902 (as cited by Rolfe 1931, p. 28) as one of the early USGS Atlas Folio series. During the middle of the next decade, ISGS mappers began to assume responsibility for much of the geological mapping in the state, and the USGS phased out its direct mapping assistance, publishing its last Folio in 1926. ISGS mappers included U. S. Grant (Northwestern University), T. E. Savage (University of Illinois), J. A. Udden (Augustana College) (Figure 1-8), and Stuart Weller (University of Chicago), who were on contract to the Survey. The mappers were hired by Bain to help jump-start geological mapping and the stratigraphic effort. Dirt roads were used for transportation (Figure 1-9). J. Marvin Weller (Fisher 1963, p. 9), describing the early field work, noted that it

> was done on foot from camps established usually in some farmers' orchards. From them, the country was mapped as far as it was convenient to walk. Then the tents were moved by farm wagon to another site. Later, livery rigs were used, ordinarily a two-horse surrey, which enlarged the area of operations. Automobiles did not enter the picture until 1917. The first was a model T Ford. . . .

History of Investigations

Figure 1-8 Johan A. Udden, field geologist, traveling on mule, circa early 1900s. (Photograph from the Illinois State Geological Survey collection.)

Figure 1-9 Sangamon Soil near Red Bud, Randolph County, during the days of dirt roads and travel by horse and buggy, 1914. (Photograph by E. W. Shaw; Illinois State Geological Survey collection.)

Figure 1-10 Field work with improved means of transportation—including Model T Fords at Campbell Hill, August 2, 1925. Shown, from left to right, are G. Moulton, A. C. Noé, P. J. Sedgwick, T. Root, and J. McCormack. (Photograph from the M. M. Leighton collection.)

The use of Model Ts is depicted in Figure 1-10. The infrastructure for this new transportation had its share of problems even then (Figure 1-11). Stuart Weller (1906a, 1906b) published both a revised geological map of Illinois and the first map of the geological structures of the state. The structure map was a rudimentary small-scale map showing six principal lines of deformation. Obviously, structural geology in Illinois was in its infancy. In 1907, Frank W. DeWolf, on loan to the ISGS from the USGS, began to publish the results of the first of a series of field investigations that extended across southern Illinois. His structural contour map on the base of Coal No. 5 in the *Year-Book for 1907* was one of the earliest of such publications relating to Illinois geology. DeWolf succeeded Bain as the director of the ISGS in 1909 when Bain resigned. Statewide compilations of structure maps appear to have been initiated by J. M. Weller in 1926 when he mapped the base of the New Albany Shale, the base of the Pennsylvanian, and the base of the No. 6 Coal, all with 100-foot contour intervals. The published compilations (Weller 1936) were a remarkable set of maps and an amazing feat for that time, notwithstanding the fact that the maps were limited by the amount of subsurface data available.

Surficial Geology

Also expanding during this period from 1905 to 1930 was work on glacial geology. M. M. Leighton (1923), a student of T. C. Chamberlin, used weathering (especially depth of leaching), geomorphic form, and other criteria to differentiate and classify the various drift sheets (Figure 1-12). In 1923, M. M. Leighton succeeded Frank DeWolf as the third chief of the Illinois State Geological Survey. Leighton (1926), noting a soil zone, was the first to recognize an older loess lying below the Peoria loess and resting on the weathered Sangamon zone. This loess location became known as the classic Farm Creek section.

Figure 1-11 Infrastructure failures happened even in 1927. State geologists examine the damage after a small bus broke through a wooden bridge. (Photograph from the Illinois State Geological Survey collection.)

While Bradley (in Worthen et al. 1870, p. 229) alluded to the formation of a glacial outlet from Lake Michigan, it was Ekblaw and Athy (1925) in their article on the Kankakee torrent who described the erosional process resulting from glacial meltwaters, their velocities, and their power to erode.

Stratigraphy

During this period, G. H. Cady focused on coal and Pennsylvanian stratigraphy; Stuart Weller, on the Mississippian System; and T. E. Savage, on Silurian stratigraphy. Savage (1920, 1925) also contributed to Devonian and Ordovician stratigraphy. In 1923, C. F. Bassett worked out the complete succession of Devonian strata (J. M. Weller 1944a). Stuart Weller, in 1920, and J. E. Lamar, in 1925, were the first to give an account of the entire Chesterian Series in southwestern Illinois (Willman 1982). Details of the stratigraphic efforts in this and later periods are described by Willman et al. (1975).

Udden (1912) described repetitive sedimentary cycles in the Peoria area, apparently the first to recognize cyclic deposition of the Pennsylvanian rocks of Illinois.

Growing interest in the geology of the state parks is reflected in an old photograph of a field excursion at Starved Rock State Park (Figure 1-13).

Paleontology

After the recognition of microfossils during the late 1800s, emphasis slowly shifted to fossilized spores and pollen (palynomorphs) in Pennsylvanian studies. Sellards (1902) and R. Thiessen (Phillips et al. 1973) noted the occurrence of fossilized spores in the Illinois Basin. During the early 1900s, paleontology also evolved into a useful tool for determining paleoclimates and paleoenvironments. White (1913), USGS, concluded from his studies of the Pennsylvanian Period flora that the climate during that time was tropical to subtropical.

Figure 1-12 Left to right: Pleistocene geologists M. M. Leighton and Harold Moses in the field in Illinois, 1922. (Photograph from the M. M. Leighton collection.)

Figure 1-13 Public geological field trip to Starved Rock. The attendees, circa 1920, are dressed for the occasion. (Photograph from the Illinois State Geological Survey collection.)

Mineral Resource Investigations

Coal

Knowledge of Illinois' mineral resources rapidly increased and proved vital to the state's industrialization. From the new Survey's beginning, investigations on coal geology and utilization were emphasized in the program developed by H. F. Bain and his successor, F. W. DeWolf. At the request of Director Bain, Bement (1910) calculated the original resources of the Illinois coal field at just over 201 billion tons (about 182 billion metric tonnes). Together with his colleagues in the Department of Mining Engineering at the University of Illinois and the U.S. Bureau of Mines, DeWolf initiated the Illinois Mining Investigations series (DeWolf 1931). The series resulted in 33 publications from 1913 to 1930, eight of them on coal resources in the eight mining districts and the remainder on mining practices, coal chemistry, strippable coal possibilities, low-sulfur coals, and the use and marketing of coals (Illinois State Geological Survey 1993, p. 51–52). G. H. Cady played a distinguished role as sole author of five of the district publications and two others on low-sulfur coals and coal-stripping possibilities.

Oil

With the discovery of the Casey oil field in 1905 (Figure 1-6), the first oil boom in Illinois was in full swing. Drillers M. L. Benedum and J. C. Trees were two of the greatest wildcatters in the area, and, in part, they used information from geologists or geological observations. In general, however, drillers remained skeptical of the worth of geological information (Miller 1981). This attitude was about to change. Director H. F. Bain (1906) arranged for W. S. Blatchley from Indiana to review the new discoveries in Illinois. Blatchley (1906) summarized those in Clark, Cumberland, and Crawford Counties. In 1908, following the discovery of additional fields (such as the Main oil field in 1906 and the Lawrence County field in 1907), Bain hired R. S. Blatchley, son of W. S. Blatchley, to study and report on the rapidly expanding developments. R. S. Blatchley and others collected "skeleton logs" (drillers' logs), often traveling from well to well on horseback. The collection of the Blatchley logs marked the informal beginning of the now extensive and widely used records collection at the ISGS (A. Faber, personal communication 2003). R. S. Blatchley's (1913) report on the southeastern oil fields along the La Salle Anticlinorium illustrates the advances made in geological knowledge since 1905 (Crockett and Helpingstine 1996). Production of the southeastern Illinois oil fields peaked in 1910 at 33 million barrels per year; production and industry activity declined steadily thereafter (Preston 1986). A few geologists carried on the ISGS studies started by the Blatchleys, including L. A. Mylius (1927) and A. H. Bell in a series of reports starting in 1926 (Willman et al. 1968).

Fluorspar, Fire Clays, and Other Industrial Minerals

Demand for fluorspar, fire clays, and other industrial minerals continued to grow. In response to increasing uses for fluorspar and expanding markets stimulated by World War I, investigators undertook further studies of the geology and fluorspar deposits of Hardin and Pope Counties (S. Weller et al. 1920). Because of the popularity of brick pavements and brick houses, early studies of fire clays focused on brick materials (Purdy and De Wolf 1907, Rolfe et al. 1908, St. Clair 1917, Parmelee and Shroyer 1921). Additionally, Rolfe was instrumental in initiating the University of Illinois' ceramics program, serving for eight years as the first head of what later became the Department of Ceramics Engineering (Langenheim 2001).

With the hiring of J. E. Lamar in 1920, the ISGS focus switched to the other industrial mineral resources in Illinois—sands and gravels, aggregates, and limestone. Krey and Lamar (1925) published an extensive report on Illinois' limestone resources and outcrops that revealed possible quarry sites and included descriptions of abandoned quarries. Lamar (1928) also published an exceptionally thorough economic study of the silica sand industry in the Ottawa, Illinois, region. That classic report also described the properties of the St. Peter Sandstone that favored its widespread use as silica glass sand, molding sand, and abrasives and its use in chemical industries.

Engineering Geology

In 1927, engineering geology was first formally recognized in Illinois as a distinct branch of geology. Chief Leighton appointed George E. Ekblaw to head the Survey's engineering geology program, which initially focused on dam sites and highway construction (Ekblaw 1953, Bergstrom 1980).

Groundwater Supplies

Groundwater supplies became critical in rapidly expanding metropolitan areas. In the Chicago region, investigators included C. B. Anderson, C. B. Williams, F. Thwaites, and G. E. Ekblaw. Only limited scientific work was conducted on groundwater location in glacial drift until Warren (1927) recorded wells in a buried river valley at Sullivan, Illinois.

Geology and the Economy: 1930 to 1955

During the period from 1930 to 1955, Illinois experienced the Great Depression, the second oil boom, World War II, and economic expansion. In 1930, the year after the stock market crash, M. M. Leighton broadened the scope of energy and mineral investigations to assist the State in restoring the economy. He thought that a program on mineral resources should include research on both the nature of the state's mineral resources and their utilization. Leighton's vision included multidisciplinary teams of geologists, chemists, physicists, and engineers. In 1931, he obtained the backing of the "captains of industry" and convinced legislators and the governor of the wisdom of this approach. In the teeth of the Depression, he added seven key staff members to conduct research on both the nature and utilization of Illinois mineral resources (Bergstrom 1980, Goodwin and Bergstrom 1988). This duality in the character of natural resources research continues to this day. The year 1931 marked the change of the ISGS from an agency focused on geological field surveys of the state's mineral resources to a multidisciplinary research organization working to expand the state's economy. Results of this new program were soon to follow (Figure 1-14).

Mineral Resource Investigations

Coal

Work on the utilization of coal progressed rapidly following the addition of chemists, a physicist, and engineering staff at the ISGS during the early 1930s. Smoke abatement had become a major environmental issue. Physicist R. J. Piersol carried out intensive experiments on coal briquettes from 1933 to 1936; he found that coal briquettes produced significantly less smoke than corresponding lump coals (Piersol 1936). This discovery was an important advance in the effort to develop smokeless fuels (Hays 1980) and a forerunner of environmental geology.

Of great importance, G. H. Cady (1935) published a new classification for Illinois coals, recognizing them as bituminous and rejecting earlier suggestions that they were subbituminous. Bituminous coal was widely recognized as having a greater heating value and better burning qualities than subbituminous or lower-ranked coals and, thus, had greater economic value. The classification was subsequently accepted by the American Society of Testing Materials and the American Institute of Mining and Metallurgical Engineers with whom Cady had been working diligently for a number of years. The classification as bituminous proved to be a significant advantage for Illinois coal in the marketplace (Leighton 1955, Simon 1974) and was mutually respected by both science and industry.

Cady (1952) published the first comprehensive inventory of the reserves, occurrence, distribution, and character of Illinois coal beds (Figure 1-15). That publication, the culmination of Cady's long and distinguished career, followed many years of mapping and detailed investigations of Illinois coals and a year's intensive work by a dozen or so scientists at the ISGS, led by J. A. Simon.

Oil and Gas

During the same period, other research investigations led to Illinois' second oil boom (Figure 1-16). At a meeting of the Western Society of Engineers in Chicago in 1930, Alfred H. Bell displayed a map he had developed that showed areas favorable for exploration in the central part of the Il-

Figure 1-14 Pictured are two important scientists who joined the Illinois State Geological Survey in 1933 at the height of the Great Depression: petrographer Ralph E. Grim (**a**), who later became known as the Father of Clay Mineralogy, and Frank H. Reed (**b**), chief chemist, who was hired to lead the Survey's new geochemical program. These men are just two of a number of outstanding hires who helped the Survey find new uses for Illinois' minerals and create badly needed jobs. Photographs were taken in 1940 and 1941, respectively. (Photographs from the Illinois State Geological Survey collection.)

linois Basin (Bell 1938, p. 6; 1941, p. 780). This now famous map stimulated industry activity in the Illinois Basin. Then the Pure Oil Company's Chief Geologist, Theron Wasson (1938a, 1938b), successfully applied geology and geophysics to discover the Clay City field in 1937. This discovery, in the area mapped by Bell as having the "best possibilities," began the second Illinois oil boom in 1937. Bell (1941) observed that the reflection seismograph proved to be the key that unlocked the door to hidden structures lying in the drift-mantled, deeper portions of the Illinois Basin where surface mapping of bedrock was impossible. Wasson's involvement, which pushed Pure Oil Company toward the Illinois Basin ahead of other oil companies, led to Pure Oil Company's early dominance in the Basin.

Figure 1-15 J. Norman Payne in the office with G. H. Cady (right), head of the Illinois State Geological Survey Coal Section, working on coal resources in Illinois in 1940. (Photograph from the Illinois State Geological Survey collection.)

Figure 1-16 A. H. Bell, Head of the Illinois State Geological Survey Oil and Gas Section, with distinguished oil and gas industry leaders, including A. I. Levorsen (far right) at a petroleum conference in Robinson, Illinois, 1941. Bell's famous 1929 map showed favorable areas for oil exploration in Illinois and focused industry's attention on this area. (Photograph from the Illinois State Geological Survey collection.)

Figure 1-17 Stuck in the mud at Salem, Illinois, as drilling crews and others moved in to explore for oil. This phenomenon was a common occurrence, as problems with infrastructure continued on Illinois roads in the late 1930s. (Photograph from the Illinois State Geological Survey collection.)

Figure 1-18 In 1936, a Carter Oil Company geophysical truck and shot hole crew joined other companies in applying new geophysical methods to oil exploration in the deep Illinois Basin. (Photograph from the Illinois State Geological Survey collection.)

Figure 1-19 The Salem Oil Field, shown circa 1940. The discovery well was drilled and completed in 1938. By June 1939, the field was the second largest in the United States. (Photograph from the Illinois Oil and Gas Association.)

Figure 1-20 This clipping from the November 12, 1941, *Mattoon Gazette* highlights the oil field development in Illinois. (Image reproduced with permission of the *Journal Gazette/Times-Courier.*)

Companies such as the Carter Oil Company, The Texas Company (Texaco), and others soon followed Pure Oil Company's lead (Preston 1986) and were also successful (Figures 1-17 and 1-18). Large oil discoveries were made from 1937 to 1939, including the Louden, Salem, and New Harmony fields and numerous smaller fields. Geology was now known to be of vital importance, along with geophysics, in oil and gas exploration in Illinois. To further the geological profession, the Illinois Geological Society was formed in 1939 with 115 charter members (Crockett and Helpingstine 1996).

Production during the second oil boom peaked in 1940 at 147 million barrels per year (Preston 1986) (Figure 1-19). For a short time that year, Illinois ranked second in the nation in daily oil production (Miller 1981). Only Russia and Venezuela exceeded Illinois' oil production during the first 10 months of 1940 (Crockett and Helpingstine 1996) (Figure 1-20). Additional anecdotal stories about early oil exploration in the Illinois Basin are available elsewhere (Preston 1986).

After 1940, Illinois' production entered a period of decline, only partially offset by the implementation of sec-

ILLINOIS OIL YIELD TOPS IRAQ AND IRAN

Urbana, Ill.—(INS)—One "I" betters two when it comes to oil production, for Illinois oil fields are outmatching barrel-for-barrel the combined output of Iraq and Iran, near east fields coveted by the totalitarian powers, Dr. M. M. Leighton, Illinois state geological survey chief, said today.

During the first 10 months of 1941 Illinois' oil output was approximately 135,000,000 barrels—nearly twice the oil consumption for 1939 in the British Isles.

Dr. Leighton said that, outside of the United States, only two nations exceed Illinois in oil output. They are Russia with 212,919,000 barrels in 1940 and Venezuela with 185,000,000 barrels. Illinois is the fourth oil producing state in the union, trailing Texas, California and Oklahoma.

Russia's annual output for 1940, however, was dwarfed by United States output for that year of 1,351,847,050 barrels.

Illinois' output was over twice that of all South American countries combined with Venezuela excepted.

The state geological survey reported that Illinois' probable 1941 oil output will be from four to five times as large as 1938 imports of the German Reich.

ondary recovery technologies and by discoveries such as the Marine field in 1943. Lowenstam and DuBois (1946) and Lowenstam (1948b) were the first to describe this new Silurian field as a buried reef and to note that such reefs formed traps for hydrocarbons, triggering additional exploration for Silurian reefs. During the late 1940s, reef-drape fields became a subsidiary play. The reef-drape effect was produced by differential compaction of weaker strata overlying the knob-like feature and providing a domal effect that trapped hydrocarbons (Bristol 1974). During 1953 to 1954, a sub-unconformity play was pursued successfully on the Sangamon Arch where truncated wedges of porous units were onlapped by New Albany Shale (Whiting 1956, Whiting and Stevenson 1965).

Frederick Squires, a petroleum engineer hired to undertake research on secondary recovery, began his extensive ISGS career investigating repressuring operations (Bell and Squires 1932) and waterflooding methods (Squires 1934) in the southeastern oil fields. After new field discoveries in 1937 to 1939, Squires turned his attention to those fields. An article on waterflooding of Illinois' oil sands by Squires and Bell (1943) called attention to the increased oil this method recovered. By 1944, the initial impact of waterfloods on Illinois oil production was apparent (Figure 1-21). Squires continued to publish on secondary recovery into the 1950s and initiated the annual reporting of waterflooding statistics in 1950. His research and publications greatly stimulated the industry to recover millions of barrels of additional oil through waterflooding.

Clays and Clay Minerals

Ralph E. Grim (Figure 1-14a), hired in 1931 as a petrographer to study Illinois clays and clay minerals, undertook microscopic studies of clay components. In 1934, W. F. Bradley joined Grim, and together they worked at identifying and characterizing clay minerals. Clark et al. (1937) published an article on the use of x-ray diffraction to identify minerals in clays, and, in the same year, Grim et al. (1937) discovered and characterized illite, a new clay mineral. Bradley (1945) provided diagnostic criteria for identifying clay minerals. During the 1930s and into the 1950s, Grim focused on the characteristics of various clay minerals in Illinois, especially their properties when used in ceramics and molding sands and their use as catalysts, as a source of alumina, and as fillers in paper products. Bradley, beginning in the late 1930s and continuing through the 1950s, focused on the structures, crystal chemistry, and thermal behavior of clay minerals (Figure 1-22). Grim's (1953) classic book, *Clay Mineralogy,* discusses the concept, classification, and nomenclature of clay minerals; their physical and chemical characteristics and properties; and their origin and occurrence. Because of the quality and quantity of his firsts in the investigation of clay minerals, Grim became known as the "Father of Clay Mineralogy" (White 1990).

Mineral Economics and Industrial Minerals

Noteworthy during this period was the 1931 appointment of Walter Voskuil as a mineral economist, the first of any state geological survey. Voskuil and Eich (1932) initiated the annual reporting of Illinois mineral statistics. From 1930 to 1955, Voskuil wrote a number of reports on the state's economic development and its competitive position

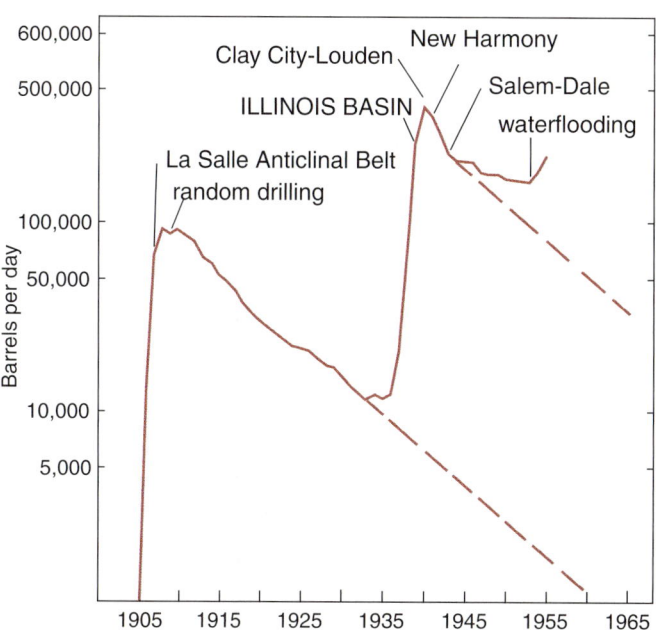

Figure 1-21 Major peaks in oil production, 1905 to 1965, reflect the introduction of new ideas into the Illinois Basin (Dickey 1958). (Graph published with the permission of the Tulsa Geological Society.)

Figure 1-22 William Bradley, clay mineralogist, investigates the properties of Illinois clay minerals in his laboratory. Photograph taken during the late 1930s or early 1940s. (Photograph from the M. M. Leighton collection.)

and markets in coal, oil and gas, and fluorspar. His reports were widely used by industry, government, and the public sector. In the 1930s, as the United States struggled out of the Great Depression, the USGS and ISGS revived field studies in the Fluorspar District (Hubbert and Weller 1934, Currier 1937a, Hatmaker and Davis 1938). Around this time also, articles by Bastin (1931) and by Currier (1937b) addressed the origin of the fluorspar deposits. Lamar and Willman (1938) described diverse uses of limestone and dolomite and the chemical and physical specifications of stone for most applications. These descriptions were very useful for industry (Willman 1982) (Figure 1-23).

Figure 1-23 Steam locomotives were in use hauling stone from the Lehigh Stone Company quarry in 1940. (Photograph from the Illinois State Geological Survey collection.)

Figure 1-24 M. King Hubbert (center) and two field party members doing field research with resistivity equipment in southern Illinois in the early 1930s. (Photograph courtesy of the Society of Exploration Geophysicists.)

Groundwater Resources

In 1931, M. M. Leighton, curious as to the potential of the new field of geophysics, hired M. King Hubbert to investigate the applicability of geophysical methods as a simple and inexpensive prospecting tool for several specific problems, such as locating water supplies in Pleistocene gravel deposits. Hubbert's work was the beginning of geophysical research at the ISGS. His analysis indicated that electrical earth resistivity (EER) was the most promising approach. His main success was in locating major water-bearing gravel deposits. The sharp boundaries of these deposits could be detected because their specific resistivity was higher than that of the adjacent materials (Hubbert 1932) (Figure 1-24). Additional studies in groundwater location followed. Case studies for 11 municipalities (Workman and Leighton 1937) increased confidence in the reliability of EER. M. Buhle, who joined the ISGS in 1938, worked with others to move the method into routine practice (Buhle and Brueckmann 1964). The EER method is still in use today by the ISGS.

One monumental event in particular gave impetus to groundwater resource studies in Illinois during this period: the publication of the map of the state's bedrock surface (Horberg 1950). This map, *Bedrock Topography of Illinois,* published at 1:500,000 scale, showed the extent and nature of buried bedrock valleys. Horberg concluded that (1) large undiscovered groundwater resources occur within the glacial deposits filling major bedrock valleys throughout the state; (2) the thickest and most continuous aquifers are within the present Mississippi, Illinois, and Wabash Valleys, which in part coincide with preglacial valleys; (3) important aquifers also are present in large buried valleys such as the Mahomet, Princeton, Shabbonna, Troy, Ticona, and Carthage Bedrock Valleys; and (4) although glacial aquifers are commonly present in the drift outside bedrock valleys, within the Wisconsin drift sheet, the most prolific sources generally are present along these valleys. With this publication, Horberg provided vision, direction, and priorities for groundwater exploration work for years to come (Figure 1-25).

Impact of World War II

Shortly after the attack on Pearl Harbor in 1941, M. M. Leighton requested an appropriation to further assist in the development of Illinois' energy and water resources needed for the war effort (Figure 1-26). This appropriation resulted in the use of geophysical well logs from the oil and gas industry as an aid in delineating groundwater supplies, completing water wells (Bays and Folk 1942), and mapping coals and other key beds (Taylor et al. 1944, Payne and Cady

History of Investigations

1944). The years 1943 to 1944 marked the beginning of the extensive use of geophysical logs outside the oil industry for subsurface geology. Those years also marked an overall increase in subsurface work (Workman 1944) (Figure 1-27).

As the war effort increased the demand for fluorspar, the ISGS, USGS, and U.S. Bureau of Mines undertook additional studies in the Fluorspar District of southern Illinois. This work is found in bibliographic citations for O. M. Bishop, 1947 to 1949 (Willman et al. 1968, J. M. Weller et al. 1952). Interest in the Galena Zinc-Lead District was also renewed, leading to the discovery of two new ore bodies (Leighton 1945, Willman et al. 1946) and the establishment of an ISGS office in Galena. Intensive investigations continued from 1947 to 1954 (Bradbury 1955, Bradbury et al. 1956).

Also notable during the war years were special efforts by ISGS chemists and engineers, Frank Reed, H. W. Jackman, P. W. Henline, and others, on blending Illinois coal with eastern coals to produce metallurgical coke (Figure 1-28). Blends were coked successfully in the Koppers plant at Granite City, Illinois, in 1944. These efforts were summarized by Reed et al. (1947). By the end of World War II, over 2 million car miles of freight transportation on the east-west lines were saved by substituting Illinois coal for a portion of eastern coking coal (Leighton 1966, Bergstrom 1980). As noted by Goodwin and Bergstrom (1988), work on blending Illinois coal resulted not only in saving freight car miles during the war but also in adding a market for coal not previously available to Illinois.

Significantly, during the war years, the federal government brought attention to the research at ISGS on synthesizing aromatic fluorine compounds for possible industrial applications. This work had been initiated soon after F. H. Reed joined the ISGS as head chemist in 1931 (Reed and Finger 1936). Samples of one of the chemicals synthesized by G. C. Finger were requested by the federal government when it was discovered that the chemical could not be obtained from any other source (Figure 1-29). The chemical was furnished to the Manhattan Project, which developed the atomic bomb. One of the Manhattan Project engineers wrote that the loan "enabled us to fulfill our most critical requirements for it [came] at a time we could ill afford [delay]" (Leighton 1966).

Figure 1-26 A field trailer laboratory used by the Illinois State Geological Survey during the war years of 1941 to 1945 for investigations of the potential for new coal resources to aid in the World War II effort. Shown, left to right, are geologists W. Pullen, A. Eddings, and T. Karlstrom. (Photograph from the Illinois State Geological Survey collection.)

Figure 1-25 Geologists from the Illinois State Geological Survey are shown testing the new Kelly water well in Champaign County in 1950. The well was drilled in response to demand for additional municipal water supplies in the area. (Photograph from the Illinois State Geological Survey collection.)

Figure 1-27 Lewis Workman, head of subsurface studies, in his office/laboratory at the Illinois State Geological Survey during the mid-1940s. (Photograph from the Illinois State Geological Survey collection.)

Fundamental Geological Investigations

Surficial Geology

The period 1930 to 1955 marked the remarkable progress of ISGS investigations of surficial materials (Figure 1-30). Four publications stand out:

1. A seminal paper by Leighton and MacClintock (1930) introduced the concept of a weathering profile with four "horizons." Leighton applied this concept to the Farm Creek section using his practical eye for weathering profiles to expand on his 1926 interpretation of the Farm Creek section. Leighton (1931b) recognized that the youthful profile on the late Sangamon (later Farmdale) loess developed when cold temperate or subarctic temperatures prevailed, signaling the beginning of the Wisconsin glacial epoch.
2. In a second paper, Leighton (1933) and Kay and Leighton (1933) divided the Wisconsin Stage into four substages—Iowa, Tazewell, Cary, and Mankato—a classification that endured for more than 25 years.
3. A classic study by G. D. Smith (1942) on loess deposits strongly supported the eolian origin of loess (Shimek 1896). Shortly after Smith's study and based in part on his work, Leighton and Willman (1950) successfully defended the eolian theory against the loessification theory of Russell (1944) and Fisk (1944).
4. In the fourth publication, following the physiographic diagrams of Fenneman (1938), M. M. Leighton et al. (1948) delineated and described physiographic units in Illinois. This basic classification of physiographic divisions in Illinois has remained.

Bedrock Geology

Lowenstam (1948a) found that the fauna of reef and off-reef strata of Silurian age were distinctly different, leading directly to the new field of paleoecological studies. L. L. Sloss et al. (1949) at Northwestern University introduced facies mapping of subsurface units (i.e., the areal mapping of different types of sedimentary rocks segregated into distinct, genetically related deposits). This delineation greatly facilitated exploration for petroleum traps and deposits of low-sulfur coal.

Figure 1-29 Glen Finger, head of fluorine chemistry, in his laboratory in 1954. During the 1940s, geologists at the Illinois State Geological Survey laboratories synthesized a fluorine compound needed in the war effort to speed the production of the first atomic bomb. (Photograph from the Illinois State Geological Survey collection.)

Figure 1-28 Paul Henline pulls a coke charge from the coke oven as his assistant cools the charges with a hose, resulting in a large amount of steam. Both are wearing protective gear. This procedure was part of a successful project to determine the suitability of blending Illinois coal with eastern coals to make coke. (Photograph from the Champaign-Urbana *News-Gazette,* September 29, 1944, obtained from the University of Illinois Archives. Photograph reproduced by permission of The News-Gazette, Inc. Permission does not imply endorsement.)

Figure 1-30 Three distinguished Pleistocene geologists in the field at a loess outcrop during a Mississippi Valley field conference in 1949. The geologists are shown assessing the origin of these loess deposits. From left to right: J. Harlan Bretz, University of Chicago; Morris M. Leighton, Illinois State Geological Survey; and A. Trowbridge, Iowa Geological Survey. (Photograph from the Illinois State Geological Survey collection.)

Wanless and Weller (1932), in a classic paper, expanded the earlier work of Udden (1912) that recognized similar cyclicity throughout the Illinois Basin in Pennsylvanian strata. Wanless and Weller promulgated the cyclothem concept that described these cycles, normally placing the base of each cycle at a disconformity beneath a channel sandstone. Each cyclothem was generally composed of a terrestrial to marine succession grading upwards to or interbedded with the terrestrial succession. Each cyclothem included an underclay and coal roughly in the middle of a sequence. The cyclothems proved helpful to the geologists in correlating coal beds and determining lateral facies equivalents.

In the area of paleontology, a paper by Schopf (1936) was the earliest detailed study of spores in Illinois, and his synopsis of Paleozoic fossil spores (Schopf et al. 1944) became a classic. A subsequent treatise by Kosanke (1950) on Pennsylvanian spores was the first major basin-wide pollen spore correlation in North America (Figure 1-31).

Among the early, more detailed correlation charts prepared to depict the succession of strata and show the lateral equivalency of various Pennsylvanian units were those by Wanless (1939) and Moore (1944). The Wanless paper was the first in a series of correlation papers, one for each period, with Illinois geologists contributing to all of them. Also in this period, for the first time, individual investigators began to piece together the tectonic history of the Illinois Basin as structural features were identified and mapped (Weller 1936; Weller and Bell 1936, 1937; Clark and Royds 1948). J. M. Weller et al. (1945) compiled a statewide geological map that added considerable detail to the earlier map of his father's (S. Weller 1906a).

Significant progress was made in defining one of the marked unconformities in Illinois when Siever (1951) and Wanless (1955) documented further the early work of Cady (1921), who described the sub-Pennsylvanian surface as a major Phanerozoic unconformity.

Engineering Geology

Engineering geology flourished during this period and helped to put in place the state's first system of paved roads (Ekblaw 1930, 1932, 1953). Other construction projects involving engineering geology included those for dams, reservoirs, and recreational lakes, mainly during the late 1930s and 1940s; for foundation conditions throughout the state; for Chicago subways commencing in 1938 (Peck 1940); and for mined caverns for liquid petroleum gas (LPG) storage in the early 1950s. Ralph B. Peck and Karl Terzaghi were major geotechnical contributors during this period, establishing a soil strength testing laboratory in Chicago during the late 1930s and giving birth to the science of soil mechanics (Terzaghi 1943). Peck (1948) reviewed the history of building foundations in Chicago, wrote about the engineering properties of Chicago's subsoils (Peck 1954), and provided information on observed and computed settlements (Peck and Uyanik 1955). These works became standard references for foundation conditions in Chicago.

M. M. Leighton retired as the chief of the ISGS in 1954 and was succeeded by John C. Frye from the Kansas State Geological Survey.

Geology and Societal Issues: 1955 to 1980

The period 1955–1980 included post-World War II recovery, the Korean and Vietnam conflicts, and the Arab oil embargo. The period also was marked by growing societal concerns for the environment and the initiation of geological and geochemical programs to assess the extent of pollution and contamination and the nature of geological hazards in Illinois. This effort was aided materially by the shift at the Illinois State Geological Survey during this period from wet chemical methods of analysis to highly sophisticated equipment capable of measuring chemicals in

Figure 1-31 Robert Kosanke, a distinguished palynologist, is shown during the late 1950s. Kosanke photographed coal samples and spore and pollen samples to help date and correlate strata. (Photograph from the Illinois State Geological Survey collection.)

parts per billion rather than parts per million and also to detect small amounts of radioactive material (Figure 1-32). Additionally, the implementation of geological protocol, the stratigraphic code, was especially important in formalizing and standardizing geological nomenclature.

Groundwater Investigations

G. B. Maxey, who led the ISGS Groundwater Section during the mid-1950s, significantly expanded its groundwater resource studies (Figure 1-33). One outstanding early study was by Pryor (1956), who used several thousand electrical logs to map the distribution of sandstone aquifers of Pennsylvanian age in White County. Pryor was one of the first to estimate on a regional level the quality of water in sandstone aquifers in Illinois by calculating the sodium chloride equivalent in parts per million (milligrams per liter) from resistivity values obtained from the electrical logs.

Following Horberg's revelations of buried bedrock valleys, studies of groundwater resources blossomed in Illinois. From 1954 to 1958, eight groundwater reports, completed for the various Agricultural Extension Districts, focused on groundwater supplies for agriculture. These reports together covered the entire state. R. E. Bergstrom effectively led the team, serving as senior author of three of the reports (Bergstrom et al. 1955, Bergstrom 1956, Bergstrom and Zeizel 1957).

After Horberg's (1950) elucidation of buried bedrock valleys and investigations that confirmed their importance, scientific curiosity was generated about how the boundaries of these features could be located more precisely. McGinnis and Kempton (1961) and McGinnis et al. (1963) conducted integrated studies in the Troy Bedrock Valley that involved seismic refraction, gravity surveys, and existing well information to map bedrock valleys filled with glacial deposits (Figures 1-34, 1-35, and 1-36). Using a compilation of earlier observations to develop their exploration concepts, the ISGS team developed a system for mapping these buried valleys. After a series of field experiments, Cartwright (1968) added yet another geophysical tool, temperature surveys, for delineating groundwater flow.

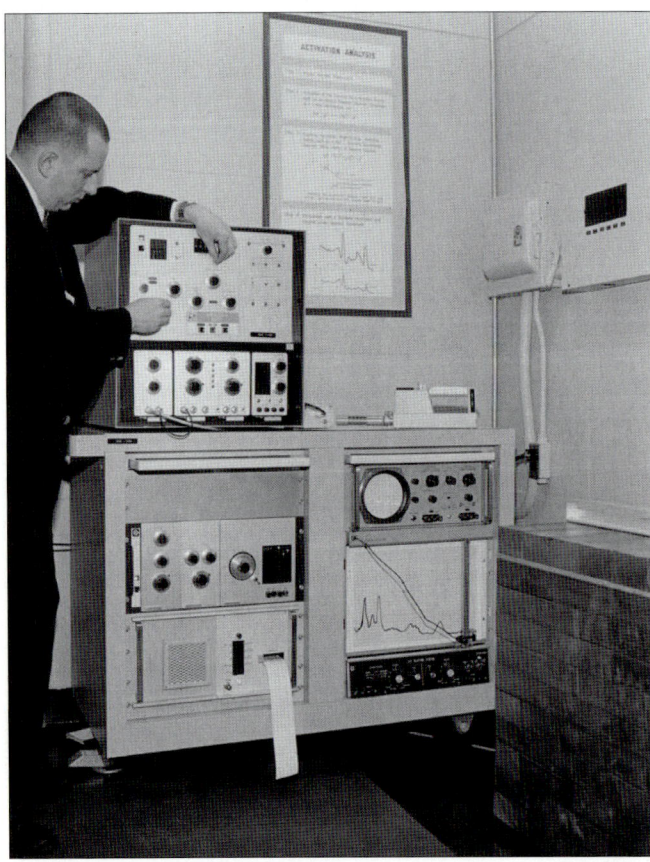

Figure 1-32 Geochemists Gus Ruch and Neil Shimp led the way in implementing new instrumental methods of geochemical analysis at the Illinois State Geological Survey. Here Gus Ruch is shown in 1966 operating the new neutron activation analysis machine to detect trace elements in geochemical samples. (Photograph from the Illinois State Geological Survey collection.)

Figure 1-33 George Burke Maxey, shown in 1961, led the expanded Illinois State Geological Survey groundwater program to locate and protect this vital resource. (Photograph from the Illinois State Geological Survey collection.)

Expanding cooperation between the ISGS and Illinois State Water Survey (ISWS) resulted in the first of a number of ISGS-ISWS Cooperative Groundwater Reports (Suter et al. 1959) as the greater Chicago area commanded additional attention. Additionally, T. A. Prickett, at the ISWS, who had noted the similarities between electrical flow and water flow, developed scaled electrical models to replicate water flow (Prickett and Lonnquist 1971). His technique became widely used and added an additional capability to both Surveys.

Mineral Resources

Coal

As the economy and coal demand grew, and as coal companies undertook massive surface mining of coals, W. H. Smith and others devoted major effort during the period from 1957 to 1968 to identify and map surface-minable coal reserves. This work led to a reevaluation of coal reserves (Simon and Smith 1969, Treworgy et al. 1978) and to further coal development (Figure 1-37).

Figure 1-35 Merlyn Buhle operates resistivity equipment (a commutator) to detect possible shallow water-bearing sand channels in 1961. (Photograph from the Illinois State Geological Survey collection.)

Figure 1-34 A blast is tripped using dynamite in a shot hole in the search for buried stream valleys with the potential for water-bearing sands, circa 1960. (Photograph from the Illinois State Geological Survey collection.)

Figure 1-36 John Kempton, Pleistocene geologist, uses field notes to verify and update bedrock surface maps, 1961. (Photograph from the Illinois State Geological Survey collection.)

Low-sulfur (less than 1.3%) coal had been recognized early in the development of Illinois Basin coal resources (Cady 1920). Hopkins (1968) and Gluskoter and Simon (1968) expanded on this work in companion studies and developed predictive discovery tools. These scientists used the presence of 20 feet (about 6 m) or more of gray shale overlying coals, mappable from borehole and well log records, to delineate favorable areas for low-sulfur coal. This work built on that of Cady (1920), whose studies demonstrated that low-sulfur coal in Illinois is directly overlain by thick gray shale about 25 feet (about 7.6 m) or more in thickness.

Johnson (1972) developed a depositional model to explain the origin of low-sulfur coal, noting that the wedges of gray shale were crevasse splay deposits that shielded coal-forming peat from the invasion of sulfur-bearing marine waters, thus preventing transfer of sulfate ions to the coal to form sulfides. Later, others (e.g., Palmer et al. 1979, Jacobson 1983) applied Johnson's model in mapping low-sulfur coal. Treworgy and Jacobson (1985) broadened the concept to locate low-sulfur coal resources by including wedge deposits of other nonmarine sediments above coal.

Oil and Gas

Oil and Gas Exploration. Independent producers became especially active during this period—branching out and finding smaller fields and improving recovery from existing fields—as existing oil and gas developments reached maturity (Howard 1963, Crockett and Helpingstine 1996). P. A. Dickey (1958) perhaps best summed up oil and gas exploration in Illinois at this stage of development, noting that each big wave of discovery and production could be attributed to the influence of a new idea (Figure 1-21).

Impacts of the Arab Oil Embargo. The 1973–1974 Arab oil embargo prompted investigations to ensure future stable energy supplies for Illinois. The coal and water resources needed for one or more plants to convert coal to synthetic energy fuels were studied (Smith and Stall 1975). Simon (1977a) detailed this alternative and others in an extensive report on long-range resource planning. Additionally, New Albany Shale investigations, which had started as oil shale studies during the 1950s (Lamar et al. 1956, Armon and Rees 1960), attracted further work as a possible unconventional gas source (Chou and Dickerson 1979, Cluff and Dickerson 1980). Crude oil price increases that followed the embargo and the subsequent removal of price controls resulted in increased drilling for oil in smaller, higher-risk "plays" (oil-producing strata) and a small increase in the state's oil production, which peaked again during the middle 1980s.

Industrial Minerals

Following World War II, demand for fluorine in fluorspar continued. Chemists at the ISGS developed fungicides containing fluorine to control fungus and mold on vinyl and cotton. This technology was used during the Korean and Vietnam conflicts for control of jungle rot on boots (R. H. Shiley and D. R. Dickerson 2003, personal communication). During the late 1950s, some emphasis was shifted from fungicides to plant regulators, or herbicides (Finger et al. 1959). In all, Finger and staff published 13 papers from 1944 to 1965 on preparing these and other aromatic fluorine compounds (Shiley et al. 1978). Work on fluorine chemistry and fluorspar deposits dwindled rapidly as foreign competition increased, and interest in metallic min-

Figure 1-37 (**a**) A large stripping shovel removes overburden to lay bare the underlying coal for mining, 1970s. (**b**) The size of the stripping bucket is emphasized in this 1970s photograph of a school class standing in it. The size of coal mining equipment greatly increased over the years, leading to greater efficiency and productivity of coal mining. (Photographs from the M. W. Leighton collection.)

Figure 1-38 Mining silica sand from the St. Peter Sandstone in the mid-1950s at the Ottawa Silica Sand Company, Ottawa, Illinois. (Photograph from the Illinois State Geological Survey collection.)

Figure 1-39 Herb Glass, clay mineralogist and Pleistocene geologist, shown during the early 1960s preparing samples for x-ray diffraction analysis to determine the mineral content of Pleistocene deposits. (Photograph from the Illinois State Geological Survey collection.)

erals also faded. Scientific investigations by Lamar (1957, 1967) and others in industrial minerals increased, keeping pace with the increasing need for construction materials (Figure 1-38). Pursuing limestone utilization, Harvey (1970) and other researchers from 1971 to 1974 (Raymond Martin 2001) published key papers on the sorptive properties of carbonate rocks to determine those most suited to sulfur dioxide sorption in the flue gas scrubbers coming into use. The thinking was, and still is, that scrubbers might be a key to renewed use of high-sulfur Illinois coal.

Mineral Economics

In 1959, H. E. Risser succeeded Voskuil as the head of the ISGS Mineral Economics Group. Risser introduced changes in the analytical approach to mineral economics as he projected industry trends and analyzed economic impacts. Following methods established by M. King Hubbert, Risser (1960) accurately forecast the inability of domestic oil and gas supplies to meet domestic demands and brought further attention to the concept that mineral resources are finite. His 1973 energy report—a visionary document for its time—pointed out that the domestic production capacity was not sufficient to meet future needs for fuels and energy, especially considering further restrictions on coal production (e.g., the U.S. Environmental Protection Agency's [U.S. EPA] regulations limiting sulfur emissions, mine subsidence, and acid mine wastes) (Risser 1973).

Glacial Geology
The Stratigraphic Code

Development of a new protocol, the stratigraphic code of the American Stratigraphic Commission, influenced in major ways how investigations on glacial geology were conducted from 1955 to 1980 (Frye and Richmond 1958). In 1958, ISGS adopted a new stratigraphic policy that adhered to this code (Willman et al. 1958). Following field work to define type sections and reference sections in accordance with the code, and in an effort to avoid conflicting usage of terms that had grown over the years, Frye and Willman (1960, 1975) and Willman and Frye (1970) developed a new, fourfold classification for Pleistocene deposits that provided for time-stratigraphic, rock-stratigraphic, soil-stratigraphic, and morpho-stratigraphic units. Also, during the 1960s, Glass et al. (1964, 1968) found clay and other minerals to be useful for Pleistocene stratigraphic correlation and as indicators of the origin of loess deposits (Figure 1-39). Then, based on available information, J. A. Lineback (1979) compiled the lithostratigraphic map, *Quaternary Deposits of Illinois,* at 1:500,000 scale.

Carbon Dating

Carbon-14 dating had a major impact on glacial geology research in Illinois during the last half of the twentieth century. The carbon dating method was developed by Professor W. H. Libby, University of Chicago, during the post-World War II years (Libby 1952). In recognition, Libby was awarded the Nobel Prize in Chemistry in 1960. J. C. Frye (1968) recognized the significance of radiocarbon dating for glacial stratigraphy and decided to establish the ISGS Radiocarbon Dating Laboratory. R. R. Ruch, ISGS chemist, set up the laboratory in 1968.

Bedrock Geology
Stratigraphy and Structure

Willman et al. (1975) captured, in ISGS Bulletin 95, *Handbook of Illinois Stratigraphy,* the period's major contributions to Illinois stratigraphy. This comprehensive collection included stratigraphic descriptions and historical correlation and classification charts for all Illinois strata. Willman et al. (1967) had previously compiled a new geological map of Illinois and, in the 1975 bulletin, compared the map with a number of previous geological maps of Illinois. New stratigraphic concepts introduced in Illinois at this time included Lineback's (1968) identification of the first turbidites to be recognized in the Illinois Basin. Lineback (1969) also called attention to a Mississippian carbonate bank facing a deeper water trough where sediments were deposited in a "starved basin," that is, a basin in which the rate of basin fill was less than the rate of basin subsidence.

The evolving thinking on the structure and tectonic history of Illinois during the period is reflected in a series of papers (Swann and Bell 1958, Swann 1968, Atherton 1971, Bristol and Buschbach 1971). Additionally, H. H. Damberger (1971, 1974), by showing that the rank of Illinois coal changed systematically in the Illinois Basin laterally and with depth, calculated the depth to which Illinois coals had been buried. Damberger found that depth had amounted to as much as about 1 to 1.9 miles (1.6 to 3 km) in parts of Illinois before post-Pennsylvanian erosion removed much of the overlying deposits.

Of major impact on the study of stratigraphy and tectonics was the recognition by L. L. Sloss (1963, 1988) with others (Sloss et al. 1949) of major sequences that were distinguished by the unconformities that bound them. Sloss found that major widespread unconformities, whether created by tectonics or eustasy (the rise and fall of sea level), were identifiable discontinuities in the stratigraphic record that could be readily used to subdivide geologic history during Phanerozoic time. Sloss (1963) called the packages of strata between interregional unconformities "sequences" and gave them Indian names. He named the bounding unconformities after the overlying sequence (e.g., sub-Absaroka, sub-Kaskaskia, and sub-Tippecanoe). The stratigraphic history of Illinois according to Sloss sequences has been traced by Swann (1968), Buschbach (1971), Sloss (1988), Collinson et al. (1988), and Leighton et al. (1991).

In a classic paper, Bristol and Howard (1971) mapped the sub-Absaroka surface, showing the strata exposed beneath this major widespread unconformity in Illinois. They used more than 53,000 electric logs accumulated by the ISGS over the years—a priceless data bank. Also, the widespread Devonian unconformity (sub-Kaskaskia) was documented with cross sections and subcrop maps by Collinson and Atherton (1975) and later Kolata (1991), expanding on earlier work by Weller (1944a, 1944b) and by Workman (1944). Buschbach (1961) and Willman and Buschbach (1975) mapped the sub-Tippecanoe unconformity below the St. Peter Sandstone.

Paleontology

During the late 1950s and into the 1970s, Charles Collinson led the expansion of conodont research and applied this knowledge to precise and detailed zonations and correlations of bedrock strata in Illinois (Collinson et al. 1981). R. A. Peppers, a student of Kosanke at the University of Illinois, significantly added to Kosanke's palynological research, especially in two publications using spores to detail Pennsylvanian correlations (Peppers 1964, 1970) (Figure 1-40). Lois Kent curated the ISGS collection of fossils (Figure 1-41).

Figure 1-40 Russell Peppers examines spores and pollen using a microscope circa 1965. (Photograph from the Illinois State Geological Survey collection.)

Figure 1-41 Lois Kent, curator, catalogues fossils in the Illinois State Geological Survey fossil collection, 1965. (Photograph from the Illinois State Geological Survey collection.)

History of Investigations

Environmental and Engineering Geology

J. E. Hackett coined the term "environmental geology" in 1963, a few years after the ISGS and the ISWS opened the northeastern Illinois office in Naperville in 1959 (R. E. Bergstrom, February 9, 1983, letter to W. C. Ackerman; Hackett 1967; Bergstrom 1992). Hackett, who was assigned to head ISGS activities in the northeastern office, found that many of his visitors needed information about issues such as the disposal of solid and liquid wastes, including landfill siting, utilization of old quarries for other purposes, land use planning and conflicts in land use, and foundation conditions (Larsen and Hackett 1965). According to Bergstrom (1992), Chief John Frye "saw the timeliness of employing geology... to the solution of many environmental problems and established environmental geology as one of the Survey's main research and service programs," thus ushering in a new era of geology in Illinois. Spurred by federal and state legislation, environmental geology expanded and took its present form. Principal environmental laws having direct impact on this field are listed in Table 1-1.

Table 1-1 Selected federal and state environmental legislation of the past 40 years.

Year	Legislation
1963	Clean Air Act
1965	Water Pollution Control Act Solid Waste Disposal Act
1969	National Environmental Policy Act, as amended
1970	Clean Air Act, as amended
1972	Federal Insecticide, Fungicide, Rodenticide Act (FIFRA)
1974	Energy Reorganization Act (Nuclear Regulatory Commission established); Safe Drinking Water Act
1976	Resource Conservation and Recovery Act (RCRA) Toxic Substances Control Act (TSCA)
1977	Clean Water Act, as amended; Clean Air Act, as amended; Surface Mining Control and Reclamation Act (SMCRA)
1980	Solid Waste Disposal Act Amendments; Low-Level Radioactive Waste Policy Act, as amended; Comprehensive Environmental Response, Compensation, and Liability Act (CERCLA) (Superfund), as amended
1982	Nuclear Waste Policy Act, as amended
1984	Hazardous and Solid Waste Amendments
1987	Illinois Groundwater Protection Act
1990	Clean Air Act, as amended Illinois Pesticide Act
1997	Illinois Pesticide Act, as amended

Geology for Planning

Environmental geology had its roots in groundwater studies and also in urban geology—geological studies for use in urban planning. Studies included the work by J Harlen Bretz (1939, 1955) on the Chicago region. "Geology for planning" began in earnest under the new ISGS environmental geology program during the 1960s. J. E. Hackett (1966) published the first geology for planning report applying the new approach to the Chicago metropolitan area, and, in 1967, he described the concepts and principles of these studies. The reports grew to include information on groundwater availability, sands and gravels for construction purposes, conditions affecting construction, and, especially, the adequacy of locations for landfills; solid, liquid, and hazardous waste disposal sites; septic tanks; and the surface application of wastes. Approximately 20 or so geology for planning reports for communities, regions, and counties were produced by ISGS authors and others during this time.

Engineering Geology

Large engineering projects included those for gas storage in mined caverns during the 1950s and storage in petroleum reservoirs or saline aquifers in later decades (Witherspoon et al. 1962, Buschbach and Bond 1974). The Deep Tunnel Project in Chicago, now called TARP (Tunnel and Reservoir Plan), was another major undertaking. Following a long history of contamination problems in the Chicago region and downstream in the Illinois River, the courts ordered the Metropolitan Sanitary District of Greater Chicago to restrict water use from Lake Michigan. This restriction forced the District to initiate a massive program to build large-diameter sewers, a holding system, and new sewage treatment plants (Harza Engineering and Bauer Engineering 1968a, 1968b). The system ultimately included more than 100 miles (161 km) of large-diameter tunnels in Chicago area bedrock. Buschbach and Heim (1972) and Buschbach et al. (1982) published the geological criteria for the tunnels and summarized the massive geotechnical effort that included more than 3,000 boreholes, 170,000 feet (51,816 m) of core, geophysical logging of 110 boreholes, groundwater testing, and 420 miles (676 km) of seismic reflection data. The Illinois State Geological Survey began testing the strength of Illinois rock materials at the end of the 1970s (Figure 1-42).

Geological Hazards

During this period, research directions were influenced by societal concerns about threats to the safety and security of Illinois citizens from earthquakes and coastal erosion.

Studies were begun during the 1970s to determine the potential of known fault zones for reactivation (Kolata and Buschbach 1976, Kolata et al. 1978) as the Nuclear Regulatory Agency posed questions about the tectonic stability of areas for the construction of nuclear power plants. Questions focused mainly on the New Madrid Seismic Zone in the northern Mississippi Embayment area of Missouri, Arkansas, and Tennessee and the zone's possible extension into southern Illinois and Kentucky. The historic New Madrid (1811–1812) series of major earthquakes (magnitudes greater than 7.0) and continuing seismicity in the area were of major concern. Beginning in 1977, investigators from seven states undertook numerous structural studies of the area (Buschbach 1986). The multi-year studies were supported by the National Research Council, coordinated by Buschbach, and undertaken in cooperation with the USGS.

High Lake Michigan lake levels during the mid-1970s caused damage from erosion and wave impact (especially during storms) on beaches, residences, and public property (Collinson et al. 1974, 1975) (Figure 1-43). DuMontelle et al. (1971) conducted landslide investigations along the Illinois River Valley that added information to that previously developed by G. E. Ekblaw.

Groundwater Contamination

Responding to complaints of leaking landfills, the ISGS and University of Illinois Department of Geology together initiated a series of investigations between 1967 and 1972 on the hydrogeology of and water quality in and near landfill sites (Hughes 1967). Cartwright and McComas (1969) and Cartwright and Sherman (1972) used geophysical surveys to identify and delineate plumes of leaking contaminants. Those researchers also developed many methods later found to be useful in monitoring landfills. After the Illinois Department of Public Health's promulgation of rules and regulations for refuse disposal sites, Hughes and Cartwright (1972) set out scientific and administrative criteria for shallow waste disposal. Then Cartwright and others (Cartwright et al. 1976, 1977; Johnson and Cartwright 1980) established the principles of contaminant migration through porous media, through glacial tills, and in the unsaturated zone (Figure 1-44).

Environmental Geochemistry

Changes in the use of environmental geochemistry illustrate the shift to the modern era that occurred in Illinois during the late 1960s and 1970s. The focus shifted

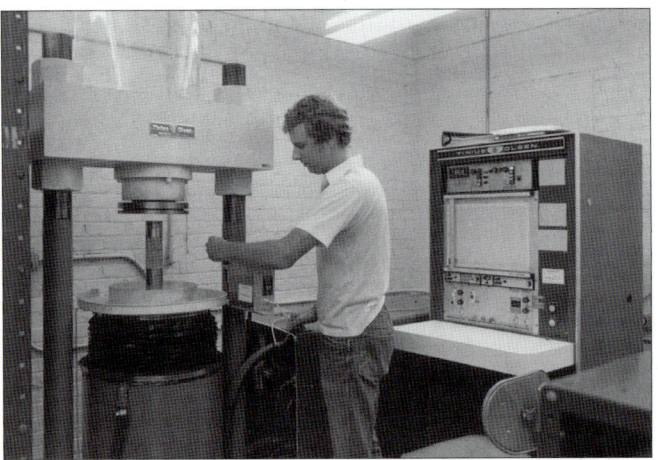

Figure 1-42 Robert Bauer, engineering geologist, in the Illinois State Geological Survey laboratory in 1979 testing the compressive strength of rock materials. He later set up the first geological engineering testing laboratory for the Illinois State Geological Survey. (Photograph from the Illinois State Geological Survey collection.)

Figure 1-43 Storm damage at Winthrop Harbor, Illinois, north of the Zion Nuclear Power Station, 1973. (Photograph by Charles Collinson.)

Figure 1-44 Keros Cartwright, hydrogeologist, measuring pore water pressure in clay using a pressure transducer in the Illinois State Geological Survey's mobile laboratory, 1970. (Photograph from the Illinois State Geological Survey collection.)

from using geological science for economic development to applying it to environmental issues. During the 1960s, as geologists began to confront societal issues of air and groundwater pollution, chemists and geochemists set about to detect and quantitatively measure in natural settings the concentrations of those elements and compounds posing an environmental threat.

The environmental geochemistry studies initially focused on pyrite in coal and air pollutants (Gluskoter 1965, Gluskoter and Simon 1968, Ruch et al. 1974, Gluskoter et al. 1977) (Figure 1-45) and contaminants in bottom sediments of Lake Michigan (Shimp et al. 1970, Ruch et al. 1970, Schleicher and Kuhn 1970, Lineback and Gross 1972). Chief J. A. Simon (1977a) reported that generally no Illinois coal could meet the U.S. EPA regulations limiting emissions to 1.2 pounds (0.54 kg) of sulfur dioxide per million British Thermal Units (BTUs) in new power plants without emission controls. (Simon had replaced Frye when he retired in 1974 to become executive director of the Geological Society of America.)

During the mid-1970s, geochemical interest shifted to the fate of contaminants leaking from landfills (Griffin et al. 1976, 1977). During the late 1970s, Griffin et al. (1978, 1980) refocused their attention to the examination of coal waste pollutants. Those studies led to the chemical characterization of leachates from coal waste.

At about the same time, J. P. Gibb (1976) verified the migration of hazardous wastes from land disposal sites in Illinois. His study, mass media publicity (e.g., *Illinois Times*, October 13–19, 1978, p. 4), and the involvement of the state legislature caused another shift in research emphasis to hazardous leachates during the late 1970s. Gibb's study led also to the use of mass spectrometer-gas chromatograph methodology to detect and measure contaminants in parts per billion (micrograms per liter) rather than parts per million (milligrams per liter) (Griffin and Chian 1979).

Isotope Geochemistry

Following the establishment of the ISGS Radiocarbon Dating Laboratory in 1968 and the hiring of D. D. Coleman in 1970, the field of isotope geochemistry in Illinois began one of its major growth periods for studies of both radioisotopes (Coleman et al. 1972) and stable isotopes. Coleman (1976) used isotopic differences to distinguish bacterial drift gas from deep thermogenic oil and gas field gas (Figure 1-46). In a series of papers, Coleman and others went on to "fingerprint" gases originating from sources including coals and leakage gas from underground storage (Coleman et al. 1977, Coleman and Meents 1978,

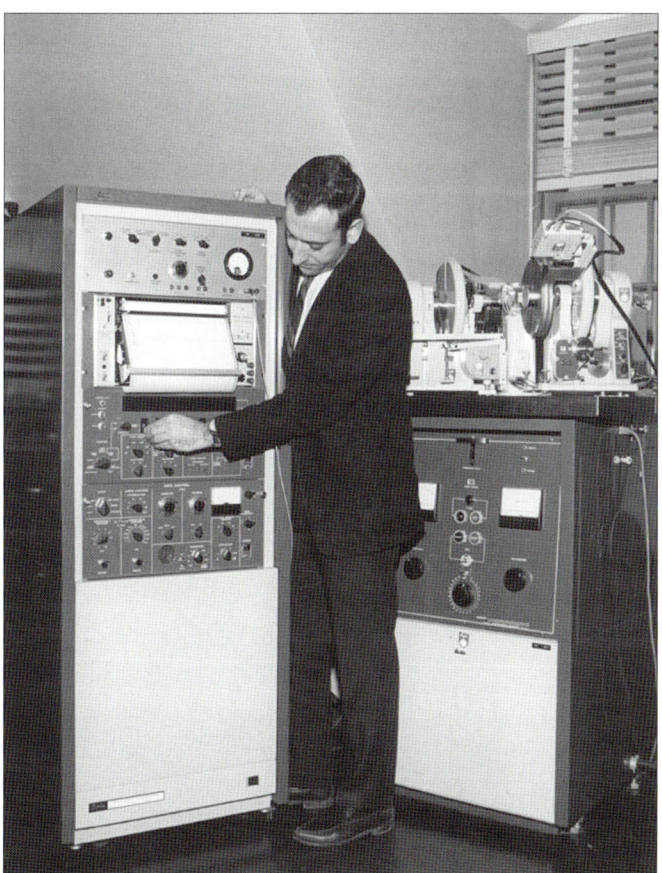

Figure 1-45 Harold Gluskoter, coal geologist, uses the Phillip x-ray diffractometer to determine the mineral content of Pennsylvanian coal samples, circa late 1970s. (Photograph from the Illinois State Geological Survey collection.)

Figure 1-46 Dennis Coleman operates a hybrid scintillation spectrometer to measure the radiocarbon content (^{14}C) in samples prepared in his laboratory at the Illinois State Geological Survey, 1971. (Photograph from the Illinois State Geological Survey collection.)

Coleman 1979) Isotopic fingerprinting was used to identify sources of leaking gases in life-threatening situations and in resolving ownership issues for certain gas accumulations.

In 1980, Simon summarized the accomplishments of the ISGS and looked toward the future during a symposium observing the 75th anniversary of the Survey (Simon 1982).

Balancing Environmental and Economic Concerns: 1980 to 2005

The period from 1980 to 2005 was marked by continuing environmental concerns and swings between economic growth and recession. Geological investigations were conducted to support both a healthy environment and a secure economy. Many investigations during this period await the test of time or are reported elsewhere in this volume; hence, this section is condensed. R. E. Bergstrom, ISGS Chief from 1981 to 1983, was succeeded by M. W. Leighton, son of former chief M. M. Leighton. After M. W. Leighton's retirement in 1994, he was succeeded by W. W. Shilts, from the Geological Survey of Canada.

Environmental Concerns

Coal Emissions

Major efforts during the 1980s and early 1990s were dedicated to finding ways to remove sulfur from coal before combustion. In 1983, R. E. Bergstrom, Neil Shimp, and Carl Kruse helped to establish the Center for Research on Sulfur in Coal (CRSC), later to become the Illinois Clean Coal Institute (ICCI) in Carterville. Funding for research grants came mainly from the Coal Development Board. During the early 1990s, further research on pre-combustion removal of sulfur from coal was limited because of industry's success in developing emission controls for coal pollutants and because of the difficulties encountered in pre-combustion removal of sulfur from coal owing to the complex nature of coal itself. However, of the various methods tried, two cleaning methods showed considerable promise. One, initiated by Latif Khan in the 1980s, represented significant improvements in the froth flotation process for the cleaning of fine coal; that process was licensed for commercialization (Khan 2004). The second coal cleaning method was initiated by Massoud Rostam-Abadi in 1990 in a program to develop activated carbon as a sorbent (Figure 1-47). That program led in 2001 to the successful testing of carbon-based sorbents for removing mercury emissions at a power plant scale (Illinois State Geological Survey 2002, p. 25; Rostam-Abadi, personal communication, September 19, 2003).

Delineation and Protection of Groundwater Resources

Monitoring Procedures and Special Investigations to Manage Wastes. Using temperature surveys, EER surveys, and other studies that successfully mapped both heat and resistivity plumes of groundwater flow from contamination sites, Keros Cartwright and his team, during the 1980s, developed monitoring procedures for landfills and hazardous waste sites (Johnson and Cartwright 1980, Gilkeson and Cartwright 1982, Cartwright 1983, Mehnert et al. 1987) and sampling and testing procedures for monitoring wells (Schuller et al. 1981, Herzog et al. 1988, Herzog 1994). These procedures are now widely used. The movement of contaminants from solid, radioactive, and hazardous waste sites prompted further investigations on ways to contain them. Experimental trench covers were constructed and evaluated at the Sheffield low-level radioactive waste site (Cartwright et al. 1982, Herzog et al. 1982, Johnson et al. 1983b). A field-scale liner was also constructed and evaluated (Cartwright and Krapac 1990, Toupiol et al. 2002, Willingham et al. 2004). Frank et al. (2005) reported that the experiment proved that a liner made of earth materials could meet U.S. EPA regulations.

Groundwater Contamination Potential. Beginning in 1984, ISGS investigators undertook a series of statewide studies in attempts to delineate areas having the

Figure 1-47 Massoud Rostam-Abadi, chemical engineer at the Illinois State Geological Survey laboratory, explains his team's efforts in clean coal technologies, 1995. (Photograph from the Illinois State Geological Survey collection.)

History of Investigations

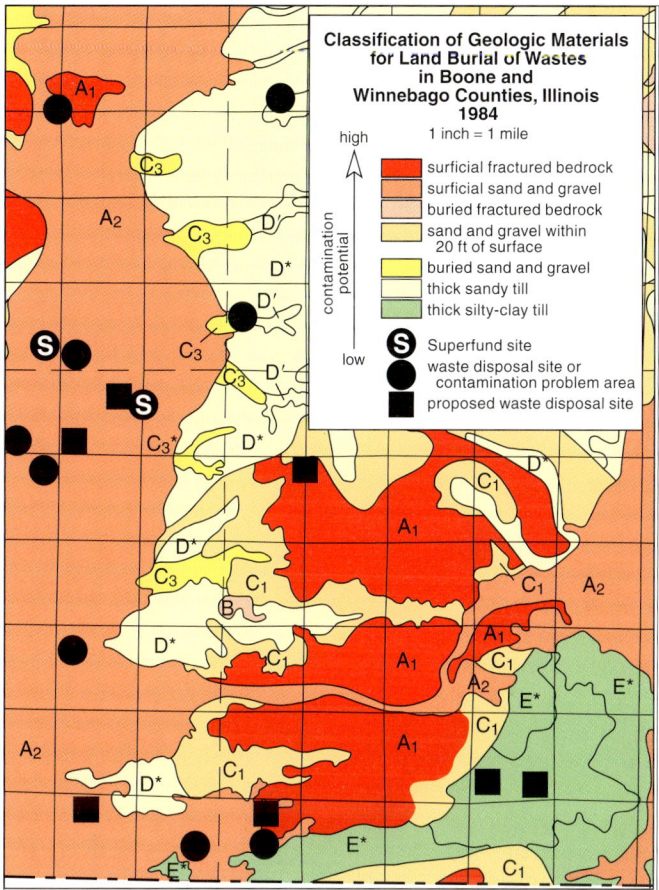

Figure 1-48 This map of Boone and Winnebago Counties marked a milestone in surficial geological mapping. It summarizes the nature of near-surface earth materials and their suitability for land burial of waste. Note the location of some Superfund sites and other waste disposal facilities in the surficial sands and gravels and fractured bedrock, which have higher contamination potential (Berg et al. 1984).

potential to contaminate shallow aquifers (Figure 1-48). Berg and Kempton (1984) also developed a map of the state (1:500,000 scale) showing the potential for contamination of shallow aquifers from the land burial of municipal wastes. This map was followed by a series of maps and reports on aquifer recharge and sensitivity to contamination (Keefer et al. 1990; McKenna and Keefer 1991; Keefer 1993, 1995; Berg 2001).

Environmental Geochemistry

During the 1980s, work on hazardous waste issues intensified. Wilsonville, in Macoupin County, became a field laboratory for ISGS scientists studying the migration of industrial organic chemicals from hazardous waste sites. T. M. Johnson et al. (1983a) reviewed the failure mechanisms that led to the migration of hazardous wastes there. Fracture-prone zones in soils or paleosols were determined to be major factors in the migration of hazardous wastes (Follmer 1984). A report by an 11-member team (Griffin et al. 1984) detailed the findings and summarized the case. During this time, batch-type procedures for estimating soil adsorption of chemicals, later to be adopted by other laboratories, were developed by W. R. Roy et al. (1984, 1991). Significantly, J. B. Risatti (1992), who set up a microbial geochemical laboratory at the ISGS in 1982, the first at any state geological survey, found that anaerobic bacteria naturally degraded the highly toxic compounds polychlorinated biphenyls (commonly known as PCBs).

During the late 1980s and early 1990s, teams of up to 15 scientists from the ISWS and ISGS jointly investigated the degree of pollution in private wells (McKenna et al. 1989, Barnhardt et al. 1992, Schock et al. 1992, Mehnert et al. 1995). Another aspect under study was the impact of pesticides and herbicides at other point sources, namely, agrichemical facilities. A team led by W. R. Roy contributed to the *Agrichemical Facility Site Contamination Study* (Illinois Department of Agriculture and Illinois State Geological Survey 1993). This publication led to state regulations for agrichemical handling at these sites. Studies of agrichemical point sources continued into the twenty-first century (Roy 2001, Roy et al. 2001, Roy and Krapac 2006) and included swine pits (Krapac et al. 2002). These studies were extended to include karst areas (Panno et al. 1996, 2003) (Figure 1-49) and nonpoint sources of contaminants (e.g., floods and field runoff).

Continuing to command attention during this period were airborne toxins from coal combustion (Cahill et al. 1982, Demir et al. 1996), environmental pollution in Illinois lakes and streams (Cahill and Steele 1986, Cahill and Autrey 1987, Cahill 2001) (Figure 1-50), and radon in Illinois homes (Cahill et al. 1994). Investigations also were conducted on other issues involving radioactive materials,

Figure 1-49 Samuel Panno, geochemist, explores Illinois Caverns, Monroe County, in southern Illinois. He is collecting water samples for chemical analysis to determine the amount and extent of groundwater contaminants within this cave in karst country, 1995. (Photograph by Joel M. Dexter.)

Figure 1-50 Richard Cahill, geochemist, samples groundwater seeps for contaminants, Middle Fork watershed, Vermillion County, 1994. (Photograph by Joel M. Dexter.)

such as high readings of uranium in deep groundwater supplies and the disposal of low-level radioactive wastes. More recently, Warner et al. (2003) called attention to elevated concentrations of arsenic in some groundwater supplies. All of these investigations were conducted to better understand the extent, transport, and fate of pollutants in the environment and the nature of radioactive materials in Illinois. These investigations helped determine the role that science could and should play in addressing societal issues related to groundwater protection and remediation, radiation protection, surface water quality, and clean air. Importantly, a number of these studies found evidence of contamination or contamination potential in excess of standards set by the U. S. EPA.

Isotope Geochemistry

Aided by major equipment acquisitions from Governor James Thompson's Build Illinois program in 1986 (Figure 1-51) and by the addition of a highly sophisticated isotope ratio mass spectrometer funded by the National Science Foundation in 1995, ISGS geochemists led by D. D. Coleman, C. L. Liu, and K. C. Hackley assembled an environmental isotope laboratory. The laboratory was capable of analyzing two radiogenic isotopes (radiocarbon and tritium) and five stable isotopes (carbon, oxygen, hydrogen, nitrogen, and sulfur). Since the 1970s, Coleman, Liu, and Hackley, have developed a "tool kit" of isotopes ($\delta^{13}C$ ratios, ^3H, ^{14}C, and $\delta^{18}O$ ratios) for fingerprinting substances. As a result, landfill gas moving in the subsurface can now be recognized (Coleman et al. 1990, 1995), and plumes of landfill gas and leachate can be detected (Hackley et al. 1996). Panno et al. (2001) used $^{15}N/^{18}O$ isotope ratios to determine the sources of nitrate contamination in springs in karst areas.

Geological Hazards and Engineering Geology

Considerable emphasis was also placed on another issue of societal concern, geological hazards, such as landslides (Figure 1-52), karst terrane (Figure 1-53), mine subsidence,

Figure 1-52 Pig barn on unstable land, 1985. (Photograph by Paul DuMontelle.)

Figure 1-51 Jack Liu in the radiocarbon laboratory age-dating water samples, 1989. (Photograph by Joel M. Dexter.)

Figure 1-53 Karst topography in southwestern Illinois presents threats to new construction and potential groundwater contamination problems, circa mid-1990s. (Photograph by Joel M. Dexter.)

flooding, and earthquakes. Landslide investigations included inventories of their occurrence in Illinois (Killey and DuMontelle 1984, Killey et al. 1984, Su and Stohr 1991, Su et al. 1993). Karst studies dealt with areas subject to collapse and contamination (Panno et al. 1994) and their mapping in Illinois (Weibel and Panno 1997).

Mine Subsidence. Mine subsidence studies included reporting case studies (Hunt 1980, Bauer and Hunt 1982), providing information to legislators and homeowners on mine subsidence insurance (DuMontelle et al. 1980, 1981), and quantifying the impacts of subsidence on crops and farmlands (Trent et al. 1996). This latter study, an eight-year effort, was the first of its kind in the United States. It was initiated in 1985 by the Illinois Coal Association and the Illinois Farm Bureau, involved investigators at a number of Illinois institutions, and was carried out under the direction of P. B. DuMontelle and R. A. Bauer, both at the ISGS, in cooperation with the U. S. Bureau of Mines.

Flooding. The Great Flood of 1993 along the Mississippi and Illinois Rivers led to rapid investigations by a team of eight ISGS geologists (Figure 1-54). The scientists studied the geological controls of flooding, levee failure, and flood impacts on erosion, deposition, groundwater recharge and contamination, and ground instability. The report, *The Great Flood of 1993,* by Chrzastowski et al. (1994) won the Geological Society of America's John C. Frye Award in 1997. Coastal erosion and flooding in the Chicago area that accompanied high lake levels in the mid to late 1980s were documented by C. C. Collinson, M. J. Chrzastowski, C. W. Shabica, and others (Figures 1-55 and 1-56). Especially notable was the side-scan sonar investigation of Chicago's shore defense structures by Chrzastowski and Schlee (1988) and Chrzastowski (1989). This study, the first of its kind, identified the causes of underwater damage that were inducing failure in the shore defense structures (Figure 1-57). The results proved useful in Chicago's design and construction of new shoreline defenses.

Additionally, the ISGS provided technical information to the City of Chicago consultants immediately after the

Figure 1-55 Lake Michigan flooding during high lake levels, 1987. (Photograph by Charles Collinson.)

Figure 1-56 Disruptions of storm defense structures testify to the energy of the storm waves along the Lake Michigan shoreline. (Photograph by Joel M. Dexter.)

Figure 1-54 Shown is the impact of the Great Flood of 1993, a major historic flooding event of the Mississippi River. (Photograph by Joel M. Dexter.)

Figure 1-57 Mike Chrzastowski, coastal geologist, examining side-scan sonar recordings, 1990. (Photograph by Joel M. Dexter.)

Chicago flood of 1992 supporting the idea that the flooded tunnels beneath the city could be rapidly dewatered without inducing failure in either the tunnels or the foundations of nearby structures. The rapid flooding resulted during construction work when workers drove a piling into a freight tunnel under the Chicago River. The tunnel was connected to a network of other freight tunnels containing electrical lines under downtown Chicago. Some power failures resulted, and other power was shut off to avoid electrical damage or explosions. Business in the downtown area was quickly shut down as water rushed into 26 basements of downtown buildings, some having their electrical and mechanical systems beneath 35 feet (about 11 m) of water. Economic losses were mounting as some major stores were losing sales of $1 million to $1.5 million per day. Costs were roughly $1 billion in damages and $0.5 billion in lost revenues (R. A. Bauer, personal communication, March 30, 2006). The decision to dewater rapidly after the leak was repaired saved the huge economic losses from mounting still further (Illinois State Geological Survey 1992b, p. 19)

Earthquakes. In 1992, mindful of the potential for another New Madrid-type earthquake, state geologists working with the Central U.S. Earthquake Consortium (CUSEC) of state emergency managers organized to assist CUSEC in addressing selected earthquake preparedness issues. Two examples of their investigations are soil amplification maps for estimating earthquake ground motions (Bauer et al. 2001) and a microzonation map of the Carbondale area classifying earth materials by their potential to amplify ground motions during earthquakes (Illinois State Geological Survey 2002).

Environmental Geology and Infrastructure. In 1989, P. B. DuMontelle and C. J. Stohr made an initial visit to the Illinois Department of Transportation (IDOT) offices in Springfield where DuMontelle made a presentation. The two men left the outline of a proposal with IDOT to provide IDOT with geological, hydrogeological, and environmental information pertinent to IDOT's infrastructure projects throughout Illinois. These efforts were aided by Nicholas Schneider and R. A. Bauer and approved by Chief Morris W. Leighton. The resulting long-term relationship between a state geological survey and a state department of transportation is unique to Illinois. Since then, other state surveys have consulted Illinois to assist them in proposing such programs to their own transportation departments. During the 20 years of its existence, the environmental site assessment program has provided assessments of thousands of road, rail, and airport projects. These assessments are used to assess environmental risk, improve worker and public safety, and avoid delays in construction and the costs associated with such delays. In 1993, through the efforts of

Figure 1-58 Colin Treworgy, coal geologist, shown in 1995 examining logs and maps as part of an investigation of Illinois' minable coal resources. (Photograph by Joel M. Dexter.)

M. V. Miller and N. Schneider, a wetlands component of the program was added to conduct research into the hydrogeology and geochemistry of wetlands that may be impacted by IDOT work and provide input into IDOT decisions on wetland mitigation and restoration. The importance of wetlands in Illinois geology is further discussed in Chapter 21.

Economic Concerns

Coal Resources and Coal By-products

Work begun in the 1970s by C. G. Treworgy and others culminated in an updated assessment of deep-minable coal resources (Treworgy and Bargh 1982). In 1993, the ISGS began a cooperative study with the USGS to determine the amount of coal that is available after considering environmental, legal, and technical restrictions on future mining in Illinois (Figure 1-58). Colin Treworgy et al. (1997) gave the new demonstrated reserve base for coal in Illinois as 105 billion tons (95.3 billion tonnes), the second largest in the nation and the largest for bituminous coal. The accessible or available resource base was estimated to be 70 billion tons (63.5 billion tonnes) and the recoverable reserves to be 38 billion tons (34.5 billion tonnes). Results of the coal availability studies, conducted under the supervision of Treworgy, were published throughout the course of the investigations beginning in 1993; the final report was published by Korose et al. (2003).

In 1995, researchers at the ISGS began to experiment with using fly ash from Illinois coals for manufacturing bricks (Hughes et al. 1995). Testing at bench and pilot plant scales was conducted by M.-I. M. Chou and others in 2001 and 2002 (Figure 1-59). Subsequently, several successful commercial-scale test runs using various fly ash sources were conducted (Chou et al. 2003, 2005), lending support to the commercial viability of manufacturing bricks using fly ash.

Figure 1-59 Mei-In Melissa Chou, Illinois State Geological Survey Coal Section, 2002, examines bricks made using a coal fly ash by-product component in their manufacture. (Photograph by Joel M. Dexter.)

Figure 1-60 Donald Oltz, head of the Illinois State Geological Survey Oil and Gas Section, shown here circa 1992 lecturing on improved oil recovery in Illinois. (Photograph by Joel M. Dexter.)

Oil and Gas Investigations

Crockett and Helpingstine (1996) traced a number of developments in oil and gas exploration during this period. Emphasis was on the development of "plays," the favorable configurations for oil and gas entrapment (Crockett et al. 1988; Whitaker 1988; Howard and Whitaker 1988, 1990; Lasemi and Grube 1995). The most recent discovery in the Geneva Dolomite "fairway" (a favorable play area) used technologies previously reviewed for independent oil and gas producers through the technology transfer workshops sponsored by the Petroleum Technology Transfer Council. This organization was conceived in 1993 in industry meetings with the U.S. Department of Energy and the state geological surveys and was implemented in 1994. The Midwest Regional Office was hosted by the ISGS from 1994 to 2007. New technologies were used by Ceja Petroleum Inc., including three-dimensional seismic technologies, horizontal drilling, and underbalanced drilling (D. G. Morse 2003, personal communication). A large petroleum discovery was made in 2002 in the Geneva Dolomite under Stephen A. Forbes State Park. Initial production there was up to 3,000 barrels a day, the highest production of any well in Illinois during the last 50 years (Seyler et al. 2003).

A network of structural-stratigraphic cross sections was constructed across the Illinois Basin by J. D. Treworgy and S. T. Whitaker and others during the late 1980s and early 1990s to help portray the geology of the Basin and identify and visualize plays (J. D. Treworgy et al. 1997). Descriptions of the modes of occurrence of oil and gas in the Illinois Basin were summarized in a publication produced jointly by the Surveys of Illinois, Indiana, and Kentucky (Zuppann et al. 1988) and in the American Association of Petroleum Geologists Memoir 51 (Leighton et al. 1991). Especially relevant were those articles on oil and gas production and recovery estimates (Mast and Howard 1991); on hydrocarbon reservoir distribution (Howard 1991); on the relationship between types of petroleum traps and different structural and stratigraphic styles (Seyler and Cluff 1991); and on the future potential of the region (Oltz et al. 1991b).

Secondary Recovery and Improved and Enhanced Recovery in Illinois. During the late 1980s, new research was initiated on improved and enhanced oil recovery. The target of this research was mobile oil that had been left behind during field development, especially in the productive Mississippian reservoirs. With a four-year, 4.5-million dollar U.S. Department of Energy contract in 1988, D. F. Oltz, head of the ISGS Oil and Gas Section, assembled a team to investigate this aspect (Oltz et al. 1991a, Oltz 1994) (Figure 1-60). Fourteen fields producing from Mississippian sandstones were studied. The majority of reports on those studies described compartmentalization of the reservoirs and concluded that additional significant mobile oil reserves could be added with infill drilling and/or well-designed waterflood programs. Principal authors of the various reports included H. E. Leetaru, S. T. Whitaker, J. P. Grube, E. O. Udegbunam, B. G. Huff, R. J. Rice, D. F. Oltz, S. K. Sim, J. Xu, B. J. Seyler, and D. G. Morse.

Oil Shales, Oil and Gas Source Rocks, Maturation, and Migration. During the late 1980s, the New Albany Shale was studied for its potential as an unconventional gas source (Cluff et al. 1981, Cluff and Dickerson 1982). These studies were updated by Hasenmueller and Comer (1994) and continued by the Illinois Basin Consortium (2000). Modern work on source rocks and their hydrocarbon potential, especially the principal source rock, the New Albany Shale, was introduced by Barrows and Cluff (1984) and

Chou and Dickerson (1985). Hatch et al. (1991) reviewed the results of their in-depth investigation of the geochemistry of Illinois Basin oils and hydrocarbon source rocks. Cluff and Byrnes (1991) assessed the maturation of organic matter and timing of oil generation. Bethke et al. (1991) presented models of paleohydrologic flow and concluded that long-range migration of oil occurred late in the Basin's history after sediments had fully compacted during a period of regional groundwater flow. The models suggested that both early and late structures could be possible targets for oil exploration and that some might be flushed or only partially full.

Industrial Minerals

Realizing the need to maintain a resource base for the Illinois economy, Mikulic and Goodwin (1985) examined the impact of urban encroachment on dolomite quarries in the Chicago area (Figure 1-61). They noted drastic reductions in the availability of these resources in northeastern Illinois. As an alternative to quarry pits, Baxter (1980, 1989) suggested underground mining of limestone and dolomite as a more efficient use of space.

An insightful review of the stone industry (Mikulic 1989) and an annotated bibliography of nonfuel industrial minerals and metals (Goodwin et al. 1990) provided an extensive reference collection for future investigations. Looking to future development of aggregate resources in Illinois, Mikulic (1990) developed a cross section of the Paleozoic rocks of northeastern Illinois.

During the late 1990s, Lasemi and Norby (2001) found a new application for sequence stratigraphy. They used depositional cycles in the Mississippian Salem Limestone to predict the amount and quality of limestone reserves in quarrying operations in the St. Louis area. During the 1990s, the demand for construction aggregate increased significantly even as coal and oil and gas production declined. As a result, industrial minerals became the leading

Figure 1-61 This photograph, taken circa 1985, is of a tunnel through a wall in Thornton Quarry where quarrying operations were conducted right up to and beneath roads and utility lines. Space for quarries appropriate to the construction industry had become severely restricted in the Chicago area. (Photograph from the Illinois State Geological Survey collection.)

Figure 1-62 Subhash B. Bhagwat, mineral economist, is shown here circa 1998. Bhagwat was honored and recognized internationally as the recipient of two J. William Fulbright awards. The first was as Senior Specialist to lecture in Germany in 2002. The second award, in 2005, honored him as Senior Fellow to work in India on his economics research for the management of water resources in that country. (Photograph by Joel M. Dexter.)

mineral commodity in value—even ahead of coal and petroleum. That lead continued into the twenty-first century (Lasemi et al. 2003, 2008).

Mineral Economics

The transition from statistical reporting to analytical studies continued (Bhagwat and Robare 1982; Bhagwat 1982, 1987, 1993). S. B. Bhagwat (Figure 1-62) looked for ways to reduce costs to make Illinois coal more competitive. As a result of a request by Chief M. W. Leighton, Bhagwat investigated the cost and benefits of geological mapping in Boone and Winnebago Counties (Bhagwat and Berg 1991). Leighton thought that quantitative economic analysis could help make the case for the National Geologic Mapping Act then under consideration by Congress and for increased state support for mapping. In his studies with Berg, Bhagwat found that geological mapping had the potential to reduce risks and costly errors in the siting of landfills, hazardous waste disposal, and septic tanks. Their publication (Bhagwat and Berg 1991) was a breakthrough in the application of economic principles to assess the benefits and costs of geological mapping.

Bhagwat introduced the concept of geological mapping as a public good and found the resulting economic benefits of geological mapping far outweighed the costs (Illinois State Geological Survey 1992a). Under Chief William W. Shilts, Bhagwat and Ipe (2000) expanded the study to include the cost-benefit analysis of detailed geological mapping, using Kentucky's statewide mapping data as a case in point. This benchmark study showed that the value of geological mapping was 25 to 29 times the cost of the mapping.

In another first, during the 1990s, these researchers applied the economic concepts of demand, prices, and scarcity rents to water resources (Ipe and Bhagwat 2003). Additionally, Bhagwat (2003) projected the U.S. energy scenario for high-sulfur coals, such as those in Illinois, to the year 2020. The study projected that, in order to thrive in the future, the Illinois coal industry needed improved productivity, efficiency, and technologies to reduce the existing preference for western coals.

Basic Geological Investigations

Topographic Mapping and Physiographic Expression

In 1985, a revolutionary and popular satellite image map of Illinois was produced (Dahlberg et al. 1985) (Figure 1-63). In 1987, after a century of topographic mapping in Illinois, the USGS, with the cooperation of the ISGS, completed first-time coverage of all of the state with topographic maps at a 1:24,000 scale (Leighton 1988). Following C. C. Abert's (1996) computer-generated *Shaded Relief Map*

Figure 1-63 Donald Luman and Christopher Stohr, with portions of the Illinois satellite image map, 1986. (Photograph by Joel M. Dexter.)

of Illinois (1:500,000 scale), Luman et al. (2003) used USGS Digital Elevation Models (DEMs) to produce *Illinois Surface Topography*, a striking three-dimensional color simulation of a shaded relief map. These new relief maps represent a major evolution from earlier, mostly hand-drawn physiographic landforms or relief maps of Fenneman (1938), Lobeck (1950), and Bier (1980).

Surficial Geology

Glacial Processes. Glacial processes are described in detail in Chapter 12, Quaternary Period. These processes include flooding and formation of glacial lakes, processes of glacial scour and erosion, the power of sub-ice channels, ice-margin deposits, the thickness, weight, and pressures exerted by moving glacial ice, and the isostatic rebound of the Earth's crust following retreat of the glaciers. All of these processes have intriguing histories. Follmer (1996) reviewed the evolution in loess concepts and the processes that formed and modified them.

Detailed Mapping of Surficial Geology. New mapping efforts began during the 1980s by the ISGS, USGS, and others in northern and northeastern Illinois. The identification of paleosols and new subsurface control from borings and water wells, radiocarbon dating, and mineralogical and lithological analyses (Glass and Killey 1987) led to further map revisions in the area (Kempton et al. 1985). Based on new detailed mapping of surficial geology by R. C. Berg, B. B. Curry, L. R. Follmer, A. K. Hansel, W. H. Johnson, J. P. Kempton, and others during the late 1980s and the 1990s, Hansel and Johnson (2000) developed a new version of the statewide map, *Surficial Deposits in Illinois*. During that same period, Hansel and Johnson (1996) proposed a major modification to the basic Quaternary stratigraphic framework established by Frye and Willman

Figure 1-64 William Shilts, who served as the chief of the Illinois State Geological Survey from 1995 to 2008, led an expanded geological mapping program at the Survey. Here he is pictured on a geological science field trip to southern Illinois, circa 1998. (Photograph by Joel M. Dexter.)

Figure 1-65 Ardith Hansel and Richard Berg, Quaternary geologists, examine an exposure of glacial materials at a sand and gravel pit, 1998. (Photograph by Joel M. Dexter.)

Computer Visualization. Truly three-dimensional mapping in a time-stratigraphic sense was initiated in 1993.

The USGS-ISGS joint mapping project for the Champaign Quadrangle successfully developed the methodology for three-dimensional, surficial geological mapping (Soller et al. 1999). During 1995–1996, Chief W. W. Shilts (Figure 1-64) introduced a newer and more vigorous three-dimensional approach that used computer visualizations of the geology to depths of 1,000 feet (305 m). The visualizations were based on information gained through the use of multidisciplinary teams, deep capability drilling rigs, geophysical and seismic information, and digitized geological information from borings and from surface mapping (Figure 1-65). Pilot mapping was conducted in the Villa Grove Quadrangle and Vincennes Quadrangle in 1996 and has since been expanded to other parts of Illinois. The study produced a series of quadrangle maps and three-dimensional illustrations (Lasemi and Berg 2001).

Bedrock Geology

After a lull in bedrock mapping during the 1970s, the USGS began funding a cooperative geological mapping program, COGEOMAP, in 1985 in response to criticism that USGS-supported geological mapping had been steadily decreasing and that much of the nation still was not mapped. The new program supported bedrock mapping in southern Illinois, and the ISGS component of the program was led by W. J. Nelson and others. In 1993, federal funding for COGEOMAP was replaced by STATEMAP, authorized by the 1992 National Cooperative Geologic Mapping Act and supported by new annual appropriations. Data gained from this and other mapping were incorporated into the statewide map, *Bedrock Geology of Illinois* (1:500,000 scale), compiled by Dennis R. Kolata (2005) to commemorate the ISGS centennial year.

Stratigraphy: 1980 to 2005. Sequence stratigraphy and seismic stratigraphy were found to be effective by a number of investigators (e.g., Weibel 1991, 1996; Witzke et al. 1996; Leetaru 1996, 2000; Lasemi et al. 1998; Kolata et al. 1998, 2001; Lasemi and Norby 2001; Nelson et al. 2002). Sequence stratigraphy and seismic stratigraphy were important in determining geological correlations, in understanding the dynamics of stratigraphic development, and in using new stratigraphic results to predict resources of economic value.

Other significant developments include the chemical fingerprinting and correlation of K-bentonites in the Ordovician System of Illinois and adjacent areas (Kolata et al. 1983). K-bentonite beds, occurring over widespread areas, were shown to be gigantic ash falls related to global tectonics (Kolata et al. 1996). In a giant leap forward, Kolata then

in 1970. The new classification showed the intertonguing relationships between glacial deposits (e.g., tills) and proglacial deposits (loess, outwash, lacustrine, and eolian). Importantly, the new classification was based on a dynamic framework of episodes driven by climatic changes.

Three-dimensional Geological Mapping. J. P. Kempton (1981) introduced a new mapping technique for surficial deposits in order to better meet societal needs, especially in identifying areas susceptible to groundwater contamination and in delineating sand and gravel deposits as resources for construction purposes. Initially called three-dimensional mapping, it later became known as stack-unit mapping (Berg and Kempton 1988). E. D. McKay and his staff introduced three-dimensional computer visualization of stacked surficial materials through a series of county assistance programs during the early 1990s.

led an effort to incorporate this information to unravel the history of the regional geology of the two Ordovician carbonate platforms, the Galena and the Lexington Platforms in Illinois and Indiana-Kentucky, respectively, and the Sebree Trough in between them, as the team advanced new sequence stratigraphic concepts in this area.

From the stratal patterns, Kolata et al. (2001) mapped the extent of Middle Ordovician subtidal, sediment-starved platform carbonates, noting the evidence of hardgrounds and flooding or "omission" surfaces. These geologists recognized that the relative sea-level rise outpaced carbonate accumulation at times on both of the platforms as well as sedimentation in the Sebree Trough. Drowning of widespread areas by relative sea-level rise led to starved areas of deposition, not only in the Sebree Trough but in the adjacent platform areas as well. Kolata et al. (2001) thus introduced the new concepts of hardgrounds and flooding surfaces into Illinois geology and described another type of starved basin in the area.

Cyclothems. The cyclothem concept was initially developed by Wanless and Weller (1932). In their review, Langenheim and Nelson (1992) traced the subsequent modification and recognition of the cyclothem concept. During the development of the 1992 review, Weibel (1991), using modern sequence stratigraphy, modified the concept. Weibel recognized that the base of each cycle commenced with the marine transgression surface. This surface provided a more practical mappable horizon than the unconformity that normally was found at the base of a channel sandstone beneath or within the terrestrial portion of each cycle. Thus, Weibel provided a more meaningful basis to understand, describe, and map the stratigraphic succession within these cyclical Pennsylvanian beds.

Paleontology. Working in the 1970s and 1980s using coal ball concretions and spore assemblages, respectively, Phillips and Peppers (1984) described the changing patterns of Pennsylvanian age coal swamp vegetation. They discovered alternating trends of wetter and drier pulses within the Pennsylvanian. Peppers (1996) established correlations between the Illinois Basin and other coal basins inside and outside the United States. Norby (1991) succinctly summarized the most important successions of biostratigraphic zones in Paleozoic strata used for Basin, regional, and intercontinental zonations.

Structural Features and Styles. J. D. Treworgy (1981) published a compendium of structural features in Illinois. Bertagne and Leising (1991) displayed striking reflection seismic lines across the Rough Creek Graben showing its structural style. Notably, Nelson (1991, 1995) applied the concept of structural styles to the various major structural features in Illinois, organizing and cataloguing them and

Figure 1-66 John McBride examining geophysical records for deep structures in the Illinois Basin in 1996. (Photograph by Joel M. Dexter.)

explaining their origins and development (see also Chapter 4, Structural Features).

Syntheses of the Tectonic History of the Illinois Basin. Sloss (1988), in a remarkable compendium article for the *Geology of North America,* documented the tectonic evolution of the craton, including the Illinois Basin, in Phanerozoic time. In 1991, Kolata and Nelson (1991) provided a comprehensive tectonic history of the Basin. Utilizing previous reports and both old and new geophysical and geological data, those researchers described the tectonic history for each of the major Sloss sequences, graphically illustrated the timing of structural activity, and directly related the tectonic history of the Illinois Basin to plate tectonics. M. W. Leighton and Kolata (1991) and Leighton (1996) summarized the profound effects of plate tectonic interactions on the evolution of the supposedly stable interior cratonic basins, including the Illinois Basin. Those geologists noted the reach of stress fields and the impact during divergence, convergence, and collision of tectonic plates.

Pratt et al. (1989) and Heigold (1991) piqued the interest of geologists in the underpinnings of the Illinois Basin with their articles on major Proterozoic basement features in the eastern midcontinent and in the Illinois Basin, respectively. Heigold's interpretation of new, high-quality, deep seismic data released by the United Seismic Data Brokers began a major revolution in effort and thought devoted to deep seismic reflection data in the Illinois Basin. Pratt et al. (1992) described widespread Precambrian layered sequences in the U. S. midcontinent, naming one the Centralia sequence.

John C. McBride, following Heigold at the ISGS, proceeded aggressively to acquire and reprocess additional industry seismic reflection data, ultimately building a network of seismic lines covering large portions of the Illinois Basin (Figure 1-66). The new data affected the understanding of the deep subsurface, much as the acquisition of geophysical well logs for the public domain that began in 1942 to 1944 affected subsurface mapping and groundwater investigations. Kolata and Hildenbrand (1997) brought together the results from a number of the investigations on the underpinnings of the Illinois Basin and on neotectonism. During the 1990s and early 2000s, investigators detected high-angle reverse faults in Precambrian basement rocks that were similar in style to structures produced by the rise of the ancestral Rocky Mountains (McBride and Nelson 1999). The researchers noted multiple episodes of fault displacements under a variety of stress fields (Marshak et al. 2003). Additionally, a new idea was proposed suggesting the existence of a Proterozoic caldera in east-central Illinois and west-central Indiana that, along with the New Madrid Rift System, shaped the configuration of the Illinois Basin by the cumulative effects of subsidence (McBride et al. 2003).

Neotectonics and Seismic Risk. After nuclear power plant construction was suspended in the United States in 1984, Buschbach (1986) issued a final result summary for the National Research Council-sponsored studies of seismic risk. Scientific curiosity about earthquakes and seismic risk, however, aroused by work in the 1970s and early 1980s, generated new studies on neotectonics, led mainly by W. J. Nelson's mapping efforts in southern Illinois (Figure 1-67). These studies, plus the ideas stimulated by the articles in the American Association of Petroleum Geologists' Memoir 51 (Leighton et al. 1991) led to further studies that covered both neotectonics and the nature of crust beneath the Phanerozoic section—the "underpinnings" of the Illinois Basin. As a part of these new studies, Nelson et al. (1997, 1999) and McBride and Nelson (2001) carried out extensive field mapping and high resolution seismic studies in the extension of the Mississippi Embayment into southern Illinois. These studies were conducted to better determine the nature of the faulting and its timing in that area. Results suggested that faulting of the Paleozoic bedrock and younger sediments of the northern Mississippi Embayment are more pervasive and younger than previously thought.

Groundwater Resources

Buried Bedrock Valleys

Notable contributions during this period were those by Kempton et al. (1991), who updated knowledge of the Mahomet Bedrock Valley, and by Herzog et al. (1994a), who issued an updated map (1:500,000 scale) of the buried bedrock surface of Illinois, based on the data accumulated since Horberg's (1950) original mapping (Figure 1-68).

Regional Groundwater Studies

The last ISGS reports on groundwater resources before the modern era were completed in 1966. Fifteen years later,

Figure 1-67 John Nelson, Illinois State Geological Survey geologist, takes field notes while mapping in southern Illinois during the 1990s. (Photograph by Joel M. Dexter.)

Figure 1-68 Beverly L. Herzog taking field notes, 1992. (Photograph by Joel M. Dexter.)

interest was renewed in regional studies to identify and delineate additional groundwater resources. The scattering of reports during the 1980s included Cooperative Report 8 (Kempton et al. 1982), which evaluated sand and gravel aquifers in east-central Illinois. During the 1990s, work focused on groundwater resources began in earnest with the publication of three cooperative reports on McLean and Tazewell Counties (Kempton and Visocky 1992, Herzog et al. 1994b, 1994c; Wilson et al. 1998), another on the confluence of the Mackinaw and Mahomet Bedrock Valleys (Wilson et al. 1995), and a report on Danville's groundwater supply (Larson et al. 1997). Other investigations of groundwater resources dealt with northwestern Illinois (T. H. Larson et al. 1993, D. R. Larson et al. 1995) and central Illinois (D. R. Larson et al. 2003).

Groundwater Modeling

University of Illinois researcher Craig Bethke (1985) and Bethke et al. (1991) significantly advanced groundwater flow modeling. He developed regional quantitative computer models of both compaction- and topographic-driven flow in the Illinois Basin. He calculated the strength of groundwater flow under both conditions and found that topographic relief provides a much stronger drive than compaction forces.

SUMMARY OF PAST INVESTIGATIONS

In considering the history of research on Illinois geology and the total contributions of so many investigators, one is struck by the number and quality of new earth science concepts, ideas, and methods that have proven essential to the fundamental understanding of Illinois geology. Their application to the betterment of the state's economy and environmental well-being have evolved over lengthy periods of time. These major contributions, as reported on the previous pages, stand out:

1. The recognition over time of the major bounding unconformities and the stratigraphic sequences they bound, the value of sequence stratigraphy in deciphering the geological events that shaped and controlled them, and the application of these concepts to predict the occurrence of the state's natural resources.
2. The development of structural concepts from the early days of geological investigations to the present, leading to the current understanding of the variety of structural styles found in the Illinois Basin and to evolving syntheses of the Basin's tectonic history.
3. The ideas and concepts of the many and varied processes that yielded multiple types of glacial deposits—beginning with the early efforts in glacial geology during the mid-1880s and continuing to this day—and the resultant mapping that helped locate industrial mineral and groundwater resources.
4. The concepts that led to the mapping of favorable areas for oil and gas occurrence and the definition of the many oil and gas plays in the Illinois Basin.
5. The understanding of the origin of Illinois' low-sulfur coals and the methods employed to map and predict them.
6. The concepts, ideas, and geophysical methods that led to the understanding, development, and exploration of the state's groundwater resources.
7. The notable concepts of cyclic deposition and the origin of cyclothems that occurred from the early 1900s to the 1990s, thus aiding not only the mapping of Illinois' many coal beds, but also the understanding of their origin and development.
8. The use of geochemistry to help characterize the nature of the state's mineral resources and identify new uses for them, beginning in the 1930s and then expanding to methods to fingerprint the sources of gases and to detect and locate contaminants in Illinois water supplies, soils, and subsurface.
9. The growth of environmental geology from the 1960s on to help resolve issues of land use and planning and to find ways to help protect the environment and the earth and water resources.
10. The numerous successful applications of engineering geology, commencing with information to support the building of roads, highways, and dams in the 1920s and evolving into studies of foundation conditions for building projects, both large and small, as well as enhancing the understanding and mitigation of the impacts from flooding, mine subsidence, earthquakes, landslides, and other earth hazards.

Many other significant concepts, ideas, and methods that are not listed here were introduced and used in the many and varied geological investigations. Not to be forgotten are contributions of individual researchers and organizations not mentioned within this chapter because of space limitations. Also to be remembered are the major steps taken to build the extensive databases required for ongoing geological investigations. The need for data preservation was recognized early on, as the collections of geological and paleontological samples were introduced. These collections were followed by the collection of well logs, field data, photographs, water and geochemical samples, and geophysical data (Figure 1-69). These collections still find major uses in the conduct of continuing and far-ranging geological, geochemical, geophysical, and engineering investigations. Means to store, maintain, and protect these collections of basic data have been implemented over time. The need and value of publishing the results of investigations and placing them in the public domain—and the need for funds to do so—were recognized early on by investigators and administrators alike. With the advent of computers, ways were found to enhance the storage and preservation of basic data in readily accessible forms. Still later, Geographic Information Systems (GIS) have been developed to allow further manipulation of both surface and subsurface data for comparative purposes and further analyses. Computerization has improved the portrayal and visualization of geological results in two and three dimensions.

Overall, one is humbled by the impressive quantity and quality of the thought and effort that have gone into unraveling the state's geology and into the means of its portrayal, retention, and display.

THE NEXT 100 YEARS: WHAT DOES HISTORY SUGGEST?

Over the past 100 years, the manner in which geological investigations have been conducted has become more complex, moving increasingly from studies conducted predominantly by single individuals to multidisciplinary investigations. Additionally, the investigations have become more multi-institutional and more team-oriented. Funding sources have changed radically. More changes can be expected in the future in the ways studies are conducted and funded. New efforts in combining resource and environmental investigations from their conception can be expected with increasing participation by the public, government, and private sectors. Yet there will always be a role for the individual and for individual thought.

More effort can be expected on the management of carbon and hydrogen cycles from "cradle to grave" as those carbon-based energy resources are unearthed, developed, and utilized. Coal bed methane and other unconventional gas sources (e.g., the New Albany Shale) may be exploited to degrees that are unknown today. A mix of energy sources will begin to replace conventional oil, gas, and coal sources. Already experiments are under way to sequester carbon dioxide emissions from coal-fired power plants and ethanol plants by pumping the gas deep underground into saline aquifers or coal seams or into oil and gas reservoirs. The goal is to design, construct, and operate new, highly efficient coal-fired power plants that will produce hydrogen for fuel cells, generate electricity, and reduce greenhouse

Figure 1-69 Anne Faber, Geological Records Unit, helps maintain the collection of massive amounts of subsurface well log data used by industry, researchers, and the general public, 1999. (Photograph by Joel M. Dexter.)

gases and toxic emissions to the atmosphere: a near zero emissions power plant. New sorbents and nanotechnology are visualized as playing major roles in limiting toxic emissions. Nanotechnology, proliferating by leaps and bounds, may find application in as yet unforeseen developments in utilizing Illinois' vast resources. New technologies to deal with coal wastes are being developed, taking advantage of mine mouth power plants to produce useful by-products such as bricks and ceramics from mixtures of coal fly ash and natural clays in the area. If confidence is restored in nuclear power generation, that energy source may be revived.

Increased use of three-dimensional seismic studies in Illinois may lead to additional oil and gas discoveries and will provide a better understanding of the subsurface and of the heterogeneities in oil and gas reservoirs. Such knowledge should lead to further improved oil recoveries. By the end of the next 100 years, three-dimensional mapping of glacial deposits of the state, aided by high resolution shallow seismic data, should be nearly completed, adding significantly to the knowledge about these materials. This information should be able to assist land use planning, including the siting of facilities (airports, highways and roads, high-speed transit systems, buildings, factories, schools, utilities); the location of resources needed for their construction; and location and protection of groundwater resources. New studies will be made of other contaminants in surface waters and groundwater that are now entering or will be entering the systems as a result of the use of drugs, hormones, or other chemicals with potential toxicological effects. Strategies for managing and pricing water resources will evolve over time.

Scientific investigations leading to new advances will develop from ultra-deep drill holes in the vast, deep areas of the Illinois Basin that have not yet been penetrated by the drill bit—the last frontier for exploration in Illinois. Their locations will be selected from high-quality, high-resolution, deep seismic reflection data. The drill holes will serve as linchpins to tie together regional seismic lines to enhance the three-dimensional interpretations of the underpinnings of the Illinois Basin. Additionally, underground development of space will occur more quickly as challenges to prime farmland become even greater, as urban encroachment continues to overtake and limit existing quarries, and as the cost of transporting construction aggregate continues to rise. New, high-speed methods of data acquisition, manipulation, visualization, and communication will allow greater and faster access by investigators, the public, government, and private sectors to technological developments. These are but a few of the many possible scenarios that may occur in the next 100 years. Ongoing work at the end of the next century will somewhat resemble today's investigations, just as today's studies resemble some of the investigations being carried out 100 years ago. But also, just as we experience looking back on the previous century, in 2105, when we look back on today, many new avenues will be apparent that could not have been foreseen "back then."

References

Abert, C. C., 1996, Shaded relief map of Illinois: Illinois State Geological Survey, Illinois Map 6, 1:500,000.

Alden, W. C., 1902, Description of the Chicago District, Illinois-Indiana: Chicago Folio (Riverside, Chicago, Des Plaines, and Calumet Quadrangles): Washington, D.C., U.S. Geological Survey, Geological Atlas Folio 81, 14 p.

Andrews, E. B., 1867, Observations upon the glacial drift beneath the bed of Lake Michigan, as seen in the Chicago tunnel: American Journal of Science and Arts, ser. 2, v. 43, no. 127, p. 75–77.

Andrews, E. B., 1869, On some remarkable relations and characters of the western boulder drift: American Journal of Science and Arts, ser. 2, v. 48, art. 18, p. 172–179.

Armon, W. J., and O. W. Rees, 1960, Chemical evaluation of Illinois oil shales: Illinois State Geological Survey, Circular 307, 22 p.

Atherton, E., 1971, Tectonic development of the Eastern Interior Region of the United States, in Background materials for symposium on future petroleum potential of NPC Region 9 (Illinois Basin, Cincinnati Arch, and northern part of Mississippi Embayment): Illinois State Geological Survey, Illinois Petroleum 96, p. 29–43.

Bain, H. F., 1904, Lead and zinc deposits of Illinois: Washington, D.C., U.S. Geological Survey, Bulletin 225, p. 202–207.

Bain, H. F., 1905, The fluorspar deposits of southern Illinois: Washington, D.C., U.S. Geological Survey, Bulletin 255, p. 505–511.

Bain, H. F., 1906, June 20, 1906 Letter of Transmittal of Bulletin 2, in W. S. Blatchley, The petroleum industry of southeastern Illinois: Illinois State Geological Survey, Bulletin 2, 109 p.

Bain, H. F., 1931, The initiation of the State Geological Survey: Illinois State Geological Survey, Bulletin 60, p. 29–33.

Barnhardt, M. L., E. Mehnert, C. Ray, and S. C. Schock, eds., 1992, Characterization of the study areas for the pilot study: Agricultural chemicals in rural private wells in Illinois: Illinois State Geological Survey and Illinois State Water Survey, Cooperative Groundwater Report 15, 114 p.

Barrows, M. H., and R. M. Cluff, 1984, New Albany Shale Group (Devonian-Mississippian) source rocks and hydrocarbon generation in the Illinois Basin, in G. Demaison and R. J. Murris, eds., Petroleum geochemistry and basin evaluation: American Association of Petroleum Geologists, Memoir 35, p. 111–138.

Bastin, E. S., 1931, The fluorspar deposits of Hardin and Pope Counties, Illinois: Illinois State Geological Survey, Bulletin 58, 116 p.

Bauer, R. A., and S. R. Hunt, 1982, Profile, strain, and time characteristics of subsidence from coal mining in Illinois, in S. S. Peng and M. Harthill, eds., Proceedings, Workshop on Surface Subsidence Due to Underground Mining, 1981: Morgantown, West Virginia, West Virginia University, Department of Mining and Engineering, p. 207–217.

Bauer, R. A., J. Kiefer, and N. Hester, 2001, Soil amplification maps for estimating earthquake ground motions in the central U.S.: Engineering Geology, v. 62, p. 7–17.

Baxter, J. W., 1980, Factors favoring expanded underground mining of limestone in Illinois: Mining Engineering, v. 32, no. 10, October 1980, p. 1497–1504.

Baxter, J. W. 1989, Possible underground mining of limestone and dolomite in central Illinois, in Proceedings of the 23rd Forum on the Geology of Industrial Minerals, held May 11–14, 1987, North Aurora, Illinois: Illinois Geological Survey, Illinois Mineral Notes 102, p. 21–28.

Bays, C. A., and S. H. Folk, 1942, Geophysical surveys in water wells: Illinois Well Driller, v. 12, no. 4, p. 7–8.

Bell, A. H., 1938, Current developments in oil and gas: Illinois State Geological Survey, Circular 23-B, p. 1–12.

Bell, A. H., 1941, Role of fundamental geologic principles in the opening of the Illinois Basin: Economic Geology, v. 36, no. 8, p. 774–785.

Bell, A. H., and F. Squires, 1932, Preliminary summary of results obtained from a survey of re-pressuring operations in the southeastern Illinois oilfield: Illinois State Geological Survey, Illinois Petroleum 23, 22 p.

Bement, A., 1910, The Illinois coal field: Illinois State Geological Survey, Bulletin 16, p. 182–202.

Berg, R. C., 2001, Aquifer sensitivity classification for Illinois using depth to uppermost aquifer material and aquifer thickness: Illinois State Geological Survey, Circular 560, 14 p.

Berg, R. C., and J. P. Kempton, 1984, Potential for contamination of shallow aquifers from land burial of municipal wastes: Illinois State Geological Survey, Land Use Planning Map, 1:500,000.

Berg, R. C., and J. P. Kempton, 1988, Stack-unit mapping of geologic materials in Illinois to a depth of 15 meters: Illinois State Geological Survey, Circular 542, 23 p.

Berg, R. C., J. P. Kempton, and A. N. Stecyk, 1984, Geology for planning in Boone and Winnebago Counties: Illinois State Geological Survey, Circular 531, 69 p.

Bergstrom, R. E., 1956, Groundwater geology in western Illinois, north part—A preliminary geologic report: Illinois State Geological Survey, Circular 222, 24 p.

Bergstrom, R. E., 1980, Illinois State Geological Survey: Its history and activities: Illinois State Geological Survey, Educational Series 12, 37 p.

Bergstrom, R. E., 1992, Environmental geology and early contributions by Fritiof Fryxell, in D. A. Schroeder and R. C. Anderson, eds., Earth interpreters: Augustana College Library Publications 36, p. 35–47.

Bergstrom, R. E., J. W. Foster, L. F. Selkregg, and W. A. Pryor, 1955, Groundwater possibilities in northeastern Illinois—A preliminary geologic report: Illinois State Geological Survey, Circular 198, 24 p.

Bergstrom, R. E., and A. J. Zeizel, 1957, Ground-water geology in western Illinois, south part—A preliminary geologic report: Illinois State Geological Survey, Circular 232, 28 p.

Berkson, A., and M. D. Wiant, eds., 2001, Discover Illinois archaeology: Urbana, Illinois, Illinois Association for Advancement of Archaeology and Illinois Archaeological Survey Joint Publication, 26 p.

Bertagne, A. J., and T. C. Leising, 1991, Interpretation of seismic data from the Rough Creek Graben of western Kentucky and southern Illinois, in M. W. Leighton, D. R. Kolata, D. F. Oltz, and J. J. Eidel, eds., Interior cratonic basins: Tulsa, Oklahoma, American Association of Petroleum Geologists, Memoir 51, p. 199–208.

Bethke, C. M., 1985, Compaction-driven groundwater flow and heat transfer in intracratonic sedimentary basins and genesis of the Upper Mississippi Valley mineral district: University of Illinois at Urbana-Champaign, Ph. D. dissertation, 125 p.

Bethke, C. M., J. D. Reed, and D. F. Oltz, 1991, Long-range petroleum migration in the Illinois Basin, in M. W. Leighton, D. R. Kolata, D. F. Oltz, and J. J. Eidel, eds., Interior cratonic basins: American Association of Petroleum Geologists, Memoir 51, p. 455–472.

Bhagwat, S. B., 1982, Cost of surface mining coal in Illinois, in D. H. Graves, ed.: Proceedings, 1982 Symposium of Surface Mining Hydrology, Sedimentology and Reclamation, December 5–10, 1982, Lexington, Kentucky, University of Kentucky, p. 111–114.

Bhagwat, S. B., 1987, Analysis of coal surface mining equipment in Illinois: Transactions of the Society of Mining Engineers, v. 280, p. 2015–2019.

Bhagwat, S. B., 1993, What influenced the price of crude oil in the U.S.?—An analysis of the 1971–1990 period: Energy modeling: Optimizing information and resources: Proceedings of the Institute of Gas Technology Conference, Chicago, Illinois, June 7–8, 1993, 9 p. (Illinois State Geological Survey, Reprint 1993-M.)

Bhagwat, S. B., 2003, High-sulfur coals in the U.S. energy scenario in 2020: Society of Mining Engineering, Inc. (SME) Annual Meeting, Cincinnati, Ohio, 2003, SME Preprint #03-159, 11 p.

Bhagwat, S. B., and R. C. Berg, 1991, Benefits and costs of geologic mapping programs in Illinois: Case study of Boone and Winnebago Counties and its statewide applicability: Illinois State Geological Survey, Circular 549, 40 p.

Bhagwat, S. B., and V. C. Ipe, 2000, Economic benefits of detailed geologic mapping to Kentucky: Illinois State Geological Survey, Special Report 3, 39 p.

Bhagwat, S. B., and P. Robare, 1982, Cost of underground coal mining in Illinois: Illinois State Geological Survey, Illinois Mineral Notes 84, 14 p.

Bier, J. A., 1980, Landforms of Illinois [map]: Illinois State Geological Survey, 1:1,000,000.

Blatchley, R. S., 1913, The oil fields of Crawford and Lawrence Counties: Illinois State Geological Survey, Bulletin 22, 442 p.

Blatchley, W. S., 1906, The petroleum industry of southeastern Illinois: Illinois Geological Survey, Bulletin 2, 109 p.

Bradbury, J., 1817, Travels in the interior of America, in the years 1809, 1810, and 1811; including a description of Upper Louisiana, together with the states of Ohio, Kentucky, Indiana and Tennessee, with the Illinois and Western Territories: London, Sherwood, Neely, and Jones, 364 p. plus map.

Bradbury, J. C., 1955, Geochemical prospecting in the zinc-lead district of northwestern Illinois: Illinois State Geological Survey, Report of Investigation 179, 11 p.

Bradbury, J. C., R. M. Grogan, and R. J. Cronk, 1956, Geologic structure map of the northwestern Illinois zinc-lead district: Illinois State Geological Survey, Circular 214, 7 p.

Bradley, W. F., 1945, Diagnostic criteria for clay minerals: American Mineralogist, v. 30, no. 11–12, p. 704–713.

Bretz, J H., 1939, Geology of the Chicago region, Part 1, General: Illinois State Geological Survey, Bulletin 65, 118 p.

Bretz, J H., 1955, Geology of the Chicago region, Part 2, The Pleistocene: Illinois State Geological Survey, Bulletin 65, 132 p. plus maps.

Bristol, H. M., 1974, Silurian pinnacle reefs and related oil production in southern Illinois: Illinois State Geological Survey, Illinois Petroleum 102, 98 p.

Bristol, H. M., and T. C. Buschbach, 1971, Structural features of the Eastern Interior Region of the United States, in Background materials for Symposium on Future Petroleum Potential of NPC Region 9 (Illinois Basin, Cincinnati Arch, and northern part of Mississippi Embayment, Champaign, Illinois, March 11–12, 1971): Illinois State Geological Survey, Illinois Petroleum 96, p. 21–28.

Bristol, H. M., and R. H. Howard, 1971, Paleogeologic map of the sub-Pennsylvanian Chesterian (Upper Mississippian) surface in the Illinois Basin: Illinois State Geological Survey, Circular 458, 14 p.

Buhle, M. B., and J. E. Brueckmann, 1964, Electrical earth resistivity surveying in Illinois: Illinois State Geological Survey, Circular 376, 51 p.

Buschbach, T. C., 1961, The morphology of the sub-St. Peter surface of northeastern Illinois: Transactions of the Illinois State Academy of Science: v. 54, no. 1–2, p. 83–89.

Buschbach, T. C., 1971, Stratigraphic setting of the Eastern Interior Region of the United States, in Background materials for Symposium on Future Petroleum Potential of NPC Region 9 (Illinois Basin, Cincinnati Arch, and northern part of Mississippi Embayment): Illinois State Geological Survey, Illinois Petroleum 96, p. 3–20.

Buschbach, T. C., 1986, New Madrid seismotectonic program: U.S. Nuclear Regulatory Commission, NUREG/CR 4632, 61 p.

Buschbach, T. C., and D. C. Bond, 1974, Underground storage of natural gas in Illinois, 1973: Illinois Petroleum 101, 71 p.

Buschbach, T. C., R. T. Cyrier, and G. E. Heim, 1982, Geology and deep tunnels in Chicago, in R. F. Legget, ed., Geology under cities: Reviews in Engineering Geology, v. 5, p. 41–54.

Buschbach, T. C., and G. E. Heim, 1972, Preliminary geologic investigations of rock tunnel sites for flood and pollution control in the greater Chicago area: Illinois State Geological Survey, Environmental Geology Notes 52, 35 p.

Cady, G. H., 1920, Low-sulphur coal in Illinois: Transactions of the American Institute of Mining and Metallurgical Engineers, v. 63, p. 641–643; with discussion, p. 644–648.

Cady, G. H., 1921, Coal resources of District IV: Illinois State Geological Survey, Illinois Coal Mining Investigations, Bulletin 26, 247 p.

Cady, G. H., 1935, Classification and selection of Illinois coals: Illinois State Geological Survey, Bulletin 62, 354 p.

Cady, G. H., 1952, Minable coal reserves of Illinois: Illinois State Geological Survey, Bulletin 78, 138 p.

Cahill, R. A., 2001, Final report: Assessment of sediment quality in Peoria Lake: Results from the chemical analysis of sediment core samples collected in 1998, 1999, and 2000: Illinois State Geological Survey, Open File Report 2001-4, 58 p.

Cahill, R. A., and A. D. Autrey, 1987, Measurement of ^{210}Pb, ^{137}Cs, organic carbon and trace elements in sediments of the Illinois and Mississippi rivers, in International Symposium on Nuclear Analytical Chemistry: Proceedings: Journal of Radioanalytical and Nuclear Chemistry, v. 110, no. 1, p. 197–205.

Cahill, R. A., R. H. Shiley, and N. F. Shimp, 1982, Forms and volatilities of trace and minor elements in coal: I, A comparison of three pyrolysis units: Illinois State Geological Survey, Environmental Geology Notes 102, 22 p.

Cahill, R. A., and J. D. Steele, 1986, ^{137}Cs as a tracer of recent sedimentary processes in Lake Michigan, in Proceedings, 4th International Symposium on Paleolimnology, 1985: Hydrobiologia, v. 143, no. 1, p. 29–35.

Cahill, R. A., J. D. Steele, and L. R. Smith, 1994, Indoor radon exposure, in Technical Report of the Critical Trends Assesment Project, v. 6, Sources of Environmental Stress, p. 61–68, ILENR/RE-EA-94/05(6).

Cartwright, K., 1968, Temperature prospecting for shallow glacial and alluvial aquifers in Illinois: Illinois State Geological Survey, Circular 433, 41 p.

Cartwright, K., 1983, Detecting and monitoring contaminated groundwater, in Papers for and Summary of a Workshop on Groundwater Resources and Contamination in the United States, March 14 and 15, 1983: Washington, D.C., National Science Foundation, Division of Policy Research and Analysis, Report 83-12, p. 173–224.

Cartwright, K., R. A. Griffin, and R. H. Gilkeson, 1976, Migration of landfill leachate through unconsolidated porous media, in Advances in groundwater hydrology: Proceedings of Symposium held at Chicago, Illinois, 1976, American Water Resources, p. 215–227.

Cartwright, K., R. A. Griffin, and R. H. Gilkeson, 1977, Migration of landfill leachate through glacial tills: Ground Water, v. 15, no. 4, p. 294–305.

Cartwright, K., S. J. Klein, T. M. Johnson, A. G. Devine, and B. L. Herzog, 1982, A study of trench covers to minimize infiltration at waste disposal sites: Task III report: Review of practices used in trench cover design and construction: Report prepared for U.S. Nuclear Regulatory Commission, Division of Waste Management, Urbana, Illinois: Illinois State Geological Survey, 92 p.

Cartwright, K., and I. G. Krapac, 1990, Construction and performance of a long-term earthen liner experiment, in R. Bonaparte, ed., Waste containment systems: Construction, regulation, and performance (GSP 26): Proceedings of a Symposium in conjunction with the American Society of Civil Engineers National Convention, 1990, p. 135–155.

Cartwright, K., and M. R. McComas, 1969, Geophysical surveys in the vicinity of landfills in northeastern Illinois, Appendix F, in Hydrogeology of solid waste sites in northeastern Illinois: An interim report on a Solid Waste Demonstration Project, U.S. Public Health Service, Bureau of Solid Waste Management, p. 114–128.

Cartwright, K., and F. B. Sherman Jr., 1972, Electrical resistivity surveying in landfill investigations: Proceedings of the 10th Annual Engineering Geology and Soils Engineering Symposium, p. 77–92.

Chamberlin, T. C., 1877, Geology of eastern Wisconsin: Geology of Wisconsin, Survey of 1873–1877: Wisconsin Geological Survey, v. 2, part 2, p. 91–405.

Chamberlin, T. C., 1878, On the extent and significance of the Wisconsin kettle moraine: Transactions of the Wisconsin Academy of Science, Arts and Letters, v. 4, p. 201–234.

Chamberlin, T. C., 1882, Preliminary paper on the terminal moraine of the second glacial epoch: Washington, D.C., U.S. Geological Survey, 3rd Annual Report, p. 291–402.

Chamberlin, T. C., 1894, Glacial phenomena of North America, in J. Geikie, ed., The Great Ice Age (3rd ed.): New York, D. Appleton and Co., p. 724–775.

Chamberlin, T. C., 1895, The classification of American glacial deposits: Journal of Geology, v. 3, p. 270–277.

Chamberlin, T. C., and R. D. Salisbury, 1885, Preliminary paper on the Driftless Area of the upper Mississippi Valley: Washington, D.C., U.S. Geological Survey, 6th Annual Report, p. 199–322.

Chicago Census Records, 2002, Chicago census facts. http://www.censusrecords.net/cites/Chicago_census.htm.

Chou, M.-I. M., S.-F. J. Chou, V. Patel, J. W. Stucki, and F. Botha, 2003, Commercialization of fired bricks with fly ash from Illinois coals (Phase IV); Final report to the Illinois State Department of Commerce and Community Affairs, Illinois Coal Development Board, and Illinois Clean Coal Institute, 2003: Illinois State Geological Survey, 28 p.

Chou, M.-I. M., S.-F. J. Chou, V. Patel, J. W. Stucki, and F. Botha, 2005, Commercial production of fired bricks with Illinois coal fly ash and bottom ash (Phase V); Final report to the Illinois State Department of Commerce and Community Affairs, Illinois Coal Development Board, and Illinois Clean Coal Institute, 2005: Illinois State Geological Survey, 29 p.

Chou, M.-I. M., and D. R. Dickerson, 1979, Pyrolysis of Eastern gas shale: Effects of temperature and atmosphere on the production of light hydrocarbons, in H. Barlow, ed., Proceedings of the Third Eastern Gas Shales Symposium, October 1–3, 1979, p. 211–223.

Chou, M.-I. M., and D. R. Dickerson, 1985, Organic geochemical characterization of the New Albany Shale Group in the Illinois Basin: Organic Geochemistry, v. 8, no. 6, p. 413–420.

Chrzastowski, M. J., 1989, Sidescan-sonar examination of deteriorated revetments and bulkheads along Chicago's lake front, in O. T. Mazoon, H. Converse, D. Miner, L. T. Tobin, and D. Clark, eds., Coastal Zone '89: Proceedings of the Sixth Symposium on Coastal and Ocean Management, American Society of Civil Engineers, Charleston, South Carolina, July 11–14, 1989, p. 3931–3944.

Chrzastowski, M. J., M. M. Killey, R. A. Bauer, P. B. DuMontelle, A. L. Erdmann, B. L. Herzog, J. M. Masters, and L. R. Smith, 1994, The Great Flood of 1993: Geologic perspectives on flooding along the Mississippi River and its tributaries in Illinois: Illinois State Geological Survey, Special Report 2, 45 p.

Chrzastowski, M. J., and J. S. Schlee, 1988, Preliminary sidescan-sonar investigation of shore-defense structures along Chicago's northside lake front: Wilson Avenue groin to Ohio Street beach: Illinois State Geological Survey, Environmental Geology Notes 128, 32 p.

Clark, G. L., R. E. Grim, and W. F. Bradley, 1937, Notes on the identification of minerals in clays by x-ray diffraction: Zeitschrift für Krystallographie und Mineralogie, v. 96, p. 322–324.

Clark, S. K., and J. S. Royds, 1948, Structural trends and fault systems in Eastern Interior Basin: American Association of Petroleum Geologists Bulletin 32, no. 9, p. 1728–1749.

Cluff, R. M., and A. P. Byrnes, 1991, Lopatin analyses of maturation and petroleum generation in the Illinois Basin, in M. W. Leighton, D. R. Kolata, D. F. Oltz, and J. J. Eidel, eds., Interior cratonic basins: Tulsa, Oklahoma, American Association of Petroleum Geologists, Memoir 51, p. 425–454.

Cluff, R. M., and D. R. Dickerson, 1980, Natural gas potential of the New Albany Shale Group (Devonian-Mississippian) in southeastern Illinois: Society of Petroleum Engineers and U.S. Department of Energy Symposium on Unconventional Gas Recovery, Pittsburgh,

PA, SPE/DOE Paper 8924, p. 21–28.

Cluff, R. M., and D. R. Dickerson, 1982, Natural gas potential of the New Albany Shale Group (Devonian-Mississippian) in southeastern Illinois: Society of Petroleum Engineers Journal, v. 22, no. 2, p. 291–300.

Cluff, R. M., M. L. Reinbold, and J. A. Lineback, 1981, The New Albany Shale Group of Illinois: Illinois State Geological Survey, Circular 518, 83 p.

Coleman, D. D., 1976, Isotopic characterization of Illinois natural gas: Urbana-Champaign, University of Illinois, Department of Geology, Ph.D. dissertation, 175 p.

Coleman, D. D. 1979, The origin of drift gas as determined by radiocarbon dating, in R. Berger and H. E. Suess, eds., Radiocarbon dating: Proceedings of the 9th International Conference, 1976, Los Angeles and La Jolla, California, p. 365–387.

Coleman, D. D., L. J. Benson, and P. J. Hutchinson, 1990, The use of isotopic analysis for identification of landfill gas in the subsurface, in Proceedings of the Government Refuse Collection and Disposal Association 13th Annual Landfill Gas Symposium: Silver Spring, Maryland, Government Refuse Collection and Disposal Association, p. 213–229.

Coleman, D. D., C.-L. Liu, D. R. Dickerson, and R. R. Frost, 1972, Improvement in trimerization of acetylene to benzene for radiocarbon dating with a commercially available vanadium oxide catalyst: Proceedings of the 8th International Conference on Radiocarbon Dating, Wellington, New Zealand, 1972, p. B50–B62.

Coleman, D. D., C.-L. Liu, K. C. Hackley, and S. R. Pelphrey, 1995, Isotopic identification of landfill methane: Environmental Geosciences, v. 2, no. 2, p. 95–103.

Coleman, D. D., and W. F. Meents, 1978, Chemical and isotopic composition of natural gases in Paleozoic rocks in Illinois, in R. E. Zartman, ed., Short Papers of the Fourth International Conference: Geochronology, Isotope Geology, 1978: Reston, Virginia, U.S. Geological Survey, Open-File Report 78-1710 p. 73–74.

Coleman, D. D., W. F. Meents, C.-L. Liu, and R. A. Keogh, 1977, Isotopic identification of leakage gas from underground storage reservoirs: A progress report: Illinois State Geological Survey, Illinois Petroleum 111, 10 p.

Collinson, C., and E. Atherton, 1975, Devonian System, in H. B. Willman, E. Atherton, T. C. Buschbach, C. Collinson, J. C. Frye, M. E. Hopkins, J. A. Lineback, and J. A. Simon, eds., Handbook of Illinois stratigraphy: Illinois State Geological Survey, Bulletin 95, p. 104–123.

Collinson, C., J. W. Baxter, R. D. Norby, H. R. Lane, and P. D. Brenkle, 1981, Mississippian stratotypes: Illinois State Geological Survey, Field guidebook in conjunction with the 15th annual meeting of the North-Central Section, Geological Society of America, 56 p.

Collinson, C., P. L. Drake, and C. K. Anchor, 1975, Inventory of physical characteristics of the Illinois shore north of Chicago: Report prepared for Illinois Coastal Zone Management Program: Illinois State Geological Survey, 72 p.

Collinson, C., J. A. Lineback, P. B. DuMontelle, D. C. Brown, R. A. Davis Jr., and C. E. Larsen, 1974, Coastal geology, sedimentology, and management: Chicago and northshore: Illinois State Geological Survey, Guidebook 12, 55 p.

Collinson, C., M. L. Sargent, and J. R. Jennings, 1988, Illinois Basin region, in L. L. Sloss, ed., Sedimentary cover—North American craton: U.S.: The Geology of North America, v. D-2: Boulder, Colorado, The Geological Society of America, p. 383–426.

Crockett, J. E., and D. K. Helpingstine, 1996, Milestones in the history of oil development in Illinois: Illinois State Geological Survey, unpublished manuscript, available from the Illinois State Geological Survey Library.

Crockett, J. E., B. Seyler, and S. J. Whitaker, 1988, Buckhorn Consolidated Field, Illinois, in C. W. Zuppann, B. D. Keith, and S. J. Keller, eds., Geology and petroleum production of the Illinois Basin, v. 2: Illinois and Indiana-Kentucky Geological Societies, p. 51–53.

Currier, L. W., 1937a, Geologic factors in the interpretation of fluorspar reserves in the Kentucky-Illinois field: Washington, D.C., U.S. Geological Survey, Bulletin 886-B, p. 5–14.

Currier, L. W., 1937b, Origin of the bedding replacement deposits of fluorspar in the Illinois field: Economic Geology, v. 32, no. 3, p. 364–386.

Dahlberg, R. E., D. E. Luman, and A. Warren, 1985, Satellite image map of Illinois: Illinois State Geological Survey, 1:500,000.

Damberger, H. H., 1971, Coalification pattern of the Illinois Basin: Economic Geology and the Bulletin of Economic Geologists, v. 66, no. 3, p. 488–494.

Damberger, H. H., 1974, Coalification patterns of Pennsylvanian coal basins of the eastern United States, in R. R. Dutcher, P. A. Hacqebard, J. M. Schopf, and J. A. Simon, eds., Carbonaceous materials as indicators of metamorphism: Geological Society of America, Special Paper 153, p. 53–74.

Demir, I., R. R. Ruch, R. A. Cahill, J. M. Lytle, and K. K. Ho, 1996, Washability of air toxics in marketed Illinois coals: American Chemical Society, Division of Fuel Chemistry, Preprints of Papers, v. 41, no. 3, p. 769–776.

DeWolf, F. W., 1907, Coal investigations in the Saline-Gallatin field, Illinois, and adjoining area: in H. F. Bain, Year-book for 1907: Illinois State Geological Survey, Bulletin 8, p. 211–299.

DeWolf, F. W., 1931, The State Geological Survey during the period 1909–1923, in Papers presented at the quarter centennial of the Illinois State Geological Survey: Illinois State Geological Survey, Bulletin 60, p. 35–43.

Dickey, P. A., 1958, Oil is found with ideas: Tulsa Geological Society Digest, v. 26, p. 84–101.

DuMontelle, P. B., S. C. Bradford, R. A. Bauer, and M. M. Killey, 1981, Mine subsidence in Illinois: Facts for the homeowner considering insurance: Illinois State Geological Survey, Environmental Geology Notes 99, 24 p.

DuMontelle, P. B., N. C. Hester, and R. E. Cole, 1971, Landslides along the Illinois River Valley south and west of La Salle and Peru, Illinois: Illinois State Geological Survey, Environmental Geology Notes 48, 16 p.

DuMontelle, P. B., R. E. Yarbrough, and R. S. Pocreva, compilers, 1980, Review of underground mining practices in Illinois as related to aspects of mine subsidence with recommendations for legislation: Illinois Institute of Natural Resources, Document 80/10, 150 p.

Ekblaw, G. E., 1930, Cause and prevention of potential rock falls north of Savanna, Illinois: Transactions of the Illinois Academy of Science (1929), v. 22, p. 450–454.

Ekblaw, G. E., 1932, Landslides near Peoria: Transactions of the Illinois Academy of Science (1931), v. 24, no. 2, p. 350–353.

Ekblaw, G. E., 1953, Twenty-five years of engineering geology in Illinois: Transactions of the Illinois Academy of Science, v. 46, p. 716.

Ekblaw, G. E., and L. F. Athy, 1925, Glacial Kankakee torrent in northeastern Illinois: Geological Society of America Bulletin, v. 36, no, 2, p. 417–427.

Emmons, S. F., 1893, Fluorspar deposits of southern Illinois: Transactions of the American Institute of Mining and Metallurgical Engineers, v. 21, p. 31–53.

Fenneman, N. M., 1938, Physiography of the eastern United States: New York, McGraw-Hill Book Co., Inc., 714 p.

Finger, G. C., M. J. Gortatowski, R. H. Shiley, and R. H. White, 1959,

Aromatic fluorine compounds, VIII, Plant growth regulators and intermediates: Journal of the American Chemical Society, v. 81, no. 1, p. 94–101.

Fisher, D. J., 1963, The seventy years of the Department of Geology, University of Chicago: Chicago, Illinois, The University of Chicago, 147 p.

Fisk, H. N., 1944, Geological investigations of the alluvial valley of the lower Mississippi River: Mississippi River Commission, U.S. Army Corps of Engineers, Vicksburg, Mississippi, p. 63.

Follmer, L. R., 1984, Soil: An uncertain medium for waste disposal, in Municipal and industrial waste, Seventh Annual Madison Waste Conference Proceedings, September 11–12, 1984: Madison, Wisconsin, University of Wisconsin, p. 296–311.

Follmer, L. R., 1996, Loess studies in central United States—Evolution of concepts: Engineering Geology, v. 45, no. 1–4, p. 287–304.

Foster, J. W., 1857, On the geologic position of the deposits in which occur the remains of fossil elephant of North America: Proceedings of the American Association for the Advancement of Science, v. 10, part 2, p. 148–169.

Frank, T. E., I. G. Krapac, T. D. Stark, and G. D. Strack, 2005, Long-term behavior of water content, density, and degree of saturation in a composted earthen liner: Journal of Geotechnical and Geoenvironmental Engineering, v. 131, p. 800–808.

Frye, J. C., 1968, Development of Pleistocene stratigraphy in Illinois, in R. E. Bergstrom, ed., The Quaternary of Illinois: A symposium in observance of the centennial of the University of Illinois: Urbana-Champaign, University of Illinois, College of Agriculture, Special Publication 14, p. 3–10.

Frye, J. C., and G. M. Richmond, 1958, Note 20—Problems in applying standard stratigraphic practice in nonmarine Quaternary deposits: American Association of Petroleum Geologists Bulletin, v. 42, no. 8, p. 1979–1983.

Frye, J. C., and H. B. Willman, 1960, Classification of the Wisconsinan Stage in the Lake Michigan glacial lobe: Illinois State Geological Survey, Circular 285, 16 p.

Frye, J. C., and H. B. Willman, 1975, Quaternary system, in H. B. Willman, E. Atherton, T. C. Buschbach, C. Collinson, T. C. Frye, M. E. Hopkins, J. A. Lineback, and J. A. Simon, Handbook of Illinois stratigraphy: Illinois State Geological Survey, Bulletin 95, p. 211–239.

Fryxell, F., 1962, Memorial to Carl Leland Horberg (1910–1955): Geological Society of America Bulletin, v. 73, no. 12, p. 99–107.

Gibb, J. P., 1976, Field verification of hazardous industrial waste migration from land disposal sites, in Residual management by waste disposal: Proceedings of the Hazardous Waste Research Symposium, p. 94–101.

Gilkeson, R. H., and K. Cartwright, 1982, The application of surface geophysical methods in monitoring network design, in D. M. Neilsen, ed., Proceedings of the Second National Symposium on Aquifer Restoration and Groundwater Monitoring, May 26–28, 1982, Columbus, Ohio, National Water Well Association, p. 169–183.

Glass, H. D., J. C. Frye, and H. B. Willman, 1964, Record of Mississippi River diversion in the Morton Loess of Illinois: Transactions of the Illinois Academy of Science, v. 57, no. 1, p. 24–27.

Glass, H. D., J. C. Frye, and H. B. Willman, 1968, Clay mineral composition, a source indicator of Midwest loess, in R. E. Bergstrom, ed., The Quaternary of Illinois: A symposium in observance of the centennial of the University of Illinois: Urbana-Champaign, University of Illinois, College of Agriculture, Special Publication 14, p. 35–40.

Glass, H. D., and M. M. Killey, 1987, Principles and applications of clay mineral composition in Quaternary stratigraphy: Examples from Illinois, USA, in J. J. M. Van der Meer, ed., Tills and glaciotectonics, Rotterdam, The Netherlands, A. A. Balkema, p. 117–125. (Illinois State Geological Survey, Reprint 1988-C.)

Gluskoter, H. J., 1965, Electronic low-temperature ashing of bituminous coal: Fuel, v. 44, no. 4, p. 285–291.

Gluskoter, H. J., 1982, Coal geology—Who needs it? in Perspectives in Geology: Invited Papers: Illinois State Geological Survey, Circular 525, p. 7–11.

Gluskoter, H. J., R. R. Ruch, W. G. Miller, R. A. Cahill, G. B. Dreher, and J. K. Kuhn, 1977, Trace elements in coal: Occurrence and distribution: Illinois State Geological Survey, Circular 499, 154 p.

Gluskoter, H. J., and J. A. Simon, 1968, Sulfur in Illinois coals: Illinois State Geological Survey, Circular 432, 28 p.

Goodwin, J. H., and R. E. Bergstrom, 1988, Illinois State Geological Survey, in A. A. Socolow, ed., The state geological surveys—A history: Association of American State Geologists, p. 103–116.

Goodwin, J. H., D. G. Mikulic, and J. W. Baxter, 1990, Industrial minerals and metals publications of the Illinois State Geological Survey through December 1989: Illinois State Geological Survey, Illinois Minerals 104, 33 p.

Griffin, R. A., K. Cartwright, N. F. Shimp, J. D. Steele, W. A. White, G. M. Hughes, and R. H. Gilkeson, 1976, Attenuation of pollutants in municipal landfill leachate by clay minerals: Part 1, Column leaching and field verification: Illinois State Geological Survey, Environmental Geology Notes 78, 34 p.

Griffin, R. A., and E. S. K. Chian, 1979, Attenuation of water-soluble polychlorinated biphenyls by earth materials: Illinois State Geological Survey, Environmental Geology Notes 86, 97 p.

Griffin, R. A., R. R. Frost, A. K. Au, G. D. Robinson, and N. F. Shimp, 1977, Attenuation of pollutants in municipal landfill leachate by clay minerals: Part 2, Heavy-metal adsorption: Illinois State Geological Survey, Environmental Geology Notes 79, 47 p.

Griffin, R. A., R. E. Hughes, L. R. Follmer, C. J. Stohr, W. J. Morse, T. M. Johnson, J. K. Bartz, J. K. Steele, K. Cartwright, M.M. Killey, and P. B. DuMontelle, 1984, Migration of industrial chemicals and soil-waste interactions at Wilsonville, Illinois, in Land Disposal of Hazardous Waste: Proceedings of the Tenth Annual Research Symposium, 1984: Washington, D.C., U.S. Environmental Protection Agency EPA-600/9-84-007, p. 61–77.

Griffin, R. A., R. M. Schuller, J. J. Sulloway, S. J. Russell, W. F. Childers, and N. F. Shimp, 1978, Solubility and toxicity of potential pollutants in solid coal wastes, in Symposium Proceedings: Environmental Aspects of Fuel Conversion Technology III, 1977: Washington, D.C., U.S. Environmental Protection Agency, U.S. EPA-600/7-78-063, p. 506–558.

Griffin, R. A., R. M. Schuller, J. J. Sulloway, N. F. Shimp, W. F. Childers, and R. H. Shiley, 1980, Chemical and biological characterization of leachates from coal solid wastes: Illinois State Geological Survey, Environmental Geology Notes 89, 99 p.

Grim, R. E., 1953, Clay mineralogy: New York, McGraw-Hill Book Company, Inc., 384 p.

Grim, R. E., R. H. Bray, and W. F. Bradley, 1937, The mica in argillaceous sediments: American Mineralogist, v. 22, no. 7, 813–829.

Hackett, J. E., 1966, An application of geologic information to land use in the Chicago metropolitan region: Illinois State Geological Survey, Environmental Geology Notes 8, 6 p.

Hackett, J. E., 1967, Geology and physical planning, in Water Geology and the Future Conference, Water Resources Research Center, Indiana University, Bloomington, Indiana, April 26–27, 1967, p. 83–90. (Illinois Geological Survey, Reprint 1967-I.)

Hackley, K. C., C.-L. Liu, and D. D. Coleman, 1996, Environmental isotope characteristics of landfill leachates and gases: Ground Water, v. 34, no. 5, p. 827–836.

Hall, J., 1843a, Geology of New York—Part 4, with geologic map of middle and western states: Albany, New York, Carrol and Cook, 685 p.

Hall, J., 1843b, Notes, explanatory of a section from Cleveland, Ohio to the Mississippi River, in a southwest direction, with remarks upon the identity of the western formations with those of New York *in* Reports of the first, second, and third meetings of the Association of American Geologists and Naturalists: Transactions of the Association, p. 267–293.

Hansel, A. K., and W. H. Johnson, 1996, Wedron and Mason Groups: Lithostratigraphic reclassification of deposits of the Wisconsin Episode, Lake Michigan lobe area: Illinois State Geological Survey, Bulletin 104, 116 p.

Hansel, A. K., and W. H. Johnson, 2000, Surficial deposits of Illinois (digitally adapted by B. J. Stiff): Illinois State Geological Survey, Open File Series 2000-7, 1:500,000.

Harvey, R. D., 1970, Petrographic and mineralogic characteristics of carbonate rocks related to sorption of sulfur oxides in flue gases: Illinois State Geological Survey, Environmental Geology Notes 38, 31 p.

Harza Engineering Company and Bauer Engineering Inc., 1968a, Pollution and flood control—A program for Chicagoland: Report for Metropolitan Sanitary District of Greater Chicago, 30 p.

Harza Engineering Company and Bauer Engineering Inc., 1968b, Chicagoland deep tunnel system for pollution and flood control—First construction zone definite project report: Report prepared for Metropolitan Sanitary District of Greater Chicago, 176 p.

Hasenmueller, N. R., and J. B. Comer, eds., 1994, Gas potential of the New Albany Shale (Devonian and Mississippian) in the Illinois Basin: Illinois Basin Studies 2 and GRI-92/0391: Prepared by the Illinois Basin Consortium for the Gas Research Institute, Des Plaines, Illinois.

Hatch, J. R., J. B. Risatti, and J. D. King, 1991, Geochemistry of Illinois Basin oils and hydrocarbon source rocks, *in* M. W. Leighton, D. R. Kolata, D. F. Oltz, and J. J. Eidel, eds., Interior cratonic basins: Tulsa, Oklahoma, American Association of Petroleum Geologists, Memoir 51, p. 403–423.

Hatmaker, P., and H. W. Davis, 1938, The fluorspar industry of the United States with special reference to the Illinois-Kentucky district: Illinois State Geological Survey, Bulletin 59, 128 p.

Hays, R. G., 1980, State science in Illinois—The scientific surveys, 1850–1978: Carbondale, Southern Illinois University Press, and London and Amsterdam, Emmons, Feffer and Simons, Inc., 257 p.

Heigold, P. C., 1991, Crustal character of the Illinois Basin, *in* M. W. Leighton, D. R. Kolata, D. F. Oltz, and J. J. Eidel, eds., Interior cratonic basins: Tulsa, Oklahoma, American Association of Petroleum Geologists, Memoir 51, p. 247–261.

Hennepin, L., 1698, A new discovery of a vast country in America, extending above four thousand miles between New France and New Mexico with a description of the Great Lakes, cataracts, rivers, plants, and animals: London, Printed for M. Bentley, J. Tonson, H. Bonwick, T. Goodwin, and S. Manship, 243 p. (Available in the rare book room, Library, University of Illinois at Urbana-Champaign.)

Herron, W. H., 1909, Report of the co-operative topographic Survey of Illinois *in* H. F. Bain, Year-book for 1908: Illinois State Geological Survey, Bulletin 14, p. 31–182.

Herzog, B. L. 1994, Slug tests for determining hydraulic conductivity of natural geologic deposits, *in* D. E. Daniel and S. J. Trautwein, eds., Hydraulic conductivity and waste contaminant transport in soil: West Conshohocken, Pennsylvania, ASTM, Special Technical Publication 1142, p. 95–110.

Herzog, B. L., K. Cartwright, T. M. Johnson, and J. H. Harris, 1982, A study of trench covers to minimize infiltration at waste disposal sites: Task I report: Illinois State Geological Survey, Contract/Grant Report, 1981-5, 245 p.

Herzog, B. L., S.-F. J. Chou, J. R. Valkenburg, and R. A. Griffin, 1988, Changes in volatile organic chemical concentrations after purging slowly recovering wells: Groundwater Monitoring Review, v. 8, no. 4, p. 177–187.

Herzog, B. L., B. J. Stiff, C. A. Chenoweth, K. L. Warner, J. B. Sieverling, and C. Avery, 1994a, Buried bedrock surface of Illinois (3rd ed.): Illinois State Geological Survey, Illinois Map 5, 1:500,000.

Herzog, B. L., S. D. Wilson, D. R. Larson, E. C. Smith, T. H. Larson, and M. L. Greenslate, 1994b, Hydrogeology and groundwater availability in southwest McLean and southeast Tazewell Counties: Part I, Aquifer characterization: Illinois State Geological Survey and Illinois State Water Survey, Cooperative Groundwater Report 17, 70 p.

Herzog, B. L., S. D. Wilson, D. R. Larson, E. C. Smith, T. H. Larson, and M. L. Greenslate, 1994c, Hydrogeology and groundwater availability in southwest McLean and southeast Tazewell Counties: Part I, Aquifer characterization: Illinois State Geological Survey and Illinois State Water Survey, Cooperative Groundwater Report 17a, 143 p.

Hobbs, H., 1999, Origin of the Driftless Area by subglacial drainage—A new hypothesis, *in* D. M. Mickelson and J. W. Attig, eds., Glacial processes past and present: Boulder, Colorado, Geological Society of America, Special Paper 337, p. 93–102.

Hopkins, M. E., 1968, Harrisburg (No. 5) Coal reserves of southeastern Illinois: Illinois State Geological Survey, Circular 431, 25 p.

Horberg, C. L., 1950, Bedrock topography of Illinois: Illinois Geological Survey, Bulletin 73, 111 p.

Howard, R. H., 1963, Wapella East oil pool, DeWitt County, Illinois—A Silurian reef: Illinois State Geological Survey, Circular 349, 15 p.

Howard, R. H., 1991, Hydrocarbon reservoir distribution in the Illinois Basin, *in* M. W. Leighton, D. R. Kolata, D. F. Oltz, and J. J. Eidel, eds., Interior cratonic basins: American Association of Petroleum Geologists, Memoir 51, p. 299–327.

Howard, R. H., and S. T. Whitaker, 1988, Hydrocarbon accumulation in a paleovalley at the Mississippian-Pennsylvanian unconformity near Hardinville, Crawford County, Illinois: A model paleogeomorphic trap: Illinois State Geological Survey, Illinois Petroleum 129, 26 p.

Howard, R. H., and S. T. Whitaker, 1990, Fluvial-estuarine valley fill at the Mississippian-Pennsylvanian unconformity, Main Consolidated Field, Illinois, *in* The collection: Casebooks in earth sciences: New York, Springer-Verlag, p. 319–341. (Illinois State Geological Survey, Reprint 1990N.)

Hubbert, M. K., 1932, Results of earth resistivity survey on various geologic structures in Illinois: American Institute of Mining and Metallurgical Engineers, Technical Publication 463, 23 p.

Hubbert, M. K., and J. M. Weller, 1934, Location of faults in Hardin County, Illinois, by the earth-resistivity method, *in* Geophysical prospecting: Transactions of the American Institute of Mining and Metallurgical Engineers, v. 110, p. 40–48.

Huff, B. G., and J. H. Goodwin, 1999, History of oil and gas production in Illinois: Illinois State Geological Survey, Geobit 8, 4 p.

Hughes, G. M., 1967, Selection of refuse disposal sites in northeastern Illinois: Illinois State Geological Survey, Environmental Geology Notes 17, 26 p.

Hughes, G. M., and K. Cartwright, 1972, Scientific and administrative criteria for shallow waste disposal: Civil Engineering, v. 42, no. 3, p. 70–73.

Hughes, R. E., G. B. Dreher, D. M. Moore, M. Rostam-Abadi, and T. Fiocchi, 1995, Brick manufacture with fly ash from Illinois coals: Technical report: Washington, D.C., U.S. Department of Energy, DOE/PC/92521-T239. NTIS.

Hunt, S. R., 1980, Surface subsidence due to coal mining in Illinois: University of Illinois at Urbana-Champaign, Ph.D. dissertation, 134 p.

Hutchins, T., 1778, A topographical description of Virginia, Pennsylvania, Maryland, and North Carolina, comprehending the Rivers Ohio, Kenhawa, Siota, Cherokee, Wabash, Illinois, Mississippi: London, England, J. Almon, 67 p.

Illinois Basin Consortium, 2000, New Albany Shale/Illinois Basin data: Workshop held at Indiana Memorial Union, Bloomington, Indiana, October 13, 2000, unnumbered pages. (CD-ROM.)

Illinois Department of Agriculture and Illinois State Geological Survey, 1993, Agrichemical facility site contamination study: Springfield, Illinois, Illinois Department of Agriculture, July, 1993, 245 p.

Illinois State Geological Survey, 1992a, Geological mapping for the future of Illinois: Illinois State Geological Survey, Special Report 1, 49 p.

Illinois State Geological Survey, 1992b, Illinois State Geological Survey efforts toward a better economy and a healthier environment: Illinois State Geological Survey, Mini-Series Report, 26 p.

Illinois State Geological Survey, 1993, Publications of the Illinois State Geological Survey, p. 51–52.

Illinois State Geological Survey, 2002, Illinois geology: Science for society: Illinois State Geological Survey, Annual Report, p. 25, 51.

Illinois State Museum, 2002, Native Americans: Springfield, Illinois. http://www.museum.state.il.us/muslink/nat_amer. Accessed October 14, 2009.

Illinois Times, October 13–19, 1978: Springfield, Illinois, p. 4.

Ipe, V. C., and S. B. Bhagwat, 2003, Water resources in Illinois: Demand, prices, and scarcity rents: Illinois State Geological Survey, Illinois Minerals 126, 11 p.

Jacobson, R. J., 1983, Murphysboro Coal, Jackson and Perry Counties: Resources with low to medium sulfur potential: Illinois State Geological Survey, Mineral Notes 85, 19 p.

Jefferson, T., 1788, Notes on the State of Virginia: Philadelphia, Pennsylvania, Pritchard & Hall, 244 p.

Johnson, D. O., 1972, Stratigraphic analysis of the interval between the Herrin (No. 6) Coal and the Piasa Limestone in southwestern Illinois: University of Illinois at Urbana-Champaign, Ph.D. dissertation, 135 p.

Johnson, M. E., 1977, Geology in American education: 1825–1860: Geological Society of America Bulletin, v. 88, no. 8, p. 1192–1198.

Johnson, T. M., and K. Cartwright, 1980, Monitoring of leachate migration in the unsaturated zone in the vicinity of sanitary landfills: Illinois State Geological Survey, Circular 514, 82 p.

Johnson, T. M., R. A. Griffin, K. Cartwright, L. R. Follmer, B. L. Herzog, W. J. Morse, P. B. DuMontelle, M. M. Killey, C. J. Stohr, and R. E. Hughes, 1983a, Hydrogeologic investigations of failure mechanisms and migration of organic chemicals at Wilsonville, Illinois, *in* D. M. Nielson, ed., Proceedings of the Third National Symposium on Aquifer Restoration and Ground-Water Monitoring, National Water Well Association, 1983, p. 413–420.

Johnson, T. M., T. H. Larson, B. L. Herzog, K. Cartwright, C. J. Stohr, and S. J. Klein, 1983b, A study of trench covers to minimize infiltration at waste disposal sites: Task II report, Laboratory evaluation and computer modeling of trench cover design: Illinois State Geological Survey, Contract/Grant Report 1983-3, 94 p.

Kay, G. F., and M. M. Leighton, 1933, Eldoran Epoch of the Pleistocene period: Geological Society of America Bulletin 44, no. 4, p. 669–674.

Keefer, D. A., 1993, Evaluation of pesticide movement in a tile-drained soil: University of Illinois at Urbana-Champaign, M. S. thesis, 119 p.

Keefer, D. A., 1995, Aquifer sensitivity to contamination by pesticide leaching in Illinois: Illinois State Geological Survey, Open File Series 1995-5s, 1:500,000.

Keefer, D. A., R. C. Berg, and W. S. Dey, 1990, Potential for aquifer recharge in Illinois (appropriate recharge area): Illinois State Geological Survey, 1:1,000,000.

Kemptom, J. P. 1981, Three-dimensional geologic mapping for environmental studies in Illinois: Illinois State Geological Survey, Environmental Geology Notes 100, 43 p.

Kempton, J. P., R. C. Berg, and L. R. Follmer, 1985, Revision of the stratigraphy and nomenclature of glacial deposits in central northern Illinois, *in* R. C. Berg, J. P. Kempton, L. R. Follmer, and D. P. McKenna, eds., Illinoian and Wisconsinan stratigraphy and environments in northern Illinois: The Altonian revised: Midwest Friends of the Pleistocene 32nd Field Conference: Illinois State Geological Survey, Guidebook 19, p. 1–19.

Kempton, J. P., W. H. Johnson, P. C. Heigold, and K. Cartwright, 1991, Mahomet Bedrock Valley in east-central Illinois: Topography, glacial drift stratigraphy, and hydrogeology, *in* W. N. Melhorn and J. P. Kempton, eds., Geology and hydrogeology of the Teays-Mahomet Bedrock Valley systems: Boulder, Colorado, Geological Society of America, Special Paper 258, p. 91–124.

Kempton, J. P., W. J. Morse, and A. P. Visocky, 1982, Hydrogeologic evaluation of sand and gravel aquifers for municipal groundwater supplies in east-central Illinois: Illinois State Geological Survey and Illinois State Water Survey, Cooperative Groundwater Report 8, 59 p.

Kempton, J. P., and A. Visocky, 1992, Regional groundwater resources in western McLean and eastern Tazewell Counties: Illinois State Geological Survey and Illinois State Water Survey, Cooperative Groundwater Report 13, 41 p.

Keyes, C. R., 1895, Synopsis of American Paleozoic echinoids: Proceedings of the Iowa Academy of Science, v. 2, p. 178–194.

Khan, L. A., 2004, Method and apparatus for froth flotation: U.S. Patent Publication No. US2004/0099575 A1, May 27, 2004.

Killey, M. M., and P. B. DuMontelle, 1984, Earthquakes in the Illinois area: Illinois State Geological Survey and Illinois Emergency Services and Disaster Agency, 4 p.

Killey, M. M., J. K. Hines, P. B. DuMontelle, and E. E. Brabb, compilers, 1984, Illinois landslide inventory map: Reston, Virginia, U.S. Geological Survey, Miscellaneous Field Studies Map MF 1691, 1:500,000.

Klem, M. J., 1904, A revision of the Paleozoic Palaeëchinoidea, with a synopsis of all known species: Transactions of the Academy of Science at St. Louis, v. 14, p. 1–98.

Kolata, D. R., 1991, Overview of sequences, *in* M. W. Leighton, D. R. Kolata, D. F. Oltz, and J. J. Eidel, eds., Interior cratonic basins: Tulsa, Oklahoma, American Association of Petroleum Geologists, Memoir 51, p. 59–73.

Kolata, D. R., compiler, 2005, Bedrock geology of Illinois: Illinois State Geological Survey, 1:500,000.

Kolata, D. R., and T. C. Buschbach, 1976, Plum River Fault Zone of northwestern Illinois: Illinois State Geological Survey, Circular 491, 20 p.

Kolata, D. R., T. C. Buschbach, and J. D. Treworgy, 1978, The Sandwich Fault Zone of northern Illinois: Illinois State Geological Survey, Circular 505, 26 p.

Kolata, D. R., and T. G. Hildenbrand, 1997, Structural underpinnings and neotectonics of the southern Illinois Basin: An overview: Seismological Research Letters, v. 68 no. 4, p. 499–510.

Kolata, D. R., W. D. Huff, and S. M. Bergström, 1996, Ordovician K-bentonites of eastern North America: Boulder, Colorado, Geological Society of America, Special Paper 313, 84 p.

Kolata, D. R., W. D. Huff, and S. M. Bergström, 1998, Nature and regional significance of unconformities associated with the Middle Ordovician Hagan K-bentonite complex in the North American midcontinent: Geological Society of America Bulletin, v. 110, no. 6, p. 723–739.

Kolata, D. R, W. D. Huff, and S. M. Bergström, 2001, The Ordovician Sebree Trough—An oceanic passage to the midcontinent United States: Geological Society of America Bulletin, v. 113, no. 8, p. 1067–1078.

Kolata, D. R., W. D. Huff, and J. K. Frost, 1983, Correlation of K-bentonites in the Decorah Subgroup of the Mississippi Valley by chemical fingerprinting, in Ordovician Galena Group of the Upper Mississippi Valley: Deposition, diagenesis, and paleoecology: 13th Annual Field Conference, Society of Economic Paleontologists and Mineralogists, Great Lakes Section, 1983, p. F1–F15.

Kolata, D. R., and W. J. Nelson, 1991, Tectonic history of the Illinois Basin, in M. W. Leighton, D. R. Kolata, D. F. Oltz, and J. J. Eidel, eds., Interior cratonic basins: Tulsa, Oklahoma, American Association of Petroleum Geologists, Memoir 51, p. 263–285.

Korose, C. P., S. D. Elrick, and R. J. Jacobson, 2003, Availability of the Colchester Coal for mining in northern and western Illinois: Illinois State Geological Survey, Illinois Minerals 127, 21 p.

Kosanke, R. M., 1950, Pennsylvanian spores of Illinois and their use in correlation: Illinois State Geological Survey, Bulletin 74, 128 p.

Kosanke, R. M., 1979, Memorial to James Morton Schopf (1911–1978): Boulder, Colorado, Geological Society of America, June 1979, 4 p.

Krapac, I. G., W. S. Dey, W. R. Roy, C. A. Smyth, S. Storment, S. L. Sargent, and J. D. Steele, 2002, Impacts of swine manure pits on groundwater quality: Environmental Pollution, v. 202, no. 2, p. 475–492.

Krey, F., and J. E. Lamar, 1925, Limestone resources of Illinois: Illinois State Geological Survey, Bulletin 46, 392 p.

Lamar, J. E., 1928, Geology and economic resources of the St. Peter Sandstone of Illinois: Illinois State Geological Survey, Bulletin 53, 175 p.

Lamar, J. E., 1957, Chemical analyses of Illinois limestones and dolomites: Illinois State Geological Survey, Report of Investigations 200, 33 p.

Lamar, J. E., 1967, Handbook on limestone and dolomite for Illinois quarry operators: Illinois State Geological Survey, Bulletin 91, 119 p.

Lamar, J. E., W. J. Armon, and J. A. Simon, 1956, Illinois oil shales: Illinois State Geological Survey, Circular 208, 22 p.

Lamar, J. E., and H. B. Willman, 1938, A summary of the uses of limestone and dolomite: Illinois State Geological Survey, Report of Investigations 49, 50 p.

Lane, N. G., 1966, New Harmony and pioneer geology: Geotimes, v. 11, no. 2, p. 18–21.

Langenheim, R. L., Jr., 2001, Geology moves on at Illinois: Benjamin C. Jillson and Charles Wesley Rolfe: Windows into the past: Department of Geology Newsletter, 2000 Year in Review, p. 14–15.

Langenheim, R. L. Jr., and W. J. Nelson, 1992, The cyclothemic concept in Illinois: A review, in R. H. Dott, ed., Eustasy: The historical ups and downs of a major geological concept: Boulder, Colorado, Geological Society of America, Memoir 180, p. 55–71.

Larsen, J. I., and J. E. Hackett, 1965, Activities in environmental geology in northeastern Illinois: Illinois State Geological Survey, Environmental Geology Notes 3, 5 p.

Larson, D. R., B. L. Herzog, and T. H. Larson, 2003, Groundwater geology of DeWitt, Piatt, and northern Macon Counties: Illinois State Geological Survey, Environmental Geology 155, 35 p.

Larson, D. R., B. L. Herzog, R. C. Vaiden, C. A. Chenoweth, Y. Xu, and R. C. Anderson, 1995, Hydrogeology of the Green River Lowland and associated bedrock valleys in northwestern Illinois: Illinois State Geological Survey, Environmental Geology 149, 20 p.

Larson, D. R., J. P. Kempton, and S. Meyer, 1997, Geologic, geophysical, and hydrologic investigations for a supplemental municipal groundwater supply, Danville, Illinois: Illinois State Geological Survey and Illinois State Water Survey, Cooperative Groundwater Report 18, 62 p.

Larson, T. H., A. M. Graese, and P. G. Orozco, 1993, Hydrogeology of the Silurian dolomite aquifer in parts of northwestern Illinois: Illinois State Geological Survey, Environmental Geology 145, 29 p.

Lasemi, Z., and R. C. Berg, 2001, Three-dimensional geologic mapping: A pilot program for resource and environmental assessment in the Villa Grove Quadrangle, Douglas County, Illinois: Illinois State Geological Survey, Bulletin 106, 117 p.

Lasemi, Z., S. B. Bhagwat, T. J. Kemmis, and I. Demir, 2003, Illinois—State activities: Annual review 2002: Mining Engineering, v. 55, no. 5, p. 69–73.

Lasemi, Z., and J. P. Grube, 1995, Mississippian "Warsaw" play makes waves in Illinois Basin: Oil and Gas Journal, v. 93, no. 2, p. 47–51.

Lasemi, Z., D. G. Mikulic, and S. D. Elrick, 2008, Annual review of Illinois' industrial minerals and coal production: Mining Engineering, v. 60, no. 5, p. 89–92.

Lasemi, Z., and R. D. Norby, 2001, Depositional cycles in the Salem Limestone of southwestern Illinois: Implications for predicting limestone quality and reserves, in R. D. Hagni, ed., Studies on ore deposits, mineral economics, and applied mineralogy: With emphasis on Mississippi Valley-type base metal and carbonatite-related ore deposits: Rolla, Missouri, University of Missouri, p. 373–383.

Lasemi, Z., R. D. Norby, and J. D. Treworgy, 1998, Depositional facies and sequence stratigraphy of a Lower Carboniferous bryozoan-crinoidal carbonate ramp in the Illinois Basin, midcontinent USA, in T. P. Burchette and V. P. Wright, eds., Carbonate ramps: London, England, The Geological Society of London, Special Publications 149, p. 369–395.

Lathrop, S. P., 1851, Mastodon in northern Illinois: American Journal of Science, ser. 2, v. 12, no. 36, p. 439.

Le Conte, J. L., 1854, Notes on some fossil suilline pachyderms from Illinois: Philadelphia Academy of Natural Science Proceedings, v. 6, p. 3–5, p. 56–57.

Leetaru, H. E., 1996, Seismic stratigraphy, a technique for improved oil recovery planning at King Field, Jefferson County, Illinois: Illinois State Geological Survey, Illinois Petroleum 151, 37 p.

Leetaru, H. E., 2000, Sequence stratigraphy of the Aux Vases Sandstone: A major oil producer in the Illinois Basin: American Association of Petroleum Geologists Bulletin, v. 84, no. 3, p. 399–422.

Leighton, M. M., 1923, The differentiation of the drift sheets in northwestern Illinois: Journal of Geology, v. 31, no. 4, p. 265–281.

Leighton, M. M., 1926, A notable type Pleistocene section: The Farm Creek exposure near Peoria, Illinois: Journal of Geology, v. 34, no. 2, p. 167–174.

Leighton, M. M., 1928, The utility of the topographic map: Journal of the Western Society of Engineers, v. 33, no. 1, January 1928, p. 25–30.

Leighton, M. M., 1931a, The State Geological Survey during the period 1923–1930, in Papers presented at the quarter centennial celebration of the Illinois State Geological Survey: Illinois State Geological Survey, Bulletin 60, p. 45–61.

Leighton, M. M., 1931b, The Peorian loess and the classification of glacial drift sheets of the Mississippi Valley: Journal of Geology, v. 39, no. 1, p. 45–53.

Leighton, M. M., 1933, The naming of the subdivisions of the Wisconsin glacial age: Science, v. 77, no. 1989, p. 168.

Leighton, M. M., 1945, The Illinois State Geological Survey in war min-

eral research: Illinois State Geological Survey, Circular 121, 23 p.
Leighton, M. M., 1955, The geological survey's response to the changing economic patterns of the past fifty years: Presentation at the 50th anniversary celebration of the Illinois State Geological Survey, Tuesday, November 11, 1955, Gregory Hall, University of Illinois, Urbana: Illinois State Geological Survey, Manuscript 257, 19 p.
Leighton, M. M., 1966, Recollections and reflections: The Illinois Geological Survey 1905–1954: Geotimes, September, p. 13–17.
Leighton, M. M., G. E. Ekblaw, and C. L. Horberg, 1948, Physiographic divisions of Illinois: Journal of Geology, v. 56, no. 1, p. 16–33.
Leighton, M. M., and P. MacClintock, 1930, Weathered zones of the drift-sheet of Illinois: Journal of Geology, v. 38, no. 1, p. 28–53.
Leighton, M. M., and H. B. Willman, 1950, Loess formations of the Mississippi Valley: Journal of Geology, v. 58, no. 6, p. 599–623.
Leighton, M. W., 1988, Recollections and reflections on topographic mapping: Illinois Mapnotes, v. 10/11, p. 2–4.
Leighton, M. W., 1996, Interior cratonic basins: A record of regional tectonic influences, in B. A. van der Pluijm and P. A. Catacosinos, eds., Basement and basins of eastern North America: Boulder, Colorado, Geological Society of America, Special Paper 308, p. 77–93.
Leighton, M. W., and D. R. Kolata, 1991, Selected interior cratonic basins and their place in the scheme of global tectonics: A synthesis, in M. W. Leighton, D. R. Kolata, D. F. Oltz, and J. J. Eidel, eds., Interior cratonic basins: Tulsa, Oklahoma, American Association of Petroleum Geologists, Memoir 51, p. 729–797.
Leighton, M. W., D. R. Kolata, D. F. Oltz, and J. J. Eidel, eds., 1991, Interior cratonic basins: Tulsa, Oklahoma, American Association of Petroleum Geologists, Memoir 51, 819 p.
Leverett, F., 1899, The Illinois glacial lobe: Washington, D.C., U.S. Geological Survey, Monograph, v. 38, no. 2, 817 p.
Libby, W. F., 1952, Radiocarbon dating: Chicago, Illinois, University of Chicago Press, 124 p.
Lineback, J. A., 1968, Turbidites and other sandstone bodies in the Borden Siltstone (Mississippian) in Illinois: Illinois State Geological Survey, Circular 425, 29 p.
Lineback, J. A., 1969, Illinois Basin—Sediment-starved during Mississippian: American Association of Petroleum Geologists Bulletin 53, no. 1, p. 112–126.
Lineback, J. A., compiler, 1979, Quaternary deposits of Illinois: Illinois State Geological Survey, 1:500,000.
Lineback, J. A., and D. L. Gross, 1972, Depositional patterns, facies, and trace element accumulation in the Waukegan Member of the Late Pleistocene Lake Michigan Formation in southern Lake Michigan: Illinois State Geological Survey, Environmental Geology Notes 58, 25 p.
Lobeck, A. K., 1950, Physiographic diagram of North America: Geographical Press, Maplewood, New Jersey, C. S. Hammond and Co., 16 p.
Locke, J., 1840, Report on the lead region of the Upper Mississippi: United States 26th Congress, 1st Session, House Executive Document 239, p. 116–159.
Lowenstam, H. A., 1948a, Biostratigraphic studies of the Niagaran inter-reef formations in northeastern Illinois: Springfield, Illinois, Illinois State Museum Scientific Papers, v. 4, 146 p.
Lowenstam, H. A., 1948b, Marine pool, Madison County, Illinois, Silurian reef producer, in J. V. Howell, ed., Structure of typical American oil fields: Tulsa, Oklahoma, American Association of Petroleum Geologists, v. 3, p. 153–188.
Lowenstam, H. A., and E. P. DuBois, 1946, Marine pool, Madison County, a new type of oil reservoir in Illinois: Illinois State Geological Survey, Report of Investigations 114, 30 p.
Luman, D. E., L. R. Smith, and C. C. Goldsmith, 2003, Illinois surface topography: Illinois State Geological Survey, Illinois Map 11, 1:500,000.
Maclure, W., 1809, Observations on the geology of the United States, explanatory of a geological map: Transactions of the American Philosophical Society, v. 6, p. 411–428, plus map.
Marshak, S., W. J. Nelson, and J. H. McBride, 2003, Phanerozoic strike-slip faulting in the continental interior platform of the United States: Examples from the Laramide orogen, midcontinent, and ancestral Rocky Mountains, in F. Storti, R. E. Holdsworth, and F. Salvini, eds., Intraplate strike-slip deformation belts: London, Geological Society of London, Special Publications 210, p. 159–184.
Mast, R. F., and R. H. Howard, 1991, Oil and gas production and recovery estimates in the Illinois Basin, in M. W. Leighton, D. R. Kolata, D. F. Oltz, and J. J. Eidel, eds., Interior cratonic basins: Tulsa, Oklahoma, American Association of Petroleum Geologists, Memoir 51, p. 295–298.
McBride, J. H., D. R. Kolata, and T. G. Hildenbrand, 2003, Geophysical constraints on understanding the origin of the Illinois Basin and its underlying crust: Tectonophysics, v. 363, p. 45–78.
McBride, J. H., and W. J. Nelson, 1999, Style and origin of mid-Carboniferous deformation in the Illinois Basin, USA—Ancestral Rockies deformation?: Tectonophysics, v. 305, no. 1–3, p. 240–273.
McBride, J. H., and W. J. Nelson, 2001, Reflection images of shallow faulting, northernmost Mississippi Embayment, north of the New Madrid seismic zone: Bulletin of the Seismological Society of America, v. 91, p. 128–139.
McChesney, J. H., 1860, Description of new species of fossils from the Palaeozoic rocks of the western states: Transactions of the Chicago Academy of Science (Extract 1859), v. 1, p. 1–76.
McGinnis, L. D., and J. P. Kempton, 1961, Integrated seismic, resistivity, and geologic studies of glacial deposits: Illinois State Geological Survey, Circular 323, 23 p.
McGinnis, L. D., J. P. Kempton, and P. C. Heigold, 1963, Relationship of gravity anomalies to a drift-filled valley system in northern Illinois: Illinois State Geological Survey, Circular 354, 23 p.
McKenna, D. P., and D. A. Keefer, 1991, Potential for agricultural chemical contamination of aquifers in Illinois: Illinois State Geological Survey, Open File Series 1991-7, 18 p.
McKenna, D. P., S. C. Schock, E. Mehnert, S. C. Mravik, and D. A. Keefer, 1989, Agricultural chemicals in rural, private water wells in Illinois: Illinois State Geological Survey and Illinois State Water Survey, Cooperative Groundwater Report 11, 109 p.
McLure, J. W., 1962, A history of the Illinois State Geological Survey, 1851–1875: Urbana-Champaign, University of Illinois, M. A. thesis, 175 p.
Mease, J., 1807, A geological account of the United States: Philadelphia, Pennsylvania, Birch & Small, 496 p.
Mehnert, E., B. L. Herzog, B. R. Hensel, J. R. Miller, and T. M. Johnson, 1987, Design of groundwater monitoring systems: Hydrogeological considerations, in Geotechnical and geohydrological aspects of waste management: Chelsea, Michigan, Lewis Publishers, p. 271–285.
Mehnert, E., S. C. Schock, M. L. Barnhardt, M. E. Caughey, S.-F. J. Chou, W. S. Dey, G. B. Dreher, and C. Ray, 1995, The occurrence of agricultural chemicals in Illinois rural private wells: Results from the pilot study: Ground Water Monitoring and Remediation, v. 15, no. 1, p. 142–149.
Merrill, G. P., 1906, Contributions to the history of North American biology, in Report of the National Museum: Smithsonian Institute Annual Report, 1904, p. 189–733.
Mikulic, D. G. 1989, The Chicago stone industry: A historical perspective, in R. E. Hughes and J. C. Bradbury, eds., Proceedings of the

23rd Forum on the Geology of Industrial Minerals, May 11–15, 1987, North Aurora, Illinois: Illinois State Geological Survey, Illinois Mineral Notes 102, p. 83–89.

Mikulic, D. G., 1990, Cross section of the Paleozoic rocks of northeastern Illinois: Implications for subsurface aggregate mining: Illinois State Geological Survey, Illinois Minerals 106, 14 p.

Mikulic, D. G., and J. H. Goodwin, 1985, Urban encroachment on dolomite resources of the Chicago area, Illinois, in J. D. Glaser and J. Edwards, eds., Proceedings of the Twentieth Forum on the Geology of Industrial Minerals, May 15–18, 1984: Baltimore, Maryland, Maryland Geological Survey, Special Publication 2, p. 125–131.

Miller, K. L., 1981, Petroleum and profits in the Prairie State, 1889–1980: Straws in the cider barrel: Illinois Historical Journal, v. 78, no. 3, p. 163–176.

Moore, R. C., 1944, Correlation of Pennsylvanian formations of North America: Geological Society of America Bulletin, v. 55, no. 6, p. 657–706.

Mylius, L. A., 1927, Oil and gas development and possibilities in east-central Illinois: Illinois State Geological Survey, Bulletin 54, 205 p.

Nelson, W. J., 1991, Structural styles of the Illinois Basin, in M. W. Leighton, D. R. Kolata, D. F. Oltz, and J. J. Eidel, eds., Interior cratonic basins: Tulsa, Oklahoma, American Association of Petroleum Geologists, Memoir 51, p. 209–243.

Nelson, W. J., 1995, Structural features in Illinois: Illinois State Geological Survey, Bulletin 100, 144 p.

Nelson, W. J., F. B. Denny, J. A. Devera, L. R. Follmer, and J. M. Masters, 1997, Tertiary and Quaternary tectonic faulting in southernmost Illinois: Engineering Geology, v. 46, no. 3–4, p. 235–258.

Nelson, W. J., F. B. Denny, L. R. Follmer, and J. M. Masters, 1999, Quaternary grabens in southernmost Illinois: Deformation near an active intraplate seismic zone: Tectonophysics, v. 305, no. 1–3, p. 381–397.

Nelson, W. J., L. B. Smith, and J. D. Treworgy, contributions by L. C. Furer and B. D. Keith, 2002, Sequence stratigraphy of the Lower Chesterian (Mississippian) strata of the Illinois Basin: Illinois State Geological Survey, Bulletin 107, 70 p.

Newberry, J. S., 1870, The surface geology of the basin of the Great Lakes and the valley of the Mississippi: The American Naturalist, v. 4, no. 4, p. 193–214.

Newberry, J. S., 1874, Geology: Surface geology: Report of the Geological Survey of Ohio, v. II, part 1, p. 1–80.

Newberry, J. S., A. H. Worthen, F. B. Meek, and L. Lewquereux, 1866, Palaeontology: Geological Survey of Illinois, v. 2, 595 p.

Norby, R. D., 1991, Biostratigraphic zones in the Illinois Basin, in M. W. Leighton, D. R. Kolata, D. F. Oltz, and J. J. Eidel, eds., Interior cratonic basins: American Association of Petroleum Geologists, Memoir 51, p. 179–194.

Oltz, D. F., 1994, Improved and enhanced oil recovery in Illinois by reservoir characterization: Final report prepared for U.S. Department of Energy: Illinois State Geological Survey, 403 p.

Oltz, D. F., H. E. Leetaru, B. Seyler, and S. T. Whitaker, 1991a, An integrated approach to reservoir characterization in the Illinois Basin: Multidisciplinary studies, in The integration of geology, geophysics, petrophysics and petroleum engineering in reservoir delineation, description and management: Proceedings of the First Archie Conference, held October 22–25, 1990, in Houston, Texas: Tulsa, Oklahoma, American Association of Petroleum Geologists, p. 38–60.

Oltz, D. F., J. A. Rupp, B. Keith, and J. Beard, 1991b, Future hydrocarbon opportunities in the Illinois Basin, in M. W. Leighton, D. R. Kolata, D. F. Oltz, and J. J. Eidel, eds., Interior cratonic basins: Tulsa, Oklahoma, American Association of Petroleum Geologists, p. 491–502.

Owen, D. D., 1844, Report of a geological exploration of part of Iowa, Wisconsin, and Illinois, made under instructions from the Secretary of the Treasury of the United States, in the autumn of the year 1839: United States [28th] Congress, 1st Session, Senate Executive Document 407, 191 p.

Owen, D. D., 1846, On the geology of the western states of North America: Quarterly Journal of the London Geological Society, v. 2, p. 433–447.

Palmer, J. E., R. J. Jacobson, and C. B. Trask, 1979, Depositional environments of strata of late Desmoinesian age overlying the Herrin (No. 6) Coal Member in southwestern Illinois, Field Trip 9, Ninth International Congress of Carboniferous Stratigraphy and Geology, Part 2: Invited papers, in J. E. Palmer and R. R. Dutcher, eds., Depositional and structural history of the Pennsylvanian System of the Illinois Basin: Illinois State Geological Survey, Guidebook 15A, p. 92–102.

Panno, S. V., K. C. Hackley, H. H. Hwang, and W. R. Kelly, 2001, Determination of the sources of nitrate contamination in karst springs using isotopic and chemical indicators: Chemical Geology, v. 179, p. 113–128.

Panno, S. V., W. R. Kelly, C. P. Weibel, I. G. Krapac, and S. L. Sargent, 2003, Water quality and agrichemical loading in two groundwater basins of Illinois' sinkhole plain: Illinois State Geological Survey, Environmental Geology 156, 36 p.

Panno, S. V., I. G. Krapac, C. P. Weibel, and J. D. Bade, 1996, Groundwater contamination in karst terrain of southwestern Illinois: Illinois State Geological Survey, Environmental Geology 151, 43 p.

Panno, S. V., C. P. Wiebel [sic], P. C. Heigold, and P. C. Reed, 1994, Formation of regolith-collapse sinkholes in southern Illinois: Interpretation and identification of associated buried cavities: Environmental Geology, v. 23, no. 3, p. 214–220.

Parmelee, C. W., and C. R. Shroyer, 1921, Further investigations of Illinois fire clays, in Year book for 1917 and 1918: Administrative reports and economic and geological papers: Illinois State Geological Survey, Bulletin 38, 149 p.

Payne, J. N., and G. H. Cady, 1944, Structure of the Herrin (No. 6) Coal beds in Christian and Montgomery Counties and adjacent parts of Fayette, Macon, Sangamon, and Shelby Counties: Illinois State Geological Survey, Circular 105, 57 p.

Peck, R. B., 1940, Sampling methods and laboratory tests for Chicago subway soils, in P. C. Rutledge, ed.: Proceedings, Purdue Conference on Soil Mechanics and Its Applications, September 2–6, 1940, Purdue University, Lafayette, Indiana, p. 140–150.

Peck, R. B., 1948, History of building foundations in Chicago: A report of an investigation: Urbana, Illinois, University of Illinois, Engineering Experiment Station Bulletin 373, 64 p.

Peck, R. B., 1954, Engineering properties of Chicago subsoils: Urbana, Illinois, University of Illinois, Engineering Experiment Station Bulletin 423, 62 p.

Peck, R. B., and M. E. Uyanik, 1955, Observed and computed settlements of structures in Chicago: Urbana, Illinois, University of Illinois, Engineering Experiment Station Bulletin 429, 60 p.

Peppers, R. A., 1964, Spores in strata of late Pennsylvanian cyclothems in the Illinois Basin: Illinois State Geological Survey, Bulletin 90, 89 p.

Peppers, R. A., 1970, Correlation and palynology of coals in the Carbondale and Spoon Formations (Pennsylvanian) of the northeastern part of the Illinois Basin: Illinois State Geological Survey, Bulletin 93, 173 p.

Peppers, R. A., 1996, Palynological correlation of major Pennsylvanian (Middle and Upper Carboniferous) chronostratigraphic boundaries in the Illinois and other coal basins: Boulder, Colorado, Geological

Society of America, Memoir 188, 111 p.

Petterchak, J. A., 1999, A chronology of Illinois history. http://www.state.il.us/hpa/lib/ILChronology.htm. Accessed December 18, 2004.

Pettijohn, F. J., 1984, Memoirs of an unrepentant geologist: Chicago, Illinois, The University of Chicago Press, 260 p.

Phillips, T. L., and R. A. Peppers, 1984, Changing patterns of Pennsylvanian coal-swamp vegetation and implications of climatic control on coal occurrence: International Journal of Coal Geology, v. 3, no. 3, p. 205–255.

Phillips, T. L., H. W. Pfefferkorn, and R. A. Peppers, 1973, Development of paleobotany in the Illinois Basin: Illinois State Geological Survey, Circular 480, 86 p.

Piersol, R. J., 1936, 1, Smokeless briquetes—Impacted without binder from partially volatilized Illinois coals; 2, Smoke index—Quantitative measurement of smoke: Illinois State Geological Survey, Report of Investigations 41, 113 p.

Pratt, T. L., R. C. Culotta, E. Hauser, D. Nelson, L. Brown, S. Kaufman, J. Oliver, and W. Hinze, 1989, Major Proterozoic basement features of the eastern midcontinent of North America revealed by recent COCORP profiling: Geology, v. 17, no. 6, p. 505–509.

Pratt, T. L., E. C. Hauser, and K. D. Nelson, 1992, Widespread buried Precambrian layered sequences in the U.S. mid-continent: Evidence for large Proterozoic depositional basins: American Association of Petroleum Geologists Bulletin 76, no. 9, p. 1384–1401.

Preston, A. F., 1986, The exploration and discovery of the Illinois Basin: Illinois State Geological Survey, manuscript 1, p. 1–30.

Prickett, T. A., and C. G. Lonnquist, 1971, Selected digital computer techniques for groundwater resource evaluation: Illinois State Water Survey, Bulletin 55, 62 p.

Pryor, W. A., 1956, Quality of groundwater estimated from electric resistivity logs: Illinois State Geological Survey, Circular 215, 15 p.

Purdy, R. C., and F. W. DeWolf, 1907, Preliminary investigations of Illinois fire clays, in Year-book for 1906: Illinois State Geological Survey, Bulletin 4, p. 129–176.

Raymond Martin, L., 2001, Bibliography and index of Illinois geology, 1966–1996: Illinois State Geological Survey, Bulletin 105, 447 p.

Reed, F. H., and G. C. Finger, 1936, Illinois fluorspar as a chemical raw material: Transactions of the Illinois Academy of Science (1935), v. 28, no. 2, p. 129–130.

Reed, F. H., H. W. Jackman, O. W. Rees, G. R. Yohe, and P. W. Henline, 1947, Use of Illinois coal for production of metallurgical coke: Illinois State Geological Survey, Bulletin 71, 132 p.

Risatti, J. B., 1992, Rates of microbial dechlorination of polychlorinated biphenyls (PCBs) in anaerobic sediments from Waukegan Harbor: Illinois Hazardous Waste Research and Information Center, Research Report 061, 36 p.

Risser, H. E., 1960, Coal in the future energy market: Illinois State Geological Survey, Circular 310, 15 p.

Risser, H. E., 1973, Trends in energy supply: Outlook for Energy Conference, 1972, Upper Midwest Council, p. 13–20. (Illinois State Geological Survey, Reprint 1973-E.)

Risser, H. E., and R. L. Major, 1968, History of Illinois mineral industries: Illinois State Geological Survey, Educational Series 10, 30 p.

Rolfe, C. W., 1931, Investigations previous to the founding of the present State Geological Survey: Illinois Geological Survey, Bulletin 60, p. 23–28.

Rolfe, C. W., R. C. Purdy, A. N. Talbot, and I. O. Baker, 1908, Paving brick and paving-brick clays of Illinois: Illinois State Geological Survey, Bulletin 9, 316 p.

Roy, W. R., 2001, The environmental fate and movement of organic solvents in water, soil, and air (sec. 17.1) and Fate-based management of organic solvent-containing wastes (sec. 17.2), in G. Wypych, ed., Handbook of solvents: Toronto, Ontario, Canada, Chemtec Publishing, p. 1149–1162, p. 1162–1169.

Roy, W. R., C. C. Ainsworth, R. A. Griffin, and I. G. Krapac, 1984, Development and application of batch adsorption procedures for designing earthen landfill liners: Seventh Annual Madison Waste Conference, Municipal and Industrial Waste, 1984: Madison, Wisconsin, University of Wisconsin, p. 390–398.

Roy, W. R., and I. G. Krapac, 2006, Guidance for conducting site assessments at retail agrichemical facilities: Springfield, Illinois, Illinois Department of Agriculture, 75 p.

Roy, W. R., I. G. Krapac, S.-F. J. Chou, and R. A. Griffin, 1991, Batch-type procedures for estimating soil adsorption of chemicals: Washington, D.C., U.S. Environmental Protection Agency, Technical Resource Document, U.S. EPA/530-SW-87-006-F.

Roy, W. R., I. G. Krapac, S.-F. J. Chou, and F. W. Simmons, 2001, Pesticide storage and release in unsaturated soil in Illinois, USA: Journal of Environmental Science and Health, v. 36, no. 3, p. 245–260.

Ruch, R. R., H. J. Gluskoter, and N. F. Shimp, 1974, Occurrence and distribution of potentially volatile trace elements in coal: A final report: Illinois State Geological Survey, Environmental Geology Notes 72, 96 p.

Ruch, R. R., E. J. Kennedy, and N. F. Shimp, 1970, Distribution of arsenic in unconsolidated sediments from southern Lake Michigan: Illinois State Geological Survey, Environmental Geology Notes 37, 16 p.

Russell, R. J., 1944, Lower Mississippi Valley loess: Geological Society of America Bulletin, v. 55, p. 1–40.

Salisbury, R. D., 1893, Distinct glacial epochs and the criteria for their recognition: Journal of Geology, v. 1, p. 61–84.

Savage, T. E., 1920, The Devonian formations of Illinois: American Journal of Science, ser. 4, v. 49, part 12, p. 169–182.

Savage, T. E., 1925, Correlation of the Maquoketa and Richmond rocks of Iowa and Illinois: Transactions of the Illinois Academy of Science, v. 17, p. 233–247.

Schleicher, J. A., and J. K. Kuhn, 1970, Phosphorus content in unconsolidated sediments from southern Lake Michigan: Illinois State Geological Survey, Environmental Geology Notes 39, 15 p.

Schock, S. C., E. Mehnert, M. E. Caughey, G. B. Dreher, W. S. Dey, S. D. Wilson, C. Ray, S.-F. J. Chou, J. Valkenburg, J. M. Gosar, J. R. Karny, M. L. Barnhardt, W. F. Black, M. R. Brown, and V. J. Garcia, 1992, Pilot study: Agricultural chemicals in rural private wells in Illinois: Illinois State Geological Survey and Illinois State Water Survey, Cooperative Groundwater Report 14, 80 p.

Schopf, J. M., 1936, Spores characteristic of Illinois Coal No. 6: Transactions of the Illinois Academy of Science, v. 28, no. 2, p. 173–176.

Schopf, J. M., L. R. Wilson, and R. Bentall, 1944, An annotated synopsis of Paleozoic fossil spores and the definition of generic groups: Illinois State Geological Survey, Report of Investigations 91, 72 p.

Schuller, R. M., J. P. Gibb, and R. A. Griffin, 1981, Recommended sampling procedure for monitoring wells: Ground Water Monitoring Review, v. 1, no. 1, p. 42–46.

Sellards, E. H., 1902, On the fertile fronds of *Crossotheca* and *Myriotheca*, and on the spores of other Carboniferous ferns, from Mazon Creek, Illinois: American Journal of Science, ser. 4, v. 14, no. 81, art. 22, p. 195–202.

Seyler, B., and R. M. Cluff, 1991, Petroleum traps in the Illinois Basin, in M. W. Leighton, D. R. Kolata, D. F. Oltz, and J. J. Eidel, eds., Interior cratonic basins: Tulsa, Oklahoma, American Association of Petroleum Geologists, Memoir 51, p. 361–401.

Seyler, B., J. P. Grube, and Z. Lasemi, 2003, The origin of prolific reservoirs in the Geneva Dolomite (Middle Devonian), west-central

Illinois Basin: Illinois State Geological Survey, Illinois Petroleum 158, 36 p.

Shaver, R. H., ed., 1987, A field guide and recollections: The David Dale Owen years to the present: A sesquicentennial commemoration of service by the Geological Survey: Bloomington, Indiana, Indiana Department of Natural Resources, Geological Survey, Special Report 44, p. 1–30; esp. fig. 1, p. 6.

Shepard, C. U., 1838, Geology of upper Illinois: American Journal of Science, v. 34, no. 1, p. 134–161.

Shiley, R. H., D. R. Dickerson, and G. C. Finger, 1978, Aromatic fluorine chemistry at the Illinois State Geological Survey: Research notes, 1934–1976: Illinois State Geological Survey, Circular 501, 114 p.

Shimek, B., 1896, A theory of the loess: Proceedings of the Iowa Academy of Science, v. 3, p. 82–89.

Shimp, N. F., H. V. Leland, and W. A. White, 1970, Distribution of major, minor, and trace constituents in unconsolidated sediments from southern Lake Michigan: Illinois State Geological Survey, Environmental Geology Notes 32, 19 p.

Siever, R., 1951, The Mississippian-Pennsylvanian unconformity in southern Illinois: American Association of Petroleum Geologists Bulletin, v. 35, no. 3, p. 542–581.

Silliman, B., 1819, New locality of fluorspar, or fluat of lime, and of galena, or sulphuret of lead: American Journal of Science, ser. 1, v. 1, art. 3, p. 52–53.

Simon, J. A. 1974, Memorial to Gilbert Haven Cady, 1882–1970: Boulder, Colorado, Geological Society of America, Memorials, v. 3, p. 42–52.

Simon, J. A., 1977a, Coal: Illinois' major fuel resource: Panel Chairman's Report to the Technical Committee of the Illinois Energy Commission, 166 p.

Simon, J. A., 1977b, Memorial to Hubert Elias Risser, 1914–1974: Boulder, Colorado, Geological Society of America, Memorials, v. 6, 4 p.

Simon, J. A., 1982, The Illinois State Geological Survey—The next quarter century, in Perspectives in geology: Illinois State Geological Survey, Circular 525, p. 1–5.

Simon, J. A., and W. H. Smith, 1969, An evaluation of Illinois coal reserve estimates: Proceedings of the Illinois Mining Institute, p. 57–68.

Sloss, L. L., 1963, Sequences in the cratonic interior of North America: Geological Society of America Bulletin, v. 74, p. 93–114.

Sloss, L. L., 1988, Tectonic evolution of the craton in Phanerozoic time, in L. L. Sloss, ed., Sedimentary cover—North American craton: U.S.: The Geology of North America, v. D-2: Boulder, Colorado, The Geological Society of America, p. 25–51.

Sloss, L. L., W. C. Krumbein, and E. C. Dapples, 1949, Integrated facies analysis, in C. R. Longwell, chairman, Sedimentary facies in geologic history: Boulder, Colorado, Geological Society of America, Memoir 39, p. 91–123.

Smith, G. D., 1942, Illinois loess—Variations in its properties and distributions—A pedologic interpretation: Urbana-Champaign, University of Illinois, Agricultural Experiment Station Bulletin 490, p. 139–184.

Smith, W. H., and J. B. Stall, 1975, Coal and water resources for coal conversion in Illinois: Illinois State Geological Survey and Illinois State Water Survey, Cooperative Resources Report 4, 79 p.

Soller, D. R., S. D. Price, J. P. Kempton, and R. C. Berg, 1999, Three-dimensional geologic maps of Quaternary sediments in east-central Illinois: Prepared in cooperation with the Illinois State Geological Survey: Reston, Virginia, U.S. Geological Survey, Geological Investigations Series, Map I-2669, 3 sheets, 1:500,000.

Squires, F., 1934, Accidental floods in Illinois and Indiana oil sands: Illinois-Indiana Petroleum Association and Illinois State Geological Survey, 2nd Annual Petroleum Conference, p. 27–35.

Squires, F., and A. H. Bell, 1943, Water flooding of oil sands in Illinois: Illinois State Geological Survey, Report of Investigations 89, 101 p.

St. Clair, S., 1917, Clay deposits near Mountain Glen, Union County, Illinois, in Year book for 1916: Administrative reports and economic and geological papers: Illinois State Geological Survey, Bulletin 36, 15 p.

Su, W.-J., L. R. Follmer, and K. Ghiassi, 1993, Landslides in the New Madrid seismic zone: Along the Mississippi River from Chester to East St. Louis, Illinois: Final Technical Report: Illinois State Geological Survey, 100 p.

Su, W.-J., and C. J. Stohr, 1991, Landslides in the New Madrid seismic zone: Landslide inventory and risk assessment in Illinois along the Ohio and the Mississippi rivers from Olmsted to Chester, Illinois: Final Technical Report: Illinois State Geological Survey, 137 p.

Suter, M., R. E. Bergstrom, H. F. Smith, G. H. Emrich, W. C. Walton, and T. E. Larson, 1959, Preliminary report on the ground-water resources of the Chicago region: Illinois: Illinois State Geological Survey and Illinois State Water Survey, Cooperative Ground-water Report 1, 89 p.

Swann, D. H., 1968, A summary geologic history of the Illinois Basin, in Geology and petroleum production of the Illinois Basin, A symposium: Illinois and Indiana-Kentucky Geological Societies, p. 3–22.

Swann, D. H., and A. H. Bell, 1958, Habitat of oil in the Illinois Basin, in L. G. Weeks, ed., Habitat of oil: Tulsa, Oklahoma, American Association of Petroleum Geologists, p. 447–472.

Taylor, E. F., M. W. Pullen Jr., P. K. Sims, and J. N. Payne, 1944, Methods of subsurface study of the Pennsylvanian strata encountered in rotary drill-holes, in Progress reports on subsurface studies of the Pennsylvanian system in the Illinois Basin: Illinois State Geological Survey, Report of Investigation 93, p. 9–21.

Terzaghi, K., 1943, Theoretical soil mechanics: New York, John Wiley and Sons, 528 p.

Thompson, M. D., 1988, The Illinois State Museum—Historical sketch and memoirs: Springfield, Illinois, Illinois State Museum Society, 204 p.

Thwaites, F. T., 1927, The development of the theory of multiple glaciation in North America: Transactions of the Wisconsin Academy of Sciences, Arts and Letters, v. 23, p. 41–164.

Toupiol, C., T. W. Willingham, A. J. Valocchi, C. J. Werth, I. G. Krapac, T. D. Stark, and D. E. Daniel, 2002, Long-term tritium transport through a field-scale compacted soil liner: Journal of Geotechnical and Geoenvironmental Engineering, v. 128, no. 8, p. 640–650.

Trent, B. A., R. A. Bauer, P. J. DeMaris, and N. Kawamura, 1996, Findings and practical applications from the Illinois Mine Subsidence Research Program: Illinois Mine Subsidence Research Program XII, 146 p.

Treworgy, C. G., and M. H. Bargh, 1982, Deep-minable coal resources of Illinois: Illinois State Geological Survey, Circular 527, 62 p.

Treworgy, C. G., L. E. Bengal, and A. G. Dingwell, 1978, Reserves and resources of surface-minable coals in Illinois: Illinois State Geological Survey, Circular 504, 44 p.

Treworgy, C. G., and R. J. Jacobson, 1985, Paleoenvironments and distribution of low-sulfur coal in Illinois, in Ninth International Congress on Carboniferous Stratigraphy and Geology, May 17–26, 1979, Washington, D.C., and Urbana-Champaign, Illinois, v. 4, Economic geology: Coal, oil and gas: Carbondale, Illinois, Southern Illinois University Press, p. 349–359.

Treworgy, C. G., E. I. Prussen, M. A. Justice, C. A. Chenoweth, M. H. Bargh, R. J. Jacobson, and H. H. Damberger, 1997, Illinois coal reserve assessment and database development: Final report: Illinois State Geological Survey, Open File Series 1997-4, 105 p.

Treworgy, J. D., 1981, Structural features in Illinois: A compendium: Illinois State Geological Survey, Circular 519, 22 p.

Treworgy, J. D., S. T. Whitaker, and Z. Lasemi, 1997, Structural cross section of the Paleozoic rocks in Illinois, Wayne County to Stephenson County: Illinois State Geological Survey, Illinois Map 7, 1:500,000.

Udden, J. A., 1912, Geology and mineral resources of the Peoria Quadrangle: Washington, D.C., U.S. Geological Survey, Bulletin 506, 103 p.

Ulrich, E. O., and C. Schuchert, 1902, Paleozoic seas and barriers in eastern North America: New York State Museum, Bulletin 52, p. 633–663.

U.S. Census Bureau, 2005, Resident population and apportionment of the U.S. House of Representatives, 1810–2000. http://www.census.gov/dmd/www/resapport/states/illinois.pdf. Accessed October 22, 2009.

Voskuil, W. H., and A. R. Eich, 1932, Illinois mineral industry in 1932—A preliminary statistical summary and economic review: Illinois State Geological Survey, Report of Investigations 25, 49 p.

Wanless, H. R., 1939, Pennsylvanian correlations in the Eastern Interior and Appalachian coal fields: Baltimore, Maryland, Geological Society of America, Special Paper 17, 130 p.

Wanless, H. R., 1955, Pennsylvanian rocks of Eastern Interior Basin: American Association of Petroleum Geologists Bulletin, v. 49, no. 9, p. 1753–1820.

Wanless, H. R., and J. M. Weller, 1932, Correlation and extent of Pennsylvanian cyclothems: Geological Society of America Bulletin, v. 4, no. 4, p. 1003–1016.

Warner, K. L., A. Martin Jr., and T. L. Arnold, 2003, Arsenic in Illinois ground water—Community and private supplies: Reston, Virginia, U.S. Geological Survey, Water-Resources Investigations Report 03-4103, 12 p.

Warren, W. P. D., 1927, Wells in buried river valley at Sullivan, Illinois: Associated State Engineering Societies Bulletin 2, v. 2, p. 88–97.

Wasson, T., 1938a, Recent oil discoveries in southeastern Illinois: American Association of Petroleum Geologists Bulletin, v. 22, no. 1, p. 71–78.

Wasson, T., 1938b, Oil exploration in the eastern portion of the Illinois Basin: Illinois State Geological Survey, Circular 23-B, p. 13–19.

Weibel, C. P., 1991, Application of cyclothemic-based sequence stratigraphy to Upper Pennsylvanian strata, east-central Illinois, in C. P. Weibel, ed., Sequence stratigraphy in mixed clastic-carbonate strata, Upper Pennsylvanian, east-central Illinois: Great Lakes Section, Society of Economic Paleontologists and Mineralogists, 21st Annual Field Conference, p. 1–25.

Weibel, C. P., 1996, Applications of sequence stratigraphy to Pennsylvanian strata in the Illinois Basin, in B. J. Witzke, G. A. Ludvigson, and J. Day, eds., Paleozoic sequence stratigraphy: Views from the North American craton: Boulder, Colorado, Geological Society of America, Special Paper 306, p. 331–339.

Weibel, C. P., and S. V. Panno, 1997, Karst terrains and carbonate bedrock in Illinois: Illinois State Geological Survey, Illinois Map 8, 1:500,000.

Weller, J. M., 1936, Geology and oil possibilities of the Illinois Basin: Illinois State Geological Survey, Illinois Petroleum 27, 19 p.

Weller, J. M. 1939, Progress in geological mapping of Illinois, 1839–1939: Transactions of the Illinois State Academy of Science, v. 32, no. 2, p. 173–174.

Weller, J. M., 1944a, Devonian system in southern Illinois, in (1940) Symposium on Devonian Stratigraphy: Illinois State Geological Survey, Bulletin 68, p. 89–102.

Weller, J. M., 1944b, Devonian correlations in Illinois and surrounding states: A summary, in (1940) Symposium on Devonian Stratigraphy: Illinois State Geological Survey, Bulletin 68, p. 205–213.

Weller, J. M., and A. H. Bell, 1936, The geology and oil and gas possibilities of parts of Marion and Clay Counties, with a discussion of the central portion of the Illinois Basin: Illinois State Geological Survey, Report of Investigation 40, 54 p.

Weller, J. M., and A. H. Bell, 1937, Illinois Basin: American Association of Petroleum Geologists Bulletin, v. 21, no. 6, p. 771–788.

Weller, J. M., R. M. Grogan, and F. E. Tippie, 1952, Geology of the fluorspar deposits of Illinois: Illinois State Geological Survey, Bulletin 76, 147 p.

Weller, J. M., L. E. Workman, G. H. Cady, A. H. Bell, J. E. Lamar, and G. E. Ekblaw, 1945, Geologic map of Illinois: Illinois State Geological Survey, 1:500,000.

Weller, S., 1906a, The geological map of Illinois: Illinois Geological Survey, Bulletin 1, 24 p.

Weller, S., 1906b, Geological structure of the state, in The petroleum industry of southeastern Illinois: Illinois Geological Survey, Bulletin 2, p. 21–22.

Weller, S., 1920, The Chester Series in Illinois: Journal of Geology, v. 28, no. 4, p. 281–303; no. 5, p. 395–416.

Weller, S., C. Butts, L. W. Currier, and R. D. Salisbury, 1920, The geology of Hardin County and the adjoining part of Pope County: Illinois State Geological Survey, Bulletin 41, 402 p.

Whitaker, S. T., 1988, Silurian pinnacle reef distribution in Illinois: Model for hydrocarbon exploration: Illinois State Geological Survey, Illinois Petroleum 130, 32 p.

White, D., 1913, Physiographic conditions attending the formation of coal, in D. White and R. Thiessen, eds., The origin of coal: U.S. Bureau of Mines, Bulletin 38, p. 52–84.

White, G. W., 1967, Accounts of Illinois geology by John Bradbury in 1817: Transactions of the Illinois Academy of Science, v. 60, no. 4. p. 337–339.

White, G. W., 1973, History of investigation and classification of Wisconsinan drift in north-central United States: Boulder, Colorado, Geological Society of America, Memoir 136, p. 3–34.

White, G. W., and B. O. Slanker, 1962, Early geology in the Mississippi Valley: An exhibition of selected works held in the University of Illinois Library: University of Illinois, Urbana, 26 p.

White, W. A., 1990, Ralph Early Grim, 1902–1989: Clay Minerals, v. 25, no. 1, p. 1–2

Whiting, L. L., 1956, Geology and history of oil production in the Decatur-Mt. Auburn-Springfield area, Illinois: Illinois State Geological Survey, Circular 211, 17 p.

Whiting, L. L., and D. L. Stevenson, 1965, The Sangamon Arch: Illinois State Geological Survey, Circular 383, 20 p.

Whittlesey, C., 1867, Fresh-water glacial drift of the northwestern states: Smithsonian Contributions to Knowledge, v. XV, art. 3, 32 p.

Wilber, C. D., 1861, *Mastodon giganteus*—Its remains in Illinois: Transactions of the Illinois Natural History Society, v. 1, ser. 1, p. 59–64.

Willingham, T. W., C. J. Werth, A. J. Valocchi, I. G. Krapac, C. Toupiol, T. D. Stark, and D. E. Daniel, 2004, Evaluation of multi-dimensional transport through a field-scale compacted soil liner over a 12-year period: Journal of Geotechnical and Geoenvironmental Engineering, v. 130, p. 887–895.

Willman, H. B., 1982, Memorial to John Everts Lamar, 1897–1979: Boulder, Colorado, Geological Society of America, Memorials, v. 12, 7 p.

Willman, H. B., E. Atherton, T. C. Buschbach, C. Collinson, J. C. Frye, M. E. Hopkins, J. A. Lineback, and J. A. Simon, 1975, Handbook of

Illinois stratigraphy: Illinois State Geological Survey, Bulletin 95, 261 p.

Willman, H. B., and T. C. Buschbach, 1975, Ordovician System, *in* H. B. Willman, E. Atherton, T. C. Buschbach, C. Collinson, J. C. Frye, M. E. Hopkins, J. A. Lineback, and J. A. Simon, eds., Handbook of Illinois stratigraphy: Illinois State Geological Survey, Bulletin 95, p. 47–87.

Willman, H. B., and J. C. Frye, 1970, Pleistocene stratigraphy of Illinois: Illinois State Geological Survey, Bulletin 94, 204 p.

Willman, H. B., J. C. Frye, J. A. Simon, K. E. Clegg, D. H. Swann, E. Atherton, C. Collinson, J. A. Lineback, and T. C. Buschbach, 1967, Geologic map of Illinois: Illinois State Geological Survey, 1:500,000.

Willman, H. B., R. R. Reynolds, and P. Herbert Jr., 1946, Geological aspects of prospecting and area for prospecting in the zinc-lead district of northwestern Illinois: Illinois State Geological Survey, Report of Investigations 116, 48 p.

Willman, H. B., J. A. Simon, B. M. Lynch, and V. A. Langenheim, 1968, Bibliography and index of Illinois geology through 1965: Illinois Geological Survey, Bulletin 92, 373 p.

Willman, H. B., D. H. Swann, and J. B. Frye, 1958, Stratigraphic policy of the Illinois State Geological Survey: Illinois State Geological Survey, Circular 249, 14 p.

Wilson, S. D., J. P. Kempton, and R. B. Loft, 1995, The Sankoty-Mahomet aquifer in the confluence area of the Mackinaw and Mahomet Bedrock Valleys, central Illinois: Illinois State Geological Survey and Illinois State Water Survey, Cooperative Groundwater Report 16, 64 p.

Wilson, S. D., G. S. Roadcap, B. L. Herzog, D. R. Larson, and D. Winstanley, 1998, Hydrogeology and groundwater availability in southwest McLean and southeast Tazewell Counties, Part 2: Aquifer modeling and final report: Illinois State Geological Survey and Illinois State Water Survey, Cooperative Groundwater Report 19, 138 p.

Witherspoon, P. A. Jr., T. D. Mueller, and R. W. Donovan, 1962, Evaluation of underground gas-storage conditions in aquifers through investigations of groundwater hydrology: Journal of Petroleum Technology, v. 14, no. 5, p. 555–561.

Witzke, B. J., G. A. Ludvigson, and J. Day, eds., 1996, Paleozoic sequence stratigraphy: Views from the North American craton: Boulder, Colorado, Geological Society of America, Special Paper 306, 446 p.

Workman, L. E., 1944, Subsurface geology of the Devonian System in Illinois, *in* (1940) Symposium on Devonian Stratigraphy: Illinois State Geological Survey, Bulletin 68, p. 189–199.

Workman, L. E., and M. M. Leighton, 1937, Search for ground-waters by the electrical resistivity method: American Geophysical Union Transactions, 18th Annual Meeting, part 2, p. 403–409.

Worthen, A. H., H. M. Bannister, F. H. Bradley, H. A. Green, J. S. Newberry, and L. Lesquereux, 1870, Geology and palaeontology: Geological Survey of Illinois, v. 4, 545 p.

Worthen, A. H., G. C. Broadhead, E. T. Cox, O. St. John, and F. B. Meek, 1875, Geology and palaeontology: Geological Survey of Illinois, v. 6, 577 p.

Worthen, A. H., H. Engelmann, H. C. Freeman, H. M. Bannister, and F. B. Meek, 1868, Geology and palaeontology: Geological Survey of Illinois, v. 3, 605 p.

Worthen, A. H., J. Lindahl, C. Wachsmuth, F. Springer, E. O. Ulrich, and O. Everett, 1890, Geology and palaeontology (two volumes): Geological Survey of Illinois, v. 8, 878 p.

Worthen, A. H., J. Shaw, and F. B. Meek, 1873, Geology and palaeontology: Geological Survey of Illinois, v. 5, 658 p.

Worthen, A. H., O. St. John, and S. A. Miller, 1883, Geology and palaeontology: Geological Survey of Illinois, v. 7, 373 p.

Worthen, A. H., J. D. Whitney, L. Lesquereux, and H. Engelmann, 1866, Geology: Geological Survey of Illinois, v. 1, 507 p.

Zuppann, C. W., B. D. Keith, and S. J. Keller, eds., 1988, Geology and petroleum production of the Illinois Basin, volume 2: Joint Publication of the Illinois and Indiana-Kentucky Geological Societies, 272 p.

Overview of Illinois Geology 2

Dennis R. Kolata

Introduction

The geology of Illinois is as robust as any place else on Earth. A casual observer, however, might look out over the subtle landscape that covers most of the state—flat and open agricultural fields alternating with urban areas—and be unaware of the numerous extraordinary and diverse geological features lying beneath the surface. A trained eye might notice recent geological landforms—perhaps the low hills that are glacial moraines, the subtle rises and depressions that are knobs and kettles, or the round sinkholes that signal karst topography. But for the most part, an understanding of Illinois geology does not come easily. The bulk of the state's geology lies far below the landscape surface. To see it requires visits to sand and gravel pits, quarries, road cuts, construction sites, and natural outcrops along streams, river banks, and lakeshores. Much of the hidden geology of the state must be interpreted using drilled core sample descriptions and geophysical techniques that are capable of remotely sensing subsurface geological features.

The geological record in Illinois reveals a 1.5-billion-year-long history of remarkable changes. The most ancient terrane—buried beneath several thousand feet of sedimentary rocks—consists of granite and rhyolite that speak of volcanic activity and perhaps mountain building. Above these "basement" rocks lie a succession of limestones, sandstones, and shales, locally containing a diverse assemblage of fossilized marine organisms that indicate long periods of inundation by warm, shallow seas filled with life. The abundant and widespread coal deposits in the state reveal a time of lush marshes, rivers, deltas, and coastal swamps. Relatively recent episodes of glaciation are apparent from the surficial deposits of peat, clay, silt, sand, and gravel that mantle much of the state. The stark contrast between past and present climate, geography, life-forms, and environments is clearly evident at Thornton Quarry on the far south side of Chicago where coral-rich reef rocks, formed in a tropical sea 400 million years ago, are being mined to produce the aggregate that is necessary to support our modern lifestyle (Figure 2-1). Most of the changes

Figure 2-1 Aerial view of Thornton Quarry and Interstate 80 in southern Cook County. The quarry exposes a 400-million-year-old reef composed of abundant fossils, including brachiopods, corals, trilobites, mollusks, and echinoderms. The reef formed in a warm shallow marine environment during the Silurian Period. (Photograph by Joel M. Dexter.)

that took place over the last 1.5 billion years were the result of extremely slow, yet relentless processes. Some changes, though, were rapid and catastrophic.

Plate Tectonics

One of the most important breakthroughs in understanding geological history was made in the mid-1960s when geologists discovered the dynamics of plate tectonics. The Earth's outer shell, the lithosphere, consists of the crust and upper part of the mantle that is composed of a patchwork of rigid plates. A key finding was that new ocean crust is being produced in sea-floor spreading centers. This discovery was made by determining the age and polarity of magnetic lineations in the bands of rock that bracketed these ridges. Another finding was that old crust was being consumed in subduction zones. The zones form a type of conveyer system that moves the plates over the Earth's surface. Many plates consist of continents, made up largely of relatively low density granitic rocks and denser basaltic ocean crust. The way in which plates move relative to one another can be determined by examining the boundaries between them.

Three major boundary types are recognized: divergent (plates moving apart at spreading centers), convergent (plates coming together with subduction zones), and transform plate boundaries (plates moving laterally past each other). Plate boundaries are fundamentally important because they determine the locations of ocean basins, mountain ranges, earthquake belts, volcanic systems, and other large-scale features on the Earth's surface. Even though Illinois is situated within the continental interior, far from the current plate margins, its geology has been greatly influenced by the movement of the North American plate and its interaction with neighboring tectonic plates. Plate tectonic theory provides a unifying explanation for past climates, sea level fluctuations, and structural deformation in Illinois. Major aspects of Illinois' tectonic history are discussed in Chapter 3, and neotectonic features are described in Chapter 5.

Geological Structures

Where exposed in quarries, road cuts, and natural exposures, the bedrock of Illinois appears to be lying flat, but actually it is tilted. Much of the state is surrounded by broad, subtle arches including the Wisconsin Arch and Kankakee Arch on the north, Cincinnati Arch on the east, Pascola Arch on the south, and Ozark Dome and Mississippi River Arch on the west (Figure 2-2). The regional tilt of the beds is obvious on the statewide cross sections (Figures 2-3 and 2-4). The strata dip and thicken away from the arches toward the center of the Illinois Basin in southern Illinois and adjacent parts of Indiana and Kentucky. The Illinois Basin, which underlies the southern two-thirds of Illinois, is an oval depression that covers approximately 110,000 square miles (284,899 km^2). It contains about 120,000 cubic miles (500,000 km^3) of Cambrian through Permian sedimentary rocks and a wealth of resources, including groundwater, coal, oil and gas, and industrial minerals and metals (Kolata et al. 1992). The outline of the Basin is arbitrarily marked by the −500-foot (−152-m) contour on top of the Ordovician Kimmswick ("Trenton") Limestone (Buschbach and Kolata 1991) (Figure 2-2).

The Illinois Basin is characterized by numerous geological structures, ranging in magnitude from small, localized faults and folds to extensive, highly deformed areas that have had a major influence on the overall geometry of the Basin. Periods of structural deformation can be correlated with the major plate tectonic events (Figure 2-5). One of the most significant structural features is the New Madrid Rift System, which underlies the southern part of the Illinois Basin (Figure 2-2). This structure consists of the Reelfoot Rift (Ervin and McGinnis 1975) and Rough Creek Graben (Soderberg and Keller 1981).

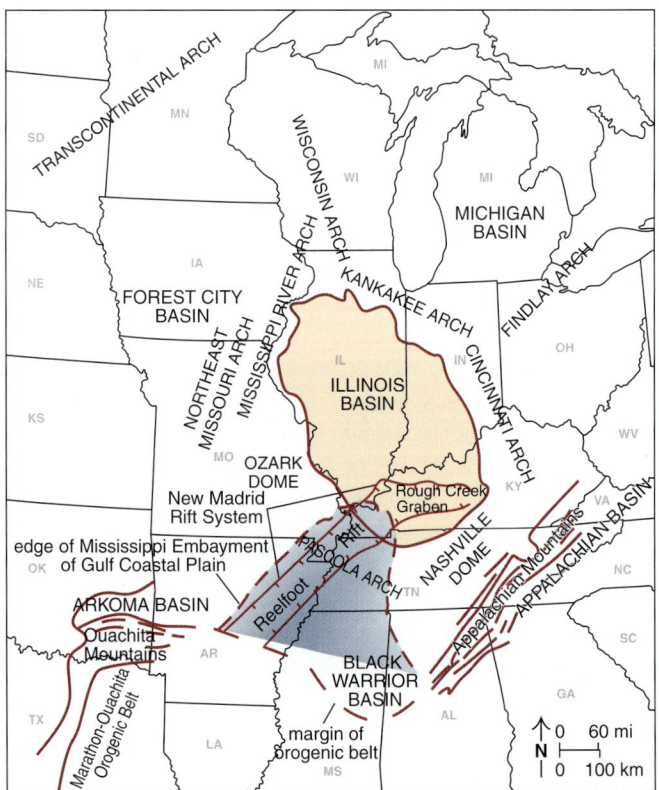

Figure 2-2 Major structural features in the midcontinental United States (Buschbach and Kolata 1991). Light shading shows the areal extent of the Illinois Basin based on the −500-foot (−152-m) contour on top of the Ordovician Kimmswick ("Trenton") Limestone. Blue area is the Mississippi Embayment of the Gulf Coastal Plain.

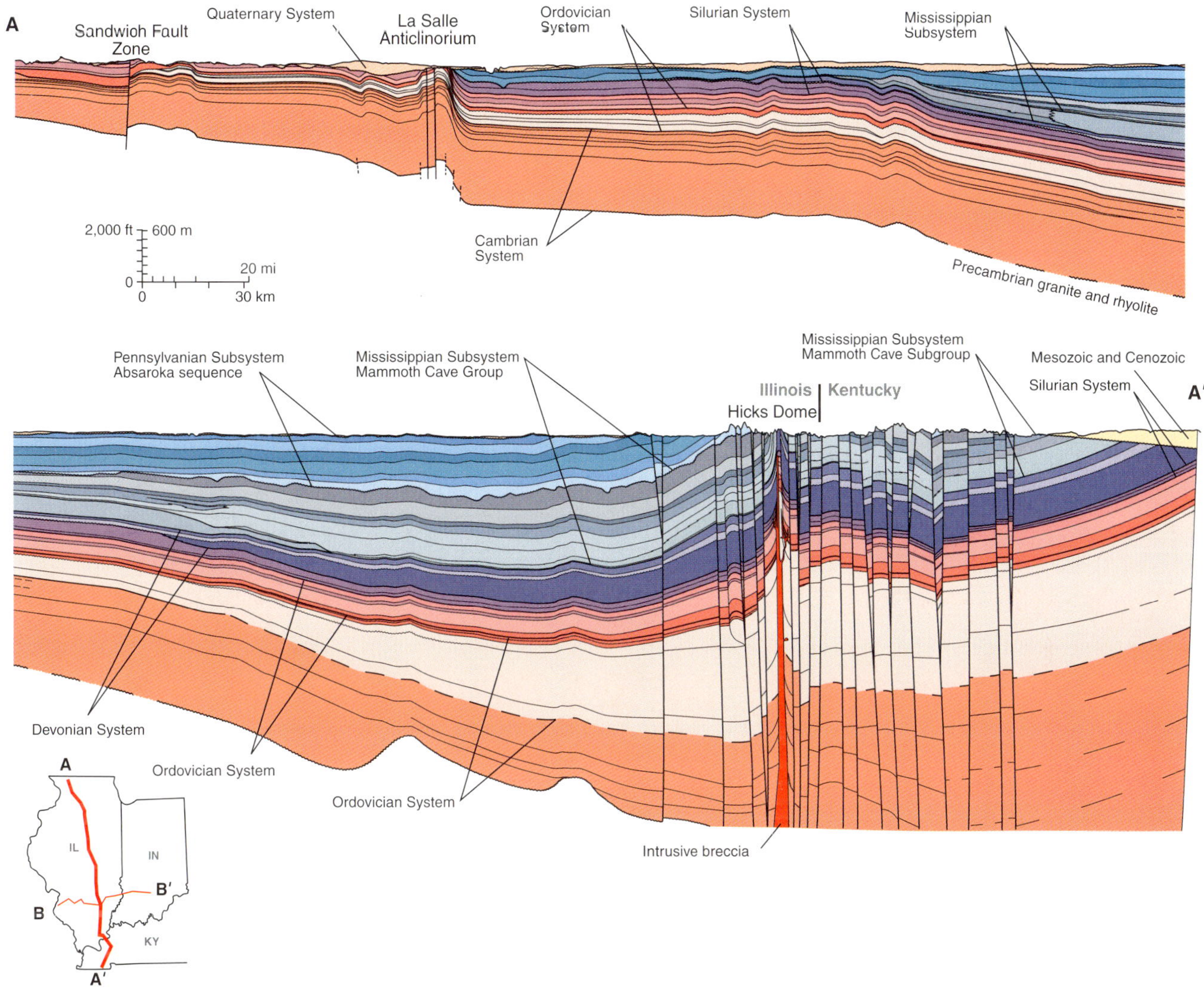

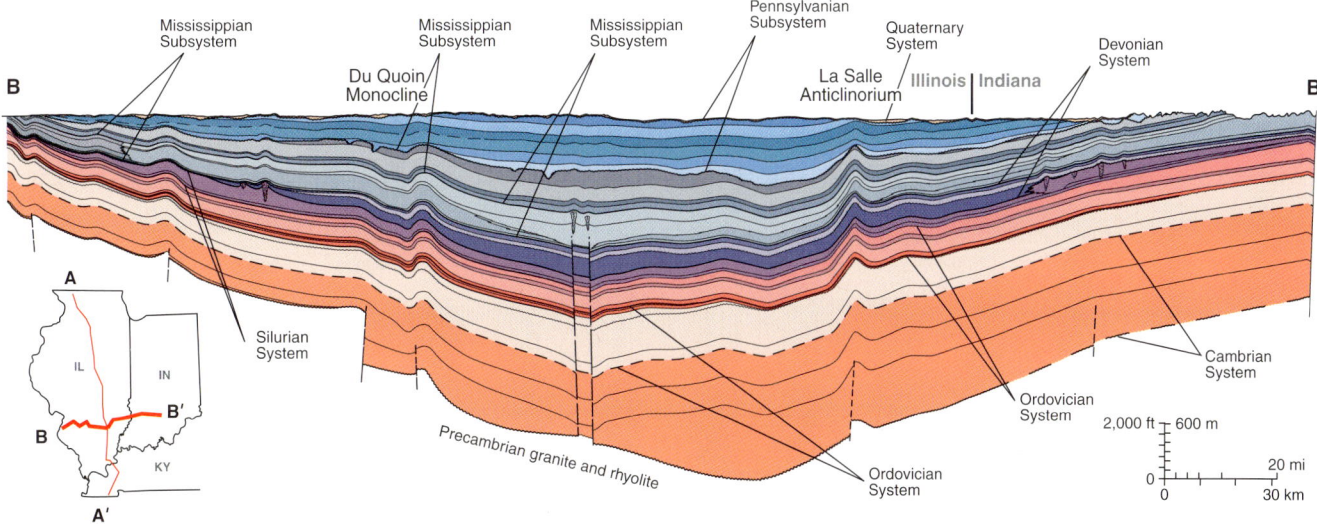

Figure 2-3 North-south cross section showing primary rock types and selected stratigraphic formations in the subsurface of Illinois and western Kentucky (Whitaker et al. 1992, Treworgy and Whitaker 1997, Kolata 2005).

Figure 2-4 West-east cross section showing primary rock types and selected stratigraphic formations in the subsurface of Illinois and eastern Indiana (Treworgy and Whitaker 1990, Whitaker and Treworgy 1990, Kolata 2005).

Other major geological structures in the Illinois Basin include (1) the La Salle Anticlinorium and Du Quoin Monocline, both of which delimit the Fairfield Basin; (2) the Cottage Grove, Rough Creek-Shawneetown, Lusk Creek, and Wabash Valley Fault Systems; (3) the Fluorspar Area Fault Complex; (4) the Cap au Grès Faulted Flexure; (5) the Ste. Genevieve Fault Zone; and (6) the Sangamon Arch (Figure 2-6). The Western Shelf and its southern extension, the Sparta Shelf, stand out as relatively flat regions in the Basin that underwent little structural deformation. Notable structures that occur north of the Basin include the Sandwich and Plum River Fault Zones. Some of the most intensely deformed areas within the state include the Des Plaines and Glasford Disturbances, which are thought to be ancient impact features. Hicks Dome and Omaha Dome are roughly circular structures that lie over deep igneous intrusions that occurred during latest Paleozoic time. These and many other structures are discussed in detail in Chapter 4.

Most Paleozoic sequences in the Illinois Basin thicken in the region of the New Madrid Rift System, suggesting that the processes operating there were linked to Basin

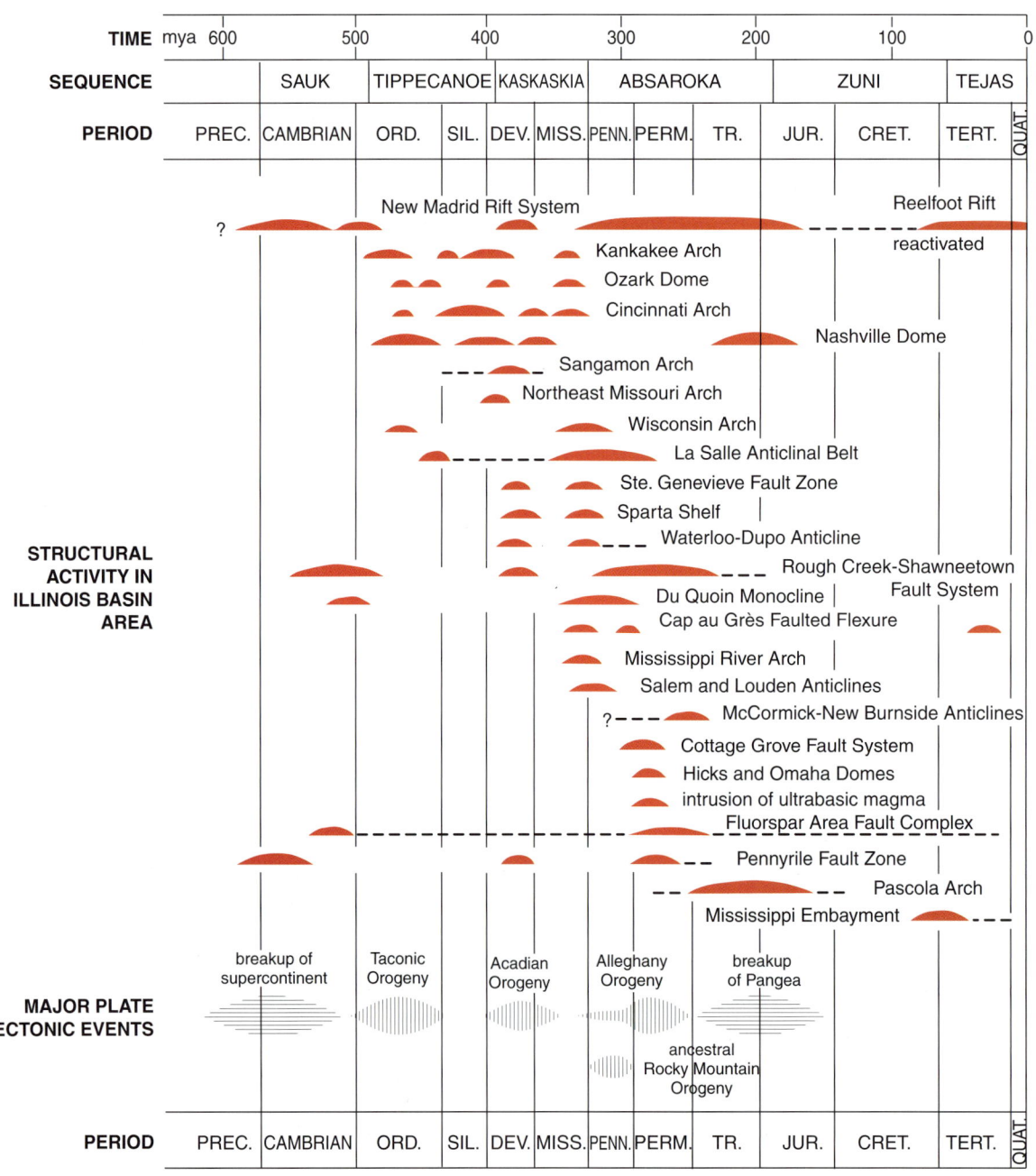

Figure 2-5 Diagram showing the correlation between structural activity (red) and major plate tectonic events (striped) in Illinois (modified from Kolata and Nelson 1991). Dashed lines indicate suspected periods of structural deformation. Abbreviations: mya, million years ago; PREC., Precambrian; ORD., Ordovician; SIL., Silurian; DEV., Devonian; MISS., Mississippian; PENN., Pennsylvanian; PERM., Permian; TR., Triassic; JUR., Jurassic; CRET., Cretaceous; TERT., Tertiary; QUAT., Quaternary.

subsidence. Structural and stratigraphic evidence indicates that the Illinois Basin likely began to form during the late Precambrian to Early Cambrian time as a rift along the southern margin of the continent. Rifting ended, however, before sea-floor spreading began, and the southeastern portion of the continent remained attached to the North American plate. Thermal cooling and isostatic adjustments in the crust followed, causing the rift and surrounding region to continue subsiding throughout most of the Paleozoic Era. Thickness patterns in the Late Cambrian Mt. Si-

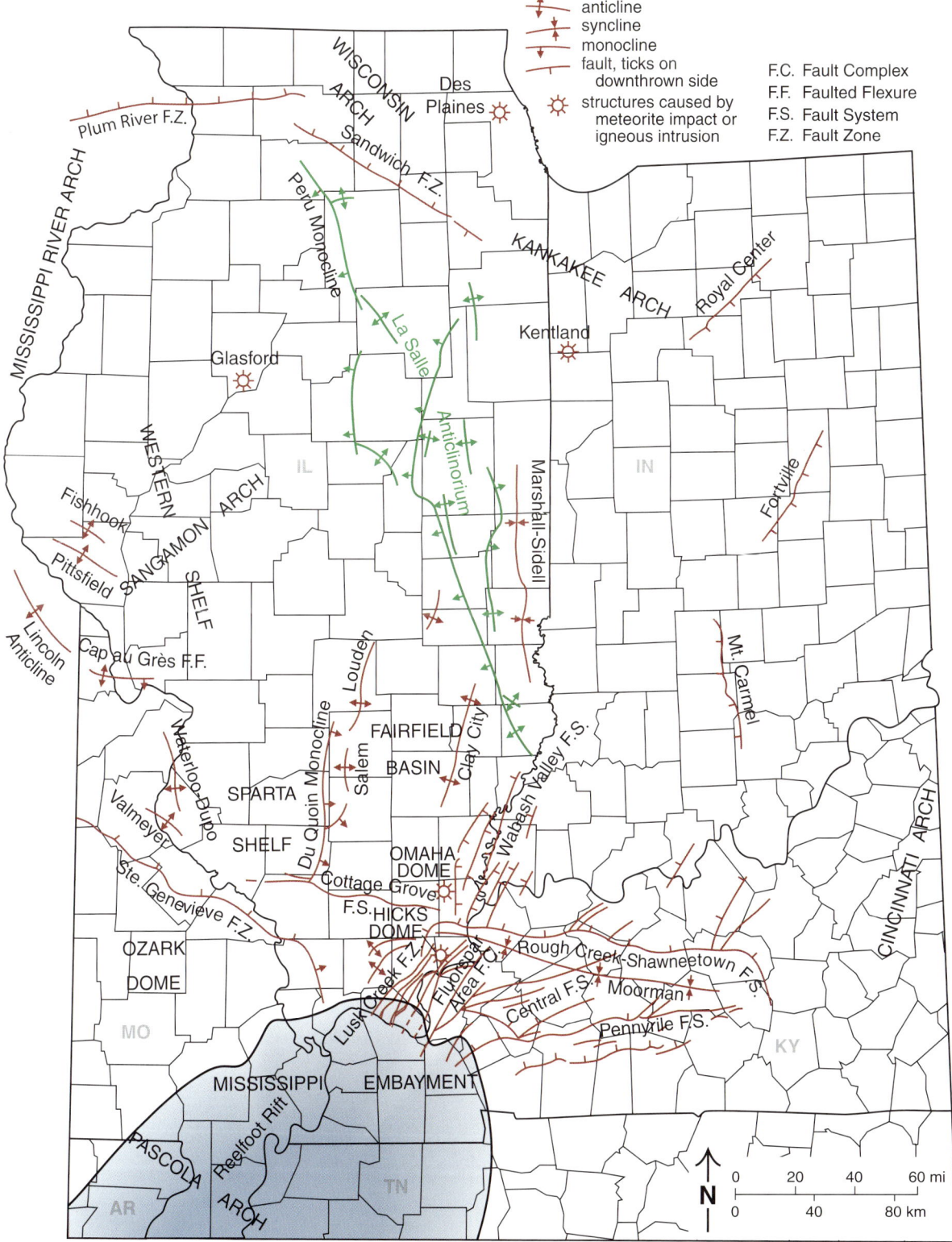

Figure 2-6 Major structural features in Illinois and adjacent states (Buschbach and Kolata 1991, Nelson 1995). Shaded area is the Mississippi Embayment of the Gulf Coastal Plain.

mon Sandstone suggest that there may have been a separate and discrete area of subsidence in east-central Illinois before or concurrent with the formation of the New Madrid Rift System.

THE ROCK RECORD AND TIME

Illinois is underlain by a thick succession of sedimentary rocks and surficial deposits that accumulated in somewhat horizontal layers, called strata, one atop the other over hundreds of millions of years. Stratigraphy is the science by which geologists decipher geological history, using two fundamental categories of information: (1) earth materials and (2) time or geological age (Figure 2-7). Earth materials, also called the rock record, are described by their content, attributes, and physical limits. The classification scheme used most often to define these materials is referred to as lithostratigraphy, which consists of a hierarchy of lithostratigraphic units. The concept of time in the geologic record is commonly expressed as either relative or absolute time. Relative time is the order in which a sequence of past events occurred and is documented and displayed in chronostratigraphic units. Absolute time is the time in numbers of years in which an event occurred or will occur and is expressed in geochronometric units. Somewhat more abstract are diachronic units, which are a means of comparing the spans of time represented by earth materials that have different age boundaries at different localities. Definitions and examples of these classification schemes are given in the following paragraphs.

Lithostratigraphic Units

A lithostratigraphic unit is a body of strata that is delimited and distinguished by composition and stratigraphic position. For example, the Middle Devonian Grand Tower Limestone (Figure 2-8) is noticeably different in composition from the strata above and below it, making the unit relatively easy to map in the field. The basic unit in lithostratigraphic classification is the formation, which can be subdivided locally into members and beds. Over time, geologists working in Illinois have identified and formally named several hundred formations, members, and beds. In Illinois, formations can be combined to form groups or supergroups.

The supergroups, groups, and formations that make up the lithostratigraphic succession are commonly bounded by unconformity surfaces. These unconformity surfaces formed during periods of erosion or nondeposition. Some unconformities are localized and represent small intervals of time; others span great distances and formed over long periods. The major unconformities divide the stratigraphic record into discrete sedimentary packages that are thought to represent global changes in sea level that lasted tens of millions of years. Sloss (1963) noted the presence of six major unconformity-bounded packages of rock in North America. He referred to these packages as sequences and named them after North American Native American tribes. Parts of the six primary sequences (Sauk, Tippecanoe, Kaskaskia, Absaroka, Zuni, and Tejas) are present in Illinois and are useful in interpreting the Paleozoic and Mesozoic sedimentary history (Figures 2-8 and 2-9).

Chronostratigraphic, Geochronometric, and Diachronic Units

Two kinds of time, relative and absolute, are used to reconstruct geological history. Relative time is based on the order of a sequence of events. In any succession of sedimentary rocks, the layers, or strata, were deposited from bottom (oldest) to top (youngest). The relative age of any two layers can be determined by their relative positions in the rock layers, assuming that the layers were not overturned after deposition. Fossils are important in determining the relative age of rocks, particularly over the last half billion years, because they record the gradual and irreversible change of life-forms during this time. Particular fossils occur at predictable points in rock layers; for example, fossil trilobites occur in the stratigraphic record before mastodons; therefore, trilobites are relatively older. Across wide areas of Illinois, the relative age of a given sedimentary layer can be determined by noting the presence of key fossils and tying their relative ages to the geological time scale. Relative time is expressed in chronostratigraphic units (Figures 2-7, 2-8, and 2-9).

Chronostratigraphic units, also known as time-stratigraphic units, represent all of the rocks, regardless of composition, that were formed during a specific interval of geological time. The boundaries are of the same age and time value everywhere. During field work, geologists commonly recognize boundaries in the field by the first appearance in the rock succession of a particular fossil or by a key bed,

MATERIAL UNITS	TIME UNITS		
LITHOSTRATIGRAPHIC UNITS	CHRONOSTRATIGRAPHIC UNITS	GEOCHRONOMETRIC UNITS	DIACHRONIC UNITS
Supergroup	Eonothem	Eon	
Group	Erathem	Era	
Formation	System	Period	Episode
Member	Series	Epoch	Phase
Bed	Stage	Age	Span

Figure 2-7 Stratigraphic classification categories used in this book.

known as a marker bed, such as a volcanic ash layer. The primary unit in chronostratigraphic classification is the system, which represents a time span of Earth history great enough to be used worldwide. Systems are further subdivided into series and stages.

In contrast to relative time, absolute time is the actual time measured in years before the present. Absolute or numerical age is determined from the decay rate of radioactive isotopes or other time-dependent properties that are measurable in the rocks. Absolute time is expressed in geochronometric units (Figures 2-7 and 2-10). The primary unit is the geological period, which is the time (in millions of years before the present) during which a geological system is formed. For example, the Devonian System, which is a chronostratigraphic unit, started and stopped forming at the same time everywhere worldwide. The Devonian System formed entirely during the Devonian Period, which occurred between 417 and 354 million years ago (U.S. Geological Survey 2005). The geochronometric time units eon, era, period, epoch, and age correspond to but are distinguished from the chronostratigraphic units eonothem, erathem, system, series, and stage, respectively (Figure 2-7).

Following a long-established tradition, geologists have found it convenient to group the systems and periods in both the chronostratigraphic and geochronometric time scales into four major units reflecting profound changes that occurred in living things. These time units (era and erathem) are ordered oldest to youngest: Cryptozoic (origin of life: simple organisms), Paleozoic (ancient life: including trilobites, cephalopods, corals, fishes, amphibians, many plants), Mesozoic (intermediate life: reptiles), and Cenozoic (recent life: mammals) (Figures 2-8 and 2-10). These units are further grouped into the Proterozoic and Phanerozoic eonothems or eons. The Phanerozoic Eon (revealed life) constitutes the last half billion years and is the time during which multicellular organisms left a detailed fossil record and built up complex and diverse ecosystems. Little is known about the Proterozoic and earlier history of Illinois.

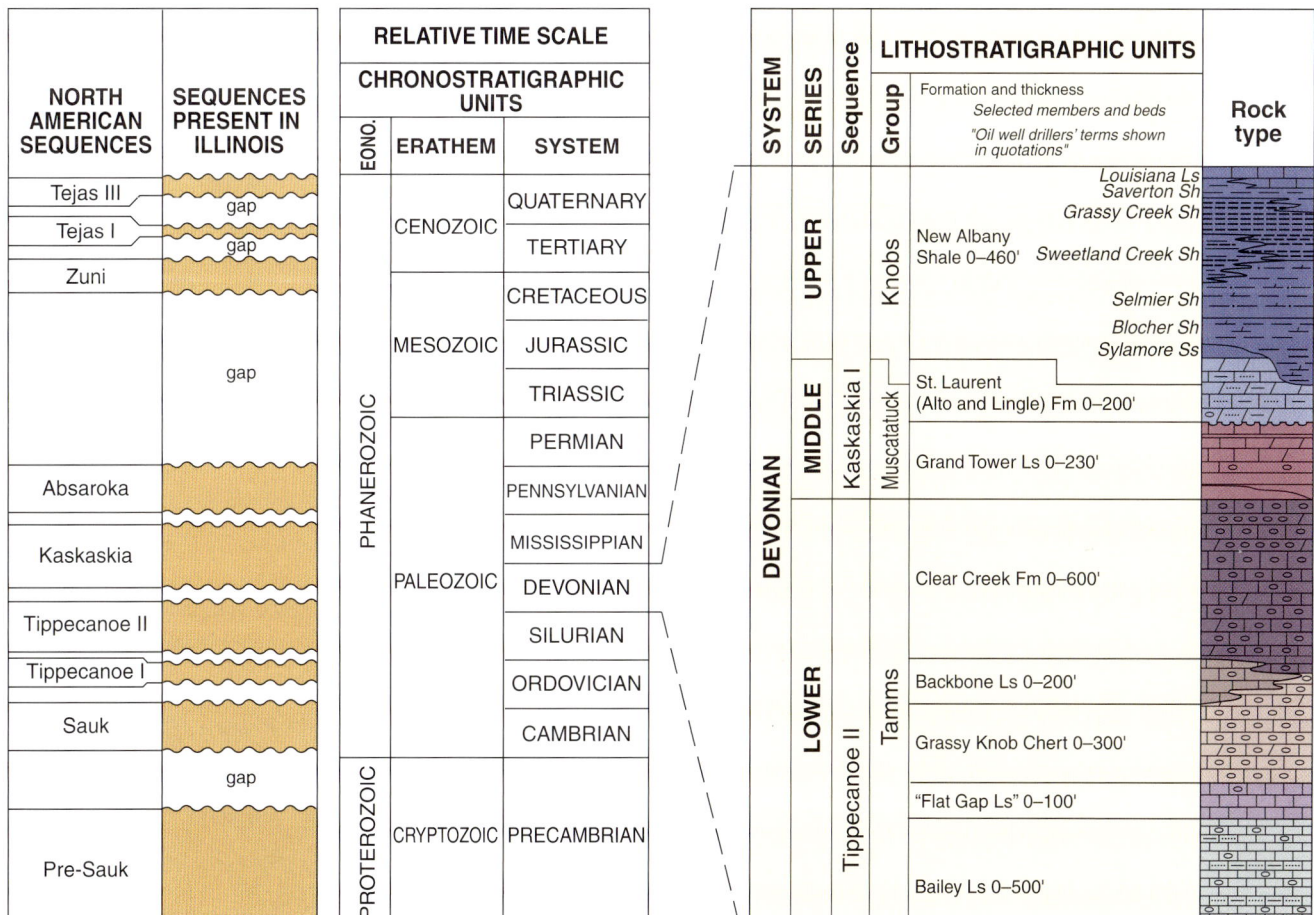

Figure 2-8 Named North American stratigraphic sequences (Sloss 1963) that are present in Illinois (shaded) correlated with chronostratigraphic units of the relative time scale. In North America, the Pennsylvanian and Mississippian are regarded as subsequences of the Carboniferous System. Subsequences are shown for Tejas and Tippecanoe. Major gaps in the stratigraphic record are due to erosion or non-deposition. The Devonian System is used as an example of the lithostratigraphic classification scheme used for material units in Illinois. Drawing is not to scale. Abbreviations: EONO., eonothem; Fm, formation; Ls, limestone; Sh, shale; Ss, sandstone.

The approach to interpreting the geological history of the Quaternary System in Illinois is slightly different from that used for the bedrock systems (Figure 2-11). Abundant radiocarbon age data collected from the midcontinental United States, including Illinois, show that the boundaries of Quaternary deposits are highly diachronous; that is,

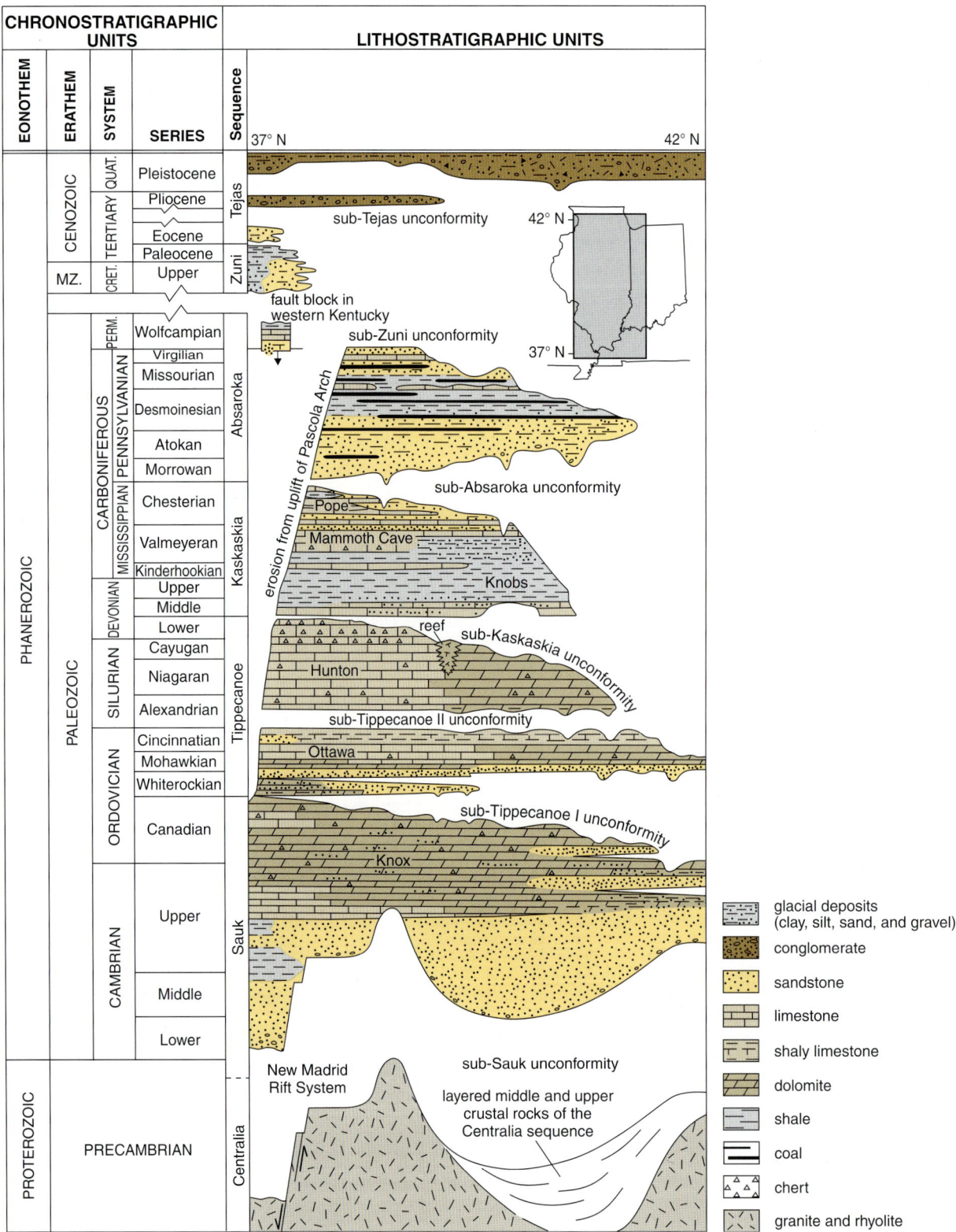

Figure 2-9 Diagrammatic cross section of Illinois representing a wide swath (shaded on inset map) from the Wisconsin border on the north (far right side of diagram) to the southern tip of the state (modified from Kolata 1991). The cross section shows that the stratigraphic record is most complete in south-central Illinois. Abbreviations: CRET., Cretaceous; MZ., Mesozoic; PERM., Permian; QUAT., Quaternary.

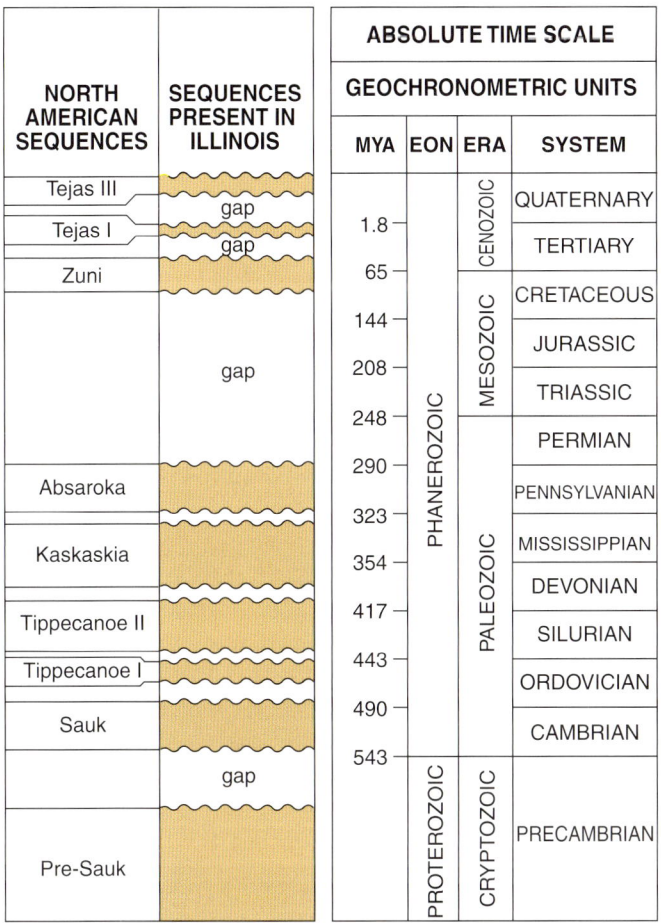

Figure 2-10 Named North American stratigraphic sequences (Sloss 1963) that are present in Illinois (shaded) correlated with geochronometric units of the absolute time scale. In North America, the Pennsylvanian and Mississippian are regarded as subsystems of the Carboniferous System. Abbreviation: MYA, million years ago.

their age differs greatly from one location to another as a result of the advance and retreat of the glaciers. Because these deposits can be more accurately dated and tend to lie closer to the surface, their diachronous characteristics are useful to study and interpret the geological events that caused them. The time-transgressive nature of Quaternary deposits is represented by diachronic units, which begin and end at different times, depending on location. The basic unit in diachronic classification is an episode, which may be subdivided into phases and spans.

BEDROCK AND SURFICIAL DEPOSITS

Much of the current understanding of the bedrock and surficial deposits in Illinois has been driven by a desire (1) to locate economically valuable mineral resources such as water, coal, oil, natural gas, lead, zinc, fluorite, crushed stone, and sand and gravel; (2) to site and construct foundations, dams, highways, tunnels, and underground facilities for natural gas storage, flood and pollution control, and

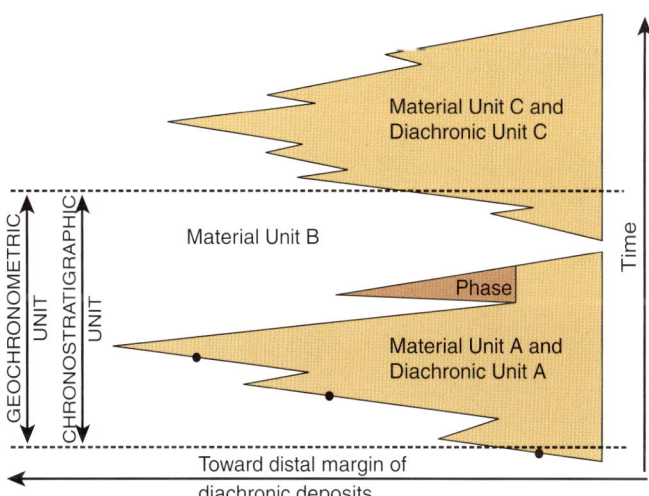

Figure 2-11 Comparison of geochronometric, chronostratigraphic, and diachronic units (modified from the North American Commission on Stratigraphic Nomenclature 1983). Colored areas show the areal extent of material units that define diachronic units A and C. Plotted against time and distance, it is obvious that the boundaries of geochronometric and chronostratigraphic units are synchronous, whereas those of diachronic units differ depending on location.

waste disposal; (3) to identify potential geological hazards such as landslides and earthquake-sensitive sediments and faults; and (4) to protect the environment by locating areas sensitive to contamination. An understanding of the thickness, physical properties, and distribution of bedrock and surficial deposits is fundamental to understanding the economic, engineering, and environmental aspects of Illinois geology.

Generally speaking, earth materials present beneath the surface of Illinois, from the top down, are soil, surficial deposits, and bedrock. The soil is a weathered zone, typically a few feet thick, that developed in surficial sediments or directly on bedrock. Surficial deposits, in places as much as 500 feet (152 m) thick, are largely of the Quaternary System and include peat, clay, silt, sand, and gravel deposited by wind, water, glacial ice, or some combination of these processes. The underlying sedimentary bedrock consists mainly of Paleozoic rocks, mostly marine in origin, that vary in thickness from 2,000 feet (610 m) in northern Illinois to about 20,000 feet (6,096 m) in southern Illinois. These rocks overlie Precambrian igneous rocks, and, in parts of east-central Illinois, possibly Precambrian sedimentary and volcaniclastic rocks (McBride et al. 2003). Dolomite and limestone make up more than half of the Paleozoic rocks; sandstones, almost one-quarter. The remainder is shale, siltstone, chert, coal, and minor amounts of anhydrite (Buschbach and Kolata 1991). Most Paleozoic sequences thicken southward toward centers of deposition in southern Illinois and adjacent parts of western Kentucky and southeastern Missouri. The sedimentary record tends to be less complete in the areas of uplift (arches and domes) because of erosion or lack of deposition and tends to be more complete in the subsiding basins. The distribution of units at the bedrock surface in Illinois is shown on the bedrock geology map (Figure 2-12). Most of the major subdivisions (series) of the Paleozoic systems are present in Illinois except for the Lower and Middle Cambrian and most of the Permian. The distribution of Paleozoic systems at the bedrock surface is largely controlled by local geological structure, unconformities, and topography. The Paleozoic succession and geological history are discussed in Chapters 7 through 10.

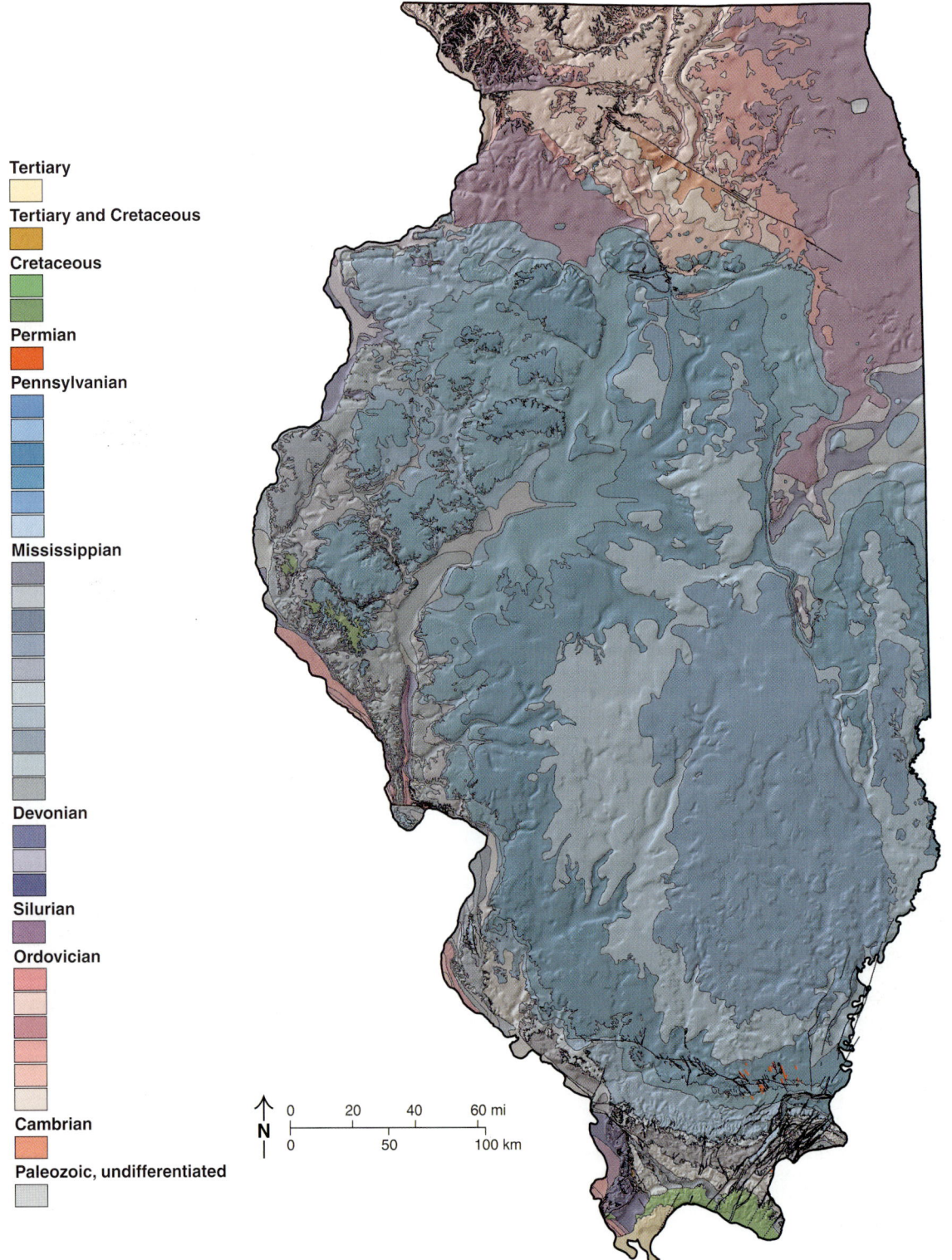

Figure 2-12 Bedrock geology of Illinois (Kolata 2005).

Overview

Mesozoic rocks are poorly represented in Illinois. The Triassic, Jurassic, and lower Cretaceous rocks are absent. In southernmost Illinois, marine, deltaic, and nearshore sediments of Late Cretaceous and Tertiary age overlap Paleozoic strata (Figure 2-12). Outliers of Late Cretaceous sands and gravels also occur in limited parts of western Illinois. The Tertiary is represented by the Paleocene, Eocene, and Pliocene Series; the Oligocene and Miocene are absent. Mesozoic and Cenozoic rocks are described in Chapter 11.

Approximately 2.5 million years ago, prior to the buildup of continental ice sheets and the advance of glaciers into Illinois, the land surface was rugged, marked by steep hills, abrupt bluffs, and deep valleys. That ancient land surface had been shaped by millions of years of erosion before being covered by glacial deposits and can still be seen today in the unglaciated areas of northwestern and far southern Illinois (Figure 2-13). Each advance of the glaciers diverted rivers, filled river and stream valleys, and deposited thick layers of clay, silt, sand, and gravel on the ancient bedrock surface. The configuration of this buried bedrock surface is revealed on the bedrock topography map (Figure 2-14), which was compiled with information from thousands of drill holes throughout the state.

Glacial drift, deposited during at least six glacial episodes during the past 2.5 million years, mantles the bedrock surface throughout Illinois, except for the unglaciated areas in southernmost and northwestern Illinois. Glacial outwash and alluvium are present along major stream courses. Clay, silt, sand, and gravel were transported into Illinois by wind, water, and ice from the northwest, northeast, and east, shaping most of the present topographic surface of the state (Figure 2-15). Because of the moderate to thick, fertile soils that developed in the glacial deposits, Illinois is one of the richest agricultural regions of the world. A description of Quaternary deposits and landforms and an account of the geological history during glacial times are presented in Chapter 12. Climate changes during the Quaternary Period are discussed in Chapter 13.

Modern Landscapes

The state of Illinois is 390 miles (628 km) long and 210 miles (388 km) wide; it covers an area of 57,918 square miles

Figure 2-13 Camel Rock, Garden of the Gods, Shawnee National Forest, in the unglaciated region of southern Illinois. (Photograph by Joel M. Dexter.)

Figure 2-14 Bedrock topography of Illinois (Abert 2005).

(150,007 km²), making it the 25th largest of the 50 states. Illinois is bordered on the west by the Mississippi River and on the south-southeast by the Wabash and Ohio Rivers, and its northeast corner lies along the Lake Michigan shoreline. About 90% of Illinois lies in the central plains, a landscape of gently rolling plains formed by glacial processes, in the stable interior of the North American continent.

Distinct physiographic regions are defined by the state's modern landscape (Figure 2-16). Most of Illinois is situated in the south-central part of the Central Lowlands Province, all of which was glaciated except the Wisconsin Driftless Section. Other major provinces that extend into the state include the Ozark Plateaus in southwestern Illinois, Interior Low Plateaus in southern Illinois, and the Coastal Plain Province in the southern tier of counties, all of which lie outside the southern glacial boundary (Leighton et al. 1948).

The Till Plains Section, mantled with unsorted and unconsolidated glacial deposits, covers about four-fifths of the state and is subdivided into the Rock River Hill Country, Green River Lowland, Galesburg Plain, Bloomington Ridged Plain, Kankakee Plain, Ancient Illinois Floodplain, Griggsville Plain, Springfield Plain, and Mt. Vernon Hill Country (Figure 2-16). These subdivisions are based on subtle landform differences (Leighton et al. 1948). The till plains in most of Illinois are in a youthful stage of erosion. In contrast, the maturely eroded Lincoln Hills Section of western Illinois and the Dissected Till Plains Section in adjacent parts of northern Missouri and southern Iowa occur on older glacial deposits. In those places where the ice front remained stationary for long periods of time (tens to hundreds of years), glacial sediment accumulated in a series of broad end moraines up to 100 feet (30.5 m) high and as much as 5 miles (8 km) wide. Some of the more prominent end moraines occur in the Bloomington Ridged Plain and in the Wheaton Morainal Country of the Great Lakes Section in northeastern Illinois (Figure 2-16). Moraines are discussed in Chapter 12.

The relief over most of the state is moderate to slight, and the mean elevation is about 600 feet (183 m) above sea level. The highest point in the state, 1,241 feet (378 m), is Charles Mound in north-central Jo Daviess County. The lowest point, 268 feet (82 m), is at the junction of the Ohio and Mississippi Rivers in southern Alexander County.

NATURAL RESOURCES, LAND USE, AND ENVIRONMENTAL HAZARDS

Illinois has a wealth of natural resources that are vitally important to the economic development and well-being of the state. Some of the more obvious geological resources include (1) industrial minerals (Chapter 17) such as limestone, sand, and gravel needed for building purposes (e.g., highways, dams, bridges, and buildings); (2) fuels such as coal (Chapter 14), oil and natural gas (Chapter 15) for energy; (3) water for domestic, agricultural, and industrial purposes as well as sustainable wetlands (Chapters 18 through 21); and (4) soils (Chapter 22) for the production of food. Although not mined presently, Illinois also has significant deposits of lead, zinc, and fluorite (Chapter 16).

In a sense, the geology that surrounds us is a "double-edged sword." Geological materials are essential for nearly all activities, including the provision of water, food, shelter, transportation, and energy. These geological materials, however, can also present serious natural hazards, such as landslides, subsidence and collapse of limestone solution

Figure 2-15 Surface topography of Illinois (Luman et al. 2003).

cavities, earthquakes, and natural radiation. Earth materials can contribute to or lessen the impacts of flooding. Many hazards and unwanted results are created by human activities, including contamination of soil and groundwater, subsidence of the land surface over abandoned underground coal mines, erosion due to poor agricultural or construction practices or shoreline modification, and degradation of the environment from surface mining. Chapters 22 through 30 focus on examples of natural and human-induced hazards and land-use issues.

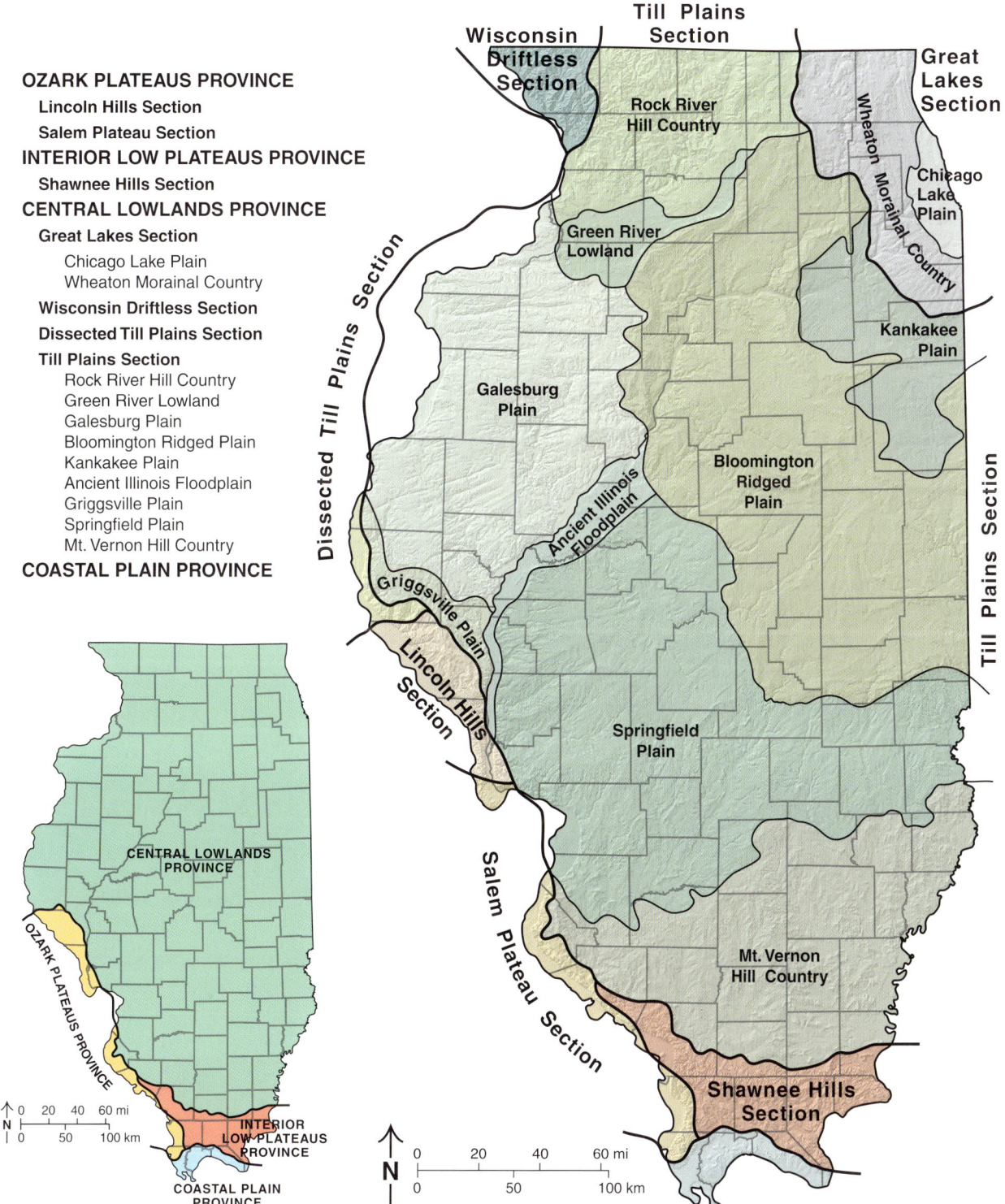

Figure 2-16 Physiographic provinces in Illinois (Phillips 2002) overlaid on the surface topography map of Illinois (Luman et al. 2003). Note province and section boundaries may extend beyond the state's borders.

Table 2-1 County location and geological information for several popular sites in Illinois. Site numbers correspond to numbered locations.

Site no.	Site name	County	Geological information
1	Apple River Canyon State Park	Jo Daviess	Ordovician dolomite bluffs; Apple River
2	Glacial Park	McHenry	Glacial moraines
3	Illinois Beach State Park	Lake	Ancient beach ridges, Lake Michigan shoreline
4	Mississippi Palisades State Park	Carroll	Silurian dolomite exposed in bluffs of the Mississippi River
5	Rock River Hill Country, Lowden and Castle Rock State Parks	Ogle	Ordovician dolomite and sandstone exposures
6	Green River Lowland	Lee, Bureau, Whiteside, Henry	Quaternary sand dunes
7	Thornton Quarry and adjacent road cuts	Cook	Silurian reef
8	Buffalo Rock, Starved Rock, Matthiessen State Parks	La Salle	Ordovician St. Peter Sandstone exposures; bluffs and ravines of the Illinois River and its tributaries
9	Kankakee River State Park	Kankakee	Evidence of the catastrophic floodwaters
10	Moraine View State Recreation Area	McLean	Glacial moraines
11	Eagle Creek State Park	Shelby	Glacial moraines
12	Fox Ridge State Park	Coles	Glacial moraines
13	Pere Marquette State Park and nearby bluffs	Jersey	Loess deposits and dipping beds of Mississippian limestone; Mississippi River
14	Illinois Caverns State Natural Area	Monroe	Caves in Mississippian limestone
15	Pine Hills area, Shawnee National Forest	Jackson	Natural exposures of Devonian, Mississippian, and Pennsylvanian rocks; Mississippi River bluffs
16	Cache River State Natural Area	Pulaski, Johnson	Site of ancient abandoned Ohio River valley
17	Bell Smith Springs	Pope	Scenic canyons in Pennsylvanian Caseyville sandstone
18	Garden of the Gods, Shawnee National Forest	Saline	Scenic ridge of Pennsylvanian Caseyville sandstone
19	Cave-In-Rock State Park	Hardin	Cave in Mississippian St. Louis Limestone along the Ohio River bluff

GEOLOGICAL ATTRACTIONS IN ILLINOIS

Numerous interesting landscapes and geological features are present in Illinois, some of which are accessible in the approximately 130 state parks and state recreation, wildlife, and natural areas (Illinois Department of Natural Resources 2009). Many of the state parks have visitor centers, interpretive programs, or posted information about the local geology. Several of the more popular sites and their locations are presented in Table 2-1.

REFERENCES

Abert, C. C., 2005, Shaded relief map of the bedrock surface of Illinois: Illinois State Geological Survey, GIS database layer, 1:500,000.

Buschbach, T. C., and D. R. Kolata, 1991, Regional setting of the Illinois Basin, in M. W. Leighton, D. R. Kolata, D. F. Oltz, and J. J. Eidel, eds., Interior cratonic basins: Tulsa, Oklahoma, American Association of Petroleum Geologists, Memoir 51, p. 29–55.

Ervin, C. P., and L. D. McGinnis, 1975, Reelfoot Rift—Reactivated precursor to the Mississippi Embayment: Geological Society of America Bulletin, v. 86, no. 9, p. 1287–1295.

Illinois Department of Natural Resources, 2009, Parks and recreation. http://www.dnr.state.il.us/lands/landmgt/PARKS/index.htm. Accessed October 14, 2009.

Kolata, D. R., 1991, Overview of sequences—Illinois Basin, *in* M. W. Leighton, D. R. Kolata, D. F. Oltz, and J. J. Eidel, eds., Interior cratonic basins: Tulsa, Oklahoma, American Association of Petroleum Geologists, Memoir 51, p. 59–73.

Kolata, D. R., compiler, 2005, Bedrock geology of Illinois: Illinois State Geological Survey, Illinois Map 14, 1:500,000.

Kolata, D. R., B. D. Keith, and J. A. Drahovzal, 1992, Illinois Basin Consortium program plan: Illinois State Geological Survey, Illinois Basin Studies, 21 p.

Kolata, D. R., and W. J. Nelson, 1991, Basin-forming mechanisms of the Illinois Basin, *in* M. W. Leighton, D. R. Kolata, D. F. Oltz, and J. J. Eidel, eds., Interior cratonic basins: Tulsa, Oklahoma, American Association of Petroleum Geologists, Memoir 51, p. 287–294.

Leighton, M. M., G. E. Ekblaw, and L. Horberg, 1948, Physiographic Divisions of Illinois: Illinois State Geological Survey, Report of Investigations 129, 18 p.

Luman, D. E., L. R. Smith, and C. C. Goldsmith, 2003, Illinois surface topography: Illinois State Geological Survey, Illinois Map 11, 1:500,000.

McBride, J. H., D. R. Kolata, and T. G. Hildenbrand, 2003, Geophysical constraints on understanding the origin of the Illinois Basin and its underlying crust: Tectonophysics, v. 363, no. 1–2, p. 45–78.

Nelson, W. J., 1995, Structural features in Illinois: Illinois State Geological Survey, Bulletin 100, 144 p.

North American Commission on Stratigraphic Nomenclature, 1983, North American stratigraphic code: American Association of Petroleum Geologists Bulletin, v. 67, no. 5, p 841–875.

Phillips, A. C., 2002, Geomorphic summary of the Illinois River basin, Contract Report DACW25-98-D-0017, U.S. Army Corps of Engineers, Rock Island District, Rock Island, Illinois: Illinois State Geological Survey.

Sloss, L. L., 1963, Sequences in the cratonic interior of North America: Geological Society of America Bulletin, v. 74, no. 2, p. 93–114.

Soderberg, R. K., and G. R. Keller, 1981, Geophysical evidence for deep basin in western Kentucky: American Association of Petroleum Geologists Bulletin, v. 65, no. 2, p. 226–234.

Treworgy, J. D., and S. T. Whitaker, 1990, 3 o'clock cross section in the Illinois Basin: Wayne County, Illinois, to Switzerland County, Indiana: Illinois State Geological Survey, Open File Report 1990-3, 1" = 400' vertical; 1:250,000 horizontal.

Treworgy, J. D., and S. T. Whitaker, 1997, Structural cross section of the Paleozoic rocks in Illinois, Wayne County to Stephenson County: Illinois State Geological Survey, Illinois Map 7, 1:250,000, two sheets; booklet, 12 p.

U.S. Geological Survey, 2005, The geologic time scale. http://vulcan.wr.usgs/glossary/geo_time_scale.html. Accessed October 22, 2009.

Whitaker, S. T., and J. D. Treworgy, 1990, 9 o'clock cross section in the Illinois Basin: Wayne County, Illinois, to St. Clair County, Illinois: Illinois State Geological Survey, Open File Report 1990-4, 1" = 400' vertical; 1:250,000 horizontal.

Whitaker, S. T., J. D. Treworgy, and M. C. Noger, 1992, 6 o'clock cross section in the Illinois Basin, Wayne County, Illinois, to Gibson County, Tennessee: Illinois State Geological Survey, Open File Report 1992-10, 1" = 400' vertical; 1:250,000 horizontal.

Tectonics and Structural Geology of Illinois

Dennis R. Kolata

The structural framework of Illinois, like that of the rest of the North American continent, is largely the result of external forces created by the motion of tectonic plates. Driven by processes operating deep in the Earth's interior, the tectonic plates have undergone a long history of collision, divergence, and sub-crustal heating. As a result, the underpinnings of Illinois consist of assorted blocks of continental crust bounded by ancient faults and zones of weakness. Plate tectonic interactions operating over the past 1.5 billion years have reactivated these crustal elements, influencing the formation of basins and arches, sedimentation rates, facies relationships, regional subsurface fluid flow systems, ancient and modern river systems, and earthquake activity. These interactions are also responsible for the geological structures in Illinois, including high-angle faults, force folds, reverse faults, detached thrust folds, and strike-slip deformation as well as local intrusions of magma.

The three chapters contained in this part of the book include a brief review of Illinois' tectonic history, an overview of the major structural features, and a discussion of neotectonics, the tectonic events and processes that have occurred during post-Miocene time (approximately the past 5 to 6 million years).

Photograph by Joel M. Dexter.

Image on previous page: Alto Pass, Shawnee National Forest, Union County, Illinois.

Tectonic History 3

Dennis R. Kolata and W. John Nelson

INTRODUCTION

Plate tectonic theory provides insight into the driving forces that have shaped the long and varied geological history of Illinois. In brief, the theory recognizes that the upper part of the Earth's crust is composed of rigid tectonic plates (slabs) that slide laterally over a layer of weaker, viscous rock in the underlying upper mantle known as the asthenosphere (Figure 3-1). Powered by the Earth's internal heat energy and moving at a fraction of an inch per year, the plates can flex up or down and break apart or collide with one another. Where plates collide, compressional stresses tend to deform the plate margins, forming mountain ranges. Geological structures situated within plate interiors, far from the margins, are subject to reactivation during plate collisions. The far-field effects of the compressional stress commonly result in offsets along fault zones, uplift of basement rocks, and renewed subsidence of sedimentary basins. In contrast, where plates move apart, extensional stresses take over, commonly resulting in rift zones, characterized by long narrow fractures in oceanic or continental crust (Figure 3-1). Rifts commonly form over regions where there is an upwelling of basaltic magma and resultant thinning of the crust.

Even though Illinois is situated within the stable interior of the North American plate, the state has not been immune to the far-reaching effects of plate tectonic interactions. Basin subsidence, sedimentation, formation of geological structures, migration of groundwater and hydrocarbons, and contemporary earthquake activity in Illinois are a direct result of these processes.

Because Illinois lies far removed from ancient and modern deformed plate margins, the succession of sedimentary rocks is mostly flat-lying or gently tilted, relatively thin, and punctuated by major unconformities. Seaward from Illinois, along the continental margins, geological structures consist largely of deep, sediment-filled troughs and highly deformed fold-and-thrust belts.

The stable interior of North America—the North American craton—is a collage of Precambrian tectonic terranes that formed by lateral accretion more than a billion years ago (Sims et al. 1987, Karlstrom et al. 1999). Outside the ancient Canadian Shield are successively younger terranes (Figure 3-2). Illinois lies within the Eastern

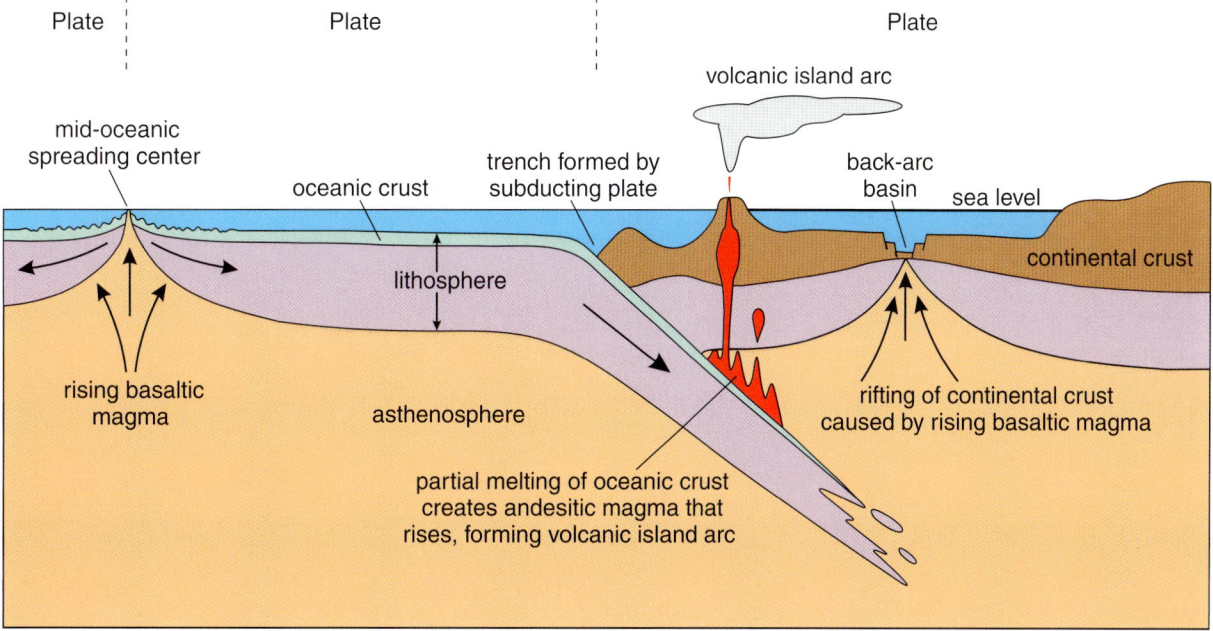

Figure 3-1 Diagrammatic cross section of magma rising from the Earth's asthenosphere, causing oceanic and continental plates to move laterally. On the left side, a mid-oceanic spreading center creates oceanic crust that forms two diverging plates. At center, oceanic crust collides with less dense continental crust causing the oceanic crust to be pushed under the continent (subduction). Above the subduction zone, a deep oceanic trench forms. The sinking oceanic crust eventually melts, forming andesitic magma that rises into the overlying continental crust, creating an explosive volcanic island arc. On the right, tensional stress from the subduction processes causes the continental crust to break up, forming a back-arc basin.

Granite-Rhyolite Province, a complex of basement rocks that extends eastward from Missouri and eastern Iowa into maritime Canada. Published uranium-lead zircon age measurements indicate that the province formed 1.48 to 1.44 billion years ago (Bickford et al. 1981, Van Schmus and Bickford 1981, Hoppe et al. 1983, Thomas et al. 1984, Van Schmus et al. 1987). The granites and rhyolites appear to have been formed by the partial melting of pre-existing crust (Nelson and DePaolo 1985; Bowring et al. 1988, 1991; Van Schmus 1991). The Grenville Province accreted to the southeastern margin of the Eastern Granite-Rhyolite Province approximately one billion years ago.

Through time, large areas of the North American craton subsided to become basins as the surrounding areas

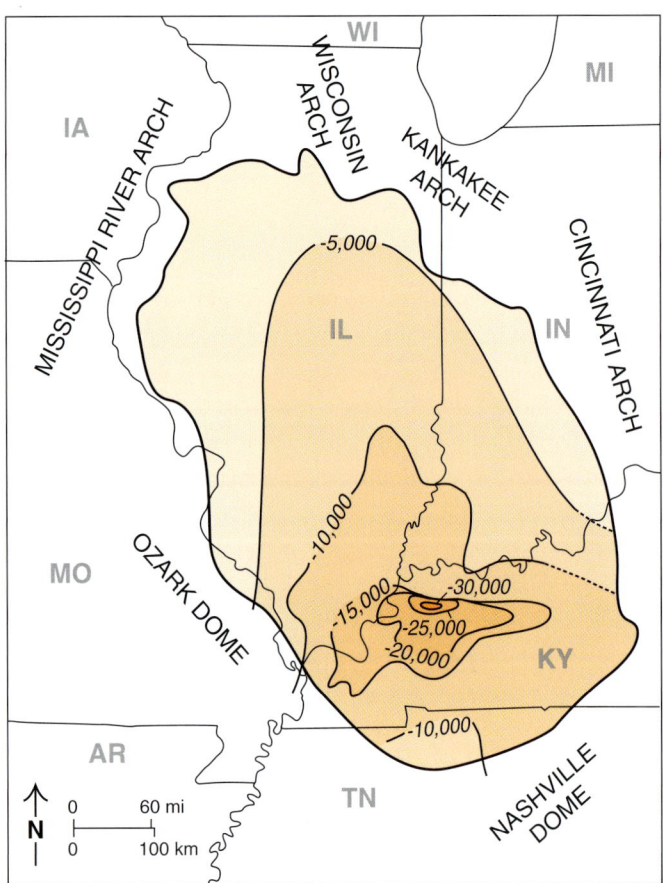

Figure 3-3 Illinois Basin (shaded) surrounded by prominent, long-lived arches and domes (modified from Buschbach and Kolata). The depth to the Precambrian basement is shown. Contour interval is 5,000 feet (1,524 m).

rose and became arches and domes. The Illinois Basin (also called the Eastern Interior Basin) is one of several large cratonic basins that developed on the Precambrian crust of North America (Leighton and Kolata 1991). The Basin is bounded by a series of prominent arches and domes (Figure 3-3). The Illinois Basin encompasses most of Illinois, contiguous portions of southwestern Indiana and western Kentucky, and small areas of Tennessee and Missouri (Collinson et al. 1988). The Illinois Basin covers an area of about 110,000 square miles (284,899 km^2). The Basin contains approximately 120,000 cubic miles (500,000 km^3) of Paleozoic rocks, consisting mainly of limestone, dolomite, sandstone, shale, and siltstone. Maximum thickness is about 30,000 feet (9,144 m) in a small area of western Kentucky (Goetz et al. 1992).

The Geological Evolution of Illinois

At least five major tectonic episodes occurred during late Precambrian and Phanerozoic time in the areas of Illinois that were characterized by distinct layering patterns that formed during periods of regional subsidence or uplift.

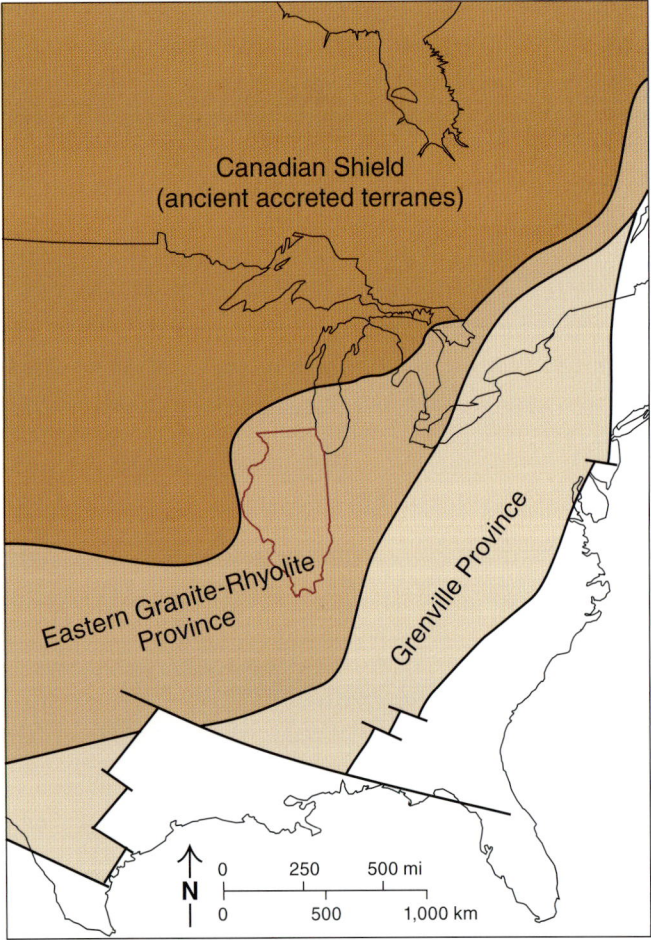

Figure 3-2 The central part of the North American plate showing major Precambrian terranes relative to Illinois. Precambrian basement rocks consist of laterally accreted terranes. The Canadian Shield, the oldest part of the continent, comprises mainly igneous and metamorphic rocks ranging from 1.5 to 2.5 billion years old. The 1.4- to 1.5-billion-year-old Eastern Granite-Rhyolite Province, which underlies most of Illinois, was formed by partial melting of older terranes. Outside of the Eastern Granite-Rhyolite Province is the 1.0- to 1.3-billion-year-old Grenville Province. The outer boundary of the Grenville Province is a rifted margin formed during several episodes of post-Grenville crustal extension (Karlstrom et al. 1999).

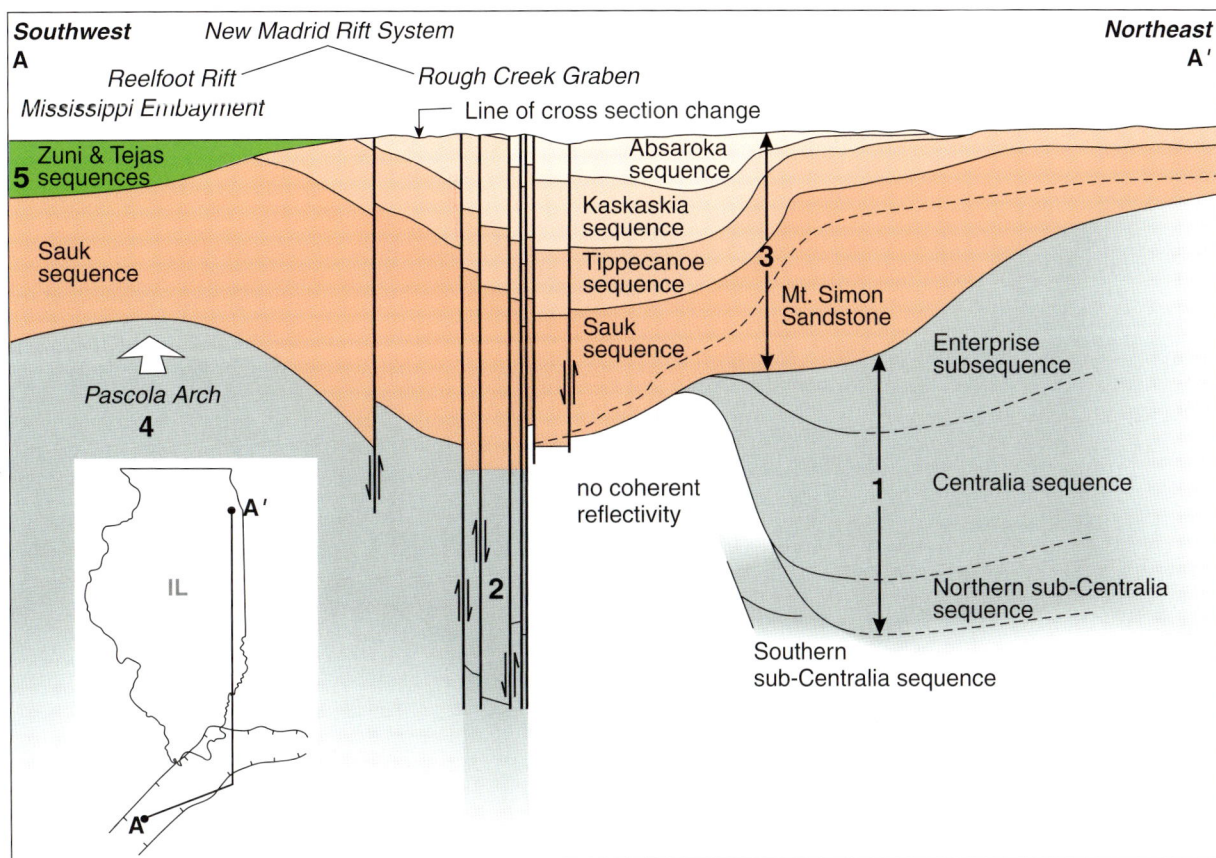

Figure 3-4 Southwest-northeast cross section (inset map shows location) showing the five main structural features formed during the major tectonic episodes in the geologic history of Illinois. **(1)** The Centralia sequence, which includes the Enterprise subsequence and sub-Centralia sequences, is centered in east-central Illinois and west-central Indiana. The sequence formed during Precambrian time. Note that the thickness pattern of the Mt. Simon Sandstone coincides with the underlying Centralia. **(2)** Crustal stretching and extension resulted in formation of the New Madrid Rift System. Half-grabens subsided rapidly, initially filling with as much as 10,000 feet (3,000 m) of probable Early and Middle Cambrian sediments, locally over 25,000 feet (7,600 m) thick. **(3)** A broad, slowly subsiding cratonic basin developed over the rift system but spread far beyond. The Paleozoic Sauk, Tippecanoe, Kaskaskia, and Absaroka sequences were deposited. Some ancient faults within the rift system were reactivated during the middle to late Paleozoic. **(4)** The Pascola Arch uplifted between Permian and Late Cretaceous time, creating the present structural configuration of the Illinois Basin. Sediment eroded from the crest of the arch was transported by rivers to the southern margin of the continent and deposited. **(5)** The Pascola Arch sank back within the rift system by the Late Cretaceous. The Mississippi Embayment subsequently subsided during the Late Cretaceous and Tertiary Periods and filled with the Mesozoic and Cenozoic Zuni and Tejas sequences.

These episodes (Figure 3-4) consist of (1) Precambrian to Late Cambrian subsidence in east-central Illinois, forming the proto-Illinois Basin containing the Centralia sequence including the Enterprise subsequence and the northern and southern sub-Centralia sequences; (2) late Precambrian to Early Cambrian faulting and rapid subsidence, forming a thick succession of strata in the Rough Creek Graben; (3) Paleozoic subsidence of the Illinois Basin forming the Sauk, Tippecanoe, Kaskaskia, and Absaroka sequences; (4) Late Pennsylvanian to Late Cretaceous uplift and erosion of the Pascola Arch in southern Illinois and adjacent parts of Missouri and Kentucky; and (5) Late Cretaceous and Tertiary subsidence centered in parts of the New Madrid Rift System, forming the Zuni and Tejas sequences of the Mississippi Embayment. For the most part, these tectonic episodes can be related to the broader forces of compression and extension (pulling apart) that occurred within the Laurentia/North American plate during late Precambrian and Phanerozoic time (Figure 3-5). Figure 2-5 (Chapter 2, Overview) shows the correlation between tectonic activities and structural features. The plate tectonic events that influenced the geology of Illinois are discussed next.

Precambrian to Late Cambrian Subsidence

Long before the Illinois Basin was formed, a similar large structural depression developed in the same general area sometime between 1.48 billion and 500 million years ago. This proto-Illinois Basin lies within Precambrian rocks as yet untouched by the drill bit, but evidence for it has been observed on regional seismic reflection profiles from east-central Illinois and west-central Indiana (Figure

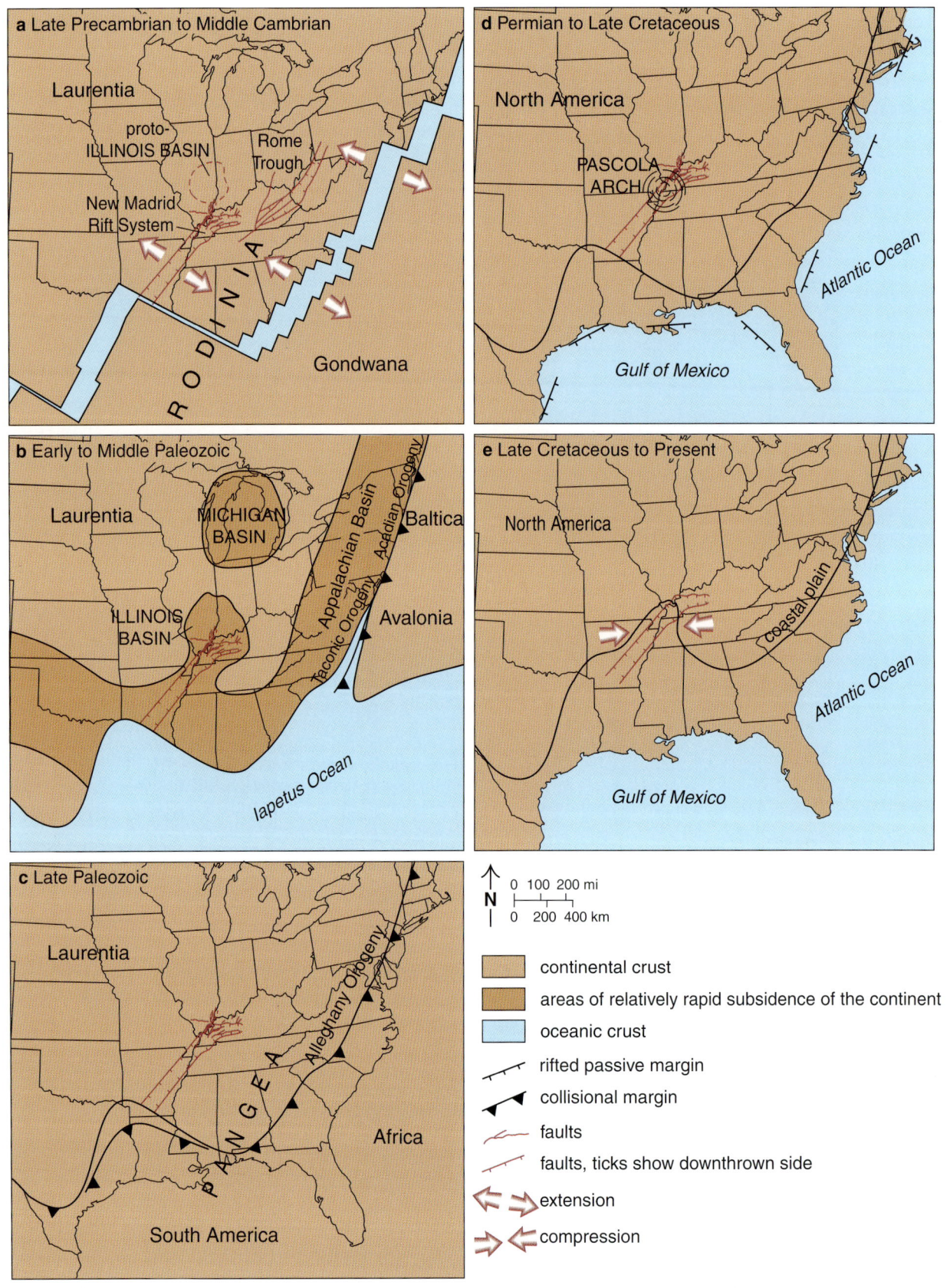

3-5a). The profiles reveal a prominent succession of reflective rocks that lie below the Cambrian Mt. Simon Sandstone (Figure 3 6) (Pratt et al. 1989, McBride and Kolata 1999, McBride et al. 2003). These rocks display a distinct, layered configuration—or fabric—and appear to sag into a depression on the surface of the granite and rhyolite basement rocks. This bowl-shaped succession has conspicuous pinch-out boundaries on the west and south (Figure 3-6a); the northern and eastern limits of the basin have not yet been defined. Pratt et al. (1989) referred to this seismic facies as the Centralia sequence. McBride and Kolata (1999) subdivided the Centralia into an upper part, a succession of unconformity-bound depositional sequences they called the Enterprise subsequence, and a lower part characterized by less coherent layering. The geometry shown by seismic reflection strongly suggests a succession of layered rocks, possibly sedimentary or volcaniclastic in composition. The thickness pattern of the Mt. Simon Sandstone generally coincides with the underlying Centralia (Figure 3-6b), suggesting that the subsidence that affected and accommodated both units was centered in east-central Illinois and west-central Indiana (McBride and Kolata 1999, McBride et al. 2003).

Figure 3-5 Periods of major tectonic plate activity that influenced the geology of Illinois (modified from Kolata and Nelson 1991b): Many tectonic events are simplified and condensed. **(a)** Late Precambrian through Middle Cambrian extension resulted in breakup of the Rodinia supercontinent and rapid sedimentation in the New Madrid Rift System and Rome Trough. The proto-Illinois Basin containing the Centralia sequence probably started to form prior to this rifting event. **(b)** Early to middle Paleozoic collision of Laurentia with the continental plates of Baltica and Avalonia (highly generalized). During Paleozoic time, the southeastern margin of Laurentia and the Illinois and Michigan Basins subsided rapidly and filled with a thick succession of sediment. The Taconic Orogeny occurred during the Ordovician in the southern and central Appalachians and involved collisions with Baltica and several microcontinents. The Acadian Orogeny took place during latest Silurian through Early Mississippian time in the northern Appalachians between New York and Newfoundland and involved collision with Avalonia. **(c)** Late Paleozoic convergence of Laurentia, Africa, and South America, forming the supercontinent Pangea. The Alleghany Orogeny resulted from these collisions. **(d)** Permian to Late Cretaceous breakup of the Pangea supercontinent and uplift and erosion of the Pascola Arch. The breakup resulted in reorganization of plates forming the present North American plate. Among other changes, Florida detached from Africa and became part of North America. **(e)** Subsidence occurring from Late Cretaceous to the present in the New Madrid Rift System has led to formation of the Mississippi Embayment of the Gulf Coastal Plain. A push away from the mid-Atlantic spreading center is presently causing widespread compressive stress in the interior of the North American plate. These stresses are reactivating ancient faults within the New Madrid Rift System, causing earthquake activity in southeastern Missouri, southern Illinois, and eastern Kentucky.

The cause of the subsidence is unclear. A rifting origin, commonly postulated for the initial subsidence in cratonic basins, appears to be unlikely because the Centralia sequence is elliptical rather than linear and is bounded by a series of arcuate faults. The sequence does not have the high gravity and high magnetic readings that would suggest basaltic igneous rocks, which typically form the floor of a rift system (Bickford et al. 1986). The overall dimensions and geometry of the Centralia sequence are similar to that of a volcanic depression, such as the Yellowstone caldera complex of northwestern Wyoming (McBride et al. 2003). Such calderas can be many tens of miles across and typically form from the rapid and catastrophic ejection of volcanic ash and gases, followed by collapse of the magma chamber roof, producing a large depression. Outcrops of the Eastern Granite-Rhyolite Province in the nearby St. Francois Mountains of east-central Missouri reveal the deeply penetrating, arcuate faults that are typical of a caldera. Volcanic rocks accumulated along these faults during repeated episodes of caldera collapse (Kisvarsanyi 1980, Sides et al. 1981). Numerous eruptions occurred in the Missouri calderas over a long period of geological time, producing a complex of overlapping calderas. The three-dimensional geometry of the Centralia assemblage of east-central Illinois and west-central Indiana bears a strong resemblance to the Missouri caldera systems. Rhyolitic caldera systems such as these have been associated with intraplate hot spots generated by deep, long-lived mantle plumes (Miller and Smith 1999). As molten rock rises from the mantle into the overlying lithosphere, it sometimes reaches the Earth's surface and forms calderas. A hot spot origin of the Eastern Granite-Rhyolite Province has been suggested by Kisvarsanyi (1980). Subsidence continued in east-central Illinois and west-central Indiana during the deposition of the Cambrian Mt. Simon Sandstone, perhaps in response to thermal cooling that occurred after hot spot formation. After deposition of the Mt. Simon Sandstone, subsidence accelerated in the region of the New Madrid Rift System.

Development of the New Madrid Rift System

The North American continent was part of a supercontinent called Rodinia that broke apart during latest Precambrian time between approximately 800 and 550 million years ago (Figure 3-5a). Major rift systems developed along the margins of North America as the supercontinent slowly pulled apart. Some rifts started to form but stopped before the plates separated. One such failed rift, the New Madrid Rift System, developed along the southern margin of the North American continent. The rift system consists of the

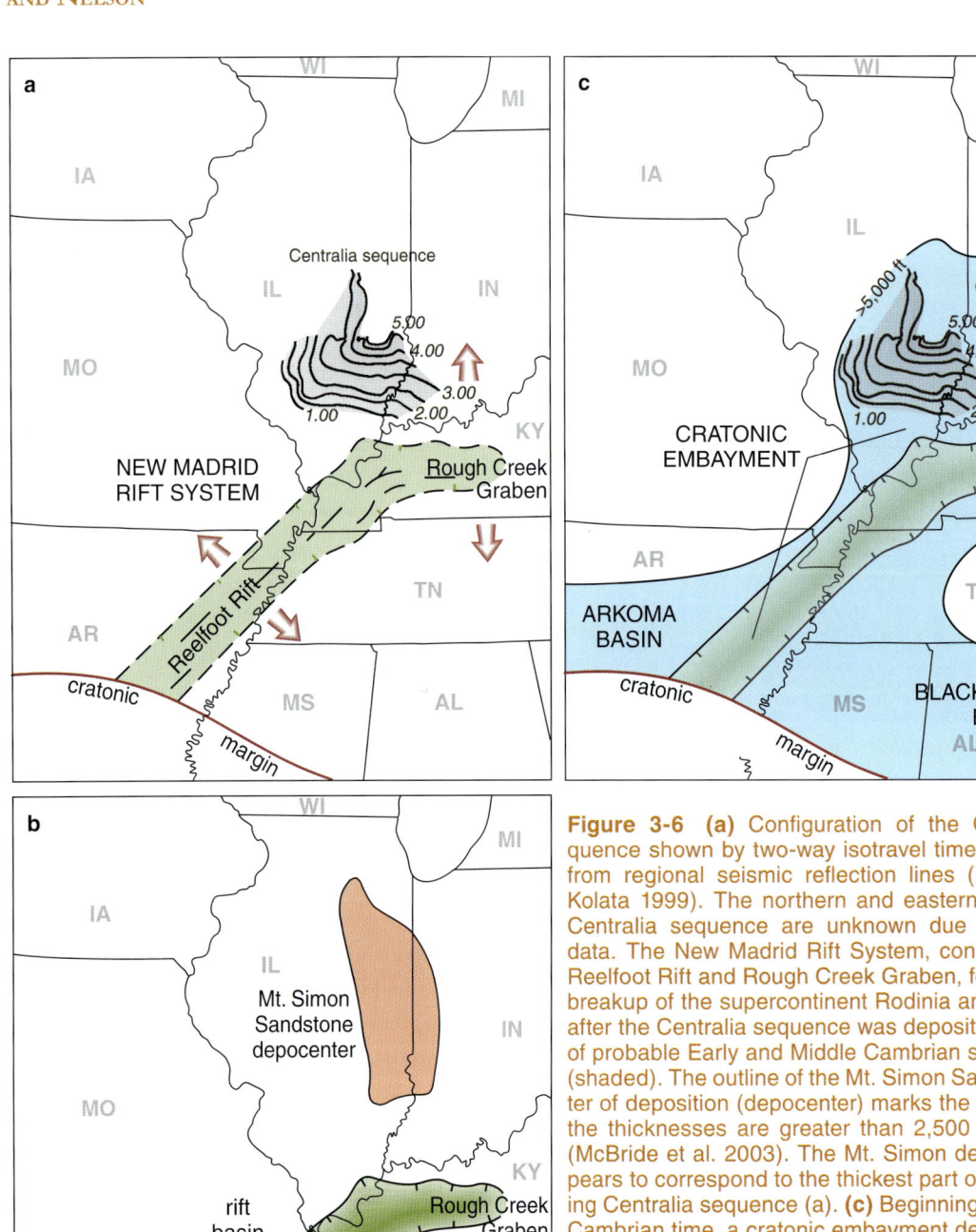

Figure 3-6 (a) Configuration of the Centralia sequence shown by two-way isotravel times determined from regional seismic reflection lines (McBride and Kolata 1999). The northern and eastern limits of the Centralia sequence are unknown due to a lack of data. The New Madrid Rift System, consisting of the Reelfoot Rift and Rough Creek Graben, formed during breakup of the supercontinent Rodinia and most likely after the Centralia sequence was deposited. (b) Areas of probable Early and Middle Cambrian sedimentation (shaded). The outline of the Mt. Simon Sandstone center of deposition (depocenter) marks the region where the thicknesses are greater than 2,500 feet (762 m) (McBride et al. 2003). The Mt. Simon depocenter appears to correspond to the thickest part of the underlying Centralia sequence (a). (c) Beginning during latest Cambrian time, a cratonic embayment developed over the region of the Centralia sequence and the New Madrid Rift System. The area continued to subside for the remainder of the Paleozoic. The structure of the Centralia sequence's base is shown by the two-way isotravel time (seconds) contours. Configuration of the basin corresponds to the region where Paleozoic rocks are in excess of 5,000 feet (1,524 m) thick. The basin formed a broad southwest-plunging trough (shaded) that extended to the cratonic margin where it merged with the Arkoma Basin in central Arkansas and the Black Warrior Basin in northern Mississippi.

northeast-trending Reelfoot Rift, extending from what was the cratonic margin in central Arkansas into southern Illinois and curving abruptly to the east into western Kentucky, forming the Rough Creek Graben (Figure 3-6a). Seismic reflection profiles suggest that the rift system consists of a series of downfaulted and tilted crustal blocks (Howe and Thompson 1984, Nelson and Zhang 1991, Bertagne and Leising 1991, Kolata and Nelson 1997). Some downfaulted blocks formed localized narrow basins, called grabens, that are filled with as much as 10,000 feet (3,000 m) of sediment. The exact age range of the sedimentary rocks that lie on the floor of the rift system is not known, but, based on fossils in samples from deep wells in western Kentucky, portions are known to be Middle Cambrian.

Paleozoic Structural Deformation and Subsidence of the Illinois Basin

By Late Cambrian time, crustal stretching and the faulting caused by this extension had largely ceased within the New Madrid Rift System. The tectonic setting gradually changed from a rift basin to a broad, slowly subsiding cratonic embayment centered over the rift but spreading far beyond it. For the remainder of Paleozoic time, the Illinois Basin was a broad trough plunging southwest and extending from central Illinois to the continental margin in Arkansas (Figure 3-6c). The sedimentary record during this time indicates that Illinois and surrounding areas fluctuated between marine and terrestrial environments. Subsidence rates were greatest, and most Paleozoic stratigraphic sequences were thickest, within and adjacent to the rift system. The present configuration of the Illinois Basin resulted from the cumulative effects of subsidence in the rift system and, to a lesser extent, in the area of the Centralia sequence in east-central Illinois and west-central Indiana (McBride et al. 2003) (Figure 3-6c). Initial rapid subsidence, caused by thermal cooling and contraction within and adjacent to the rift system, gradually gave way to relatively slow subsidence rates by latest Middle Ordovician time (Heidlauf et al. 1986, Treworgy et al. 1989, Kolata and Nelson 1997). A discussion of the stages of Paleozoic subsidence and structural deformation in Illinois follows.

Ordovician through Early Devonian Deformation

Geological and geophysical evidence suggests that subsidence of the Illinois Basin after Middle Ordovician time was largely caused by downwarping of the Earth's crust beneath the New Madrid Rift System. Dense rock, originating from the mantle and emplaced in the lower part of the crust during formation of the rift, weighted down the Basin floor and caused it to subside (Braile et al. 1986; Kolata and Nelson 1991a, 1991b, 1997). In other words, the dense rock created instability in the crust, causing it to sink into the mantle, forming a depression at the Earth's surface. Subsidence rates were relatively slow due to the rigidity of the crust, but periodically they increased. As pointed out by DeRito et al. (1983), the most likely mechanism responsible for the increased subsidence was a global increase in the geothermal gradient or an increase in regional stress related to the collision of tectonic plates. Accordingly, both mechanisms could have decreased the viscosity of the lithosphere, activating subsidence of the dense rock in the crust, causing the rift and surrounding areas of the Illinois Basin to subside. The extension of the Illinois Basin far to the north of the rift system into east-central Illinois and west-central Indiana suggests that some subsidence continued in the area of the Centralia sequence. The two discrete areas of subsidence apparently linked together, creating the present Basin fill pattern (Figure 3-6c).

Beginning during the Early Ordovician, the eastern margin of the North American plate began to converge with several tectonic plates that would later form northern Europe. A subduction zone developed offshore of the North American plate, deforming and metamorphosing much of the rifted passive margin. By the Late Ordovician, further compression of the plate margin resulted in mountainous uplift, extensive volcanic activity, and ultramafic igneous intrusions in eastern North America, all part of the Taconic Orogeny (Figure 3-5b). Volcanic ash was carried westward by winds from supervolcanoes in the Taconic orogenic belt into the midcontinent, including Illinois (Kolata et al. 1986, 1987). A likely source of the ash was a volcanic island arc system associated with subduction (Figure 3-1). As uplift and erosion accelerated, large volumes of sediment were shed from the uplifted mountains westward across the craton, forming a large delta in New York and Pennsylvania (not shown). The shales and siltstones of the Upper Ordovician Maquoketa Group in Illinois are the distal submarine deposits of the delta. The Taconic Orogeny continued into the Early Silurian. The plate tectonic processes that affected the eastern margin of the continent are very much like those illustrated in Figure 3-1.

During Ordovician and Early Silurian time, structural deformation in Illinois was relatively minor. Subtle examples of suspected Ordovician deformation can be seen in places. For example, disturbed bedding and breccias in the Middle Ordovician Pecatonica Formation along the axis of the Peru Monocline near Utica, Illinois, suggest possible deformation within the La Salle Anticlinorium during this time.

The thinning of the Upper Ordovician Scales Shale over the Peru Monocline in the same region suggests uplift along the monocline's axis (Kolata and Graese 1983). The Kankakee and Wisconsin Arches of northern Illinois and the Ozark Dome of eastern Missouri and southwestern Illinois were also uplifted during the Ordovician. The deformation presumably formed in response to distant compressive stresses associated with the onset of the Taconic Orogeny (Figure 3-5b).

From Middle Silurian through Early Devonian time, the Ozark Dome, Northeast Missouri Arch, and Sangamon Arch (Figure 3-7a) were tectonically uplifted; folding occurred on the Waterloo-Dupo Anticline, and faulting occurred in the Plum River Fault Zone (Nelson and Marshak 1996).

Middle Devonian through Early Mississippian Deformation

Middle to Late Devonian deformation in Illinois was centered along discrete fault zones, mostly oriented east-west to southeast-northwest. Much of the faulting, folding, and igneous activity was concurrent with, and perhaps in response to, the Acadian Orogeny, which occurred along the North American plate margin from New York through eastern Newfoundland (Figure 3-5b). Displacements of up to 1,000 feet (305 m) occurred on some Illinois fault zones. Some structures, such as the Ozark Dome and Plum River Fault Zone, continued to undergo mild deformation. Significant areas of new or reactivated deformation occurred in the Rough Creek-Shawneetown Fault System, Pennyrile Fault System, Ste. Genevieve Fault System, Waterloo-Dupo Anticline, Lincoln Anticline, Cap au Grès Faulted Flexure, and Media Anticline (Kolata and Nelson 1991a, Nelson and Marshak 1996). In addition, explosive igneous activity resulting in volcanic vents or pipes took place near the Ste. Genevieve Fault System at Farmington in eastern Missouri (Figure 3-7b). The volcanic structures formed here are called the Avon diatremes. By the Late Devonian and Early Mississippian (Kinderhookian Series), uplift and erosion of highlands in the Acadian Orogen of New England and eastern Canada (Figure 3-5b) caused large volumes of sediment to be shed westward across the continent, forming an extensive delta. In Illinois, the shales and siltstones of the New Albany Shale and Borden Siltstone represent the distal basin deposits of the delta.

Early Mississippian through Late Permian Deformation

Beginning during the Early Mississippian (Osagian Series), the South American and African plates converged on the southern and southeastern margins of North America. By Late Permian, all of the continents became tightly joined, forming Pangea (Figure 3-5c). Illinois underwent significant deformation as a result of these plate tectonic interactions (Figure 3-7c). The orientation and intensity of compressive stresses apparently changed as the continents, microcontinents, and island arcs gradually and sequentially converged during the late Paleozoic. The relationship is not entirely clear between specific cratonic deformation and the various mountain-building events that led to assembly of the supercontinent, but the general sequence of tectonic events and the resulting deformation in and around Illinois appear to include increased subsidence, widespread deformation, changes in compressive stresses, and local extension.

Increased Subsidence. During the Early Mississippian (Osagean Series), subsidence of the Illinois Basin abruptly increased (Treworgy et al. 1989; Kolata and Nelson 1991a, 1991b). The southern part subsided as much as 1,200 feet (366 m) (Craig and Varnes 1979). Initially, local water depths probably exceeded 1,000 feet (305 m) (Lineback 1969) before sedimentation rates caught up with subsidence rates. Abnormal, dysaerobic conditions developed (Ausich et al. 1979), which extinguished many life-forms, and carbonate deposition ceased. Subsequently, a submarine fan of Borden Siltstone spread along the western margins of the Basin and was succeeded by the deposition of siliceous and fossiliferous limestone to the east-southeast of the submarine fan. The remainder of the Early Mississippian (Valmeyeran Series) generally was tectonically stable, and widespread fossiliferous carbonate rocks were deposited across the midcontinent.

Widespread Deformation. Beginning during Late Mississippian (Chesterian) time, widespread deformation occurred throughout Illinois, peaking during the Early Pennsylvanian (Morrowan and Atokan) and continuing intermittently through the remainder of the Pennsylvanian (Figure 3-7c). High-angle reverse faults in the Precambrian basement propagated upward, forming monoclines and asymmetrical anticlines in the overlying Paleozoic sedimentary cover (McBride and Nelson 1999). Notable structures are the Lincoln Anticline and Cap au Grès Faulted Flexure, Waterloo-Dupo Anticline, Ste. Genevieve Fault System, Louden Anticline, Du Quoin Monocline, La Salle Anticlinorium, and probably the Sandwich Fault Zone (Figure 3-7c; see also Chapter 4, Structural Features). The Late Mississippian also marked the beginning of an influx of clay, silt, sand, and conglomerate into Illinois from eastern North America, heralding the Alleghany Orogeny (Figure 3-5c) and the rise of the Appalachian Mountains. These siliciclastic sediments were deposited intermittently throughout the remainder of the Paleozoic. The latest

Tectonic History

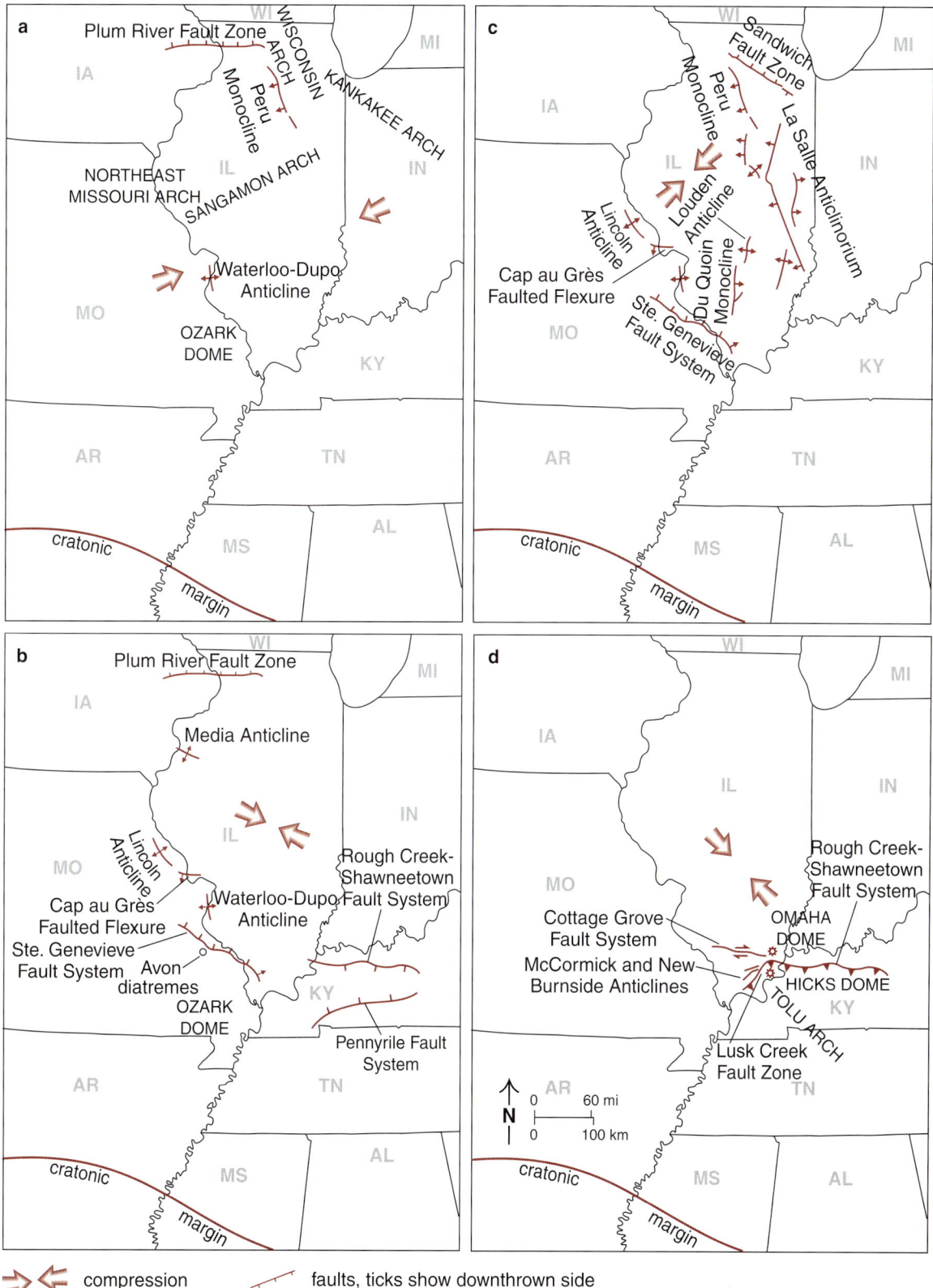

Figure 3-7 Geological structures that underwent compressional deformation **(a)** during Ordovician through Early Devonian time, **(b)** during Middle and Late Devonian time, **(c)** during Late Mississippian through Middle Pennsylvanian time, and **(d)** during Late Pennsylvanian through Early Permian time (modified from Kolata and Nelson 1991a). Arrows show inferred orientation of the regional compressive stress field.

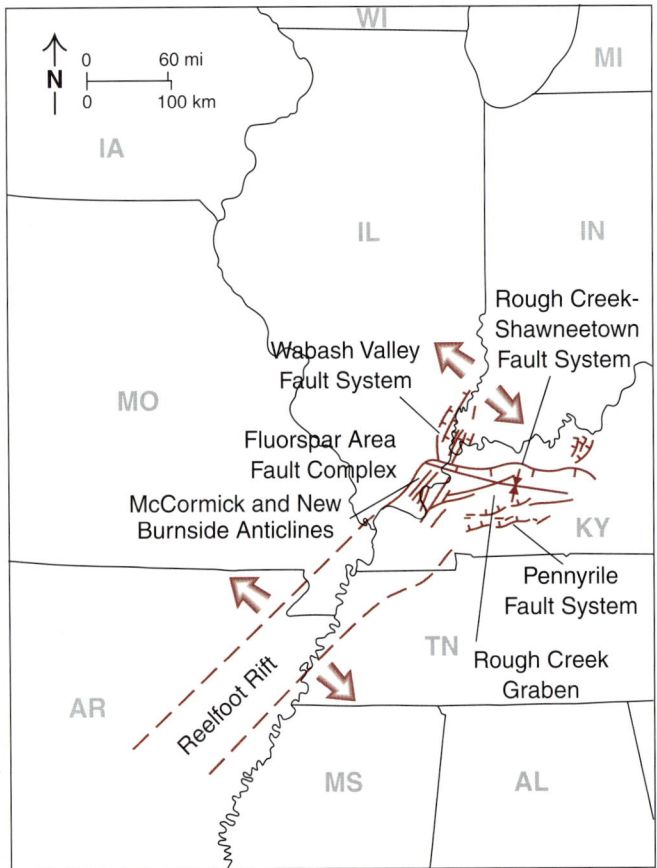

Figure 3-8 Geological structures that underwent extensional deformation during Late Permian through early Mesozoic time (modified from Kolata and Nelson 1991a). Arrows show inferred orientation of the regional tensional stress field.

nant compressive stresses in Illinois changed to a north-northwest orientation, which was consistent with stresses emanating from the Appalachian Orogeny (Figure 3-7d). During this episode of deformation, several structures were formed or reactivated in southern Illinois including (1) the basement-rooted, high-angle reverse faults of the Rough Creek-Shawneetown Fault System and Lusk Creek Fault Zone; (2) the northeast-trending McCormick and New Burnside Anticlines, the underpinnings of which are interpreted to be sub-horizontal detachment structures or thrust faults called décollements; (3) the east-west-trending Cottage Grove Fault System; and (4) the doming and igneous intrusions at the Tolu Arch, Omaha Dome, and Hicks Dome and the formation of igneous dikes along tension fractures in the Cottage Grove Fault System (Kolata and Nelson 1991a, McBride and Nelson 1999). The igneous intrusions have been dated as Early Permian in age (Zartman et al. 1967, Reynolds et al. 1997).

Extension. The period of crustal compression occurring during the Late Mississippian to Early Permian (Figure 3-7d) was followed by an episode of extension that

Mississippian time was marked by a major drop in sea level. Across Illinois, deep valleys were eroded into the exposed Mississippian sediments prior to the advance of Early Pennsylvanian seas.

The orientation and compressive-block style of many structures in Illinois suggest that these may in part be due to the far-field effects of the ancestral Rocky Mountain Orogeny (McBride and Nelson 1999), perhaps operating in concert with ongoing plate collisions along the eastern (Alleghany Orogeny) and southeastern (Ouachita Orogeny) margins of the continent (Kluth 1986). The ancestral Rocky Mountain Orogeny, which peaked during the Early Pennsylvanian (Morrowan and Atokan), is thought to have involved subduction of oceanic lithosphere beneath North America along a northeast-dipping subduction zone situated at the southwestern margin of the continent in the region of east-central Mexico (Ye et al. 1996). The compressive stress from the orogeny caused northeast-southwest-oriented crustal shortening from west Texas through Colorado and eastward, apparently into Illinois.

Change in Compressive Stresses. During the latest Pennsylvanian and Early Permian time, the domi-

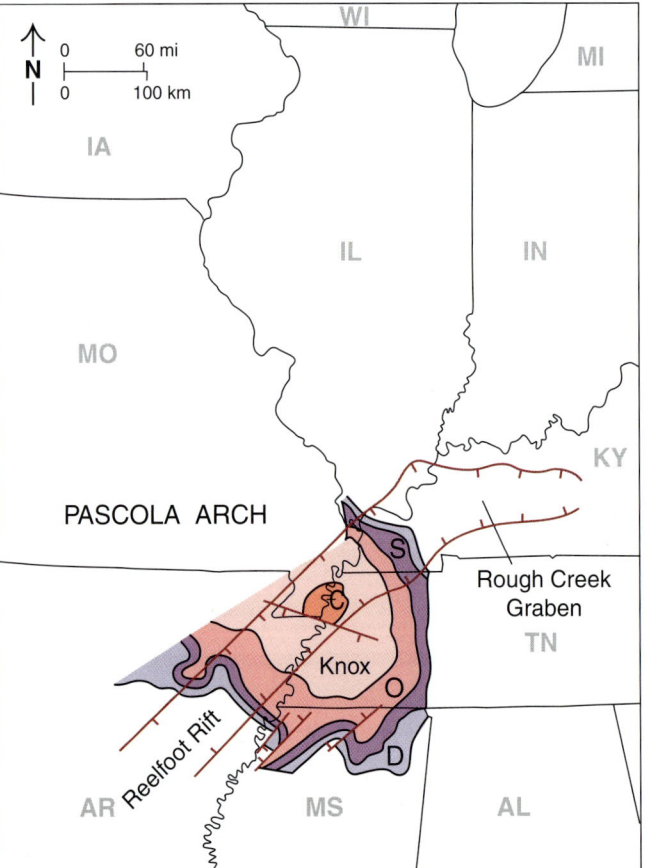

Figure 3-9 Sub-Cretaceous geological map showing eroded crest of the Pascola Arch (modified from Kolata and Nelson 1991a). Cambrian (€) rocks are surrounded by concentric belts of eroded and truncated Ordovician (O), Silurian (S), and Devonian (D) rocks.

Tectonic History

to possibly as much as 15,000 feet (4,572 m) of Paleozoic strata were eroded from the crest of the arch (Marcher and Stearns 1962), probably during the early Mesozoic breakup of the Pangea supercontinent (Figure 3-5d), although stratigraphic evidence is meager. The arch may have resulted from thermotectonic doming caused by the reactivation of the Reelfoot Rift during this episode of crustal breakup and extension (Kolata and Nelson 1991a, 1991b). Also during this interval, wide areas of the midcontinent underwent significant erosion, and hundreds of feet of late Paleozoic sediments were removed.

Late Cretaceous and Tertiary Subsidence of the Mississippi Embayment

By Late Cretaceous time, approximately 65 million years ago, the Pascola Arch subsided and was submerged in a shallow seaway that extended northward into southernmost Illinois from the expanding Gulf of Mexico, forming the Mississippi Embayment (Figure 3-10). The embayment, positioned over the Reelfoot Rift, continued to subside and

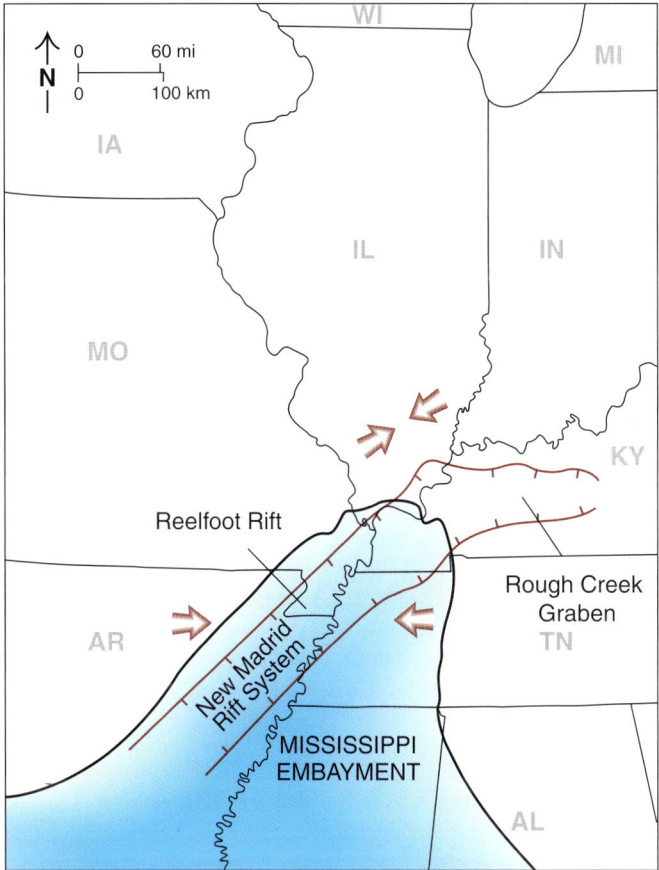

Figure 3-10 Mississippi Embayment (shaded) resulted from Late Cretaceous to Holocene subsidence of the ancient Reelfoot Rift (modified from Kolata and Nelson 1991a). Arrows show inferred orientation of the regional compressive stress field.

probably coincided with the breakup of the Pangea supercontinent during Triassic and Jurassic times (Kolata and Nelson 1991a) (Figures 3-5d and 3-8). Northeast-trending normal faults developed in the Fluorspar Area Fault Complex, along the McCormick and New Burnside Anticlines, and in the Wabash Valley Fault System. In addition, the crustal block situated south and southeast of the Rough Creek-Shawneetown and Lusk Creek Fault Systems sank back to the position it had before the Late Mississippian. Locally, some of the igneous dikes emplaced during the Early Permian underwent faulting and displacement.

Uplift and Erosion of the Pascola Arch

The Pascola Arch, centered over the Reelfoot Rift in southeastern Missouri and western Tennessee, was uplifted approximately 250 to 100 million years ago between Permian and Late Cretaceous time, structurally closing the southern end of the Illinois Basin and creating its present configuration. Broad, concentric belts of eroded and truncated Ordovician, Silurian, and Devonian rocks surround a Cambrian center (Figure 3-9). From 8,000 feet (2,438 m)

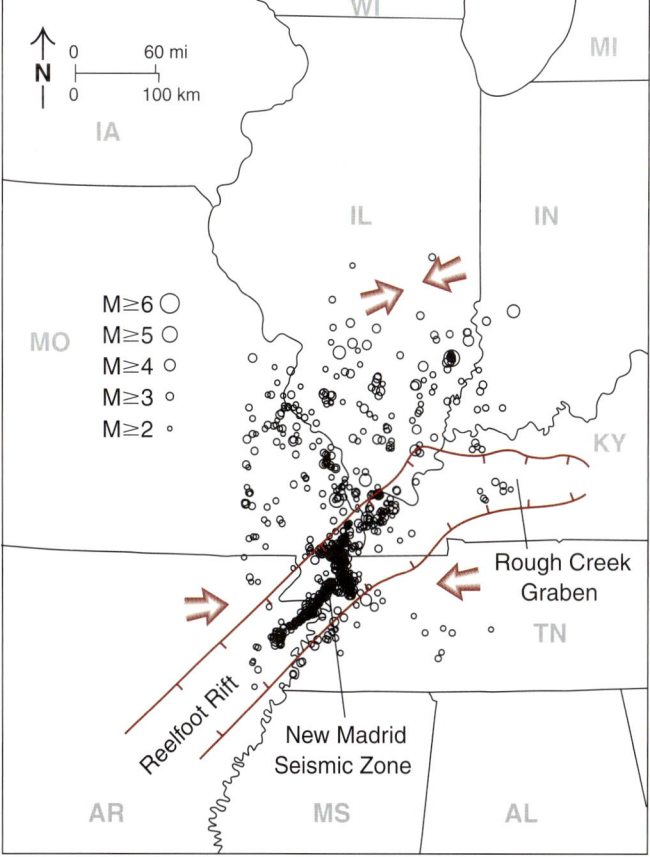

Figure 3-11 Earthquake epicenters recorded between 1974 and 1994 (Center for Earthquake Research and Information 2003). Diameter of circles indicates magnitude (M) of earthquake. Present-day compressive stress (red arrows) within Illinois is oriented in an east-west direction. Arrows show inferred orientation of the regional compressive stress field.

receive siliciclastic sediment during Paleocene, Eocene, and probably Pliocene and Pleistocene times. The Embayment's overall trough-shaped geometry and sediment thickness patterns are suggestive of the Paleozoic Illinois Basin, although not extending into central Illinois and Indiana.

During the past 100 million years, in Illinois and the surrounding regions of the midcontinent, stress has changed from tension to horizontal compression that is oriented east-west to northeast-southwest (Sbar and Sykes 1973, Zoback and Zoback 1980, Nelson and Bauer 1987). This stress field is probably related to the westward drift of the North American plate away from the mid-Atlantic spreading center and concurrent collision with the Pacific plate (Irving 1977) (Figure 3-5e). The stress appears to be reactivating ancient faults within the Reelfoot Rift, causing numerous earthquakes in the New Madrid Seismic Zone (Figure 3-11) (Crone et al. 1985, Zoback 1992, Liu et al. 1992) and in southern Illinois (McBride et al. 1997).

Conclusion

The long and varied geological history of Illinois is the direct result of global-scale tectonic processes. Plate tectonic interactions operating over geological time have been a major influence on the formation of the basins, arches, folds, and fault zones; accumulations of sedimentary rocks; intrusion of igneous rocks; overall uplift and erosion of the continent; and contemporary earthquake activity. Furthermore, Illinois and surrounding regions of the midcontinent are still undergoing tectonic stress, a factor to reckon with during the siting and construction of certain facilities, mine planning, and deep drilling activities.

References

Ausich, W. I., T. W. Kammer, and N. G. Lane, 1979, Fossil communities of the Borden (Mississippian) Delta in Indiana and northern Kentucky: Journal of Paleontology, v. 53, no. 5, p. 1182–1196.

Bertagne, A. J., and T. C. Leising, 1991, Interpretation of seismic data from the Rough Creek Graben of western Kentucky and southern Illinois, in M. W. Leighton, D. R. Kolata, D. F. Oltz, and J. J. Eidel, eds., Interior cratonic basins: Tulsa, Oklahoma, American Association of Petroleum Geologists, Memoir 51, p. 199–208.

Bickford, M. E., K. L. Harrower, W. J. Hoppe, B. K. Nelson, R. L. Nusbaum, and J. J. Thomas, 1981, Rb-Sr and U-Pb geochronology and distribution of rock types in the Precambrian of Missouri and Kansas: Geological Society of America Bulletin, v. 92, no. 6, p. 323–341.

Bickford, M. E., W. R. Van Schmus, and I. Zietz, 1986, Proterozoic history of the midcontinent region of North America: Geology, v. 14, no. 6, p. 492–496.

Bowring, S. A., R. A. Arvidson, and F. A. Podosek, 1988, The Missouri gravity low: Evidence for a cryptic suture? (abs.): Geological Society of America, Abstracts with Programs, v. 20, no. 2, p. 91.

Bowring, S. A., T. B. Housh, C. E. Isachsen, and F. A. Podosek, 1991, Trace element and Nd isotopic constraints on the evolution of Earth's earliest crust (abs.): EOS, v. 72, no. 17 (suppl.), p. 296.

Braile, L. W., W. J. Hinze, G. R. Keller, E. G. Lidiak, and J. L. Sexton, 1986, Tectonic development of the New Madrid Rift Complex, Mississippi Embayment, North America: Tectonophysics, v. 131, no. 1–2, p. 1–21.

Center for Earthquake Research and Information, 2003, Online catalog of earthquakes in the central U.S.: Memphis, Tennessee, University of Memphis. http://www.ceri.memphis.edu/seismic/index.html. Accessed November 11, 2009.

Collinson, C., M. L. Sargent, and J. R. Jennings, 1988, Illinois Basin region, in L. L. Sloss, ed., Sedimentary cover—North American Craton: U.S.: The Geology of North America, v. D-2: Boulder, Colorado, Geological Society of America, p. 383–426.

Craig, L. C., and K. L. Varnes, 1979, History of the Mississippian System—An interpretive summary, in L. C. Craig and C. W. Connor, eds., Paleotectonic investigations of the Mississippian System in the United States—Part II, Interpretive summary and special features of the Mississippian System: Reston, Virginia, U.S. Geological Survey, Professional Paper 1010, p. 371–406.

Crone, A. J., F. A. McKeown, S. T. Harding, R. M. Hamilton, D. P. Russ, and M. D. Zoback, 1985, Structure of the New Madrid seismic source zone in southeastern Missouri and northeastern Arkansas: Geology, v. 13, no. 8, p. 547–550.

DeRito, R. F., F. A. Cozzarelli, and D. S. Hodge, 1983, Mechanism of subsidence of ancient cratonic rift basins: Tectonophysics, v. 94, no. 1–4, p. 141–168.

Goetz, L. K., J. G. Tyler, R. L. Macarevich, D. Brewster, and J. Sonnad, 1992, Deep gas play probed along Rough Creek Graben in Kentucky part of southern Illinois Basin: Oil and Gas Journal, v. 90, no. 38, p. 97–101.

Heidlauf, D. T., A. T. Hsui, and G. deV. Klein, 1986, Tectonic subsidence analysis of the Illinois Basin: Journal of Geology, v. 94, no. 6, p. 779–794.

Hoppe, W. J., C. W. Montgomery, and W. R. Van Schmus, 1983, Age and significance of Precambrian basement samples from northern Illinois and adjacent states: Journal of Geophysical Research, v. 88, no. B9, p. 7276–7286.

Howe, J. R., and T. L. Thompson, 1984, Tectonics, sedimentation and hydrocarbon potential of the Reelfoot Rift: Oil and Gas Journal, v. 82, no. 46, p. 179–190.

Irving, E., 1977, Drift of the major continental blocks since the Devonian: Nature, v. 270, no. 5635, p. 304–309.

Karlstrom, K. E., S. S. Harlan, M. L. Williams, J. Mclelland, J. W. Geissman, and K.-I. Åhäll, 1999, Refining Rodinia: Geologic evidence for the Australia-western U.S. connection in the Proterozoic: GSA Today, v. 9, no. 10, p. 1–7.

Kisvarsanyi, E. B., 1980, Granitic ring complexes and Precambrian hot-spot activity in the St. Francois terrane, midcontinent region, United States: Geology, v. 8, no. 1, p. 43–47.

Kluth, C. F., 1986, Plate tectonics of the ancestral Rocky Mountains, in J. A. Peterson, ed., Paleotectonics and sedimentation in the Rocky Mountain Region, United States: Tulsa, Oklahoma, American Association of Petroleum Geologists, Memoir 41, p. 353–369.

Kolata, D. R., J. K. Frost, and W. D. Huff, 1986, K-bentonites of the Ordovician Decorah Subgroup, Upper Mississippi Valley—Correlation by chemical fingerprinting: Illinois State Geological Survey, Circular 537, 30 p.

Kolata, D. R., J. K. Frost, and W. D. Huff, 1987, Chemical correlation of K-bentonite beds in the Middle Ordovician Decorah Subgroup, Upper Mississippi Valley: Geology, v. 15, no. 3, p. 208–211.

Kolata, D. R., and A. M. Graese, 1983, Lithostratigraphy and depositional environments of the Maquoketa Group (Ordovician) in northern Illinois: Illinois State Geological Survey, Circular 528, 49 p.

Kolata, D. R., and W. J. Nelson, 1991a, Tectonic history of the Illinois Basin, in M. W. Leighton, D. R. Kolata, D. F. Oltz, and J. J. Eidel, eds., Interior cratonic basins: Tulsa, Oklahoma, American Association of Petroleum Geologists, Memoir 51, p. 263–285.

Kolata, D. R., and W. J. Nelson, 1991b, Basin-forming mechanisms of the Illinois Basin, in M. W. Leighton, D. R. Kolata, D. F. Oltz, and J. J. Eidel, eds., Interior cratonic basins: Tulsa, Oklahoma, American Association of Petroleum Geologists, Memoir 51, p. 287–292.

Kolata, D. R., and W. J. Nelson, 1997, Role of the Reelfoot Rift/Rough Creek Graben in the evolution of the Illinois Basin, in R. W. Ojakangas, A. B. Dickas, and J. C. Green, eds., Middle Proterozoic to Cambrian rifting, central North America: Boulder, Colorado, Geological Society of America, Special Paper 312, p. 287–298.

Leighton, M. W., and D. R. Kolata, 1991, Selected interior cratonic basins and their place in the scheme of global tectonics—A synthesis, in M. W. Leighton, D. R. Kolata, D. F. Oltz, and J. J. Eidel, eds., Interior cratonic basins: Tulsa, Oklahoma, American Association of Petroleum Geologists, Memoir 51, p. 729–797.

Lineback, J. A., 1969, Illinois Basin—Sediment-starved during Mississippian: Tulsa, Oklahoma, American Association of Petroleum Geologists Bulletin, v. 53, no. 1, p. 112–126.

Liu, L., M. D. Zoback, and P. Segall, 1992, Rapid intraplate strain accumulation in the New Madrid Seismic Zone: Science, v. 257, no. 5077, p. 1666–1669.

Marcher, M. V., and R. G. Stearns, 1962, Tuscaloosa Formation in Tennessee: Geological Society of America Bulletin, v. 73, no. 11, p. 1365–1386.

McBride, J. H., and D. R. Kolata, 1999, Upper crust beneath the central Illinois Basin, United States: Geological Society of America Bulletin, v. 111, no. 3, p. 375–394.

McBride, J. H., D. R. Kolata, and T. G. Hildenbrand, 2003, Geophysical constraints on understanding the origin of the Illinois Basin and its underlying crust: Tectonophysics, v. 363, no. 1–2, p. 45–78.

McBride, J. H., and W. J. Nelson, 1999, Style and origin of mid-Carboniferous deformation in the Illinois Basin, USA—Ancestral Rockies deformation?: Tectonophysics, v. 305, no. 1–3, p. 249–273.

McBride, J. H., M. L. Sargent, and C. J. Potter, 1997, Investigating possible earthquake-related structure beneath the southern Illinois basin from seismic reflection: Seismological Research Letters, v. 68, no. 4, p. 641–649.

Miller, D. S., and R. B. Smith, 1999, P and S velocity structure of the Yellowstone volcanic field from local earthquake and controlled-source tomography: Journal of Geophysical Research, v. 104, no. B7, p. 15105–15121.

Nelson, B. K., and D. J. DePaolo, 1985, Rapid production of continental crust 1.7 to 1.9 b.y. ago: Nd isotopic evidence from the basement of the North American mid-continent: Geological Society of America Bulletin, v. 96, no. 6, p. 746–754.

Nelson, K. D., and J. Zhang, 1991, A COCORP deep reflection profile across the buried Reelfoot Rift, south-central United States: Tectonophysics, v. 197, no. 2–4, p. 271–293.

Nelson, W. J., and R. A. Bauer, 1987, Thrust faults in southern Illinois Basin—Result of contemporary stress?: Geological Society of America Bulletin, v. 98, no. 3, p. 302–307.

Nelson, W. J., and S. Marshak, 1996, Devonian tectonism of the Illinois Basin region, U.S. continental interior, in B. A. van der Pluijm and P. A. Catacosinos, eds., Basement and basins of eastern North America: Boulder, Colorado, Geological Society of America, Special Paper 308, p. 169–180.

Pratt, T. L., R. Culotta, E. C. Hauser, K. D. Nelson, L. Brown, S. Kaufman, J. Oliver, and W. Hinze, 1989, Major Proterozoic basement features of the eastern midcontinent of North America revealed by recent COCORP profiling: Geology, v. 17, no. 6, p. 505–509.

Reynolds, R. L., M. B. Goldhaber, and L. W. Snee, 1997, Paleomagnetic and $^{40}Ar/^{39}Ar$ results from the Grant intrusive breccia and comparison to the Permian Downeys Bluff sill—Evidence for Permian igneous activity at Hicks Dome, southern Illinois Basin: Reston, Virginia, U.S. Geological Survey, Bulletin 2094-G, 16 p.

Sbar, M. L., and L. R. Sykes, 1973, Contemporary compressive stress and seismicity in eastern North America—An example of intra-plate tectonics: Geological Society of America Bulletin, v. 84, no. 6, p. 1861–1882.

Sides, J. R., M. E. Bickford, R. D. Shuster, and R. L. Nusbaum, 1981, Calderas in the Precambrian terrane of the St. Francois Mountains, southeastern Missouri: Journal of Geophysical Research, v. 86, no. B11, p. 10349–10364.

Sims, P. K., E. B. Kisvarsanyi, and G. B. Morey, 1987, Geology and metallogeny of Archean and Proterozoic basement terranes in the northern midcontinent, U.S.A.—An overview: Reston, Virginia, U.S. Geological Survey, Bulletin 1815, 51 p.

Thomas, J. J., R. D. Shuster, and M. E. Bickford, 1984, A terrane of 1,350- to 1,050-m.y.-old silicic volcanic and plutonic rocks in the buried Proterozoic of the mid-continent and in the Wet Mountains, Colorado: Geological Society of America Bulletin, v. 95, no. 10, p. 1150–1157.

Treworgy, J. D., M. L. Sargent, and D. R. Kolata, 1989, Tectonic subsidence history of Illinois Basin (abs.): American Association of Petroleum Geologists Bulletin, v. 73, no. 8, p. 1040–1041.

Van Schmus, W. R., 1991, Age and crustal history of the midcontinent region in the United States (abs.): EOS, v. 72, no. 17 (suppl.), p. 297.

Van Schmus, W. R., and M. E. Bickford, 1981, Proterozoic chronology and evolution of the midcontinent region, North America, in A. Kröner, ed., Precambrian plate tectonics, v. 4, Development in Precambrian geology: Amsterdam, The Netherlands, Elsevier Publishing Company, p. 261–296.

Van Schmus, W. R., M. E. Bickford, and I. Zietz, 1987, Early and middle Proterozoic provinces in the central United States, in A. Kröner, ed., Proterozoic lithospheric evolution: Washington, D.C., American Geophysical Union, Geodynamics Series, v. 17: Boulder, Colorado, Geological Society of America, 273 p.

Ye, H., L. Royden, C. Burchfiel, and M. Schuepbach, 1996, Late Paleozoic deformation of interior North America—The greater ancestral Rocky Mountains: American Association of Petroleum Geologists Bulletin, v. 80, no. 9, p. 1397–1432.

Zartman, R. E., M. R. Brock, A. V. Heyl, and H. H. Thomas, 1967, K-Ar and Rb-Sr ages of some alkalic intrusive rocks from central and eastern United States: American Journal of Science, v. 265, no. 10, p. 848–870.

Zoback, M. L., 1992, State of stress in the Earth's crust (abs.): Geological Society of America, Abstracts with Programs, v. 24, no. 7, p. A40.

Zoback, M. L., and M. Zoback, 1980, State of stress in the conterminous United States: Journal of Geophysical Research, v. 85, no. B11, p. 6113–6156.

4 Structural Features

W. John Nelson

INTRODUCTION

Structural geology examines how rock units are deformed after they are deposited or formed. In Illinois, nearly all of the rocks at or near the surface are sedimentary rocks that were originally laid down in more or less horizontal layers. Thus, any significant departure from the horizontal reflects deformation and is a geological structure.

Geological structures can be economically important. For example, oil and natural gas are less dense than water, so they seek to rise; as a result, they can be trapped in certain types of geological structures. The search for favorable structures is a key component of petroleum exploration. Many types of mineral deposits—including the fluorite, lead, and zinc formerly mined in Illinois—are concentrated in minable quantities only along specific structures, notably faults. Structure also determines where other mineral deposits, such as coal, limestone, and clay, are close enough to the surface to be mined economically. Finally, structural features, such as those in southern Illinois, provide some of the state's most spectacular scenery.

On the negative side, faults and fractures weaken strata, causing rock falls and letting water and gas into mines. Faults also have troublesome engineering properties, providing a weak substrate for foundations. Because earth movements along faults are the primary cause of earthquakes, a thorough understanding of structural geology is crucial to assess seismic hazards.

Four classes of geological structures are discussed:

1. large, regional basins, arches, and domes such as the Illinois Basin and the Ozark Dome;
2. smaller-scale folds, particularly the anticlines and domes that host many of the state's oil fields;
3. faults and fault zones—fractures along which movement has taken place and, in some cases, may still be in progress; and
4. localized structures, including some of the state's most interesting features: the unique Hicks Dome, two meteorite impact sites, and domes produced by compaction over buried hills and reefs.

Figure 2-5 (Chapter 2, Overview) shows the correlation between periods of structural deformation and major plate tectonic events.

STRUCTURES

Regional Basins, Domes, Arches, and Shelves

The largest geological structures in Illinois are the broad, regional uplifts and downwarps labeled as basins, domes, arches, and shelves. *Basins* are large depressions in the Earth's crust that contain the thickest accumulations of sedimentary rocks. Basins are the best places to find oil and natural gas because (1) the greatest amounts of organic-rich, marine sediments accumulated there, (2) these sediments were buried deeply enough to generate hydrocarbons, and (3) the geological opportunities for trapping oil and gas in reservoir rocks generally are most numerous. *Domes* are roughly circular, uplifted areas. Many domes, such as the Ozark Dome, were uplifted repeatedly through geological time and remain high today. *Arches* are elongated uplifts; they commonly separate one basin from another. *Shelves* are structural terraces where strata lie flat and at a level intermediate between basins and domes or arches.

Basins, domes, arches, and shelves similar to those in Illinois are characteristic of the entire midwestern United States as well as other regions of the world within the interiors of continental plates. In terms of plate tectonic theory, the origins of many of these structures are rather obscure and remain a subject for debate among geologists. Among the favored mechanisms are these:

- The intrusion of dense igneous rocks into the lower crust promoted sinking, forming a basin in the buoyant crust riding on the semi-fluid mantle of the Earth's interior.
- The intrusion of large amounts of granite and similar igneous rocks of low density caused areas to rise.
- Tectonic plate movements stretched, thinned, and partially rifted areas of the crust, forming a basin. These processes apparently occurred in the southern part of the Illinois Basin and in the Michigan Basin.
- The stresses of mountain-building plate collisions, such as those that produced the Appalachian Mountains, were transmitted inland and warped or buckled the crust.

Basins, domes, arches, and shelves that still exist at the bedrock surface are evident on the *Bedrock Geology of Illinois* map (Kolata 2005). Other such structures are older and either have been buried by younger rocks or have been transformed by later crustal movements. Structures that are in existence today are described first.

Structures That Are Still Evident at the Bedrock Surface

Illinois Basin

Also called the Eastern Interior Basin, the Illinois Basin underlies most of central and southern Illinois, southwestern Indiana, and part of western Kentucky (Figure 4-1; see Illinois Basin cross sections, Figures 2-3 and 2-4). The Basin area is roughly 110,000 square miles (285,000 km^2) and contains an estimated 120,000 cubic miles (500,000 km^3) of sedimentary rocks (Kolata et al. 1992). The maximum depth to the Precambrian "basement" rocks may reach 30,000 feet (9,100 m) in western Kentucky. The southern part of the Basin overlies part of an ancient rift (Reelfoot Rift and Rough Creek Graben) where, during the rifting process, the Earth's crust thinned and subsided beginning in Cambrian time. From Cambrian through Middle Pennsylvanian time, the Illinois Basin area was generally a trough or embayment open on the south toward the deep ocean. Continental plate collision in Middle Pennsylvanian time closed off the ocean and formed the Ouachita Mountains, but Illinois remained a shallow, intercontinental seaway at least into early Permian time. Uplift of the Pascola Arch, mostly during the Mesozoic, upturned the rocks at the southern end of the Illinois Basin and gave it its present structure.

Fairfield Basin

In Illinois, the deeper, oil-productive part of the Illinois Basin is known as the Fairfield Basin (Figure 4-1). The Fairfield Basin is bounded on the west by the Du Quoin Monocline, on the east by the La Salle Anticlinorium, and on the south by the Cottage Grove and Rough Creek-Shawneetown Fault Systems (Figure 4-2). Maximum depth to the Precambrian rocks is about 15,000 feet (4,600 m). The Fairfield Basin took its form during and after Pennsylvanian deposition.

Moorman Syncline (or Trough)

Located largely in Kentucky, the Moorman Syncline is the intricately faulted and deeply downwarped portion of the Illinois Basin lying south of the Rough Creek-Shawneetown Fault System (Figure 4-2). This syncline is largely the product of post-Pennsylvanian downwarping.

Ozark Dome

Located almost entirely in Missouri, with its northeastern flank barely extending into Illinois, the Ozark Dome (or uplift) coincides with the modern Ozark Plateaus physiographic division. Between 1,380 and 1,480 million years ago, a tremendous amount of granite and rhyolite was emplaced in the central Ozark region. These rocks formed a remarkably long-lived uplift because (1) intrusion of the granite uplifted the region, (2) the rocks are highly resistant to erosion, and (3) granite and rhyolite have relatively low density, producing buoyant crust that rides high on the Earth's mantle. The sedimentary rock succession is very thin on the Ozark Dome and is absent in the highest area, the St. Francois Mountains. Fission-track data by Charles W. Naeser (U.S. Geological Survey, written communication 2004) suggest that the Ozark Dome was buried beneath about 2 to 3 miles (3.2 to 4.8 km) of sediment during the early Mesozoic. By Late Jurassic or Cretaceous time, most of this overburden had been removed as a consequence of uplift and erosion. The Ozark Dome is bordered by faults in many places.

Kankakee Arch

The Kankakee Arch constitutes the divide between the Illinois Basin and the Michigan Basin (Figure 4-1) and formed initially during Ordovician time (Atherton 1971).

Wisconsin Arch

The southern part of the Wisconsin Arch extends into northern Illinois, separating the Michigan Basin on the east from the Forest City Basin in Iowa on the west.

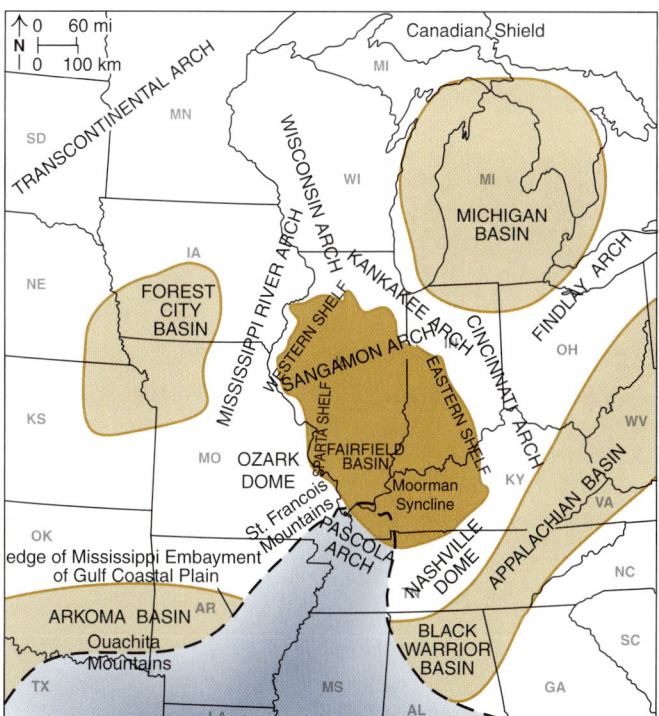

Figure 4-1 Map showing selected regional structural features in and around Illinois. Darker brown identifies the Illinois Basin.

The Wisconsin Arch began forming late in Cambrian time (Paull and Paull 1977) and, like the Kankakee Arch, remains structurally high today.

Mississippi River Arch

Separating the Illinois and Forest City Basins, the Mississippi River Arch developed after Early Pennsylvanian time (Bunker et al. 1985).

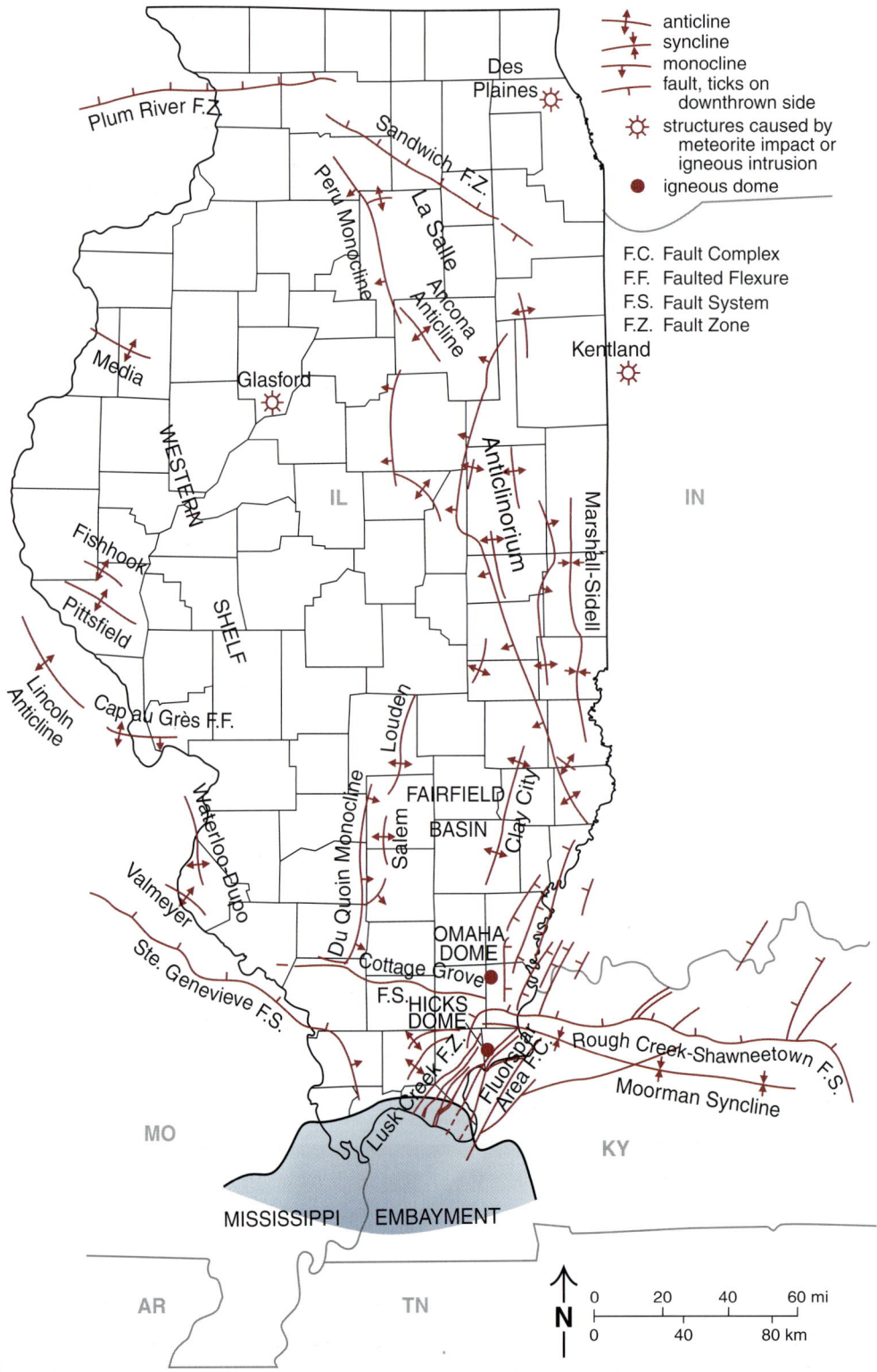

Figure 4-2 Some of the folds, faults, and other structures in and near Illinois (Nelson 1991, 1995; Buschbach and Kolata 1991).

Western Shelf and Sparta Shelf

A broad area of nearly flat-lying sedimentary rocks in western Illinois forms a structural terrace between the Fairfield Basin and the Ozark Dome. Western Shelf is the inclusive term for this area, and Sparta Shelf refers to the portion south of Bond and Madison Counties.

Mississippi Embayment

The Mississippi Embayment is a northward extension of the Gulf Coastal Plain into southernmost Illinois. The Embayment developed as a structural downwarp during Late Cretaceous and Tertiary time and was filled with marine, deltaic, and coastal sediments. The axis of the Embayment overlies the ancient Reelfoot Rift, which may have been rejuvenated as the Gulf of Mexico opened (Ervin and McGinnis 1975). Another theory is that the Embayment sank in response to stresses from the Laramide Orogeny, which produced the Rocky Mountains (Harrison and Schultz 2002). Widespread eruptive and intrusive igneous activity in Arkansas, Mississippi, and northern Louisiana accompanied the initial submergence of the Embayment.

Ancient Arches

Sangamon Arch

A broad arch developed in central Illinois during Devonian time, forming a shoal or string of islands through the Devonian seaway. Silurian and older rocks were slightly bulged upward and eroded along the crest of the arch, which ran roughly from Champaign through Springfield, Illinois, to Louisiana, Missouri (Whiting and Stevenson 1965). After the Devonian, the Sangamon Arch became inactive and was buried by younger strata.

Pascola Arch

The Pascola Arch was essentially a southeastward projection off the Ozark Dome (Grohskopf 1955, Marcher and Stearns 1962, Harrison and Schultz 2002). Its northeastern flank lies near the southern tip of Illinois and is responsible for closing off the southern end of the Illinois Basin. The arch probably developed in response to Ouachita and Appalachian mountain building during Late Pennsylvanian and Permian time (Harrison and Schultz 2002). The crest of the arch became deeply eroded and then sank as the Mississippi Embayment formed during Late Cretaceous time (Marcher and Stearns 1962, Stearns and Marcher 1962).

Smaller-Scale Folds

Folds that are smaller than arches and basins include anticlines (upward folds), synclines (downward folds), and step-like monoclines. Anticlines are of greatest interest to economic geologists and are more likely to be named than synclines because the anticlines often provide structural traps for oil and gas.

The folds discussed here have many features in common. As shown by Figure 4-2, most of those in the Fairfield Basin run north-south, or nearly so. Folds on the Western Shelf (and in Missouri and eastern Iowa) strike dominantly northwest-southeast. Some folds are nearly linear, but many are arcuate or strongly curved in map view, and, in some cases, they branch. Most anticlines are asymmetrical, having one limb in which the strata are gently inclined and one in which their incline is relatively steep. The steep flanks typically overlie faults in the Precambrian basement rocks. On the Western Shelf, where the depth to basement rocks is less, faults commonly reach the surface and cut the crests of folds. Deep seismic reflection profiles in the Fairfield Basin indicate that high-angle reverse faults underlie the steep limbs of structures such as the Du Quoin Monocline, Salem Anticline, and La Salle Anticlinorium. In most cases, these deep-seated faults die out upward in the lower part of the Paleozoic succession (McBride 1998, McBride and Nelson 1999).

Monoclines and anticlines of Illinois are comparable to folds of the Colorado Plateau and the central and northern Rocky Mountains in the western United States. Such structures have been called drape folds because the strata drape over the buried fault like a carpet over a stair. A more technical term is fault-propagation fold, which implies that the fault began at depth and propagated itself toward the surface.

Most drape or fault-propagation folds in Illinois developed during or shortly after the Pennsylvanian Period. This timing is shown by angular unconformities between Pennsylvanian and older rocks and by the abrupt thinning of Pennsylvanian strata on anticlines. When anticlines rose during deposition of a particular layer, that layer is thin or absent. Structures of similar age and style occur throughout the southern midcontinent west of Illinois, an area encompassing Missouri, Kansas, Oklahoma, west Texas, and the southern Rocky Mountains in New Mexico, southern Colorado, and eastern Utah. Because such features were first recognized in the modern Rocky Mountains, but are older than the mountains, the event that formed them is called the ancestral Rocky Mountain Orogeny (Melton 1925, Ver Wiebe 1930, McBride and Nelson 1999).

La Salle Anticlinorium

An anticlinorium is a compound anticline and is also referred to as an anticlinal belt. The La Salle Anticlinorium runs north-northwest from Lawrenceville through Champaign, Bloomington, and La Salle toward Dixon. An eastern branch heads north-northeast from west of Champaign toward Kankakee. The belt is composed of curving and

Figure 4-3 The upward arching of the La Salle Anticlinorium is visible near its axis at Matthiessen State Park. The view shown here is of the anticlinorium's steeper west side where glacial till directly overlies Ordovician Galena-Platteville dolomite. The Pennsylvanian and Silurian rocks have been eroded away at this location. (Photograph by Joel M. Dexter.)

branching anticlines, domes, monoclines, and synclines. In most places, the steeply dipping flanks of folds face west. Glacial deposits hide most of the structure, except in stream cuts along the Illinois River and its tributaries in La Salle County (Figure 4-3). At Tuscola in Douglas County, Silurian and Devonian limestones are exposed in a large quarry at the crest of the fold. The total uplift at Tuscola is 2,500 feet (760 m), the largest of any fold in Illinois. The La Salle Anticlinorium is chiefly known through seismic reflection surveys and the logs of tens of thousands of oil and gas test holes drilled along its trend. An estimated 750 million barrels (120 billion L) of oil have been produced from fields in structural traps along the La Salle Anticlinorium (Nelson 1995).

Approximately half of the total uplift on the La Salle Anticlinorium took place during Late Mississippian to Early Pennsylvanian time (Clegg 1965, 1970). The result is an angular unconformity between the Pennsylvanian and older rocks that is most pronounced in La Salle County, where Pennsylvanian rocks lie directly on folded Ordovician strata. Upward movements continued intermittently throughout Pennsylvanian time, causing coal beds and other rock intervals to thin on anticlines and thicken in synclines (Jacobson 1985). Additional movements took place after Pennsylvanian time, folding the youngest Pennsylvanian strata.

An example of a fold within the La Salle Anticlinorium is the Ancona Anticline (Figure 4-4). This fold straddles the La Salle-Livingston County line in northern Illinois. Unlike most anticlines of the La Salle Anticlinorium, the Ancona has a steeper northeast flank. Seismic profiles suggest the northeast flank is faulted at depth, but those faults do not reach the surface. Like several large domes along

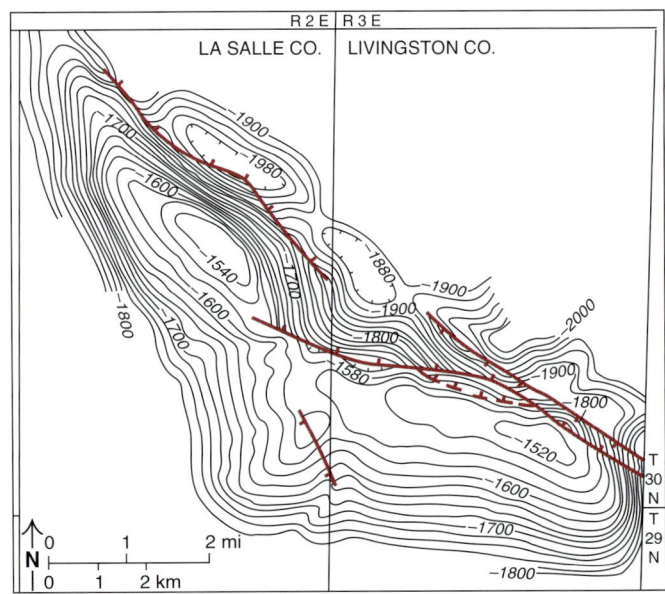

Faults, ticks on downthrown side

Figure 4-4 Structure map of the Ancona Anticline, part of the La Salle Anticlinorium in southern La Salle County and northwestern Livingston County (Nelson 1995). Contours represent elevation on top of the basal Cambrian Mt. Simon Sandstone in feet below mean sea level. Faults affect Cambrian rocks but die out in younger strata. The geometry of the fold suggests it overlies a block of Precambrian basement rock that was uplifted and tilted (Nelson 1995). Contour interval is 20 feet.

the trend of the La Salle, the Ancona structure has been developed as a gas storage field. Such structural traps are used to store natural gas from the Gulf Coast that has been pumped into the Cambrian Mt. Simon Sandstone during the summer; the natural gas can later be withdrawn during the winter when demand is greatest.

Seismic surveys show that high-angle reverse faults displace Precambrian and lower Paleozoic rocks beneath many folds in the La Salle Anticlinorium. These faults die out as they extend upward and do not reach the surface. Given its time of origin and geometry, the La Salle Anticlinorium is considered to be a typical ancestral Rocky Mountains structure (McBride 1998, McBride and Nelson 1999).

Du Quoin Monocline

The branched, step fold known as the Du Quoin Monocline separates the Fairfield Basin from the Western (Sparta) Shelf (Figure 4-2). Like the La Salle Anticlinorium, the Du Quoin Monocline was active throughout Pennsylvanian time, the greatest uplift taking place during the Early Pennsylvanian. As a result, the relief on Mississippian and older rocks is nearly twice as great, more than 1,000 feet (300 m), as on Middle Pennsylvanian coals. Both branches of the monocline are faulted at depth, and there is evidence for two episodes of movement. Early reverse faulting during Early Pennsylvanian time raised the western side of the fold. Later, normal faulting during or after Late Pennsylvanian time lowered the western side (Nelson 1995, McBride 1998, McBride and Nelson 1999).

Anticlines in the Fairfield Basin

Most anticlines in the Fairfield Basin have sinuous trends that run more or less north-south. The largest are the Salem and Louden Anticlines (Figure 4-2), both of which have more than 200 feet (61 m) of closure, the vertical distance between the structure's highest and lowest closed structural contour. The Clay City Anticline is more subtle, plunging gently southward and topped by a series of small domes. These three anticlines and many lesser ones provide structural traps for oil and natural gas fields. The Salem, Louden, and Clay City Consolidated fields all are Illinois "giants," having cumulative production in excess of 300 million barrels (47.7 billion L).

Cap au Grès Faulted Flexure

The Cap au Grès Faulted Flexure (Figure 4-2), north of St. Louis, Missouri, is closely associated with the larger Lincoln fold in northeastern Missouri. Together, the Lincoln and Cap au Grès constitute a belt of sinuous, branching monoclines and anticlines accompanied by domes (Harrison and Schultz 2002). The south or southwest flank is steep. In places, bedding along the steep flank is vertical or even overturned, and faults reach the surface. The Cap au Grès structure was active from Late Devonian through Pennsylvanian time, and the largest movements took place between Late Mississippian to Early Pennsylvanian (Rubey 1952). Detailed study of small folds, faults, and fractures indicates two episodes of deformation, the first involving north-south compression, and the second, larger episode entailing northeast-southwest compression (Harrison and Schultz 2002). As Rubey (1952) first deduced, the Cap au Grès fold probably overlies large reverse faults in the Precambrian basement rocks. Evidence from gravity (Mateker 1959) and magnetic (Douthit 1959) surveys supports this conclusion.

Other Anticlines

The Valmeyer and Waterloo-Dupo Anticlines in southwestern Illinois both trend northwest and have steep southwest flanks. The Waterloo-Dupo fold is a continuation of the Cap au Grès Faulted Flexure (Harrison and Schultz 2002) and forms the trap for oil fields producing from the Middle Ordovician Kimmswick ("Trenton") Limestone, the oldest formation from which oil has been produced in Illinois. Farther north on the Western Shelf, the Pittsfield Anticline was the site of Illinois' first commercial gas discovery, in 1886, at a depth of 265 feet (81 m) in Silurian dolomite. The gas field was exhausted by 1930. The Media Anticline in Henderson County (Figure 4-2) is unusual in that most of its uplift took place during Late Devonian or very early in Mississippian time. This timing is shown by the marked thinning of the Devonian New Albany Shale from the flank to the crest of the fold; younger Mississippian limestone is only slightly folded (Nelson 1995).

Faults

Faults and fault zones are concentrated in southern Illinois (Figure 4-2), although the Plum River and Sandwich Fault Zones of northern Illinois are among the longest faults in the state. Faults in Illinois are of three main types:

- high-angle normal faults, resulting from crustal extension;
- high-angle reverse faults, resulting from crustal compression involving the Precambrian basement; many are related to ancestral Rocky Mountain-building processes; and
- strike-slip faults, resulting from horizontal shearing stresses.

Several fault zones in Illinois combine more than one type of fault and bear evidence of two or more episodes of movement under different stress fields. Once a fault has formed, it remains a zone of weakness in the Earth's crust. Generally, it is easier for new stresses to reactivate old faults than to create new faults. Some faults in southern Illinois have undergone at least four episodes of displacement. The principal times of tectonic activity include the Cambrian, Devonian, Pennsylvanian, Permian through Cretaceous, and, in southernmost Illinois, the late Tertiary and Quaternary.

Current tectonic activity and its relationship to earthquakes in Illinois are discussed in Chapter 5 (Neotectonics) and Chapter 27 (Earthquakes).

Plum River Fault Zone

The Plum River Fault Zone extends 112 miles (180 km) west-southwest from near Byron, Illinois, into east-central Iowa. The zone varies from a few hundred feet to nearly a mile wide; the net vertical offset to the north is 100 to 400 feet (30 to 120 m) down (Kolata and Buschbach 1976). Domes flank the south side of the fault zone, and a syncline flanks the north side in Illinois (Figure 4-5). Bunker et al. (1985) concluded that the largest movements took place between Middle Devonian and Pennsylvanian time, and slight adjustments occurred after Pennsylvanian time.

Mapping by Templeton (1951) in Ogle County shows many northwest-trending en echelon (in steplike arrangement) normal faults and extensional fractures and a few en echelon northeast-trending thrust or reverse faults. Also, east-trending faults bear horizontal polished surfaces called slickensides. As Bunker et al. (1985) noted, the degree of rock shattering along the Plum River is out of proportion with the relatively small vertical displacements, which suggests that the Plum River Fault Zone underwent an episode of right-lateral, strike-slip motion (Marshak et al. 2003).

Sandwich Fault Zone

The Sandwich Fault Zone runs about 85 miles (137 km) from near Joliet to Mt. Morris, Illinois. The timing and style of faulting are poorly known because the zone is largely concealed by glacial drift. The youngest rocks that are clearly broken are Silurian; Pleistocene glacial deposits are not faulted. Small outliers of Pennsylvanian rock occur near the fault zone, but whether they are affected is not clear (Kolata et al. 1978). The zone comprises many steeply dipping, normal faults and a few small reverse-and-thrust faults (Templeton 1951, Kolata et al. 1978, Nelson 1995). Along the middle segment of the zone, displacement is as great as 800 feet (240 m) down to the northeast. Near both ends, the southwest side is downthrown. A narrow, upthrown fault slice of Cambrian strata at Oregon, Illinois, strikes northwest and is slightly oblique to the overall trends of the zone (Figure 4-6). The Sandwich Fault Zone may have undergone either strike-slip movement or more than one episode of up-and-down faulting (Marshak et al. 2003).

Wabash Valley Fault System

The broad fracture system known as the Wabash Valley Fault System straddles the Illinois-Indiana border and is composed mainly of high-angle normal faults that strike

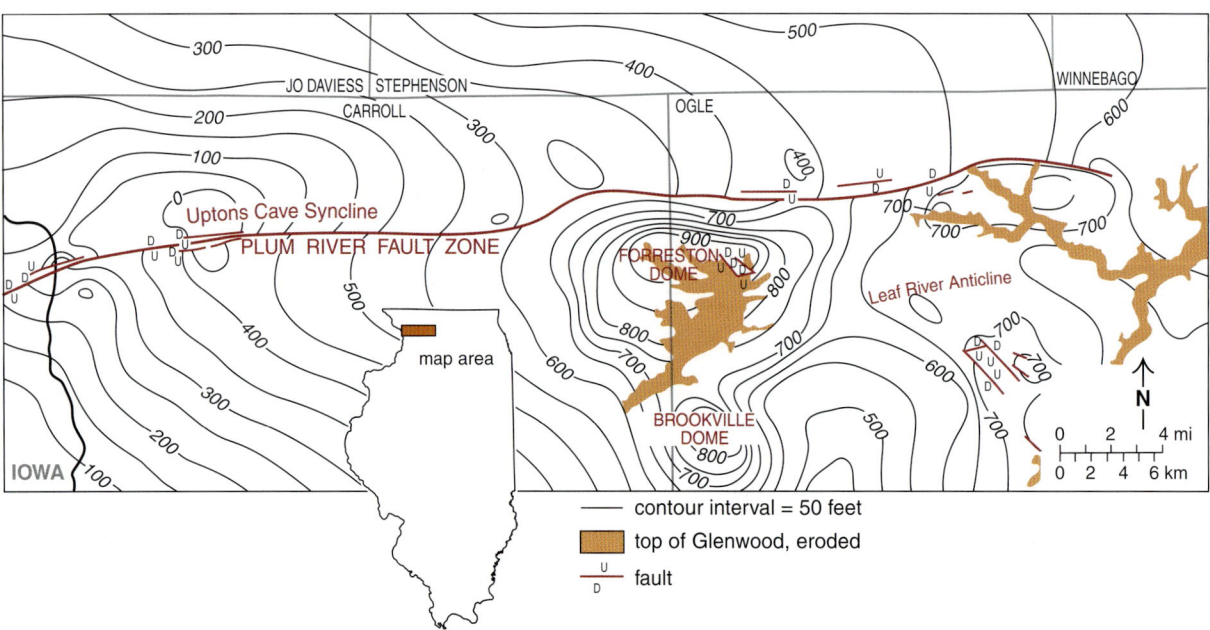

Figure 4-5 Geologic map of the eastern end of the Plum River Fault Zone (Kolata and Buschbach 1976). The combination of en echelon normal faults oriented northwest, en echelon thrust faults oriented northeast, and east-west fractures bearing horizontal slickensides imply an element of strike-slip faulting. Abbreviations: U, upthrown; D, downthrown.

north-south and north-northeast to south-southwest. Gently arched and tilted fault blocks provide structural traps for oil and natural gas and disrupt underground coal mining. The fault system is about 55 miles (88 km) long and as wide as 30 miles (48 km); it displaces strata as much as 480 feet (146 m) (Bristol and Treworgy 1979, Ault et al. 1980). Major movement occurred after Pennsylvanian time, although abrupt changes in the thickness and character of Upper Mississippian and Lower Pennsylvanian rocks indicate earlier activity (Nelson et al. 2002).

The Wabash Valley experiences moderate earthquake activity in present times, and deformed stream sediments indicate that strong earthquakes occurred here within the last 10,000 years (Munson et al. 1992, 1997; Obermeier 1992; Obermeier et al. 1998). Large faults in Precambrian and Cambrian rocks beneath the Wabash Valley may be ancestral to the Wabash Valley Fault System (Sexton et al. 1986, Bear et al. 1997). However, earthquakes can originate in deeply buried structures that are not directly related to near-surface faults (Hildenbrand and Ravat 1997, McBride et al. 2002). These topics are addressed more thoroughly in Chapter 5, Neotectonics.

Cottage Grove Fault System

Approximately 70 miles (110 km) long and more than 10 miles (16 km) wide in places, this complex fracture zone trends slightly north of west across southern Illinois (Figure 4-7). The Cottage Grove Fault System contains these components:

- A steeply dipping, sinuous, strike-slip "master fault" with maximum vertical displacement of about 200 feet (60 m) (Figure 4-7). Seismic reflection profiles, deep drilling, and coal mine exposures show that the master fault typically consists of high-angle reverse faults that form "positive flower structures" (Nelson and Krausse 1981, Marshak et al. 2003, Duchek et al. 2004).
- Numerous subsidiary faults that strike northwest, oblique to the master fault. Most are high-angle normal faults, but some display clear evidence of strike-slip movement. In the eastern part of the system, the faults are accompanied by ultramafic igneous dikes, nearly vertical sheetlike intrusions of igneous rocks.
- Anticlines that lie mostly south of the master fault and trend either parallel or slightly oblique to the latter in en echelon fashion.

The youngest displaced rocks are Upper Pennsylvanian; the dikes are dated as Early Permian. Given that the dikes parallel and follow subsidiary faults, the faulting probably was under way by Early Permian time (Nelson and Lumm

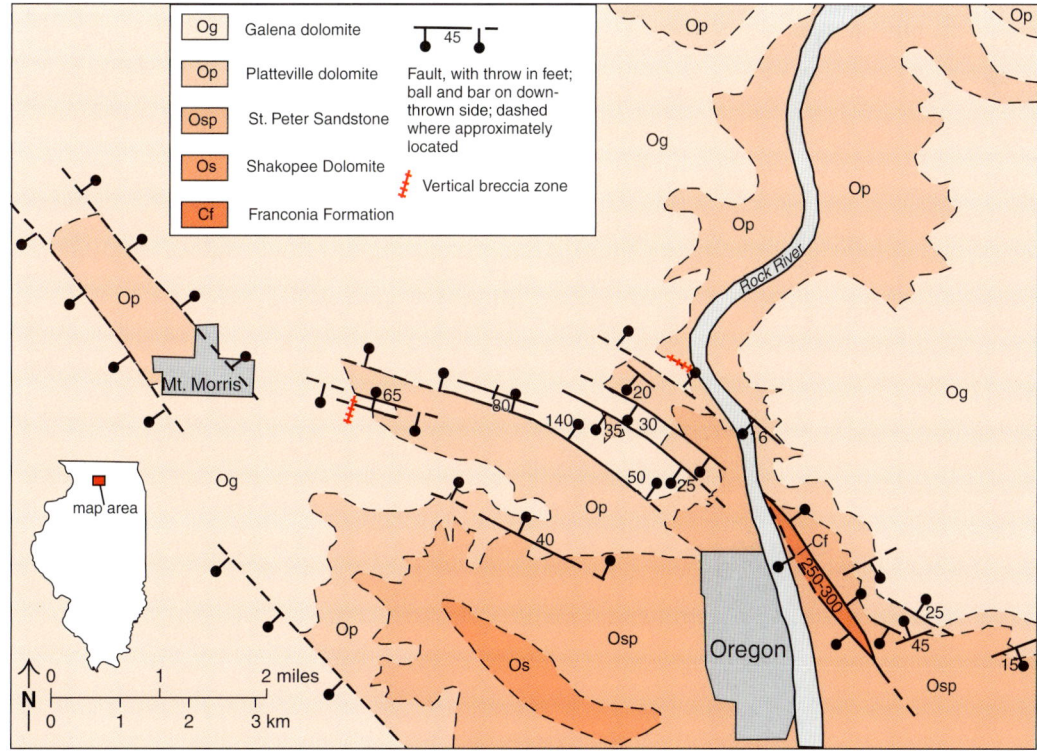

Figure 4-6 Geologic map of the Oregon area, Ogle County, northern Illinois, showing part of the Sandwich Fault Zone. The map is based primarily on data from Templeton (1951). The narrow, upthrust slice of Franconia east of the Rock River at Oregon is a curious feature that suggests either strike-slip faulting or multiple episodes of faulting.

1985). Quaternary deposits along the fault system are not deformed. The Cottage Grove is generally interpreted to be a right-lateral fault (north side moved eastward) with a strong element of north-south compression. The amount of horizontal movement has not been measured but is probably less than 1,000 feet (300 m) (Duchek et al. 2004).

Ste. Genevieve Fault System

The Ste. Genevieve Fault System lies in Missouri, but enters Illinois at Grand Tower in Jackson County, trending southeast and dying out into a monocline. The Ste. Genevieve follows and partially defines the border between the Illinois Basin and the Ozark Dome (Figure 4-2). This intricate system is composed of steeply dipping normal and reverse faults that exhibit dogleg bends and form a braided pattern in map view. As shown in Figure 4-8, two episodes of displacement are documented (Weller and St. Clair 1928, Nelson and Lumm 1985, Harrison and Schultz 2002). The first episode was during Middle to Late Devonian, when the south or southwest side of the zone was downthrown as much as 1,000 feet (300 m) in Illinois and more than 2,000 feet (610 m) in Missouri. This episode involved extension and normal faulting and a component of left-lateral strike slip (Nelson and Marshak 1996, Harrison and Schultz 2002). The second episode, during latest Mississippian and Pennsylvanian time, raised the southwestern block as much as 3,000 feet (915 m) along high-angle reverse faults, forming a sharp monocline with the flank vertical or overturned.

Rough Creek-Shawneetown Fault System

One of the largest fault systems in the Midwest, the Rough Creek-Shawneetown, is situated mostly in western Kentucky, trending westward 130 miles (209 km) to enter Illinois at Shawneetown on the Ohio River. Southeast of Harrisburg, the fault zone bends abruptly to the southwest and merges with the Lusk Creek Fault Zone, which displays the same structural style. In map view, the Rough Creek-Shawneetown presents an intricate, braided fracture pattern. Steeply dipping normal and reverse faults outline narrow "slices" of upthrown and downdropped rocks, many highly sheared. Although the net displacement across the system is small, individual faults commonly have throw measured in thousands of feet (Nelson and Lumm 1987; Potter et al. 1995, 1997). Narrow slices of rock much older than those on either side are upthrust within the fault zone. For example, at the "Horseshoe upheaval" near the Saline-Gallatin County line, Upper Devonian black shale

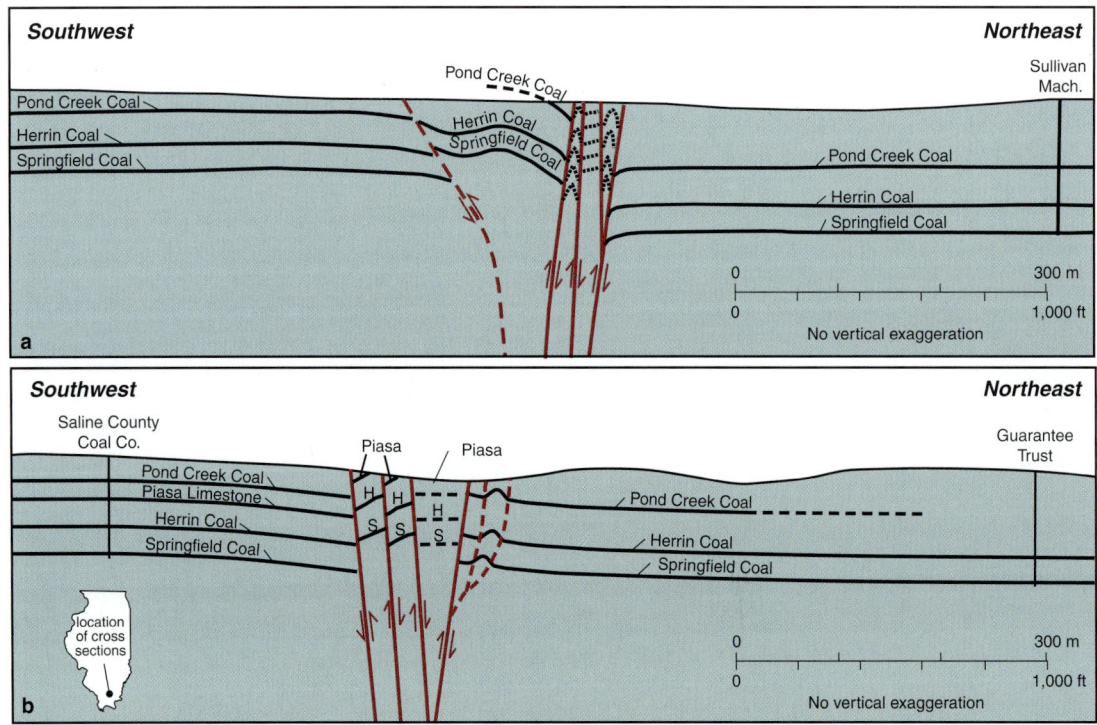

Figure 4-7 Two cross sections of the Cottage Grove Fault System in eastern Williamson County. These are "positive flower structures": a series of narrow, upthrown slices outlined by high-angle reverse faults and characteristic of strike-slip faults that formed under compression. Vertical black lines are boreholes. (a) Attila, Sec. 33, T8S, R4E; drawing is based mostly on unpublished ISGS field notes by G. H. Cady (1920). (b) Bethel Church, Sec. 35, T8S, R4E and Sec. 2, T9S, R4E; drawing is based on observations by W. J. Nelson and unpublished ISGS field notes by G. H. Cady, H. R. Wanless, and M. W. Fuller (1932).

Structural Features

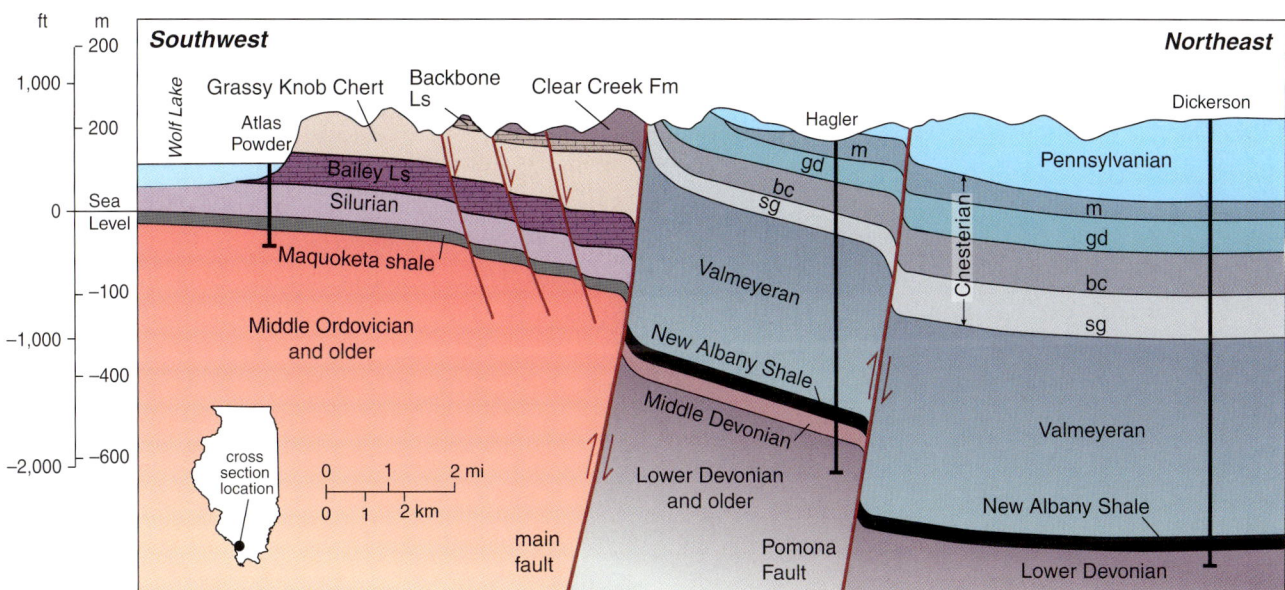

Figure 4-8 Cross section of the Ste. Genevieve Fault System in southern Jackson and northern Union Counties, based on data from boreholes (shown) and outcrop mapping (Desborough 1961, Devera 1993, Devera and Nelson 1995). The main fault is a steep reverse fault with about 2,000 feet (610 m) of throw. Normal faults southwest of the main fault may intersect the latter at depth. The Pomona Fault, which runs parallel to the main fault, is a reverse fault with the southwest side upthrown near the surface. However, Middle Devonian strata are present on the southwest but absent northeast of the Pomona Fault, implying that during Middle to Late Devonian time the northeast side was raised. Notice also that Valmeyeran and Chesterian strata thin markedly southwest of the Pomona Fault. Such thinning suggests uplift of the southwestern side during the Mississippian. Abbreviations: m, top of the Menard Limestone; gd, top of the Glen Dean Limestone; bc, top of the Beech Creek (Barlow) Limestone; sg, top of the Ste. Genevieve Limestone; Fm, Formation; Ls, Limestone.

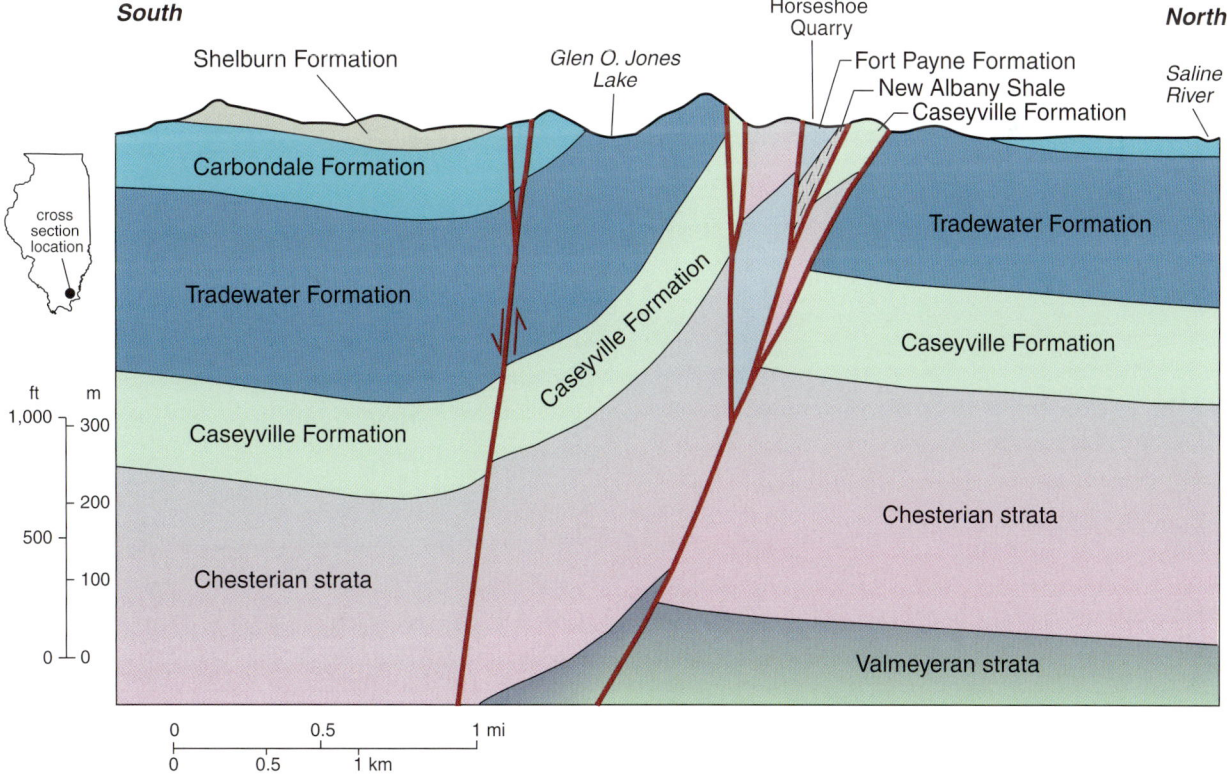

Figure 4-9 Cross section of the Rough Creek-Shawneetown Fault System at the "Horseshoe upheaval" southeast of Harrisburg on the Saline-Gallatin County line (modified from Nelson and Lumm 1986). A slice of steeply tilted Devonian New Albany Shale and Lower Mississippian Fort Payne Formation is 3,500 feet (1,070 m) above its position on either side of the fault zone. Although the fault geometry resembles a "positive flower structure" due to strike-slip faulting (compare with Figure 4-7), the Horseshoe upheaval appears to be better explained as the product of two or more episodes of up-and-down motion.

and Lower Mississippian cherty limestone are jammed between Pennsylvanian rocks on either side, 3,500 feet (1,070 m) above their normal elevation (Figure 4-9).

The Rough Creek-Shawneetown system originated during the Cambrian Period as the north side of a failed rift, the Rough Creek Graben. At this time, the south side of the system was downthrown along normal faults having as much as 8,000 feet (2,400 m) of displacement (Soderberg and Keller 1981; Bertagne and Leising 1991; Potter et al. 1995, 1997). During the Devonian Period, the southern side of the zone was downdropped again (Freeman 1951, Nelson and Marshak 1996). Major earth movements took place after Pennsylvanian time, most likely during the Permian as the Appalachian Mountains were forming. The Rough Creek Graben appears to be tectonically quiet today.

Although geologists such as Heyl (1972) postulated strike-slip faulting of the system, large horizontal movements can be ruled out because Pennsylvanian paleochannels in Kentucky are not offset. Multiple episodes of vertical movement provide a better explanation for the observed structure. Upthrust slices such as the "Horseshoe upheaval" may be the product of early reverse faulting that raised the southern block, followed by normal faulting during which slices of older rocks were sheared off and wedged high in the fault zone (Nelson and Lumm 1985).

Fluorspar Area Fault Complex

The complex array of fractures in the Illinois-Kentucky Fluorspar District includes high-angle normal, reverse, and strike-slip faults having various orientations. The fault complex is found in Hardin, southeastern Pope, and Massac Counties in Illinois and the adjacent area of Kentucky. The Fluorspar Area Fault Complex is bounded on the north by the Rough Creek-Shawneetown Fault System and on the northwest by the Lusk Creek Fault Zone. Eastern and southern boundaries are in Kentucky. Northeast-striking faults that outline horsts and grabens dominate (Baxter et al. 1963, 1967; Baxter and Desborough 1965; Hook 1974; Trace 1974; Trace and Amos 1984; Nelson 1991; Kolata and Nelson 1991; Bradbury and Baxter 1992; Potter et al. 1995, 1997). These faults have undergone multiple episodes of movement, beginning with Cambrian rifting and continuing locally in the New Madrid area to the present day. A major episode of faulting, together with igneous intrusion and emplacement of ore minerals, took place during the Permian. In Massac and Pulaski Counties, activity continued during Tertiary and Pleistocene time (Nelson et al. 1997, 1999). These faults are directly in line with the active New Madrid Seismic Zone to the southwest (see Chapter 5, Neotectonics). The Illinois-Kentucky Fluorspar District contains the largest deposits of fluorite in the United States, along with plentiful lead, zinc, barite, silver, and other valuable commodities (see Chapter 16, Lead, Zinc, and Fluorite Mining, and Chapter 17, Industrial Minerals).

Localized Structures

Hicks Dome

Although intimately related to the Fluorspar Area Fault Complex, Hicks Dome in Hardin County has so many unusual and unique features that it merits separate discussion. This feature is both a topographic and structural dome, forming a bull's-eye pattern on a geological map (Figure 4-10). The dome is about 10 miles (16 km) in diameter, and rocks at its apex are uplifted 4,000 feet (1,200 m). Middle Devonian rocks at the center are surrounded concentrically by younger rocks out to Pennsylvanian on the rim. Faults radiate from the center and surround the dome concentrically; other northeast-trending faults crosscut it. Dikes of Lower Permian ultramafic rock, such as peridotite, alnoite, and carbonatite, radiate from the dome in a bow-tie pattern (see Figure 4-10). Also present are diatremes, or explosion breccias, which contain a mixture of angular fragments of sedimentary and other rocks in a groundmass of igneous material (Baxter and Desborough 1965, Baxter et al. 1967, Zartman et al. 1967, Bradbury and Baxter 1992). Igneous rocks of Hicks Dome, in fact, are similar to those found in many of the world's diamond-bearing districts. Companies have explored for diamonds at Hicks Dome, apparently without success.

The presence of igneous rocks suggests that Hicks Dome was formed by intrusion at depth. Geophysical surveys, however, do not indicate a large mass of igneous rock beneath the dome (McGinnis and Bradbury 1964). Seismic reflection surveys show that the internal structure is highly disrupted and that doming extends down into the Precambrian basement rocks (Potter et al. 1995, 1997). Hicks Dome is best explained as the product of multiple small intrusions and explosive igneous activity within the Earth's crust (Bradbury and Baxter 1992).

Meteorite Impact Structures

Two probable buried meteorite impact structures, the Des Plaines and Glasford disturbances, occur within the state. A third, at Kentland, Indiana, is only a few miles outside of Illinois. The Des Plaines structure (Figure 4-2), in the northwestern suburbs of Chicago, is approximately 5 miles (8 km) in diameter. It is covered with glacial drift but is well-known from drilling records, including cored test holes for the Chicago deep tunneling project. Pulverized Ordovician St. Peter Sandstone at the center of the structure is more than 800 feet (240 m) above its expected depth as a result of rebound of the central area immediately

Structural Features

after impact. This center is surrounded by highly brecciated older rocks in a mosaic of fault blocks. Outward from the center is a ring-shaped trough containing Mississippian dolomite and Pennsylvanian shale and coal, rocks that elsewhere have been eroded from the Chicago region (Emrich and Bergstrom 1962).

The Glasford structure, southwest of Peoria, appears as a simple, circular dome about 2.5 miles (4 km) across in Silurian through Pennsylvanian rocks. Below the Silurian, drilling shows the Maquoketa Group (Upper Ordovician) to be about 100 feet (30 m) thicker than normal. The impact in Late Ordovician time formed a crater, which was filled with thicker than normal Maquoketa Shale, which apparently filled the actual crater. Below the Maquoketa, core drilling reveals a highly jumbled, shattered mixture of Cambrian and Ordovician rocks (Buschbach and Ryan 1963, Buschbach and Bond 1974).

Both the Des Plaines and Glasford structures are remote from major faults; no igneous rocks are known in the region. The geometry of both structures is consistent with an impact origin. More conclusively, shatter cones have been identified in drill cores from both structures. These distinctive, cone-shaped, striated fractures are known only from large meteorite impacts and from nuclear test sites. Distinctive crystal fractures in quartz grains from the St. Peter Sandstone at the apex of the Des Plaines structure provide additional support for its impact origin (McHone et al. 1986).

The Des Plaines structure is Pennsylvanian or younger, but occurred before Pleistocene time. The Glasford structure can be dated much more precisely to the Late Ordovician. All older rocks are brecciated, whereas younger rocks are merely domed, apparently as a result of crustal rebound.

Buried Hills and Reefs

A final category of local structure, of special interest to petroleum exploration, is a dome, which is formed by compaction over buried hills and reefs. As sediments are deposited over a buried knob of resistant rock, they bulge upward, creating a dome that can trap oil and natural gas.

When Cambrian sediments were deposited in Illinois, the Precambrian surface was hilly. Knobs of resistant rock, particularly rhyolite porphyry, rose as much as 1,500 feet (460 m) above the surrounding landscape (Dake and Bridge 1932, Nelson 1995). The tops of such hills remained above the sea through much of Late Cambrian and, in some cases, into Ordovician time. Eventually the hills were buried, but, as the sediments were compacted, they draped over the tops of the hills and formed structural domes. In some

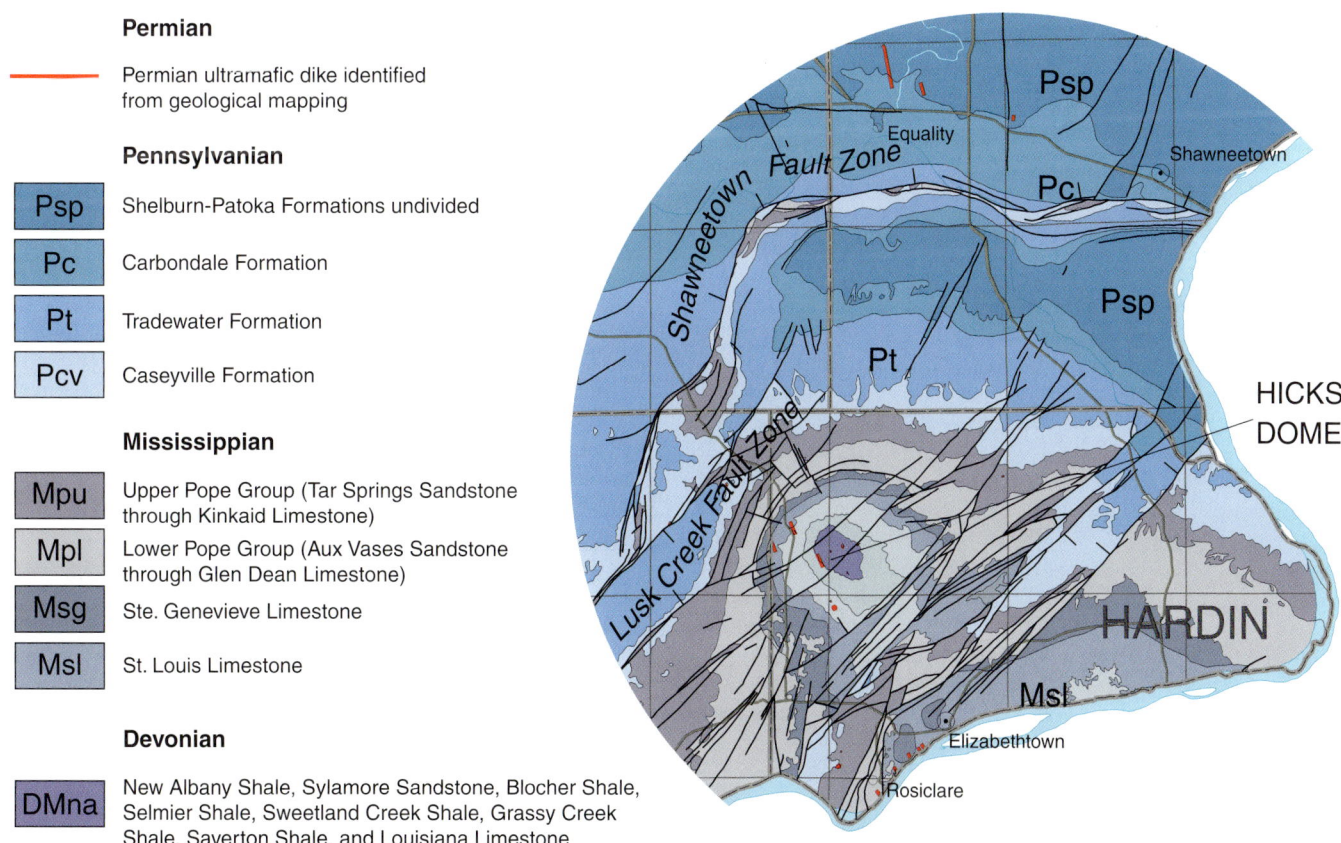

Figure 4-10 Geological map of Hicks Dome (Bradbury and Baxter 1992, Kolata 2005).

cases, drape folding extends upward to include Pennsylvanian strata. Thus, oil operators can find domes by shallow drilling and often complete wells in multiple pay zones, the rock layers that produce oil and gas.

Silurian pinnacle reefs also produce domes. The massive reef core resists compaction, much like a buried knob of rhyolite. Hundreds of Silurian reefs are scattered across Illinois and neighboring states. Nearly 30 Illinois reefs are productive, yielding close to 100 million barrels (15.9 billion L) of oil. Both the reefs themselves and the younger, permeable rocks draped over them can trap hydrocarbons. A further discussion of pinnacle reefs appears in Chapter 8, Silurian and Lower Devonian.

CONCLUSION

Although lacking spectacular exposures like those of the Appalachian and Rocky Mountains, Illinois has a wide variety of structural features that reflect a long history of tectonic activity. These structures played a key role in localizing economic mineral deposits, especially oil, gas, and fluorspar and associated minerals.

REFERENCES

Atherton, E., 1971, Tectonic development of the Eastern Interior Region of the United States, *in* Background materials for symposium on future petroleum potential of NPC Region 9 (Illinois Basin, Cincinnati Arch, and northern part of Mississippi Embayment): Illinois State Geological Survey, Illinois Petroleum 96, p. 29–43.

Ault, C. H., D. M. Sullivan, and G. F. Tanner, 1980, Faulting in Posey and Gibson Counties, Indiana: Indiana Academy of Science Proceedings, v. 89, p. 275–289.

Baxter, J. W., and G. A. Desborough, 1965, Areal geology of the Illinois Fluorspar District, Part 2—Karbers Ridge and Rosiclare Quadrangles: Illinois State Geological Survey, Circular 385, 40 p.

Baxter, J. W., G. A. Desborough, and C. W. Shaw, 1967, Areal geology of the Illinois Fluorspar District, Part 3—Herod and Shetlerville Quadrangles: Illinois State Geological Survey, Circular 413, 41 p.

Baxter, J. W., P. E. Potter, and F. L. Doyle, 1963, Areal geology of the Illinois Fluorspar District, Part 1—Saline mines, Cave in Rock, De koven, and Repton Quadrangles: Illinois State Geological Survey, Circular 342, 44 p.

Bear, G. W., J. A. Rupp, and A. J. Rudman, 1997, Seismic interpretation of the deep structure of the Wabash Valley Fault System: Seismological Research Letters, v. 68, no. 4, p. 624–640.

Bertagne, A. J., and T. C. Leising, 1991, Interpretation of seismic data from the Rough Creek Graben of western Kentucky and southern Illinois, *in* M. W. Leighton, D. R. Kolata, D. F. Oltz, and J. J. Eidel, eds., Interior cratonic basins: Tulsa, Oklahoma, American Association of Petroleum Geologists, Memoir 51, p. 199–208.

Bradbury, J. C., and J. W. Baxter, 1992, Intrusive breccias at Hicks Dome, Hardin County, Illinois: Illinois State Geological Survey, Circular 550, 23 p.

Bristol, H. M., and J. D. Treworgy, 1979, The Wabash Valley Fault System in southeastern Illinois: Illinois State Geological Survey, Circular 509, 19 p.

Bunker, B. J., G. A. Ludvigson, and B. J. Witzke, 1985, The Plum River Fault Zone and the structural and stratigraphic framework of eastern Iowa: Iowa Geological Survey, Technical Information Series No. 13, 123 p.

Buschbach, T. C., and D. C. Bond, 1974, Underground storage of natural gas in Illinois, 1973: Illinois State Geological Survey, Illinois Petroleum 101, 71 p.

Buschbach, T. C., and D. R. Kolata, 1991, Regional setting of the Illinois Basin, *in* M. W. Leighton, D. R. Kolata, D. F. Oltz, and J. J. Eidel, eds., Interior cratonic basins: Tulsa, Oklahoma, American Association of Petroleum Geologists, Memoir 51, p. 29–55.

Buschbach, T. C., and R. Ryan, 1963, Ordovician explosion structure at Glasford, Illinois: American Association of Petroleum Geologists Bulletin, v. 47, no. 12, p. 2015–2022.

Clegg, K. E., 1965, The La Salle Anticlinal Belt and adjacent structures in east-central Illinois: Illinois Academy of Science Transactions, v. 58, no. 2, p. 82–94.

Clegg, K. E., 1970, The La Salle Anticlinal Belt in Illinois, *in* W. H. Smith, R. B. Nance, M. E. Hopkins, R. G. Johnson, and C. W. Shabica, eds., Depositional environments in parts of the Carbondale Formation, western and northern Illinois: Francis Creek Shale and associated strata, and Mazon Creek biot: Illinois State Geological Survey, Guidebook 8, p. 106–110.

Dake, C. L., and J. Bridge, 1932, Buried and resurrected hills of central Ozarks: American Association of Petroleum Geologists Bulletin, v. 16, no. 7, p. 629–652.

Desborough, G. A., 1961, Geology of the Pomona Quadrangle, Illinois: Illinois State Geological Survey, Circular 320, 16 p.

Devera, J. A., 1993, Geologic map of the Wolf Lake Quadrangle, Jackson and Union Counties, Illinois: Illinois State Geological Survey, Illinois Geologic Quadrangle map, IGQ 13, 1:24,000.

Devera, J. A., and W. J. Nelson, 1995, Geologic map of the Cobden Quadrangle, Jackson and Union Counties, Illinois: Illinois Geologic Quadrangle map, IGQ 16, 1:24,000.

Douthit, T. D. N., 1959, Magnetic survey of the Cap au Grès Faulted Flexure: St. Louis, Missouri, St. Louis University, M.S. thesis, 111 leaves.

Duchek, A. B., J. H. McBride, W. J. Nelson, and H. E. Leetaru, 2004, The Cottage Grove Fault System (Illinois Basin): Late Paleozoic transpression along a Precambrian crustal boundary: Geological Society of America Bulletin, v. 116, no. 11/12, p. 1465–1484.

Emrich, G. H., and R. E. Bergstrom, 1962, Des Plaines disturbance, northeastern Illinois: Geological Society of America Bulletin, v. 73, no. 8, p. 959–968.

Ervin, C. P., and L. D. McGinnis, 1975, Reelfoot Rift: Reactivated precursor to the Mississippi Embayment: Geological Society of America Bulletin, v. 86, no. 9, p. 1287–1295.

Freeman, L. B., 1951, Regional aspects of Silurian and Devonian stratigraphy in Kentucky: Lexington, Kentucky, Kentucky Geological Survey, Series IX, Bulletin 6, 565 p.

Grohskopf, J. G., 1955, Subsurface geology of the Mississippi Embayment of southeast Missouri: Missouri Geological Survey and Water Resources, v. 37, 2nd ser., 133 p.

Harrison, R. W., and A. Schultz, 2002, Tectonic framework of the southwestern margin of the Illinois Basin and its influence on neotectonism and seismicity: Seismological Research Letters, v. 73, no. 5, p. 698–731.

Heyl, A. V., 1972, The 38th parallel lineament and its relationship to ore deposits: Economic Geology, v. 67, no. 7, p. 879–894.

Hildenbrand, T. G., and D. Ravat, 1997, Geophysical setting of the Wabash Valley Fault System: Seismological Research Letters, v. 68, no. 4, p. 567–585.

Hook, J. W., 1974, Structure of the fault systems in the Illinois-Kentucky Fluorspar District: Lexington, Kentucky, Kentucky Geological Survey, Series X, Special Publication 22, p. 77–86.

Jacobson, R. J., 1985, Coal resources of Grundy, La Salle, and Livingston Counties, Illinois: Illinois State Geological Survey, Circular 536, 58 p.

Kolata, D. R., compiler, 2005, Bedrock geology of Illinois: Illinois State Geological Survey, Illinois Map 14, 1:500,000.

Kolata, D. R., and T. C. Buschbach, 1976, Plum River Fault Zone of northwestern Illinois: Illinois State Geological Survey, Circular 491, 20 p.

Kolata, D. R., T. C. Buschbach, and J. D. Treworgy, 1978, The Sandwich Fault Zone of northern Illinois: Illinois State Geological Survey, Circular 505, 26 p.

Kolata, D. R., B. D. Keith, and J. A. Drahovzal, 1992, Illinois Basin Consortium program plan: Illinois State Geological Survey, Illinois Basin Studies, 21 p.

Kolata, D. R., and W. J. Nelson, 1991, Tectonic history of the Illinois Basin, in M. W. Leighton, D. K. Kolata, D. F. Oltz, and J. J. Eidel, eds., Interior cratonic basins: Tulsa, Oklahoma, American Association of Petroleum Geologists, Memoir 51, p. 263–285.

Marcher, M. V., and R. G. Stearns, 1962, Tuscaloosa Formation in Tennessee: Geological Society of America Bulletin, v. 73, no. 11, p. 1365–1386.

Marshak, S., W. J. Nelson, and J. H. McBride, 2003, Phanerozoic strike-slip faulting in the continental interior platform of the United States—Examples from the Laramide orogen, midcontinent, and ancestral Rocky Mountains, in F. Storti, R. E. Holdsworth, and F. Salvani, eds., Intraplate strike-slip deformation belts: London, England, Geological Society of London, Special Publication 210, p. 159–184.

Mateker, E. J., 1959, Some gravity and magnetic interpretation problems: St. Louis, Missouri, St. Louis University, M.S. thesis, 165 leaves.

McBride, J. H., 1998, Understanding basement tectonics of an interior cratonic basin: Southern Illinois Basin, USA: Tectonophysics, v. 293, no. 1–2, p. 1–20.

McBride, J. H., T. G. Hildenbrand, W. J. Stephenson, and C. J. Potter, 2002, Interpreting the earthquake source of the Wabash Valley Seismic Zone (Illinois, Indiana, and Kentucky) from seismic-reflection, gravity, and magnetic-intensity data: Seismological Research Letters, v. 73, no. 5, p. 660–686.

McBride, J. H., and W. J. Nelson, 1999, Style and origin of mid-Carboniferous deformation in the Illinois Basin, USA—Ancestral Rockies deformation?: Tectonophysics, v. 305, no. 1–3, p. 249–273.

McGinnis, L. D., and J. C. Bradbury, 1964, Aeromagnetic study of the Hardin County area, Illinois: Illinois State Geological Survey, Circular 363, 12 p.

McHone, J. H., M. L. Sargent, and W. J. Nelson, 1986, Shatter cones in Illinois—Evidence for meteoritic impacts at Glasford and Des Plaines (abs.): Abstracts and Program for the 49th Annual Meeting of the Meteoritical Society: Houston, Texas, Lunar and Planetary Institute, p. G-3.

Melton, F. A., 1925, The ancestral Rocky Mountains of Colorado and New Mexico: Journal of Geology, v. 33, no. 1, p. 84–89.

Munson, P. J., C. A. Munson, N. K. Bleuer, and M. D. Labitzke, 1992, Distribution and dating of prehistoric earthquake liquefaction in the Wabash Valley of the central U.S.: Seismological Research Letters, v. 63, no. 3, p. 337–342.

Munson, P. J., S. F. Obermeier, C. A. Munson, and E. R. Hajic, 1997, Liquefaction evidence for Holocene and latest Pleistocene seismicity in the southern halves of Indiana and Illinois—A preliminary overview: Seismological Research Letters, v. 68, no. 4, p. 521–536.

Nelson, W. J., 1991, Structural styles of the Illinois Basin, in M. W. Leighton, D. R. Kolata, D. F. Oltz, and J. J. Eidel, eds.: Tulsa, Oklahoma, American Association of Petroleum Geologists, Memoir 51, p. 209–243.

Nelson, W. J., 1995, Structural features in Illinois: Illinois State Geological Survey, Bulletin 100, 144 p.

Nelson, W. J., F. B. Denny, J. A. Devera, L. R. Follmer, and J. M. Masters, 1997, Tertiary and Quaternary tectonic faulting in southernmost Illinois: Engineering Geology, v. 46, no. 3–4, p. 235–258.

Nelson, W. J., F. B. Denny, L. R. Follmer, and J. M. Masters, 1999, Quaternary grabens in southernmost Illinois: Deformation near an active intraplate seismic zone: Tectonophysics, v. 305, no. 1–3, p. 381–397.

Nelson, W. J., and H.-F. Krausse, 1981, The Cottage Grove Fault System in southern Illinois: Illinois State Geological Survey, Circular 522, 65 p.

Nelson, W. J., and D. K. Lumm, 1985, Ste. Genevieve Fault Zone, Missouri and Illinois: Illinois State Geological Survey, Contract/Grant Report 1985-3, 94 p.

Nelson, W. J., and D. K. Lumm, 1986, Geology of the Rudement Quadrangle, Saline County, Illinois: Illinois State Geological Survey, Illinois Geologic Quadrangle map 3, 1:24,000.

Nelson, W. J., and D. K. Lumm, 1987, Structural geology of southeastern Illinois and vicinity: Illinois State Geological Survey, Circular 538, 70 p.

Nelson, W. J., and S. Marshak, 1996, Devonian tectonism of the Illinois Basin region, U.S. continental interior, in B. A. van der Pluijm and P. A. Catacosinos, eds., Basement and basins of eastern North America: Boulder, Colorado, Geological Society of America, Special Paper 308, p. 169–180.

Nelson, W. J., L. B. Smith, and J. D. Treworgy, 2002, Sequence stratigraphy of the lower Chesterian (Mississippian) strata of the Illinois Basin: Illinois State Geological Survey, Bulletin 107, 70 p.

Obermeier, S. F., 1998, Liquefaction evidence for strong earthquakes of Holocene and latest Pleistocene ages in the states of Indiana and Illinois, USA: Engineering Geology, v. 50, no. 3–4, p. 227–254.

Obermeier, S. F., P. J. Munson, C. A. Munson, J. R. Martin, A. D. Frankel, T. L. Youd, and E. C. Pond, 1992, Liquefaction evidence for strong Holocene earthquake(s) in the Wabash Valley of Indiana-Illinois: Seismological Research Letters, v. 63, no. 3, p. 321–335.

Paull, R. K., and R. A. Paull, 1977, Geology of Wisconsin and upper Michigan—Including parts of adjacent states: Dubuque, Iowa, Kendall-Hunt Publishing Co., 232 p.

Potter, C. J., J. A. Drahovzal, M. L. Sargent, and J. H. McBride, 1997, Proterozoic structure, Cambrian rifting, and younger faulting as revealed by a regional seismic reflection network in the southern Illinois Basin: Seismological Research Letters, v. 68, no. 4, p. 537–552.

Potter, C. J., M. B. Goldhaber, P. C. Heigold, and J. A. Drahovzal, 1995, Structure of the Reelfoot–Rough Creek rift system, Fluorspar Area Fault Complex, and Hicks Dome, southern Illinois and western Kentucky—New constraints from regional seismic reflection data: Reston, Virginia, U.S. Geological Survey, Professional Paper 1538-Q, 19 p

Rubey, W. W., 1952, Geology and mineral resources of the Hardin and Brussels Quadrangles (in Illinois): Washington, D.C., U.S. Geological Survey, Professional Paper 218, 179 p.

Sexton, J. L., L.W. Braile, W. J. Hinze, and M. J. Campbell, 1986, Seismic reflection profiling studies of a buried Precambrian rift beneath the Wabash Valley fault zone: Geophysics, v. 51, no. 3, p. 640–660.

Soderberg, R. K., and G. R. Keller, 1981, Geophysical evidence for deep basin in western Kentucky: American Association of Petroleum Geologists Bulletin, v. 65, no. 2, p. 226–234.

Stearns, R. G., and M. V. Marcher, 1962, Late Cretaceous and subsequent structural development of the northern Mississippi Embayment area: Geological Society of America Bulletin, v. 73, no. 11, p. 1387–1394.

Templeton, J. S., 1951, The geology and mineral resources of the Oregon Quadrangle: Illinois State Geological Survey, unpublished manuscript no. 3, available from the Illinois State Geological Survey Library, 99 p.

Trace, R. D., 1974, Illinois-Kentucky Fluorspar District: Kentucky Geological Survey, Series X, Special Publication 22, p. 58–76.

Trace, R. D., and D. H. Amos, 1984, Stratigraphy and structure of the western Kentucky Fluorspar District: Reston, Virginia, U.S. Geological Survey, Professional Paper 1151-D, 41 p.

Ver Wiebe, W. A., 1930, Ancestral Rocky Mountains: American Association of Petroleum Geologists Bulletin, v. 14, no. 6, p. 765–788.

Weller, S. C., and S. St. Clair, 1928, Geology of Ste. Genevieve County, Missouri: Missouri Bureau of Geology and Mines, v. 22, 2nd ser., 352 p.

Whiting, L. L., and D. L. Stevenson, 1965, The Sangamon Arch: Illinois State Geological Survey, Circular 383, 20 p.

Zartman, R. E., M. R. Brock, A. V. Heyl, and H. H. Thomas, 1967, K-Ar and Rb-Sr ages of some alkalic intrusive rocks from central and eastern United States: American Journal of Science, v. 265, no. 10, p. 848–870.

Neotectonics 5

W. John Nelson and John H. McBride

INTRODUCTION

At first glance, Illinois appears to be rather quiet tectonically. Despite the occasional present-day tremors that occur within Illinois and the great historical New Madrid earthquakes of 1811 and 1812 in the Missouri "bootheel" area, geologists have traditionally viewed Illinois as part of the "stable craton," deep within the continent and far removed from dynamic areas such as the West Coast. The New Madrid earthquakes have generally been seen as anomalies.

Earth scientists now recognize that Illinois has active faults and a recent history of large earthquakes. Four discoveries support that view:

1. Surface faults in southernmost Illinois have undergone large movements during the past 2 million years and in one place possibly underwent very small movements within 5,000 years of the present.
2. Large earthquakes (magnitude 6 to 7.5) liquefied river sediments in large areas of Illinois during the past 12,000 years.
3. Active tectonic stress is forming new faults and fractures and causes roof failure in underground mines.
4. Some earthquakes in Illinois are centered on zones of weakness in the crust that are hundreds of millions of years old.

This chapter on neotectonics, or "new tectonics," addresses recent and ongoing earth movements, particularly those associated with earthquakes. This area is one of many where geological processes have a direct impact on human activities.

QUATERNARY FAULTS AT THE SURFACE

Although faults are present throughout Illinois (see Chapter 3, Tectonic History, and Chapter 4, Structural Features), most show no signs of recent activity. Until the mid-1990s, no definitive evidence had been found for faulting in Illinois since the Cretaceous Period (65 million years ago). Recent investigations at the southern tip of the state show that some faults there underwent large movements during the Pleistocene. The best examples are in Massac County, near Metropolis and Joppa (Figure 5-1), where the main activity took place during and before the Illinois Episode of the Quaternary Period, more than 125,000 years ago.

The only other place in Illinois where a geologist has suggested Quaternary faulting may exist lies just north of St. Louis. Rubey (1952) mapped a unit that he named the Grover Gravel of late Tertiary age on the uplands of Calhoun County. The unit is composed of well-rounded pebbles, cobbles, and boulders of purple, red, white, and black quartzite and chert. The map shows the gravel ranging from 0 to 20 feet (6 m) in thickness. Rubey concluded that the Grover Gravel was higher on the north side of the Cap au Grès Faulted Flexure than on the south side, indicating post-Tertiary movement on the Cap au Grès. During recent mapping of the Brussels and Winfield Quadrangles by Seid (2008a, 2008b), no gravel exposures were seen in place. Gravel was found along upland streams coming off the Dividing Ridge in Calhoun County, but the gravel's elevation could not be determined. On the maps by Rubey (1952) and Harrison (1997), the Tertiary gravel unit is overmapped (i.e., overestimated), which makes the unit appear much thicker than it probably is. The distribution, thickness, and elevation of the Grover Gravel are so poorly constrained that post-Tertiary deformation cannot be supported.

Southernmost Illinois

Numerous faults displace Quaternary sediments in the Mississippi Embayment at the southern tip of Illinois. Although suspected by some early geologists, those faults were not documented until the late 1990s. Early reports of Quaternary deformation (Shaw 1915, Ross 1963) lacked well-defined evidence, and closer inspection of some reported sites indicated that landsliding or other non-tectonic processes had been at work (Kolata et al. 1981). Detailed geological mapping at a scale of 1:24,000 and a dedicated program of drilling, trenching, and geophysical studies were required to prove Quaternary tectonic faulting in the northern Mississippi Embayment.

The Quaternary faults of southernmost Illinois are part of the Fluorspar Area Fault Complex (Figure 4-2, Chapter 4, Structural Features; see also Chapter 13, Quaternary Paleoclimate). These faults originated as part of a "failed rift" (Reelfoot Rift) and have been active repeatedly since Cambrian time. The New Madrid Seismic Zone lies within the Reelfoot Rift.

Most Quaternary faults in southern Illinois trend northeast or north-northeast (Figure 5-1), following the dominant trend of the Fluorspar Area Fault Complex (Figure 5-2). A few strike north-south to slightly west of north; at Round Knob, they strike northwest. All dip vertically or nearly so. Typically, these faults form grabens, downthrown blocks that range from a few feet to perhaps a quarter mile (about 400 m) wide. Although the rocks dip steeply at the edges of grabens, key beds on either side of the dropped block lie at nearly equal elevations. This characteristic makes it easy for Quaternary faults to escape detection, given the scarcity of outcrops and the wide spacing of boreholes in the Mississippi Embayment area. Several grabens were discovered by following up on reports of unusual strata or abnormally great depth to bedrock in a single well. A railroad cut yielded a lucky exposure at Kelley (Figure 5-1).

Faults displace the Mounds Gravel (late Tertiary to early Pleistocene) or younger units at nine sites, shown in Figure 5-1 and Table 5-1.

Barnes Creek, located about 5 miles (8 km) north of Metropolis, is the only known place in Illinois where tectonic faults may displace Holocene sediments. The creek was artificially straightened and deepened to drain fields for agriculture, providing a ready-made trench. Many faults are exposed that displace the Metropolis Formation (Pleistocene); offsets range from a few inches to more than 90 feet (27.4 m) (Nelson et al. 1997, 1999; McBride et al. 2002). A few fractures continue through the Metropolis into younger Holocene sediments radiocarbon dated at 5,000 to 7,000 years (Science Applications International Corporation 2002). Some of these clay-lined fractures exhibit a very small (a few inches) offset of layers in Holocene sediment. Although Science Applications International Corporation (2002) suggested tectonic displacement, it is possible that ordinary soil processes, such as settling and compaction, produced the offsets. No faults enter the youngest sediments at Barnes Creek; those sediments yield radiocarbon dates of 200 to 5,000 years (Science Applications International Corporation 2002).

Wisconsinan (late Pleistocene) faulting probably took place at Mallard Creek, northeast of Metropolis (Figure 5-3). A stream-bank exposure, improved with a backhoe, showed multiple faults that offset the McNairy (Cretaceous) and Metropolis Formations. Faults also offset a younger gravel in which the Sangamon Soil is developed. The younger gravel was dropped into a small graben about

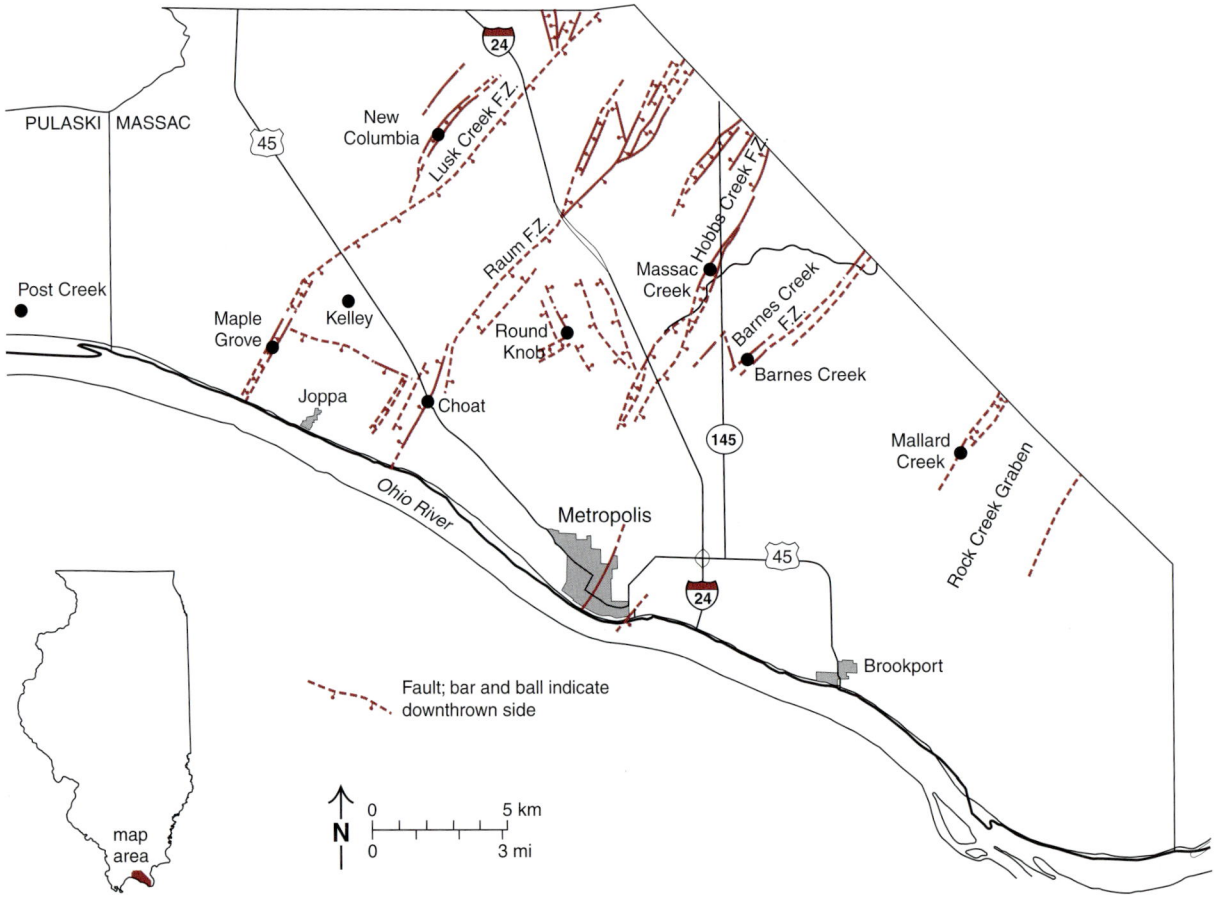

Figure 5-1 Map of Massac County and a portion of Pulaski County showing sites where faults affect Quaternary sediments. A solid line means fault accurately located; a dashed line means approximately located or inferred. Abbreviation: F.Z., Fault Zone.

NEOTECTONICS

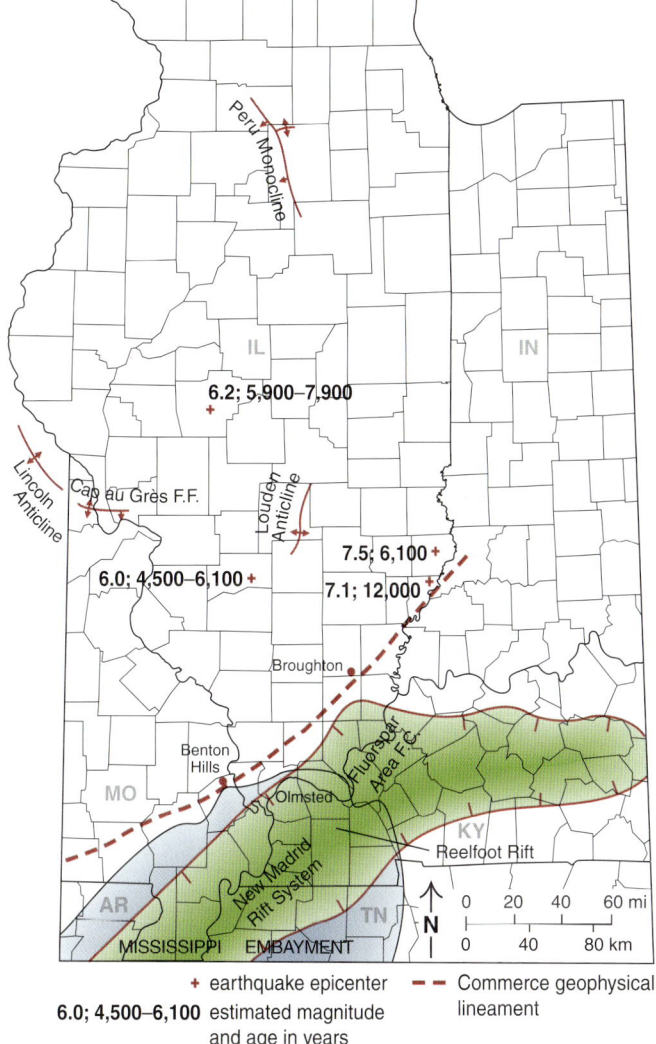

Figure 5-2 Map of Illinois and portions of surrounding states showing neotectonic features. Red crosses indicate epicenters of prehistoric earthquakes inferred from paleo-liquifaction studies (see text for explanation). Abbreviations: F.C., fault complex; F.F., faulted flexure.

3 feet (0.9 m) wide and 3 feet (0.9 m) deep. The youngest sediments at Mallard Creek, presumably Holocene, are not faulted. No organic matter was found that could be radiocarbon dated (Nelson et al. 1999).

Kelley, in Massac County, is a second site where Wisconsinan tectonic movements may have taken place (Figure 5-4). Trenching in a railroad cut revealed that the Mounds Gravel, Metropolis Formation, and Loveland Silt were dropped into a north-trending graben. The Loveland, a loess unit of Illinoian age, is confined to the deepest part of the graben. The overlying Roxana and Peoria Silts, both loesses of Wisconsin age, thicken and sag into the graben, but are not faulted.

The largest known late Tertiary to Quaternary displacement on any fault in the Mississippi River valley (including the New Madrid Seismic Zone) is at Massac Creek about 6 miles (9.6 km) north of Metropolis. The Illinois State Geological Survey drilled two continuously cored test holes into the graben (Figure 5-5). The first hole was drilled in 1995. The deeper hole, drilled in 1999, shows that the Mounds Gravel is downthrown a phenomenal 450 feet (137.4 m) compared with its elevation in a nearby gravel pit. Overlying the Mounds is 385 feet (117.4 m) of sediment, largely clay, silt, and fine sand that contain abundant peat and organic matter. Fossil pollen from the two cores was examined by two pollen specialists from the U.S. Geological Survey in Reston, Virginia (Ronald Litwin, written communication 1995; Norman Frederiksen, written communication 1996, 2000). Their conclusions differ somewhat, but samples from the first hole were dated as no older than Miocene and possibly as young as Pliocene. Pollen from a depth of 333 feet (101.5 m) in the deeper boring "probably represents some temperate part of the Quaternary," and the assemblage was described as representing "temperate forest; the local environment of deposition was a pond, marsh, or other wet place" (Frederiksen, written communication 2000). Evidently the graben subsided over a long period, forming a swampy lowland in which sediment continuously accumulated (Nelson 1996; Nelson et al. 1997, 1999; McBride and Nelson 2001; McBride et al. 2002).

Tectonic activity is not the only force that deforms geologically young materials. Landslides and karst activity

Table 5-1 Faults displacing the Mounds Gravel or younger units.

Structure	Trend	Configuration	Youngest unit affected	Age
Barnes Creek	Various	Multiple grabens	Metropolis Formation and possibly younger alluvium	Wisconsinan; possible Holocene
Choat	North-northeast	Multiple faults	Metropolis Formation	Illinoian
Kelley	North-south	Graben	Loveland Silt	Sangamonian to early Wisconsinan
Mallard Creek	North-northeast	Graben	Sangamon Geosol	Wisconsinan
Maple Grove	North-northeast	Graben	Metropolis Formation	Illinoian
Massac Creek	North-northeast	Complex graben	Unnamed	Quaternary
New Columbia	Northeast	Graben	Mounds Gravel	Late Tertiary to early Quaternary
Post Creek	North to northeast	Graben	Mounds Gravel	Late Tertiary to early Quaternary
Round Knob	Northwest	Underlain	Metropolis Formation	Illinoian

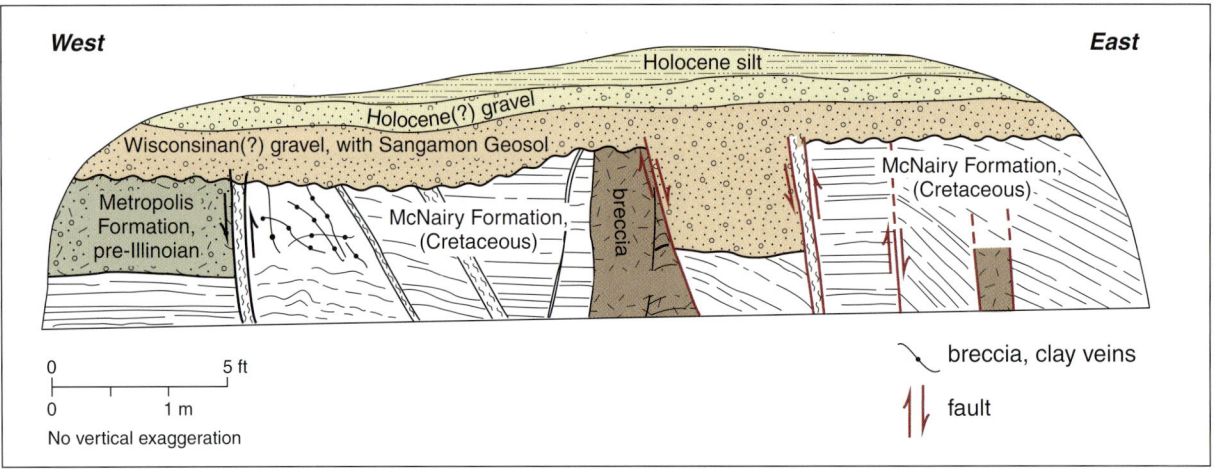

Figure 5-3 Sketch of the bank of Mallard Creek, Massac County, showing faults displacing Quaternary sediments (Nelson et al. 1999, Denny and Nelson 2005). Location shown on Figure 5-1.

are common in southern Illinois and elsewhere (see Chapter 22, Soils, and Chapter 28, Karst Terrane). Kolata et al. (1981) documented several cases of karst and landslide disturbance in southernmost Illinois. Distinguishing tectonic structures from nontectonic structures can be difficult, but, typically, Quaternary tectonic faults

- are directly in line with mapped tectonic faults in bedrock and have the same trend;
- extend directly into bedrock, as shown by seismic reflection surveys and drilling;
- include reverse or strike-slip offsets, which generally are incompatible with gravity-driven earth failure;
- show repeated episodes of movement over long spans of geological time; and
- occur where necessary conditions for karst or landslide failure do not exist (e.g., no limestone bedrock and no steep slopes).

Benton Hills and Commerce Geophysical Lineament

Widespread Holocene tectonic faulting took place within a few miles of the Illinois border in the Benton Hills of southeastern Missouri (Figure 5-2). Detailed geological mapping, extensive geophysical surveys, drilling, and trenching revealed a complex array of faults, most of which trend northeast (Harrison and Schultz 1994, Harrison 1999). Many faults displace Peoria Silt (Woodfordian or latest Pleistocene), and several offset sediments that have yielded Holocene radiocarbon dates. One site yields evidence for movement during the New Madrid earthquakes of 1811 and 1812 (Harrison et al. 1999, 2002; Harrison and Schultz 2002). The Benton Hills lie along a larger feature called the Commerce geophysical lineament (Figure 5-2). Defined by gravity and magnetic anomalies (Langenheim and Hildenbrand 1997), the lineament extends from north-

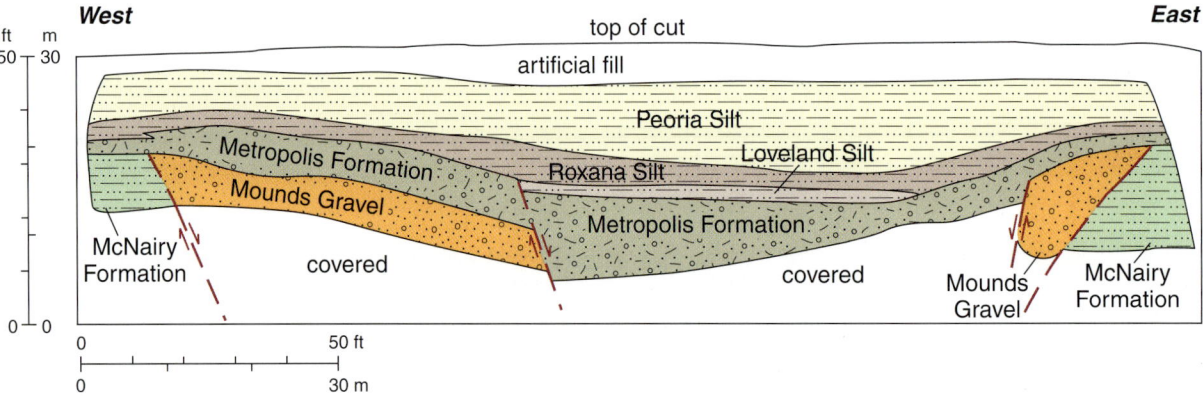

Figure 5-4 Cross section of Kelley site, Massac County, based on trench exposures. Location shown on Figure 5-1.

NEOTECTONICS

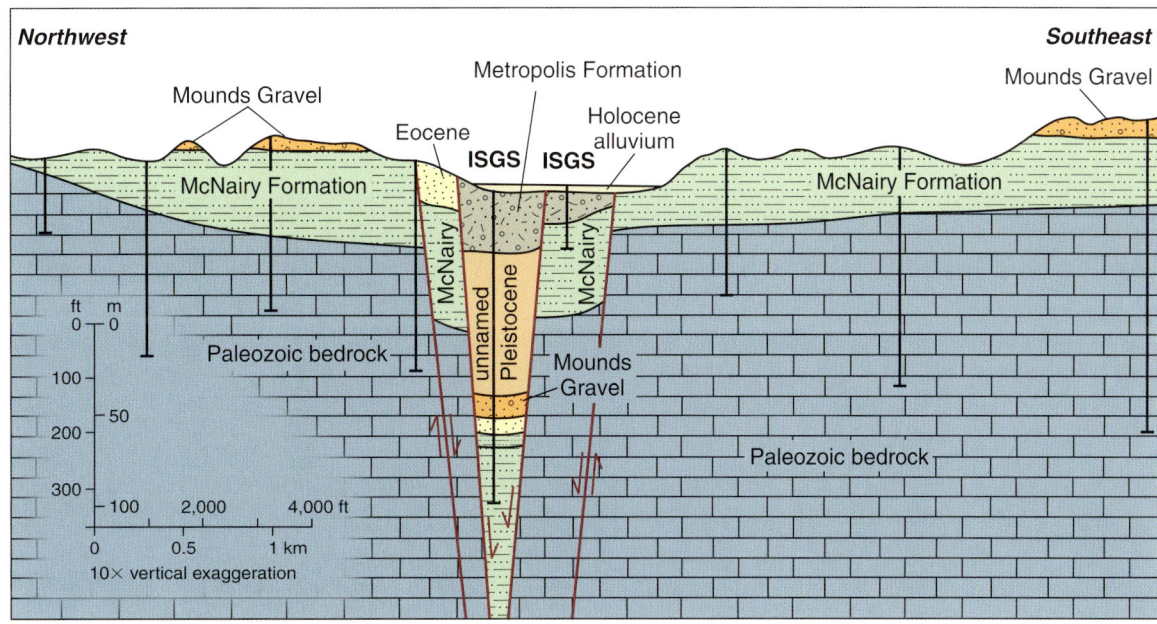

Figure 5-5 Cross section of Massac Creek graben, Massac County. Location shown on Figure 5-1.

eastern Arkansas to near Vincennes, Indiana, and lies parallel to the Reelfoot Rift and New Madrid Seismic Zone. The lineament may be an ancient, deep-seated fracture zone into which dense igneous rocks intruded. About a dozen earthquake epicenters, including several in southern Illinois, fall along the Commerce geophysical lineament (Harrison and Schultz 1994, 2002; Langenheim and Hildenbrand 1997; Nelson et al. 1997; Hildenbrand et al. 2002; Odum et al. 2002).

Earthquake Liquefaction

One of the most dramatic effects of the great New Madrid earthquakes of 1811–1812 was soil liquefaction. Conditions were ideal for liquefaction: a high water table and thick, loose, water-saturated river sands capped by clay-rich topsoil. Water and sand erupted violently from thousands of fissures, forming "sand blows" that are still visible today on the ground and from the air (Fuller 1912, Stewart and Knox 1993, Wesnousky and Leffler 1994). (See also Chapter 27, Earthquakes.)

Sand blows from prehistoric earthquakes in the New Madrid area can be studied in trenches that reveal, in cross section, irregular "sand dikes" that cut upward through silt and clay layers and, in some cases, the cone-shaped deposit of the sand blow itself. Using radiocarbon age-dating techniques and cultural artifacts, geologists can reconstruct the history of the New Madrid Seismic Zone and estimate how often great earthquakes occur (Kelson et al. 1992, Wesnousky and Leffler 1994). Studies of New Madrid and other historic earthquakes indicate that magnitude 6 is about

Figure 5-6 Sand blow liquefaction feature, Cahokia Creek, east of Roxana, Madison County. (Photograph by Robert A. Bauer, October 2007.)

the lower limit of energy needed to liquefy water-saturated sand (Wesnousky and Leffler 1994). The larger the quake, the larger and more numerous are the sand blows and the affected area.

Hundreds of sand dikes, similar to those of the New Madrid area, occur along river banks in central and southern Illinois, southwestern Indiana, and eastern Missouri (Figure 5-6). Investigators have concluded that, within the last 10,000 years, earthquakes larger than any recorded in historic times have taken place within the region. Initial studies in the Wabash River valley (Munson et al. 1992, Obermeier et al. 1992) have suggested the occurrence of a single, large earthquake with an epicenter near Vincennes, Indiana, between 2,500 and 7,500 years ago. Later studies over a wider area indicated additional events (Obermeier 1996, 1998; Hajic and Wiant 1997; Munson et al. 1997; Tuttle et al. 1999):

- A quake of about magnitude 7.5 centered about 15 miles (24 km) west of Vincennes, Indiana, about 6,100 years ago (Figure 5-2).
- A quake of magnitude 7.1 located about 25 miles (40 km) southwest of Vincennes about 12,000 years ago.
- An event of magnitude 6.2 or greater near Springfield, Illinois, 5,900 to 7,900 years ago.
- A quake of magnitude 6.0 or greater, centered between Carlyle and Centralia, Illinois, either 4,500 or 6,100 years ago (radiocarbon dating results differ).

STRESS, FRACTURES, AND MINE ROOF CONTROL

Mine roof control is one of the greatest concerns in underground mining. Mine roof falls account for nearly half of the injuries suffered by underground coal miners. Control of the mine roof, therefore, ranks high among the costs incurred by mining companies, and many mines have been abandoned prematurely due to difficulty in controlling ground movement. Miners in Illinois have long known that roof falls and fractures in some mines followed certain preferred trends, but the causes behind these trends were a mystery until recently. Within the last 20 years, instrumental studies have supplied the answer: tectonic stress.

Tectonic Stress Field

Tectonic stress near the Earth's surface is the product of continental plate motions. Hence, stress orientation remains consistent across large areas. Maps based on hundreds of stress measurements throughout North America show that the principal maximum compressive stress axis (hereafter, σ_1) is generally oriented northeast to east-northeast throughout the eastern and central United States. These stresses apparently result from the westward motion of the North American plate away from the ever-widening Atlantic Ocean (Sbar and Sykes 1973, Zoback and Zoback 1980, Gough 1984, Gough and Gough 1987). Measurements taken in Illinois show that σ_1 is nearly east-west in the southern part of the state; σ_1 averages about N60° E elsewhere in Illinois (Figure 5-7) (Nelson and Bauer 1987, 1991).

Various methods are used to measure stress. Stress deforms any opening in the earth, such as a borehole or mine, and instruments measure the deformation. The direction of maximum shortening is σ_1. In a vertical borehole that is under stress, fractures develop parallel to this σ_1; these fractures can be recorded by various means. One method

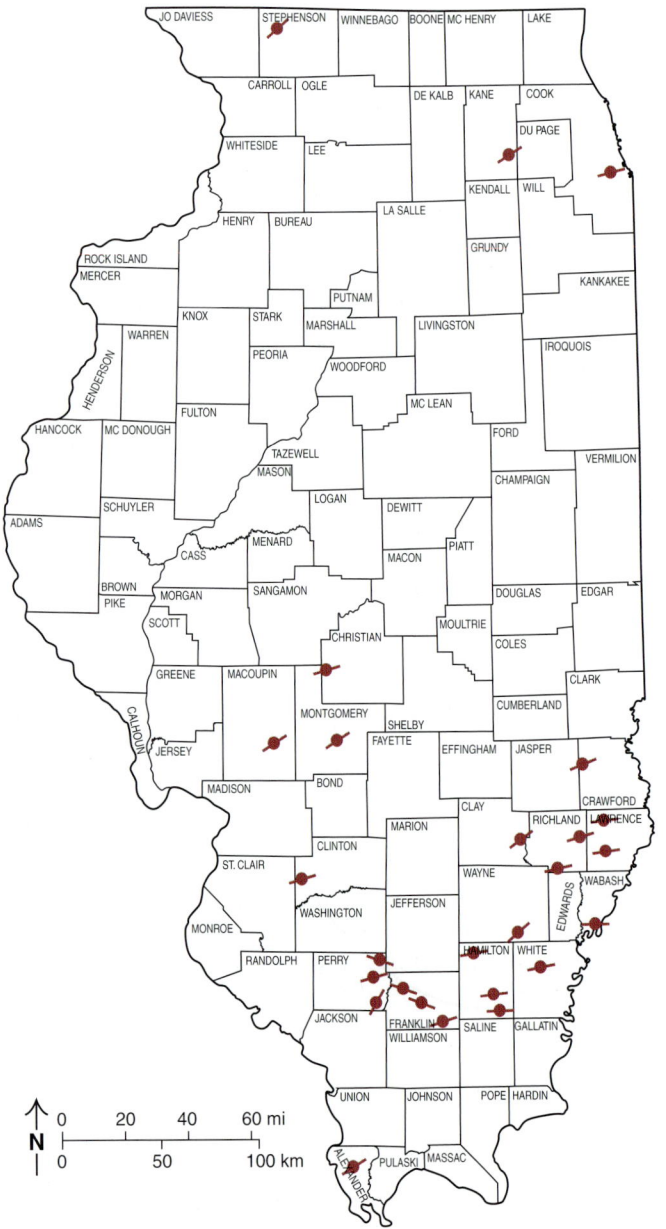

Figure 5-7 Map showing orientations of principal compressive stress axes in Illinois. (Map by Robert A. Bauer.)

of determining σ_1 is analysis of seismograph records from earthquakes, called focal plane analysis. The basic direction in which earthquake waves propagate is related to the orientation of the stress field and the movement it induces on faults in the Earth's crust.

Fractures

Several sets of rock fractures in Illinois are so closely correlated with σ_1 as to strongly suggest those fractures are of recent origin. These fracture sets include the following:

- Joints in the shales that overlie coal seams, as observed in mines (Figure 5-8). These parallel, planar, vertical fractures run consistently parallel to σ_1. Joints in southeastern Illinois run east-west; in the rest of the state, they trend N50° E to N70° E (Nelson and Bauer 1987, 1991). In southwestern Indiana, joints in roof shales dominantly trend east-northeast, except near the Ohio River, where they run north-northeast, parallel with nearby large faults (Ault et al. 1985).
- Small thrust faults (Figure 5-9) in southern Illinois and western Kentucky (Nelson and Bauer 1987, 1991) and in Indiana (Ault et al. 1985) strike north-south to north-northwest, perpendicular to σ_1. The thrusts, which have throws of a few feet at the most, appear unrelated to older fault systems.
- Open, vertical crevices (Figure 5-10) in coal mine roofs run parallel to σ_1. These fractures contain little or no mineralization, suggesting that they formed quite recently. In Macoupin County, south of Springfield, such fractures are associated with a strike-slip fault that extends more than 3 miles (5 km) through mine workings (Nelson and Bauer 1991, Nelson 1995).

All of these types of fractures seem to be confined to coal seams and enclosing rocks, especially roof shales. They

Figure 5-8 Joints (fractures) in black shale overlying the Herrin Coal (Krausse et al. 1979, Nelson 1983). These planar, vertical, evenly spaced fractures consistently run parallel to the principal compressive stress axis (σ_1). The metal plates under the heads of the roof bolts are about 6 inches (15 cm) square, and the header boards are likely 2- × 8-inch (5- × 20-cm) planks about 2 feet (0.6 m) long.

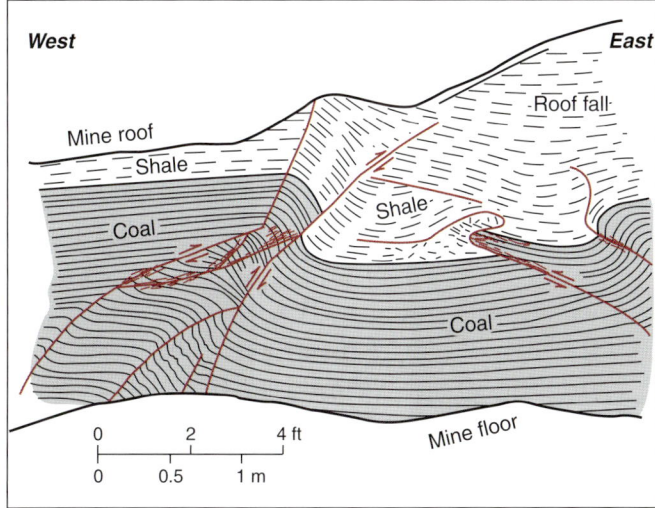

Figure 5-9 Sketch of small thrust faults in underground coal mine, Saline County (Nelson 1995).

Figure 5-10 Crevice in limestone roof of mine near Crown Fault, Macoupin County (Nelson and Bauer 1991).

have not been observed in thick layers of competent rocks, such as sandstone and limestone, which suggests that the stress field currently is capable of inducing failure only in relatively weak rocks. This observation is in line with measured values of rock strength versus the magnitude of current stress in Illinois.

Roof Failure

At many past and present Illinois mines, roof falls occur predominantly in headings driven perpendicular to σ_1 (Blevins 1982, Ingram and Molinda 1988), which generally means north-south entries in southeastern Illinois and northwest-southeast entries elsewhere in the state. Once mine operators realized that such roof failures were caused by stress, they responded by changing mine layouts to avoid the dangerous direction, which markedly reduced roof falls. New mines now are typically planned with main entries driven at about 45° to either side of σ_1.

The typical mode of stress failure in Illinois coal mines is by "kink zone" (Figure 5-11). Kinks develop only in fissile, brittle rocks, especially shales. These linear or slightly sinuous zones of failure nearly always run at a right angle to σ_1 (Figure 5-12). Layers of shale on either side of the kink zone snap and bend downward. Close examination reveals that the roof is not merely sagging; the broken rock layers are literally being jammed together horizontally.

Effects were more dramatic at the Inland No. 2 Mine near McLeansboro, one of the deepest mines in the state; coal was 930 feet (283.5 m) deep at the shaft. When mining north and south, falls extending 2 feet (0.6 m) to 4 feet (1.2 m) into the roof often occurred immediately after removing the coal. In an effort to relieve the stress, the company used a chain cutter to make vertical slots 6 inches (15 cm) wide down the center of the headings. Crushed rock began to dribble from these slots as soon as they were cut. Within 24 hours, the slots narrowed to 4.5 to 4.75 inches (11.4 to 12 cm). After a week, they were 3 inches (7.6 cm) wide, and, after 10 days, only 1 inch (2.5 cm) wide at the top. The company abandoned the slot-cutting and resorted to changing the mining direction, which produced a marked improvement (Blevins 1982).

EARTHQUAKES AND SUBSURFACE GEOLOGICAL STRUCTURES

Investigating the causes of earthquakes in Illinois is not easy, unlike in California, where active faults such as the San Andreas are exposed at the surface and where their movements may be observed directly (Iacopi 1971).

Figure 5-11 Roof fall in laminated Dykersburg Shale in the Kerr-McGee Company Galatia Mine. Note the "kink zone" in the roof. (Photograph by Frank Chase.)

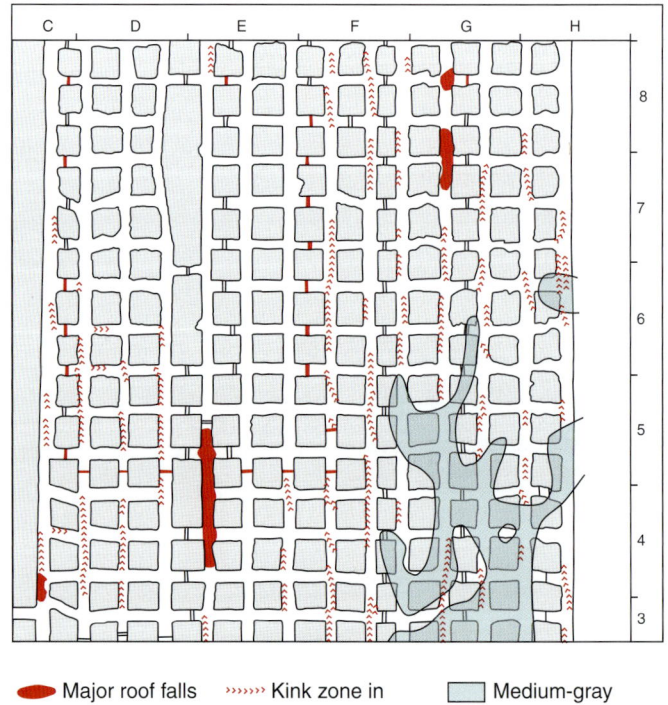

Figure 5-12 Map of part of an underground mine, showing preferential orientation of kink zones (Krausse et al. 1979).

NEOTECTONICS

Although faults can be seen at the surface in a few places in Illinois, most are deeply buried by soil and glacial deposits. Mapping the state's faults requires subsurface data, such as well records. Because most earthquakes take place at depths of many miles, below the bottoms of the deepest wells, investigating such earthquakes requires the use of techniques that probe deeper into the Earth's crust.

The New Madrid Seismic Zone (Figure 5-13) is one of the most studied seismic zones in the world (Johnston and Schweig 1996). An array of investigative techniques has been brought to bear there, bringing to light the active geological structures hidden beneath the Mississippi River alluvial plain. The demonstrated potential for devastating earthquakes provides great incentive for thorough study of the New Madrid area.

Earthquake epicenters in the New Madrid Seismic Zone are highly concentrated (Figure 5-14). A narrow, linear belt of epicenters runs northeast through northeastern Arkansas into the "Bootheel" region of Missouri, where it intersects a shorter and broader band of intense activity trending slightly west of north. Another band of epicenters runs northeast from the "Bootheel" into westernmost Kentucky and southernmost Illinois. These linear bands of epicenters represent ancient, deeply buried faults that originated as part of the Reelfoot Rift during Cambrian time and that are being reactivated today under tectonic stress caused by continental plate movements (Johnston and Schweig 1996).

In contrast to the New Madrid Seismic Zone, epicenters in Illinois are highly dispersed (Figure 5-13). Aside from a higher concentration of events in the southern third of the state, no particular trends are evident (Nuttli 1979), although statistical studies suggest some alignments (Amorèse 2003). The absence of any obvious alignment of seismicity, which would be indicative of faults, provides few clues to the investigator.

To investigate the causes of earthquakes in southern Illinois, the Illinois State Geological Survey during the early 1990s embarked on a program of procuring seismic reflection data from the petroleum industry. This effort garnered thousands of miles of industry seismic profiles, which revealed many previously unsuspected, deeply buried faults in undrilled lower Paleozoic and Precambrian rocks. (Seismic reflection is an exploration method

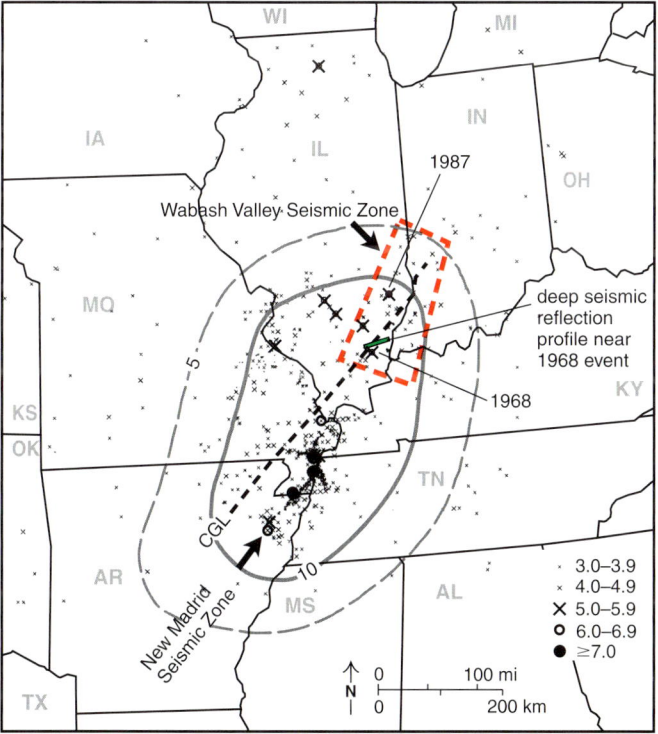

Figure 5-13 Central U.S. earthquakes from 1810 through 1995 having body-wave magnitude of at least 3.0. Large X represents event of body-wave magnitude 5.0 or greater in Illinois. Contour lines indicate seismic hazard surrounding the New Madrid Seismic Zone, measured as peak acceleration (percent gravity) with a 10% probability of being exceeded within 50 years (Frankel 1995). Abbreviation: CGL, Commerce geophysical lineament. Sources: Nuttli and Brill (1981) and instrumental data from the Central Mississippi Earthquake Catalog, June 29, 1974, through December 31, 1995 (compiled by the St. Louis University Regional Seismic Network). Epicenter base map courtesy of Timothy H. Larson, Illinois State Geological Survey.

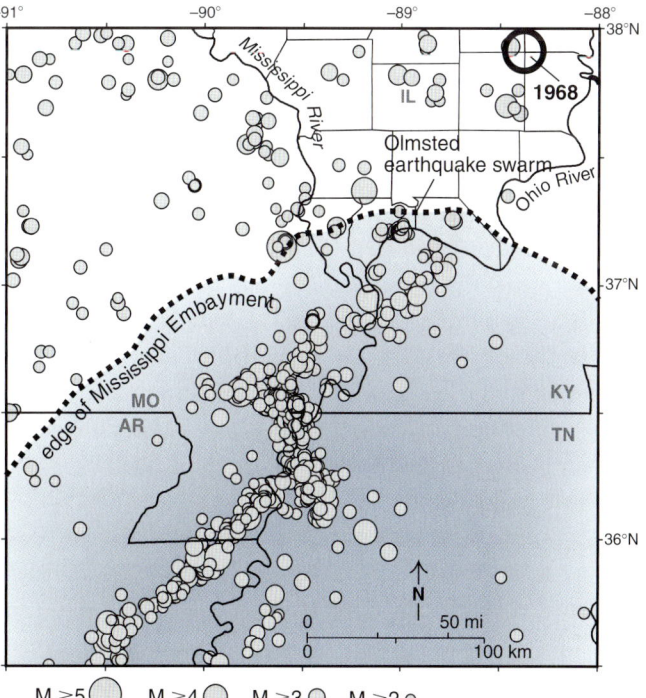

Figure 5-14 Map of instrumentally recorded earthquake epicenters in the New Madrid Seismic Zone and surrounding regions, 1974 to 2001, of body-wave magnitude (M) 2.0 or greater (Saint Louis University and Central Mississippi Valley Earthquake Bulletins, Tennessee Earthquake Information Center Earthquake Bulletin, and continuous monitoring by Saint Louis University and the Center for Earthquake Research and Information, University of Memphis).

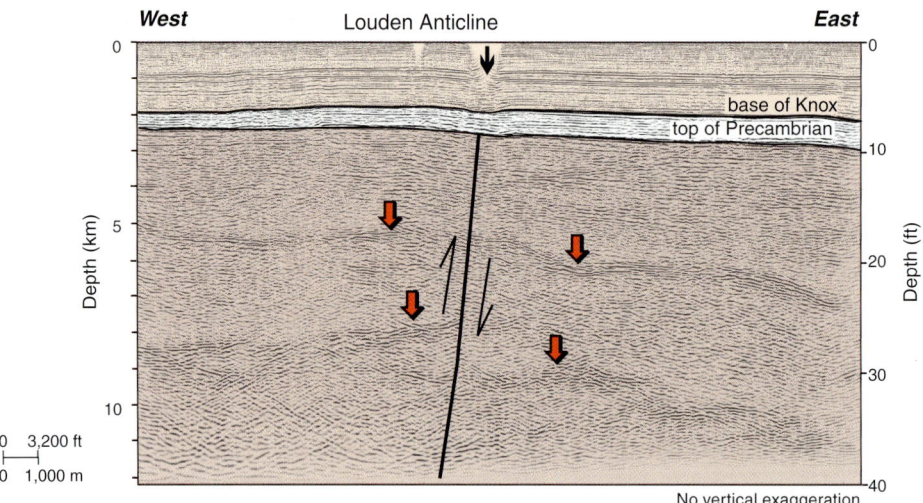

Figure 5-15 Reprocessed seismic reflection section over the Louden Anticline. Vertical red arrows indicate the inferred offset of Precambrian stratal markers.

in which sound waves are generated that penetrate the Earth and reflect back to the surface from geological boundaries such as rock layers and faults. The reflected energy is recorded on an array of geophones and processed to yield a profile that resembles a geological cross section.) The seismic reflection profiles across many oil-producing anticlines and monoclines in southern Illinois reveal deep-seated faults. Other profiles unveil fold and fault zones that have no near-surface expression. Some of these structures are closely associated with recent earthquakes.

A long-standing paradigm for the North American craton is that many earthquakes result from reactivation of old faults (Marshak and Paulsen 1996). This theoretical framework has been especially true for the northern Mississippi Embayment and southern Illinois Basin, which have been tectonically active through more than half a billion years of Earth's history. For example, the Louden Anticline hosts one of the largest oil fields in the Illinois Basin. Industry seismic profiles, such as the one reproduced here (Figure 5-15), show that the flank of the Louden Anticline is underlain by deep-seated reverse faults (McBride 1997, 1998; McBride and Kolata 1999; McBride and Nelson 1999). Although the Louden structure is not seismically active, large faults such as those shown are capable of generating earthquakes if they are oriented favorably to the tectonic stress field. Analysis of seismic reflection profiles across other large folds in Illinois, including the Du Quoin Monocline, Salem Anticline, and La Salle Anticlinorium, indicates that these structures also are underlain by large faults that probably do not reach the surface (McBride 1997, 1998; McBride and Nelson 1999; McBride and Kolata 1999).

A segment of the La Salle Anticlinorium in northern Illinois may be seismically active. Larson (2002) located the epicenters of three earthquakes along or very close to the Peru Monocline, a large northwest-striking fold at the northern end of the La Salle trend. The focal mechanism of a 1972 earthquake of magnitude 4.5 is consistent with strike-slip motion along a concealed fault lying parallel to the monocline.

The Wabash River valley is locally known in Illinois as "earthquake country." The Wabash Valley Fault Zone contains northeast-trending normal and strike-slip faults that follow the axis of the valley (Figure 5-16). Although these near-surface faults are prime suspects for seismicity (Mitchell et al. 1991), earthquakes seldom correlate with mapped faults (Kim 2003). Earthquake focal mechanisms in the Wabash River valley present a combination of north-northeast–trending dextral strike-slip and reverse faulting (Taylor et al. 1989), like many seismically active faults of the surrounding region (Langer and Bollinger 1991). These mechanisms are consistent with a maximum horizontal compressive stress that trends just north of east in southern Illinois and Indiana.

The largest earthquake recorded to date in Illinois (magnitude 5.5) took place near Broughton in Hamilton County on November 9, 1968 (Gordon et al. 1970, Stauder and Nuttli 1970, Herrmann 1973, Heigold and Larson 1990). In an effort to understand the focal mechanism of this earthquake, a seismic profile that passed close to the epicenter was reprocessed; that is, the raw data were reanalyzed for additional information (McBride et al. 2002). On the reprocessed profile, the earthquake focus, or point of initial rupture, closely matches a series of strong, apparently west-dipping reflections in the middle crust (Figure 5-16). Such strong and deep reflectors are more typical of mountain belts such as the Appalachian Mountains than of the North American midcontinent. Thus, the deep reflectors at Broughton possibly represent part of a buried Precambrian mountain belt. Large, gently dipping thrust faults associated with mountain belts are prime candidates for reactivation.

Dipping reflections occur occasionally on deep seismic reflection profiles in southeastern Illinois, but only one (Figure 5-16) seems to be spatially associated with a known earthquake focus, as far as we know. This particular reflection is enigmatic since it disappears off the bottom of the record (middle, Figure 5-17) and thus continues to an unknown depth. The area of the Earth's crust near the 1968 earthquake is characterized by zones of dipping reflections arriving over distinct time intervals. For example, a shallow reflection arrives within the crust above the deeper reflection (top, Figure 5-17). Elsewhere in southeastern Illinois, enigmatic dipping reflections disappear off the bottom of seismic records (bottom, Figure 5-17). The meaning of these mysterious features is not clear and invites further study.

Faults that do not reach the surface, like those at Broughton, Peru, and Loudon, are said to be "blind." Blind faults are unnerving because they can generate destructive quakes with no warning. The catastrophic 1980 magnitude 7.7 El Asnam (Algeria) and the 1974 magnitude 6.7 Northridge (California) events are infamous examples of earthquakes on blind faults. The recognition of large, active blind faults in Illinois heightens the importance of continued research.

Another area where seismic reflection profiles have been analyzed to investigate earthquakes is southernmost Illinois (Stephenson et al. 1999, McBride and Nelson 2001, McBride et al. 2002) at the northern end of the New Madrid Seismic Zone (Wheeler 1997) (Figure 5-14). During February 1984, a swarm of more than 100 earthquakes, the largest having magnitude 3.6, took place north of Olmsted in Pulaski County (Stauder et al. 1984, Stover 1988). This area lies within the Mississippi Embayment, where outcrops and wells are few and no faults are exposed. A seismic reflection profile near Olmsted (Figure 5-18) indicates a complex array of fractures affecting Paleozoic bedrock and also Cretaceous and Cenozoic strata of the Mississippi Embayment. Faults interpreted on the seismic profile outline an intricate graben, similar to those found in the Fluorspar Area Fault Complex northeast of Olmsted. The 1984 earthquake swarm thus appeared to have been centered

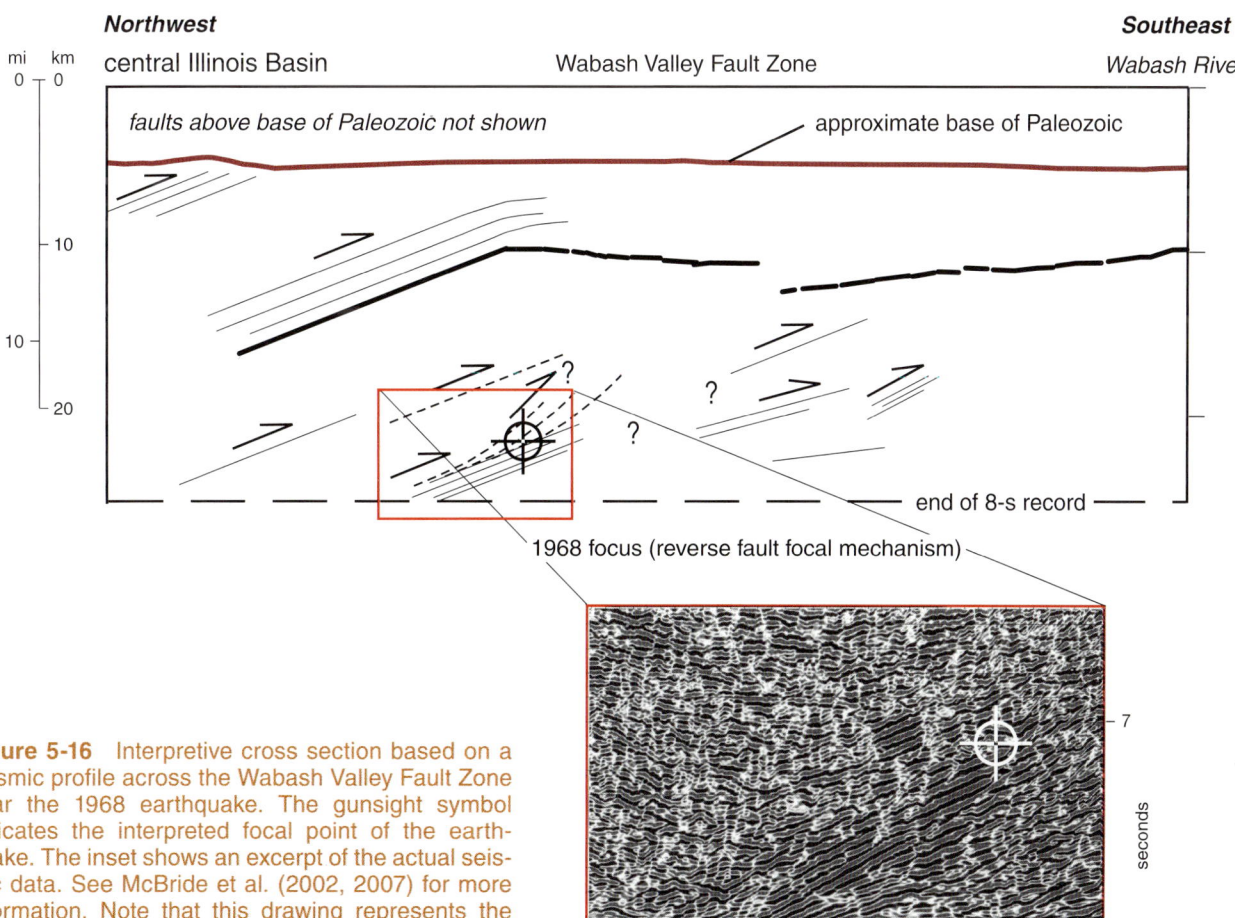

Figure 5-16 Interpretive cross section based on a seismic profile across the Wabash Valley Fault Zone near the 1968 earthquake. The gunsight symbol indicates the interpreted focal point of the earthquake. The inset shows an excerpt of the actual seismic data. See McBride et al. (2002, 2007) for more information. Note that this drawing represents the authors' interpretation, which needs to be tested and refined by further data.

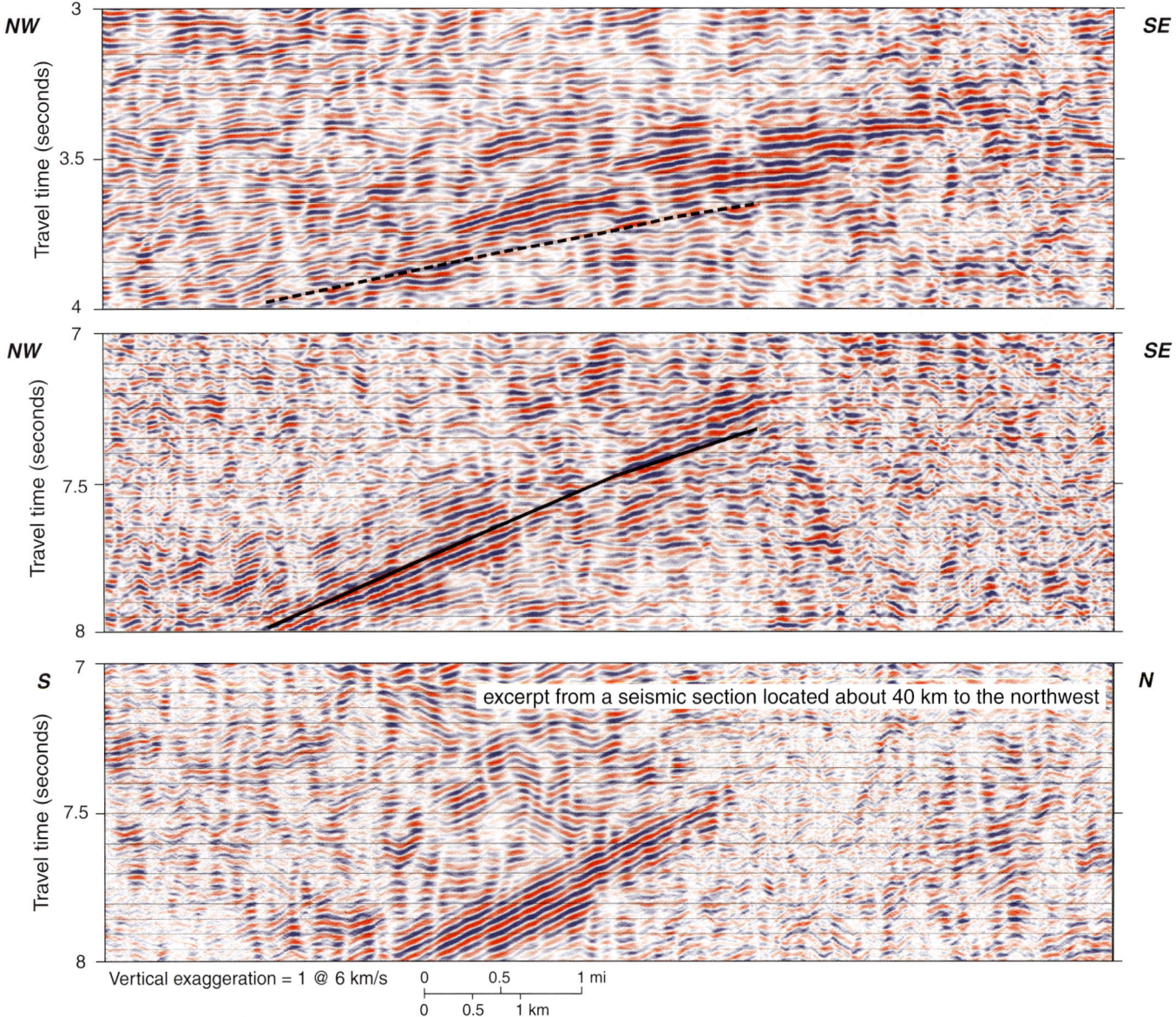

Figure 5-17 Excerpts from seismic records showing dipping reflections (unmigrated) from southeastern Illinois. Top panel shows a shallow dipping reflection directly above the focal region for the 1968 earthquake (Figure 5-16). Middle panel is the excerpt for the focal region as shown in Figure 5-16. The dashed line in the upper panel represents the expected position of the reflection in the middle panel (solid line) if the latter were a multiple reflection ("double bounce") of the former. Because the dashed line in the upper panel does not match any arrivals, the reflection in the middle panel is interpreted as a primary arrival. The lower panel provides an example of another enigmatic reflection in southeastern Illinois, not known to be associated with an earthquake, that dips off the bottom of the record. See McBride et al. (2002) for more explanation.

along faults that directly link the Fluorspar District with the New Madrid Seismic Zone.

CONCLUSION

The geological record belies the old notion that Illinois is part of the "stable craton" and has been tectonically quiet since the Appalachian Mountains rose. The North American tectonic plate is continuously under stress, causing ground failure in man-made openings and occasional movement along ancient lines of weakness. Immediately south of Illinois, one of the largest series of earthquakes in world history occurred less than 200 years ago. Shocks nearly as large rocked Illinois within the last 10,000 years, literally liquefying the earth. Recent seismological research in Illinois suggests that deeply buried pre-existing structures, known as "blind faults," have the potential to rejuvenate and cause earthquakes. Therefore, identifying and mapping such structures are important. Although these findings should not cause alarm, they remind us to plan and prepare for the motions of our unsteady earth. Information on seismic hazards and their mitigation can be found in Chapter 27, Earthquakes.

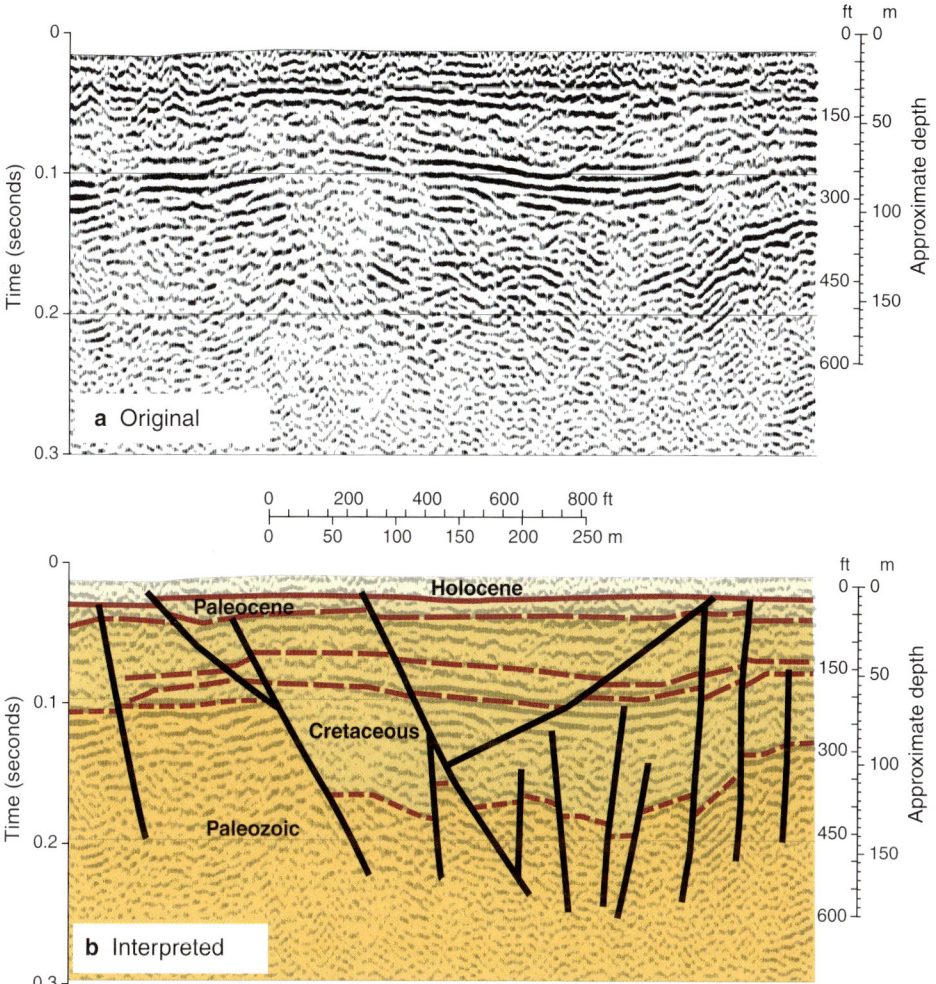

Figure 5-18 (a) Original and (b) interpreted high-resolution seismic profile from Ballard County, Kentucky, just east of the Ohio River near Olmsted, Illinois (McBride et al. 2003).

REFERENCES

Amorèse, D., 2003, A new approach for associating earthquakes with geological structures: Application to epicenters in southern Illinois and southeastern Missouri: Geophysical Research Letters, v. 30, no. 1689, doi:10.1029/2003GL017247.

Ault, C. H., D. Harper, C. R. Smith, and M. A. Wright, 1985, Faulting and jointing in and near surface mines of southwestern Indiana: Washington, D.C., U.S. Nuclear Regulatory Commission, NUREG/CR-4117, 27 p.

Blevins, C. T., 1982, High lateral stresses at Inland Mine No. 2: Proceedings of the Illinois Mining Institute, p. 13–20.

Denny, F. B., and W. J. Nelson, 2005, Bedrock geology of Paducah NE Quadrangle, Massac and Pope Counties, Illinois: Illinois State Geological Survey, Illinois Geologic Quadrangle Paducah NE-BG, 1;24,000.

Frankel, A., 1995, Mapping seismic hazard in the central and eastern United States: Seismological Research Letters, v. 66, no. 4, p. 8–21.

Fuller, M. L., 1912, The New Madrid earthquake: Washington, D.C., U.S. Geological Survey, Bulletin 494, 119 p.

Gordon, D. W., T. J. Bennett, R. B. Herrmann, and A. M. Rogers, 1970, The south-central Illinois earthquake of November 9, 1968: Macroseismic studies: Bulletin of the Seismological Society of America, v. 60, no. 3, p. 953–971.

Gough, D. I., 1984, Mantle upflow under North America and plate dynamics: Nature, v. 311, no. 5985, p. 428–433.

Gough, D. I., and W. I. Gough, 1987, Stress near the surface of the earth: Annual Review of Earth and Planetary Sciences, v. 15, p. 545–566.

Hajic, E. R., and M. D. Wiant, 1997, In search of ancient earthquakes: The Living Museum, v. 58, no. 4, p. 12–13.

Harrison, R. W., 1997, Bedrock geologic map of the St. Louis 30' × 60' Quadrangle, Missouri and Illinois: Reston, Virginia, U.S. Geological Survey, Map I-2533.

Harrison, R. W., 1999, Geologic map of the Thebes Quadrangle, Illinois and Missouri: Reston, Virginia, U.S. Geological Survey, Geologic Quadrangle Map GQ-1779, 1:24,000.

Harrison, R. W., D. Hoffman, J. D. Vaughn, J. R. Palmer, C. L. Wiscombe, J. P. McGeehin, W. J. Stephenson, J. K. Odum, R. A. Williams, and S. L. Forman, 1999, An example of neotectonism in a continental interior—Thebes Gap, midcontinent, United States: Tectonophysics, v. 305, no. 1–3, p. 399–417.

Harrison, R. W., J. R. Palmer, D. Hoffman, J. D. Vaughn, J. E. Repetski, N. O. Frederiksen, and S. L. Forman, 2002, Geologic map of the Scott City 7.5-minute quadrangle, Scott and Cape Girardeau Counties, Missouri: Reston, Virginia, U.S. Geological Survey, Geologic Investigations Series Map I-2744, 2 sheets plus 12 p. booklet.

Harrison, R. W., and A. Schultz, 1994, Strike-slip faulting at Thebes gap, Missouri and Illinois: Implications for New Madrid tectonism:

Tectonics, v. 13, no. 2, p. 246–257.

Harrison, R. W., and A. Schultz, 2002, Tectonic framework of the southwestern margin of the Illinois Basin and its influence on neotectonism and seismicity: Seismological Research Letters, v. 73, no. 5, p. 698–731.

Heigold, P. C., and T. H. Larson, 1990, Seismicity of Illinois: Illinois State Geological Survey, Environmental Geology Notes 133, 21 p.

Herrmann, R. B., 1973, Surface-wave generation by the south central Illinois earthquake of November 9, 1968: Bulletin of the Seismological Society of America, v. 63, no. 6, p. 2121–2134.

Hildenbrand, T. G., J. H. McBride, and D. Ravat, 2002, The Commerce geophysical lineament and its possible relation to Mesoproterozoic igneous complexes and large earthquakes in the central Illinois Basin: Seismological Research Letters, v. 73, no. 5, p. 640–659.

Iacopi, R., 1971, Earthquake country: How, why and where earthquakes strike in California (3rd ed.): Menlo Park, California, Lane Publishing Co., 160 p.

Ingram, D. K., and G. M. Molinda, 1988, Relationship between horizontal stresses and geologic anomalies in two coal mines in southern Illinois: [Washington, D.C.], U.S. Bureau of Mines, Report of Investigations 9189, 18 p.

Johnston, A. C., and E. S. Schweig, 1996, The enigma of the New Madrid earthquakes of 1811–1812: Annual Review of Earth and Planetary Sciences, v. 24, p. 339–384.

Kelson, K. I., R. B. VanArsdale, G. D. Simpson, and W. R. Lettis, 1992, Assessment of the style and timing of surficial deformation along the central Reelfoot scarp, Lake County, Tennessee: Seismological Research Letters, v. 63, no. 3, p. 349–356.

Kim, W.-Y., 2003, The 18 June 2002 Caborn, Indiana, earthquake: Reactivation of ancient rift in the Wabash Valley Seismic Zone?: Bulletin of the Seismological Society of America, v. 93, no. 5, p. 2201–2211.

Kolata, D. R., J. D. Treworgy, and J. M. Masters, 1981, Structural framework of the Mississippi Embayment of southern Illinois: Illinois State Geological Survey, Circular 516, 38 p.

Krausse, H. F., H. H. Damberger, W. J. Nelson, S. R. Hunt, C. T. Ledvina, C. G. Treworgy, and W. A. White, 1979, Roof strata of the Herrin (No. 6) coal member in mines of Illinois: Their geology and stability: Summary report: Illinois State Geological Survey, Illinois Mineral Notes, v. 72, 54 p.

Langenheim, V. E., and T. G. Hildenbrand, 1997, Commerce geophysical lineament—Its source, geometry, and relation to the Reelfoot Rift and New Madrid Seismic Zone: Geological Society of America Bulletin, v. 109, no. 5, p. 580–595.

Langer, C. J., and G. A. Bollinger, 1991, The southeastern Illinois earthquake of 10 June 1987: The later aftershocks: Bulletin of the Seismological Society of America, v. 81, no. 2, p. 423–445.

Larson, T. H., 2002, The earthquake of 2 September 1999 in northern Illinois: Intensities and possible neotectonism: Seismological Research Letters, v. 73, no. 5, p. 732–738.

Marshak, S., and T. Paulsen, 1996, Midcontinent U.S. fault and fold zones: A legacy of Proterozoic intracratonic extensional tectonism?: Geology, v. 24, no. 2, p. 151–154.

McBride, J. H., 1997, Variable deep structure of a midcontinent fault and fold zone from seismic reflection: La Salle deformation belt, Illinois Basin: Geological Society of America Bulletin, v. 109, no. 11, p. 1502–1513.

McBride, J. H., 1998, Understanding basement tectonics of an interior cratonic basin, southern Illinois Basin, USA: Tectonophysics, v. 293, no. 1–2, p. 1–20.

McBride, J. H., T. G. Hildenbrand, W. J. Stephenson, and C. J. Potter, 2002, Interpreting the earthquake source of the Wabash Valley Seismic Zone (Illinois, Indiana, and Kentucky) from seismic-reflection, gravity, and magnetic-intensity data: Seismological Research Letters, v. 73, no. 5, p. 660–686.

McBride, J. H., and D. R. Kolata, 1999, Upper crust beneath the central Illinois Basin, United States: Geological Society of America Bulletin, v. 111, no. 3, p. 375–394.

McBride, J. H., H. E. Leetaru, R. A. Bauer, B. E. Tingey, and S. E. A. Schmidt, 2007, Deep faulting and structural reactivation beneath the southern Illinois Basin: Precambrian Research, v. 157, p. 289–313, doi: 10.1016/j.precamres. 2007.02.020.

McBride, J. H., and W. J. Nelson, 1999, Style and origin of mid-Carboniferous deformation in the Illinois Basin, USA—Ancestral Rockies deformation?: Tectonophysics, v. 305, no. 1–3, p. 249–273.

McBride, J. H., and W. J. Nelson, 2001, Seismic reflection images of shallow faulting, northernmost Mississippi Embayment, north of the New Madrid Seismic Zone: Bulletin of the Seismological Society of America, v. 91, no. 1, p. 128–139.

McBride, J. H., W. J. Nelson, and W. J. Stephenson, 2002, Integrated geological and geophysical study of Neogene and Quaternary-age deformation in the northern Mississippi Embayment: Seismological Research Letters, v. 73, no. 5, p. 597–627.

McBride, J. H., A. J. M. Pugin, W. J. Nelson, T. H. Larson, S. L. Sargent, J. A. Devera, F. B. Denny, and E. W. Woolery, 2003, Variable post-Paleozoic deformation detected by seismic reflection profiling across the northwestern "prong" of New Madrid Seismic Zone: Tectonophysics, v. 368, no. 1–4, p. 171–191.

Mitchell, B. J., O. W. Nuttli, R. B. Herrmann, and W. Stauder, 1991, Seismotectonics of the central United States, in D. B. Slemmons, E. R. Engdahl, M. D. Zoback, and D. D. Blackwell, eds., Neotectonics of North America: Boulder, Colorado, Geological Society of America, Decade Map, v. 1, p. 245–260.

Munson, P. J., C. A. Munson, N. K. Bleuer, and M. D. Labitzke, 1992, Distribution and dating of prehistoric earthquake liquefaction in the Wabash Valley of the central U.S.: Seismological Research Letters, v. 63, no. 3, p. 337–342.

Munson, P. J., S. F. Obermeier, C. A. Munson, and E. R. Hajic, 1997, Liquefaction evidence for Holocene and latest Pleistocene seismicity in the southern halves of Indiana and Illinois: A preliminary overview: Seismological Research Letters, v. 68, no. 4, p. 521–536.

Nelson, W. J., 1983, Geologic disturbances in Illinois coal seams: Illinois State Geological Survey, Circular 530, 47 p.

Nelson, W. J., 1995, Structural features in Illinois: Illinois State Geological Survey, Bulletin 100, 144 p.

Nelson, W. J., 1996, Geologic map of the Reevesville Quadrangle, Illinois: Illinois State Geological Survey, Illinois Geologic Quadrangle map, IGQ 17, 1:24,000.

Nelson, W. J., and R. A. Bauer, 1987, Thrust faults in southern Illinois Basin—Result of contemporary stress? Geological Society of America Bulletin, v. 98, no. 3, p. 302–307.

Nelson, W. J., and R. A. Bauer, 1991, Coping with tectonic stress in the Illinois Basin coal field, in D. C. Peters, ed., Geology in coal resource utilization: Fairfax, Virginia, TechBooks, p. 321–334.

Nelson, W. J., F. B. Denny, J. A. Devera, L. R. Follmer, and J. M. Masters, 1997, Tertiary and Quaternary tectonic faulting in southernmost Illinois: Engineering Geology, v. 46, no. 3–4, p. 235–258.

Nelson, W. J., F. B. Denny, L. R. Follmer, and J. M. Masters, 1999, Quaternary grabens in southernmost Illinois: Deformation near an active intraplate seismic zone: Tectonophysics, v. 305, no. 1–3, p. 381–397.

Nuttli, O. W., 1979, Seismicity of the central United States, in A. W. Hatheway and C. R. McClure Jr., eds., Geology in the siting of nuclear power plants: Boulder, Colorado: Geological Society of America, Reviews in Engineering Geology, v. 4, p. 67–93.

Nuttli, O. W., and K. G. Brill Jr., 1981, Earthquake source zones in the central United States determined from historical seismicity, in N. Barstow, K. G. Brill, O. W. Nuttli, and P. W. Pomeroy, eds., An approach to seismic zonation for siting nuclear electric power generating facilities in the eastern United States: Washington, D.C., U.S. Nuclear Regulatory Commission, NUREG/CR-1577, p. 98–143.

Obermeier, S. F., 1996, Use of liquefaction-induced features for paleoseismic analysis—An overview of how seismic liquefaction features can be distinguished from other features and how their regional distribution and properties of source sediment can be used to infer the location and strength of Holocene paleo-earthquakes: Engineering Geology, v. 44, no. 1–4, p. 1–76.

Obermeier, S. F., 1998, Liquefaction evidence for strong earthquakes of Holocene and latest Pleistocene ages in the states of Indiana and Illinois, USA: Engineering Geology, v. 50, no. 3–4, p. 227–254.

Obermeier, S. F., P. J. Munson, C. A. Munson, J. R. Martin, A. D. Frankel, T. L. Youd, and E. C. Pond, 1992, Liquefaction evidence for strong Holocene earthquake(s) in the Wabash Valley of Indiana-Illinois: Seismological Research Letters, v. 63, no. 3, p. 321–335.

Odum, J. K., W. J. Stephenson, R. A. Williams, J. A. Devera, and J. R. Staub, 2002, Near-surface faulting and deformation overlying the Commerce geophysical lineament in southern Illinois: Seismological Research Letters, v. 73, no. 5, p. 687–697.

Ross, C. A., 1963, Structural framework of southernmost Illinois: Illinois State Geological Survey, Circular 351, 28 p.

Rubey, W. W., 1952, Geology and mineral resources of the Hardin and Brussels Quadrangles (in Illinois): Washington D.C., U.S. Geological Survey, Professional Paper 218, 179 p.

Sbar, M. L., and L. R. Sykes, 1973, Contemporary compressive stress and seismicity in eastern North America: An example of intra-plate tectonics: Geological Society of America Bulletin, v. 84, no. 6, p. 1861–1882.

Science Applications International Corporation, 2002, Seismic investigation report for siting of a potential on-site CERCLA waste-disposal facility at the Paducah gaseous diffusion plant, Paducah, Kentucky: San Diego, California, DOE/R/07-2038 & D1, pages unnumbered.

Seid, M. J., 2008a, Bedrock geology of Brussels Quadrangle, Calhoun and Jersey Counties: Illinois State Geological Survey, Illinois Preliminary Geologic Map, IPGM Brussels-BG, 2 sheets.

Seid, M. J., 2008b, Bedrock geology of Winfield Quadrangle, Calhoun County: Illinois State Geological Survey, Illinois Preliminary Geologic Map, IPGM Winfield-BG, 2 sheets.

Shaw, E. W., 1915, Quaternary deformation in southern Illinois and southeastern Missouri (abs.): Geological Society of America Bulletin, v. 26, p. 67–68.

Stauder, W., R. Herrmann, J. Chulick, M. J. Mascarenas, V. John, P. L. Leu, T. C. Shin, H. Yepes, and C. Finn, 1984, Central Mississippi Valley Earthquake Bulletin, No. 39, First Quarter 1984: St. Louis, Missouri, Department of Earth and Atmospheric Sciences, St. Louis University, 90 p.

Stauder, W., and O. W. Nuttli, 1970, Seismic studies: South central Illinois earthquake of November 9, 1968: Bulletin of the Seismological Society of America, v. 60, no. 3, p. 973–981.

Stephenson, W. J., J. K. Odum, R. A. Williams, T. L. Pratt, R. W. Harrison, and D. Hoffman, 1999, Deformation and Quaternary faulting in southeast Missouri across the Commerce geophysical lineament: Bulletin of the Seismological Society of America, v. 89, no. 1, p. 140–155.

Stewart, D., and R. Knox, 1993, The earthquake that never went away: Marble Hill, Missouri, Gutenberg-Richter Publications, 221 p. plus slide set.

Stover, C. W., 1988, United States earthquakes, 1984: Reston, Virginia, U.S. Geological Survey, Bulletin 1862, 179 p.

Taylor, K. B., R. B. Herrmann, M. W. Hamburger, G. L. Pavlis, A. Johnston, C. Langer, and C. Lam, 1989, The southeastern Illinois earthquake of 10 June 1987: Seismological Research Letters, v. 60, no. 3, p. 101–110.

Tuttle, M., J. Chester, R. Lafferty, K. Dyer-Williams, and R. Cande, 1999, Paleoseismicity study northwest of the New Madrid Seismic Zone: U.S. Nuclear Regulatory Commission, NUREG/CR-5730, 96 p. plus appendices.

Wesnousky, S. G., and L. M. Leffler, 1994, A search for paleoliquefaction and evidence bearing on the recurrence behavior of the great 1811–12 New Madrid earthquakes: Reston, Virginia, U.S. Geological Survey, Professional Paper 1538-H, 42 p.

Wheeler, R. L., 1997, Boundary separating the seismically active Reelfoot Rift from the sparsely seismic Rough Creek Graben, Kentucky and Illinois: Seismological Research Letters, v. 68, no. 4, p. 586–598.

Zoback, M. L., and M. Zoback, 1980, State of stress in conterminous United States: Journal of Geophysical Research, v. 85, no. B11, p. 6113–6156.

Geological History and the Stratigraphic Record of Illinois

Dennis R. Kolata

Illinois is underlain by a thick succession of sedimentary rocks consisting primarily of limestone, shale, sandstone, and siltstone, most of which is overlain by unlithified deposits of clay, silt, sand, and gravel. The composition, thickness, distribution, and fossil content of these rocks and sediments, as well as the regional and interregional unconformities that form their boundaries, reveal important clues about the geological history of Illinois during the past 500 million years. The vertical succession of rock layers are like the pages of a book with younger historical events stacked upon older ones. Each new drill hole, quarry, mine shaft, and excavation is a passageway to this ancient archive that provides insight to the geological history of Illinois.

This part of the book presents the geological history chronologically, from oldest to youngest, beginning with the Precambrian and concluding with the Quaternary Period. Many of the chapters are organized according to cratonic sequences, each of which reflects major episodes in the rise and fall of sea level on the stable interior (craton) of North America during the past 570 million years and represents depositional and tectonic phases in the geological history. Parts of six primary sequences are present in Illinois, including (oldest to youngest) the Sauk, Tippecanoe, Kaskaskia, Absaroka, Zuni, and Tejas Sequences. The Quaternary history is marked by recurring episodes of continental glaciation revealed in a complex series of glacial tills, fluvial sands and gravels, paleosols, and lacustrine and aeolian deposits. Quaternary depositional and erosional processes have played a major role in shaping the modern landscape of Illinois.

Photograph by Joel M. Dexter.

Image on previous page: Caseyville Formation sandstone at Garden of the Gods Recreation Area, Gallatin County.

The Precambrian Crust 6

John H. McBride, Dennis R. Kolata, Michael L. Sargent, and Thomas G. Hildenbrand

INTRODUCTION

Little is known of Illinois' oldest rocks, commonly referred to as the "Precambrian basement" or "crystalline crust." Deposited from 4,500 to 544 million years ago (mya), the Precambrian rocks lie at great depth, where historically there has been little economic incentive to drill. Geological cross sections of the Midwest (e.g., Figure 6-1) typically show the Precambrian as a uniform mass that simply disappears off the bottom of the diagram, a *terra incognita* not greatly different from the outer edges of early medieval maps of the world. The structure, composition, and origins of the deeply buried Precambrian crust of the Midwest have thus remained mostly beyond the geological frontier. This chapter briefly summarizes the general knowledge of the Precambrian in Illinois and then discusses an exceptional area in east-central Illinois where the coincidence of deep drill-hole, seismic reflection, and gravity- and magnetic-field data provides a unique "window" into the deep Earth beneath Illinois.

PRECAMBRIAN BASEMENT OF ILLINOIS

Because the Precambrian strata in Illinois are so deeply buried, knowledge of these rocks is limited to approximately 40 drill holes and a small amount of remote sensing data acquired from seismic surveys and from studying Earth's gravitational and magnetic fields. Most of the Precambrian drill holes have brought up only small fragments of rock. A few precious cores of solid rock have been obtained by oil and utility companies from Precambrian rocks at great expense and have been made available to geologists for study (Sargent 1991) (Figure 6-2). Published descriptions of the geology of the Precambrian in and around Illinois are available (Bradbury and Atherton 1965, Lidiak 1996, Van Schmus et al. 1996), and the overall shape of the buried surface of the Precambrian rocks can be visualized from a structural contour map (Figure 6-3). The map was constructed using data from deep drill holes and from regional seismic reflection surveys. The depth to the top of the Precambrian ranges from 1,000 feet (about 300 m) below sea level in northernmost Illinois to almost 20,000 feet (6,100 m) below in the extreme southeastern part of the state. The relatively smooth surface shown on such deep structural contour maps is due largely to a lack of data and thus is somewhat misleading. Structural relief is approximately 5,000 feet (1,500 m) at the edge of the Reelfoot Rift and Rough Creek Graben in southern Illinois (Wheeler 1997). Many borings have been drilled by the petroleum industry on elevated basement structures, such as anticlines (convex-upward folds) and buried Precambrian hills; therefore, in many places, the actual depth to basement is probably deeper and more complex than shown (Bradbury and Atherton 1965). For example, actual rock outcrops of the basement rock in the St. Francois Mountains of southeastern Missouri (Figure 6-4) and a small number of closely spaced deep drill holes in Illinois reveal that the Precambrian surface probably has as much as 500 to 1,000 feet (150 to 300 m) of relief in places over distances of a mile or less (Sargent 1991).

Precambrian basement rocks in Illinois, as in other parts of the Midwest, consist primarily of granite plutons, granodiorite, and rhyolite (light-colored [felsic] igneous rocks rich in silica, sodium, aluminum, and potassium) and are assigned to the Eastern Granite-Rhyolite Province (Lidiak 1996), as shown on a simplified reconstruction of

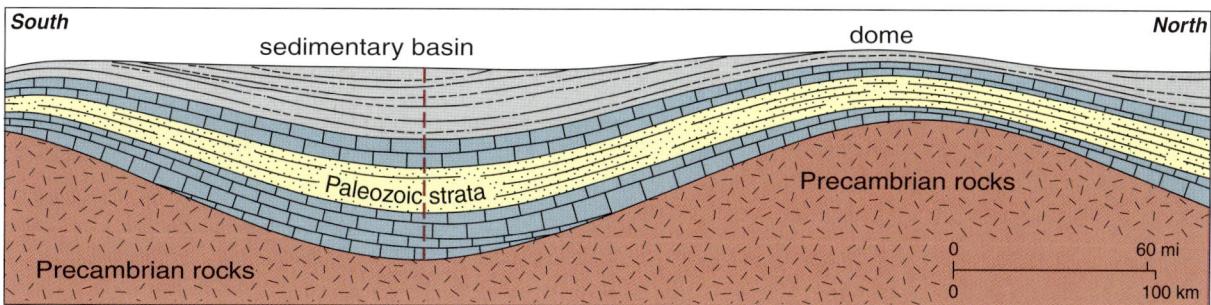

Figure 6-1 Typical diagrammatic geological cross section through a sedimentary basin in the central United States.

Figure 6-2 Polished surface of granite cut from a drill core in Precambrian basement rocks from the St. Jacob Oil Pool, Madison County, Illinois, depth 5,014 feet (1,528 m) (Buschbach and Kolata 1991, Wheeler 1997). (Photograph by Dennis R. Kolata.)

A question critical to the state's economy is whether or not sedimentary deposits, which may host petroleum accumulations, lie within or on top of the Eastern Granite-Rhyolite Province beneath the better known sedimentary Paleozoic strata. This question has led petroleum exploration companies to target the deeply buried Precambrian rocks in the east-central part of the state, south of Springfield, where petroleum production has been ongoing for well over a century. Research geologists have also focused on this area due to the abundance and availability of geophysical data (seismic reflection profiles and gravity- and the continents (Figure 6-5). The drill core shown in Figure 6-2 is an example of the granite that makes up much of this province underpinning Illinois. Part of a vast igneous belt stretching from northern Mexico to eastern Québec (Karlstrom et al. 1999) (Figures 6-4 and 6-5), the Eastern Granite-Rhyolite Province beneath Illinois is thought to represent a thin veneer (a mile or so thick) on Earth's approximately 24-mile-thick crust. Geological age measurements (based on the radioactive decay of uranium isotopes to lead) indicate that the province formed 1.4 to 1.5 billion years ago (bya) (Van Schmus et al. 1996). The origins of the granite and rhyolite, and of materials that lie even deeper, are not well understood.

Geochemical studies based on rare earth elements (samarium and neodymium) show the presence of a major geological boundary cutting diagonally across North America, including Illinois (Figure 6-4) (Van Schmus et al. 1996). Results of those studies suggest that the granites and rhyolites situated northwest of the boundary formed from the remelting of ancient lower crust that formed before about 1.6 bya. Southeast of the boundary, in Illinois, the primary age of the lower crust is younger than this, and the Eastern Granite-Rhyolite Province in the upper crust is slightly younger still (Van Schmus et al. 1996). This age difference means that the Precambrian basement beneath southern Illinois, southeast of the limit of the older crust (Figure 6-4), belongs to the Middle Proterozoic Era (900 mya to 1.6 bya). Furthermore, many geologists have interpreted the geochemically defined boundary to have once marked the very edge of the older part of the ancient continent called Laurentia (Figure 6-5).

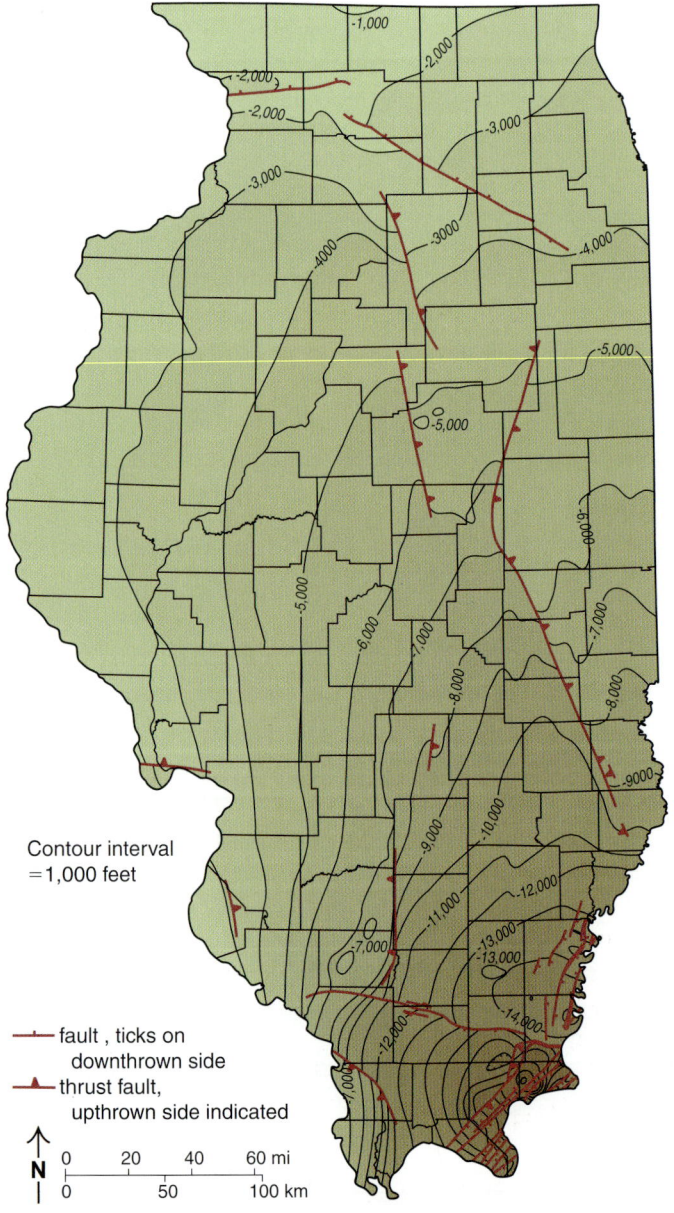

Figure 6-3 Configuration of the base of the Upper Cambrian Mt. Simon Sandstone in Illinois (modified from Buschbach and Kolata 1991). Throughout most of the state, this surface marks the top of Precambrian granite or rhyolite, but, in east-central Illinois, the Precambrian may include sedimentary or volcaniclastic rocks.

Precambrian

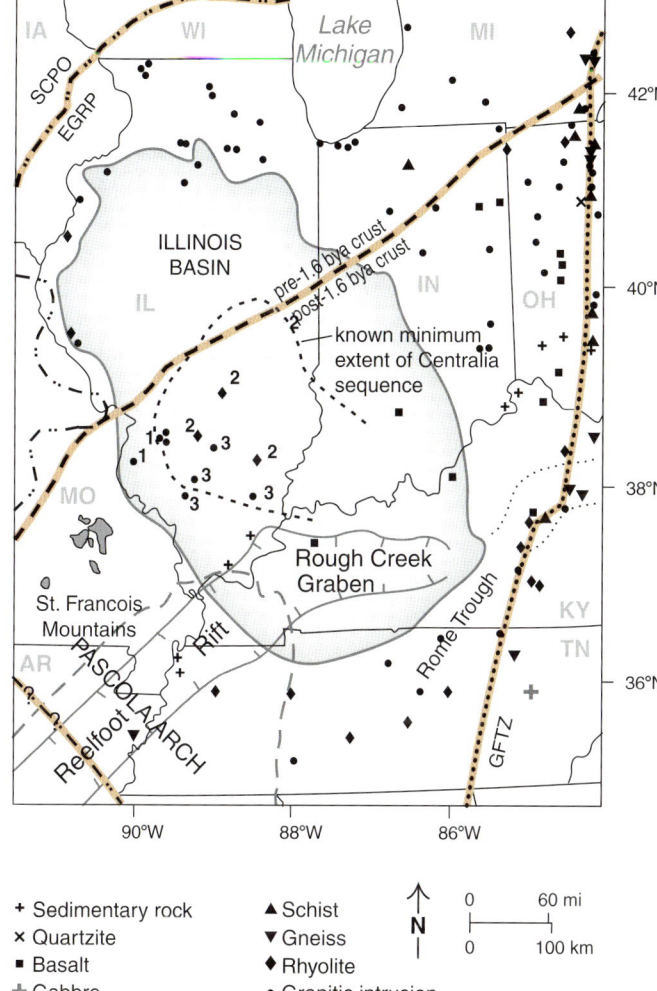

- + Sedimentary rock
- × Quartzite
- ■ Basalt
- + Gabbro
- ▲ Schist
- ▼ Gneiss
- ◆ Rhyolite
- ● Granitic intrusion

Figure 6-4 Major basement rocks encountered in drill holes of the central United States (based mainly on Lidiak 1996) and principal basement provinces (Van Schmus et al. 1996). The thick dashed line indicates the inferred southeastern limit of crust more than 1.6 billion years ago (bya) (Van Schmus et al., 1996). The dot-dashed line represents the boundary between the southern central plains orogen (SCPO) and the Eastern Granite-Rhyolite Province (EGRP). Abbreviation: GFTZ, Grenville Front Tectonic Zone. Numbered drill holes: **1**, phaneritic intrusive igneous rock; **2**, aphanitic extrusive igneous rock; **3**, granophyric-textured rock.

Figure 6-5 Paleogeographic reconstruction of the ancient supercontinent Laurentia including North America at about 1.6 to 1.3 billion years ago with continents shown in equal-area projection in North American coordinates (Karlstrom et al. 1999). Eastern Granite-Rhyolite Province and juvenile crust are shown in fine stipple. An outline of Illinois has been superimposed for reference. Brown shading represents crust older than 1.5 billion years. Light tan shading with random dashed pattern represents crust that was later accreted.

magnetic-field data), which can be integrated with drill-hole information in order to reveal fascinating insights into the deeply buried Precambrian here (Heigold 1991, Drahovzal 1997, McBride 1999). These geophysical and geological data have served to open an intriguing window into the Precambrian in this part of Illinois.

East-Central Illinois Precambrian Basement Revealed
Deep Seismic Reflection Profiles

Since the first successful petroleum exploration well was drilled in the Illinois Basin in 1886, over 4 billion barrels of oil and an estimated 4 trillion cubic feet (about 113 billion m³) of natural gas have been produced (Mast and Howard 1991). Beginning in 1937, production climbed rapidly as a result of the successful application of a new exploration technique—seismic reflection profiling. This subsurface exploration method generates sound waves, which pass through and reflect off geological boundaries

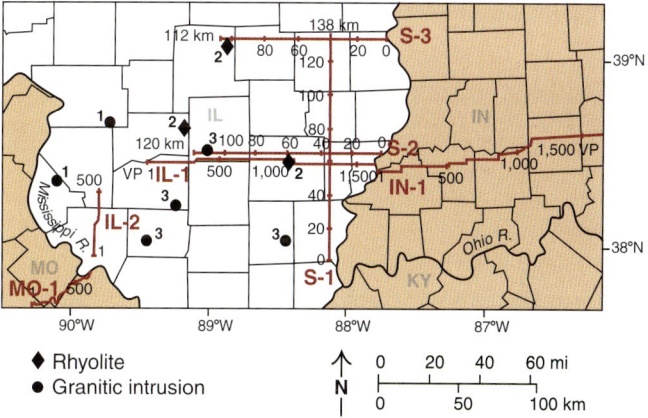

- ◆ Rhyolite
- ● Granitic intrusion

Figure 6-6 Map of east-central Illinois showing a subset of the seismic lines available to the Illinois State Geological Survey (e.g., McBride 1999, McBride and Kolata 1999, McBride et al. 2003) and publicly available profiles from the Consortium for Continental Reflection Profiling (Pratt et al. 1989). Circles and diamonds indicate drill holes for which petrographic analyses were performed by Sargent (1993). Numbered drill holes: **1**, phaneritic intrusive igneous rock; **2**, aphanitic extrusive igneous rock; **3**, granophyric-textured rock.

(rock layers and fault surfaces that act as "reflectors") to be received and recorded back at the surface by an array of geophones. The data are interpreted as a geological cross section through the Earth. The vertical axis is usually given in units of two-way travel time, which can be thought of as an approximate proxy for depth below the Earth's surface (Figures 6-6 and 6-7). A rough estimate of depth can be made by multiplying the travel time by the speed of sound, measured in seconds (s) in solid rock. For the deeper compacted rocks beneath Illinois, travel time can range from 15,000 feet/s (4,600 m/s) to 20,000 feet/s (6,100 m/s) (Heigold and Oltz 1991). Because the time is doubled during reflection back to the surface, it is necessary to divide the time result by 2. Conveniently, for Precambrian rocks, time measures can be roughly converted to depth in kilometers merely by multiplying the travel time by 3. Over the years, major oil companies have generated tens of thousands of miles of seismic reflection exploration data. Until recently, almost none of these data had been published or released to academic researchers.

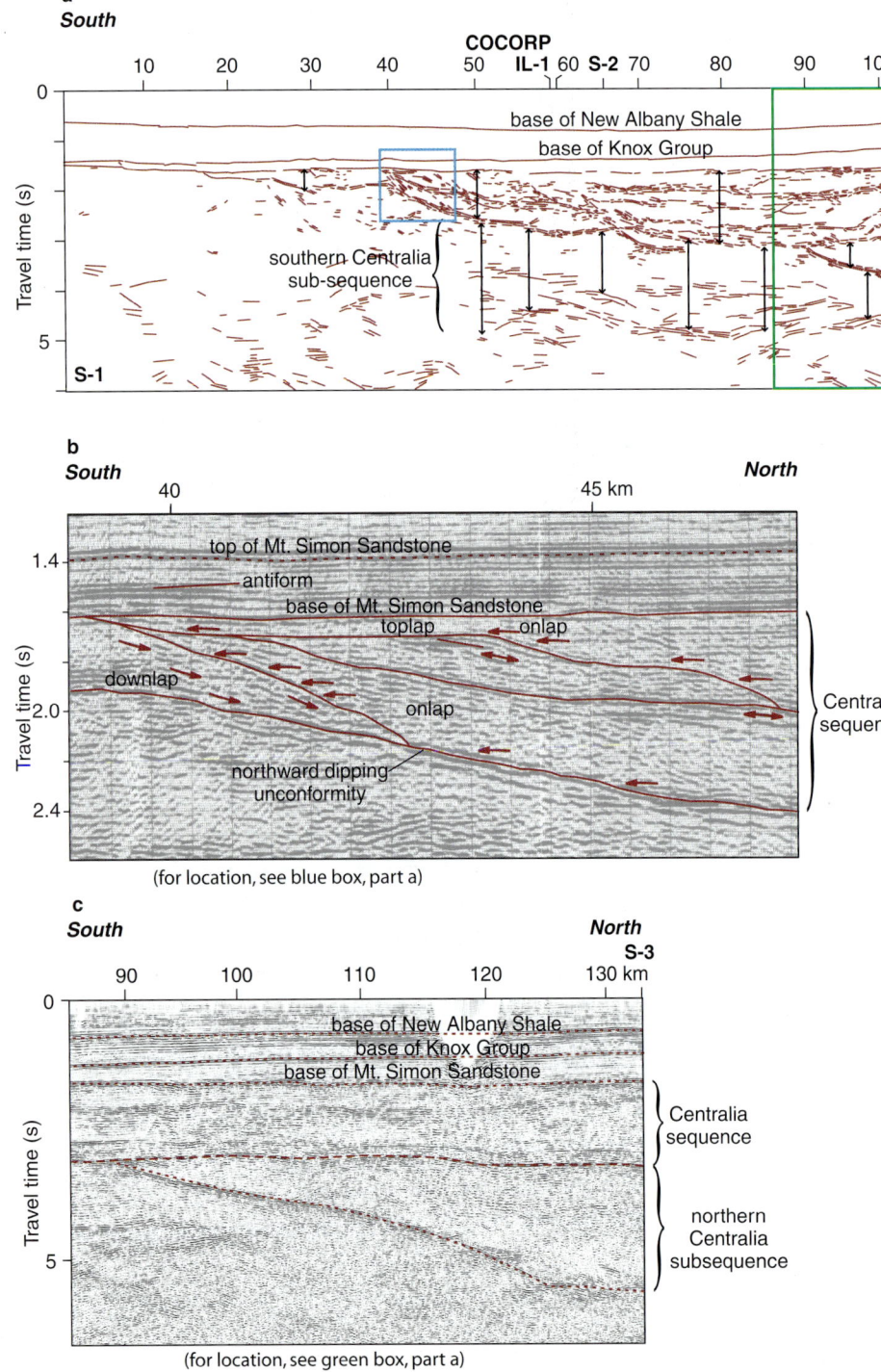

Figure 6-7 (a) Interpretation of reprocessed seismic reflection line S-1 for the upper 6 seconds (s); prominent interpreted faults and seismic sequences are noted. The Paleozoic portion of the record (the base of Mt. Simon Sandstone and higher) is represented by the most prominent stratigraphic markers only. Double-ended arrows indicate upper and lower boundaries of the Centralia sequence. Vertical exaggeration is 2:1 using conversion velocity of 6.2 km/s (20,300 feet/s). (b) An excerpt of the migrated seismic reflection profile S-1 focused on the upper Precambrian subsequence (indicated by bracket) with a seismic stratigraphy interpretation. Horizontal arrows indicate interpreted horizontal truncations or onlap; dipping arrows indicate dipping truncations or downlap. (c) An excerpt of the northern part of seismic reflection profile S-1 showing northward-thickening sub-Mt. Simon Sandstone sequences. Note especially the strong dipping reflection forming the lower boundary of the northern Centralia subsequence. The seismic data for this figure and all other displays of the S profiles are provided by Seismic Exchange, Inc.; in all cases, the interpretation is that of the authors. These sections have been modified from McBride et al. (2003).

Precambrian

A major breakthrough came in the late 1980s, when 239 miles (385 km) of deep reflection profiles were acquired across Indiana and Illinois (lines IL-1, IL 2, and IN-1 on Figure 6-6) by the Consortium for Continental Reflection Profiling (COCORP) based at Cornell University. Actual images of Precambrian basement structure began to emerge (Pratt et al. 1989). For the first time geologists could "see" into the basement rocks beneath the Midwest, and what they saw was a major surprise. Instead of a blank region of no reflectors, as had been expected for a basement of just granitic rock (Figure 6-1), a broad expanse of highly reflective rock layers appeared beneath a large area of east-central Illinois and west-central Indiana. These stunning results immediately caused geologists to dismiss the idea that the Eastern Granite-Rhyolite Province had no internal structure—in fact, the Precambrian basement was found to be more structurally complex and interesting than the better studied Paleozoic sedimentary rocks of the overlying Illinois Basin, which had been explored for decades. Scientists from Cornell University coined the term "Centralia sequence" to identify the internally reflective Precambrian basement (Pratt et al. 1992).

A second breakthrough came several years later when a loose network of several hundred miles of petroleum industry seismic reflection profiles was released to the Illinois State Geological Survey (profiles S-1, S-2, S-3 on Figure 6-6). Reanalysis and mapping (McBride and Kolata 1999, McBride et al. 2003) of the profile network revealed the three-dimensional shape and the fine-scale internal structure of the Precambrian Centralia sequence that was first identified by the single straight-line COCORP transect. The higher resolution of the industry profile network led to the discovery that the Centralia sequence consists of three sequences that are highly coherent, layered, unconformity-bounded and that overlie or may be incorporated within the Eastern Granite-Rhyolite Province beneath the Paleozoic Illinois Basin (Figure 6-7a). What was even more exciting was that these sequences were found to be much thicker than originally thought and that they extended down into middle crustal depths of about 10.5 miles (16.9 km) below the surface (Figure 6-7b). These sequences are bounded by strong, laterally continuous reflectors (Figure 6-7c) that are mappable over distances of more than 124 miles (200 km) and are expressed as broad bowl-shaped packages that become narrower with depth (Figures 6-7a and 6-8).

The accumulating information from seismic reflection profiles challenged long-standing beliefs about the Precambrian basement beneath Illinois. The seismic profiles revealed that Precambrian rocks are actually layered and possess a coherent structure. Thus, what had historically been thought to be a homogeneous, featureless mass (e.g., Figure 6-1) is in fact marked by previously unmapped deeply buried basins, dipping and offset reflectors, and extensive angular unconformities (Figure 6-7c). Three-dimensional mapping of the seismic sequences represented by a contour map (Figure 6-8) reveals a broad, bowl-shaped succession of rocks (Centralia sequence) that reach a maximum depth of roughly 6 miles (10 km) and a thickness of about 3.7 miles (6 km), based on an assumed seismic compressional wave velocity of about 20,000 feet/s (6 km/s) (Figure 6-7). Normal faults progressively disrupt the sequences with depth along their outer margins (McBride and Kolata 1999). The overall distribution of the sequences mimics that of the overlying Cambrian Mt. Simon Sandstone unit (Figures 6-8 and 6-9) (McBride et al. 2003), which is particularly evident when the travel-time isopach contours of the Enterprise subsequence, the uppermost unit of the Centralia sequence, are plotted on the thickness map of the Mt. Simon Sandstone (Figure 6-10). This relationship sug-

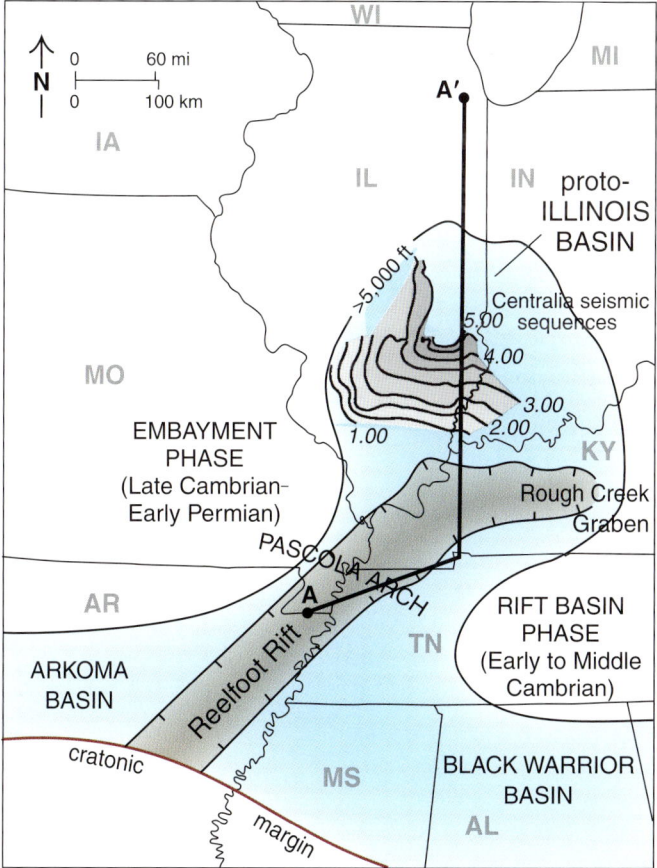

Figure 6-8 Map showing the location of the rift basin and subsequent formation of the proto-Illinois Basin centered over the rift junction. Blue shading indicates where Paleozoic strata are presently thicker than 5,000 feet (1,500 m) (modified from Kolata and Nelson 1991). Also shown is the simplified structural contour map (in seconds travel time) for the base of the Precambrian seismic reflection sequences as mapped from the seismic lines shown in Figure 6-7 and other proprietary seismic lines (McBride and Kolata 1999, McBride et al. 2003).

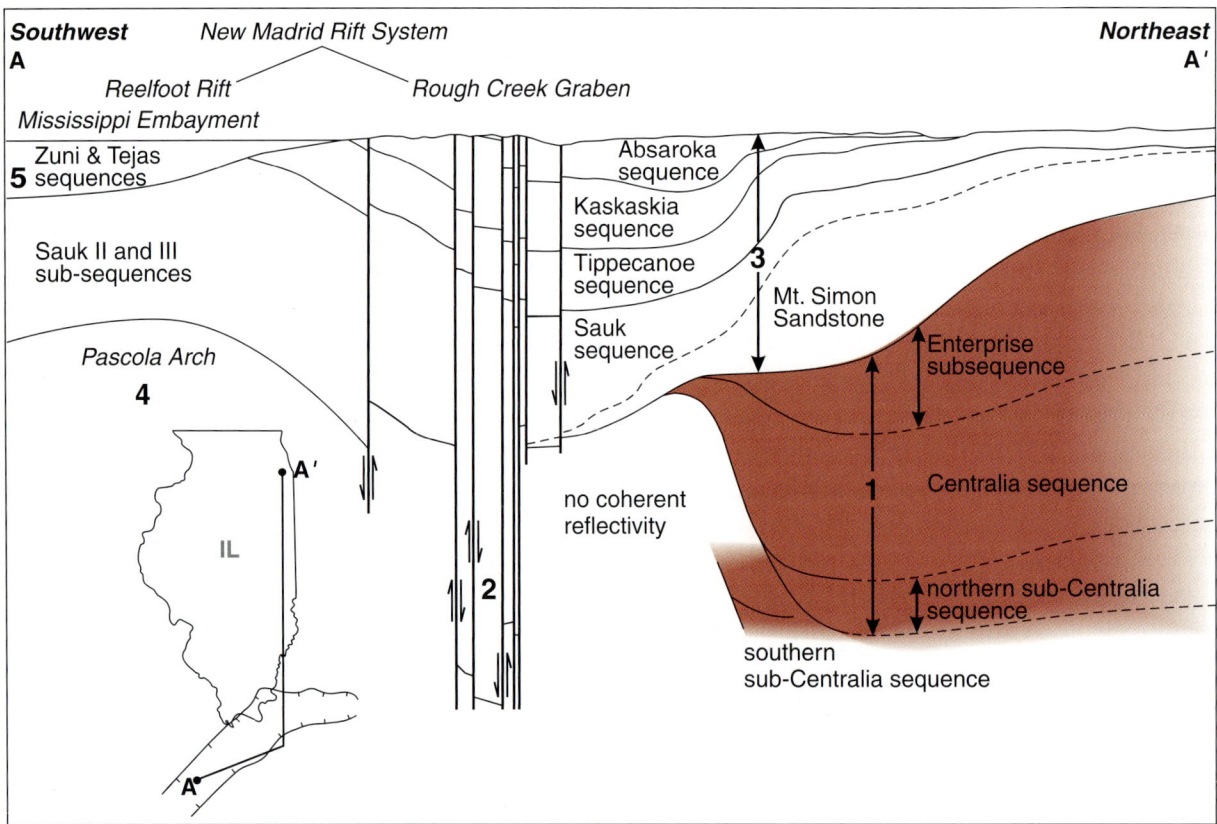

Figure 6-9 Schematic cross section of Illinois Basin Paleozoic succession and known Precambrian sequences along the A–A′ profile in Figure 6-8. Faults that formed the Rough Creek Graben continued to be active during deposition of Paleozoic sequences. Numbers refer to the chronology of geological events. This diagram has been modified from McBride et al. (2003).

gests that an episode of localized subsidence was under way before deposition of the Paleozoic stratigraphic succession, which fills most of the well-known Paleozoic Illinois Basin and which is centered farther south over the Reelfoot Rift and Rough Creek Graben (Figure 6-9).

Geopotential-Field Data

Seismic reflection profiles aid in the interpretation of regional geometrical patterns in Precambrian rocks but provide little direct information about rock composition. A means of determining the compositional variations in rock type is needed to interpret the geology between the scattered drill holes and guide the understanding of the seismic data. Maps depicting variations in Earth's gravitational and magnetic fields (geopotential maps) (Figures 6-11 and 6-12) are used by geophysicists to infer the composition of rocks that are deeply buried and thus are unavailable for direct examination. The gridded gravity anomaly map (Figure 6-11) provides information about lateral changes in rock density in the subsurface by depicting small changes in acceleration due to Earth's gravity. Gravity "highs" represent rock that is denser than the surrounding rock or "average" crust. Magnetic anomaly (intensity or magnetic-field strength) maps (Figure 6-12) show lateral changes in the

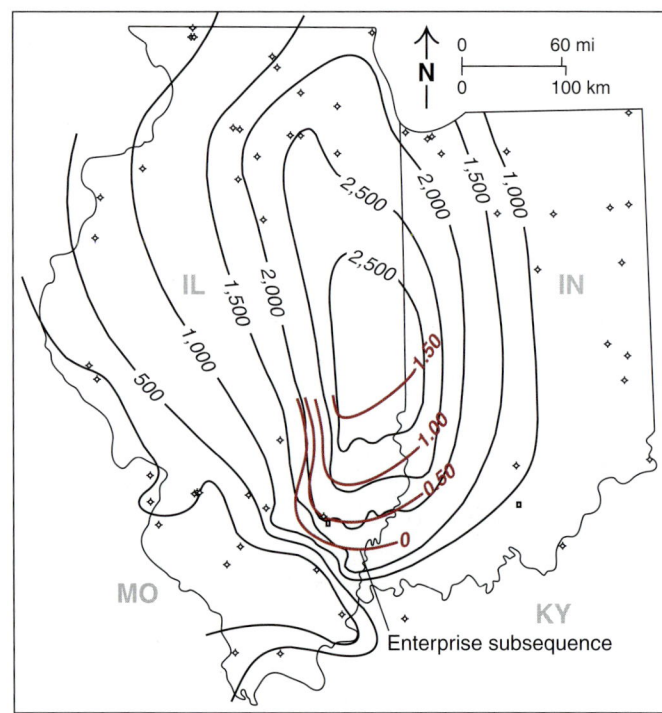

Figure 6-10 Thickness map of the Cambrian Mt. Simon Sandstone overlain with travel-time isopach contours of the Enterprise subsequence, the uppermost unit of the Centralia sequence (modified from McBride et al. 2003). Also shown are wells that penetrated the Mt. Simon. The Illinois Basin Paleozoic sequences are further explained in Chapters 7 through 10.

magnetization of subsurface rocks. Magnetic intensity changes indicate lateral changes in the amount of magnetic minerals (e.g., magnetite) that are present in subsurface rocks. Mafic and ultramafic igneous rocks (rich in olivine, pyroxene, amphibole, and biotite mica) usually produce gravity and magnetic highs (e.g., representing basalt or gabbro). Areas of thick sedimentary rocks (e.g., sandstone and limestone) commonly produce lows. Some of the highest magnetic values occur over iron ore deposits. Granite is typically less magnetic than mafic igneous rocks and, by comparison, most sedimentary rocks lack significant amounts of magnetic minerals. The combination of gravity and magnetic maps provides a powerful tool to approximate the basic compositional properties of Earth's crust.

Progress has been made in understanding the geometry, distribution, and probable composition of the deeply buried Precambrian rocks of east-central Illinois by integrating drill-hole information, seismic reflection profiles, and geopotential-field data. This integration has been accomplished by posting the locations of drill holes that have penetrated Precambrian basement and seismic lines (S-1, S-2, and S-3) on the geopotential anomaly maps (Figures 6-11 and 6-12). The most important observation from the geopotential-field data is that the upper crustal sequences do not, as a whole, correspond to high magnetic or to high gravity values, as might be expected for large bodies of mafic igneous rocks, such as those found in the Midcontinent Rift System in Iowa and Minnesota (Figure 6-12) (Lidiak et al. 1985). This fact is especially clear from the broad regional perspective, as shown on the regional magnetic-intensity map for the central Midwest (Figure 6-12a). On the gravity and magnetic maps centered over southern Illinois (Figures 6-11 and 6-12), there is not a high proportion of "hot" grid colors (pinks and reds) within the area of the Precambrian layered seismic sequences. Instead, areal extents of these seismic sequences correspond to subdued intensities on the gravity and magnetic maps (Figures 6-11 and 6-12). To the south and west, the known boundaries of the Precambrian seismic sequences match an overall change in magnetic anomaly

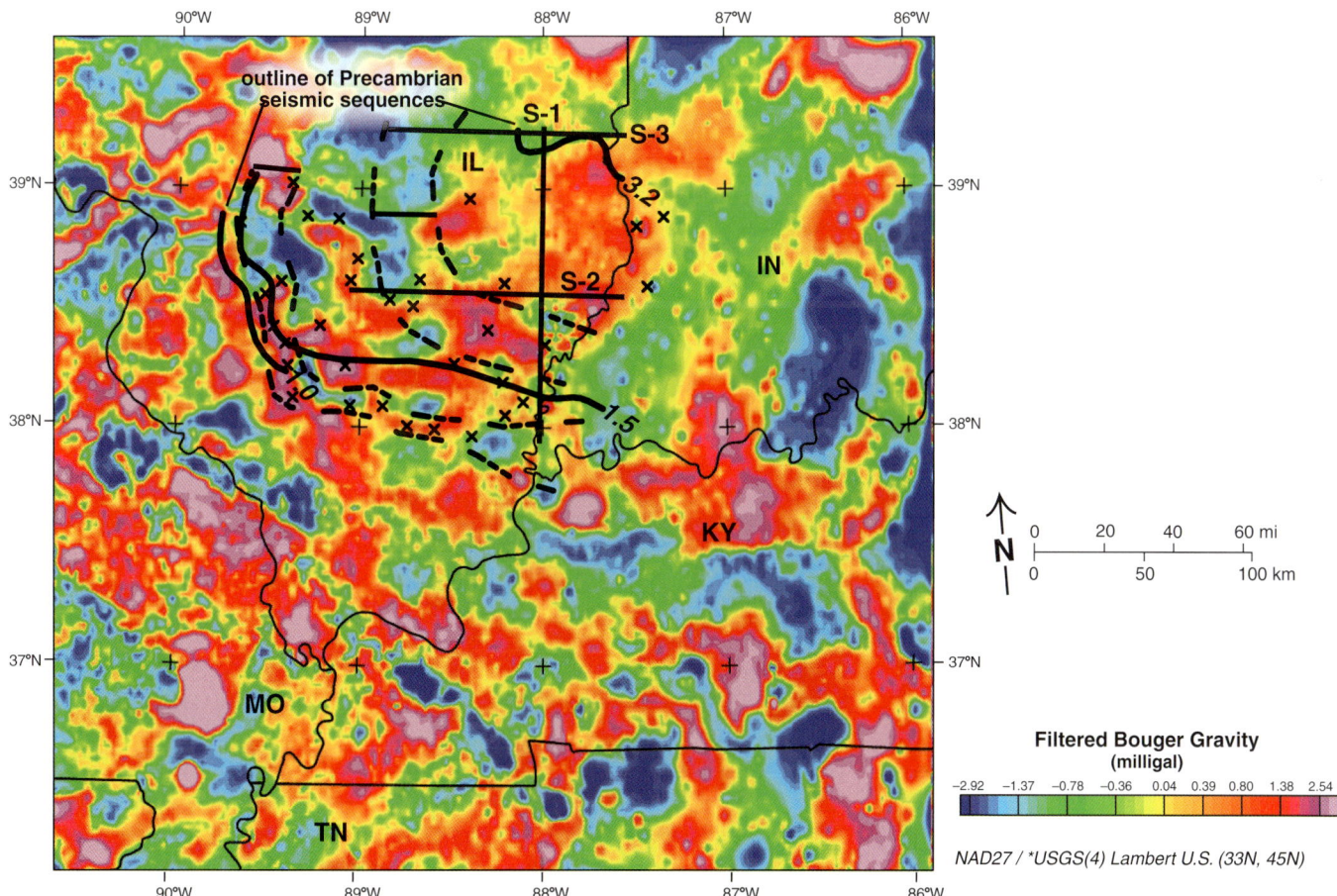

Figure 6-11 Residual Bouguer gravity computed by removing a 2,000-m (6,563-feet) upward continuation; that is, by removing the expression of the field as if it were observed 2,000 m (6,563 feet) above the ground. Three key seismic reflection lines are shown (Figure 6-6), plus selected travel-time contours (heavy solid lines) for the Centralia sequence (Figure 6-8). Curved dashed lines show two interpreted rings of positive geopotential field anomalies. A different version of this map is presented in McBride et al. (2003). Abbreviations: x, peak of positive anomaly observed from magnetic intensity (Figure 6-12b); milligal, 0.001 of a gal (1 gal = 1 cm/s^2).

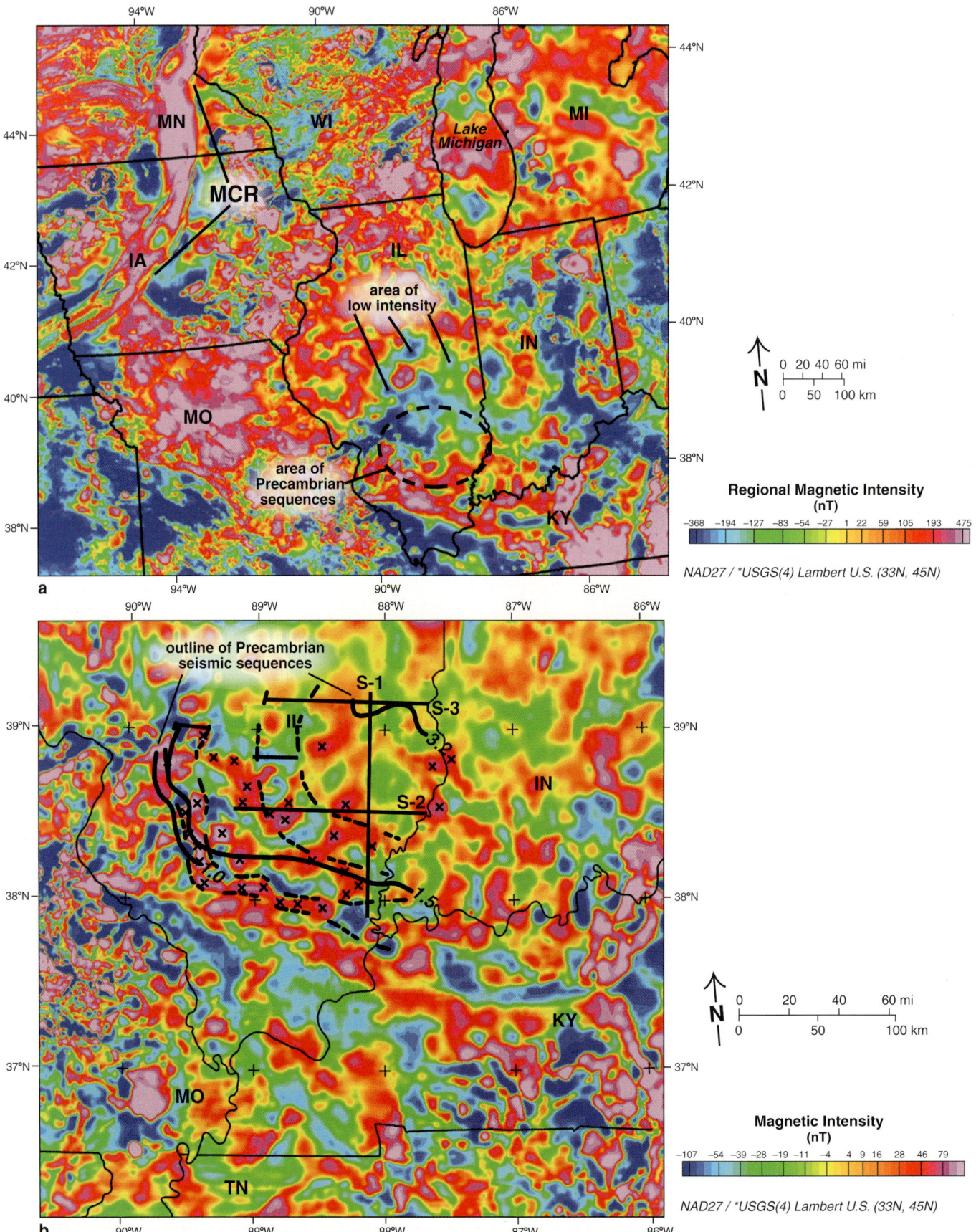

Figure 6-12 (a) Gridded regional magnetic intensity map for central United States. (b) Reduced-to-pole magnetic intensity with the 2,000-m (6,563 feet) upward continuation subtracted to produce a residual map. Reduced-to-pole means that the magnetic data have been normalized to show the shape of the anomalies as if they were at the Earth's pole. A different version of this map is presented in McBride et al. (2003). Abbreviations: MCR, midcontinent rift system; nT, nanoteslas.

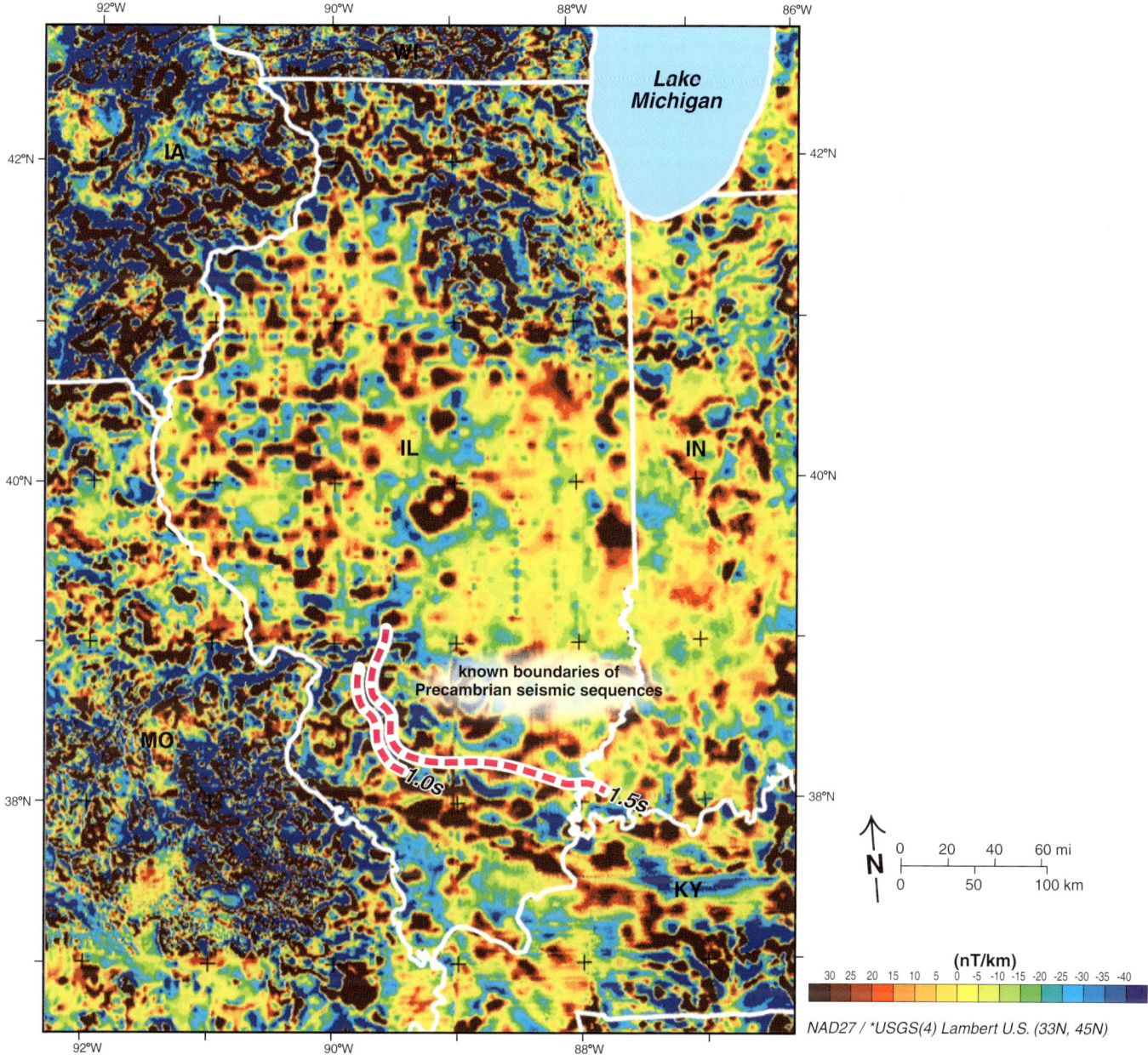

Figure 6-13 First vertical derivative (analogous to the slope of the contoured values) of reduced-to-pole magnetic intensity for the central United States midcontinent showing western and southern boundaries of the sub-Mt. Simon sequences using the shallowest travel-time contours from Figure 6-8. For a detailed discussion of the methods used to produce these maps and their significance, see Hildenbrand and Ravat (1997), Hildenbrand et al. (2002), and McBride et al. (2003). Abbreviation: nT, nanoteslas.

wavelength and intensity as depicted on a "slope" anomaly map (Figure 6-13). However, to the northeast, a much larger area of subdued geopotential field values extends well beyond the area for which seismic reflection data are available (see area of low variation on map dominated by mostly yellow in Figure 6-13). Thus, the visualized portions of the Precambrian sequences might represent a small part of what could ultimately be an enormous feature of the upper crust below the Paleozoic in east-central Illinois and continuing to the north and east (Figure 6-13).

A second observation is that the western and southern boundaries of the Precambrian seismic sequences are marked by two approximately concentric curved alignments of small, localized magnetic anomaly highs (Figure 6-12b). These highs are matched by a less well-defined alignment of small gravity anomaly highs (Figure 6-11). The positive magnetic anomalies constituting this outer ring mostly correspond to the centers of gravity highs ("x" symbols in Figures 6-11 and 12b), which suggests mafic igneous sources (e.g., basalt or gabbro) (Lidiak et al. 1985). The inner ring of quasi-circular anomalies is less well-defined and does not show such a distinct one-to-one correspondence of prominent magnetic- and gravity-anomaly highs. Previous analyses of magnetic and gravity anomalies in the Illinois Basin region (Hildenbrand and Ravat 1997, Hildenbrand et al. 2002) indicate that the sources of the anomalies are small, isolated mafic igneous rock bodies that lie mostly below the base of the basement seismic sequences (below 4 to 9 miles [6.4 to 14.5 km]). In summary, the geopotential-field data and information from drill-hole penetrations below the Mt. Simon Sandstone do not support the presence of a large basaltic complex beneath central Illinois; however, the geopotential data indicate that small, deeply buried, mafic igneous rock bodies have intruded lower-density, less-magnetic crust.

Direct Sampling from Drill-Hole Penetrations

Over the past 15 years, drilling samples and geophysical data for the basement rocks beneath Illinois and nearby states have been restudied, especially in southwestern Ohio, southern Indiana, and northern Kentucky where the Paleozoic sedimentary section is relatively shallow and easier to reach by drilling (e.g., Shrake et al. 1991). Drill holes in some of these areas have encountered rocks that are Precambrian (i.e., beneath the basal Cambrian Mt. Simon Sandstone) but sedimentary, not igneous as had previously been expected. These rocks, which could act as oil or gas reservoirs, have been interpreted as occupying a large sedimentary basin analogous to the younger Paleozoic Illinois Basin (Shrake et al. 1991, Dean and Baranoski 2002). These exciting results have prompted a re-examination by the Illinois State Geological Survey of Precambrian rocks recovered from drill holes in Illinois. For the central part of the Illinois Basin (Figure 6-4), pre-Mt. Simon sedimentary strata are not well-known. A few drill-hole penetrations of arkosic sandstone have been tentatively assigned to the Precambrian in southern Illinois (Buschbach and Kolata 1991, Lidiak 1996), but no supporting geological age determinations are available.

Studies of rock samples from drill holes that penetrate the base of the Mt. Simon Sandstone in the Illinois Basin can be used to constrain the interpretation of seismic reflection profiles (e.g., Figure 6-7). Thin sections produced from cuttings or available core from eight of the drill holes labeled in Figure 6-6 were studied and described (Sargent 1993, Sargent and Lasemi 1993). Three distinct petrographic textures are represented that are indicative of igneous rock: coarse-grained phaneritic (grains are visible) rocks, aphanitic (grains are invisible) rocks, and granophyres (Figures 6-4 and 6-6). The granophyres, which typically have a micrographic texture (an intergrowth pattern of two minerals), contain medium-grained quartz, potassium feldspar, sodic plagioclase, and traces of amphibole and opaque minerals (Figure 6-14). These rocks show micrographic intergrowths of quartz and potassium feldspar that cannot be grown authigenically in a sediment; that is, they cannot crystallize at the same time as the formation of the rock in which they occur in a sediment and probably could not have survived a single cycle of erosion and deposition. Furthermore, the porphyritic (an igneous rock with crystals embedded in a finer groundmass of minerals) and spherulitic (a minute spherical crystalline grain having a radiated structure, observed in some glassy volcanic rocks) textures of the aphanitic rocks are also distinctly igneous. They would have been destroyed by weathering if they had been part of a sediment. This examination thus showed that the Precambrian rocks, where penetrated by drilling beneath the southern Illinois Basin in Illinois, are clearly igneous in composition, not sedimentary.

Toward a More Coherent View of Illinois' Precambrian Basement

The challenges to understanding Illinois' Precambrian rocks have historically been that they do not crop out anywhere in the state and that only a tiny fraction of the thousands of drill holes in Illinois have penetrated these basement rocks. Only by integrating seismic profile interpretations, geopotential-field data, and scattered drill-hole information can a coherent view of this vast geological province beneath the state begin to be pieced together.

Seismic reflection profiles reveal at least three highly coherent, stratified Precambrian basement sequences (Centralia sequences) beneath the Paleozoic succession of east-central Illinois that continue down to roughly 10.5 miles (17 km) in depth. The interiors of these sequences display evidence for unconformity-bounded depositional sequences (Figure 6-7b) within rocks that the drill-hole data indicate to be non-sedimentary basement. Indeed, reflection terminations and configurations (Figure 6-7b) are similar to the patterns formed in river deltas that are typically imaged along many passive continental margins such as the Gulf of Mexico and the Atlantic Ocean. Such patterns suggest a sedimentary or volcaniclastic (sediments derived from volcanic products) succession filling a basin. Although the information from several drill-hole penetrations in and around east-central Illinois indicate non-sedimentary granite-rhyolite textures and compositions (Figure 6-14), as just explained, only one or two of the observed reflective basement sequences have actually been tested by drilling. This lack of data leaves open the possibility that part of the Precambrian basement in between the few base-

PRECAMBRIAN

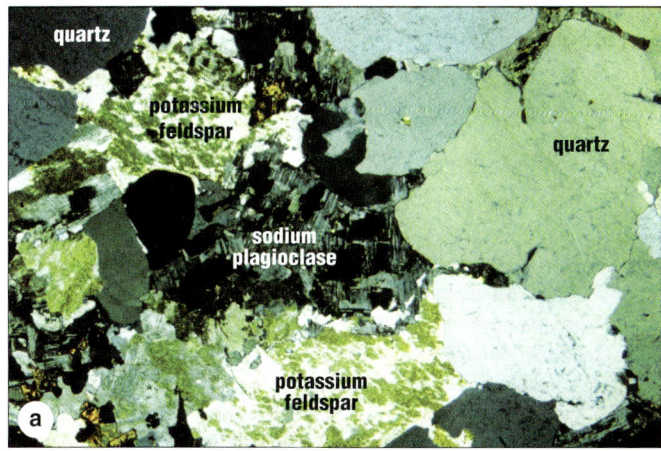

Figure 6-14 Photomicrographs of petrographic thin sections using cross-polarized light from sub-Mt. Simon Sandstone rocks in **(a)** a core sample from Madison County, Illinois, and **(b)** drill-hole cuttings from Marion County, Illinois. These sections show unweathered components of igneous rocks, which indicate, at least for the immediate area of the drill holes, that the Precambrian rocks are not sedimentary. (Photomicrographs by Michael L. Sargent.)

ment drill-hole penetrations (Figure 6-6) may be sedimentary or volcaniclastic. McBride et al. (2003) suggested that chaotic mixtures of volcanic rocks and sediments would be expected from the explosive interaction of magma and wet sediments in a Precambrian sedimentary basin, as has been documented from the rifted margins of Western Australia (Planke et al. 2000).

In general, the gravitational- and magnetic-field data (Figures 6-11 and 6-12) show overall low intensity values, meaning relatively low density and low magnetic character. Geophysical maps (Figure 6-12) indicate that the deeper parts of the seismic sequences correspond to subdued geopotential field values, which implies, together with the felsic igneous rocks from drill holes, that the sequences are lacking in high-density, highly magnetized rocks relative to the surrounding region. These results preclude a large mafic igneous component to the crust. This key informa-

tion suggests that there is something anomalous about the basement here—that it perhaps includes some sedimentary component that has not yet been revealed by drilling.

Illinois' Precambrian basement, the geological age of which is at least three times that of the overlying Illinois Basin succession, likely has a complex history involving more than one episode of formation. The broad bowl shape and rich reflective structure of the Precambrian layered seismic sequences have been explained by an episode of explosive (or extrusive) volcanism that originally produced the Eastern Granite-Rhyolite Province (McBride et al. 2003). Such extrusive structures, known as calderas (a bowl- or basin-shaped volcanic depression), are well-documented in the nearby St. Francois Mountains in southeastern Missouri (Figure 6-4). The rift-like geometries seen in parts of the basement (Figure 6-7a), however, also suggest an episode of crustal extension.

The Precambrian structure appears to have possibly governed the much later subsidence of the Paleozoic Illinois Basin. The location of the deepest parts of the Precambrian sequences well north of the Rough Creek Graben (Figures 6-8, 6-9, and 6-10) implies that the Basin's subsidence history was influenced by processes operating separately in two discrete regions (Figures 6-8 and 6-9): (1) the area of the Precambrian seismic (Centralia) sequences and (2) the Reelfoot Rift and Rough Creek Graben in southern Illinois and adjacent areas. Tectonic subsidence curves from deep wells in the Basin (Heidlauf et al. 1986, Kolata and Nelson 1991) indicate an initial episode of rapid subsidence associated with late Precambrian to Early Cambrian rifting, followed by thermal contraction of the lithosphere and subsidence that lasted into the Middle Ordovician. For the remainder of the Paleozoic, subsidence was localized over the Reelfoot Rift and the Rough Creek Graben (Kolata and Nelson 1997). The anomalously thick succession of Mt. Simon Sandstone rests over the deep Precambrian sequences (Figure 6-9). The meaning of this spatial correspondence remains unresolved, but it may be that the deep sequences governed the earliest Paleozoic subsidence of the Basin prior to the effects of the Rough Creek Graben (Figure 6-10). Thus, the present configuration of the Illinois Basin resulted from the cumulative effects of the Precambrian subsidence in east-central Illinois and the rift and graben to the south (Figures 6-8 and 6-9).

CONCLUSION

The state of knowledge of the Precambrian rocks of Illinois depends heavily on geophysical remote sensing, guided wherever possible by actual rock samples. The geophysical and geological data needed for understanding Precambrian structure and history are most plentiful for the southern half of the state. Because geochemical modeling suggests that the center of Illinois is crossed by a major crustal boundary, which has been interpreted to mark an ancient active continental margin of Precambrian age (Figure 6-4), crustal tectonic and magmatic activity would, therefore, be expected to have been more intense in the southern half of the state. This activity could account for the complexity in basement structure that is observed there. A coherent view of the deeply buried rocks just south of the geochemical boundary has emerged in which the Precambrian, far from being the homogeneous mass that is typically shown in textbooks (Figure 6-1), is actually highly structured and diverse. Drill-hole samples show multiple types of felsic igneous Precambrian rocks, and seismic data reveal discrete Precambrian sequences of apparently layered deposits that point to a complex and multiphase history. Geopotential data (Figures 6-11 and 6-12) show an anomalous basement beneath southern Illinois and provide clues that suggest a concentration of volcaniclastic and/or sedimentary strata within or superimposed on the Precambrian Eastern Granite-Rhyolite Province. As in the past, exploratory drilling for resources and new geophysical surveying will likely reveal further secrets of the deeply buried Precambrian rocks of Illinois.

References

Bradbury, J. C., and E. Atherton, 1965, The Precambrian basement of Illinois: Illinois State Geological Survey, Circular 382, 13 p.

Buschbach, T. C., and D. R. Kolata, 1991, Regional setting of the Illinois Basin, in M. W. Leighton, D. R. Kolata, D. F. Oltz, and J. J. Eidel, eds., Interior cratonic basins: Tulsa, Oklahoma, American Association of Petroleum Geologists, Memoir 51, p. 29–55.

Dean, S. L., and M. T. Baranoski, 2002, Ohio Precambrian: 2, Deeper study of Precambrian warranted in western Ohio: Oil and Gas Journal, v. 100, no. 30, p. 37–40.

Drahovzal, J. A., 1997, Proterozoic sequences and their implications for Precambrian and Cambrian geologic evolution of western Kentucky: Evidence from seismic-reflection data. Seismological Research Letters, v. 68, no. 4, p. 553–566.

Heidlauf, D. T., A. T. Hsui, and G. deV. Klein, 1986, Tectonic subsidence analysis of the Illinois Basin: Journal of Geology, v. 94, no. 6, p. 779–794.

Heigold, P. C., 1991, Crustal character of the Illinois Basin, in M. W. Leighton, D. R. Kolata, D. F. Oltz, and J. J. Eidel, eds., Interior cratonic basins: Tulsa, Oklahoma, American Association of Petroleum Geologists, Memoir 51, p. 247–261.

Heigold, P. C., and D. F. Oltz, 1991, Seismic expression of the stratigraphic succession, in M. W. Leighton, D. R. Kolata, D. F. Oltz, and J. J. Eidel, eds., Interior cratonic basins: Tulsa, Oklahoma, American Association of Petroleum Geologists, Memoir 51, p. 169–178.

Hildenbrand T. G., J. H. McBride, and D. Ravat, 2002, The Commerce geophysical lineament and its possible relation to Mesoproterozoic igneous complexes and large earthquakes in the central Illinois Basin: Seismological Research Letters, v. 73, no. 5, p. 660–686.

Hildenbrand, T. G., and D. Ravat, 1997, Geophysical setting of the Wabash Valley Fault System: Seismological Research Letters, v. 68, no. 4, p. 567–585.

Karlstrom, K. E., S. S. Harlan, M. L. Williams, J. McLelland, J. W. Geissman, and K.-I. Åhäll, 1999, Refining Rodinia: Geologic evidence for the Australia-western U.S. connection in the Proterozoic: GSA Today, v. 9, no. 10, p. 1–7.

Kolata, D. R., and W. J. Nelson, 1991, Tectonic history of the Illinois Basin, in M. W. Leighton, D. R. Kolata, D. F. Oltz, and J. J. Eidel, eds., Interior cratonic basins: Tulsa, Oklahoma, American Association of Petroleum Geologists, Memoir 51, p. 263–285.

Kolata, D. R., and W. J. Nelson, 1997, Role of the Reelfoot Rift/Rough Creek Graben in the evolution of the Illinois Basin, in R. W. Ojakangas, A. B. Dickas, and J. C. Green, eds., Middle Proterozoic to Cambrian rifting, central North America: Boulder, Colorado, Geological Society of America, Special Paper 312, p. 287–298.

Lidiak, E. G., 1996, Geochemistry of subsurface Proterozoic rocks in the eastern midcontinent of the United States: Further evidence for a within-plate tectonic setting, in B. A. van der Pluijm and P. A. Catacosinos, eds., Basement and basins of eastern North America: Boulder, Colorado, Geological Society of America, Special Paper 308, p. 45–66.

Lidiak, E. G., W. J. Hinze, G. R. Keller, J. E. Reed, L. W. Braile, and R. W. Johnson, 1985, Geologic significance of regional gravity and magnetic anomalies in the east-central midcontinent, in W. J. Hinze, ed., The utility of regional gravity and magnetic anomaly maps: Tulsa, Oklahoma, Society of Exploration Geophysicists, p. 287–307.

Mast, R. F., and R. H. Howard, 1991, Oil and gas production and recovery estimates in the Illinois Basin, in M. W. Leighton, D. R. Kolata, D. F. Oltz, and J. J. Eidel, eds., Interior cratonic basins: Tulsa, Oklahoma, American Association of Petroleum Geologists, Memoir 51, p. 295–298.

McBride, J. H., 1999, Without firing a shot: Seismic exploration of the Illinois Basin: Geotimes, v. 44, no. 5, p. 19–23.

McBride, J. H., and D. R. Kolata, 1999, Upper crust beneath the central Illinois Basin, United States: Geological Society of America Bulletin, v. 111, no. 3, p. 375–394.

McBride, J. H., D. R. Kolata, and T. G. Hildenbrand, 2003, Geophysical constraints on understanding the origin of the Illinois Basin and its underlying crust: Tectonophysics, v. 363, no. 1–2, p. 45–78.

Planke, S., P. A. Symonds, E. Alvestad, and J. Skogseid, 2000, Seismic volcanostratigraphy of large-volume basaltic extrusive complexes on rifted margins: Journal of Geophysical Research v. 105, no. B8, p. 19335–19351.

Pratt, T., R. Culotta, E. Hauser, D. Nelson, L. Brown, S. Kaufman, J. Oliver, and W. Hinze, 1989, Major Proterozoic basement features of the eastern midcontinent of North America revealed by recent COCORP profiling: Geology, v. 17, no. 6, p. 505–509.

Pratt, T. L., E. C. Hauser, and K. D. Nelson, 1992, Widespread buried Precambrian layered sequences in the U.S. midcontinent: Evidence for large Proterozoic depositional basins: American Association of Petroleum Geologists Bulletin, v. 76, no. 9, p. 1384–1401.

Sargent, M. L., 1991, Sauk Sequence: Cambrian System through Lower Ordovician Series, in M. W. Leighton, D. R. Kolata, D. F. Oltz, and J. J. Eidel, eds., Interior cratonic basins: Tulsa, Oklahoma, American Association of Petroleum Geologists, Memoir 51, p. 75–85.

Sargent, M. L., 1993, Structural and tectonic implications of pre-Mt. Simon strata—or a lack of such—in the western part of the Illinois Basin (abs.): Boulder, Colorado, Geological Society of America, Abstracts with Programs, v. 25, no. 3, p. 78.

Sargent, M. L., and Z. Lasemi, 1993, Tidally dominated depositional environment for the Mt. Simon Sandstone in central Illinois (abs.):

Boulder, Colorado, Geological Society of America, Abstracts with Programs, v. 25, no. 3, p. 78.

Shrake, D. L., R. W. Carlton, L. H. Wickstrom, P. E. Potter, B. H. Richard, P. J. Wolfe, and G. M. Sitler, 1991, Pre-Mount Simon basin under the Cincinnati Arch: Geology, v. 19, p. 139–142.

Van Schmus, W. R., M. E. Bickford, and A. Turek, 1996, Proterozoic geology of the east-central midcontinent basement, *in* B. A. van der Pluijm and P. A. Catacosinos, eds., Basement and basins of eastern North America: Boulder, Colorado, Geological Society of America, Special Paper 308, p. 7–32.

Wheeler, R. L., 1997, Boundary separating the seismically active Reelfoot Rift from the sparsely seismic Rough Creek Graben, Kentucky and Illinois: Seismological Research Letters, v. 68, no. 4, p. 586–598.

7 Cambrian and Ordovician Systems
(Sauk Sequence and Tippecanoe I Subsequence)
Dennis R. Kolata

INTRODUCTION

The rocks of the Cambrian and Ordovician Systems underlie most of Illinois and include the entire Sauk sequence and part of the overlying Tippecanoe sequence. These rocks are an important source of material to make cement, construction aggregate, glass sand, and agricultural lime. Locally, they contain oil and gas, and their potential for bearing hydrocarbons is considered to be promising. In the northern part of the state, these rocks form aquifers that supply a major amount of the groundwater in the region. Therefore, an understanding of the thickness and distribution of these rocks and their chemical and physical characteristics is critical to the well-being of Illinois citizens.

The Sauk sequence in Illinois consists of Cambrian and Lower Ordovician rocks deposited approximately 543 to 490 million years ago (mya). The sequence is named for Sauk County in southwestern Wisconsin, where it rests on the Precambrian Baraboo Quartzite and lies unconformably beneath the Middle Ordovician St. Peter Sandstone (Sloss et al. 1949). The Sauk sequence (Figures 7-1 and 7-2) occurs in the subsurface throughout Illinois, and the upper part is exposed at the bedrock surface locally in north-central Illinois. The Sauk constitutes at least half of the total sedimentary cover in the state and consists primarily of sandstone and dolomite with lesser amounts of limestone, shale, siltstone, and arkose.

The Sauk sequence is formally subdivided into three subsequences (Sloss 1988): (1) Sauk I, from latest Precambrian to Early Cambrian, includes the pre-Mt. Simon Sandstone layered sequences and parts of the sedimentary fill in the New Madrid Rift System; (2) Sauk II, from Middle through early Late Cambrian, includes the Mt. Simon Sandstone and Eau Claire and Bonneterre (southern Illinois) Formations; and (3) Sauk III, from mid-Late Cambrian through Early Ordovician, includes all sedimentary rocks above the Eau Claire and Bonneterre Formations and below the sub-Tippecanoe unconformity.

Overlying the Sauk is the Tippecanoe sequence, which consists of Middle Ordovician through Lower Devonian rocks. The sequence is named for Tippecanoe County, northwestern Indiana, where it rests on the Lower Ordovician Knox Group and is overlain by the Middle Devonian Muscatatuck Group. The Tippecanoe sequence includes the Middle and Upper Ordovician Tippecanoe I subsequence and the Silurian through earliest Devonian Tippecanoe II subsequence (Sloss 1988). The Tippecanoe I subsequence is marked by prominent unconformities at top and bottom and is the most widely distributed subsequence in the North American craton. The subsequence is present in the subsurface throughout the state and is exposed in numerous quarries, road cuts, and natural exposures in north-central Illinois and along the bluffs of the Mississippi River and its tributaries in west-central and southwestern Illinois. The Tippecanoe I subsequence is a major source of dolomite aggregate where it occurs at the bedrock surface, and, in some urbanized parts of northeastern Illinois, the subsequence is mined underground.

Additional details regarding Sauk and Tippecanoe I subsequence stratigraphy can be found in publications by Willman and Buschbach (1975), Templeton and Willman (1963), Willman and Kolata (1978), Kolata and Graese (1983), and Buschbach (1964). The Tippecanoe II subsequence is discussed in Chapter 8, Silurian and Lower Devonian.

GEOLOGICAL HISTORY

During late Precambrian time, most of the continents were assembled in a single supercontinent known as Rodinia that was situated over the south pole (Scotese and McKerrow 1990). Approximately 800 million years ago, the supercontinent began to break into smaller tectonic plates. One of these plates was Laurentia, which consisted largely of the present-day North American continent and Greenland. By about 525 million years ago, Laurentia was mainly surrounded by newly formed ocean basins (Rankin et al. 1989, Thomas 1989). From 800 million to 525 million years ago, Laurentia apparently had drifted from a position deep in the southern hemisphere to the equator, situated along a line extending roughly from Hudson Bay to the Gulf of Mexico (Scotese and McKerrow 1990) (Figure 7-3). If this interpretation of paleogeography is correct, the climate in Illinois during Sauk deposition can be assumed to have been tropical to subtropical.

Unlike later times during the Paleozoic Era, there were no collisional tectonics in North America during most of Cambrian time. Consequently, there were no mountain belts or volcanic arcs. The continent was encircled by flat

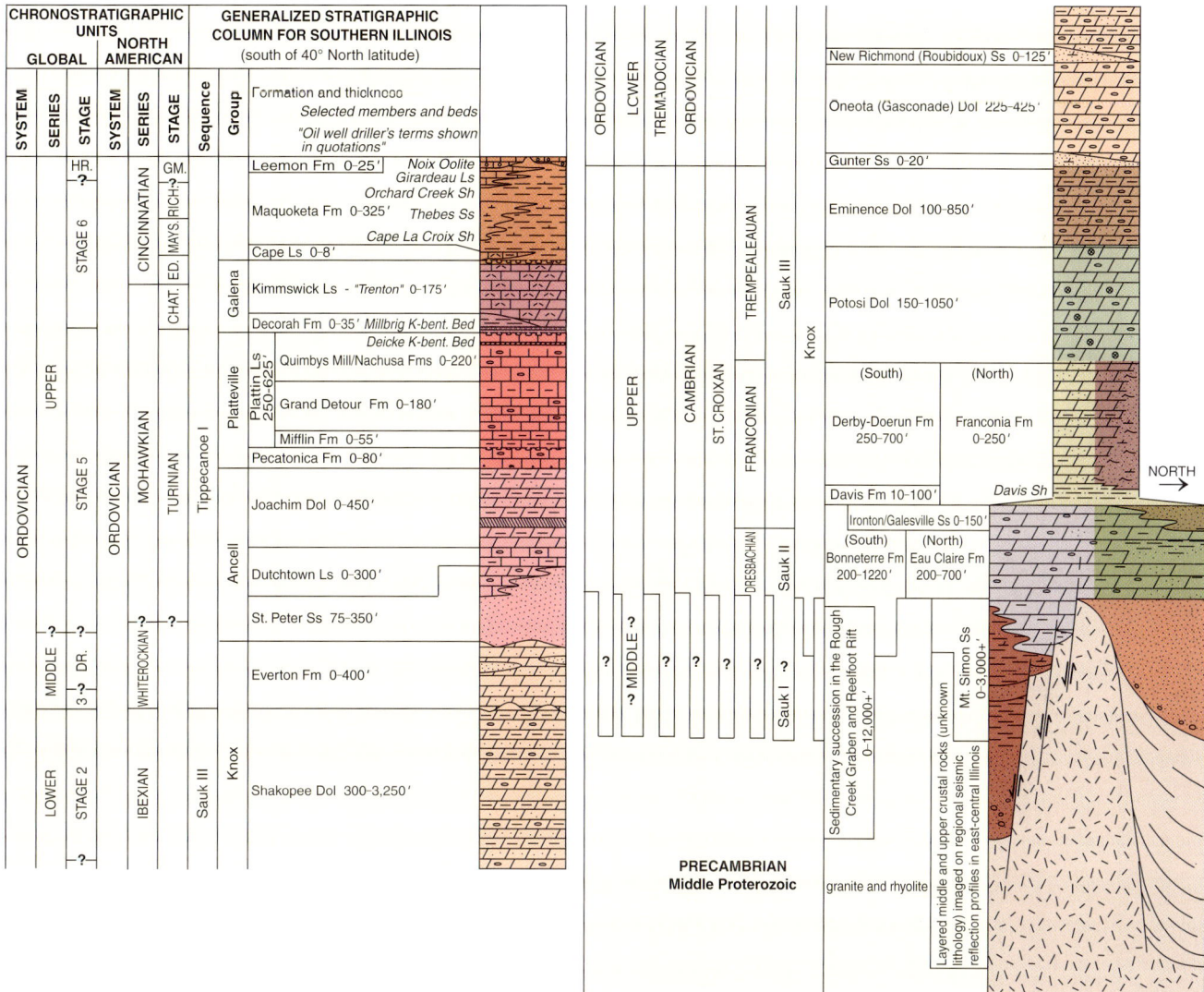

Figure 7-1 Generalized stratigraphic column of the Sauk sequence and Tippecanoe I subsequence in southern Illinois, south of 40° N latitude (Kolata 2005). Abbreviations: CHAT., Chatfieldian; DR., Darriwilian; ED., Edenian; GM., Gamachian; HR., Hirnantian; K-bent., K-bentonite; MAYS., Maysvillian; RICH., Richmondian; Dol, Dolomite; Fm, Formation; Ls, Limestone; Mbr, Member; Sh, Shale; Ss, Sandstone. Lithologic symbols are explained in Appendix I.

coastlines and received uniform bands of similar sediments (Palmer 1960). Three broad, shoreline facies belts (Figure 7-4) developed on the North American continent during the deposition of the Sauk sequence: (1) an inner detrital belt of sediment situated adjacent to exposed parts of the Canadian Shield, (2) a middle carbonate belt, and (3) an outer siliciclastic belt along the margins of the continent (Palmer 1960). When sea level rose and the seas progressively onlapped the continent, the carbonate facies belt broadened, and the inner detrital belt shifted shoreward. When sea level dropped, the inner detrital belt shifted seaward. During Cambrian and Early Ordovician times, the many episodes of rising and falling seas caused the facies to shift laterally across Illinois many times. As a result, elements of the inner detrital and carbonate belts are found in the vertical succession of geological units in Illinois. Another consequence of these migrating facies belts is that the overall character of the Sauk sequence changes from north to south in Illinois. Above the basal Mt. Simon Sandstone, the sequence is dominantly dolomite and limestone in the southern part of Illinois and shale, siltstone, sandstone, and dolomite in the northern part.

Two notable structural features developed in response to plate tectonic interactions in Illinois during Precambrian to Late Cambrian time that influenced the Sauk sequence deposition. The first structural feature is the Illinois Basin, which formed in east-central Illinois and west-central Indiana, apparently over a pre-existing caldera complex and/or rift (McBride et al. 2003). The Illinois Basin is the center of the thickest deposits of Mt. Simon Sandstone in the midcontinental United States. The second structural feature is the New Madrid Rift System (includes the Reelfoot Rift

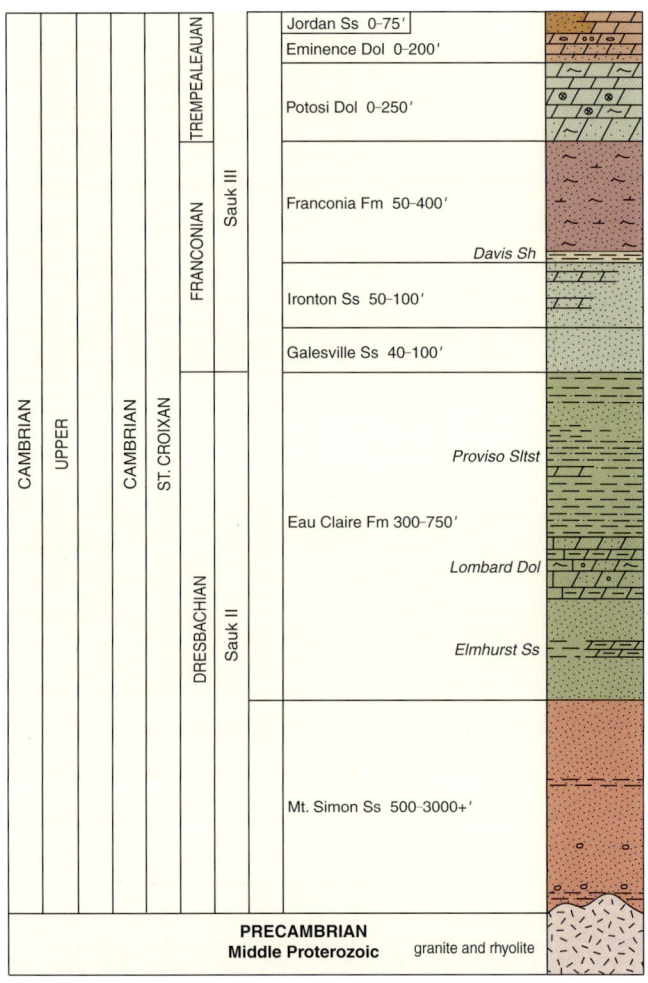

Figure 7-2 Generalized stratigraphic column of the Sauk Sequence and Tippecanoe I subsequence in northern Illinois, north of 40° N latitude (Kolata, 2005). Abbreviations: DR., Darriwilian; K-bent., K-bentonite; MAYS., Maysvillian; WH., Whiterockian; Dol, Dolomite; Fm, Formation; Ls, Limestone; Mbr, Member; Sh, Shale; Sltst, Siltstone; Ss, Sandstone. Lithologic symbols are explained in Appendix I.

and Rough Creek Graben), one of several failed rifts that formed along the margins of the North American craton as the late Precambrian supercontinent broke apart. For more details about these structures, see Chapter 4, Structural Features.

During the more than 70 million years in which the Sauk sequence was deposited, life evolved from relatively simple single-celled organisms to a diverse array of multicellular forms. The soft-bodied metazoans, known as the Ediacaran fossils, which dominated the late Precambrian seas, were succeeded by locally abundant and widespread skeletonized organisms beginning in Early Cambrian time. By the end of the Cambrian, nearly all animal phyla had evolved. Unfortunately, because the environments were not conducive to fossil preservation and because of a lack of present-day outcrops, fossils of the Sauk sequence in Illinois are not particularly abundant or well-known. The most common fossils of this time include inarticulate brachiopods, trilobites, hyolithids, and stromatolites, all of which are indicative of a marine environment. Trilobite fossils

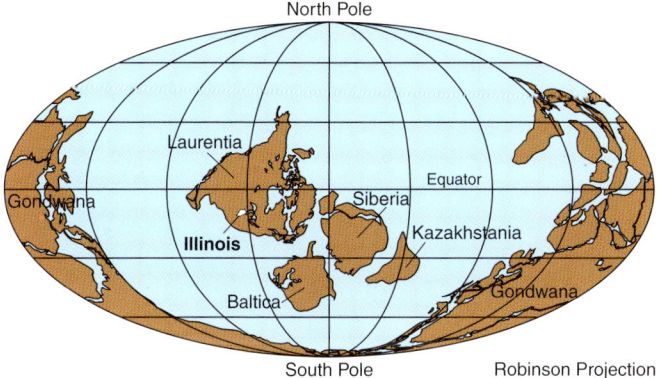

Figure 7-3 Late Precambrian paleogeography (Scotese and McKerrow 1990). Laurentia included the Canadian Shield and the United States, Greenland, Mexico, Yucatan, northern Ireland, and northern Scotland. Baltica consisted of Scandinavia, Poland, and part of Russia. The core continents of Gondwana included South America, Africa, Madagascar, India, Antarctica, and Australia. (Image used with the permission of the Geological Society of London.)

are a particularly important key to understanding the relative ages of the rocks. An inarticulate brachiopod and a hyolithid recovered from drill core in Grundy County are shown in Figure 7-5.

The North American continent remained in an equatorial position during deposition of the Tippecanoe I subsequence. The most significant change from the Sauk to the Tippecanoe was the convergence of the North American plate with several microplates, resulting in uplift of the eastern (present-day coordinates) continental margin and the formation of mountains that shed large quantities of detrital sediments. A wedge of sediment slowly expanded westward into Illinois during the Late Ordovician. The distribution of volcanic ash observed in beds such as the Middle Ordovician Millbrig and Deicke K-bentonite Beds suggests that the ash originated from supervolcanic eruptions in an arc situated east of South Carolina (Kolata et al. 1996, 1998).

During the Late Ordovician, sea level rose, and more of the North American continent was submerged than at any other time during the Paleozoic Era. The proto-Illinois Basin, centered over the New Madrid Rift System, continued to subside more rapidly than surrounding regions. A broad southwest-plunging trough centered over the rift remained the Basin's center of deposition (depocenter) during deposition of the Tippecanoe I subsequence.

The Ordovician Period was a time of major faunal diversification. This episode in the history of life was unique in its taxonomic, ecological, and biogeographical aspects. Many of the invertebrates that evolved at this time dominated the marine realm for the next 250 million years. The fossil record is exceptionally well preserved in the Middle and Upper Ordovician Tippecanoe I subsequence of Illinois and surrounding regions of the midcontinental United States.

SEQUENCE STRATIGRAPHY

Sub-Sauk Sequence Surface

Illinois and the surrounding regions underwent a long period (about a billion years) of subaerial and coastal erosion prior to the deposition of the Sauk sequence. Throughout much of the state, the approximately 520-million-year-old Upper Cambrian Mt. Simon Sandstone rests unconformably on 1.4- to 1.5-billion-year-old early and middle Proterozoic igneous rocks of the Eastern Granite-Rhyolite Province (Bickford et al. 1986, Lidiak 1996).

The sub-Sauk surface is confined to the subsurface in Illinois and has been penetrated by about 40 drill holes within the state. Consequently, little is known about its configuration. Based on the small number of closely spaced drill holes in Illlinois and exposures of the unconformity in the nearby St. Francois Mountains of Missouri, as much as 500 to 1,000 feet (150 to 300 m) of local relief is likely in the sub-Sauk surface in Illinois (Sargent 1991). Seismic reflection profiles and limited drill-hole data reveal a sub-Sauk surface marked by steep-sided hills and knobs. As a result, the overlying Cambrian and Ordovician strata were unevenly deposited and compacted (Nelson 1995).

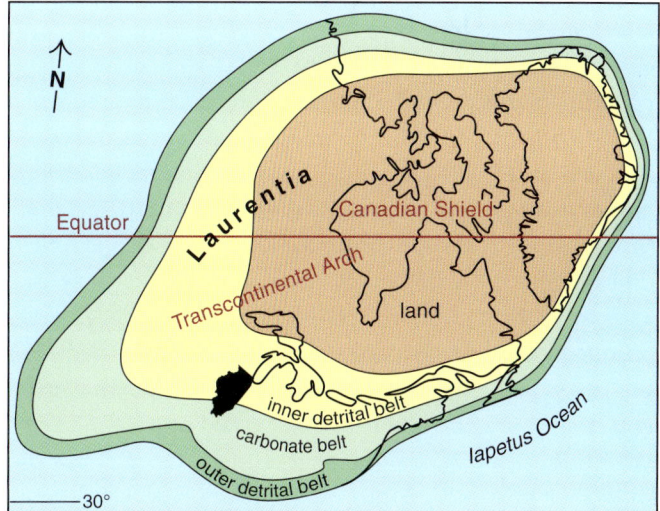

Figure 7-4 Paleogeography of Laurentia during the deposition of the Cambrian through the Early Ordovician Sauk sequence (modified from Scotese and McKerrow 1990). The continent straddled the equator and was surrounded by a passive margin with flat coastlines that received uniform bands of similar sediment. The carbonate and inner detrital facies belts episodically shifted across what is now Illinois during the rising and falling of sea level.

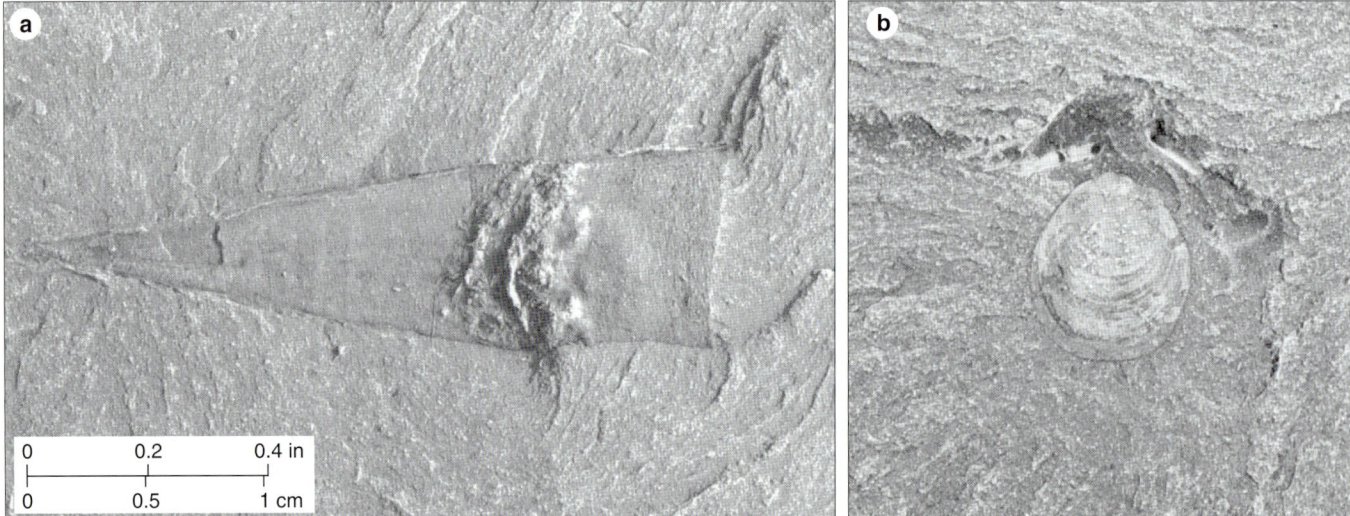

Figure 7-5 Cambrian hyolithid (a) and inarticulate brachiopod (b) fossils from the Lombard Dolomite Member of the Eau Claire Formation in Grundy County, Illinois (Section 20, T34 N, R8E). Recovered from a drill core at a depth of 1,785 feet (547 m).

Sauk Sequence Stratigraphy

Sauk I Subsequence: Pre-Mt. Simon Layered Sequences

Rocks of the Sauk I subsequence are confined for the most part to the passive margin regions of the North American craton (Sloss 1988). However, deep seismic reflection profiles surveyed in east-central Illinois and west-central Indiana reveal layered reflective rocks beneath the Sauk II Mt. Simon Sandstone (Figures 7-1 and 7-6; see also Chapter 6, Precambrian) (Pratt et al. 1989, McBride and Kolata 1999, McBride et al. 2003). The composition and age of these rocks are unknown, but their stratigraphic position and the presence of internal unconformities suggest that they may be part of the Sauk I subsequence. The uppermost unit, known as the Enterprise subsequence (not shown on generalized stratigraphic column), may be as much as 12,000 feet (3,658 m) thick and is characterized by well-defined depositional features and truncations that are similar to those observed in known seismic reflection profiles of sedimentary and volcaniclastic rocks (see Chapter 6, Precambrian, and McBride et al. 2003). Some seismic reflection clusters strongly resemble prograding wedge-shaped strata imaged on passive continental margins, suggesting that the Enterprise could be a sedimentary or volcaniclastic succession that is more recent than the 1.4-billion-year-old granite and rhyolite on which it rests. The Enterprise predates the overlying Late Cambrian Mt. Simon Sandstone, which is thought to be about 520 million years old. It is also possible that the Enterprise may represent a series of unconformity-bound sequences that are older than the Sauk I subsequence.

The Reelfoot Rift and Rough Creek Graben, structural elements of the New Madrid Rift System, also contain thick sedimentary packages that probably include Sauk I deposits (Figure 7-6). These rocks are preserved in down-dropped fault blocks that presumably formed during the late Precambrian to Early Cambrian breakup of the Rodinia supercontinent. The precise age of the rift succession in southern Illinois and adjacent regions of Kentucky and Missouri is unknown, but, based on indirect evidence, it is thought to have formed during the Early and Middle Cambrian (Sloss 1988).

Sauk II Subsequence: Middle and Late Cambrian (Dresbachian Stage)

Except for minor intervals, sea level gradually rose during deposition of the Sauk II subsequence. Basal sandy siliciclastic sediments, derived from the North American craton, progressively onlapped the continent. The presence of the diagnostic Middle Cambrian trilobite *Baltagnostus* sp., recovered from a deep drill hole in the Rough Creek Graben of western Kentucky, indicates that the initial deposition of the Sauk II sequence took place within the New Madrid Rift System (Schwalb 1982). By Dresbachian (early Late Cambrian) time, the Sauk II subsequence covered Illinois and wide areas of the midcontinental United States. A blanket of coarse arkose—the basal Mt. Simon Sandstone—covered most of the exposed and weathered granite and rhyolite terrane, except for high-standing hills and buttes. Meanwhile, erosion of exposed igneous and metamorphic terranes in central Canada continued to supply siliciclastic sediment to the midcontinent. As a result of the long transport distances and multiple cycles of erosion and redeposition in high-energy intertidal zones, the arkose was succeeded by mature, well-sorted, rounded, pure quartz sandstone—the upper Mt. Simon Sandstone. Analyses of sedimentary features in drill cores indicate that the Mt. Simon was deposited in marine environments ranging from subtidal to upper tidal flats (Sargent and Lasemi 1993). According to Palmer (1960), the Mt. Simon was deposited in the inner detrital belt (Figure 7-4).

Mt. Simon Sandstone. The Mt. Simon consists primarily of fine- to coarse-grained, partly pebbly, poorly sorted quartzose to arkosic sandstone. In some locations, the unit contains graded cross-stratified beds of coarse to very coarse sand and beds of red and green micaceous shales (Buschbach 1975). The high degree of rounding of the grains suggests that much of the sand passed through at least one and possibly several cycles of erosion, transportation, and deposition (Willman et al. 1975). The Mt. Simon underlies all of Illinois except in scattered localities where it failed to cover hills and buttes on the Precambrian surface. The Mt. Simon is part of a vast sheet of basal Cambrian sandstone that covers wide areas of the midcontinental United States. In most of Illinois, the Mt. Simon generally ranges from 500 to 2,000 feet (153 to 610 m) thick, but, in east-central Illinois, data from seismic reflection profiles suggest that the Mt. Simon may be thicker than 3,000 feet (914 m) (Figure 7-1). The Mt. Simon correlates with the Lamotte Sandstone of Missouri (Workman and Bell 1948). The contact with the overlying Eau Claire and Bonneterre Formations is conformable.

Throughout Illinois, the Mt. Simon Sandstone lacks fossils that can be used for age assignment, but the upper part is considered to be Late Cambrian in age on the basis of stratigraphic position and physical tracing to the type section near Eau Claire. The overlying Eau Claire and Bonneterre Formations contain Late Cambrian fossils. In the Illinois Basin, particularly in east-central Illinois where it is unusually thick, the lower part of the Mt. Simon is probably older than Late Cambrian.

In the northern part of the state, the upper part of the Mt. Simon is an important aquifer. Lower in the formation, the water contains more dissolved solids and generally is not suitable for drinking south of Cook County. Locally in central Illinois, sandstone reservoirs within the Mt. Simon are used for the temporary underground storage of natural gas (Buschbach and Bond 1974).

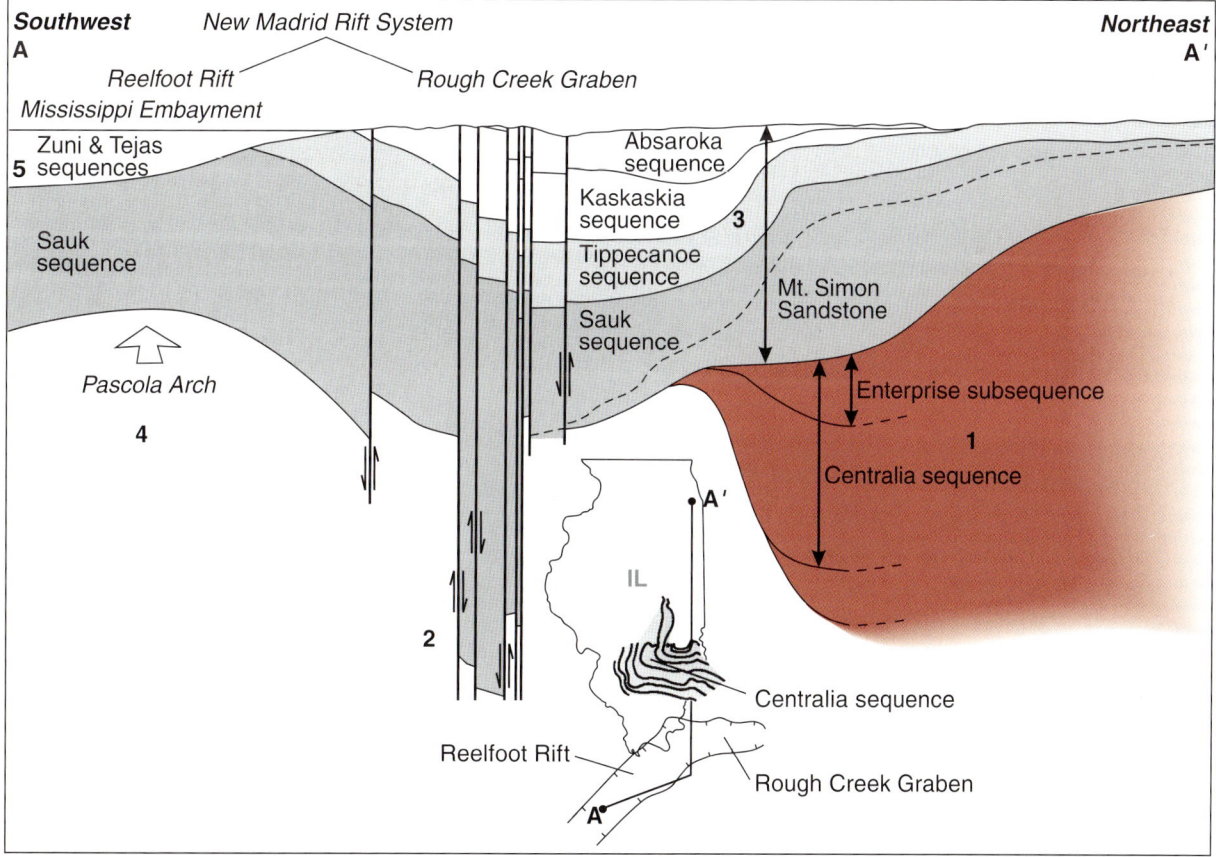

Figure 7-6 Diagrammatic southwest-northeast cross section of Illinois and adjacent parts of Indiana, Kentucky, and Missouri showing the primary tectono-stratigraphic successions that occur in Illinois. These include (1) the late Precambrian Centralia sequence, (2) late Precambrian through Middle Cambrian New Madrid Rift System succession, (3) Paleozoic Illinois Basin stratigraphic succession, (4) Mesozoic uplift and erosion of the Pascola Arch, and (5) subsidence and deposition within the Mississippi Embayment.

Eau Claire and Bonneterre Formations. In northern Illinois, the Mt. Simon Sandstone is overlain by the Eau Claire Formation, which consists mainly of dolomitic siltstones and sandstones with interbedded dolomite and shale. The formation is particularly well exposed near Eau Claire, Wisconsin, from which the name is derived. Southward, the Eau Claire becomes increasingly more dolomitic and limey, grading to dominantly dolomite and limestone in the southern part of Illinois where it is commonly referred to as the Bonneterre Formation, named for the town of Bonneterre in southeastern Missouri where the formation is well exposed. The Bonneterre carbonate rocks are assigned to the Knox Group (Figure 7-1). The Eau Claire and Bonneterre Formations underlie virtually all of Illinois but are not exposed within the state. The formations range in thickness from about 200 feet (61 m) in western Illinois (Buschbach 1975) to more than 1,200 feet (370 m) in southernmost Illinois (Sargent 1991) (Figures 7-1 and 7-2). In contrast to the underlying Mt. Simon, which is thickest in east-central Illinois, the Eau Claire and Bonneterre thicken toward their depocenter in western Kentucky. The formations grade abruptly from oolitic limestone and dolomite immediately north of the New Madrid Rift System to shale with interbeds of oolitic carbonates within the rift (Palmer 1989). Locally in the southern part of Illinois, where the Mt. Simon Sandstone was not thick enough to cover the tops of buttes and hills on the Precambrian surface, the Bonneterre rests directly on Precambrian rocks. The Eau Claire in northeastern Iowa contains *Cedaria, Crepicephalus, Aphelaspis,* and *Dunderbergia* Zone trilobite faunas (McKay (1988). The *Crepicephalus* and *Aphelaspis* Zones are also present in the Bonneterre Formation in eastern Missouri (Thompson 1995), confirming that those rocks are Dresbachian in age. Hyolithids and inarticulate brachiopods are locally common in northern Illinois drill cores (Figure 7-5).

In Illinois, faunal zones have not been differentiated, and the Sauk II and III contact is tentatively placed at the top of the Eau Claire Formation. In Wisconsin and in the northern parts of Iowa and Illinois, the contact is disconformable and sharp (McKay 1988, Sargent 1991). Local relief on the surface of the Eau Claire is up to 20 feet (6.1 m) in northeastern Iowa, but the magnitude of the disconformity appears to decrease southward between the type area of west-central Wisconsin and Iowa (McKay 1988). Palmer (1981) reported that the disconformity likely formed during a continent-wide but short-lived lowering of worldwide sea level. In response, the inner clastic belt shifted southward, depositing the quartz sands that became the Galesville Sandstone, Ironton Sandstone, and Franconia Formation across northern Illinois. Presumably, the amount of erosion at the top of the Eau Claire was greater shoreward in Wisconsin.

Sauk III Subsequence: Late Cambrian (Franconian and Trempealeauan Stages)

The renewed transgression of the seas during the Late Cambrian resulted in the deposition of mature quartz sandstones across the northern part of the interior of the North American craton, followed by widespread deposition of carbonate rocks. The Ironton and Galesville Sandstones and the Franconia Formation are assigned to the Franconian Stage; the Potosi Dolomite, Eminence Dolomite, and Jordan Sandstone are assigned to the Trempealeauan Stage of the St. Croixan Series (Figures 7-1 and 7-2). These assignments are based primarily on faunal studies in Wisconsin and Minnesota, but are provisional assignments in Illinois where diagnostic fossils are lacking.

Ironton and Galesville Sandstones. The Ironton and Galesville Sandstones are named for towns in southern Wisconsin where they are well exposed. In those areas, the coarser-grained Ironton overlies the finer-grained Galesville. The Ironton and Galesville Sandstones constitute the most productive bedrock aquifer in northern Illinois. The Ironton is locally fossiliferous and contains the *Elvinia* Zone trilobite fauna, marking the base of the Franconian Stage. The two formations extend southward into the subsurface of northern Illinois but are not easily distinguished in drilling samples. They are combined as a single formation in most subsurface studies in Illinois and range in total thickness from 150 to 200 feet (46 to 61 m) in northern Illinois. In central Illinois, the sandstones grade to dolomite that is assigned to the upper part of the Eau Claire and Bonneterre Formation and to the Knox Group (Buschbach 1975).

Franconia and Derby-Doerun Formations. This stratigraphic package consists of a northern siliciclastic- and glauconite-rich facies known as the Franconia Formation and a southern relatively pure dolomite facies, the Derby-Doerun Formation. The Franconia is named for exposures in southeastern Minnesota, and the Derby-Doerun is named for outcrops in southeastern Missouri. In the lower part of the Franconia and Derby-Doerun Formations is a widespread and persistent shaly unit called the Davis Shale Member. The Franconia in northern Illinois consists of fine-grained, glauconitic, argillaceous, dolomitic sandstone with interbeds of red and green shale (Buschbach 1975). The Franconia ranges in thickness from 50 feet (15 m) in northernmost Illinois to 400 feet (120 m) in west-central Illinois. The Franconia sands and glauconite grade southward in central Illinois to slightly glauconitic dolomite of the Derby-Doerun Formation. As in the underlying carbonate rocks in southern Illinois, the Derby-Doerun carbonate facies is assigned to the Knox Group. Franconia and Derby-Doerun equivalents probably reach thicknesses of about 700 feet (210 m) in southernmost Illinois based on the general pattern of thickening and lateral transition into carbonate-dominated facies observed in other formations in the Knox Group (Sargent 1991). The Franconia/Derby-Doerun is conformable with the underlying Ironton-Galesville and the overlying Potosi Dolomite. Franconia Formation rocks are the oldest rocks exposed in Illinois and are present in outcrops near Ashton in Lee County and Oregon in Ogle County. Fragments of the trilobites *Dikelocephalus, Illaenurus,* and *Saukiella* have been identified in the Franconia Formation of northern Illinois (Willman and Templeton 1951).

Potosi Dolomite. By Trempealeauan time, the middle carbonate facies belt shifted northward again, and the Potosi and Eminence Dolomites blanketed most of Illinois, except where covered by the Jordan Sandstone, a facies of the Eminence confined to the northwesternmost part of the state. The Potosi Dolomite, named for a town in east-central Missouri where it is well exposed, is relatively pure. It overlies the glauconitic sandstone of the Franconia Formation in northern Illinois and, in southern Illinois, is nearly indistinguishable from the underlying Derby-Doerun dolomite. Pre-Tippecanoe erosion locally removed the Potosi in parts of northeastern Illinois (Buschbach 1975). The Potosi Dolomite consists of fine-grained dolomite containing traces of fine glauconite grains near the top and bottom. Small cavities lined with drusy quartz are a diagnostic feature of the dolomite in outcrop and subsurface samples. The Potosi ranges in thickness from 100 feet (30 m) or less in northern Illinois to more than 1,000 feet (300 m) in the southern part of the state. This dolomite is exposed in a few quarries in Ogle and Lee Counties (Willman and Templeton 1951) (Figures 7-7 and 7-8). Willman and Templeton (1951) noted the occurrence of cryptozoon algal structures and the trilobite *Saukiella* in several outcrops in north-central Illinois.

Eminence Dolomite. The Eminence, named for a locality in southeastern Missouri, is the uppermost unit of the Cambrian System in most of Illinois. The Eminence is a sandy, fine- to medium-grained dolomite that contains oolitic chert and thin beds of sandstone. A thin sandstone is present locally at the base of the formation. The Eminence conformably overlies the relatively pure Potosi Dolomite. In northwestern Illinois, the Eminence grades laterally into the Jordan Sandstone, which extends north and west through Iowa, Wisconsin, and Minnesota.

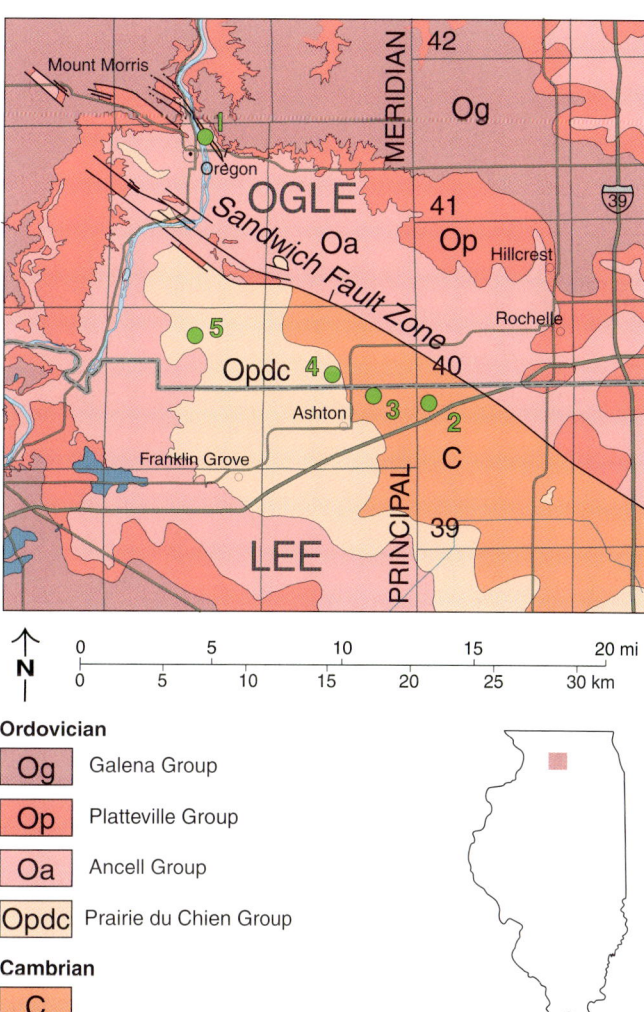

Figure 7-7 Bedrock geology in parts of eastern Ogle and Lee Counties showing the Sandwich Fault Zone and associated Ashton Anticline (Kolata 2005). Numbers refer to key outcrops of Cambrian and Lower Ordovician rocks: **1** and **2**, Franconia Formation; **3**, Potosi Dolomite; **4** and **5**, Oneota Dolomite.

Sauk III Subsequence: Early Ordovician (Ibexian Series)

The transition from the Cambrian to the Ordovician in Illinois was uneventful, except for a minor drop in sea level that interrupted deposition in northern and western Illinois. Carbonates dominated the remainder of the Sauk III deposition in northern Illinois, except for the basal Ordovician Gunter Sandstone and the New Richmond (Roubidoux) Sandstone. The middle carbonate belt persisted in the southern part of the Illinois Basin, and sedimentation continued with little or no apparent interruption during the deposition of the post-Mt. Simon Sauk II subsequence and all of the Sauk III subsequence.

Except for the regressive episodes that resulted in the deposition of the Gunter and New Richmond Sandstones, Early Ordovician Sauk sedimentation in Illinois continued

Figure 7-8 Fifteen-foot (4.6-m)-thick exposure of Potosi Dolomite near Ashton, Lee County, Illinois (NE¼ NE¼ NE¼ Section 23, T22N, R11E). Location 3 in Figure 7-7. (Photograph by Dennis R. Kolata.)

to be dominated by carbonate deposition. Sauk deposition ended during the Early Ordovician with the retreat of the sea, followed by widespread exposure and significant erosion of the North American interior.

Gunter Sandstone. The Gunter Sandstone, the oldest Ordovician rock unit in Illinois, is named for a locality in central Missouri. The unit consists mainly of fine- to medium-grained, rounded quartz sand that is widespread but discontinuous in Illinois. Locally, beneath parts of La Salle and McLean Counties in north-central Illinois, the Gunter is up to 25 feet (7 m) thick (Buschbach 1975). Unconformities are suspected at the top and bottom of this unit based on its sharp contacts and patchy distribution. The Gunter is exposed in a small number of outcrops and quarries in parts of Lee and Ogle Counties (Willman and Templeton 1951). Although regarded as a formation by Willman and Buschbach (1975), the relative thinness and patchy distribution of the Gunter argue for assignment as the basal member of the overlying Oneota Dolomite.

Oneota (Gasconade) Dolomite. The Oneota (Gasconade) Dolomite is named for the Oneota River in northeastern Iowa, where the formation is well exposed. The Oneota consists of relatively pure, medium- to coarse-grained, cherty dolomite that is massively bedded and contains minor amounts of sand. The chert is generally white and chalky, can be oolitic and sandy, and occurs as beds, lenses, nodules, and irregularly shaped bodies with distinctive branching shapes (Buschbach 1975). One of the most distinguishing characteristics of the Oneota is the coarseness of the dolomite grains. The Oneota ranges in thickness from less than 100 feet (30 m) in northern Illinois to more than 400 feet (122 m) in the southern part of the state, where its boundaries are difficult to distinguish in the subsurface. The Oneota is exposed in several quarries near the Ashton Anticline in parts of Ogle and Lee Counties (Figures 7-7 and 7-9) where algal domes and a sparse fauna consisting largely of gastropods have been reported (Willman and Templeton 1951). The Oneota is equivalent to the Gasconade Dolomite, which is exposed in southeastern Missouri (Buschbach 1975, Sargent 1991).

New Richmond (Roubidoux) Sandstone. The New Richmond (Roubidoux) Sandstone is named for a town in west-central Wisconsin, where the formation is well exposed. In Illinois, it consists of relatively pure, fine- to medium-grained, rounded, somewhat well-sorted quartz sand that is locally cross-bedded. The New Richmond also contains a moderate amount of sandy, fine-grained dolomite that contains oolitic chert (Buschbach 1975). The New Richmond is generally about 50 feet (15 m) thick in northern Illinois but 100 to 125 feet (30 to 38 m) thick in a linear belt extending through the subsurface from La Salle County through Jersey County. The formation terminates abruptly on the southeast side of this linear belt. The New Richmond rests unconformably on the Oneota and is gradational with the overlying Shakopee Dolomite. The New Richmond is well exposed in natural outcrops along Franklin Creek in Lee County and in the bluffs of the Fox River in La Salle County (Willman and Templeton 1951) (Figure 7-10).

Shakopee Dolomite. The Shakopee Dolomite, named for a town in southeastern Minnesota, consists of pure to argillaceous, fine- to medium-grained dolomite that is interbedded with lesser amounts of medium-grained sandstone, siltstone, and shale. The formation contains nodules and lenses of chert that locally are sandy, oolitic, or both. Depositional structures are locally abundant and include cross-bedded sandstone and oolites, laminated dolomite,

Cambrian and Ordovician

Sub-Tippecanoe Surface

A long period of erosion, including late Ibexian and much of Whiterockian time, separates the Sauk from the Tippecanoe sequence over much of the central midcontinent (Witzke 1980). The amount of time represented by the unconformity is much greater in northern Illinois where, locally, rocks as old as the Cambrian Franconia Formation are overlain by the basal unit of the Tippecanoe sequence, the St. Peter Sandstone of presumed Ordovician Turinian age (Figure 7-12). Farther south, in the Illinois Basin, the basal St. Peter Sandstone is late Whiterockian in age and rests either on the early Whiterockian Everton Formation or on Knox carbonate rocks of Ibexian age (Shaw 1999). Detailed study of subsurface data from closely spaced drill holes in east-central Indiana reveals a surface of locally occurring steep-sided, ovate, butte-like, erosional remnants with vertical relief of about 150 feet (46 m) (Keller and Abdulkareem 1980). Data from wells in northern Illinois just a few hundred feet apart show relief of 200 feet (61 m) (Buschbach 1964). The unconformity is exposed at some excavations in La Salle County in northern Illinois (Figure 7-13).

Tippecanoe I Subsequence

Tippecanoe I Subsequence Stratigraphy

The Middle and Late Ordovician Tippecanoe I subsequence was deposited during a major transgressive cycle that began with deposition of the Everton Formation followed by a succession of mixed carbonates and siliciclastics of the Ancell Group and widespread carbonate rocks of the Platteville and Galena Groups (Figures 7-1 and 7-2). Siliciclastic rocks of the Maquoketa Group, derived in large part from the Taconic Orogeny, make up the final phase of Tippecanoe I deposition.

Tippecanoe I: Middle Ordovician (Whiterockian and Mohawkian Series)

The initial deposition of the Tippecanoe I subsequence was restricted to the continental margin and the Illinois Basin depocenter, centered over the New Madrid Rift System. Sandy carbonate rocks of the Whiterockian Everton Formation were deposited on the Sauk sequence within the Basin depocenter and gradually spread northward on the more deeply eroded Sauk surface.

Everton Formation. The Everton Formation, named for Everton, Arkansas, is confined to the southern part of the Illinois Basin. Early Whiterockian in age (Shaw 1999), the Everton consists of a lower fine-grained quartz sand-

Figure 7-9 Thirty-foot (9.1-m)-thick exposure of Oneota Dolomite near Ashton, Lee County, Illinois (SE¼ SE¼ SW¼ Section 17, T22N, R11E). Location 4 in Figure 7-7. (Photograph by Dennis R. Kolata.)

mud cracks, and intraformational conglomerate. In the Shakopee of Illinois, fossils are neither diverse nor abundant. They typically consist of gastropods, cephalopods, and algal domes and mats. In much of the northern two tiers of Illinois counties, the Shakopee Dolomite, New Richmond Sandstone, and Oneota Dolomite were removed by erosion during Early Ordovician time prior to the deposition of the Tippecanoe sequence (Buschbach 1975). The Shakopee thickens southward from its erosional feather edge in northern Illinois to about 500 feet (152 m) in central Illinois. Farther south, where the underlying New Richmond Sandstone pinches out, the Shakopee is not readily distinguished from the underlying carbonate rocks of the Knox Group. The overall thickening trend suggests that Shakopee equivalent rocks may be more than 3,000 feet (910 m) in southernmost Illinois. The Shakopee is well exposed in quarries and natural outcrops in La Salle, Kendall, Lee, and Ogle Counties (Figure 7-11).

stone and an upper sandy dolomite that locally contains interbeds of sandstone, shale, anhydrite, and limestone. In southernmost Illinois and adjacent parts of Missouri, Indiana, and Kentucky, the Everton is over 400 feet (120 m) thick but thins northward to about 39° N latitude where it pinches out (Collinson et al. 1988). The dramatic basinward thickening of the Everton and the lack of a well-developed physical break at the base suggest that the Everton may be conformable with the underlying Sauk sequence in the Basin depocenter of southernmost Illinois and western Kentucky (Rexroad et al. 1982, Norby et al. 1986, Shaw et al. 1988, Norby 1991, Shaw 1999). The magnitude of the break diminishes basinward (Norby et al. 1986, Shaw et al. 1988). Deposition appears to have been continuous within the New Madrid Rift System well into Whiterockian time (Sargent 1991). An unconformity marks the top of the Everton; however, conodont biostratigraphic data from southwestern Indiana suggest that only a slight time gap, or no time gap, occurred between the deposition of the St. Peter Sandstone and the underlying Everton Formation (Rexroad et al. 1982, Norby 1991) in that region. The Everton is not exposed in Illinois, but is well exposed in nearby eastern Missouri.

Ancell Group. The Ancell Group, named for outcrops near Ancell in southeastern Missouri, lies atop the Everton Formation in southern Illinois and Early Ordovician or Late Cambrian rocks in northern Illinois. The St. Peter Sandstone is a blanket sandstone that covers wide areas of the midcontinental United States. A relatively thick, 80-mile (129-km)-wide, linear belt of the St. Peter extends from Chicago, Illinois, to Kansas City, Missouri (Nunn 1986), separating relatively pure carbonates of the Dutchtown Limestone and Joachim Dolomite on the south side from sandstone, siltstone, shale, and argillaceous dolomite of the Glenwood Formation on the north (Figure 7-14).

The St. Peter Sandstone is a supermature quartz sand (Figure 7-15) that is well-sorted, pure (more than 99%) quartz and is characterized by large- and small-scale trough cross-strata as well as tabular to concave-upward sets of cross-strata. The St. Peter is most commonly between 100 (30.5 m) and 200 feet (61 m) thick but in places can range from a few feet to more than 600 feet (180 m) (Willman and

Figure 7-10 New Richmond Sandstone in the west bluff of the Fox River viewed from a bridge on County Route 22 at Sheridan, La Salle County, Illinois (SE¼ SE¼ SE¼ Section 6, T35N, R5E). (Photograph by Dennis R. Kolata.)

CAMBRIAN AND ORDOVICIAN

Figure 7-11 Thirty-five-foot (10.1-m)-thick exposure of Shakopee Dolomite along the former Chicago, Rock Island, and Pacific Railroad west of Utica, La Salle County, Illinois (SE¼ SE¼ SE¼ Section 7, T33N, R2E). (Photograph by Dennis R. Kolata.)

Figure 7-12 Geology of the sub-Tippecanoe surface (Willman et al. 1967). Throughout most of Illinois, the Tippecanoe sequence rests on the Shakopee Dolomite or New Richmond Sandstone. In northernmost Illinois, the Tippecanoe rests on rocks as old as the Franconia Formation.

Buschbach 1975). The St. Peter is diachronous; that is, the age of the St. Peter Sandstone is different in different places. It is late Whiterockian in age in southern Illinois and early Turinian in northern Illinois (Shaw 1999). At some localities, the Kress Member occurs at the base of the St. Peter, filling depressions on the Tippecanoe I/Sauk III unconformity. The Kress Member consists of a coarse conglomerate of reworked Sauk chert and residuum of shale and sandstone (Figure 7-13). The Tonti Sandstone Member forms the major part of the widespread blanket sands that extend over most of the midcontinent. The succeeding Starved Rock Sandstone Member is confined to the thick linear belt of sandstone and is thought to have formed as a barrier island complex (Fraser 1976, Nunn 1986) (Figure 7-14).

By early Mohawkian (Turinian Stage) time, the Tonti Sandstone Member had progressively onlapped the irregular sub-Tippecanoe surface throughout the midcontinent, including Illinois. The pure quartz sands of the St. Peter were deposited in an advancing shoreline dominated by eolian dune and beach processes.

Because of its purity, the St. Peter Sandstone is mined for glass making and other uses, most recently at Oregon in Ogle County and at Ottawa in La Salle County. The St. Peter Sandstone is also a locally productive aquifer in northern Illinois. The formation is well exposed in north-central Illinois, particularly at Starved Rock, Matthiessen, and Buffalo Rock State Parks and on Split Rock on the

Illinois and Michigan Canal State Trail between Utica and La Salle (Figure 7-16) in La Salle County and Castle Rock State Park in Ogle County.

Figure 7-13 Unconformable contact between the Kress Member of the St. Peter Sandstone (above line) and the Shakopee Dolomite at an excavation along the Illinois River for the Starved Rock Lock and Dam near La Salle, La Salle County, Illinois (NW¼ NE¼ NW¼ Section 22, T33N, R2E). (Photograph by Joel M. Dexter.)

The Glenwood Formation lies above the Tonti Sandstone Member and to the north of and in facies relationship with the linear belt of the Starved Rock Sandstone Member of the St. Peter Sandstone (Figure 7-14). The Glenwood Formation consists of sandstone, shale, and argillaceous dolomite and typically is 25 to 75 feet (8 to 23 m) thick where present. The rocks are bioturbated but, for the most part, are unfossiliferous. The Glenwood Formation was deposited in a restricted lagoon on the north side of the barrier island complex (Starved Rock Sandstone Member) and extends back into southeastern Minnesota (Fraser 1976). The Dutchtown Limestone and overlying Joachim Dolomite lie above the Tonti Sandstone Member and to the south of and in facies relationship with the Starved Rock Sandstone Member (Willman and Buschbach 1975). The Dutchtown is as much as 300 feet (91 m) thick in southeastern Illinois and consists mainly of dark gray, organic-rich, argillaceous, locally fossiliferous (mollusks, ostracodes, and conodonts) lime mudstone. The Dutchtown is confined to the subsurface in Illinois. The formation thins north of its depocenter in western Kentucky and is absent north of

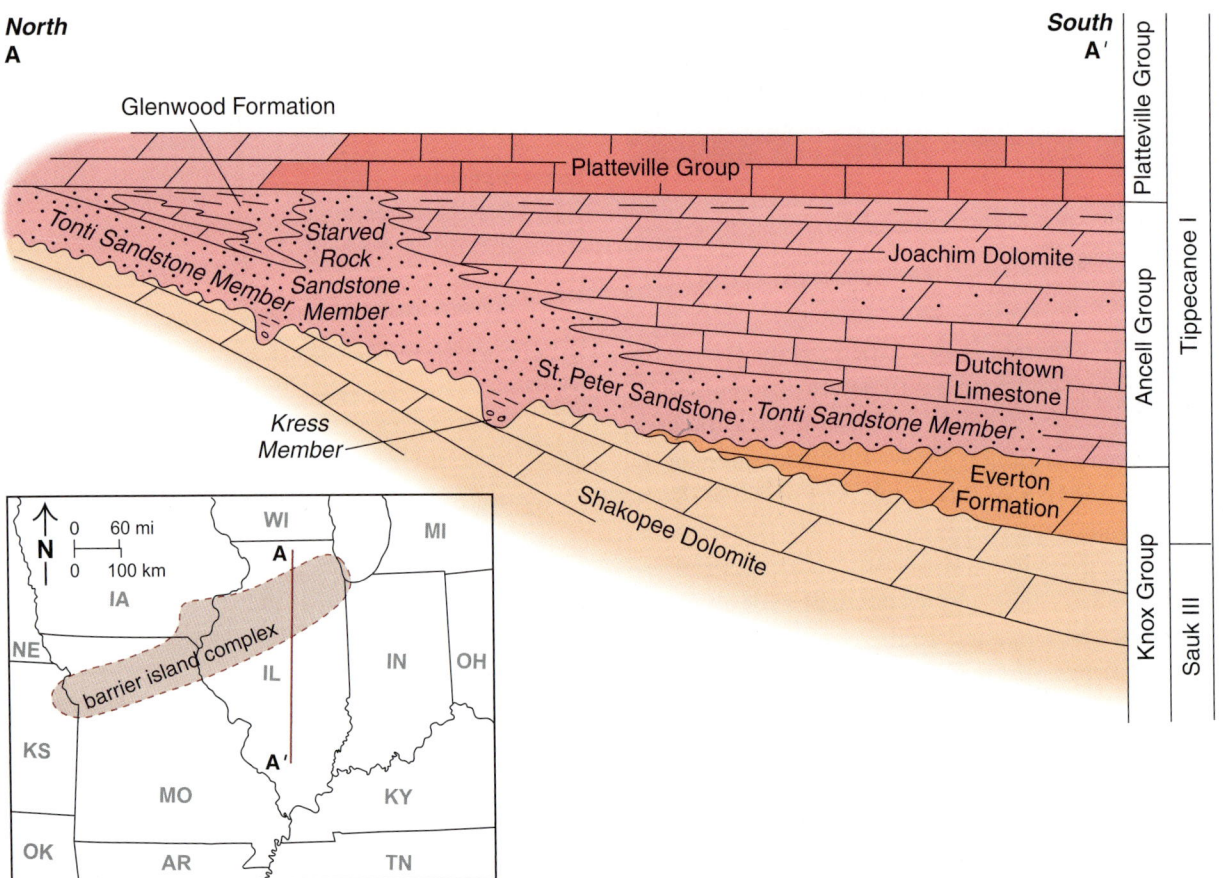

Figure 7-14 Diagrammatic north-south cross section showing facies of the Ancell Group (Templeton and Willman 1963). The St. Peter Sandstone consists of the lower, widespread Tonti Sandstone Member and overlying Starved Rock Sandstone Member, which form a barrier island complex extending from near Chicago, Illinois, to Kansas City, Missouri (Nunn 1986). The Shakopee Dolomite is assigned to the Prairie du Chien Group in northern Illinois.

Jackson County, Illinois. The Joachim Dolomite overlies the Dutchtown Limestone in southern Illinois and the St. Peter Sandstone in central Illinois. The Joachim consists of silty, sandy, and argillaceous dolomite with interbedded limestone, sandstone, and shale as well as anhydrite and gypsum (Shaw 1999). The beds are locally laminated and have mud cracks and ripple marks, indicating that deposition occurred in peritidal to shallow subtidal environments. Except for several outcrops in Calhoun County, the Joachim is confined to the subsurface in Illinois.

Like the St. Peter Sandstone, the Dutchtown Limestone and Joachim Dolomite are time-transgressive from south (oldest) to north (youngest). Based on conodont biostratigraphy, both units are assigned to the upper Whiterockian Series in the White County Superior-Ford well in southern Illinois (Shaw 1999). In central Illinois, the Joachim is thought to be early Turinian (Blackriveran).

Platteville Group. By Turinian time, during the deposition of the Tippecanoe, the advancing seas covered the barrier island complex, and a blanket of fossiliferous carbonate rocks of the Platteville Group (referred to as Black River Limestone in the subsurface of Illinois) was deposited across the midcontinent. The Platteville Group, named for the city of Platteville in southwestern Wisconsin, conformably overlies the Ancell Group in Illinois. Based on the relative amount of disseminated clay, the Platteville of Illinois is subdivided into five formations and 24 members (Templeton and Willman 1963, Willman and Kolata 1978). The abundance of lime mudstone, wackestone, ooids, and coated grains in the Platteville Group and its equivalents suggests that deposition occurred in relatively warm water. Many formations in the Platteville are remarkably uniform and continuous over hundreds of miles, indicating extensive, shallow seas, lack of a shelf-slope break, and relatively low rates of net sedimentation (Kolata et al. 2001). Open marine conditions are indicated by the abundant and diverse fauna consisting of brachiopods, bryozoans, corals, mollusks, echinoderms, trilobites, and ostracods (Sloan 1987). In northern Illinois and southern Wisconsin, echinoderms alone are represented by at least seven classes and 27 species (Kolata 1975).

Figure 7-16 Fifty-foot (15.2-m)-thick exposure of St. Peter Sandstone at Split Rock along the former Chicago, Rock Island, and Pacific Railroad west of Utica, La Salle County, Illinois (NW¼ SW¼ NE¼ Section 13, T33N, R1E). The St. Peter Sandstone dips westward from the Peru Monocline. Pennsylvanian age sandstone overlies the St. Peter along a pronounced unconformity at the top of the outcrop just to the right of the picture. Horizontal beds of Shakopee Dolomite (Figure 7-14) are exposed along the railroad between Split Rock and Utica east of this outcrop. (Photograph by Dennis R. Kolata.)

Figure 7-15 Well-rounded quartz sand grains of St. Peter Sandstone. (Photograph from the Illinois State Geological Survey collection.)

Figure 7-17 Characteristic fossils of the Ordovician Platteville and Galena Groups: **a,** trilobite *Gabriceraurus* sp.; **b,** trilobite *Thaleops ovata*; **c,** trilobite *Encrinurus* sp.; **d,** crinoid *Cupulocrinus gracilis*; **e,** colonial coral *Foerstephyllum* sp.; **f,** ostracod *Eoleperditia fabulites*; **g,** brachiopod *Dalmanella* sp.; **h,** brachiopod *Sowerbyella punctostriata*; **i,** brachiopod *Campylorthis deflecta*; **j,** brachiopod *Hesperorthis concava*; **k,** horn coral *Streptelasma* sp.; **l,** trepostome bryozoan; **m,** brachiopod *Strophomena plattinensis*; **n,** blue-green algae *Receptaculites oweni*; **o,** clam *Vanuxemia* sp.; **p,** snail *Maclurites* sp.; **q, r,** brachiopod *Opikina minnesotensis*; **s,** snail *Hormotoma major*; **t,** cephalopod *Richardsondoceras* sp.; **u,** snail *Ectomaria* sp.; **v,** snail *Phragmolites* sp.; **w,** snail *Tetranota* sp.; **x,** snail *Lophospira* sp.; **y,** snail *Clathrospira* sp. Fossils are from the Platteville Group, except g, h, k, n, and s, which are from the Galena Group. All specimens are shown approximately life size. (Photographs by Dennis R. Kolata.)

In northern Illinois, the Platteville Group is largely bluish gray to buff, very fine- to fine-grained, partly cherty, argillaceous, fossiliferous dolomite. Locally, as near Dixon in Lee County, the Platteville consists of limestone and dolomite-mottled limestone. Platteville rocks of northern Illinois are highly bioturbated and contain abundant and diverse faunas. Common fossils include brachiopods (*Öpikina, Strophomena, Hesperorthis, Campylorthis deflecta,* and *Rostricellula*), bryozoans (numerous trepostome species), mollusks (cephalopods, *Endoceras* and *Richardsondoceras;* snails, *Phragmolites, Maclurites, Ectomaria, Clathrospira, Tetranota,* and *Lophospira;* and clams, *Vanuxemia*), trilobites (*Thaleops, Gabriceraurus, Isotelus,* and *Encrinurus*), corals (*Foerstephyllum* and *Streptelasma*), and echinoderms (*Cupulocrinus* and *Calceocrinus*). Many of these fossils are shown in Figure 7-17. In southern Illinois and adjacent parts of Missouri, the Plattin (=Platteville) is not as richly fossiliferous and contains peritidal features including laminations, dessication cracks, algal mats, fenestral fabrics, and intraformational breccias. The Platteville thickens southeastward from about 50 feet (15 m) in western Illinois to about 700 feet (213 m) in southernmost Illinois and western Kentucky along a linear trend that corresponds to the late Precambrian-Early Cambrian New Madrid Rift System (Kolata et al. 2001), which also coincides with the depocenter of the underlying Tippecanoe and Sauk strata. The Deicke K-bentonite Bed, an air-fall ash bed derived from volcanic eruptions in the Taconic orogenic belt, is present in the upper part of the Platteville Group over a wide area of southern and central Illinois (Kolata et al. 1996, 1998, 2001).

The top of the Platteville is characterized by a widespread hardground surface that is pitted and locally encrusted with phosphatic grains and iron-rich minerals (Figure 7-18). There is a lack of subaerial exposure features, such as karst, suggesting that the surface formed in submarine conditions (Kolata et al. 1998). The surface marks a significant change from the relatively pure lime mud–dominated carbonate rocks of the Platteville to the shales and argillaceous wackestones and packstones of the overlying Decorah Formation and the packstones and grainstones that characterize the Kimmswick "Trenton" Limestone of the Galena Group in southern Illinois. The surface also marks a significant change in the shelly faunas (Sloan 1987).

Galena Group. The Galena Group, named for outcrops near Galena in Jo Daviess County, overlies Platteville Group rocks (Figure 7-18). The Galena and its equivalents extend far onto the Canadian Shield, recording one of the most extensive episodes of marine inundation in the history of the North American continent. Two dominant facies are recognized: (1) a lime mudstone-wackestone facies with variable amounts of disseminated clay and interbedded shale is present in northern Illinois and into Wisconsin, Iowa, and Minnesota; and (2) a relatively pure lime packstone-grainstone facies is present in the subsurface of central and southern Illinois and is exposed in numerous outcrops in southwestern Illinois and eastern Missouri. The latter is commonly referred to as the Kimmswick Limestone (in outcrop) or "Trenton" limestone (in the subsurface), the oldest oil-producing formation in Illinois. The Galena of Illinois contains abundant and diverse faunas. Common fossils include algae (*Receptaculites* and *Ischadites*), brachiopods (*Rafinesquina, Sowerbyella, Dalmanella, Pionodema,* and *Pseudolingula*), bryozoans (numerous trepostomes), mollusks (cephalopods; snails, *Hormotoma* and *Maclurites;* and clams, *Vanuxemia*), trilobites (*Illaenus*), corals (*Streptelasma*), and echinoderms.

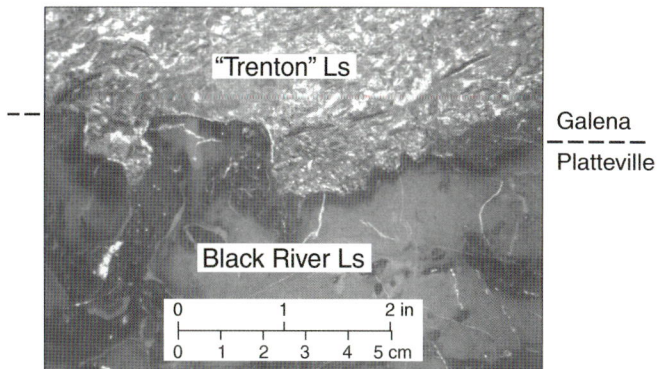

Figure 7-18 Sculpted hardground omission surface between lime mudstone of the Platteville Group (=Black River Limestone) and fossiliferous grainstone of the Kimmswick ("Trenton") Limestone in drill core from White County, Illinois. (Photograph by Dennis R. Kolata.)

In northern Illinois and southern Wisconsin, the Galena is extensively dolomitized. The Galena ranges in thickness from over 250 feet (76 m) in northern Illinois to less than 50 feet (15 m) in southern Illinois. Several very thin but persistent, altered volcanic ash beds are present in the Galena (Willman and Kolata 1978). Like the Platteville and Decorah, the top of the Galena throughout Illinois is capped by a prominent, locally sculpted, pyrite- and phosphate-rich hardground omission surface. The Galena is exposed in numerous road cuts (Figure 7-19) and quarries in north-central and northwestern Illinois where it is mined primarily for aggregate (Figure 7-20). In southern Illinois, the Kimmswick is quarried as a source of high-calcium limestone. The Platteville, Decorah, and Galena together form one of the most important bedrock aquifers in the northern part of the state.

Decorah Formation. The Decorah Formation, assigned to the Galena Group, overlies the Platteville Group in the northern and western parts of Illinois (Herbert 1949). The formation consists largely of fossiliferous shale and ar-

Figure 7-19 Fifteen-foot (4.6-m)-thick exposure of the Wise Lake Formation of the Galena Group in road cut along Interstate 39 immediately south of the Kishwaukee River near Rockford, Illinois. (Photograph by Dennis R. Kolata.)

gillaceous lime mudstone, wackestone, and packstone and reaches a maximum thickness of about 30 feet (9 m) in westernmost Illinois. The widespread Millbrig K-bentonite Bed, which marks the base of the Chatfieldian Stage (Leslie and Bergström 1995), occurs either on the hardground surface at the top of Platteville or within 5 feet (1.5 m) of the Platteville in the overlying Decorah Formation shale (Kolata et al. 1996, 1998, 2001). The Decorah forms a clastic wedge that thins southeastward away from siliciclastic sources on the Transcontinental Arch in southern Minnesota. Lead and zinc ore were once mined in northwestern Illinois from the Decorah as well as from the upper part of the Platteville and the lower part of the Galena (Chapter 16, Lead, Zinc, and Fluorite).

Tippecanoe I, Late Ordovician (Cincinnatian Series)

Graphic correlation methods based on scaled ranges of conodont species suggest that the Mohawkian and Cincinnatian Series boundary is near the Dunleith and Wise Lake Formation contact at the approximate level of the Dygerts K-bentonite Bed (not shown on generalized stratigraphic column) (Sweet 1987). This boundary is also the contact between the Edenian and Chatfieldian Stages. Cincinnatian strata of northern Illinois, therefore, include the Wise Lake and Dubuque Formations in the upper part of the Galena

Figure 7-20 Dolomite of the Galena Group overlain by shale of the Maquoketa Group at the Vulcan Materials quarry near Sycamore, Illinois (NE¼ NE¼ Section 12, T41N, R4E). (Photograph by Dennis R. Kolata.)

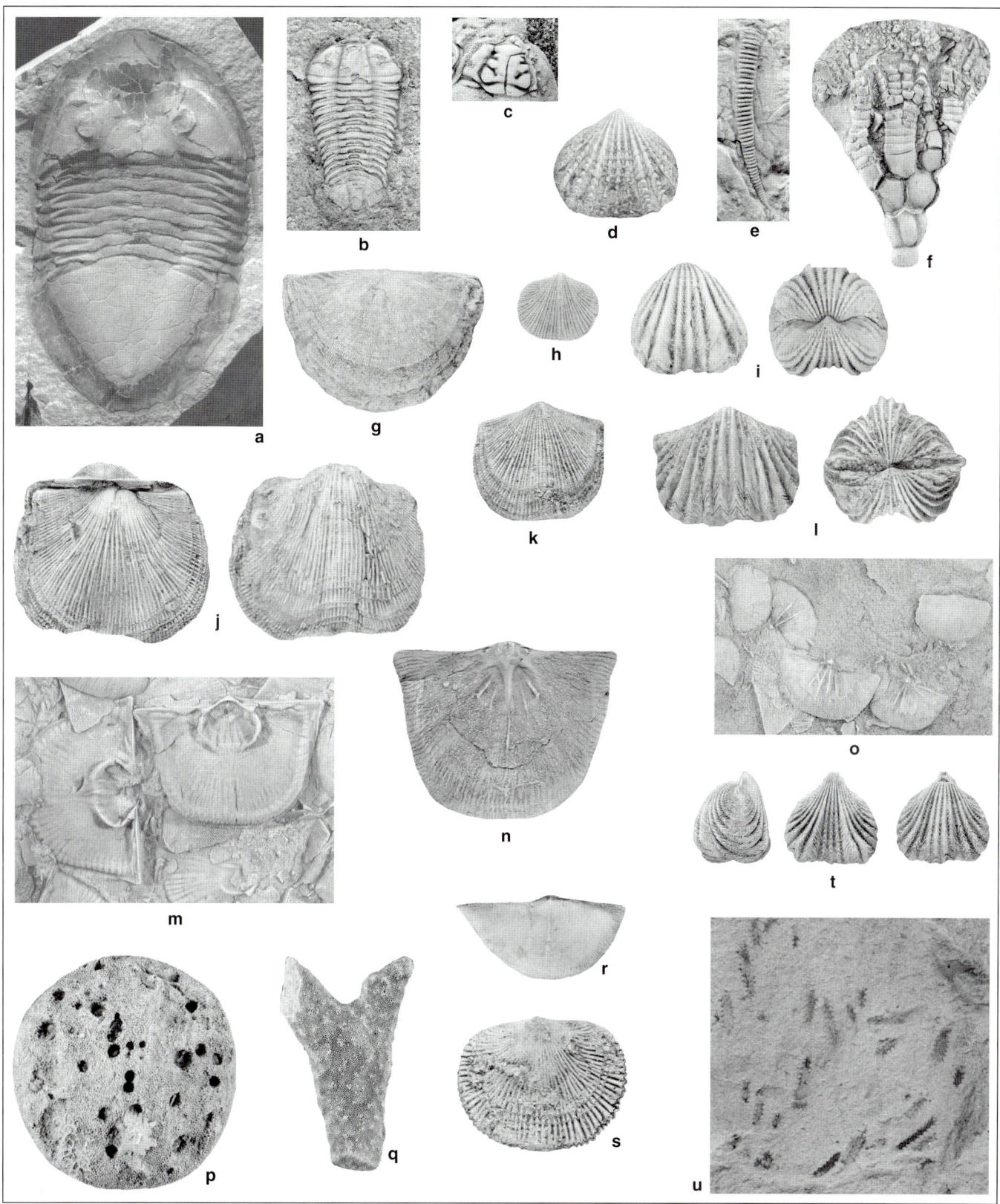

Figure 7-21 Characteristic fossils of the Maquoketa Group: **a,** trilobite *Isotelus iowensis;* **b,** calymenid trilobite; **c,** trilobite *Gravicalymene;* **d,** brachiopod *Lepidocyclus* sp.; **e,** tentaculitoid *Tentaculites* sp.; **f,** crinoid *Cupulocrinus angustatus;* **g,** brachiopod *Strophomena* sp.; **h,** brachiopod *Diceromyonia* sp.; **i,** brachiopod *Lepidocyclus* sp.; **j,** brachiopod *Hebertella* sp.; **k,** brachiopod *Glyptorthis* sp.; **l,** brachiopod *Platystrophia* sp.; **m,** brachiopod *Strophomena* sp.; **n,** brachiopod *Öpikina* sp.; **o,** brachiopod *Thaerodonta* sp.; **p,** trepostome bryozoan *Prasopora* sp.; **q,** trepostome bryozoan; **r,** brachiopod *Megamyonia unicostata;* **s,** brachiopod *Plaesiomys* sp.; **t,** brachiopod *Hypsiptycha* sp.; and **u,** graptolites. All specimens are shown approximately life size.

Group and all of the Maquoketa Group (southern Illinois) (Figures 7-1 and 7-2).

During late Chatfieldian time, the detrital wedge shed from the Taconic Highlands, centered in what is now Massachusetts, New York, and Vermont, began to encroach on the Illinois Basin. Progressively increasing quantities of fine siliciclastics, forming the Maquoketa Group, were shed in a westward direction, eventually covering Illinois and reaching the Upper Mississippi Valley region by the late Cincinnatian. During times of low detrital input, deposition of carbonate rocks outpaced that of shale. Faunal diversity and abundance increased up-section in the Maquoketa, suggesting a gradual change from cool, oxygen-poor, phosphate-rich, density-stratified oceanic waters to well-mixed, oxygen-rich, clear seas with abundant and diverse marine faunas.

Maquoketa Group. The Maquoketa Group, named for outcrops along the Little Maquoketa River near Dubuque, Iowa, overlies the Galena Group. The Maquoketa shales rest on the prominent hardground at the top of the Galena, the abrupt change in lithology indicating a significant change in depositional environments but not necessarily a long hiatus. The Maquoketa is part of a wedge of mainly siliciclastic rocks that thickens from about 200 feet (61 m) in northwestern Illinois to over 300 feet (91 m) in southeastern Illinois, gradually thickening to nearly 6,000 feet (1,829 m) in the Appalachian basin in Pennsylvania (Kolata and Graese 1983). The Maquoketa is the first of a series of Paleozoic siliciclastic wedges that had an evident eastern source.

The Maquoketa is characterized by several local and regional facies. In northern Illinois, the lower unit is the Scales Shale, which is dark olive-gray, laminated to highly bioturbated, locally organic-rich, and dolomitic. Typically at the base, known as the depauperate zone, are thin layers of phosphatic pellets, nodules, and diminutive gastropods, bivalves, cephalopods, and brachiopods (Kolata and Graese 1983). In northern Illinois, the Scales Shale is overlain by yellowish brown, fossiliferous packstone and grainstone of the Fort Atkinson Limestone, which is overlain in turn by the greenish gray, fossiliferous, dolomitic shale of the Brainard Shale. The Scales Shale typically has sparse faunas of little diversity, but the Brainard and Fort Atkinson contain rich and diverse faunas (Figure 7-21). Common fossils include brachiopods (*Hebertella, Strophomena, Diceromyonia, Glyptorthis, Lepidocyclus, Megamyonia, Platystrophia, Plaesiomys, Hypsiptycha, Öpikina,* and *Thaerodonta*), bryozoans (numerous trepostome genera including *Prasopora*), trilobites (*Isotelus, Gravicalymene,* and *Flexicalymene,* mollusks (cephalopods, snails, and clams), tentaculitoids (*Tentaculites*), graptolites, and echinoderms (*Cupulocrinus*).

Locally, as in northwestern Illinois, the Fort Atkinson grades to shale, and the Maquoketa is not differentiated into formations (Kolata and Graese 1983). In southern Illinois, the Maquoketa succession consists of (in ascending order) the Cape La Croix Shale, Thebes Sandstone, Or-

Figure 7-22 Angular unconformity between the Late Ordovician Brainard Shale of the Maquoketa Group and overlying Silurian dolomite in a quarry near Hillside, Cook County, Illinois (SE¼, NE¼ Section 17, T39 N, R12E). (Photograph by Dennis R. Kolata.)

chard Creek Shale, Girardeau Limestone, and Noix Oolite. The top of the Maquoketa in Illinois is marked by a widespread unconformity that developed during the drop in sea level associated with Late Ordovician glaciation (Kolata and Graese 1983). Pre-Silurian (Tippecanoe II) erosion created as much as 150 feet (46 m) of relief on the Maquoketa surface in parts of northeastern Illinois and northwestern Indiana (Figure 7-22). In some parts of the midcontinent, including northern Illinois where the Maquoketa is thickest, the upper 10 feet (3 m) or less consists of red shale interbedded with flattened spheroids of geothite and hematite that apparently developed by soil-forming processes during a long period of exposure in a tropical environment (Kolata and Grease 1983). The unit is referred to as the Neda Formation.

CONCLUSION

During the Cambrian and Ordovician, a span of approximately 100 million years, significant geological and biological changes took place on Earth. The primary driving forces for this and subsequent changes were the movement and interaction of global tectonic plates. By the dawn of Cambrian time, Laurentia, the Paleozoic continent consisting largely of North America and Greenland, was positioned along the equator. It was surrounded by passive margins with flat coastlines and flanked by newly formed oceans, including the Iapetus Ocean (Figure 7-4). Exposed and weathered Precambrian crystalline rocks near the center of the continent (i.e., the Canadian Shield) supplied sediment to the shallow continental seas, forming a detrital belt that surrounded the land mass. Beyond this detrital belt was a middle belt of carbonate sediments and an outer concentric belt of fine detrital sediments that extended over the continental margin. With the passing of time and fluctuations in sea level, these facies shifted laterally across Illinois, producing a stratigraphic succession dominated by marine sandstones, carbonate rocks, and shales.

By the Early Ordovician, Laurentia began to converge with Scandinavia, the southern British Isles, and fragments of the Gondwana continent. The Iapetus Ocean, which formed during the Cambrian, now began to be consumed in subduction zones. During the Middle and Late Ordovician, the eastern margin of Laurentia underwent significant deformation, resulting from the collision of tectonic plates. Cambrian and Ordovician continental shelf rocks and various incoming microcontinental and oceanic terranes were deformed into an assemblage of elevated thrust sheets that shed detrital sediment (i.e., Late Ordovician Maquoketa Group) across the southeastern half of Laurentia spreading as far west as present-day Iowa.

Life at the beginning of the Cambrian consisted of relatively simple, soft-bodied organisms, but, by the end of the Early Cambrian, organisms comparable in morphology and organization to present-day animals had evolved. Shelly faunas dominated by trilobites, hyolithids, and inarticulate brachiopods were abundant and widespread during the Cambrian. Although most phyla evolved during the Cambrian, life-forms underwent their most dramatic diversification during the Early Ordovician. The Ordovician faunal radiation resulted in global transition in dominance from the Cambrian evolutionary faunas (trilobites, hyolithids, and inarticulate brachiopods) to faunas (articulate brachiopods, bryozoans, molluscs, and echinoderms) that would persist through Phanerozoic time. Also, community structure dramatically increased in complexity. These tectonic and evolutionary events are well documented in the rock record of Illinois.

REFERENCES

Bickford, M. E., W. R. Van Schmus, and I. Zietz, 1986, Proterozoic history of the midcontinent region of North America: Geology, v. 14, no. 6, p. 492–496.

Buschbach, T. C., 1964, Cambrian and Ordovician strata of northeastern Illinois: Illinois State Geological Survey, Report of Investigations 218, 90 p.

Buschbach, T. C., 1975, Cambrian System, in H. B. Willman, E. Atherton, T. C. Buschbach, C. Collinson, J. C. Frye, M. E. Hopkins, J. A. Lineback, and J. A. Simon, eds., Handbook of Illinois stratigraphy: Illinois State Geological Survey, Bulletin 95, p. 34–46.

Buschbach, T. C., and D. C. Bond, 1974, Underground storage of natural gas in Illinois—1973: Illinois State Geological Survey, Illinois Petroleum 101, 71 p.

Collinson, C. W., M. L. Sargent, and J. R. Jennings, 1988, Illinois Basin region, in L. L. Sloss, ed., Sedimentary cover—North American craton: U.S.: The Geology of North America, v. D-2: Boulder, Colorado, Geological Society of America, 506 p.

Fraser, G. S., 1976, Sedimentology of a Middle Ordovician quartz arenite-carbonate transition in the Upper Mississippi Valley: Geological Society of America Bulletin, v. 87, no. 6, p. 833–845.

Herbert, P. Jr., 1949, Stratigraphy of the Decorah Formation in western Illinois: University of Chicago, Ph.D. dissertation: Illinois State Geological Survey, manuscript PH-1.

Keller, S. J., and T. F. Abdulkareem, 1980, Post-Knox unconformity—Significance at Unionport gas-storage project and relationship to petroleum exploration in Indiana: Bloomington, Indiana, Indiana Geological Survey, Occasional Paper OP31, 19 p.

Kolata, D. R., 1975, Middle Ordovician echinoderms from northern Illinois and southern Wisconsin: Paleontological Society Memoir 7: Journal of Paleontology, v. 49, no. 3, suppl., 74 p.

Kolata, D. R., compiler, 2005, Bedrock geology of Illinois: Illinois State Geological Survey, Illinois Map 14, 1:500,000.

Kolata, D. R., and A. M. Graese, 1983, Lithostratigraphy and depositional environments of the Maquoketa Group (Ordovician) in northern Illinois: Illinois State Geological Survey, Circular 528, 49 p.

Kolata, D. R., W. D. Huff, and S. M. Bergström, 1996, Ordovician K-bentonites of eastern North America: Boulder, Colorado, Geological Society of America, Special Paper 313, 84 p.

Kolata, D. R., W. D. Huff, and S. M. Bergström, 1998, Nature and regional significance of unconformities associated with the Middle Ordovician Hagan K-bentonite complex in the North American midcontinent: Geological Society of America Bulletin, v. 110, no. 6, p. 723–739.

Kolata, D. R., W. D. Huff, and S.M. Bergström, 2001, The Ordovician Sebree Trough—An oceanic passage to the midcontinent United States: Geological Society of America Bulletin, v. 113, no. 8, p. 1067–1078.

Leslie, S. A., and S. M. Bergström, 1995, Revision of the North American late Middle Ordovician standard stage classification and timing of the Trenton transgression based on K-bentonite Bed correlation, in J. D. Cooper, M. L. Droser, and S. C. Finney, eds., Ordovician odyssey: Short papers for the Seventh International Symposium on the Ordovician System: Fullerton, California, Pacific Section Society for Sedimentary Geology (SEPM), Book 77, p. 49–54.

Lidiak, E. G., 1996, Geochemistry of subsurface Proterozoic rocks in the eastern midcontinent of the United States: Further evidence for a within-plate tectonic setting, in B. A. van der Pluijm and P. A. Catacosinos, eds., Basement and basins of Eastern North America: Boulder, Colorado, Geological Society of America, Special Paper 308, p. 45–66.

McBride, J. H., and D. R. Kolata, 1999, Upper crust beneath the central Illinois Basin, United States: Geological Society of America Bulletin, v. 111, no. 3, p. 375–394.

McBride, J. H., D. R. Kolata, and T. G. Hildenbrand, 2003, Geophysical constraints on understanding the origin of the Illinois Basin and its underlying crust: Tectonophysics, v. 363, no. 1–2, p. 45–78.

McKay, R. M., 1988, Stratigraphy and lithofacies of the Dresbachian (Upper Cambrian) Eau Claire Formation in the subsurface of eastern Iowa, in G. A. Ludvigson and B. J. Bunker, eds., New perspectives on the Paleozoic history of the Upper Mississippi Valley: Des Moines, Iowa, Iowa Department of Natural Resources, Geological Survey Bureau, Guidebook 8, p. 33–53.

Nelson, W. J., 1995, Structural features in Illinois: Illinois State Geological Survey, Bulletin 100, 144 p.

Norby, R. D., 1991, Biostratigraphic zones in the Illinois Basin, in M. W. Leighton, D. R. Kolata, D. F. Oltz, and J. J. Eidel, eds., Interior cratonic basins: Tulsa, Oklahoma, American Association of Petroleum Geologists, Memoir 51, p. 179–194.

Norby, R. D., M. L. Sargent, and R. L. Ethington, 1986, Conodonts from the Everton Dolomite (Middle Ordovician) in southern Illinois (abs.): Geological Society of America, Abstracts with Programs, v. 18, no. 3, p. 258.

Nunn, J. R., 1986, The petrology and paleogeography of the Starved Rock Sandstone in southeastern Iowa, western Illinois and northeastern Missouri: Iowa City, Iowa, University of Iowa, M.S. thesis, 144 p.

Palmer, A. R., 1960, Some aspects of the early Upper Cambrian stratigraphy of White Pine County, Nevada and vicinity: Intermountain Association of Petroleum Geologists, 11th Annual Field Conference, Guidebook to the Geology of East-central Nevada, p. 53–58.

Palmer, A. R., 1981, Subdivision of the Sauk Sequence, in M. E. Taylor, ed., Short papers for the Second International Symposium on the Cambrian System: Reston, Virginia, U.S. Geological Survey, Open-File Report 81-743, p. 160–162.

Palmer, J. R., 1989, Late Upper Cambrian shelf depositional facies and history, southern Missouri, in J. M. Gregg, J. R. Palmer, and V. E. Kurtz, eds., Field Guide to the Upper Cambrian of southeastern Missouri—Stratigraphy, sedimentology, and economic geology: Rolla, Missouri, Department of Geology and Geophysics, University of Missouri, p. 1–24.

Pratt, T. L., R. Culotta, E. C. Hauser, K. D. Nelson, L. Brown, S. Kaufman, J. Oliver, and W. Hinze, 1989, Major Proterozoic basement features of the eastern midcontinent of North America revealed by recent COCORP profiling: Geology, v. 17, no. 6, p. 505–509.

Rankin, D. W., A. A. Drake Jr., L. Glover III, R. Goldsmith, L. M. Hall, D. P. Murray, N. M. Ratcliffe, J. F. Read, D. T. Secor Jr. et al., 1989, Pre-orogenic terranes, in R. D. Hatcher Jr., W. A. Thomas, and G. W. Viele, eds., The Appalachian-Ouachita orogen in the United States: The Geology of North America, v. F-2: Boulder, Colorado, Geological Society of America, p. 7–100.

Rexroad, C.B., J. B. Droste, and R. L. Ethington, 1982, Conodonts from the Everton Dolomite and the St. Peter Sandstone (lower Middle Ordovician) in a core from southwestern Indiana: Bloomington, Indiana, Indiana Geological Survey, Occasional Paper 39, 17 p.

Sargent, M. L., 1991, Sauk sequence: Cambrian System through Lower Ordovician Series, in M. W. Leighton, D. R. Kolata, D. F. Oltz, and J. J. Eidel, eds., Interior cratonic basins: Tulsa, Oklahoma, Association of Petroleum Geologists, Memoir 51, p. 75–86.

Sargent, M. L., and Z. Lasemi, 1993, Tidally dominated depositional environment for the Mt. Simon Sandstone in central Illinois (abs.): Geological Society of America, Abstracts with Programs, v. 25, no. 3, p. 78.

Schwalb, H. R., 1982, Geologic-tectonic history of the area surrounding the northern end of the Mississippi Embayment: University of Missouri at Rolla Journal, no. 5, p. 31–42.

Scotese, C. R., and W. S. McKerrow, 1990, Revised world maps and introduction, in W. S. McKerrow and C. R. Scotese, eds., Palaeozoic palaeogeography and biogeography: London, The Geological Society of London, Memoir 12, p. 1–21.

Shaw, T. H., 1999, Early-Middle Ordovician lithostratigraphy, biostratigraphy, and depositional environments of the Illinois Basin: New York, The City University of New York, Ph. D. dissertation, 929 p.

Shaw, T. H., R. D. Norby, and M. L. Sargent, 1988, Lower Ordovician conodonts from the Shakopee Dolomite (upper Prairie du Chien Group) in southwestern Illinois (abs.): Geological Society of America, Abstracts with Programs, v. 20, no. 5, p. 388.

Sloan, R. E., ed., 1987, Middle and Late Ordovician lithostratigraphy and biostratigraphy of the Upper Mississippi Valley: St. Paul, Minnesota, Minnesota Geological Survey, Report of Investigations 35, 232 p.

Sloss, L. L., 1988, Tectonic evolution of the craton in Phanerozoic time, in L. L. Sloss, ed., Sedimentary cover—North American Craton: U.S.: The Geology of North America, v. D-2L: Boulder, Colorado, Geological Society of America, p. 25–51.

Sloss, L. L., W. C. Krumbein, and E. C. Dapples, 1949, Integrated facies analysis: Boulder, Colorado, Geological Society of America, Memoir 39, p. 91–123.

Sweet, W. C., 1987, Distribution and significance of conodonts in Middle and Upper Ordovician strata of the Upper Mississippi Valley region, in R. E. Sloan, ed., Middle and Late Ordovician lithostratigraphy and biostratigraphy of the Upper Mississippi Valley: St. Paul, Minnesota, Minnesota Geological Survey, Report of Investigations 35, p. 167–172.

Templeton, J. S., and H. B. Willman, 1963, Champlainian Series (Middle Ordovician) in Illinois: Illinois State Geological Survey, Bulletin 89, 260 p.

Thomas, W. A., 1989, The Appalachian-Ouachita orogen beneath the Gulf Coastal Plain between the outcrops in the Appalachian and Ouachita Mountains, in R. D. Hatcher, Jr., W. A. Thomas, and G. W. Viele, eds., The Appalachian-Ouachita orogen in the United States: The Geology of North America: v. F-2, Boulder, Colorado, Geological Society of America, p. 537–553.

Thompson, T. L., 1995, The stratigraphic succession in Missouri: Rolla, Missouri, Missouri Department of Natural Resources, Division of Geology and Land Survey, v. 40, 190 p.

Willman, H. B., E. Atherton, T. C. Buschbach, C. Collinson, J. C. Frye, M. E. Hopkins, J. A. Lineback, and J. A. Simon, 1975, Handbook of Illinois stratigraphy: Illinois State Geological Survey, Bulletin 95, 261 p.

Willman, H. B., and T. C. Buschbach, 1975, Ordovician System, *in* H. B. Willman, E. Atherton, T. C. Buschbach, C. Collinson, J. C. Frye, M. E. Hopkins, J. A. Lineback, and J. A. Simon, eds., Handbook of Illinois stratigraphy: Illinois State Geological Survey, Bulletin 95, p. 47–87.

Willman, H. B., J. C. Frye, J. A. Simon, K. E. Clegg, D. H. Swann, E. Atherton, C. Collinson, J. A. Lineback, and T. C. Buschbach, 1967, Geologic map of Illinois: Illinois State Geological Survey, 1:500,000.

Willman, H. B., and D. R. Kolata, 1978, The Platteville and Galena Groups in northern Illinois: Illinois State Geological Survey, Circular 502, 75 p.

Willman, H. B., and J. S. Templeton, 1951, Cambrian and Lower Ordovician exposures in northern Illinois: Illinois State Academy of Science Transactions, v. 44, p. 109–125.

Witzke, B. J., 1980, Middle and Upper Ordovician paleogeography of the region bordering the Transcontinental Arch, *in* T. D. Fouch and E. R. Magathan, eds., Paleozoic paleogeography of west-central United States: Denver, Colorado, Society of Economic Paleontologists and Mineralogists, Rocky Mountain Paleogeography Symposium 1, p. 1–18.

Workman, L. E., and A. H. Bell, 1948, Deep drilling and deeper oil possibilities in Illinois: Tulsa, Oklahoma, American Association of Petroleum Geologists Bulletin, v. 32, no. 11, p. 2041–2062.

8 Silurian System and Lower Devonian Series (Tippecanoe II Subsequence)

Donald G. Mikulic, Joanne Kluessendorf, and Rodney D. Norby

INTRODUCTION

The rocks of the Tippecanoe II subsequence have contributed considerable scientific, economic, and aesthetic value to Illinois. As defined by Sloss (1963, 1982, 1988), the Tippecanoe II subsequence (444 to 398 million years ago) includes basal Silurian through Lower Devonian rocks. Silurian rocks and fossils were among the earliest to be studied scientifically in Illinois, and they formed the basis of several classic studies of international importance. Since the 1830s, Silurian dolomite deposits have been an economically important, major source of construction materials for northeastern, northwestern, and west-central Illinois and surrounding areas. Additionally, the Silurian rocks of central Illinois have served as an important hydrocarbon reservoir since the 1940s. Exposures of Silurian rocks also contribute greatly to the scenic beauty and popularity of some of the largest parks in the state.

GEOLOGICAL HISTORY AND SETTING

During the deposition of the Tippecanoe II subsequence, the portion of the North American plate (Laurentia) that is now Illinois was located in a subtropical latitude and was moving northward from about 25° south of the equator (Figure 7-4) (Scotese et al. 1979). Near the end of the Ordovician (and the end of the Tippecanoe I subsequence), glaciation had developed at the south pole, producing a major drop in sea level. As a result, most of Laurentia became emergent and was subjected to various amounts of erosion. When glaciation declined at the beginning of the Silurian (444 million years ago and the beginning of the Tippecanoe II subsequence), shallow seas transgressed across this area, depositing shallow-water carbonate sediments. Primary controls on the depositional character of these rocks were the late Early Silurian to Late Silurian development or reactivation of three depositional basins: the Michigan Basin in the northeast, the East-Central Iowa Basin in the northwest, and the Illinois Basin in central and southern Illinois. The Late Silurian-Early Devonian deposition in each of the three basins, although related, is distinctive enough so that each has its own rock record, necessitating individual descriptions of each basin to provide an overview of the Silurian in Illinois.

At the beginning of the Wenlock (428 million years ago), numerous reefs began to develop and, in conjunction with the subsidence of the Illinois Basin and surrounding basins, prominent shelf-edge carbonate banks began to develop, yielding paleogeographic and lithologic complexity. Toward the end of the Silurian, the worldwide changes in sea level (eustacy), cratonic uplift, and local tectonic events caused a drop in sea level, ending deposition for most of the area and exposing the Silurian rocks to significant erosion for millions of years. Only in southern Illinois did Early Devonian marine sedimentation continue, infilling some of the remaining Silurian topography. The entire area was not completely flooded again until the Middle Devonian (398 million years ago).

ROCK DISTRIBUTION AND FEATURES

The character and distribution of Tippecanoe II subsequence rocks in Illinois and surrounding areas have been described in several reports (Willman 1973, Becker 1974, Willman and Atherton 1975, Collinson and Atherton 1975, Droste and Shaver 1987, Mikulic 1991, Thompson 1993, Berg and Masters 1994). The Silurian and Lower Devonian rocks of this subsequence are not as well-known as most of those in the Illinois Paleozoic. This lack of knowledge reflects a general scarcity of exposures and limited subsurface information for much of the state. Also, most Silurian rocks are carbonates that were extensively dolomitized, which destroyed many of their primary characteristics and resulted in a superficially featureless appearance. As a result, the Silurian rocks are among the most difficult to study of the Paleozoic rocks in the region. Only the Chicago area has enough available information to provide a good local understanding of these strata. Despite their superficially similar appearance, the Silurian rocks had a complex depositional history that is described here within a sequence stratigraphic framework rather than as individual units.

Silurian rocks occur throughout the state except in the north-central part and portions of northwestern and west-central Illinois, where they have been removed by erosion. Throughout most of central and southern Illinois, they are buried deeply by younger Paleozoic rocks. The Silurian rocks are found at the bedrock surface only in the northeastern and northwestern parts of the state and in limited areas in the west-central and southwestern parts.

In northeastern Illinois, natural exposures of these rocks are limited in extent because the bedrock surface is usually buried under Quaternary sediments. Quarrying in this region, however, has produced some of the most extensive exposures of the Silurian in the Midwest. In areas lacking Quaternary sediments, such as the Driftless Area of northwestern Illinois and a few locations along the Illinois and Mississippi Rivers to the south, Silurian rocks are exposed in natural bluffs.

Silurian rocks in Illinois consist primarily of carbonates, such as dolomite and limestone. Locally, however, the basal strata commonly are more clastic because they incorporate reworked shaly sediments of the underlying Ordovician rocks. Upper Silurian rocks in central and southern Illinois also have a high clastic content. The thickness of Silurian rocks varies greatly because of post-Silurian erosion. The maximum average thickness generally ranges from 500 to 600 feet (150 to 180 m), although in some central Illinois reefs, it is up to 1,000 feet (305 m) thick. In contrast to the rocks of most other systems, Silurian rocks do not thicken appreciably southward into the Illinois Basin.

Lower Devonian rocks occur only in the southern one-third of Illinois and are exposed only in the extreme southwestern part of the state where they primarily crop out in bluffs along the Mississippi River. These rocks have an unusually siliceous composition, with chert or cherty limestone predominating. In contrast to the Silurian, the thickness of Lower Devonian rocks increases markedly to the south, ranging from zero at their northern erosional edge to over 1,200 feet (370 m) at the state's southern boundary.

SEQUENCE BOUNDARIES

The Tippecanoe II subsequence is part of a Sloss sequence, which reflects depositional history at the cratonic scale (Sloss 1963, 1982, 1988). Sequence stratigraphy subdivides the Sloss sequences into smaller intervals related to cycles of sedimentation called depositional sequences.

At least four prominent depositional sequence boundaries characterize the Tippecanoe II subsequence in Illinois: sub-Tippecanoe II, sub-Offerman, sub-Brandon Bridge, and sub-Kaskaskia unconformities. The sub-Tippecanoe II unconformity is the Ordovician-Silurian boundary (Figure 8-1). This depositional break is related to a glacio-eustatically controlled regression at the end of the Ordovician, followed by transgression during the early Silurian (Berry and Boucot 1973). This depositional sequence boundary also marks the end of Late Ordovician clastic deposition and the onset of predominantly shallow-water carbonate deposition, which characterizes much of the Silurian rock record for the region.

The sub-Offerman (within the Kankakee) and the sub-Brandon Bridge (basal Joliet) sequence boundaries (Kluessendorf and Mikulic 1996, Mikulic and Kluessendorf 1998) are prominent markers within the Silurian (Figure 8-1). They represent brief regressions or breaks in deposition that may be related to weak pulses of glaciation.

The sub-Kaskaskia sequence boundary may be a response to a combination of both sea-level change and tectonic events, including moderate uplift of the North American craton and regional arches plus some subsidence of area basins (Johnson 1996, Shaver 1996). Ross and Ross (1996) suggested a similar scenario but thought that the larger role was played by eustacy. Sloss (1988), however, thought that tectonic events may have been the overriding factor and cited evidence of differential vertical tectonic motions that were both regional and local in scale. The sub-Kaskaskia sequence boundary, as currently used by the Illinois State Geological Survey, marks the contact between Silurian and Middle Devonian rocks throughout much of the state and between Early Devonian and Middle Devonian rocks in southern Illinois.

The sub-Kaskaskia unconformity that caps the Tippecanoe II subsequence in Illinois has been placed at two different horizons, which has led to confusion. Sloss (1963, 1982, 1988) placed the top of the Tippecanoe at the top of the Gedinnian Stage (earliest Devonian stage that has been replaced by and is approximately equal to the Lochkovian Stage, Figure 8-1c). However, earlier, Sloss et al. (1949) had used the top of the Clear Creek Formation (or top of the Lower Devonian) as the top of the Tippecanoe sequence. Willman and Atherton (1975) used the top of the Clear Creek Formation (Sloss et al. 1949) instead of the horizon used later by Sloss (1963). Most authors in a publication by Leighton et al. (1991) followed the placement of the unconformity (and sequence boundary) of Willman and Atherton (1975) and used the top of the Lower Devonian (Clear Creek) as the top of their Tippecanoe sequence. For the sake of conformity with other related papers in this volume, the local unconformity at the base of the Dutch Creek Sandstone Member of the Grand Tower Limestone is used in this chapter as the top of the Tippecanoe II subsequence (see Chapter 9, Middle Devonian–Mississippian). With additional study, however, the widely accepted, modified boundary between the Tippecanoe and the Kaskaskia sequences suggested by Sloss (1963, 1982, 1988) is likely to be reapplied to the Illinois sequences.

Depositional Sequence 1

The oldest Tippecanoe II subsequence rock units represent the initial transgression of Silurian seas and the burial of the eroded Ordovician surface. Areas of exten-

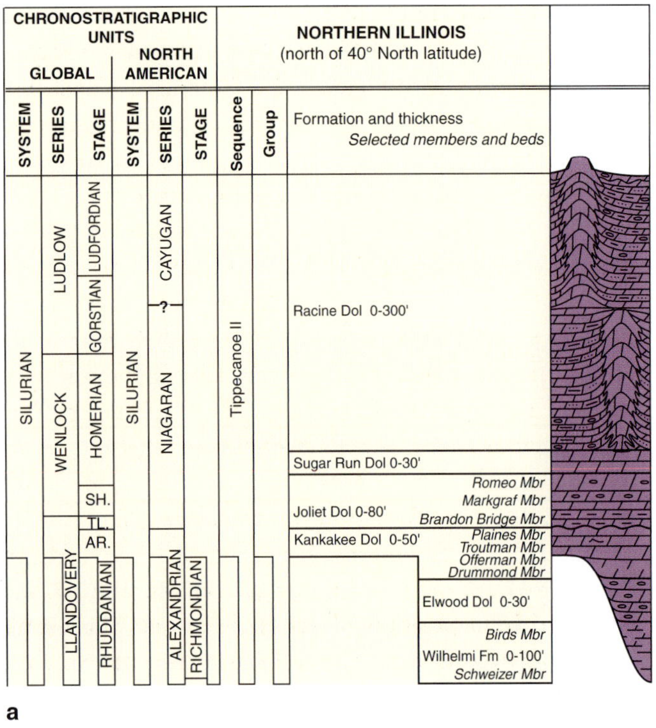

Figure 8-1 Silurian stratigraphic sections for (a) northern Illinois, (b) northwestern Illinois, and (c) Silurian and Lower Devonian stratigraphic section for central and southern Illinois. Abbreviations: ALEX., Alexandrian; AR., Aeronian; CAY., Cayugan; GR., Gorstian; HM., Homerian: LD., Ludfordian; LUD., Ludlow; LLAND., Llandovery; RH., Rhuddanian; SH., Sheinwoodian; TL., Telychian; Wen., Wenlock; Dol, Dolomite; Fm, Formation; Ls, Limestone; Mbr, Member. Lithologic symbols are explained in Appendix I.

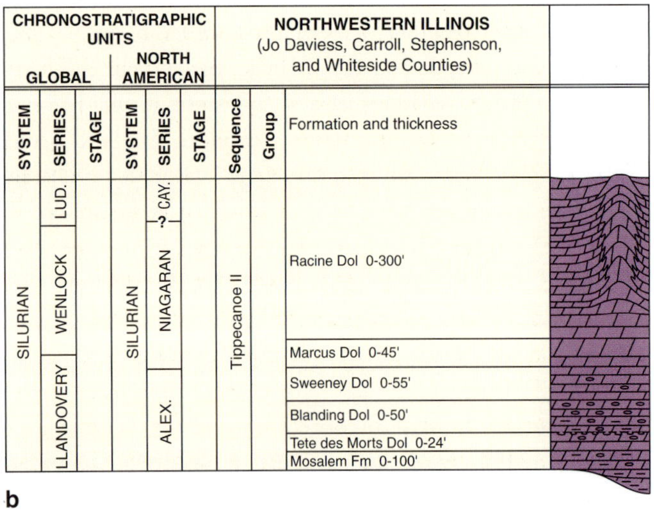

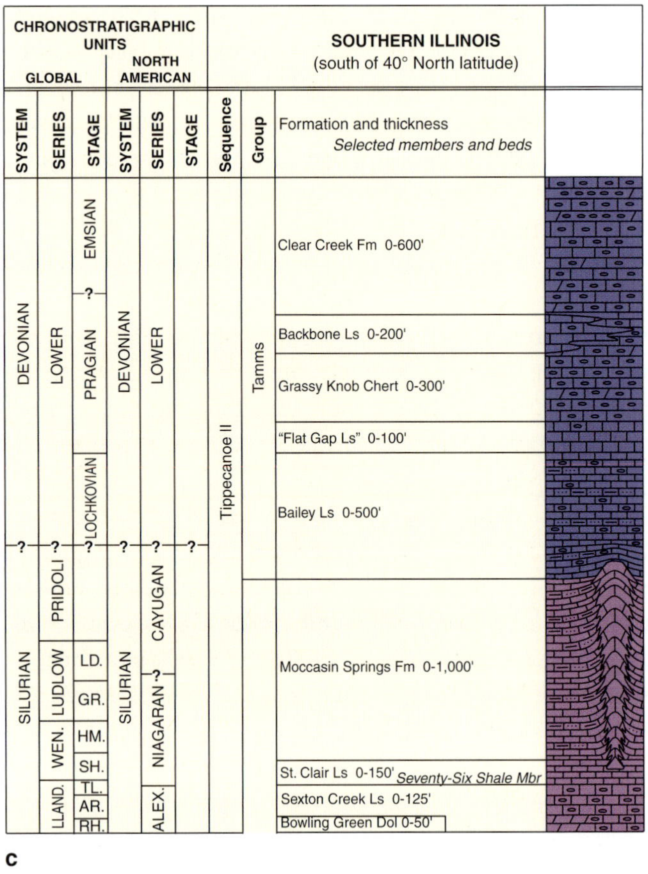

sive topographic relief occurred at that time as the underlying clastic sediments of the Ordovician Maquoketa Group were eroded into a series of broad hills and valleys, some with more than 100 feet (30.5 m) of relief. This topography is not distributed uniformly but seems to be more typical of the northern part of the state.

The basal strata of these units are highly argillaceous, containing a considerable amount of reworked Ordovician sediment. Clastic content gradually diminishes upward as carbonate content increases and as progressively more of the old Ordovician surface was buried. The final burial of the highest parts of the Ordovician surface produced the region's earliest clean (relatively pure) carbonates, such as the Middle to Early Silurian Drummond Member of the Kankakee Dolomite.

The primary rock units of this sequence in northeastern Illinois (part of the Michigan Basin) are the basal argillaceous Wilhelmi Formation, which is overlain by the highly cherty Elwood Dolomite and capped by the high-purity Drummond Member of the Kankakee Dolomite. The maximum combined thickness of this depositional sequence is 140 feet (43 m) (Figure 8-1a). Most of this interval contains typical marine fossils; a conspicuous zone of the brachiopod *Platymerella* is found near the top of the Elwood. In northwestern Illinois (part of the East-Central Iowa Basin), the same time interval is represented by the basal argillaceous Mosalem Formation, which is overlain by the Tete des Morts Dolomite, having a combined thickness of as much as 120 feet (37 m) (Figure 8-1b).

In west-central Illinois, a 30-foot (9-m)-thick clean carbonate unit, the Bowling Green Dolomite, is equivalent to this interval (Figure 8-1c). The Bowling Green typically succeeds Ordovician carbonates, such as the Noix Oolite and Leemon Formation (not shown), which did not provide a source of clastic material to the early Silurian sediments. In central and southern Illinois, few, if any, rocks appear to represent this time interval. The rocks of the overlying sequence apparently blanket the Ordovician surface as they do consistently to the south in Missouri, Arkansas, and Oklahoma, suggesting that this entire region was emergent during earliest Silurian deposition, although little evidence exists of the pre-Silurian large-scale erosional topography that is found in northern Illinois. It has been suggested that the lower Mosalem and Wilhelmi may be Ordovician in age (Loydell et al. 2002, Kleffner et al. 2005), although conclusive proof is lacking. Most evidence continues to point to a Silurian age for these rocks.

Depositional Sequence 2

The second depositional sequence occurs between the sub-Offerman sequence boundary and the sub-Brandon Bridge sequence boundary. This interval represents a change from the high purity carbonates of the Drummond and related units to more argillaceous carbonates of the Offerman Member of the Kankakee Dolomite. Although generally thin (less than 40 feet [12 m]) compared with the previous sequence, this sequence also grades upward into a high-purity carbonate. In northeastern Illinois (Figure 8-1a), this interval is represented by the Kankakee Dolomite (excluding the Drummond Member); in northwestern Illinois (Figure 8-1b), by the much thicker Blanding and Sweeny Dolomites (possibly including part of the overlying Marcus Dolomite); and in southern Illinois, by the Sexton Creek Limestone (Figure 8-1c). In most areas, the upper strata of this sequence are characterized by an abundance of the brachiopods *Pentamerus* (Figure 8-2) and *Stricklandia*.

Depositional Sequence 3

The most prominent depositional sequence boundary within the Silurian is marked by the sub-Brandon Bridge sequence boundary (Kluessendorf and Mikulic 1996). Historically, this feature has been described as the contact between the Alexandrian Series (as defined by Savage 1908) and the overlying Niagaran Series.

The base of the Brandon Bridge Member of the Joliet Dolomite marks the base of depositional sequence 3 in northeastern Illinois. The Brandon Bridge is a highly argillaceous dolomite up to 30 feet (9 m) thick, colored conspicuously red and green. In the rest of the state, equivalent strata, such as the Seventy-Six Shale Member at the base of the St. Clair Limestone in southern Illinois (Figure 8-1c) and possibly part of the Marcus Dolomite in northwestern Illinois (Figure 8-1b), are much thinner. In all three areas, these strata contain the most diverse and prolific microfossil fauna (conodonts and foraminiferans) known from the Silurian rocks of the state. This interval is recognized worldwide, and its top marks the approximate base of the Wenlock Series of the Silurian System. The Brandon Bridge Member is overlain conformably by the Markgraf and Romeo Members of the Joliet Dolomite in the northeastern part of the state (Figure 8-1a), parts of the Marcus Dolomite and Racine Dolomite in the northwest (Figure 8-1b), and the St. Clair Limestone in the rest of Illinois (Figure 8-1c). In general, these strata represent lagoonal deposition succeeded by deposition of echinoderm-rich grainstones in shoal environments (e.g., the Romeo and the St. Clair). The development of these grainstones marks a significant change in the overall depositional character of the Silurian.

Depositional characteristics of earlier Silurian rocks show little evidence for significant basin development in Illinois. Other than accommodation of the localized pre-Silurian erosional topography, these earlier rock units are similar in thickness, texture, and depositional features, suggesting that none of the basins in this region was active. Following Romeo and St. Clair deposition, three distinct basins (Michigan, Illinois, and East-Central Iowa) can be recognized in the Silurian rocks of Illinois, as their depositional environments became more complex and variable. Most of this change can be attributed to the development of reefs and carbonate banks along the margins of slowly subsiding basins and to the appearance of reefs on the surrounding shelf areas.

In northeastern Illinois, at the southwestern edge of the Michigan Basin, depositional sequence 3 is followed by the Sugar Run and Racine Dolomites, which constitute a section of reefs and argillaceous non-reef dolomite more than 300 feet (90 m) thick (Figure 8-1a). A similar thickness of Racine Dolomite comprising carbonate banks occurs in northwestern Illinois (Figure 8-1b) along the southeastern border of the East-Central Iowa Basin. Both of these basins were filled with younger Silurian hypersaline carbonates or evaporites, which are not represented in Illinois, with the possible exception of the exposed strata in the Morrison area that resemble the Anamosa Member of the Gower Formation in Iowa (Mikulic and Hickerson 1994).

In central and southern Illinois, which encompasses much of the Illinois Basin, the Moccasin Springs Formation represents most, if not all, of the post-St. Clair Limestone

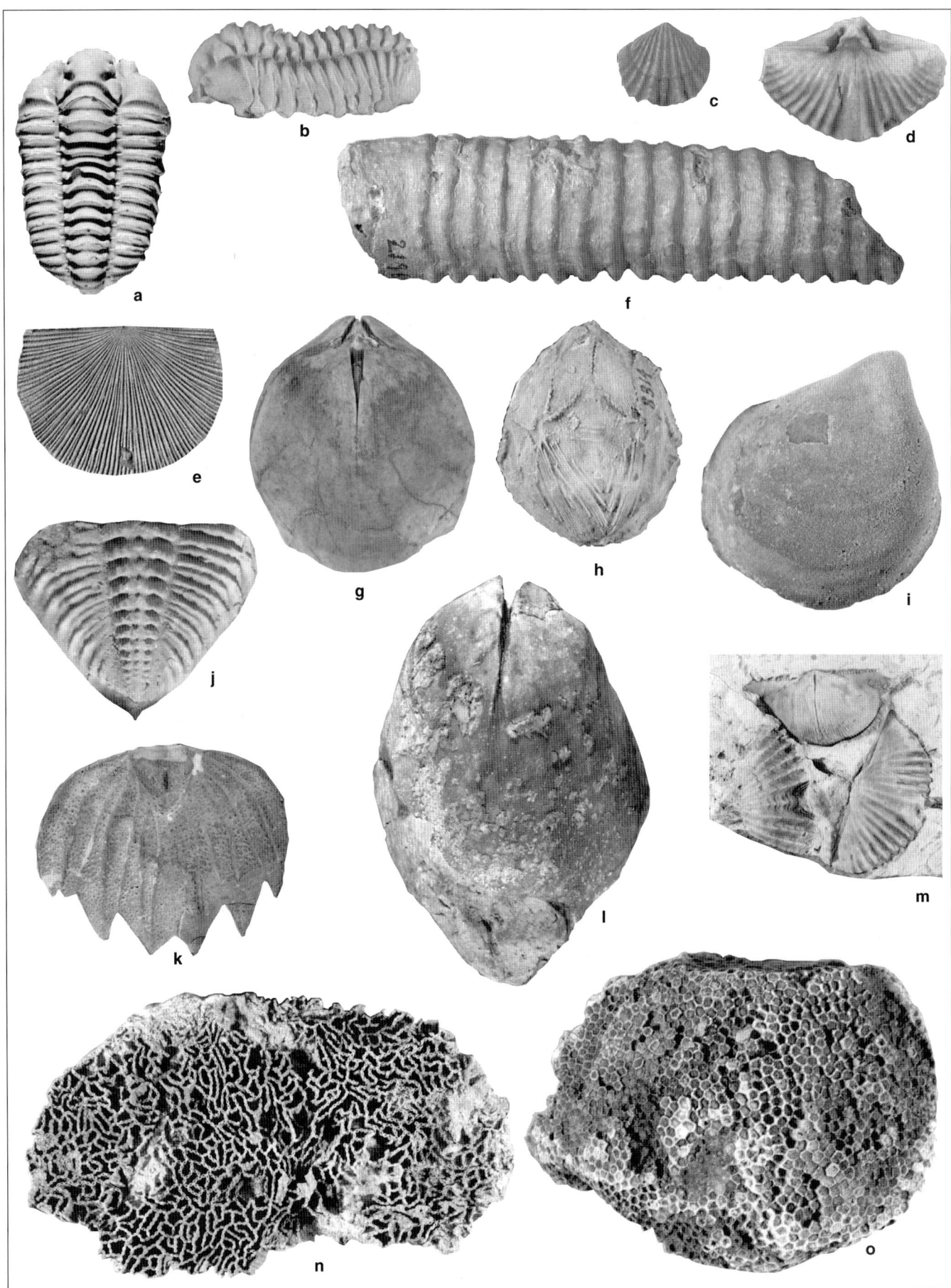

Silurian deposition (Figure 8-1c). The Moccasin Springs Formation exhibits two main lithologies, primarily representing either shelf-edge carbonate buildups or non-reef basinal deposits. The former consists mostly of localized carbonate banks and reefs, ranging up to 1,000 feet (300 m) in thickness. These structures generally form a high rim along the northern margins of the basin along a shelf-slope break. To the north are the shelf deposits of the Racine Dolomite, and to the south are the basinal non-reef strata of the Moccasin Springs Formation, comprising limestone and argillaceous limestone. These shelf and basin deposits are considerably thinner, from 200 to 400 feet (60 to 120 m), than the reefs and banks, which are up to 1,000 feet (300 m) thick. Alternative interpretations have claimed a ramp configuration for the north-south trend of the Illinois Basin (Whitaker 1988), but convincing evidence is lacking (Mikulic 1991).

Unlike the Racine Dolomite in northern Illinois, the Moccasin Springs Formation does not appear to have been truncated by pre-Middle Devonian erosion in much of the southern area of its distribution, where it is probably conformable with the overlying Bailey Limestone. The Moccasin Springs is poorly dated but appears to range from Wenlock through Ludlow (based on conodonts and graptolites; Ross 1962, Norby 1990), possibly extending through Pridoli (based on graptolites; Berry and Boucot 1970). Proposals have been made to include the overlying Bailey Limestone in the Silurian (Becker and Droste 1978, Droste and Shaver 1987) despite its historical assignment to the Early Devonian. Evidence for this interpretation is very limited, however, and part of the confusion results from misidentification of conodonts that have been reported from overlying strata (Norby 1990). Acritarch fossils indicate that the entire Bailey is Early Devonian (Norby 1990, Mikulic 1991).

Lower Devonian rocks are confined to southern Illinois and, except for exposures in a small area of the southwestern part of the state, are otherwise known only from subsurface information. These strata, which are overlain conformably by sub-Kaskaskia rocks in the center of the Illinois Basin, suggest that marine deposition was continuous during much of this time. These rocks, of considerable thickness at the southern tip of the state, consist of a number of conspicuously chert-rich units (Figure 8-1c). The basal Bailey is 200 to 500 feet (60 to 150 m) thick and is primarily a cherty limestone. It has been interpreted as a carbonate turbidite resulting from the exposure and erosion of Silurian platform deposits circling the Illinois Basin following a late Silurian fall in sea level (Carozzi and Banaee, 1984). Overlying the Bailey are the "Flat Gap Limestone" (0 to 100 feet [0 to 30.5 m] thick), the Grassy Knob Chert (200 to 300 feet [60 to 90 m] thick), the Backbone Limestone (0 to 200 feet [0 to 60 m] thick), and the Clear Creek Formation (300 to 600 feet [90 to 180 m] thick). Much of the Bailey, Backbone, and Clear Creek contain normal marine faunas. The conspicuously cherty nature of the Clear Creek has been interpreted as being due to hydrothermal silicification of carbonates rather than to primary deposition of siliceous sediments (Berg and Masters 1994).

Scientific Importance

The rocks and fossils of Illinois have long played an important role in the study of the Silurian geology and paleontology of North America. Although early reports on these rocks date back to the 1830s, little was published until the 1860s, when some of the earliest geological studies of the region included descriptions of Silurian fossils from the Chicago area. At the time, new quarries provided opportunities for amateur naturalists to assemble collections of local fossils that attracted the attention of professional scientists (Mikulic and Kluessendorf 1999). Most of these fossils were from Silurian reefs, which provided some of the most diverse fossil biotas known in the country. Other localities in Illinois are also famous as sources of Silurian fossils. For example, the old quarries at Grafton, Illinois, have long produced complete specimens of the trilobite *Gravicalymene celebra,* and specimens from there can be found in museums throughout the world (Mikulic and Kluessendorf 2000). Figure 8-2 illustrates some of the characteristic fossils of the Silurian and Lower Devonian.

In the 1940s and 1950s, Illinois became famous for geological studies of its Silurian reefs (Lowenstam 1948, 1957). In a classic paper on Marine reef in Madison County, Lowenstam (1948) was one of the first to recognize the role of these ancient structures as oil reservoirs. Eventually, Illinois Silurian reefs would become internationally famous, and the Thornton reef would become a textbook example of an ancient reef. Millions of motorists have viewed the

Figure 8-2 Characteristic fossils of the Silurian Period and the Early Devonian Epoch. **(a, b)** trilobite *Gravicalymene celebra,* ×1.25, Joliet Dolomite; **(c, d)** spiriferid brachiopods, ×1.5; **(e)** brachiopod *Schuchertera* sp., ×1.8; **(f)** cephalopod *Dawsonoceras annulatum,* ×0.7, Racine Dolomite; **(g)** brachiopod *Amphigenia curta,* ×1.4; **(h)** cystoid *Caryocrinites* sp., ×1.2, Racine Dolomite; **(i)** bivalve *Amphicoelia neglecta,* ×1.0, Racine Dolomite; **(j)** trilobite pygidium *Glyptambon gassi,* ×2.0, Racine Dolomite; **(k)** trilobite pygidium *Arctinurus occidentalis,* ×0.9, Racine Dolomite; **(l)** brachiopod *Pentamerus* sp., ×0.9, Marcus Dolomite; **(m)** brachiopod *Eodevonaria arcuata* (upper) and spiriferid brachiopods (lower), ×1.1; **(n)** colonial coral *Halysites* sp., ×0.6; **(o)** colonial coral *Favosites* sp., ×0.6, Louisville Limestone. Fossils c, d, e, g, and m are from the Lower Devonian Clear Creek Formation.

steeply dipping flank strata of Thornton reef in road cuts along Interstate 80, one of the busiest highways in the nation (Figure 8-3).

The name Alexandrian, which covers nearly the same time interval as the internationally accepted Llandovery Series, was defined by Savage (1908) using lower Silurian rocks exposed in Alexander County.

Economic Importance

The primary economic importance of Silurian dolomite rocks in Illinois lies in their use as construction materials. In 2001, annual production of crushed stone and lime from these rocks was valued at almost $250 million, ranking second only to coal as the state's most valuable natural resource. In fact, Silurian rocks have been the single largest source of stone construction materials used in the state since the early 1800s and now account for well over half of all its crushed stone production. Most of this industry has been centered on the Chicago metropolitan area, which, as one of the largest urban areas in the United States, has an enormous need for these materials. This need has been met largely by the ready availability and low transportation costs of its local high-quality Silurian stone products. Other regions, including the Rock Island area and east-central Illinois, currently produce and use large amounts of Silurian stone as well.

Although use of Silurian rocks has long been important, the character of products produced from them has changed dramatically over the last 170 years, following advances in technology and building methods (Mikulic 1989). Throughout most of the nineteenth century, lime and dimension stone were the main construction materials used in erecting buildings, bridges, sidewalks, and nearly everything else now made of concrete. The geological features of the Silurian rocks, including their hardness, purity, and nature of the bedding, made them the best stone materials available. Lime production in and around Chicago, as well as around Port Byron, made Illinois one of the largest lime-producing centers in the Midwest during the late 1800s.

A product that was even more important than lime was the building stone quarried in the Lemont-Joliet and Grafton areas. The Lemont-Joliet area was the main source of construction stone used in Chicago both before and after the fire of 1871. With the development of a rail system in the mid-nineteenth century, this stone was shipped widely throughout the region. Because of their location at the confluence of the Illinois and Mississippi Rivers, the great quarries at Grafton could ship their stone cheaply, and Grafton stone was used extensively in the St. Louis area (Mikulic and Kluessendorf 2000). Thousands of individuals were employed in the Silurian-based lime and building stone industry of this time period.

For a variety of reasons, such as the introduction of portland cement, rising labor costs, and the importation of stone materials from outside the area, an abrupt and profound change in the use of stone in Illinois occurred around 1900. The building stone and lime industries largely disappeared, only to be replaced by an even greater demand for crushed stone used as aggregate in concrete (Mikulic 1989). The Silurian rocks of the region were well suited for this new use, and they remain the basis for the modern stone industry.

Stone is not the only valuable resource produced from Silurian rocks of the state. Beginning in the 1940s in central Illinois, these rocks were discovered to be an important reservoir for oil. In 2003, approximately 10% of Illinois' annual oil production of 11,000,000 barrels was produced from Silurian rocks or Silurian reef-controlled structures in the state (Bryan Huff and Beverly Seyler, personal communication 2003).

In addition to their value in terms of economic products, Silurian rocks are important in northern Illinois as a major shallow aquifer, supplying water to many small communities and rural homeowners throughout the northeastern and some northwestern portions of the state. The character of the Silurian rocks has also played an important role in the construction of a number of major engineering projects, such as the Sanitary and Ship Canal and the Tunnel and Reservoir Plan (known as TARP) of the Metropolitan Water Reclamation District of Greater Chicago.

The Lower Devonian Clear Creek Formation also is the source of tripoli, a microcrystalline silica used as an abrasive, filler, or extender. Although the overall value of

Figure 8-3 Aerial photograph of the Material Service Corporation quarry at Thornton, Cook County, Illinois, one of the oldest and largest producers of stone products from Silurian rocks in the state. (Photograph by Joel M. Dexter.)

Silurian and Lower Devonian

Figure 8-4 Silurian rock exposures in Mississippi Palisades State Park, Savannah, Carroll County, Illinois. (Photograph by Joel M. Dexter.)

this product in the state is small, Illinois ranks as the largest producer in the nation.

Aesthetic Value

The Silurian rocks of Illinois contribute to the scenery and recreational appeal of the state. Many of the most prominent state parks and other sites owe much of their scenic beauty to cliffs and outcrops of Silurian rocks. Examples include Pere Marquette State Park near Grafton, Mississippi Palisades State Park north of Savannah (Figure 8-4), Kankakee River State Park near Kankakee, the Illinois and Michigan Canal Heritage Corridor, and the valleys along the Fox, DuPage, Des Plaines, and Kankakee Rivers. Charles Mound, the highest point in Illinois, and many other mounds in the Driftless Area of Jo Daviess County owe their very existence to their capping Silurian rocks. One of the most prominent geological sites of the state, the Thornton Quarry in Cook County (one of the ten largest quarries in the United States), is crossed on Interstate 80 by tens of thousands of commuters and travelers daily and is an internationally famous Silurian reef (Figure 8-3).

Silurian rocks also lend character to many prominent cultural features, for example, the Illinois State Capitol, the Eads Bridge, the Rock Island Arsenal, the Joliet Penitentiary, the locks along the Illinois and Michigan Canal in Will County, and the Chicago Water Tower. These structures were built totally or in part from Silurian stone quarried in Illinois. Moreover, the use of local building stone imparts a distinctive architectural flavor to a number of historical communities, such as Batavia, Grafton, Joliet, Lockport, and Lemont. The Lower Devonian rocks of Illinois have a more limited distribution, but are also well exposed in scenic outcrops along the Mississippi River. Lower Devonian rocks form several of the highest hills in southern Illinois, such as Bald Knob and the scenic bluffs of the La Rue-Pine Hills in western Union County (W. John Nelson, personal communication 2004).

References

Becker, L. E., 1974, Silurian and Devonian rocks in Indiana southwest of Cincinnati Arch: Bloomington, Indiana, Indiana Geological Survey, Bulletin 50, 83 p.

Becker, L. E., and J. B. Droste, 1978, Late Silurian and Early Devonian sedimentologic history of southwestern Indiana: Bloomington, Indiana, Indiana Geological Survey, Occasional Paper 24, 14 p.

Berg, R. B., and J. M. Masters, 1994, Geology of microcrystalline silica (tripoli) deposits, southernmost Illinois: Illinois State Geological Survey, Circular 555, 89 p.

Berry, W. B. N., and A. J. Boucot, 1970, Correlation of the North American Silurian rocks: Boulder, Colorado, Geological Society of America, Special Paper 102, 289 p.

Berry, W. B. N., and A. J. Boucot, 1973, Glacio-eustatic control of Late Ordovician-Early Silurian platform sedimentation and faunal changes: Geological Society of America Bulletin, v. 84, p. 275–283.

Carozzi, A. V., and J. Banaee, 1984, Bailey Limestone (Lower Devonian) of southwestern Illinois: A carbonate turbidite: Transactions of the Illinois State Academy of Sciences, v. 77, no. 3–4, p. 271–282.

Collinson, C., and E. Atherton, 1975, Devonian System, in H. B. Willman, E. Atherton, T. C. Buschbach, C. Collinson, J. C. Frye, M. E. Hopkins, J. A. Lineback, and J. A. Simon, eds., Handbook of Illinois stratigraphy: Illinois State Geological Survey, Bulletin 95, p. 104–123.

Droste, J. B., and R. H. Shaver, 1987, Upper Silurian and Lower Devonian stratigraphy of the central Illinois Basin: Bloomington, Indiana, Indiana Geological Survey, Special Report 39, 29 p.

Johnson, M. E., 1996, Stable cratonic sequences and a standard for Silurian eustasy, in B. J. Witzke, G. A. Ludvigson, and J. Day, eds., Paleozoic sequence stratigraphy: Views from the North American craton: Boulder, Colorado, Geological Society of America, Special Publication 306, p. 203–211.

Kleffner, M. A., S. M. Bergström, and B. Schmitz, 2005, Revised chronostratigraphy of the Ordovician/Silurian boundary interval in eastern Iowa and northeastern Illinois based on $\delta^{13}C$ chemostratigraphy, in G. A. Ludvigson and B. J. Bunker, eds., Facets of the Ordovician geology of the Upper Mississippi Valley region: Iowa City, Iowa, Iowa Geological Survey, Guidebook Series No. 24, p. 47–50.

Kluessendorf, J., and D. G. Mikulic, 1996, An early Silurian sequence boundary in Illinois and Wisconsin, in B. J. Witzke, G. A. Ludvigson, and J. Day, eds., Paleozoic sequence stratigraphy: Views from the North American craton: Boulder, Colorado, Geological Society of America, Special Paper 306, p. 177–186.

Leighton, M. W., D. R. Kolata, D. F. Oltz, and J. J. Eidel, eds., 1991, Interior cratonic basins: Tulsa, Oklahoma, American Association of Petroleum Geologists, Memoir 51, 819 p.

Lowenstam, H. A., 1948, Marine pool, Madison County, Illinois, Silurian reef producer, in Structure of typical American oil fields, v. 3: Tulsa, Oklahoma, American Association of Petroleum Geologists, p. 153–188.

Lowenstam, H. A., 1957, Niagaran reefs in the Great Lakes area, in H. S. Ladd, ed., Treatise on marine ecology and paleoecology: Boulder, Colorado, Geological Society of America, Memoir 67, v. 2, p. 215–248.

Loydell, D. K., A. Mallett, D. G. Mikulic, J. Kluessendorf, and R. D. Norby, 2002, Graptolites from near the Ordovician-Silurian boundary in Illinois and Iowa: Journal of Paleontology 76, p. 134–137.

Mikulic, D. G., 1989, The Chicago stone industry: A historical perspective, in R. E. Hughes and J. C. Bradbury, eds., Proceedings of the 23rd Forum on the Geology of Industrial Minerals held May 11–15, 1987, North Aurora, Illinois: Illinois State Geological Survey, Industrial Mineral Note 102, p. 83–89.

Mikulic, D. G., 1991, Tippecanoe II subsequence: Silurian System through Lower Devonian Series, in M. W. Leighton, D. R. Kolata, D. F. Oltz, and J. J. Eidel, eds., Interior cratonic basins: Tulsa, Oklahoma, American Association of Petroleum Geologists, Memoir 51, p. 101–107.

Mikulic, D. G., and W. Hickerson, 1994, Trilobites of the Silurian Racine Formation of northwestern Illinois, in B. J. Bunker, ed., Paleozoic stratigraphy of the Quad-cities region, east-central Iowa, northwestern Illinois: Iowa City, Iowa, Geological Society of Iowa, Guidebook 59, p. 17–21

Mikulic, D. G., and J. Kluessendorf, 1998, Sequence stratigraphy and depositional environments of the Silurian and Devonian rocks of southeastern Wisconsin: Fall Field Conference for the Joint Meeting of the Society of Economic Paleontologists and Mineralogists, Great Lakes Section and the Michigan Basin Geological Society, 84 p.

Mikulic, D. G., and J. Kluessendorf, 1999, The classic Silurian reefs of the Chicago area: Geological Society of America-North Central Section, Geological Field Trip 4: Illinois State Geological Survey, Guidebook 29, 42 p.

Mikulic, D. G., and J. Kluessendorf, 2000, Silurian geology and the history of the stone industry at Grafton, Illinois, in R. D. Norby and Z. Lasemi, eds., Paleozoic and Quaternary geology of the St. Louis Metro East area of western Illinois: 63rd Tri-State Geological Field Conference: Illinois State Geological Survey, Guidebook 32, p. 39–46.

Norby, R. D., 1991, Biostratigraphic zones in the Illinois Basin, in M. W. Leighton, D. R. Kolata, D. F. Oltz, and J. J. Eidel, eds., Interior cratonic basins: Tulsa, Oklahoma, American Association of Petroleum Geologists, Memoir 51, p. 179–194.

Ross, C. A., 1962, Silurian monograptids from Illinois: Palaeontology, v. 5, p. 59–72.

Ross, C. A., and J. R. P. Ross, 1996, Silurian sea-level fluctuations, in B. J. Witzke, G. A. Ludvigson, and J. Day, eds., Paleozoic sequence stratigraphy: Views from the North American craton: Boulder, Colorado, Geological Society of America, Special Paper 306, p. 187–192.

Savage, T. E., 1908, On the lower Paleozoic stratigraphy of southwestern Illinois: American Journal of Science, ser. 4, v. 25, p. 431–443.

Scotese, C. R., R. K. Bambach, C. Barton, R. Van der Voo, and A. M. Ziegler, 1979, Paleozoic base maps: Journal of Geology, v. 87, p. 217–277.

Shaver, R. H., 1996, Silurian sequence stratigraphy in the North American craton, Great Lakes area, in B. J. Witzke, G. A. Ludvigsen, and J. Day, eds., Paleozoic sequence stratigraphy: Views from the North American craton: Boulder, Colorado, Geological Society of America, Special Paper 306, p. 193–202.

Sloss, L. L., 1963, Sequences in the cratonic interior of North America: Geological Society of America Bulletin, v. 74, p. 93–114.

Sloss, L. L., 1982, The midcontinent province: United States, in A. R. Palmer, ed., Perspectives in regional geological synthesis, Planning for The Geology of North America: Boulder, Colorado, Geological Society of America, D-NAG Special Publication 1, p. 27–39.

Sloss, L. L., 1988, Tectonic evolution of the craton in Phanerozoic time, in L. L. Sloss, ed., Sedimentary cover—North American craton: U.S.: The Geology of North America, v. D-2: Boulder, Colorado, Geological Society of America, p. 25–51.

Sloss, L. L., W. C. Krumbein, and E. C. Dapples, 1949, Integrated facies analysis: Boulder, Colorado, Geological Society of America, Memoir 39, p. 91–123.

Thompson, T. L., 1993, Paleozoic succession in Missouri, Part 3, Silurian and Devonian Systems: Rolla, Missouri, Missouri Department of Natural Resources, Division of Geology and Land Survey, Report of Investigations 70, 228 p.

Whitaker, S. T., 1988, Silurian pinnacle reef distribution in Illinois: Model for hydrocarbon exploration: Illinois State Geological Survey, Illinois Petroleum 130, 32 p.

Willman, H. B., 1973, Rock stratigraphy of the Silurian System in northeastern and northwestern Illinois: Illinois State Geological Survey, Circular 479, 55 p.

Willman, H. B., and E. Atherton, 1975, Silurian System, in H. B. Willman, E. Atherton, T. C. Buschbach, C. Collinson, J. C. Frye, M. E. Hopkins, J. A. Lineback, and J. A. Simon, eds., Handbook of Illinois stratigraphy: Illinois State Geological Survey, Bulletin 95, p. 87–104.

Middle Devonian Series through Mississippian System (Kaskaskia Sequence)   9

Joseph A. Devera, W. John Nelson, and Rodney D. Norby

INTRODUCTION

Rocks of the Kaskaskia sequence (Middle Devonian through Late Mississippian) underlie nearly all of Illinois south of Interstate 80 but are covered by Pennsylvanian bedrock and Pleistocene glacial deposits through most of this area. The rocks are best exposed along the bluffs of the Mississippi River, extending from Bettendorf, Iowa, southward to Grand Tower in Jackson County, Illinois. Many outcrops are also found in the Shawnee Hills region of southern Illinois and along the Illinois River and its tributaries in western Illinois. Outcrops along the Mississippi River bluffs in Illinois contain part of the world standard section for the Mississippian System, now officially designated as the Mississippian Subsystem of the Carboniferous System.

Devonian and Mississippian rocks are of great economic importance in Illinois. More than 80% of all the oil and gas produced in the Illinois Basin came from reservoir rocks of this age, and the Upper Devonian New Albany Shale is the principal petroleum source rock in the Basin. Mississippian rocks are also the leading source of limestone quarried in the state.

In Illinois, the Kaskaskia sequence ranges in age from Middle Devonian through Late Mississippian, from about 390 to 295 million years before present (Figure 9-1). The Kaskaskia sequence is dominated by carbonate rock but also contains siliciclastics—shale, siltstone, and sandstone. These rocks were deposited in the Illinois Basin over a period of about 60 million years. The Kaskaskia sequence lies above the sub-Kaskaskia unconformity and below the sub-Absaroka unconformity.

Rocks of this sequence are exposed near the Kaskaskia River in southwestern Illinois (Sloss et al. 1949). The Kaskaskia sequence is thickest, about 3,800 feet (1,158 m), and most complete in southern Illinois counties (Gallatin, Saline, eastern Williamson, northern Johnson, and Pope). In this area, subsidence during deposition of the Kaskaskia sequence was more rapid than elsewhere, resulting in nearly continuous sedimentation and maximum preservation of Middle Devonian through Upper Mississippian strata beneath the sub-Absaroka unconformity. Northward, individual units of the Kaskaskia sequence become thinner, and the gaps in the geological record become greater. Notably, the sub-Absaroka erosion surface is more deeply incised to the north. In part of northeastern Illinois, the Kaskaskia sequence is entirely absent, and the rocks of the Absaroka (Pennsylvanian strata) directly overlie Silurian and older rocks (Figure 9-2).

Sloss (1988) divided the Kaskaskia sequence into two subsequences. The Kaskaskia I subsequence comprises the Middle Devonian Series through the Mississippian Kinderhookian Series, including rocks of the Grand Tower Limestone in southern Illinois through the Chouteau Limestone (Figure 9-1). The Kaskaskia II subsequence contains the Valmeyeran Series through upper Chesterian Series, including formations from the Meppen Limestone through the Kinkaid Limestone (Figure 9-1).

GEOLOGICAL HISTORY

The Kaskaskia sequence was greatly influenced by two mountain-building events, the Acadian Orogeny and the Alleghany Orogeny. Collision of the Laurentian (North America) and Baltican (Europe) paleocontinents caused upwarping in Illinois, regression of the sea, and erosion of topographic high points prior to deposition of the Kaskaskia sequence. This erosion produced the sub-Kaskaskia unconformity (Figure 9-2), which cuts across older rocks, except in southernmost Illinois. With the ensuing sea-level rise, carbonate deposition (Grand Tower Limestone) shows a deepening-upward cycle of sediment in the basal part of the Kaskaskia sequence.

During the Middle Devonian, the Illinois Basin was positioned 10 to 20° south of the equator (Droste and Shaver 1983). The continent continued to move northward toward the equator throughout deposition of the Kaskaskia sequence. By Mississippian time, the Illinois Basin was 5° south of the equator. A warm, equatorial climate is indicated by the dominance of warm-water, limestone-producing fossil organisms in the rocks.

Carbonate production was interrupted numerous times during the deposition of the Kaskaskia sequence, beginning in late Middle Devonian time, when mud reached Illinois from the Catskill Delta in the eastern United States (Collinson et al. 1967). By Late Devonian time, the Illinois Basin was covered with poorly oxygenated seas west of the Catskill Delta. The New Albany Shale, deposited during this time, is part of a vast deposit of black, organic-rich Late Devonian shale covering much of North America.

Figure 9-3 shows the unit's distribution in the eastern United States. Such shales are prolific source rocks for petroleum and natural gas.

In Pike, Calhoun, and Jersey Counties, carbonate production resumed as a result of local clearing of the seawater in latest Devonian and earliest Mississippian time (Kinderhookian), as evidenced by limestones (Louisiana and Horton Creek = "Glen Park" limestone) that straddle this boundary. However, dark muds were still being deposited in the eastern portion of the Illinois Basin. The first phase of Kaskaskia deposition, the Kaskaskia I subsequence, came to a close at the end of Kinderhookian time as sea level dropped and the upper part of the Chouteau Limestone was eroded in the shallowest areas of the Illinois Basin.

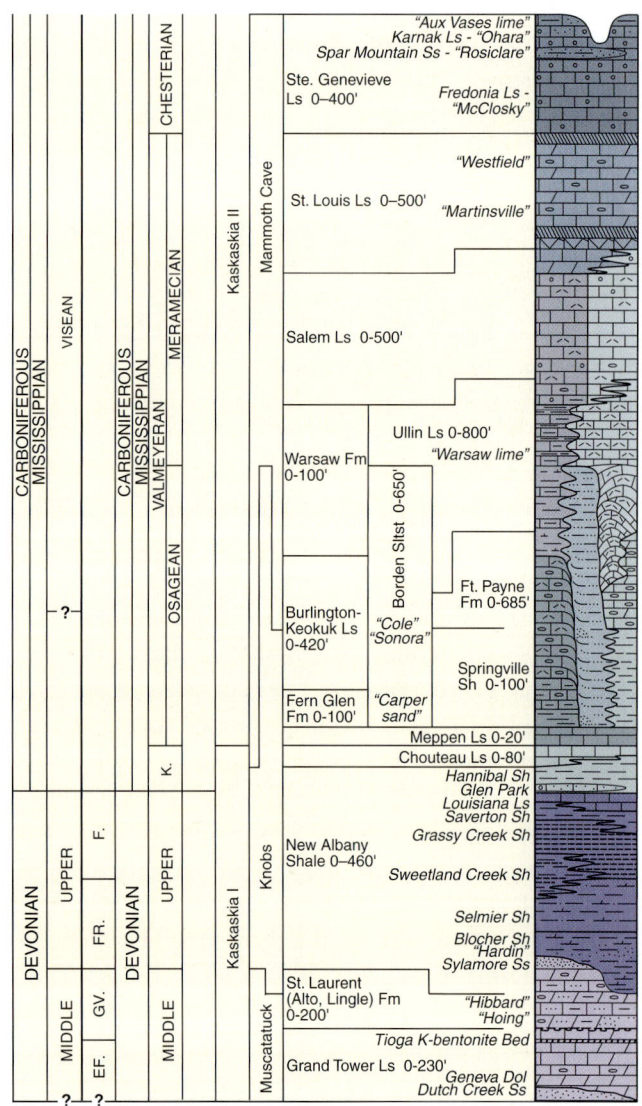

Figure 9-1 Stratigraphic column of the Kaskaskia sequence (Kolata 2005). Unit thicknesses in the text may vary from the column due to scale and because the text is based on more recent, detailed research. Abbreviations: CH., Chesterian; EF., Eifelian; F., Famennian; FR., Frasnian; GV, Givetian; K., Kinderhookian; Muscat., Muscatatuck; Dol, Dolomite; Fm, Formation; Ls, Limestone; Sh, Shale; Sltst, Siltstone; Ss, Sandstone; NE, northeast; NW, northwest. Lithologic symbols are explained in Appendix I.

Deposition of the Kaskaskia II subsequence resumed with a transgression of the sea. The central part of the Illinois Basin in southeastern Illinois subsided rapidly and was covered by deep water while western Illinois remained a shallow, stable shelf. A large crinoidal bank developed on this shelf in western Illinois, forming the Burlington and Keokuk Limestones. The deeper part of the Illinois Basin became a "starved basin" (Lineback 1969), subsiding more rapidly than the rate of sedimentation. Siliceous limestone of the Fort Payne was deposited in this part. Meanwhile, a large delta prograded—built outward—into the Illinois Basin from the east, depositing submarine fans (Borden Siltstone) and prodeltaic muds (Springville Shale) (Swann et al. 1965, Lineback 1966, Ausich et al. 1979). By mid-Valmeyeran time, the Illinois Basin was largely filled with sediment: carbonate bank on the west, Borden submarine fan to the north and east, and Fort Payne siliceous limestone to the southeast.

During Late Valmeyeran time, subsidence slowed, but sedimentation continued, causing the ocean covering the Illinois Basin to become progressively more shallow. This process commenced with the deposition of the shallow, subtidal cross-bedded grainstones and packstones of the Ullin Limestone and its equivalents. The upward progression is from the cyclic, shallow subtidal deposits of the Salem Limestone, through subtidal and supratidal dolomite and anhydrite of the St. Louis Limestone, and finally oolite shoals and exposure surfaces in the Ste. Genevieve Limestone (Cluff and Lineback 1981).

During Chesterian time, the final phase of the Kaskaskia sequence, a shallow sea covered most of Illinois (Figure 9-4). The Illinois Basin was bordered on the east by the Cincinnati Arch, on the northwest by the Transcontinental Arch, and on the southwest by the Ozark Dome. The Illinois Basin was open to the south, facing the deep Ouachita Trough. Other arms of the shallow sea extended westward across the southern midcontinent and northeast-

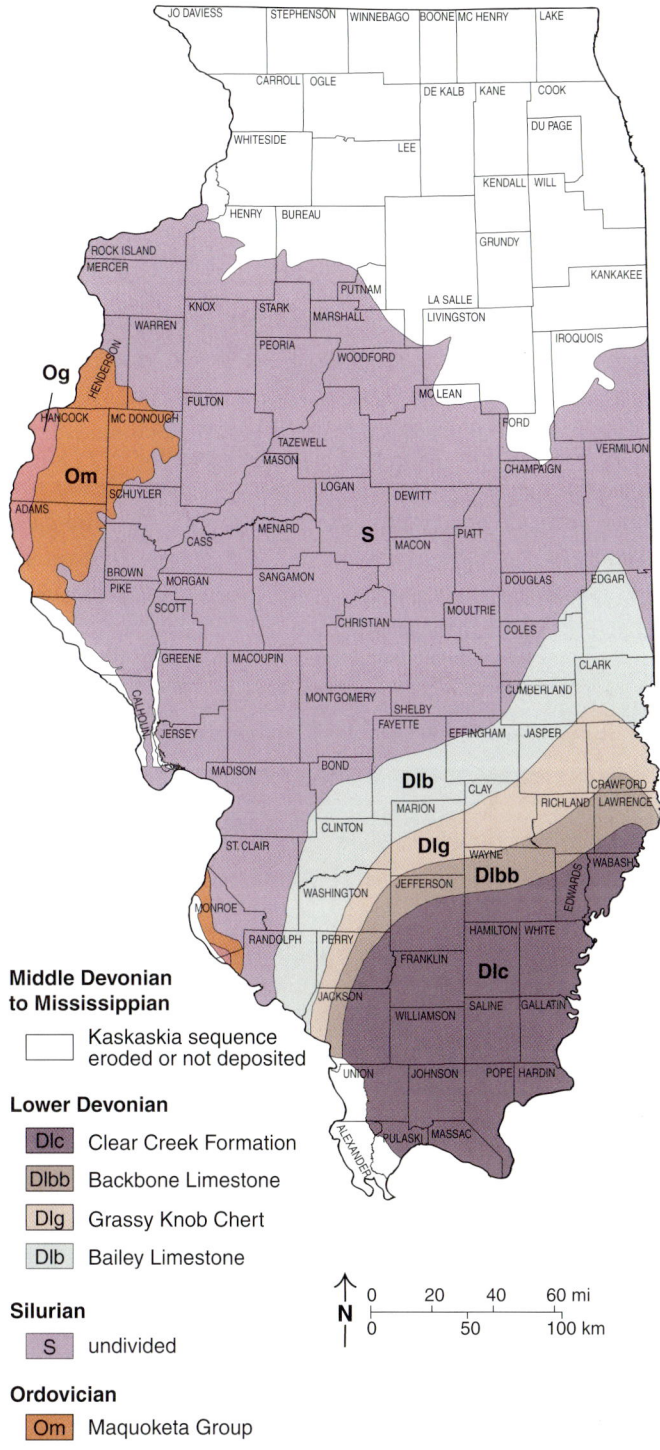

Figure 9-2 Geological map of the sub-Kaskaskia surface (Willman et al. 1967, Kolata 2005).

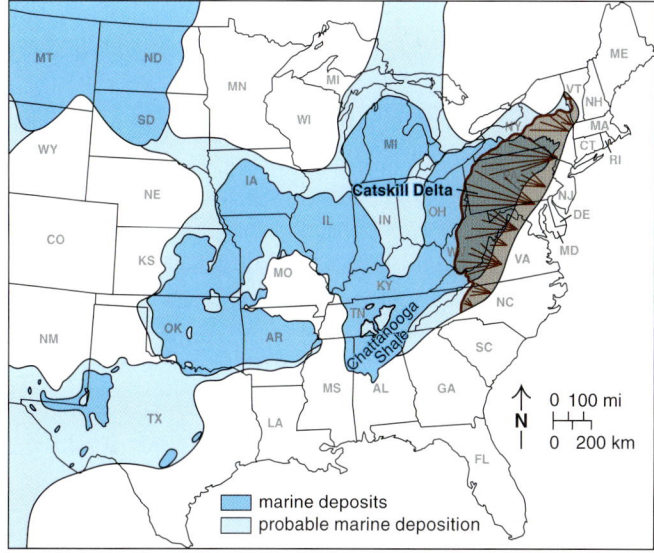

Figure 9-3 Distribution of the New Albany Shale and equivalent strata and the relationship of the Illinois Basin to the Catskill Delta.

ward into the Appalachian Basin. The primary source of cratonic sediment came from the Canadian Shield (Swann 1963) and, to a lesser extent, the Transcontinental Arch and Ozark Dome. The Appalachian Mountains are also a sediment source. The Chesterian Series ended with a global drop in sea level, ending the Kaskaskia sequence.

Kaskaskia I Subsequence

The Kaskaskia I subsequence consists of the Middle Devonian through Kinderhookian Series (Grand Tower Limestone through Chouteau Limestone).

Sub-Kaskaskia Unconformity

A major regression of the sea at the end of Early Devonian time exposed most of the sub-Kaskaskia surface. This surface is an unconformity throughout most of Illinois, except in the deepest part of the Illinois Basin to the south, where outcrops and well records indicate continuous sedimentation. Shallower areas in Illinois, such as the Sparta Shelf, Sangamon Arch, and Kankakee Arch, were exposed and then deeply eroded (Figure 9-5). In western Illinois, channels in Silurian rocks as deep as 50 feet (15 m) are filled with Middle Devonian sandstone and limestone (Meents and Swann 1965). During the Middle Devonian, the short-lived Sangamon Arch separated the Illinois Basin from the Iowa Basin (Whiting and Stevenson 1965, Nelson 1995).

Middle Devonian Series

The Middle Devonian Series in Illinois is equivalent to the Eifelian (lower Middle Devonian) and Givetian (upper Middle Devonian) Stages, which are global chronostratigraphic units (Collinson et al. 1967). These units (Figure 9-1) are defined by the presence of conodonts—microscopic tooth parts of a primitive swimming vertebrate—and invertebrate fossils.

Life flourished in the Middle Devonian seas of Illinois. This seascape was dotted with small organic "patch reefs" dominated by stromatoporoids, corals, and calcareous algae. Fish were rapidly evolving; in fact, the Devonian has been called the "Age of Fishes." Middle Devonian strata contain a cornucopia of brachiopods, stick-like crinoids, lacy bryozoans, trilobites, bivalves, and snails. Straight nautiloid cephalopods (an octopus-like animal in an elongate, conical shell) and placoderm fish were the early predators of sea life. Most of these animals lived in and around the reef community, harbored from the waves by an encrusting, coralline ridge. Behind the ridge, species adapted to quieter water conditions, and delicate branching organisms thrived. Figure 9-6 shows some of the characteristic fossils of Devonian and Mississippian time.

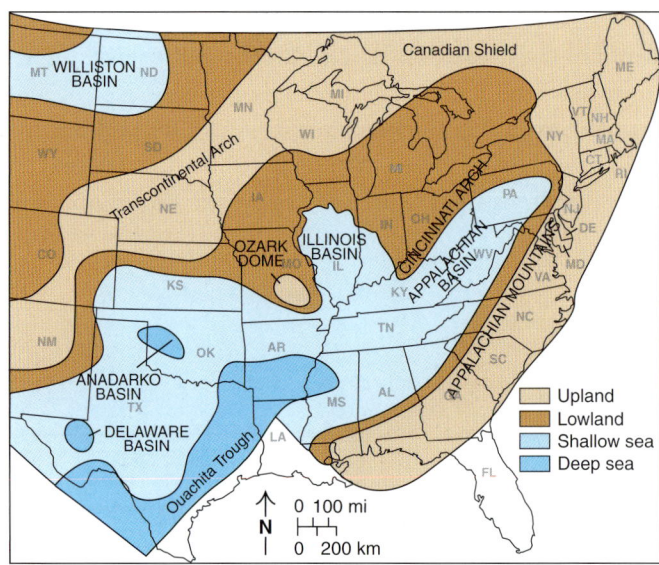

Figure 9-4 Sediment sources and generalized geography during Chesterian time.

On land, vascular plants came into being and eventually developed into the first primitive forests. These early plants were precursors to plants of the vast coal-forming swamps of the Pennsylvanian, and thin layers of Devonian coal formed on the north side of the Sangamon Arch (Peppers and Damberger 1969).

Distribution and General Stratigraphy

Lower Middle Devonian (Eifelian) strata south of the Sangamon Arch are assigned to the Grand Tower Limestone. Rocks northwest of the arch are assigned to the Wapsipinicon Limestone. The correlative (and directly contiguous) unit in Indiana and Kentucky is the Jeffersonville Limestone.

Sedimentation was continuous from Early Devonian into Middle Devonian times in the southern part of the Illinois Basin. Evidence includes the lack of erosion between the Lower Devonian (Clear Creek Formation) and Middle Devonian (Grand Tower Limestone) rocks and fossil evidence that the Grand Tower straddles the boundary between the Lower and Middle Devonian (Devera and Fraunfelter 1988).

The Grand Tower Limestone is composed of three members and one bed in Illinois. The Dutch Creek Sandstone occurs in some places at the base. Two laterally equivalent facies overlie the Dutch Creek: the Geneva Dolomite in east-central Illinois and an unnamed bioclastic facies that makes up the bulk of the Grand Tower in southern Illinois (Figure 9-1). The Tioga K-bentonite Bed, a widespread volcanic ash bed, occurs in the upper part of both the Geneva Dolomite and the southern bioclastic facies.

MIDDLE DEVONIAN–MISSISSIPPIAN

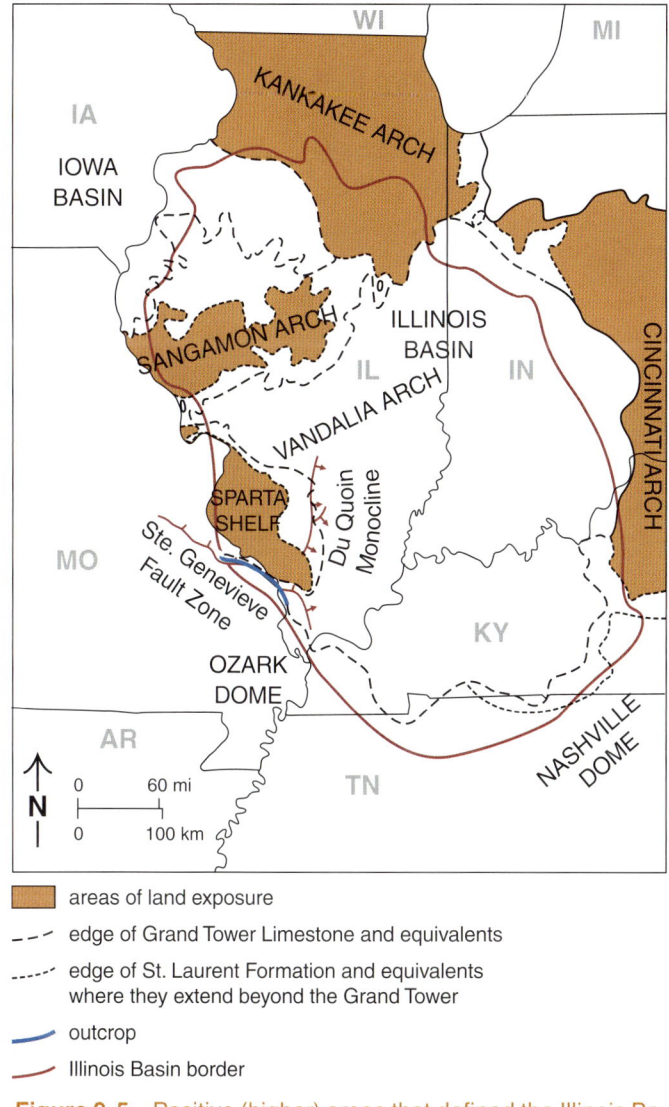

Figure 9-5 Positive (higher) areas that defined the Illinois Basin during early deposition of the Kaskaskia sequence.

The Dutch Creek Sandstone is a white, fossiliferous, calcareous-cemented, medium-grained, quartz arenite with frosted grains. The sand was probably recycled from the St. Peter Sandstone (Ordovician) and older units that were exposed to erosion around the margins of the Illinois Basin during Early Devonian time (Summerson and Swann 1970). Tidal laminations and fossils in southern Illinois indicate that deposition occurred on a shallow shelf. Among characteristic fossils are the trilobites *Odontocephalus* and *Eldredgeops (Phacops)*, the brachiopods *Amphigenia* and *Eodevonaria*, the tabulate coral *Pleurodictyum problematicum*, bivalves, conularids, and rugose corals (Figure 9-6). The sandstone grades upward into bioclastic facies of the Grand Tower Limestone in the south and the Geneva Dolomite in the north.

The bioclastic facies of the Grand Tower Limestone crops out in southwestern Illinois. This deepening-upward carbonate succession is unlike the typical shallowing-upward successions common to most limestones. The lower part contains white, cross-bedded, coarse, crinoidal grainstones with disseminated quartz grains. The middle part contains abundant encrusting bryozoans and a calcareous algae called *Asphaltinoides grandtowerensis* (Devera 1987). These encrusting organisms stabilized the sediment and set the stage for overlying patch-reef development. The dominant patch-reef organisms were stromatoporoids and the dome-shaped colonial corals *Hexagonaria* sp. and *Favosites* sp. Continued deepening of the sea yielded a thick package of lime mudstones at the top.

The Geneva Dolomite consists of a dark brown, vuggy, dolostone found in the subsurface in east-central Illinois and west-central Indiana. Its diverse and abundant fauna suggests a normal marine setting. The upper part of the Grand Tower Limestone above the Geneva contains variegated lime mudstones and laminated dolostones with bird's-eye structures, sporadic quartz grains, and brecciated laminae. In Indiana, this facies is called the Vernon Fork Member (Droste and Shaver 1975). Its environment was probably similar to modern carbonate flats and sebkhas, which are shallower and more restricted than the depositional environment of the Geneva Dolomite.

North of the Sangamon Arch, the lower part of the Wapsipinicon Limestone (Otis Member; not shown on generalized stratigraphic column) is a fine-grained limestone that contains a sparse open-marine fauna, indicating a direct connection and circulation between the Iowa Basin and the Illinois Basin. On the northern flank of the arch, the Otis Member thins and becomes sandy at the base (Willman et al. 1975). The upper Pinicon Ridge Member (not shown) of the Wapsipinicon contains fine-grained limestone, dolostone, solution collapse breccias, and a sparse fauna of stromatolites and ostracodes as well as burrows and fish remains. These features suggest that the Iowa Basin became isolated, shallow, and highly restricted during this time (Anderson 1998).

Renewed structural uplift and a drop in sea level at the close of early Middle Devonian time exposed much of the Sparta Shelf and Sangamon Arch (Devera and Fraunfelter 1988). The result is a disconformity between the Grand Tower Limestone and the overlying St. Laurent (Alto and Lingle) Formation. The Ste. Genevieve Fault Zone became active, elevating the Sparta Shelf relative to the Ozark Dome. As a consequence, thick Middle Devonian strata are preserved (or were formerly present) south of the fault zone, whereas Middle and Upper Devonian rocks are greatly thinned or absent on the Sparta Shelf in the north. In places on the shelf, Lower Mississippian strata lie directly on Ordovician rocks (Weller and St. Clair 1928, Nelson and Lumm 1985, Nelson and Marshak 1996). The Avon

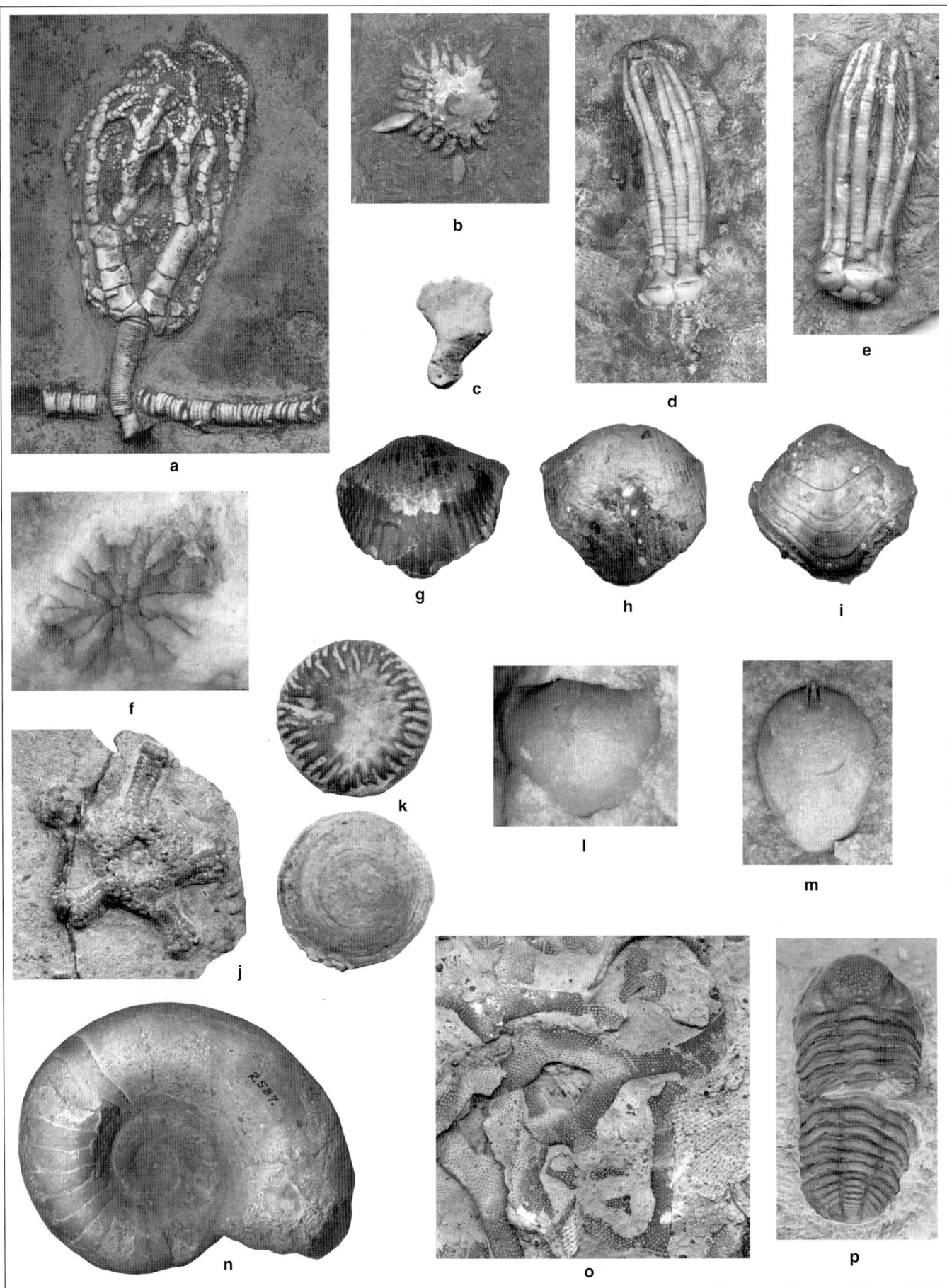

(Missouri) diatremes erupted concurrently with this fault activity (Figure 3-7b). Eruptions through the pipe-like diatremes carried rocks to the surface from great depths in the Earth's crust and mantle. In some cases, rocks that line the vent may fall downward during periods of activity. The Avon diatremes occur in an area where no Devonian rocks presently exist, having been eroded away. The presence of rock fragments containing Middle Devonian fossils within the diatremes is evidence that Devonian layers originally extended across this area and well up onto the northeastern flank of the Ozark Dome when the diatremes were active.

The upper Middle Devonian (Givetian) unit south of the Sangamon Arch is the St. Laurent Formation (Nelson et al. 1995); north of the arch is the Cedar Valley Limestone.

The St. Laurent Formation is lithologically diverse and contains complex facies associations between siliciclastic and carbonate rocks. It is composed of argillaceous limestone, dolostone, cherty limestone, shale, siltstone, thin quartz sandstones, and a local oolitic packstone. The index fossil *Microcyclus discus,* a small button-shaped coral (Figure 9-6k), occurs in the basal, gray-brown, argillaceous limestone beds. Other fossils associated with the lower part of the formation are the trilobite *Eldredgeops (Phacops),* spiriferid brachiopods, and silicified rugose corals. Chert, glauconite, and interbedded dark brown shale are also common. The upper part of the St. Laurent ranges from a dolostone and calcareous shale to a dark cherty lime mudstone at the top. These dark siliceous limestones contain the conodont *Polygnathus cristata.* The presence of this conodont places the upper St. Laurent in the upper Middle Devonian (Collinson et al. 1967). Eastward in the deeper part of the Illinois Basin, the middle to upper St. Laurent laterally interfingers with the Blocher Shale Member of the New Albany Shale.

Figure 9-6 Some characteristic fossils of Devonian and Mississippian time: **(a)** crinoid *Onychocrinus* sp., ×0.8; **(b)** crinoid *Pterotocrinus* sp., ×1.0; **(c)** crinoid wing plate of *Pterotocrinus* sp., ×1.1; **(d)** crinoid *Phanocrinus* sp., ×1.1; **(e)** crinoid *Phanocrinus* sp., ×1.1; **(f)** tabulate coral *Pleurodictyum problematicum,* ×1.2; **(g)** brachiopod *Anthracospirifer increbescens,* ×1.2; **(h)** brachiopod *Spirifer increbescens,* ×1.2; **(i)** brachiopod *Composita subquadrata,* ×1.2; **(j)** brittle star *Cholaster peculiaris,* ×1.9; **(k)** base and septal view of the button coral *Microcyclus discus,* ×2.8; **(l)** brachiopod *Eodevonaria* sp., ×2.2; **(m)** brachiopod *Amphigenia curta,* ×1.3; **(n)** cephalopod *Endolobus* sp., ×0.9; **(o)** bryozoan *Cystodictya* sp., ×1.2; **(p)** trilobite *Eldredgeops (Phacops) cristata* Hall, found in the lower Middle Devonian, near the upper part of the Grand Tower Formation. Fossils f, l, and m are from the Dutch Creek Member of the Grand Tower Limestone, lower Middle Devonian. Fossil k is an index fossil from the base of the St. Laurent Formation, upper Middle Devonian. Fossils a, b, c, d, and e are from the Renault through Golconda limestones, lower Chesterian. Fossils g, h, and i are from the Clore Formation, upper Chesterian. Fossil j is from the middle Chesterian.

North of the Sangamon Arch, the Cedar Valley Limestone lies unconformably on the Wapsipinicon Limestone. The Cedar Valley has patchy occurrences of fine- to medium-grained, well-rounded, quartz sandstone ("Hoing" sandstone) at the base (Figure 9-1). Overlying and laterally equivalent to the sandstone is a brownish gray, fossil packstone that is clay-rich and has carbonaceous debris. Its bedding is irregular to brecciated. The brachiopod *Independatrypa independensis* is an index fossil of the lower Cedar Valley Limestone (Stainbrook 1941). Occurring in the upper Cedar Valley Limestone are coralline patch reefs or biostromes dominated by rugose and tabulate corals, the colonial coral *Hexagonaria profunda,* and stromatoporoids. Dolomite also becomes more prevalent upward. Shallowing-upward cycles include lithologies that represent open marine, subtidal environments to shallow, restricted marine environments. The Cedar Valley Limestone is unconformably overlain by the Sylamore Sandstone and Sweetland Creek Shale (Upper Devonian).

Upper Devonian Series

The Upper Devonian Series is divided into Frasnian (lower) and Famennian (upper) Stages based on conodont ranges. The base of the Upper Devonian is at the top of the St. Laurent in southern Illinois and at the base of the Sylamore Sandstone in central and western Illinois (Collinson et al. 1967). The Upper Devonian in Illinois is dominated by dark-colored, organic-rich, marine basinal shales that were primarily derived from fine prodeltaic mud of the Catskill Delta. Except for conodonts, the shales contain few fossils. Only shallow marine shales of upper Famennian age contain macrofossils. The lack of large fossils reflects a devastating mass extinction of marine life that occurred near the end of the Frasnian Stage (Stanley 1989). Among the hardest hit were patch-reef animals such as corals and stromatoporoids and filter-feeding organisms such as brachiopods and bryozoans. Trilobites nearly went extinct and never regained their previous diversity.

General Stratigraphy

Except for the Blocher Shale at the base (Mid-Devonian) and the Hannibal Shale at the top (Mississippian), most of the New Albany Shale is Upper Devonian in age (Figure 9-1). The New Albany Shale is dominantly shale, but minor sandstone beds occur near the base, and thin carbonate beds occur in the upper part. The New Albany has been divided into nine members (Willman et al. 1975, Cluff et al. 1981, Devera and Hasenmueller 1991).

The Blocher Shale at the base of the New Albany Shale is a brownish gray to black, finely laminated, organic-rich, calcareous shale. The Sylamore Sandstone (Hardin equivalent in Missouri) occurs at the base of the New Albany in the shallow areas of the northwestern part of the Illinois Basin where the Blocher is absent (Figure 9-7). This thin, patchy sandstone consists of well-rounded, fine- to medium-grained quartz grains.

The Selmier Shale conformably overlies the Blocher in the deeper part of the Illinois Basin. The Selmier Shale consists of alternating beds of poorly laminated, brownish black and indistinctly bedded, dark greenish gray shale. The Selmier Shale is thickest, 200 feet (61 m), in Hardin County and is absent on the Sparta Shelf in the St. Louis Metro East area. Conformably overlying the Selmier in southern Illinois is the Grassy Creek Shale, which has the highest concentration (4% to 15%) of organic carbon of any member of the New Albany Shale (Cluff et al. 1981). The Sweetland Creek Shale in western Illinois is equivalent to the Grassy Creek Shale (Figure 9-7). The Sweetland Creek is composed of bioturbated greenish gray shales and olive-black laminated shale that rarely contains burrows.

On the western and northwestern side of Illinois, the Saverton Shale overlies the Grassy Creek and Sweetland Creek Shales. The Saverton is a bluish to greenish gray, silty partly calcareous shale that grades into the Louisiana Limestone. The Louisiana Limestone is a thin unit confined to a small area running westward from Fayette to Calhoun Counties. The Horton Creek Formation, formerly the Glen Park, is confined locally to west-central Illinois where it overlies the Louisiana Limestone and the Saverton Shale

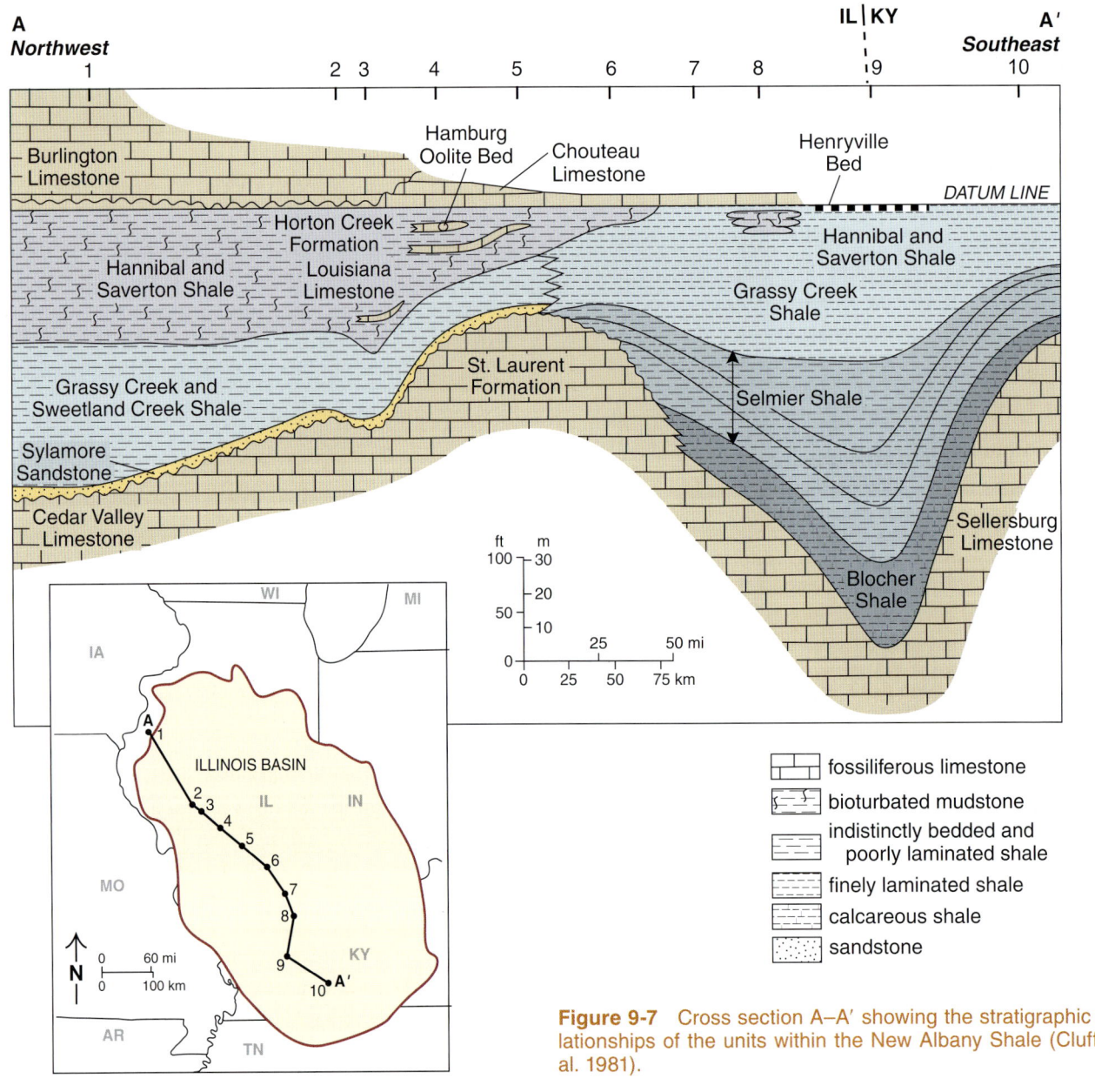

Figure 9-7 Cross section A–A' showing the stratigraphic relationships of the units within the New Albany Shale (Cluff et al. 1981).

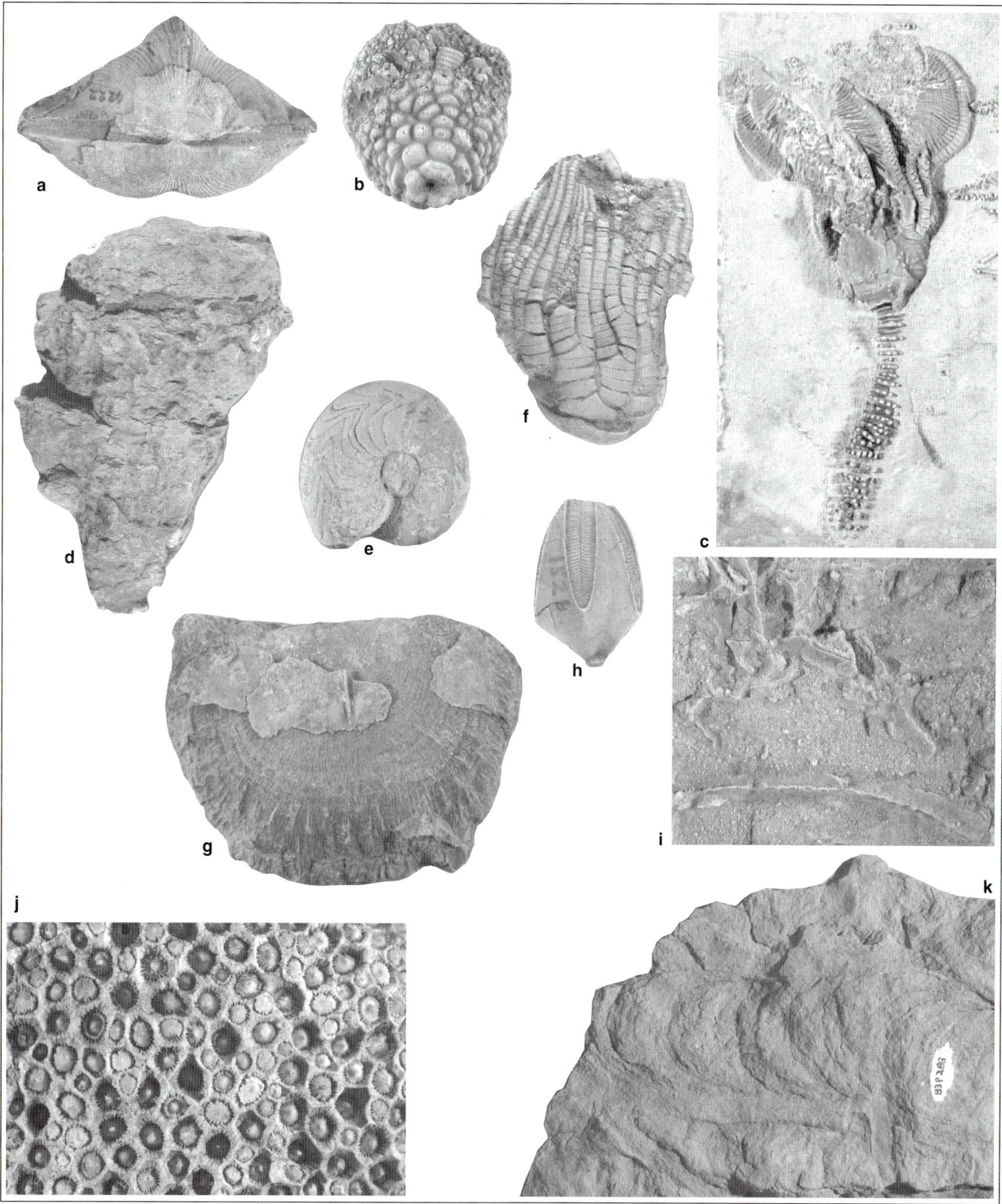

Figure 9-8 Some characteristic fossils of the Mississippian: **(a)** brachiopod *Spirifer logani*, ×0.7; **(b)** crinoid *Talarocrinus* sp.,×1.0; **(c)** crinoid *Platycrinites penicillus*, ×0.8; **(d)** trace fossil *Conostichus* ichnosp., ×0.9; **(e)** cephalopod *Muensteroceras* sp., ×0.9; **(f)** crinoid *Zeacrinites wortheni*, ×1.0; **(g)** brachiopod *Marginirugus magnus*, ×0.9; **(h)** blastoid *Pentremites* sp., 1.0; **(i)** bryozoan *Prismopora serrulata*, ×0.8; **(j)** colonial coral *Acrocyathus* sp., ×0.5; **(k)** trace fossil *Rhizocorallium* ichnosp., ×0.8. Fossils c, d, f, h, i, and k are all from Chesterian formations. Fossils a and b are from the Burlington-Keokuk, and fossil e is from the Rockford Limestone in Indiana, which is equivalent to the Chouteau in Illinois. Fossil g is from the Warsaw Formation, and fossil i is from the Glen Dean Limestone within the Chesterian. Fossil j is from the base of the St. Louis Limestone.

farther south. The Hannibal Shale (basal Kinderhookian) at the top of the New Albany Shale is a greenish gray, silty shale that is highly bioturbated (Cluff et al. 1981).

Chouteau Limestone

The Chouteau Limestone of uppermost Kinderhookian age is as thick as 80 feet (24 m) in Calhoun County, thinning to a feather edge in southernmost Illinois. The Chouteau Limestone consists of irregular beds of light brown to greenish gray fine-grained or lithographic lime mudstone. The ammonoid *Protocanites lyoni* and *Muensteroceras* sp. (Figure 9-8) and the trilobites *Phillipsia* sp. and *Proetides* sp. are rare but diagnostic. Calcite-filled geodes 1 to 8 inches (2.5 to 20 cm) in diameter and gray chert nodules occur in the limestone. The Chouteau conformably overlies the Hannibal, except in southwestern Illinois, where the Hannibal is absent and the Chouteau overlies Devonian and older strata (Willman et al. 1975). The Chouteau Limestone is unconformably overlain by rocks of the Kaskaskia II subsequence.

Kaskaskia II Subsequence

The Kaskaskia II subsequence comprises the Valmeyeran and Chesterian Series (Meppen Limestone through Kinkaid Limestone).

Valmeyeran Series Versus Osagean and Meramecian Series

The Valmeyeran (Weller and Sutton 1940) is the middle series of the Mississippian System in Illinois and Indiana. The previously named Osagean and Meramecian Series, both named in Missouri, were combined in Illinois and Indiana because the boundary was difficult to identify in those states. The U. S. Geological Survey and most other state surveys continue to use Osagean and Meramecian. The Valmeyeran takes its name from the old historic location of Valmeyer at the base of the bluff of the Mississippi River in Monroe County, Illinois. Most units of the lower part of the Valmeyeran (Osagean equivalent) are well exposed in Dennis Hollow leading eastward from historic Valmeyer. The upper Valmeyeran is present, but not easily accessible, in the bluffs north and south of present-day Valmeyer.

The upper boundary of the Valmeyeran has been moved several times (Swann 1963). Maples and Waters (1987) proposed moving the boundary down to the base of the Ste. Genevieve. Brenckle et al. (1988) agreed with this proposal, but recommended additional study. Nelson et al. (2002) were the first at the Illinois State Geological Survey to officially use the base of the Ste. Genevieve as the top of the Valmeyeran and base of the Chesterian. Although geologists today have more means at their disposal to recognize the Osagean-Meramecian boundary in Illinois (Kammer et al. 1990), the Valmeyeran remains the official series name.

Valmeyeran rocks were deposited over an interval of approximately 18 million years and consist predominantly of limestone; siltstone and shale are found near the middle of the series in some areas. The series attains a maximum thickness of over 1,800 feet (549 m) in the subsurface of southeastern Illinois (Atherton et al. 1975). Microscopic conodonts and foraminifera are the most useful fossils for subdividing the Valmeyeran. Specialists recognize about six zones of conodonts (Collinson et al. 1962, 1971; Lane et al. 1980) and five to eight zones of foraminifers (Mamet and Skipp 1970, Baxter and Brenckle 1982).

Regional Setting and Controls on Deposition

At the beginning of the Valmeyeran deposition, a wide, relatively deep seaway occupied southern Illinois. This seaway opened southward into the eastern end of the Ouachita Trough (Gutschick and Sandberg 1983; Lasemi et al. 1998, 2003) (Figure 9-9a). The seaway had been subsiding since Late Devonian time, but the subsidence ended or slowed considerably early in the Valmeyeran (Lasemi et al. 1998). The sedimentation rate slowed markedly during Late Devonian and Kinderhookian time, when organic-rich shales and thin limestones were deposited. The previously sediment-starved Illinois Basin of the early Valmeyeran provided the space to accommodate new sediments. By the middle of the Valmeyeran, the Illinois Basin began to fill rapidly with sediment. Lineback (1969) estimated that the Illinois Basin seaway was 900 to 1,000 feet (274 to 305 m) deep in southern Illinois.

Surrounding the deep seaway were shallow areas that extended into northern Illinois, Iowa, and Missouri. Carbonate sediments were deposited in these shelf areas, most thickly on the north and northwest, where the seaway was more shallow (Lineback 1981). These deposits represent the eastern portion of what is known as the Burlington Shelf (Lane 1978, Lineback 1981), a carbonate bank that formed along the north side of the Ouachita Trough from New Mexico through Oklahoma, northern Arkansas, Illinois, and into Indiana (Figure 9-9a). The shelf entered southwestern Illinois in Monroe County, continued north, and then curved gently eastward across central Illinois. As the carbonate bank grew, the thin, phosphatic Springville Shale accumulated in the Illinois Basin. The Springville Shale was succeeded by the thick Borden Siltstone, which coursed into central Illinois from the east as part of a large submarine fan called the Borden Delta (Figure 9-9b). South of the Borden Delta Complex, the dark-colored, cherty, silica-rich limestone of the Fort Payne Formation was deposited. The Fort Payne in turn was overlain by crinoid- and bryozoan-rich carbonate sediments of the Ullin Limestone, which intergrades laterally with the Fort Payne (Lineback 1966; Lasemi et al. 1994, 1998, 2003). Finally, clean light-colored bioclastic to oolitic carbonate sediments of the Salem and St. Louis Limestones largely filled the Illinois Basin by the end of the Valmeyeran (Lineback 1966).

Distribution and General Stratigraphy

In Illinois, Valmeyeran rocks crop out or lie beneath glacial sediments from Henderson and Warren Counties on the north to Madison County on the south (Figure 9-10). Farther south, the outcrop belt follows the Mississippi River and swings eastward to the Ohio River in Hardin County. Valmeyeran rocks also subcrop in east-central Illinois. The Valmeyeran Series attains its maximum thickness of 1,800 feet (549 m) in southeastern Illinois, thinning northward to less than 800 feet (244 m) in central Illinois as the individual units thin. Farther north, the upper Valmeyeran is truncated by erosion and thins to a feather edge (Atherton et al. 1975). Stratigraphy of the Valmeyeran is depicted in Figure 9-1. Fossil zonation has been summarized by Norby (1991).

Meppen through Keokuk Limestones

The oldest Valmeyeran unit in southern Illinois, the Meppen Limestone (Figure 9-1), is generally a buff-colored, fine-grained, dolomitic limestone or dolomite. The Meppen crops out in a small area of western Illinois and attains a maximum thickness of 20 feet (7 m) (Atherton et al. 1975). The overlying Fern Glen, Burlington, and Keokuk formations are more widespread through a region extending southward from Henderson County to Monroe County. The Fern Glen is a 30- to 40-foot-thick (9- to 12-m-thick) unit of red-to-green, clay-rich limestone and calcareous shale. The unit grades upward and laterally into the light gray, coarse-grained, crinoidal Burlington Limestone, which is up to 200 feet (61 m) thick. The Burlington becomes very cherty and finer grained toward the south.

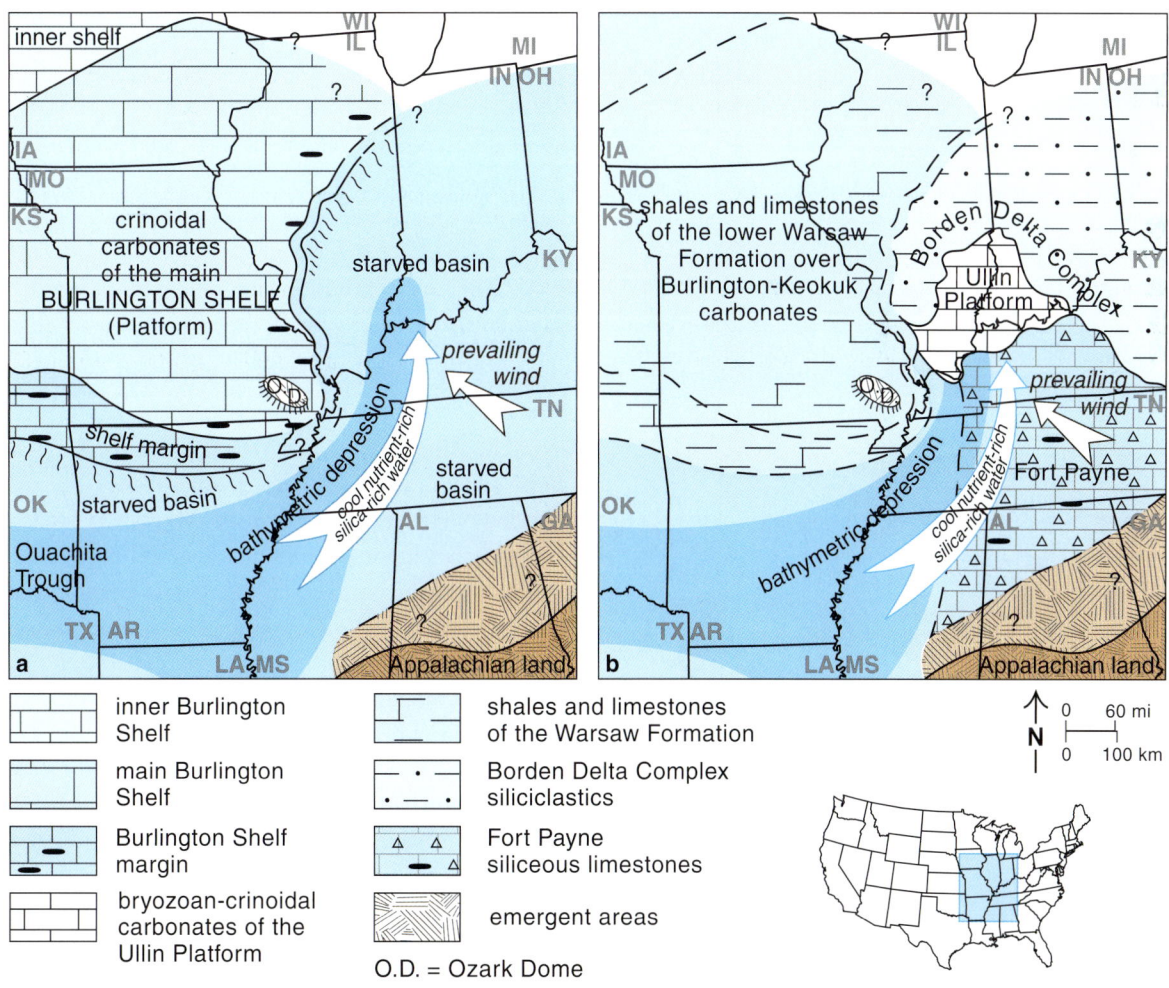

Figure 9-9 Paleogeography during early (a) and later (b) Valmeyeran time (Lasemi et al. 2003).

Large spiriferid brachiopods, *Spirifer grimesi,* are common in white crinoidal packstone and grainstone of the lower Burlington. The large echinoid *Melonechinus multiporus* (Figure 9-11), although not common (but highly prized by fossil collectors), seems to be restricted to formations from the Burlington Limestone to the St. Louis Limestone. The upper Burlington contains more glauconite, numerous blastoids, and the crinoid *Dizygocrinus.* The Keokuk Limestone is largely crinoidal limestone that is similar to the Burlington, containing interbeds of clay-rich dolomite, fine-grained limestone, and calcareous gray shale. Siliceous sponge spicules and chert nodules are abundant in the lower part. The Keokuk reaches 60 to 80 feet (18 to 24 m) thick (Atherton et al. 1975). Because of their lithologic similarity, many geologists treat the Burlington and Keokuk as a single unit.

Springville, Borden, Warsaw, Ullin, and Fort Payne Formations

In central to southeastern Illinois, a thin veneer of greenish gray clay-rich shale, the Springville Shale, forms the basal Valmeyeran (Figure 9-12). Outcrops are restricted to two small areas of southern Illinois between Mountain Glen and Jonesboro, Illinois. The Springville ranges up to 100 feet (30.5 m) thick and grades laterally and vertically to the Borden Siltstone. The Borden ranges up to 650 feet (198 m) thick in east-central and south-central Illinois, but does not crop out in Illinois. The Springville Shale represents prodelta deposits; that is, it was deposited in deep water off shore in advance of the delta slope. The Borden Siltstone comprises the subaqueous portion of the delta proper, including delta front and delta top sediments.

Dark brownish gray, fine-grained, siliceous, cherty limestone of the Fort Payne Formation overlies the Springville Shale in southern Illinois. The Fort Payne thins northward from more than 600 feet (183 m) thick in Pope County. The Fort Payne grades laterally and vertically into the lower part of the Ullin Limestone. The lower Ullin is mostly fine-grained limestone that is somewhat siliceous and cherty. The upper Ullin is light-colored, fine- to coarse-grained cross-bedded limestone that is very rich in bryozoan and crinoid debris. The Ullin is thickest, about 800 feet (244 m), in Hamilton County in south-central Illinois. Where the Fort Payne is thickest, the Ullin is thinnest and vice versa (Lineback 1966; Lasemi et al. 1994, 1998, 2003). Added together, thickness is more or less uniform, although the reasons for this are unclear.

The Warsaw Formation overlies the Keokuk Limestone in western Illinois, grading eastward into the uppermost Borden Siltstone. To the east, the Warsaw is predominantly a clay-rich siltstone up to 300 feet (91 m) thick. On the west, the lower part of the Warsaw is predominantly gray shale with thin beds of clay-rich to dolomitic limestone and abundant geodes (avidly sought by collectors). The upper Warsaw is mostly thick-bedded, shaly limestone composed of fenestrate bryozoans and crinoid fragments. In western Illinois, the Warsaw is thinner than 100 feet (30.5 m). The Sonora formation is a sandy facies of the upper Warsaw and Salem, found in Adams and Hancock Counties in west-central Illinois.

The Warsaw is noted for whole, articulated crinoid specimens. *Tricoelocrinus woodmani* is known only in this unit. Common species include *Dizygocrinus gorbyi, Abrotocrinus coreyi, Eratocrinus coxanus,* and *Synbathocrinus* sp. The blastoid

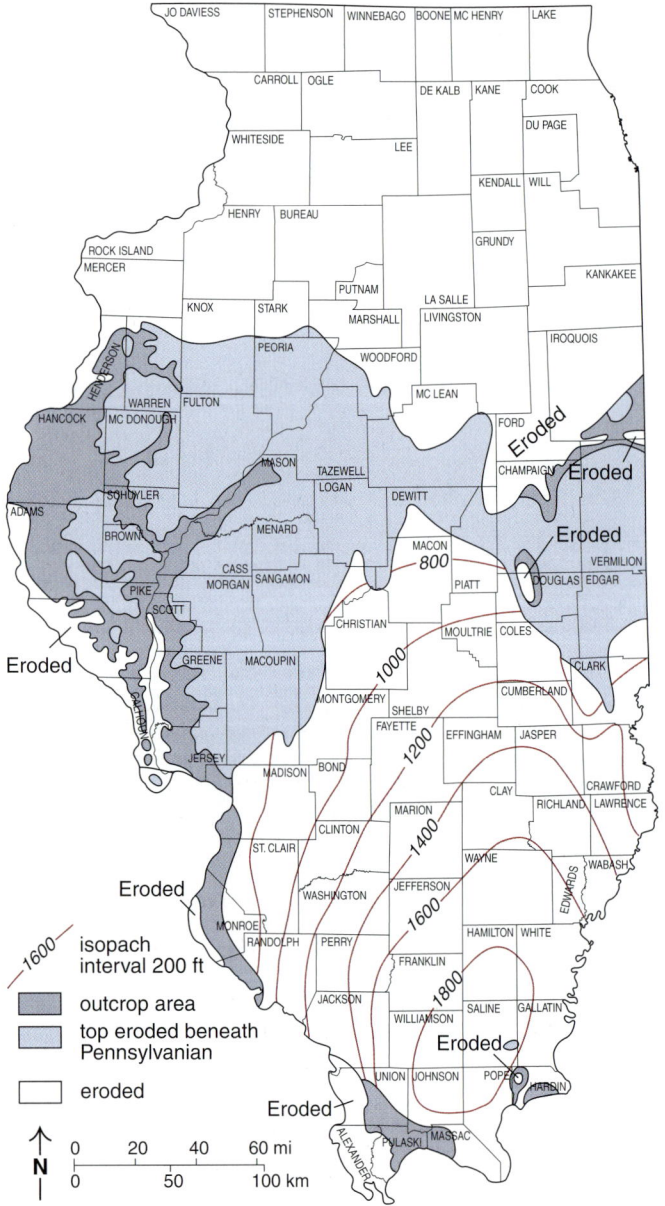

Figure 9-10 Thickness of the Valmeyeran Series and outcrop areas where Chesterian rocks are present (Atherton et al. 1975).

Metablastus sp. is very common in the Warsaw Formation; productid brachiopods, particularly *Marginirugus magnus*, are abundant in thin layers 20 feet (6 m) above the base of the Warsaw. The most abundant fossil in the Warsaw is the "corkscrew" bryozoan *Archimedes* sp.; in fact, the Warsaw was called "the Archimedes limestone" in many early reports (Figure 9-13).

Salem Limestone

The Salem Limestone overlies the Warsaw Formation and Ullin Limestone and consists predominantly of a granular limestone with abundant abraded grains of fossil fragments, carbonate-coated grains, and oolites. Minor amounts of dolomite, dolomitic limestone, clay-rich limestone, and silica-rich limestone are commonly present. The

Figure 9-11 *Melonechinus multiporus* Norwood and Owen occurs in the St. Louis Limestone. The fossil is typically found as disarticulated plates. This specimen shows two large, beautifully preserved echinoids with many small spines intact in a shaly facies of the St. Louis Limestone. The fossil was found in riprap on the Mississippi River. (Photograph by Dennis R. Kolata.)

Salem is quarried extensively for limestone aggregate. The Salem Limestone is distributed over the southern half of Illinois and ranges up to 500 feet (152 m) in southeastern Illinois. The most noted index fossil is the microfossil *Globoendothyra baileyi,* a foraminiferid. Other fossils include bryozoans, corals, brachiopods, crinoids, and echinoids, found mainly as disarticulated fragments deposited under high-energy conditions of the shallow shelf, including sand waves and shoals. The upper part of the Salem grades laterally into the lower part of the St. Louis Limestone, especially in southwestern Illinois (Lineback 1972, Baxter and Brenckle 1982).

St. Louis Limestone

The St. Louis Limestone is the youngest Valmeyeran unit. It is typically very fine-grained, cherty limestone (lime mudstone) with interbeds of dolomite, dolomitic limestone, granular limestone, and anhydrite (mainly in the lower part). The anhydrite is represented by limestone breccia in outcrop. The St. Louis is typically 200 to 300 feet (61 to 91 m) thick, but reaches 500 feet (152 m) in southeastern Illinois.

The colonial rugose coral *Acrocyathus* sp. (Figure 9-8j) is abundant in the lower St. Louis and forms a mappable bed in southwestern Illinois from Alton in Madison County to Prairie Du Rocher in Randolph County and also locally in southern Illinois. Vertebrate fossils occur locally as phosphatic bone beds near St. Louis, Missouri, in the middle to upper part of the formation. Teeth of pavement sharks or of shell-crushing sharks also occur throughout the unit. Many species of bryozoans are known, particularly the flat, branching *Cystodictya* sp. (Figure 9-6o), which occurs in great abundance in shaly green partings of the upper St. Louis. Brachiopods, mollusks, trilobites, crinoid stems, and whole crinoids also occur in argillaceous layers. The crinoid *Platycrinites penicillus* (Figure 9-8c), although not restricted to this unit, is commonly found as whole crowns in thin, green, shaly layers of the upper St. Louis Limestone.

Chesterian Series

The uppermost division of the Mississippian in North America is the Chesterian Series, named after Chester, Illinois, where pioneer geologist Amos Worthen first described these rocks in 1860. The series has been redefined several times, evolving from a lithostratigraphic unit (based on rock description) to a chronostratigraphic unit (denoting an epoch of Earth's history) (Weller 1913, 1920; Weller and Sutton, 1940; Swann 1963). Swann (1963) placed the Valmeyeran-Chesterian boundary within the Renault Limestone. Maples and Waters (1987), Brenckle et al. (1988), and Nelson et al. (2002) place the boundary at the base of the Ste. Genevieve; their placement is accepted here. The Chesterian Series in Illinois is composed of numerous alternating beds of limestone, sandstone, and shale. Thus, the Chesterian contrasts with older Mississippian strata, which are largely limestone, and with overlying Pennsylvanian rocks, which are mostly sandstone and shale. Geologists divide the Chesterian into many formations (Figure 9-1). In general, formations of interbedded sandstone and shale alternate with formations of limestone or of interbedded limestone and shale. Some limestone units are nearly continuous throughout the Illinois Basin. Others are more localized and grade laterally to sandstone and shale. Formation names change from place to place to reflect these lithological variations.

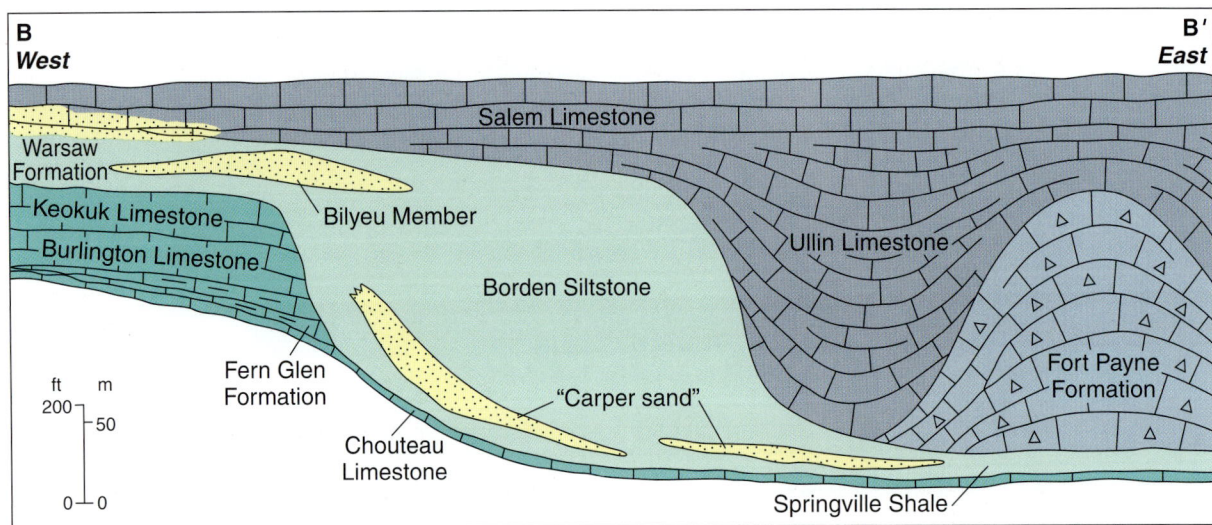

Figure 9-12 Generalized cross section showing the complex stratigraphic relationships of the Burlington Limestone, Keokuk Limestone, Springville Shale, Borden Siltstone, Ullin Limestone, and Fort Payne Formation (Lineback 1968).

Distribution and General Stratigraphy

Chesterian rocks are confined to the southern half of Illinois (Figure 9-10). Outcrops extend discontinuously along the Mississippi River bluffs from Alton southward through Chester to Grand Tower and continue eastward through Anna, Vienna, Golconda, and Cave-in-Rock along the southern flank of the Shawnee Hills. In the subsurface, Chesterian rocks extend northward to approximately Springfield, Clinton, Tuscola, and Paris. From south to north, Chesterian rocks are truncated with angular unconformity beneath basal Pennsylvanian strata. This erosion surface reflects the regional tilting of the land surface that took place during this time. Thus, Chesterian rocks thin from as much as 1,800 feet (549 m) in Johnson County (far southern Illinois) to a feather edge in central Illinois. Near the feather edge, only the oldest Chesterian rocks remain; north of this line, the Pennsylvanian directly overlies Valmeyeran and older rocks.

The Ste. Genevieve Limestone consists of alternating fine- and coarse-grained carbonate rocks and minor sandstone, siltstone, and shale (Figure 9-1). Fine-grained intervals are composed of wackestone and microcrystalline dolomite. Coarse-grained intervals are largely oolitic and skeletal grainstone. Oolitic limestone tends to occur as lenses that have flat bottoms and convex tops and are elongate in map view. Individual lenses are one-half mile (0.8 km) to several miles long, one-eighth to three-fourths mile (0.2 to 1.2 km) wide, and up to 50 feet (15 m) thick. The lenses commonly occur in northeast-trending, parallel swarms (mega-ripples) and are interpreted as tidal bars analogous to those of the modern Bahama Banks (Carr 1973, Cluff and Lineback 1981, Choquette and Steinen 1980, Gibson 2001). These oolite tidal bars in the Ste. Genevieve are highly sought by the petroleum industry because they commonly form good reservoir rocks (Cluff 1986). The Ste. Genevieve has yielded more than 1 billion barrels of oil. The Ste. Genevieve also contains a bryozoan and algal mud-mound facies (Lasemi et al. 2003). Interbeds of sandstone, siltstone, and shale become thicker and more numerous toward the northwest. The Ste. Genevieve is commonly 150 to 200 feet (46 to 61 m) thick, locally exceeding 400 feet (122 m) in southernmost Illinois. This unit becomes thin or absent in places near the flank of the Ozark Dome as a result of tectonic uplift during or shortly following the deposition of the Ste. Genevieve (Weller and Sutton 1940).

The remainder of the Chesterian comprises formations of limestone and shale that alternate with formations of sandstone and shale (Figure 9-1). Many of the limestone units are uniform in thickness across large areas and, thus, are very useful for subsurface correlation and structural mapping. The Downeys Bluff (lower Paint Creek Formation), Beech Creek ("Barlow") Limestone, Glen Dean Limestone, Vienna Limestone, Menard Limestone, and Kinkaid Limestone all are widely mapped marker units. To generalize, most limestones of the lower Chesterian resemble the Ste. Genevieve. They are light-colored, coarsely granular, oolitic, and fossiliferous (grainstones and packstones). Progressing upward, the limestones tend to become darker and finer grained and possess greater clay content (argillaceous wackestones and lime mudstones).

Formations of sandstone and shale vary laterally more than do units of limestone and shale. For example, the Aux Vases Sandstone is as thick as 160 feet in southwestern Illinois, forming bold cliffs at Prairie du Rocher in Randolph County. This unit thins eastward to isolated lenses of sandstone less than 20 feet (6 m) thick near the Kentucky and Indiana borders. The Yankeetown Sandstone and Degonia

Figure 9-13 Central support column for bryozoan colony, *Archimedes wortheni*, Warsaw Formation, Mississippian System, Hancock County, Illinois. (Photograph by Dennis R. Kolata.)

Sandstone also are well developed on the west but change to thin shale or mudstone eastward. Conversely, the Bethel, Cypress, Hardinsburg, and Waltersburg sandstones are thick on the east and pinch out or change to shale toward the west. Also, within each formation, sandstone bodies can be highly lenticular. It is not unusual for a sandstone 50 to 100 feet (30.5 m) thick to change to shale entirely in less than a mile (1.6 km). Many Chesterian sandstone bodies are tidal sand bars (Potter 1963, Cole and Nelson 1995, Seyler 1998), and others are incised valley fills (Morse 2001, Nelson et al. 2002).

Chesterian sandstones are dominantly white to light gray, buff, or greenish gray and very fine to medium grained. Most are at least 90% quartz; feldspar and rock fragments are minor components. Especially in the lower Chesterian, some sandstones contain glauconite, calcite cement, and marine fossils. Siltstone and shale are various shades of gray, green, olive, and in some places red. Nonbedded mudstones variegated in red, green, and gray are useful horizon markers and are significant in interpreting Chesterian deposition.

Important fossils include the crinoid genera *Pterotocrinus, Talarocrinus, Taxocrinus, Phanocrinus, Onychocrinus, Platycrinites,* and *Zeacrinites* (some of these are shown in Figures 9-6 and 9-8). Various *Pterotocrinus* spp. possess distinctive "wing plates" (Figure 9-6c) that are diagnostic for specific formations from the Renault through Kinkaid (Sutton 1934, Gutschick 1965, Chestnut and Ettensohn 1988). *Dichocrinus* sp. and *Talarocrinus* sp. occur mainly from the Downeys Bluff Limestone through the Ridenhower Formation, the latter being an index fossil for the lower Gasperian Stage (not shown) (Strimple 1977). The stemless crinoid *Agassizocrinus* is found in most of the higher energy carbonates from the Renault to the Kinkaid Limestone. The *Pentremites* sp. blastoids (Figure 9-8h) are common throughout the Chesterian but become abundant in the lower third of the series.

The bryozoan *Prismopora serrulata* is characteristic of the Haney and Glen Dean Limestones (Willman et al. 1975), whereas *Archimedes* sp. is common throughout the Chesterian. The brachiopods *Anthracospirifer increbescens* (Figure 9-6g, h) and *Composita subquadrata* (Figure 9-6i) are abundant in the Elviran Stage (upper Chesterian; Figure 9-1). Fossil plants are found in shale associated with thin coal in the Cypress, Hardinsburg, Tar Springs, Waltersburg, Palestine, Tygett, and Degonia formations. Trace fossils such as *Conostichus* ichnosp. and *Rhizocorallium* ichnosp. are common in the upper units of these sandstone formations (Figure 9-8d, k), which suggests a marine environment. Vertebrate fossils, including lungfish and microsaurs, have been found in red mudstone of the Kinkaid. Microsaurs were small four-legged reptiles that resembled a dwarfed brontosaurus-like dinosaur.

Sources of Sediment

Several sources of detrital sediment were available during Chesterian time (Figure 9-4). The Transcontinental Arch, a broad peninsula or isthmus that extended from northern New Mexico to Minnesota and beyond, supplied sand to the Illinois Basin chiefly during early Chesterian time, as shown by thickness and facies patterns of the Spar Mountain, Aux Vases, and Yankeetown Sandstones (Willman et al. 1975, Leetaru 2000, Nelson et al. 2002). The Ozark Uplift, although closer to Illinois, apparently was not a large contributor of Chesterian sediment. The newly rising Appalachian Mountains shed copious volumes of sediment, most of which was trapped along the shoreline close to the mountain front. The one sandstone that may have an Appalachian source is the Big Clifty Sandstone of the Golconda Formation (Figure 9-1). This sandstone is thickest on the southeastern margin of the Illinois Basin in Kentucky, changing northwestward to shale (Fraileys Shale Member) with thin and isolated sandstone lenses in eastern Illinois (Swann 1963, 1964; Treworgy 1988; Nelson et al. 2002).

The principal source of Chesterian sand was the southern part of the Canadian Shield, including the northern Great Lakes region and southern Ontario. Swann (1963) proposed that the ancient "Michigan River system" flowed generally southwestward, transporting sediment from the Canadian Shield into Illinois. He visualized the Michigan River building a series of deltas that advanced and retreated and shifted laterally in response to sea-level changes and tectonic uplift. Of the postulated rivers and their coastal plain, essentially nothing remains. This model is likely too simplistic, but no new hypothesis has arisen to replace it. The key problem is absence of evidence because the area where the purported delta lay now has no Mississippian rocks. However, the observed distribution of units such as the Bethel, Cypress, Hardinsburg, Tar Springs, and Palestine Sandstones is best explained by a northeastern provenance of the sediment.

Tectonic Activity

Continental collision between southeastern North America and Africa began to form the Appalachian Mountains during Late Mississippian time (see Chapter 3, Tectonic History, and Chapter 4, Structural Features). Projected inland, these forces initiated movements in Illinois, particularly along pre-existing faults. The margins of the Ozark Uplift rose, creating angular unconformities in lower Chesterian strata (Weller and Sutton 1940). Other active areas included the ancient Reelfoot Rift and its northeastern extension, the Wabash Valley Fault System. Narrow, northeast-trending, fault-bounded depressions here served as pathways for sediment movement and deposition (Nelson et al. 2002). It was Sullivan (1972) who first recognized "stacking" of thick sandstone bodies into the narrow, linear to slightly sinuous "West Baden clastic belt" along the Wabash Valley. The younger Hardinsburg Sandstone forms a similar clastic belt in the same area (Potter 1962, 1963; Potter et al. 1958).

Sea-Level Changes

The Chesterian rocks of Illinois consist of several sequences. Each is a defined body of rock bounded above and below by unconformities. Largest are the major sequences, such as the Tippecanoe and Kaskaskia (Sloss 1988). Smaller sequences of alternating marine and non-marine deposits occur in the Chesterian. Their bounding unconformities represent times when the sea temporarily withdrew from Illinois, and the region was subjected to erosion and soil formation with little or no new sediment being laid down.

Nelson et al. (2002) divided the lower Chesterian of Illinois (Ste. Genevieve through Glen Dean) into 11 sequences. Applying the same principles to the upper Chesterian yields 9 more sequences, for a total of 20 in the entire Chesterian (Figure 9-14).

The unconformities that delimit the Chesterian sequences take two forms. The more striking form is the incised valley, a deep channel cut into underlying rock layers (Figure 9-15). Most such valleys are filled with sandstone, although some contain shale or even limestone. As mapped from borehole data, most incised valleys are slightly sinuous and trend northeast to south-southwest, reflecting the flow out of the Illinois Basin toward the Ouachita Trough. Channels are typically 20 to 50 feet (6 to 15 m) deep, but some exceed 100 feet (30.5 m). A valley filled with Bethel Sandstone in western Kentucky is as deep as 250 feet (76 m), cutting into the St. Louis Limestone (Reynolds and Vincent 1967, Sedimentation Seminar 1969).

Other Chesterian unconformities are defined by ancient soils known as paleosols. These commonly take the form of mudstones that are mottled and variegated in red, green, and gray. Other paleosols, developed in siltstone, sandstone, or limestone, reveal features such as calcite concretions, caliche or "hardpan," root traces of land plants, fractures caused by repeated wetting and drying, and even development of sand dunes. All are evidence that the sea floor was exposed to the air, allowing weathering and growth of land vegetation to proceed (Nelson et al. 2002).

Deposits in the lower part of a Chesterian sequence (Figure 9-15) reflect the initial rise of sea level that followed valley cutting and soil formation. As already noted, incised valleys were largely filled with sand by the rivers that cut them. As sea level rose, drowned valleys became estuaries subject to tides and supporting brackish to marine organisms. Areas between valleys often became swamps where

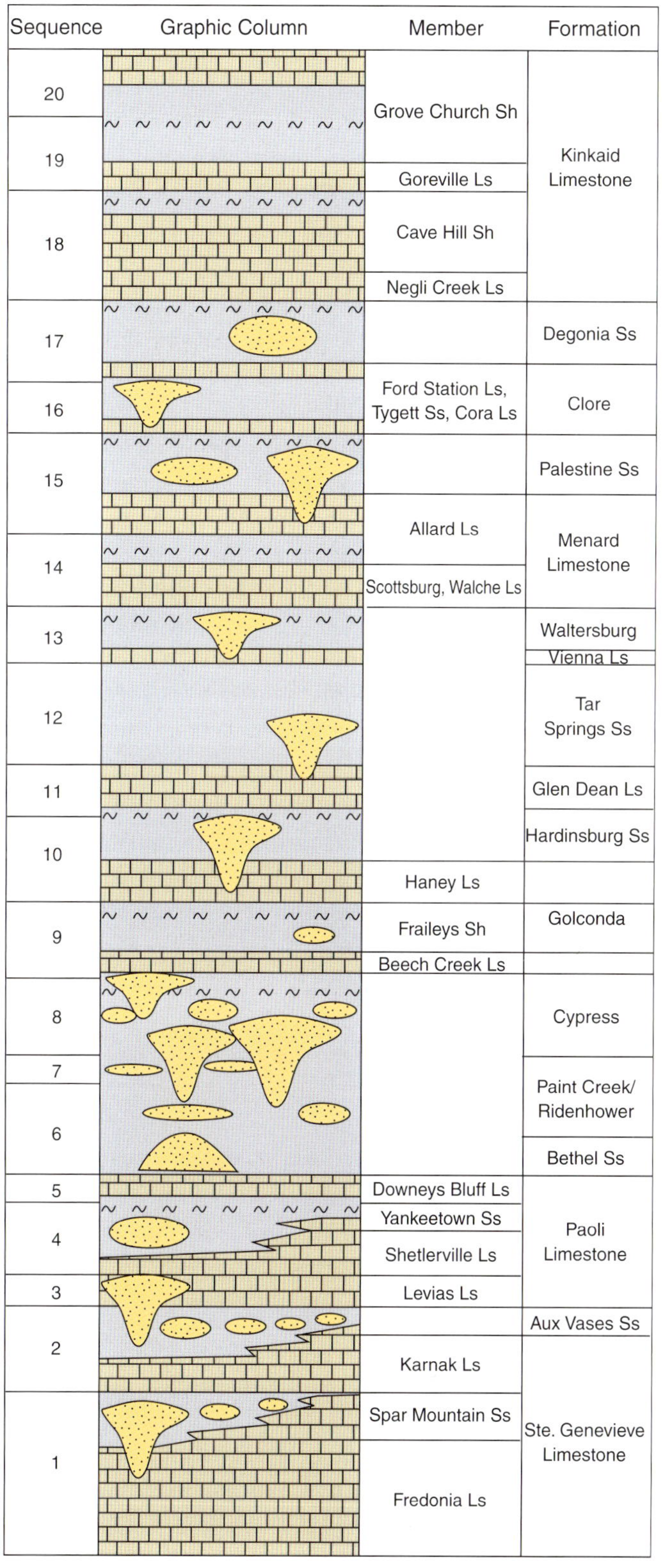

Figure 9-14 Diagram of Chesterian sequences. Abbreviations: Ls, Limestone; Sh, Shale; Ss, Sandstone.

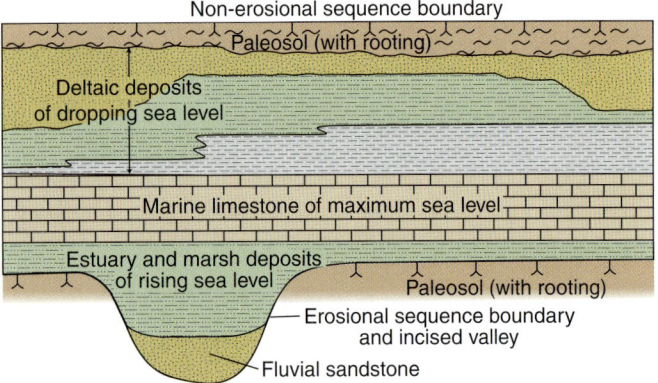

Figure 9-15 Diagram of a typical Chesterian sequence.

patchy peat deposits (now thin coal beds) developed. As the entire Illinois Basin area became submerged, the deposits changed to limestone or sandstone. Tidal and longshore currents commonly resulted in numerous lens-shaped sand bars trending northeast to southwest (Treworgy 1988, Leetaru 2000, Nelson et al. 2002). As sea level peaked and began to fall, deltas built out from the advancing shoreline, typically represented by sediments that coarsened from shale at the base to sandstone at the top. With further retreat of the sea, a new episode of valley incision and soil formation ensued.

Nelson et al. (2002) found that the 11 lower Chesterian sequences are traceable (with local gaps) throughout the Illinois Basin. Al-Tawil (1998) documented the same number of sequences in Lower Chesterian rocks of the northern Appalachian Basin. Such widespread continuity seems to rule out merely local processes, such as fault movements or episodes of delta construction and abandonment. Furthermore, similar styles of sedimentation are observed in Upper Mississippian and Pennsylvanian rocks in many parts of the world, indicating that global sea-level changes were at work. The probable cause: episodes of glaciation in the southern hemisphere that began in the Middle Mississippian and continued through the Early Permian (Hambrey and Harland 1981, Crowell 1983, Caputo and Crowell 1985, Veevers and Powell 1987, Ross and Ross 1988, Frakes et al. 1992). Each glacial episode locked water in the polar ice cap and caused sea level to drop; glacial melting raised sea level.

Economic Geology

The economic geology of Illinois is discussed thoroughly later in this volume. However, it should be mentioned here that more than 80% of the oil and gas produced in Illinois has come out of Chesterian rocks (Howard 1991). The Ste. Genevieve, Aux Vases, and Cypress are the most prolific producers, but all of the Chesterian sandstones have yielded hydrocarbons in one area of Illinois or another. The Ste. Genevieve Limestone is also quarried extensively for use as agricultural lime, concrete aggregate, road gravel, and other uses. Other Chesterian limestone units support quarries, and Chesterian sandstone has been quarried on a small scale for use as building stone.

Conclusion

Chesterian sedimentation was brought to a close in Illinois by a dramatic drop of sea level, one of the largest in Earth's history. This change left the entire sea floor exposed to erosion; it was deeply scoured by southwest-flowing rivers. At the same time, climatic changes took place, and there was a great upsurge of tectonic activity. The Applachian Mountains began to grow, as many faults and folds in Illinois and elsewhere deformed Chesterian and older strata. Signaling major changes in global conditions, the resulting unconformity separates the Kaskaskia sequence from the overlying Absaroka sequence.

References

Al-Tawil, A., 1998, High-resolution sequence stratigraphy of Late Mississippian carbonates in the Appalachian Basin: Blacksburg, Virginia, Virginia Polytechnic Institute and State University, Ph.D. dissertation, 109 p.

Anderson, W. I., 1998, Iowa's geologic past: Three billion years of change: Iowa City, Iowa, University of Iowa Press, 424 p.

Atherton, E., C. Collinson, and J. A. Lineback, 1975, Mississippian System, in H. B. Willman, E. Atherton, T. C. Buschbach, C. Collinson, J. C. Frye, M. E. Hopkins, J. A. Lineback, and J. A. Simon, Handbook of Illinois stratigraphy: Illinois State Geological Survey, Bulletin 95, p. 123–163.

Ausich, W. I., T. W. Kammer, and N. G. Lane, 1979, Fossil communities of the Borden (Mississippian) Delta in Indiana and northern Kentucky: Journal of Paleontology, v. 53, no. 5, p. 1182–1196.

Baxter, J. W., and P. L. Brenckle, 1982, Preliminary statement on Mississippian calcareous foraminiferal successions of the midcontinent (U.S.A.) and their correlation to western Europe: Newsletters on Stratigraphy, v. 11, no. 3, p. 136–153.

Brenckle, P. L., J. F. Baesemann, F. J. Woodson, J. W. Baxter, J. L. Carter, C. Collinson, H. R. Lane, R. D. Norby, and C. B. Rexroad, 1988, Comment and reply on "redefinition of the Meramecian/Chesterian boundary (Mississippian)": Geology, v. 16, no. 5, p. 471, 472.

Caputo, M. V., and J. C. Crowell, 1985, Migration of glacial centers across Gondwana during the Paleozoic Era: Geological Society of America Bulletin, v. 96, no. 8, p. 1020–1036.

Carr, D. D., 1973, Geometry and origin of oolite bodies in the Ste. Genevieve Limestone (Mississippian) in the Illinois Basin: Indiana Geological Survey, Bulletin 48, 81 p.

Chestnut, D. R. Jr., and F. R. Ettensohn, 1988, Hombergian (Chesterian) echinoderm paleontology and paleoecology, south-central Kentucky: Ithaca, New York, Paleontological Research Institution, Bulletins of American Paleontology, v. 95, no. 330, 102 p.

Choquette, P. W., and R. P. Steinen, 1980, Mississippian non-supratidal dolomite, Ste. Genevieve Limestone, Illinois Basin: Evidence for mixed-water dolomitization: Tulsa, Oklahoma, SEPM Society for Sedimentary Geology, SEPM Special Publication 28, p. 163–196.

Cluff, R. M., 1986, Application of modern sand models to oil and gas exploration, Mississippian Ste. Genevieve Limestone, Illinois Basin, *in* B. Seyler, ed., Aux Vases and Ste. Genevieve formations: A core workshop and field trip guidebook: Illinois Geological Society and Illinois State Geological Survey, 67 p.

Cluff, R. M., and J. A. Lineback, 1981, Middle Mississippian carbonates of the Illinois Basin: Illinois Geological Society and Illinois State Geological Survey, 88 p.

Cluff, R. M., M. L. Reinbold, and J. A. Lineback, 1981, The New Albany Shale Group of Illinois: Illinois State Geological Survey, Circular 518, 83 p.

Cole, R. D., and W. J. Nelson, 1995, Stratigraphic framework and environments of deposition of the Cypress Formation in the outcrop belt of southern Illinois: Illinois State Geological Survey, Illinois Petroleum 149, 47 p.

Collinson, C. C., L. E. Becker, G. W. James, J. W. Koenig, and D. H. Swann, 1967, Illinois Basin, *in* D. H. Oswald, ed., International Symposium on the Devonian System: Calgary, Alberta, Canada, Alberta Society of Petroleum Geologists, v. 1, p. 940–962.

Collinson, C., C. B. Rexroad, and T. L. Thompson, 1971, Conodont zonation of the North American Mississippian, *in* W. C. Sweet and S. M. Bergstrom, eds., Symposium on Conodont Biostratigraphy: Geological Society of America Memoir, v. 127, p. 353–394.

Collinson, C., A. J. Scott, and C. B. Rexroad, 1962, Six charts showing biostratigraphic zones, and correlations based on conodonts from the Devonian and Mississippian rocks of the Upper Mississippi Valley: Illinois State Geological Survey, Circular 328, 32 p.

Crowell, J. C., 1983, Ice ages recorded on Gondwanan continents: Transactions of the Geological Society of South Africa, v. 86, p. 237–263.

Devera, J. A., 1987, *Asphaltinoides insertae sedis,* a new genus from the Devonian of Illinois: Journal of Paleontology, v. 61, no. 6, p. 1274–1278.

Devera, J. A., and G. H. Fraunfelter, 1988, Middle Devonian paleogeography and tectonic relationships east of the Ozark Dome, southeastern Missouri, southwestern Illinois and parts of southwestern Indiana and western Kentucky, *in* N. J. McMillan, A. F. Embry, and D. J. Glass, eds., Devonian of the World: Proceedings of the Second International Symposium on the Devonian System, Calgary, Canada, v. II, Sedimentation, p. 179–196.

Devera, J. A., and N. R. Hasenmueller, 1991, Kaskaskia sequence: Middle and Upper Devonian Series through Mississippian Kinderhookian Series, *in* M. W. Leighton, D. R. Kolata, D. F. Oltz, and J. J. Eidel, eds., Interior cratonic basins: Tulsa, Oklahoma, American Association of Petroleum Geologists, Memoir 51, p. 113–123.

Droste, J. B., and R. H. Shaver, 1975, Jeffersonville Limestone (Middle Devonian) of Indiana: Stratigraphy, sedimentation, and relation to Silurian reef-bearing rocks: American Association of Petroleum Geologists Bulletin, v. 59, no. 3, p. 393–412.

Droste, J. B., and R. H. Shaver, 1983, Atlas of early and middle Paleozoic paleogeography of the southern Great Lakes area: Indiana Geological Society, Special Report 32, 32 p.

Frakes, L. A., J. E. Francis, and J. L. Sykto, 1992, Climate modes of the Phanerozoic: Cambridge, England, Cambridge University Press, 274 p.

Gibson, A. C., 2001, Three-dimensional geometries and porosity trends of subsurface ooid shoal hydrocarbon reservoirs in the Mississippian Ste. Genevieve Formation of the Illinois Basin, USA: University of Illinois at Urbana-Champaign, M. S. thesis, 38 p.

Gutschick, R. C., 1965, *Pterotocrinus* from the Kinkaid Limestone (Chester, Mississippian) of Illinois and Kentucky: Journal of Paleontology, v. 39, no. 4, p. 636–646.

Gutschick, R. C., and C. A. Sandberg, 1983, Mississippian continental margins of the conterminous United States, *in* D. J. Stanley and G. T. Moore, eds., The shelfbreak: Critical interface on continental margins: Tulsa, Oklahoma, Society of Economic Paleontologists and Mineralogists, Special Publication 33, p. 79–96.

Hambrey, M. J., and W. B. Harland, 1981, Earth's pre-Pleistocene glacial record: New York, Cambridge University Press, 1004 p.

Howard, R. H., 1991, Hydrocarbon reservoir distribution in the Illinois Basin, *in* M. W. Leighton, D. R. Kolata, D. F. Oltz, and J. J. Eidel, eds., Interior cratonic basins: Tulsa, Oklahoma, American Association of Petroleum Geologists, Memoir 51, p. 299–327.

Kammer, T. W., P. L. Brenckle, J. L. Carter, and W. I. Ausich, 1990, Redefinition of the Osagean-Meramecian boundary in the Mississippian stratotype region: Palaios, v. 5, no. 5, p. 414–431.

Kolata, D. R., compiler, 2005, Bedrock geology of Illinois: Illinois State Geological Survey, Illinois Map 14, 1:500,000.

Lane, H. R., 1978, The Burlington Shelf (Mississippian, north-central United States): Geologica et Palaeontologica, v. 12, p. 165–176.

Lane, H. R., C. A. Sandberg, and W. Ziegler, 1980, Taxonomy and phylogeny of some lower Carboniferous conodonts and preliminary standard post-*Siphonodella* zonation: Geologica et Palaeontologica, v. 14, p. 117–164.

Lasemi, Z., R. D. Norby, and J. D. Treworgy, 1998, Depositional facies and sequence stratigraphy of a Lower Carboniferous bryozoan-crinoidal carbonate ramp in the Illinois Basin, mid-continent USA, *in* T. P. Burchette and V. P. Wright, eds., Carbonate ramps: London, England, The Geological Society of London, Special Publication 149, p. 369–395.

Lasemi, Z., R. D. Norby, J. E. Utgaard, W. R. Ferry, R. J. Cuffey, and G. R. Dever, Jr., 2003, Mississippian carbonate buildups and development of cool-water-like carbonate platforms in the Illinois Basin, midcontinent, U.S.A., *in* W. M. Ahr, P. M. Harris, W. A. Morgan, and I. D. Somerville, eds., Permo-Carboniferous carbonate platforms and reefs: Society of Sedimentary Geology, Special Publication 78, p. 69–95.

Lasemi, Z., J. D. Treworgy, R. D. Norby, J. P. Grube, and B. G. Huff, 1994, Waulsortian mounds and reservoir potential of the Ullin Limestone ("Warsaw") in southern Illinois and adjacent area in Kentucky: Illinois State Geological Survey, Guidebook 25, 65 p.

Leetaru, H. E., 2000, Sequence stratigraphy of the Aux Vases Sandstone: A major oil producer in the Illinois Basin: American Association of Petroleum Geologists Bulletin, v. 84, no. 3, p. 399–422.

Lineback, J. A., 1966, Deep-water sediments adjacent to the Borden Siltstone (Mississippian) delta in southern Illinois: Illinois State Geological Survey, Circular 401, 48 p.

Lineback, J. A., 1968, Turbidites and other sandstone bodies in the Borden Siltstone (Mississippian) in Illinois: Illinois State Geological Survey, Circular 425, 23 p.

Lineback, J. A., 1969, Illinois Basin—Sediment-starved during Mississippian: American Association of Petroleum Geologists Bulletin 53, no. 1, p. 112–126.

Lineback, J. A., 1972, Lateral gradation of the Salem and St. Louis Limestones (Middle Mississippian) in Illinois: Illinois State Geological Survey, Circular 474, 23 p.

Lineback, J. A., 1981, The eastern margin of the Burlington-Keokuk (Valmeyeran) carbonate bank in Illinois: Illinois State Geological Survey, Circular 520, 24 p.

Mamet, B., and B. Skipp, 1970, Lower Carboniferous calcareous Foraminifera: Preliminary zonation and stratigraphic implications for the Mississippian of North America: Sixth International Congress on Carboniferous Stratigraphy and Geology, Sheffield, 1967: Compte Rendu, v. 3, p. 1129–1146.

Maples, C. G., and J. A. Waters, 1987, Redefinition of the Meramecian/Chesterian boundary (Mississippian): Geology, v. 15, p. 647–651.

Meents, W. F., and D. H. Swann, 1965, Grand Tower Limestone (Devonian) of southern Illinois: Illinois State Geological Survey, Circular 389, 34 p.

Morse, D. G., 2001, Sedimentology, diagenesis and trapping style, Mississippian Tar Springs Sandstone, Inman East Consolidated field, Gallatin County, Illinois: Illinois State Geological Survey, Illinois Petroleum 157, 67 p.

Nelson, W. J., 1995, Structural features in Illinois: Illinois State Geological Survey, Bulletin 100, 144 p.

Nelson, W. J., J. A. Devera, and J. M. Masters, 1995, Geology of the Jonesboro 15-minute Quadrangle, southwestern Illinois: Jonesboro, Mill Creek, Ware, and McClure 7.5-minute Quadrangles: Illinois State Geological Survey, Bulletin 101, 57 p.

Nelson, W. J., and D. K. Lumm, 1985, Ste. Genevieve Fault Zone, Missouri and Illinois: Illinois State Geological Survey, Contract/Grant Report 1985-3, 94 p.

Nelson, W. J., and S. Marshak, 1996, Devonian tectonism of the Illinois Basin region, U.S. continental interior, in B. A. van der Pluijm and P. A. Catacosinos, eds., Basement and basins of eastern North America: Boulder, Colorado, Geological Society of America, Special Paper 308, p. 169–180.

Nelson, W. J., L. B. Smith, and J. D. Treworgy, 2002, Sequence stratigraphy of the lower Chesterian (Mississippian) strata of the Illinois Basin: Illinois State Geological Survey, Bulletin 107, 70 p.

Norby, R. D., 1991, Biostratigraphic zones in the Illinois Basin, in M. W. Leighton, D. R. Kolata, D. F. Oltz and J. J. Eidel, eds., Interior cratonic basins: Tulsa, Oklahoma, American Association of Petroleum Geologists, Memoir 51, p. 179–194.

Peppers, R. A., and H. H. Damberger, 1969, Palynology and petrography of a Middle Devonian coal in Illinois: Illinois State Geological Survey, Circular 445, 36 p.

Potter, P. E., 1962, Late Mississippian sandstones of Illinois: Illinois State Geological Survey, Circular 340, 36 p.

Potter, P. E., 1963, Late Paleozoic sandstones of the Illinois Basin: Illinois State Geological Survey, Report of Investigations 217, 92 p.

Potter, P. E., E. Nosow, N. M. Smith, D. H. Swann, and F. H. Walker, 1958, Chester cross-bedding and sandstone trends in Illinois Basin: American Association of Petroleum Geologists Bulletin, v. 42, no. 5, p. 1013–1046.

Reynolds, D. W., and J. K. Vincent, 1967, Western Kentucky's Bethel channel—The largest continuous reservoir in the Illinois Basin, in W. D. Rose, ed., Proceedings of the Technical Sessions, Kentucky Oil and Gas Association, Twenty-ninth Annual Meeting, June 3–4, 1965: Lexington, Kentucky, Kentucky Geological Survey, Series X, Special Publication 14, p. 19–30.

Ross, C. A., and J. R. P. Ross, 1988, Late Paleozoic transgressive-regressive deposition, in C. K. Wilgus, B. S. Hastings, H. Posamentier, J. Van Wagoner, C. A. Ross, and C. G. St. C. Kendall, eds., Sea-level changes: An integrated approach: Tulsa, Oklahoma, Society of Economic Paleontologists and Mineralogists (SEPM) Special Publication 42, p. 227–247.

Sedimentation Seminar, 1969, Bethel Sandstone (Mississippian) of western Kentucky and south-central Indiana, a submarine-channel fill: Kentucky Geological Survey, Series X, Report of Investigations 11, 24 p.

Seyler, B., 1998, Geologic and engineering controls on Aux Vases Sandstone reservoirs in Zeigler field, Illinois: Illinois State Geological Survey, Illinois Petroleum 153, 79 p.

Sloss, L. L., 1988, Tectonic evolution of the craton in Phanerozoic time, in L. L. Sloss, ed., Sedimentary cover—North American craton: U.S.: The Geology of North America, v. D-2: Boulder, Colorado, Geological Society of America, p 25–51.

Sloss, L. L., W. C. Krumbein, and E. C. Dapples, 1949, Integrated facies analysis, in C. R. Longwell, ed., Sedimentary facies in geologic history: New York, Geological Society of America, Memoir 39, p. 91–123.

Stainbrook, M. A., 1941, Biotic analysis of Owen's Cedar Valley Limestones: Pan-American Geologist, v. 75, no. 5, p. 321–327.

Stanley, S. M., 1989, Earth and life through time (2nd ed.): New York, W. H. Freeman and Company, 689 p.

Strimple, H. L., 1977, Chesterian (upper Mississippian) and Morrowan (Lower Pennsylvanian) crinoids of northeastern Oklahoma and northwestern Arkansas, in R. K. Sutherland and W. L. Manger, eds., Upper Chesterian-Morrowan stratigraphy and the Mississippian-Pennsylvanian boundary of northeastern Oklahoma and northwestern Arkansas: Norman, Oklahoma, Oklahoma Geological Survey, Guidebook 18, p. 171–176.

Sullivan, D. M., 1972, Subsurface stratigraphy of the West Baden Group in Indiana: Indiana Geological Survey, Bulletin 47, 31 p.

Summerson, C. H., and D. H. Swann, 1970, Patterns of Devonian sand on the North American craton and their interpretation: Geological Society of America Bulletin, v. 81, no. 2, p. 469–490.

Sutton, A. H., 1934, Evolution of *Pterotocrinus* in the eastern interior basin during the Chester Epoch: Journal of Paleontology, v. 8, no. 4, p. 393–416.

Swann, D. H., 1963, Classification of Genevievian and Chesterian (Late Mississippian) rocks of Illinois: Illinois State Geological Survey, Report of Investigations 216, 91 p.

Swann, D. H., 1964, Late Mississippian rhythmic sediments of Mississippi Valley: American Association of Petroleum Geologists Bulletin, v. 48, no. 5, p. 637–658.

Swann, D. H., J. A. Lineback, and E. Frund, 1965, The Borden Siltstone (Misssissippian) delta in southwestern Illinois: Illinois State Geological Survey, Circular 386, 20 p.

Treworgy, J. D., 1988, The Illinois Basin—A tidally and tectonically influenced ramp during mid-Chesterian time: Illinois State Geological Survey, Circular 544, 20 p.

Veevers, J. J., and C. M. Powell, 1987, Late Paleozoic glacial episodes in Gondwanaland reflected in transgressive-regressive sequences in Euramerica: Geological Society of America Bulletin, v. 98, no. 4, p. 475–487.

Weller, J. M., and A. H. Sutton, 1940, Mississippian border of Eastern Interior Basin: American Association of Petroleum Geologists Bulletin, v. 24, no. 5, p. 765-858.

Weller, S., 1913, Stratigraphy of the Chester Group in southwestern Illinois: Transactions of the Illinois Academy of Science, v. 6, p. 118–129.

Weller, S., 1920, The Chester Series in Illinois: Journal of Geology, v. 28, no. 4, p. 281–303; no. 5, p. 395–416.

Weller, S., and S. St. Clair, 1928, Geology of Ste. Genevieve County, Missouri: Missouri Bureau of Geology and Mines, v. 22, 2nd ser., 352 p.

Whiting, L. L., and D. L. Stevenson, 1965, The Sangamon Arch: Illinois State Geological Survey, Circular 383, 20 p.

Willman, H. B., E. Atherton, T. C. Buschbach, C. Collinson, J. C. Frye, M. E. Hopkins, J. A. Lineback, and J. A. Simon, 1975, Handbook of Illinois stratigraphy: Illinois State Geological Survey, Bulletin 95, 261 p.

Willman, H. B., J. C. Frye, J. A. Simon, K. E. Clegg, D. H. Swann, E. Atherton, C. Collinson, J. A. Lineback, and T. C. Buschbach, 1967, Geologic map of Illinois: Illinois State Geological Survey, 1:500,000.

Pennsylvanian Subsystem and Permian System (Absaroka Sequence)

10

W. John Nelson and Russell J. Jacobson

INTRODUCTION

The term "Coal Measures" or "Pennsylvania Series" was introduced by Williams (1891, p. 83) in reference to the coal-bearing rocks in Pennsylvania. The Pennsylvanian was a subdivision of the Carboniferous System, defined by Conybeare and Phillips (1822) in England and Wales. Weller (1906) was the first to use "Pennsylvanian" in Illinois, although "Pennsylvanian," "Coal Measures," and "Upper Carboniferous" were used interchangeably in early reports. North American geologists gradually adopted the Pennsylvanian as a system, recognizing that Mississippian and Pennsylvanian rocks differ in fundamental ways and are separated by a regional unconformity. By international agreement, the Pennsylvanian was reclassified in 2004 as the Pennsylvanian Subsystem of the Carboniferous System (Heckel 2004).

The Pennsylvanian Subsystem in the United States commonly is divided into five series: Morrowan (oldest), Atokan, Desmoinesian, Missourian, and Virgilian (Figure 10-1). Also, the Pennsylvanian is divided into Lower (equivalent to Morrowan), Middle (Atokan and Desmoinesian), and Upper (Missourian and Virgilian) Series (Peppers 1996).

Pennsylvanian rocks contain the bituminous coal resources of Illinois, the largest of any state (Chapter 14, Coal). They also have yielded 13% of Illinois' cumulative oil and gas production (Chapter 15, Oil and Gas) and are an important source of limestone and of clay and shale used in brick-making (Chapter 17, Industrial Minerals).

The Permian Period was named by British geologist R. I. Murchison in 1841. The period takes its name from the city of Perm near the Ural Mountains of Russia, near which rocks of this age are extensively exposed. Permian sedimentary rocks are widespread in the southern U.S. midcontinent, especially in west Texas, where the North American standard Permian section was assembled. The only known Permian rocks in Illinois are igneous rocks in the southern part of the state, but evidence suggests that sedimentary layers of this age formerly covered much of Illinois. The cyclic character of Permian strata, which is comparable to underlying Pennsylvanian rocks, is not consistent with deposition in a small basin (Kehn et al. 1982).

The Absaroka sequence was named for the Absaroka Mountains in Wyoming (Sloss et al. 1949). Originally the sequence included Pennsylvanian rocks, but the upper limit was not specified. Sloss (1963) defined the upper contact as the unconformity that underlies Middle Jurassic strata in the Rocky Mountains and Gulf Coast. The sequence is divided into Absaroka I, II, and III subsequences (Sloss 1988). The Absaroka I extends from latest Mississippian, approximately 330 million years before present, through Wolfcampian (Early Permian) time 268 million years ago. The rock record of the Absaroka sequence in Illinois is entirely Absaroka I, comprising Pennsylvanian sedimentary rocks and Lower Permian intrusive igneous rocks (Figure 10-1). This chapter covers Absaroka I only; Chapter 11 deals with Absaroka II and III and younger sequences identified by Sloss (1963).

GEOLOGICAL HISTORY

Near the end of Mississippian time, glaciation in the southern hemisphere caused the sea to withdraw from the Illinois Basin. Rivers eroded a series of deep, parallel, southwest-trending valleys into the exposed sea floor. The ancestral Rocky Mountain Orogeny commenced around the same time, having impacts as far east as Illinois (McBride and Nelson 1999). Many faults and folds developed in Illinois, strongly influencing sedimentation.

As the sea level rose in early Morrowan time, incised valleys of Illinois were drowned and filled with sediment. Initial sediments included large volumes of pebbly, quartz-rich sand carried by rivers from the northeastern United States and southeastern Canada. Apparently derived from older Paleozoic sedimentary rocks, these deposits constitute the Caseyville Formation (Nelson 1989). Sea level rose and fell repeatedly in concert with tectonic activity in Illinois, producing multiple episodes of valley cutting and filling.

By Atokan time, sediments had accumulated in most of Illinois, and the drowned-valley topography was gradually leveling. Precambrian crystalline rocks were being eroded in the northern and central Appalachian Mountains, supplying mica, feldspar, rock fragments, and clay to Illinois sediments. Peat-forming swamps, which developed along

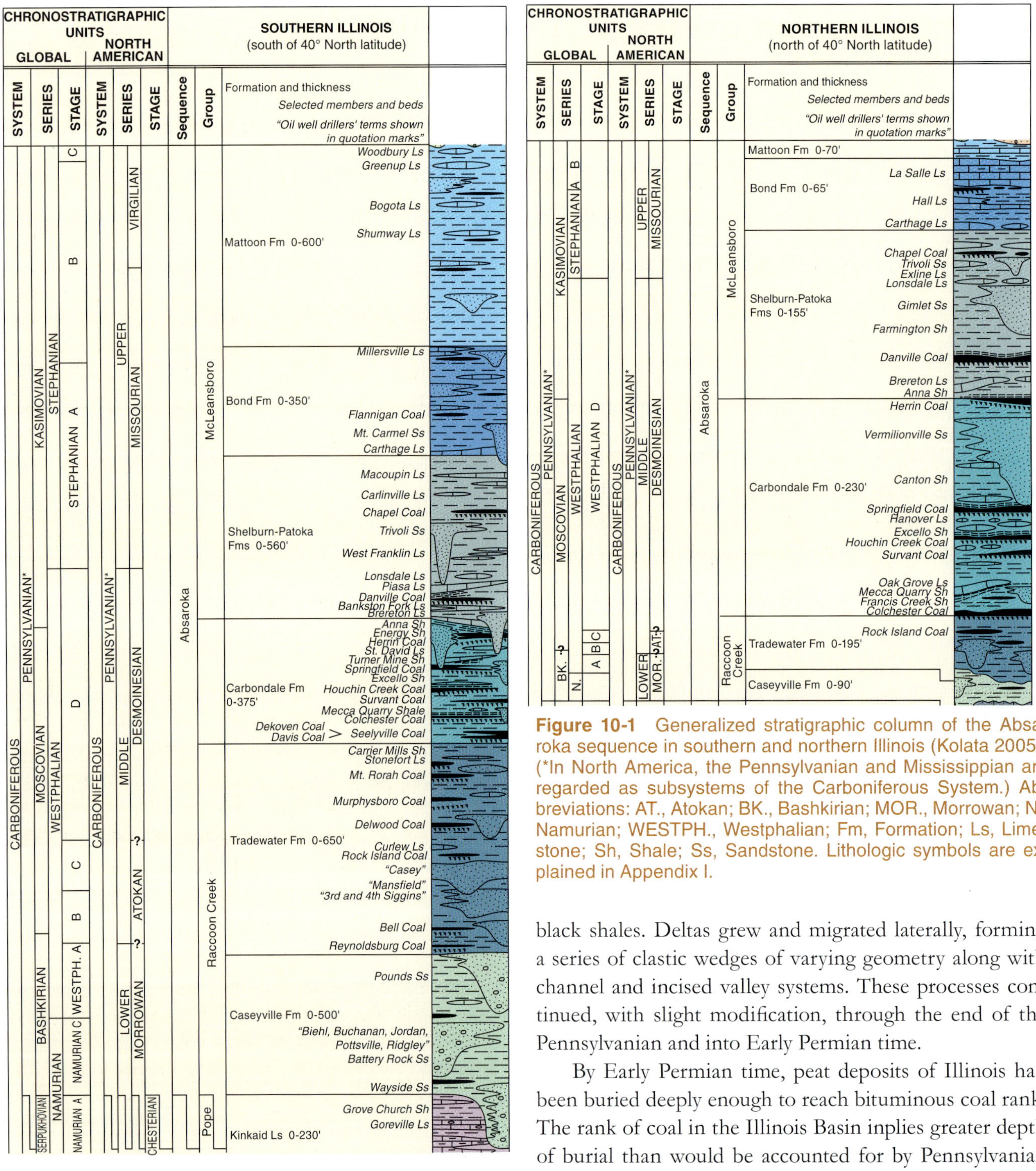

Figure 10-1 Generalized stratigraphic column of the Absaroka sequence in southern and northern Illinois (Kolata 2005). (*In North America, the Pennsylvanian and Mississippian are regarded as subsystems of the Carboniferous System.) Abbreviations: AT., Atokan; BK., Bashkirian; MOR., Morrowan; N., Namurian; WESTPH., Westphalian; Fm, Formation; Ls, Limestone; Sh, Shale; Ss, Sandstone. Lithologic symbols are explained in Appendix I.

the deltaic coastline on each sea-level rise, were becoming more extensive. Occasionally the sea water became clear enough to allow limestone deposition.

By Desmoinesian (late Middle Pennsylvanian) time, the topography of Illinois and adjacent areas was nearly level. Tectonic activity continued, but more slowly than previously. These conditions promoted the formation of extremely widespread beds of peat (coal), limestone, and black shales. Deltas grew and migrated laterally, forming a series of clastic wedges of varying geometry along with channel and incised valley systems. These processes continued, with slight modification, through the end of the Pennsylvanian and into Early Permian time.

By Early Permian time, peat deposits of Illinois had been buried deeply enough to reach bituminous coal rank. The rank of coal in the Illinois Basin inplies greater depth of burial than would be accounted for by Pennsylvanian rocks alone (Damberger 1971). Several thousand feet of the latest Pennsylvanian and Permian sediments were subsequently eroded away except in a few places in western Kentucky, where they are preserved in downdropped fault slices. The Alleghany Orogeny culminated as North America and Africa collided. The midcontinental crust deformed in response, with many faults being reactivated and new faults forming, along with intrusion of ultramafic igneous rocks in southeastern Illinois and adjacent Kentucky.

Pennsylvanian Subsystem

Extent and Thickness

Pennsylvanian rocks underlie about two-thirds of Illinois, including most of the area south of Interstate 80. They are the youngest sedimentary bedrock in the state. Pennsylvanian strata are extensively mantled by Pleistocene glacial deposits and, in western Illinois, by Cretaceous and Tertiary sand and gravel. The Pennsylvanian rocks are best exposed in the Shawnee Hills of southern Illinois, where they form some of the state's most scenic topography and provide the setting for a national forest and several state parks. Outcrops also are numerous along the Illinois River and its tributaries, from Pittsfield (Pike County) and Macomb (McDonough County) on the south to La Salle (La Salle County) and the Quad Cities (Rock Island and Moline, Illinois, and Davenport and Bettendorf, Iowa) on the Mississippi River in northwestern Illinois and southeastern Iowa. Elsewhere, Pennsylvanian outcrops are small and widely scattered in stream banks and artificial exposures, such as mines and highway and railroad cuts.

The largest contiguous area of thick Pennsylvanian rocks is in the Fairfield Basin of southeastern Illinois where they attain a thickness of about 2,500 feet (760 m) southeast of Fairfield near the corner of Hamilton, Wayne, and White Counties (Willman et al. 1975). However, more than 4,000 feet (1,200 m) of Pennsylvanian strata are present in narrow, downfaulted blocks along the Ohio River between Gallatin County, Illinois, and Union County, Kentucky (Palmer 1976). The youngest known Pennsylvanian rock in the state is shale of late Virgilian age, overlying the Woodbury Limestone (not shown on generalized stratigraphic column) along Webster Branch in southern Cumberland County (Weibel et al. 1989). The Pennsylvanian thins markedly north and west of the Fairfield Basin, where Lower Pennsylvanian rocks are largely absent, and most or all of the Upper Pennsylvanian has been eroded.

General Character

The Pennsylvanian of Illinois is composed mostly (90 to 95%) of sandstone, siltstone, and shale with thin but widely traceable beds of limestone, coal, black fissile shale, and underclay (Figure 10-1). Sandstone is dominant in the lower part (Caseyville and Tradewater Formations); the proportions of shale and siltstone increase upward. Although rare in the Caseyville and uncommon in the Tradewater, limestone is common in younger Pennsylvanian formations, where it constitutes 5 to 10% of the rock volume and includes several widespread units of quarry stone. Coal accounts for only 1 to 2% of total rock volume and is concentrated in the Middle Pennsylvanian (Kosanke et al. 1960).

Pennsylvanian rocks originated as sediments within and bordering an extensive shallow seaway that covered much of the eastern and central United States (Figure 10-2). This sea advanced and retreated many times. At its maximum extent, the sea covered all of Illinois except possibly the Wisconsin Arch in the far north. The crest of the Ozark Dome (in Missouri) also remained above sea level. At the maximum retreat of the sea, all of Illinois was coastal plain, and the shoreline lay in Oklahoma. Specific environments of deposition in Illinois included open shallow marine waters (mostly limestone); restricted lagoons (black shale); deltas, bays, estuaries, and coastal plains (gray shale, siltstone, and sandstone); freshwater lakes (some limestone); ancient soils (underclay); and vast coastal swamps (coal).

Fossils

Pennsylvanian rocks of Illinois contain abundant and diverse fossils, including some that are unique in the world and are highly sought by collectors (Figures 10-3 and 10-4). Most common are marine invertebrates, especially brachiopods, gastropods, and pelecypods (Figure 10-4). Corals, cephalopods, trilobites, bryozoans, and worms are present but less prevalent than in older rocks (Newberry et al. 1866, Worthen et al. 1875, Moore 1944, Wanless 1958). Marine fossils are found chiefly in limestones, although many shale units are fossiliferous. Microscopic fossils such as fusulinids (Dunbar and Henbest 1942, Thompson et al. 1959, Thompson and Shaver 1964), ostracods (tiny shelled crustaceans) (Cooper 1946), and plant spores found in coal (Peppers 1996) are key to identifying and correlating Pennsylvanian strata from place to place.

Plant fossils are locally abundant and well preserved, especially in shale overlying coal beds (Read and Mamay

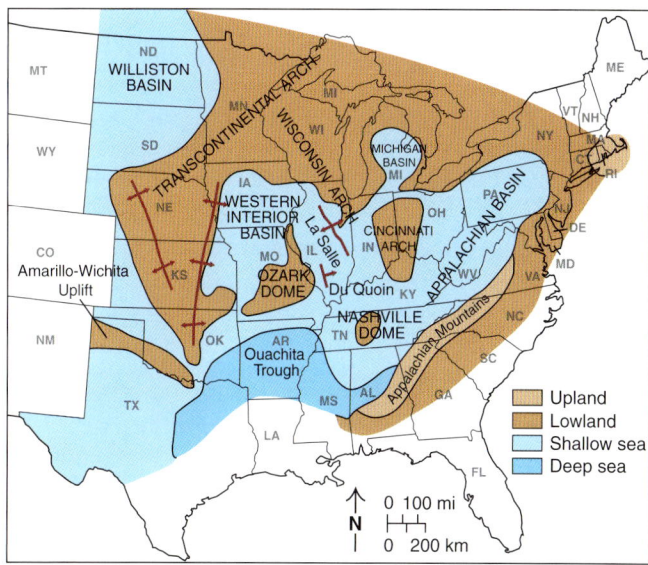

Figure 10-2 Regional setting of Illinois during Atokan (early Middle Pennsylvanian) (McKee and Crosby 1975).

Figure 10-3 Characteristic Pennsylvanian plant fossils: **(a)** articulate plant foliage *Annularia* sp., ×0.7; **(b)** tree fern *Asterotheca* sp., ×0.8; **(c)** seed fern *Neuropteris* sp., ×0.7; **(d)** seed fern *Sphenopteris* sp., ×1.5; **(e)** articulate plant stem *Calamites* sp., ×0.6; **(f)** lycopod tree bark *Lepidodendron* sp., ×0.6. (Photographs by Rodney D. Norby and Dennis R. Kolata.)

1964, Jennings 1990) (Figure 10-3). Some coal seams in Illinois contain coal balls, which are masses of fossil peat replaced by calcite and other minerals (Figure 10-5). When sliced and examined under the microscope, coal balls provide an intimate view of the vegetation that went into forming coal (Phillips et al. 1976).

The most famous fossils of Illinois are the soft-bodied organisms of the Mazon Creek fauna (Baird et al. 1985; Earth Science Club of Northern Illinois 1989, 1990; Johnson and Richardson 1966, 1970; Richardson 1956; Shabica 1970; Shabica and Hay 1997; Wittry 2006). These are preserved in siderite (iron carbonate) concretions in the Francis Creek Shale Member, overlying the Colchester Coal in north-central Illinois. Fossils can be collected from stream cuts, but most come from spoil banks of abandoned surface mines in Grundy and Will Counties. The Mazon Creek fauna include many animals rarely preserved as fossils, including forms found nowhere else: a variety of insects, spiders, scorpions, small crustaceans, worms, jellyfish, amphibians, fish, and Tully monster—Illinois' state fossil (Figure 10-6).

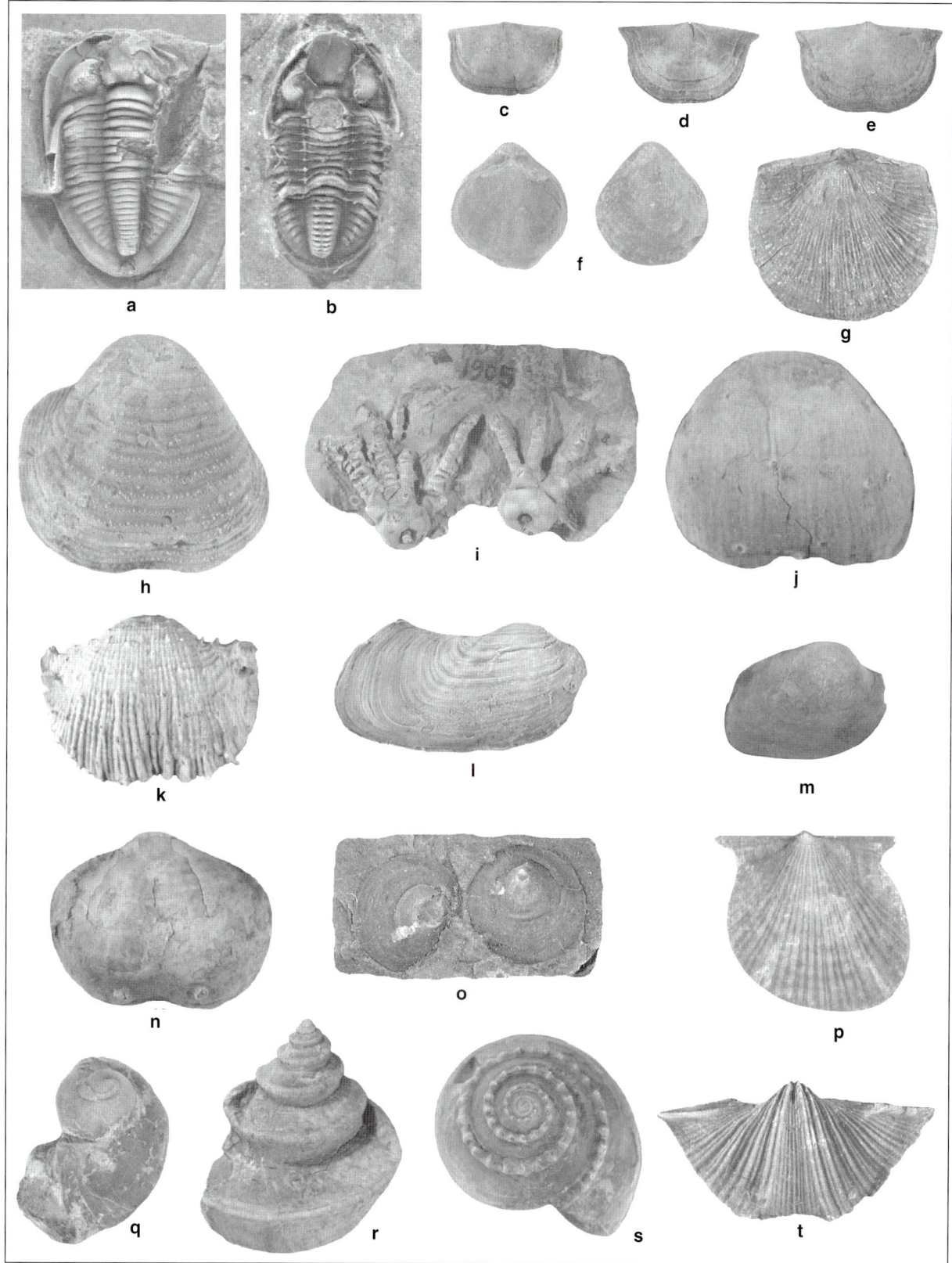

Figure 10-4 Characteristic Pennsylvanian animal fossils: **(a)** trilobite *Ameura sauki,* ×2.5; **(b)** trilobite *Ditomopyge* sp., ×4; **(c)** brachiopod *Mesolobus mesolobus,* ×1.3; **(d)** brachiopod *Neochonetes granulifer,* ×1.1; **(e)** brachiopod *Neochonetes* sp., ×1.3; **(f)** brachiopod *Composita* sp., ×1.5; **(g)** brachiopod *Derbyia* sp., ×1.0; **(h)** brachiopod *Dictyoclostus* sp., ×1.2; **(i)** crinoids *Endelocrinus fayettensis* sp., ×0.7; **(j)** brachiopod *Marginifera* sp., ×4; **(k)** brachiopod *Antiquatonia* sp., ×0.8; **(l)** pelecypod *Chaenomya* sp., ×0.8; **(m)** pelecypod *Cardiomorpha* sp., ×2.0; **(n)** brachiopod *Marginifera* sp., ×4.0; **(o)** brachiopods *Orbiculoides newberryi,* ×1.8; **(p)** pelecypod *Dunbarella* sp., ×1.0; **(q)** gastropod *Natacopsis* sp., ×1.7; **(r)** gastropod *Worthenia* sp., ×1.0; **(s)** gastropod *Trepospira* sp., ×1.5; **(t)** brachiopod *Neospirifer* sp., ×0.8. (Photographs by Rodney D. Norby and Dennis R. Kolata.)

Controls on Deposition

Climate

Clues to ancient climate can be gleaned from a variety of sources. The anatomy of fossil plants, geometry and character of coal beds, mineral content of rocks, and character of ancient soils are especially relevant to interpreting the Pennsylvanian climate because certain types of plants, peat deposits, minerals, and soils are characteristic of specific climates.

These factors indicate that during the Late Mississippian, the climate of the eastern and central United States was dry, seasonal, and tropical (Cecil et al. 1985). Further evidence is that Upper Mississippian limestones formed escarpments and capped plateaus, as they do today in arid regions such as the western United States (Bristol and Howard 1971).

A profound shift to an ever wet, tropical climate took place at the onset of the Pennsylvanian. Due to plate tectonic movements, North America and western Europe were then located close to the equator. After partial drying during the Atokan, ever wet conditions returned in force during the Desmoinesian, fostering maximum coal development (Schopf 1975, Phillips and Peppers 1984, Cecil et al. 1985, DiMichele and Phillips 1996).

Figure 10-6 The Illinois state fossil: Tully monster, *Tullimonstrum gregarium*, ×0.7. The animal is of unknown phylum and was obtained from Mazon Creek nodules. (Photograph by Rodney D. Norby and Dennis R. Kolata.)

At the end of the Desmoinesian, the climate became markedly drier, inducing mass extinction of coal swamp vegetation and markedly reducing coal bed thickness (Phillips and Peppers 1984, Cecil et al. 1985, DiMichele and Phillips 1996). A likely explanation of Late Pennsylvanian drying is continental convergence and the rise of the Appalachian and Ouachita Mountains, restricting the movement of humid sea air (Phillips and Peppers 1984).

Tectonic Setting

Continents were converging during the late Paleozoic, with profound consequences. As mentioned, North America and Europe lay close together, within the equatorial belt. South America and Africa, on the south, drifted northward. Eventual collision of continental plates created new mountain ranges, including the Appalachians and Ouachitas. By Permian time, all of the continents had united into a single landmass, Pangea.

During Morrowan time (Early Pennsylvanian), southern Illinois occupied a shallow embayment nearly isolated from the Appalachian Basin. Through Middle Pennsylva-

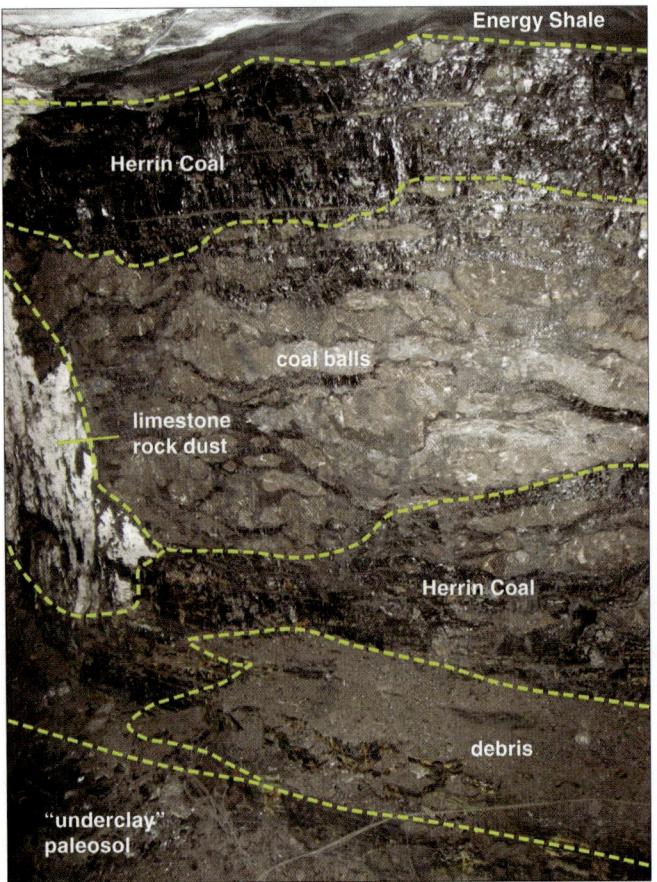

Figure 10-5 Coal balls shown in place within the Herrin Coal at the Galatia Mine, Saline County, Illinois.

nian time, the seaway expanded and the Ozark and Nashville Domes and Cincinnati Arch became islands (Figure 10 2). By Late Pennsylvanian time, these areas of uplift were largely submerged, and the seaway extended continuously from the midcontinent to the Appalachians (McKee and Crosby 1975).

Huge volumes of sediment were transported through Illinois into the deep, rapidly subsiding Ouachita Trough during Morrowan and Atokan time. Collision between North and South America, commencing in the early Desmoinesian, closed off the trough and thrust up the Ouachita Mountains (Figure 10-7). The Appalachian Mountains underwent uplift throughout the Pennsylvanian, especially during the Morrowan and Atokan and in latest Pennsylvanian to Permian time (McKee and Crosby 1975).

Tectonic activity was widespread throughout the North American craton during the Pennsylvanian Era (see Chapter 3, Tectonic History). In Illinois, many north- and northwest-trending monoclines and anticlines developed. Examples of these folds include the La Salle Anticlinorium and the Salem, Louden, and Clay City Anticlines, all of which hold large oil fields (see Chapter 4, Structural Features). Such folds generally overlie faults in Precambrian rocks. Faulting and folding in Illinois began shortly before the onset of Pennsylvanian sedimentation, producing angular unconformities between Pennsylvanian and older rocks. Earth movements continued intermittently throughout the Pennsylvanian and into Permian time (Kolata and Nelson 1991, Nelson 1995). This activity apparently is a product of the ancestral Rocky Mountain Orogeny, which affected a vast area of the continental interior (McBride and Nelson 1999).

Earth movements profoundly influenced Pennsylvanian sedimentation in Illinois. Areas that sank received thick sediment deposits; areas that rose received little or no sediment and were subjected to weathering, erosion, and soil formation. The trends of ancient rivers and the thickness and distribution of peat (coal) deposits were particularly sensitive to structural movements (Palmer et al. 1979, Jacobson 1985, Greb 1989, Archer and Greb 1995).

Sediment Sources

The dominant source of sediment to the Illinois Basin during the Pennsylvanian lay in the northern Appalachian Mountains and the Canadian Shield. This source is shown by paleocurrent measurements, trends of channels and incised valleys, and reconstruction of deltaic systems (Potter and Siever 1956a, 1956b; Potter and Pryor 1961; Potter 1963; Pryor and Sable 1974; Wanless and Wright 1978). The Ozark Dome and remnants of the Transcontinental Arch were minor sediment sources (Figure 10-2). The central

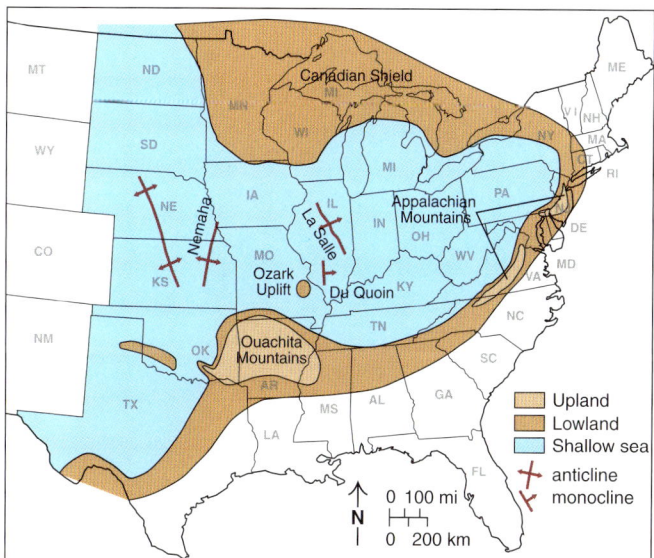

Figure 10-7 Regional setting of Illinois during Missourian (early Late Pennsylvanian) time (McKee and Crosby 1975).

and southern Appalachians and, later, the Ouachita Mountains yielded large volumes of eroded sediment, but most of this sediment was trapped in foreland basins adjoining the mountain fronts.

Cyclothems and Eustasy

Pennsylvanian rocks in the midcontinent contain many thin, widely traceable units that seem to recur in regular order. Udden (1912) was the first to describe cycles of sedimentation in Middle Pennsylvanian rocks around Peoria, Peoria County, Illinois. He described four cycles, each composed of coal overlain first by marine shale and limestone, and then sandstone or shale, and capped by the underclay—the rooted soil horizons that underlie coal beds—of the next coal bed. (The fourth cycle lacked coal, but otherwise resembled the other three.) Each cycle seemed to represent an episode of advance and retreat of the sea. Further mapping in Illinois confirmed Udden's observations (Weller 1930, 1931; Wanless 1931). Wanless and Weller (1932, 1944) coined the word *cyclothem,* derived from the Greek words *cyclos,* meaning circle, and *thema,* meaning deposit.

Geologists differed on the causes of cyclicity. Udden (1912, p. 49) suggested "recurrent interruption of a progressive submergence." J. M. Weller (1930) advocated repeated episodes of tectonic uplift and subsidence. Wanless and Shepard (1936) proposed that glacial advance and retreat in the polar regions, driven by climatic cycles, raised and lowered sea level worldwide (eustasy). Ferm (1970, 1975) maintained that cyclicity was more apparent than real and that repetition of strata could be explained by successive episodes of delta building.

Since the time of Wanless and Shepard (1936), Pennsylvanian age glacial deposits have been documented in southern continents (Ross and Ross 1985, Saunders and Ramsbottom 1986, Veevers and Powell 1987). Most geologists today accept glacially induced sea-level changes as an important control on Pennsylvanian sedimentation. Other processes, particularly tectonic activity and short-term climate change, also were highly influential.

The cyclothem concept was a predecessor to sequence stratigraphy, which arose in the 1970s as a tool in petroleum exploration (Payton 1977, Van Wagoner et al. 1990). Sequences are packages of sedimentary rocks that are bounded by unconformities or that can be traced laterally to conformable or gradational contacts (correlative conformities). Sequences are interpreted as being the product of recurrent rises and falls of relative sea level. Sequences have been identified in rocks of many geological ages from many parts of the world.

Weibel (1996, 2002) and Nadon (1999) recognized cyclothems as mappable rock units bounded by unconformities. Their concepts merge traditional cyclothems with sequence stratigraphy. Unconformities that bound cyclothems represent lowstands, when sea level was lowest and Illinois was subjected to erosion, weathering, and soil development. Underclays (e.g.), the rooted claystone layers below coal seams) and incised valleys (e.g., the Galatia channel) carved by rivers flowing across the exposed coastal plain, are the result (Figure 10-8). As sea level began to rise, incised valleys were flooded and became estuaries where sediment (largely sand) was deposited. At the same time, the coastal plain became a vast swamp where peat (coal) formed. Continued rise of the sea drowned the swamp, and marine sediments were laid on top of the peat. Near the peak sea level (highstand), deltas began to advance across the region, typically laying coarser sediments on top of finer ones. As sea level fell, the prograding delta plain became exposed again, repeating the cycle.

Stratigraphy

Classifying Pennsylvanian Rocks

Pennsylvanian rocks in Illinois are difficult to classify. Many coal, limestone, and marine shale units are distinctive and widely traceable but are too thin to serve as formations (North American Commission on Stratigraphic Nomenclature 1983). Coal and limestone beds prominent in one part of the Illinois Basin may be absent in other areas. The dominant shale, siltstone, and sandstone do not change greatly in character from the base of the Pennsylvanian to its top. Although definite vertical trends are recognized, the changes are gradual. The same rocks undergo abrupt and dramatic changes when traced laterally. Hence, several different classification schemes have arisen over the years (Figure 10-9).

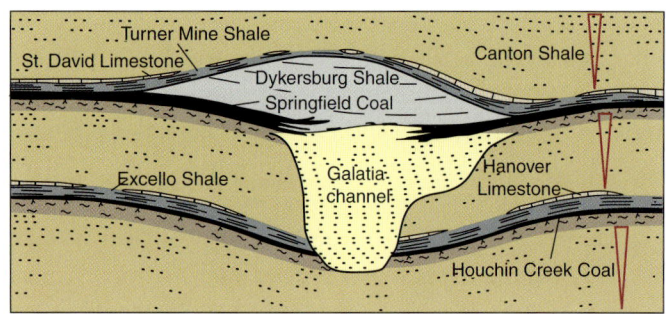

Figure 10-8 Middle and Upper Pennsylvanian depositional sequences, exemplified by Houchin Creek and Springfield Coals in southeastern Illinois. A typical cyclothem comprises, in ascending order: (a) coal; (b) black fissile shale such as Excello and Turner Mine; (c) marine limestone such as Hanover and St. David; (d) gray shale, siltstone, and sandstone coarsening upward (symbolized by tall triangles); and (e) rooted underclay beneath the next coal. Incised valleys, typified by the Galatia channel, developed in areas of maximum subsidence and are filled with sandstone that grades upward to siltstone and shale. Coal thickens toward channel margins because of greater subsidence, but is absent within the channel, which persisted during peat formation. With continued subsidence, the peat swamp along channel margins was drowned and buried by a thick lens of non-marine gray shale in an estuarine setting.

Early geologists such as Shaw and Savage (1912) broke the Pennsylvanian into a few large formations. Those units were too large to be of much use for mapping small areas, given their typical low relief and gentle structure. Beginning in the 1930s, many geologists used cyclothems as formations. Difficulties arose because many cyclothems are too thin to show on maps, their boundaries can be difficult to locate, and geologists disagree on where to place their boundaries. Kosanke et al. (1960) dropped cyclothemic formations and proposed seven larger units, some old and some new. This classification was based on overall changes in lithology, coupled with key limestone and coal layers used as boundaries. The thinner key units were classified as members or beds. A revised version of the classification of Kosanke et al. (1960) is used today (Tri-State Committee on Correlation of the Pennsylvanian System in the Illinois Basin 2001) (Figures 10-1 and 10-9). Problems continue because several of the key boundary units do not persist throughout the Illinois Basin. Thus, the current chart should be considered a progress report, not a final determination.

Basal Unconformity

The base of the Absaroka sequence is a substantial unconformity that separates Pennsylvanian from older rocks throughout Illinois and neighboring states. The Mississippian-Pennsylvanian hiatus is smallest in Johnson County in southern Illinois, where some of the oldest Pennsylvanian rocks in Illinois overlie the youngest preserved Mississippian unit, the Grove Church Shale (Jennings and Fraunfelter 1986, Weibel and Norby 1992). Pennsylvanian rocks overlie Upper Mississippian in southern Illinois, progressively overlapping older Paleozoic strata to the north. At the northern end of the Illinois Basin (from about Moline, Illinois, eastward through La Salle and Ottawa toward Kankakee), Pennsylvanian strata directly overlie Ordovician or Silurian strata, reflecting a time gap as great as 150 million years.

The unconformity is the result of a worldwide drop in sea level approximately 330 million years ago, exposing the entire Illinois Basin area to valley incision (Saunders and Ramsbottom 1986). Because many tectonic structures were active during this episode of erosion, the unconformity is angular in places. A good place to see an angular unconformity is Matthiessen State Park southeast of La Salle. There, horizontal Pennsylvanian coal and underclay overlie Middle Ordovician dolomite that dips 40° to 50° west.

Detailed subsurface mapping (Siever 1951, Bristol and Howard 1971) revealed a series of southwest-trending incised valleys on the sub-Pennsylvanian surface (Figure 10-10). These valleys originated in the southeastern Canadian Shield and northern Appalachians, traversed the Illinois Basin area in a dendritic pattern, and drained into the Ouachita Trough (Sedimentation Seminar 1978, Archer and Greb 1995). In some areas, large blocks slumped from the valley walls during incision (Bristol and Howard 1974).

Caseyville Formation

The Caseyville Formation (Morrowan) rocks are the oldest Pennsylvanian rocks in the state (Figure 10-1). The type section is in the bluffs of the Ohio River between Gentry Landing and the mouth of the Saline River in Hardin County, Illinois (Lee 1916). The Caseyville is largely confined to the southeastern part of the Illinois Basin. Farther north, the formation is restricted to deep paleovalleys and an outlier near the Quad Cities (Bettendorf and Davenport, Iowa, and Moline and Rock Island, Illinois) (Willman et al. 1975, Wanless 1975, Ludvigson and Swett 1987). These outliers truncate rocks as old as Devonian and are overlain unconformably by rocks ranging from Atokan to Desmoinesian in age.

The thickness of the Caseyville was controlled by regional subsidence and by sub-Pennsylvanian paleotopogra-

Figure 10-9 Changing classification of Pennsylvanian rocks in Illinois is shown. Italics indicate that the unit is a member. Abbreviations: Fm, Formation; Ls, Limestone.

phy. The Caseyville thickens southward regionally, attaining its maximum thickness locally, over 600 feet (182 m), in the Evansville paleovalley in western Kentucky (Greb et al. 1992). The Caseyville thickens in paleovalleys and thins on the intervening divides. The oldest strata are confined to the valley bottoms; younger beds overlap divides. Local thickness changes as great as 300 feet (90 m) are recorded.

Cliff-forming massive and cross-bedded sandstone is typical of the Caseyville (Figure 10-11). Caseyville sandstone is quartz arenite (95 to 100% quartz) that commonly

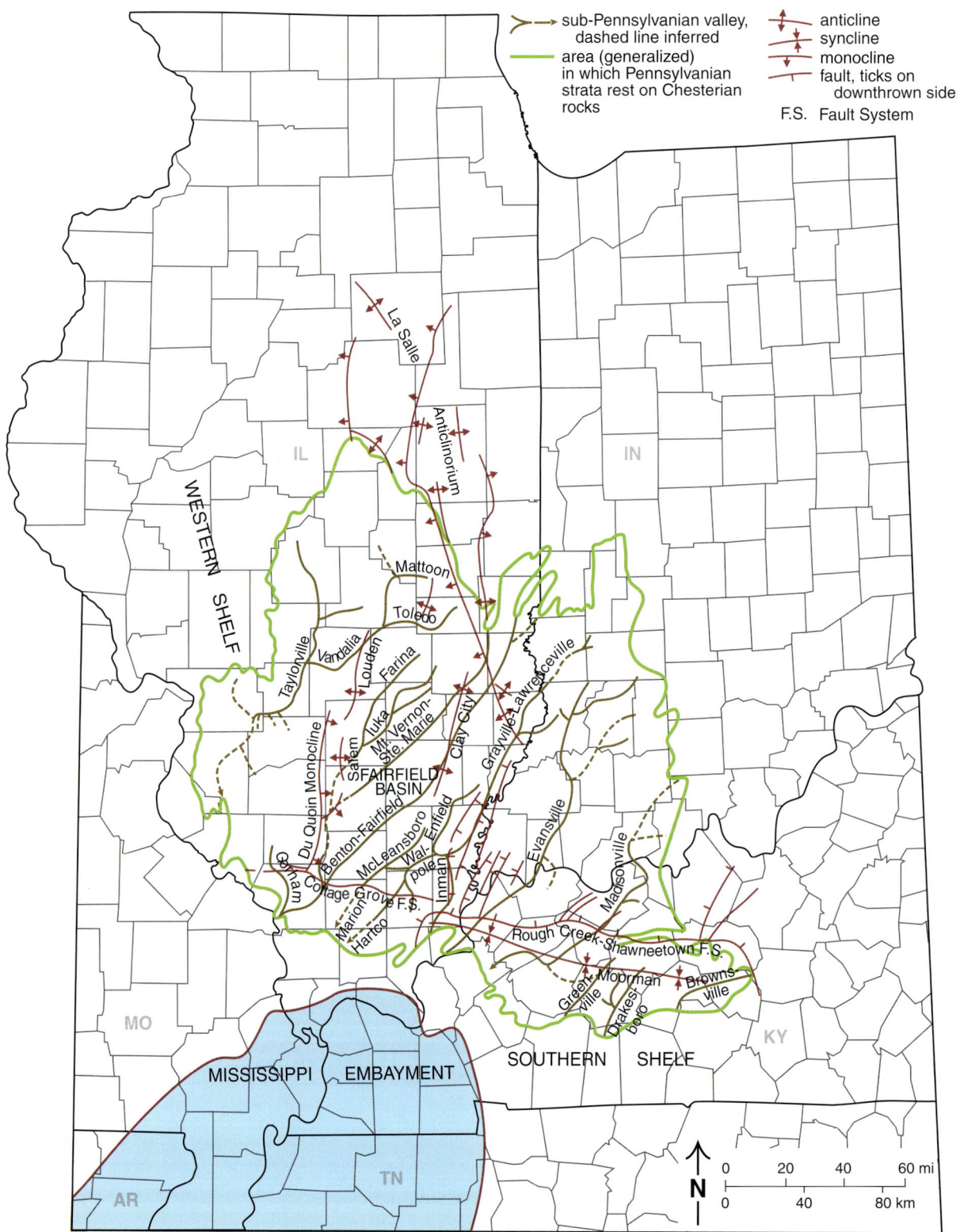

Figure 10-10 Map of Illinois showing sub-Pennsylvanian valleys in relation to structural features. The paleocurrent toward the southwest is indicated by tributary pattern and sedimentary structures, such as cross-bedding (Bristol and Howard 1971).

contains well-rounded quartz pebbles as large as 1 inch (2.5 cm) in diameter. The change to less mature sandstones (more mica, feldspar, rock fragments, and clay matrix) in the overlying Tradewater is gradational (Potter and Glass 1958, Nelson 1989, Nelson et al. 1991). The Caseyville also contains thin-bedded sandstone, mudstone, shale, siltstone, conglomerate, underclay, coal, and limestone. Caseyville shales are gray to black and contain freshwater, marine, and brackish water fossils (Devera et al. 1987, Devera 1989, Greb 1989). Coal beds are discontinuous, rarely more than 2 feet (0.6 m) thick, and generally shaly. Rare limestones in the upper half of the formation are shaly to sandy, are iron-rich, and contain marine to brackish water fossils (Wanless 1939).

Lower Pennsylvanian quartz arenite and quartz-pebble conglomerate similar to the Caseyville are found throughout the eastern United States. Grain size and paleocurrent trends indicate that rivers and estuaries transported materials from the northeastern United States and southern Canada through the Illinois Basin toward the Ouachita Trough (Sedimentation Seminar 1978, Rice 1984, Nelson 1989, Archer and Greb 1995). The maturity of Caseyville (and equivalent) sand and gravel implies it was largely recycled from older Paleozoic sandstone and conglomerate (Potter and Siever 1956a, 1956b; Potter and Glass 1958).

Depositional sequences that fill incised valleys 100 to 150 feet (30 to 45 m) deep occur throughout the formation. Many valleys are filled with pebbly sandstone, the lower part having fluvial aspects and the upper part locally exhibiting estuarine features, such as herringbone cross-bedding (Nelson 1989). Some valleys are entirely filled with dark gray brackish water to marine shale (Greb et al. 1992, Kvale and Barnhill 1994). Where valley fill is sandstone, it is commonly capped by paleosol and thin coal. Overlying these units are highstand deposits of dark gray shale and siltstone, which are locally marine, but in most places brackish. Estuarine indicators include bidirectional cross-bedding, tidal rhythmites, body fossils (uncommon), and common to abundant ichnofossils (Devera 1989, Kvale and Barnhill 1994, Archer and Greb 1995).

Although some individual Caseyville sequences are traceable for many miles using well logs, correlation of Caseyville units across the Illinois Basin is not possible. Multiple sandstone sequences commonly become "stacked" and cannot be differentiated. Shaly sequences 15 to 60 feet (4.6 to 18 m) thick have the greatest lateral continuity. In general, facies are more continuous down depositional dip (northeast-southwest) than along strike; that is, facies are confined to the paleovalleys, which generally flowed toward the southwest.

Figure 10-11 (a) Sandstone of the Caseyville Formation forms scenic bluffs at Garden of the Gods Recreation Area in Gallatin County. (b) "Streets" in Giant City State Park, Jackson County, formed by widened joints in massive sandstone of the lower Tradewater Formation. (Photographs by Joel M. Dexter.)

Tradewater Formation

The Tradewater Formation of Atokan and lower Desmoinesian (Middle Pennsylvanian) age (Figure 10-1) records a transition to more "cyclic" sedimentation concurrent with the climax of the Ouachita Orogeny. Pennsylvanian rocks gradually advanced across the shelves of the Illinois Basin, filling valleys and making a flatter landscape. Tectonic folding and faulting continued to impact sedimentation in Illinois.

The Tradewater Formation (Lee 1916) was named for the Tradewater River in western Kentucky, where it includes strata from the top of the Caseyville to the base of the Davis Coal. Northward, where the Davis and Dekoven Coals merge to become the Seelyville Coal, the top of the Tradewater is mapped at the base of the Seelyville (Jacobson 1987, Tri-State Committee on Correlation of the Pennsylvanian System in the Illinois Basin 2001). Westward and northwestward in Illinois, the Davis and Seelyville Coals become thin, discontinuous, and difficult to identify (Chapter 14, Coal). In these areas, the top of the Tradewater is raised to the base of the widely continuous Colchester Coal (Tri-State Committee on Correlation of the Pennsylvanian System in the Illinois Basin 2001).

Thickness changes are more gradual in the Tradewater than in the Caseyville because the Tradewater was deposited on a flatter surface. The Tradewater reaches more than 600 feet (180 m) in southern Illinois, thinning toward the north, dramatically so west of the Du Quoin Monocline, which was actively rising during Middle Pennsylvanian time. In most of western and northern Illinois, the Tradewater varies from a few feet to about 100 feet (30.5 m) thick.

In southern Illinois, the lower half (Atokan) is mostly sandstone interbedded with lesser amounts of siltstone and shale. Some lower Tradewater sandstones form scenic bluffs in the state parks in the southern portion of the state (Figure 10-11). Lower Tradewater sandstone bodies are generally thinner and less deeply incised than those of the Caseyville. However, multiple sandstone units commonly are "stacked" to produce continuous sandstone intervals as thick as 200 feet (60 m). As a consequence, correlation of lower Tradewater depositional sequences using well log information is difficult even over short distances. Some lower Tradewater sandstones are quartz arenites that contain scattered quartz granules and small pebbles, like those of the Caseyville. Upward through the formation, the proportions of feldspar, mica, rock fragments, and interstitial clay increase (Potter and Siever 1956a, 1956b; Potter and Glass 1958). Limestone is uncommon; coal beds are discontinuous, shaly, and rarely thick enough to mine.

The upper Tradewater of southern Illinois contains more shale and siltstone, less sandstone, and more numerous and widely traceable limestone and coal units than does the lower Tradewater. Shale and siltstone are various shades of gray and contain plentiful mica and finely divided carbonaceous matter. Sandstones of the upper Tradewater (and all younger Pennsylvanian units) are lithic arenites rich in mica, feldspar, carbonaceous debris, rock fragments, and clay matrix (Potter and Siever 1956a, 1956b; Potter and Glass 1958). Regional well-log correlation of depositional sequences becomes easier, especially near the top of the Tradewater. Several thin coal, limestone, and black fissile shale units can be correlated across many counties.

In western and northern Illinois, the Tradewater is much thinner and differs markedly from that formation in southern Illinois. Widespread but lenticular quartzose sandstone (Babylon Sandstone Member; not shown on generalized stratigraphic column) occurs at the base (Wanless 1957) and is overlain by interbedded shale, claystone, marine limestone, and coal beds that are minable in some locales. In places, multiple underclays (ancient soil horizons) are superimposed, forming a continuous succession as thick as 50 feet (15 m) (Cheltenham Clay; not shown; formerly mined for making ceramics) and providing evidence that many areas of western Illinois remained above sea level well into Desmoinesian time.

Carbondale Formation

The Carbondale Formation (Figure 10-1) accounts for more than 90% of Illinois' coal reserves and past and present production. The Colchester, Springfield, and Herrin Coals are extensively mined, ranking among the most widespread Carboniferous coal beds in the world (Greb et al. 2003). The formation was named by Shaw and Savage (1912) for Carbondale in Jackson County. Formation boundaries have changed over the years. Currently the Davis or Seelyville Coal marks the base, and the base of the Brereton Limestone is the top (Tri-State Committee on Correlation of the Pennsylvanian System in the Illinois Basin 2001). The Brereton lies within a few feet above the Herrin Coal in most of Illinois. The Carbondale is 250 to 375 feet (75 to 120 m) thick in most of the Fairfield Basin and 150 to 250 feet (45 to 75 m) thick on most of the Western Shelf and in northern Illinois (Willman et al. 1975). Abrupt thinning along the Du Quoin Monocline, La Salle Anticlinorium, and other anticlines reflects active uplift of these structures during Carbondale deposition (see Chapter 4, Structural Features).

Highly continuous coal, black shale, and limestone beds are characteristic of the Carbondale. Many such units extend not only through most of the Illinois Basin, but also into the Western Interior coal field of Missouri, Iowa, eastern Kansas, and northeastern Oklahoma (Wanless 1939, 1955; Wanless and Wright 1978). Coal seams vary from a feather edge to 13 feet (4 m) thick; the Springfield and Herrin Coals are 4 to 8 feet (1.2 to 2.4 m) thick across large regions (see Chapter 5, Neotectonics).

Black shale layers 2 to 5 feet (0.6 to 1.5 m) thick overlie most coal beds in the Carbondale (Figure 10-8). Called "slate" by miners, these brittle, highly fissile shales contain so much carbonaceous matter that they will burn. Pyrite and siderite concretions are common. Black shales vary from nearly barren to abundantly fossiliferous; nearly all fossils are flattened along bedding planes. The inarticulate brachiopods and bivalves are best preserved. Conodonts, fish scales and spines, and poorly preserved cephalopods are common in places. Swimmers or floaters dominate the fauna; evidence of bottom-dwelling life is rare. Like the Devonian New Albany Shale, Pennsylvanian black shales were deposited in extremely quiet water where the substrate and part of the water column had very low oxygen content. Peat was the probable source of the carbon. Geologists disagree on how these unusual shales formed. Heckel (1977) argued that they formed in deep, stratified water during maximum rises of sea level, but Zangerl and Richardson (1963) presented convincing evidence that at least some black shales accumulated in shallow, restricted lagoons that may have been covered by floating mats of vegetation.

Limestone beds lie directly over many black shales and also occur beneath coal beds (Figure 10-8). Although the limestone beneath the Herrin Coal exceeds 10 feet (3 m) thick in places, most Carbondale limestones are thinner than 2 feet (0.6 m). Generally, limestone is buff or gray to nearly black, fine-grained, and shaly (wackestone and lime mudstone). Marine fossils are abundant, particularly large brachiopods.

Underclay is massive claystone or mudstone that lies directly under nearly all Pennsylvanian coal beds. Blocky or hackly fracture, carbonate and siderite nodules, and fossil root traces testify that underclays are ancient soils (Hughes et al. 1987). These soils developed during episodes of low sea level, forming the substrate upon which peat formed. Soils also can be developed in rock types other than mudstone, including siltstone, sandstone, and limestone. Where limestone has been subjected to soil formation, it takes on a highly nodular or "bouldery," brecciated structure riddled with clay-filled stringers and fractures.

Gray shale, siltstone, and sandstone make up the bulk of the Carbondale Formation. These rocks are arranged in three characteristic patterns (Figure 10-12):

- Upward-coarsening sequences, commonly 25 to 75 feet (8 to 23 m) thick. Dark gray shale grades upward in turn to gray silty shale, siltstone, and sandstone. Such sequences record a prograding delta or shoreline.
- Fining-upward sequences as thick as 100 feet (30.5 m) filling channels or valleys. The fill is typically sandstone containing basal conglomerate and grading upward to siltstone and shale. In map view, such channels have sinuous to strongly meandering and dendritic patterns with flow toward the south and southwest (Hopkins 1968, Treworgy and Jacobson 1985).
- Clastic wedges as thick as 100 feet (30.5 m) separating coal beds from overlying black shale and limestone. These wedges are dominantly silty medium-gray shale that contains fossil plants and siderite nodules. This gray shale formed in the fresh to brackish water of estuaries and bays. Examples include the Francis Creek Shale (not shown in Figure 10-1) overlying Colchester Coal, the Dykersburg Shale (not shown in Figure 10-1) overlying the Springfield Coal, and the Energy Shale above the Herrin Coal (see Chapter

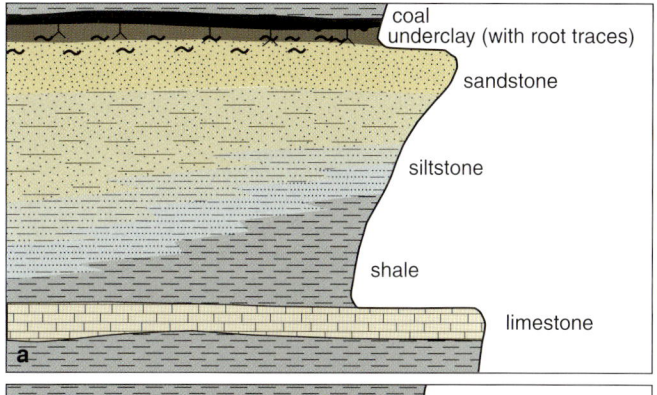
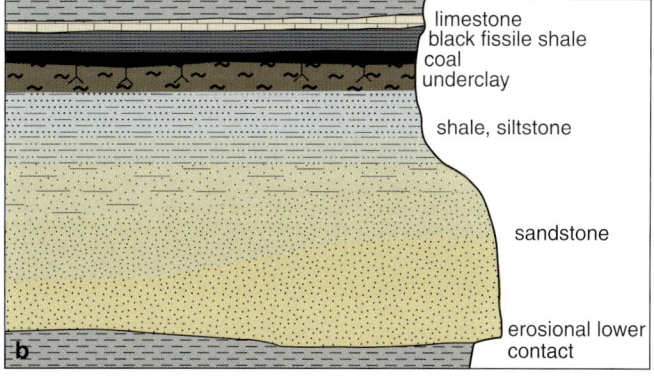
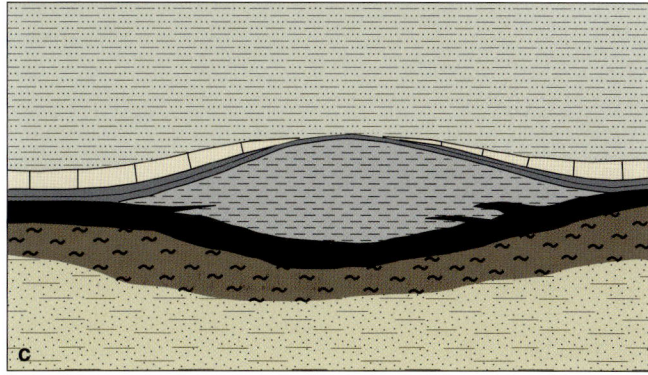

Figure 10-12 Three characteristic patterns of Carbondale Formation rocks: **(a)** upward-coarsening sequence, commonly the product of an advancing delta as sea level drops; **(b)** fining-upward sequence, the filling of an incised valley or channel; and **(c)** clastic wedge of gray shale, intervening between coal and black shale or limestone.

14, Coal). Coal overlain by such gray shale is typically low in sulfur.

McLeansboro Group

The McLeansboro Group (DeWolf 1910) comprises all Pennsylvanian rocks above the base of the Brereton Limestone (Figure 10-1). Like other Pennsylvanian units, the McLeansboro has been reclassified periodically (Figure 10-9). Currently the group contains four formations: Shelburn (oldest), Patoka, Bond, and Mattoon (Tri-State Committee on Correlation of the Pennsylvanian System in the Illinois Basin 2001). The McLeansboro Group differs from the Carbondale in two primary ways:

1. Coal beds rarely are thick enough to mine, most being thinner than 2 feet (0.6 m). Only the Danville Coal, near the base (Figure 10-1), contains significant resources.
2. Limestone beds are thicker and more numerous in the McLeansboro than in the Carbondale. Several are 10 feet (3 m) thick or thicker over large areas; the Millersville Limestone attains 50 feet (15 m). McLeansboro limestones are an important source of quarry stone in central Illinois where older carbonate rocks are absent or deeply buried.

Black, fissile shale units persist as useful markers throughout the McLeansboro Group. The bulk of the McLeansboro is composed of gray shale, siltstone, and sandstone arranged into sequences as is the Carbondale. The top of the McLeansboro is eroded throughout Illinois. Maximum thickness is approximately 1,200 feet (370 m) in Jasper County (Treworgy and Bargh 1982).

Shelburn Formation. The Shelburn Formation was named by Cumings (1922) for a small town south of Terre Haute, Indiana. The formation was extended into Illinois by the Tri-State Committee on Correlation of the Pennsylvanian System in the Illinois Basin (2001), designating rocks formerly included in the lower part of the Modesto Formation (Kosanke et al. 1960). The upper boundary is the top of the West Franklin Limestone Member in eastern Illinois or the equivalent Exline and Scottville Limestones (not shown in Figure 10-1) in western Illinois. Because these limestones are absent or discontinuous in many areas of Illinois, identifying the top of the Shelburn can be problematic.

The Shelburn Formation occupies the Fairfield Basin, lapping onto the Eastern and Western Shelves and northward into La Salle and Bureau Counties. Erosional outliers are present west of the Illinois River. Where fully preserved (top not eroded), the Shelburn ranges from about 150 feet (45 m) thick in northern and western Illinois to 275 feet (80 m) in southeastern Illinois. The formation thins to as little as 100 feet (30.5 m) on the Louden Anticline in central Illinois, reflecting contemporaneous uplift (Tri-State Committee on Correlation of the Pennsylvanian System in the Illinois Basin 2001).

The most widespread sandstone is the Anvil Rock Sandstone Member (not shown in Figure 10-1 because of scale; it falls between the Brereton and Bankston Fork Limestones), which fills channels eroded more than 100 feet (30.5 m) into underlying strata (Hopkins 1958, Potter and Simon 1961). Among limestones, the Brereton, Bankston Fork, Piasa, and West Franklin are thickest and most extensive. All are commonly 5 feet (1.5 m) to more than 10 feet (3 m) thick; the West Franklin locally reaches 25 feet (8 m) and comprises as many as three limestone benches each 5 to 10 feet (1.5 to 3 m) thick. The Danville Coal, which ranges up to about 6 feet (2 m) thick, has been mined in several areas of Illinois. Other coal beds are mostly less than 1 foot (0.3 m) thick, although Treworgy and Bargh (1982) mapped resources of the Jamestown or Hymera Coals in east-central Illinois (see Chapter 14, Coal).

Patoka Formation. The Patoka Formation was defined by Wier and Gray (1961) and Wier (1961, 1965) in southwestern Indiana and was extended into Illinois by the Tri-State Committee on Correlation of the Pennsylvanian System in the Illinois Basin (2001). The upper boundary is the base of the Carthage (formerly Shoal Creek) Limestone Member, which occurs virtually throughout the Illinois Basin (Wanless 1939, 1956; Tri-State Committee on Correlation of the Pennsylvanian System in the Illinois Basin 2001). The Patoka occurs in the Fairfield Basin and extends part way onto the adjacent shelves, attaining a thickness of 300 feet (90 m) in southeastern Illinois and thinning to as little as 35 feet (10 m) on the northwest.

Shale and sandstone compose more than 85% of the Patoka. The Trivoli Sandstone, near the base, is the most extensive and commonly ranges from 20 to 80 feet (6 to 24 m) thick (Andresen 1961). Like many Pennsylvanian sandstones, the Trivoli fills channels or valleys incised into older rocks. Other named units within the Patoka include the Chapel, Womac (not shown), and New Haven Coals (not shown), all of which are overlain by black, fissile shale and marine limestone (see Chapter 14, Coal). The coal beds are rarely thick enough to support even small, local mines.

Bond Formation. Named for Bond County in southwestern Illinois (Kosanke et al. 1960), the Bond Formation extends from the base of the Carthage Limestone to the top of the Millersville (also called Livingston) Limestone (Figure 10-1). The Bond is found throughout the Fairfield Basin; its eroded margin extends a short distance onto the

Western Shelf and into north-central Illinois. Where fully preserved, the formation varies from less than 150 feet (45 m) to just over 300 feet (90 m) thick, being thickest on the south (Willman et al. 1975).

The Carthage Limestone is 6 to 8 feet (2 to 2.5 m) thick across large areas and locally attains 20 feet (6 m). The La Salle Limestone, best developed near its namesake city in north-central Illinois, reaches 30 feet (9 m) thick. The Millersville (Livingston) Limestone locally reaches 50 feet (15 m) thick, but more commonly consists of two or three limestone beds separated by shale or mudstone (Willman et al. 1975). All of these limestones contain abundant and diverse marine fossils and are important sources of quarry stone. Coal beds are widespread but thin and economically insignificant, although the Flannigan and Bristol Hill Coals were formerly mined in Crawford County of eastern Illinois (Nance and Treworgy 1981).

Mattoon Formation. Named by Kosanke et al. (1960), the Mattoon Formation is largely confined to the Fairfield Basin and an area east of the La Salle Anticlinorium. The top of the formation is eroded everywhere in Illinois. Maximum thickness is a little over 600 feet (180 m) in Jasper County; the youngest preserved strata are in southern Cumberland County (Willman et al. 1975; Weibel et al. 1989). In fault blocks in western Kentucky, the Mattoon is completely preserved and is approximately 1,500 feet (460 m) thick (Greb et al. 1992).

The Mattoon Formation has been little studied because of the scarcity of its outcrops. Kosanke et al. (1960) described it as dominated by shale and sandstone and containing thin discontinuous coal beds and widespread limestones. Using well logs, Weibel (1996) and Weibel et al. (1989) found that the sandstone component is similar to that of Shelburn through Bond Formations. The most widely traceable beds are marine limestones and shales, especially the black shales. Several coal beds (not named on stratigraphic column) in the Mattoon Formation have been mined in Illinois, including Friendsville (being mined in 2003 near Mt. Carmel), Belle Rive and Opdyke (near Mt. Vernon), Oconee, Shelbyville, and Trowbridge (formerly mined near Shelbyville and elsewhere), and the Calhoun Coal (several counties in eastern Illinois) (Nance and Treworgy 1981).

Permian System

No Permian sedimentary rocks are preserved in Illinois, but they once probably covered much of the state. The Permian was a time of widespread tectonic faulting and folding throughout the state and igneous activity in southern Illinois (Figure 10-13).

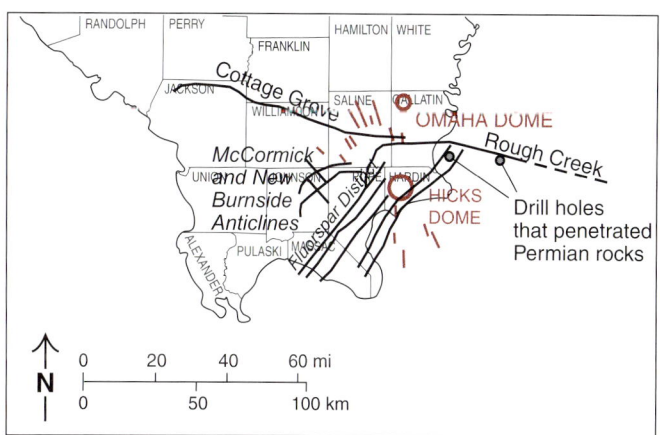

Figure 10-13 Permian rocks and structural features in southern Illinois. Maroon indicates igneous rocks; black lines are faults.

A core drilled in Kentucky, 15 miles (24 km) east of Shawneetown, Illinois, yielded 400 feet (120 m) of sandstone and shale interbedded with limestone containing marine fossils, including Permian fusulinids (Kehn et al. 1982). These rocks occupy a narrow fault slice. The base of the Permian is approximately 1,900 feet (580 m) above the Herrin Coal; by comparison, the youngest preserved rocks in the Mattoon Formation in Illinois are about 1,200 feet (370 m) above the Herrin. Also in Kentucky and less than 4 miles (6.4 km) from the Illinois state line, a dry oil test hole penetrated 2,100 feet (640 m) of strata overlying the Carthage Limestone and an estimated 2,550 feet (780 m) overlying the Herrin Coal. This hole, also drilled into a downfaulted block, likely contains the youngest Paleozoic rocks in the Illinois Basin (Palmer 1976).

The cored Permian rocks described by Kehn et al. (1982) are essentially similar to the upper part of the McLeansboro Group in Illinois. They are also similar to the next nearest Lower Permian rocks in eastern Kansas and in the Dunkard Basin of southeastern Ohio, northern West Virginia, and southwestern Pennsylvania. It is likely that Permian rocks originally were continuous from the Dunkard Basin across Illinois and Kentucky into the midcontinent.

The stratigraphic placement—or rank—of coal in Illinois requires that Pennsylvanian strata were once more deeply buried than at present. Coal rank is highest south of the Rough Creek Fault System in western Kentucky and far southeastern Illinois. Although igneous and hydrothermal activity in the faulted area may be partially responsible for the high rank, deeper burial here is also strongly implied (Damberger 1971, Greb et al. 1992). Also, Clegg (1955) reported that the character of natural coke along an igneous dike in Williamson County indicates that the coal was at or near its present rank when intrusion took place. Given that the dikes are now known to be Permian, that means the coal reached its present rank during the Permian.

Permian igneous rocks are found in far southeastern Illinois (Figure 10-13). A series of dikes, diatremes, and small stocks radiate along a northwest-southeast axis from Hicks Dome in Hardin County. A dike is a tabular body of intrusive igneous rock that crosscuts layering of the host rock. A diatreme is a small, roughly circular intrusion, and a stock is a larger body of intrusive rock that may be circular or irregularly outlined, crosscutting layered rocks. Hicks Dome, a major structural upheaval, is the product of explosive igneous activity at depth (Bradbury and Baxter 1992). Another swarm of northwest-trending igneous dikes accompanies the Cottage Grove Fault System north of Hicks Dome. Omaha Dome, northeast of Harrisburg, is an oil-productive structure that is the product of intrusion by a series of igneous sills.

These rocks in southeastern Illinois are ultramafic, classified as peridotite, alnoite, lamprophyre, and carbonatite. They are composed principally of olivine, brown mica, and pyroxene and lesser quantities of magnetite, apatite, garnet, and other minerals (Clegg and Bradbury 1956). Many intrusions contain fragments of sedimentary rocks altered by heat. Because the presumed source of ultramafic magma is the Earth's mantle, the fractures that served as feeders apparently penetrated the entire crust. Radiometric dating of igneous rocks using potassium-argon and rubidium-strontium analysis (Zartman et al. 1967, Nelson and Lumm 1987) yields Early Permian ages.

During the Permian, tectonic stresses from the Alleghany Orogeny were transferred inland, reactivating many faults, particularly in the south. The Rough Creek and Cottage Grove Fault Systems and many faults in the Fluorspar District were active at this time (Figure 10-13) as were many ancestral Rocky Mountain structures, including the Du Quoin Monocline, La Salle Anticlinorium, Louden and Salem Anticlines, and many others (Kolata and Nelson 1991).

It is difficult to segue from the Permian to the Mesozoic (Chapter 11) in Illinois. The Permian is presently represented by a small number of igneous intrusions. There are no Triassic or Jurassic rocks in the state, and the Cretaceous record is confined to the two latest (out of 12) stages. There was a long hiatus of almost 200 million years for which no direct record exists of the geological happenings in Illinois.

Conclusion

Following deposition of Lower Permian sediments in part of the Illinois Basin, the region was subjected to a long period of erosion prior to the deposition of the oldest Mesozoic strata, which are covered in the following chapter.

References

Andresen, M. J., 1961, Geology and petrology of the Trivoli Sandstone in the Illinois Basin: Illinois State Geological Survey, Circular 316, 31 p.

Archer, A. W., and S. F. Greb, 1995, An Amazon-scale drainage system in the Early Pennsylvanian of central North America: Journal of Geology, v. 103, no. 6, p. 611–628.

Baird, G. C., C. W. Shabica, J. L. Anderson, and E. S. Richardson, Jr., 1985, Biota of a Pennsylvanian muddy coast—Habitats within the Mazonian delta complex, northeast Illinois: Journal of Paleontology, v. 59, no. 2, p. 253–281.

Bradbury, J. C., and J. W. Baxter, 1992, Intrusive breccias at Hicks Dome, Hardin County, Illinois: Illinois State Geological Survey, Circular 550, 23 p.

Bristol, H. M., and R. H. Howard, 1971, Paleogeologic map of the sub-Pennsylvanian Chesterian (Upper Mississippian) surface in the Illinois Basin: Illinois State Geological Survey, Circular 458, 14 p.

Bristol, H. M., and R. H. Howard, 1974, Sub-Pennsylvanian valleys in the Chesterian surface of the Illinois Basin and related Chesterian slump blocks, in G. Briggs, ed., Carboniferous of the southeastern United States: Boulder, Colorado, Geological Society of America, Special Paper 148, p. 315–336.

Cecil, C. B., R. W. Stanton, S. G. Neuzil, F. T. Dulong, L. F. Ruppert, and B. S. Pierce, 1985, Paleoclimate controls on Late Paleozoic sedimentation and peat formation in the Central Appalachian basin (U.S.A.): International Journal of Coal Geology, v. 5, no. 1–2, p. 195–230.

Clegg, K. E., 1955, Metamorphism of coal by peridotite dikes in southern Illinois: Illinois State Geological Survey, Report of Investigations 178, 18 p.

Clegg, K. E., and J. C. Bradbury, 1956, Igneous intrusive rocks in Illinois and their economic significance: Illinois State Geological Survey, Report of Investigations 197, 19 p.

Conybeare, W. D., and W. Phillips, 1822, Outlines of the geology of England and Wales: London: W. Phillips, 470 p. [Reprinted 1978, New York, Arno Press.]

Cooper, C. L., 1946, Pennsylvanian ostracodes of Illinois: Illinois State Geological Survey, Bulletin 70, 177 p.

Cumings, E. R., 1922, Nomenclature and description of the geological formations of Indiana, in W. N. Logan, E. R. Cumings, C. A. Malott, S. S. Visher, W. M. Tucker, and J. R. Reevs, eds., Handbook of Indiana geology: Bloomington, Indiana, Indiana Department of Conservation Publication 21, part 4, p. 403–570.

Damberger, H. H., 1971, Coalification pattern of the Illinois Basin: Economic Geology, v. 66, p. 488–494.

Devera, J. A., 1989, Ichnofossil assemblages and associated lithofacies of the Lower Pennsylvanian (Caseyville and Tradewater Formations), southern Illinois, in J. C. Cobb, coordinator, Geology of the Lower Pennsylvanian in Kentucky, Indiana, and Illinois: Illinois Basin Consortium, Illinois Basin Studies, v. 1, p. 57–83.

Devera, J. A., C. E. Mason, and R. A. Peppers, 1987, A marine shale in the Caseyville Formation (Lower Pennsylvanian) in southern Illinois (abs.): Boulder, Colorado, Geological Society of America, Abstracts with Programs, v. 19, no. 4, p. 220.

DeWolf, F. W., 1910, Studies of Illinois coal—Introduction: Illinois State Geological Survey, Bulletin 16, p. 178–181.

DiMichele, W. A., and T. L. Phillips, 1996, Climate change, plant extinctions and vegetational recovery during the Middle–Late Pennsylvanian transition—The case of tropical peat-forming environments in North America, in M. B. Hart, ed., Biotic recovery from mass extinction events: London, England, The Geological Society of London, Special Publication 102, p. 201–221.

Dunbar, C. O., and L. G. Henbest, 1942, Pennsylvanian Fusulinidae of Illinois: Illinois State Geological Survey, Bulletin 67, 218 p.

Earth Science Club of Northern Illinois, 1989, Keys to identify Pennsylvanian fossil animals of the Mazon Creek area: Downers Grove, Illinois, Earth Science Club of Northern Illinois, 125 p.

Earth Science Club of Northern Illinois, 1990, Keys to identify Pennsylvanian fossil plants of the Mazon Creek area: Downers Grove, Earth Science Club of Northern Illinois, 60 p.

Ferm, J. C., 1970, Allegheny deltaic deposits, in J. P. Morgan, ed., Deltaic sedimentation, modern and ancient: Society of Economic Paleontologists and Mineralogists, Special Publication 15, p. 246–255.

Ferm, J. C., 1975, Pennsylvanian cyclothems of the Appalachian Plateau, a retrospective view, in E. D. McKee and E. J. Crosby, coordinators, Paleotectonic investigations of the Pennsylvanian System in the United States: Reston, Virginia, U.S. Geological Survey, Professional Paper 853, part. 2, p. 57–64.

Greb, S. F., 1989, Structural controls on the formation of the sub-Absaroka unconformity in the U.S. Eastern Interior basin: Geology, v. 17, no. 10, p. 889–892.

Greb, S. F., W. M. Andrews, C. F. Eble, W. DiMichele, C. B. Cecil, and J. C. Hower, 2003, Desmoinesian coal beds of the Eastern Interior and surrounding basins: The largest tropical peat mires in Earth history, in M. A. Chan and A. W. Archer, eds., Extreme depositional environments: Boulder, Colorado, Geological Society of America, Special Paper 370, p. 127–150.

Greb, S. F., D. A. Williams, and A. D. Williamson, 1992, Geology and stratigraphy of the Western Kentucky Coal Field: Kentucky Geological Survey, Series XI, Bulletin 2, 77 p.

Heckel, P. H., 1977, Origin of phosphatic black shale facies in Pennsylvanian cyclothems of mid-continent North America: American Association of Petroleum Geologists Bulletin, v. 61, no. 7, p. 1045–1068.

Heckel, P. H., 2004, Newsletter on Carboniferous stratigraphy: UIGS Subcommission on Carboniferous Stratigraphy, v. 22, p. 1–3.

Hopkins, M. E., 1958, Geology and petrology of the Anvil Rock Sandstone of southern Illinois: Illinois State Geological Survey, Circular 431, 24 p.

Hopkins, M. E., 1968, Harrisburg (No. 5) Coal reserves of southeastern Illinois: Illinois State Geological Survey, Circular 431, 25 p.

Hughes, R. E., P. J. DeMaris, W. A. White, and D. K. Cowin, 1987, Origin of clay minerals in Pennsylvanian strata of the Illinois Basin: Proceedings of 1985 International Clay Conference, Clay Minerals Society, Bloomington, Indiana, p. 97–104.

Jacobson, R. J., 1985, Coal resources of Grundy, La Salle, and Livingston Counties, Illinois: Illinois State Geological Survey, Circular 536, 58 p.

Jacobson, R. J., 1987, Stratigraphic correlations of the Seelyville, Dekoven, and Davis Coals of Illinois, Indiana, and western Kentucky: Illinois State Geological Survey, Circular 539, 27 p.

Jennings, J. R., 1990, Guide to Pennsylvanian fossil plants of Illinois: Illinois State Geological Survey, Educational Series 13, 75 p.

Jennings, J. R., and G. H. Fraunfelter, 1986, Preliminary report on macropaleontology of strata above and below the upper boundary of the type Mississippian: Transactions of the Illinois State Academy of Science, v. 79, no. 3–4, p. 253–261.

Johnson, R. G., and E. S. Richardson, Jr., 1966, A remarkable Pennsylvanian fauna from the Mazon Creek area, Illinois: Journal of Geology, v. 74, no. 5, p. 626–631.

Johnson, R. G., and E. S. Richardson, Jr., 1970, Fauna of the Francis Creek Shale in the Wilmington area, in W. H. Smith, R. B. Nance, M. E. Hopkins, R. G. Johnson, and C. W. Shabica, Depositional environments in parts of the Carbondale Formation—Western and northern Illinois: Illinois State Geological Survey, Guidebook 8, p. 53–60.

Kehn, T. M., J. G. Beard, and A. D. Williamson, 1982, Mauzy Formation, a new stratigraphic unit of Permian age in western Kentucky: Reston, Virginia, U.S. Geological Survey, Bulletin 1529-H, p. H73–H86.

Kolata, D. R., compiler, 2005, Bedrock geology of Illinois: Illinois State Geological Survey, Illinois Map 14, 1:500,000.

Kolata, D. R., and W. J. Nelson, 1991, Tectonic history of the Illinois Basin, in M. W. Leighton, D. R. Kolata, D. F. Oltz, and J. J. Eidel, eds., Interior cratonic basins: Tulsa, Oklahoma, American Association of Petroleum Geologists, Memoir 51, p. 263–285.

Kosanke, R. M., J. A. Simon, H. R. Wanless, and H. B. Willman, 1960, Classification of the Pennsylvanian strata of Illinois: Illinois State Geological Survey, Report of Investigations 214, 84 p.

Kvale, E. P., and M. L. Barnhill, 1994, Evolution of Lower Pennsylvanian estuarine facies within two adjacent paleovalleys, Illinois Basin, Indiana, in R. W. Dalrymple, R. Boyd, and B. A. Zaitlin, eds., Incised-valley systems: Tulsa, Oklahoma: Society for Sedimentary Geology (SEPM), Special Publication 51, p. 191–207.

Lee, W., 1916, Geology of the Kentucky part of the Shawneetown Quadrangle: Lexington, Kentucky, Kentucky Geological Survey, series 4, v. 4, part 2, 73 p.

Ludvigson, G., and K. Swett, 1987, Pennsylvanian strata at Wyoming Hill and Wildcat Den State Park, Muscatine County, Iowa, in G. R. McCormick, ed., Guidebook for the Fifty-first Annual Tri-State Geological Field Conference: Iowa City, Iowa, University of Iowa, p. C1–C20.

McBride, J. H., and W. J. Nelson, 1999, Style and origin of mid-Carboniferous deformation in the Illinois Basin, USA—Ancestral Rockies deformation?: Tectonophysics, v. 305, no. 1–3, p. 249–273.

McKee, E. D., and E. J. Crosby, coordinators, 1975, Paleotectonic investigations of the Pennsylvanian System in the United States: Reston, Virginia, U.S. Geological Survey, Professional Paper 853, 3 v., 541 p.

Moore, R. C., 1944, Correlation of Pennsylvanian formations of North America: Geological Society of America Bulletin, v. 55, no. 6, p. 657–706.

Nadon, G. C., 1999, Ice house sequence stratigraphy—The constraints of glacial-eustasy on Pennsylvanian sedimentation (abs.): Boulder, Colorado, Geological Society of America, Abstracts with Programs, v. 31, no. 7, p. 182.

Nance, R. B., and C. G. Treworgy, 1981, Strippable coal resources of Illinois, Part 8, Central and southeastern counties: Illinois State Geological Survey, Circular 515, 32 p.

Nelson, W. J., 1989, The Caseyville Formation (Morrowan) of the Illinois Basin: Regional setting and local relationships, in J. C. Cobb, coordinator, Geology of the Lower Pennsylvanian in Kentucky, Indiana, and Illinois: Illinois Basin Consortium, Illinois Basin Studies 1, p. 84–95.

Nelson, W. J., 1995, Structural features in Illinois: Illinois State Geological Survey, Bulletin 100, 144 p.

Nelson, W. J., J. A. Devera, R. J. Jacobson, C. P. Weibel, L. R. Follmer, M. H. Riggs, S. P. Esling, E. D. Henderson, and M. S. Lannon, 1991, Geology of the Eddyville, Stonefort, and Creal Springs Quadrangles, southern Illinois: Illinois State Geological Survey, Bulletin 96, 85 p.

Nelson, W. J., and D. K. Lumm, 1987, Structural geology of southeastern Illinois and vicinity: Illinois State Geological Survey, Circular 538, 70 p.

Nelson, W. J., C. B. Trask, R. J. Jacobson, H. H. Damberger, A. D. Williamson, and D. A. Williams, 1991, Absaroka sequence—Pennsylvanian and Permian Systems, in M. W. Leighton, D. R. Kolata, D.

F. Oltz, and J. J. Eidel, eds., Interior cratonic basins: American Association of Petroleum Geologists, Memoir 51, p. 143–164.

Newberry, J.S., A. H. Worthen, F. B. Meek, and L. Lesquereux, 1866, Palaeontology: Geological Survey of Illinois, v. 2, 595 p.

North American Commission on Stratigraphic Nomenclature, 1983, North American stratigraphic code: American Association of Petroleum Geologists Bulletin, v. 67, no. 5, p. 841–875.

Palmer, J. E., 1976, Geologic map of the Grove Center quadrangle, Kentucky–Illinois, and part of the Shawneetown quadrangle, Kentucky: Reston, Virginia, U.S. Geological Survey, Geologic Quadrangle Map GQ-1314, 1:24,000.

Palmer, J. E., R. J. Jacobson, and C. B. Trask, 1979, Depositional environments of strata of late Desmoinesian age overlying the Herrin (No. 6) Coal Member in southwestern Illinois, in J. E. Palmer and R. R. Dutcher, eds., Depositional and structural history of the Pennsylvanian System of the Illinois Basin: Field trip 9, Ninth International Congress of Carboniferous Stratigraphy and Geology, part 2: Invited papers, p. 92–102.

Payton, C. E., ed., 1977, Seismic stratigraphy—Applications to hydrocarbon exploration: Tulsa, Oklahoma, American Association of Petroleum Geologists, Memoir 26, v. 11, 516 p.

Peppers, R. A., 1996, Palynological correlation of major Pennsylvanian (Middle and Upper Carboniferous) chronostratigraphic boundaries in the Illinois and other coal basins: Boulder, Colorado, Geological Society of America, Memoir 188, 111 p.

Phillips, T. C., M. J. Avcin, and D. Berggren, 1976, Fossil peat of the Illinois Basin—A guide to the study of coal balls of Pennsylvanian age: Illinois State Geological Survey, Educational Series 11, 39 p.

Phillips, T. L., and R. A. Peppers, 1984, Changing patterns of Pennsylvanian coal-swamp vegetation and implications of climatic control on coal occurrence: International Journal of Coal Geology, v. 3, no. 3, p. 205–255.

Potter, P. E., 1963, Late Paleozoic sandstones of the Illinois Basin: Illinois State Geological Survey, Report of Investigations 217, 92 p.

Potter, P. E., and H. D. Glass, 1958, Petrology and sedimentation of the Pennsylvanian sediments in southern Illinois: Vertical profile: Illinois State Geological Survey, Report of Investigations 204, 60 p.

Potter, P. E., and W. A. Pryor, 1961, Dispersal centers of Paleozoic and later clastics of the upper Mississippi Valley and adjacent areas: Geological Society of America Bulletin, v. 72, no. 8, p. 1195–1250.

Potter, P. E., and R. Siever, 1956a, Sources of basal Pennsylvanian sediment in the Eastern Interior Basin: Part I, Cross-bedding: Journal of Geology, v. 64, no. 3, p. 225–244.

Potter, P. E., and R. Siever, 1956b, Sources of basal Pennsylvanian sediment in the Eastern Interior Basin: Part II, Sedimentary petrology: Journal of Geology, v. 64, no. 4, p. 317–335.

Potter, P. E., and J. A. Simon, 1961, Anvil Rock Sandstone and channel cutouts of Herrin (No. 6) Coal in west-central Illinois: Illinois State Geological Survey, Circular 314, 12 p.

Pryor, W. A., and E. G. Sable, 1974, Carboniferous of the Eastern Interior Basin, in G. Briggs, ed., Carboniferous of the southeastern United States: Boulder, Colorado, Geological Society of America, Special Paper 148, p. 281–313.

Read, C. B., and S. H. Mamay, 1964, Upper Paleozoic floral zones and floral provinces of the United States: Washington D.C., U.S. Geological Survey, Professional Paper 454-K, 33 p.

Rice, C. L., 1984, Sandstone units of the Lee Formation and related strata in eastern Kentucky: Reston, Virginia, U.S. Geological Survey, Professional Paper 1151-G, 53 p.

Richardson, E. S., Jr., 1956, Pennsylvanian invertebrates of the Mazon Creek area, Illinois: Chicago, Illinois, Natural History Museum, Fieldiana: Geology, v. 12, no. 1–4, p. 3–76.

Ross, C. A., and J. R. P. Ross, 1985, Late Paleozoic depositional sequences are synchronous and worldwide: Geology, v. 13, no. 3, p. 194–197.

Saunders, W. B., and W. H. C. Ramsbottom, 1986, The mid-Carboniferous eustatic event: Geology, v. 14, no. 3, p. 208–212.

Schopf, J. M., 1975, Pennsylvanian climate in the United States, in Paleotectonic investigations of the Pennsylvanian System in the United States: Reston, Virginia, U.S. Geological Survey, Professional Paper 853, p. 23–31.

Sedimentation Seminar, 1978, Sedimentology of the Kyrock Sandstone (Pennsylvanian) in the Brownsville paleovalley, Edmonson and Hart Counties, Kentucky: Lexington, Kentucky, Kentucky Geological Survey, Series X, Report of Investigations 21, 24 p.

Shabica, C. W., 1970, Depositional environments in the Francis Creek Shale, in W. H. Smith, R. B. Nance, M. E. Hopkins, R. G. Johnson, and C. W. Shabica, eds., Environments in parts of the Carbondale Formation—Western and northern Illinois: Illinois State Geological Survey, Guidebook Series 8, p. 43–52.

Shabica, C. W., and A. A. Hay, eds., 1997, Richardson's guide to the fossil fauna of Mazon Creek: Chicago, Illinois, Northeastern Illinois University, 308 p.

Shaw, E. W., and T. E. Savage, 1912, Murphysboro-Herrin folio: Washington D.C., U.S. Geological Survey, Geological Atlas Folio 185, 15 p.

Siever, R., 1951, The Mississippian-Pennsylvanian unconformity in southern Illinois: American Association of Petroleum Geologists Bulletin, v. 35, no. 3, p. 542–581.

Sloss, L. L., 1963, Sequences in the cratonic interior of North America: Geological Society of America Bulletin, v. 74, no. 2, p. 93–114.

Sloss, L. L., 1988, Tectonic evolution of the craton in Phanerozoic time, in L. L. Sloss, ed., Sedimentary cover—North American craton: U.S.: The Geology of North America, v. D-2: Boulder, Colorado, Geological Society of America, p. 25–51.

Sloss, L. L., W. C. Krumbein, and E. C. Dapples, 1949, Integrated facies analysis, in C. R. Longwell, chairman, Sedimentary facies in geologic history: New York, Geological Society of America, Memoir 39, p. 91–123.

Thompson, M. L., and R.H. Shaver, 1964, Early Pennsylvanian microfaunas of the Illinois Basin: Transactions of the Illinois State Academy of Sciences, v. 57, no. 1, p. 3–23.

Thompson, M. L., R. H. Shaver, and E. A. Riggs, 1959, Early Pennsylvanian fusulinids and ostracods of the Illinois Basin: Journal of Paleontology, v. 33, no. 5, p. 770–792.

Treworgy, C. G., and M. H. Bargh, 1982, Deep-minable coal resources of Illinois: Illinois State Geological Survey, Circular 527, 62 p.

Treworgy, C. G., and R. J. Jacobson, 1985, Paleoenvironments and distribution of low-sulfur coal in Illinois, in A. T. Cross, ed., Economic geology, coal, oil, and gas: Compte Rendu, 9th Congrès International de Stratigraphie et de Géologie du Carbonifère, v. 4: Carbondale, Illinois, Southern Illinois University Press, p. 372–382.

Tri-State Committee on Correlation of the Pennsylvanian System in the Illinois Basin, 2001, Toward a more uniform stratigraphic nomenclature for rock units (formations and groups) of the Pennsylvanian System in the Illinois Basin: Illinois Basin Consortium, Illinois Basin Consortium Study 5, 26 p.

Udden, J. A., 1912, Geology and mineral resources of the Peoria Quadrangle: Washington D.C., U.S. Geological Survey, Bulletin 506, 103 p.

Van Wagoner, J. C., R. M. Mitchum, K. M. Campion, and V. D. Rahmanian, 1990, Siliciclastic sequence stratigraphy in well logs, cores, and outcrops; Concepts for high-resolution correlation of time and facies: Tulsa, Oklahoma, American Association of Petroleum Geologists, Methods in Exploration Series, no. 7, 55 p.

Veevers, J. J., and C. M. Powell, 1987, Late Paleozoic glacial episodes in Gondwanaland reflected in transgressive-regressive depositional sequences in Euramerica: Geological Society of America Bulletin, v. 98, no. 4, p. 475–487.

Wanless, H. R., 1931, Pennsylvanian cycles in western Illinois: Illinois State Geological Survey, Bulletin 60, p. 179–193.

Wanless, H. R., 1939, Pennsylvanian correlations in the Eastern Interior and Appalachian coal fields: Boulder, Colorado, Geological Society of America, Special Paper 17, 130 p.

Wanless, H. R., 1955, Pennsylvanian rocks of Eastern Interior Basin: American Association of Petroleum Geologists Bulletin, v. 39, no. 9, p. 1753–1820.

Wanless, H. R., 1956, Classification of the Pennsylvanian rocks of Illinois as of 1956: Illinois State Geological Survey, Circular 217, 14 p.

Wanless, H. R., 1957, Geology and mineral resources of the Beardstown, Glasford, Havana, and Vermont Quadrangles: Illinois State Geological Survey, Bulletin 82, 233 p.

Wanless, H. R., 1958, Pennsylvanian faunas of the Beardstown, Glasford, Havana, and Vermont Quadrangles: Illinois State Geological Survey, Report of Investigations 205, 59 p.

Wanless, H. R., 1975, Illinois Basin region, in E. D. McKee and E. J. Crosby, coordinators, Paleotectonic investigations of the Pennsylvanian System in the United States, Part 1—Introduction and regional analyses of the Pennsylvanian system: Reston, Virginia, U.S. Geological Survey, Professional Paper 853-E, p. 71–95.

Wanless, H. R., and F. P. Shepard, 1936, Sea level and climatic changes related to late Paleozoic cycles: Geological Society of America Bulletin, v. 47, no. 8, p. 1177–1206.

Wanless, H. R., and J. M. Weller, 1932, Correlation and extent of Pennsylvanian cyclothems: Geological Society of America Bulletin, v. 4, no. 4, p. 1003–1016.

Wanless, H. R., and J. M. Weller, 1944, Eastern Interior Region, in Correlation of Pennsylvanian formations of North America: Geological Society of America Bulletin 55, p. 657–706.

Wanless, H. R., and C. R. Wright, 1978, Paleoenvironmental maps of Pennsylvanian rocks, Illinois Basin and northern Midcontinent region: Boulder, Colorado, Geological Society of America, Map and Chart Series 23, 32 p.

Weibel, C. P., 1996, Applications of sequence stratigraphy to Pennsylvanian strata in the Illinois Basin, in B. J. Witzke, G. A. Ludvigson, and J. Day, eds., Paleozoic sequence stratigraphy: Views from the North American craton: Boulder, Colorado, Geological Society of America, Special Paper 306, p. 331–339.

Weibel, C. P., 2002, Evolution of the boundaries used to map Pennsylvanian cyclothemic sequences in Illinois, USA, in L. V. Hills, C. M. Henderson, and E. W. Bamber, eds., Carboniferous and Permian of the world: Calgary, Alberta, Canada, Canadian Society of Petroleum Geologists, Memoir 19, p. 239–251.

Weibel, C. P., R. L. Langenheim Jr., and D. A. Willard, 1989, Cyclic strata of the Late Pennsylvanian outlier, east-central Illinois, in J. D. Vineyard and W. K. Wedge: Geological Society of America 1989 Field Trip Guidebook, Missouri Division of Geology and Land Survey, Special Publication 5, p. 141–169.

Weibel, C. P., and R. D. Norby, 1992, Paloepedology and conodont biostratigraphy of the Mississippian-Pennsylvanian boundary interval, type Grove Church Shale area, southern Illinois: Norman, Oklahoma, Oklahoma Geological Survey, Circular 94, p. 39–53.

Weller, J. M., 1930, Cyclical sedimentation of the Pennsylvanian Period and its significance: Journal of Geology, v. 38, no. 2, p. 97–135.

Weller, J. M., 1931, The conception of cyclical sedimentation during the Pennsylvanian Period: Illinois State Geological Survey, Bulletin 60, p. 163–177.

Weller, S., 1906, The geological map of Illinois: Illinois State Geological Survey, Bulletin 1, 26 p.

Wier, C. E., 1961, Stratigraphy of the Middle and Upper Pennsylvanian rocks in southwestern Indiana: Indiana Geological Survey, unpublished manuscript, 172 p. (Available from the Geology Department library, Indiana University, Bloomington, Indiana.)

Wier, C. E., 1965, Stratigraphy of the Middle and Upper Pennsylvanian rocks in southwestern Indiana: Indiana Geological Survey, unpublished manuscript, 194 p. (Available from the Geology Department library, Indiana University, Bloomington, Indiana.)

Wier, C. E., and H. H. Gray, 1961, Geologic map of the Indianapolis 1° × 2°-degree quadrangle, Indiana and Illinois, showing bedrock and unconsolidated deposits: Bloomington, Indiana, Indiana Geological Survey, Regional Geologic Map, 1:250,000.

Williams, H. S., 1891, Correlation papers—Devonian and Carboniferous: Washington D.C., U.S. Geological Survey, Bulletin 80, 279 p.

Willman, H. B., E. Atherton, T. C. Buschbach, C. Collinson, J. C. Frye, M. E. Hopkins, J. A. Lineback, and J. A. Simon, 1975, Handbook of Illinois stratigraphy: Illinois State Geological Survey, Bulletin 95, 261 p.

Wittry, J., 2006, The Mazon Creek fossil flora: Downers Grove, Illinois, Earth Science Club of Northern Illinois (ESCONI), 154 p.

Worthen, A. H., G. C. Broadhead, E. T. Cox, O. St. John, and F. B. Meek, 1875, Geology and palaeontology: Geological Survey of Illinois, v. 6, 577 p.

Zangerl, R., and E. S. Richardson, Jr., 1963, The paleoecological history of two Pennsylvanian black shales: Chicago, Illinois, Natural History Museum, Fieldiana—Geology Memoirs, v. 4, 352 p.

Zartman, R. E., M. R. Brock, A. V. Heyl, and H. H. Thomas, 1967, K-Ar and Rb-Sr ages of some alkalic intrusive rocks from central and eastern United States: American Journal of Science, v. 265, no. 10, p. 848–870.

11 Mesozoic and Tertiary Eras

W. John Nelson

INTRODUCTION

This chapter covers a long span of Earth's history during which many geological events occurred in Illinois, but for which little direct evidence remains. Mesozoic and Tertiary strata occur only as scattered erosional remnants in western and southernmost Illinois. To deduce what was happening in Illinois during this time, rocks of other U.S. regions must be considered.

The Mesozoic includes the upper part of the Absaroka sequence (Chapter 10, Pennsylvanian and Permian) and most of the Zuni sequence (Sloss 1963, 1988; Sloss et al. 1949). The Tertiary comprises the uppermost part of the Zuni and the overlying Tejas sequence, which extends through the Quaternary to the present day. Because the record of these sequences in Illinois is so fragmentary, this chapter is organized according to the U.S. Geological Survey, Geologic Names Committee (2007).

MESOZOIC ERA

The Mesozoic (from Greek meaning "middle life") Era was the period of Earth's history between approximately 250 and 65 million years ago when the dinosaurs flourished. The era is popularly known as the "Age of Reptiles" and is divided into three periods: Triassic (oldest), Jurassic, and Cretaceous. Only the Upper Cretaceous rocks are preserved in Illinois (Figure 11-1).

Triassic Period

As outlined in Chapters 3 (Tectonic History) and 4 (Structural Features), all of the world's continents came together to form a single giant landmass, Pangea, near the end of the Paleozoic. The shallow seas that previously covered Illinois and surrounding regions withdrew during the Permian. Situated close to the equator and isolated from ocean breezes by the newly formed Appalachian and Ouachita Mountains (Figure 11-2), Illinois undoubtedly had a torrid desert climate as the Permian passed into the Triassic; erosion, rather than deposition, prevailed.

During Late Triassic time, Pangea began to break apart (Chapter 3, Tectonic History). Asia and Australia drifted westward away from the Americas as Europe and Africa drifted eastward, creating the Atlantic Ocean. The Gulf of Mexico began to develop. A series of fault-bounded basins along the Atlantic and Gulf Coastal Plains (Figure 11-2) were filled with thick successions consisting largely of red, terrestrial sediments and basaltic lava flows (King 1977, Byerly 1991, Salvador 1991).

Many faults in southern Illinois probably became active. Normal faults in the Wabash Valley Fault Zone may have developed in response to Gulf Coast rifting. Normal faulting also took place in the Rough Creek-Shawneetown Complex and Fluorspar Area Fault Complex some time during the Mesozoic; the Triassic is the most logical time for such faulting to have occurred (Nelson and Lumm 1987, Kolata and Nelson 1991).

In the western United States, the ancestral Rocky Mountains were being worn down. Streams and wind carried and deposited red clay, silt, sand, and gravel. The east-

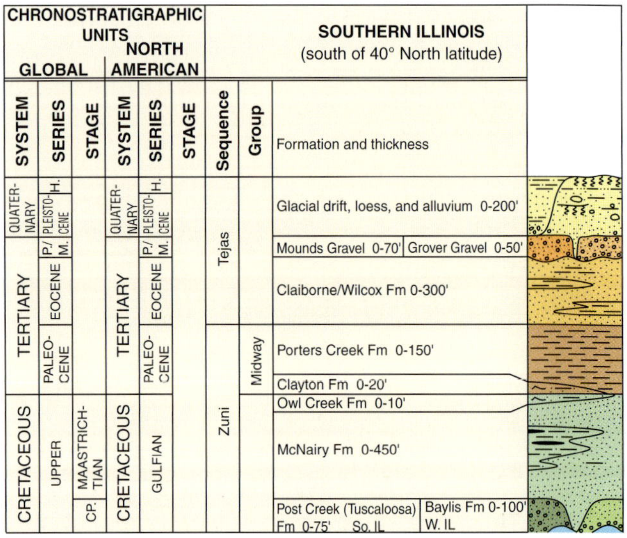

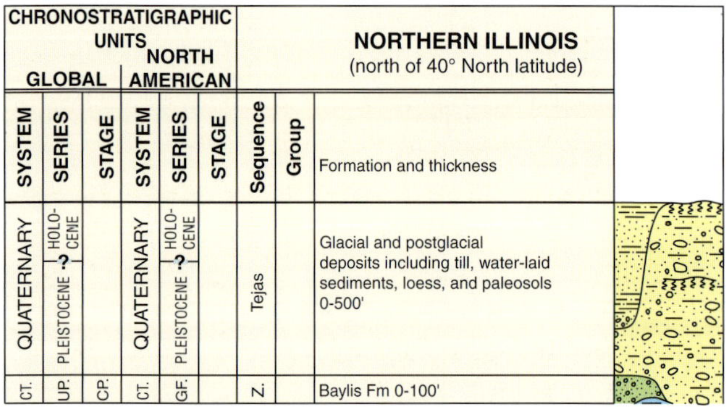

Figure 11-1 Stratigraphic column of the Mesozoic and Tertiary strata of Illinois (Kolata 2005). Abbreviations: CP., Campanian; CT., Cretaceous; GF., Gulfian; H., Holocene; M., Miocene; P., Pliocene; Z., Zuni; Fm, Formation; So., southern; UP., Upper; W., western.

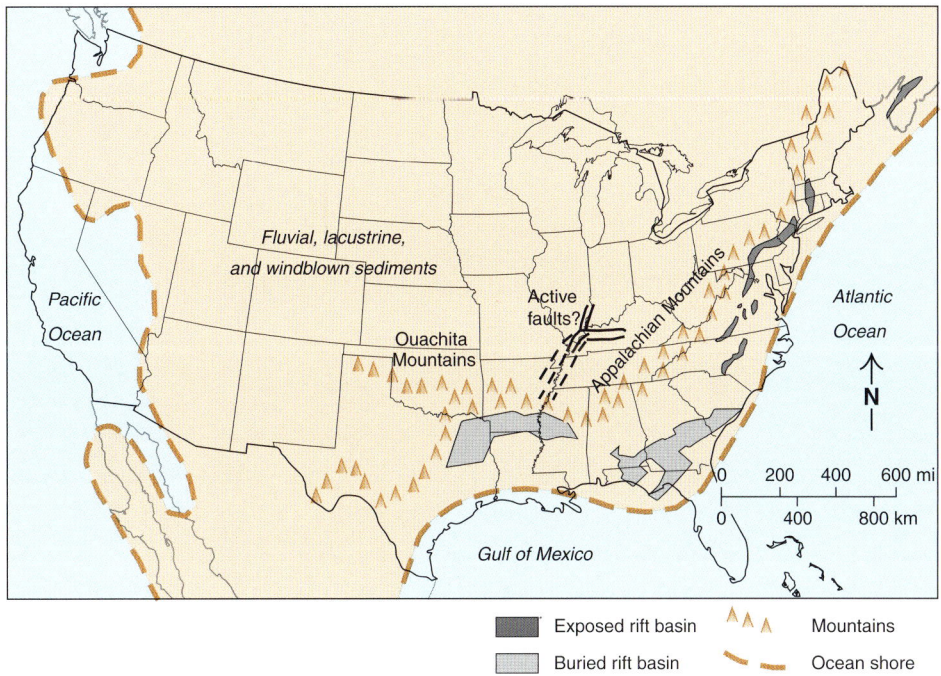

Figure 11-2 Map of the United States during the Late Triassic Period. Buried rift locations are from Salvador (1991); other features are from Dunbar (1960).

ernmost approach of the sea into what is now central Utah took place during Early Triassic time (Dunbar 1966). Terrestrial sedimentary rocks, such as the Chinle Formation that hosts Arizona's Petrified Forest, typify Triassic rocks of the western United States. Triassic red beds are extensive in the Rocky Mountains and western Great Plains, but have been entirely eroded farther east.

Jurassic Period

Rifting along the Atlantic and Gulf Coastal Plains ended early in the Jurassic Period as the Gulf became a true ocean basin (Figure 11-3). One of the world's greatest salt deposits, the Louann Salt, formed along the Gulf Coast during this time, evidence of a prolonged hot and dry climate that evaporated vast quantities of sea water. Younger Jurassic deposits in the Gulf include marine limestone, marl, and anhydrite, which extend in the subsurface as far north as southern Arkansas and central to southern Mississippi.

In the western United States, Jurassic rocks are largely dune sands and fluvial red beds, such as the Navajo and Entrada Sandstones of Utah's canyonlands and the dinosaur-bearing Morrison Formation. At times, shallow seas advanced eastward across what is now the Great Plains. The Jurassic rocks nearest to Illinois are at Fort Dodge, Iowa, and in central Michigan (Figure 11-3). Both locations contain beds of gypsum, a rock that typically forms through evaporation of sea water (Fisher et al. 1988, Bunker et al. 1988, Catacosinos et al. 1991). Given Illinois' position between areas of marine Jurassic rocks, it is likely that the sea entered Illinois at least briefly during the Jurassic.

Cretaceous Period

Profound geological changes continued during the Cretaceous. The Atlantic Coast became a "passive margin," characterized by quiet tectonic conditions, and the scene of gradual sediment accumulation. In contrast, vigorous tectonic, volcanic, and igneous activity took place along the Gulf Coastal Plain (Figure 11-4). Volcanoes erupted in Mississippi and northern Louisiana, probably during the middle part of the Cretaceous, as indicated by igneous rocks in central Arkansas that have been dated radiometrically at 86 to 106 million years (Byerly 1991). The Gulf Coast volcanism of Cretaceous time appears to have been localized and almost entirely in the subsurface; it is poorly understood. Geophysical data and samples from deep wells (Heigold 1976, McGinnis et al. 1976, Hildenbrand et al. 1977) indicate that more buried Cretaceous igneous bodies occur farther north in Arkansas, western Tennessee, southeastern Missouri, and southernmost Illinois. Large, nearly circular magnetic and gravity highs coincide with southern Illinois' two microcrystalline silica (tripoli) districts—Elco in Alexander County and Wolf Lake in Union County (Figure 11-5). The mafic igneous rocks are dense (high gravity) and have relatively high magnetism. They aren't the only rocks having such properties but are the most common cause of gravity and magnetic highs. Berg and Masters (1994) thus proposed that southern Illinois tripoli depos-

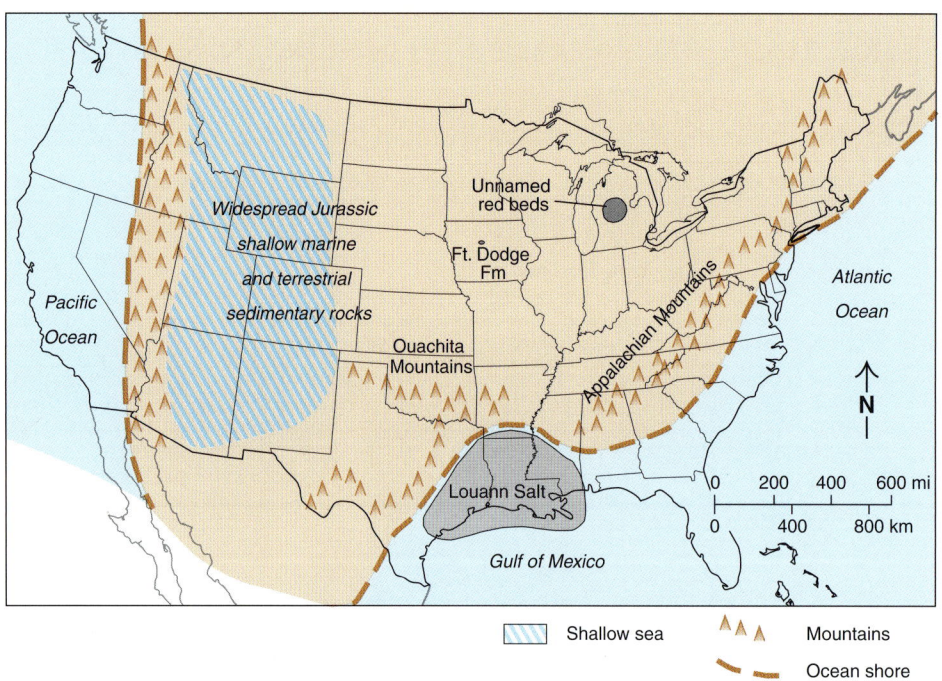

Figure 11-3 Map of the United States during Middle Jurassic time. (Source: W. John Nelson.)

its are the product of hydrothermal activity derived from deep-seated magma bodies.

By Cretaceous time, the Pascola Arch (Figure 11-4), which formed during Late Pennsylvanian and Permian times (Chapter 4, Structural Features), had been deeply eroded and began to subside (Marcher and Stearns 1962, Stearns and Marcher 1962, Harrison and Schultz 2002). The arch was an extension of the Ozark Dome running southeast from Missouri into western Kentucky and Tennessee, closing off the southern end of the Illinois Basin (Figure 11-4).

Late Cretaceous faulting took place in and near southernmost Illinois. Seismic and borehole data near Olmsted portray complex faulting and abrupt variation in the thickness of the McNairy Formation between adjacent fault blocks (McBride et al. 2003). A Cretaceous fault-bounded

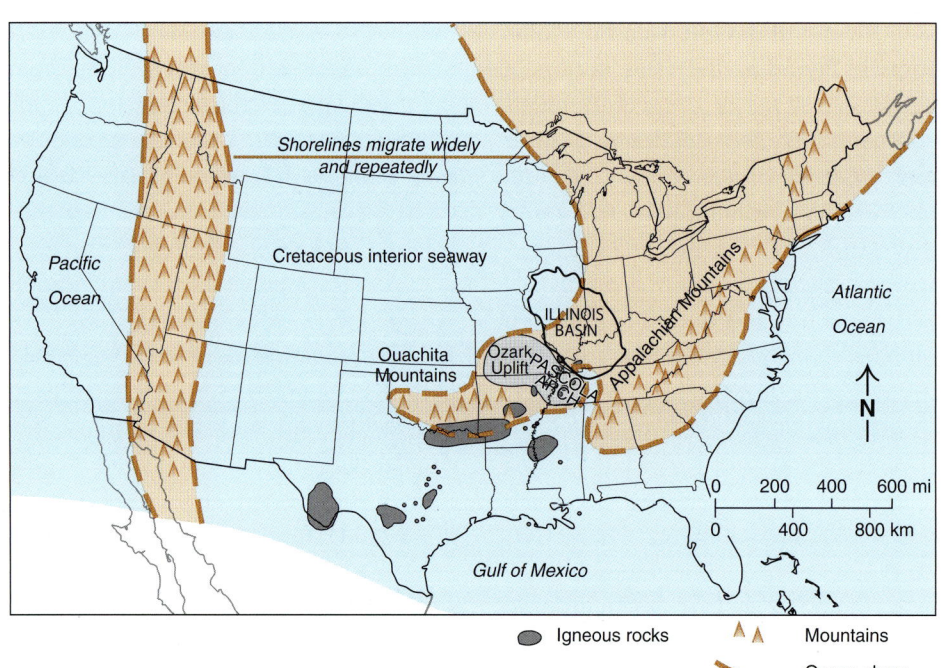

Figure 11-4 Map of the United States during mid-Cretaceous time. The extent of igneous rocks is from Byerly (1991).

depression near Marble Hill, Missouri, has yielded dinosaur fossils. Vertebrae of a hadrosaur were unearthed there in 1942 as a water well was being dug. Recent excavations have yielded more hadrosaur remains, teeth of a tyrannosaur-like dinosaur (*Albertosaurus*?), and bones of crocodiles, turtles, and smaller vertebrates (Gilmore and Stewart 1945; R. J. Jacobson, personal communication, 2003).

Meanwhile, during the Late Cretaceous, a series of tectonic plates collided and accreted to the western margin of North America, producing mountain ranges. A shallow seaway through the midcontinent linked the Gulf of Mexico with the Arctic Ocean (Figure 11-4). Rising western mountains shed vast quantities of sediment into the sea-

way. Rocks of the western coastal plain are magnificently exposed today in the Rocky Mountains, but deposits of the eastern coastal plain have been eroded or buried by glacial deposits. Outliers of Upper Cretaceous sand and gravel survive near Pittsfield and Quincy, Illinois, as hilltop erosional remnants covering 80 to 100 square miles (207 to 259 km²) of the divide between the Illinois and Mississippi Rivers (Figures 11-1 and 11-5). These sand and gravel deposits, called the Baylis Formation, apparently were deposited along the eastern margin of the Cretaceous interior seaway. Formerly a continuous sheet, the Baylis was reduced by erosion to scattered remnants prior to Pleistocene glaciation. These deposits comprise as much as 100 feet (30.5 m) of loess or weakly cemented white, fine- to medium-grained quartz sand with lenses of clay. A basal gravel of chert and quartz pebbles, commonly cemented by iron oxide, is present (Frye et al. 1964).

Although no fossils have been found in these deposits, Frye et al. (1964) inferred they were Cretaceous in age because their mineralogical and stratigraphic relationships agree with Cretaceous strata farther west. The Windrow Formation of northeastern Iowa, southeastern Minnesota, and western Wisconsin consists of sediments that closely resemble the Baylis (Thwaites and Twenhofel 1921, Andrews 1958). The Windrow is established as being Cretaceous in age by fossil angiosperms (flowering trees) collected southeast of Minneapolis-St. Paul, Minnesota (Winchell 1888, p. 43–45; Lesquereux 1895).

Additional clues as to the former extent of Cretaceous rocks come from fossils in glacial drift. Worthen (1890) reported sharks' teeth, echinoids, ammonoids, and belemnites in drift at numerous sites in eastern Iowa, Missouri, and near Alton, Petersburg, and Warsaw, Illinois (Figure 11-5). Cobban (1983) refined the age of some of Worthen's ammonites as mid-Turonian (early Late Cretaceous). Woodward and Thomas (1895) described Cretaceous Radiolaria and Foraminifera from glacial deposits in south Chicago and Calumet City, Illinois. These microscopic marine fossils match those of various Cretaceous rocks in the western United States. Hence, Cretaceous marine shales appear to have extended east of the Mississippi River prior to Pleistocene glaciation. These shales presumably overlie the Baylis and Windrow Formations, which, by inference, are Turinian or older.

A structural trough called the Mississippi Embayment developed, extending from the Gulf of Mexico up to southern Illinois. Marine and shoreline sediments accumulated in the Embayment. At the base is the Post Creek (previously called Tuscaloosa) Formation. Highly variable in thickness, the Post Creek rests on the irregular, deeply weathered surface of Paleozoic bedrock. Gravel is the dominant material:

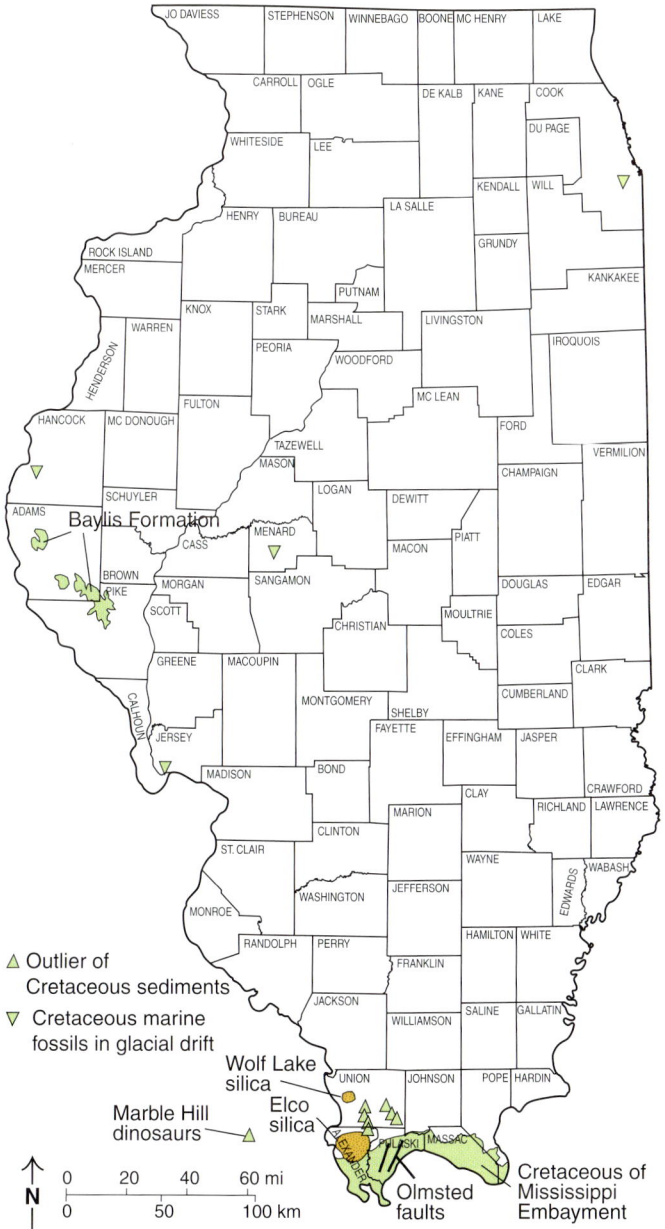

Figure 11-5 Location of Cretaceous sediments and fossils in Illinois. (Source: W. John Nelson.)

white to light gray pebbles and small cobbles of chert along with lesser amounts of quartz, sandstone, and quartzite partially cemented by iron oxide (Figure 11-6). The Post Creek also includes beds of quartz sand, clays and silts of various colors, and ancient silicified soils that contain root traces. Fossil pollen indicates these rocks are Campanian in age (the next-to-last of 12 Cretaceous subdivisions) (Harrison and Litwin 1997).

Overlying the Post Creek is the McNairy Formation, a unit of interbedded dark gray clay and silt and white (fresh) to orange and red (weathered), very fine to medium sand (Figure 11-7a). Abundant muscovite mica is characteristic of the the McNairy. Common are fine, rhythmic laminations arranged in bundles that imply tidal neap and spring cycles. Bundles of thick laminations represent spring tide, which occurs twice monthly at full and new moon when tidal range is greatest. Bundles of thinner laminations correspond to neap tide, when sun and moon are at right angles and tidal range is lowest. Thin beds of impure lignite are present locally, as are gravel lenses (mostly near the base).

Figure 11-6 Post Creek Formation: **(a)** coarse, light-colored chert gravel contains cobbles more than a foot (0.3 m) in diameter; **(b)** silcrete, an ancient, silicified soil bearing numerous root traces. Both outcrops occur about 5 miles (8 km) west of Tamms in Alexander County, where the Post Creek overlies Devonian bedrock. (Photographs by Joseph A. Devera.)

Figure 11-7 **(a)** Fine-grained, light-colored sand of the Mc Nairy Formation in a pit near the Mississippi River at Fayville, south of Thebes. Cross-bedding and burrows, as seen here, are common features. **(b)** Highly burrowed, clay-rich, glauconitic sand of the Clayton Formation in a ravine north of Olmsted. (Photographs by John M. Masters.)

Fossils include leaf imprints, petrified wood, oysters (Figure 11-8c), and *Ophiomorpha,* the burrows of marine brine shrimp (Olive 1980). The Maastrichtian (latest Cretaceous) age of the McNairy is based on fossil pollen (Olive 1980, Harrison and Litwin 1997). The McNairy thickens southward from a feather edge to more than 200 feet (61 m) at the southern state border (Pryor and Ross 1962, Ross 1963, Kolata et al. 1981). Outcrops of McNairy can be observed in Pulaski County at the Post Creek cutoff and in ravines along the Ohio River north of Olmsted, as well as alongside the river road south of Thebes in Alexander County. Outliers of brightly colored sand, silt, and clay, containing Cretaceous petrified wood, occur north of the Mississippi Embayment in the bedrock uplands of Union County and northern Alexander County (Nelson et al. 1995).

The McNairy Formation in southern Illinois represents a variety of fluvial, deltaic, and marginal-marine depositional settings at the head of the Mississippi Embayment. Much of the sediment was carried to the Embayment by rivers draining westward from the Appalachian Mountains (Potter and Pryor 1961). The diagnostic mica and other minerals came from metamorphic rocks in the Blue Ridge and Piedmont Provinces.

The youngest Cretaceous deposit in southern Illinois is the thin (0 to 10 feet [0 to 3 m]) Owl Creek Formation. Where present, the Owl Creek is distinguished from the underlying McNairy by the presence of glauconite (a dark green to black mineral formed in marine sediments) and abundant U-shaped burrows (Pryor and Ross 1962, Kolata et al. 1981).

TERTIARY PERIOD

The name "Tertiary" derives from the dawn of geological study, when Earth's history was divided into only four periods. Although "Primary" and "Secondary" have fallen into disuse, the terms "Tertiary" and "Quaternary" survive. The Tertiary is presently defined as the period from approximately 65 million years ago to 2 million years ago and is divided into five stages: Paleocene (oldest), Eocene, Oligocene (not shown in Figure 11-1), Miocene, and Pliocene. The Tertiary, like the Mesozoic, is poorly represented in Illinois, being restricted to the Mississippi Embayment area and to scattered remnants elsewhere in the state.

The oldest Tertiary unit in the Mississippi Embayment is the Clayton Formation (Paleocene). Less than 20 feet (6 m) thick, the Clayton consists of clayey sand and sandy clay that is extensively burrowed and contains abundant dark green to black glauconite (Figure 11-7b). Chert pebbles are common near the base. Fossils include foraminifera, bryozoans, annelids, bivalves, gastropods (Figure 11-8b), crabs, lobsters, sharks (Figure 11-8a), rays, ratfish, crocodiles, and turtles (Cope 1999). These indicate shallow marine to brackish water environments. The Clayton overlies the Owl Creek or McNairy with an unconformity.

Overlying the Clayton is the Porters Creek Formation, a unit of dark gray or olive-gray marine clay up to about 150 feet (45 m) thick. The clay is massive, somewhat silty and glauconitic, and breaks with conchoidal fracture (Figure 11-9). It contains Radiolaria, Foraminifera, sharks' teeth, fish scales, and molds and casts of bivalves and pelecypods (Pryor and Ross 1962, Kolata et al. 1981). The Porters Creek is mined as a source of clay for making cat litter and other absorbents (Figure 11-9). Porters Creek and Clayton

Figure 11-8 Selected Cretaceous and Tertiary fossils: **(a)** shark's tooth *(Lamna cuspidata),* Clayton Formation, Paleocene Series, Tertiary System, Pulaski County; **(b)** gastropod *(Turritella* sp.), Clayton Formation, Paleocene Series, Tertiary System, Pulaski County; and **(c)** oyster *(Exogyra costata),* Cretaceous System, Alexander County. (Photographs by Dennis R. Kolata.)

Figure 11-9 Porters Creek Formation clay in pits near Olmsted. **(a)** Note the blocky, conchoidal fracture and lack of well-defined bedding. **(b)** Shown is a small normal fault, which is probably a tectonic feature or a product of landsliding. (Photographs by John M. Masters.)

Figure 11-10 Eocene strata: **(a)** cross-bedded sand of the Wilcox Formation at Moses gravel pit near Mounds, Pulaski County. (Photograph by John M. Masters.) **(b)** Steeply tilted sand and gravel of Eocene(?) age south of Jonesboro, Union County. Well-rounded white, gray, and black chert pebbles resemble those of the Wilcox Formation, but are much larger. (Photograph from Nelson et al. 1995.)

exposures can be viewed in clay pits and gullies along the Ohio River north of Olmsted.

The Eocene Wilcox Formation, which overlies the Porters Creek with an unconformity, contains sand that is white to orange and fine to coarse grained. Small pebbles of highly polished white quartz and gray to black chert are characteristic. Silt and clay of the Wilcox are colored gray, pink, and yellowish orange. Although outcrops are scarce, the Wilcox is exposed from time to time in clay and gravel pits. Drilling indicates Eocene sediments may be as thick as 250 feet (75 m) beneath modern river sediments at the southern tip of the state at Cairo. These may include not only the Wilcox, but also the younger Jackson and Claiborne Formations as mapped in adjacent western Kentucky (Olive 1980).

Eocene strata also occur north of the Embayment as ridge-top outliers and deposits downdropped into grabens and sinkholes. These deposits resemble the Wilcox, but contain polished chert cobbles up to 6 inches (15 cm) in diameter (Figure 11-10). Clay deposits near Alto Pass, mined during the early twentieth century for making pottery (St. Clair 1920, Parmelee and Schroyer 1921, Lamar 1948), are thought to be Eocene in age based on fossil pollen collected from the clay pits (D. J. Nichols 1993, personal communication, cited in Nelson and Devera 1995; A. T. Cross 1984, personal communication, and D. J. Nichols 1993, personal communication, cited in Nelson et al. 1995).

No Oligocene strata have been identified in Illinois.

MESOZOIC AND TERTIARY

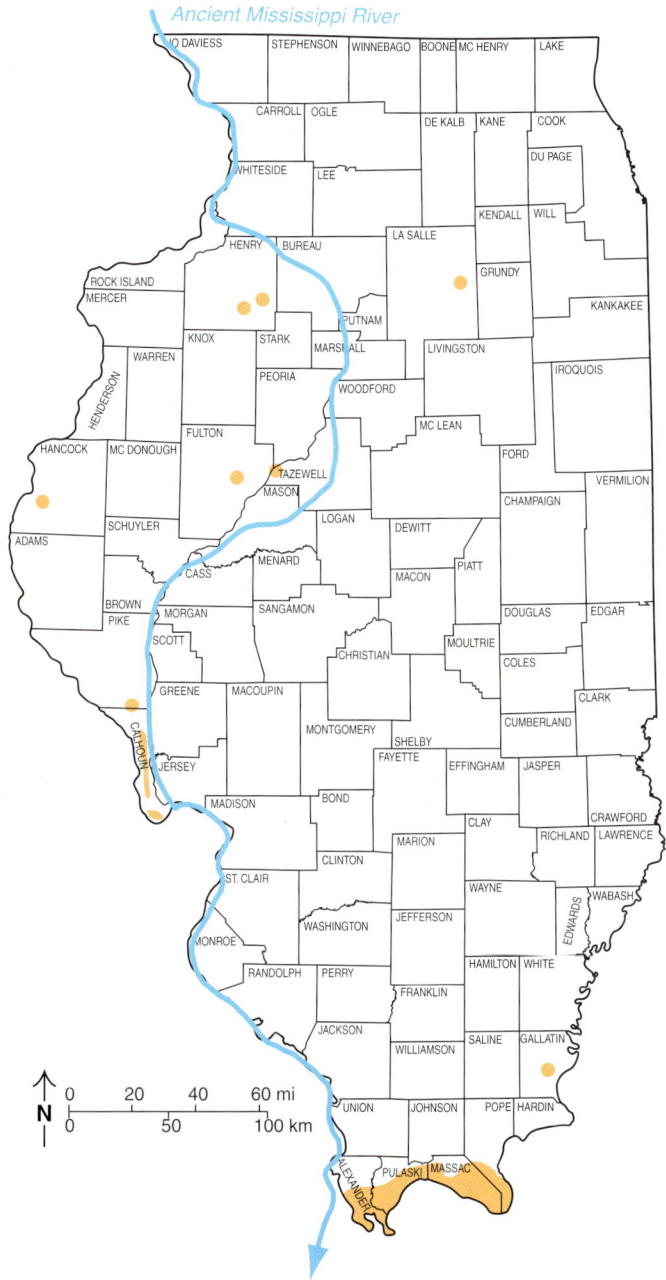

Figure 11-11 Distribution of the Mounds and Grover Gravels in Illinois. (Source: W. John Nelson.)

Figure 11-12 Mounds Gravel: **(a)** Large-scale trough cross-bedding in a quarry at Black Powder Hollow, south of Thebes. Staff is 1 m (3.3 feet) long. **(b)** A closer view shows rounding of pebbles and a poor degree of sorting. Scale is in centimeters at left, inches at right. (Photographs by John M. Masters.)

Distinctive brown chert gravel of late Miocene(?) to early Pleistocene age is widely scattered in Illinois (Figure 11-11). This unit is called Mounds Gravel in southern Illinois and Grover Gravel in western Illinois. Both are yellowish to reddish brown and are composed dominantly of chert pebbles up to several inches in diameter, with small quartz pebbles and a few fragments of sandstone and other rock types. Pebbles bear a characteristic bronze-colored patina of iron oxide. Lenses of red to brown, cross-bedded sand are interstratified with the gravel (Figure 11-12). Ranging up to about 50 feet (15 m) thick, the Mounds Gravel is quarried in many small pits and is used for surfacing roads and as ornamental gravel.

Mounds Gravel is identical and contiguous to the widespread Lafayette Gravel of the southeastern United States. Potter (1955a, 1955b) interpreted the Lafayette as alluvial fan and braided river deposits derived from weathered cherty Paleozoic limestone and older Cretaceous gravels. The Mounds occupies several distinct terrace levels that represent successive stages of fluvial downcutting and deposition. The basic principle was this: As the rivers eroded their channels downward through time, the rate of erosion and sediment deposition was not constant. Periods of stability alternated with episodes of accelerated erosion caused by such things as tectonic uplift and climate changes. Each time a river cut its channel to a lower level, the previous floodplain became a terrace. Through time, as the valleys became more deeply incised, they tended to become narrower, leaving successive terraces as a series of steps. The age of the Mounds Gravel is poorly constrained as late Miocene(?) to early Pleistocene, based on fossil pollen from two sites in Kentucky (Olive 1980).

The Grover Gravel, which closely resembles the Mounds, is scattered in northern and western Illinois (Figure 11-11). Its pebbles and boulders of pink and purple quartzite, red hematitic chert, and other rocks were derived from Precambrian exposures in Wisconsin and Minnesota and distributed by the ancestral Mississippi River (Horberg 1950, Rubey 1952).

Conclusion

Deposition of the Mounds and Grover Gravels was the final episode of sedimentation in Illinois prior to the onset of continental glaciation, which ushered in the Pleistocene Epoch (Chapter 12, Quaternary Period).

References

Andrews, G. W., 1958, Windrow Formation of Upper Mississippi Valley region, a sedimentary and stratigraphic study: Journal of Geology, v. 66, no. 6, p. 597–624.

Berg, R. B., and J. M. Masters, 1994, Geology of microcrystalline silica (tripoli) deposits, southernmost Illinois: Illinois State Geological Survey, Circular 555, 89 p.

Bunker, B. J., B. J. Witzke, W. L. Watney, and G. A. Ludvigson, 1988, Phanerozoic history of the central midcontinent, United States, in The Geology of North America, v. D-2, Sedimentary cover—North American craton: Boulder, Colorado, Geological Society of America, p. 243–260.

Byerly, G. R., 1991, Igneous activity in The Geology of North America, v. J, The Gulf of Mexico Basin: Boulder, Colorado, Geological Society of America, p. 91–108.

Catacosinos, P. A., P. A. Daniels, Jr., and W. B. Harrison 1991, Structure, stratigraphy, and petroleum geology of the Michigan Basin, in M. W. Leighton, D. R. Kolata, D. F. Oltz, and J. J. Eidel, eds., Interior cratonic basins: Tulsa, Oklahoma, American Association of Petroleum Geologists, Memoir 51, p. 561–601.

Cobban, W. A., 1983, Molluscan fossil record from the northeastern part of the Upper Cretaceous seaway, Western Interior: Reston, Virginia, U.S. Geological Survey, Professional Paper 1253-A, p. 1–25.

Cope, K. H., 1999, Paleontology and paleoecology of the fauna of the Clayton Formation (Paleocene) in southern Illinois: Carbondale, Illinois, Southern Illinois University, M.S. thesis, 92 p.

Dunbar, C. O., 1960, Historical geology (2nd ed.): New York, John Wiley & Sons, 500 p.

Fisher, J. H., M. W. Barratt, J. B. Droste, and R. H. Shaver, 1988, Michigan basin, in The Geology of North America, v. D-2, Sedimentary cover—North American craton: Boulder, Colorado, Geological Society of America, p. 361–382.

Frye, J. C., H. B. Willman, and H. D. Glass, 1964, Cretaceous deposits and the Illinoian glacial boundary in western Illinois: Illinois State Geological Survey, Circular 364, 28 p.

Gilmore, C. W., and D. R. Stewart, 1945, A new sauropod dinosaur from the Upper Cretaceous of Missouri: Journal of Paleontology, v. 19, no. 1, p. 23–29.

Harrison, R. W., and R. J. Litwin, 1997, Campanian coastal-plain sediments in southeastern Missouri and southern Illinois: Significance to the early geologic history of the northern Mississippi Embayment: Cretaceous Research, v. 18, no. 5, p. 687–696.

Harrison, R. W., and A. Schultz, 2002, Tectonic framework of the southwestern margin of the Illinois Basin and its influence on neotectonism and seismicity: Seismological Research Letters, v. 73, no. 5, p. 698–731.

Heigold, P. C., 1976, An aeromagnetic survey of southwestern Illinois: Illinois State Geological Survey, Circular 495, 28 p.

Hildenbrand, T. G., M. F. Kane, and W. Stauder, 1977, Magnetic and gravity anomalies in the northern Mississippi Embayment and their spacial relation to seismicity: Reston, Virginia, U.S. Geological Survey, Miscellaneous Field Studies, map MF-914, 2 sheets.

Horberg, C. L., 1950, Preglacial gravels in Henry County, Illinois: Transactions of the Illinois State Academy of Science, v. 43, p. 171–175.

King, P. B., 1977, The evolution of North America: Princeton, New Jersey, Princeton University Press, 197 p.

Kolata, D. R., 2005, Bedrock geology of Illinois: Illinois State Geological Survey, Illinois Map 14, 1:500,000.

Kolata, D. R., and W. J. Nelson, 1991, Tectonic history of the Illinois Basin in M. W. Leighton, D. R. Kolata, D. F. Oltz, and J. J. Eidel, eds., Interior cratonic basins: Tulsa, Oklahoma, American Association of Petroleum Geologists, Memoir 51, p. 263–285.

Kolata, D. R., J. D. Treworgy, and J. M. Masters, 1981, Structural framework of the Mississippi Embayment of southern Illinois: Illinois State Geological Survey, Circular 516, 38 p.

Lamar, J. E., 1948, Clay and shale resources of extreme southern Illinois: Illinois State Geological Survey, Report of Investigations 128, 107 p.

Lesquereux, L., 1895, Cretaceous fossil plants from Minnesota, in L. Lesquereux, A. Woodward, B. W. Thomas, C. Schuchert, E. O. Ulrich, and N. H. Winchell, Paleontology: Minnesota Geological and Natural History Survey, The Geology of Minnesota, v. 3, part 1, p. 1–22.

Marcher, M. V., and R. G. Stearns, 1962, Tuscaloosa Formation in Tennessee: Geological Society of America Bulletin, v. 73, no. 11, p. 1365–1386.

McBride, J. H., A. J. M. Pugin, W. J. Nelson, T. H. Larson, S. L. Sargent, J. A. Devera, F. B. Denny, and E. W. Woolery, 2003, Variable post-Paleozoic deformation detected by seismic reflection profiling across the northwestern "prong" of New Madrid Seismic Zone: Tectonophysics, v. 368, no. 1–4, p. 171–191.

McGinnis, L. D., P. C. Heigold, C. P. Ervin, and M. Heidari, 1976, The gravity field and tectonics of Illinois: Illinois State Geological Survey, Circular 494, 28 p.

Nelson, W. J., and J. A. Devera, 1995, Geologic map of the Cobden Quadrangle, Illinois: Illinois State Geological Survey, Illinois Geological Quadrangle map, IGQ 16, 1:24,000.

Nelson, W. J., J. A. Devera, and J. M. Masters, 1995, Geology of the

Jonesboro 15-minute Quadrangle, southwestern Illinois: Jonesboro, Mill Creek, Ware, and McClure 7.5-minute Quadrangles: Illinois State Geological Survey, Bulletin 101, 57 p.

Nelson, W. J., and D. K. Lumm, 1987, Structural geology of southeastern Illinois and vicinity: Illinois State Geological Survey, Circular 538, 70 p.

Olive, W. W., 1980, Geologic maps of the Jackson Purchase region, Kentucky: Reston, Virginia, U.S. Geological Survey, Miscellaneous Investigations Series, Map I-1217, and booklet, 11 p.

Parmelee, C. W., and C. R. Schroyer, 1921, Further investigations of Illinois fire clays *in* Year book for 1917 and 1918: Administrative reports and economic and geological papers: Illinois State Geological Survey, Bulletin 38, 149 p.

Potter, P. E., 1955a, The petrology and origin of the Lafayette gravel, Part 1, Mineralogy and petrology: Journal of Geology, v. 63, no. 1, p. 1–38.

Potter, P. E., 1955b, The petrology and origin of the Lafayette gravel, Part 2, Geomorphic history: Journal of Geology, v. 63, no. 3, p. 115–132.

Potter, P. E., and W. A. Pryor, 1961, Dispersal centers of Paleozoic and later clastics of the Upper Mississippi Valley and adjacent areas: Geological Society of America Bulletin, v. 72, no. 8, p. 1195–1249.

Pryor, W. A., and C. A. Ross, 1962, Geology of the Illinois parts of the Cairo, La Center, and Thebes Quadrangles: Illinois State Geological Survey, Circular 332, 39 p.

Ross, C. A., 1963, Structural framework of southernmost Illinois: Illinois State Geological Survey, Circular 351, 27 p.

Rubey, W. W., 1952, Geology and mineral resources of the Hardin and Brussels Quadrangles (in Illinois): Washington D.C., U.S. Geological Survey, Professional Paper 218, 179 p.

Salvador, A., 1991, Triassic-Jurassic, *in* The Geology of North America, v. J, Gulf of Mexico Basin: Boulder, Colorado, Geological Society of America, p. 131–180.

Sloss, L. L., 1963, Sequences in the cratonic interior of North America: Geological Society of America Bulletin, v. 74, no. 2, p. 93–114.

Sloss, L. L., 1988, Tectonic evolution of the craton in Phanerozoic time, *in* L. L. Sloss, ed., Sedimentary Cover—North American Craton: U.S.: The Geology of North America, v. D-2: Boulder, Colorado, Geological Society of America, p. 25–51.

Sloss, L. L., W. C. Krumbein, and E. C. Dapples, 1949, Integrated facies analysis: *in* C. R. Longwell, chairman, Sedimentary facies in geologic history: Geological Society of America, Memoir 39, p. 91–123.

St. Clair, S., 1920, Clay deposits near Mountain Glen, Union County, Illinois *in* Year book for 1916: Administrative reports and economic and geological papers: Illinois State Geological Survey, Bulletin 36, p. 71–83.

Stearns, R. G., and M. V. Marcher, 1962, Late Cretaceous and subsequent structural development of the northern Mississippi Embayment area: Geological Society of America Bulletin, v. 73, no. 11, p. 1387–1394.

Thwaites, F. T., and W. H. Twenhofel, 1921, Windrow Formation; An upland gravel formation of the Driftless and adjacent areas of the Upper Mississippi Valley: Geological Society of America Bulletin, v. 32, p. 293–314.

U.S. Geological Survey, Geological Names Committee 2007, Divisions of geologic time—Major chronostratigraphic and geostratigraphic units: Reston, Virginia, U.S. Geological Survey Fact Sheet 2007–3015, 2 p.

Winchell, N. H., 1888, The geology of Minnesota: St. Paul, Minnesota, Minnesota Geological and Natural History Survey, v. 2, 695 p.

Woodward, A., and B. W. Thomas, 1895, The microscopical fauna of the Cretaceous, in Minnesota, with additions from Nebraska and Illinois (foraminifera, radiolarian, coccoliths, rhabdoliths), *in* L. Lesquereux, A. Woodward, B. W. Thomas, C. Schuchert, E. O. Ulrich, and N. H. Winchell, Paleontology: Minnesota Geological and Natural History Survey, The Geology of Minnesota, v. 3, part 1, p. 23–54.

Worthen, A. H., 1890, Drift deposits of Illinois: Geological Survey of Illinois, v. 8, part 1, p. 1–24.

12 Quaternary Period

Ardith K. Hansel and E. Donald McKay III

INTRODUCTION

The modern Illinois landscape is largely a product of the most recent geological period, the Quaternary, which began approximately 2.6 million years ago and continues to the present. This period of geological time has been characterized by repeated climatic fluctuations between cold glacial episodes and warm interglacial episodes, each lasting tens of thousands of years. Continental ice sheets have come and gone in North America many times. What is now Illinois has experienced extreme ecological shifts between subarctic tundra conditions at some times and warm intervals with subtropical vegetation at others. Worldwide, the buildup and decay of continental ice sheets were accompanied by global changes in sea level, ocean current patterns, the level of the land surface, and by plant and animal migrations and extinctions.

During the glacial episodes in North America, great continental ice sheets formed in Canada and spread outward from their centers (Figure 12-1). The largest of these ice sheets reached its southernmost extreme in Illinois. More than 90% of the state was covered by glacial ice at least once, and the remaining 10% of the state was affected by glacial meltwater, dust storms, and a climate more frigid than that of today. During the warm interglacial episodes, including the current one that began in Illinois about 12,500 years ago (as determined by radiocarbon dating), more stable landscape conditions allowed vegetation and soils to develop.

Lobes of glacial ice likely entered Illinois during six or more glaciations (Johnson 1986). Regional ice flow was from the northwest, northeast, and east during the earliest (pre-Illinois Episode) glaciations and from the northeast and east during the later Illinois and Wisconsin Episode glaciations (Figure 12-1). The northwestern-sourced lobes of ice are thought to have come from snow accumulation centers in the western part of the Canadian Shield (Precambrian bedrock) region of Canada west of Hudson Bay (the Keewatin center). The northeastern- and eastern-sourced lobes are attributed to accumulation centers in the Canadian Shield region east of Hudson Bay (the Labrador center) (Fulton 1989) (Figure 12-1). During the glaciations, the basins of the Great Lakes were deepened and widened by glacial erosion and became increasingly important routes for ice flowing southward and southwestward from the Labrador accumulation center. The ice lobe streaming through the trough that would become Lake Michigan was largely responsible for the modern landscape of Illinois, and the glacial deposits it left are composed predominantly of finely ground Paleozoic rocks (limestone, dolomite, and shale) eroded from the lake basin.

The record of Quaternary time in Illinois consists of a fairly thick succession of glacial and related sediments, referred to as drift, in which weathered zones (soils) developed during warm interglacial times. These sediments average about 100 feet (30.5 m) in thickness. They are 250 to 500 feet (76 to 152 m) thick in some bedrock valleys in northern and central Illinois (Piskin and Bergstrom 1975). One reason the Quaternary deposits are so thick is that, during each of the major glacial episodes, Illinois was at the southern limit of the ice sheet. In this end region, glacial

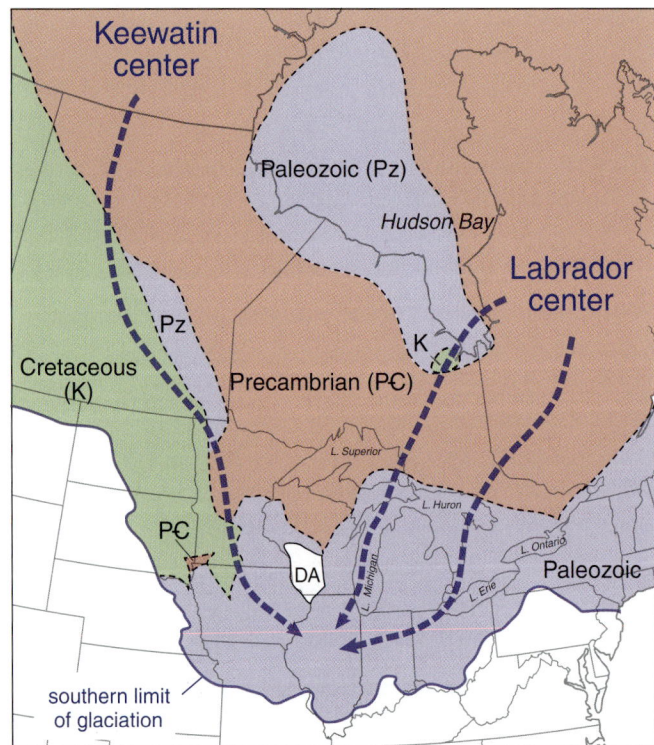

Figure 12-1 Source areas (Keewatin and Labrador centers of accumulation) and flow paths of ice lobes that entered Illinois from the northwest, northeast (Lake Michigan lobe), and east (Huron-Erie lobe). Glacial limit, Driftless Area (DA), and bedrock types encountered along flow paths: Precambrian igneous, metamorphic, and sedimentary rocks; Paleozoic carbonate and shale; and Cretaceous shale and sandstone. Based on Prest (1984) and Fulton (1989) and the Geological Survey of Canada (1970) geological map.

deposition was greater than erosion. Only the northeastern quarter of the state was glaciated during the most recent glacial episode. Consequently, older drift and soils farther west and south were not destroyed by glacial erosion (Johnson and Hansel 1999). The Quaternary glacial stratigraphic record is a series of offlapping drift sheets that generally pinch out northeastward toward the Lake Michigan basin beneath successively younger drift sheets.

Quaternary Landforms

The present landscape of Illinois reflects its glacial makeover during the Quaternary Period (Figure 12-2). As the glaciers flowed into Illinois, carrying ground rock and soil debris that had been eroded along their flowpaths, they decapitated old bedrock landforms, smeared on new layers of sediment, and built new ridges, particularly along their edges. Old river valleys were filled with sediment, and new valleys were cut by glacial meltwater. In some areas, rivers changed course or flow direction.

The unglaciated areas of Illinois (northwest, west, and south) were not buried by glacial sediment, although thin windblown and waterborne Quaternary drift was deposited. These areas are more rugged than glaciated areas (Figure 12-2), and their surface topography generally parallels that of the shallowly buried bedrock surface, which has significant local relief. Bedrock outcrops are common along river bluffs, stream channels, and road cuts.

Each time the glaciers retreated from Illinois during the Quaternary Period, a new landscape emerged from beneath the glacier. It is important to remember that glacier retreat refers to a general withdrawal of an ice cover caused by ice melting. No reversal of ice flow is involved. There are two general types of ice withdrawal, and both occurred at different times and places in Illinois. One type of retreat simply involves the gradual recession of the active ice margin because the volume of ice flowing in is not enough to replace that lost to melting. With this type of retreat, the ice margin at times melted back in a stepped fashion, pausing long enough for the forward movement of the ice to deliver huge piles of sediment to the stationary ice margin. The second type of glacier retreat involves a large-scale stoppage of ice movement to an area of the glacier. The resultant "stagnant" or "dead" ice slowly disintegrates (melts down), and the suite of landforms produced is much different from those constructed in active ice retreat.

Generally speaking, the glaciers tended to smooth the Illinois landscape, forming broad plains underlain by till and lake sediments. New broad, low-relief, curvilinear ridges (end moraines) were created where sediment was delivered to the glacier's leading edge during active ice flow conditions. Major river valleys leading away from the melting ice lobes were cut by meltwater draining from the glaciers. Locally, these meltwater channels formed narrow gorges where they breached end moraines. As the glaciers receded from Illinois, they left behind a low-relief, poorly drained landscape with shallow lakes and wetlands. Gradually, rill-like streams began to form on the end moraine slopes, and water that accumulated in the low-relief plains between end moraines drained into tributaries to the major valleys. Stream erosion progressively increased the local relief as streams lengthened and deepened their valleys. In the area most recently deglaciated, even today, after heavy rains, ephemeral shallow lakes appear in flat Illinois farm fields, but, in the oldest part of the glacial landscape, streams are deeper, closer together, and drain virtually all of the land.

In contrast to the large-scale plains and the low-relief ridges formed by the glaciers' smoothing processes are small-scale Quaternary landforms that resulted from deposition (1) in contact with melting glacier ice, particularly during stagnant ice disintegration (kames, kettles, eskers, esker fans, and ice-walled lake plains), (2) in glacial and postglacial streams and lakes (outwash terraces, lake plains), and (3) in dune fields. The ice-contact landforms often have stepped slopes where collapse occurred when the ice melted. These ice-contact landforms are generally more apparent than the larger-scale plains and end moraines, but some, such as ice-walled lake plains, are subtle, low-relief features. In some areas, erosion beneath the glacier sculpted the landscape and reformed the topography into subtle streamlined landforms called drumlins. In Illinois, many of the glacial landforms are partially concealed because they were later buried by a blanket of windblown silt (loess).

Quaternary Studies in Illinois

Although the surface drift in Illinois was originally thought to have been deposited from icebergs adrift in a sea that later vanished (Worthen 1866), mapping of the drift sheets and the end moraines that are so characteristic of the Illinois landscape soon led to the conclusion that the sediments were the products of continental glaciation (Chamberlin 1883). Early studies focused on the use of end moraines to map ice movement. Weathering and the degree of stream dissection were used to estimate the relative ages of drift sheets. By the turn of the twentieth century, scholars recognized the widespread preservation of buried soils, and the concept of multiple glacial and interglacial episodes was firmly established (Salisbury 1893, Chamberlin 1894, Leverett 1899).

Figure 12-2 Surface topography map. The land surface reflects the influence of glaciers and rivers on the Illinois landscape (Luman et al. 2003).

Further subdivision of the Quaternary sediment record in Illinois during the first half of the twentieth century was based primarily on the relative size of end moraines, contrasts in morphology and drainage, the configuration of moraine fronts, and stratigraphic relationships among buried soils (paleosols), windblown silts (loesses), and deposits directly from glaciers (tills) (e.g., Leverett 1909; Leighton 1926, 1933; Leighton and Willman 1950). During the second half of the twentieth century, studies focused on stratigraphy and classification, characterization of lithostratigraphic units using mineral content and grain size, and refinement of the chronology of Quaternary events using radiocarbon and other dating methods (e.g., Leighton 1957; Frye and Willman 1960; Frye et al. 1968; Willman and Frye 1970; Johnson et al. 1972; Lineback 1979a, 1979b; Follmer et al. 1979; Wickham 1979a, 1979b; Berg et al. 1985; Johnson et al. 1985; Johnson 1986; Wickham et al. 1988; Curry 1989; Hansel and Johnson 1992, 1996). Another important focus of Illinois Quaternary research during the second half of the twentieth century was the development of applied Quaternary studies for environmental geology and, particularly, groundwater applications (e.g., Frye 1967, Hackett and McComas 1969, Kempton 1981, Berg et al. 1984).

A more recent trend in Illinois Quaternary studies has emphasized the study of glacial sediments to reconstruct ice-margin fluctuations and glacial history and to identify proglacial (beyond the limits of the glacier) fluvial sediment facies that serve as important drift aquifers (e.g., Johnson and Hansel 1990; Hansel and Johnson 1992, 1996; Hansel et al. 1999; Curry et al. 1999; McKay et al. 2008). Studies since 1992 have focused on detailed (1:24,000 scale, where 1 inch on the map represents 2,000 feet on the ground) three-dimensional mapping (e.g., Lasemi and Berg 2001, Berg et al. 2002, Hansel et al. 2004). This detailed scale allows geologists to portray complexities at the surface and in the subsurface more accurately and to contribute to understanding of the continuity and character of glacial deposits, particularly those that are buried. Recent innovations in shallow seismic reflection geophysics have yielded new insights into subsurface mapping of glacial deposits (Pugin et al. 2003).

Quaternary Sediments and Glacial Sequences

As a glacier advances across a landscape, it creates different sedimentary environments, including the proglacial, ice-marginal, and subglacial depositional environments. In the proglacial environment, sedimentation originates dominantly from glacial meltwater, streams and lakes, and sand or dust storms. The resultant deposits are glacial lake, glacial stream, and glacial windblown (loess) sediments (Figure 12-3 a, b, c).

Near the ice margin, rocks, sand, silt, and clay that were incorporated into the glacier as it advanced melt out on top of or adjacent to ice and slide or flow to low spots, generally resulting in somewhat chaotic deposits. Diamictons (unsorted or poorly sorted sediment with a wide range of particle sizes) are commonly interbedded with water-sorted sediments (Figure 12-3d). Glacially induced folds and faults are common in these materials.

In the subglacial environment (the environment beneath the glacier), sedimentation is dominated by debris that has melted out of the glacier and sediment that has been deformed by ice movement. This subglacial sediment, commonly called till, is generally fairly uniform diamicton in Illinois (Figure 12-3e), although locally till can contain fine lenses and channel-shaped bodies of waterlain sediment (Figure 12-3f). Subglacial drainage of meltwater localized in channels can also deposit thick gravel, sand, and sediments that form eskers, kames, and other landforms after the glacier melts.

In Illinois, a typical glacial sequence, from the base upward, consists of proglacial, subglacial, and ice-marginal sediments. When the ice margin passes back and forth over an area several times during a glaciation, multiple glacial sequences can be preserved. These sequences are often incomplete because of subglacial erosion. Figure 12-4 shows multiple glacial sequences that resulted from ice-margin fluctuations in northeastern Illinois during the last glacial episode. Glacial sequences of multiple glacial episodes are common in Quaternary successions in Illinois. Paleosols (buried soils) of interglacial episodes are helpful to differentiate deposits of different glacial episodes. Figure 12-5 shows tills of three glacial episodes and soils of three interglacial episodes preserved at Tuscola Quarry, Douglas County, east-central Illinois.

Quaternary sediments occur as distinctly different sequences in different parts of the state. Beyond the southern limit of glaciation, Quaternary upland sequences are proglacial sediments, mainly loess and windblown sand. Soils developed in these sediments during interstadial phases (warmer, ice-free intervals within a glacial episode) and during interglacial episodes when the landscape supported vegetation. In unglaciated areas, lowlands consist of proglacial stream or lake deposits interbedded with interglacial lake, river, and slope deposits. Sequences within, but near, a glacial limit consist of paleosols and loesses intertongued with ice-marginal and subglacial deposits. Upglacier, away from the glacial limit, deposits may be preserved that record multiple sequences from multiple glacial advances. Some of the thickest and most varied Quaternary successions occur over and within deep valleys that had been cut into the bedrock by preglacial rivers.

CLASSIFICATION OF QUATERNARY MATERIALS

As with older sedimentary bedrock, the Quaternary deposits of Illinois have been assigned to lithostratigraphic (rock succession) units (Frye and Willman 1960; Frye et al. 1968; Willman and Frye 1970; Willman et al. 1971; Johnson et al. 1971, 1985; Lineback et al. 1974; Wickham 1979a, 1979b; Berg et al. 1985; Hansel and Johnson 1996). The formation is the basic unit. Groups (several formations) and subdivisions of formations called members and tongues have been defined, the latter two particularly in northeastern Illinois, by Hansel and Johnson (1996). Qua-

Figure 12-3 Quaternary sediments: **(a)** glacial lake sediments above the Farmdale Geosol at Charleston Quarry, Coles County; **(b)** glacial-fluvial sand and gravel beneath till at Wedron Quarry, La Salle County; **(c)** loess bluff near Edwardsville, Madison County; **(d)** ice-marginal sediments including diamictons interbedded with water-sorted sands and silts at sand and gravel pit near Lemont, Cook County; **(e)** uniform diamicton interpreted to be Lemont till at the Chicago-O'Hare Reservoir, Cook County; **(f)** channel fill of sand, gravel, and silt in Tiskilwa till at Wedron Quarry, La Salle County. (Photographs a, b, e, and f by Ardith K. Hansel; photograph c by Joel M. Dexter; photograph d by Leon R. Follmer.)

ternary lithostratigraphic units are defined on the basis of their physical properties (e.g., color, grain size, lithology, mineralogy, and bedding) and stratigraphic position. Most of the Quaternary lithostratigraphic units in Illinois can be

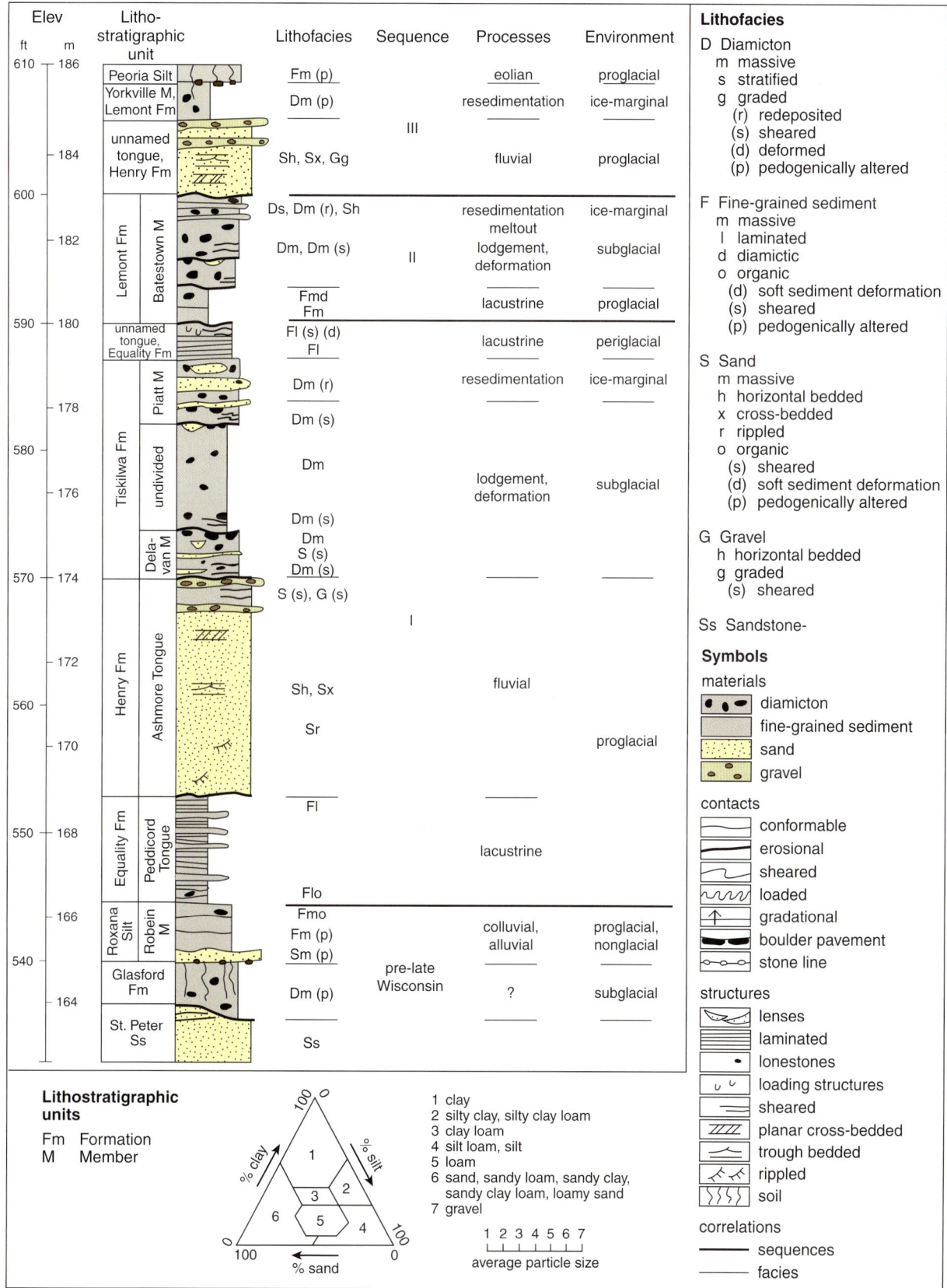

Figure 12-4 Generalized succession of Quaternary deposits overlying the St. Peter Sandstone at Wedron Quarry, La Salle County. Three glacial sequences of the Wisconsin Episode are shown (Johnson and Hansel 1990, Hansel et al. 2004).

mapped over areas of many square miles. They are lithologically distinct units of till and ice-marginal sediment, outwash, loess, and lake sediment. Formal units based on buried soils are known as pedostratigraphic units or geosols. These units, which are zones of weathering, are important markers in the Quaternary successions.

Figure 12-6 shows the major lithostratigraphic and pedostratigraphic units for the glacial and interglacial episodes of the Quaternary Period in Illinois. Episodes, subepisodes, and phases are diachronic temporal (time) units; that is, they begin and end at different times in different places, just like the ice-margin fluctuations that created them (Hansel and Johnson 1996, Johnson et al. 1997, Karrow et al. 2000). Episodes are based on events interpreted from lithostratigraphic and pedostratigraphic units; the ages and time-transgressive nature of these units are well demonstrated by laboratory age determinations. Because of correlation uncertainties and a paucity of age determinations of the oldest Quaternary deposits, the glacial and interglacial episodes that predate the Illinois Episode are referred to collectively as the pre-Illinois Episode. The latest part of this oldest time, based on the Yarmouth Geosol, is referred to as the Yarmouth Episode.

PREGLACIAL EVENTS

There is no evidence for glaciation in Illinois during the first half of the Quaternary Period, and, in fact, soil and sediment evidence from that preglacial time is rare. For millions of years prior to the Quaternary Period, the exposed bedrock surface of Illinois was weathered and eroded. Today, the topographic expression of the bedrock surface (Figures 12-7 and 12-8), now mostly buried, is the principal record of preglacial time.

Broad bedrock uplands and regional slopes were formed mainly before glaciation as were many of the broad benches and low areas on the bedrock surface (Horberg

Figure 12-5 Tuscola Quarry in Douglas County showing tills, from top down, of Wisconsin, Illinois, and pre-Illinois glacial episodes; modern, Sangamon, and Yarmouth interglacial episode soils; and a preglacial soil developed in bedrock. Soil B horizons are oxidized to strong reddish brown. (Photograph by Ardith K. Hansel.)

QUATERNARY PERIOD

1950a, Willman and Frye 1970). The preglacial bedrock surface was significantly modified during the middle and late Quaternary by glaciers and by rivers carrying glacial meltwater (Leverett 1899; Horberg 1946, 1950a; Willman

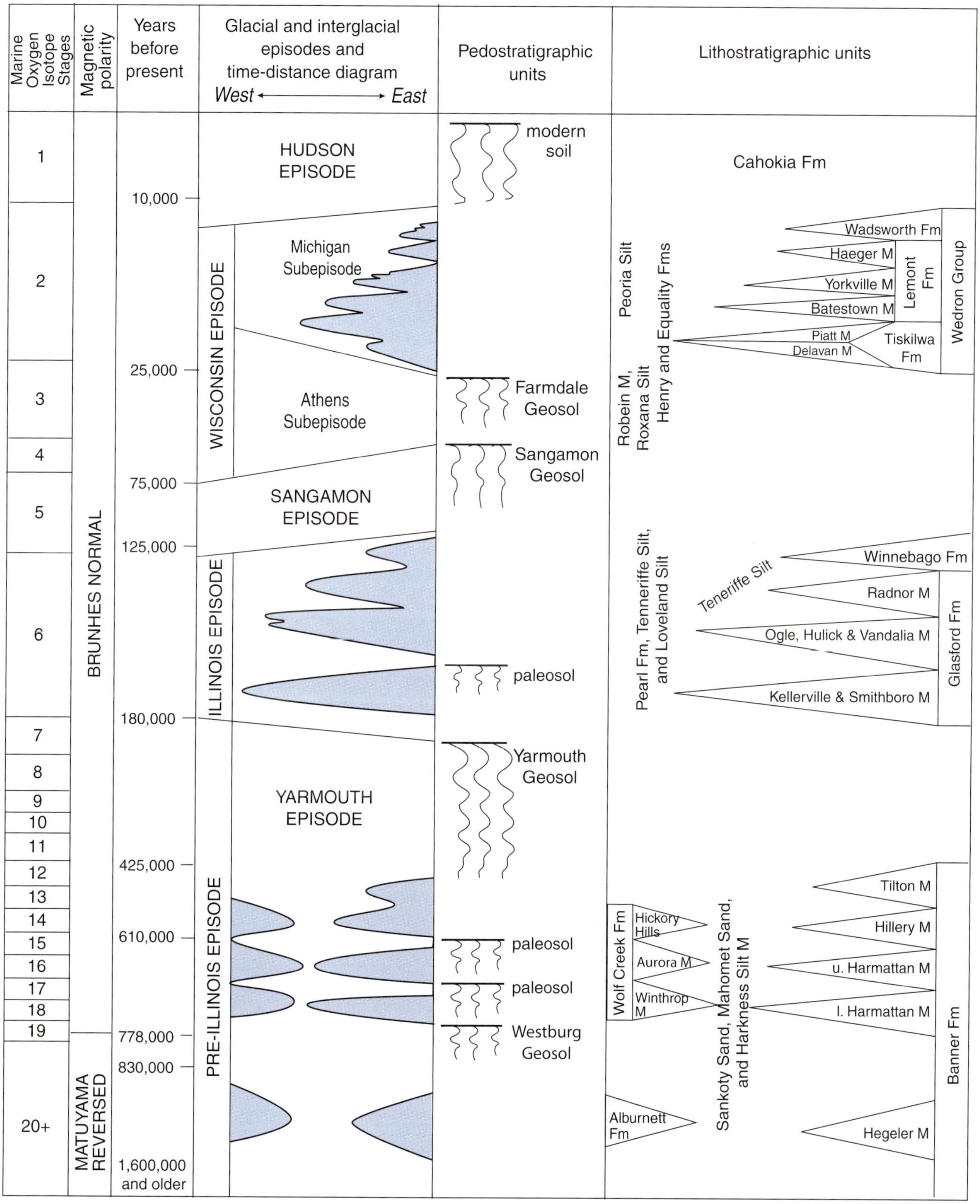

Marine Oxygen Isotope Stages	Magnetic polarity	Years before present	Glacial and interglacial episodes and time-distance diagram West ← → East	Pedostratigraphic units	Lithostratigraphic units
1	BRUNHES NORMAL	10,000	HUDSON EPISODE	modern soil	Cahokia Fm
2		25,000	WISCONSIN EPISODE — Michigan Subepisode		Peoria Silt; Robein M, Roxana Silt, Henry and Equality Fms — Wadsworth Fm, Haeger M, Yorkville M, Batestown M (Lemont Fm), Piatt M, Delavan M (Tiskilwa Fm) — Wedron Group
3			Athens Subepisode	Farmdale Geosol	
4		75,000		Sangamon Geosol	
5		125,000	SANGAMON EPISODE		
6		180,000	ILLINOIS EPISODE	paleosol	Pearl Fm, Tenneriffe Silt, and Loveland Silt; Teneriffe Silt — Winnebago Fm, Radnor M, Ogle, Hulick & Vandalia M, Kellerville & Smithboro M — Glasford Fm
7				Yarmouth Geosol	
8			YARMOUTH EPISODE		
9					
10					
11					
12		425,000			
13			PRE-ILLINOIS EPISODE		Wolf Creek Fm — Hickory Hills, Aurora M, Winthrop M; Sankoty Sand, Mahomet Sand, and Harkness Silt M — Tilton M, Hillery M, u. Harmattan M, l. Harmattan M — Banner Fm
14		610,000			
15				paleosol	
16				paleosol	
17					
18					
19		778,000		Westburg Geosol	
	MATUYAMA REVERSED	830,000			Alburnett Fm — Hegeler M
20+		1,600,000 and older			

Figure 12-6 Timetable of Quaternary glacial and interglacial events and primary lithostratigraphic and pedostratigraphic units on which they are based (Willman and Frye 1970; McKay 1979, 1986; Wickham 1979a, 1979b; Hallberg 1980, 1986; Berg et al. 1985; Hajic 1986; Johnson 1986; Miller et al. 1994; Hansel and Johnson 1996; McKay et al. 2008). Abbreviations: Fm, Formation; M, Member; l, lower; u, upper.

and Frye 1970; Kempton et al. 1991; Herzog et al. 1994). Many of the bedrock valleys are deeply incised. Some are occupied by modern rivers, for example, the Mississippi, Illinois, and Rock Rivers. Others, such as the Mahomet-Teays, Mackinaw, Princeton, and Troy Bedrock Valleys (Figure 12-7), are partly or completely filled with Quaternary sediment (Figure 12-8). The ages of the bedrock valleys are estimated from and limited by the ages of the sediments that fill them. The absence of preglacial soils, alluvial deposits, and the earliest Quaternary deposits from the deepest valleys indicates that the valleys were deepened at some time during the Quaternary Period. Extensive infilling of the deep bedrock valleys with middle Quaternary (so-called "Kansan age") glacial deposits indicates that the deep valley incision predated the first extensive glaciation of Illinois (Horberg 1945, Willman and Frye 1970, Kempton et al. 1991).

Several investigators have suggested that most of the deep bedrock valleys originally formed as ice-marginal streams. For instance, the ancient Mississippi Valley in central Illinois (Figure 12-8) may have developed along the margins of the earliest glacial lobes from the west and the east (Trowbridge 1921; Horberg 1945, 1950a) (Figure 12-7).

Early, Pre-Illinois Episode Events

Glaciation in Illinois began more than 800,000 years ago (Figure 12-6). The landforms, deposits, and paleosols of those earliest, pre-Illinois Episode glaciations and interglacials, however, are the least known. They are (1) discontinuously preserved, (2) intensely weathered, (3) deeply buried, (4) deeply eroded in the small area where they are the surficial units, and (5) largely undated. Pre-Illinois deposits and soils are important, not only because they contain records of very old events and climate, but also because the deposits fill many of the large bedrock valleys and comprise some of the state's most significant sand and gravel aquifers (Chapter 18, Aquifers).

Landscape

Deposits of the earliest glaciations in Illinois occur at the land surface beyond the limit of Illinois Episode sediments in western Pike, Adams, and Hancock Counties in western Illinois (Figure 12-9). The landscape of that area has a well-integrated, dendritic (branching) stream network and few remnants of original glacial landforms. In northwestern Illinois, outwash high on the landscape in the Driftless Area (Figures 12-1 and 12-8) of Jo Daviess County may have been deposited during one or more early glacial episodes (Willman and Frye 1969), but the absolute

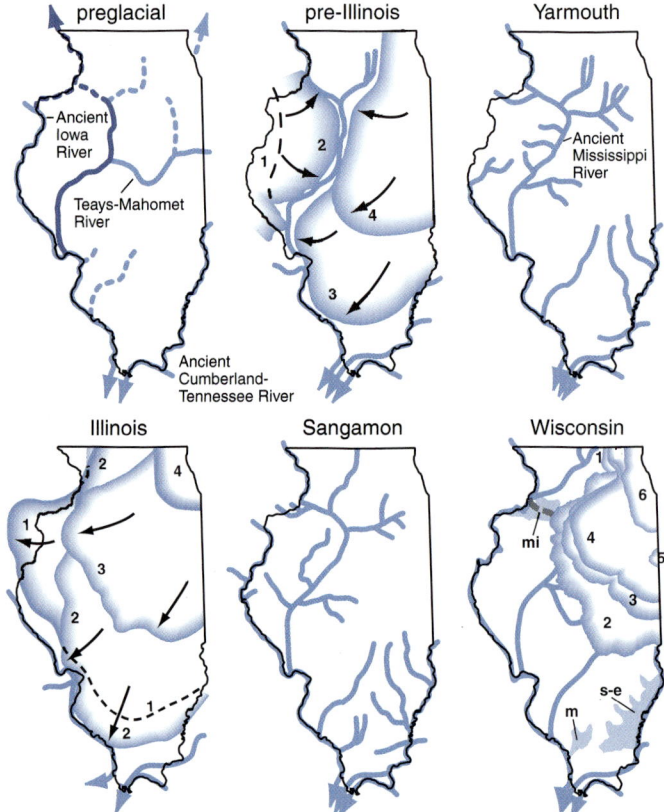

Figure 12-7 Quaternary drainage history, including glacial Lake Milan (mi), Lake Muddy (m), Lake Saline-Embarras (s-e), and the inferred maximum ice-margin positions in Illinois during the pre-Illinois, Illinois, and Wisconsin Episodes. Numbers indicate the sequence of ice-margin advances. Based on data from Horberg (1950b), Willman and Frye (1970), Curry (1998), Hansel and Johnson (1992, 1996), Herzog et al. (1994), and McKay et al. (2008).

age of the outwash has not been determined. The deposits are thin and scattered on a deeply eroded bedrock upland. Pre-Illinois Episode drift is not known to have extended beyond the margin of Illinois Episode drift elsewhere in Illinois, but it is widespread in Iowa, Missouri, Kansas, and Nebraska.

Sediment and Paleosol Record

Deposits from the earliest glaciations are found in the subsurface in places in the southern two-thirds of the state where they are thick, complex, and interbedded with fluvial deposits and paleosols. West of the Illinois River, these glacial deposits were derived from the Keewatin ice-dispersal center and are included in the Wolf Creek and Alburnett Formations (Hallberg et al. 1980, Hallberg 1986) (Figures 12-1 and 12-6). The diamictons from the Keewatin ice-dispersal center tend to be rich in cobbles of limestone, greenstone, igneous rocks, dark red granite of unknown source, and red argillite similar to the well-known pipestone (catlinite) of South Dakota (Horberg 1956). These diamictons contain approximately equal amounts of dolomite and calcite in their silt fraction and considerably more expandable clay minerals than illite in their clay fraction (Willman et al. 1963, Hallberg et al. 1980, Killey 1996).

East of the Illinois River in central, eastern, and southern Illinois, the earliest glacial deposits are included in the Banner Formation. These are thought to be deposits of the northeastern-source Lake Michigan glacial lobe and the eastern-source Huron-Erie glacial lobe derived from one or more Labrador dispersal centers (Johnson et al. 1971, Johnson 1986) (Figures 12-1 and 12-6). Diamicton units from a northeastern-source dispersal center, brought into Illinois via the Lake Michigan glacial lobe, contain cobbles of salmon-colored rhyolite porphyry of unknown source, purple quartzite, and more dolomite than limestone (Horberg 1956). They contain slightly more garnet than epidote in their fine sand fraction, considerably more dolomite than calcite in their silt fraction, and abundant illite in their clay fraction (Willman et al. 1963; Johnson et al. 1971, 1972). Diamicton units from an eastern source, the Huron-Erie glacial lobe (Figure 12-1), tend to have more limestone than dolomite cobbles, considerably more garnet than epidote in their fine sand fraction, and more calcite than dolomite in their silt fraction (Willman et al. 1963). Eastern-source diamicton units also tend to have higher magnetic susceptibility than those from the other source areas (Johnson 1986). Diamicton units from both Lake Michigan and Huron-Erie lobe sources contain a small number of distinctive cobbles of jasper quartzite conglomerate derived from the bedrock that crops out north of Lake Huron.

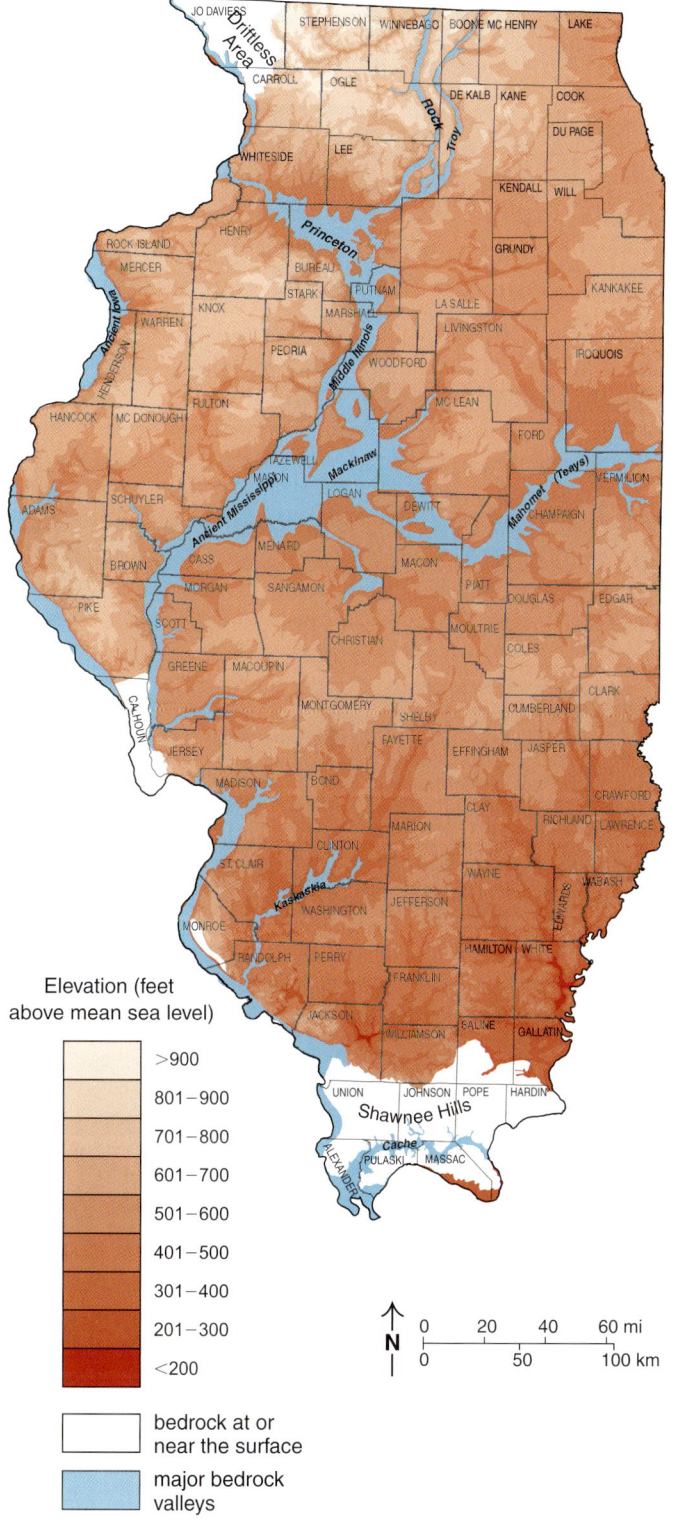

Figure 12-8 Bedrock topography of Illinois. Major bedrock valleys are shown (Herzog et al. 1994).

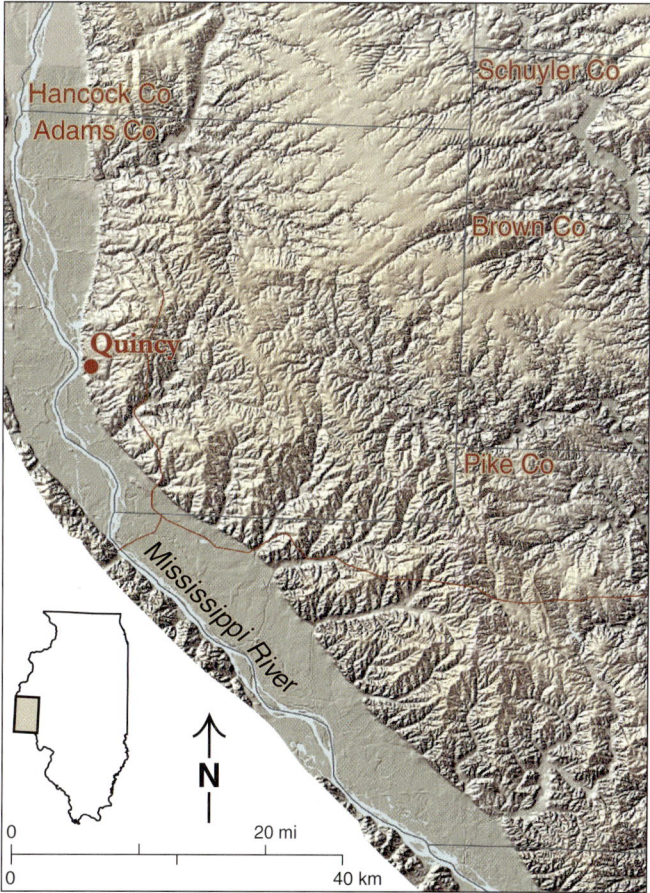

Figure 12-9 Dissected landscape underlain by pre-Illinois Episode drift in western Illinois (Luman et al. 2003).

Many bedrock valleys, particularly those in central Illinois, are filled with pre-Illinois Episode glacial-fluvial sand and gravel, diamicton, and fine-grained river and lake deposits. As much as 200 feet (61 m) of sand and gravel (the Mahomet Sand Member of the Banner Formation) fills the buried Mackinaw Valley and, in some places, overlies older alluvium (Kempton et al. 1991) (Figure 12-10). Portions of the ancient Mississippi Valley are filled with more than 200 feet (61 m) of reddish brown sand (the Sankoty Sand Member of the Banner Formation) (Horberg 1950b), which is a very important source of groundwater.

Glacial and Interglacial History of Early Pre-Illinois Episode Events

Early workers studying old deposits and paleosols in Nebraska, Kansas, and Iowa identified two early glaciations and two interglacial intervals: the Nebraskan (oldest), Afton, Kansan, and Yarmouth stages (Chamberlin 1894, Calvin 1897, Shimek 1909). This terminology was used in Illinois until the 1970s (Willman and Frye 1970). Detailed study of those deposits and soils, however, has demonstrated that they are complex and are widely miscorrelated in their type areas in Kansas, Nebraska, and Iowa (Boellstorff 1978; Hallberg and Boellstorff 1978). Consequently, the names Nebraskan, Aftonian, Kansan, and Yarmouthian have been replaced by the more general and inclusive term, *pre-Illinoian age* (e.g., Wickham 1979a, 1979b) and, more recently, *pre-Illinois Episode* (Johnson et al. 1997, Karrow et al. 2000), a term that recognizes the diachronic nature of these deposits. This volume uses the name Yarmouth Episode to refer to that part of the late pre-Illinois Episode when the Yarmouth Geosol formed.

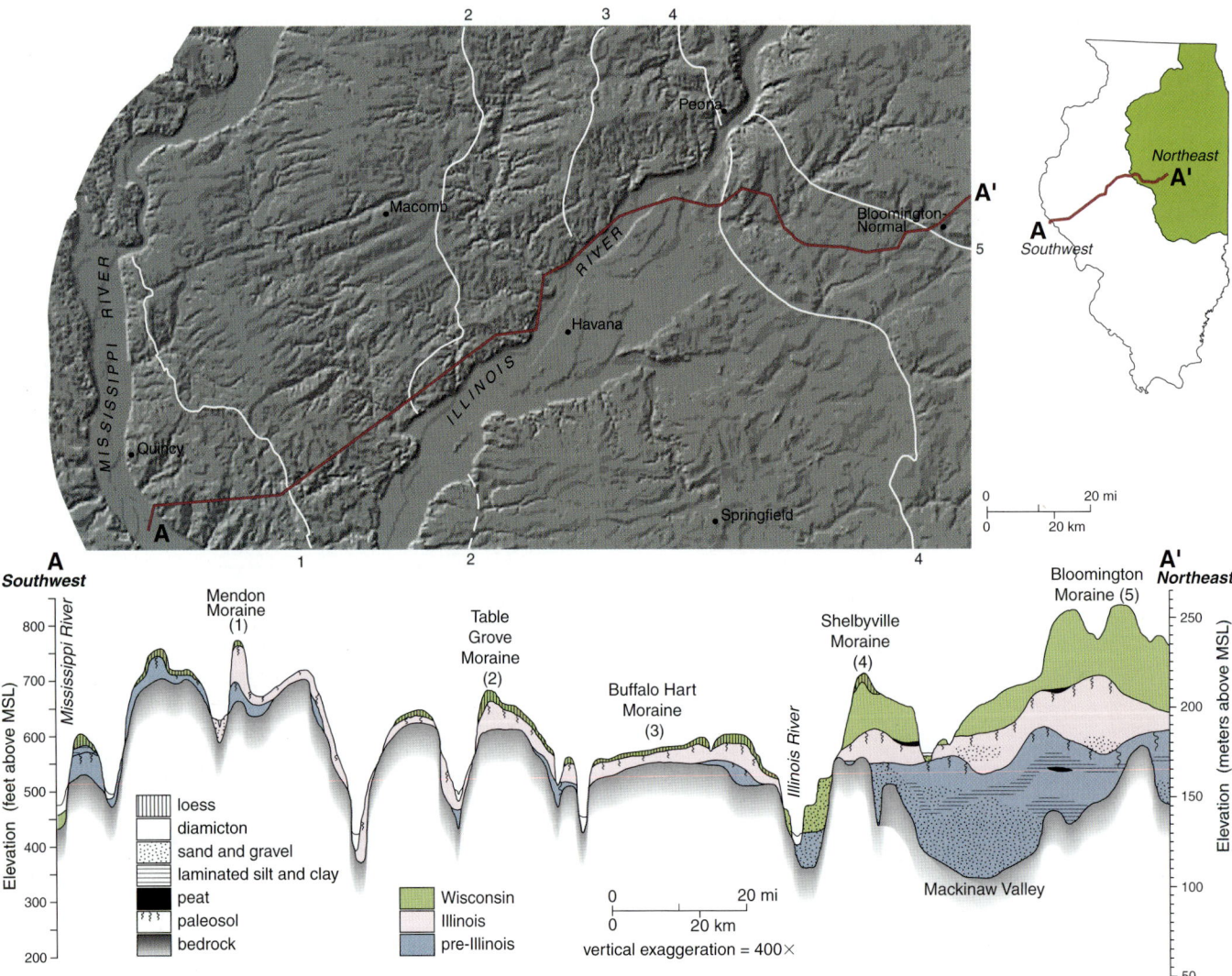

Figure 12-10 Generalized cross section from the Mississippi River in western Illinois, near Quincy, to the central part of the Wisconsin Episode till plain near Bloomington, showing Illinois and pre-Illinois Episode deposits that are thin and widespread in western Illinois, but thick and deeply buried in bedrock valleys of central Illinois. The cross section is based on borehole and outcrop data (Horberg 1950b, 1953; Wanless 1957; Willman and Frye 1970; Piskin and Bergstrom 1975; Lineback 1979b; Wickham 1979b; Herzog et al. 1995; bedrock surface data from Herzog et al. 1994). Abbreviation: MSL, mean sea level.

The preglacial and early glacial records of Illinois, like the geological records of other central states and of the deep ocean (Chapter 13, Quaternary Paleoclimate), are more complex than early researchers supposed. In recent decades, cores of ocean sediments have been shown to contain widely correlated variations in stable oxygen isotopes and other factors that reflect global changes in temperature and glacial ice volume associated with many glacial-interglacial cycles during the Quaternary. Marine Oxygen Isotope Stages (OIS) based on those records are used widely to correlate Quaternary records (Imbrie et al. 1984) using a convention whereby each odd-numbered OIS is a warm (interglacial) climate cycle and each even-numbered OIS is a cold (glacial) cycle (Figure 12-6). Studies in the Midwest during the past 25 years have revealed evidence for many climatic shifts between glacial and nonglacial conditions; the evidence suggests significantly more climatic shifts than previously thought but fewer than implied by the ocean record.

No evidence of the earliest Quaternary climate shifts has been recognized in the state's sediments or paleosols. Illinois, located far from past accumulation centers for continental ice sheets, was apparently situated poorly for the preservation of a complete record of early Quaternary climate fluctuations. Ice sheets of lesser extent did not flow south far enough to reach Illinois or the headwaters of its rivers and more extensive ice sheets that reached Illinois eroded deeply in some areas of the state.

The age of glacial deposits thought to be the earliest in Illinois is beginning to be better known. Pre-Illinois Episode deposits (Alburnett and Wolf Creek Formations), which constitute the thickest portion of the glacial cover west of the Illinois River (Horberg 1956), have been approximately dated in Iowa by relating them to a recorded reversal (Brunhes-Matuyama reversal) of the Earth's magnetic field at about 778,000 years before present (BP) (Hallberg 1986) (Figure 12-6). At that Iowa location, the Alburnett Formation underlies magnetically reversed sediments. In an outcrop in Hancock County, Illinois, a fossil mammal not known to exist anywhere in sediments older than 830,000 years BP was found in magnetically reversed sediments that underlie the Wolf Creek Formation (Miller et al. 1994). These findings suggest that the Wolf Creek Formation is younger than about 778,000 years BP and that the Alburnett Formation is older than about 778,000 and possibly older than about 830,000 years BP. These ages are also consistent with the fission-track age of 610,000 years BP for the Pearlette 'O' (Lava Creek B) volcanic ash bed where it occurs between tills of the Wolf Creek and Alburnett Formations in central Iowa (Hallberg and Boellstorff 1978, Hallberg 1986, Wickham 1979b). Thus, in western Illinois, the oldest pre-Illinois Episode glacial deposits (Alburnett Formation) are likely older than 778,000 years BP, and the younger pre-Illinois Episode deposits (Wolf Creek Formation) are likely younger than 610,000 years BP.

Near Danville in east-central Illinois, the lowermost glacial unit (Hegeler Member) of five pre-Illinois Episode Banner Formation diamicton units (Johnson 1964, 1976, 1986; Johnson et al. 1971) underlies magnetically reversed silt, which indicates that, as in western Illinois, some pre-Illinois Episode glacial sediment in eastern Illinois is older, but most is younger, than about 778,000 years BP (Figure 12-6).

Another age estimate comes from loess deposits in southern Illinois along the Mississippi River near Thebes in Alexander County (Figure 12-11). There, pre-Illinois Episode loess that has been correlated with the Crowley's Ridge loess of the lower Mississippi Valley (Porter and Bishop 1990) underlies the Loveland Silt (Illinois Episode). Based on the degree of soil development in several paleosols at the Thebes site, Grimley et al. (2003) have suggested that the pre-Illinois Episode loess was deposited during OIS 12, prior to approximately 430,000 years BP. It is not known to which early glacial advance this loess may correlate.

Deposits of the most widespread pre-Illinois Episode glacial advance in Illinois originated from the Lake Michigan glacial lobe (Johnson et al. 1971) and are present in much of central and southern Illinois as far south as Franklin County. These deposits, previously referred to as "Eastern Kansan till," are tentatively correlated to the Harmattan Member of the Banner Formation (Johnson 1986). Evidence for early glaciation is sparse in northern Illinois, where younger deposits rest directly on bedrock. Some recent work in southern Wisconsin suggests the Winslow till in that area, previously thought to correlate to the Illinois Episode (Willman and Frye 1970), underlies magnetically reversed sediments and may be pre-Illinois Episode in age and older than 778,000 BP (Miller 2000).

The deepest bedrock channel of the Mahomet Bedrock Valley, one of the deepest and largest bedrock valleys in Illinois (Figure 12-8), predates the Mahomet Sand Member but postdates at least one early Quaternary diamicton unit (Kempton et al. 1991). This unnamed diamicton may correlate to the West Lebanon till of western Indiana (Bleuer 1976, 1980), the Hegeler Member (lowermost diamicton) of the Banner Formation in east-central Illinois, and diamicton of the Alburnett Formation in western Illinois (Johnson 1986) (Figure 12-6), which would suggest that the deep erosion of the Mahomet Valley occurred after 778,000 BP.

Three paleosols (from oldest to youngest, the Westburg, Franklin, and Dysart paleosols) occur within the succession of Alburnett and Wolf Creek Formation diamictons in their type area in Iowa (Hallberg 1980, 1986). Several weathered zones described within the Alburnett and Wolf Creek diamictons in western Illinois likely correlate with those in Iowa, indicating that multiple glacial and interglacial cycles predating the Illinois Episode are recorded in western Illinois, but only one of the western Illinois paleosols has been correlated to a named paleosol—Westburg Geosol—in Iowa (Wickham 1979b) (Figure 12-6).

Along the Mississippi River near Collinsville in Madison County in southwestern Illinois, the Quaternary succession includes a pre-Illinois Episode loess (Burdick Member) at the base, an unnamed paleosol A, a diamicton (Omphghent Member), an upper loess (Maryville Member), and an unnamed paleosol B, thus containing a record of two or three glacial episodes and two interglacial episodes that predate the Illinois Episode (McKay 1979, 1986). About 40 miles (64 km) to the northwest, at the Pancake Hollow Section in Calhoun County, four units of loess and loess-derived sediment (units 1, 2, 3, and 4) and three unnamed paleosols record at least one, and perhaps as many as three, episodes of interglacial climate that predate the Illinois Episode (Hajic 1986).

Pre-Illinois Episode deposits in western Illinois are capped by a very strongly developed paleosol, the Yarmouth Geosol (Leverett 1898, 1899; Hallberg and Baker 1980; Woida and Thompson 1993). That paleosol probably records the cumulative weathering of several interglacial episodes during a long hiatus in glacial deposition after the youngest pre-Illinois Episode diamicton in Illinois was deposited. That hiatus, the Yarmouth Episode, ended with burial of the Yarmouth Geosol by Illinois Episode loess or diamicton. The average thickness of the soil profile of the Yarmouth Geosol in western Illinois is more than twice that of the Sangamon Geosol, and the time required for its formation may have been three or four times longer than for the Sangamon Geosol (Willman and Frye 1970). Soil profile analysis and a comparison of physical and chemical properties of the Yarmouth, Sangamon, and modern soils in loess near Thebes indicate that the time required for the formation of the Yarmouth Geosol was more than 3.5 times the estimated 50,000 years required to form the Sangamon Geosol at the site (Grimley et al. 2003). According to this analysis, the Yarmouth Geosol at Thebes represents cumulative weathering of about 180,000 years, beginning about 430,000 years BP and spanning three interglacial and

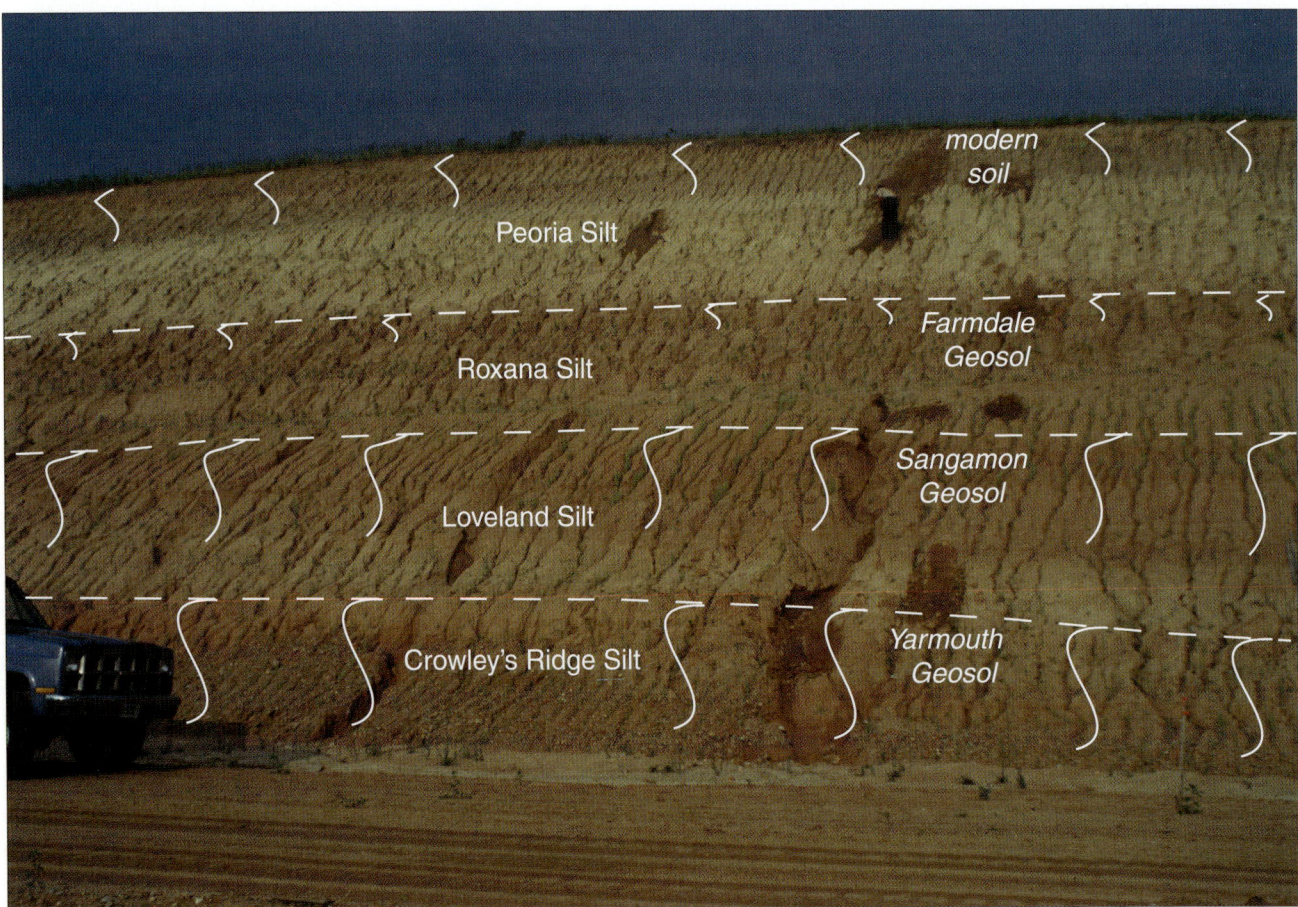

Figure 12-11 Thebes road cut, formerly exposed along Illinois Route 3 along the Mississippi River (Alexander County). (Photograph by Leon R. Follmer; annotation by David A. Grimley.)

two glacial episodes (OIS 7 through OIS 11; Figure 12-6) (Grimley et al. 2003).

Illinois Episode

The Illinois Episode was originally named for deposits of the Lake Michigan glacial lobe in Illinois (Leverett 1899). It likely correlates to marine OIS 6 (McKay et al. 2008) (Figure 12-6). Deposits of this episode are widespread, locally thick, intensely weathered in the upper part, and nearly everywhere buried by younger loess or tills. The relationships of Illinois Episode to pre-Illinois Episode deposits and paleosols were first worked out in western Illinois and adjacent eastern Iowa where these deposits overlap (Leverett 1899). The distinctive glacial landscape and sediment record of the Illinois Episode suggest that the suite of glacial processes that produced them differed markedly from those that prevailed in Illinois later during the Wisconsin Episode (Leighton and Brophy 1961, Willman and Frye 1970, Stiff and Hansel 2004).

Landscape

At its maximum extent, Illinois Episode glacial ice in Illinois was more widespread than the ice of any earlier or later glaciation, covering nearly 90% of the state (Figures 12-7 and 12-12). The landscape this glaciation produced, the "Illinoian till plain," is distinguished as much by its flatness as by its unique assemblage of glacial landforms, referred to as the "ridged drift." Farthest south, near the southern limit of glaciation, thin Illinois Episode drift deposits mantle, yet do not conceal, the underlying, undulating, bedrock surface as it rises southward into the Shawnee Hills in Jackson and Williamson Counties. To the north,

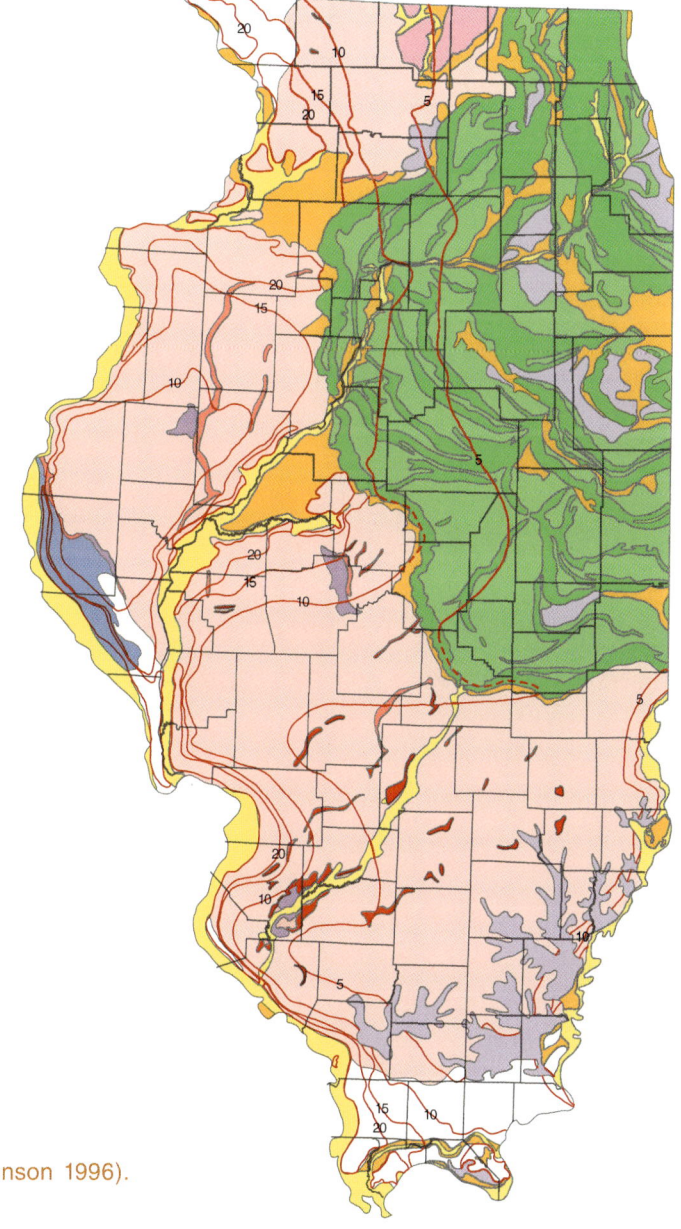

Figure 12-12 Quaternary deposits of Illinois (Hansel and Johnson 1996). Abbreviations: Fm, Formation; Mbr, Member.

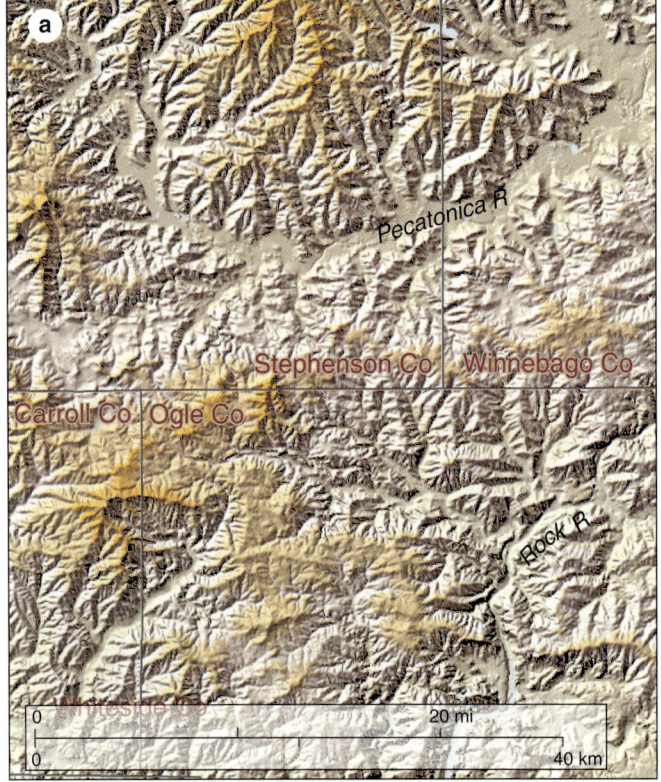

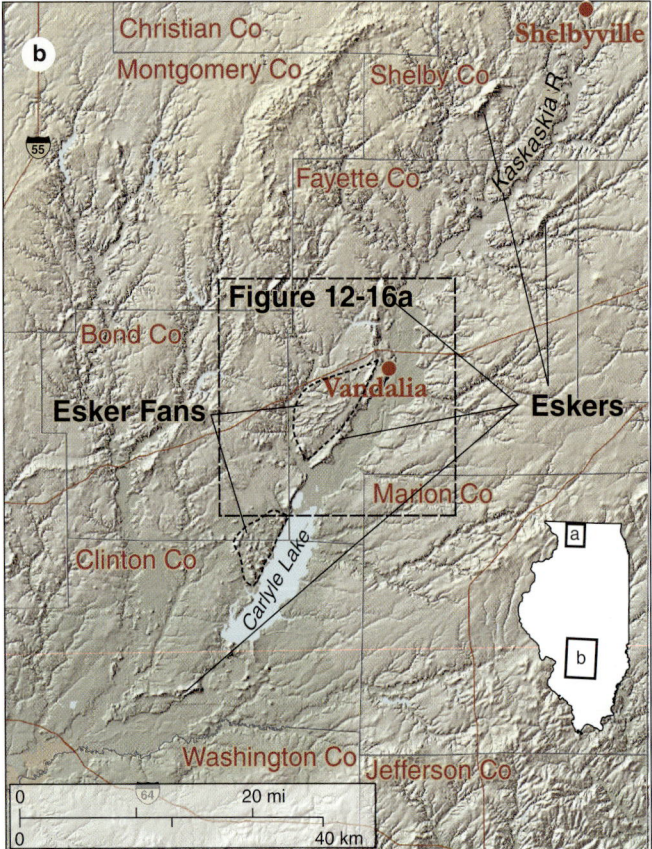

Figure 12-13 (a) Dissected Illinois Episode landscape of northwestern Illinois; (b) ice-contact drift ridges (kamic moraine, kames, and eskers), outwash fans, and lake plains in the flat till plain area (Kaskaskia River watershed) of south-central Illinois (Luman et al. 2003).

in the broad Kaskaskia River watershed, the ridged drift region includes broad expanses of flat till plain interrupted by linear valleys and glacial ridges, irregular gravel hills, eskers, esker fans, and lake plains (Figures 12-12 and 12-13b).

The mode of origin of these features has been controversial since they were first described more than a century ago. Leverett (1899) considered them to be segments of the end moraine of the Huron-Erie lobe. Leighton and Brophy (1961) interpreted the ridged drift and associated linear and trellis valleys as evidence of widespread stagnation of a portion of the Illinois Episode glacier but did not conclude that the ice was necessarily from a Huron-Erie source. On the basis of differences in heavy mineral suites in the drift on the east and west sides of the ridged drift complex, Willman et al. (1963) interpreted the ridges as marking an interlobate ice-contact zone between a Lake Michigan glacial lobe to the west and a Huron-Erie glacial lobe to the east. Research into the origin of landforms and sediments on this glacial landscape is ongoing.

In western Illinois, a series of strikingly aligned linear stream valleys (Figure 12-14) were interpreted as having been cut by meltwater flowing in crevasses in a stagnant Illinois Episode glacier (Leighton and Brophy 1961). The orientation of these valleys and their association with drumlin fields suggest to the authors that they are related to an Illinois Episode Lake Michigan glacial lobe and were formed by subglacial processes.

In northwestern Illinois, surficial Illinois Episode deposits beyond the Wisconsin Episode margin have been deeply eroded (Figure 12-13a). Few depositional landforms remain, and widespread Wisconsin Episode erosion greatly modified the landscape, leaving scant evidence of the Sangamon Geosol on much of the landscape (Follmer and Kempton 1985, Kempton et al. 1985).

End moraines, glacial landforms that dominate the Wisconsin Episode landscape, are few, narrow, and discontinuous in the area of Illinois Episode glaciation. Along the outer limit of Illinois Episode drift, a terminal moraine (Mendon Moraine) is well developed only in Adams County in western Illinois. Elsewhere along the limit, discontinuous moraine segments occur in a few places. Within the area of the Illinoian till plain, two significant end moraines are present west of the Illinois River (Figures 12-10 and 12-12). The westernmost moraine, the Table Grove Moraine, is thought to mark the limit of a readvance of active ice during the Illinois Episode (Willman and Frye 1970). The younger Buffalo Hart Moraine also marks a readvance. North and south of the areas where they were defined (their type localities), both of the moraines are discontinuous and difficult to trace and correlate.

Glacial-fluvial landforms are widespread on the Illinoian till plain. Several large eskers are present along the Kaskaskia River in the Vandalia area (Figure 12-13b). Another, along the Leaf River in northwestern Ogle County, was named the Adeline esker (Leighton and Brophy 1961). Lake plains have been mapped along glacial margins in Sangamon and McDonough Counties where glacial readvances overrode lake sediments (Figure 12-12). A significant outwash deposit (Pearl Formation) occurs in the lower Kaskaskia River valley (Figure 12-12), but, elsewhere in major valleys, Illinois Episode valley train deposits were buried by younger deposits or were eroded by Wisconsin Episode meltwater.

Drumlins and long, low-relief, linear ridges are present in Boone, Winnebago, Knox, Warren, and McDonough Counties (Figures 12-14a and 12-15) and also in Macoupin, Montgomery, and Christian Counties east of the Illinois River valley. These landforms may be subglacial bedforms (flutes) that are generally oriented parallel to the direction of ice flow.

Sediment and Paleosol Record

Illinois Episode diamictons, like most others in Illinois, consist mainly of sand, silt, and clay. They also contain pebbles, cobbles, and boulders made up of a variety of erratic (transported from other places) and local rock types. Based on lithology, mineralogy, and stratigraphic position, these diamictons have been defined and classified as members of the Glasford Formation (Willman and Frye 1970) and Winnebago Formation (Kempton et al. 1985) (Figure 12-6).

Although the Glasford Formation is widespread throughout the state (Figure 12-12), tracing and correlating diamictons over long distances have been challenging, leading to regional definition of several members. In western and west-central Illinois, the Glasford Formation, as originally defined by Willman and Frye (1970), included three diamicton members: the Kellerville (oldest), Hulick, and Radnor (youngest). In southern and east-central Illinois, three diamicton members also were recognized: the Smithboro (oldest), Vandalia, and Radnor Members (Jacobs and Lineback 1969, Willman and Frye 1970, Johnson et al. 1972). In northern Illinois, a similar threefold diamicton classification included the Winslow (oldest), Ogle, and Sterling Members. The Winslow Member has since been shown to be pre-Illinois Episode in age (Miller 2000). Lineback (1979b) informally subdivided the Glasford Formation of western and west-central Illinois into as many as eight individual members based on clay mineralogy, carbonate mineralogy, and grain size. Mineralogic and grain-size properties have also been used to define a number

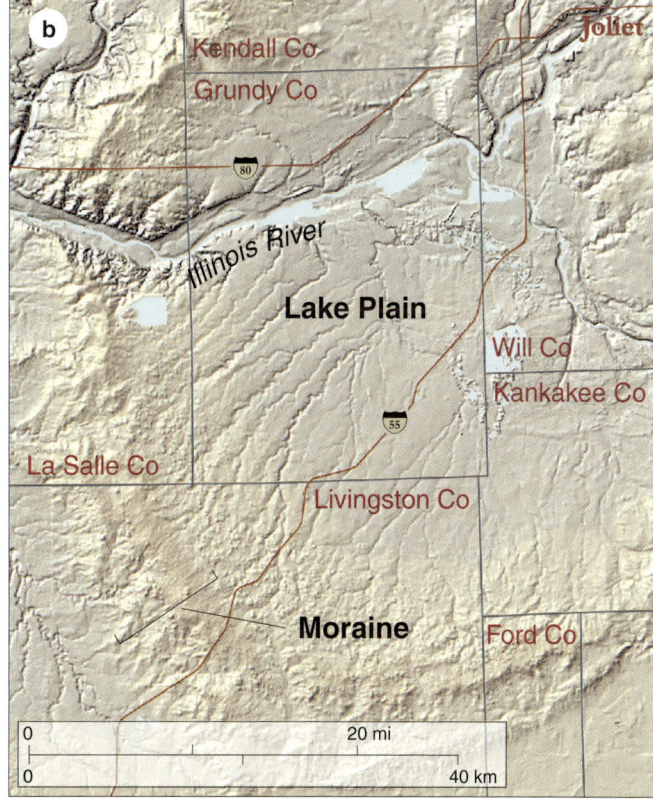

Figure 12-14 (a) Loess-covered drumlin area of western Illinois with linear stream valleys and trellis drainage pattern; (b) Wisconsin Episode landscape in northeastern Illinois showing asymmetric end moraines of the lobate, double-crested Marseilles morainic system that grade into till plain and lake plain in the up-ice direction (Luman et al. 2003).

Figure 12-15 Oblique aerial view looking west-southwest along the axis of a glacial ridge (flute) on the Illinois till plain in Section 5, T7N, R1W, McDonough County. (Photograph by J. Wickham.)

of other Illinois Episode stratigraphic units in studies of northern and southern Illinois (McKay 1979, Kettles 1980, Fleeger 1980, Hartline 1981, Kempton et al. 1985).

Reliance on mineralogical and grain-size analyses for stratigraphic unit distinction and definition has led to differentiation of some units that are difficult to distinguish and trace in the field. Consequently, stratigraphic names have proliferated as multiple units have been named for the lateral variations in the composition of the debris transported and deposited during single glacial advances.

The oldest Glasford Formation diamictons in southern and western Illinois (Smithboro and Kellerville Members, respectively) are similar in composition and stratigraphic position (Figure 12-6). They are likely parts of a single till sheet (Lineback 1980). Both are silty, clay-rich, and contain relatively large amounts of expandable clay minerals and small amounts of carbonate, a composition reflecting erosion and incorporation of weathered deposits from the underlying Yarmouth Geosol. The next younger unit is mainly sandy diamicton (Hulick, Vandalia, and Ogle Members). This sandy diamicton is widespread and contains large amounts of carbonate and illite from deep erosion of Paleozoic bedrock units in the Lake Michigan trough. The sandy diamicton is overlain by silty to clayey diamicton of the Radnor Member and its equivalents (Sterling, Lee Center, and Esmond Members), which are also rich in carbonate and illite.

The relationship of the Winnebago Formation diamictons (Figure 12-6) to those of the Glasford Formation remains uncertain. In northern Illinois, the older diamicton members of the Winnebago Formation are thought to range in age from Kellerville equivalents to Radnor equivalents and the younger part of the Winnebago Formation (the Clinton, Argyle, and Nimtz Members; not shown) in Boone and Winnebago Counties may be among the youngest Illinois Episode glacial deposits in the state (Kempton et al. 1985, Johnson 1986).

The upper surface of Illinois Episode deposits beyond the limit of Wisconsin Episode glaciation is usually marked by the deeply weathered Sangamon Geosol, which has been traced beneath Wisconsin Episode deposits 50 to 75 miles (80.5 to 121 km) northeast of the Shelbyville Moraine in central Illinois (Figure 12-10). Beyond that distance the soil is missing, and Illinois Episode deposits were deeply eroded. They are discontinuous and rarely preserved in the Chicago area.

Illinois Episode loess (Loveland Silt; Figure 12-6) was deposited and is locally preserved near the valleys of the ancient Mississippi River and the modern Mississippi, Ohio, Wabash, and Missouri Rivers. Where it interfingers with Illinois Episode diamicton units, most of the Illinois Episode loess occurs beneath the oldest diamicton (Kellerville Member) (Leighton and Willman 1950, Willman and Frye 1970). In most areas, little Illinois Episode loess occurs between drift units, and only thin loess overlies the uppermost Illinois Episode diamictons.

Glacial History

During the Illinois Episode, glacial ice entered Illinois mainly from the north via the Lake Michigan glacial lobe (Leverett 1899, Johnson 1986). At its maximum extent, the ice margin reached as far south as 37° 35′ N latitude in southern Illinois, 8 to 10 miles (13 to 16 km) south of the city of Carbondale on the north slope of the Shawnee Hills (Figure 12-8). That limit is farther south than was reached by any other continental glacier in the northern hemisphere. Illinois Episode ice crossed the ancient Mississippi River (in the position of the modern Illinois River), advanced farther west than pre-Illinois Episode ice from the Labrador ice-dispersal center, and continued westward across the ancient Iowa River (in the position of the modern Mississippi River) (Figures 12-7 and 12-8). The ice margin advanced 22 miles (35 km) into what is now southeastern Iowa, temporarily diverting the ancient Mississippi River (Leighton and Brophy 1961). In northwestern Illinois, the ice encroached upon but, like earlier glaciers, did not cover the topographically high Driftless Area (Figure 12-8).

The glacial retreat that followed the initial advance of the Illinois Episode is marked by a rarely observed paleosol (Pike Soil of Willman and Frye 1970) (Figure 12-6). An extensive advance of the Lake Michigan lobe ice and perhaps the Huron-Erie lobe ice then moved far into the state, depositing the diamicton of the Vandalia and Hulick Members of central and western Illinois and the Ogle Member of northern Illinois. The few discontinuous end moraines, extensive hummocky sand and gravel hills, eskers, esker fans, crevasse-fill deposits, and possible crevasse traces indicate that a significant area of ice in south-central Illinois stagnated rather than retreated under active flow conditions (Leighton and Brophy 1961) (Figures 12-12, 12-13b, and 12-16). A readvance of ice then deposited the Radnor

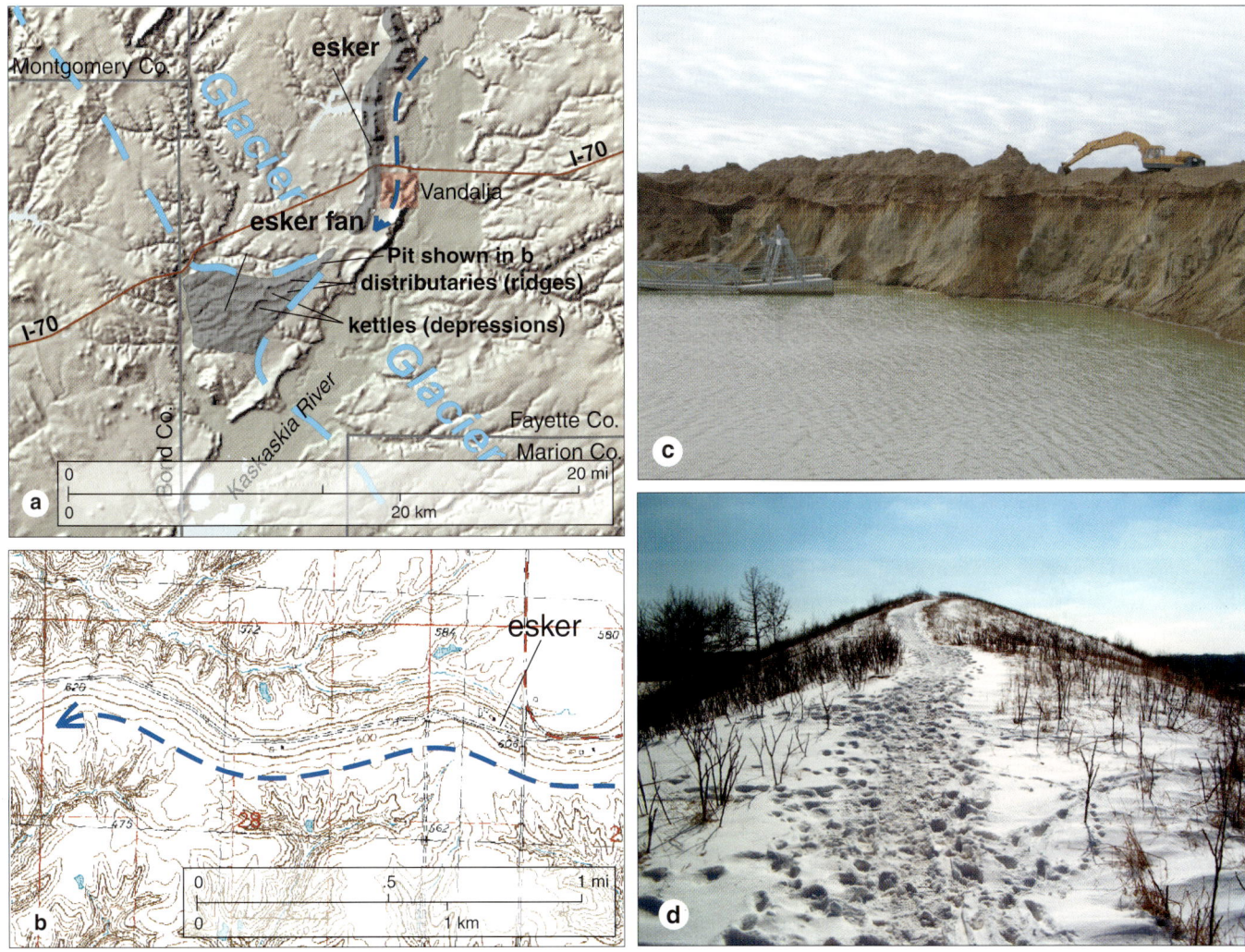

Figure 12-16 (a) Shaded relief map of surface topography showing Illinois Episode eskers and an associated esker fan (dark gray shading) that formed near the margin (dashed blue line) of the Illinois Episode glacier when it stood southwest of Vandalia. Meltwater flowing southwestward (blue arrow) beneath the glacier in the esker conduit (tunnel) discharged near the glacier margin, depositing fan sediments (sand and gravel) on and around blocks of glacier ice. When all the ice melted, the sediment-filled conduit and distributary stream channels remained as ridges. Melting of buried ice blocks formed depressions (kettles) in the fan surface. The esker shown is the best preserved of several that parallel the Kaskaskia River valley in south-central Illinois, where they are mined extensively for sand and gravel. (b) Topographic map showing a portion of an Illinois Episode esker located near Exeter in northern Scott County, Illinois. This esker, which rises up to 40 feet (12 m) above the surrounding till plain, was formed by meltwater flowing westward (right to left) during the middle part of the Illinois Episode. The contour interval on the Florence 7.5-minute Quadrangle map shown here is 10 feet (3 m). (c) Backhoe removing Illinois Episode overburden as dredge mines sand and gravel in the esker fan shown in Figure 12-16a, Fayette County. (Photograph by E. Donald McKay.) (d) Snow-covered Wisconsin Episode esker located along Deerpath Trail, Glacial Park, in McHenry County, Illinois. (Photograph by Wayne T. Frankie 2007.)

diamicton and equivalents (the Sterling, Esmond, and Lee Center Members and perhaps also the Belvidere Member; not shown) in central and northern Illinois.

Each of these Illinois Episode glacial advances blocked and temporarily diverted the ancient Mississippi River from its course through the middle of the state into a course along the western edge of the state. With each retreat, the river was reestablished closely, but not exactly, along its ancient course, where it remained until late in the Wisconsin Episode, when it was permanently diverted to its present course (Figure 12-8).

Although Lineback (1979b) suggested that several paleosols occur within the succession of Illinois Episode deposits and indicated that the Illinois Episode spans several glacial-interglacial cycles, most other researchers have been more conservative. Johnson (1986) suggested that the several observed paleosols in different locations may correlate with the Pike paleosol and, therefore, that the Illinois Episode may consist of two glacial episodes and one interglacial episode, spanning OIS 8, 7, and 6 from about 280,000 years BP to about 132,000 years BP.

Age estimates for the Loveland Silt, based on atmospheric accumulation of isotopic beryllium (^{10}Be), however, fall between 190,000 and 170,000 years BP (Markewich et al. 1998). Other age determinations, based on infrared stimulated luminescence of the loess, indicate an age range of about 180,000 to 140,000 years BP for the Loveland Silt of the Missouri and Mississippi River valleys (Forman and Pierson 2002). Both ^{10}Be and luminescence age estimates support an interpretation that the Loveland Silt and, therefore, the overlying Kellerville diamicton and younger Illinois Episode deposits were deposited during one glacial climatic cycle, OIS 6 (Figure 12-6). Recent optically stimulated luminescence dates from fluvial sand between and beneath Illinois Episode tills in the ancient Mississippi Bedrock Valley in Marshall and Putnam Counties confirm that the Illinois Episode tills were deposited during OIS 6 (McKay and Berg 2008).

SANGAMON EPISODE

The Sangamon Episode is the interglacial interval between the Illinois and Wisconsin glacial episodes and is commonly referred to as the "last interglacial." The Sangamon is represented in the geological record of Illinois mainly by the weathering profile and soil development of the Sangamon Geosol (Figure 12-17) (Leverett 1899; Follmer 1979, 1983; Curry and Follmer 1992). It is generally agreed that the Sangamon Episode correlates with OIS 5 and in Illinois transgresses into OIS 4 and OIS 3. The sediment record of the Sangamon Episode is not widespread

Figure 12-17 Exposure of the Sangamon Geosol in Putnam County where the paleosol is developed in Illinois Episode fluvial deposits overlying Illinois Episode glacial till. The paleosol is overlain by early Wisconsin Episode loess, the Roxana Silt. (Photograph by Richard C. Berg.)

but is important because it contains information about climate during the last interglacial, the last time Earth's climate resembled present postglacial conditions for a period of tens of thousands of years (Chapter 13, Quaternary Paleoclimate).

Paleosol and Sediment Record

The Sangamon Geosol (Leverett 1898) is a widely recognized stratigraphic marker. In its type area, the Sangamon Geosol includes Mollisols (dark grassland soils), Alfisols (light forest soils), and Ultisols (very strongly developed soils) that formed in Illinois and Sangamon Episode sediments, including diamicton of the Vandalia Member and colluvial sediment (Berry Clay Member) of the Glasford Formation. In northern Illinois, the Sangamon Geosol occurs in the upper part of the Winnebago Formation. Throughout the state, this paleosol is overlain by Wisconsin Episode silt (loess) or glacial deposits.

Compared with the modern soil, Sangamon Geosol profiles generally have more strongly developed subsoil B horizons in which clay-sized material accumulates; the profiles are thicker, and minerals are slightly more weathered. These soil profiles have been described as ranging from (1) reddish brown to dark-gray soil profiles developed in place under drainage conditions ranging from good to poor, to (2) accretion-gley (wetland) profiles developed under poorly drained conditions in flat areas or depressions into which local sediment accumulated. In all cases, the soil parent material is leached of calcite and dolomite, usually to a depth of 6 to 10 feet (1.8 to 3 m), and oxidation and mottling commonly extend much deeper.

Geological and Climatic History

The Sangamon Episode, represented by the Sangamon Geosol, began as the post-Illinois Episode climate warmed sufficiently to allow soil formation and ended when the soil was buried by younger sediment. Thus, the Sangamon Episode boundaries are highly time-transgressive. In its type area, the Sangamon Geosol is estimated to have begun development 135,000 to 125,000 years BP and ended as burial by the Roxana Silt began 65,000 to 55,000 years BP (Curry and Follmer 1992). Grimley et al. (2003), in their examination of paleosols in loess in southwestern Illinois near Thebes (Figure 12-11), estimated 50,000 years were required to form the Sangamon Geosol. They suggested that it formed mostly during OIS 5, but continued to develop during OIS 4 and part of OIS 3.

A number of studies during the past 30 years have examined pollen and macroscopic plant and animal remains preserved in lake and bog sediments in south-central Illinois in order to reconstruct late Illinois, Sangamon, and Wisconsin Episode climate (Grüger 1972, Saunders and King 1986, Curry and Baker 2000, Teed 2000). Those studies document the interglacial climate from the end of the Illinois Episode (OIS 6), about 132,000 years BP, through the early part of the Wisconsin Episode, about 60,000 years BP. Climate records in these sediments are of sufficient resolution to permit reconstruction of several significant intervals, characterized by different temperature and moisture conditions, during the nonglacial episode (Curry and Baker 2000). Charcoal preserved in the sediment may indicate variation in the frequency of prairie and forest fires (Teed 2000). During the warmest part of the Sangamon Episode, south-central Illinois was home to the giant tortoise *Geochelone crassiscutata*, which could not tolerate freezing temperatures; its presence indicates a very warm climate year-round (Saunders and King 1986). During this time, shallow lakes were home to a microscopic crustacean, the ostracode *Heterocypris punctata*, found today only in subtropical climates (Curry and Baker 2000). These warmest interglacial temperatures of the Sangamon Episode gradually cooled, giving way to the glacial climate of the Wisconsin Episode.

WISCONSIN EPISODE

The deposits of the last glacial episode are better known than those of earlier episodes because they are (1) less weathered and dissected by erosion, (2) less deeply buried by loess, (3) closer to the land surface and more accessible for study, and (4) at least partially within the range of radiocarbon dating. Although the Wisconsin Episode glaciers advanced only into the northeastern quarter of Illinois, beyond the glacial margin the Wisconsinan loess accumulated across the upland landscape, and waterlain sediment was deposited in major glacial meltwater channels and dammed tributaries (the Henry Formation and Equality Formation, respectively) (Figure 12-12). As a result, sediment of the Wisconsin Episode is thick, generally 30 to 260 feet (9 to 79 m) in northeastern Illinois in and adjacent to most major meltwater channels. Over most of the rest of the state, loess forms a thin veneer, 2 to 12 feet (0.61 to 3.6 m) thick, over older sediments and bedrock.

Landscape

The landscape of the Wisconsin glaciation in Illinois is characterized by a series of subparallel end moraines separated by lake plains and by low-relief till plains dissected by a few large meltwater channels (Figures 12-12, 12-18a, and 12-19). With the exception of the Iroquois Moraine, formed by Huron-Erie lobe ice flowing into Illinois from the east (Figure 12-18), the moraines were formed by Lake Michigan lobe ice flowing into Illinois from the northeast (Johnson et al. 1986, Hansel and Johnson 1992). Five sublobes of the Lake Michigan lobe were active in Illinois during the Wisconsin Episode (inset in Figure 12-18a). More than 30 end moraines have been mapped (Ekblaw 1960, Willman and Frye 1970); maps of these exceptionally developed moraines are often included in Quaternary geology texts (e.g., Sugden and John 1976, Bowen 1978, Menzies 1995, Ehlers 1996). Most of the end moraines are composed predominantly of till and appear to have formed at the ice margin during the last interval of till deposition (Mickelson et al. 1983, Hansel and Johnson 1992, Johnson and Hansel 1999). Some end moraines consist of a single crest and generally are less than 3 miles (5 km) wide. Others form broader, multiple-crested morainic systems ranging from 5 to 12 miles (8 to 19 km) wide. The composite forms are attributed to superposition during multiple intervals of moraine formation.

End moraine height varies from about 30 to 100 feet (9 to 30.5 m). The end moraines are noticeably asymmetrical

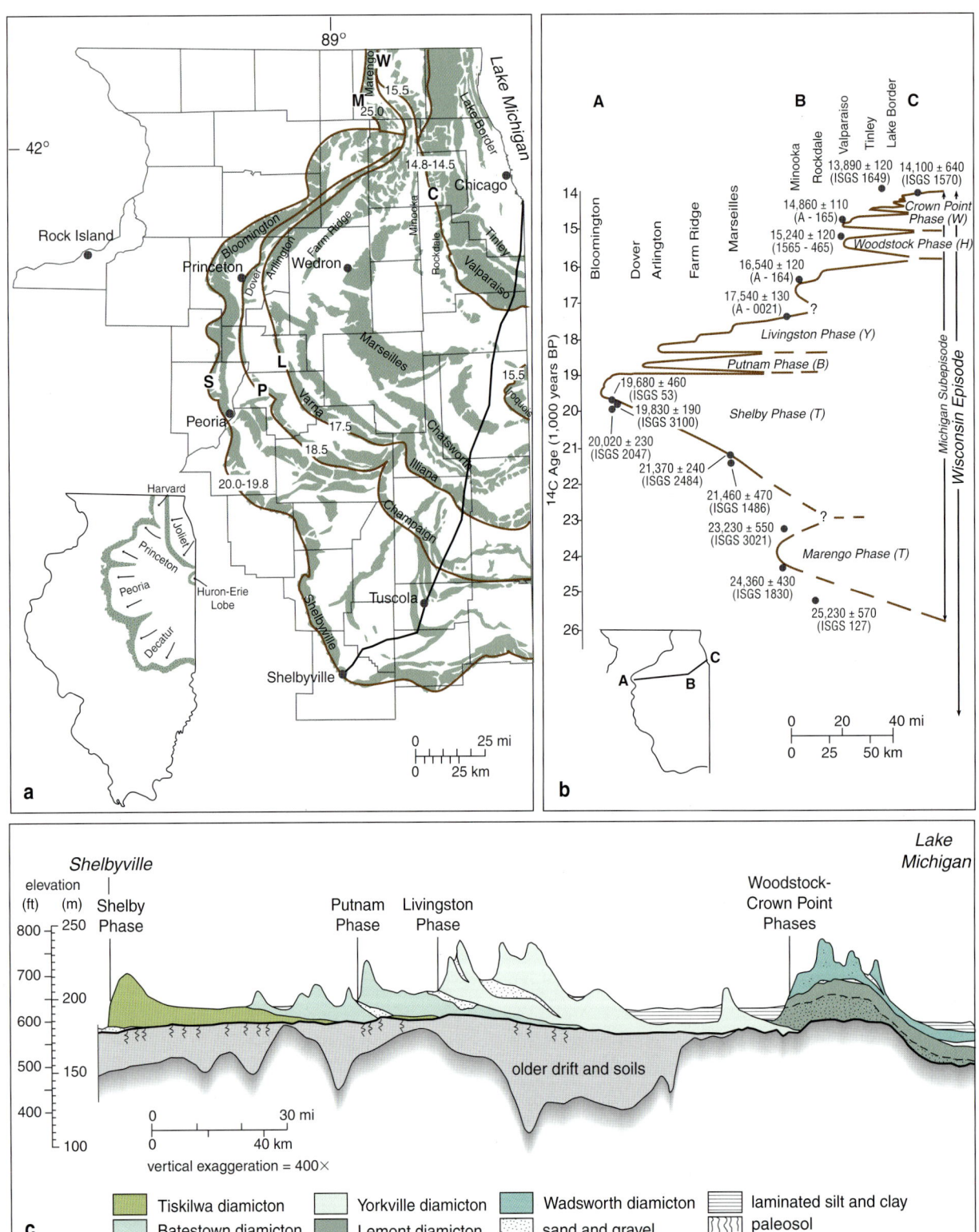

Figure 12-18 Spatial and temporal relationships of end moraines, ice-margin fluctuations, and tills of the Wisconsin Episode Lake Michigan lobe in Illinois. **(a)** Location of end moraines, sublobes (inset map), and maximum ice-margin positions during the Marengo (M), Shelby (S), Putnam (P), Livingston (L), Woodstock (W), and Crown Point (C) Phases of the Michigan Subepisode with estimated radiocarbon ages (in 1,000 years BP). **(b)** Time-distance diagram showing glacial phases for the Lake Michigan lobe in Illinois from the Bloomington Moraine to Lake Michigan and the material units on which they are based, including the Tiskilwa Formation (T), Batestown Member (B), Yorkville Member (Y), Haeger Member (H), and Wadsworth Formation (W). Radiocarbon ages and alphanumeric borehole identifications are shown. Modified from Hansel and Johnson (1992) and Hansel et al. (2004). **(c)** Generalized cross section from Shelbyville to Lake Michigan showing offlapping diamicton units of the Wedron Group (Figure 12-6). Bedrock surface data are from Herzog et al. (1994); Quaternary geology is from interpretation of geologic records at the Illinois State Geological Survey. Modified from Hansel et al. (1999).

Quaternary Period

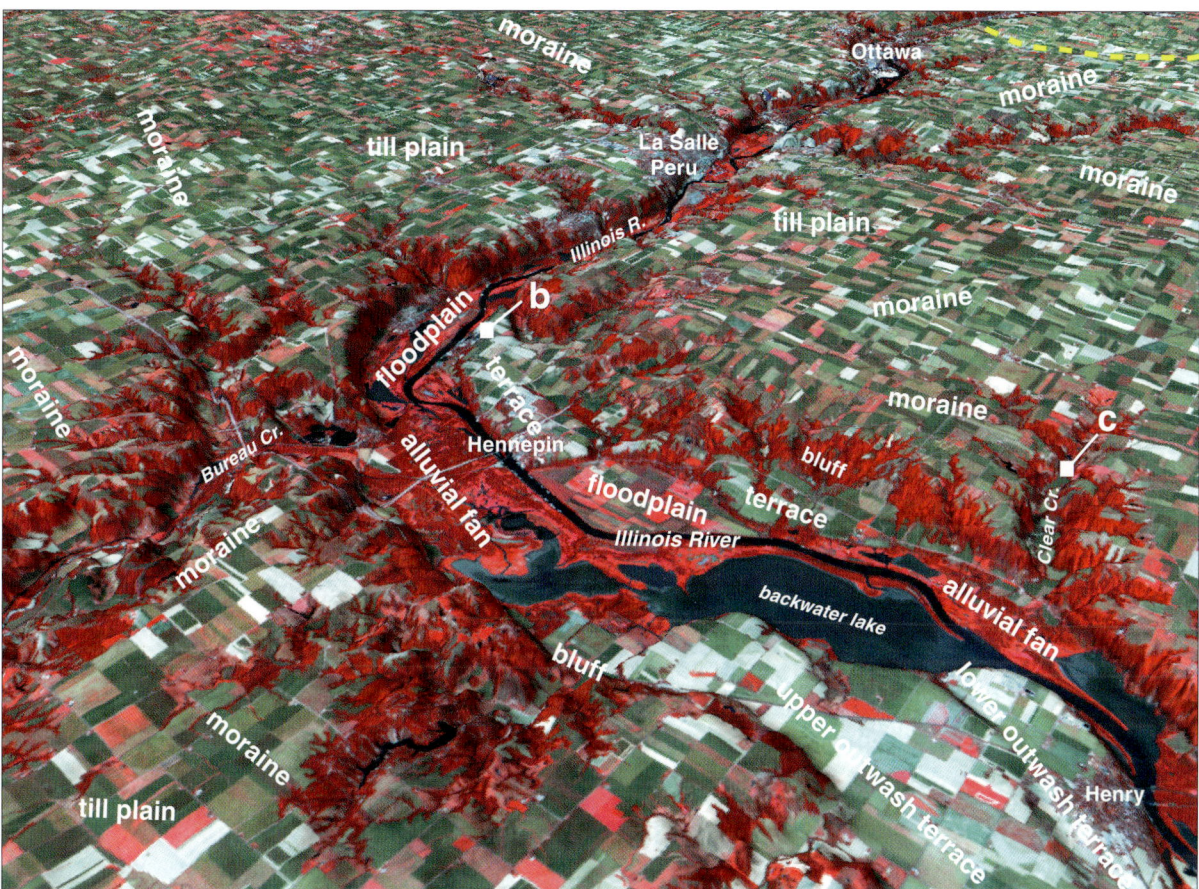

Figure 12-19 (a) False-color infrared satellite imagery has been draped on a digital representation of surface topography to create a three-dimensional image of the deeply incised Illinois River valley in Bureau, Putnam, La Salle, and Marshall Counties in north-central Illinois. Red colors indicate vegetation (mostly trees). Steep tree-covered slopes along the valley and rectangular field patterns are apparent as are a variety of landforms. This landscape was formed largely during the Wisconsin Episode of glaciation. Slightly more than 20,000 years ago, the Wisconsin glacier reached its westernmost limit west of the present Illinois River, and its margin began to retreat toward Lake Michigan as it melted. Moraines on the upland mark former stillstands of the glacier front during that retreat. (b) Meltwater from the glacier incised the Illinois River valley into drift and rock and deposited thick, coarse outwash. Periodically, large floods made major changes in the valley, carving channels and terraces into these deposits, and the floods continued to impact the valley even after the glacier retreated into the Lake Michigan Basin about 14,000 years ago. A series of particularly large floods around 16,000 years ago deepened and widened the valley and carved the main outwash deposits into terraces that are the major landforms on the valley floor today, Putnam County. The modern Illinois River has inherited a channel that lies hundreds of feet below the surrounding upland till plain and is more than 100 feet (30.5 m) below the highest terraces in these deposits. Tributaries to the modern river continue to erode into thick drift deposits in Putnam County (c) that underlie the till plain, and the river is constricted in places by alluvial fans deposited at the mouths of these tributaries. Sand and gravel deposits (b) that underlie the terraces in the valley are used widely as an important source of groundwater and are mined extensively for aggregate. (False-color image by Donald E. Luman; photographs by E. Donald McKay.)

in transverse cross section, steeper in the down-ice direction, and more gentle in the up-ice direction, where they grade into till or lake plain (Figures 12-18c and 12-14b). In map view, some of the end moraines are broadly arc-shaped, whereas others are lobate (tightly curved). Hansel and Johnson (1992) interpreted six of the broadly arc-shaped moraines to mark positions of advance or readvance of the order of tens of miles (Figures 12-18b, c); most of the others were attributed to minor readvances or recessional ice-marginal stillstands. Overall, except for the Valparaiso and Marseilles morainic complexes in northeastern Illinois, the landscape of the Wisconsin Episode lacks many of the features that are typical of a glaciated landscape (Johnson and Hansel 1999, Hansel et al. 1999). Outwash plains are generally absent or are poorly developed; where present, they are generally associated with advance and readvance end moraines as opposed to ice-marginal stillstands. Kettles and hummocky topography are generally lacking, even in the end moraines. Glacial landforms that are common farther north in Wisconsin and Michigan, such as kames, ice-walled lake plains, eskers, and tunnel valleys, are uncommon, and drumlins are absent.

Sediment Record

The Wisconsin Episode glacial succession in Illinois consists of a series of offlapping drift sheets. Except in the Valparaiso and Woodstock Moraine areas of northeastern Illinois, the drift sheets are composed predominantly of till that is uniform in appearance, physical properties, and composition. Reworked ice-marginal sediments also are present in places. In cross section, the drift sheets are shingled, and they pinch out beneath successively younger drift sheets in the up-ice direction toward the Lake Michigan basin (Figure 12-18c). Locally, tongues of proglacial sediment (outwash sand and gravel and/or lake clay and silt) separate the till units of different drift sheets and, less commonly, till beds of the same unit (Figures 12-4 and 12-18).

Proglacial sediment also occurs beyond the Wisconsin Episode glacial limit. Sand and gravel (assigned to the Henry Formation) accumulated in outwash plains proximal to the Marengo, Bloomington, and Shelbyville Moraines and as valley train within the Illinois Valley in Mason County; the ancient Mississippi Valley in Bureau, Lee, Whiteside, and Henry Counties; and the Wabash Valley along the southeastern border of Illinois (Figure 12-12). Sand dunes on many of these outwash deposits (Figure 12-20) and on outwash deposits in Kankakee and Iroquois Counties record prevailing westerly winds during the late Wisconsin Episode and early postglacial (Hudson) episodes. Fine-grained proglacial sediment (assigned to the Equality Formation) accumulated in lakes that formed (1) between older moraines and an ice margin; (2) in low areas between moraines, which were flooded periodically by glacial meltwater; and (3) in tributary valleys dammed by aggradation in major meltwater channels, particularly the Wabash Valley (Figure 12-12).

With the exception of the Haeger Member of the Lemont Formation (Figure 12-6), the Wisconsin Episode tills are relatively fine grained. Silt and clay in the matrix each generally range from 30 to 50%, and sand is 10 to 40% (Johnson 1976, Johnson and Hansel 1999). Illite is the dominant clay mineral (generally 65 to 80% in unaltered samples); it was derived predominantly from Paleozoic shale from the Lake Michigan basin and northern Illinois (Willman and Frye 1970, Glass and Killey 1987). Gravel clast lithologies also are dominated by Paleozoic carbonate and shale, and locally the till is enriched with older Quaternary deposits, most notably organic-rich material that was eroded and incorporated as the glacier advanced.

Wisconsin Episode Lake Michigan lobe diamictons in Illinois are assigned to the Wedron Group (Figurer 12-6). The Wedron Group includes the reddish gray to grayish brown clay loam to loam Tiskilwa Formation diamicton; the Lemont Formation, which contains the gray to grayish brown silt loam Batestown Member diamicton; the gray silty clay to silty clay loam Yorkville Member diamicton; the light gray to gray sandy loam Haeger Member diamicton; and the gray silty clay to silty clay loam Wadsworth Formation diamicton (Figure 12-6). In the Chicago area, Wadsworth Formation diamicton overlies gravelly silt loam diamicton of the undivided Lemont Formation. Locally, the Tiskilwa Formation contains lower and upper members (Delavan and Piatt Members, respectively) that are grayer and less clayey than type Tiskilwa diamicton, the red hue of which generally has been attributed to a Lake Superior source area (Wickham and Johnson 1981). Tongues of proglacial sorted sediments regionally occur between the diamicton units of the Wedron Group (Figures 12-4, 12-6, 12-18c); some are named and are important aquifers (e.g., see Hansel et al. 2004). Huron-Erie lobe diamicton (Trafalgar Formation) occurs in Illinois in the Iroquois Moraine area (Figure 12-18a). The Trafalgar Formation is gray, loam to silt loam diamicton, and, unlike the Lake Michigan lobe diamicton units, contains more garnet than epidote in the fine sand fraction, a characteristic attributed to an eastern source area, indicating a Labrador ice-dispersal center in the Grenville Province of the Canadian Shield (not shown) (Johnson et al. 1986).

Glacial History

Illinois appears to have been free of ice during the early part of the Athens Subepisode (Figure 12-6) of the Wis-

QUATERNARY PERIOD

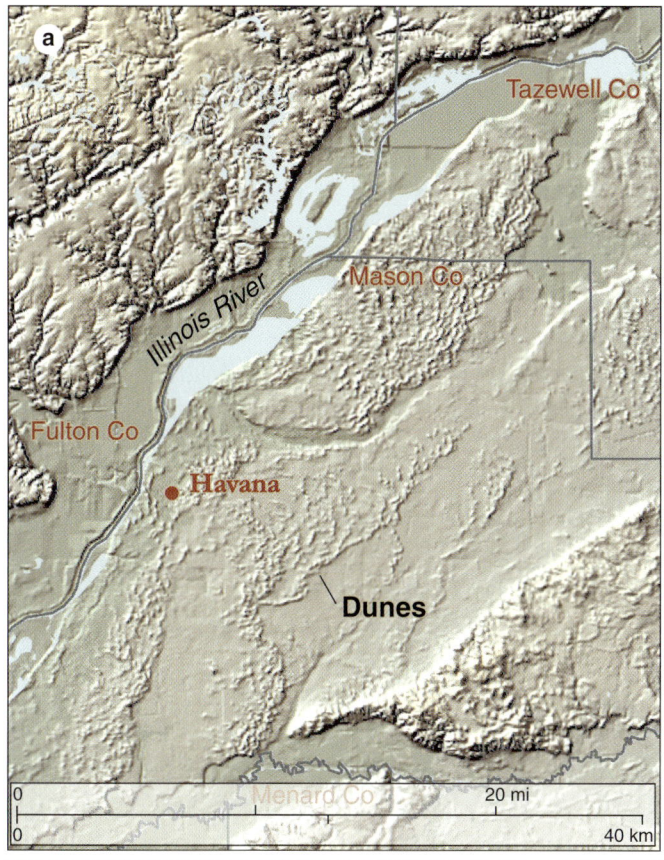

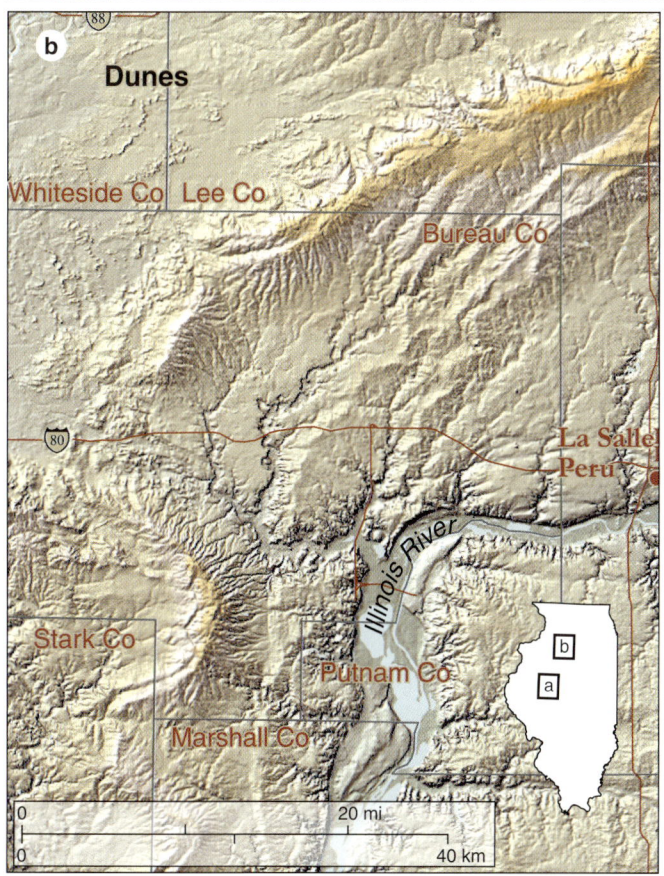

Figure 12-20 (a) Sand dunes on Wisconsin Episode outwash in the Illinois River valley near Havana; (b) Big Bend area of the Illinois River valley (Luman et al. 2003).

consin Episode (Hansel and Johnson 1996, Johnson et al. 1997, Stiff and Hansel 2004). Tills of the Winnebago Formation, formerly attributed to this time period (Frye et al. 1968, Willman and Frye 1970), now are interpreted to be older than the last interglacial soil, the Sangamon Geosol, because remnants of the soil have been found developed in them (Berg et al. 1985, Curry 1989). Therefore, Winnebago Formation tills are now interpreted to be Illinois Episode deposits (Berg et al. 1985, Hansel and Johnson 1996) (Figure 12-6). The Athens Subepisode, between about 55,000 and 28,000 radiocarbon years BP, was characterized by cool climate conditions and loess deposition (Roxana Silt, Figure 12-6). Based on its distribution and mineralogic composition, the immediate source of the loess is clearly linked to the ancient Mississippi River floodplain (Smith 1942; McKay 1977, 1979; Johnson and Follmer 1989; Follmer 1996). A definite upper Mississippi basin glacial source for the sediments from which the Roxana loess was derived has not been identified.

Deposition of the Roxana loess was followed by a cool interval (the Farmdale Phase, beginning about 28,000 radiocarbon years BP) during which a distinctive, organic-rich soil (the Farmdale Geosol, Figure 12-6) developed in Illinois. Development of this soil, which is best known for its organic horizon, was short-lived. The soil and the coniferous trees that grew in it were buried between about 26,000 and 20,000 radiocarbon years BP, generally by proglacial sediment (such as loess, outwash, or lake sediment) of the advancing Lake Michigan glacial lobe.

During the later part of the Wisconsin Episode (the Michigan Subepisode) in Illinois, the Lake Michigan lobe ice advanced out of the lake basin. Radiocarbon ages from wood from the Farmdale Geosol date the beginning of this advance at about 26,000 radiocarbon years BP. Initially, during the Marengo Phase, when the north-south–trending Marengo Moraine formed, the lobe margin likely was parallel to the Lake Michigan basin (Figure 12-18a, b). Later, interference by Huron-Erie lobe ice to the east resulted in a "westward bulge" of the lobe during the Shelby, Putnam, and Livingston Phases (Wickham et al. 1988). The lobe margin reached the last glacial maximum position during the Shelby Phase about 20,000 radiocarbon years BP and had melted back from Illinois by about 13,500 radiocarbon years BP (Figure 12-18a, b).

The advance of the lobe margin into the valley of the ancient Mississippi River between Princeton and Peoria during the Shelby Phase resulted in the formation of glacial Lake Milan in the section of the valley between Rock Island and Princeton (Anderson 1968) (Figure 12-7). This glacial lake, fed by upper Mississippi drainage and meltwater from the Lake Michigan lobe, soon spilled over a former drainage divide (the Andalusia sill), creating a new channel about 20,350 radiocarbon years BP (Curry 1998) that diverted the Mississippi River to its present-day course along the western margin of Illinois.

The advance of Lake Michigan glacial lobe ice into the valleys of the ancient Mississippi River and its tributaries (including the glacial Wabash River, which drained to the ancient Mississippi River via the Ohio Valley) resulted in widespread sedimentation in the major meltwater valleys (the Mississippi, Illinois, and Wabash Valleys) and initiation of the largest loess-producing event on the North American continent (Frye et al. 1965). This loess, the Peoria Silt, is distinct from the older Roxana Silt. The Roxana loess is generally reddish brown and more leached (except where thick); the Peoria loess is very pale brown to light gray and is calcareous. Deposition of the Peoria loess in Illinois began about 25,000 radiocarbon years BP and ceased by about 12,500 radiocarbon years BP (McKay 1977).

The coldest part of the Michigan Subepisode in Illinois was between about 21,000 and 16,000 radiocarbon years BP, when a narrow zone of permafrost developed along the margin of the Lake Michigan ice lobe (Johnson 1990). Johnson (1990) suggested that the duration of permafrost formation was likely limited to less than 1,000 years and that the zone likely migrated during ice-margin fluctuations. On the basis of fossil ice-wedge casts and patterned ground, Johnson (1990) concluded that the permafrost was more discontinuous southward and extended as far south as 38° 30' N latitude. The paucity of radiocarbon ages for wood between 19,500 and 15,500 radiocarbon years BP is consistent with an interpretation of full-glacial climate and permafrost on a landscape with limited woody vegetation (King 1979; Baker et al. 1989; Hansel and Johnson 1992, 1996).

Readvances on the order of 50 miles (80.5 km) or more occurred during the Putnam and Livingston Phases (Figure 12-18a, b). Those readvances resulted in offlapping glacigenic sequences (Figure 12-18c) that generally are marked by a distinct change in till lithology. The lithologic changes are attributed to a combination of source-area changes caused by ice-margin fluctuations and to source-material change due to subglacial erosion. In some key sections, for example, Wedron Quarry (Johnson and Hansel 1990) (Figure 12-4), Tuscola Quarry (Hansel et al. 1999) (Figure 12-5), and Clear Creek (Patterson et al. 2003), unconformities separate lithologically distinct tills in the same glacigenic sequence. Those unconformities are interpreted to be subglacial erosion surfaces that developed between times of till deposition during ice margin advance and during ice margin retreat. The abrupt differences in till lithology reflect changes in the lithology of debris in the ice during a single glacial advance and retreat event (Johnson and Hansel 1999).

Proglacial lakes formed in low areas between the ice margin and older end moraines as the ice melted back from the last glacial maximum position. The lakes eventually overtopped low spots in the moraines to form new drainage patterns linked to major meltwater channels (Figure 12-4). At the end of the Livingston Phase, the Lake Michigan lobe margin retreated into the present lake basin at Chicago and Milwaukee for the first time during the deglaciation (Mickelson et al. 1983, Hansel and Johnson 1992). This meltback is recorded by proglacial fluvial and lacustrine sediments between the clayey Yorkville till of the Livingston Phase and sandy Haeger till and laterally equivalent silty Lemont till of the Woodstock Phase (Hansel and Johnson 1992, 1996). The meltback likely corresponds to a regional ice recession in the eastern and northern Great Lakes region about 16,000 radiocarbon years BP, referred to by Karrow et al. (2000) as the Erie Phase.

During the Woodstock Phase, the Lake Michigan lobe ice margin readvanced about 30 miles (48 km) to the Woodstock and Valparaiso Moraines in Illinois, to the Darien Moraine in Wisconsin, and to a position near the Valparaiso Moraine in Indiana. Regionally, ice lobes (the Green Bay, Lake Michigan, Saginaw, and Huron-Erie lobes) apparently advanced out of the Great Lakes basins at this time to abut adjacent lobes, resulting in compressive flow and stacking of debris-rich ice (Mickelson et al. 1983, Johnson et al. 1986, Hansel and Johnson 1992). The end moraines that formed during the Woodstock Phase in Illinois, and also in Indiana, Wisconsin, and Michigan, are characterized by uneven, or hummocky, topography and ice-contact drift (Mickelson et al. 1983). Thick proglacial fluvial sediment underlies and is inset into the sandy till of the Woodstock Phase.

Terraces in the Fox, Des Plaines, Kankakee, and Illinois Valleys are consistent with the interpreted release of tremendous volumes of meltwater during the Woodstock Phase, about 15,500 radiocarbon years BP (Figure 12-21). Those floods may relate, in part, to the catastrophic drainage of subglacial lakes that possibly formed in the Great Lakes basins. On the basis of broad, scoured channels and streamlined bars in the Kankakee and Fox River valleys, those flood events have been referred to as the Kankakee and Fox River torrents, respectively (Ekblaw and Athy 1925, Willman and Payne 1942).

At the close of the Woodstock Phase, the margin of the Lake Michigan ice lobe again retreated into the lake basin at Chicago and Milwaukee before readvancing about 30 miles (48 km) to the Valparaiso Moraine complex during the Crown Point Phase (about 14,500 radiocarbon years ago). In Illinois and Indiana, ice of that advance overrode the position of the earlier Woodstock ice margin, except in

Figure 12-21 Starved Rock, Starved Rock State Park, La Salle County, was carved by the same forces that sculpted the Illinois River valley: massive glacial meltwater torrents from ancestral Lake Michigan and the Great Lakes. These erosional events are recorded in inclined benches that are visible on the rock of St. Peter Sandstone. The upper erosional bench was formed by glacial meltwater torrents, and the lower bench was carved by later meltwater events, possibly floodwaters coming through the Chicago Outlet. (Photograph by Joel M. Dexter.)

the area of the Haeger Member till in McHenry County, Illinois. In the Valparaiso Moraine complex, the ice deposited thick clayey till and ice-marginal reworked diamicton over the sandy and silty till of the Woodstock Phase. The Tinley and Lake Border Moraines represent minor readvances and/or stillstands during the Crown Point Phase (Figure 12-18a, b).

About 14,000 radiocarbon years ago, glacial Lake Chicago formed as the glacial margin retreated to the Chicago area and into the lake basin. The lake drained via low spots in the Valparaiso and Tinley Moraines southwest of Chicago (Chicago Outlet) to the lower Des Plaines and Illinois Valleys. When the Lake Michigan ice lobe retreated north of isostatically depressed lower outlets at the north end of the basin, the lake level fell in the Lake Michigan basin, and the Chicago Outlet was abandoned.

Based on the sediment record and landforms, the Wisconsin Episode glaciation in Illinois was characterized by wet-based, fast-moving ice that deposited thick sheets of fairly uniform till in a relatively short time period (Willman and Frye 1970; Mickelson et al. 1981, 1983; Begét 1986; Johnson and Hansel 1990, 1999; Hansel and Johnson 1992; Clark 1992, 1994; Mickelson and Colgan 2004). Sedimentological studies of key stratigraphic sections (Johnson and Hansel 1990, Hansel and Johnson 1992, Hansel et al. 1999, Patterson et al. 2003) indicate that net deglaciation from the last glacial maximum was characterized by a fluctuating active ice margin with multiple ice-margin meltbacks and readvances (during the Putnam, Livingston, Woodstock, and Crown Point Phases) on the order of 30 to 50 miles (48 to 80.5 km) (Figure 12-18). After the glacier receded from Illinois (about 13,500 radiocarbon years ago), the Lake Michigan lobe ice margin continued to fluctuate farther north on an even larger scale during the Port Huron and Two Rivers Phases. Readvances were on the order of tens to hundreds of miles (Hansel et al. 1985, Hansel and Johnson 1992). By about 12,500 radiocarbon years ago, loess deposition had ceased in Illinois, and the landscape stabilized, allowing vegetation and soils to develop.

At the end of the last glacial episode, climate and vegetation changes were accompanied by the extinction of many large mammals. Fossils of many of these mammals—including the American mastodon, woolly mammoth, giant beaver, Jefferson's ground sloth, and stag moose—have been found in Illinois (Figure 12-22). Sometime during the last 25,000 years, humans entered and settled Illinois. They are thought by some (e.g., Styles 2004) to have played a role in the extinction of some of the large mammals.

POSTGLACIAL HUDSON EPISODE

The present interglacial episode, also known as the postglacial Hudson Episode (Hansel and Johnson 1996, Johnson et al. 1997) in Illinois, is characterized by a warm stable climate and development of the modern soil. Deposits of this most recent time interval are the alluvium that is accumulating in stream valleys (the Cahokia Formation), dune and beach sand (the Henry Formation), fine-grained lake sediment (the Equality Formation), peat (the Grayslake Peat), and colluvium (the Peyton Formation). At various times during the late glacial and postglacial episodes (periodically prior to about 11,000 radiocarbon years BP and between about 6,000 and 4,000 radiocarbon years BP), ancestral Lake Michigan drained via the Des Plaines Valley and Cal-Sag Channel through the Valparaiso and Tinley Moraines southwest of Chicago to the Illinois Valley (Hansel et al. 1985). Some erosion and sedimentation in the outlet channels and in the Des Plaines and Illinois Valleys likely date to these times of higher lake levels in the Great Lakes, although the discharge probably was less than that supplied by the late glacial Kankakee and Fox River torrent events (Hajic 1989).

CONCLUSION

Quaternary research is very active in Illinois, and much remains to be learned. Exciting new technologies that promise to extract more information from the geological record are also becoming available. Detailed isotopic analyses of plant and animal fossils, sediments, and cave formations offer high-resolution measures that document

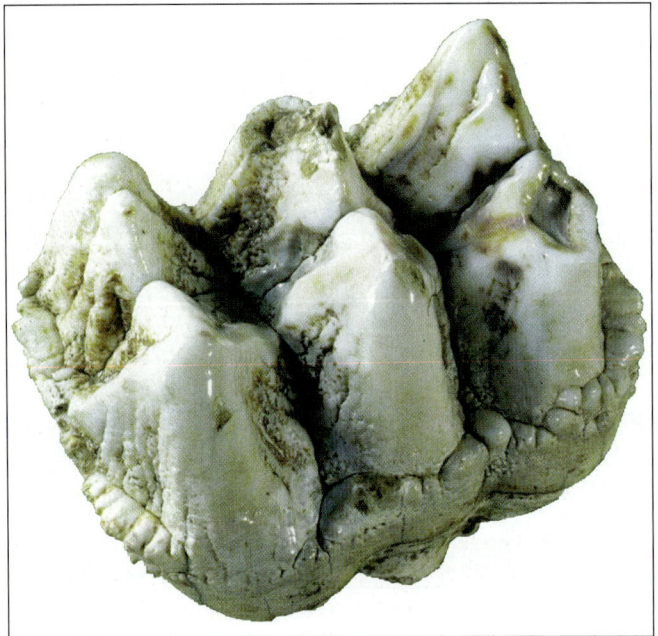

Figure 12-22 Mastodon molar, *Mammut americanum,* Urbana, Champaign County, Equality Formation. (Photograph by Dennis R. Kolata and Rodney D. Norby 2004).

dramatic changes in past climate. Luminescence, radiometric, and cosmogenic dating of materials older than achievable with radiocarbon are providing ages for previously undatable materials and yielding a better understanding of the ages and correlation of those older deposits. Advanced computing is allowing the construction and visualization of detailed three-dimensional models of glacial deposits.

Quaternary deposits in the state are important societal resources. They underlie the foundations of homes, industries, and transportation infrastructure. They are the materials shaken or liquefied by earthquakes, and they contain much of the groundwater that Illinois residents require. Geological mapping at large scale (1:24,000 or 1 inch on the map represents 2,000 feet on the ground) is under way and requires detailed field examination of small areas. Such mapping is leading to discovery of new exposures of glacial deposits, delineation of aquifers for water supply, and maps of some deposits that can be mined for aggregate and others that can safely host waste disposal sites. Tested methods, such as drilling and seismic profiling, are being applied more systematically and widely to glacial deposits. Continuous cores, longer than 300 feet (91 m) in many cases, are giving geologists their first looks at some buried sediments and ancient soils. Seismic reflection profiling is facilitating better correlations of those materials between boreholes and, thereby, more accurate three-dimensional models.

The Quaternary Period was a time of dramatic climate change, and, as awareness of the importance of climate change to human society rises, the relevance of the recent geological past should also become more apparent. Detailed studies of the natural climate signature recorded in Quaternary deposits are providing a glimpse of climate change that can be extended back thousands of years before instrumental records. These geological climate records, which are being studied in greater detail, may contain significant lessons about the impact of past climates on Illinois landscape, soils, water bodies, and life. Stay tuned: the history of the Quaternary in Illinois is still being written.

References

Anderson, R. C., 1968, Drainage evolution in the Rock Island area, western Illinois and eastern Iowa, in R. E. Bergstrom, ed., The Quaternary of Illinois—A Symposium in observance of the centennial of the University of Illinois: Urbana-Champaign, University of Illinois, College of Agriculture, Special Publication 14, p. 11–18.

Baker, R. G., A. E. Sullivan, G. R. Hallberg, and D. G . Horton, 1989, Vegetational changes in western Illinois during the onset of late Wisconsinan glaciation: Ecology, v. 70, no. 5, p. 1363–1376.

Begét, J. E., 1986, Modeling the influence of till rheology on the flow and profile of the Lake Michigan lobe, southern Laurentide Ice Sheet, U.S.A.: Journal of Glaciology, v. 32, no. 111, p. 235–241.

Berg, R. C., J. P. Kempton, L. R. Follmer, and D. P. McKenna, 1985, Illinoian and Wisconsinan stratigraphy and environments in northern Illinois: The Altonian revised: Illinois State Geological Survey, Guidebook 19, 177 p.

Berg, R. C., J. P. Kempton, and A. N. Stecyk, 1984, Geology for planning in Boone and Winnebago Counties: Illinois State Geological Survey, Circular 531, 69 p.

Berg, R. C., E. D. McKay III, D. A. Keefer, R. A. Bauer, P. D. Johnstone, B. J. Stiff, A. Pugin, C. P. Weibel, T. H. Larson, W.-J. Su, and G. T. Homrighous, 2002, Three-dimensional geologic mapping for transportation planning in central-northern Illinois—Data selection, map construction, and model development, in L. H. Thorliefson and R. C. Berg, eds., Three-dimensional geological mapping for groundwater applications: Ottawa, Ontario, Canada, Geological Survey of Canada, Open File 1449, p. 13–17.

Bleuer, N. K., 1976, Remnant magnetism of Pleistocene sediments of Indiana: Indiana Academy of Science Proceedings, v. 85, p. 277–294.

Bleuer N. K., 1980, Correlation of pre-Wisconsinan tills of the Lake Michigan lobe and Huron-Erie lobe through the Teays Valley fill (abs.): Boulder, Colorado, Geological Society of America, Abstracts with Programs, v. 12, no. 5, p. 219.

Boellstorff, J. D., 1978, Proposed abandonment of pre-Illinoian Pleistocene stage terms (abs.): Boulder, Colorado, Geological Society of America, Abstracts with Programs, v. 10, no. 6, p. 247.

Bowen, D. Q., 1978, Quaternary geology: Oxford, Pergamon Press, 221 p.

Calvin, S., 1897, Synopsis of the drift deposits of Iowa: American Geologist, v. 19, no. 4, p. 270–272.

Chamberlin, T. C., 1883, Geology of Wisconsin: Madison, Wisconsin, Wisconsin Geological and Natural History Survey, v. 1, p. 1–300.

Chamberlin, T. C., 1894, Glacial phenomena of North America, in J. Geikie, The Great Ice Age and its relation to the antiquity of man (3rd ed.): London, Edward Stanford, p. 724–774.

Clark, P. U., 1992, Surface form of the southern Laurentide Ice Sheet and its implications to ice-sheet dynamics: Boulder, Colorado, Geological Society of America Bulletin, v. 104, no. 5, p. 595–605.

Clark, P. U., 1994, Unstable behavior of the Laurentide Ice Sheet over deforming sediment and its implications for climate change: Quaternary Research, v. 41, no. 1, p. 19–25.

Curry, B. B., 1989, Absence of Altonian glaciation in Illinois: Quaternary Research, v. 31, no. 1, p. 1–13.

Curry, B. B., 1998, Evidence at Lomax, Illinois, for mid-Wisconsin (~40,000 yr B.P.) position of the Des Moines Lobe and for diversion of the Mississippi River by the Lake Michigan Lobe (20,350 yr B.P.): Quaternary Research, v. 50, no. 2, p. 128–138.

Curry, B. B., and R. G. Baker, 2000, Palaeohydrology, vegetation, and climate since the late Illinois Episode (~130 ka) in south-central Illinois: Palaeogeography, Palaeoclimatology, and Palaeoecology, v. 155, no. 1–2, p. 59–81.

Curry, B. B., and L. R. Follmer, 1992, The last interglacial-glacial transition in Illinois: 123–25 ka, in P. U. Clark and P. D. Lea, eds., The last interglacial-glacial transition in North America: Boulder, Colorado, Geological Society of America, Special Paper 270, p. 71–88.

Curry, B. B., D. A. Grimley, and J. A. Stravers, 1999, Quaternary geology, geomorphology, and climatic history of Kane County, Illinois: Illinois State Geological Survey, Guidebook 28, 40 p.

Ehlers, J., 1996, Quaternary and glacial geology: New York, John Wiley and Sons, 578 p.

Ekblaw, G. E., 1960, Glacial map of northeastern Illinois (rev. ed.): Illinois State Geological Survey, 1:1,650,000.

Ekblaw, G. E., and L. F. Athy, 1925, Glacial Kankakee torrent in northeastern Illinois: Boulder, Colorado, Geological Society of America,

Bulletin, v. 36, p. 417–428.

Fleeger, G. M., 1980, Pre-Wisconsinan till stratigraphy in the Avon, Canton, Galesburg, and Maquon 15-minute Quadrangles, western Illinois: Urbana-Champaign, Illinois, University of Illinois, M. S. thesis, 90 p.

Follmer, L. R., 1979, A historical review of the Sangamon soil, in L. R. Follmer, E. D. McKay, J. A. Lineback, and D. L. Gross, eds., Wisconsinan, Sangamonian, and Illinoian stratigraphy in central Illinois: Illinois State Geological Survey, Guidebook 13, p. 79–91.

Follmer, L. R., 1983, Sangamon and Wisconsinan pedogenesis in the midwestern United States, in S. C. Porter, ed., The Late Pleistocene: Late-Quaternary environments of the United States, v. 1: Minneapolis, Minnesota, University of Minnesota Press, p. 138–144.

Follmer, L. R., 1996, Loess studies in central United States—Evolution of concepts: Engineering Geology, v. 45, no. 1–4, p. 287–304.

Follmer, L. R., and J. P. Kempton, 1985, A review of the Esmond Till Member, in R. C. Berg, J. P. Kempton, L. R. Follmer, and D. P. McKenna, eds., Illinoian and Wisconsinan stratigraphy and environments in northern Illinois: The Altonian revised: Illinois State Geological Survey, Guidebook 19, p. 139–155.

Follmer, L. R., E. D. McKay, J. A. Lineback, and D. L. Gross, 1979, Wisconsinan, Sangamonian, and Illinoian stratigraphy in central Illinois: Illinois State Geological Survey, Guidebook 13, 139 p.

Forman, S. L., and J. Pierson, 2002, Late Pleistocene luminescence chronology of loess deposition in the Missouri and Mississippi River valleys, United States: Palaeogeography, Palaeoclimatology, Palaeoecology, v. 186, no. 1–2, p. 25–46.

Frankie, W. T., J. J. Miner, S. E. Benton, G. E. Pociask, E. T. Plankell, A. J. Stumpf, and R. J. Jacobson, 2007, Guide to the geology of Moraine Hills, Glacial Park, and Volo Bog areas, McHenry and Lake Counties, Illinois: Illinois State Geological Survey, Geological Science Field Trip Guidebook 2007A, 52 p.

Frye, J. C., 1967, Geological information for managing the environment: Illinois State Geological Survey, Environmental Geology Notes 18, 12 p.

Frye, J. C., and H. B. Willman, 1960, Classification of the Wisconsinan Stage in the Lake Michigan glacial lobe: Illinois State Geological Survey, Circular 285, 16 p.

Frye, J. C., H. B. Willman, and R. F. Black, 1965, Outline of glacial geology of Illinois and Wisconsin, in H. E. Wright Jr. and D. G. Frey, eds., The Quaternary of the United States: Princeton, New Jersey, Princeton University Press, p. 43–61.

Frye, J. C., H. B. Willman, M. Rubin, and R. F. Black, 1968, Definition of Wisconsinan Stage: Washington D.C., U.S. Geological Survey, Bulletin 1274-E, p. E1–E22.

Fulton, R. J., 1989, Foreword, in R. J. Fulton, ed., Quaternary geology of Canada and Greenland: Ottawa, Ontario, Canada, Geological Survey of Canada, Geology of Canada, no. 1, p. 1–11. (Also Geological Society of America, The Geology of North America, v. K-1, p. 1–11.)

Geological Survey of Canada, 1970, Geological map of Canada, in R. J. W. Douglas, ed., Geology and economic minerals of Canada: Ottawa, Ontario, Canada, Economic Geology Report 1, Map 1250A, 1:500,000.

Glass, H. D., and M. M. Killey, 1987, Principles and applications of clay mineral composition in Quaternary stratigraphy—Examples from Illinois, USA, in J. J. M. van der Meer, ed., Tills and glaciotectonics: Rotterdam, The Netherlands, A. A. Balkema, p. 117–125.

Grimley, D. A., L. R. Follmer, R. E. Hughes, and P. A. Solheid, 2003, Modern, Sangamon and Yarmouth soil development in loess of unglaciated southwestern Illinois: Quaternary Science Reviews, v. 22, no. 2–4, p. 225–244.

Grüger, E., 1972, Late Quaternary vegetation development in south-central Illinois: Quaternary Research, v. 2, no. 2, p. 217–231.

Hackett, J. E., and M. R. McComas, 1969, Geology for planning in McHenry County: Illinois State Geological Survey, Circular 438, 31 p.

Hajic, E. R., 1986, Pre-Wisconsinan loesses and paleosols at Pancake Hollow, west-central Illinois, in R. W. Graham, B. W. Styles, J. J. Saunders, M. D. Wiant, E. D. McKay, T. R. Styles, and E. J. Hajic, eds., Quaternary records of southwestern Illinois and adjacent Missouri: Illinois State Geological Survey, Guidebook 23, p. 91–98.

Hajic, E. R., 1989, Late Pleistocene and Holocene landscape evolution, depositional subsystems, and stratigraphy in the lower Illinois River valley and adjacent central Mississippi River valley: Urbana-Champaign, Illinois, University of Illinois, Ph.D. dissertation, 342 p.

Hallberg, G. R., 1980, Pleistocene stratigraphy in east-central Iowa: Iowa City, Iowa, Iowa Geological Survey, Technical Information Series, no. 10, 168 p.

Hallberg, G. R., 1986, Pre-Wisconsinan glacial stratigraphy of the central plains region in Iowa, Nebraska, Kansas, and Missouri: Quaternary Science Reviews, v. 5, p. 11–15.

Hallberg, G. R., and R. G. Baker, 1980, Reevaluation of the Yarmouth type area, in G. R. Hallberg, ed., Illinoian and Pre-Illinoian stratigraphy of southeast Iowa and adjacent Illinois: Iowa City, Iowa, Iowa Geological Survey, Technical Information Series, no. 11, p. 111–150.

Hallberg, G. R., and J. D. Boellstorff, 1978, Stratigraphic "confusion" in the region of the type areas of Kansan and Nebraskan deposits (abs.): Boulder, Colorado, Geological Society of America, Abstracts with Programs, v. 10, no. 6, p. 255–256.

Hallberg, G. R., N. C. Wollenhaupt, and J. T. Wickham, 1980, Pre-Wisconsinan stratigraphy in southeast Iowa, in G. R. Hallberg, ed., Illinoian and Pre-Illinoian stratigraphy of southeast Iowa and adjacent Illinois: Iowa City, Iowa, Iowa Geological Survey, Technical Information Series, no. 11, p. 1–110.

Hansel, A. K., R. C. Berg, A. C. Phillips, and V. Gutowski, 1999, Glacial sediments, landforms, paleosols, and 20,000-year-old forest bed in east-central Illinois: Illinois State Geologic Survey, Guidebook 26, 31 p.

Hansel, A. K., and W. H. Johnson, 1992, Fluctuations of the Lake Michigan lobe during the late Wisconsin subepisode, in A.-M. Robertsson, B. Ringberg, U. Miller, and L. Brunnberg, eds., Quaternary stratigraphy, glacial morphology and environmental changes: Uppsala, Sweden, Sveriges Geologiska Undersökning, p. 133–144. (Illinois State Geological Survey, Reprint Series 1993-F.)

Hansel, A. K., and W. H. Johnson, 1996, Wedron and Mason Groups: Lithostratigraphic reclassification of deposits of the Wisconsin Episode, Lake Michigan Lobe area: Illinois State Geological Survey, Bulletin 104, 116 p.

Hansel, A. K., D. M. Mickelson, A. F. Schneider, and C. E. Larsen, 1985, Late Wisconsinan and Holocene history of the Lake Michigan basin, in P. F. Karrow and P. E. Calkin, eds., Quaternary evolution of the Great Lakes: St. John's, Newfoundland, Canada, Geological Association of Canada, Special Paper 30, p. 39–53.

Hansel, A. K., B. J. Stiff, and M. L. Barnhardt, 2004, Three-dimensional geologic mapping in rapid-growth areas: A case study from Lake County, northeastern Illinois, in R. C. Berg, H. Russell, and L. H. Thorliefson, convenors, Three-dimensional geologic mapping for groundwater applications: Illinois State Geological Survey, Open File Series 2004-8, p. 23–27.

Hartline, L. E., 1981, Illinoian stratigraphy of the Bond County region of west central Illinois: Urbana-Champaign, Illinois, University of Illinois, M. S. thesis, 104 p.

Herzog, B. L., B. J. Stiff, C. A. Chenoweth, K. L. Warner, J. B. Sieverling, and C. Avery, 1994, Buried bedrock surface of Illinois (3rd ed.): Illinois State Geological Survey, Illinois Map 5, 1:500,000.

Herzog, B. L., S. D. Wilson, D. R. Larson, E. C. Smith, T. H. Larson, and M. L. Greenslate, 1995, Hydrogeology and groundwater availability in southwest McLean and southeast Tazewell Counties: Part 1, Aquifer characterization: Illinois State Geological Survey, Cooperative Groundwater Report 17, 70 p.

Horberg, C. L., 1945, A major buried valley in east-central Illinois and its regional relationships: Journal of Geology, v. 53, no. 5, p. 349–359.

Horberg, C. L., 1946, Preglacial erosion surfaces in Illinois: Journal of Geology, v. 54, no. 3, p. 179–192.

Horberg, C. L., 1950a, Bedrock topography of Illinois: Illinois State Geological Survey, Bulletin 73, 111 p.

Horberg, C. L., 1950b, Geology in L. Horberg, M. Suter, and T. E. Larson, Groundwater in the Peoria region: Illinois State Geological Survey, Bulletin 75, p. 13–49.

Horberg, C. L., 1953, Pleistocene deposits below the Wisconsin drift in northeastern Illinois: Illinois State Geological Survey, Report of Investigations 165, 61 p.

Horberg, C. L., 1956, Pleistocene deposits along the Mississippi Valley in central-western Illinois: Illinois State Geological Survey, Report of Investigations 192, 39 p.

Imbrie, J., J. D. Hays, D. G. Martinson, A. McIntyre, A. C. Mix, J. J. Morley, N. G. Pisias, W. L. Prell, and N. J. Shackleton, 1984, The orbital theory of Pleistocene climate: Support from a revised chronology of the marine $d^{18}O$ record, in A. L. Berger, ed., Milankovitch and climate, v. 1: Norwell, Massachusetts, Reidel Publishing Company, p. 269–305.

Jacobs, A. M., and J. A. Lineback, 1969, Glacial geology of the Vandalia, Illinois, region: Illinois State Geological Survey, Circular 442, 24 p.

Johnson, W. H., 1964, Stratigraphy and petrography of Illinoian and Kansan drift in central Illinois: Illinois State Geological Survey, Circular 378, 38 p.

Johnson, W. H., 1976, Quaternary stratigraphy in Illinois: Status and current problems, in W. C. Mahaney, ed., Quaternary stratigraphy of North America: Stroudsburg, Pennsylvania, Dowden, Hutchinson and Ross, p. 161–196.

Johnson, W. H., 1986, Stratigraphy and correlation of the glacial deposits of the Lake Michigan Lobe prior to 14 ka B.P.: Quaternary Science Reviews, v. 5, p. 17–22.

Johnson, W. H., 1990, Ice-wedge casts and relict patterned ground in central Illinois and their environmental significance: Quaternary Research, v. 33, no. 1, p. 51–72.

Johnson, W. H., and L. R. Follmer, 1989, Source and origin of the Roxana Silt and middle Wisconsinan midcontinent glacial activity: Quaternary Research, v. 31, no. 3, p. 319–331.

Johnson, W. H., L. R. Follmer, D. L. Gross, and A. M. Jacobs, 1972, Pleistocene stratigraphy of east-central Illinois: Illinois State Geological Survey, Guidebook 9, 97 p.

Johnson, W. H., D. L. Gross, and S. R. Moran, 1971, Till stratigraphy of the Danville region, east-central Illinois, in R. P. Goldthwait, ed., Till—A Symposium: Columbus, Ohio, Ohio State University Press, p. 184–216.

Johnson, W. H., and A. K. Hansel, 1990, Multiple Wisconsinan glacigenic sequences at Wedron, Illinois: Journal of Sedimentary Petrology, v. 60, no. 1, p. 26–41.

Johnson, W. H., and A. K. Hansel, 1999, Wisconsin Episode glacial landscape of central Illinois: A product of subglacial deformation processes?, in D. M. Mickelson and J. W. Attig, eds., Glacial processes past and present: Boulder, Colorado, Geological Society of America, Special Paper 337, p. 121–135.

Johnson, W. H., A. K. Hansel, E. A. Bettis III, P. F. Karrow, G. J. Larson, T. V. Lowell, and A. F. Schneider, 1997, Late Quaternary temporal and event classifications, Great Lakes region, North America: Quaternary Research, v. 47, no. 1, p. 1–12.

Johnson, W. H., A. K. Hansel, B. J. Socha, L. R. Follmer, and J. M. Masters, 1985, Depositional environments and correlation problems of the Wedron Formation (Wisconsinan) in northeastern Illinois: Illinois State Geological Survey, Guidebook 16, 91 p.

Johnson, W. H., D. W. Moore, and E. D. McKay III, 1986, Provenance of late Wisconsinan (Woodfordian) till and origin of the Decatur sublobe, east-central Illinois: Geological Society of America Bulletin, v. 97, no. 9, p. 1098–1105.

Karrow, P. F., A. Dreimanis, and P. J. Barnett, 2000, A proposed diachronic revision of late Quaternary time-stratigraphic classification in the eastern and northern Great Lakes area: Quaternary Research, v. 54, no. 1, p. 1–12.

Kempton, J. P., 1981, Three-dimensional geologic mapping for environmental studies in Illinois: Illinois State Geological Survey, Environmental Geology Notes 100, 43 p.

Kempton, J. P., R. C. Berg, and L. R. Follmer, 1985, Revision of the stratigraphy and nomenclature of glacial deposits in central northern Illinois, in R. C. Berg, J. P. Kempton, L. R. Follmer, and D. P. McKenna, eds., Illinoian and Wisconsinan stratigraphy and environments in northern Illinois: The Altonian revised: Midwest Friends of the Pleistocene 32nd Field Conference: Illinois State Geological Survey, Guidebook 19, p. 1–19.

Kempton, J. P., W. H. Johnson, P. C. Heigold, and K. Cartwright, 1991, Mahomet Bedrock Valley in east-central Illinois: Topography, glacial drift stratigraphy, and hydrogeology, in W. H. Melhorn and J. P. Kempton, eds., Geology and hydrogeology of the Teays-Mahomet Bedrock Valley system: Boulder, Colorado, Geological Society of America, Special Paper 258, p. 91–124.

Kettles, I. M., 1980, Till stratigraphy of the Vandalia-Effingham-Marshall region, east-central Illinois: Urbana-Champaign, Illinois, University of Illinois, M.S. thesis, 124 p.

Killey, M. M., 1996, Easternmost extent of western-source pre-Illinois sediments in Illinois (abs.): Geological Society of America, Abstracts with Programs, v. 28, no. 6, p. 49.

King, J. E., 1979, Pollen analysis of some Farmdalian and Woodfordian deposits, central Illinois, in L. R. Follmer, E. D. McKay, J. A. Lineback, and D. L. Gross, eds., Wisconsinan, Sangamonian, and Illinoian stratigraphy in central Illinois: Illinois State Geological Survey, Guidebook 13, p. 109–113.

Kolata, D. R., and R. D. Norby, 2004, Illinois fossils: Illinois State Geological Survey, poster.

Lasemi, Z., and R. C. Berg, 2001, Three-dimensional geological mapping: A pilot program for resource and environmental assessment in the Villa Grove Quadrangle, Douglas County, Illinois: Illinois State Geological Survey, Bulletin 106, 117 p.

Leighton, M. M., 1926, A notable type Pleistocene section: The Farm Creek exposure near Peoria, Illinois: Journal of Geology, v. 34, no. 2, p. 167–174.

Leighton, M. M., 1933, The naming of the subdivisions of the Wisconsin glacial age: Science, v. 77, no. 1989, p. 168.

Leighton, M. M., 1957, The Cary-Mankato-Valders problem: Journal of Geology, v. 65, no. 1, p. 108–111.

Leighton, M. M., and J. A. Brophy, 1961, Illinoian glaciation in Illinois: Journal of Geology, v. 69, no. 1, p. 1–31.

Leighton, M. M., and H. B. Willman, 1950, Loess formations of the Mississippi Valley: Journal of Geology, v. 58, no. 6, p. 599–623.

Leverett, F., 1898, The weathered zone (Sangamon) between the Iowan loess and Illinoian till sheet: Journal of Geology, v. 6, no. 2, p. 171–181.

Leverett, F., 1899, The Illinois glacial lobe: Washington, D.C., U.S. Geological Survey, Monograph 38, 817 p.

Leverett, F., 1909, Weathering and erosion as time measures: American Journal of Science, 4th ser., v. 27, p. 349–368.

Lineback, J. A., 1979a, Quaternary deposits of Illinois: Illinois State Geological Survey, 1:500,000.

Lineback, J. A., 1979b, The status of the Illinoian glacial stage, in L. R. Follmer, E. D. McKay, J. A. Lineback, and D. L. Gross, eds., Wisconsinan, Sangamonian, and Illinoian stratigraphy in central Illinois: Illinois State Geological Survey, Guidebook 13, p. 69–78.

Lineback, J. A., 1980, The Glasford Formation of western Illinois, in G. R. Hallberg, ed., Illinoian and Pre-Illinoian stratigraphy of southwest Iowa and adjacent Illinois: Iowa City, Iowa, Iowa Geological Survey, Technical Information Series, no. 11, p. 181–184.

Lineback, J. A., D. L. Gross, and R. P. Meyer, 1974, Glacial tills under Lake Michigan: Illinois State Geological Survey, Environmental Geology Notes 69, 48 p.

Luman, D. E., L. R. Smith, and C. C. Goldsmith, 2003, Illinois surface topography: Illinois State Geological Survey, Illinois Map 11, 1:500,000.

Markewich, H. W., D. A. Wysocki, M. J. Pavich, E. M. Rutledge, H. T. Millard, Jr., F. J. Rich, P. B. Maat, M. Rubin, and J. P. McGeehin, 1998, Paleopedology plus TL, ^{10}Be, and ^{14}C dating as tools in stratigraphic and paleoclimatic investigations, Mississippi River valley, U.S.A.: Quaternary International, v. 51/52, p. 143–167.

McKay, E. D. III, 1977, Stratigraphy and zonation of Wisconsinan loesses in southwestern Illinois: Urbana-Champaign, University of Illinois, Ph.D. dissertation, 242 p.

McKay, E. D., 1979, Stratigraphy of Wisconsinan and older loesses in southwestern Illinois, in J. D. Treworgy, E. D. McKay, and J. T. Wickham, eds., Geology of western Illinois, Tri-State Geological Field Conference 43: Illinois State Geological Survey, Guidebook 14, p. 37–67.

McKay, E. D., 1986, Illinoian and older loesses and tills at the Maryville section, in R. W. Graham, B. W. Styles, J. J. Saunders, M. D. Wiant, E. D. McKay, T. R. Styles, and E. R. Hajic, eds., Quaternary records of southwestern Illinois and adjacent Missouri: Illinois State Geological Survey, Guidebook 23, p. 21–30.

McKay, E. D. III, and R. C. Berg, 2008, Optical ages spanning two glacial-interglacial cycles from deposits of the ancient Mississippi River, north-central Illinois (abs.): North-Central Section, 42nd Annual Meeting, 24–25 April 2008: Geological Society of America, Abstracts with Programs, v. 40, no. 5, p. 78.

McKay, E. D. III, R. C. Berg, A. K. Hansel, A. J. Stumpf, and T. J. Kemmis, 2008, Quaternary deposits and history of the ancient Mississippi River valley, north-central Illinois: Illinois State Geological Survey, Guidebook 35, 106 p.

Menzies, J., ed., 1995, Past glacial environments—Sediments, forms and techniques: Glacial environments, v. 2: Oxford, United Kingdom, Butterworth-Heinemann Ltd., 598 p.

Mickelson, D. M., L. J. Acomb, and C. R. Bentley, 1981, Possible mechanism for rapid advance and retreat of the Lake Michigan lobe between 13,000 and 11,000 years BP (abs.): Annals of Glaciology, v. 2, p. 185–186.

Mickelson, D. M., L. Clayton, D. S. Fullerton, and H. W. Borns Jr., 1983, The Late Wisconsin glacial record of the Laurentide Ice Sheet in the United States, in S. C. Porter, ed., Late-Quaternary environments of the United States, v. 1, The Late Pleistocene: Minneapolis, Minnesota, University of Minnesota Press, p. 3–37.

Mickelson, D. M., and P. M. Colgan, 2004, The southern Laurentide Ice Sheet, in A. R. Gillespie, S. C. Porter, and B. F. Atwater, eds., The Quaternary Period in the United States: Developments in Quaternary Science, v. 1, New York, Elsevier, p. 1–16.

Miller, B. B., R. W. Graham, A. V. Morgan, N. G. Miller, W. D. McCoy, D. F. Palmer, A. J. Smith, and J. J. Pilny, 1994, A biota associated with Matuyama-age sediments in west-central Illinois: Quaternary Research, v. 41, no. 3, p. 350–365.

Miller, J. W., 2000, A magnetically reversed till in southern Wisconsin (abs.): Geological Society of America, Abstracts with Programs, v. 32, no. 4, p. A-53.

Patterson, C. J., A. K. Hansel, D. M. Mickelson, D. J. Quade, E. A. Bettis III, P. M. Colgan, E. D. McKay, and A. J. Stumpf, 2003, Contrasting glacial landscapes created by ice lobes of the southern Laurentide Ice Sheet, day 1: The Lake Michigan lobe in Illinois, in D. J. Easterbrook, ed., Quaternary geology of the United States: Reno, Nevada, Desert Research Institute, INQUA 2003 Field Guide Volume, p. 135–139.

Piskin, K., and R. E. Bergstrom, 1975, Glacial drift in Illinois: Thickness and character: Illinois State Geological Survey, Circular 490, 35 p.

Porter, D., and S. Bishop, 1990, Soil and lithostratigraphy below the Loveland/Sicily silt, Crowley's Ridge, Arkansas: Journal of the Arkansas Academy of Sciences, v. 44, p. 86–90.

Prest, V. K., 1984, The Late Wisconsinan glacier complex, in R. J. Fulton, ed., Quaternary stratigraphy of Canada—A Canadian contribution to IGCP Project 24: Ottawa, Ontario, Canada, Geological Survey of Canada, Paper 84-10, p. 21–36; Map 1584A, 1:7,500,000.

Pugin, A. J. M., T. H. Larson, T. C. Young, S. Sargent, and R. S. Nelson, 2003, Extensive geophysical mapping of the buried Teays-Mahomet bedrock buried valley, Illinois: Symposium on the Application of Geophysics to Engineering and Environmental Problems, April 6–10, 2003, San Antonio, Texas, p. 1121–1133. (CD-ROM.)

Salisbury, R. D., 1893, Distinct glacial epochs and the criteria for their recognition: Journal of Geology, v. 1, no. 1, p. 61–84.

Saunders, J. J., and J. E. King, 1986, Stratified Illinoian-Sangamon pollen, plant macrofossil, invertebrate and vertebrate record at Hopwood Farm, in R. W. Graham, B. W. Styles, J. J. Saunders, M. D. Wiant, E. D. McKay, T. R. Styles, and E. R. Hajic, eds., Quaternary records of southwestern Illinois and adjacent Missouri: Illinois State Geological Survey, Guidebook 23, p. 1–6.

Shimek, B., 1909, Aftonian sands and gravels in western Iowa: Geological Society of America Bulletin, v. 20, p. 399–408.

Smith, G. D., 1942, Illinois loess—Variations in its properties and distribution—A pedologic interpretation: Urbana-Champaign, University of Illinois, Agricultural Experiment Station Bulletin 490, p. 139–184.

Stiff, B. J., and A. K. Hansel, 2004, Quaternary glaciations in Illinois, in J. Ehlers and P. Gibbard, eds., Quaternary glaciations—Extent and chronology, Part II: North America: New York, Elsevier, Developments in Quaternary Science, v. 2, p. 71–82.

Styles, B. W., 2004, Changes—Dynamic Illinois environments: The Living Museum, v. 66, no. 2–3, p. 6–13.

Sugden, D. E., and B. S. John, 1976, Glaciers and landscape—A geomorphological approach: New York, John Wiley and Sons, 376 p.

Teed, L. R., 2000, A >130,000-year-long pollen record from Pittsburg basin, Illinois: Quaternary Research, v. 54, no. 2, p. 264–274.

Trowbridge, A. C., 1921, The erosional history of the Driftless Area: Iowa City, Iowa, University of Iowa Studies in Natural History, v. 9, no. 3, 127 p.

Wanless, H. R., 1957, Geology and mineral resources of the Beardstown, Glasford, Havana, and Vermont Quadrangles: Illinois State Geological Survey, Bulletin 82, 233 p.

Wickham, J. T., 1979a, Glacial geology of north-central and western Champaign County, Illinois: Illinois State Geological Survey, Circular 506, 30 p.

Wickham, J. T., 1979b, Pre-Illinoian till stratigraphy in the Quincy, Illinois, area, *in* J. D. Treworgy, E. D. McKay, and J. T. Wickham, eds., Geology of western Illinois: Tri-State Geological Field Conference 43: Illinois State Geological Survey, Guidebook 14, p. 69–70.

Wickham, S. S., and W. H. Johnson, 1981, The Tiskilwa Till, a regional view of its origin and depositional processes: Annals of Glaciology, v. 2, p. 176–182.

Wickham, S. S., W. H. Johnson, and H. D. Glass, 1988, Regional geology of the Tiskilwa Till Member, Wedron Formation, northeastern Illinois: Illinois State Geological Survey, Circular 543, 35 p.

Willman, H. B., and J. C. Frye, 1969, High-level glacial outwash in the Driftless Area of northwestern Illinois: Illinois State Geological Survey, Circular 440, 23 p.

Willman, H. B., and J. C. Frye, 1970, Pleistocene stratigraphy of Illinois: Illinois State Geological Survey, Bulletin 94, 204 p.

Willman, H. B., H. D. Glass, and J. C. Frye, 1963, Mineralogy of glacial tills and their weathering profiles in Illinois, Part I, Glacial tills: Illinois State Geological Survey, Circular 347, 55 p.

Willman, H. B., A. B. Leonard, and J. C. Frye, 1971, Farmdalian lake deposits and faunas in northern Illinois: Illinois State Geological Survey, Circular 467, 12 p.

Willman, H. B., and J. N. Payne, 1942, Geology and mineral resources of the Marseilles, Ottawa, and Streator Quadrangles: Illinois State Geological Survey, Bulletin 66, 388 p.

Woida, K., and M. L. Thompson, 1993, Polygenesis of a Pleistocene paleosol in southern Iowa: Geological Society of America Bulletin, v. 105, no. 11, p. 1445–1461.

Worthen, A. H., 1866, Physical features, general principles and surface geology, *in* A. H. Worthen, ed., Geology: Champaign, Illinois, Geological Survey of Illinois, v. 1, p. 1–39.

13 Quaternary Paleoclimate

B. Brandon Curry, Hong Wang, Samuel V. Panno, and Keith C. Hackley

INTRODUCTION

What is climate? To most, it is the long-term measurement of practical environmental parameters that affect humans, such as wind speed, temperature, precipitation, the annual number of drought or frost-free days, and so on. These data are desired not only in terms of annual averages but also as seasonal averages and as maximum and minimum values.

Studies of paleoclimate—the climate of the geological past—comprise two equally important, but completely independent parts. First is the estimated value of a parameter, such as precipitation or temperature; this value commonly is referred to as a climatic proxy. Second is the geochronology of the record—the age of the events—and the time period between events. Today's investigators of paleoclimate strive to improve upon both climatic proxies and geochronological methods. Investigators are attempting to determine the causes of climate change in order to prepare for the effects of future climate change.

In the absence of recorded measurements of environmental parameters, paleoclimatologists interpret other records that function as proxies for such measurements. For these interpretations to be valid, two important criteria need to be met. First, the records must be layered so that a time series can be established. The layering may be concentric, such as in tree rings or stalagmites, or tabular, such as layers of lake sediment, ice, or loess. Second, the basis of the proxy record must be firmly established. For example, tree-ring widths may be correlated to effective moisture (the balance between annual precipitation and evaporation), based on historical data. The known relationship can then be applied to measurements of the proxy beyond the historical record, permitting an interpreted record of climate change. Known broadly as the "Method of Modern Analogs," this is an application of the basic geology tenet "the present is the key to the past."

Interpretation of a proxy record of climate change is only as good as the geochronology—the record of when the changes took place. Additionally, it is important to determine whether the changes correspond with cyclical phenomena such as sunspots (solar activity), El Niño–La Niña cycles, or Milankovitch cycles. Many of the recent advancements in understanding climate change are a result of improvements in dating methods. Radiocarbon dating, for example, has been used since the 1950s, but revolutionary breakthroughs in atomic particle accelerator technologies have resulted in radiocarbon ages that have greater precision using much smaller samples. Other dating methods used for Quaternary samples include luminescence dating of sediment (Forman and Pierson 2002), uranium-series dating of speleothems (Dorale et al. 1998, Richards and Dorale 2003), beryllium-10 (^{10}Be) inventories of paleosols (Curry and Pavich 1996), and amino acid racemization of snail shells in loess records (Miller et al. 1994).

History of Climate Change Studies

More than 500,000 years of climate change have been interpreted from the Quaternary fossils, soils, and sediment that mantle Illinois (Table 13-1). Evidence of climate change was observed by the first scientists as they described and mapped the surficial deposits of the midwestern United States by canoe and on horseback near the end of the nineteenth century. One of the most unusual and exciting of their discoveries was the layers of well-preserved conifer debris, organic-rich soil, and weathered sediment sandwiched by unweathered glacial "boulder clay" (Figure 13-1) (Leverett 1898, 1899). Although moraines had provided evidence for local ice margins, this evidence indicated that the glaciers had waxed and waned and that vegetation had regained a foothold between the glaciations.

These early scientific observations in Illinois were important for at least two reasons. First, evidence of glacial ice was found as far south as Carbondale (37° 20' N latitude), the southernmost reach of the great Laurentide ice sheet that once covered much of Canada and the northern United States (Geikie 1894, Dyke et al. 2003). Second, the evidence for climate change was overwhelming. Scientists were confronted with the need to explain how the layers of Quaternary sediment and fossils were deposited under climate conditions ranging from glacial to warm temperate.

Since their discovery, the buried organic-rich deposits and soils observed by Leverett and others have been studied meticulously by several generations of geologists (e.g., Leighton 1926, Willman and Frye 1970, Follmer et al. 1979, Wang et al. 2003). Until about the 1980s, Illinois' Quaternary stratigraphic succession was considered by most to

Table 13-1 The paleoclimate in Illinois over the past 500,000 years.

Years ago[1]	Climatic conditions	Comments
500,000 to 480,000	Mild, interglacial	Imbrie et al. (1984).
480,000 to 425,000	Intense, glacial	Glaciers probably entered Illinois from the north.
425,000 to 370,000	Intense, interglacial	
370,000 to 338,000	Mild, glacial	Ice may not have reached Illinois.
338,000 to 275,000	Moderate, interglacial	Martinson et al. (1987).
275,000 to 244,000	Mild, glacial	Ice may not have reached Illinois.
244,000 to 190,000	Intense, interglacial	
190,000 to 130,000	Intense, glacial	The Lake Michigan lobe of Laurentide ice sheet deposited Illinois Episode glacial deposits as far south as Carbondale, Illinois (latitude 37° 20' N).
130,000 to 127,000	Postglacial	Boreal forests grew under a cool, moist climate at Raymond Basin (Curry and Baker 2000, Curry et al. 2002).
127,000 to 122,000	Interglacial	Early Sangamon Episode climate was wetter than today.
122,000 to 77,000	Interglacial	Middle Sangamon Episode climate was like modern Illinois; some intervals had nonfreezing winters.
77,000 to 71,000	Interglacial	Middle Sangamon Episode continued; climate was much wetter than today; some intervals had nonfreezing winters.
71,000 to 55,000	Interglacial	Interglacial conditions at Raymond Basin; late Sangamon Episode; climate unlike any known today but best described as prairie-like (Curry and Baker 2000, Curry et al. 2002).
55,000 to 33,000	Glacial onset	Glacial ice entered Mississippi River drainage system but did not yet reach Illinois; Illinois climate was cool to cold and dry.
33,000 to 29,400	Glacial	Lake Michigan lobe ice flowed from the lake basin to its maximum position in northeastern Illinois, forming the Marengo Moraine (Hansel and Johnson 1992, Curry et al. 1999).
29,400 to 23,700	Glacial	Lake Michigan lobe retreated from the Marengo Moraine and advanced to its southernmost position at 39° 20' N latitude, forming the Shelbyville Moraine (Hansel and Johnson 1996). In the unglaciated part of the state, the climate was cool to cold and slightly drier than today (Wang et al. 2000). Permafrost occurred in many deposits to about as far south as East St. Louis, Illinois (Johnson 1990).
23,700 to 16,800	Glacial	Lake Michigan lobe remained active in northeastern Illinois (Hansel and Johnson 1992, 1996; Curry et al. 1999). The St. Louis, Missouri, area was cooler and drier than today (Johnson 1990, Wang et al. 2000, Curry 2008, Curry and Yansa 2004). Depending on the location of the Lake Michigan lobe in the Chicago area, the vegetation varied from spruce forests to tundra (Curry et al. 1999, Curry and Yansa 2004). The area between the Lake Michigan lobe and St. Louis likely contained patches of permafrost (Johnson 1990).
16,800 to 15,900	Glacial retreat	The Lake Michigan lobe retreated from Illinois. Spruce forests invaded the new glacial soil and quickly replaced the tundra vegetation (Curry 2008). Further south, Illinois remained cooler and drier than today. This period also marked the end of loess deposition, a semi-continuous process that had begun about 55,000 years ago.
15,900 to 14,700	Postglacial	Dense boreal forests grew in northeastern Illinois; the climate of the remainder of the state is largely unknown.
14,700 to 11,800		Mixed boreal-deciduous tree forests developed in northeastern Illinois (Curry et al. 1999, Curry 2008).
11,800	Abruptly warmer	Sudden climate change drives spruce from many places in northeastern Illinois. This is the age of the Pleistocene-Holocene boundary.
11,800 to 9,500	Moist, warm	Growth of forests was dominated by oak, black ash, and elm.
9,500 to 6,300	Somewhat drier	Patches of prairie developed in northeastern Illinois (Nelson et al. 2006); elm and ash become much less important.
6,300 to 4,000	Drier than today	Maximum prairie development in northeastern Illinois; overall conditions were drier than today (Nelson et al. 2006).
4,000 to present	Modern	Modern climate of Illinois established.

[1] The ages from 500,000 to 300,000 years ago come from work by Imbrie et al. (1984); from 300,000 to 130,000 years ago, from Martinson et al. (1987); and from 130,000 to 55,000 years ago, from Dorale et al. (1998), Dorale (2004), and Curry et al. (2002). Ages from 33,000 to the present were calibrated from radiocarbon years. Calibration was done online using CALIB 3.0 (Stuiver and Reimer 1993). The period from 130,000 to 55,000 years ago is based on data from Raymond Basin, Montgomery County, Illinois.

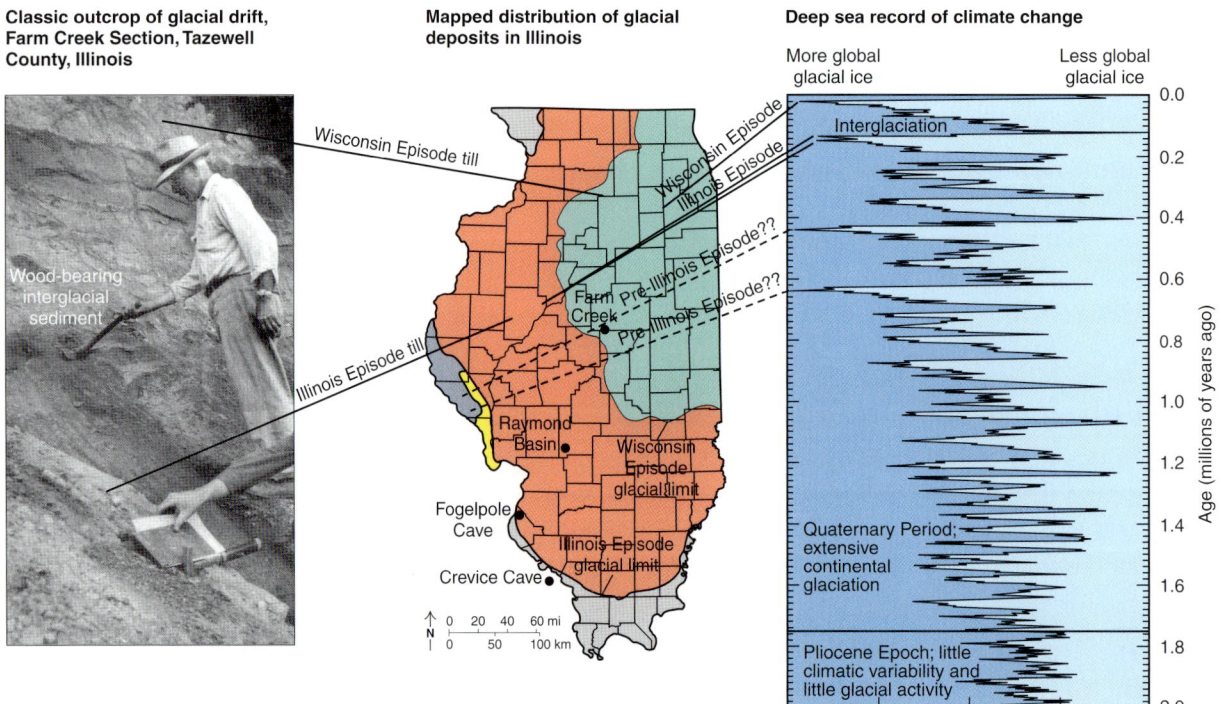

Figure 13-1 Correlation of the classic succession of glacial drift deposits at Farm Creek with the mapped distribution of those deposits in Illinois (Lineback 1979) and the deep sea $^{18}O:^{16}O$ record of benthic foraminifera (Zachos et al. 2001). M. M. Leighton is shown in the photograph. (Photograph by K. P. Oakley, British Museum, July 1950.) The locations of Crevice Cave, Missouri, and Raymond Basin and Fogelpole Cave, Illinois, are discussed in the chapter text.

represent a classic model of climate change: drift-soil-drift (Morrison and Frye 1965, Bowen 1985). Based on studies of fossils and the chemistry of deep sea sediment, ice cores, and speleothems (cave formations), it is now known that the classic theory of drift-soil-drift succession in Illinois contains only fragments of the state's full geological history (Figure 13-1).

Investigations of climate change are rapidly evolving and becoming increasingly complex. These investigations are spurred on largely by scientists trying to understand the mechanisms and implications of global warming—an urgent topic that enters into the realm of public policy and global politics. Evidence of global warming is manifest in the 0.5°C increase in mean global temperature over the past 50 years (Karl and Trenberth 2003) and its effect on many environments, such as shrinking Alaskan glaciers (Cubasch et al. 2001, Arendt et al. 2002). Excellent historical treatments of climate change are presented by Imbrie and Imbrie (1979), Cronin (1999), Bradley (1999), Alley (2000), and Cubasch et al. (2001). This chapter briefly discusses (1) a few of the mechanisms of climate change (such as insolation), (2) evidence of climate changes in the chemistry of the fossils contained in deep sea sediment and in polar ice, and (3) global climate models. The chapter concludes with reviews of three current paleoclimate studies in Illinois.

GATHERING EVIDENCE OF CLIMATE CHANGE
Insolation

Proposed in the early 1940s, but initially ignored by students of glacial sediments and soils, was a theory of long-term cyclical changes in insolation, the amount of solar light and heat energy received at the Earth's surface. The cyclicity, caused by changes in the Earth's orbital geometry, the wobble of its axis, and the timing of the solstices, eventually became known as "Milankovitch cycles," named for the Serbian scientist who first determined values of insolation at different latitudes on the Earth's surface and showed how those values changed over time (Imbrie and Imbrie 1979). One of the implications of the theory was that many more Quaternary glacial cycles were suggested than were indicated by the terrestrial glacial record. Today, most, if not all, paleoclimatologists think that Milankovitch's theory is a vital part of the explanation of long-term climate change (Cooperative Holocene Mapping Project Members 1988).

One of the greatest challenges facing the field of global climate change is understanding the mechanisms that transport heat from the equator to the poles, including the combined effects of enormous floods, rising mountain chains, and configurations of the continents that deflect or alter currents in the ocean and atmosphere. These mechanisms are but some of those proposed to control the long-term seasonal patterns of precipitation and temperature in Illinois. Another challenge in Illinois is correlating terrestrial records of environmental change with the large amount of well-known, more accurately dated oceanic and polar ice records and determining why these records do not always agree. For example, little is known about the global mechanisms that promoted the southern Laurentide ice sheet to advance into Illinois about 30,000 years ago, a time for which there is no evidence for pronounced global cooling.

Changes in Fossil Chemistry

Studies of deep sea sediment have been important in understanding the timing and magnitude of climate change over long stretches of geological time. Perhaps the most telling were the results of geochemical studies of microfossils that indirectly indicated changes in the volume of global moisture stored in glacial ice (Shackleton 1987). As water from the ocean was displaced to ice sheets, ocean levels lowered by more than 400 feet (122 m) in some regions. As this displacement occurred, evaporation caused naturally occurring stable isotopes of oxygen to become fractionated. During evaporation, water molecules containing the lighter isotope of oxygen, ^{16}O, evaporate more readily than water molecules containing the heavier isotope, ^{18}O. The vapor and, therefore, precipitation become progressively more enriched in the ^{16}O isotope, leaving the world's oceans with a relatively greater concentration of ^{18}O.

These changes in the oceanic isotopic composition were recorded by Foraminifera, one-celled organisms that form protective shells (called "tests") of calcium carbonate ($CaCO_3$) from the dissolved ions in seawater. The dissolved carbonate ions, in chemical equilibrium with the ocean water, reflect its isotopic composition. Analysis of layers upon layers of sediment with foraminifera tests in relation to an ever-improving chronological framework provided scientists with a history of the world's ice volume (Shackleton 1987, Cronin 1999). The deep sea record of the ratio of ^{18}O to ^{16}O, expressed by geochemists by $\delta^{18}O$ notation (Bradley 1999), shows evidence for more than 30 glacial episodes over the past 1.8 million years. Illinois' continental record of climate change shows evidence for only three or four episodes of glaciation (Wilman and Frye 1970, Grimley et al. 2003). In north-central North America, there is evidence for nine glaciations (Roy et al. 2004). Although Illinois' record of climate change tells only part of the story (Figure 13-1), records for other continents, such as the Chinese loess-paleosol records (Ding et al. 1994, Rutter et al. 1996), appear to be more complete.

Today, the key evidence for climate change is considered by many to be the $\delta^{18}O$ records of ice cored from Antarctica (the Vostok cores) and Greenland (the GISP, GISP2, and GRIP cores) (Bradley 1999). The completeness and chronology of these records, based on annual layer counts and models of ice-flow velocity, are also considered to be the standard to which other records are compared. Other important parameters have been measured in the ice, such as the concentration of the "greenhouse" gases methane and carbon dioxide. Present-day research on the ice cores focuses on the period from about 135,000 years ago to the present; emphasis is on understanding the details of the last deglaciation.

Global Climate Models

Global circulation models are an additional tool used by paleoclimatologists (Kutzbach 1987, Bader et al. 2008). The models are computer-generated solutions to complex formulas that attempt to map complexly interrelated features such as ice sheets and vegetation types, precipitation and temperature values, and the direction and relative intensity of ocean and wind currents. Most variables are interdependent and form feedback loops. For example, a growing glacier surface area reflects increasingly greater amounts of the Sun's energy back into the atmosphere, reducing the insolation in the upper latitudes and promoting colder temperatures and, given a souce of moisture, the further growth of the ice sheet. Some equations take several months to solve on the most sophisticated of supercomputers. One drawback of the models is that their solutions are portrayed in large 500 × 500-mile (805 × 805-km) grids (10° longitude and latitude) and thus lack the degree of detail that many scientists need. Despite the many drawbacks of global circulation model output, the models remain a valuable tool for evaluating the sensitivity of climatic parameters and provide touchpoints for the direction of future research.

RECENT STUDIES USING CLIMATIC PROXIES

Climatic proxies are fossil or chemical records that indirectly record changes in temperature, precipitation, or some other parameter that might be linked to climate. Most proxies yield relative data, that is, evidence of the direction in which climate was changing (e.g., drier, warmer, more variable). Only in special circumstances are the results quantitative (e.g., the mean annual temperature increased 3°C over 10 years). Of utmost importance is the establishment of a modern database that forms the basis for inter-

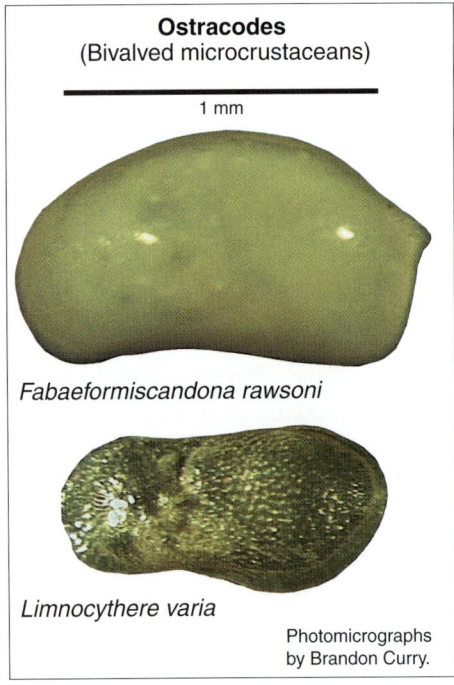
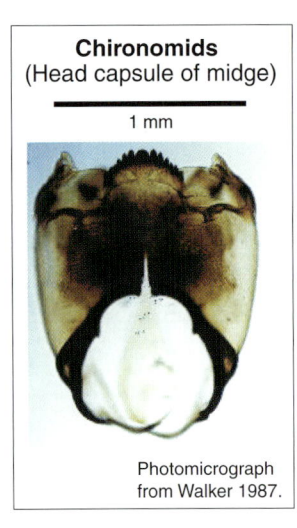
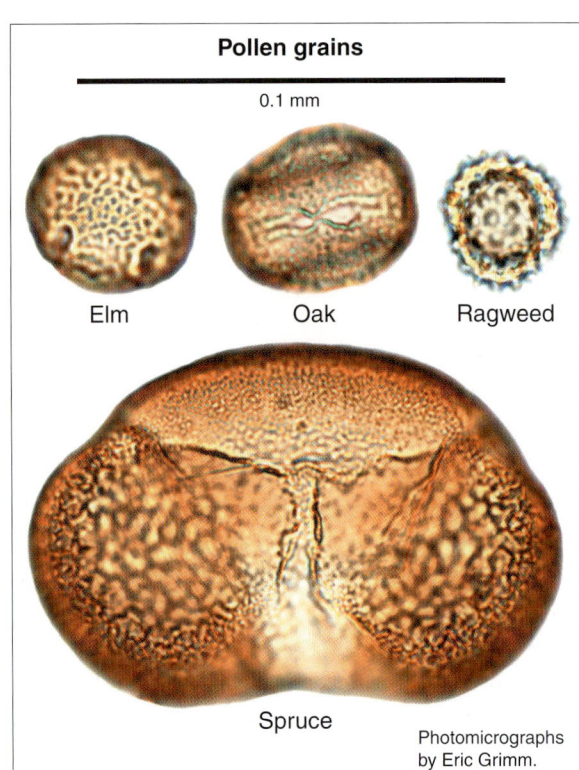
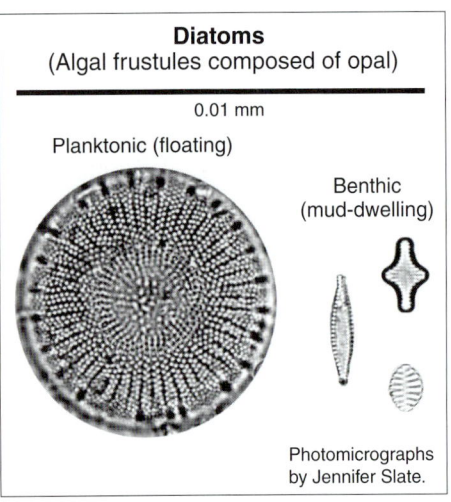
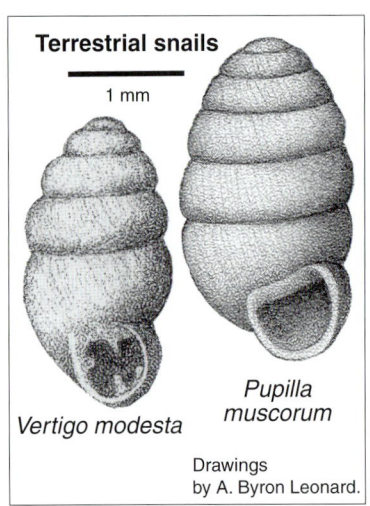

Figure 13-2 Examples of climatically sensitive fossils commonly found in Illinois, including ostracodes, diatoms, chironomids, pollen, plant macrofossils, and gastropods (snails). The lacustrine microfossils (diatoms, chironomids, and ostracodes) are sensitive to the depth, temperature, and chemistry of the water they inhabit. The relationship between lake conditions and climate needs to be established for the fossils to be used in paleoclimatic studies (Warner 1990, Curry 2003). Pollen grains and plant macrofossils are commonly preserved in the Quaternary fossil record. The climatic ranges of plant and vegetation types are well-known (MacDonald 1990; http://www.ncdc.noaa.gov/paleo/pollen.html). For example, abundant spruce pollen is indicative of cool, wet conditions, such as those experienced during late glacial conditions in northeastern Illinois 13,000 radiocarbon years ago; abundant ragweed pollen is indicative of environmental disturbance caused by soil conditioning by early European settlers (King 1981).

preting past climate change. An often-used methodology employs fossil pollen preserved in layers of lake or bog sediment. An important resource is the North American Pollen Database, which contains data from published records of Quaternary pollen successions in addition to the pollen assemblages from more than 1,000 modern lakes and wetlands (http://www.ncdc.noaa.gov/paleo/pollen.html). Sophisticated multivariate statistical analyses are used to find the best fit between the assemblage of fossil and modern pollen spectra (Webb et al. 1987). Similar data sets are available for other fossils, such as seeds, needles, rootlets (collectively known as "plant macrofossils"), snails (gastropods) and clams (pelecypods) (Baker 1920), and microscopic aquatic organisms such as diatoms (freshwater algae that form siliceous skeletons) (Laird et al. 2003), ostracodes (microcrustaceans that form valves of calcium carbonate) (Curry 2003, Forester et al. 2006), and chironomids (midge larvae) (Walker 1987). Figure 13-2 shows some of these fossils. The more proxies that can be employed during an investigation of a site or core, the more viable is the paleoclimatic interpretation.

Coarse reconstructions of climate are refined by interpreting the physical characteristics and mineralogy of sediment and soil. For example, abundant fine sand in an otherwise silty lake or wetland deposit might have been deposited under relatively arid conditions when stabilizing vegetation was suppressed and erosion washed or blew in sand. Such assertions are more tenable in the presence of supporting evidence, such as the presence of a microfossil that prefers relatively shallow or more saline water, conditions that would be expected to develop under drier conditions.

Another type of climatic proxy is chronological variation in the chemical composition of minerals or organic matter, including stable isotope composition and trace element concentrations. The use of proxies is illustrated using the following examples from paleoclimate studies in Illinois.

Climatic Reconstruction from Paleosol and Loess Successions

Thick windblown silt (loess) deposits occur in the bluffs adjacent to major rivers in north-central North America. The silt was blown from bare, unvegetated bars of glacially derived alluvium (outwash) deposited in the floodplains of major river valleys such as the Mississippi River. Loess deposits are thickest adjacent to the river valley. At the Keller Farm section at Keller Farm near East St. Louis, Illinois, geologists (Wang et al. 2000, 2003) found a unique exposure of loess that contains several dark bands that represent the organic upper layer (A-horizon) of buried soils (Figures 13-3 and 13-4). Typically, loess is a massive deposit with subtle variations in sediment character. The Keller Farm section paleosols are remarkably well preserved and include rhizoliths, which are calcite-rich concretions that formed around roots and rootlets while the soil was forming.

At Keller Farm, carbon isotopes were used for two independent purposes. First, radiocarbon ages of organic carbon in A-horizons provided evidence of the frequency of soil development and the rate of loess deposition. Second, stable isotopes of carbon were used to interpret changes in vegetation (Wang et al. 2000). Stable carbon isotopes, measured as the ratio of carbon-13 to carbon-12 ($^{13}C:^{12}C$) in each sample, are compared against an international standard. Some grasses yield greater $^{13}C:^{12}C$ than trees do, which is reflected in the organic matter of their soils. Because North American grasses tolerate a drier climate than does forest vegetation, the $^{13}C:^{12}C$ values are a proxy for the balance between precipitation and evaporation.

The $^{13}C:^{12}C$ profile during the last glaciation suggests that in the East St. Louis area the vegetation changed every 500 to 1,000 years. The dark bands in the loess were likely formed under a forest of spruce trees; the loess was deposited under grass. The number of years between soil-forming events corresponds with long-term El Niño–Southern Oscillation (ENSO) cycles (Wang et al. 2000). These cycles are driven by ocean currents and winds that move warm masses of water to either the east Pacific (setting up El Niño) or the west Pacific (setting up La Niña). Global circulation modeling suggests that, during El Niño events, the glacial climate of Illinois included warm, wet summers and warm, dry winters; during La Niña events, the climate was cold with dry summers and snowy winters. The regional ENSO-influenced climate may have influenced the advance and retreat of the southern margin of the Laurentide ice sheet, which was then located more than 87 miles (140 km) northeast of Keller Farm.

Climatic Reconstruction from Speleothems

Speleothems include stalactites, stalagmites, and flowstone in caves. Because of the manner in which they form, speleothems also contain proxy records of the local history of the terrestrial climate and vegetation. Formation of speleothems begins with movement of slightly acidic rainwater and snow melt through the soil zone where the water dissolves carbon dioxide (CO_2) in soil and becomes more acidic. The slightly acid soil water percolates through the ground and dissolves carbonate minerals. As this mineralized groundwater drips into caves, some of the dissolved carbonate minerals begin to precipitate and crystallize (Figure 13-5). The soil zone contains CO_2 with an isotopic

Figure 13-3 The Keller Farm section (near East St. Louis, Illinois) exposed 39 paleosols developed in loess (Peoria Silt) from the last glaciation. The inset shows details from a shallow trench at the bottom of the section.

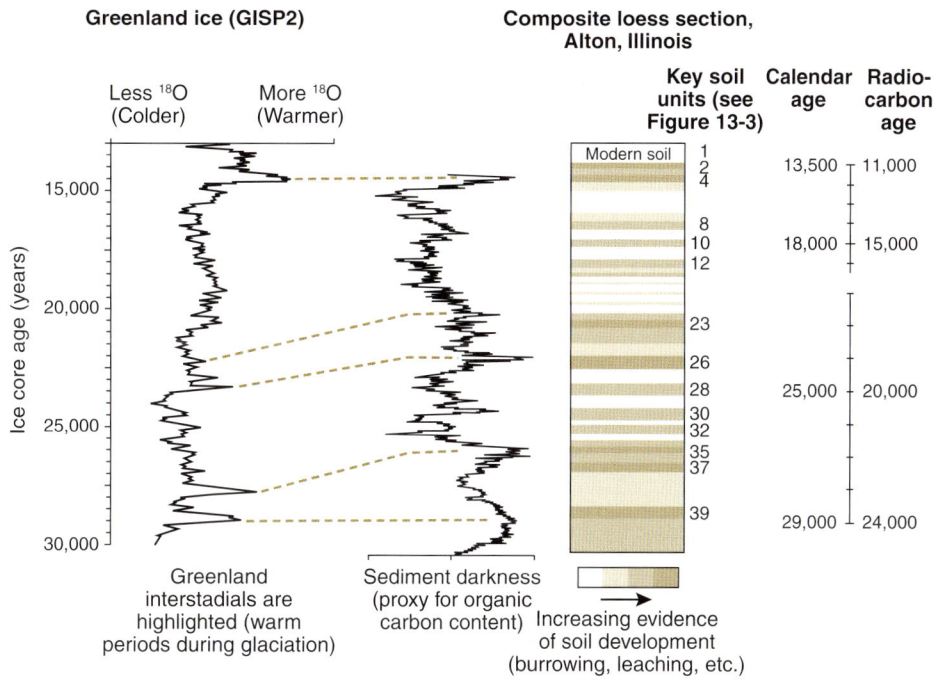

Figure 13-4 Correlation (dashed lines) of the $^{18}O:^{16}O$ record from GISP2 (an ice core from Greenland; Grootes et al. 2003) with sediment darkness and soil development in the Peoria Silt at the Keller Farm section near East St. Louis, Illinois. The cyclicity of soil formation (about 1,000 years) is evidence that climate was impacted by long-term development of El Niño–La Niña cycles in the equatorial Pacific region (Wang et al. 2003).

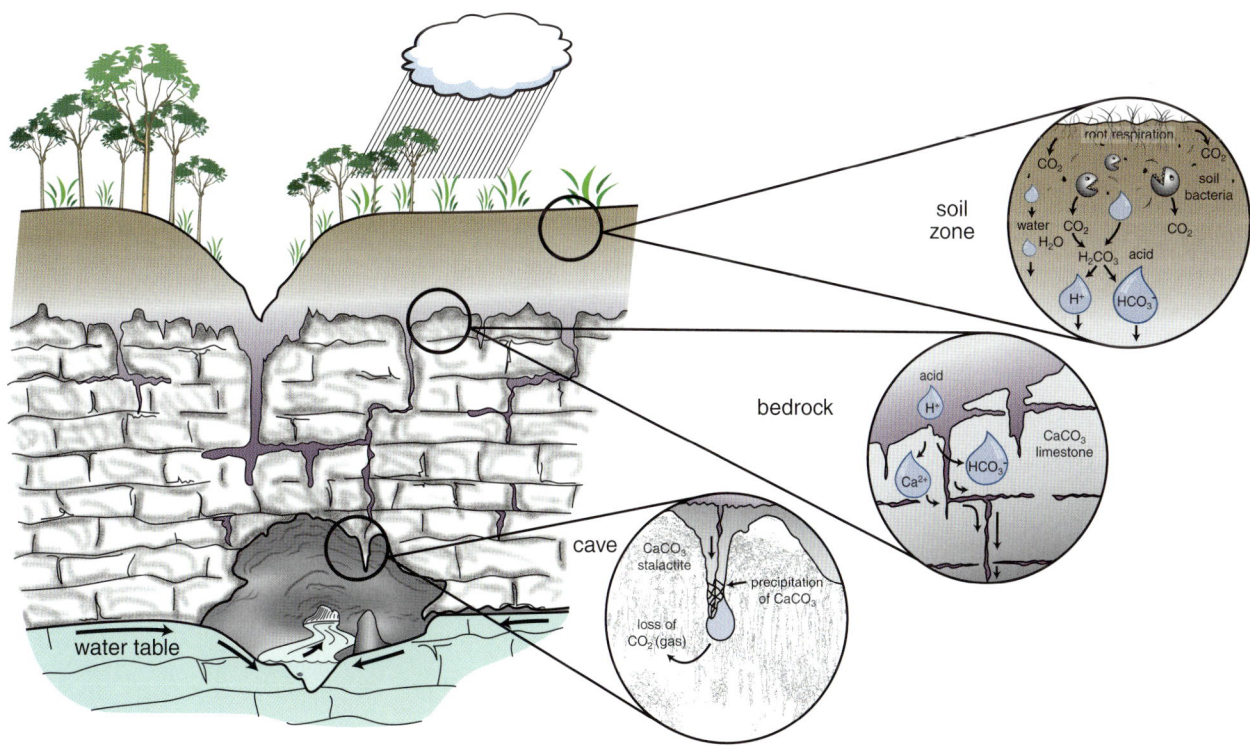

Figure 13-5 Schematic drawing of how the chemistry of a speleothem reflects the chemistry of infiltrating soil water (Panno 2004b). The CO_2, generated from root respiration and decaying vegetation, mixes with water to form relatively weak carbonic acid. The acidic soil water eventually migrates into the bedrock and dissolves some of the limestone or dolomite. Once mineral-enriched soil water finds its way into a cave, the lower gas pressure in the cave (same as atmospheric pressure) causes CO_2 (under a higher pressure in the soil zone) to come out of solution (degassing in a manner similar to that of opening a bottle of soda pop). As the gas escapes, the water becomes less acidic and can no longer hold as much dissolved limestone or dolomite in solution as it could when it was under pressure. The water then drips or flows into the cave and deposits minute crystals of calcite or, in some cases, other carbonate minerals. These minerals form in discrete layers, most easily seen in stalagmites (Panno et al. 2004b).

signature representative of the carbon inherited from surface vegetation and the oxygen from rainwater and snow melt. This CO_2 mixes with the CO_2 generated by dissolution of the carbonate bedrock to form a mix of carbon isotopes and oxygen isotopes. The degree of mixing is generally constant. Changes in the $^{13}C:^{12}C$ and $^{18}O:^{16}O$ ratios tend to be diagnostic of vegetation types and average annual temperature of the surface, respectively. The calcite layers of the stalagmite can yield records of $^{13}C:^{12}C$ and $^{18}O:^{16}O$ values that may be attributed to the changes in vegetation and precipitation characteristics (Ford 1997).

To determine the history of vegetation and climate change, stalagmite bands must be age dated. The most accurate and precise dating method is uranium (U)-series dating, which involves sophisticated analysis of trace amounts of ^{238}U and its decay products measured from individual bands in the speleothems. Uranium dating techniques of carbonate speleothems have greatly improved during the past two decades with the development of new instrumentation. Prior to the late-1980s, alpha-spectrometry was used to measure the radioactive elements associated with uranium decay for dating purposes (Edwards et al. 1986). New mass spectrometry techniques, such as thermal ionization mass spectrometry and, more recently, inductively coupled plasma mass spectrometry, have improved the precision and rapidity of isotopic measurements, reduced errors, and greatly decreased required sample size (Richards and Dorale 2003).

Our example of a speleothem (Figure 13-6) comes from Fogelpole Cave near Waterloo, in Monroe County, Illinois. Uranium-series dating indicates that the speleothem formed during the interval from about 80,000 to 31,000 years ago (Zhou et al. 2005). The layers of calcite are interbedded with thin laminae of silt; these appear as dark banded areas in Figure 13-6. The most prominent silt layers date at 77,500, 71,000, 60,000, 55,000, and 50,000 years ago. Speleothem growth ceased at about 31,000 years ago when the caves in this region became choked with silty sediment (Panno et al. 2004a). Layers near the top and on the side of the stalagmite, dating from about 60,000 to 31,000 years, are especially silty. The inferred increase in sediment flux through the cave suggests increased erosion and climate change associated with the advancing Laurentide ice sheet during the last glaciation. The 60,000-year uranium-series age is especially important because the errors of the ages of the layers that bracket the silty zone are small (less than 1,000 years). A similar age of 55,000 years for the onset of the last glaciation was determined on a speleothem record from Missouri (Dorale et al. 1998). The same value was determined by other less reliable methods based on the

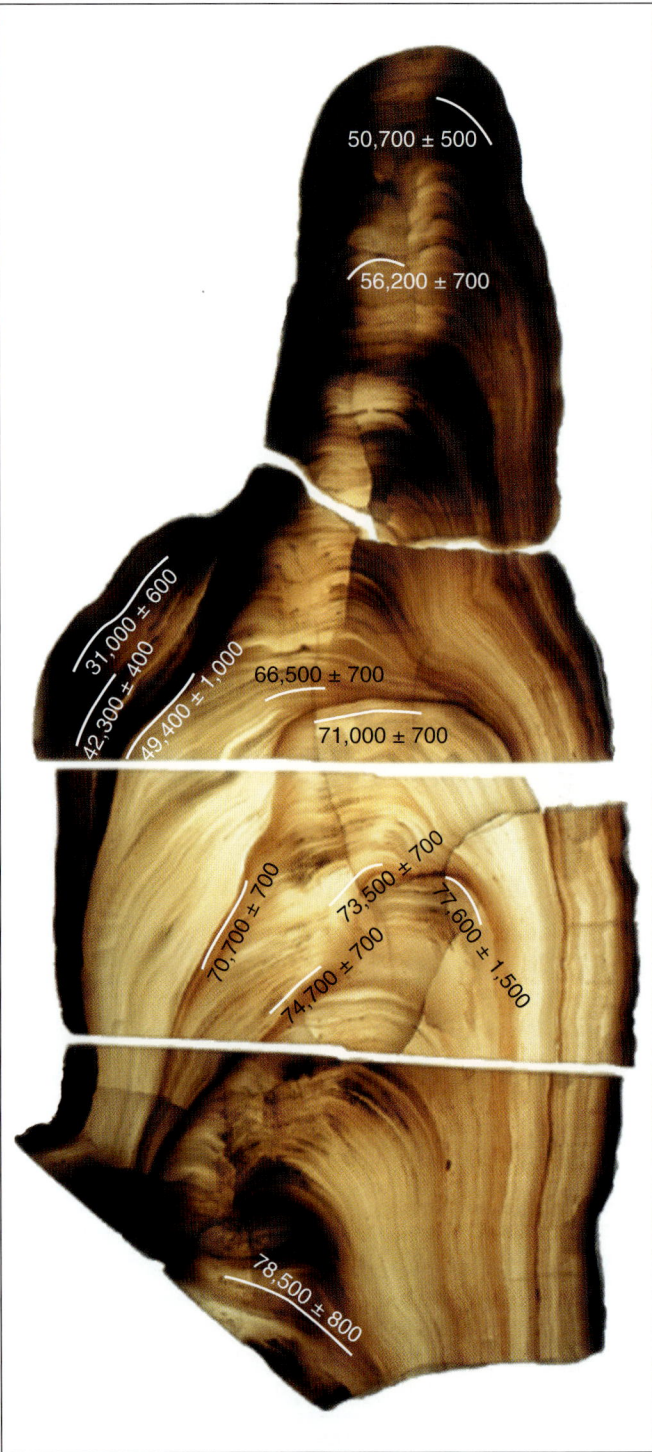

Figure 13-6 Uranium-series ages of banded calcite constituting a stalagmite from Fogelpole Cave, Monroe County, Illinois (Zhou et al. 2005). The bands were made visible by transmitting light through a 0.5-inch-thick (1.27-cm-thick) slab. The darker bands contain silt and clay deposited by muddy floods. Note the shoulder of stalagmite growth dating from about 50,000 to 31,000 years ago. Silt and clay bands are especially thick in the shoulder region of the stalagmite; contemporaneous silty subterranean lake deposits have been observed elsewhere in Fogelpole Cave (Panno et al. 2004a). Older thin layers of sediment that cover local unconformities in the layered calcite date from about 77,600 and 71,000 years ago.

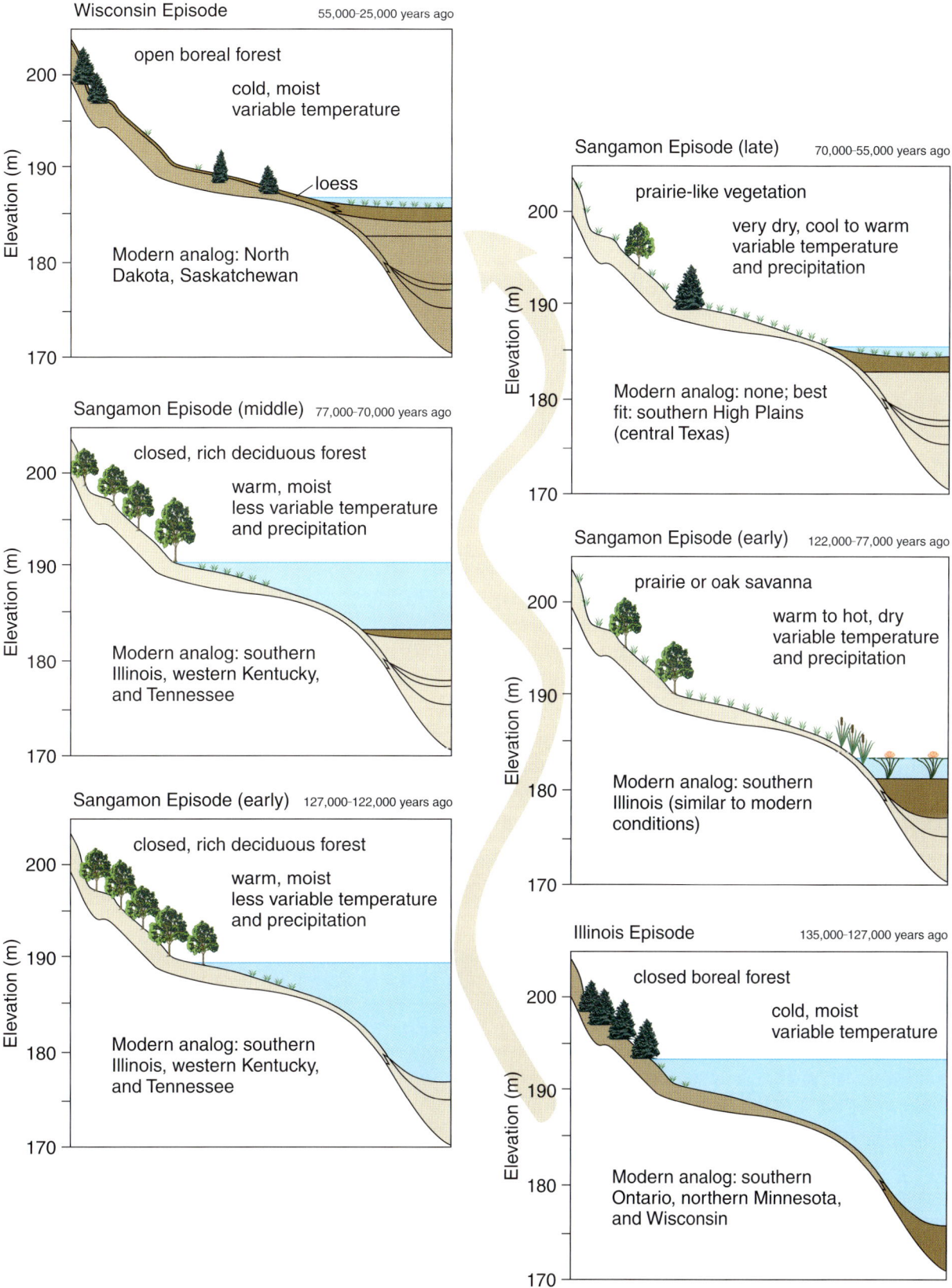

Figure 13-7 Schematic diagram of changes in vegetation, lake levels, and general climate in south-central Illinois from about 135,000 to 25,000 years ago (Zhu and Baker 1995, Dorale et al. 1998, Curry and Baker 2000, Curry et al. 2002). The dark brown layers highlighted in each panel are the layers from which pollen and ostracodes were analyzed and are the basis for the reconstructed environment. An open forest has openings between stands of trees where shrubs and grass typically grow. A closed forest is dense enough that no non-arboreal elements can grow between the stands of trees.

Correlation of Proxy Records of Vegetation from a Speleothem and Pollen Succession

Speleothems also can be used to correlate $^{13}C:^{12}C$ profiles of calcite with pollen records of paleovegetation. An outstanding uranium-series–dated $^{13}C:^{12}C$ profile from Crevice Cave, Missouri (Dorale et al. 1998), near Chester, Randolph County, Illinois, correlated with a pollen record from Raymond Basin, Montgomery County, Illinois (Zhu and Baker 1995), indicates significant climate change during the last interglaciation (Figure 13-7). The pollen record comes from analysis of a 49-foot-long (15-m-long) core taken near the center of Raymond Basin, a drained 2-mile (3.3-km)-wide kettle lake on the Illinoian till plain. Subsamples that yielded predominantly spruce, pine, and fir pollen suggest the vegetation was a boreal forest similar to that found today in southern Ontario and northern Minnesota and Wisconsin (Curry and Baker 2000) (Figure 13-7). According to the correlation with the Crevice Cave speleothems, such vegetation existed during the waning of the next-to-last glaciation (Illinois Episode) 135,000 to 127,000 years ago and during the onset of the last glaciation (Wisconsin Episode) 55,000 to 25,000 years ago. The interpretation of cool conditions is supported by the occurrence of *Limnocythere friabilis,* an ostracode common in near-glacier lake deposits in northeastern Illinois (Curry and Yansa 2004). During the intervening interglacial period (Sangamon Episode), the vegetation changed from dominantly prairie (122,000 to 77,000 years ago and 71,000 to 55,000 years ago) to deciduous forests (127,000 to 122,000 years ago and 77,000 to 71,000 years ago). Prairie vegetation is indicated by a pollen assemblage dominated by ragweed, goosefoot, and grass. Forests of deciduous trees are indicated by pollen of oak, hickory, walnut, black ash, and sweet gum (Zhu and Baker 1995). Some intervals of the Sangamon interglaciation included warm winters, judging from the remains of *Geochelone crassiscutata,* a giant tortoise discovered at Hopwood Farm about 18.6 miles (30 km) east-southeast of Raymond Basin (King and Saunders 1986). The warm winters are implied because this large creature could not burrow to avoid subfreezing temperatures. Another warm-winter indicator from the Sangamon records of Raymond Basin and other records from central Illinois is *Heterocypris punctata,* a freshwater ostracode with a modern distribution no farther north than the Gulf Coast (Curry and Baker 2000). One of the more intriguing aspects of these records is the indication that the warmest and wettest conditions Illinois experienced over the past 180,000 years occurred from about 77,000 to 71,000 years ago when moisture from the Gulf of Mexico was drawn across Illinois on its way to supplying moisture to the growing Laurentide ice sheet in Canada (Curry et al. 2002, Dorale 2004).

CONCLUSION

Climate change in Illinois has been documented through a few important studies of proxy records dating from 135,000 years ago to the present, including those of the physical record of glacial sediment and paleosols, speleothems, and fossils contained in lake sediment. The viability of interpretations increases as the number of proxies increases and with improved accuracy and precision of the accompanying record of event ages. In addition to the investigations of loess and paleosol successions, speleothems, lake sediment, and fossils mentioned in this chapter, exciting new research is being conducted at the Illinois State Geological Survey and other research institutions on topics such as fire and drought frequency over the past 15,000 years (Nelson et al. 2006), grain size variation and paleo-wind conditions during the late Wisconsin glaciation (Wang et al. 2006), and the history and environments of deglaciation (Curry and Yansa 2004, Curry 2008).

REFERENCES

Alley, R. B., 2000, The two-mile time machine: Ice cores, abrupt climate change, and our future: Princeton, New Jersey, Princeton University Press, 229 p.

Arendt, A. A., K. A. Echelmeyer, W. D. Harrison, C. S. Lingle, and V. B. Valentine, 2002, Rapid wastage of Alaska glaciers and their contribution to rising sea level: Science, v. 297, no. 5580, p. 382–386.

Bader, D. C., C. Covey, W. J. Gutowski Jr., I. M. Held, K. E. Kunkel, R. L. Miller, R. T. Tokmakian, and M. H. Zhang, 2008, Climate models: An assessment of strengths and limitations: A report by the U.S. Climate Change Science Program and the Subcommittee on Global Change Research: Washington, D.C., Department of Energy, Office of Biological and Environmental Research, 124 p.

Baker, F. C., 1920, The life of the Pleistocene or glacial period: Urbana-Champaign, University of Illinois, Bulletin 17, no. 41, 476 p.

Bowen, D. Q., 1985, Quaternary geology: A stratigraphic framework for multidisciplinary work: New York, Pergamon Press, 237 p.

Bradley, R. S., 1999, Paleoclimatology: Reconstructing climates of the Quaternary (2nd ed.): San Diego, California, Harcourt Academic Press, 613 p.

Cooperative Holocene Mapping Project Members, 1988, Climatic changes of the last 18,000 years: Observations and model simulations: Science, v. 241, no. 4869, p. 1043–1052.

Cronin, T. M., 1999, Principles of paleoclimatology: New York, Columbia University Press, 560 p.

Cubasch, U., G. A. Meehl, G. J. Boer, R. J. Stouffer, M. Dix, A. Noda, C. A. Senior, S. Raper, and K. S. Yap, 2001, Projections of future climate change, *in* J. Houghton et al., eds., Climate change 2001; The scientific basis; Contribution of Working Group 1 to the Third Assessment Report of the Intergovernmental Panel on Climate

Change: New York, Cambridge University Press, p. 525–582.

Curry, B. B., 2003, Linking ostracodes to climate and landscape, *in* L.E. Park and A.J. Smith, eds., Bridging the gap: Trends in the ostracode biological and geological sciences: The Paleontological Society Papers, v. 9, p. 223–246.

Curry, B. B., ed., 2008, Deglacial history and paleoenvironments of northeastern Illinois: Illinois State Geological Survey, Open File Series 2008-1, 175 p.

Curry, B. B., and R. G. Baker, 2000, Paleohydrology, vegetation, and climate since the late Illinois Episode (~130 ka) in south-central Illinois: Palaeogeography, Palaeoclimatology, and Palaeoecology, v. 155, no. 1–2, p. 59–81.

Curry, B. B., J. A. Dorale, and R. A. Henson, 2002, Function-fitting vegetation proxy record profiles of the Sangamon and Wisconsin Episodes from Missouri and Illinois (abs.): Geological Society of America, Abstracts with Programs, v. 34, no. 6, p. 199.

Curry, B. B., and L. R. Follmer, 1992, The last interglacial-glacial transition in Illinois: 123–25 ka, m, *in* P. U. Clark and P. D. Lea, eds., The last interglacial-glacial transition in North America: Boulder, Colorado, Geologic Society of America, Special Paper 270, p. 71–88.

Curry, B. B., D. A. Grimley, and J. A. Stravers, 1999, Quaternary geology, geomorphology, and climatic history of Kane County, Illinois: Illinois State Geological Survey, Guidebook 28, 40 p.

Curry, B. B., and M. J. Pavich, 1996, Absence of glaciation in Illinois during Marine Isotope Stages 3 through 5: Quaternary Research, v. 46, no. 1, p. 19–26.

Curry, B. B., and C. H. Yansa, 2004, Evidence of stagnation of the Harvard sublobe (Lake Michigan lobe) in northeastern Illinois, USA, from 24 000 to 17 600 BP and subsequent tundra-like ice-marginal paleoenvironments from 17 600 to 15 700 BP: Géographie physique et Quaternaire, v. 58, p. 305–321.

Ding, Z., Z. Yu, N. W. Rutter, and T. Liu, 1994, Towards an orbital time scale for Chinese loess deposits: Quaternary Science Reviews, v. 13, no. 1, p. 39–70.

Dorale, J. A., 2004, Forest-grassland oscillations at Crevice Cave, Missouri, during the last interglacial-glacial cycle: A regional response to global scale forcing: American Quaternary Association Program and Abstracts for the 18th Biennial Meeting, Lawrence, Kansas, June 26–28, 2004, p. 15–16.

Dorale, J. A., R. L. Edwards, E. Ito, and L. A. González, 1998, Climate and vegetation history of the midcontinent from 75 to 25 ka: A speleothem record from Crevice Cave, Missouri, USA: Science, v. 282, no. 5395, p. 1871–1874.

Dyke, A. S., A. Moore, and L. Robertson, 2003, Deglaciation of North America: Ottawa, Ontario, Geological Survey of Canada, Open File 1574. (CD-ROM.)

Edwards, R. L., J. H. Chen, and G. J. Wasserburg, 1986, ^{238}U-^{234}U-^{230}Th-^{232}Th systematics and the precise measurement of time over the past 500,000 years: Earth and Planetary Science Letters, v. 81, no. 2–3, p. 175–192.

Follmer, L. R., E. D. McKay, J. A. Lineback, and D. L. Gross, 1979, Wisconsinan, Sangamonian, and Illinoian stratigraphy in central Illinois: Illinois State Geological Survey, Guidebook 13, p. 19–21.

Ford, D., 1997, Dating and paleo-environmental studies of speleothems, *in* C. Hill and P. Forti, eds., Cave minerals of the world (2nd ed.): Huntsville, Alabama, National Speleological Society, Inc., p. 271–284.

Forester, R. M., A. J. Smith, D. F. Palmer, and B. B. Curry, 2006, North American Non-Marine Ostracode Database "NANODe", Version 1: Kent, Ohio, Kent State University.

Forman, S. L., and J. Pierson, 2002, Late Pleistocene luminescence chronology of loess deposition in the Missouri and Mississippi River valleys, United States: Palaeogeography, Palaeoclimatology, Palaeoecology, v. 186, nos. 1–2, p. 25–46.

Geikie, J., 1894, The great ice age (3rd ed.): London, Edward Stanford, 853 p.

Grimley, D. A., L. R. Follmer, R. E. Hughes, and P. A. Solheid, 2003, Modern, Sangamon and Yarmouth soil development in loess of unglaciated southwestern Illinois: Quaternary Science Reviews, v. 22, no. 2–4, p. 225–244.

Grootes, P. M., M. Stuiver, J. W. C. White, S. Johnsen, and J. Jouzel, 1993, Comparison of oxygen isotope records from the GISP2 and GRIP Greenland cores: Nature, v. 366, p. 552–554.

Hansel, A. K., and W. H. Johnson, 1992, Fluctuations of the Lake Michigan lobe during the late Wisconsin subepisode, *in* A.-M. Robertsson, B. Ringberg, U. Miller, and L. Brunnberg, eds., Quaternary stratigraphy, glacial morphology, and environmental changes: Uppsala, Sweden, Sveriges Geologiska Undersökning, v. 81, p. 133–144. (Illinois State Geological Survey, Reprint Series 1993-F.)

Hansel, A. K., and W. H. Johnson, 1996, Wedron and Mason Groups: Lithostratigraphic reclassification of deposits of the Wisconsin Episode, Lake Michigan Lobe area: Illinois State Geological Survey, Bulletin 104, 116 p.

Imbrie, J., J. D. Hays, D. G. Martinson, A. McIntyre, A. C. Mix, J. J. Morley, N. G. Pisias, W. L Prell, and N. J. Shackleton, 1984, The orbital theory of Pleistocene climate: Support from a revised chronology of the marine record, *in* A. Berger, J. Imbrie, J. Hays, G. Kukla, and B. Saltzman, eds., Milankovitch and climate: Dordrecht, The Netherlands, Reidel, p. 269–305.

Imbrie, J., and K. P. Imbrie, 1979, Ice ages: Solving the mystery. Short Hills, New Jersey, Enslow Publishers, 224 p.

Johnson, W. H., 1990, Ice-wedge casts and relict patterned ground in central Illinois and their environmental significance: Quaternary Research, v. 33, no. 1, p. 51–72.

Karl, T. R., and K. E. Trenberth, 2003, Modern global climate change: Science, v. 302, no. 5651, p. 1719–1723.

King, J. E., 1981, Late Quaternary vegetational history of Illinois: Ecological Monographs, v. 51, no. 1, p. 43–62.

King, J. E., and J. J. Saunders, 1986, Geochelone in Illinois and the Illinoian-Sangamonian vegetation of the type region: Quaternary Research, v. 25, no. 1, p. 89–99.

Kutzbach, J. E., 1987, Model simulations of the climatic patterns during the deglaciation of North America, *in* W. F. Ruddiman and H. E. Wright Jr., eds., North America and adjacent oceans during the last deglaciation: The geology of North America, v. K-3: Boulder, Colorado, The Geological Society of America, p. 425–446.

Laird, K. R., B. F. Cumming, S. Wunsam, J. Rusak, R. J. Oglesby, S. C. Fritz, and P. R. Leavitt, 2003, Lake sediments record large-scale shifts in moisture regimes across the northern prairies of North America during the past two millenia: Proceedings of the National Academy of Sciences, v. 100, no. 5, p. 2483–2488.

Leighton, M. M., 1926, A notable type Pleistocene section: The Farm Creek exposure near Peoria, Illinois: Journal of Geology, v. 34, no. 2, p. 167–174.

Leverett, F., 1898, The weathered zone (Sangamon) between the Iowan loess and Illinoian till sheet: Journal of Geology, v. 6, no. 2, p. 171–181.

Leverett, F., 1899, The Illinois glacial lobe: Washington, D.C., U.S. Geological Survey, Monograph, v. 38, no. 2, 817 p.

Lineback, J. A., 1979, Quaternary deposits of Illinois: Illinois State Geological Survey, 1:500,000.

MacDonald, G. M., 1990, Palynology, *in* B.G. Warner, ed., Methods in Quaternary ecology: Geoscience Canada, v. 5, p. 37–52.

Martinson, D. G., N. G. Pisias, J. D. Hays, J. Imbrie, T. C. Moore Jr., and N. J. Shackleton, 1987, Age dating and the orbital theory of the Ice Ages: Development of a high-resolution 0 to 300,000-year chronostratigraphy: Quaternary Research, v. 27, no. 1, p. 1–29.

Miller, B. B., J. E. Mirecki, and L. R. Follmer, 1994, Pleistocene molluscan faunas from central Mississippi Valley loess sites in Arkansas, Tennessee, and southern Illinois: Southeastern Geology, v. 34, no. 2, p. 89–96.

Morrison, R. B., and J. C. Frye, 1965, Correlation of the Middle and Late Quaternary successions of the Lake Lahontan, Lake Bonneville, Rocky Mountain (Wasatch Range), southern Great Plains, and eastern Midwest areas: Reno, Nevada, Nevada Bureau of Mines Report 9, 45 p.

Nelson, D. M., F. S. Hu, E. C. Grimm, B. B. Curry, and J. E. Slate, 2006, The influence of aridity and fire on Holocene prairie communities in the eastern prairie peninsula: Ecology 87:2523–2536.

Panno, S. V., B. B. Curry, H. Wang, K. C. Hackley, C.-L. Liu, C. Lundstrom, and J. Zhou, 2004a, Climate change in southern Illinois, USA, based on age and $\delta^{13}C$ of organic matter in cave sediments: Quaternary Research, v. 61, no. 3, p. 301–313.

Panno, S. V., S. E. Greenberg, C. P. Weibel, and P. K. Gillespie, 2004b, Guide to the Illinois Caverns State Natural Area: Illinois State Geological Survey, GeoScience Education Series 19, 116 p.

Richards, D. A., and J. A. Dorale, 2003, U-series chronology and environmental applications of speleothems, in B. Bourdon, G. M. Henderson, C. C. Lundstrom, and S. P. Turner, eds., Uranium-series geochemistry: Washington, D.C., The Mineralogical Society of America, Reviews in Mineralogy and Geochemistry, v. 52, p. 407–460.

Roy, M., P. U. Clark, R. W. Barendregt, J. R. Glasmann, and R. J. Enkin, 2004, Glacial stratigraphy and paleomagnetism of late Cenozoic deposits of the north-central United States: Geological Society of America Bulletin, v. 116, no. 1–2, p. 30–41.

Rutter, N. W., Z. Ding, and T. Liu, 1996, Long paleoclimate records from China: Geophysica, v. 32, p. 7–34.

Shackleton, N. J., 1987, Oxygen isotopes, ice volume and sea level: Quaternary Science Reviews, v. 6, no. 3–4, p. 183–190.

Stuiver, M., and P. J. Reimer, 1993, Extended ^{14}C database and revised CALIB 3.0 ^{14}C age calibration program: Radiocarbon, v. 35, no. 1, p. 215–230.

Walker, I. R., 1987, Chironomidae (Diptera) in paleoecology: Quaternary Science Reviews, v. 6, p. 29–40.

Wang, H., L. R. Follmer, and J. C. Liu, 2000, Isotope evidence of paleo-El Niño-Southern Oscillation cycles in the loess-paleosol record in the central United States: Geology, v. 28, no. 9, p. 771–774.

Wang, H., R. E. Hughes, J. D. Steele, S. W. Lepley, and J. Tian, 2003, Correlation of climate cycles in middle Mississippi Valley loess and Greenland ice: Geology, v. 31, no. 9, p. 179–182.

Wang, H., J. A. Mason, and W. L. Balsam, 2006, The importance of both geological and pedological processes in control of grain size and sedimentation rates in Peoria Loess: Geoderma, v. 136, p. 388–400.

Warner, B. G., ed., 1990. Methods in Quaternary ecology: Geoscience Canada 5, 170 p.

Webb, T. III, P. J. Bartlein, and J. E. Kutzbach. 1987. Climatic change in eastern North America during the past 18,000 years; Comparison of pollen data with model results, in W. F. Ruddiman and H. E. Wright Jr., eds., North America and adjacent oceans during the last deglaciation, Geology of North America, K-3: Boulder, Colorado, Geological Society of America, p. 447–462.

Willman, H. B., and J. C. Frye, 1970, Pleistocene stratigraphy of Illinois: Illinois State Geological Survey, Bulletin 94, 204 p.

Zachos, J., M. Pagani, L. Sloan, E. Thomas, and K. Billups, 2001, Trends, rhythms, and aberrations in global climate 65 Ma to present: Science, v. 292, no. 5517, p. 686–693.

Zhou, J., C. C. Lundstrom, B. Fouke, S. Panno, K. Hackley, and B. Curry, 2005, Geochemistry of speleothem records from southern Illinois: Development of $(^{234}U)/(^{238}U)$ as a proxy for paleoprecipitation: Chemical Geology, v. 221, no. 1–2, p. 1–20.

Zhu, H., and R.G. Baker, 1995, Vegetation and climate of the last glacial-interglacial cycle in southern Illinois, USA: Journal of Paleolimnology, v. 14, no. 3, p. 337–354.

Mineral Resources of Illinois

Subhash B. Bhagwat

Long before the French explorers came to Illinois, the indigenous people of the area were using Illinois minerals. They formed pottery from local clay, shaped tools from flint, and carved fluorspar into trinkets. They also mined and carved galena and produced small amounts of lead. Salt was produced from mineral springs. Evidence exists to show that minerals were traded down the Mississippi River.

In the 1670s, Joliet and Marquette reported the occurrence of coal and other minerals in Illinois. The construction of the Illinois and Michigan Canal in 1848 and the building of railroads promoted mineral production, especially coal. The industrialization that occurred between the Civil War and World War I gave rise to large iron and steel centers around Chicago and St. Louis. Coal production reached its all-time high of 90 million tons (81.6 million tonnes) toward the end of World War I, and production of other minerals also flourished. Illinois' mining industry has contributed significant quantities of zinc, lead, clay, silica sand, fluorspar, tripoli, and building stone to the state's economy. Crude oil production in Illinois began in significant quantities in 1906, reaching its peak at almost 150 million barrels a year during World War II.

The increased use of automobiles and the development of the interstate highway network in the 1950s, the development of the coal industry in the western United States in the 1970s, and the residential and industrial construction boom have contributed to the value of Illinois non-fuel minerals, estimated at about $1 billion. The total value of all minerals and fuels produced in Illinois is estimated at about $2.2 billion annually.

Mineral resource extraction competes with other land uses everywhere in the United States, including Illinois. Environmental concerns further add to the difficulty in mineral extraction. Nevertheless, Illinois remains endowed with abundant mineral and fuel resources that could supply the needs of the state for generations to come. Their continued use and economic viability will depend on policies put in place today.

In this part of the book, discussion of Illinois mineral resources is divided into chapters on coal; oil and gas; zinc, lead, and fluorite; and industrial minerals.

Photograph by Joel M. Dexter.

Image on previous page: Oil pump at Salem Field,
Marion County, Illinois.

Coal 14

W. John Nelson, Russell J. Jacobson, Scott D. Elrick, Gary B. Dreher, and William R. Roy

INTRODUCTION

The first written record of coal in the New World is from Illinois. A map from Father Louis Hennepin's 1668 expedition down the Illinois River shows a "cole mine" at what is now Ottawa (*Coal Report* 1954). Whether this "mine" was merely an outcrop or a place where Native Americans dug coal is not recorded.

Coal has played a key role in the industrial growth of Illinois. Settlers first mined Illinois coal in 1810 along the Big Muddy River near Murphysboro, shipping the fuel down the Mississippi River on flatboats. By the 1840s, shaft mines were operating near Belleville. The 1840 U.S. Census records coal mining in 19 counties.

From 1833 to 2004, about 6.1 billion tons (5.5 billion metric tonnes) of coal have been mined in Illinois (Coal Reports 1882 to 2008). This quantity amounts to a cubic block of coal more than a mile on a side. A train loaded with this much coal would have to be long enough to encircle Earth 21 times at the equator. Six billion tons (5.4 billion tonnes) of coal could generate 9.3 trillion megawatts of electrical energy, enough to serve every household in the United States for 6.25 years. Yearly production since 2001 has averaged between 30 and 35 million tons (between 27 and 32 million tonnes) (*Coal Report* 2008).

Through 1954, Illinois ranked third in the United States (behind Pennsylvania and West Virginia) in all-time bituminous coal production. Illinois' rank has fallen to eighth in recent years due to enormous growth in output from western states, where low-sulfur coal seams as thick as 100 feet (30.5 m) are mined in gigantic open pits (*Keystone Coal Industry Manual* 2003). Mining thinner seams in Illinois is more labor intensive; the higher sulfur content of most Illinois coal also places it at a disadvantage. Even though near its lowest point for a century, coal production provides nearly $1 billion a year to the economy of Illinois.

Seventy-three of the 102 counties in Illinois have produced coal at one time or another. Currently, Illinois coal production is near a 100-year low, and only 11 counties reported coal production in 2007. The three leading counties were Saline with 10.9 million tons (9.9 million tonnes), Macoupin with 4.5 million tons (4.1 million tonnes), and White with 2.5 million tons (2.3 million tonnes). The ranking of total coal production since 1882 by counties is (1) Franklin, (2) Perry, (3) Williamson, (4) St. Clair, (5) Macoupin, (6) Christian, (7) Saline, and (8) Fulton. Each of these counties has produced more than 300 million tons (272 million tonnes) of coal (Figure 14-1).

Coal-bearing rocks underlie 37,000 square miles (95,830 km^2), or 68%, of the state, and more than 211 billion tons (191 billion tonnes) of coal are estimated to remain in place (Jacobson and Korose 2003). Illinois is second among all states for total coal resources and first for bituminous coal. Available coal resources in Illinois contain as much energy as Saudi Arabia's oil (Office of Coal Development, Illinois Department of Commerce and Economic Opportunity 2006). Coal makes up nearly 85% of total fuel resources in the United States, and, at present rates of consumption, coal resources should last more than 250 years.

Electricity generation is by far the largest use for U.S. coal, and coal-fired power plants provide 52% of the nation's electricity. By 2020, electricity consumption is expected to grow by 35% in the United States and by 70% worldwide.

ORIGIN OF COAL

Illinois coal originated as in-place peat deposits in freshwater to brackish water swamps that occupied a vast coastal lowland bordering the shallow Pennsylvanian sea. During the Pennsylvanian age (Chapter 10, Pennsylvanian and Permian), Illinois lay near the equator, and forest plants (nearly all extinct today) flourished in the tropical climate. Giant ancestors of present-day club mosses, horsetails, ferns, conifers, and cycads dominated these forests. Dense undergrowth included ferns, fernlike plants, and club mosses. The lack of annual growth rings in coal-forming trees points to rapid growth and a year-round hot, wet climate typical of modern tropical lowlands (White and Thiessen 1913, Jacobson 2000).

Under these conditions, plant debris rapidly accumulated, forming mats of peat many feet to tens of feet thick. Highly acidic, oxygen-poor swamp water retarded decay. As the Illinois Basin subsided, the swamps were drowned, and the peat was covered with sediment. With continued subsidence and burial over time, peat gradually was transformed into coal. Pressure, heat, and time expelled water and volatile matter and reconstituted the carbon compounds, changing peat first to lignite, then to subbituminous coal, and finally to bituminous coal. Illinois coal apparently reached its deepest burial and its present rank by Early Permian time (Damberger 1971, Clegg 1955).

COAL CHARACTERISTICS
Thickness and Continuity of Coal Beds

Most coal beds in the state are highly continuous, extending through large areas of Illinois, Indiana, and western Kentucky; some are correlated with coal fields in other states. Such continuity is a big advantage for planning and operating mines. Although faults and other discontinuities occur in Illinois (Nelson 1981, 1983), they are less prevalent here than in most other U.S. coal fields.

Coal seams in Illinois range from a mere streak to as much as 13 feet (4 m) thick. The lower minable limits are about 1.5 feet (0.46 m) for surface mining, 2.5 feet (0.76 m) for historic underground mining, and 4 feet (1.2 m) for modern underground mining. Most production takes place in beds that are 5 to 10 feet (1.5 to 3 m) thick. Underground miners appreciate Illinois' "high coal," which allows them to walk upright.

Rank

Rank specifies the degree to which original plant material was altered during coalification. Rank is vital in determining the best uses of a particular coal. Increasing heat and pressure drive out moisture and gaseous components, which increases the proportion of carbon and, thus, the energy value of the coal. From lowest to highest, coal rank progresses through peat, lignite, subbituminous, bituminous, semi-anthracite, and finally anthracite.

Illinois coal is classified as high-volatile bituminous. The most abundant rank of coal in the United States, bituminous coal has a moisture content of less than 20% (by weight) and a heating value of 10,500 to 14,000 BTU/lb (/0.45 kg). (A British Thermal Unit is the amount of energy required to raise the temperature of one pound of water by 1°F.) Illinois bituminous coal is compact, black, and banded, typically having alternating layers of shiny (vitrain) and dull (durain, clairain, and fusain) material.

The heating value of Illinois coal increases with depth at rates ranging from about 60 BTU/lb per 100 feet (30.5 m) of depth in southern Illinois to 150 BTU/lb per 100 feet (30.5 m) in northwestern Illinois. Such an increase is in line with increases with depth observed in other coal fields (Damberger 1971), although the actual values vary from one field to another. Heating value increases from about 11,000 BTU/lb in northwestern Illinois to nearly 15,000 BTU/lb in the southeast (Figure 14-2). Moisture content drops from around 20% in the northwest to less than 5% in the southeast. As Damberger (1971) concluded, the rank change probably reflects the increasing original depth of burial toward the southeast, combined with heat flow from Permian igneous intrusions in southeastern Illinois. Note that rank abruptly decreases across the La Salle Anticlinorium in east-central Illinois. This structure actively rose during and after Pennsylvanian time, so coal there was buried less deeply.

Ash

Ash, the incombustible matter in coal, is derived from three major sources: (1) the nonflammable portion of plant matter, analogous to wood ash; (2) mineral sediment (such

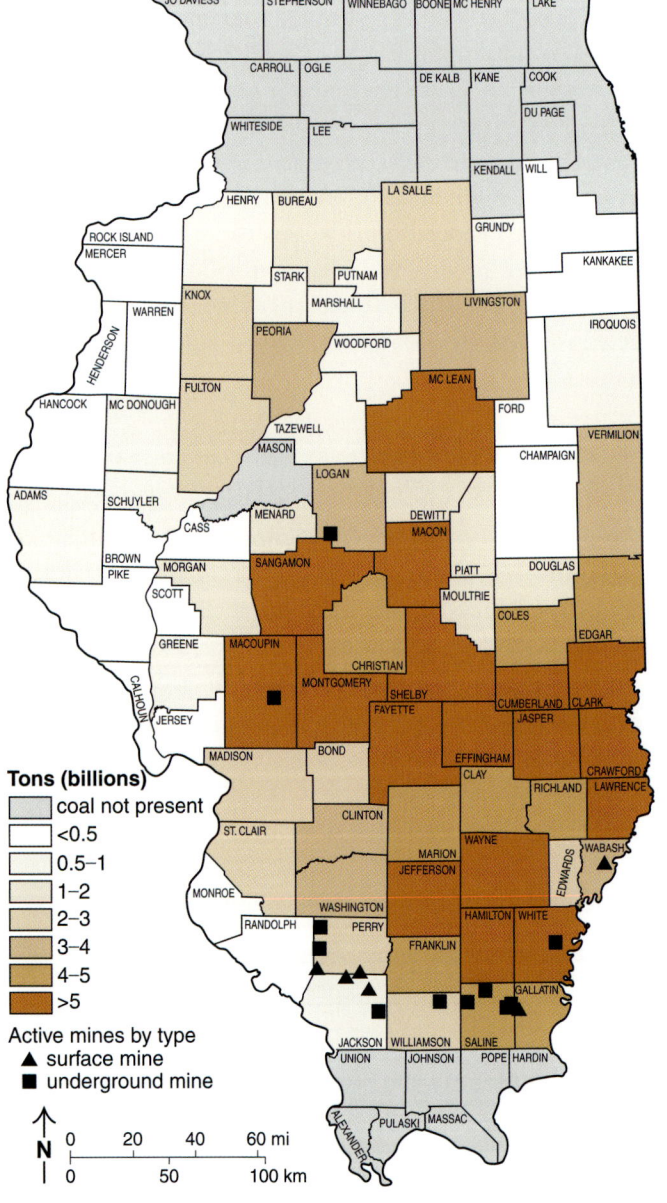

Figure 14-1 Coal reserves by county and active mines in 2008 (Elrick 2008).

as clay, silt, and sand) that was deposited with the peat and occurs both finely dispersed through the coal and as distinct rock layers and partings; and (3) mineral matter introduced (generally by groundwater) after the peat (or coal) was buried. The most common components of ash are quartz (silica), clay, pyrite (iron sulfide), and calcite (calcium carbonate).

A primary objective of coal preparation is to reduce ash content, but this objective is only partially met because so much ash is microscopic. Preparation is most effective at removing rock layers and minerals introduced after burial, including the out-of-seam dilution caused by the roof and floor rock that are unavoidably excavated with the coal.

Coal mined in Illinois typically contains 6 to 14% ash; 10% is average, and less than 6% is remarkable. Seams thick enough to mine seldom contain more than 14% ash, although values as high as 28% have been reported. The distribution of ash can be irregular and unpredictable, although the Springfield and Herrin Coals have remarkably consistent ash contents.

Sulfur

Sulfur produces noxious and corrosive fumes when coal is burned. Environmental laws mandate the reduction of sulfur emissions, requiring power plants to install expensive scrubbers when burning high-sulfur coal. Thus, low-sulfur coal from the West (as low as 0.3% sulfur) is more economical to use than Illinois coal (3% to 5% sulfur).

Sulfur occurs in three forms: organic, pyritic, and sulfate. Organic sulfur, present in the original plant matter, is locked up in molecules of coal and, for practical purposes, cannot be removed. Organic sulfur in Illinois coal varies from 0.4% to 3.0%. Pyritic sulfur occurs as iron sulfide (pyrite) and other sulfide minerals. These generally occur as visible bands, lenses, and vein fillings that can be removed in the preparation plant. However, some pyritic sulfur is microscopic and cannot be removed economically. Pyritic sulfur content varies from almost none to more than 5% of Illinois coal. Sulfate sulfur, which results from the weathering of pyrite, averages about 0.2% and is largely extracted during preparation (Gluskoter and Simon 1968). Total sulfur, the sum of the three varieties, ranges from 0.5% to more than 7% in Illinois coal. Coal having less than 2.5% sulfur is found in a few well-defined areas and represents about 6% of the state's resources (Jacobson and Korose 2003).

Sulfur content is closely related to the type of rock overlying the coal. Much sulfur in Illinois coal came from seawater and marine sediments. Peat buried by gray shale was shielded from this sulfur source (Gluskoter and Simon 1968, Hopkins 1968, Gluskoter and Hopkins 1970, Allgaier and Hopkins 1975, Jacobson 1983). Coal beds overlain by marine rocks, such as black shale or limestone, contain 3% or more of sulfur. Nearly all low-sulfur coal in Illinois is overlain by thick (20 feet [6 m] or greater) gray shale or siltstone. Fossils and sedimentary features indicate that these overlying rocks were deposited in fresh or brackish water. This relationship between gray shale and low-sulfur coal holds for the Murphysboro Coal in Jackson County, the Colchester Coal in northeastern Illinois, the Springfield Coal in southeastern Illinois, and the Herrin Coal in several areas.

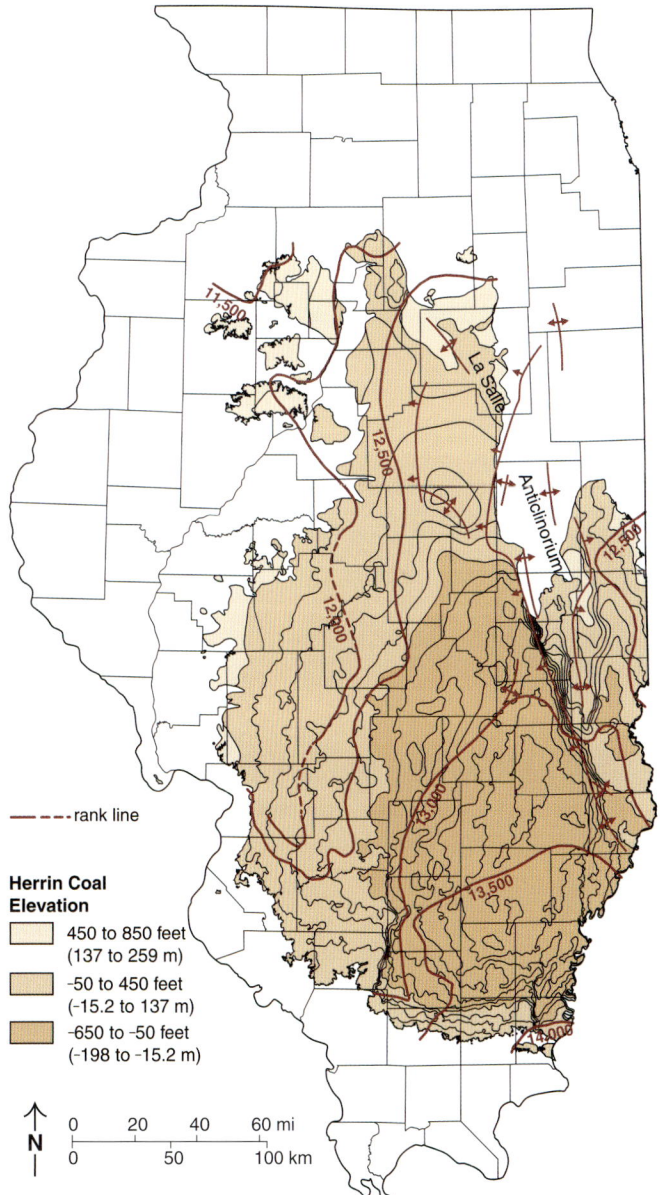

Figure 14-2 Herrin coal ranks (Damberger 1971): less than 13,000 BTU/lb (/0.45 kg), high-volatile C bituminous; 13,000 to 14,000 BTU/lb, high-volatile B bituminous; greater than 14,000 to nearly 15,000 BTU/lb, high-volatile A bituminous.

Minor and Trace Elements

Coal contains many elements in concentrations too small to detect without special equipment. Most minor and trace elements occur in quantities not greatly different from those in the average rock. Only boron, chlorine, and selenium are significantly enriched. Of the three, chlorine presents a special concern because it contributes to fouling and corrosion of steam boilers. Gluskoter and Rees (1964) found that the chlorine content of Illinois coal increases with depth and is closely tied to the salinity of groundwater.

Zinc and cadmium are enriched in coal samples from several counties in western Illinois. These elements were emplaced chiefly as mineral veins during or after coalification. Zinc and cadmium constitute as much as 0.2% of whole coal and 1.5% or more of mine waste. In such quantities, zinc and cadmium could be recovered as a by-product of coal mining, given favorable economics (Hatch et al. 1976; Cobb et al. 1979, 1980).

Other elements that occur in Illinois coal include arsenic, beryllium, cobalt, chromium, copper, fluorine, lead, lithium, manganese, mercury, molybdenum, nickel, phosphorus, tin, thorium, uranium, and vanadium (Demir et al. 1997). All are measured in parts per million (milligrams per kilogram). Because essentially all coal shipped from Illinois mines is cleaned, concentrations of most of these elements are further reduced during preparation. None occurs in large enough concentrations to represent a health hazard or to deter the use of Illinois coal (Gluskoter et al. 1977, Harvey et al. 1983).

COAL RESOURCES

Definition and Mapping

Coal resources include all coal in the ground that meets specified depth and thickness limits. *Coal reserves* refer only to coal that can be mined at a profit under current economic conditions. Government agencies generally map resources; private coal companies establish reserves. For many decades, the Illinois State Geological Survey (ISGS) has defined resources for underground mining as being all coal thicker than 28 inches (0.71 m), regardless of depth. All coal thicker than 18 inches (0.46 m) and less than 150 feet (46 m) deep is considered strippable and is counted as resources. These values approximate historic and current limits for mining bituminous coal in the eastern and central United States. To map resources, geologists compile information on coal depth and thickness from mines, outcrops, and boreholes. Cored coal test borings provide the best borehole information. Electric, gamma-ray, density, and other geophysical logs from oil and gas test holes also enable geologists to map coal in areas of Illinois where no mining or coal test drilling has taken place.

Illinois has the largest bituminous coal resources (211 billion tons [191 billion tonnes]), including the largest strippable bituminous coal resources (22 billion tons [19.9 billion tonnes]) of any state (Jacobson and Korose 2003). Illinois has the third largest total coal resources of any state and is second only to Montana in terms of demonstrated reserves (53% of total, or about 111 billion tons [101 billion tonnes]).

Development Potential and Availability

Beyond basic resource measurement, the ISGS seeks to identify coal deposits that have the best potential for mining by considering economic, cultural, and political factors. For example, coal underlying densely populated areas, large bodies of water, state and county parks, cemeteries, and interstate highways cannot be surface mined (Treworgy et al. 1978). These surface features also restrict underground mining because subsidence results from longwall mining or pillar extraction. Areas that have been densely drilled for oil and gas may be impractical to mine because a barrier of coal must remain around each well (Treworgy and Bargh 1982).

The size of the resource block is another important consideration. The cost of opening a mine increases as shallow reserves become exhausted, or as the depth to the coal seam increases. Thus, most coal companies need substantial tonnage to recoup the investment of opening a mine. In assessing surface-minable resources, Treworgy et al. (1978) identified contiguous blocks of coal containing at least 6 million tons (5.4 million tonnes). For underground mining, Treworgy and Bargh (1982) mapped blocks of 25 million tons (22.7 million tonnes) and larger. Those scientists further classified Illinois' underground coal resources into four categories, ranging from high development potential (comparable in depth and thickness to coal currently being mined) to restricted development potential (coal difficult to mine because of surface features or dense petroleum drilling). Subsequent studies (e.g., Treworgy et al. 1999, Jacobson and Korose 2003) classified development potential similarly.

According to those standards, about 46% of the state's resources have high development potential. Approximately 11%, or 22 billion tons (19.9 billion tonnes), of Illinois coal resources are classified as surface minable. About 6.1 billion tons (5.5 billion tonnes) are defined as having high development potential for surface mining. Interestingly, this amount is nearly equal to all the coal mined in Illinois from 1833 to the present.

Coal Seams

The first State Geologist, Amos H. Worthen, numbered coal seams in Illinois with the oldest, or lowest, being No. 1. Coal beds of neighboring states were numbered in similar fashion, but the numbers do not match. Over time, Illinois coal seams acquired names taken from places where they were mined or exposed (Figure 14-3). Although names are preferred for formal geological reports, the numbers persist throughout the mining industry: Rock Island (No. 1), Colchester (No. 2), Houchin Creek (No. 4), Springfield (No. 5), Herrin (No. 6), and Danville (No. 7). Currently, 48 coal beds in Illinois have names, but only 25 have been mined. The 12 seams that contain significant resources are described herein from oldest to youngest.

Rock Island Coal

Mined prior to 1970 near the Quad Cities (Rock Island, Davenport, Bettendorf, and Moline) and Monmouth, Galesburg, and Peoria, the Rock Island Coal occurs as elongate, lenticular bodies that mostly trend east-west or northeast-southwest. The seam thickens from a streak on the margins to as much as 8 feet (2.4 m) along the central axes of its lenses. The Rock Island Coal was commonly 4 to 5 feet (1.2 to 1.5 m) thick where mined (Searight and Smith 1969). As Wanless (1965) deduced, these coal bodies probably represent peat that accumulated in estuaries (drowned river valleys) as sea level began to rise. Marine limestone directly overlies the coal, producing excellent roof conditions in underground mines. County averages for the Rock Island Coal were 10,290 to 11,470 BTU/lb, 7.9 to 10.4% ash, and 4.4 to 5.5% sulfur (Cady 1935).

The Litchfield and Assumption Coals, which were mined during the early twentieth century near their namesake cities, correlate with the Rock Island Coal. On analysis, the Assumption Coal yielded 11,600 BTU/lb, 8.9% ash, and 2.3% sulfur (Cady 1935). Current estimates of original resources for the Rock Island, Litchfield, and Assumption seams are nearly 1.5 billion tons (1.4 billion tonnes) of which 200 million tons (181 million tonnes) are deemed minable (Searight and Smith 1969, Treworgy and Bargh 1982).

Murphysboro Coal

The Murphysboro seam yielded the first coal mined in Illinois and supported large underground mines at Murphysboro and Carbondale from the 1880s through the 1940s. Surface mining of that seam continues today. Although "patches" of thick Murphysboro Coal occur widely in southern Illinois, reserves have been mapped only in Jackson, Perry, and western Williamson Counties, where the seam is 1 to 7 feet (0.30 to 2.23 m) thick (Jacobson 1983). Coal mined near Murphysboro contained 5 to 7% ash and 0.6 to 2.0% sulfur and yielded 12,300 to 12,700 BTU/lb. All are excellent values for midwestern coal. Some of this coal was used to make coke. At Carbondale, ash and sulfur contents were higher, and the heating value was slightly lower. Jacobson (1983) showed that thick, low-sulfur Murphysboro Coal underlies thick, non-marine silty gray shale along the flanks of an ancient river channel now filled with sandstone. This setting is nearly identical to that of other low-sulfur coal deposits in Illinois and provides a model for new reserve exploration. Although thick coal near Murphysboro is largely mined out, substantial

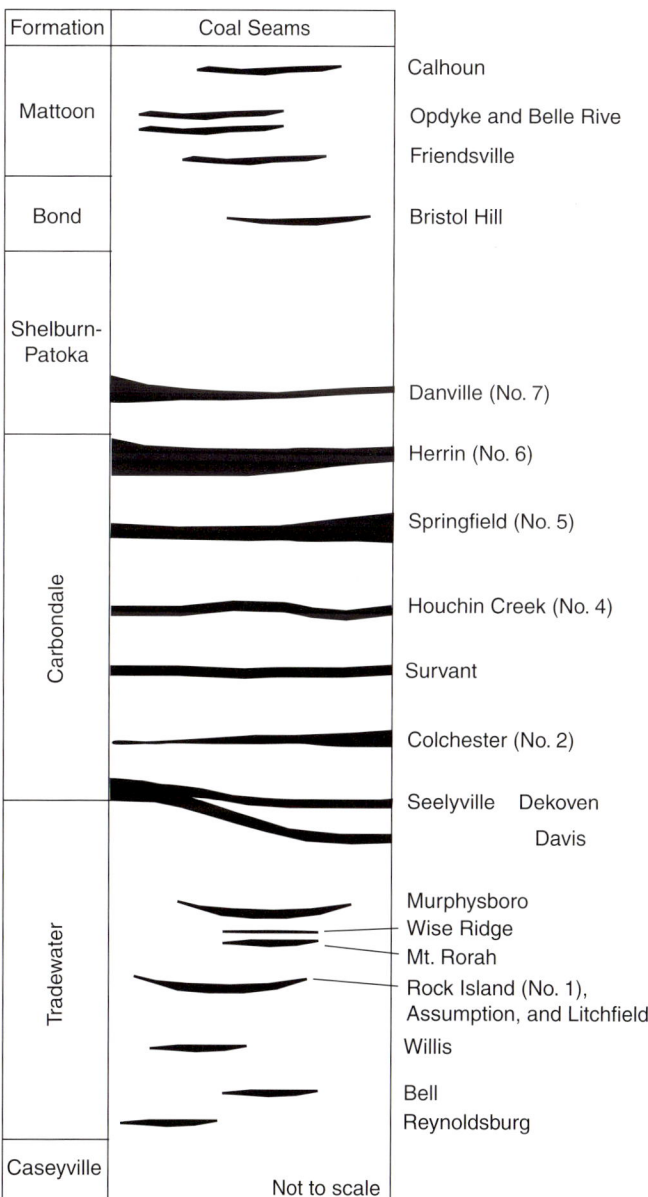

Figure 14-3 Significant coal seams mentioned in this chapter. Not shown is the Jamestown Coal, which is only a few inches thick in most places in the state and has never been mined in Illinois. (Source: Coal Section, Illinois State Geological Survey.)

resources of low-sulfur coal extend northward. Jacobson (1983) estimated 650 million tons (590 million tonnes) of coal in place, of which approximately 160 million tons (145 million tonnes) are strippable.

Dekoven and Davis Coals

The Dekoven and Davis seams commonly are 10 to 25 feet (3 to 7.6 m) apart and are surface mined in tandem in Saline and Gallatin Counties of southeastern Illinois. Jacobson (1993) mapped 4.1 billion tons (3.7 billion tonnes) of resources in those two counties. Additional resources, most too deep for surface mining, have been mapped in Franklin, Williamson, White, Hamilton, Wabash, and Edwards Counties, all in southeastern Illinois. Nearly all mining in these two seams in Illinois has been at the surface, but they have been mined underground nearby in Kentucky, and a highwall drift mine in the Davis was recently opened in Gallatin County. Where mined, the Dekoven averages 3 to 3.5 feet (0.9 to 1.07 m) thick, and the Davis averages 3.5 to 4 feet (1.07 to 1.2 m). Both seams contain 8 to 13% ash and 3 to 5% sulfur; heating values range from 11,900 to 12,800 BTU/lb. The Dekoven and Davis merge northward to become the Seelyville Coal (Jacobson 1987). Current estimates of in-place resources are nearly 6 billion tons (5.4 billion tonnes) for the Dekoven Coal and almost 9.6 billion tons (8.7 billion tonnes) for the Davis Coal (Korose et al. 2002, Jacobson and Korose 2003).

Seelyville Coal

Mined extensively in Indiana, the Seelyville Coal provides an untapped resource of 7.5 billion tons (6.8 billion tonnes) in east-central Illinois (Treworgy 1981, Treworgy and Bargh 1982). Borehole data indicate that the seam is 3.5 to 9 feet (1.07 to 2.74 m) thick beneath an area of 1,900 square miles (4,921 km^2), although the coal contains interbeds of shale in places. The few available analyses indicate a high-sulfur coal having ash and heating values comparable with other Illinois Basin coal. One factor that has deterred mining the Seelyville is its depth, which is as much as 1,500 feet (457 m). Although well within the range of current technology, mining costs increase with depth; to date, no Illinois mines have operated much below 1,000 feet (305 m). Substantial Seelyville resources at less than 800 feet (244 m) depth occur close to the Indiana border.

Colchester Coal

The Colchester Coal may have the largest original extent of any coal seam in the world (Figure 14-4). It correlates with the Whitebreast Coal of Iowa; the Croweburg Coal of Missouri, Kansas, and Oklahoma; the Princess No. 6 of eastern Kentucky; the Lower Kittanning of Ohio and Pennsylvania; and the No. 6 Block Coal of West Virginia (Wanless 1957, Peppers 1996). The Colchester occurs throughout the Illinois Basin but is thick enough to mine only in parts of northern and western Illinois.

The earliest large-scale mining in Illinois was in this seam in northern Illinois, especially around Wilmington, Braidwood, La Salle, and Spring Valley. Proximity to the Chicago market was a great advantage. Production peaked around 1905 at 12 to 15% of the state's total. Many operators practiced a form of longwall mining to maximize recovery in a coal seam that was only 2 to 4 feet (0.6 to 1.2 m) thick. As rail shipping costs decreased, mining Colchester Coal underground in northern Illinois became unprofitable in competition with much thicker coal in southern Illinois. The last deep mine closed in the early 1950s; surface mining continued for another two decades. Extensive surface mining and shallow underground mining in the Colchester also took place between Peoria and Rushville. The last surface mine in the Colchester, located south of Macomb, closed in 2003.

Korose et al. (2003) mapped 18.5 billion tons (16.8 billion tonnes) of Colchester resources; 0.5 billion tons (0.45 billion tonnes) have been mined. Resources lie on the northern and western margin of the coal field from Kankakee through La Salle and Peoria and southward to Alton. In most places, the coal is 1.5 to 3.5 feet (0.46 to 1.07 m) thick; locally it exceeds 4 feet (1.2 m) (Figure 14-4). Elsewhere in Illinois, the Colchester is too deep for surface mining and too thin for underground mining.

The Colchester Coal has low ash content (5 to 10%). Its heating value varies from 10,400 to 11,700 BTU/lb, which is somewhat lower than that of coal from southern Illinois but in line with regional trends. As with other Illinois coal beds, the sulfur content of the Colchester correlates with the type of rock overlying the seam. Where the black, fissile, marine Mecca Quarry Shale Member lies directly on or close to the coal, sulfur content is 3% or higher. Low-sulfur Colchester Coal (0.7 to 2.5%) occurs in parts of northern Illinois where the thick, gray Francis Creek Shale Member overlies the coal (Gluskoter and Hopkins 1970). The Francis Creek is noteworthy for its fossils of soft-bodied organisms and plants, the Mazon Creek fauna (Chapter 11, Mesozoic and Tertiary).

Survant Coal

Called the Lowell, Shawneetown, and No. 2A Coal in early reports, the Survant Coal is highly persistent but is too thin to mine in most of Illinois. The Survant thickens eastward to 8 feet (2.4 m) and has been mined in several

Indiana counties that border Illinois (Wier 1973, Shaver et al. 1986). In Illinois, deep-minable resources have been mapped near Lawrenceville and Mt. Olive, where the coal reaches a thickness of 3 to 4 feet (0.9 to 1.2 m) (Treworgy and Bargh 1984). Unmapped resources of Survant Coal probably exist, but development potential is low.

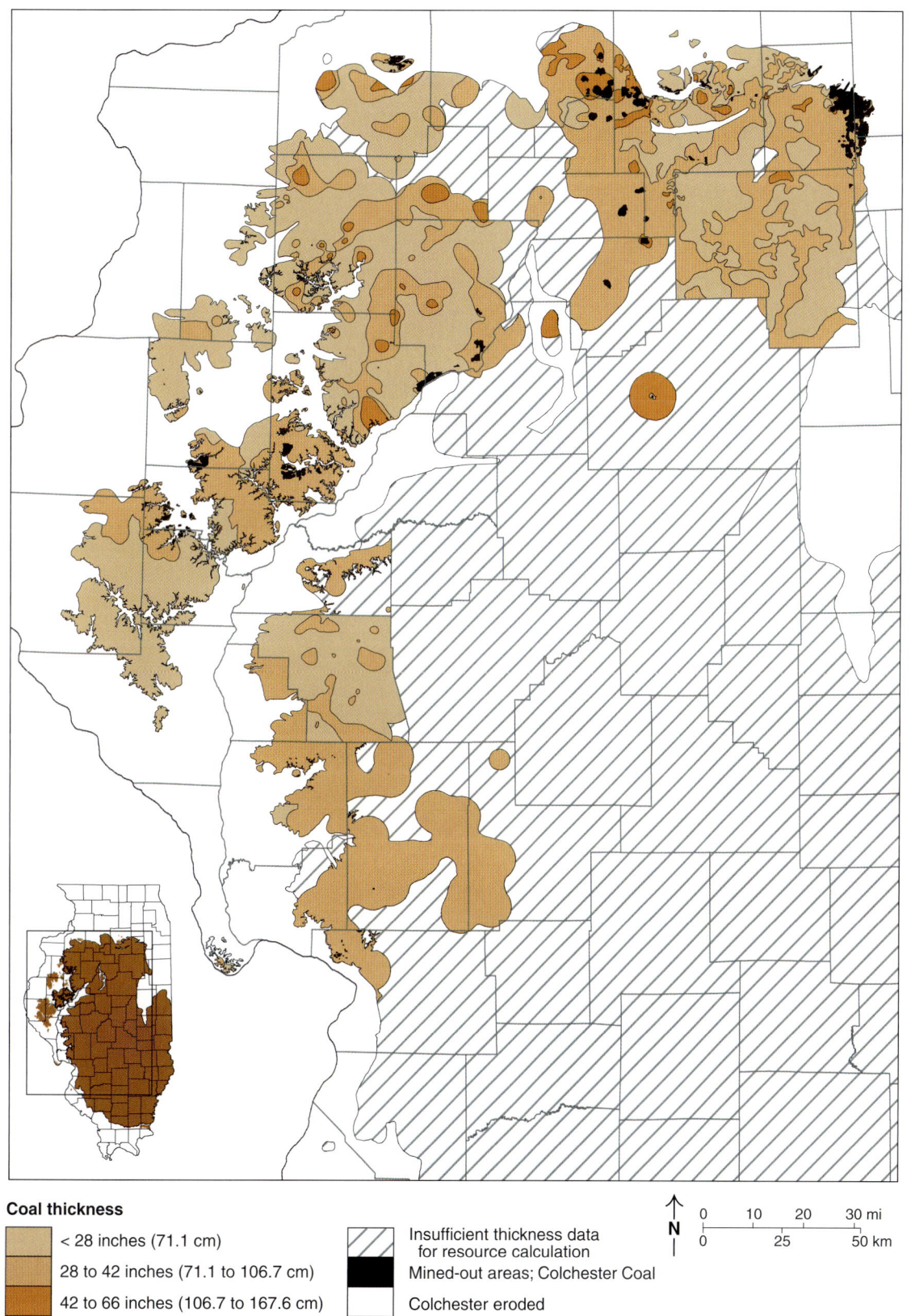

Figure 14-4 Resources of Colchester Coal, which may have the largest original extent of any coal seam in the world. The earliest large-scale mining in Illinois was in the Colchester Coal (Korose et al. 2003).

Houchin Creek Coal

The Houchin Creek Coal is another seam that is regionally extensive, yet attains minable thickness only locally. Small underground mines near Soperville (Knox County), Summum (Fulton County), and Greenville (Bond County) formerly exploited this coal. The Houchin Creek Coal also was formerly surface mined near Wilmington along with the underlying Colchester Coal and southwest of Harrisburg along with the Davis and Dekoven Coals. Total historic production is probably less than 10 million tons (9 million tonnes). Treworgy and Bargh (1982) mapped 514 million tons (466 million tonnes) of deep-minable Houchin Creek Coal resources; Treworgy et al. (1978) identified about 70 million tons (63.5 million tonnes) of surface-minable Houchin Creek Coal. Reserves are scattered around the state but lie chiefly in the region between Springfield and St. Louis and in the northeast corner of the coal field (Treworgy and Bargh 1984). In much of Illinois, this coal has not been mapped.

The Houchin Creek generally ranges from less than 1 foot (0.3 m) to about 2.5 (0.76 m) feet thick. Locally it exceeds 3 feet (0.9 m); maximum reported thickness is 5 feet (1.5 m) in Grundy County in northeastern Illinois (Treworgy et al. 1978). The coal is universally overlain by black, fissile marine shale that is in turn topped by a thin marine limestone. Analyses from Knox County show 7 to 9% ash, 3 to 4% sulfur, and 10,800 to 11,300 BTU/lb.

Springfield Coal

The Springfield Coal is the most important commercial coal bed in the Illinois Basin and ranks second among Illinois coal seams in cumulative production and remaining resources. It has been mined underground extensively in and around Springfield, yet large resources remain in central Illinois. One underground mine near Elkhart in Logan County is currently active. The coal in this region is 4.5 to 6 feet (1.37 to 1.8 m) thick, is overlain by black marine shale, and has a high sulfur content (3 to 5%). Abundant claystone dikes or "horsebacks" in the seam add much waste material to the mine-run coal and contribute to roof failures (Damberger 1970).

The second major mining area extends from Harrisburg through Eldorado, McLeansboro, Grayville, and Mt. Carmel in southeastern Illinois. The thickest coal, up to 8 feet (2.4 m), flanks the Galatia channel, an ancient river that meandered through the swamp while the Springfield peat was forming. Coal near the channel is overlain by the Dykersburg Shale, a gray, silty shale deposited in fresh to brackish water (Figure 14-5). Where the Dykersburg is thicker than 20 feet (6 m), the Springfield Coal consistently contains less than 2.5% sulfur (Hopkins 1968). Several large underground mines are currently exploiting this reserve. Mining problems encountered near the Galatia channel include shale "splits" (lenses of rock) within the coal, "rolls" (sharp undulations in the seams), or small channels that partially cut out the coal, and unstable roof where the Dykersburg Shale is thinly laminated (Nelson 1983). Away from the Galatia channel, the coal thins to 4 or 5 feet (1.2 to 1.5 m), its sulfur content increases, and the Dykersburg becomes absent or lenticular. A vast tract of essentially unexploited Springfield Coal extends northward to Paris and Pana and westward to Centralia and Marion (Hopkins 1968; Treworgy and Bargh 1982, 1984).

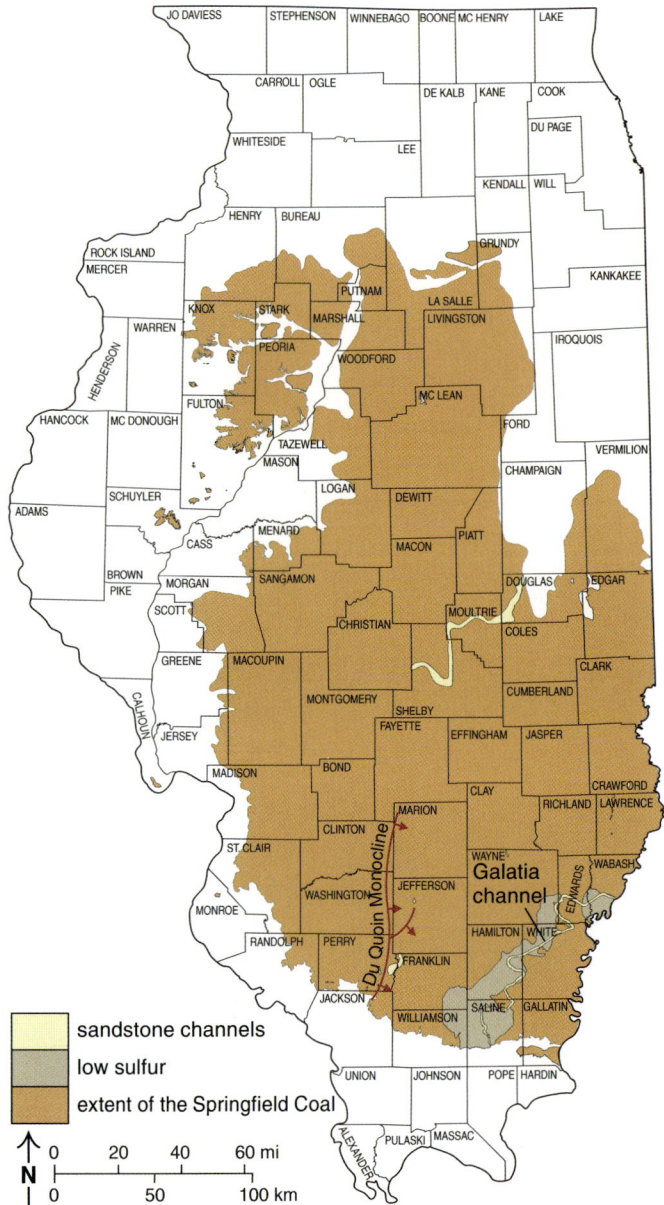

Figure 14-5 Areas of low-sulfur (less than 3%) Springfield Coal along with channels where coal is thickest (Treworgy et al. 1999).

Two more sizeable blocks of Springfield Coal have been mapped. One extends west and southwest from Peoria, and the other lies west of the Du Quoin Monocline in western Illinois. Small-scale underground mining and large-scale surface mining have taken place in both areas.

Remaining resources of Springfield Coal are estimated at 63 billion tons (57 billion tonnes), or about 31% of the state's total. Of this amount, about 8 billion tons (7.3 billion tonnes) are surface minable, and 1.4 billion tons (1.3 billion tonnes) are thought to be low in sulfur (Treworgy et al. 1999).

Herrin Coal

The Herrin Coal is the most extensively mined seam in Illinois. The seam accounts for about 40% of the remaining resources and almost 75% of those having high development potential. The Herrin is known as the No. 11 Coal in western Kentucky and correlates with the Lexington Coal in Missouri and the Mystic Coal in Iowa (Wanless 1939, Peppers 1996). The Herrin is absent or too thin to mine in Indiana where the overlying and thicker Hymera (VI) Coal is mined.

Herrin Coal is thickest bordering the Walshville Channel, the course of an ancient river that flowed through the swamp while peat accumulated (Figure 14-6). Low-sulfur coal is overlain by non-marine gray shale (the Energy Shale), which forms large lobe-shaped bodies as thick as 100 feet (30.5 m) flanking the Walshville Channel (Gluskoter and Simon 1968, Gluskoter and Hopkins 1970, Allgaier and Hopkins 1975, Nelson 1987). The relationship is precisely the same as the Springfield Coal with the Dykersburg Shale and the Colchester Coal with the Francis Creek Shale. Drawbacks of mining near the channel include "splits" of rock and "rolls." A split is when part of the coal seam splits from the main seam. A roll is when the coal seam thins because of the protrusion of overlying rock or an upheaval of the seam floor. The Energy Shale typically forms a weak mine roof and deteriorates rapidly when exposed to air (Krausse et al. 1979; Nelson 1983, 1987).

The "Quality Circle" (Figure 14-6), an area of about 250 square miles (647 km²) encompassing Herrin, Marion, West Frankfort, Benton, and Mt. Vernon in southern Illinois, contained the thickest coal ever mined in Illinois as well as its largest deposit of low-sulfur coal, much of which was suitable for blending to make metallurgical coke. The coal averaged 8 feet (2.4 m) thick and locally reached 13 feet (4 m) thick. Sulfur content ranged from 0.5 to 2.5%, averaging about 1.5%. The "Quality Circle" is essentially mined out.

Other deposits of low-sulfur Herrin Coal include the Troy District in Madison County east of St. Louis, Missouri, and the Hornsby District south of Springfield, Illinois (Figure 14-6). Difficult roof control in thick, moisture-sensitive gray shale deters mining in both areas (Nelson 1987). The largest area of gray shale roof is in east-central Illinois. Except in Vermilion County south of Danville and near Murdock in Douglas County, little mining has taken place in this tract. Discouraging factors include thinner coal, thick splits of shale, difficult roof conditions, and (toward the south) depth.

Huge resources of high-sulfur Herrin Coal lie in the Fairfield Basin of southeastern Illinois and on the Western Shelf from Macoupin and Christian Counties southward. The coal is consistently 6 to 8 feet (1.8 to 2.4 m) thick on

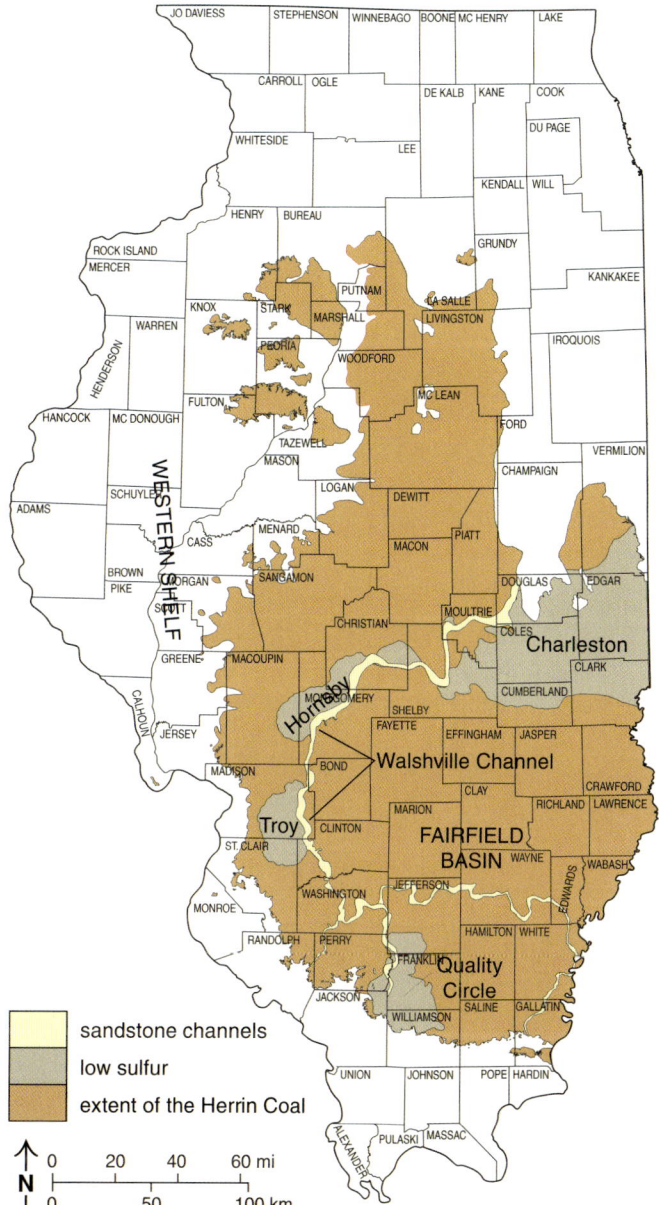

Figure 14-6 Areas of low-sulfur (less than 3%) Herrin Coal along with channels where coal is thickest (Treworgy et al. 2000).

the Western Shelf and generally is thinner in the Fairfield Basin. Smaller patches of minable Herrin Coal occur near Spring Valley, La Salle, and Streator in northern Illinois (historic underground mining) and west of the Illinois River (recently active surface mining). In all of these areas, the coal is overlain in turn by lenses of Energy Shale, black Anna Shale, Brereton Limestone, and sandstone or shale. The Brereton Limestone is the most competent unit and the key to roof stability in deep mines (Krausse et al. 1979). Claystone dikes or "horsebacks" and "white top," in which the upper part of the seam is riddled with small clay veins, are common in some areas, mainly on the Western Shelf. These features contaminate the coal and weaken the roof (Damberger 1970).

A total of 78.9 billion tons (71.6 billion tonnes) of original in-place resources have been mapped, including 8.7 billion tons (7.9 billion tonnes) of strippable coal and 9.5 billion tons (8.6 billion tonnes) of low-sulfur coal (Treworgy et al. 2000, Jacobson and Korose 2003).

Jamestown or Hymera Coal

The Jamestown Coal is an important commercial bed in Indiana, where it is called the Hymera (VI), and in western Kentucky, where it is called the Paradise (No. 12). The Jamestown Coal is widespread in Illinois but typically is only a few inches thick. Borehole data show that the seam thickens markedly near the Indiana border. Korose et al. (2002) mapped 3.6 billion tons (3.3 billion tonnes) of deep-minable Jamestown Coal in Clark, Crawford, and Lawrence Counties. The seam has never been mined in Illinois, and no analytical data are available. The counties where the Jamestown is minable have the potential for multiseam underground operations.

Danville Coal

The Danville Coal, the youngest seam mined widely in Illinois, is developed best around Danville. A belt of resources has been mapped southward along the Indiana border from Vermilion to Lawrence Counties. Recent exploration reportedly has outlined some relatively low-sulfur Danville Coal in Clark County (Korose et al. 2002). Danville Coal having sulfur content lower than 2.4% is widespread in contiguous areas of Indiana (Wier 1973, Harper 1985). Low-sulfur Danville Coal, currently being mined in southwestern Indiana, is overlain by thick gray shale and siltstone containing abundant fossil land plants, which fits the usual correlation of non-marine gray shale roof with low-sulfur coal.

Northern Illinois, particularly from La Salle southeastward toward Pontiac and near Sparland, Colfax, and Bloomington, is the location of historic mining and resources of the Danville Coal. The quality of the northern reserve is lower: 12 to 15% ash, 3 to 4% sulfur, and only 9,600 to 10,3000 BTU/lb. Statewide in-place resources total 19.6 billion tons (17.8 billion tonnes) (Korose et al. 2002).

Other Coals

Numerous coal beds other than those described add small resources to the statewide total, including the Reynoldsburg, Bell, Willis, Mt. Rorah, and Wise Ridge Coals in the Tradewater Formation and the Bristol Hill, Friendsville, Belle Rive, Opdyke, and Calhoun Coals in the McLeansboro Group (Nance and Treworgy 1981; Treworgy and Bargh 1982, 1984). All have been mined in the past by surface mining or from small slope and drift mines. Some minor coals are highly lenticular; others are widely traceable but attain minable thickness only locally. These minor coals rarely exceed 4 feet (1.2 m) in thickness and are minable in areas ranging from a few square miles to a few tens of square miles.

Coal Mining

Surface Mining

Surface mining, also known as strip mining, takes place where coal seams lie relatively close to the surface. The soil and rock overlying the coal, known as overburden, are excavated to expose the coal, which is then removed and loaded into trucks. The generally flat landscape of Illinois coupled with thick, shallow, and widely continuous coal seams make the state especially favorable for surface mining.

Some of the world's earliest surface mining took place in Illinois. Horse-drawn scrapers uncovered coal at Grape Creek south of Danville as early as 1866. The nation's first steam shovel for coal mining appeared near Danville in 1885. Ever larger shovels and draglines fostered rapid growth of Illinois surface mining around 1915 (Bottomley 1944; Hollingsworth 1963). Surface mining accounted for 30% of Illinois coal production in 1950, peaked at 59.2% in 1972, and subsequently declined over a 30-year period to about 20% of total output (*Coal Reports* 1972 to 2002). Depletion of surface-minable coal, increased reclamation costs, and advances in underground mining technology account for the change. After more than a century, less than 1% of Illinois has been affected by surface mining.

The enormous stripping shovels, draglines, and bucket-wheel excavators used in Illinois surface mines from the 1950s through the 1980s were among the largest mobile machines that ever operated on land (Chenoweth et al.

2008) (Figure 14-7). "Big Paul," the stripping shovel at the River King Mine, had a 70-cubic-yard bucket that moved 100 tons (91 tonnes) of material at each bite. Draglines having buckets as large as 220 cubic yards (about 325 tons [295 tonnes]) have been built (Crowell 1995). Recently the trend has turned toward smaller equipment such as bulldozers, scrapers, smaller shovels, and front-end loaders. Such machines cost less to purchase and operate and are better suited for mining smaller tracts of coal in compliance with reclamation laws.

Surface mining has many advantages over underground mining. Surface mining recovers 85 to 95% of the coal compared with 40 to 90% from underground mines. Productivity (tons mined per worker per day) is much higher in surface mines, which also are safer places in which to work. Accident rates are typically half those in underground mines; roof falls, methane and coal dust explosions, and black lung disease in workers are unknown in surface mines.

Most early surface miners did not restore mined land. By the 1930s, some enlightened Illinois companies practiced voluntary reclamation. They planted trees, stocked pit lakes with fish, and converted mined land to grazing (Bristow 1944). Of the 67,000 acres (27,114 ha) mined in 1954, about 12,000 acres (4,856 ha) were planted in trees, 9,000 acres (3,642 ha) in grasses and legumes, and nearly 10,000 acres (4,047 ha) in pasture (*Coal Report* 1954). Some abandoned surface mines became state parks and wildlife refuges.

State laws requiring reclamation first took effect in 1962. Since that time, federal, state, and, in some cases, county governments have imposed increasingly stringent land restoration laws. Currently, rock and soil layers must be replaced in their original order, and the original contour of the land must be restored. Achieving higher crop yields after mining than before mining is not unusual, and some coal companies have agricultural subsidiaries to farm reclaimed mines. No longer an afterthought, reclamation today is an integral process of surface mining. See Chapter 30, Surface Mine Reclamation, for additional information.

Underground Mining

Underground mining takes place where coal is too deep—typically 100 feet (30.5 m) or more—for economical surface mining. In drift mines, tunnels are driven directly into coal exposed in the side of a hill or the highwall of

Figure 14-7 Surface-mining equipment: **(a)** Peabody Coal Company's River King Mine in St. Clair County, about 1958. The bucket-wheel excavator shown in the background is removing glacial deposits while the stripping shovel at center removes bedrock that has been blasted. The drill in the foreground is drilling shot holes in the bedrock. (Photograph from Illinois State Geological Survey collection and Chenoweth et al. 2008.) **(b)** The scene at Vigo Coal Company's Friendsville Mine near Mt. Carmel in 2003. Bulldozers, scrapers, front-end loaders, and trucks are more versatile in small mines than are the giant earth movers of the past. Equipment in the background is removing and stockpiling overburden. The bulldozer and loader in the foreground have uncovered the coal seam. (Photograph by Scott D. Elrick.) **(c)** Loading coal from the Davis and Dekoven seams at Jader Coal's No. 4 Mine in Gallatin County in 2003. Coal trucks of 100 tons (90.7 tonnes) or larger capacity are the largest machines in use and might not be found on a typical construction job site. (Photograph by Christopher P. Korose.)

a surface mine. The highwall is the vertical rock wall on the advancing side of a surface pit. Drift mining, primarily along the bluffs of rivers and streams, is prevalent in mountainous country such as eastern Kentucky and was formerly common in Illinois. Slope mines use inclined tunnels to reach coal that lies below valley bottoms. Slopes are best suited for relatively shallow mining, but have been driven to coal as deep as 1,000 feet (305 m) in Illinois. Shaft mines access the coal via vertical entrances fitted with hoists or elevators.

Room-and-Pillar Mining

Coal is extracted by either room-and-pillar or longwall methods. In room-and-pillar mining, a series of tunnels called entries, rooms, and crosscuts are driven in the coal seam, typically intersecting at right angles to form a pattern similar to that of city streets (Figure 14-8). Pillars of coal are left to support the overburden. Generally, only 40 to 50% of the coal is recovered by room-and-pillar mining. In some cases, miners extract some of the pillars in the final phase of mining, allowing the roof to collapse in as they retreat; using this technique, 60 to 70% of the coal can be recovered.

The traditional method of room-and-pillar mining is called conventional mining. First, the coal is undercut by cutting a deep slot beneath the seam so that the coal breaks down when blasted. Originally a pick was used; cutting machines became commonplace by 1915. Next, a set of holes is drilled in the coal and loaded either with explosives or compressed-air cartridges. After blasting, the coal is loaded into pit cars or onto a conveyor belt for transport to the surface. By the 1980s, this type of conventional mining was largely phased out.

All room-and-pillar mining in Illinois today is done by continuous mining machines (continuous miners). Introduced in the late 1940s, the continuous mining machine is equipped with rotating drums armed with carbide bits to cut and load the coal in one operation (Figure 14-9). Shuttle cars carry coal from the continuous miner to conveyor belts that transport it out of the mine.

Rock falls are the leading hazard in underground mining. In early mining, the mine roof was supported with timbers. Roof bolting is the primary method of roof control today (Figure 14-10). Developed in the late 1940s—and pioneered by Illinois operators—roof bolting reinforces the rock and ties weak strata to strong strata while leaving the entries open for moving workers and equipment. Roof bolting machines install steel bolts several feet long into holes drilled in the roof. The bolts are anchored by expansion shells or epoxy resin. Timbers, wooden cribs, and even steel arches are installed in weak, fractured roof where bolts alone do not suffice.

Although continuous mining entails a larger investment, the method loads more coal in greater safety by a smaller crew than conventional mining methods do. Currently, many Illinois operations are converting to super sections, where a crew of 10 to 12 operates two continuous miners, three or four shuttle cars, and two roof bolters.

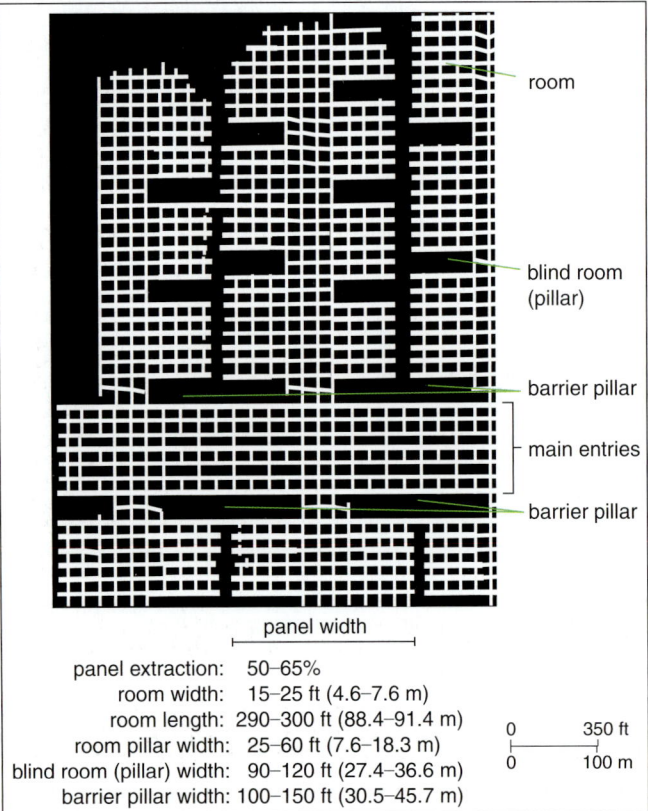

Figure 14-8 Modern room-and-pillar underground mining system (Bauer 2006).

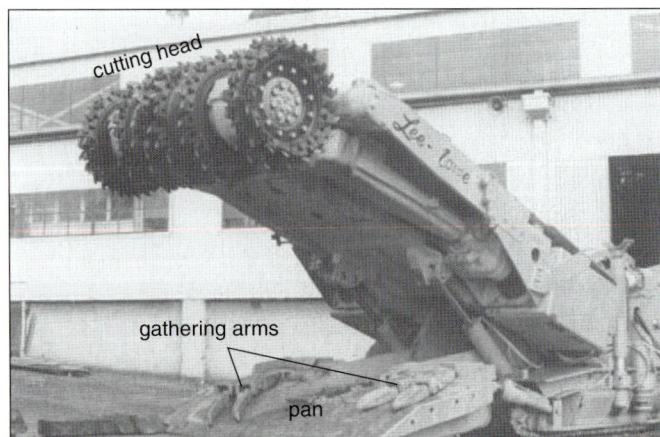

Figure 14-9 Continuous miner used for room-and-pillar underground mining. The rotating head armed with carbide steel bits breaks the coal, which falls onto the pan below, where gathering arms feed it to a conveyor. (Photograph from the Illinois State Geological Survey collection.)

COAL

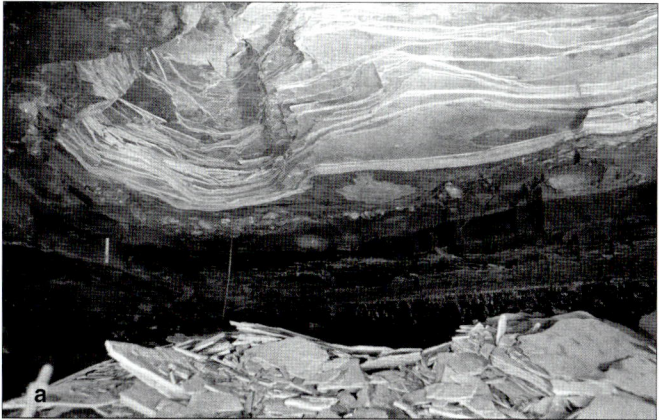

A single-unit continuous miner can load up to 2,000 tons (1,814 tonnes) of raw coal per 8-hour shift, and a super section can achieve 2,500 to 3,000 tons (2,268 to 2,722 tonnes).

Longwall Mining

Longwall mining was developed to enhance coal recovery, especially in thin seams. After setting temporary roof supports, the miners remove all of the coal along a face hundreds to thousands of feet long (Figure 14-11a). Roof supports advance with the working face; the mined-out area is allowed to collapse behind. Longwall mining enables 85 to 90% coal recovery. Pillars are left only along entries used for transportation and ventilation.

A primitive form of longwall mining was practiced in northern Illinois until the 1940s. The mine plan resembled the rim and spokes of a wheel (Figure 14-11b). From a central pair of shafts, miners drove radiating entries leading to a circular longwall face, where conventional mining (either by hand or by machine) progressed outward in all directions. As coal was removed, workers built walls of waste rock to help support the overburden.

Developed in Europe, modern longwall mining was first used in southern Illinois in 1965 (Evans 1965). Longwall panels are rectangular and outlined by gate roads (Figure 14-11a). A shearer with rotating drums fitted with carbide bits cuts the coal, which falls onto conveyor belts for transport out of the mine (Figure 14-12). A row of massive hydraulic jacks or shields supports the roof above the shearer. As coal is mined, the shields advance, allowing the overburden to cave in behind. In 2000, 6 longwall units among 18 active mines produced nearly one-third of all the coal mined in Illinois (*Coal Report* 2000).

The chief drawback to longwall mining is that it causes the ground surface to subside. Subsidence damages structures on the surface and disrupts field drainage in flat farmland (DuMontelle et al. 1981, Bauer 2008), which mandates higher payments to surface landowners. Thus, longwall coal companies frequently purchase the surface rights as well as the mineral rights. Generally longwall mining beneath such features as highways, railroads, and buildings is not permitted.

Figure 14-10 (a) Rock fall at Peabody Coal Company No. 10 mine, Christian County, in 1974. Thin-bedded sandstone over shale. (b) Heavy bolting with resin-anchored head rebar roof bolts was needed to anchor the black shale above the coal. (c) Freeman Coal Mining Company, Crown No. 1 Mine, near Farmersville, Montgomery County, in 1960. This photograph shows roof bolts still anchored in limestone even after shale and coal were removed. The roof failed despite being bolted. (Photographs from the Illinois State Geological Survey collection and Chenoweth et al. 2008.)

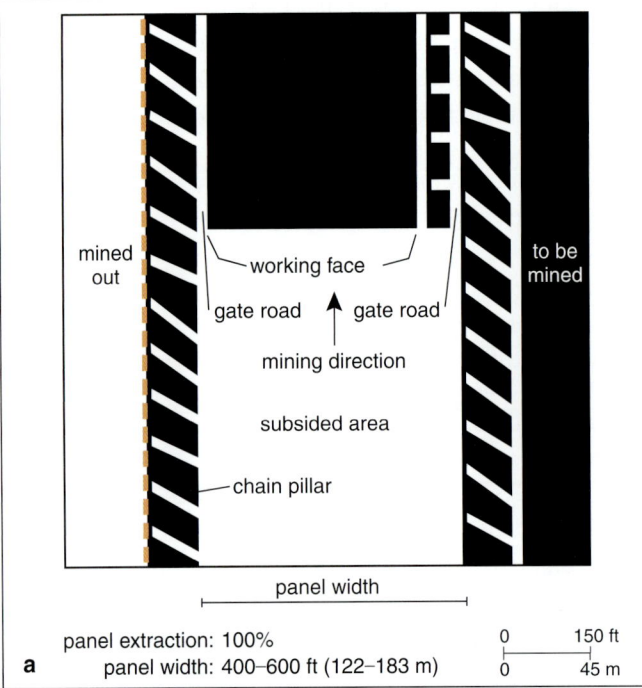

Figure 14-11 (a) Modern and (b) historic longwall mine layouts (Bauer 2006).

Figure 14-12 Longwall face equipment. The shearer cuts the coal, which falls onto the conveyor belt. The shields, or hydraulic roof supports, advance after each pass of the shearer, allowing the roof to collapse behind. (Photograph from the Illinois State Geological Survey collection and Chenoweth et al. 2008.)

All underground mines must be ventilated to provide fresh air and remove dangerous gases, especially methane, which is emitted by all coal seams and is explosive in concentrations above 5%. Huge electric fans at the surface draw fresh air throughout the workings. Stoppings (walls), doors, curtains, and other structures channel air currents underground. Air flow is monitored and remedied, if necessary, at every working place before the start of every shift. Automatic methane detectors shut down equipment when gas exceeds 1%. White powdered limestone, called rock dust, is applied throughout the mine to dilute and cover coal dust, which can be explosive in addition to being a long-term health hazard. Water sprays at the working face reduce the chance of igniting sudden outbursts of methane and further serve to keep down coal dust. Together these measures have greatly reduced the incidence of explosions and fires, which caused so many mining tragedies in the past.

Auger and Highwall Mining

When overburden becomes too thick for profitable surface mining but economics do not favor underground mining, coal companies may turn to auger and highwall mining. In the older method of augering, drills as large as 60 inches (1.5 m) in diameter bore horizontal holes as long as 280 feet (85.3 m) but typically 100 to 200 feet (30.5 to 61 m), recovering substantial tonnages of coal at a lower cost and with greater safety than by underground mining (Guthrie 1952). Although widespread in Appalachian coal fields, only a few auger mines operated in Illinois.

Highwall miners, a recent development, are basically remote-controlled continuous miners that feed a belt conveyor. They are guided by television cameras or ground-penetrating radar. Driving sets of parallel entries 1,000 feet (305 m) or more into a hillside, a crew of three to four workers can mine 100,000 to 120,000 tons (90,718 to 108,862 tonnes) of raw coal per month (Sanda 1992). Like augers, highwall miners are most common in the Appalachian coal fields but recently have been employed in southern Indiana and southern Illinois.

Figure 14-13 Surface preparation facilities of the Galatia Mine. Rock and pyrite are removed from the coal in the preparation plant (center). Coal is stored in the four huge silos to the right. The unit train of more than 100 cars is automatically loaded in the tower to the far left. More than 7 million tons (6.4 tonnes) of coal were shipped from the mine in 2001, which is an all-time record for a single mine in Illinois. (Photograph by Scott D. Elrick.)

COAL PREPARATION

Coal, as mined, contains a large portion of rock and other waste. Coal cleaning or preparation takes place at the preparation plant (Figure 14-13), also called the tipple, because originally pit cars were tipped to unload. In the early days, workers (often boys) removed waste by hand as the coal passed on a moving conveyor. Modern coal preparation, however, is totally mechanized. First, the coal is crushed and screened to remove large pieces of rock. Then it passes through a series of tanks filled with water and chemicals that are constantly stirred and aerated to form a froth. Coal floats with the froth and is skimmed off the top; denser rock, pyrite, and other mineral wastes sink and are taken out the bottom. The procedure is similar to that of ore concentrators used at metal mines. Washed coal is dried and loaded into railroad cars, trucks, or barges for shipment. Waste is placed in approved storage areas.

COAL TRANSPORTATION

The first coal mined in Illinois (1810) was floated down the Mississippi on a flatboat, but railroads quickly became the dominant mode of transportation. Railroad and coal industries grew side by side. According to the 1954 *Coal Report*, railroads hauled more than 99% of all coal produced in the state after 1902. That picture has changed dramatically. Today nearly half of the coal mined in Illinois travels by barge, about one-third by railroad, and one-sixth by truck. Barges operating on the Illinois, Kaskaskia, Ohio, and Mississippi Rivers convey coal to customers as far away as Florida. Truck shipment is used mainly by smaller mines for nearby customers. Railroads have fought to maintain their share of the traffic by introducing unit trains of 100 cars or more that shuttle from mine to power plant. Another strategy is to bring the utility closer to the mine. Mine-mouth power plants, connected to "captive" mines via short conveyor belts, operated at several Illinois locations from the 1960s through the 1990s. A new mine-mouth power station is under construction in Logan County, and a major project is being built in Washington County.

COAL MINING TRENDS

From the late 1800s through World War I, Illinois coal production grew vigorously, reaching an all-time high of nearly 90 million tons (81.7 tonnes) in 1918 (Figure 14-14). Output declined sharply during the Depression, rebounding during World War II. The market slumped again during the 1950s as railways converted from steam to diesel (steam locomotives burned as much as 20 million tons [18.1 million tonnes] of Illinois coal annually) and as industry and homeowners shifted from coal furnaces to those using oil and natural gas.

Beginning around 1960, increased use of coal by electricity-generating plants revitalized the industry. Annual production averaged close to 60 million tons (54.4 million tonnes) from about 1965 to 1995. Today, 70 to 80% of Illinois coal is used to make electricity. The remainder includes 2.5 to 3 million tons (2.3 to 2.7 million tonnes) of coal used to make coke for commercial use, a few million tons used by other industries, and a small tonnage for domestic use. The recent decline to 33 million tons (30 million tonnes) during the 2000–2002 period mainly reflects the loss of market to low-sulfur coal from enormous open pit mines in the western United States.

Initially, all coal mining in Illinois took place underground. Surface mining gained a sizeable share around 1920, peaked in the 1960s, and then fell to about 20% (Figure 14-14). The number of mines and miners has dropped dramatically since the early twentieth century. Figure 14-14 greatly understates the number of early mines, because thousands of "local" mines are not included. When the state started keeping track in 1882 there were 704 mines, and, through World War II, the tally dropped below 700 only once. The record was set in 1935 at 1,350 mines, of which 182 large shipping mines produced more than 90% of the total tonnage. Why did the number of mines increase during the Great Depression? Small local mines provided jobs and supplemented farm income and required little or no capital investment. After World War II, the industry began to consolidate through merger and acquisition, leading to fewer operators who ran larger, more productive mines. The process continues today. Only 18 mines (11 underground, 7 surface) were active in Illinois in 2008.

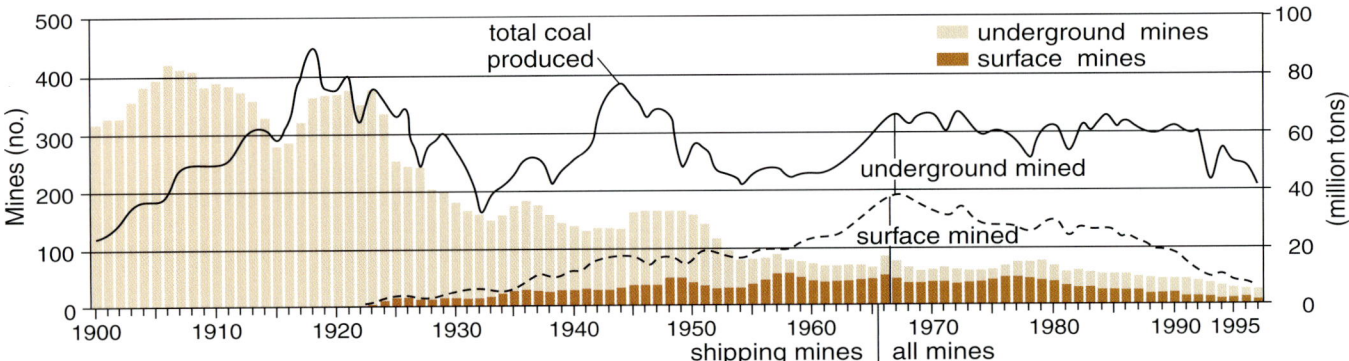

Figure 14-14 Trends in Illinois coal production as shown by number of mines and tonnage (Illinois Department of Natural Resources, Office of Mines and Minerals). One ton is equal to 0.91 tonne.

Several of the world's largest coal mines have operated in Illinois. In 1924, the New Orient mine at West Frankfort set a single-shift tonnage record that stood until recently—some 18,000 tons (16,329 tonnes), reportedly loaded by hand by nearly 2,000 miners. The Peabody No. 10 underground mine south of Springfield, which closed in 1995, produced 168 million tons (152.4 million tonnes) over its lifetime and mined out an area of 40 square miles (103.6 km²). The Captain Mine in Perry County, a surface mine, turned out 132 million tons (120 million tonnes) before it closed in 1998. The American Coal Company's Galatia Mine in Saline County produced more than 7.5 million tons (6.8 million tonnes) by longwall mining during 2000.

The number of mine workers in Illinois peaked in 1923 at 103,000 men, 92,000 of whom worked underground (*Coal Report* 1954). This number has plummeted to about 3,500 today. Mechanization means much more coal is produced by fewer workers. The average miner shoveled about 2 tons (1.8 tonnes) per day during the 1880s, 4 tons (3.6 tonnes) around 1912, 8 tons (7.3 tonnes) in 1935, and 16 tons (14.5 tonnes) by 1955. By 2000, average daily output ranged from 25 to 40 tons (22.7 to 36.3 tonnes) per miner in deep mines and 80 tons (72.6 tonnes) in surface mines (*Coal Report* 2000).

Another trend that thankfully continues is the trend toward safer mines. Between 1881 and 1954, there were 8,441 fatalities and 81,000 injuries resulting in 30 days or more lost time in Illinois mines (*Coal Report* 1954). Disasters at Braidwood, Zeigler, Royalton, Centralia, and West Frankfort are stamped into local memory. The worst single tragedy in Illinois occurred in 1909 at the Cherry Mine in Bureau County, when a dripping oil lamp set fire to a carload of hay intended for mine mules. The blaze spread to timbers in the shaft and then to the coal. Their escape cut off, 256 miners died. Appalling as these accidents were, the day-to-day toll from roof falls, mishaps with machinery, and the like was greater. Since the 1980s, it is not unusual to go through a year without a single fatality in Illinois mines. Many advances are to be credited for this trend, including improved roof support techniques, better ventilation, regular safety training, and advanced technology.

Coal Products and By-Products

Coke

Coke is made by heating coal in the absence of oxygen at temperatures as hot as 2,000°F. Many tons of coal are heated in huge, airtight coke ovens, where volatile matter is driven off, and the carbon and ash fuse. The coke product is dull gray, hard, and porous. Metallurgical coke is used as fuel, dispersant, and reducing agent for smelting iron ore in blast furnaces. Chemical coke, which has different properties, is used for smelting ore in electric furnaces and manufacturing products such as calcium carbide.

Good metallurgical coke must have low ash, sulfur, and phosphorus contents. Also, it must be highly porous yet physically strong in order to maintain the flow of reducing gases in the furnace. These properties can be determined only by thorough testing. Most Illinois coal is not suitable for making metallurgical coke. Low-sulfur Illinois coal, particularly the Herrin and Springfield Coals from southern Illinois, produce coke that is adequate for iron smelting when blended with higher-rank coal from the Appalachian coal fields. The requirements for chemical coke are less rigorous, and more varieties of Illinois coal are suitable either alone or in blends (Reed et al. 1947; Jackman et al. 1959; Jackman and Helfinstine 1967, 1968).

Coal Bed Methane

Methane is a natural product of coalification and is trapped in all coal beds. This colorless, odorless gas is both a hazard to underground miners and a resource to be tapped. With the longevity of petroleum deposits in question, attention has turned to coal bed methane and other "unconventional" energy sources. Coal bed methane has been exploited successfully elsewhere in the United States and in other nations but has been barely tapped in Illinois. Gas from abandoned underground mines has been recovered for a number of years. More recently, exploration and development wells have been drilled to extract methane from virgin coal. Commercial production is now under way in Saline County.

Future coal bed methane potential in Illinois lies in drilled wells producing methane from multiple seams. The potential for economic production depends on gas content, permeability, development of appropriate well-completion techniques, and, of course, the price of natural gas.

Archer and Kirr (1984) estimated that the Illinois Basin contains 21 trillion cubic feet (0.59 trillion m^3) of coal bed methane, a volume comparable with that of the actively producing Black Warrior Basin in Alabama. Illinois resource numbers need to be defined better through new testing and production case histories. Based on reconnaissance testing, Demir et al. (2002) have suggested that gas content of Illinois coals may be 50 to 100% greater than previous estimates have indicated.

Coal differs from conventional natural gas reservoirs in being both the source of the gas and the gas reservoir. The composition and burial history of the coal affect its composition, gas content, and storage capacity. Storage capacity is a key parameter as most gas in coal beds is adsorbed or held as extremely thin layers within coal, rather than freely occupying pore space, as occurs in sandstone. Regularly spaced natural fractures, termed cleats, are important pathways for gas to reach a producing well. Various completion techniques, such as hydraulic fracturing and cavitation, are employed to widen cleats and stimulate gas flow. Pressure on the coal must be reduced by pumping water out to free adsorbed gas. Removing and disposing of this water are significant economic factors. Potable water can be routed to surface streams, but water that contains high concentrations of dissolved solids must be pumped back into the ground through costly separate wells.

Coal Gasification and Liquefaction

Gas from coal dates back to the "gaslight era" of the 1890s. Basically, to retrieve coal gas, pulverized coal is exposed to hot steam plus air or oxygen in carefully regulated quantities. The complex carbon molecules of coal break down, releasing hydrogen, carbon monoxide, and other flammable gases. Using modern technology, coal gasification is emerging as a clean energy source. Coal gasification plants in Indiana and Florida are operational (U.S. Department of Energy 2005). The Wabash River project at West Terre Haute, Indiana, and the plant in Tampa, Florida, both use Illinois Basin coal. Coal slurry is combined with oxygen at high temperature and pressure, oxidizing the slurry as the fluid ash is drained off the bottom. The resulting gas is treated with additional coal slurry, and then the sulfur is extracted for sale. Finally, the clean synthetic gas is burned to generate electricity (e.g., Los Alamos National Laboratory 2005).

Extracting liquid products from coal also has a long history. Coal tar obtained from coke ovens has been used for purposes such as roofing and road surfacing and as a chemical feedstock, notably for dyes (Stutzer 1940). During World War II, when its access to liquid petroleum was blocked, Germany made gasoline from coal. Spurred by government grants and tax credits, advanced forms of coal liquefaction are coming to fruition in the United States.

Coal Waste and By-Products

Coal mining and combustion generate materials that can harm the environment. Laws mandating safe and proper disposal and reclamation have mitigated such problems greatly. Better yet, creative technology enables recovery of useful by-products and uses one form of waste material to neutralize another.

The coarse rock waste rejected during coal preparation, known as gob, formerly was simply piled on the surface, and volcano-shaped cones of gob dot the old coal-producing district in northern Illinois. These mounds erode to a lunar landscape, contaminate streams with acidic runoff, and may catch fire spontaneously. Coal companies today dispose of gob at specific sites where it is graded, covered with soil, and revegetated.

Slurry is the fine-grained waste from a coal preparation mixed with water. Slurry is commonly pumped into impoundments, where the solids settle out and the water is reused. Slurry contains pyrite that oxidizes to produce sulfuric acid. Mixing alkaline products such as limestone with slurry neutralizes the acid so the site will support vegetation when reclaimed (Dreher et al. 1996, Green 1995, Roy et al. 1997). Operators today increasingly tend to pump slurry into abandoned underground mines in preference to surface impoundment.

Slurry from old mines contains a high proportion of coal, in some cases 50% or more. Several Illinois operations practice carbon recovery by mining old slurry ponds for fuel. Slurry can be blended with coal or, in some cases, used in raw form, particularly in fluidized-bed combustion.

Combustion Waste

The best-known noxious products from burning coal are the airborne contaminants emitted into the atmosphere during coal combustion. Environmental laws now strictly regulate emissions from power plants and other coal-burning facilities. One way to meet emission standards in Illinois is to install scrubbers that remove sulfur at the stack. The sulfur is neutralized in solid form with powdered limestone; in some cases, sulfur and sulfur compounds can be captured and sold.

Solid waste from the combustion of coal includes slag, bottom ash, and fly ash. Slag is a glassy material composed of mineral matter that melted in the furnace. Bottom ash is the solid, granular ash that settles to the bottom of the furnace. Fly ash is the fine material that goes up the stack. Electrostatic precipitators and fabric filters collect fly ash from flue gases. Power plants in Illinois produce about 3 million tons (2.7 million tonnes) of fly ash each year.

Crushed bottom ash and slag are routinely used as road cinders and in roofing tiles. Slag and ash also contain valuable metals that can be recovered. Among them are germanium, which is used in the semiconductor industry (Machin and Witters 1956), and mercury (Ruch et al. 1971).

Most fly ash is buried at specially designed disposal sites (Roy et al. 1981, 1984; Roy and Griffin 1984). Recently, science and industry have been looking at ways to put fly ash to use. Fly ash is used in several portland cement and lightweight concrete products, structural fills, embankments, and road bases. Fly ash serves either as inert fill or as cement to improve cohesion and stability. About 95,570 tons (86,700 tonnes) of fly ash were used in highway construction in Illinois in 2001. In a commercial demonstration, high-carbon Illinois fly ash was used to make portland cement (Bhatty et al. 2002). Other uses include mineral filler in asphalt paving and as a component of bricks and blocks (Chou et al. 2002).

Fluidized-bed combustion is a process in which pulverized coal is mixed with crushed limestone or dolomite to capture sulfur. Jets of air agitate the mixture into a suspension of red-hot particles that flow like a fluid. The heat produces the steam to drive electrical generators (Dreher et al. 1996, U.S. National Energy Technology Laboratory 2005). Because fluidized-bed combustion inherently produces low emissions, it is well suited for use in urban settings. Fluidized-bed combustion also enables the burning of very low grades of coal, including coal recovered from slurry ponds. This ash by-product is strongly alkaline, creating a disposal issue, but one creative solution is to use it to neutralize acidic slurry during reclamation of slurry impoundments.

FUTURE OF ILLINOIS COAL

As this chapter is written, the immediate future of Illinois coal looks bright. Increasing scarcity of petroleum and natural gas is driving up the price of coal and stimulating commercial interest in unconventional energy resources, such as coal bed methane and synthetic liquid fuels derived from coal. Major underground mine complexes are being planned to support mine-mouth power generation and synthetic fuel production. One of these projects by itself would double current annual coal production in Illinois. At the same time, small operators are mining blocks of high-quality coal previously considered too small to be worthwhile. For the first time, coal bed methane from virgin coal is being commercially extracted in Illinois.

Clean coal technology promises further breakthroughs for Illinois coal. As more power plants install scrubbers, the use of Illinois coal becomes increasingly attractive. Transportation costs and total energy usage are slashed by using the coal at or near its source, especially in mine-mouth facilities. Fluidized-bed combustion is rapidly advancing as a way to produce clean energy using "dirty" coal. Gasification and liquefaction also provide clean-burning fuels for a wide range of uses. Such processes bypass the old roadblocks to cleaning the coal before it is used. Instead, sulfur and other contaminants are captured during combustion. In many cases, formerly harmful waste materials are becoming salable by-products.

REFERENCES

Allgaier, G. J., and M. Hopkins, 1975, Reserves of the Herrin (No. 6) Coal in the Fairfield Basin in southeastern Illinois: Illinois State Geological Survey, Circular 489, 31 p.

Archer, P. L., and J. N. Kirr, 1984, Pennsylvanian geology, coal, and coalbed methane resources of the Illinois Basin—Illinois, Indiana, and Kentucky, in C. T. Rightmire, G. E. Eddy, and J. N. Kirr, eds., Coalbed methane resources of the United States: Tulsa, Oklahoma, American Association of Petroleum Geologists, Studies in Geology 17, p. 105–134.

Bauer, R. A., 2006, Mine subsidence in Illinois: Facts for homeowners: Illinois State Geological Survey, Circular 569, 20 p.

Bauer, R. A., 2008, Planned coal mine subsidence in Illinois: A public information booklet: Illinois State Geological Survey, Circular 573, 19 p.

Bhatty, J. I., J. Gajda, and F. M. Miller, 2002, Converting Illinois coal prep wastes into supplementary portland cements: Illinois Clean Coal Institute, Final Technical Report. http://www.icci.org/02final/02battyw.pdf. Accessed August 25, 2006.

Bottomley, J. A., 1944, History and development of strip mining in Illinois: Proceedings of the Illinois Mining Institute, p. 90–100.

Bristow, J. W., 1944, Land reclamation accomplishments of Illinois strip mining companies: Proceedings of the Illinois Mining Institute, v. 52, p. 103–109.

Cady, G. H., 1935, Classification and selection of Illinois coals: Illinois State Geological Survey, Bulletin 62, 354 p.

Chenoweth, C., A. R. Myers, and J. M. Obrad, 2008, Photographic history of coal mining practices in Illinois: Illinois State Geological Survey, Circular 572, 178 p.

Chou, M.-I. M., S.-F. J. Chou, V. Patel, J. Stucki, and J. Wu, 2002, Commercialization of fired brick with fly ash from Illinois coals: Final Technical Report, Illinois Clean Coal Institute. http://www.icci.org/reports.php. Accessed November 9, 2009.

Clegg, K. E., 1955, Metamorphism of coal by peridotite dikes in southern Illinois: Illinois State Geological Survey, Report of Investigations 178, 18 p.

Coal Reports, 1882 to 2008: Springfield, Illinois, Illinois Bureau of Labor Statistics (1882–1910), State Mining Board (1911–1916), Department of Mines and Minerals (1917–1996), and Office of Mines and Minerals (1997 to 2008).

Cobb, J. C., J. M. Masters, C. G. Treworgy, and R. J. Helfinstine, 1979, Abundance and recovery of sphalerite and fine coal from mine waste in Illinois: Illinois State Geological Survey, Illinois Minerals Note 71, 11 p.

Cobb, J. C., J. D. Steele, C. G. Treworgy, and J. F. Ashby, 1980, The abundance of zinc and cadmium in sphalerite-bearing coals in Illinois: Illinois State Geological Survey, Illinois Minerals Note 74, 28 p.

Crowell, D. L., 1995, History of the coal mining industry in Ohio: Ohio Department of Natural Resources, Division of Geological Survey, Bulletin 72, 203 p.

Damberger, H. H., 1970, Clastic dikes and related impurities in Herrin (No. 6) and Springfield (No. 5) Coals of the Illinois Basin, in W. H. Smith, R. B. Nance, M. E. Hopkins, R. G. Johnson, and C. W. Shabica, eds., Depositional environments in parts of the Carbondale Formation—Western and northern Illinois: Illinois State Geological Survey, Guidebook 8, p. 111–119.

Damberger, H. H., 1971, Coalification pattern of the Illinois Basin: Economic Geology, v. 66, no. 3, p. 488–494.

Demir, I., R. E. Hughes, J. M. Lytle, R. R. Ruch, P. J. DeMaris, and C.-L. Chou, 1997, Mineralogical and chemical composition of inorganic matter from Illinois coals: Illinois Clean Coal Institute, Final Technical Report. http://www.icci.org/97final/findemir.htm. Accessed August 25, 2006.

Demir, I., D. G. Morse, S. D. Elrick, C. A. Chenoweth, R. J. Jacobson, and R. J. Finley, 2002, Delineation of the coalbed methane resources of Illinois—A commodity to support economic viability of Illinois Coal, Final Contract Report, Office of Coal Development, Illinois Department of Commerce and Community Affairs, December 31, 2002, 162 p.

Dreher, G. B., W. R. Roy, and J. D. Steele, 1996, Laboratory studies on the codisposal of fluidized-bed combustion residue and coal slurry solid: Illinois State Geological Survey, Environmental Geology 150, 28 p.

DuMontelle, P. B., S. C. Bradford, R. A. Bauer, and M. M. Killey, 1981, Mine subsidence in Illinois: Facts for the homeowner considering insurance: Illinois State Geological Survey, Environmental Geology Notes 99, 24 p.

Elrick, S. D., 2008, Coal industry in Illinois (rev.): Illinois State Geological Survey, Illinois Map 13, 1:500,000.

Evans, M. A., 1965, Longwall in 1965: Illinois Mining Institute Proceedings, p. 59–82.

Gluskoter, H. J., and M. E. Hopkins, 1970, Distribution of sulfur in Illinois coals: Illinois State Geological Survey, Guidebook 8, p. 89–95.

Gluskoter, H. J., and O. W. Rees, 1964, Chlorine in Illinois coal: Illinois State Geological Survey, Circular 372, 23 p.

Gluskoter, H. J., R. R. Ruch, W. G. Miller, R. A. Cahill, G. B. Dreher, and J. K. Kuhn, 1977, Trace elements in coal: Occurrence and distribution: Illinois State Geological Survey, Circular 499, 154 p.

Gluskoter, H. J., and J. A. Simon, 1968, Sulfur in Illinois coals: Illinois State Geological Survey, Circular 432, 28 p.

Green, W. P., 1995, Plant response to coal slurry solids amended with fluidized bed combustion waste: Urbana-Champaign, University of Illinois, M.S. thesis, 100 leaves.

Guthrie, R. W., 1952, Auger mining in Illinois coal: Illinois Mining Institute Proceedings, p. 47–52.

Harper, D., 1985, Coal mining in Vigo County, Indiana: Indiana Geological Survey, Special Report 34, 67 p.

Harvey, R. D., R. A. Cahill, C.-L. Chou, and J. D. Steele, 1983, Mineral matter and trace elements in the Herrin and Springfield Coals, Illinois Basin coal field: Illinois State Geological Survey, Contract/Grant Report 1983-4, 162 p.

Hatch, J. R., H. J. Gluskoter, and P. C. Lindahl, 1976, Sphalerite in coals from the Illinois Basin: Economic Geology, v. 71, no. 3, p. 613–624.

Hollingsworth, J. A., 1963, History of the development of strip mining machinery: Proceedings of the Illinois Mining Institute, p. 64–84.

Hopkins, M., 1968, Harrisburg (No. 5) Coal reserves of southeastern Illinois: Illinois State Geological Survey, Circular 431, 25 p.

Jackman, H. W., R. L. Eissler, and R. J. Helfinstine, 1959, Coke from medium-volatile and Illinois coals: Illinois State Geological Survey, Circular 278, 24 p.

Jackman, H. W., and R. J. Helfinstine, 1967, A survey of the coking properties of Illinois coals: Illinois State Geological Survey, Circular 412, 27 p.

Jackman, H. W., and R. J. Helfinstine, 1968, Drying and preheating coals before coking, Part 2, Coal blends: Illinois State Geological Survey, Circular 434, 23 p.

Jacobson, R. J., 1983, Murphysboro Coal, Jackson and Perry Counties: Resources with low to medium sulfur potential: Illinois State Geological Survey, Illinois Minerals Notes 85, 19 p.

Jacobson, R. J., 1987, Stratigraphic correlations of the Seelyville, Dekoven, and Davis Coals of Illinois, Indiana, and western Kentucky: Illinois State Geological Survey, Circular 539, 27 p.

Jacobson, R. J., 1993, Coal resources of the Dekoven and Davis Members (Carbondale Formation) in Gallatin and Saline Counties, southeastern Illinois: Illinois State Geological Survey, Circular 551, 41 p.

Jacobson, R. J., 2000, Depositional history of the Pennsylvanian rocks in Illinois: Illinois State Geological Survey, GeoNote 2, 12 p.

Jacobson, R. J., and C. P. Korose, 2003, Coal geology of Illinois, in 2003 Keystone Coal Industry Manual, Coal age: Chicago, Primedia Business Magazines and Media, p. 503–514.

Keystone Coal Industry Manual, 2003, Coal age: Chicago, Illinois, Primedia Business Magazines and Media, 731 p.

Korose, C. P., S. D. Elrick, and R. J. Jacobson, 2003, Availability of the Colchester Coal for mining in northern and western Illinois: Illinois State Geological Survey, Illinois Minerals 127, 21 p.

Korose, C. P., C. G. Treworgy, R. J. Jacobson, and S. D. Elrick, 2002, Availability of the Danville, Jamestown, Dekoven, Davis, and Seelyville Coals for mining in selected areas of Illinois: Illinois State Geological Survey, Illinois Minerals 124, 44 p.

Krausse, H.-F., H. H. Damberger, W. J. Nelson, S. R. Hunt, C. T. Ledvina, C. G. Treworgy, and W. A. White, 1979, Roof strata of the Herrin (No. 6) Coal Member in mines of Illinois: Their geology and stability: Illinois State Geological Survey, Illinois Minerals Note 72, 54 p.

Los Alamos National Laboratory, 2005. www.lanl.gov. Accessed November 9, 2009.

Machin, J. S., and J. Witters, 1956, Germanium in fly ash and its spectrochemical determination: Illinois State Geological Survey, Circular 216, 13 p.

Nance, R. B., and C. G. Treworgy, 1981, Strippable coal resources of Illinois, Part 8, Central and southeastern counties: Illinois State Geological Survey, Circular 515, 32 p.

Nelson, W. J., 1981, Faults and their effect on coal mining in Illinois: Illinois State Geological Survey, Circular 523, 38 p.

Nelson, W. J., 1983, Geologic disturbances in Illinois coal seams: Illinois State Geological Survey, Circular 530, 47 p.

Nelson, W. J., P. J. DeMaris, and R. A. Bauer, 1987, The Hornsby district of low-sulfur Herrin Coal in central Illinois (Christian, Macoupin, Montgomery, and Sangamon Counties): Illinois State Geological Survey, Circular 540, 40 p.

Office of Coal Development, Illinois Department of Commerce and Economic Opportunity, 2006, Facts and figures. http://www.commerce.state.il.us/dceo. Accessed September 30, 2010.

Peppers, R. A., 1996, Palynological correlation of major Pennsylvanian (Middle and Upper Carboniferous) chronostratigraphic boundaries in the Illinois and other coal basins: Boulder, Colorado, Geological Society of America, Memoir 188, 111 p.

Reed, F. H., H. W. Jackman, O. W. Rees, G. R. Yohe, and P. W. Henline, 1947, Use of Illinois coal for production of metallurgical coke: Illinois State Geological Survey, Bulletin 71, 132 p.

Roy, W. R., G. B. Dreher, J. D. Steele, R. G. Darmody, D. Tungate, W. E. Giles, and S. C. Pfifer, 1997, Direct revegetation of coal slurry after amendment with FBC residues: Illinois Clean Coal Institute, Final Technical Report. http://www.icci.org/97final/finalroy.htm. Accessed October 5, 2003.

Roy, W. R., and R. A. Griffin, 1984, Illinois Basin coal fly ashes: 2, Equilibria relationships and qualitative modeling of ash-water reactions: Environmental Science and Technology, v. 18, p. 739–742.

Roy, W. R., R. A. Griffin, D. R. Dickerson, and R. M. Schuller, 1984, Illinois Basin coal fly ashes, 1, Chemical characterization and solubility: Environmental Science and Technology, v. 18, no. 10, p. 734–739.

Roy, W. R., R. G. Thiery, R. M. Schuller, and J. J. Suloway, 1981, Coal fly ash: A review of the literature and proposed classification system with emphasis on environmental impacts: Illinois State Geological Survey, Environmental Geology Notes 96, 69 p.

Ruch, R.R., H. J. Gluskoter, and E. J. Kennedy, 1971, Mercury content of Illinois coals: Illinois State Geological Survey, Environmental Geology Notes 43, 15 p.

Sanda, A.P., 1992, Addington builds a better mousetrap: Coal [now Coal Age], v. 97, no. 10, p. 54–59.

Searight, T. K., and W. H. Smith, 1969, Strippable coal reserves of Illinois, Part 5, Mercer, Rock Island, Warren, and parts of Henderson and Henry Counties: Illinois State Geological Survey, Circular 439, 22 p.

Shaver, R. H., et al., 1986, Compendium of Paleozoic rock-unit stratigraphy in Indiana—A revision: Bloomington, Indiana, Indiana Geological Survey, Bulletin 59, 203 p.

Stutzer, O. (translated and revised by A. C. Noe), 1940, Geology of coal: Chicago, Illinois, The University of Chicago Press, 461 p.

Treworgy, C. G., 1981, The Seelyville Coal: A major unexploited seam in Illinois: Illinois State Geological Survey, Illinois Minerals Notes 80, 11 p.

Treworgy, C. G., and M. H. Bargh, 1982, Deep-minable coal resources of Illinois: Illinois State Geological Survey, Circular 527, 62 p.

Treworgy, C. G., L. E. Bengal, and A. G. Dingwell, 1978, Reserves and resources of surface-minable coal in Illinois: Illinois State Geological Survey, Circular 504, 44 p.

Treworgy, C. G., C. P. Korose, C. A. Chenoweth, and D. L. North, 1999, Availability of the Springfield Coal for mining in Illinois: Illinois State Geological Survey, Illinois Minerals 118, 43 p.

Treworgy, C. G., C. P. Korose, and C. L. Wiscombe, 2000, Availability of the Herrin Coal for mining in Illinois: Illinois State Geological Survey, Illinois Minerals 120, 54 p.

Treworgy, J. D., and M. H. Bargh, 1984, Coal resources of Illinois. Illinois State Geological Survey, 5 maps, 1:500,000.

U.S. Department of Energy, 2005, www.fossil.energy.gov. Accessed November 9, 2009.

U.S. National Energy Technology Laboratory, 2005, http://www.netl.doe.gov. Accessed September 30, 2010.

Wanless, H. R., 1939, Pennsylvanian correlations in the Eastern Interior and Appalachian coal fields: Baltimore, Maryland, Geological Society of America, Special Paper 17, 130 p.

Wanless, H. R., 1957, Geology and mineral resources of the Beardstown, Glasford, Havana, and Vermont Quadrangles: Illinois State Geological Survey, Bulletin 82, 233 p.

Wanless, H. R., 1965, Environmental interpretation of coal distribution in the eastern and central United States: Proceedings of the Illinois Mining Institute, p. 19–36.

White, D., and R. Thiessen, 1913, The origin of coal: U.S. Bureau of Mines, Bulletin 38, 390 p.

Wier, C. E., 1973, Coal resources of Indiana: Bloomington, Indiana, Indiana Geological Survey, Bulletin 42-I, 40 p.

Oil and Gas Geology 15

Bryan G. Huff and Beverly Seyler

HISTORY

Hydrocarbon production in Illinois began in 1853 when marsh or drift gas was produced from two water wells drilled near Champaign. This gas came from rotting vegetation buried in the glacial deposits. At the time, people knew little about where gas or oil came from or how to search for it.

In the early 1860s, several holes drilled in Clark County produced enough oil for the name Oilfield to be given to a small town there, even though commercial-scale production in the county did not begin until 40 years later. The search for oil and gas in the county began in earnest in 1866 when the Clark County Petroleum and Mining Company established its headquarters at Marshall. Natural gas seeps near Oilfield led the company's owners to think that commercial quantities of oil and gas existed there. However, because well casing technology did not yet exist, water from the upper layers of earth flowed into and filled the wells; the heavier water prevented most of the oil in deeper layers from seeping out of the rocks.

Farther west, near Litchfield, holes drilled during the late 1860s to search for coal leaked oil and water into the workings of a mine, and, for several years, people skimmed oil off the water and sold the oil. By the early 1880s, natural gas also had been discovered in the area and was being piped to Litchfield for domestic use. Continued drilling eventually established oil production, and, in 1889, area wells produced 1,460 barrels of oil. By 1902, when production ceased, the wells had produced only 6,576 barrels of oil. (A barrel equals 42 U.S. gallons, or 158.9 L, and is a standard unit of volume measurement in the petroleum industry.)

Commercial oil and gas production began in Illinois around 1904, although early oil field operations (Figure 15-1) were primitive. Commercial production refers to

Figure 15-1 Oil field operations in the early 1900s near Bridgeport, Illinois. The oil steamer in the foreground was used to separate oil from water, and the wooden derricks in background were used to drill and service wells. (Photograph by R. S. Blatchley, Illinois State Geological Survey, Oil and Gas Section collection.)

oil produced for refinement into petroleum products and profitable redistribution. Figure 15-2 shows the total annual oil production in Illinois from 1905 to 2003. The large increases in production around the years 1906, 1939, and 1957 reflected the implementation of new technology and innovative exploration or field development techniques that had been developed elsewhere and successfully applied to petroleum fields in Illinois. The development of new well casing technologies at the turn of the nineteenth century made it possible to prevent water from flowing into oil wells from overlying rock layers, a problem in very early oil wells. When this technology was applied to Illinois wells, commercial petroleum production in Illinois became possible.

The recognition that oil and gas accumulated in anticlines (crests of folds in rock layers) was a key element in the commercial development of shallow oil and gas reservoirs from 1904 to 1910 in areas of Crawford and Lawrence Counties. Mapping surface and near-surface features, such as coal beds in shallow Pennsylvanian rocks, allowed the geographic location of anticlines to be defined and predicted in what is now known as the La Salle Anticlinorium (Figure 2-6). As a result, numerous shallow reservoirs were discovered. Annual oil production jumped from 181,000 barrels in 1905 to 33 million barrels by 1910, when Illinois became the third leading oil-producing state in the United States. The drop in production beginning in 1911 (Figure 15-2) shows that all of the easily discovered anticlines—those with surface or near-surface expression—had been drilled. By 1936, with few new discoveries to replace reserves, the state's total oil production dropped to less than 4.5 million barrels.

It took another major technological development, reflection seismology, to find deeply buried and subtle, hidden anticlines in Illinois. This seismic technology was developed in the late 1930s and is still in use today. Seismic exploration uses sensitive microphones called geophones to record sound waves from controlled ground-level dynamite blasts or other energy sources as the waves reflect off the tops of successive, dense rock layers (Figure 15-3). The reflected energy is recorded and processed to form an image of the rock layers based on the time it takes the reflected sound to reach the geophones.

The application of seismic exploration techniques brought about a huge increase in Illinois oil production during the late 1930s and early 1940s. Hundreds of new anticlines and other structural oil traps were found and drilled in southern Illinois. During 1940, the state's total oil production rose to 147.6 million barrels, the historical high for Illinois (Figure 15-2). Some of the largest oil fields in Illinois were discovered during this period, including

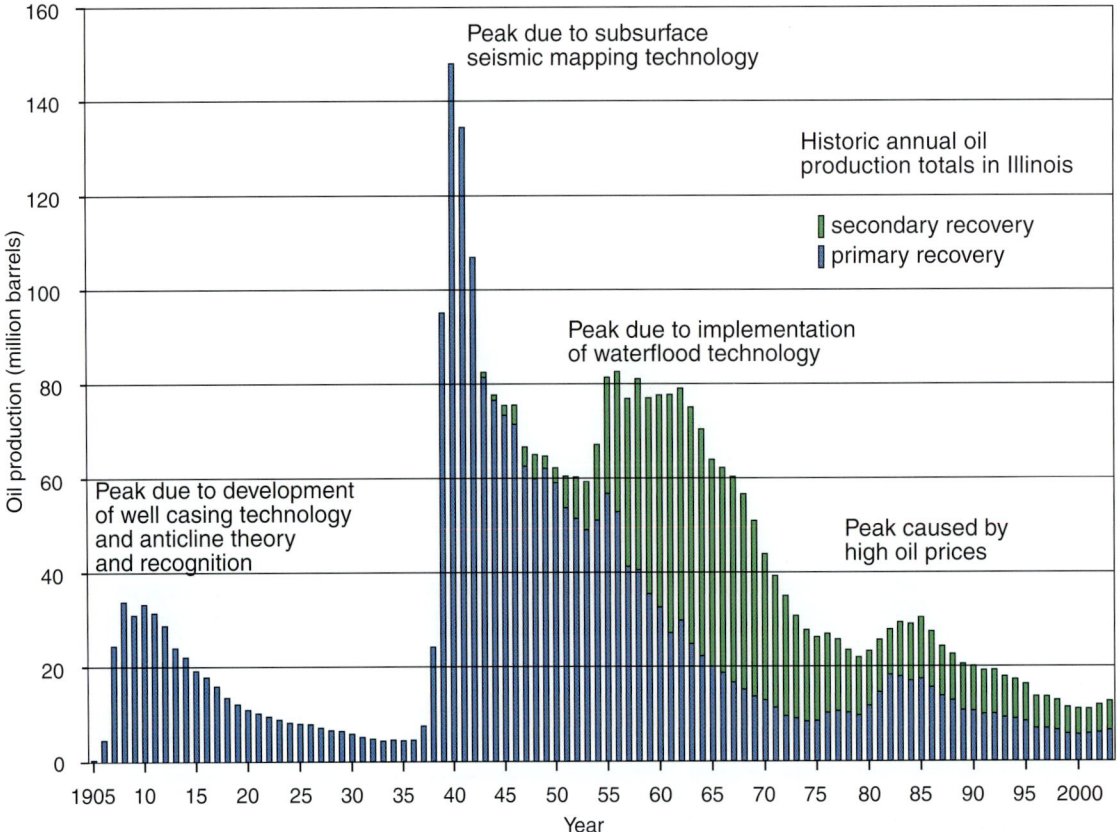

Figure 15-2 Illinois annual oil production 1905–2003 (in barrels). Most peaks in the curve correspond with the introduction of new exploration and development techniques.

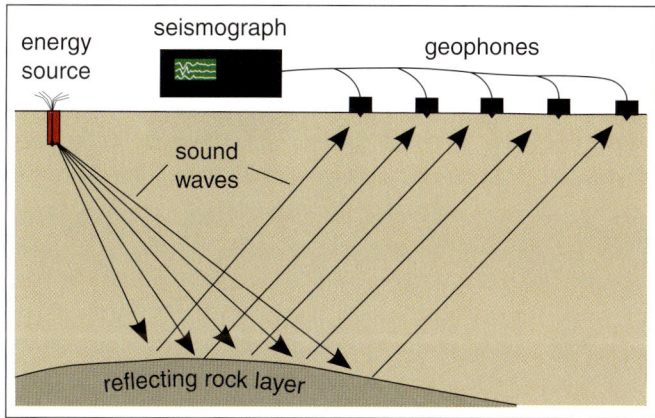

Figure 15-3 Geophones record sound waves reflected off dense rock layers (Killey and Larson 2004). Sound waves are generated by controlled ground-level dynamite blasts or from other energy sources.

Clay City, Salem, and Louden Fields. This timely increase played a major role in meeting the energy needs of the United States during World War II. Oil production declined after the 1940 peak as the size and number of fields being discovered decreased, and the volume of newly discovered oil failed to replace production from depleted older fields. Production in Illinois continued to decline until the early 1950s (Figure 15-2).

Two new technologies, developed during the 1950s, helped producers force more oil out of newly discovered and existing fields. The first process, called hydraulic fracturing, uses powerful pumps at the surface to inject a fluid, commonly with the consistency of a milkshake, into the oil-producing reservoir rocks. The pressure exerted by the fluid is great enough to fracture the rocks around the well, and sand grains injected with the fluid keep the cracks propped open once the pumping stops. The newly opened fractures make reservoir rocks more porous and permeable, allowing oil to flow more easily into the well.

In the second technique, called waterflooding, water is injected into the reservoir rocks to maintain reservoir pressure as the oil is withdrawn and to sweep the oil out of the reservoir rocks and toward the producing well. In the most commonly used design, the five-spot pattern, water is pumped into the reservoir rocks at four wells arrayed around a central producing well (Figure 15-4). This water pushes or flushes oil to the producing well.

Using such techniques to maintain reservoir pressure and to drive the oil out of the reservoir rocks is called secondary recovery or secondary production. With hydraulic fracturing and waterflooding, Illinois' total oil production rose to about 82.3 million barrels in 1957, declining from that level almost continuously since then (Figure 15-2).

Since 1973, Illinois oil producers have confronted wide swings in oil prices caused by world and national events.

(Prices are inflation-adjusted to 2007 U.S. dollars for Illinois crude; data are from the Illinois Oil and Gas Association.) The price of oil in the United States nearly doubled during the 1973–1974 Arab states oil embargo. In an effort to protect the economy from this shock, the government imposed price controls that lasted until 1978. When the controls were lifted, the price of crude oil more than doubled again to an inflation-adjusted (2007) peak of nearly $95.50 per barrel in 1980.

In response to these high oil prices, Illinois producers drilled many new wells, and production rose slightly in Illinois during the early 1980s (Figure 15-2). In 1981, however, an economic recession and increased energy costs caused energy consumption to decrease through improved efficiency and conservation. A major factor was the dramatic improvement in the fuel efficiency of U.S. automobiles. By July 1986, reduced oil demand and major additions in supply, such as from the North Sea oil fields, had driven the price of crude oil down to about $27.60 per barrel.

In 1998, the price of Illinois crude oil sank to less than $15.35 per barrel, a level at which many small, independent producers could not make a profit. Because the price of oil is determined in world commodity markets, Illinois producers, to survive, must constantly seek ways of reducing exploration and production costs and increasing the amount of oil recovery.

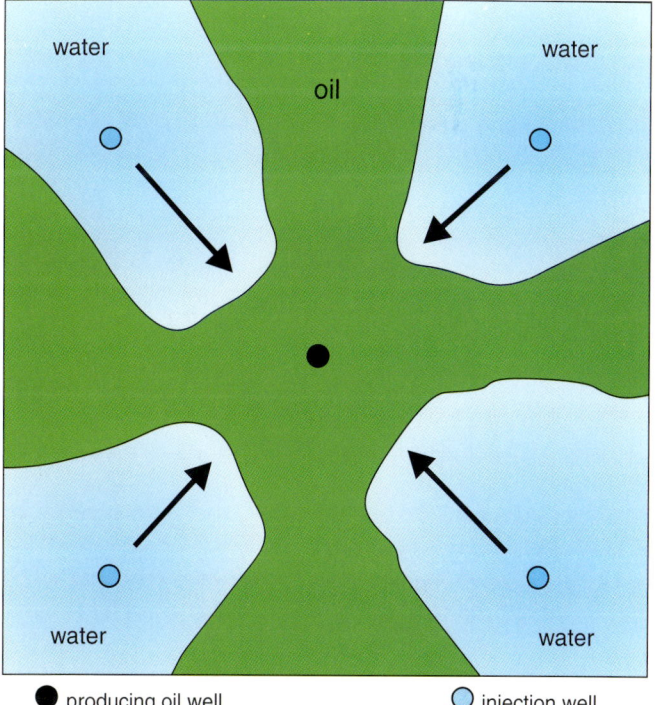

Figure 15-4 Example of a five-spot waterflood injection unit (Huff and Goodwin 2004). Water is injected into four wells surrounding a producing well, pushing oil toward the producing well.

Petroleum Geology in Illinois

The four elements needed to create a petroleum reservoir are (1) a source rock that is rich in the petroleum precursors needed to generate the hydrocarbon, (2) a porous and permeable reservoir hosting layer, (3) an overlying impermeable sealing layer, and (4) structural and/or stratigraphic conditions that trap petroleum.

Source Rocks

In Illinois, most petroleum originated from the Devonian age New Albany Shale. Minor amounts have been obtained from older rocks. The black shale intervals in the New Albany are organic-rich and have the volume and maturity necessary for the generation of hydrocarbons via geothermal cooking caused by the heat and pressure of burial over time.

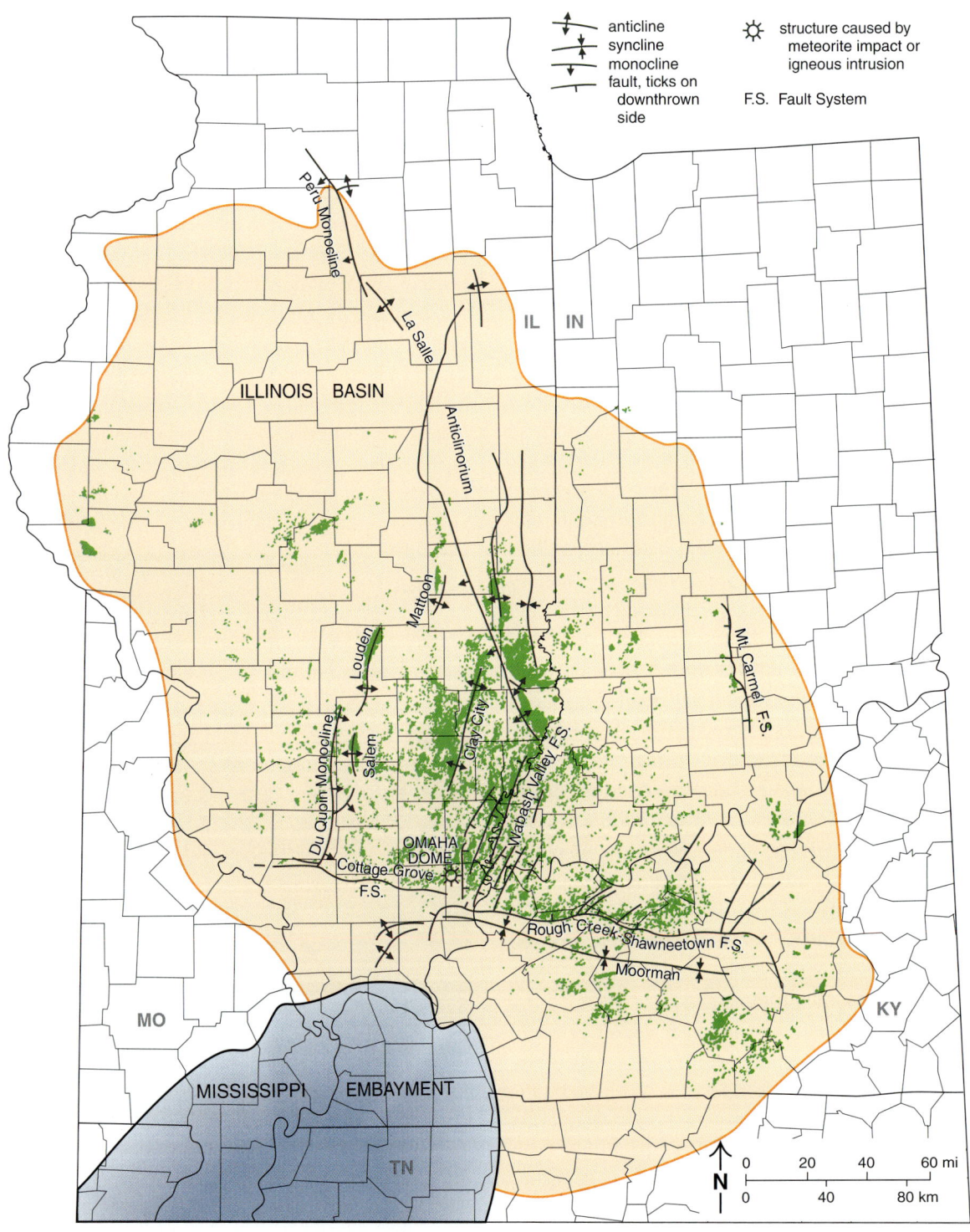

Figure 15-5 Oil fields and major structural features in the Illinois Basin (Seyler and Cluff 1991).

Oil and Gas

Reservoir Rocks

Petroleum reservoirs in Illinois (Figure 15-5) are found in Paleozoic rocks (285 to 480 million years old) containing sufficient porosity and permeability to host large volumes of fluid, primarily in Mississippian and Pennsylvanian sandstones and Ordovician, Silurian, Devonian, and Mississippian carbonates (Figure 15-6). Porosity is the volume that can be filled with fluids, usually a combination of oil, gas, and brine; porosity is measured as a percentage of total rock volume. Permeability is the capacity of a rock to transmit fluid and depends on the size and shape of the pores and the size and shape of their interconnections. Porosity in Illinois petroleum reservoirs is commonly 10 to 25% and is usually expressed in millidarcies (md).

In a reservoir, oil, salt water, and natural gas fill the open pores between the grains of the reservoir rock, much as a liquid fills the spaces between ice cubes in a glass. The weight of all of the rock layers on top of a reservoir applies pressure on the reservoir rock and on the oil, gas, and salt water within it. Therefore, reservoir pressure depends on the depth of the reservoir. Shallow reservoirs less than 1,000 feet (305 m) deep are under less pressure than those buried more deeply at 3,000 to 4,000 feet (914 to 1,219 m).

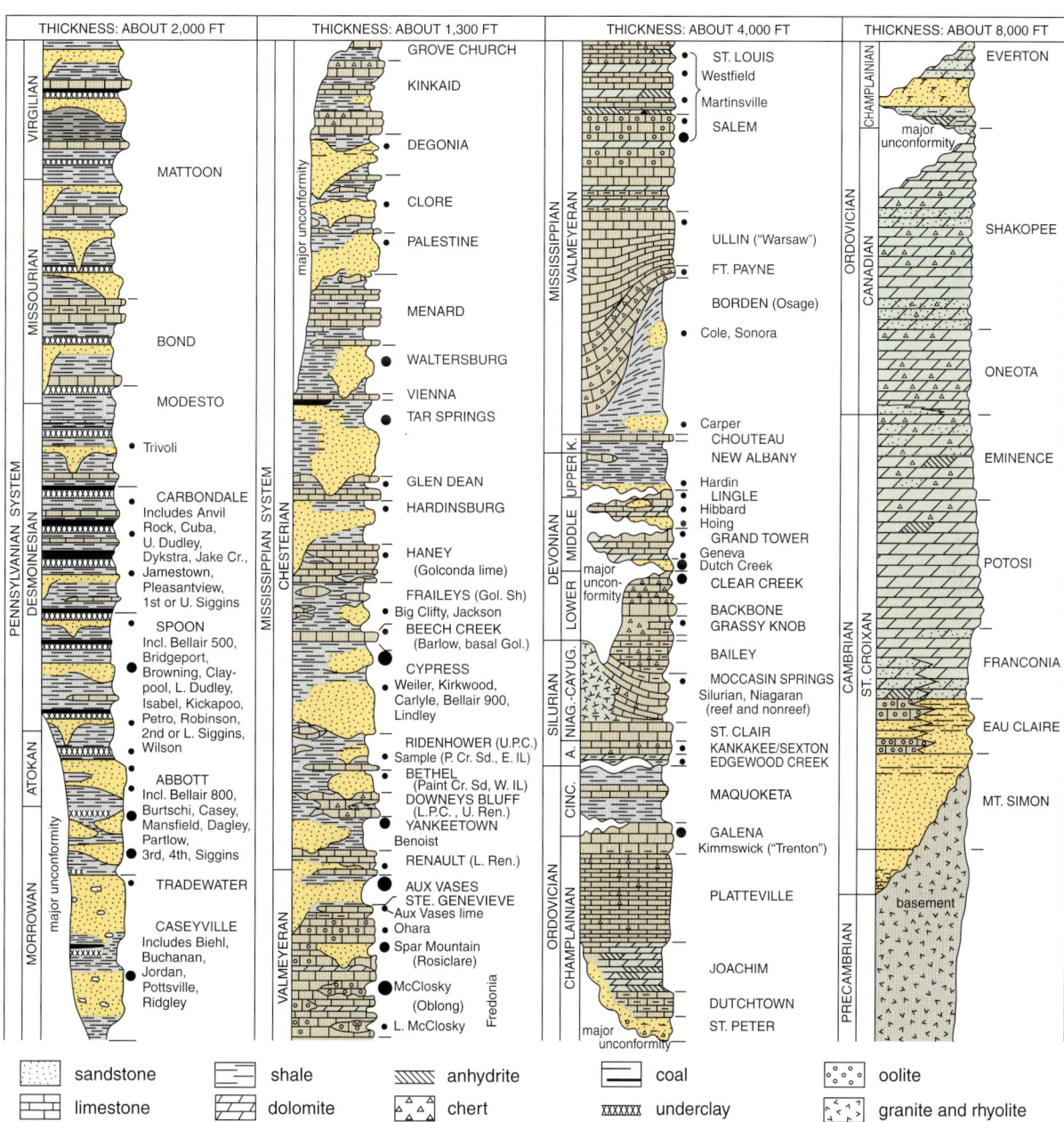

Figure 15-6 Generalized geological column of southern Illinois (Huff 1988). Size of solid dots indicates relative amount of petroleum produced from individual pay zones. Formation names are in capital letters. Abbreviations: A., Alexandrian; Cinc., Cincinnatian; Cr., Creek; E., eastern; Gol., Golconda; Incl., including; K., Kinderhookian; L., lower; Niag.-Cayug., Niagaran-Cayugan; P. Cr. Sd, Paint Creek Sand; Ren., Renault; Sh, shale; U., upper; U.P.C., Upper Paint Creek; W., western.

The shallowest reservoirs in Illinois are at about 200 feet (61 m), and the deepest are at approximately 5,400 feet (1,646 m). When a well penetrates a reservoir, the pressure in the reservoir drives the oil and gas toward the lower pressure of the open well hole. As the oil and gas are withdrawn, reducing the volume of oil between the grains of the reservoir rock, reservoir pressure falls, and, somewhat like a balloon whose neck is opened, the reservoir gradually "deflates."

As the pressure in the reservoir falls, the flow of oil into the well slows to a trickle and eventually stops. In contrast, the familiar image of the oil well gusher spewing oil high into the sky is caused by oil flowing from newly drilled reservoirs under high pressure. Oil produced under this natural driving pressure is known as the primary recovery or the primary production of a field. Almost all production from 1902 to 1950 was primary production. Using water injection techniques to maintain reservoir pressure and drive the oil out of the reservoir rocks is called secondary recovery. With hydraulic fracturing and waterflooding, Illinois' annual oil production again peaked in 1956 at about 82.3 million barrels (Figure 15-2). Waterflood operations have been widely applied and currently account for over one-half of the oil produced in the state.

Seals

A seal is an impermeable barrier that traps the hydrocarbon in the reservoir rock. A seal is most commonly formed by a change in rock type (e.g., a sandstone overlain by or pinching out into a shale or dense limestone). Changes in rock chemistry that cause pores to be filled may transform a porous reservoir rock to a seal. In Illinois, faulting that forces reservoir rock adjacent to an impermeable zone can also form a seal.

Structural and Stratigraphic Traps

There are numerous trapping mechanisms in Illinois oil fields that limit the movement of fluid through the rock and concentrate commercially viable accumulations of oil (Figure 15-7). Structural traps are caused by movements of the Earth's crust that seal the oil or gas in a reservoir. Common structural traps include anticlines, faults, and domes. Stratigraphic trapping occurs when a change in the nature, makeup, or composition of the rock creates permeability barriers, commonly when a sandstone pinches out into an updip shale or when a porous and permeable area in a limestone or dolomite transitions into a nonporous, nonpermeable rock type.

Combination stratigraphic-structural traps are the predominant reservoirs in Illinois (Figure 15-7). Although most petroleum production in Illinois is associated with anticlines, multiple stacked reservoirs have a strong stratigraphic trapping component. Significant combination stratigraphic-fault traps are another major style of combination traps and are associated with the Wabash Valley Fault System, the Mt. Carmel Fault System, and the Rough Creek-Shawneetown Fault System (Figure 2-6). Other common types of traps are unconformity pinchouts, Silurian age reefs, and drapes of porous younger strata over reefs. Most exploration taking place now and in the recent past remains directed toward structural traps. The greatest potential for discovery of future additional reservoirs, however, is in subtle stratigraphic traps.

Productive Geological Units
Cambrian Storage Reservoirs

Although no commercial petroleum has been produced from units in the Sauk Sequence or from Cambrian age rock (Figure 15-6), the reservoirs discovered in these rocks are still commercially valuable as receptacles for storing natural gas. This stored gas is needed in cities to meet peak winter demand, supplementing the gas available from pipelines coming from out of state.

Illinois is a leader in the underground storage of natural gas; gas is pumped into underground saline aquifer reservoirs at more than 37 locations. Natural gas storage is needed because of the wide seasonal fluctuation in demand for gas and the prohibitive expense of building enough pipeline capacity to meet peak winter demand. Therefore, an industry of underground gas storage reservoirs in saline aquifers has developed (Buschbach and Bond 1974). Illinois has far greater storage capacity than any other state (Vary et al. 1973), storing more than 580 billion cubic feet (16.4 billion m³) of gas, about one-third of which is working gas that can be retrieved and distributed to consumers during times of peak demand. Two-thirds is cushion gas,

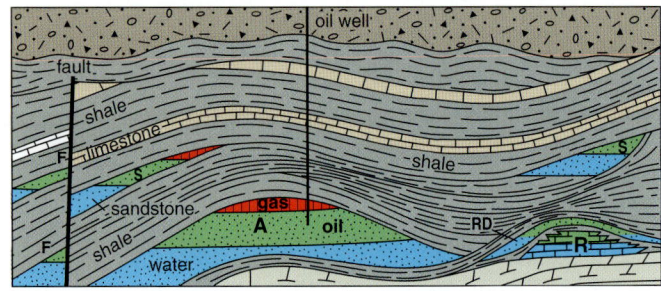

Figure 15-7 Typical petroleum traps: anticline trap (A), reef trap (R), fault (F), reef drape (RD), stratigraphic trap (S), gas (red), oil (green), and water (blue). Impervious rocks such as shale trap oil and gas in crests or upwarps of rock layers (Huff and Goodwin 2004).

that is, the gas that is needed to maintain the ability to retrieve a portion of the stored natural gas for consumption while keeping water from encroaching into producing wells during gas withdrawal.

Gas storage reservoirs have many of the same requirements as petroleum reservoirs. A reservoir layer is needed that has enough permeability and porosity to hold injected natural gas and to release the gas during the withdrawal phase. The typical gas storage reservoir rock has porosity of 15 to 25% of the total volume, almost all of which is available for gas storage. An impermeable caprock or reservoir seal overlying the reservoir strata is needed to prevent the natural gas from escaping upward, and a geological trap, such as an anticline or dome, is needed to prevent the gas from escaping horizontally.

Most gas storage in Illinois is in the Cambrian age Mt. Simon Sandstone and the Ordovician age St. Peter Sandstone. Mt. Simon Sandstone reservoirs are capped and effectively sealed against leakage by a thick sequence of impermeable shale in the Eau Claire Formation (Figure 15-6). Gas storage reservoirs in the St. Peter typically are less effectively sealed, and many St. Peter Sandstone gas storage reservoirs have been abandoned in favor of Mt. Simon reservoirs. The Mt. Simon Sandstone contains highly porous and permeable rock layers that are well suited for underground storage of natural gas in several large anticlinal structures in central and northern Illinois. A recent study (Morse and Leetaru 2005) used three-dimensional modeling of porosity data and detailed mapping to bring new insights into the depositional settings of Mt. Simon Sandstone reservoirs and pertinent information about the effective management of these reservoirs.

Ordovician Carbonate Traps

The "Trenton" formation (formally known as the Kimmswick Limestone) of the Galena Group has been a minor exploration target in Illinois. Production from Trenton reservoirs to date is controlled by structural highs, most of which occur at relatively shallow depths at the edges of the Illinois Basin in the Dupo, St. Jacob, and Waterloo Fields on the Western Shelf (Figure 2-6). Much of the Trenton production in the deeper part of the Illinois Basin occurs in the Salem and Centralia Fields. Although all of the major anticlinal structures in the Basin mapped on the Beech Creek Limestone have been tested down to the Trenton, there have been few discoveries. This lack of success does not rule out the possibility of major Trenton discoveries in the future, but it does suggest that alternative exploration strategies other than drilling present-day structural highs will need to be used.

Westfield Field in eastern Illinois, the largest Trenton reservoir in the state, is located on a structural dome in the La Salle Anticlinorium. Production is from a dolomitic limestone at the base of the Trenton that has fracture-enhanced porosity. During the 1980s and early 1990s, major oil companies targeted the Trenton for exploration in areas along the Cottage Grove Fault System and the La Salle Anticlinorium. Fractured and dolomitized Trenton rocks may be potential targets for oil accumulation as the highly productive Albion-Scipio Field in Michigan is in this type of rock. Prolific reservoirs in the Trenton were also found in dolomitized and fractured zones in grabens of the northern Appalachian Mountains of New York State during the late 1990s and early 2000s. The successful New York exploration model drilled along faulted and fractured zones in downdropped grabens. Applying the New York model to Illinois would require exploration for structural lows, a complete turnabout from the "structural high" anticlinal theory applied in early exploration in Illinois. The application of such new exploration models may lead to new discoveries in Trenton age rocks in Illinois.

Silurian Traps

Three types of traps exist within the Silurian rocks of Illinois: (1) Silurian pinnacle reefs and reef-related traps in southwestern and southeastern Illinois; (2) unconformity and structural traps associated with the paleo-Sangamon Arch in central Illinois; and (3) combination unconformity-stratigraphic traps associated with the sub-Silurian unconformity and the Sangamon Arch in the basal Silurian Kankakee Formation in west-central Illinois.

Silurian Reefs

Silurian reefs may or may not be productive; much of the reef-related production actually comes from Devonian and Mississippian reservoirs draped over the reefs. The reefs provide the structural relief necessary for oil accumulation. Seismic or gravity surveys have located many reefs by revealing the relatively dense rock at the reef core and targeting the associated compactional doming of overlying strata for exploration. Seismic expression of reefs and associated draped strata are illustrated in Figure 15-8.

The first recognized reef discovery in Illinois, and one of the most prolific oil producers, is Marine Field in Madison County. Through 2003, this 2,470-acre (999.6 ha), horseshoe-shaped reef has produced more than 13.4 million barrels of oil (Figure 15-9) from a depth of 1,700 feet (518 m). The map showing subsurface relief of the Silurian reef at Marine Field was contoured on the Devonian New Albany Shale, a consistent marker horizon commonly used for

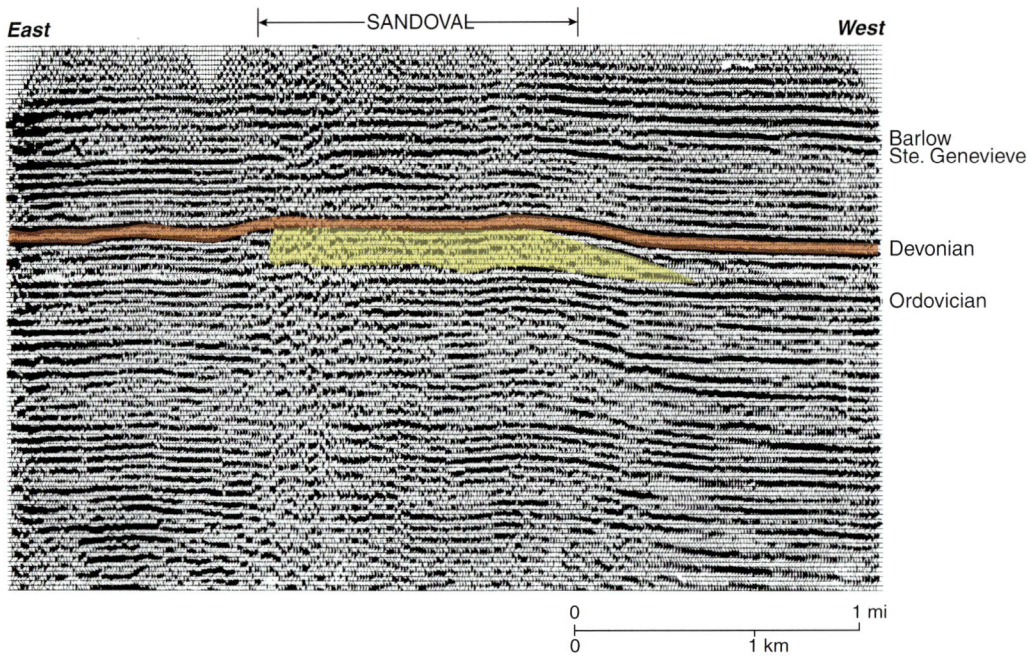

Figure 15-8 Two-dimensional seismic section showing doming in overlying strata caused by drape over the Sandoval Silurian reef (Whitaker 1988). Doming is more apparent when seismic section is viewed sideways at a low angle.

constructing structure maps that is located approximately 500 feet (150 m) above the Silurian reef rock. A diagram of an Illinois Silurian reef that is representative of Marine Field is shown in Figure 15-10. The basic components are the reef core, reef flank, detrital limestones and siliciclastics overlying the reef, and inter-reef carbonates. The reef core consists of structureless, dark bluish gray, vuggy, coarse to finely sucrosic dolomite. Most of the porosity was developed around fossil cavities enlarged by solution and along the numerous intersecting fractures that characterize the reef cores (Lowenstam 1949). The producing horizons in

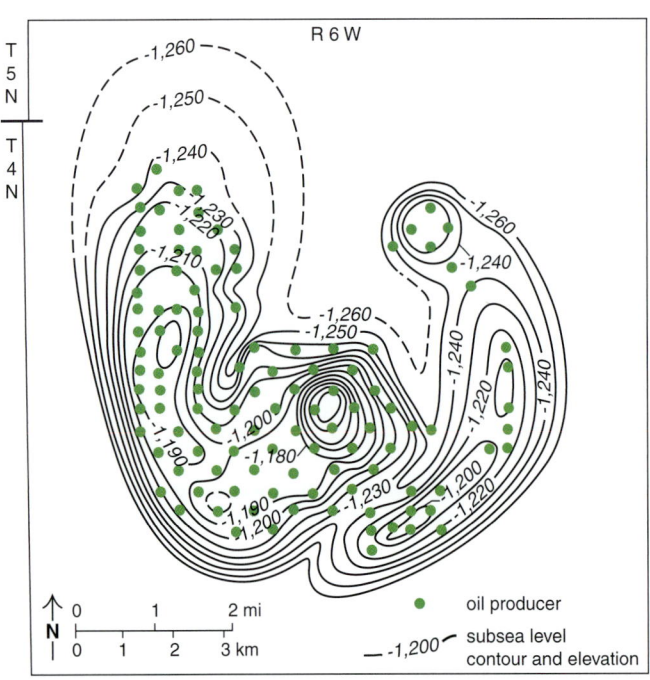

Figure 15-9 Structure map contoured on the base of the New Albany Shale shows the configuration of the underlying Marine reef. Production is from a Silurian reef in Niagaran Series rocks (Bristol 1974).

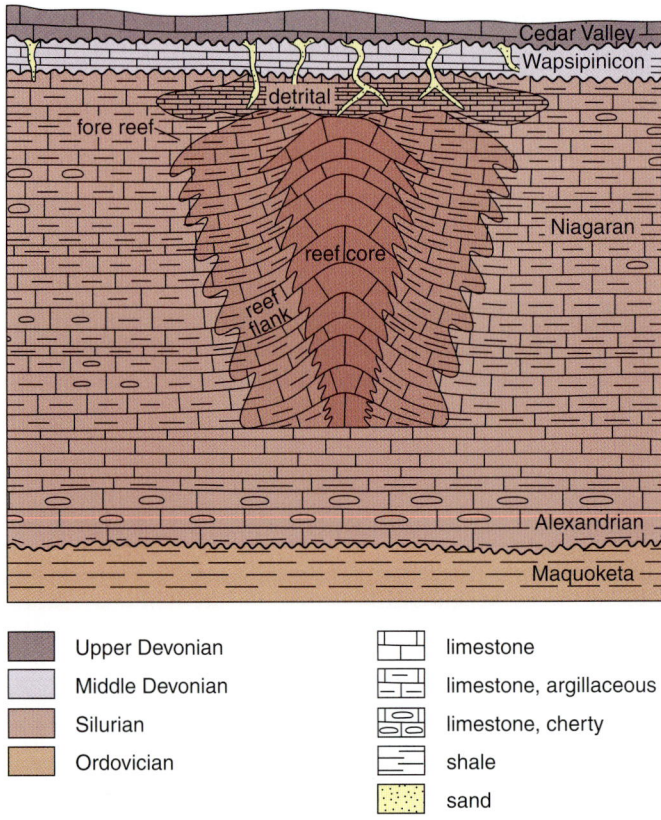

Figure 15-10 Schematic diagram of Silurian reef (Lowenstam 1948).

Marine Field are nearly all located above the reef core in an upper zone of pink, coarsely crystalline, detrital Niagaran limestone. Porosity in the producing horizons consists of fossil cavities, interskeletal spaces of colonial corals, and discontinuous fissures. Reef flank strata also may form oil reservoirs.

Unconformity and Pinchout Traps on the Sangamon Arch

The Sangamon Arch significantly influenced the distribution of petroleum reservoirs in Silurian strata in the area of Decatur, Mt. Auburn, and Springfield in central Illinois and in Adams, Brown, and Schuyler Counties in western Illinois (Figure 15-11). The Sangamon Arch (Figure 2-6) was a major upwarp that formed during Early and Middle Devonian time, resulting in intense erosion of the Silurian rocks exposed on its crest. Middle Devonian rocks lap onto the arch but do not cover it. Upper Devonian rocks in the New Albany Shale Group were then deposited on the arch (Whiting and Stevenson 1965). Erosion that took place prior to the deposition of the New Albany Shale and later tectonic movement have masked the arch on present-day structure maps. Silurian oil fields are producing from secondary porosity, which developed as a result of erosion and possible weathering. No major structural closures are present, although some production is associated with locally structurally high areas. Trapping of oil in these fields is the result of porous strata pinching out on the slope of the arch. The New Albany Shale typically seals these reservoirs.

Unconformity and Stratigraphic Traps in Western Illinois

Production from other Silurian rocks has been discovered in shallow (450 to 650 feet [137 to 198 m]) stratigraphic traps on the northwestern side of the Sangamon Arch. Included in this category are the Kellerville, Siloam, and Buckhorn Fields in Brown and Schuyler Counties, which produce from dolomitized portions of the Kankakee Formation. Subtle paleovalleys eroded into the Ordovician Maquoketa Shale Group at the Ordovician-Silurian unconformity were filled with basal Silurian Kankakee Formation (Figure 15-12) (Seyler et al. 1988). Isopach maps of a New Albany Silurian section showed that thick intervals coincided with the greatest Silurian oil production and were the result of fill in the paleovalleys. Highly porous and permeable petroleum reservoirs were formed by subsequent alteration of the carbonate fill to dolomite.

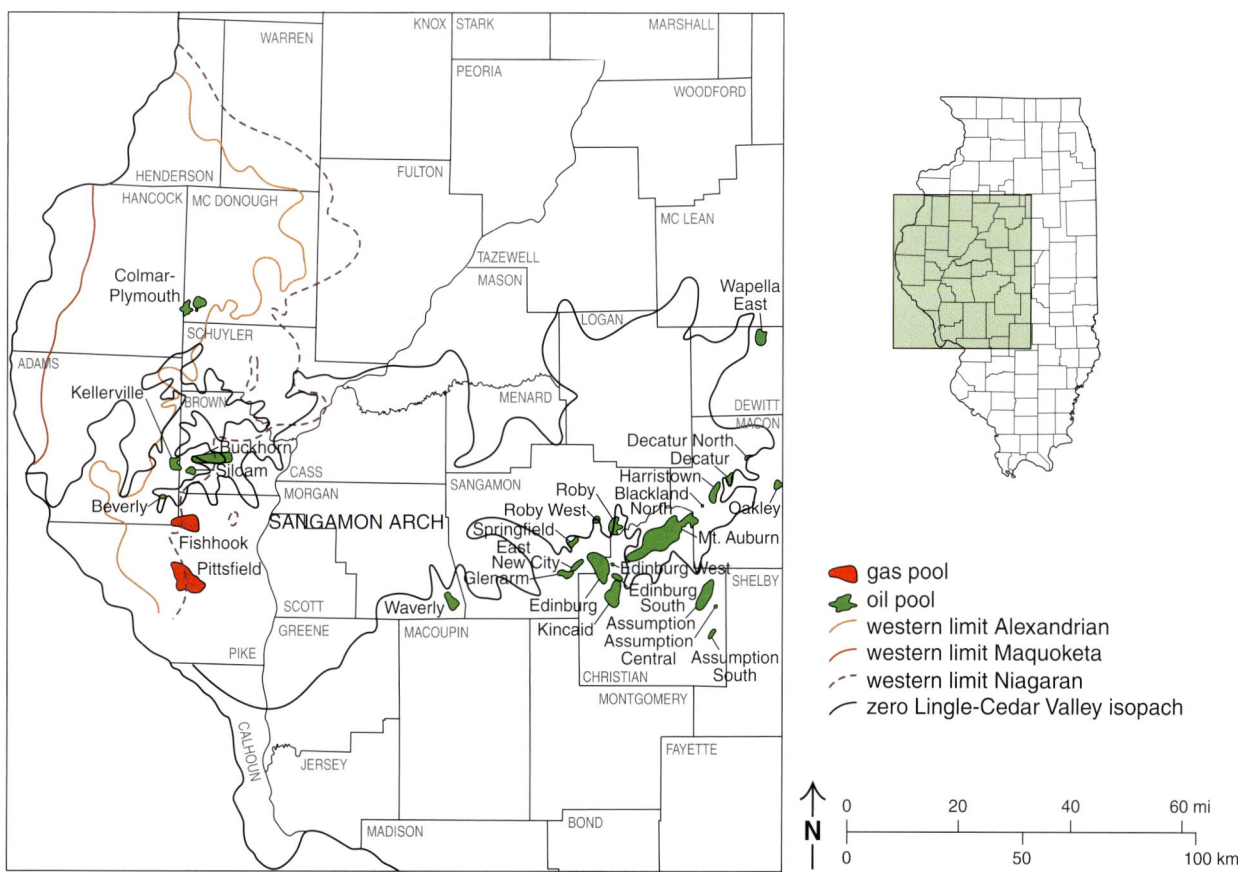

Figure 15-11   Location of the Sangamon Arch and Silurian and Devonian oil and gas fields (Whiting and Stevenson 1965).

Devonian Traps

Devonian reservoirs are generally found in structural traps, such as anticlines and the strata draped over Silurian reefs. Most Devonian production is from the Geneva Dolomite Member of the Grand Tower Limestone where the Geneva has been deposited over Silurian reefs and over other structural highs (Figure 15-6). Wells completed in the Geneva Dolomite are among the most prolific producers in Illinois; some have produced more than 500,000 barrels of oil from depths of 4,000 to 5,000 feet (1,219 to 1,524 m). Significant Devonian production has also been developed from highly porous and permeable sandstones in the Dutch Creek Sandstone Member of the Grand Tower Limestone. The sandstones and sandy limestones of the Lingle Formation and cherty carbonates of the lower Devonian also produce oil but are of minor importance in the Illinois Basin.

The Upper Devonian New Albany Shale is a widespread and areally consistent formation that is a key horizon for structural mapping in the Illinois Basin. A structure map contoured on the base of the New Albany Shale mimics the structural orientation of the Geneva Dolomite (Figure 15-13).

Pronounced structural closure (Figure 15-14), as great as 100 feet (30.5 m), is associated with Geneva Dolomite production. Much of this structure is tectonic in origin, but, in some cases, the closure is caused or enhanced by the draping of younger Middle Devonian strata over Silurian reefs (Bristol 1974), a function of different degrees of compaction. The core reef rock lithified early and/or was cemented by organisms that formed a rigid reef structure that underwent little additional compaction. In contrast, the sediments surrounding the reef were relatively unlithified carbonate muds that were more compacted than the reef (Lowenstam 1948). Structural folding, caused by the differential compaction of the fine-grained sediments flanking the rigid core of a Silurian reef, has been documented to extend upward throughout the entire overlying stratigraphic section. The structural folding can even be detected in structural highs in the overlying Pennsylvanian coals (e.g., Whitaker 1988). Many of these reefs are expressed at the ground surface and can be located by present-day radial drainage patterns (Whitaker 1988). Reef-induced paleostructure (Droste and Shaver 1980) may have influenced the deposition of the Geneva Dolomite by offering sites suitable for growth of rock structures built up by calcareous organisms and the ensuing alteration of these carbonates, resulting in improved reservoir porosity and permeability. Fractures within the Geneva beds may have resulted from differential compaction, providing enhanced flow paths in the reservoir.

In March 2002, a new field discovery well was completed immediately south of the Miletus Field in Marion County. Up to 3,000 barrels a day of oil flowed from a horizontal well bore in the Geneva Dolomite. Earlier vertical test wells nearby had produced more than 300 barrels of oil per day with a low rate of decline. Three-dimensional seismic technology was used to identify the prospect. The combined application of three-dimensional seismic and underbalanced horizontal drilling technologies in the Illinois Basin should enhance exploration and development opportunities, not only in the Middle Devonian Geneva Dolomite, but also in other structural and stratigraphic pros-

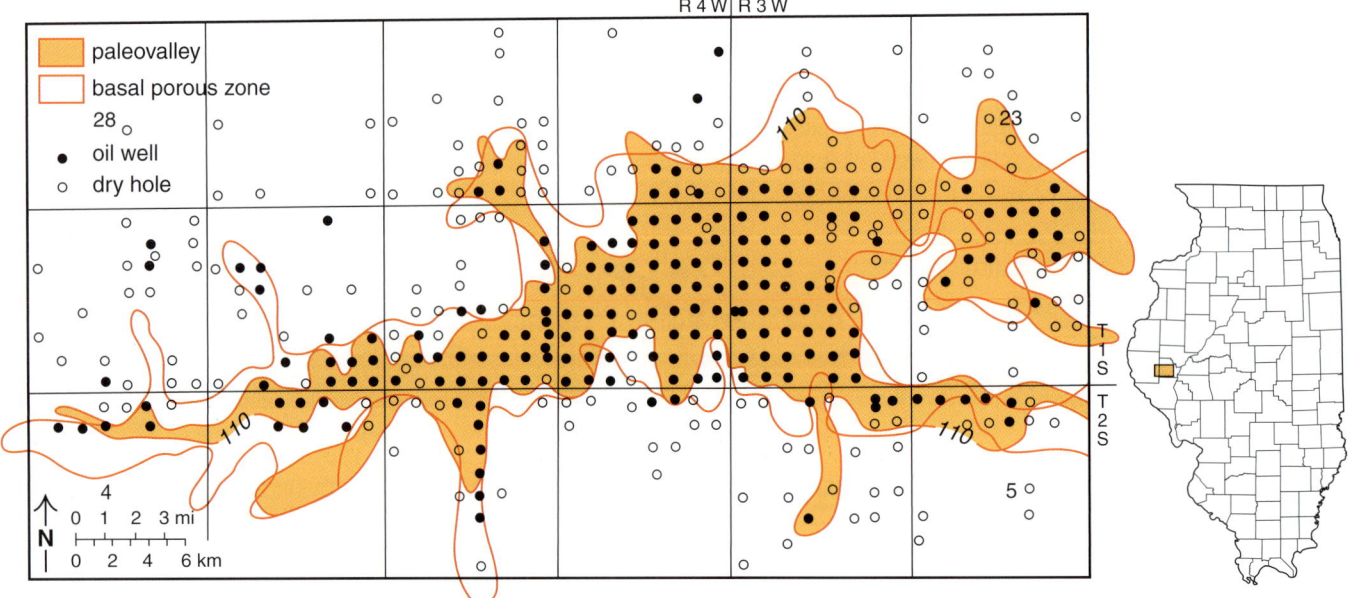

Figure 15-12 Buckhorn Consolidated Field. A paleovalley eroded into the surface of the Maquoketa shale correlates to the dolomitized porous interval of the basal Kankakee Formation (Seyler et al. 1988).

pects in other formations. These technologies had been implemented in other mature producing provinces but not until recently proved or refined for application in the Illinois Basin.

Production from an anticlinal closure at the St. James, Salem, and Centralia Fields shows that pronounced structural closure from tectonic deformation is sufficient to trap petroleum in the Geneva Dolomite. The Geneva Dolomite also produces from structural highs unrelated to Silurian reefs in the Salem, Louden, and Centralia Fields.

The Dutch Creek Sandstone Member of the Grand Tower Limestone is another prolific producing horizon, and approximately 2.4 million barrels of oil have been produced from the Texaco Silverman unit in the Aden Field at a depth of 5,324 feet (1,623 m). The Dutch Creek is a thin, laterally discontinuous, mostly fine-grained, well-sorted, and well-rounded sandstone. Dutch Creek production is confined primarily to structural highs at the Aden Consolidated, Goldengate Consolidated, and Mill Shoals Fields in Wayne County.

Mississippian Traps

Approximately 20% of the oil produced in Illinois is from Middle Mississippian (Valmeyeran) reservoir rocks. Producing horizons, in ascending order, are the Carper sand of the Borden Siltstone; the Fort Payne, Ullin, Salem, St. Louis, and Ste. Genevieve limestones; and the Aux Vases Sandstone (Figure 15-6). These units are some of the most popular drilling targets in Illinois, and most production is associated with major anticlines. However, more recent exploration for reservoirs in these units has concentrated on searching for combination structural-stratigraphic traps.

The oldest and deepest producing horizon in Middle Mississippian Valmeyeran rocks is the Carper sand. Carper reservoirs are made up of sandy siltstone or siltstone rather than sandstone. The average porosity in Carper reservoirs ranges from 13 to 18%; however, average permeability is only about 1 md (Stevenson 1964). As a result, hydraulic fracturing is generally necessary to establish commercial production. Carper reservoirs have been interpreted as offshore turbidites; that is, their morphology, stratigraphic position, and graded bedding indicate deposition by turbidity currents (Lineback 1968). Turbidity currents contain suspended matter that eventually is deposited as turbidites. Carper sands were deposited as discontinuous turbidites alternating with intervals of silt and mud, conditions that are favorable to the development of stratigraphic and combination stratigraphic-structural traps.

The Ullin Limestone (informally, the Warsaw limestone) commonly produces from combination traps in southeastern Illinois in the central part of the Fairfield Basin (Buschbach and Kolata 1991). Ullin reservoirs are unique in Illinois because they are composed of almost pure medium- to fine-grained, crinoid-bryozoan grainstone. This

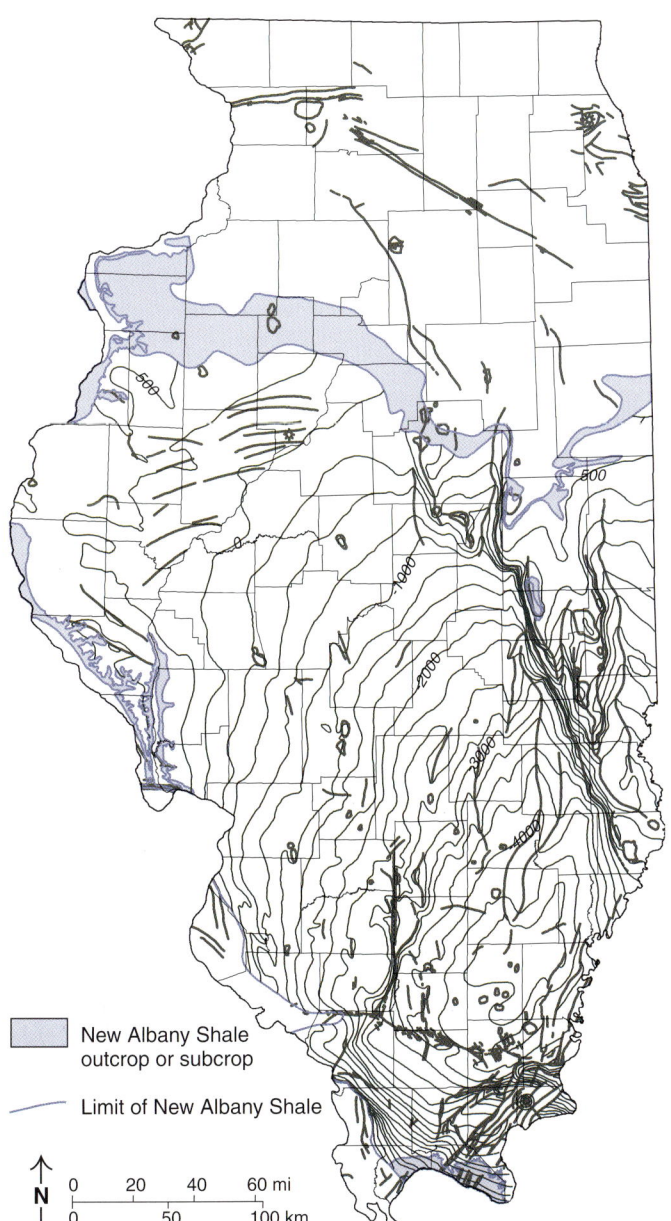

Figure 15-13 Geological structure map contoured on the base of the New Albany Shale (Stevenson et al. 1981).

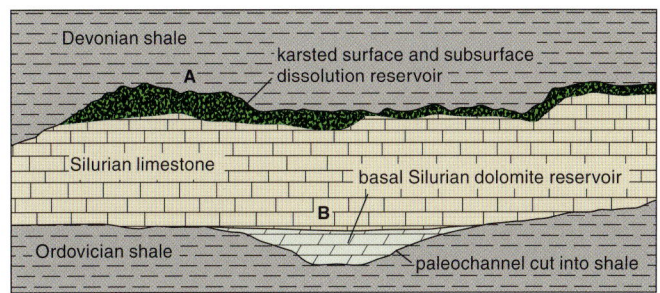

Figure 15-14 Silurian unconformity trap (A) and unconformity stratigraphic trap (B).

grainstone is usually cross-bedded and highly porous (8 to 14%) in intervals that can be up to several hundred feet thick. Reservoir seals are generally a well-cemented, bioclastic layer within the formation. Seals apparently are rare in these strata, resulting in few, but very prolific, petroleum traps in otherwise extremely high-quality reservoir rocks.

Combination stratigraphic and structural traps in the Salem Limestone are another exploration target in Illinois. Deeper drilling on structural highs that already have established production has resulted in the discovery of some Salem reservoirs. Other discoveries showed that structural highs are not essential for trapping oil in the Salem Limestone. Much of the production from these rocks is from stratigraphically trapped oil on the flanks of structures or on subtle structural noses having no closure. Most Salem reservoirs were created by the deposition of coarse-grained carbonate sand, particularly porous, oolitic grainstones located in the middle portions of a progradational cycle. Progradation is the building out of a shoreline or beach toward the water as sediments and other materials are deposited or accumulated. Carbonates advanced across the shelf as a layer of oolitic and skeletal grainstones was deposited over fine-grained carbonate mudstones. Muddy carbonates were then deposited over the carbonate grainstones. Fine-grained sediments that might include thin shales capped each cycle in most places and served as an effective reservoir seal.

The Ste. Genevieve Limestone is one of the most prolific horizons in Illinois and accounts for approximately 18% of the production in the Illinois Basin. Most of the production is from oolitic grainstone lenses, commonly known as "McClosky" oolites in the Fredonia Limestone of the Ste. Genevieve formation. Dolomitic zones in the same interval may also form Ste. Genevieve reservoirs, but those zones are less significant. The Ste. Genevieve commonly produces from combination structural-stratigraphic traps on the top and flanks of anticlines. McClosky oolite reservoirs associated with the Clay City Anticline are aligned parallel to the orientation of the anticline, reflecting their original depositional outline.

Oceanic conditions would have had to have been just right for oolites to form by the precipitation of concentric layers of calcium carbonate around grains of sand. Marine water that was close to saturation with calcium carbonate was lifted onto carbonate platforms by tidal currents and warmed, becoming supersaturated in calcium carbonate due to evaporation and increased water temperature. When carbonate sand grains (shell fragments, foraminifera, crinoids, etc.) were agitated in the supersaturated water by wave action and tidal currents, calcium carbonate precipitated onto the granular nuclei in concentric layers. As the layers built up, the oolitic grains were deposited in lenticular (lens-shaped) bars and became lithified. Cements formed where the grains touched and left open pores between grains that eventually formed high-quality reservoir rock with large amounts of primary porosity. These processes produced many highly productive reservoirs in Illinois.

McClosky oolite bars are typically lenticular, elongate, generally flat-bottomed, convex-upward deposits 13 to 25 feet (3.96 to 7.6 m) thick, 0.13 to 2 miles (0.2 to 0.6 km) wide, and up to 6 miles (1.8 m) long (Carr 1973, Cluff 1986, Choquette and Steinen 1980). The oolite bars were deposited as grainstone shoals overlain by low-porosity lime mudstones or shales that graded laterally into dense carbonate rocks that commonly form lateral seals. Several hundred million barrels of oil have been produced from these reservoirs using a combination of primary and secondary recovery techniques.

Also productive in Illinois are two other members of the Ste. Genevieve Limestone, the Spar Mountain Sandstone Member (also known as the Rosiclare Sandstone Member) and the Ohara limestone or Karnak Limestone Member. The Spar Mountain Sandstone is a siliciclastic member of the Ste. Genevieve and is not a major reservoir in Illinois, although, in some anticlinal fields (e.g., Mattoon and Cooks Mills Consolidated), it is a very prolific and productive horizon. The Karnak Limestone, the uppermost member of the Ste. Genevieve Limestone, is a persistent member throughout the Illinois Basin. Although not a major producer in Illinois, it is typically a gray, oolitic, and crinoidal limestone and may be locally productive in anticlinal structures.

Following the deposition of Ste. Genevieve carbonates, conditions changed in Illinois from those suited to deposition and accumulation of carbonate oolite and grainstone shoals to conditions favoring deposition of shallow marine siliciclastic sandstones, siltstones, and shales. The clear water, shallow marine, equatorial carbonate "factories" were overcome by an influx of terrestrial sediments during deposition of the Middle Mississippian Aux Vases Sandstone and the succeeding Chesterian and Pennsylvanian age strata. The influx of terrestrial sediments came from the Ozarks area of present-day southern Missouri and the Canadian Shield region of present-day eastern and middle Canada.

Most petroleum reservoirs in Illinois are found in Mississippian and Pennsylvanian age sandstones. Some of the more prolific sandstone reservoirs are found in the Aux Vases and Cypress formations. The Aux Vases Sandstone overlies carbonate deposits in the Ste. Genevieve formation. Several field studies show that reservoirs in the Aux Vases Sandstone across Illinois are strikingly similar in size, geometry, orientation, and sedimentary features. All indicate deposition by tidal processes that formed tidal bar complexes. Individual tidal bars reach a maximum thickness of 40 feet (12.2 m) and range between 0.5 to 1.5 miles (0.8 to 2.4 km) long and from 0.25 to 0.75 miles (0.4 to 1.2 km) wide (Seyler and Cluff 1991). Bars are convex-upward in cross section and grade laterally and vertically into interbar and tidal flat deposits. The best production is from clean, well-sorted, fine-grained, trough cross-bedded sandstones that were deposited by tidal processes in an offshore tidal complex (Seyler 1986). Aux Vases reservoirs are highly porous, permeable, and loosely cemented. A persistent problem in the Aux Vases is that geophysical logs, used to determine the amount of oil versus salt water in a reservoir, erroneously show water saturations that are very high, indicating that reservoirs contain salt water rather than oil.

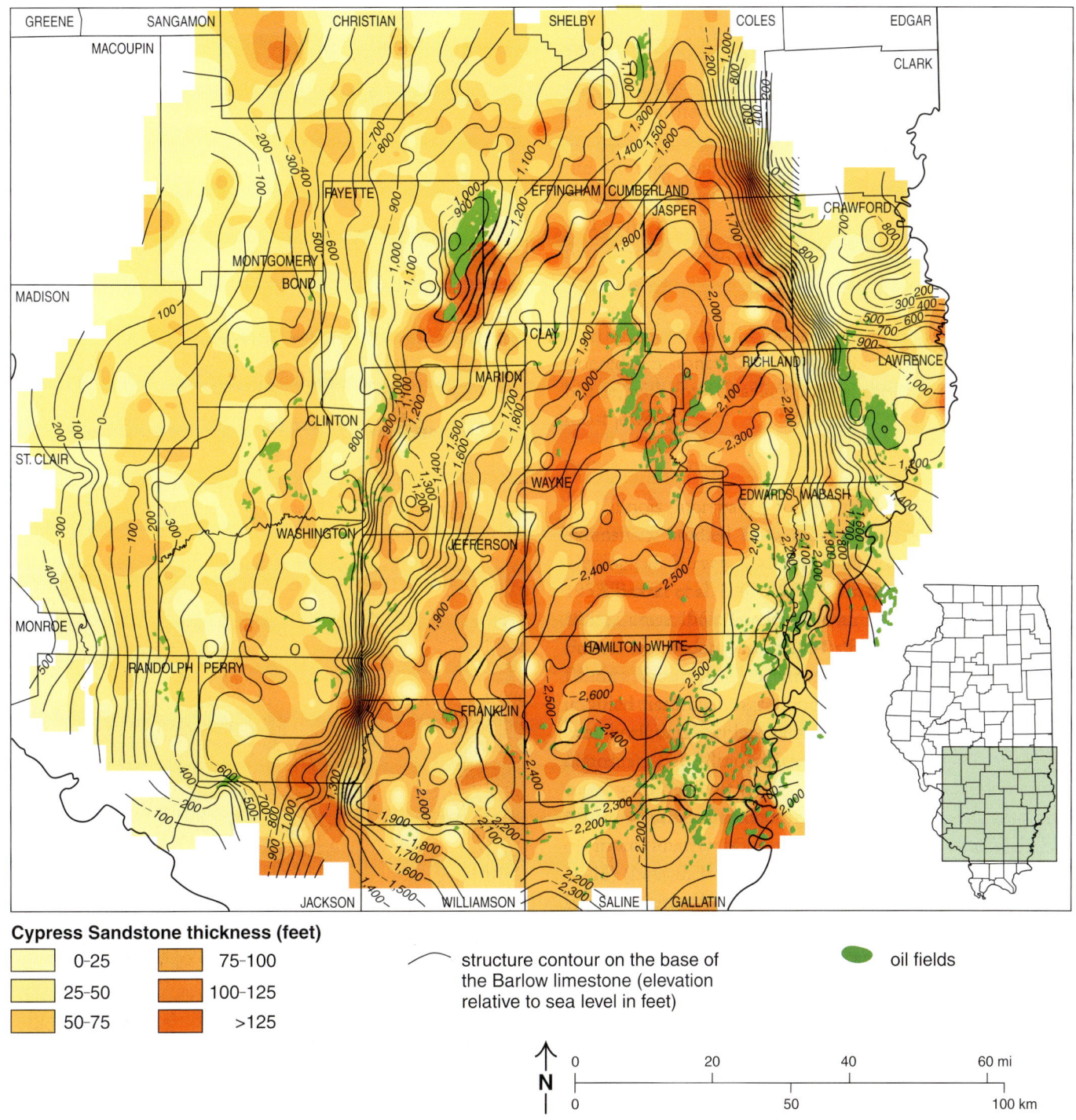

Figure 15-15 Distribution of the Cypress Sandstone in Illinois. Darker shades represent greater thickness of the Cypress Sandstone. The structure layer is contoured on the base of the Barlow limestone. Green shaded areas are oil fields producing from the Cypress Sandstone (Seyler 2001).

Seyler (1998) has shown that these high water-saturation calculations in zones that are actually filled with oil is due to a clay mineral coating on sand grains in these reservoirs. Virtually every sand grain has a thin microscopic coating of clay, which traps and holds immobile salty formation water, altering the electric currents of geophysical logs, causing misleadingly high readings for water saturation.

Sandstone reservoirs in Chesterian age rocks have been highly productive in Illinois and account for over 1 billion barrels of oil produced in the state to date. The reservoirs are primarily fine-grained, well-sorted sandstones that were cyclically deposited by repeated advances and retreats of the sea across a southward-dipping coastal plain (Swann 1963, Howard 1991). These sandstones were deposited as point bars, tidal ridges, offshore bars, and distributary channel bars in the ancient Michigan River fluvial-deltaic system. Each cycle of shoreline advance and retreat produced a new series of genetically and physically similar reservoir sandstones, shales, and limestones that became vertically stacked as the Illinois Basin subsided. Reservoirs

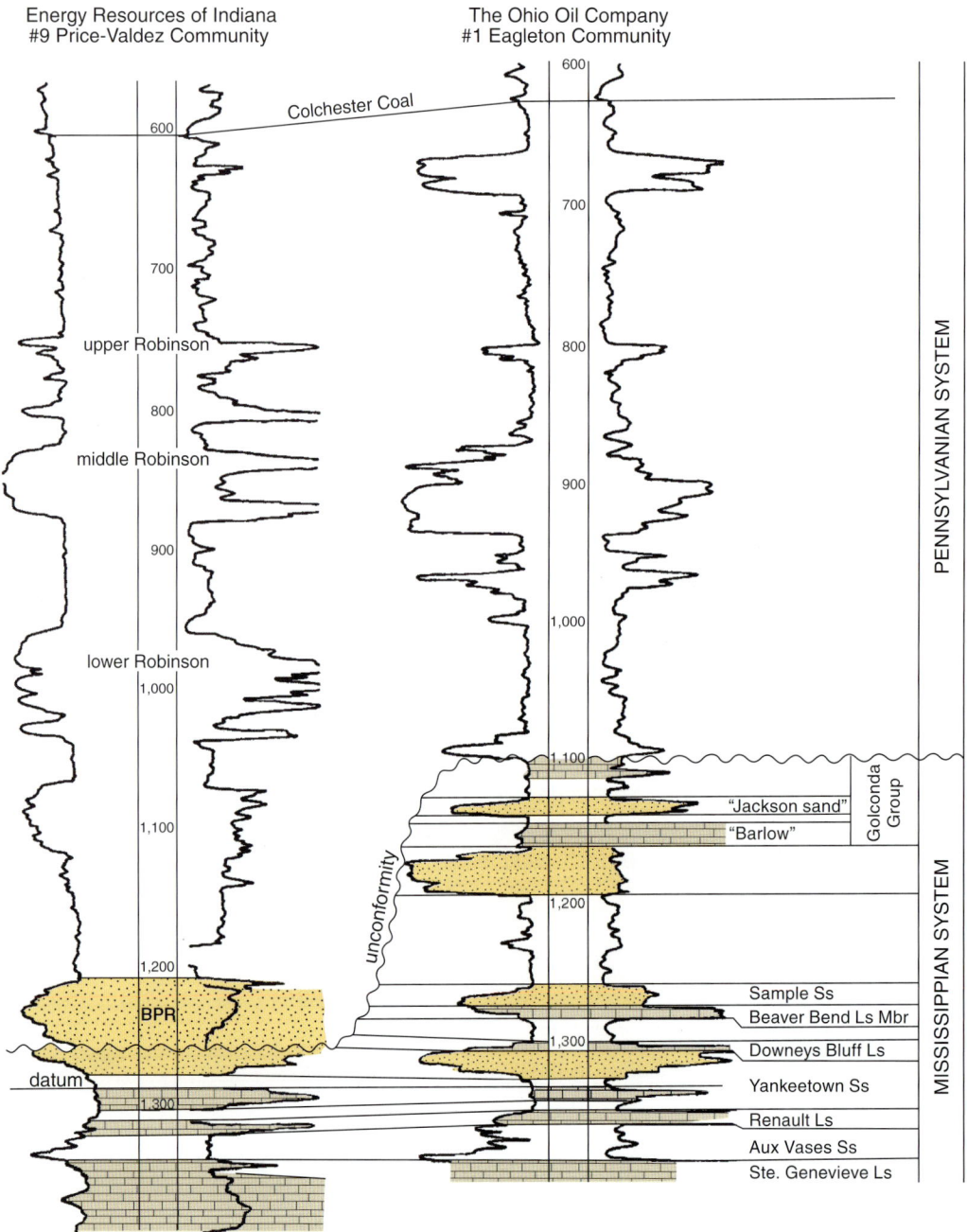

Figure 15-16 Well logs from Crawford County showing a basal Pennsylvanian channel fill reservoir filling a channel cut into Mississippian strata (Howard and Whitaker 1988). Quotation marks indicate an informal name. Abbreviations: BPR, basal Pennsylvanian reservoir; Ls, Limestone; Mbr, Member; Ss, Sandstone.

are found in the following Chesterian formations (oldest to youngest): Renault, Yankeetown, Bethel-Ridenhower-Paint Creek, Cypress, Fraileys-Jackson, Hardinsburg, Tar Springs, Waltersburg, Palestine, and Degonia-Clore (Figure 15-6). Because most of the Chesterian sandstones were deposited as discrete bars in association with impermeable shales and limestones, they can form excellent stratigraphic traps as well as structural traps on the sides and crests of anticlines and domes. Usually, the trapping mechanism is a combination of stratigraphy and structure.

The Cypress Sandstone is the most widespread and productive sandstone in Illinois, having produced almost one-third of all Illinois oil. Figure 15-15 shows the major depositional trends of the Cypress Sandstone in Illinois (Seyler 2001). Reservoirs are found in deltaic and marine tidal ridge deposits. Most production in the Cypress Sandstone is from thin, multiple stacked layers of sandstone in the upper part of the formation and not from the massive, thick sandstones in the lower Cypress. Most reservoirs are found in the periphery of the Illinois portion of the Illinois Basin.

Pennsylvanian Reservoirs

The time separating Mississippian and Pennsylvanian rocks was marked by a long period of substantial erosion. Deep valleys were eroded into Mississippian rocks, and lower Pennsylvanian sediments filled these valleys. Long, relatively narrow channel systems developed. The marine processes that dominated the deposition of Mississippian units were replaced by more terrigenously (land) and fluvially (river) dominated processes during Pennsylvanian sedimentation. Pennsylvanian reservoirs commonly consist of channel deposits laid down by ancient river systems. Features found in modern-day river systems, such as point bars, crevasse splays, and meander belt sand bodies, are common in paleochannel deposits in Pennsylvanian age rocks.

Pennsylvanian sandstone reservoirs account for 15% of the oil production in the Illinois Basin. Most Pennsylvanian reservoirs occur in the lower Pennsylvanian sandstones of the Caseyville, Tradewater, and Carbondale Formations (Figure 15-6) that were deposited by fluvial or deltaic processes. These sandstones have a high degree of lateral variation because river channel systems do not deposit laterally extensive sand bodies. A typical sequence of channel sandstones is illustrated in Figure 15-16, which shows that the base of the deposit has scoured into underlying Mississippian sediments. The bases of channel deposits are abrupt and commonly incorporate large clasts or pebbles; grain size becomes finer upward in the sequence. The basal deposits may also contain large petrified tree fern logs that were common in interfluvial regions during the Pennsylvanian. The high-energy channel-base deposits typically have good porosity and permeability and are followed in the vertical sequence by finer-grained, trough cross-bedded sandstones. The basal deposits are common reservoirs when structural conditions are suitable for trapping petroleum, particularly on the large anticlines in Clark, Lawrence, and Crawford Counties at the eastern edge of the state.

Conclusion

Illinois began producing oil and gas commercially in 1904 and, to date, has produced over 3.5 billion barrels of oil. Little natural gas is commercially produced in the state; however, Illinois relies on underground storage of natural gas in sandstone reservoirs of Ordovician and Cambrian age to meet peak seasonal gas needs.

Oil is produced from rocks of Ordovician through Pennsylvanian age at depths of from 200 to 5,400 feet (61 to 1,646 m) deep. The New Albany Shale is the source of most Illinois oil. Much of the state's oil has been produced by applying a succession of new technologies, beginning with well drilling and casing techniques, followed by waterflooding and seismic techniques, and, more recently, horizontal drilling techniques. Although Illinois may be considered a mature hydrocarbon province, new technologies—such as enhanced oil recovery through injection of carbon dioxide—should continue to help producers find and develop the vast reserves remaining in Illinois.

References

Bristol, H. M., 1974, Silurian pinnacle reefs and related oil production in southern Illinois: Illinois State Geological Survey, Illinois Petroleum 102, 98 p.

Buschbach, T. C., and D. C. Bond, 1974, Underground storage of natural gas in Illinois, 1973: Illinois State Geological Survey, Illinois Petroleum 101, 71 p.

Buschbach, T. C., and D. R. Kolata, 1991, Regional setting of the Illinois Basin, *in* M. W. Leighton, D. R. Kolata, D. F. Oltz, and J. J. Eidel, eds., Interior cratonic basins: Tulsa, Oklahoma, American Association of Petroleum Geologists, Memoir 51, p. 29–55.

Carr, D. D., 1973, Geometry and origin of oolite bodies in the Ste. Genevieve Limestone (Mississippian) in the Illinois Basin: Indiana Geological Survey, Bulletin 48, 81 p.

Choquette, P. W., and R. P. Steinen, 1980, Mississippian non-supratidal dolomite, Ste. Genevieve Limestone, Illinois Basin: Evidence for mixed-water dolomitization: Society of Economic Paleontologists and Mineralogists, Special Publication 28, p. 163–196.

Cluff, R. M., 1986, Application of modern carbonate sand models to oil and gas exploration, Mississippian Ste. Genevieve Limestone, Illinois Basin, *in* B. Seyler, ed., Aux Vases and Ste. Genevieve formations: A core workshop and field trip guidebook: Illinois Geological Society and Illinois State Geological Survey, p. 5–7.

Droste, J. B., and R. H. Shaver, 1980, Recognition of buried Silurian reefs in southwestern Indiana—Application to the Terre Haute bank: Journal of Geology, v. 88, p. 567–587.

Howard, R. H., 1991, Hydrocarbon reservoir distribution in the Illinois basin, *in* M. W. Leighton, D. R. Kolata, D. F. Oltz, and J. J. Eidel, eds., Interior cratonic basins: Tulsa, Oklahoma, American Association of Petroleum Geologists, Memoir 51, p. 299–327.

Howard, R. H., and S. T. Whitaker, 1988, Hydrocarbon accumulation in a paleovalley at the Misissippian-Pennsylvanian unconformity near Hardinville, Crawford County, Illinois: A model paleogeomorphic trap: Illinois State Geological Survey, Illinois Petroleum 129, 26 p.

Huff, B. G., 1988, Oil and gas developments in Illinois: Illinois State Geological Survey, Illinois Petroleum 154, 72 p.

Huff, B. G., and J. H. Goodwin, 2004, History of oil and gas production in Illinois: Illinois State Geological Survey, Geobit 8, 4 p.

Killey, M. M., and D. R. Larson, 2004, Illinois groundwater: A vital geologic resource: Illinois State Geological Survey, Geoscience Education Series 17, p. 43.

Lineback, J. A., 1968, Turbidites and other sandstone bodies in the Borden Siltstone (Mississippian) in Illinois: Illinois State Geological Survey, Circular 425, 29 p.

Lowenstam, H. A., 1948, Marine pool, Madison County, Illinois, Silurian reef producer: Illinois State Geological Survey, Report of Investigations 131, p. 153–188.

Lowenstam, H. A., 1949, Niagaran reefs in Illinois: Their relation to oil accumulation: Illinois State Geological Survey, Report of Investigations 145, 36 p.

Morse, D. G., and H. E. Leetaru, 2005, Reservoir characterization and three-dimensional models of Mt. Simon gas storage fields in the Illinois Basin: Illinois State Geological Survey, Circular 567, 72 p.

Seyler, B., 1986, Aux Vases and Ste. Genevieve formations: A core workshop and field trip guidebook: Illinois Geological Society, 67 p.

Seyler, B., 1998, Geologic and engineering controls on Aux Vases Sandstone reservoirs in Zeigler field, Illinois: Illinois State Geological Survey, Illinois Petroleum 153, 79 p.

Seyler, B., 2001, Regional depositional relationships of the Cypress Formation and the underlying Chesterian Strata, *in* B. Seyler, J. P. Grube, and D. G. Morse, eds., The Cypress Sandstone in Illinois: Petroleum Technology Transfer Council, Illinois State Geological Survey, and Illinois Oil and Gas Association.

Seyler, B., and R. M. Cluff, 1991, Petroleum traps in the Illinois Basin *in* M. W. Leighton, D. R. Kolata, D. F. Oltz, and J. J. Eidel, eds., Interior cratonic basins: American Association of Petroleum Geologists, Memoir 51, p. 361–401.

Seyler B., J. E. Crockett, and S. T. Whitaker, 1988, Buckhorn consolidated field, *in* Geology and petroleum production of the Illinois Basin, v. 2: Joint Publication of the Illinois and Indiana-Kentucky Geological Societies, p. 51–53.

Stevenson, D. L., 1964, Carper Sand oil production in St. James, Wilburton, and St. Paul pools, Fayette County, Illinois: Illinois State Geological Survey, Circular 362, 12 p.

Stevenson, D. L., L. L. Whiting, and R. M. Cluff, 1981, Geologic structure of the base of the New Albany Shale Group in Illinois: Illinois State Geological Survey, Illinois Petroleum 121, 2 p.

Swann, D. H., 1963, Classification of Genevievian and Chesterian (Late Mississippian) rocks of Illinois: Illinois State Geological Survey, Report of Investigations 216, 91 p.

Vary, J. A., et al. (Task Group on Underground Gas Storage Statistics), 1973, The underground storage of gas in the United States and Canada, 22nd Annual Report on Statistics, Committee on Underground Storage, American Gas Association, XU0273, December 31, 1972: Arlington, Virginia, American Gas Association, 22 p.

Whitaker, S. T., 1988, Silurian pinnacle reef distribution in Illinois: Model for hydrocarbon exploration: Illinois State Geological Survey, Illinois Petroleum 130, 32 p.

Whiting, L. L., and D. L. Stevenson, 1965, The Sangamon Arch: Illinois State Geological Survey, Circular 383, 20 p.

Lead, Zinc, and Fluorite Mining   16

Zakaria Lasemi

INTRODUCTION

Lead and zinc ores were mined for many decades in northwestern Illinois (Figure 16-1), southwestern Wisconsin, and northeastern Iowa, a region collectively referred to as the Upper Mississippi Valley Lead-Zinc District (Willman et al. 1946, Willman and Reynolds 1947, Bradbury 1959, Heyl et al. 1959). Lead was obtained from the mineral galena, or lead sulfide (PbS), and zinc was obtained from sphalerite, or zinc sulfide (ZnS). Galena and sphalerite were not useful in their dispersed natural form. The ore-bearing host rock had to be milled, and galena and sphalerite concentrates had to be sent to smelting plants, which separated the lead and zinc metals from the sulfur-bearing ore minerals.

The search for lead ore extended into southeastern Illinois and western Kentucky in the 1830s and led to the discovery of highly valuable fluorite deposits in what is now referred to as the Illinois-Kentucky Fluorspar District (e.g., Bradbury et al. 1968, Fulton and Montgomery 1994, Goldstein 1997). Fluorite, or calcium fluoride (CaF_2), also known as fluorspar in the mining industry, is a glassy mineral consisting of calcium (51%) and fluorine (49%). Along with the lead and zinc ores, great quantities of fluorite came out of the mines. Initially the fluorite was discarded because there were no known uses for it. Commercial fluorite production in Illinois did not begin until the late 1800s, when it began to be used in industry. For many years, Illinois was the largest producer of fluorite in the United States, but domestic production began to decline in the 1950s, largely due to competition from foreign sources. The closing of the last mines in southern Illinois in 1995 ended more than a century of active exploration and mining in the region (Masters 1987, Goldstein 1997, Reinertsen and Masters 1997).

ORIGIN OF LEAD-ZINC ORES AND FLUORITE

It is thought that lead and zinc ores and fluorite formed in Illinois about 270 million years ago during the early Permian Period. These minerals were once thought to have been deposited from warm or hot water that originated deep within the Earth (Currier 1923, Grogan and Bradbury 1968). Some people still think that groundwater circulating deep in the Earth along major fractures or faults brought the metals to near-surface sites of deposition, where cooling and decreased pressure led to precipitation of the ores.

Most geologists, however, now think that a regional gravity flow system drove hot, metal-rich fluids from south to north in the Illinois Basin about 270 million years ago (Bethke 1986, Rowan and Goldhaber 1996, Pitman et al. 1997). Fluid migration appears to have begun during the late Pennsylvanian and continued through the early Permian, originating from the deep part of the Illinois Basin. Tectonic uplift of the Appalachian fold belt to the east and the Ouachita fold belt to the south of the Illinois Basin provided the relief necessary to create the large-scale, gravity-driven flow system (Figure 16-2). The Cambrian-Ordovician aquifers, such as the Mt. Simon and St. Peter Sandstone, and porous and permeable dolomite provided conduits for this large-scale flow system. Metal-rich fluids migrated to shallower areas at the margin of the Basin in southeastern and northwestern Illinois where the aquifers came closer to the surface. For the Upper Mississippi Valley District, of which the northwestern Illinois Lead-Zinc

Figure 16-1 Photograph of a lead-zinc mine located north of Galena in Jo Daviess County, 1955. (Photograph from the Illinois State Geological Survey collection.)

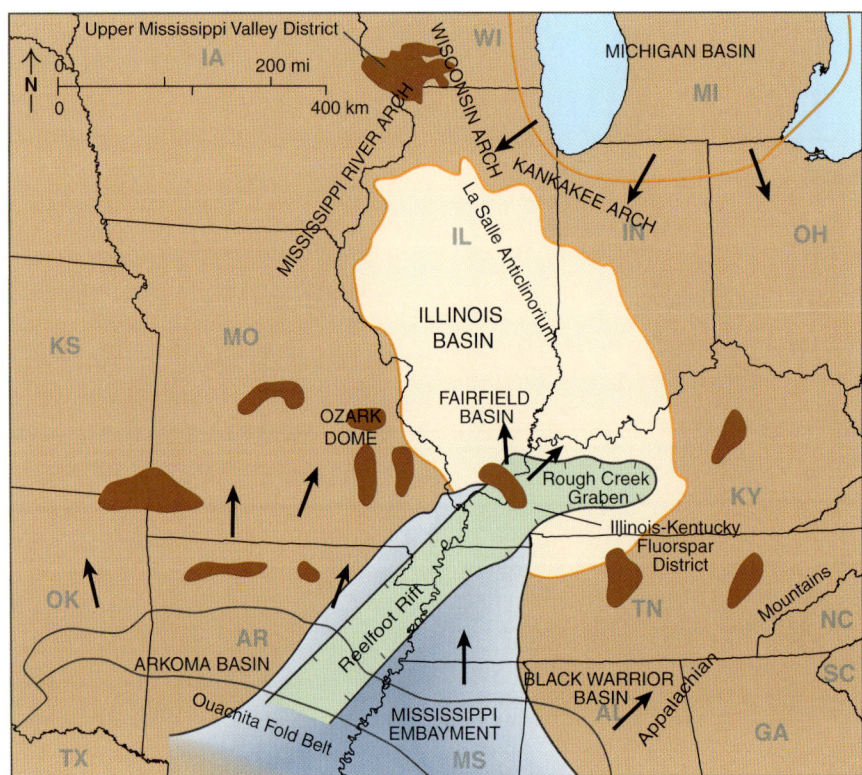

Figure 16-2 Diagram showing the paleoflow directions in the Illinois Basin area. The dominant flow component was from the south, but a smaller component was from the north. The dark brown areas are major ore districts. (Modified from Pitman et al. 1997.)

District is a part, funneling of ore-bearing solution to the site of deposition was enhanced by the existence of the La Salle Anticlinorium to the east and the Mississippi River Arch to the west of the Illinois Basin (e.g., Bethke 1986). Mineralization sites formed as fluid rose through an extensive system of faults and fractures. Upon cooling, the fluid formed fluorite and associated lead-zinc ores in the Illinois-Kentucky Fluorspar District and lead-zinc ores in the Upper Mississippi Valley District. Similar fluid flow models have been proposed for the formation of lead-zinc ores in other regions in adjacent states (Bethke 1986, Rowan and Goldhaber 1996, Pitman et al. 1997).

Formation of the ores from hot mineralizing fluids ranging from 75°C and 220°C has been demonstrated based on data from microscopic fluid inclusions trapped in the ore minerals and associated sediments. Fluid inclusions have also shown that the mineralizing fluids were highly saline (about 20% NaCl equivalent), similar to oil field brines (Hall and Friedman 1963). The source of this high salinity is unclear but may have resulted in part from the evaporative concentration of seawater (Walter et al. 1990).

The volume of the fluid involved in mineralization must have been large, as inferred from the extensive areas affected (Rowan and Goldhaber 1996). Temperatures measured from fluid inclusions in rocks of the districts suggest that southern Illinois experienced widespread igneous activity (e.g., around Hicks Dome and vicinity), and the heat from this activity likely contributed significantly to the temperature of naturally warm fluids deep in the Illinois Basin.

Lead and Zinc Deposits
Properties and Uses

Galena (PbS), from which lead is extracted, is a gray, shiny mineral that breaks into cubes or combinations of cubes (Figure 16-3). It is quite heavy (specific gravity 7.4 to 7.6). For many years, Illinois was the leading U.S. producer of lead. Major uses include batteries, solder for electronics, glass enhancement, and radiation shielding. Lead also was commonly used for plumbing, in paint, and for better performance and efficiency in gasoline, applications that have been severely restricted in recent years as the toxicity of lead has become recognized.

Zinc is extracted from the mineral sphalerite (ZnS), which can be yellow, brown, or black with a resinous luster (Figure 16-3). Major uses of zinc include the manufacturing of zinc-based alloys, such as brass and bronze, and galvanization to prevent corrosion of steel. Zinc compounds are used in batteries, and zinc oxide (ZnO) is widely used in the manufacture of paints, cosmetics, pharmaceuticals, plastics, printing inks, soap, and electrical equipment,

Lead, Zinc, and Fluorite

Figure 16-3 This specimen from Hardin County is tarnished silver-gray galena with resinous, sharp red-brown sphalerite on a very fine, oxidized, drusy quartz matrix. The largest octahedral galena crystal measures 0.4 cm (0.16 inch) across. Notice the naturally etched galena in the bottom right corner. (Photograph by Jared Freiburg.)

among others. For more information on the properties of these and other Illinois minerals, refer to ISGS Geoscience Education Series 16 (Frankie 2004) and http://www.mii.org/commonminerals.html.

Distribution of Ore Bodies

All major lead and zinc ores in Jo Daviess County in northwestern Illinois occur within an area 5 miles (8 km) wide and 15 miles (24 km) long (Figure 16-4). This area is centered on the town of Galena and extends from the Wisconsin border south to the Mississippi River. Small deposits also have been reported just west of Mt. Carroll in Carroll County and near Freeport in Stephenson County.

These northwestern Illinois lead and zinc ore deposits occur in Ordovician dolomite host rock within the Galena and Platteville Groups (Figure 16-5). Major lead ore deposits were found in the upper 150 feet (46 m) of the Galena Group, and zinc ore was mined primarily from deeper within the bedrock in the lower part of the Galena Group and the upper part of the Platteville Group (Willman et al. 1946).

Types of Deposits

Crevice and Residual Lead Deposits

Near-surface crevice and residual deposits provided the lead ores during the early phase of mining in the region. Crevice deposits occur as pods along vertical fissures and joints in dolomite (Figure 16-6), most commonly in the upper 150 feet (46 m) of the Galena Group. Typical pods are 3 feet (0.9 m) wide, 5 feet (1.5 m) high, and up to a few hundred feet long. Crevice sources are mostly solution channels where, for the most part, the galena occurs in a mixture of clay and weathered bedrock that partly or entirely fills the openings. The galena in residual deposits is concentrated in the residual soil formed by thousands of years of weathering and dissolution of near-surface, ore-bearing dolomite host rock.

"Flat" and "Pitch" Sphalerite Deposits

With few remaining near-surface deposits of galena came the need for exploration at greater depths and the discovery of deposits of another important mineral, sphalerite, a resinous mineral consisting of zinc sulfide. The zinc ore was found as pockets, irregularly shaped masses, vertical or inclined veins, or small particles scattered through the dolomite. Irregularly shaped flat and pitch bodies were the main types of deposit. A *flat* is a nearly horizontal sheet-like ore body between or parallel to the bedding planes of the host rock; a *pitch* is a similarly shaped deposit that cuts across the bedding planes (Figure 16-7). These deposits ranged from 50 to 100 feet (15 to 30.5 m) wide and up to 100 feet (30.5 m) long and occurred at a depth of about 300 feet (91 m).

Fluorite

Properties and Uses

Fluorite is a relatively soft mineral with a glassy luster that is usually white or gray, but may be purple, rose, yellow, blue, green, or, rarely, colorless. The crystal form is cubic,

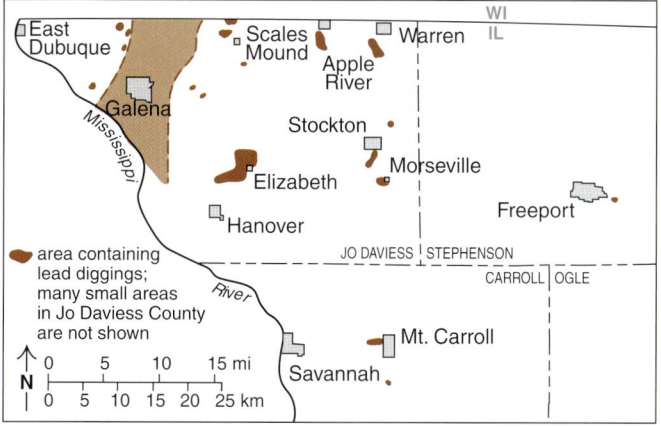

Figure 16-4 Upper Mississippi Valley Lead-Zinc District of northwestern Illinois (Willman et al. 1946). The shaded area between the colored dashed lines is the principal mineralized area.

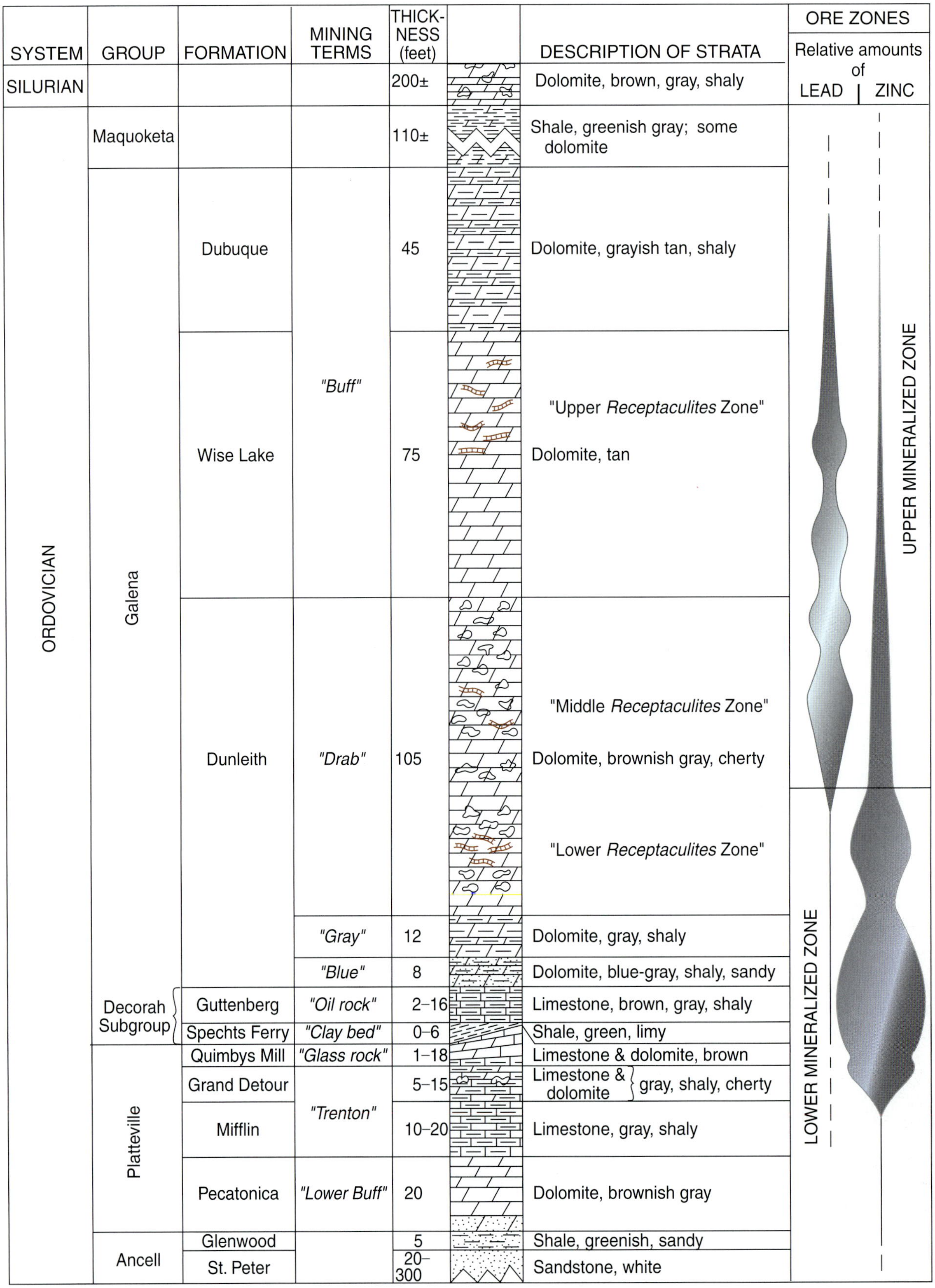

SYSTEM	GROUP	FORMATION	MINING TERMS	THICK-NESS (feet)		DESCRIPTION OF STRATA	ORE ZONES Relative amounts of LEAD / ZINC
SILURIAN				200±		Dolomite, brown, gray, shaly	
ORDOVICIAN		Maquoketa		110±		Shale, greenish gray; some dolomite	UPPER MINERALIZED ZONE
	Galena	Dubuque	"Buff"	45		Dolomite, grayish tan, shaly	
		Wise Lake	"Buff"	75		"Upper *Receptaculites* Zone" / Dolomite, tan	
		Dunleith	"Drab"	105		"Middle *Receptaculites* Zone" / Dolomite, brownish gray, cherty / "Lower *Receptaculites* Zone"	
	Decorah Subgroup		"Gray"	12		Dolomite, gray, shaly	LOWER MINERALIZED ZONE
			"Blue"	8		Dolomite, blue-gray, shaly, sandy	
		Guttenberg	"Oil rock"	2–16		Limestone, brown, gray, shaly	
		Spechts Ferry	"Clay bed"	0–6		Shale, green, limy	
	Platteville	Quimbys Mill	"Glass rock"	1–18		Limestone & dolomite, brown	
		Grand Detour	"Trenton"	5–15		Limestone & dolomite, gray, shaly, cherty	
		Mifflin	"Trenton"	10–20		Limestone, gray, shaly	
		Pecatonica	"Lower Buff"	20		Dolomite, brownish gray	
	Ancell	Glenwood		5		Shale, greenish, sandy	
		St. Peter		20–300		Sandstone, white	

Figure 16-5 Stratigraphic distribution of lead and zinc deposits (Willman et al. 1946, Reinertsen 1992a). Brown ladder-like symbols are fossil *Receptaculites*.

Lead, Zinc, and Fluorite

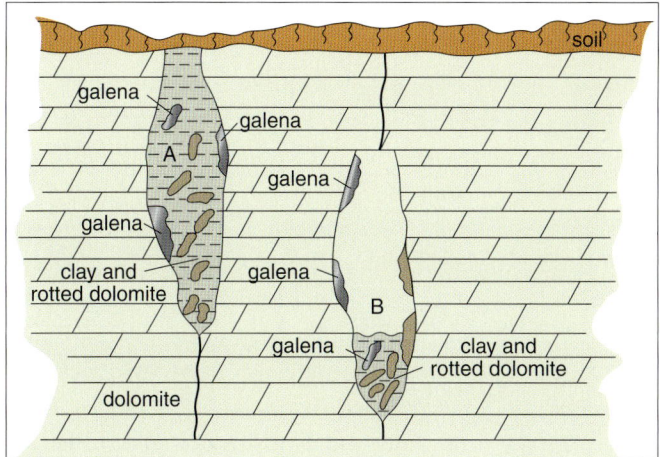

Figure 16-6 Diagrammatic cross section of galena crevice deposits (Lamar 1965).

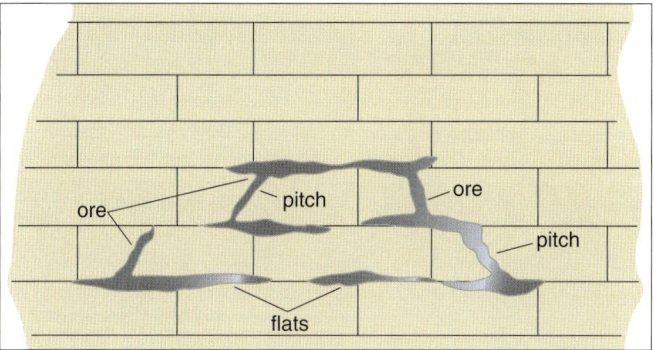

Figure 16-7 Diagrammatic cross section of a zinc ore deposit showing flats and pitches (Lamar 1965).

Figure 16-8 Translucent purple fluorite cubes are on a matrix of sharp, lustrous black sphalerite. The largest fluorite cube is 2.5 cm (1 inch) across. The specimen is from Hardin County. (Photograph by Jared Freiburg.)

and the mineral has a perfect octahedral cleavage. In most cases, fluorspar occurs as massive, interlocking crystals (Figure 16-8). The Illinois General Assembly designated fluorite as the state mineral in 1965 because of its attractiveness and because it rarely occurs in minable quantities elsewhere in the United States.

Fluorite has many uses, including applications in metallurgy and ceramics, production of highly corrosive hydrofluoric acid (HF), and fluoridation of water and toothpaste (Fulton and Montgomery 1994, Reinertsen and Masters 1997). The most important use of fluorite is in the iron and steel industry, where it is used as flux, a substance that lowers the melting temperature of a material and removes impurities. Smaller amounts of fluorite are used in the ceramic industry to manufacture glass and for enamel coatings on sinks, bathroom fixtures, stoves, refrigerators, and signs. Fluorite historically was used for making chlorofluorocarbons (e.g., Freon), which were used extensively in refrigerators and air conditioners and as aerosol propellant, but are now banned due to their damaging effects on the Earth's ozone layer. Fluorite crystals sometimes are considered gemstones, but, because they are relatively soft and brittle, with four cleavage planes, they are not suitable for most jewelry.

Distribution of Ore Bodies

Illinois fluorspar deposits and their geology have been extensively studied by geologists from the Illinois State Geological Survey and other institutions (Weller et al. 1952; Baxter et al. 1963, 1967, 1973; Baxter and Desborough 1965; Lamar 1965; Bradbury et al. 1968; Reinertsen 1992b; Fulton and Montgomery 1994). Fluorspar in southeastern Illinois (Figure 16-9) occurs in two main forms: as steeply inclined fissure fillings along faults (vein deposits) and as near-horizontal replacement bodies in limestone (bedded deposits). The Illinois-Kentucky Fluorspar District is located within one of the most intensely faulted areas of the North American midcontinent and encompasses an area of approximately 1,000 square miles (2,590 km^2) (Figure 16-10) (Nelson 1995, Baxter et al. 1989, Denny et al. 2008). Fluorite mineralization is mostly associated with the Middle Mississippian carbonates, and the largest deposits are found in the Ste. Genevieve Limestone. Moderate amounts of fluorite have also been found in the Renault Limestone above and the St. Louis Limestone below.

Types of Deposits

Vein Deposits

In the Rosiclare District (Figure 16-9) vein deposits typically occur as steeply inclined fissure fillings in a complex system of faults (Figure 16-11a). Vein deposits have been found in faults of considerable vertical movement, along which open spaces as much as 30 feet (9 m) wide were formed and later filled with fluorspar. Prolific fluorspar deposits of this type are most common along faults with 100 to 500 feet (30.5 to 152 m) of vertical displacement. Veins pinch and swell both vertically and laterally and range in thickness from a feather edge to as much as 30 feet (9 m). Vein deposits have been mined at depths greater than 800 feet (244 m). Fluorite and calcite are the main minerals in vein deposits, but minor amounts of galena, sphalerite, barite ($BaSO_4$, also known as barium sulfate or barium ore), and numerous other rare but exotic minerals also occur (Goldstein 1997).

Vein deposits are most common along more competent (very hard) rocks, such as limestone and well-cemented sandstone, that could maintain adequate open space along the fault. Because of their high competence, St. Louis and Ste. Genevieve Limestones are the most favorable rocks for vein deposits. Weaker—or less competent—blocks, such as shale, shaly limestone, and sandstone, were crushed by movement along the fault and did not create openings that were suitable for fluorite precipitation.

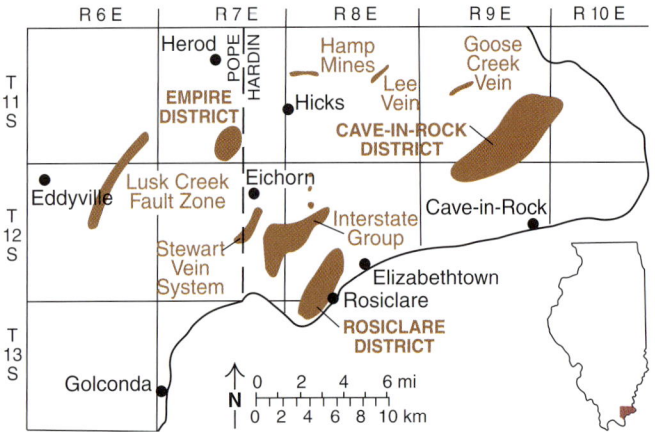

Figure 16-9 Principal fluorspar mining districts in extreme southeastern Illinois (Bradbury et al. 1968).

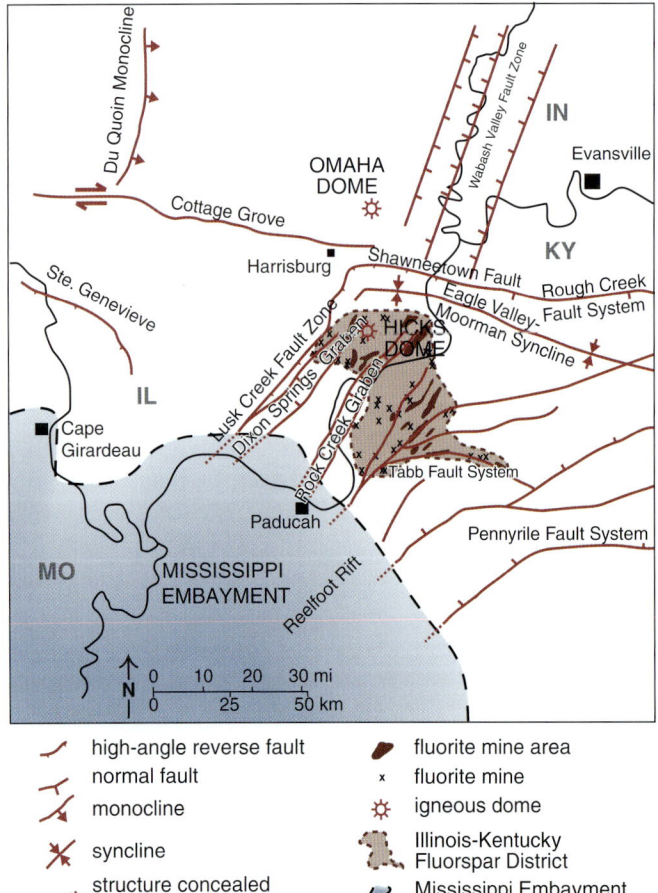

Figure 16-10 Major tectonic structures in the Illinois-Kentucky Fluorspar District (Denny et al. 2008).

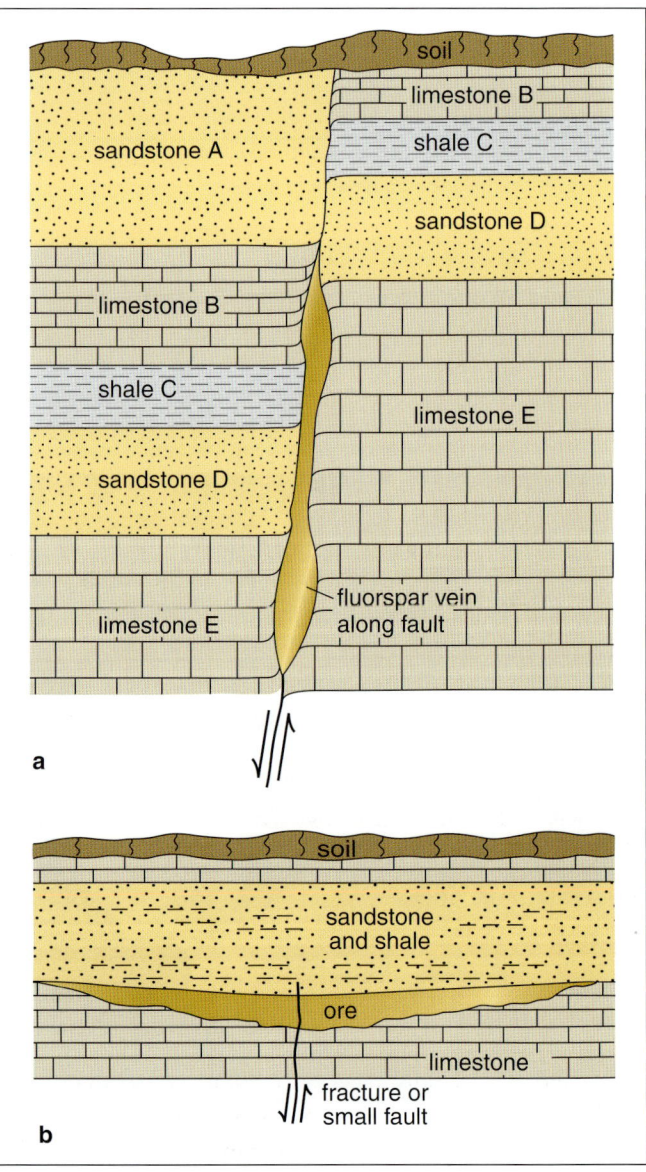

Figure 16-11 Cross section showing fluorspar (a) vein deposit along near-vertical faults and (b) horizontal, bedded replacement deposit of fluorspar (Lamar 1965).

Bedded Deposits

Typical of the Cave-in-Rock area (Figure 16-9), bedded deposits are nearly flat-lying, irregular deposits that lie parallel to the beds of the host limestone (Figure 16-11b). The deposits are commonly 5 to 15 feet (1.5 to 4.5 m) thick, 200 to 2,500 feet (61 to 762 m) long, and 50 to 300 (15 to 91 m) feet wide. Galena and sphalerite are commonly associated minerals in bedded replacement deposits.

Bedded replacement deposits occur within a relatively narrow stratigraphic interval from the base of Bethel Sandstone downward to the top of the Fredonia Limestone Member of the Ste. Genevieve Limestone (Figure 16-12). The principal deposits are found at three favored positions within this interval: at the top of Downeys Bluff Limestone, the top of the Joppa Limestone Member of the Ste. Genevieve, and the top of the Fredonia Limestone Member of the Ste. Genevieve. Less extensive bedded replacement deposits occur in the upper part of the Ste. Genevieve Limestone.

Unlike vein deposits, bedded replacement deposits are not simple void fillings but apparently were formed as a result of limestone replacement by fluorine-bearing solutions during which calcium carbonate ($CaCO_3$), the main mineral of limestone, was converted to calcium fluoride (CaF_2). Solutions moved laterally along the limestone bedding planes and possibly within porous, coarser-grained beds. Major bedded deposits all occur in limestones overlain by sandstone. Because of its porous and permeable nature, the sandstone was probably the conduit for mineralizing fluids. There is evidence that each favored interval may have been exposed to weathering and erosion prior to mineralization, thus further enhancing porosity and permeability (Grogan and Bradbury 1968, Leetaru 2000).

MINING HISTORY

The Upper Mississippi Valley Lead-Zinc District includes Jo Daviess County in northwestern Illinois and extends across the Mississippi River into Iowa and north into Wisconsin. The area is one of the nation's oldest mining districts, as mining by Native Americans dates back to at least 1658.

Lead and Zinc Mining

In northeastern Illinois, in 1659, the French learned of the existence of lead in the region from Native Americans (McClure 1951). In 1687, Father Louis Hennepin, a Catholic missionary, prepared a map showing a mine near what is now Galena, in Jo Daviess County. Actual mining by the French and the Native Americans began in 1690 when the first trading post for lead was established in Dubuque, Iowa

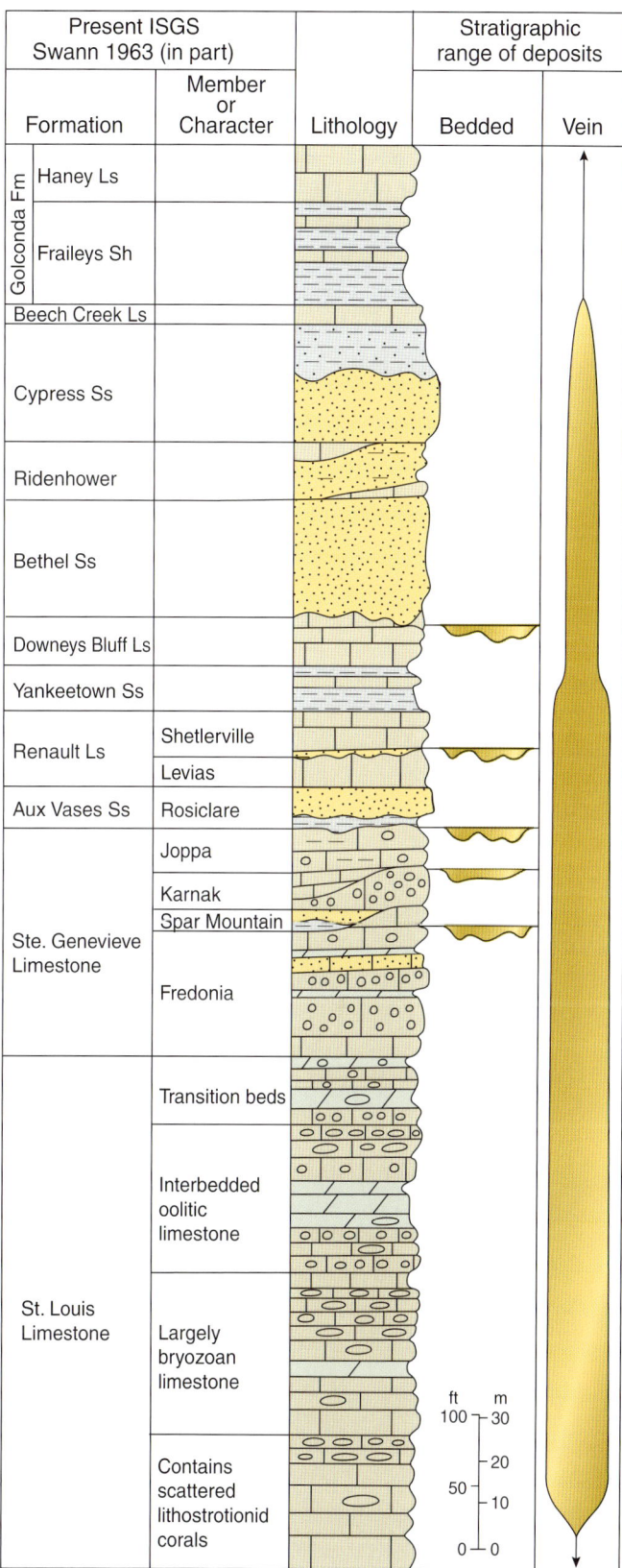

Figure 16-12 Stratigraphic distribution of fluorspar deposits (Baxter et al. 1973). Abbreviations: Ls, Limestone; Sh, Shale; Ss, Sandstone.

(Heyl et al. 1959). Although the French were the first non-natives to exploit the resources of the region, beginning in the early 1700s, English settlers later dominated the area, and, during the War of 1812, produced large amounts of lead for munitions.

American occupation of the region followed the war, and, in 1822, a mining rush began in the area. Increased incentives for lead mining tripled its production, resulting in a population explosion in Galena as the town grew from 200 in 1825 to 10,000 people in 1828 (Trowbridge and Shaw 1916, Schockel 1916, McClure 1951). Because of its good transportation infrastructure, Galena became the center of the lead-mining district in the Upper Mississippi Valley during the first half of the 1800s and led in zinc and lead mining for more than a century (Lamar 1965). In 1845, 27,000 tons (24,500 metric tonnes) of lead were produced in the area, accounting for 90% of all production in the United States, the world's leading producer at that time. Galena was the only ore mined until about 1850.

A new chapter in the mining history of Jo Daviess County began with the exploitation of the more deeply buried zinc ore deposits, discovered in the early 1850s. Exploration for lead ore also resulted in discoveries in far southeastern Illinois. Initially these mines produced lead and zinc, but fluorite, a by-product of the mining, became a more valuable resource.

In the early days of mining, between 1820 and 1865, lead ores were found in near-surface crevices or open vertical fissures in dolomite bedrock. These crevice-type deposits of the Upper Mississippi Valley District were the nation's principle source of lead ore. Production peaked from 1845 to 1847, but, following the Civil War, lead mining declined to a fraction of its former importance due to depletion of shallower deposits and changes in market conditions. Later, in the early 1850s, larger and deeper deposits of zinc ores were discovered, which revitalized mining in the district. These deeper deposits occurred in the dolomite bedrock as horizontal and inclined veins referred to as flats and pitches, respectively (Willman et al. 1946, Bradbury 1959, Lamar 1965). Commercial production of sphalerite began in 1852 when the first smelting (metal processing) plant for zinc ore was opened in La Salle, Illinois.

Fluorite Mining

Fluorite was first discovered in Illinois in a well in Rosiclare in Hardin County in 1842 (Risser and Major 1968). Although significant quantities of fluorite were recovered with the mining of lead ore, fluorite was discarded as by-product waste during the early years of mining. Major uses for fluorite in the United States were found after the 1870s,

Figure 16-13 Photograph of workers in an old fluorite mine. (Photograph from the Illinois State Geological Survey collection.)

which led to the beginnings of commercial mining in Hardin and Pope Counties in the 1880s (Figure 16-13) (Weller et al. 1952, Reinertsen and Masters 1997).

Fluorite-bearing rocks were found outcropping in many places in the Cave-in-Rock district in southeastern Illinois, attracting early prospectors to the area. Fluorspar was mined in open-pit operations during the early days, and mine tunnels (adits) also were driven horizontally into the hillsides. Increased demand for fluorite led to exploitation of deeper deposits through underground mining (Bradbury et al. 1968). Later, fluorite was mined from surface and underground mines, some up to 1,300 feet (396 m) deep, in the Illinois-Kentucky Fluorspar District, an area of about 1,000 square miles (2,590 km^2) in southeastern Illinois and western Kentucky (Figure 16-10). From 1880 through 1990, fluorspar concentrate shipments from the district totaled about 12 tons (10.9 tonnes), which is equivalent to about 36 tons (about 33 tonnes) of crude ore (Fulton and Montgomery 1994).

In 1942, Illinois became the country's leading producer of fluorite and, for many years, the state accounted for more than 50% of U.S. production. The towns of Rosiclare, Cave-in-Rock, and Elizabethtown were headquarters for this mining industry. The Rosiclare and Cave-in-Rock Mining Districts provided the majority of fluorspar mined in the Illinois-Kentucky Fluorspar District. By 1990, Illinois was the only remaining domestic producer of this mineral, and increasing costs of mining and competition from foreign sources such as China, South Africa, and Mexico eventually made fluorite mining in Illinois economically unprofitable. The last fluorspar mine was closed in December 1995 (Reinertsen and Masters 1997).

Conclusion

Lead-zinc ore deposits in the Upper Mississippi Valley district and the Illinois-Kentucky Fluorspar District are well-known throughout the world and have been the focus of many investigations that provided conceptual models for understanding the origin of similar ore deposits worldwide. Recognition of the importance of the northwestern Illinois mining district as a driving force for economic development and as a source of minerals for a variety of domestic industrial applications has all but vanished. Environmental restrictions in the 1970s required companies to reduce zinc concentrations in mine waters. These restrictions, along with rising costs, made mining economically unfeasible. The last mines closed in 1973 in Illinois and in 1978 in Wisconsin, ending almost three centuries of domestic ore production in the region (Goldstein 1997, Verzal undated).

The great mines of southeastern Illinois, which provided much of the nation's fluorite for over a century, are closed. The high cost of mining and competition from foreign sources have eliminated the prospect for renewed mining in the area for the foreseeable future. However, fluorite's legacy continues. Its attractiveness fascinates geologists and collectors, who continue to visit the old mines. Today fluorite, the official state mineral, is widely displayed in museums and at rock and mineral shows—a reminder of the important role this mineral once played in the economic development of Illinois and the nation.

There may be the possibility of renewed fluorite mining in the Illinois-Kentucky Fluorspar District. Fluorite has not been mined in Illinois since the early 1990s except for a small quantity of fluorite recovered by Hastie Mining Company as a by-product of limestone quarrying and stockpiled for future processing in Hardin County in southeastern Illinois. Since the mid-1990s, Hastie has supplied fluorite, purchased from the National Defense Stockpile (NDS), to a number of U.S. markets. With the NDS supply diminishing, and with uncertainty regarding the availability of import sources, there is renewed interest in fluorite mining in the Illinois-Kentucky Fluorspar District. Hastie owns the mineral rights to several former fluorspar properties in Illinois and plans to produce about 22,000 tons (about 20,000 tonnes) per year of acid-grade fluorspar and 33,000 tons (30,000 tonnes) per year of metallurgical grade fluorspar from its mine and quarry operations (Miller 2007).

References

Baxter, J. W., J. C. Bradbury, and N. C. Hester, 1973, A geologic excursion to fluorspar mines in Hardin and Pope Counties, Illinois: Illinois State Geological Survey, Guidebook Series 11, 28 p.

Baxter, J. W., and G. A. Desborough, 1965, Areal geology of the Illinois Fluorspar District, Part 2, Karbers Ridge and Rosiclare Quadrangles: Illinois State Geological Survey, Circular 385, 40 p.

Baxter, J. W., G. A. Desborough, and C. W. Shaw, 1967, Areal geology of the Illinois Fluorspar District, Part 3, Herod and Shetlerville Quadrangles: Illinois State Geological Survey, Circular 413, 41 p.

Baxter, J. W., E. V. Kisvarsanyi, and R. D. Hagni, 1989, Precambrian and Paleozoic geology and ore deposits of the midcontinent region: 28th International Geological Congress, IGC Field trip T147, 68 p.

Baxter, J. W., P. E. Potter, and F. L. Doyle, 1963, Areal geology of the Illinois Fluorspar District, Part 1, Saline Mines, Cave in Rock, Dekoven, and Repton Quadrangles: Illinois State Geological Survey, Circular 342, 44 p.

Bethke, C. M., 1986, Hydrologic constraints on the genesis of the Upper Mississippi Valley mineral district from Illinois Basin brines: Economic Geology, v. 81, no. 2, p. 233–249.

Bradbury, J. C., 1959, Crevice lead-zinc deposits of northwestern Illinois: Illinois State Geological Survey, Report of Investigations 210, 49 p.

Bradbury, J. C., G. C. Finger, and R. L. Major, 1968, Fluorspar in Illinois: Illinois State Geological Survey, Circular 420, 64 p.

Currier, L. W., 1923, Fluorspar deposits of Kentucky: Frankfort, Kentucky, Kentucky Geological Survey, ser. 6, v. 13, 189 p.

Denny, F. B., A. Goldstein, J. A. Devera, D. A. Williams, Z. Lasemi, and W. J Nelson, 2008, The Illinois-Kentucky Fluorite District, Hicks Dome, and Garden of the Gods in southeastern Illinois and northwestern Kentucky, in A.H. Maria and R.C. Counts, eds., From the Cincinnati Arch to the Illinois Basin: Geological Field Excursions along the Ohio River Valley: Boulder, Colorado, Geological Society of America, Field Guide 12, p. 11–24.

Frankie, W., 2004, Guide to rocks and minerals of Illinois: Illinois State Geological Survey, Geoscience Education Series 16, 68 p.

Fulton, R. B., and G. Montgomery, 1994, Fluorspar, in D. D. Carr, ed., Industrial minerals and rocks: Littleton, Colorado, Society for Mining, Metallurgy, and Exploration, p. 509–522.

Goldstein, A., 1997, The Illinois-Kentucky Fluorite District: Mineralogical Record Magazine, v. 28, no. 1, p. 3–49.

Grogan, R. M., and J. C. Bradbury, 1968, Fluorite-lead-zinc deposits of the Illinois-Kentucky mining district, in J. D. Ridge, ed., Ore deposits of the United States: New York, The American Institute of Mineral, Metallurgical, and Petroleum Engineers (AIME), p. 370–399.

Hall, W. E., and I. Friedman, 1963, Composition of fluid inclusions, Cave-in-Rock fluorite district, Illinois, and Upper Mississippi Valley zinc-lead district: Economic Geology, v. 58, no. 6, p. 886–911.

Heyl, A. V., A. F. Agnew, E. J. Lyons, and C. H. Behre, 1959, The geology of the Upper Mississippi Valley zinc-lead district: Washington, D.C., U. S. Geological Survey, Professional Paper 309, 310 p.

Lamar, J. E., 1965, Industrial minerals and metals of Illinois: Illinois State Geological Survey, Educational Series 8, 48 p.

Leetaru, H. E., 2000, Sequence stratigraphy of the Aux Vases Sandstone: A major oil producer in the Illinois Basin: American Association of Petroleum Geologists Bulletin, v. 84, no. 3, p. 399–422.

Masters, J. M., 1987, Industrial minerals and metals, in R. D. Neely and C. G. Heister, eds., The natural resources of Illinois, introduction and guide: Illinois Natural History Survey, Special Publication 6, p. 199–205.

McClure, S. S., 1951, Mineral collecting in Illinois, the northwestern zinc-lead region: Mineralogist, v. XIX, no. 5, p. 227–230.

Miller, M. M., 2007, Fluorspar, in U. S. Geological Survey 2006 Mineral Yearbook. http://minerals.usgs.gov/minerals/pubs/commodity/fluorspar/myb1-2006-fluor.pdf. Accessed October 27, 2009.

Nelson, W. J., 1995, Structural features in Illinois: Illinois State Geological Survey, Bulletin 100, 144 p.

Pitman, J. K., M. B. Goldhaber, and C. Spöetl, 1997, Regional diagenetic patterns in the St. Peter Sandstone: Implications for brine migration in the Illinois Basin: Reston, Virginia, U.S. Geological Survey Bulletin 2094-A, p. A1–A17.

Reinertsen, D. L., 1992a, Guide to the geology of the Galena area, Jo Daviess County, Illinois, Lafayette County, Wisconsin: Illinois State Geological Survey, Field Trip Guidebook 1992B, 38 p.

Reinertsen, D. L., 1992b, Guide to the geology of the Cave in Rock and Rosiclare area, Hardin County: Illinois State Geological Survey, Field Trip Guidebook 1992A, 39 p.

Reinertsen, D. L., and J. M. Masters, 1997, Fluorite—Illinois' state mineral: Illinois State Geological Survey, Geobit 4, 2 p.

Risser, H. E., and R. L. Major, 1968, History of Illinois mineral industries: Illinois State Geological Survey, Educational Series 10, 30 p.

Rowan, E. L., and M. B. Goldhaber, 1996, Fluid inclusions and biomarkers in the Upper Mississippi Valley zinc-lead district—Implications for the fluid-flow and thermal history of the Illinois Basin: Reston, Virginia, U. S. Geological Survey, Bulletin 2094-F, p. F1–F34.

Schockel, B. H., 1916, History of development of Jo Daviess County: Illinois State Geological Survey, Bulletin 26, p. 173–228.

Trowbridge, A. C., and E. W. Shaw, 1916, Geology and geography of the Galena and Elizabeth Quadrangles: Illinois State Geological Survey, Bulletin 26, p. 13–171.

Verzal, P. A., undated, A history of the Upper Mississippi Valley Lead-Zinc District. http://www.uwplatt.edu/geography/mines/history.html. Accessed August 28, 2006.

Walter, L. M., A. M. Stueber, and T. J. Huston, 1990, Br-Cl-Na systematics in Illinois Basin fluids: Constraints on fluid origin and evolution: Geology, v. 18, no. 4, p. 315–318.

Weller, J. M., R. M. Grogan, and F. E. Tippie, 1952, Geology of the fluorspar deposits of Illinois: Illinois State Geological Survey, Bulletin 76, 147 p.

Willman, H. B., and R. R. Reynolds, 1947, Geological structure of the zinc-lead district of northwestern Illinois: Illinois State Geological Survey, Report of Investigations 124, 15 p.

Willman, H. B., R. R. Reynolds, and P. Herbert, Jr., 1946, Geological aspects of prospecting and areas for prospecting in the zinc-lead district of northwestern Illinois: Illinois State Geological Survey, Report of Investigations 116, 48 p.

Industrial Minerals 17

Zakaria Lasemi, Donald G. Mikulic, Randall E. Hughes, Timothy J. Kemmis, Subhash B. Bhagwat, and Karan S. Keith

INTRODUCTION

Industrial minerals are naturally occurring, nonmetallic, nonfuel resources that are economically important and essential for industrial applications (Figure 17-1). Major industrial minerals that currently are mined in Illinois include limestone and dolomite, sand and gravel, clay, shale, industrial sand (silica sand), tripoli (microcrystalline quartz), and peat (Figure 17-2). Portland cement, lime, glass, and brick are closely related manufactured products. Fluorite, historically an important mineral resource in Illinois, is no longer mined in significant quantities in the state (see Chapter 16, Lead, Zinc, and Fluorite).

Over the last two centuries, industrial minerals have been the most important mineral resource produced in Illinois. The need for these resources is enormous. Much of the state's infrastructure, including general construction, transportation networks, and sanitation systems, is built mainly with industrial minerals and related products. Nearly 400 tons (363 metric tonnes) of aggregate are needed to construct the average modern home, and over 38,000 tons (34,473 tonnes) are needed to build one mile (1.6 km) of interstate highway (Langer and Glanzman 1993; National Stone, Sand and Gravel Association undated). Aggregate is the major component in concrete (80%) and asphalt (90%) (Tepordei 1997; National Stone, Sand and Gravel Association undated). Millions of tons of limestone are used every year in agriculture to neutralize the acids in soil. Limestone is also used for environmental remediation in coal-burning power plants and in medical and hazardous waste incinerators to reduce the emissions of toxic gases and air pollutants such as mercury.

Illinois is both a major producer and a major consumer of industrial minerals. More than a billion dollars worth of stone, sand and gravel, cement, clay, shale, silica sand, tripoli, and lime are produced or manufactured in the state every year, and together these materials represent a major segment of the Illinois economy. Production and processing of industrial minerals create a large number of jobs in Illinois—about 6,400 in the mining sector alone—as does the use of these resources in building and road construction, manufacturing, and agriculture. Crushed stone (limestone and dolomite) and sand and gravel aggregates together constitute more than 60% of the value of Illinois industrial minerals (Lasemi et al. 2008). Each year, about 80 to 90 million tons (about 73 to 82 million tonnes) of crushed stone and about 35 million tons (about 32 million tonnes) of sand and gravel are produced in Illinois from more than 150 stone quarries, underground stone mines, and sand and gravel pits.

INDUSTRIAL MINERALS BY TYPE
Limestone and Dolomite Resources

Limestone and dolomite are among the most valuable and useful rocks extracted in Illinois and across the nation (Lamar 1965, Langer and Glanzman 1993). *Limestone* is often used as a general term to include both limestone and dolomite, but the two types of carbonate rocks are quite different and are not interchangeable in many applications. Limestone consists mainly of the mineral calcite, a compound of calcium, carbon, and oxygen, or calcium carbonate ($CaCO_3$). Dolomite, or more correctly dolostone, consists mostly of the mineral dolomite, a compound of magnesium, calcium, carbon, and oxygen; its chemical formula is $CaMg(CO_3)_2$. Limestone and dolomite generally exhibit a crystalline appearance showing shiny and sparkling faces; in some cases, though, the crystals are very fine, and the rock appears dull.

Almost all limestone in Illinois formed in the warm, shallow seas that millions of years ago covered part or all of Illinois. Most limestones resulted from the accumulation of the remains of calcareous plants and sea shells (Figure 17-3). Some limestones may have formed from calcium carbonate precipitation from seawater as lime mud, which later hardened to form very fine-grained limestones. Many dolostones were originally limestones in which the calcite was replaced by dolomite when magnesium-rich water percolated through the stone. Spectacular coral reefs similar to those in modern oceans formed in areas of northeastern Illinois. Reef development was especially prolific during the Silurian Period, about 440 million years ago. The reefs, now dolomite, have been quarried for the high-quality stone used in construction in northeastern Illinois.

Limestone and Dolomite Distribution

Limestone and dolomite quarries are located where thick stone deposits occur near the surface, mainly in the northern quarter, the western side, and the southern tip of Illinois (Lamar 1967; Goodwin 1979, 1983; Goodwin and

Figure 17-1 Illinois' crushed stone quarries **(a)** and sand and gravel pits **(b)** provide essential materials for building transportation, residential, and commercial infrastructure **(c** and **d)**. (Photographs a and b are by Zakaria Lasemi. Photographs c and d are by Joel M. Dexter.)

INDUSTRIAL MINERALS

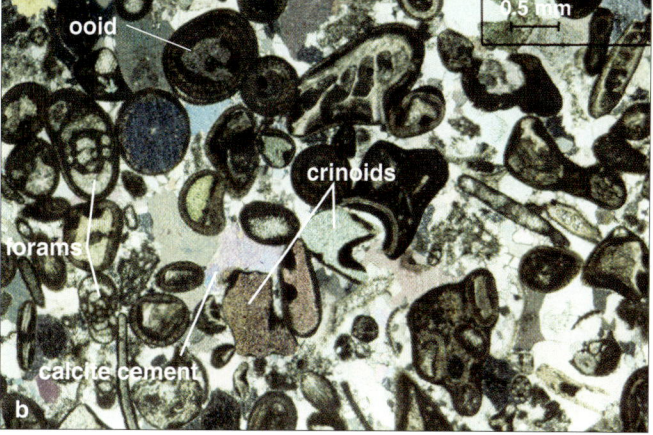

Figure 17-3 **(a)** Bedding surface of Ordovician dolomite with well-preserved brachiopod fossil remains. (Photograph by Joel M. Dexter.) **(b)** A microscopic view of Mississippian limestone showing fossil fragments, some (ooids) coated with thin bands of calcium carbonate. (Photograph by Zakaria Lasemi.)

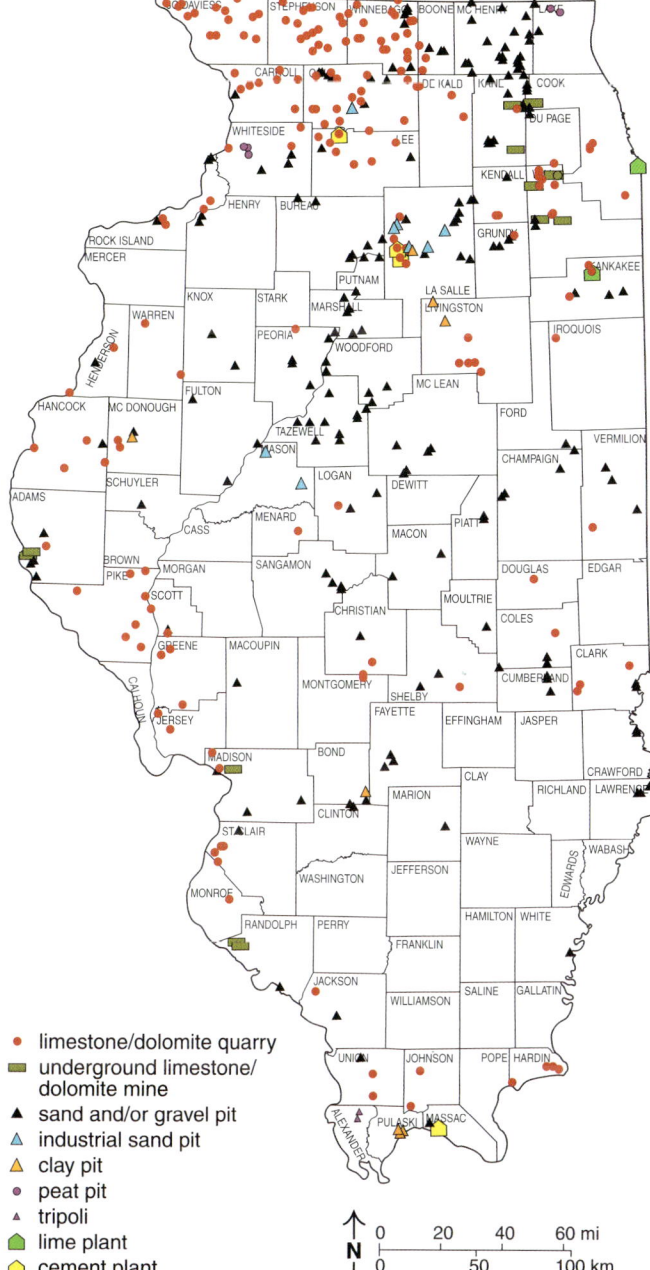

Figure 17-2 Industrial minerals extraction sites and lime and cement plants. Map was compiled by Zakaria Lasemi primarily from information provided by the Illinois Department of Natural Resources, Office of Mines and Minerals.

Baxter 1981) (Figure 17-4). Although thick limestone and dolomite deposits are present throughout most of the rest of the state, they are deeply buried in central and eastern Illinois, except where brought near the surface by structural features.

Rocks of the Ordovician and Silurian Systems are the carbonate sources in northern Illinois, where dolomites are quarried mostly for use in road construction. There, stone resources historically have been obtained from the high-quality Silurian dolomites that occur at the bedrock surface throughout most of the area (Mikulic 1990). Ordovician rocks, which underlie the Silurian, are becoming a significant source of stone as the accessible Silurian dolomite reserves are depleted. Uppermost Ordovician (Maquoketa Group) rock consists primarily of shale, up to approximately 250 feet (about 72 m) thick, which presently has no commercial value. Beneath the Maquoketa are the high-quality dolomites and limestones of the Ordovician Galena and Platteville Groups, which are up to about 300 feet (about 91 m) thick in northern Illinois (Willman and Kolata 1978). Ordovician age Kimmswick Limestone, an excellent source of high-purity limestone, was mined underground near the village of Valmeyer in Monroe County until the 1980s. The Kimmswick currently is quarried in adjacent counties of Missouri, and the potential exists for mining this unit in Alexander County in southern Illinois.

Limestones of the Mississippian System are mined in western and southern Illinois for construction aggregate, cement and lime manufacture, and other chemical purposes (Lamar 1959; Baxter 1965; Lasemi et al. 1999; Lasemi and Norby 2001, 2004). Several Mississippian strata contain high-purity, low-magnesium limestone suitable for environmental remediation and for making lime and portland cement (Lamar et al. 1956, Lamar 1966, Goodwin and Baxter 1981, Lasemi and Norby 2001). Mississippian rock units containing high-purity limestones include the upper Burlington Limestone, Ullin Limestone, upper Warsaw Formation, Salem Limestone, and Ste. Genevieve Limestone (Figure 9-1). Limited amounts of Pennsylvanian age limestones occur locally in central Illinois, where they are quarried at several locations.

Uses of Limestone and Dolomite

Limestone and dolomite are the most widely quarried rocks in Illinois, and millions of tons of stone are crushed annually for construction aggregate, road surfacing material, agricultural limestone, and lime. High-calcium limestones are also used as a scrubbing agent for pollution control in power plants and incinerators and as a major ingredient of cement, the binding agent used in concrete pavements and foundations. Portland cement is produced from low-magnesium limestone; lime (CaO or MgO) is produced from limestone or dolomite, respectively.

Limestone and dolomite are used in agriculture, glass and steel making, and paper manufacturing (Lamar 1965, Carr 1994). Ground limestone is used as pigment and filler in paints, caulks, sealants, paper, and plastics. Powdered limestone is used in coal mines for explosion prevention.

Millions of tons of pulverized limestone and dolomite are used for agricultural purposes to neutralize soil acidity and provide calcium and magnesium nutrients to plants. Powdered pure limestone is used as a calcium supplement for hogs, cattle, and chickens. For human consumption, high-purity limestone is used in antacids and calcium supplements.

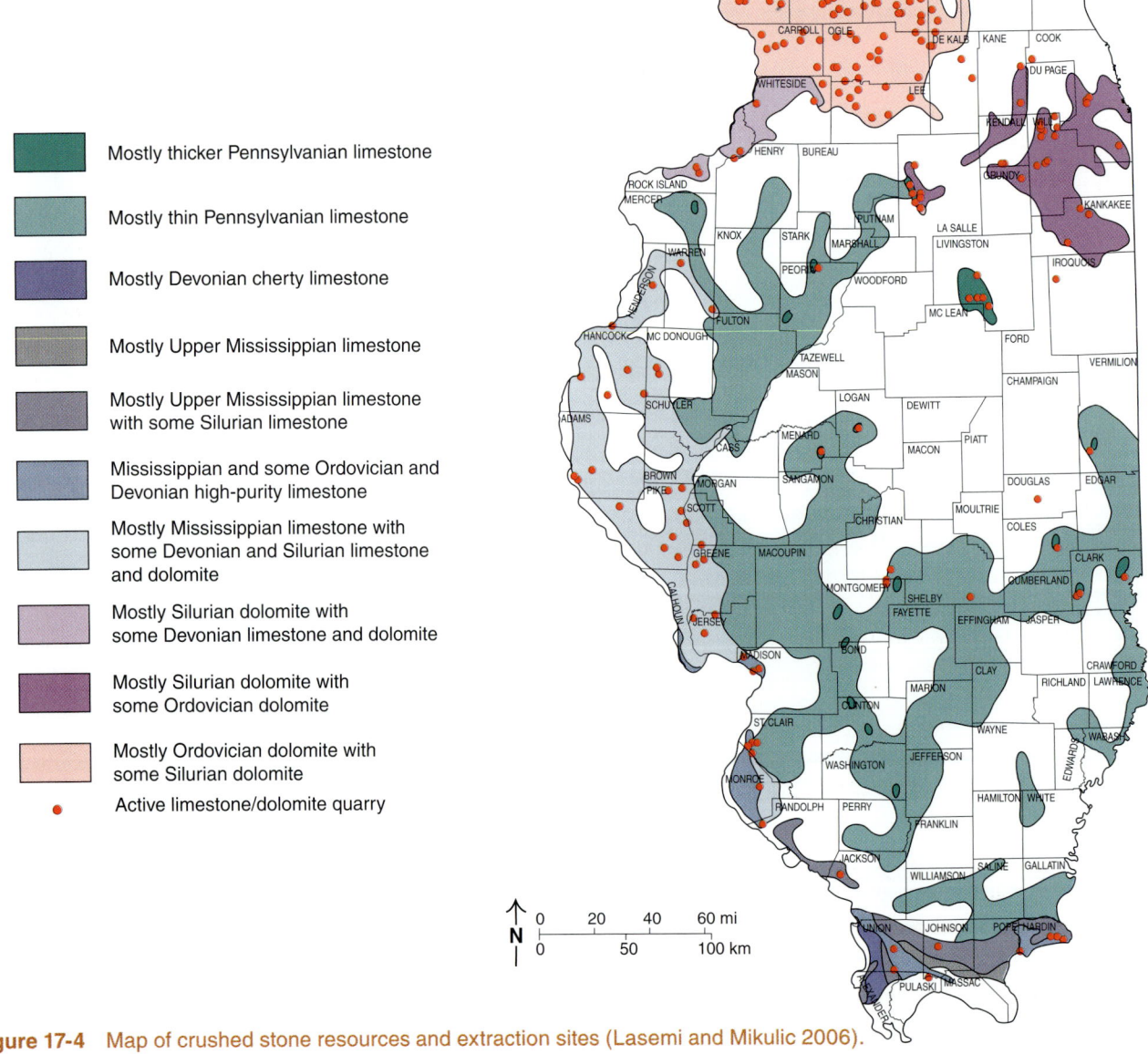

Figure 17-4 Map of crushed stone resources and extraction sites (Lasemi and Mikulic 2006).

Lime has been used in building and road construction since Roman time (National Lime Association 2005). Lime is made by heating limestone ($CaCO_3$) to a high temperature (about 825°C [about 1,517°F]) through the process called calcining in which the CO_2 is driven off and limestone is converted to calcium oxide (CaO), also called quicklime. The reaction of lime with water produces hydrated lime—calcium hydroxide, or $Ca(OH)_2$—a dry, white powder with numerous chemical, industrial, and environmental applications. Magnesia (MgO) is produced by heating dolomite to a high temperature. Both lime and magnesia require high-purity limestone and dolomite as raw materials. Lime and magnesia are widely used in the ceramic and metallurgical industries as flux to remove impurities such as silica, phosphorus, and sulfur.

Limestone is vital to the construction industry. It is a key ingredient for the manufacture of portland cement, which is made by combining low-magnesium limestone with measured amounts of clay or shale. The materials are ground, thoroughly mixed, and heated to a high temperature (about 1,450°C or 2,642°F). The resulting material, called clinker, is ground to a flourlike powder. Cement is combined along with sand and gravel or sand and crushed limestone (or dolomite) and a little water to make concrete, the nearly universal construction material. Cut and polished limestone is an excellent stone for building and for statues and monuments. Large chunks of limestone and dolomite are also used as riprap to control shoreline or stream erosion.

Limestone, lime, and dolomite are extensively used for environmental remediation (Borgwardt and Harvey 1972, Harvey et al. 1974, Boynton 1980, Bhagwat 1985, Foose and Barsotti 1999). They are used to remove mercury, sulfur oxides, and other pollutants from medical and hazardous waste incinerators, coal-fired power plants, and other industrial facilities. Limestone absorbs sulfur oxides emitted from power plants and incinerators, thus reducing acid rain. Treating industrial and mining wastewater with limestone and lime makes it less acidic and removes phosphorous and nitrogen. Municipalities use limestone for softening water and removing impurities such as lead from drinking water. Limestone is also used to treat sewage sludge.

Sand and Gravel Resources

The sand and gravel in much of Illinois was deposited by water from the melting glaciers. Huge lobes of continental ice moved into Illinois from Canada during the Pleistocene, or Ice Age, from several hundred thousand to 12,000 years ago, carrying enormous amounts of rock debris. As the glaciers melted, much of the debris was washed and sorted by the meltwaters into sand and gravel deposits (Figure 17-5). Sand deposits of more recent origin are found in larger streams and rivers and are recovered by dredging.

Principal commercial sources of sand and gravel are found in outwash plain deposits (Figure 17-5a) and valley train deposits (Figure 17-5b). Outwash plains were formed by many small meltwater streams that flowed from the glaciers and deposited sand and gravel as large apron-like formations. Modern streams have cut their courses into many of these deposits, and the remnants of the valley train deposits are now large terraces or benches that lie above present stream channels. The Fox, Rock, Illinois, Mississippi, and Wabash Rivers, for example, have such valley terraces. These deposits are used by some of the largest sand and gravel production operations in the state.

In 2007, production of construction sand and gravel totaled about 35 million tons in Illinois (Lasemi et al. 2008). Sand and gravel deposits are distributed in select locations across the state but are most abundant and of highest quality in northeastern Illinois. More than 50% of Illinois sand and gravel is produced in northern Illinois, primarily northeastern Illinois. Illinois sand and gravel production has declined since 1997, primarily because of resource depletion, the closing of several of the largest operations in the metropolitan Chicago area, and the difficulty in siting and obtaining permits for new pits.

Sand and gravel is primarily used in asphalt and concrete pavements and in commercial and residential structures. The commercial excavation of sand and gravel began in the early 1800s when small amounts of these materials were used to make dirt roads more passable in rainy weather. As the railroad system was developed, gravel was mined and sold for ballast. Only after the turn of the century, when the use of concrete in construction became common, did the production of sand and gravel in Illinois increase significantly (Figure 17-6).

Industrial Sand (Silica Sand)

Silica sand, or industrial sand, consists almost entirely of the mineral quartz (SiO_2). Commercial silica sand is produced from sandstone bedrock and some glacial deposits. Production is concentrated in two northern counties, La Salle and Ogle, where the Ordovician St. Peter Sandstone is mined (Lamar 1928, 1965). Silica sand from northern Illinois is famous for its high purity, and Illinois continues to rank first in the nation for the production of this valuable industrial mineral.

Quartz grains, originally angular, are gradually rounded by wind and water erosion. The famous St. Peter Sandstone is formed of these rounded quartz grains that were compacted by burial and cemented into rock. The St. Peter Sandstone at Starved Rock and Matthiessen State Parks near La Salle and along the highway between Dixon and Oregon can be seen in scenic bluffs and canyons. Other silica sand resources are much younger; they were deposited as windblown dune sands about 12,000 years ago, near the end of the glacial period in Illinois. These sands are currently mined in Mason County.

Unground silica sand is used primarily in the manufacture of glass, as foundry sand for making molds, and for grinding and polishing. Because the grains are rounded and withstand high temperatures without melting, large tonnages of the washed silica sand mixed with a small amount of clay are used in foundries to make molds used for metal castings of engine blocks, train wheels, and other metal products. Washed sand, because it is clean and does not dissolve in water, is used to filter impurities from drinking water. Its whiteness makes it a desirable constituent in plaster, mortar, and precast building panels. Some silica sand, ground to a fine, white powder (called ground quartz, ground silica, silica flour, or potter's flint), is used as an ingredient in paints, pottery and china, and scouring powders.

Because the coarse grains of the washed silica sand are rounded, strong, and available in uniform sizes, the petroleum industry uses thousands of tons each year in the hydraulic fracture treatment of wells to help increase oil production. In hydraulic fracturing, sand is mixed with oil, water, or other petroleum products and is forced by powerful pumps into sandstone or limestone formations that contain oil. The great pressure that is exerted opens

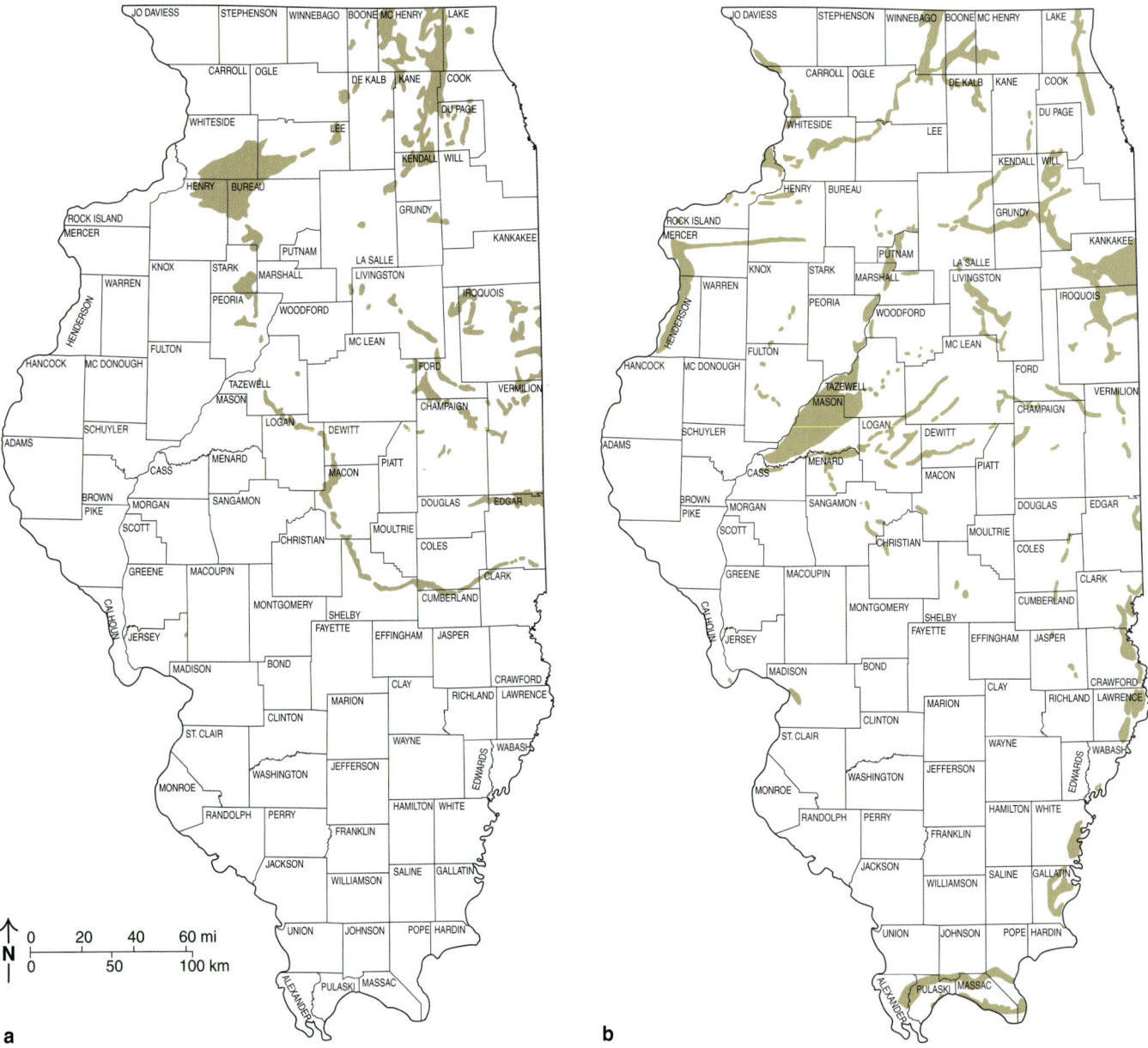

Figure 17-5 **(a)** Areas underlain by outwash plains. **(b)** Glacial outwash sand and gravel deposited in valley trains during the Wisconsin Episode of glaciation. Both maps are from Masters (1983).

fractures in the rock and pushes the liquid and sand into them. When the pressure is relieved, the sand grains serve as props to hold the fractures open. Oil can then flow more readily into the wells, and oil production is increased.

Clay and Shale

Clay is a very fine-grained material made up of a group of minerals produced by the weathering of different kinds of rocks. Illinois clay deposits are either unconsolidated surficial clays or consolidated bedrock clays and shales. The term *clay* is used in three different ways: (1) as a rock term, (2) as a group of fine-grained minerals, and (3) as a particle size. As a rock term, clay refers to a natural, earthy, fine-grained material. Mineralogically, clay refers to a group of hydrous silicate minerals comprising mainly silica, alumina, minor amounts of iron, magnesium, calcium, or sodium, and water. As a particle size term, clays are defined as particles less than 2 μm in size. The primary clay minerals important to the industrial development of Illinois include the platy clay minerals illite, kaolinite, and chlorite and the swelling clay minerals vermiculite and montmorillonite. A distinguishing feature of clay minerals is the high ratio of surface area to particle volume. This high surface area to volume ratio allows clay minerals to readily sorb water and other liquids onto their surfaces or between their layers, giving the clay important colloidal properties, such as plasticity.

The mica-like clay mineral illite was named for the state of Illinois. Illite is the most abundant clay mineral in Illinois surficial and bedrock strata and in the shales that are used for ceramics. Kaolinite is a layer silicate composed of almost pure silica, alumina, and water, making it very resistant to melting and therefore a good refractory material. Illite and quartz generally are intermediate in their tendency to melt, and iron- and magnesium-rich chlorite can melt and bond with other minerals into a steel-hard ceramic product. Many Pennsylvanian age claystones and shales have optimal proportions of clay minerals, quartz, and feldspar for use as ceramics.

Vermiculites and smectites are expandable clay minerals that have the ability to sorb liquids easily to their surface and inner-lattice structure. This property allows for shrinking and swelling of the clay. These clay minerals also readily exchange ions, such as sodium (Na^+), calcium (Ca^{2+}), and potassium (K^+) cations and phosphate (PO_4^{3-}) anions, which, in turn, significantly change the properties of the clay. When heat-activated, the smectite-rich Porters Creek clay of southern Illinois absorbs both oil- or water-based liquids, a characteristic that led to its use as an oil absorbent sweep-up compound and as pet litter. The ex-

Figure 17-6   A sand and gravel operation in northeastern Illinois. (Photograph by Joel M. Dexter.)

pandable nature of some clay minerals makes them excellent as thickeners and valuable for applications where low permeability is needed, such as in waste barriers and landfill liners. Swelling clays, however, can also create engineering problems, such as foundation cracks.

The chemical and physical processes of soil development commonly form clay minerals. Pennsylvanian age rocks that lie at the bedrock surface in Illinois contain useful clay minerals. The underclays beneath many coal seams were once the soils in which the plants grew that eventually became coal. Clay-rich silts and muds, eroded from distant upland soils, were deposited in Illinois during the Pennsylvanian Period. As the Illinois Basin subsided, the temperature and pressure rose, and the clay minerals were changed. The mineralogical and chemical compositions of the clays dictate how the materials can be used. Many Pennsylvanian age shales are mined and used to make bricks and other products. In a few places, the clays were altered to flint clay, a very fine-grained, hard clay used to make "refractory brick" to line furnaces and for other high-temperature applications. Flint clay from a particular source near present-day St. Louis has been identified as the material used by the mound builders at Cahokia to make carved figurines and ceremonial pipes (Emerson et al. 2003).

Clay materials have been used by people in Illinois since prehistoric times (Emerson et al. 2003). For more than 5,000 years, Native Americans used upland soils and river bottom sediments to create ceremonial and utilitarian ceramic ware (Figure 17-7). European settlers used native clay materials to create pottery, paving and building bricks, and structural blocks and tiles for buildings. Ceramic field drain tiles were an early innovation that made it possible to cultivate Illinois' seasonally wet soils.

Illinois clays and shales are also used for a variety of other industrial applications. Clays are used as flux that is mixed and fired with limestone to make cement. Clays that are high in alumina and low in iron, alkalis, and alkaline earth elements can be used to produce high-strength cement for construction of bridges and large buildings. These clays also are used to create vision and sound barriers to screen industrial operations along roadways and to block out transportation noise. Clays make excellent barriers to fluid transport and are used extensively as pond liners, landfill covers, and liners for waste disposal or isolation sites (Keith and Murray 1994). New products that incorporate clays include (1) underground markers for utilities, (2) "syn-soils" made from dredged sediment and composted yard wastes, (3) flowable fill, and (4) bricks made of clay, shale, and fly ash (a by-product of coal combustion).

Tripoli

Tripoli is a porous siliceous rock consisting of microcrystalline quartz (SiO_2) particles. This high-purity material has been mined for 80 years in Alexander and Union Counties from rocks of the Lower Devonian Series that are about 400 million years old (Berg and Masters 1994). Illinois has been the nation's principal producer of this material for many years.

The rocks that host tripoli, mainly the Lower Devonian Clear Creek Formation, are very siliceous and cherty limestones, which likely provided the source of silica needed for formation of tripoli. The origin of tripoli is not well understood. Two main ideas have been proposed: (1) alteration of chert and leaching of siliceous limestone through weathering (Lamar 1965) and (2) dissolution of limestone and precipitation of silica from a hot, silica-rich fluid flow system (Berg and Masters 1994). Hydrothermal events (rather than direct deposition of the silica) also may have enhanced alteration of chert and siliceous sediment to form tripoli.

Tripoli is used in the ceramic industry, in glass manufacture, and in polishing optical lenses. It is also used as paint filler and as a fine abrasive in products such as abrasive soaps. It is a filler and extender in plastics, sealants, and epoxy resins. Lower-quality tripoli is used in some portland cements to increase strength and reduce water absorption.

Peat

Peat deposits formed from decaying plant material that accumulated in low bog areas. When the last great ice sheet in Illinois melted back about 12,500 years ago, it left behind many ponds and lakes, particularly in northeastern Illinois (Lamar 1965, Hester and Lamar 1969, Masters 1987). Reeds, sedges, and mosses grew along the shores and in the shallows of these lakes and ponds. Their partially decomposed remains were preserved beneath the water, and, with gradual burial, compaction, and disintegration of the plant remains, the ponds and lakes became peat bogs. Small amounts of peat are currently mined in Lake and Whiteside Counties. Peat is used in agriculture and horticulture, primarily as a soil conditioner because of its water-holding properties and ability to lighten heavy soils.

Societal Needs and Issues

Illinois Industrial Minerals: A Rich Past

The production and use of industrial minerals have played a vital role in the state's development. Since pioneer settlement, Illinois citizens have used local sources of industrial minerals such as stone, clay, and sand and gravel to build their homes, community buildings, roads, and bridges (Figure 17-8). Cities such as Lemont, Grafton, Joliet, and Batavia developed a distinctive architectural character as a result of the local availability of high-quality building stone, which also was exported throughout the Midwest (Mikulic 1989; Mikulic and Kluessendorf 1999, 2000a, 2000b). The

Figure 17-7   Clay product manufacture began about 7,500 years ago in Illinois when Native Americans carved utilitarian and ceremonial objects from a unique claystone called flint clay. Artifacts include **(a)** a 7,000- to 7,500-year-old notched butterfly-type bannerstone; **(b)** a pair of 2,600- to 3,200-year-old plummets or net weights; and **(c)** a ceremonial platform pipe that was carved about 2,000 years ago. Bannerstones and plummets are found near the source area. Many of the pipes were never smoked and were buried with important people along the Illinois and Mississippi Rivers and at the Tremper Mound in southern Ohio. (Photographs are courtesy of K. B. Farnsworth, Illinois Transportation Research Program.) **(d)** This figurine from an excavation in the American Bottoms area is thought to have been carved about 1,250 years ago from an extremely rare variety of flint clay from nearby Missouri. These figurines were widely traded throughout the southern United States and to the north as far as Wisconsin, suggesting they were highly valued. T. E. Emerson (personal communication) has estimated from the loss of fine carving details on artifacts at some sites that the objects were used in ceremonies for a century or more. (Photograph is courtesy of the Illinois Transportation Archaeological Research Program.)

Illinois State Capitol building in Springfield is an outstanding example of the use of this high-quality building stone (Figure 17-9). As new construction techniques and architectural styles have evolved, the types of stone resources needed for construction have changed. Building stone is no longer quarried commercially in Illinois.

Geologically, stone resources are concentrated primarily in northern, southern, and western Illinois. Limited distribution of stone material and the lack of good transportation meant that most early nineteenth century communities used local sources of nearby rock, regardless of quality. As a result, hundreds of small quarries operated then. Initially, those quarries provided mainly building stone for erecting load-bearing walls and thick foundations in large buildings (Figure 17-10) and for building other durable structures such as bridges. Limestone and dolomite from these quar-

Figure 17-8 Workers at a building stone quarry in Batavia, Illinois, during the 1940s. (Photograph from the Illinois State Geological Survey collection.)

Figure 17-9 Most of the stone for the Illinois State Capitol was mined from the Lockport-Joliet area in northeastern Illinois. (Photograph by Joel M. Dexter.)

Figure 17-10 High school in Dallas City, Hancock County, built in 1895 from locally mined stone. (Photograph by Jack M. Masters.)

Figure 17-11 Lime kiln near Ullin, Pulaski County, that was in operation during the late 1800s. (Photograph from the Illinois State Geological Survey collection.)

Industrial Minerals

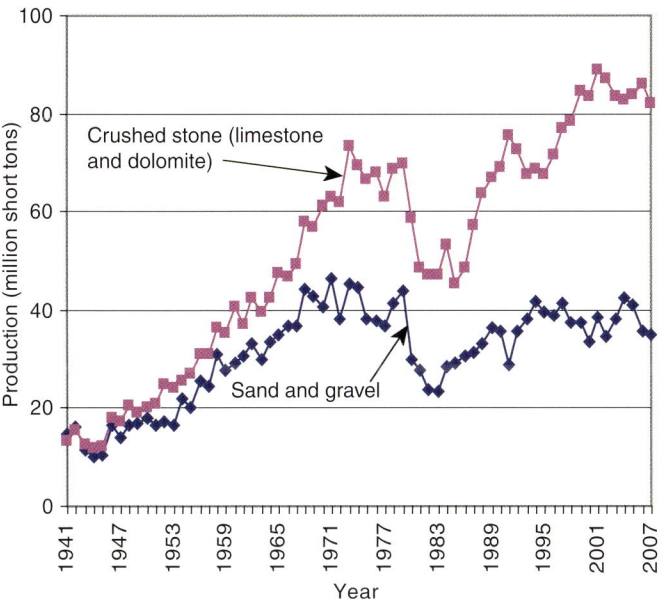

Figure 17-12 Illinois' production of crushed stone and sand and gravel, 1941 through 2007 (data are from the former U.S. Bureau of Mines and U.S. Geological Survey).

The growth of Illinois' road network also affected the stone industry. Crushed stone aggregate was being produced for road building by the mid-1800s, but demand was limited. The introduction of portland cement and the growth of automobile use beginning around 1900 created a need for good roads with concrete or asphalt pavements, which dramatically increased the need for crushed stone aggregate. Sand and gravel were also mined on a large scale for the first time as another source of aggregate. Since 1900, the need for crushed stone as aggregate has continued to increase in importance, and its production has dramatically increased in both volume and value, making it one of the most important natural resources produced in Illinois (Figure 17-12).

Value of Illinois Industrial Minerals

Industrial minerals continue to be one of the state's leading mineral resource commodities, accounting for $1.2 billion (43%) of the total value of minerals produced in Illinois in 2007. Coal ($1 billion or 36%) and oil ($602 million or 21%) are the other main mineral resources (Figure 17-13). Construction aggregates have low unit values per ton, but large tonnages are needed for construction. The cost of aggregate rises steeply as the transport distance increases, so, to be economically viable, aggregate resources must be available locally (Bhagwat 1989, 2000).

ries were also used to produce lime for making mortar (Figure 17-11). After construction of the railroad system in the mid-1800s, lime and stone production moved to locations producing the highest-quality materials. Cheaper transportation also meant that many small local quarries were abandoned.

The building stone and lime industries reached their peak during the 1890s but diminished rapidly over the next ten years. The introduction of cheap portland cement, a superior construction material, and the development of new building methods eliminated the need for foundation stone and load-bearing walls (Mikulic 1990). However, the widespread use of portland cement created an even greater need for stone as a source of crushed aggregate, a major ingredient in concrete and asphalt.

According to the U.S. Geological Survey (2003) mineral industry profile, Illinois ranked sixteenth among the 50 states in total value of nonfuel mineral production. By value, crushed stone was the Illinois' leading industrial mineral, accounting for about 46% ($573 million) of the total. Next were portland cement, 25% ($308 million); construction sand and gravel, about 14% ($176 million); and industrial sand, about 8% ($102 million). Lime, fuller's

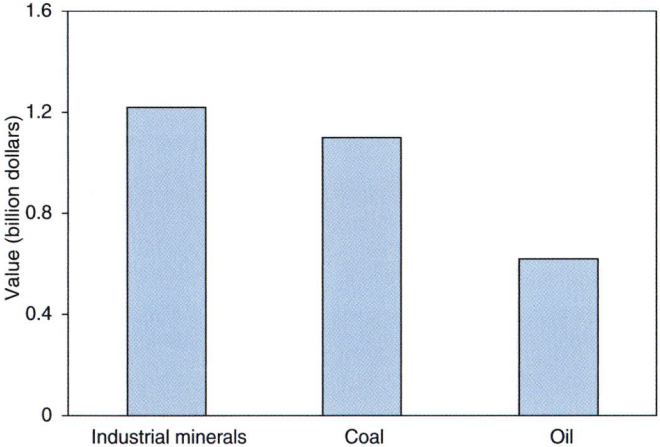

Figure 17-13 Value of Illinois fuel and nonfuel raw materials (U.S. Geological Survey, Illinois Oil and Gas Association, Annual Coal Report, 2001–2007).

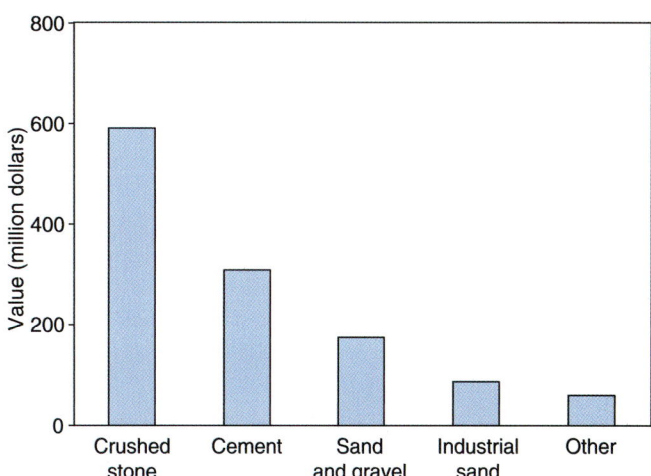

Figure 17-14 Value of Illinois industrial minerals (U.S. Geological Survey 2006); the "other" category includes lime, clay, tripoli, peat, crushed sandstone, and gemstones.

earth (absorbent clay such as that used for production of pet litter), tripoli, and other nonfuel minerals, in decreasing order, accounted for most of the remaining 7% (Figure 17-14). Among the 49 mineral-producing states, Illinois was first in the production of industrial sand and tripoli, fifth in crushed stone, and fourth in peat and fuller's earth. The state ranked ninth in cement production and remained a significant producer of construction sand and gravel and of lime (Lasemi et al. 2008).

Current Usage and Demand

Rapidly developing regions such as the St. Louis Metro East and the Chicago metropolitan area have rich deposits of limestone, dolomite, and sand and gravel, but it is becoming increasingly difficult to open new quarries as new development expands over these resources (Figure 17-15). The reserves in existing quarries are nearly depleted, and, because of permitting concerns, it is doubtful whether new surface mines will be developed in many urbanized areas in the future (Mikulic 1995). Mining for construction stone material and sand and gravel can become a source of contention between local residents, land-use planners, regulators, and aggregate companies. However, mining is a temporary land use; once the needed resources are extracted, the land can be put to other beneficial uses. Reclamation of mined-out sites can create new wildlife habitat or restore original habitats that were lost when prairies were converted to farmlands (Chapter 30, Surface Mine Reclamation). Beautiful parks and scenic lakes surrounded by residential development continue to replace areas that were once extraction sites (Figure 17-16). Underground stone mining is an increasingly attractive technique that allows both continued access to resources and ongoing expansion of residential and industrial complexes (Baxter 1980, 1989; Mikulic 1989). Underground mining also constrains noise and dust pollution and could ultimately provide temperate underground space ideal for business parks, warehouses, and cold storage (Bhagwat et al. 2004). Underground mining could safely tap deeper limestone and dolomite reserves with minimal damage to the environment, groundwater, or

Figure 17-15 This quarry (lower center) on Chicago's south side has been overrun by development of buildings and highways, which eventually resulted in its closure. (Photograph by Joel M. Dexter.)

Figure 17-16 A former sand and gravel operation in northeastern Illinois has been converted into a high-value, residential development. (Photograph by Jack M. Masters.)

Figure 17-17 An underground limestone mine in Valmeyer, Illinois, showing open rooms and pillars. Mined-out rooms such as these are often used for office space and climate-controlled storage. (Photograph by Joel M. Dexter.)

surficial landscape. Several underground mines are already in operation in Illinois, and a number of new sites are under consideration or development in the northeastern part of the state (Figure 17-17).

Conclusion

Industrial minerals continue to be one of the state's leading mineral commodities, and construction aggregates (crushed stone and sand and gravel) and portland cement account for more than 85% of the industrial minerals mined or manufactured in Illinois. Because of their importance to residential and transportation infrastructure, every dollar's worth of industrial minerals consumed in Illinois contributes $550 directly and indirectly to Illinois' gross state product.

Several ongoing challenges will need to be faced in the coming years: (1) obtaining adequate government funding for improvements to transportation infrastructure, (2) addressing the problems of supplying aggregate for the rapidly growing Chicago and metropolitan East Saint Louis areas, (3) resolving the conflicting public demands to protect the environment while obtaining needed resources, (4) and dealing with increasing opposition to mining from the general public.

The repair and maintenance of the highway system require the local availability of high-quality aggregate. With aggregate comprising approximately 80% of concrete pavements and more than 90% of asphalt, durable aggregate will continue to be in high demand throughout the state for years to come. Additionally, stringent pollution control requirements have accelerated installation of limestone-based desulfurization systems in coal-fired power plants. This trend toward increasing numbers of desulfurization units is expected to continue. Because of the importance of high-calcium limestone as a scrubbing agent, it is essential to address the issues associated with the transport, availability, and suitability of high-calcium limestone resources. Nearby sources of suitable limestone raw material must be found to feed existing and new scrubber installations for coal-fired power plants and to aid in the selection of proper resources for desulfurization systems in the future.

References

Annual Coal Report, 2001–2007, Energy Information Administration: Washington, D.C., U.S. Department of Energy, Office of Coal, Nuclear, Electricity, and Alternate Fuels, DOE/EIA 0584 (2007). http://www.eia.doe.gov/cneaf/coal/page/acr/acr.pdf. Accessed April 20, 2009.

Baxter, J. W., 1965, Limestone resources of Madison County, Illinois: Illinois State Geological Survey, Circular 390, 39 p.

Baxter, J. W., 1980, Factors favoring expanded underground mining of limestone in Illinois: Mining Engineering, v. 32, no. 10, October 1980, p. 1497–1504.

Baxter, J. W., 1989, Possible underground mining of limestone and dolomite in central Illinois, in R. E. Hughes and J. C. Bradbury, eds., Proceedings of the 23rd Forum on the Geology of Industrial Minerals, held May 11–14, 1987, North Aurora, Illinois: Illinois State Geological Survey, Illinois Minerals Notes 102, p. 21–28.

Berg, R. B., and J. M. Masters, 1994, Geology of microcrystalline silica (tripoli) deposits, southernmost Illinois: Illinois State Geological Survey, Circular 555, 89 p.

Bhagwat, S. B., 1985, The lime and limestone market for sulfur removal: Potential for 1992: Illinois State Geological Survey, Illinois Minerals Notes 90, 13 p.

Bhagwat, S. B., 1989, Model of construction aggregate demand and supply: A Chicago area case study, in R. E. Hughes and J. C. Bradbury, eds., Proceedings of the 23rd Forum on the Geology of Industrial Minerals: Illinois Geological Survey, Illinois Minerals Notes 102: 29–34.

Bhagwat, S. B., 2000, Industrial minerals in Illinois: A key to growth: Illinois State Geological Survey, informational flyer, 4 p.

Bhagwat, S. B., Z. Lasemi, and M. Dunn, 2004, Economic feasibility of underground mining of stone in the St. Louis Metro East region of Illinois, in C. M. Seeger, ed., Proceedings of the 38th Forum on the Geology of Industrial Minerals: Rolla, Missouri, Missouri Geological Survey, Report of Investigation 74, p. 47–53.

Borgwardt, R. H., and R. D. Harvey, 1972, Properties of carbonate rocks related to SO_2 reactivity: Environmental Science and Technology, v. 6, no. 4, p. 350–360. (Illinois State Geological Survey, Reprint 1972-E.)

Boynton, R. S., 1980, Chemistry and technology of lime and limestone (2nd ed.): New York, Wiley & Sons, 578 p.

Carr, D. D., senior ed., 1994, Industrial minerals and rocks (6th ed.): Littleton, Colorado, Society of Mining, Metallurgy, and Exploration, 1,196 p.

Emerson, T. E., R. E. Hughes, M. R. Hynes, and S. U. Wisseman, 2003, The sourcing and interpretation of Cahokia-style figurines in the Trans-Mississippi south and southeast: American Antiquity, v. 68, no. 2, p. 287–313.

Foose, M. P., and A. F. Barsotti, 1999, Use of limestone resources in flue-gas desulfurization power plants in the Ohio River valley, in K. S. Johnson, ed., Proceedings of the 34th Forum on the Geology of Industrial Minerals, 1998: Oklahoma Geological Survey, Circular 102, p. 273–278.

Goodwin, J. H., 1979, A guide to selecting agricultural limestone products: Illinois State Geological Survey, Illinois Minerals Notes 73, 7 p.

Goodwin, J. H., 1983, Geology of carbonate aggregate resources of Illinois: Illinois State Geological Survey, Illinois Minerals Notes 87, 12 p.

Goodwin, J. H., and J. W. Baxter, 1981, High-calcium, high-reflectance limestone resources of Illinois: Geological Society of America Bulletin, part I, v. 92, p. 621–628.

Harvey, R. D., R. R. Frost, and J. Thomas Jr., 1974, Lake marls, chalks, and other carbonate rocks with high dissolution rates in SO_2-scrubbing liquors: Illinois State Geological Survey, Environmental Geology Notes 68, 22 p.

Hester, N. C., and J. E. Lamar, 1969, Peat and humus in Illinois: Illinois State Geological Survey, Industrial Minerals Notes 37, 14 p.

Keith, K. S., and H. H. Murray, 1994, Clay liners and barriers, in D. D. Carr, ed., Industrial minerals and rocks (6th ed.): Littleton, Colorado, Society of Mining, Metallurgy, and Exploration, p. 435–452.

Lamar, J. E., 1928, Geology and economic resources of the St. Peter Sandstone of Illinois: Illinois State Geological Survey, Bulletin 53, 175 p.

Lamar, J. E., 1959, Limestone resources of extreme southern Illinois: Illinois State Geological Survey, Report of Investigations 211, 81 p.

Lamar, J. E., 1965, Industrial minerals and metals of Illinois: Illinois State Geological Survey, Educational Series 8, 48 p.

Lamar, J. E., 1966, High-purity limestones in Illinois: Illinois State Geological Survey, Industrial Minerals Notes 27, 20 p.

Lamar, J. E., 1967, Handbook on limestone and dolomite for Illinois quarry operators: Illinois State Geological Survey, Bulletin 91, 119 p.

Lamar, J. E., J. S. Machin, W. H. Voskuil, and H. B. Willman, 1956, Preliminary report on portland cement materials in Illinois: Illinois State Geological Survey, Report of Investigations 195, 34 p.

Langer, W. H., and V. M. Glanzman, 1993, Natural aggregate: Building America's future: Reston, Virginia, U. S. Geological Survey, Circular 1110, 39 p.

Lasemi, Z., and D. G. Milulic, 2006, Annual review of Illinois' industrial minerals: Mining Engineering, May, p. 86–90.

Lasemi, Z., D. G. Mikulic, and S. D. Elrick, 2008, Annual review of Illinois' industrial minerals and coal production: Mining Engineering, v. 60, no. 5, p. 89–92.

Lasemi, Z., and R. D. Norby, 2001, Depositional cycles in the Salem Limestone of southwestern Illinois: Implications for predicting limestone quality and reserves, *in* R.D. Hagni, ed., Studies on ore deposits, mineral economics, and applied mineralogy: With emphasis on Mississippi Valley-type base metal and carbonatite-related ore deposits: Rolla, Missouri, University of Missouri, p. 373–383.

Lasemi, Z., and R. D. Norby, 2004, A preliminary cross section of the Mississippian limestone resources in the St. Louis Metro East region of Illinois, *in* C. M. Seeger, ed., Proceedings of the 38th Forum on the Geology of Industrial Minerals, Rolla, Missouri: Missouri Geological Survey, Report of Investigations 74, p. 185–187.

Lasemi, Z., R. D. Norby, J. A. Devera, B. W. Fouke, H. E. Leetaru, and F. B. Denny, 1999, Middle Mississippian carbonates and siliciclastics in western Illinois: Illinois State Geological Survey, Guidebook 31, 60 p.

Masters, J. M., 1983, Geology of sand and gravel aggregate resources of Illinois: Illinois State Geological Survey, Illinois Minerals Notes 88, 10 p.

Masters, J. M., 1987, Industrial minerals and metals, *in* R. D. Neely and C. G. Heister, eds., The natural resources of Illinois: Champaign, Illinois, Illinois Natural History Survey, Special Publication 6, p. 199–205.

Mikulic, D. G., 1989, The Chicago stone industry: A historical perspective, *in* R. E. Hughes and J. C. Bradbury, eds., Proceedings of the 23rd Forum on the Geology of Industrial Minerals, held May 11–15, 1987, North Aurora, Illinois: Illinois Geological Survey, Illinois Mineral Notes 102, p. 83–89.

Mikulic, D. G., 1990, Cross section of the Paleozoic rocks of northeastern Illinois: Implications for subsurface aggregate mining: Illinois State Geological Survey, Illinois Minerals 106, 14 p.

Mikulic, D. G., 1995, Uncertain future for Chicago aggregate industry: Rock Products, v. 98, no. 8, p. 21–23, 46.

Mikulic, D. G., and J. Kluessendorf, 1999, Silurian geology and the history of the stone industry at Pere Marquette State Park and Grafton, Illinois: Illinois Association of Aggregate Producers Teachers Workshop Guidebook, Part 1, 17 p.

Mikulic, D. G., and J. Kluessendorf, 2000a, Silurian geology and the history of the stone industry at Grafton, Illinois, *in* R. D. Norby and Z. Lasemi, eds., Paleozoic and Quaternary geology of the St. Louis Metro East area of western Illinois: 63rd Annual Tri-State Geological Field Conference: Illinois State Geological Survey, Guidebook 32, p. 39–45.

Mikulic, D. G., and J. Kluessendorf, 2000b, Stop 7, Keller Quarry, *in* R. D. Norby and Z. Lasemi, eds., Paleozoic and Quaternary geology of the St. Louis Metro East area of western Illinois, 63rd Annual Tri-State Geological Field Conference: Illinois State Geological Survey, Guidebook 32, p. 76–81.

National Lime Association, 2005, Lime: The essential chemical: Arlington, Virginia. http://www.lime.org/aboutlime.html. Accessed November 10, 2009.

National Stone, Sand and Gravel Association, undated, 50 fascinating facts about stone, sand & gravel, fundamental resources for more than 5000 years: Arlington, Virginia, 6 p.

Tepordei, V. V., 1997, Natural aggregates—Foundation of America's future: Reston, Virginia, U.S. Geological Survey, Fact Sheet FS144-97, 4 p.

U.S. Geological Survey, 2006, The mineral industry of Illinois: Reston, Virginia, 8 p. http://minerals.usgs.gov/minerals/pubs/state/il.html. Accessed November 10, 2009.

Willman, H. B., and D. R. Kolata, 1978, The Platteville and Galena Groups in northern Illinois: Illinois State Geological Survey, Circular 502, 75 p.

Groundwater Resources of Illinois

Beverly L. Herzog

Groundwater resources in Illinois are extremely important to the well-being of the state's citizens and to the economy of the state as a whole. Nearly 5 million Illinois citizens, or more than one-third of the population, receive their drinking water from groundwater sources. Almost 60% of the Illinois water supply beyond Lake Michigan comes from groundwater, and about 90% of the rural population depends on groundwater to supply its water needs. Groundwater use for public, household, commercial and industrial, farming, mining, and power generation totals about 925 million gallons per day (3.5 billion L/day) pumped from more than 10,000 public supply wells and more than 440,000 household wells. To satisfy this demand, about 7,800 new wells are drilled annually. Groundwater industries, including well drillers, environmental consulting firms, and waste management firms, generate nearly $1.5 billion in sales annually. This amount does not include the direct sale of water by utilities.

Groundwater resources are inextricably linked to surface water, both in receiving recharge and in providing the base flow to streams, which keeps them flowing during dry periods and sustains aquatic ecosystems. Groundwater is essential for maintaining wetlands and supporting the special habitats they provide.

Although Illinois' groundwater resources are abundant, its aquifers are not distributed uniformly across the state. Aquifers occur from ground surface to deeper than 1,000 feet (305 m), and groundwater yield varies from location to location, from a few gallons of water per minute to more than a million gallons per day. Water quality also varies greatly with depth, geological materials, and natural protection from contamination.

The following chapters describe Illinois aquifers from a geological perspective—their geological framework, constituents, composition, spatial distribution, quality, and natural protection from contamination. This part concludes with a chapter covering the importance of geology to wetlands.

Photograph by Joel M. Dexter.

Image on previous page: Waterfall and bridge, Matthiessen State Park, La Salle County, Illinois.

Aquifers 18

David R. Larson and Beverly L. Herzog

Introduction

Understanding how the geological framework of Illinois relates to groundwater is essential to comprehending where aquifers occur and how groundwater behaves in the subsurface. By definition, an aquifer is saturated geological material that can supply sufficient water to sustain a well or spring. The geological framework dictates where aquifers occur and where they do not, how well aquifers are protected from surface activities, and how readily their water is replenished. Understanding the state's geological framework makes it easier to grasp why groundwater is available in some areas of Illinois but is difficult to obtain in others.

Groundwater within the Geological Framework

Chapters 2 through 12 describe the geological framework of Illinois in detail. Two major aspects are especially relevant to a discussion of groundwater: (1) bedrock and (2) the unconsolidated deposits that overlie the bedrock throughout most of the state. This chapter concentrates on the water-yielding characteristics of these geological materials.

Although water can move through almost all geological materials, it does so at vastly different rates. The capacity of geological materials to transmit a fluid is called permeability. If the fluid is groundwater, the capacity is called hydraulic conductivity. Groundwater readily moves through geological materials with high hydraulic conductivity; that is, with relatively large pore spaces or fractures that are well interconnected. Coarse-grained geological materials—such as sandstone, sand, and gravel—and creviced limestone and dolomite are examples of materials that have high hydraulic conductivity.

Hydraulic conductivity is a measure of the quantity of groundwater that moves in a given unit of time under a unit hydraulic gradient through an area of one square foot (0.09 m^2) of earth material oriented at right angles to the direction of groundwater flow. Hydraulic conductivity is commonly expressed in gallons (the volume of groundwater) per day (the unit of time) per square foot (the cross sectional area). A unit hydraulic gradient is a one-foot (0.3-m) vertical decline in hydraulic head over a horizontal distance of one foot (0.3 m)(Figure 18-1). Table 18-1 lists hydraulic conductivities of various aquifer and aquitard materials commonly found in Illinois.

Aquifers and Aquitards

When materials with high hydraulic conductivity are saturated, these materials are aquifers. Groundwater does not readily move through geological materials with low hydraulic conductivity; such materials form aquitards. Aquitard materials, such as shale, nonfractured limestone and dolomite, or fine-grained silts and clays, have pore spaces that are relatively small and/or poorly interconnected.

Understanding the vertical and horizontal distribution of aquifer and aquitard materials is an integral part of determining the extent and quantity of groundwater resources, managing those resources, and assessing the susceptibility of groundwater to contamination. Such knowledge is needed, for example, to evaluate the impacts of well pumping on other wells, to site waste disposal facilities to minimize their groundwater contamination potential, and

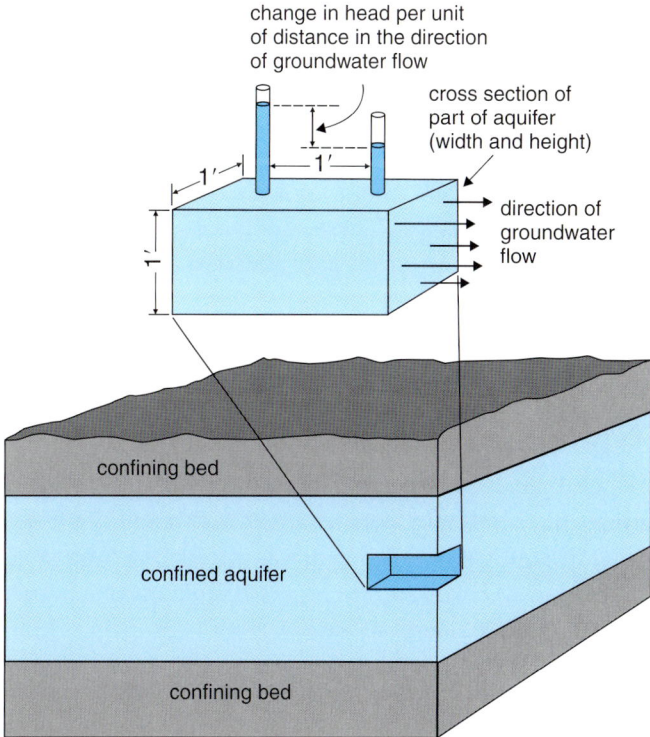

Figure 18-1 Hydraulic conductivity is the volume of groundwater that will move in a unit of time under a unit hydraulic gradient through a unit cross section area oriented at right angles to the direction of groundwater flow. A unit hydraulic gradient is a one-foot (0.3-m) vertical decline in hydraulic head over a horizontal distance of one foot (0.3 m) (Heath 1989, Killey and Larson 2004).

Table 18-1 Equivalent times for the range of hydraulic conductivity of various geological materials (Killey and Larson 2004).

Geological material	Equivalent time[1]	Hydraulic conductivity [2] (gpd/ft^2)
Gravel	0.1 to 84 seconds	1 million to 1,000
Sand	8 seconds to 1 day	10,000 to 1
Fractured limestone and dolomite	8 seconds to 10 days	1,000 to 0.1
Sandstone	14 minutes to 2.7 years	100 to 0.001
Loess	2.4 hours to 2.7 years	10 to 0.01
Silt	10 days to 100 years	0.1 to 0.01
Till	10 days to 2,740 years	0.1 to 0.000001
Clay	100 days to 2,740 years	0.01 to 0.000001
Shale	2.7 to 27,400 years	0.001 to 0.0000001

[1]The time it takes for a gallon of water to flow through one square foot of the geological material under a unit hydraulic gradient.
[2]One gallon per day per square foot is equivalent to 0.041 m/day.

to assess the effects of other land-use activities, such as the application of fertilizers and pesticides, that may harm groundwater quality (Chapter 20, Protecting Groundwater, and Chapter 29, Pollution of Natural Waters).

Hydrogeologists gather information about the vertical arrangement and regional distribution of aquifer and aquitard materials from two main sources: (1) places where bedrock or sediments are exposed at land surface, and (2) the many records of water wells, engineering boreholes, and test borings for coal, oil, and gas. These records contain drillers' logs, well-construction details, water-level data, and other information. Although local variations occur in the characteristics of geological deposits, geologists use distinctive units within them to correlate the deposits from one location to another. These correlations help in predicting the occurrence, extent, and thickness of aquifers and aquitards.

The following descriptions of typical aquifer and aquitard materials are taken from Killey and Larson (2004).

Typical Aquifer Materials

Aquifers are made up of earth materials that have relatively high hydraulic conductivities and readily transmit groundwater. Conversely, aquitards are made up of earth materials that have relatively low hydraulic conductivities, causing groundwater to move slowly through them.

Sandstone

Sandstone that is made up of well-rounded sand grains that are fairly uniform in size has mostly interconnected pore spaces. Groundwater readily flows through such sandstone. If mineral matter cements the sand grains together, pore space is reduced, decreasing the hydraulic conductivity. Some sandstone is so thoroughly cemented that groundwater moves only through fractures in the rock rather than through the spaces between grains. Other sandstone may be a mixture of very fine to very coarse grains, or the grains may be angular instead of rounded, or the pores may contain silt or clay. These conditions decrease the connections among pore spaces, thereby reducing hydraulic conductivity.

Limestone and Dolomite

Although relatively dense and typically lacking interconnected pores, many limestone and dolomite units contain numerous interconnected cracks, crevices, and solution channels. If these openings are filled with groundwater, the rocks form aquifers. Where these rocks do not have such openings, they form aquitards and restrict the flow of groundwater.

Sand and Gravel Deposits

As a result of the glaciation of Illinois (Chapter 12, Quaternary Period), deposits of sand, gravel, or their combination more commonly occur in the northern two-thirds of Illinois than in the southern third. Much of the sand and gravel was carried away from ice sheets in streams and rivers of meltwater as glacial outwash. In general, the hydraulic conductivity of these deposits is high, and they readily transmit groundwater. Some of the most productive aquifers in the state are formed by thick, saturated deposits of sand and gravel that are found in buried bedrock valleys or near the land surface in large areas of central and northwestern Illinois.

Typical Aquitard Materials

Shale

Shale is made up of clay particles stacked tightly together in parallel layers with tiny pore spaces between the particles. Although shale has relatively high porosity, the arrangement of the clay particles greatly reduces the size of the pore spaces and the interconnections among them. With its low hydraulic conductivity, shale is a barrier to groundwater flow. If the shale is weathered or fractured, the hydraulic conductivity may be greater, and groundwater may more readily move through the shale.

Coal

Coal is formed from altered and compacted plant remains that were buried in coastal swamps and lithified over time. Compared with most of the other sedimentary rocks in the state, Illinois coal beds are thin and typically not very significant as aquifers or aquitards. In some localities in southern Illinois, however, fractured coal is the only source of water.

Clay and Silt

Clay and silt found within glacial sediments have low hydraulic conductivity and also are aquitards. If the shale, clay, or silt is weathered or fractured, the hydraulic conductivity may be greater, and groundwater may then move more easily through these sediments.

Diamicton

Diamicton is a sediment composed of a relatively dense mixture of all grain sizes. The proportion of grain sizes in diamicton, however, varies considerably within Illinois. The major portion of some diamictons is clay and silt, but, in others, it is clay, silt, and sand. Pebbles, cobbles, and boulders are relatively sparse in some diamictons and more plentiful in others. Diamicton that is massive and predominantly clay and silt has low hydraulic conductivity and is a barrier to groundwater movement within glacial deposits. Weathered diamicton, diamicton with fractures, and diamicton in which sand and gravel are relatively common allow some movement of groundwater, so hydraulic conductivity is somewhat increased.

Loess

Loess is windblown dust. This fine-grained sediment has low hydraulic conductivity, and groundwater moves through it rather slowly. Openings develop in loess as it gradually weathers into soils, which allows water to move through it more readily than through the unweathered material.

Types of Aquifers

Unconfined Aquifers

Groundwater in a body of saturated, permeable earth materials in which the top of the saturated zone (the water table) is free to rise and fall is described as being unconfined. The aquifer holding this water is called an unconfined aquifer (Figure 18-2) or a water-table aquifer. An aquitard underlies an unconfined aquifer and hinders the downward movement of groundwater, causing the permeable earth materials to become saturated and form an aquifer. The permeable earth materials above the water table are not saturated, but can become so if the water table rises. As a result, the thickness of an unconfined aquifer changes with the rise and fall of the water table. The water level in a well completed in an unconfined aquifer closely approximates the water table in the aquifer adjacent to the well because the pressure on the groundwater at the water table is about equal to atmospheric pressure. Unconfined aquifers in Illinois typically occur at relatively shallow depths.

Confined Aquifers

A confined aquifer is a body of permeable earth material that is completely saturated and has aquitards (the confining units) above and below it. The groundwater in such a setting is described as being confined (Figure 18-2). Another name for a confined aquifer is an artesian aquifer. An example of a confined aquifer is saturated sand and gravel with relatively impermeable diamicton above and below it. The overlying aquitard provides the aquifer with protection from contamination. The level of protection depends on the thickness and hydraulic conductivity of the aquitard.

Because groundwater in the aquifer is confined, the pressure on the groundwater is greater than the atmospheric pressure due to the weight of the water and the overlying earth materials. Consequently, the static water level in a well completed in a confined aquifer rises above the top of the aquifer to a level where the pressure on the groundwater is balanced by atmospheric pressure. This water level marks the potentiometric surface of the aquifer (Figure 18-2). A well completed in a confined (artesian) aquifer is called an artesian well. If the pressure on the groundwater is great enough, water can flow from the well without the use of a pump. This type of well is called a flowing artesian well and can be found in scattered locations throughout northern Illinois and in Iroquois and Vermilion Counties. Before declining pressure on the groundwater stopped its flow, such wells were also found in the Chicago and Peoria areas.

Groundwater Flow

Groundwater typically does not stay in one place, but instead moves slowly in the direction of decreasing hydraulic head. Groundwater travels through the sediments and rocks along flow paths of various lengths and depths (Figure 18-3). Local flow paths are relatively short and shallow, so the amount of time groundwater spends in the subsur-

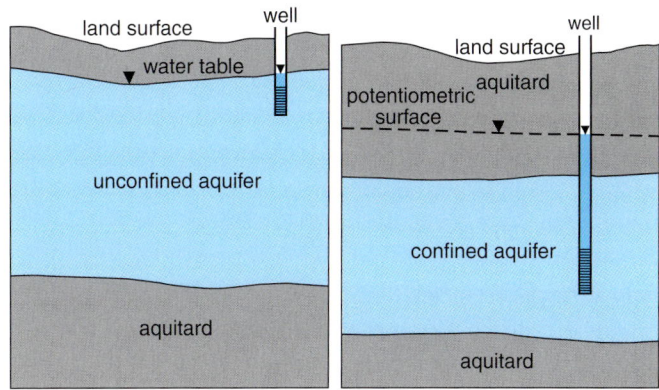

Figure 18-2 Illustration of unconfined and confined aquifers. The triangles indicate the water level in the wells, the water table, and the potentiometric surface (Killey and Larson 2004).

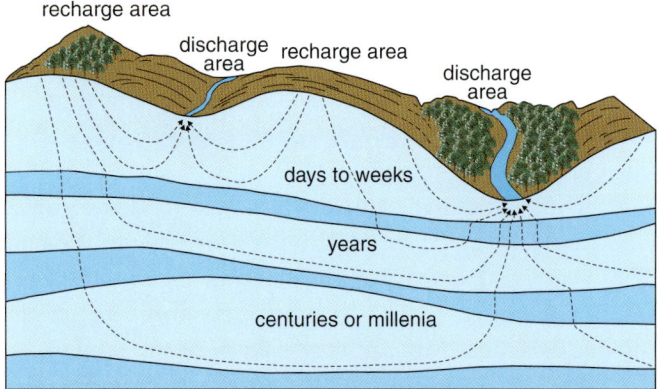

Figure 18-3 Groundwater moves through the subsurface along flow paths of various lengths and depths. The amount of time that groundwater resides in the subsurface varies from days in shallow flow systems to centuries or millennia in the deep flow systems (Killey and Larson 2004).

face is measured in days or weeks. Regional flow paths are much longer and deeper, and the length of time groundwater resides underground is measured in centuries or millennia. At the end of the flow paths, groundwater emerges at the land surface and discharges to rivers, streams, lakes, springs, and wetlands.

Unlike the surface water flowing in rivers or streams, groundwater moves very slowly because of friction between the water and the walls of the pore spaces or fracture surfaces in the sediment and rock. The amount of friction varies both laterally and vertically according to the type of sediment or rock. Where subsurface spaces are very large, such as in large crevices or in caves, groundwater flows very quickly. These features are most common in the karst areas of southwestern Illinois (Chapter 28, Karst Terrane).

Groundwater and Surface Water Interactions

The hydrologic cycle describes the circulation of water between the Earth and its atmosphere (Figure 18-4). Because groundwater and surface water are both part of the hydrologic cycle, they form an integrated resource. Groundwater discharging into rivers and streams contributes to their flow. Groundwater discharge provides the baseflow to these surface water bodies, which keeps them flowing during dry times of the year.

Springs are places where the water table intersects the land surface and groundwater flows from relatively distinct openings, such as in depressions or sinkholes. Springs can also form where an aquitard inhibits the downward movement of groundwater and forces the water to flow laterally until it intersects the land surface.

Wetlands are areas where there is enough surface water or groundwater to produce wet conditions in low areas of the landscape for at least part of a year. The wet conditions cause waterlogged (hydric) soils, which support plants and animals that thrive in wet habitats. Wetlands that are saturated or have open water during the summer are typically groundwater discharge areas, and the water table is at or near the land surface. Wetlands are covered more extensively in Chapter 21, Wetlands Geology.

Groundwater Flow between Bedrock and Glacial Deposits

Groundwater flows between glacial deposits and the bedrock beneath them. Understanding this movement is critical to understanding groundwater availability and groundwater protection. The direction of flow depends on the prevailing vertical hydraulic gradient. The hydraulic characteristics of the bedrock and the overlying glacial deposits determine how easily groundwater moves between the two. Groundwater may flow relatively freely if both have high hydraulic conductivity, for example, where sand and gravel deposits overlie creviced or fractured limestone bedrock. Groundwater flow may be restricted when there is a large contrast in hydraulic conductivity, such as diamicton (low hydraulic conductivity) overlying sandstone (high hydraulic conductivity), or sand and gravel (high hydraulic conductivity) overlying shale (low hydraulic conductivity).

BEDROCK AQUIFERS

Major bedrock aquifers, those that are capable of yielding at least 70 gallons per minute (gpm) (380 m³/day), are present under all of Illinois, although the groundwater in them is not equally suitable for all uses. Bedrock aquifers occur in bedrock units that range in age from Cambrian, the oldest, to Tertiary, the youngest. The depth to

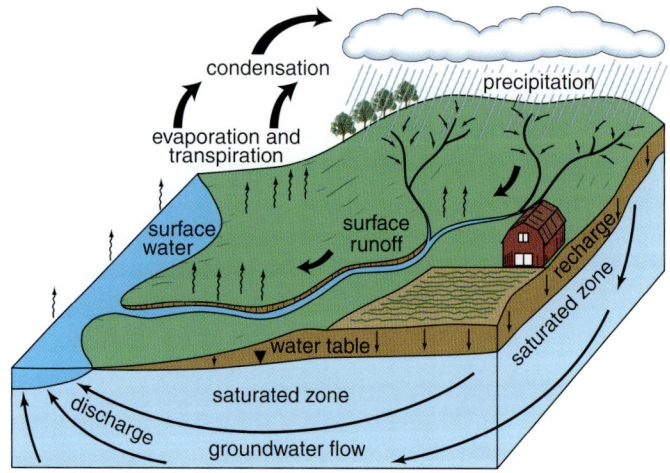

Figure 18-4 Simplified hydrologic cycle shows the circulation of water between the earth and its atmosphere (Killey and Larson 2004).

the bedrock aquifers ranges from near the land surface to thousands of feet beneath it. Bedrock aquifers commonly span several geological units. Categorizing these aquifers according to depth below land surface and identifying regions where the least mineralized water occurs in each of the categories provides a good overview of groundwater resources. The three depth categories are more than 500 feet (150 m) below land surface, within 500 feet (150 m) of land surface, and within 300 feet (90 m) of land surface. Throughout this chapter, depth and yield measurements and conversions have been generalized.

Major bedrock aquifers that are deeper than 500 feet (150 m) below the land surface underlie all of the state. They are found in rocks ranging from Cambrian and Ordovician to Pennsylvanian in age. These major bedrock aquifers can be separated into three regions based on the concentration of total dissolved solids (TDS) in the groundwater. The TDS value is a measure of the amount of dissolved matter in the water. Figure 18-5 indicates that groundwater within the deeper major bedrock aquifers underlying the northern half of Illinois typically has TDS concentrations of less than 2,500 mg/L. Water with a TDS concentration of less than 2,500 mg/L may be treatable to bring its quality up to drinking water standards. At aquifer depths of greater than about 2,000 feet (600 m) in northeastern Illinois, mineralization may be greater than 2,500 mg/L, limiting the usefulness of the very deep aquifers (Visocky et al. 1985). In the southern half of the state, the concentration of TDS in the groundwater in the deeper aquifers increases from 2,500 mg/L to 10,000 mg/L in the north to more than 10,000 mg/L in the south. Water with a TDS between 2,500 mg/L and 10,000 mg/L is designated as being potentially treatable for drinking water, although at a higher cost. The increase in TDS concentration from north to south is indicative of the highly saline water found in the center of the Illinois Basin in southeastern Illinois (Miller 1994).

Major bedrock aquifers that are deeper than 300 feet (90 m) are typically sandstone aquifers that occur in parts of the northern third of Illinois and along part of the state's southwestern border (Figure 18-6). These aquifers can yield as much as 20,000 gallons per day (gpd)/mi² (29 m³/day/km²) (Wehrmann et al. 2003). Although these aquifers commonly underlie fine-grained glacial deposits, they may directly underlie major sand and gravel aquifers such as those in buried bedrock valleys. This juxtaposition allows a direct hydraulic connection between bedrock and glacial drift aquifers. Groundwater in these aquifers is generally suitable for most uses because the TDS concentration is less than 1,500 mg/L.

Major bedrock aquifers found within 300 feet (90 m) of land surface also occur in northern Illinois and along the

Figure 18-5 Total dissolved solids content of waters in major bedrock aquifers found at depths greater than 500 feet (about 150 m) below land surface (Miller 1994).

western border. As shown in Figure 18-7, in northern Illinois, these aquifers can yield as much as 200,000 gpd/mi² (292 m³/day/km²); in western and southern Illinois, yields are commonly 50,000 to 100,000 gpd/mi² (73 to 146 m³/day/km²) (Wehrmann et al. 2003). These bedrock aquifers underlie fine-grained glacial deposits and coarse-grained outwash deposits of sand and gravel, or they may crop out at land surface. These major bedrock aquifers have a direct hydraulic connection with shallow, major sand and gravel aquifers. Groundwater in the shallow bedrock aquifers generally is suitable for most uses because the TDS concentration is less than 1,500 mg/L.

The following stratigraphic units are considered to be major bedrock aquifers in Illinois:

- The Mt. Simon Sandstone and the Elmhurst Sandstone Member (the basal member of the Eau Claire Formation), commonly called the Mt. Simon aquifer (Cambrian aquifer in northern Illinois).
- The Ironton and Galesville Sandstones (Cambrian aquifer in northern Illinois).
- The St. Peter Sandstone and Glenwood Formation, which constitute the Ancell Group (Ordovician aquifer in northern and southern Illinois).
- The Hunton Limestone Group (Silurian-Devonian aquifer in northern and southern Illinois).

These stratigraphic units and corresponding aquifers are discussed, beginning with the oldest. Information comes primarily from Smith and Stall (1975), Visocky et al. (1985), and Voelker (1989).

Cambrian-Ordovician Bedrock Aquifers

The most extensively used bedrock aquifers include the Mt. Simon Sandstone, Ironton-Galesville Sandstone, Glenwood-St. Peter Sandstone, and the Galena-Platteville dolomite. These four units are frequently grouped together as the Cambrian-Ordovician aquifer; the three sandstone units are commonly referred to as the "deep sandstone aquifer." Potable water is obtained from depths of 1,500 feet (460 m) or more in the northern third of Illinois; the average depth is about 1,300 feet (400 m). Many deep sandstone wells tap multiple sandstone intervals and have yields of more than 700 gpm (3,820 m^3/day). Some wells have produced this volume for more than 100 years.

The oldest, deepest bedrock aquifer is the Mt. Simon. Although this aquifer underlies almost all of Illinois, it is too deep and the groundwater is too mineralized to be a source of drinking water except in about the northern one-fifth of the state (line A–A′ in Figure 18-6). The depth of the Mt. Simon increases to the south and southeast because of the downward dip of the rocks into the Illinois Basin. Hydraulic conductivity of the sandstone is relatively low because the rock is moderately well cemented, although the friability and, consequently, the hydraulic conductivity of the sandstone increases to the north. The shales, siltstones, dolomites, and dolomitic sandstones of the Eau Claire Formation act as a confining unit for the Mt. Simon aquifer.

The Ironton-Galesville aquifer is composed mostly of well-sorted, fine-grained sandstone. It is the most consistently productive aquifer in the northern third of Illinois. In the remainder of the state, groundwater in the Ironton-Galesville is too highly mineralized for most uses. Groundwater in the Ironton-Galesville aquifer is confined by dolomites and shales with some sandstones that are regional in extent. The sandstones and fractured dolomite within the confining unit, however, are locally significant aquifers. These units include the Franconia Formation, the Eminence and Potosi Dolomites, and the Prairie du Chien Group.

The Glenwood-St. Peter Sandstone is another major bedrock aquifer in Illinois. The sandstone is typically very well sorted, fine to medium grained, and friable. These characteristics make it a productive aquifer. The Glenwood-St. Peter is found at relatively shallow depths in northern Illinois where it locally underlies the glacial deposits. The Glenwood-St. Peter forms the bedrock surface in parts of north-central Illinois from La Salle and Grundy Counties northward across Lee and Ogle Counties, and locally in De Kalb County and into Stephenson, Winnebago, and Boone Counties. The depth to this aquifer increases southward because the bedrock dips southward toward the

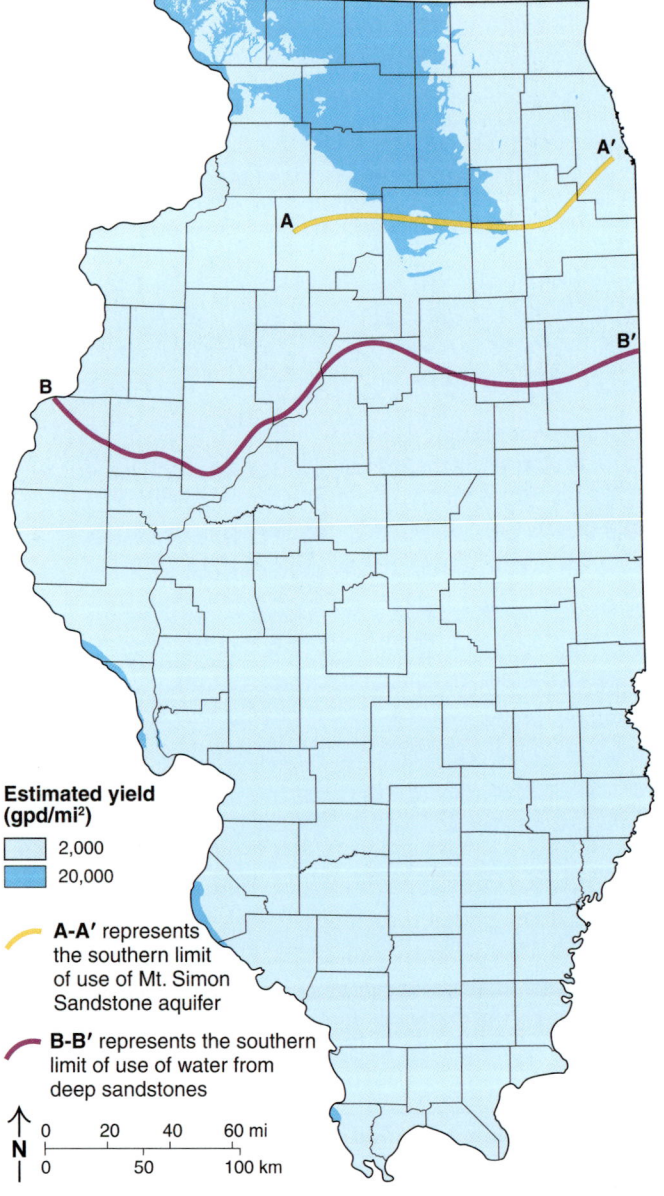

Figure 18-6 Estimated potential yield of deep bedrock aquifers (greater than 300 feet [90 m]) (Wehrmann et al. 2003) with lines of usability (Lim et al. 2002).

center of the Illinois Basin. Groundwater from this aquifer is too highly mineralized in the southern two-thirds of Illinois for most uses. Line B–B' on Figure 18-6 shows the southern limit of potable groundwater in this aquifer.

The Galena-Platteville aquifer, which is mostly fractured dolomite, overlies the Glenwood-St. Peter aquifer, and, for the most part, these two aquifers form a single hydraulic unit. Where the Galena-Platteville is not fractured, however, the dolomite is a confining unit to the Glenwood-St. Peter aquifer. Small groundwater supplies are typically obtained from the Galena-Platteville aquifer, although larger supplies are possible locally where the rock is highly fractured or where glacial deposits overlie it. The Galena-Platteville is found at relatively shallow depths in northern Illinois where it locally underlies the glacial deposits. It forms the bedrock surface in a wedge-shaped area that extends from the Mississippi River to Boone County and southeastward to parts of Lee, La Salle, De Kalb, Kendall, and Grundy Counties. Similar to the Glenwood-St. Peter, the depth to the Galena-Platteville increases southward because the bedrock dips into the Illinois Basin. Groundwater from this aquifer is too highly mineralized in the southern two-thirds of Illinois for most uses, except along the western and southwestern boundary of the state. The Glenwood-St. Peter/Galena-Platteville aquifer is overlain and confined by the shales of the Maquoketa Group.

Silurian Bedrock Aquifers

Aquifers composed of dolomite that is Silurian and Devonian in age are commonly referred to as "shallow dolomite aquifers." Rocks of Silurian age underlie almost all of the state. Because of erosion, however, these rocks are not found in the north-central to northwestern part of Illinois or in two areas near its western margin. These rocks form the bedrock surface in most of northeastern Illinois and about half of northwestern Illinois. Across the central part of the state, the thickness of these rocks increases from about 100 feet (30 m) on the west side of the state to about 700 feet (210 m) on the east side. Their thickness of 500 to 700 feet (150 to 210 m) extends southward into south-central Illinois but decreases further south to about 300 feet (90 m) along the state's southern boundary.

In northern Illinois, the Silurian rocks are predominantly dolomite with some limestone, although they may contain relatively thin, regionally continuous coarse-grained intervals (Rovey and Cherkauer 1994a, 1994b). These rocks typically are aquifers where fractures, joints, or dissolution channels (secondary permeability features) and the coarse-grained intervals are saturated. In southern Illinois, these rocks are mostly limestone with some dolomite, siltstone, shale, and chert. Groundwater is obtained from

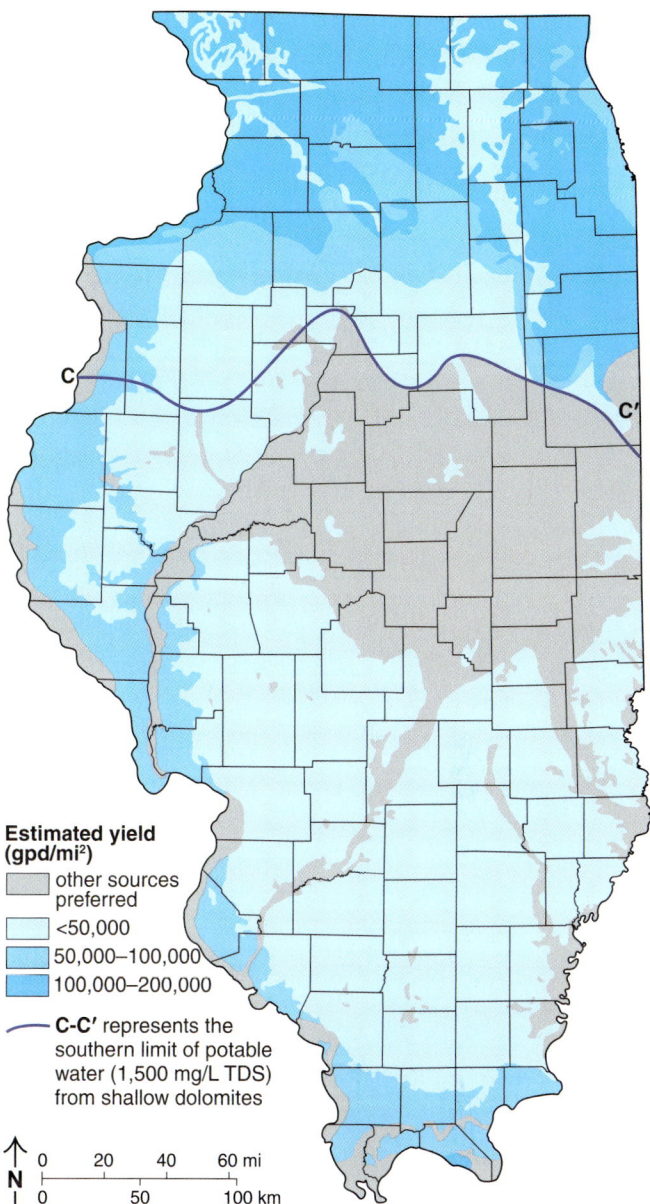

Figure 18-7 Estimated potential yield of shallow bedrock aquifers (less than 300 feet [90 m]) (Wehrmann et al. 2003) with line of usability (Lim et al. 2002). Abbreviation: TDS, total dissolved solids.

fractures, joints, or dissolution channels in these rocks. Aquifer yields range from small to locally very large and able to support municipal and industrial wells. Groundwater quality is good where the rocks are at the land surface or directly underlie the glacial deposits. The water becomes more mineralized as the depth to these rocks increases. Line C–C' on Figure 18-7 shows the southern limit of potable water in these aquifers.

Devonian Bedrock Aquifers

Rocks of Devonian age underlie the southern two-thirds of Illinois. Because of erosion, they are not present in the northern one-third of the state or in some smaller areas along the state's western boundary. Devonian rocks form the bedrock surface in some small areas along the Mississippi River valley and near the eastern edge of east-central Illinois adjacent to the Indiana state line. Thickness of these rocks is commonly about 200 feet (60 m) across central Illinois, increasing to the southeast to about 1,800 feet (550 m) at the state line.

The Devonian rocks are predominantly limestone with some dolomite and shale. These rocks form aquifers where fractures, joints, or dissolution channels are saturated. Yields are typically small to moderate, and the water quality is acceptable in the limited areas where the Devonian rocks form the land surface or directly underlie glacial deposits. Away from the outcrop areas, groundwater is highly mineralized and is unsuitable for most uses.

The black, gray, and green shales of the New Albany Shale form a confining unit for the underlying Devonian rocks.

Mississippian Bedrock Aquifers

Similar to Devonian rocks, Mississippian rocks also are found in the southern two-thirds of Illinois. Erosion has removed them from the northern one-third of the state and some areas along the Mississippi River valley. Mississippian rocks form the bedrock surface near the borders of the state from western to southern Illinois. Thickness of these rocks increases from less than 800 feet (240 m) in central Illinois to about 3,200 feet (980 m) in southern Illinois.

Mississippian rocks are predominantly limestones and dolomites with some shales, siltstones, and sandstones. Mississippian rocks form aquifers where they are fractured and aquitards where they are unfractured. These rocks are a source of water in outcrop areas or where glacial deposits directly overlie them. The yields are typically less than 25 gpm (136 m^3/day) but can be 1,000 gpm (5,450 m^3/day) or more where the rocks are highly fractured. Average depth is 250 feet (75 m). The water becomes highly mineralized away from the outcrop areas and is unsuitable for most uses.

Pennsylvanian Bedrock Aquifers

Rocks of Pennsylvanian age are found throughout Illinois, except in the northern quarter of the state from the Mississippi River valley eastward to Lake Michigan and extending southward near the Indiana state line for about 100 miles (160 km). Pennsylvanian rocks are also eroded along the state's western to southern boundaries. Throughout most of their area of occurrence, Pennsylvanian rocks are buried by Quaternary deposits but are exposed in many valleys where stream erosion has removed the Quaternary deposits.

Pennsylvanian age rocks are predominantly shale throughout the state but include some sandstone, siltstone, limestone, and coal. Thickness of these rocks increases from a few hundred feet in north-central Illinois to about 2,400 feet (730 m) in southeastern Illinois. Pennsylvanian rocks typically act as aquitards, although small water yields can be obtained from the sandstones or fractured limestones, siltstones, shale, or coal. These rocks are used as a source of water where none is available from sand and gravel aquifers and in the outcrop areas where the water quality is suitable. Wells in Pennsylvanian sandstones typically yield less than 25 gpm (136 m^3/day) and have an average depth of 170 feet (50 m). The water becomes highly mineralized away from the outcrop areas and as depth increases. The water is unsuitable for most uses.

Cretaceous-Tertiary Bedrock Aquifers

Cretaceous rocks are found in limited areas of Adams and Pike Counties in western Illinois, and Cretaceous and Tertiary rocks are found at the southern tip of the state. In Adams and Pike Counties, Cretaceous rocks are mostly clayey sand with sand and gravel prevalent at the base. Maximum thickness is about 100 feet (30 m). Although yields are typically small, these rocks form a locally important source of groundwater. In southern Illinois, the Cretaceous-Tertiary rocks are mostly sand, clay, and silt with some gravel. Maximum thickness is about 900 feet (275 m). These rocks are a local source of groundwater. Well yields in areas of greatest thickness may approach 1,000 gpm (5,450 m^3/day).

SAND AND GRAVEL AQUIFERS

Some of the most productive aquifers in Illinois are composed of bedded sand and gravel deposited by the large volumes of glacial meltwater flowing away from the continental ice sheets that periodically crossed into Illinois during the Quaternary Period (Chapter 12, Quaternary Period). The bedrock surface of Illinois, buried for the most part under as much as 400 feet (120 m) of glacial deposits, has features that are similar to those found on the present-day landscape, such as broad river valleys and adjacent uplands and hills (Figure 18-8). Two major bedrock valleys, the Mahomet Bedrock Valley from the east and the Mackinaw Bedrock Valley from the north, join in central Illinois (Figure 18-8). The Mahomet Bedrock Valley extends east to west in a large arc across the central part of Illinois from Indiana to the Illinois River valley (Figure 18-9). The Mackinaw Bedrock Valley joins the Middle Illinois Bedrock Valley that follows the present-day Illinois River valley northward where it joins the Princeton Bedrock Valley. These three bedrock valleys are the course of the ancient Mississippi River. Other major bedrock valleys in the state include the Rock River and Troy in northern Illinois, the Carthage in western Illinois, the Kaskaskia in southwestern Illinois, and the Cache in southern Illinois. Outwash sands and gravels are found as valley train deposits between diamicton units or along major stream valleys in the bedrock surface.

Outwash sands and gravels are also spread out over large areas as gently sloping deposits called outwash plains. Outwash plains are located in the Green River Lowland of north-central Illinois and in the Havana region near the Illinois River in central Illinois (Figure 18-9). These areas contain sand and gravel deposits that are locally continuous from land surface to bedrock and form some of the most prolific aquifers in Illinois. The potential yield may be as much as 400,000 gpd/mi^2 (585 m^3/day/km^2) (Wehrmann et al. 2003) or more than 500 gpm (2,725 m^3/day) (Lim et al. 2002) from individual wells.

Glacial meltwater also carried fine-grained silts and clays, which were deposited farther downstream away from the ice margin than the heavier sands and gravels or were deposited in the relatively quiet water of lakes or slowly moving streams. These fine-grained lake, or lacustrine, deposits and glacial diamicton can form confining layers for the sand and gravel aquifers or can occur as localized deposits within the aquifers themselves.

Aquifers in Buried Bedrock Valleys

Major sand and gravel aquifers in Illinois are found in buried bedrock valleys (Figure 18-8). These aquifers generally are located in the northern two-thirds of the state. The principal buried bedrock valley aquifers in the state are the Mahomet and Sankoty aquifers and aquifers within the Troy, Rock, Paw Paw, and Princeton Bedrock Valleys.

The Mahomet aquifer (Figure 18-8) trends east-west across east-central to central Illinois from Indiana to the Illinois River, underlying parts of 15 counties. Although the Mahomet aquifer is principally found in the lower half of the Mahomet Bedrock Valley, it also occupies the lower parts of its tributaries, the Pesotum, Danville, Chatsworth, Kempton, and Onarga Bedrock Valleys (not shown) (Kempton et al. 1991). The Mahomet aquifer is under confined conditions for most of its extent, but is unconfined and forms the land surface in the Havana region in central Illinois. The width of the Mahomet aquifer ranges between 3 and 6 miles (5 and 10 km), and the elevation of its top is approximately 500 feet (150 m) mean sea level (msl). Aquifer thickness averages 100 feet (30 m) but may be as much as 200 feet (60 m) thick locally. This aquifer is estimated to produce approximately 84 million gpd (318,000 m^3/day)

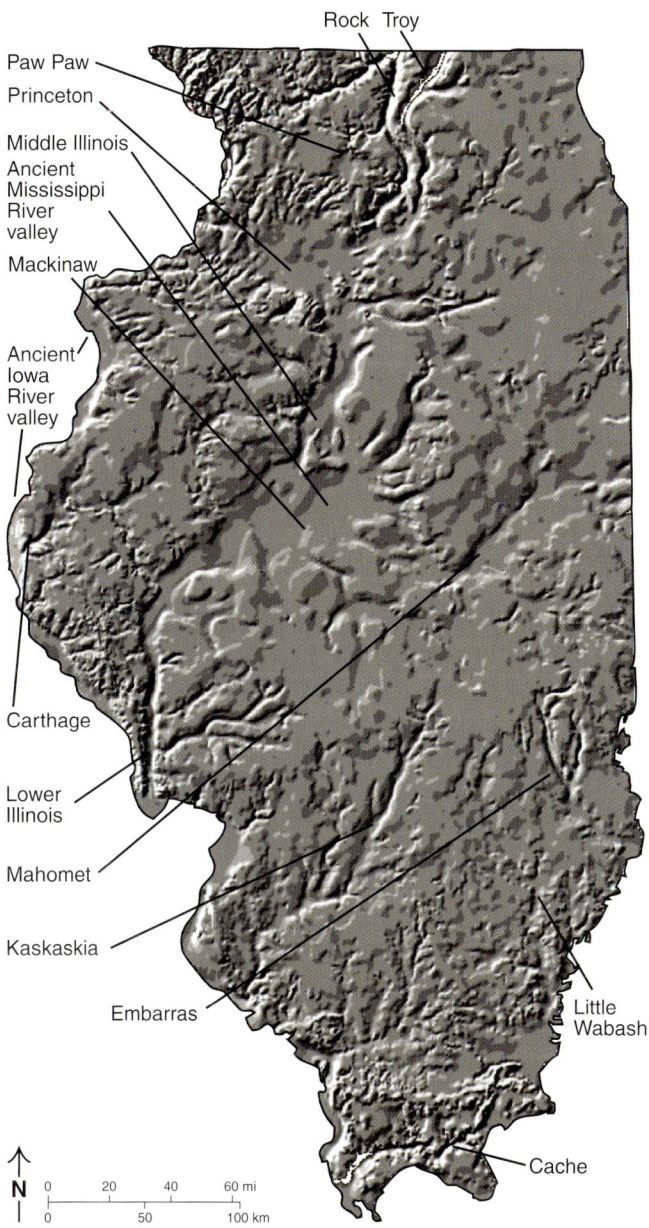

Figure 18-8 Major bedrock valleys of Illinois (Abert 1996, 2005).

and is the water source for about 800,000 people (Panno and Korab 2000). This yield translates into a potential yield of up to 200,000 gpd/mi^2 (292 m^3/day/km^2) (Wehrmann et al. 2003) or more than 500 gpm (2,000 L/min) (Lim et al. 2002) from individual wells.

The Mackinaw and Middle Illinois Bedrock Valleys of central Illinois are the course of the ancient Mississippi River. The Sankoty aquifer occupies about the lower half of these bedrock valleys (Horberg et al. 1950). Aquifer width varies from less than 2 miles (3 km) to about 10 miles (16 km); upper elevation is 500 feet (150 m) msl or lower. Aquifer thickness ranges from less than 50 feet (15 m) to about 300 feet (90 m) but averages about 100 feet (30 m). The potential yield may be as much as 400,000 gpd/mi² (585 m³/day/km²) (Wehrmann et al. 2003) or more than 500 gpm (2,725 m³/day) (Lim et al. 2002) from individual wells. The aquifer is under confined conditions where it is buried by diamicton deposited during the Wisconsin Episode. In the Illinois River valley, the diamicton is not present, so the aquifer is unconfined there.

The Troy (Vaiden et al. 2004) and Rock Bedrock Valleys (Berg et al. 1984), located in north-central Illinois, extend southward about 50 miles (80 km) from the Wisconsin state line to their confluence with the Paw Paw Bedrock Valley. These bedrock valleys are filled with a complex succession of diamicton, outwash sand and gravel deposits, and lacustrine silts and clays. The sand and gravel deposits are aquifers occupying about the lower half of each of these valleys. The aquifers vary in width from about 1 to 4 miles (2 to 6 km) to about 8 miles (13 km) in the southern reach of the Paw Paw Valley. Aquifer thickness typically ranges from less than 50 feet (15 m) to about 100 feet (30 m) but locally can exceed 100 feet (30 m). The potential yield may be as much as 400,000 gpd/mi² (585 m³/day/km²) (Wehrmann et al. 2003) or more than 500 gpm (2,725 m³/day) (Lim et al. 2002) from individual wells. It is most likely that confined conditions prevail for these aquifers, but local unconfined conditions may occur where the aquifers are hydraulically connected to overlying shallower aquifers, some of which extend to land surface.

The Princeton Bedrock Valley in northwestern Illinois, once the course of the ancient Mississippi River, extends eastward from the Mississippi River valley to where it joins the Paw Paw and Middle Illinois Bedrock Valleys (Larson et al. 1995). The Princeton Bedrock Valley aquifer, which occupies the lower half of the bedrock valley, varies in width from less than 3 miles (5 km) to about 12 miles (19 km) at the confluence of the Princeton and Paw Paw Bedrock Valleys. Aquifer thickness is typically about 100 feet (30 m) but ranges from less than 50 feet (15 m) near the margins of the bedrock valley to about 200 feet (60 m) along the valley's thalweg (its deepest part). The potential yield may be as much as 200,000 gpd/mi² (292 m³/day/km²) (Wehrmann et al. 2003) or more than 500 gpm (2,725 m³/day) (Lim et al. 2002) from individual wells. The aquifer is mostly confined. It is locally unconfined, however, where the absence of the aquitard allows a hydraulic connection with the overlying Tampico aquifer. The Tampico aquifer consists of the saturated surficial sand and gravel deposits within the Green River Lowland, a topographic feature that indicates the presence of the Princeton Bedrock Valley. The width of the Tampico aquifer varies from less than 2 miles (3 km) to about 12 miles (19 km). Aquifer thickness is typically about 50 feet (15 m) but ranges from less than 10 feet (3 m) to about 100 feet (30 m). Because the aquifer is unconfined, thickness varies with the rise and fall of the water table. The potential yield may be as much as 200,000 gpd/mi² (292 m³/day/km²) (Wehrmann et al. 2003) or more than 500 gpm (2,725 m³/day) (Lim et al. 2002) from individual wells.

The lower Illinois Bedrock Valley extends from central Illinois south-southwest to its confluence with the Mississippi River valley (Horberg 1950). The upper part of this bedrock valley originates at the confluence of the Mahomet and Mackinaw Bedrock Valleys. Although the Sankoty

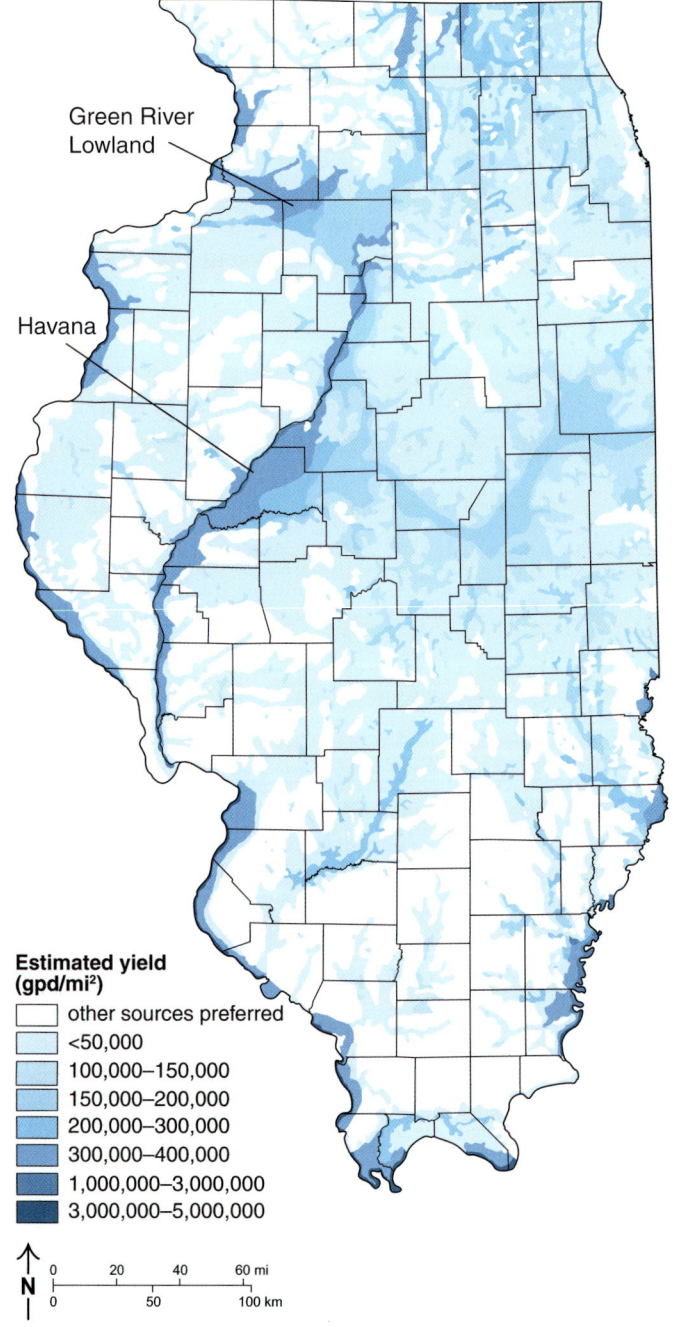

Figure 18-9 Yields of sand and gravel aquifers in Illinois (Wehrmann et al. 2003).

aquifer is found over most of the lowland, the Sankoty and Mahomet aquifers commingle along the lowland's eastern border (Walker et al. 1965). Aquifer width decreases from about 10 miles (16 km) in the northeast to about 6 miles (10 km) in the southwest. Aquifer thickness varies from less than 50 feet (15 m) to more than 150 feet (46 m). The aquifer is unconfined except where overlying moraines allow for confined conditions along the southern margin. The lower Illinois Bedrock Valley narrows along its course from about 6 miles (10 km) to about 2 miles (3 km) at its confluence with the Mississippi River valley. Depth of the bedrock valley increases from 250 feet (75 m) to more than 500 feet (150 m). The width of the aquifer in this bedrock valley varies with the width of the valley. Aquifer thickness ranges from less than 20 feet (6 m) to about 150 feet (46 m). The potential yield of this aquifer has been estimated at more than 1 million gpd/mi^2 (1,460 m^3/day/km^2) along the Illinois River and up to 400,000 gpd/mi^2 (585 m^3/day/km^2) away from the river (Wehrmann et al. 2003). Individual wells may produce more than 500 gpm (2,725 m^3/day) (Lim et al. 2002).

Other Significant Sand and Gravel Aquifers

Major sand and gravel aquifers are found within long reaches of the Mississippi River valley. The width of these aquifers varies from a mile or less where the valley is narrowest to about 10 miles (16 km) where the valley is widest. Sand and gravel occurs in the Mississippi River valley as thick outwash valley train deposits overlain by alluvial valley fill. Aquifer thickness averages about 150 feet (46 m) but ranges from about 125 to 340 feet (38 to 104 m). The potential yield of this aquifer has been estimated at more than 3 million gpd/mi^2 (4,385 m^3/day/km^2) along the Mississippi River and up to 400,000 gpd/mi^2 (585 m^3/day/km^2) away from the river (Wehrmann et al. 2003). Individual wells may produce more than 500 gpm (2,725 m^3/day) (Lim et al. 2002).

The Kaskaskia Bedrock Valley is filled mostly with fluvial and lacustrine deposits of sand, silt, and clay with minor amounts of gravel. Maximum thickness is more than 100 feet (30 m). The potential yield of this aquifer has been estimated at up to 300,000 gpd/mi^2 (440 m^3 L/day/km^2) (Wehrmann et al. 2003); individual wells may produce more than 100 gpm (545 m^3/day) (Lim et al. 2002).

The Ohio River occupies a bedrock valley that is about 2 miles (3 km) wide on average. Thickness of the sand and gravel deposits within the valley varies from 80 feet (24 m) or less to as much as 300 feet (91 m) locally. The potential yield of this aquifer has been estimated at more than 1 million gpd/mi^2 (1,460 m^3/day/km^2) along the Ohio River and up to 400,000 gpd/mi^2 (585 m^3/day/km^2) away from the river (Wehrmann et al. 2003). Individual wells may produce more than 500 gpm (2,725 m^3/day) (Lim et al. 2002).

The Cache Bedrock Valley in southern Illinois is an abandoned course of the Ohio River. The valley is filled mostly with sand and gravel outwash, the maximum thickness of which is 140 to 180 feet (43 to 55 m). The potential yield of this aquifer has been estimated locally at up to 300,000 gpd/mi^2 (440 m^3/day/km^2) away from the river (Wehrmann et al. 2003). Individual wells may produce more than 20 gpm (110 m^3/day) (Lim et al. 2002).

The Wabash River occupies a bedrock valley that averages about 5 miles (8 km) in width. It is filled with 100 to 150 feet (30 to 46 m) of outwash and alluvial deposits consisting of sand, gravel, silt, and clay in order of abundance. The potential yield of this aquifer has been estimated at more than 1 million gpd/mi^2 (1,460 m^3/day/km^2) along the Wabash River and up to 400,000 gpd/mi^2 (585 m^3/day/km^2) away from the river (Wehrmann et al. 2003). Individual wells may locally produce more than 500 gpm (2,725 m^3/day)(Lim et al. 2002). Two major tributary valleys, the Little Wabash and Embarras Bedrock Valley, are similar, but the thickness of the sediments filling the valley is not as great. Their aquifer potentially yields up to 300,000 gpd/mi^2 (440 m^3/day/km^2) (Wehrmann et al. 2003), and individual wells may produce more than 100 gpm (545 m^3/day) (Lim et al. 2002).

Linear Aquifers (Ridged Drift)

A system of elongated ridges and knolls extends from the Bloomington Morainic System (Figure 12-18) in central Illinois into southwestern Illinois (Heigold et al. 1985). The system is about 1 mile (2 km) wide and consists of as much as 113 feet (34 m) of stratified sand and gravel with some silt. One of the most prominent of these ridges parallels Illinois State Route 48 and is centered near Taylorville. Where saturated, these deposits form locally significant aquifers that are used for domestic, municipal, and industrial supplies. Potential yields have been estimated at up to 300,000 gpd/mi^2 (440 m^3/day/km^2) (Wehrmann et al. 2003), and individual wells may produce more than 100 gpm (545 m^3/day) (Lim et al. 2002).

Shallow Sand and Gravel Aquifers in Northeastern Illinois

Outwash sand and gravel occurs in a complex glacial terrane as extensive outwash plains and valley trains. Some of these deposits are buried within diamicton deposits; others locally extend from land surface to bedrock. The thickness of the sand and gravel ranges from less than 50 feet (15 m) to about 200 feet (60 m), although locally the thickness can be 250 feet (75 m). Depths to the sand and gravel deposits vary from less than 50 feet to 200 feet (15 to 61 m). Because of their great variability, estimated potential yields vary from 100,000 gpd/mi^2 (146 m^3/day/km^2) to as much as 400,000 gpd/mi^2 (585 m^3/day/km^2) in a few locations (Wehrmann et al. 2003). Individual wells can be expected to produce anywhere from about 100 gpm (545 m^3/day) to more than 500 gpm (2,725 m^3/day) (Lim et al. 2002).

Conclusion

Because groundwater is the source of drinking water for one-third of all Illinoians, including 90% of the state's rural population, and because more than one billion gallons (3.8 million m^3) of groundwater are used each day in the state, Illinois citizens clearly need an understanding of groundwater. This resource is vital to Illinoians and to the state's economic and environmental well-being. Illinois groundwater resources, although generous, are finite and are not distributed uniformly across the state. One of the reasons for this uneven distribution of groundwater resources is the variable nature of the areal extent, thickness, and other characteristics of the geological materials that constitute the aquifers. Hydrogeological mapping provides a means for illustrating and understanding the variability inherent in these geological materials and for integrating this knowledge into water resource planning. Because more land-use decisions are made each year that affect water resources, and because the demand for groundwater continues to increase, managing the state's groundwater for long-term sustainability and protecting it from various polluting activities are essential.

References

Abert, C. C., 1996, Shaded relief map of Illinois: Illinois State Geological Survey, Illinois Map 6, 1:500,000.

Abert, C. C., 2005, Shaded relief map of the bedrock surface of Illinois: Illinois State Geological Survey, GIS database layer, 1:500,000.

Berg, R. C., J. P. Kempton, and A. N. Stecyk, 1984, Geology for planning in Boone and Winnebago Counties: Illinois State Geological Survey, Circular 531, 69 p.

Heath, R. C., 1989, Basic ground-water hydrology: Reston, Virginia, U.S. Geological Survey, Water-Supply Paper 2220, 84 p.

Heigold, P. C., V. L. Poole, K. Cartwright, and R. H. Gilkeson, 1985, An electrical earth resistivity survey of the Macon-Taylorville ridged-drift aquifer: Illinois State Geological Survey, Circular 533, 23 p.

Horberg, L., 1950, Bedrock topography of Illinois: Illinois State Geological Survey, Bulletin 73, 111 p.

Horberg, L., M. Suter, and T. E. Larson, 1950, Groundwater in the Peoria region: Illinois State Geological Survey, Bulletin 75, 128 p.

Kempton, J. P., W. H. Johnson, P. C. Heigold, and K. Cartwright, 1991, Mahomet Bedrock Valley in east-central Illinois; topography, glacial drift stratigraphy, and hydrogeology, in W. N. Melhorn and J. P. Kempton, eds., Geology and hydrogeology of the Teays-Mahomet Bedrock Valley system: Boulder, Colorado, Geological Society of America, Special Paper 258, p. 91–124.

Killey, M. M., and D. R. Larson, 2004, Illinois Groundwater: A vital geologic resource: Illinois State Geological Survey, Geoscience Education Series 17, 61 p.

Larson, D. R., B. L. Herzog, R. C. Vaiden, C. A. Chenoweth, Y. Xu, and R. C. Anderson, 1995, Hydrogeology of the Green River Lowland and associated bedrock valleys in northwestern Illinois: Illinois State Geological Survey, Environmental Geology 149, 20 p.

Lim, J., B. Hacker, L. Sze, D. Splitt, S. R. Gustison, H. A. Wehrmann, H. V. Knapp, S. Sinclair, and R. J. Finley, 2002, Energy facilities screening in Illinois: Mine-mouth power plants: Springfield, Illinois, Illinois Department of Natural Resources, S-2198, Task 3, 28 p.

Miller, J. M., 1994, Major bedrock aquifers at depths greater than 500 feet of ground surface: Illinois State Geological Survey, 1:500,000.

Panno, S. V., and H. Korab, 2000, The Mahomet aquifer: The Illinois Steward, v. 9, no. 1, p. 19–21.

Rovey, C. W. II, and D. S. Cherkauer, 1994a, Relation between hydraulic conductivity and texture in a carbonate aquifer—Observations: Ground Water, v. 32, no. 1, p. 53–62.

Rovey, C. W. II, and D. S. Cherkauer, 1994b, Relation between hydraulic conductivity and texture in a carbonate aquifer—Regional continuity: Ground Water, v. 32, no. 2, p. 227–238.

Smith, W. H., and J. B. Stall, 1975, Coal and water resources for coal conversion in Illinois: Illinois State Geological Survey and Illinois State Water Survey, Cooperative Resources Report 4, 79 p.

Vaiden, R. C., E. C. Smith, and T. H. Larson, 2004, Groundwater geology of DeKalb County, Illinois, with emphasis on the Troy Bedrock Valley: Illinois State Geological Survey, Circular 563, 39 p.

Visocky, A. P., M. G. Sherrill, and K. Cartwright, 1985, Geology, hydrology, and water quality of the Cambrian and Ordovician Systems in northern Illinois: Illinois State Geological Survey and Illinois State Water Survey, Cooperative Groundwater Report 10, 136 p.

Voelker, D. C., 1989, Quality of water from public-supply wells in principal aquifers of Illinois, 1984–87: Reston, Virginia, U.S. Geological Survey, Water Resources Investigations Report 88-4111, 29 p.

Walker, W. H., R. E. Bergstrom, and W. C. Walton, 1965, Preliminary report on the ground-water resources of the Havana region in west-central Illinois: Illinois State Geological Survey and Illinois State Water Survey, Cooperative Groundwater Report 3, 61 p.

Wehrmann, H. A., S. V. Sinclair, and T. P. Bryant., 2003, An analysis of groundwater use to aquifer potential yield in Illinois: Champaign, Illinois, Illinois State Water Survey, Contract Report 2004-11, 30 p.

Geological Influences on Groundwater Quality   19

Samuel V. Panno and Keith C. Hackley

Introduction

The quality of groundwater is determined by the type and concentration of its chemical and biological constituents. These constituents, depending on their nature and concentrations, can produce water that is unsafe to drink or that has an objectionable taste or odor. Because naturally occurring substances are dissolved in or suspended as fine-grained solids in groundwater, in many cases, some water treatment is necessary before groundwater quality becomes acceptable for drinking purposes or industrial uses. This chapter discusses these naturally occurring substances, their common sources, and the general quality of the groundwater observed within the different aquifers in Illinois.

As municipal and industrial water supplies grew in importance during the early twentieth century, scientists began to study the chemical composition of groundwater and its relationship to the geological framework and associations with the host sediment or bedrock aquifers. The chemical interactions between aquifers and groundwater were identified during the first half of the century and rigorously examined during the latter half (Back and Freeze 1983). Water quality of municipal and private water supplies in Illinois has been routinely analyzed by the Illinois State Water Survey since its founding in 1895 as a means of tracing waterborne diseases (Illinois State Water Survey 2003). Today, studies of the chemical composition of groundwater are routinely performed on both public and private water supplies. The composition of groundwater is used to determine its potability and serves as a tool to determine the sources of naturally occurring and human-related contaminants.

Groundwater is derived from precipitation, such as rainfall or snow melt. Precipitation contains very low concentrations of dissolved minerals (approaching that of distilled water). The major ions—or electrically charged particles—present in rainwater and snowmelt are calcium, magnesium, sodium, silica (as $Si(OH)_4$), sulfate, chloride, and nitrogen compounds (nitrate, ammonium, and organic nitrogen). The total concentration of all constituents typically is less than 5 to 10 mg/L (parts per million) (Table 19-1). The natural sources of these ions include airborne dust (calcium, magnesium, and silica) and aerosols from seawater (sodium and chloride), which are highest near the coastal areas (Drever 1997). The majority of the remaining ions are typically from airborne particulate matter and gases from industrial emissions, automobile exhaust, and the evaporation of volatile chemicals (e.g., anhydrous ammonia fertilizers). These latter constituents are discussed in Chapter 28, Karst Terrane.

Once rainfall and snowmelt seep into the ground, the water begins its journey to the water table through the connected pores and crevices in soil, unconsolidated sediment, and rocks. The water table is the elevation at which the geological sediments and rocks become saturated with water. Permeable sediments and bedrock below the water table constitute aquifers. Aquifers supply private citizens and municipalities with potable water and are also used as a source of water for irrigation and industrial purposes throughout much of the state.

Water is a solvent and, as such, is capable of dissolving and interacting with organic and inorganic components of soils, the minerals that make up unconsolidated deposits (e.g., sand and gravel), and with various types of bedrock. Dissolution of minerals within the soil, sediment, and bedrock is a slow process that can take days, years, or eons, depending on the solubility of the materials. These materials contribute to the amount of total dissolved solids (TDS) that are present in all groundwater. Major dissolved constituents of groundwater include the cations (positively charged ions) sodium, potassium, calcium, magnesium, and silica and the anions (negatively charged ions) bicarbonate, sulfate, chloride, and nitrate (Table 19-1). When trace-level ions, such as ammonium, iron, manganese, and arsenic, are present in relatively high concentrations, they create water-quality problems. These constituents come from the dissolution of rocks such as limestone and dolomite, the weathering of less-soluble rocks such as granite, the degradation of organic debris in the soil zone (e.g., trees and leaves), and man-made contaminants (discussed in Chapter 29, Pollution of Natural Waters).

In general, the chemistry of groundwater is controlled by the type of geological materials through which the water flows and the amount of time the water is in contact with these materials. For example, precipitation percolating into a limestone aquifer would dissolve the limestone (made of calcium carbonate), and the water (now groundwater) would contain predominantly calcium and bicarbonate ions. The dominant cations and anions are used to describe the general chemical composition, or "type," of groundwater. For example, limestone aquifers contain "calcium bicarbonate type" groundwater, and deep, granitic rock eventually produces "sodium chloride type" groundwater.

AQUIFERS AND AQUITARDS: EFFECTS ON GROUNDWATER QUALITY

Aquifers in Illinois are made up of two types of material: (1) sand and gravel deposits found in association with glacial till or along stream valleys, and (2) porous and permeable bedrock units. Sand and gravel aquifers are scattered throughout the state, but the largest deposits occur in northeastern, central, and northwestern Illinois. Bedrock aquifers containing usable groundwater are located throughout the state, but potable water is found only in the bedrock aquifers of the northern third and, to a lesser extent, the western and southern parts of the state. For example, fresh groundwater with less than 1,000 mg of TDS/L can be pumped from wells over 1,000 feet (305 m) deep in the northern part of the state, but saline (salty) groundwater is encountered several hundred feet or less below the surface in southern Illinois (Lloyd and Lyke 1995). Groundwater from bedrock found in the southern two-thirds of the state is generally too saline for human use (Figure 19-1). Salinity of groundwater generally increases with depth. Bedrock aquifers in the northern portion include shallow carbonate rocks (dolomite and limestone) as well as deep sandstones that are primarily made up of quartz and silicate minerals. The bedrock aquifers in the western and southern parts of the state are primarily shallow carbonate rocks.

Sand and gravel deposits typically contain abundant clasts (i.e., sand grains, pebbles, and cobbles) of limestone and dolomite that dissolve much faster than do clasts of other composition (e.g., quartz). Consequently, calcite and dolomite, once dissolved, dominate the chemical composition of the groundwater in these aquifers, forming a calcium-magnesium bicarbonate type of groundwater (Table 19-1). Similarly, many bedrock aquifers are made of fractured limestone and dolomite, resulting in groundwater with the same general chemical composition as that found in sand and gravel aquifers. Groundwater circulating through fractured sandstone aquifers, such as the Aux Vases Sandstone in southwestern Illinois, typically forms a sodium

Table 19-1 Chemical composition[1] of averaged rainwater samples and single samples of different types of groundwater in Illinois. All parameters are reported in milligrams per liter except pH.

		Groundwater type								
	Rain[2]	Ca-Mg bicarbonate[3]	Na bicarbonate[4]	High NaCl[5]	High F[6]	High As[7]	High Mn, Fe[8]	High ammonium[9]	High sulfate[10]	Seawater[11]
pH	4.5	7.2	8.6	6.6	6.8	7.4	7.49	7.2	7.1	8.0
SC	21.1*	688	2,100	—	694	515	750	670	2,000	58,500*
Na	0.08	27.4	559	44,923	27.4	17.4	47.6	15.8	104	10,556
K	0.03	<1	<2	—	<2	1.74	1.82	3	4	380
Ca	0.24	90.1	6.0	5,740	109	67.9	82.3	82.3	279	400
Mg	0.04	33.9	3.1	2,173	20.5	32.9	41.8	36.6	96.7	1,272
SiO_2	—	19.1	8.6	15	14.8	18.1	21.8	22.7	20.1	6.43
HCO_3	—	443	1,108	61	385	479	560	480	301	140
SO_4	2.28	33.1	165	329	25.3	<0.09	16.3	0.34	981	2,649
Cl	0.15	7.4	37	85,569	25.1	1.1	41.8	0.55	8.5	18,908
F	—	0.37	—	—	1.46	0.2	0.1	—	0.28	1
NO_3	1.46	<1.9	0.26	0.0	0.09	<0.02	<0.2	0.35	<1.7	<0.5
NH_4	0.40	0.98	—	64	—	2.8	3.7	6.02	—	<0.5
Fe	—	1.89	<0.01	20	1.65	4.57	3.67	4.32	0.27	0.01
Mn	—	<0.03	<0.01	2.0	0.76	0.04	0.21	0.03	1.2	0.002
As	—	—	—	—	—	0.096	—	—	—	0.003
TDS	4.68*	661*	1,887*	142,194	611*	626*	821*	652*	1,796	34,477

[1]Abbreviations: SC, specific conductance; Na, sodium; K, potassium; Ca, calcium; Mg, magnesium; SiO_2, silicon dioxide, or silica; HCO_3, bicarbonate; SO_4, sulfate; Cl, chlorine; F, fluorine; NO_3, nitrate; NH_4, ammonia; Fe, iron; Mn, manganese; As, arsenic; TDS, total dissolved solids. Values followed by asterisk were calculated. —, data are unavailable.
[2]Average values from six active collection sites throughout Illinois from 1980 to 2003 (http://www.nadp.isws.illinois.edu).
[3]Hackley (2002); sample WRD-55, Mahomet aquifer, Champaign County.
[4]Panno (unpublished data 1995); Aux Vases Sandstone, Monroe County.
[5]Meents et al. (1952); sample 930, Ste. Genevieve Limestone, Wayne County.
[6]Panno and Hackley (2001a); Valmeyeran Series limestones, Rosiclare, Hardin County.
[7]Holm (1995); sample 93-24A, Mackinaw Valley aquifer, Tazewell County.
[8]Holm (1995); sample 93-14, Mackinaw Valley aquifer, Tazewell County.
[9]Hackley (unpublished data); Mahomet aquifer, McLean County.
[10]Hackley (2002); sample VER-94b, Mahomet aquifer, Vermilion County.
[11]Mason (1966).

GROUNDWATER QUALITY

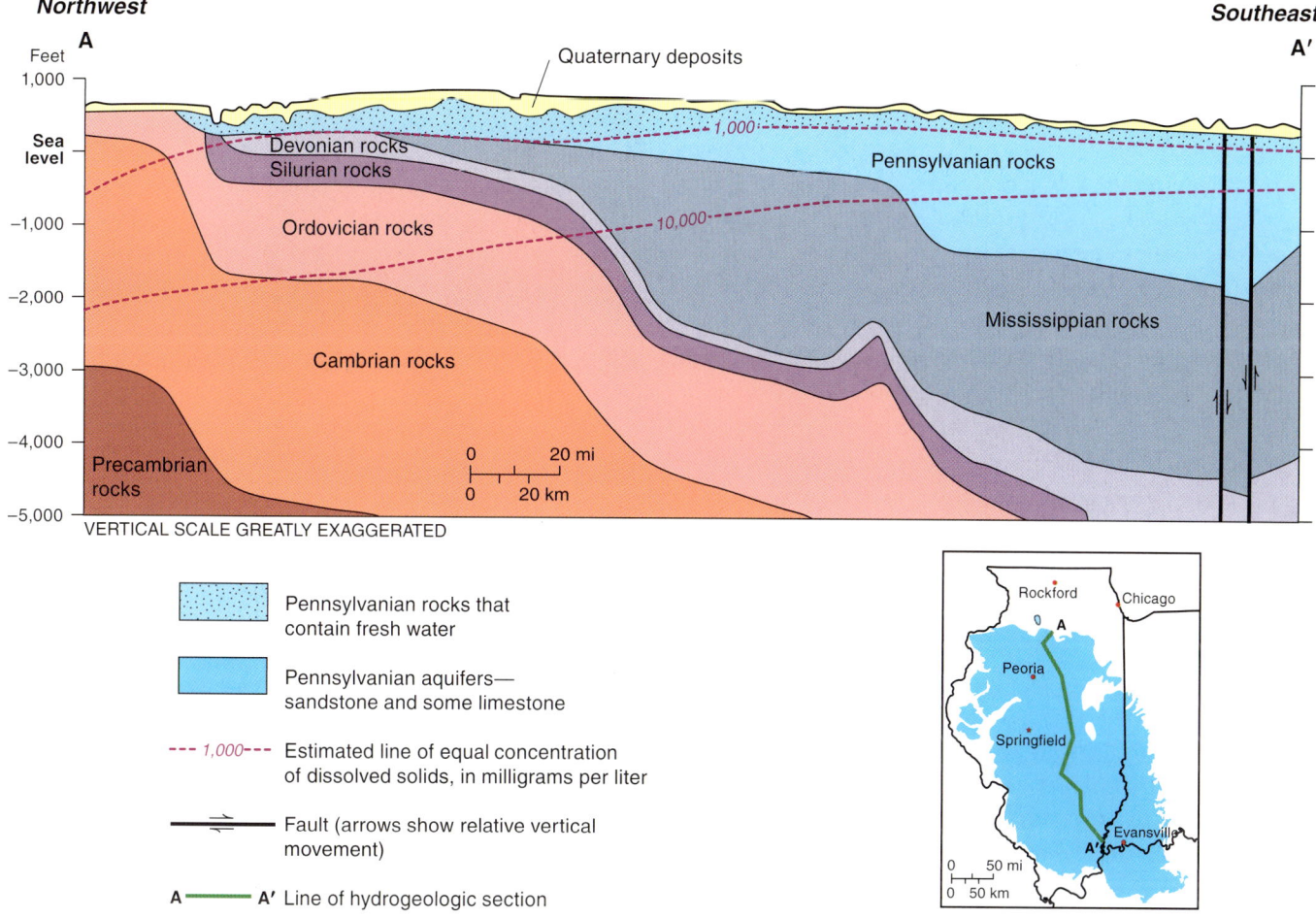

Figure 19-1 North-south cross section of Illinois showing estimated lines of equal concentration of total dissolved solids in groundwater (Lloyd and Lyke 1995).

bicarbonate type groundwater (Table 19-1). One way such groundwater is formed is for calcium bicarbonate groundwater to enter the sandstone, and, through ion exchange, calcium ions replace sodium ions on clay minerals (sodium is generally present due to degradation of feldspars in the sandstone). This process is similar to that occurring when a water softener causes calcium and magnesium to exchange positions with sodium on a resin within the softener; the result in both cases is a "soft" sodium bicarbonate water.

Aquitards are water-bearing deposits or rocks with very low permeability that do not yield water readily (e.g., clay-rich glacial till and shale) (Chapter 18, Aquifers). These deposits or rock units protect groundwater from surface-borne contaminants, isolate groundwater from the atmosphere, and confine groundwater to a single aquifer or channel it to a series of connected aquifers. The isolation of groundwater for hundreds to thousands to tens of thousands of years can affect its chemistry by fostering reducing conditions, as is discussed later in this chapter. Additionally, the interaction of groundwater with the minerals of an aquitard can alter the chemical composition of the groundwater flowing through it and the chemical composition of adjacent aquifers. For example, dilute groundwater interacting with the shale that bounds an aquifer may encounter the mineral pyrite (iron sulfide), which can be easily oxidized to form groundwater that has very high concentrations of sulfate (perhaps hundreds to thousands of milligrams per liter) (Table 19-1). Such sulfate-enriched groundwater, as it enters an underlying aquifer, can enrich the aquifer with sulfate even though there may have been no mineralogical source of sulfate in the aquifer.

CHEMICAL COMPOSITION

There are a number of constituents that can be found in groundwater, and water quality depends on the type and concentration of these constituents. For example, groundwater in some locations is immediately potable; elsewhere it may have an objectionable taste and/or odor and may even be unsafe to drink. The U.S. Environmental Protection Agency (U.S. EPA) has developed a list of acceptable concentrations of natural and man-made contaminants for municipal drinking water. These standards, the National Primary Drinking Water Regulations (NPDWR), are enforced by the U.S. EPA (1991, 2009). National Secondary Drinking Water Regulations (NSDWR) are U.S. EPA guidelines pertaining to constituents of drinking water based on cosmetic effects, taste, odor, and appearance. Both guidelines may be found on the U.S. EPA (2009) Web site.

The chemical components that are typically measured to help determine the general quality of groundwater include suspended solids, hardness, sodium, chloride, fluoride, sulfate, chemicals associated with reduction-oxidation (e.g., arsenic, iron, manganese, nitrate, and ammonium), assorted minor and trace constituents, dissolved gases (e.g., air, carbon dioxide, methane, and hydrogen sulfide), and stable and radioactive isotopes. Some of these constituents are occasionally referred to as "contaminants," but they can be naturally occurring.

Suspended Solids

Suspended solids in groundwater are occasionally observed in wells tapping aquifers that have relatively large porosity and permeability (e.g., sand and gravel, karst limestone/dolomite). Suspended solids include soil and other fine-grained material (e.g., clay, silt, and organic debris) small enough to pass through the tiny spaces between sand and gravel particles and through crevices in carbonate rock. Suspended solids can also include compounds that may precipitate in the water as it is withdrawn from the aquifer due to dramatic changes in temperature, pressure, pH, or oxygen content. Examples of dissolved constituents that can emerge from apparently clear water are calcium carbonate, which forms small white flakes when dissolved carbon dioxide is released from the water, and iron and manganese oxides, which form orange and black granules or coatings when water is exposed to oxygen. Such materials can make the water cloudy and can affect its taste unfavorably.

An extreme example of suspended solids in groundwater comes from the karst aquifers of southwestern Illinois. Karst aquifers have large crevices and conduits that can carry relatively large concentrations of suspended solids in the water. Several wells in one subdivision in southwestern Illinois intersected crevices that were in direct contact with at least one large sinkhole. During and following large rainstorms, soil washed into the sinkhole(s) and through the crevices that led to the wells. Pumps in the wells sent the sediment-laden water, now a muddy slurry, into the plumbing systems of several houses, and mud oozed from the fixtures (Figure 19-2) (Panno et al. 1996). In most cases, however, suspended solids in groundwater are not so extreme and can be removed by commercially available filters.

Figure 19-2 Examples of muddy tap water from a residential home in southwestern Illinois' karst region. Left and center: Water containing sediment from erosion of cropland adjacent to a sinkhole during and following relatively large rainfall events (1 inch [2.5 cm] or greater). Right: Fireplace ashes in water caused by dumping fireplace ashes into the sinkhole in the home's backyard. (Photograph by Joel M. Dexter.)

Rock-Water Interaction

The chemical characteristics of groundwater are determined by the interaction between groundwater and the minerals that make up the rock or sediment through which the water is moving (Figure 19-3). That is, the materials of the soil zone and the aquifer are dissolved and incorporated into the water moving through them. Some components may provide a characteristic taste to the water and some may make the water undrinkable based on taste, toxicity, or both.

Calcium, Magnesium, Bicarbonate, and Hardness

In areas where bedrock is dominated by limestone and/or dolomite or where sand and gravel aquifers contain abundant clasts of these carbonate minerals, the groundwater is typically a calcium bicarbonate type or calcium-magnesium bicarbonate type (Table 19-1; Figure 19-4). These rocks are dominated by calcite (calcium carbonate) and dolomite (calcium-magnesium carbonate), and both are soluble in slightly acidic rainwater (Table 19-1) and snowmelt and in carbon dioxide-enriched soil water (soil water made more acidic by the generation of carbon dioxide from root respiration and microbial degradation of organic debris). Carbonate minerals dissolve relatively fast when exposed to carbon dioxide-enriched waters.

Groundwater from aquifers made up of or containing carbonate rock is often referred to as "hard." Hard water contains calcium and magnesium that react with soap and produce soap scum or react with other anions to produce a solid scale or coating when the water is boiled and evaporated. Hardness and its components are not toxic, have no

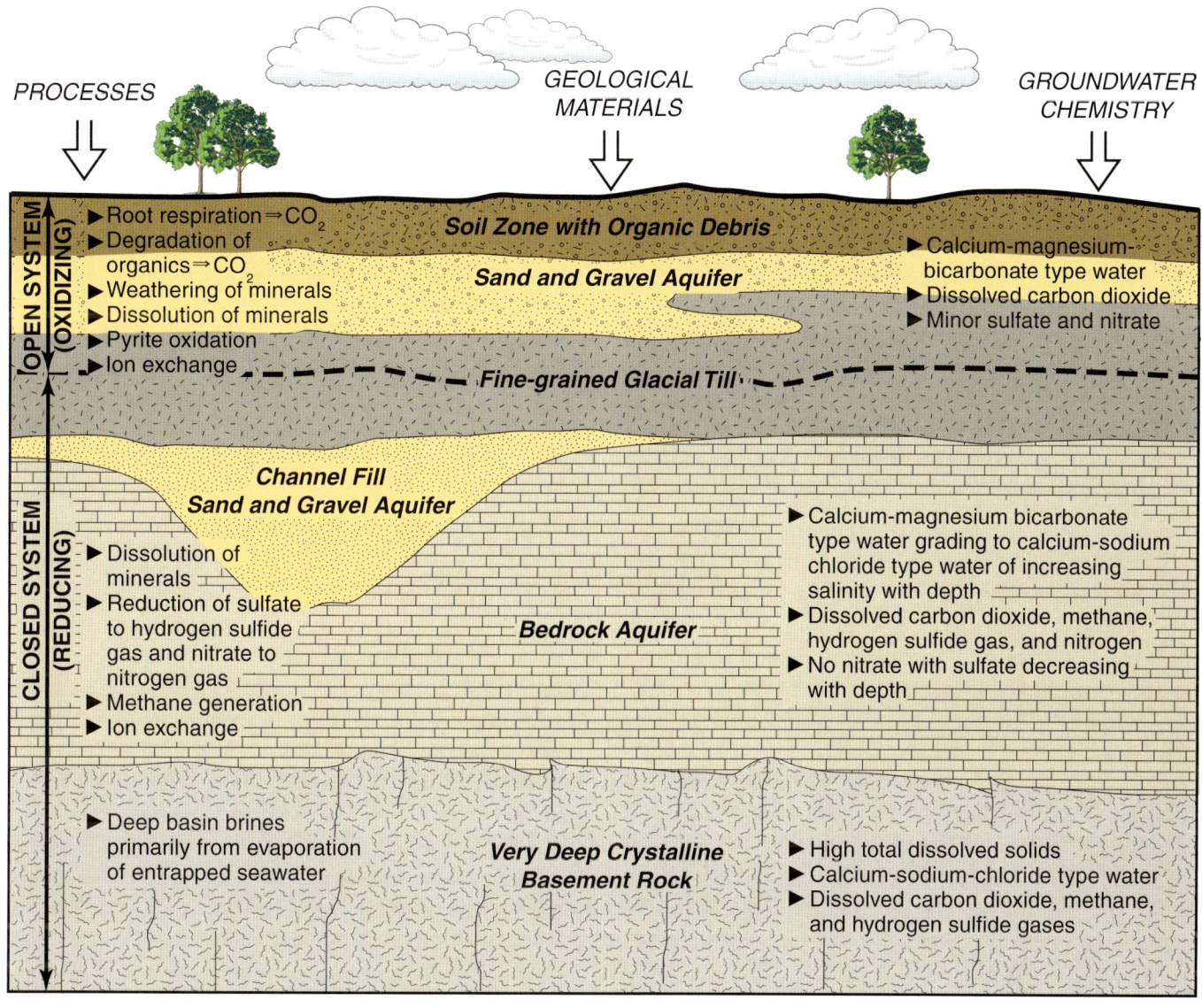

Figure 19-3 Conceptual model of the processes and chemical compositions of groundwater in Illinois. Abbreviation: CO_2, carbon dioxide.

offensive taste or odor, and are under no U.S. EPA guidelines. Chemically, hardness is defined (Freeze and Cherry 1979) as (total calcium concentration × 2.5) + (magnesium × 4.1) and is expressed as milligrams per liter, or parts per million. For example, if water contains 5 mg of calcium/L and 5 mg of magnesium/L, then the hardness is 33 mg/L. Water with a hardness value greater than 120 mg/L is considered to be hard, and greater than 180 mg/L is considered to be very hard (Lehr et al. 1980). Because the presence of these minerals controls water chemistry, groundwater from carbonate mineral-rich aquifers is typically very hard. For example, the average hardness value for the groundwater of the karst limestone aquifers of Illinois is about 230 mg/L (Panno et al. 1996, 1997). Groundwater from sand and gravel aquifers in northern Illinois has hardness values of over 300 to 400 mg/L (Panno et al. 2005a).

The scale or coating from these dissolved minerals that can be found in pots used for boiling hard water and on the bottom of hot water heaters is a deposit of carbonate minerals. In some cases, hard water may be desirable because it tends to coat the interior of metal pipes containing copper or lead or joined by lead-bearing solder, thereby protecting the metals from leaching into the water supply (Lehr et al. 1980). However, continued coating of pipes with scale may eventually (after tens of years) constrict the interior of water pipes so that the water flow slows to a trickle. Hardness can be treated easily by commercially available water softening treatment systems. Water softeners typically consist of an ion-exchange system that exchanges sodium for calcium and magnesium. What enters the water softener as a hard, calcium-magnesium bicarbonate water discharges as a soft, sodium bicarbonate water.

Figure 19-4 Groundwater discharging from a spring at Weldon Springs State Park near Clinton, Illinois. The dilute, calcium-bicarbonate groundwater that is recharged in the higher elevations to the north percolates into and flows through a sand and gravel layer in glacial till and discharges as springs in the low-lying area of the park. A pipe was inserted into this spring's discharge point and channeled to this facade; unlike its Wyoming namesake, Illinois' Old Faithful flows continually (Panno and Hackley 2001b). (Photograph by Samuel V. Panno.)

Sodium and Chloride

Water may be classified according to its salinity and/or TDS content (Table 19-2). In Illinois, most groundwater from shallow aquifers is fresh, whereas relatively deep groundwater from within the Illinois Basin is saline to briney. Water containing chloride concentrations of approximately 250 mg/L or greater may have a salty taste. This concentration is the U.S. EPA recommended limit for drinking water (U.S. EPA 1991, 2009); chloride concentrations exceeding 1,000 mg/L may be unsafe for human consumption (Lin 1977, Lehr et al. 1980, U.S. EPA 1991). Sodium concentrations are not regulated by the U.S. EPA. However, high-sodium intake may be associated with hypertension in some individuals (U.S. EPA 1991), and the sodium concentration in drinking water should be considered by those for whom dietary sodium is a concern (Haman and Bottcher 1986).

Most shallow groundwater is relatively fresh and typically has little sodium and chloride dissolved in it. Naturally occurring sodium and chloride concentrations in sand and gravel aquifers in Illinois are typically between about 1 and 15 mg/L (Panno et al. 2006a). However, sodium chloride is present in large concentrations in groundwater from deep aquifers of northern Illinois, in parts of central Illinois, and many parts of southern Illinois (Clayton et al. 1966).

Much of the bedrock in Illinois, however, is part of an underlying midcontinental sedimentary basin known as the Illinois Basin. Sedimentary basins such as this contain abundant saline groundwater that typically increases in salinity with depth (Figures 19-1 and 19-3); in the case of the Illinois Basin, salinity increases to a depth of about 1.2 miles (2 km) or so before it levels out (McIntosh et al. 2002). The source of the salinity of this groundwater has generated much interest. Walter et al. (1990) and Steuber et al. (1991, 1993) suggested that the salinity was due to the evaporation of seawater short of the precipitation of halite in Silurian and Devonian strata and a mixture of evaporated seawater and dissolution of evaporites in shallower Mississippian and Pennsylvanian strata. Recent work by Hanor and McIntosh (2006) suggests that saline groundwater of the Illinois Basin underwent extensive alteration as a result of rock-water interaction and was transformed from sodium chloride brines to calcium-sodium chloride brines with salinities up to 200,000 mg/L TDS.

Saline springs can be seen along structural features, such as along folds and faults in the Starved Rock area (the Old Salt Well) in La Salle County, in southwestern Illinois in the vicinity of oil fields near Columbia in Monroe County, and in southern Illinois near the Equality Fault Zone and the town of Equality in Saline County (Figure 19-5). Water from these springs is about as saline as seawater with about 35,000 mg/L TDS and is dominated by sodium and chloride; seawater is dominated by sodium and chloride and contains about 34,477 mg/L TDS (Table 19-1).

Saline groundwater and brines can be brought near or to the land surface by natural conditions, such as migrating up prominent fractures and/or faults in bedrock, or by anthropogenic activities, such as exploration for and exploitation of petroleum. The mixing of upward-migrating saline groundwater with fresh groundwater from shallow aquifers can make groundwater from private wells undrinkable and can present a very expensive problem for municipalities. An example of saline groundwater seeping into a freshwater aquifer can be found in east-central Illinois where the Mahomet aquifer crosses a major, roughly north-south–trending structural feature called the La Salle Anticlinorium (Figure 19-6). Saline water seeps up into the base of the aquifer from bedrock at a depth of about 300 feet (91 m) below the land surface near the Champaign County–Piatt County line. Groundwater that normally has chloride

Table 19-2 Classification of groundwater relative to total dissolved solids (TDS) contents (Hem 1986).

Water classification	TDS (mg/L)
Fresh	<1,000
Slightly saline	1,000 to 3,000
Moderately saline	3,000 to 10,000
Very saline	10,000 to 35,000
Brine	>35,000

Figure 19-5 Equality Spring in Saline County in southern Illinois is a saline spring containing over 13,000 mg/L of sodium and chloride. The spring is over 6 feet deep (2 m), is encased by railroad ties, and was used as a commercial source of salt by early settlers of Illinois. Inset photograph is of the wavy optical effects of saline groundwater on 5-inch (12.5-cm)-long electrical conductance electrode. (Photographs by Samuel V. Panno.)

concentrations less than or about 1 mg/L here exceeds 300 mg/L (Panno et al. 1994). Samples of groundwater from bedrock beneath the sand and gravel aquifer have chloride concentrations in the thousands of milligrams per liter (Panno et al. 2005a).

In addition, naturally saline springs in northern (e.g., Starved Rock State Park) and southern Illinois, usually located along fault zones or structural folds, discharge highly saline water into surface streams (Panno et al. 2005a, 2006a). Streams in areas where there are saline seeps have names such as the Saline River (Saline County) and the Salt Fork of the Vermilion River (Vermilion County). Because salt was an important commodity during early Illinois settlement by Europeans during the early 1800s, such seeps and their "salt works," initially used by Native Americans, became the nucleus of early settlements. Saline water was boiled in large iron kettles, and the residual salt was harvested and sold (Fliege 2002).

Fluoride

Fluoride typically is added to municipal water supplies because of its protective effects on teeth. However, fluoride in drinking water at concentrations greater than 3 mg/L can cause problems with tooth mottling, brown staining, and pitting in the teeth (dental fluorosis) in children with developing teeth (U.S. EPA 2009). The primary U.S. EPA drinking water standard is 4 mg/L. Fluoride concentrations greater than 4 mg/L can result in bone disease and bone and joint deformation (skeletal fluorosis) (Appelo and Postma 1994). The NSDWR for fluoride in drinking water is 2.0 mg/L (U.S. EPA 1991, 2009).

Groundwater from wells in most aquifers in Illinois has relatively low fluoride concentrations (about 0.5 mg/L) (Panno, unpublished data). Fluoride concentrations in groundwater are a function of the chemical composition of the groundwater and the presence or absence of fluorine-bearing minerals within an aquifer. Fluorine-bearing minerals include apatite (calcium phosphate), fluorite (calcium fluoride), and minerals that can exchange hydroxyl ions for fluoride ions (hornblende, mica, and clay minerals) (Hem 1986). Conrad et al. (1999) found that the greatest fluoride concentrations in Kentucky were associated with relatively deep, sodium bicarbonate-type groundwater with pH values between 7.6 and 8.0. The fluoride concentrations in Illinois aquifers are relatively low, in part, because calcium bicarbonate-type groundwater (found in limestone and sand and gravel aquifers of northern Illinois) limits the amount of the fluorite that can be dissolved (Hem 1986). Fluorite (calcium fluoride) deposits are found in southeastern Illinois within and near the town of Rosiclare, in Hardin County, site of the largest known fluorite deposits in the United States. Groundwater samples drawn from limestone for Rosiclare's municipal water supply contained 1.5 mg of fluoride/L (Panno and Hackley 2001a) (Table 19-1). Private wells that tapped into the groundwater of the Aux Vases Sandstone in southern Illinois have intersected groundwa-

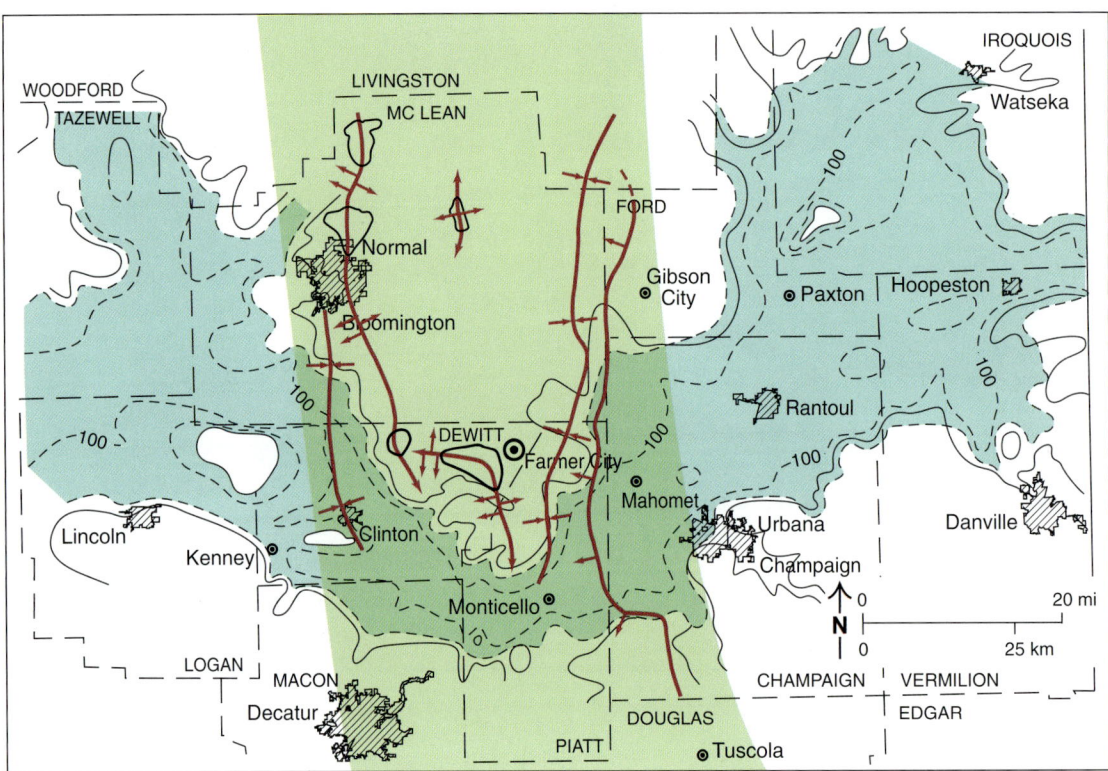

Figure 19-6 Saline groundwater can seep into a freshwater aquifer to contaminate it. Such a situation occurs in east-central Illinois where the Mahomet aquifer crosses the north-south–trending cluster of folds known as the La Salle Anticlinorium (green shading). Dashed lines are thickness contours (in feet); green shaded area is the La Salle Anticlinorium. (Modified from Panno et al. 1994.)

ter with fluoride concentrations as high as 9 mg/L (J. Bade, former sanitarian with the Monroe-Randolph Bi-County Health Department, personal communications 1997).

Sensitivity to Reduction-Oxidation

Oxygen

Oxygen, one of the most abundant elements on earth, very commonly combines with other elements. Depending on the nature of the bond, the bonded element is either reduced or oxidized. In simplest terms, reduction-oxidation (commonly shortened to redox) involves the gaining (reduction) or losing (oxidation) of electrons between elements, one of which is oxygen. The term "reduction" refers to the charge; an atom, molecule, or ion that gains an electron now has a reduced, or negative, charge. The loss of electrons through oxidation results in a positive charge. Redox reactions happen in pairs; there is no oxidation without a concurrent reduction. Redox-sensitive elements are those that either precipitate or dissolve in the presence or absence of dissolved oxygen and other electron-bearing, typically oxygen-bearing, ions (e.g., nitrate and sulfate). The abundance of dissolved oxygen (and oxygen-bearing ions) usually decreases with depth and confinement of an aquifer. Groundwater in shallow, unconfined aquifers typically has dissolved oxygen contents similar to that of surface water (about 8 mg/L). Groundwater in deeper aquifers and shallow, confined aquifers has little or no dissolved oxygen because it is readily used up by bacterial activity and other oxygen- and electron-consuming chemical reactions that occur within the aquifers (e.g., decomposition of organic matter and oxidation of pyrite); such groundwater is referred to as being under reducing conditions. Consequently, redox-sensitive elements are more readily dissolved in deeper, oxygen-depleted groundwater.

Arsenic

Arsenic is an element that is soluble in water, is redox-sensitive, and occurs naturally in minute concentrations in groundwater. The routine ingestion of water containing very low concentrations of arsenic has been linked to skin disease and cancer (Smith et al. 1992). The arsenic standard for community water supply wells has been lowered by the U.S. EPA from 50 to 10 µg/L (parts per billion) for finished drinking water (Vorman 2001). Health problems, such as arsenic-induced cancers, have resulted, in other parts of the world, from drinking water with relatively large concentrations of arsenic (e.g., Bagla and Kaiser 1996). Arsenic concentrations of shallow glacial and alluvial aquifers of Illinois range from less than 1 to as high as 107 µg/L; the greatest arsenic concentrations are in the southern part of the state and in the Mackinaw Valley aquifer (Figure 19-7) (Warner et al. 2003).

Iron and Manganese

The solubilities of iron and manganese, metals that are commonly present in groundwater, are sensitive to changes in redox conditions. The concentrations of these metals are low in groundwater from shallow aquifers because the presence of dissolved oxygen in the water causes these metals to precipitate from solution. However, in deeper aquifers where dissolved oxygen has long been consumed by biological and chemical reactions, groundwater commonly contains abundant concentrations of dissolved iron and manganese (Table 19-1).

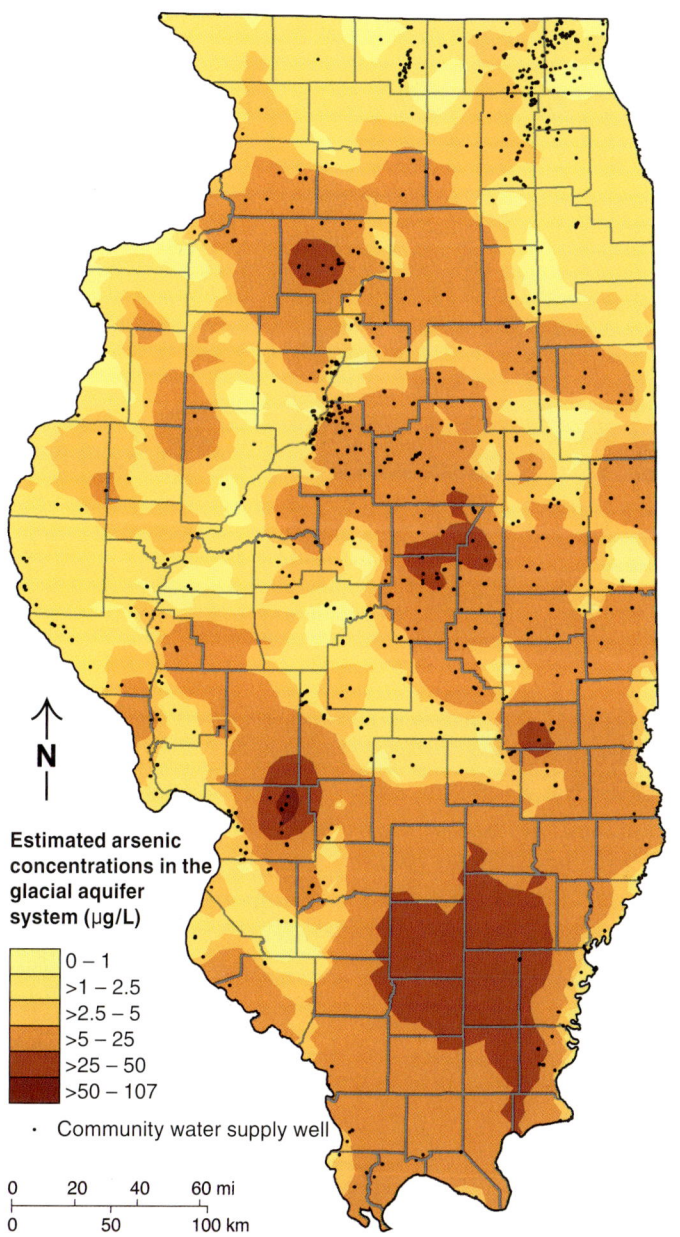

Figure 19-7 Contour map of estimated arsenic concentrations in Illinois groundwater (Warner et al. 2003).

Groundwater containing relatively high concentrations of these dissolved metals may appear to turn yellow rapidly (within minutes) in a container. Such water can form orange to black coatings on porcelain fixtures and laundry due to precipitation of manganese oxides and iron oxyhydroxides upon exposure to oxygen (Figure 19-8). The presence of these metals in water can thus be a nuisance. Iron and manganese also can precipitate on well screens and eventually plug them. Such plugging is very likely one of the causes for reduced efficiency of a municipal well in an east-central Illinois municipal well field (Panno et al. 2005b). Iron and manganese can impart a metallic taste to water. The U.S. EPA limits for iron and manganese are 0.30 and 0.05 mg/L, respectively. The U.S. EPA limit is an NSDWR, based on taste and the problems just mentioned rather than on health risk (U.S. EPA 1991). Commercially available water softeners with salt and filters can effectively remove these metals.

Nitrogen Compounds

Nitrogen compounds (nitrate, nitrite, ammonium, and organic nitrogen) are naturally occurring components of groundwater. Nitrate is the most common dissolved nitrogen compound in surface and shallow groundwater environments where the water is relatively oxidized. In general, the concentrations of these compounds are small. Naturally occurring nitrate concentrations (NO_3-N) typically are between 2.0 and 2.5 mg/L (Panno et al. 2006b). However, groundwater in some shallow aquifers in Illinois exceeds the U.S. EPA NPDWR of 10 mg of NO_3-N/L; water with concentrations at 10 mg of NO_3-N/L or more may pose health problems. The potential sources of such high-nitrate concentrations are discussed in Chapter 29, Pollution of Natural Waters. Nitrate concentrations can be naturally reduced through denitrification, a process by which denitrifying bacteria convert nitrate to nitrogen gas. As a result, nitrate in groundwater is commonly stratified or layered; the greatest concentrations occur near the top of the saturated zone and continually decrease with depth. Ammonium becomes more prevalent in the reducing conditions of deeper and confined aquifers. For example, the Mahomet aquifer of east-central Illinois has ammonium concentrations up to 6 mg/L (Table 19-1) (Hackley 2002). Roy et al. (2003) indicated that the source of ammonium in a deep sand and gravel of the Mahomet aquifer in east-central Illinois is the subsurface degradation of organic materials present in buried ancient soils (paleosols).

Sulfate

Sulfate is typically present in groundwater at concentrations in the tens of milligrams per liter, but it can be much higher. Sulfate concentrations greater than 250 mg/L is a U.S. EPA NSDWR. Drinking water with such high sulfate concentrations may have objectionable taste and odor; however, there is no evidence of ill effects from drinking such water (Centers for Disease Control, U.S. EPA 1999). In Illinois, sulfate concentrations of several hundred to 2,000 mg/L were found from the deep bedrock Cambrian and Ordovician aquifers in northeastern Illinois (Gilkeson et al. 1981) and in east-central Illinois in the unconsolidated sands and gravels of the Onarga Valley portion of the Mahomet aquifer (Hamdan 1970). Gilkeson et al. (1981) determined that the large sulfate concentrations found in the Cambrian and Ordovician aquifers were from the dissolution of marine evaporites (the calcium sulfate mineral gypsum). Sulfate in the Onarga Valley branch of the Mahomet aquifer is due to circulation of groundwater through the bedrock and the leaching of secondary sulfate minerals produced from pyrite oxidation (Table 19-1) (Panno et al. 1994, Hackley 2002).

Dissolved Gases

A variety of dissolved gases are commonly associated with groundwater (Figure 19-3): nitrogen, oxygen, carbon dioxide, methane, and hydrogen sulfide. Other gases that are often present but usually in low concentrations are the noble (inert) gases, hydrogen, chlorofluorocarbons, and radon. As water penetrates into the ground, gases typically found in the atmosphere are dissolved in the water. Thus, nitrogen is common in groundwater. Nitrogen also may be produced from microbial denitrification reactions. Due to biochemical reactions within the aquifers, oxygen is quickly consumed during oxidation of organic matter and other reduced compounds such as the sulfides found in the soils and deeper sediments. Carbon dioxide is added to the infiltrating water in the soil zone due to root respiration and oxidation of organic matter. The addition of carbon dioxide increases the carbonic acid content, lowering the water's pH, which has important impacts on the solubility of the rocks and minerals through which the groundwater flows.

In reducing environments, hydrogen sulfide and methane become more prevalent. Hydrogen sulfide results in a rotten egg odor that is sometimes present in well water. The sources of hydrogen sulfide include the bacterial reduction of naturally occurring sulfate dissolved in groundwater and decomposing organic matter within confined aquifers or in aquitards. Although the odor is objectionable, the concentrations of hydrogen sulfide gas typically found in drinking water do not pose a health problem. Hydrogen sulfide can be tasted at concentrations at or exceeding 0.05 mg/L (Lehr et al. 1980). At concentrations of 0.2 mg/L, hydrogen sulfide can be corrosive to silverware, fixtures, and plumbing (Lin 1977). Most water treatment companies can install equipment capable of removing hydrogen sulfide from domestic water supplies using oxidation, chlorination, aeration, and filtration or ion exchange with chemically active compounds (Lemley et al. 1999, Ross et al. 1999).

Figure 19-8 Falling Springs, near Dupo, in St. Clair County in southwestern Illinois, is an example of reduced groundwater. The spring discharges from a cave located 50 feet above the base of a cliff. As the groundwater is oxygenated, it precipitates manganese oxide and iron oxyhydroxide intermixed with calcite to form a dark apron on the cliff. (Photograph by Joel M. Dexter.)

Dissolved methane is quite common in Illinois groundwater (Figure 19-9). Methane occurs naturally from the microbial degradation of buried organic debris present in the glacial drift deposits throughout much of central and northeastern Illinois (Meents 1958, 1960; Coleman 1976). Many residents of Illinois have used this "drift gas" to run appliances or heat their homes (Meents 1960). Using isotope geochemistry, methane from glacial drift deposits can be distinguished from other sources, such as thermogenic methane associated with petroleum formation or generated in landfills (Coleman et al. 1977, 1988, 1995; Hackley et al. 1996). The ability to determine these differences is important when exploring possible environmental problems from methane-generating sources such as landfills.

Total Dissolved Solids

Total dissolved solids, or TDS, refers to the total mass of all dissolved constituents (as ions) present in groundwater. The U.S. EPA's NSDWR puts a suggested limit of 500 mg/L on this parameter. The TDS can easily be approximated by using a specific conductance or electrical conductance meter to measure the electrical conductivity of water (its ability to conduct electricity). Conductivity has the following relationship with TDS (Hem 1986):

$$TDS = \text{specific conductance} \times 0.59.$$

Distilled water, having almost no dissolved minerals, would have a specific conductance of near zero. Rainwater is very dilute with a low specific conductance of about 20 microsiemens per centimeter ($\mu S/cm$), and water with a large TDS load (such as seawater) has a very high value (close to 60,000 $\mu S/cm$ in the case of seawater) (Table 19-2). The specific conductance of the 500 mg/L standard would be about 850 $\mu S/cm$.

ENVIRONMENTAL ISOTOPES

Isotopes are atoms of an element that have a different number of neutrons in the nucleus. Isotopes can be stable or radioactive. Because the number of protons and the electron configuration are the same as other atoms of that element, isotopes of a particular element behave similarly during chemical reactions. However, because the atomic weights are different, the isotopes of an element react at different rates, which results in different concentrations of isotopes in the coexisting phases of a compound or in the reaction products of chemical, physical, and biological processes. This natural partitioning process is very important in the field of stable isotope geochemistry and is quite useful for determining the source of water or certain dissolved constituents and for figuring out the geochemical processes that have occurred to give the present-day chemical composition of a certain groundwater. The use of environmental isotopes can help solve groundwater quality problems by establishing the source of contaminants (natural or human-induced). The isotopes can also be used to protect groundwater by helping to identify groundwater recharge areas and groundwater flow velocities. Elements with stable isotopes that are commonly used in groundwater studies include oxygen, hydrogen, carbon, sulfur, nitrogen, and, to a lesser degree, chlorine, strontium, helium, and beryllium.

Radioactive isotopes have unstable nuclei and change into different atomic elements through spontaneous decay involving the emission of particles and energy. The rates of decay differ widely for the various radioactive isotopes, ranging from a fraction of a second to billions of years. The rate of decay is usually referred to as the half-life, the time needed for the concentration of the radioactive isotope to decrease by one-half. In hydrology, radioactive isotopes are typically used to determine the relative age of water and as tracers to help determine the source of water or of dissolved constituents in the water. Several radioactive isotopes are used in groundwater research; however, the most common are tritium (^3H), with a 12.3-year half-life, and carbon-14 (^{14}C), which has a 5,730-year half-life.

Two very good sources for information about isotope geochemistry applied to groundwater hydrology are Clark and Fritz (1997) and Kendall and McDonnell (1998). Examples of the use of isotope geochemistry applied to Illinois groundwater include the determination of the origin of high-sulfate concentrations in deep aquifers of northern Illinois (Gilkeson et al. 1981), the processes controlling the

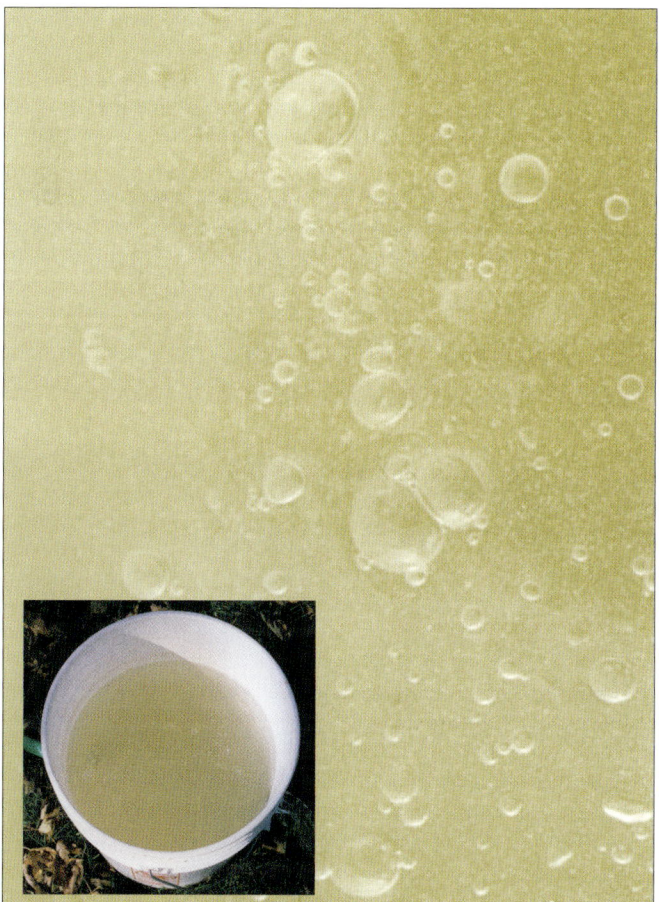

Figure 19-9 Gas (predominantly methane) bubbling out of groundwater pumped from the Mahomet aquifer in central Illinois. The largest bubbles were up to 1 inch (2.54 cm) in diameter. The presence of methane indicates reducing conditions within the aquifer. (Photographs by Samuel V. Panno.)

enrichment of radium-226 and uranium-234 in the Cambrian and Ordovician aquifers of northern Illinois (Gilkeson et al. 1983), the origin of methane in glacial drift and Paleozoic bedrock (Coleman et al. 1988), detection of the impacts of landfill leachate and gases on local groundwater chemistry (Hackley et al. 1996, 1999), determination of sources of elevated nitrate concentrations (Panno et al. 2001, Roadcap et al. 2002, Hwang et al. 2003, Hackley et al. 2007), the origin of the Illinois Basin brines and Mississippi Valley-type ore deposits (Clayton et al. 1966, Richardson et al. 1988, Stueber and Walter 1991), and the age of groundwater in a buried bedrock valley of central Illinois (Hackley 2002).

Although the concentrations of most radioactive isotopes measured in Illinois groundwater are low and do not detrimentally affect groundwater quality, dissolved radium from the deep Cambrian and Ordovician aquifers of northern Illinois is commonly present in relatively high concentrations. More than 300 wells in this area exceeded the U.S. EPA NPDWR for radium (Gilkeson et al. 1983). Radium-226 and radium-228 concentrations in groundwater ranged from 2.3 to 50.2 picocuries per liter (pCi/L); the U.S. EPA regulatory limit for the two radionuclides is 5.0 pCi/L (U.S. EPA 1991, 2009). The source of the radium in the aquifers is thought to be fine-grained materials such as the shale layers that form the bounds of the sandstone aquifers.

Conclusion

The quality of groundwater in Illinois changes with depth and location of the aquifer. Shallow surficial aquifers, such as glacially deposited sand and gravel or bedrock units near the land surface in the northern third of the state, contain relatively dilute groundwater compared with groundwater from deeper bedrock aquifers in the central and southern two-thirds of the state. Most of the aquifers in Illinois are made up of carbonate rock or contain considerable amounts of carbonate rock fragments, resulting in an almost ubiquitous problem with hard water. Many of the unlithified sand and gravel aquifers are deposited within fine-grained glacial tills and are confined aquifers that are under reducing conditions. The reducing conditions result in considerable amounts of dissolved iron and manganese and, in some cases, methane (drift gas). Most private wells in rural areas of Illinois use water softeners to extract iron and decrease the hardness of the groundwater. Groundwater pumped from wells in the central and southern parts of the state quickly become more saline as depths of wells increase and penetrate into the Pennsylvanian and deeper Paleozoic aquifers. Sodium and chloride concentrations tend to increase as depth increases.

References

Appelo, C. A. J., and D. Postma, 1994, Geochemistry, groundwater and pollution: Rotterdam, The Netherlands, A. A. Balkema, 536 p.

Back, W., and R. A. Freeze, 1983, Chemical hydrogeology, in Benchmark papers in geology 73: Stroudsburg, Hutchinson Ross Publishing Company, 416 p.

Bagla, P., and J. Kaiser, 1996, India's spreading health crisis draws global arsenic experts: Science, v. 274, no. 5285, p. 174–175.

Centers for Disease Control and Prevention, U.S. Environmental Protection Agency, 1999, Health effects from exposure to high levels of sulfate in drinking water study: Atlanta, Georgia, EPA 815-4-99-001, 25 p.

Clark, I., and P. Fritz, 1997, Environmental isotopes in hydrogeology: Boca Raton, Florida, Lewis Publishers, 328 p.

Clayton, R. N., I. Friedman, D. L. Graf, T. K. Mayeda, W. F. Meents, and N. F. Shimp, 1966, The origin of saline formation waters, 1, Isotopic composition: Journal of Geophysical Research, v. 71, no. 16, p. 3869–3882.

Coleman, D. D., 1976, Isotopic characterization of Illinois natural gas: Urbana-Champaign, University of Illinois, Department of Geology, Ph.D. dissertation, 175 p.

Coleman, D. D., C.-L. Liu, K. C. Hackley, and S. R. Pelphrey, 1995, Isotopic identification of landfill methane: Environmental Geosciences, v. 2, no. 2, p. 95–103.

Coleman, D. D., C.-L. Liu, and K. M. Riley, 1988, Microbial methane in the shallow Paleozoic sediments and glacial deposits of Illinois, U.S.A.: Chemical Geology, v. 71, p. 23–40.

Coleman, D. D., W. F. Meents, C.-L. Liu, and R. A. Keogh, 1977, Isotopic identification of leakage gas from underground storage reservoirs: A progress report: Illinois State Geological Survey, Illinois Petroleum 111, 10 p.

Conrad, P. G., D. J. Carey, J. S. Webb, J. S. Dinger, R. S. Fisher, and M. J. McCourt, 1999, Ground-water quality in Kentucky: Fluoride: Lexington, Kentucky, Kentucky Geological Survey, ser. 12, Information Circular 1, 4 p.

Drever, J. I., 1997, The geochemistry of natural waters: Surface and groundwater environments (3rd ed.): Englewood Cliffs, New Jersey, Prentice Hall, Inc., 436 p.

Fliege, S., 2002, Trails and tales of Illinois: Urbana, Illinois, University of Illinois Press, 248 p.

Freeze, R. A., and J. A. Cherry, 1979, Groundwater: Englewood Cliffs, New Jersey, Prentice-Hall, Inc., 604 p.

Gilkeson, R. H., K. Cartwright, J. B. Cowart, and R. B. Holtzman, 1983, Hydrogeologic and geochemical studies of selected natural radioisotopes and barium in groundwater in Illinois: Illinois State Geological Survey, Contract/Grant Report 1983-6, 93 p.

Gilkeson, R. H., E. C. Perry Jr., and K. Cartwright, 1981, Isotopic and geologic studies to identify the sources of sulfate in groundwater containing high barium concentrations: Illinois State Geological Survey, Contract/Grant Report 1981-4, 39 p.

Hackley, K. C., 2002, A chemical and isotopic investigation of the groundwater in the Mahomet Bedrock Valley aquifer: Age, recharge and geochemical evolution of the groundwater: Urbana-Champaign, University of Illinois, Ph.D. dissertation, 151 p.

Hackley, K. C., C.-L. Liu, and D. D. Coleman, 1996, Environmental isotope characteristics of landfill leachates and gases: Ground Water, v. 34, no. 5, p. 827–836.

Hackley, K. C., C.-L. Liu, and D. Trainor, 1999, Isotopic identification of the source of methane in subsurface sediments of an area surrounded by waste disposal facilities: Applied Geochemistry, v. 14, no. 1, p. 119–131.

Hackley, K. C., S. V. Panno, H.-H. Hwang, and W. R. Kelly, 2007, Groundwater quality of springs and wells of the sinkhole plain in southwestern Illinois: Determination of the sources of nitrate: Illinois State Geological Survey, Circular 570, 39 p.

Haman, D. Z., and D. B. Bottcher, 1986, Home water quality and safety: Gainesville, Florida, Florida Cooperative Extension Service, University of Florida, Circular 703, 15 p.

Hamdan, A. S., 1970, Ground-water hydrology of Iroquois County: Urbana-Champaign, University of Illinois, Geology Department, M.S. thesis, 72 p.

Hanor, J. S., and J. C. McIntosh, 2006, Are secular variations in seawater chemistry reflected in the compositions of basinal brines: Journal of Geochemical Exploration, v. 89, p. 153–156.

Hem, J. D., 1986, Study and interpretation of the chemical characteristics of natural water: Reston, Virginia, U.S. Geological Survey, Water-Supply Paper 2254, 263 p.

Holm, T. R., 1995, Ground-water quality in the Mahomet aquifer, McLean, Logan, and Tazewell Counties: Champaign, Illinois, Illinois State Water Survey, Contract Report 579, 42 p.

Hwang, H. H., S. V. Panno, and K. C. Hackley, 2003, Effects of urban growth on groundwater quality in Mc Henry County, Illinois based on chemical and isotopic assessment: Illinois Groundwater Consortium, Southern Illinois University, Proceedings Paper. http://orda.siuc.edu/igc/proceedings/02/contents.html. Accessed November 10, 2009.

Illinois State Water Survey, 2003, History of the Water Survey. http://www.sws.illinois.edu/chief/history.asp. Accessed November 10, 2009.

Kendall, C., and J. J. McDonnell, 1998, Isotope tracers in catchment hydrology: Amsterdam, The Netherlands, Elsevier Science B.V., 839 p.

Lehr, J. H., T. E. Gass, W. A. Pettyjohn, and J. De Marre, 1980, Significance of water-quality constituents in domestic water treatment: New York, New York, McGraw-Hill, p. 95–126.

Lemley, A., J. J. Schwartz, and L. Wagenet, 1999, Hydrogen sulfide in household drinking water; Water treatment notes: Ithaca, New York, Cornell Cooperative Extension, College of Human Ecology, Water Treatment Notes, 4 p. http://waterquality.cce.cornell.edu/publications/CCEWQ-07-HydrogenSulfide.pdf. Accessed March 31, 2010.

Lin, S. D., 1977, Tastes and odors in water supplies—A review: Champaign, Illinois, Illinois State Water Survey, Circular 127, 50 p.

Lloyd, O. B., and W. L. Lyke, 1995, Ground water atlas of the US, Segment 10, Hydrologic Investigations Atlas 730-K: Reston, Virginia, U.S. Geological Survey, 30 p.

Mason, B., 1966, Principles of geochemistry (3rd ed.): New York, John Wiley & Sons, Inc., 329 p.

McIntosh, J. C., L. M. Walter, and A. M. Martini, 2002, Pleistocene recharge to midcontinent basins: Effects on salinity structure and microbial gas generation: Geochimica et Cosmochimica Acta, v. 66, no. 10, p. 1681–1700.

Meents, W. F., 1958, Tiskilwa drift-gas area, Bureau and Putnam Counties, Illinois: Illinois State Geological Survey, Circular 253, 15 p.

Meents, W. F., 1960, Glacial drift-gas in Illinois: Illinois State Geological Survey, Circular 292, 58 p.

Meents, W. F., A. H. Bell, O. W. Rees, and W. G. Tilbury, 1952, Illinois oil-field brines: Illinois State Geological Survey, Illinois Petroleum 66, 38 p.

Panno, S. V., and K. C. Hackley, 2001a, Groundwater quality of the Rosiclare and Elizabethtown municipal wells: Illinois State Geological Survey, Open File Series 2001-2, 20 p.

Panno, S. V., and K. C. Hackley, 2001b, The origin of Weldon Springs: Illinois State Geological Survey, Open File Series 2001-3, 6 p.

Panno, S. V., K. C. Hackley, K. Cartwright, and C.-L. Liu, 1994, Hydrochemistry of the Mahomet Bedrock Valley aquifer, east-central Illinois: Indicators of recharge and ground-water flow: Ground Water, v. 32, no. 4, p. 591–604.

Panno, S. V., K. C. Hackley, H. H. Hwang, S. Greenberg, I. G. Krapac, S. Landsberger, and D. J. O'Kelly, 2005a, Database for the characterization and identification of the sources of sodium and chloride in natural waters of Illinois: Illinois State Geological Survey, Open File Series, 2005-1, 15 p.

Panno, S. V., K. C. Hackley, H. H. Hwang, S. Greenberg, I. G. Krapac, S. Landsberger, and D. J. O'Kelly, 2006a, Characterization and identification of the sources of NaCl in ground water: Ground Water, v. 44, no. 2, p. 176–187.

Panno, S. V., K. C. Hackley, H. H. Hwang, and W. R. Kelly, 2001, Determination of the sources of nitrate contamination in karst springs using isotopic and chemical indicators: Chemical Geology, v. 179, no. 1–4, p. 113–128.

Panno, S. V., K. C. Hackley, E. Mehnert, D. R. Larson, D. Canavan, and T. C. Young, 2005b, Declining specific capacity of high-capacity wells in the Mahomet aquifer: Mineralogical and biological factors: Illinois State Geological Survey, Circular 566, 51 p.

Panno, S. V., W. R. Kelly, A. Martinsek, and K. C. Hackley, 2006b, Estimating background and threshold nitrate concentrations using probability graphs: Ground Water, v. 44, no. 5, p. 697–709.

Panno, S. V., I. G. Krapac, C. P. Weibel, and J. D. Bade, 1996, Groundwater contamination in karst terrain of southwestern Illinois: Illinois State Geological Survey, Environmental Geology 151, 43 p.

Panno, S. V., C. P. Weibel, and W. B. Li, 1997, Karst regions of Illinois: Illinois State Geological Survey, Open File Series 1997-2, 42 p.

Richardson, C. K., R. O. Rye, and M. D. Wasserman, 1988, The chemical and thermal evolution of the fluids in the Cave-in-Rock fluorspar district, Illinois: Stable isotope systematics at the Deardorff mine: Economic Geology, v. 83, no. 4, p. 765–783.

Roadcap, G. S., K. C. Hackley, and H.-H. Hwang, 2002, Application of nitrogen and oxygen isotopes to identify sources of nitrate: Carbondale, Illinois, Illinois Groundwater Consortium, Southern Illinois University, Proceedings Paper, 30 p.

Ross, B., K. Parrott, and J. Woodard, 1999, Household water quality: Hydrogen sulfide in household water: Blacksburg, Virginia, Virginia Cooperative Extension, Virginia State University. http://www.ext.vt.edu/pubs/housing/356-488/356-488.html. Accessed March 27, 2008.

Roy, W. R., J. J. G. Glessner, I. G. Krapac, and T. H. Larson, 2003, Possible geological-geochemical sources of ammonium in groundwater: Preliminary results, Illinois Groundwater Consortium, Southern Illinois University, Proceedings Paper, 12 p.

Smith, A. H., C. Hopenhayn-Rich, M. N. Bates, H. M. Goeden, I. Hertz-Picciotto, H. M. Duggan, R. Wood, M. J. Kosnett, and M. T. Smith, 1992, Cancer risks from arsenic in drinking water: Environmental Health Perspectives, v. 97, p. 259–267.

Stueber, A. M., and L. M. Walter, 1991, Origin and chemical evolution of formation waters from Silurian-Devonian strata in the Illinois Basin, USA: Geochimica et Cosmochimica Acta, v. 55, no. 1, p. 309–325.

Stueber, A. M., L. M. Walter, T. J. Huston, and P. Pushkar, 1993, Formation waters from Mississippian-Pennsylvanian reservoirs, Illinois Basin, USA: Chemical and isotopic constraints on evolution and migration: Geochimica et Cosmochimica Acta, v. 57, no. 4, p. 763–784.

U.S. Environmental Protection Agency, 1991, National primary drinking water regulations, final rule: U.S. Federal Register, 40 CFR, parts 141, 142, and 143; v. 56, no. 20, p. 3526–3594.

U.S. Environmental Protection Agency, 2009, Drinking water contamination. http://www.epa.gov/safewater/contaminants/index.html.

Accessed June 23, 2009.

Vorman, J., 2001, EPA to tighten limit on arsenic in drinking water. http://americancityandcounty.com/news/government_epa_tighten_limit. Accessed March 30, 2010.

Walter, L. M., A. M. Stueber, and T. J. Huston, 1990, Br-Cl-Na systematics in Illinois basin fluids: Constraints on fluid origin and evolution: Geology, v. 18, no. 4, p. 315–318.

Warner, K. L., A. Martin, Jr., and T. L. Arnold, 2003, Arsenic in Illinois ground water—Community and private supplies: Reston, Virginia, U.S. Geological Survey, Water-Resources Investigation Report 03-4103, 12 p.

Protecting Groundwater Resources from Contamination

20

Richard C. Berg

INTRODUCTION

An important consideration for land-use planning, especially in areas of rapid population growth, is ensuring that groundwater resources are readily known and that those resources are adequately protected. Groundwater contamination problems must be prevented or eliminated to optimize the resource potential of the groundwater. The U.S. Environmental Protection Agency (1993) has placed a national priority on identifying potential contaminant sources and routes of transport and on evaluating methodologies for identifying water-bearing geological units, particularly aquifers, that are vulnerable to contamination. Only when the geological framework is well-understood and the groundwater resources are located and evaluated for their potential to become contaminated can subsequent land-use planning effectively prevent contaminants from entering aquifers. This chapter discusses the contamination potential for groundwater resources in Illinois and provides examples of maps that have been created to aid planners in protecting groundwater resources.

BACKGROUND

In Illinois, aquifer materials are mainly glacially deposited sands and gravels or sandstone and fractured and jointed limestone and dolomite bedrock, all of which are sufficiently permeable that when saturated they can yield useful quantities of groundwater to wells, springs, or streams (Illinois Groundwater Protection Act 1987). The presence of sand and gravel or highly permeable bedrock indicates aquifers or potential aquifers. Because 90% of Illinois' land area has been glaciated—in some areas more than three times—multiple former landscapes are buried beneath the present land surface, including sands and gravels from ancient rivers and streams, silts and clays from ancient lakes, and old soils. The buried sands and gravels, as well as sands and gravels on the surface, constitute glacial aquifer materials that are not uniformly distributed across the state. Beneath the glacial deposits are layers of bedrock, mainly limestone, dolomite, sandstone, and shale. Where the sandstones are not heavily cemented and where the limestone and dolomite are fractured and jointed, the bedrock may also constitute an aquifer. Because aquifers yield economically significant quantities of water and allow water and/or contaminants to travel at relatively rapid rates, they are potentially vulnerable to contamination. Fine-grained, low-permeability materials such as clay, silt, till, shale, and non-fractured, non-jointed bedrock and cemented sandstone are not considered aquifers even if these materials are water-saturated, because they will not yield sufficient quantities of water to be economically useful. They are aquitards.

The distribution of aquifers and aquitards directly controls the movement of contaminants in groundwater (Chapter 18, Aquifers). The potential for an aquifer to become contaminated largely depends on the natural protective properties of the geological materials that lie above and below it. The thicker the aquitard between an aquifer and a potential contaminant source, the less likely the aquifer is to become contaminated (Berg et al. 1984a, 1984b). Hydraulic conductivity (the ability of geological materials to transmit water) of the protective fine-grained materials is also important. For example, water moves much more slowly through clayey tills than through sandy tills (Berg 2001). The presence of low-permeability materials beneath an aquifer can also restrict further downward migration of potential contaminants into deeper aquifers. Schock et al. (1992) investigated the presence of agricultural chemicals in aquifers, and the results of this Illinois study indicated that aquifers lying within 50 feet (15 m) of the surface generally are the ones most vulnerable to potential contamination.

Many chemical and biological agents can enter into the surface soil, be transmitted into the groundwater flow system, be stored in aquifers, and eventually be pumped from wells and consumed. Examples of surface or near-surface activities that have the potential to contaminate aquifers are accidental spilling of chemicals; applying agricultural chemicals to farm fields; applying herbicides and pesticides to residential and other lawns and gardens; leaking of substances from municipal or hazardous waste landfills, septic systems, or salt storage areas; leaking of petroleum products and chemicals from storage tanks; seeping of water contaminated with road salts and other chemicals from stormwater detention areas that are improperly placed; and infiltrating metals and nitrates from surface spreading of sewage sludge and septage (Chapter 29, Pollution of Natural Waters). Each can pose serious environmental and health hazards (Soller and Berg 1992). Depending on the degree of contamination and intended future land-use activities, costly mitigation efforts are generally required to clean up contaminated groundwater and restore land so that it can be utilized for safe recreational, residential, and commercial or industrial use.

Mapping the geology in three dimensions is the key to successful development of maps that show the sensitivity of aquifers to contamination (Thorleifson et al. 2010). Mapping is the first step in assessing the suitability of geological materials for specific land uses at statewide, county, or municipal scales. When this basic geological information is combined with hydrologic data, geologists and hydrologists can best help regulators and land-use planners establish recharge protection areas and setback zones around municipal wells to restrict potentially hazardous land-use practices. Such technical information also allows decision makers to address various federal and state regulations for disinfecting water wells, designating aquifers critical for drinking as "sole source," providing protection to surface-water bodies and watersheds that may be eventual sources for groundwater recharge, developing classification systems to best separate waters for drinking from waters for industrial and other uses, and prioritizing remediation of

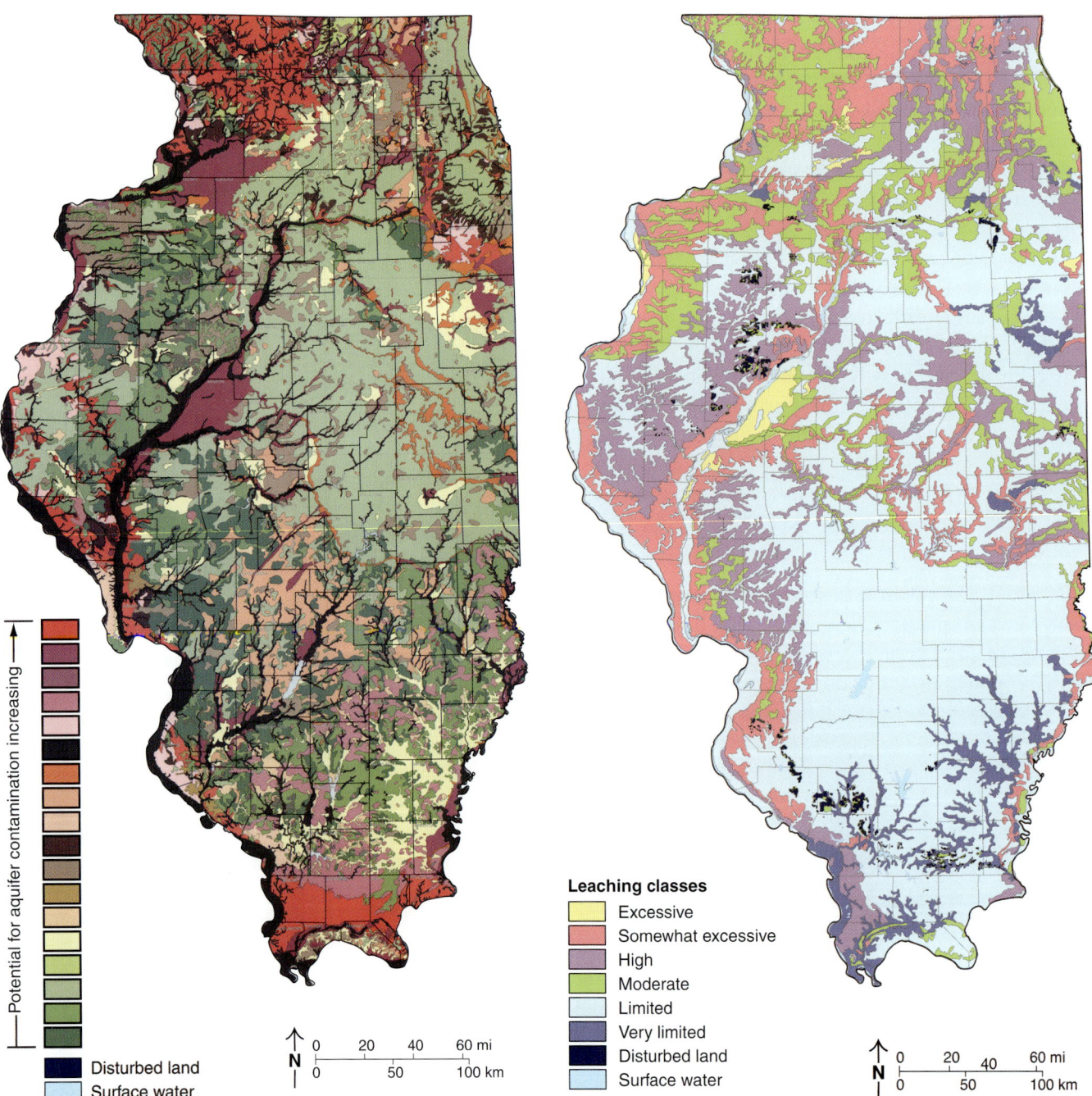

Figure 20-1 Sensitivity of aquifers within 50 feet (15.2 m) of land surface to become contaminated (Berg et al. 1984a).

Figure 20-2 Aquifer sensitivity to contamination by nitrates (Keefer 1995).

sites with contaminated groundwater based on projected risks to populations.

STATEWIDE POTENTIAL FOR GROUNDWATER CONTAMINATION

Protection of groundwater resources in Illinois requires that aquifers be accurately known and that their thickness and depth beneath the surface be delineated. Also important to determine are the thickness and extent of overlying aquitards and the ease with which aquifers and other deposits can transmit groundwater containing potential contaminants (Soller and Berg 1992). Figures 20-1, 20-2, 20-3, and 20-4 show the sensitivity of aquifers to contamination by surface or near-surface sources. These aquifer sensitivity maps show ranges in sensitivity from high, where aquifers are at the surface, to low, where aquifers are buried.

Figure 20-1 (Berg et al. 1984a) was developed for the Illinois Environmental Protection Agency to show the variability of statewide conditions for the purposes of developing groundwater protection legislation (1987 Illinois Groundwater Protection Act). The map was also used for setting priorities for future site-specific studies of hazardous waste sites. About 50% of the state has a sand and gravel or bedrock aquifer within 50 feet (15.2 m) of the surface (Figure 20-1). The red color shows the most sensitive areas where aquifer materials greater than 20 feet (6.1 m) thick are present within 5 feet (1.5 m) of the surface. These areas primarily exist (1) adjacent to Illinois' major rivers where thick sand and gravel deposits lie in floodplains and in terraces, (2) in Mason and Kankakee Counties where sand dunes are prevalent, (3) in western Lake County and McHenry County where glaciers stagnated and deposited thick sands and gravels, and (4) in extreme northwestern and southern Illinois and adjacent to the lower Illinois River valley where glacial deposits are thin or absent and sandstone or fractured limestone and dolomite are at the surface. Orange areas also reflect areas of high sensitivity where sand and gravel aquifer materials are within 20 feet (6.1 m) of land surface and less than 20 feet (6.1 m) thick. These aquifers occur as buried deposits mostly in central Illinois in Montgomery, Christian, and Fayette Counties and at the land surface, mostly in east-central Illinois in front of glacial moraines. The aquifer materials in both the red and orange map areas are highly sensitive to contaminants moving through them and getting into drinking water sources. Brown areas show sand and gravel aquifers buried at 20 to 50 feet (15.2 m) occurring as patchy deposits throughout central Illinois; bedrock aquifers buried at 20 to 50 feet (15.2 m) of the surface, occur in thin drift areas in west-central Illinois. Fine-grained materials overlying the aquifers offer moderate protection from the spreading of wastes, septic effluent, or application of agricultural chemicals. The light green map color represents areas with the lowest sensitivity, where no aquifers exist within 50 feet (15.2 m) of the surface. Aquifer sensitivity is lowest in these regions, as fine-grained materials offer reasonable protection to aquifers from most contamination sources introduced at the land surface. However, disposal of hazardous wastes or siting of a municipal landfill is not recommended since aquifers may exist deeper than 50 feet (15.2 m).

Further refinements of the state's shallow aquifer sensitivity map are shown in Figures 20-2 and 20-3 (Keefer 1995). These maps were prepared in response to Illinois'

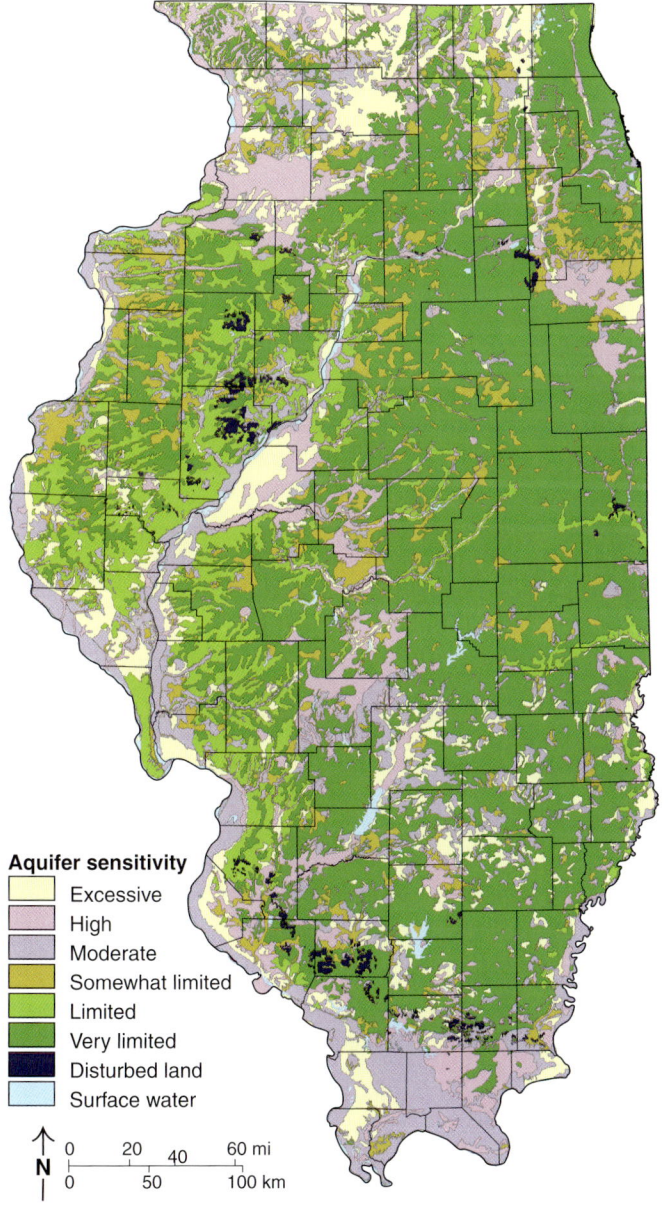

Figure 20-3 Aquifer sensitivity to contamination by pesticides (Keefer 1995).

Pesticide Management Plan and U.S. Environmental Protection Agency requirements that agricultural pesticides be regulated and based on groundwater use, value, and vulnerability. The maps incorporate information on shallow aquifers (less than 50 feet [15.2 m]) beneath the surface) as well as soil hydraulic conductivity, natural drainage class, and amount of soil organic matter. Figure 20-2 shows aquifer sensitivity to contamination by nitrates. Nitrates are found in fertilizers and travel mostly unrestricted with groundwater. Therefore, in areas with sandy surface soils overlying sandy aquifer materials or fractured bedrock, sensitivity is high, as shown by the pink and lilac areas on Figure 20-2. Areas along the major river valleys and in northwestern and southern Illinois are particularly sensitive. Areas where clayey soils overlie buried aquifers have low sensitivity.

Figure 20-3 shows aquifer sensitivity to contamination by pesticides. Unlike nitrates, when pesticides move through the soil, some of their harmful constituents are removed (adsorbed) by the organic matter and clays in the soil. Therefore, many of the high-sensitivity areas on Figure 20-2 are mapped on Figure 20-3 as less sensitive for pesticides.

In addition to their use as a basis for state-level regulatory measures, the 1995 sensitivity maps were used by the Illinois Environmental Protection Agency to evaluate potential and actual contamination problems of surface water supplies. The highest concentrations of agricultural chemicals were found to occur in rivers, streams, and lakes in the lowest aquifer sensitivity settings because of their low infiltration and high runoff.

Figure 20-4 (Keefer et al. 1990) includes information about shallow aquifers, hydraulic conductivity through surface soils, and, most importantly, information about deep aquifers (see Chapter 18, Aquifers). Figure 20-4 shows major sand and gravel aquifers at all depths and major bedrock aquifers within 300 feet (91 m) of land surface. These aquifers are the most prolific in Illinois, yielding over 100,000 gallons (380,000 L) per day of water. The deep sands and gravels of the Mahomet aquifer can be found in central Illinois, and deep bedrock aquifers occur in the northern third of the state. Green areas on Figure 20-4 lack major aquifers. Although it may be safe to dispose of hazardous and municipal wastes above deeply buried aquifers, it may still be politically unacceptable to dispose of wastes above a known high-yielding aquifer.

The 1990 sensitivity map (Figure 20-4) was used to identify areas of the state where recharge is rapid and plentiful. The map was combined with Illinois State Water Survey information showing the location of waste sites and was used by the Illinois Environmental Protection Agency to establish groundwater protection planning regions and to delineate potential problem areas of the state where more detailed studies may be warranted.

County Maps for Aquifer Sensitivity

Although Illinois' statewide aquifer sensitivity maps greatly assisted state and federal agencies with statewide

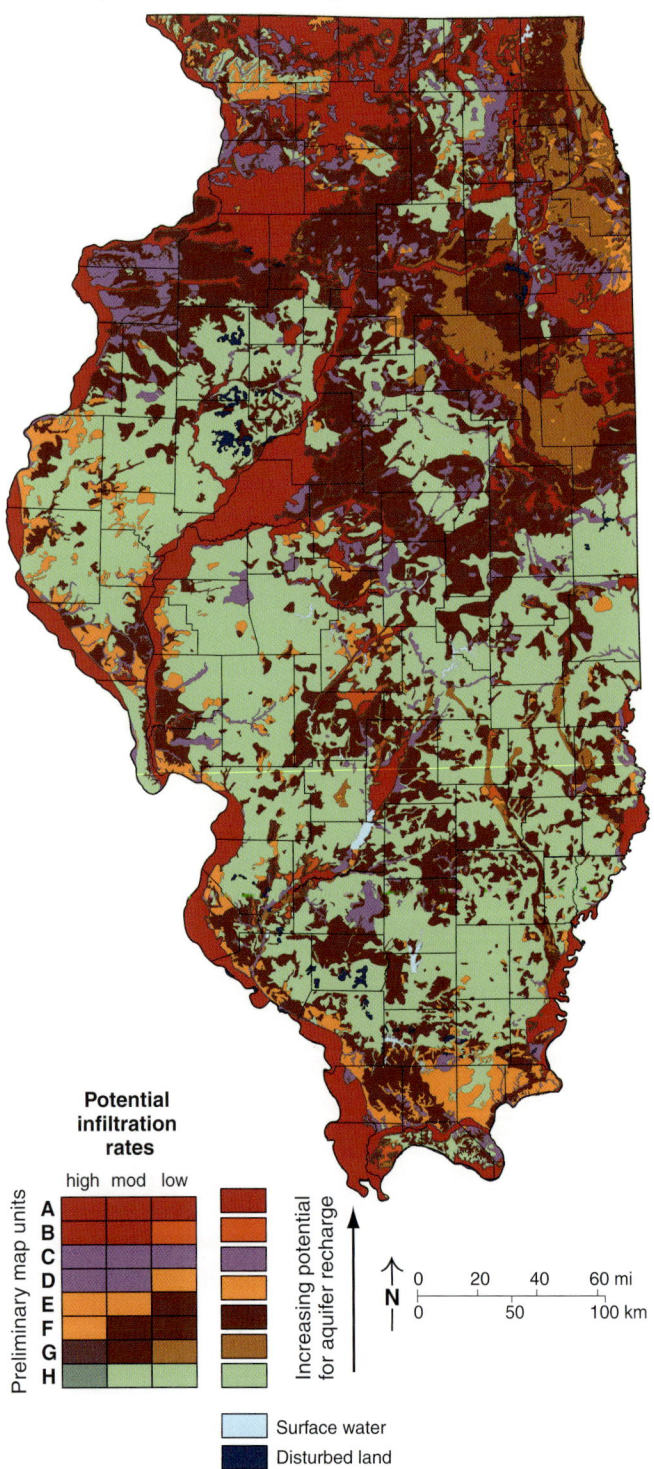

Figure 20-4 Sensitivity of shallow and deep aquifers to become contaminated (Keefer et al. 1990).

PROTECTING GROUNDWATER

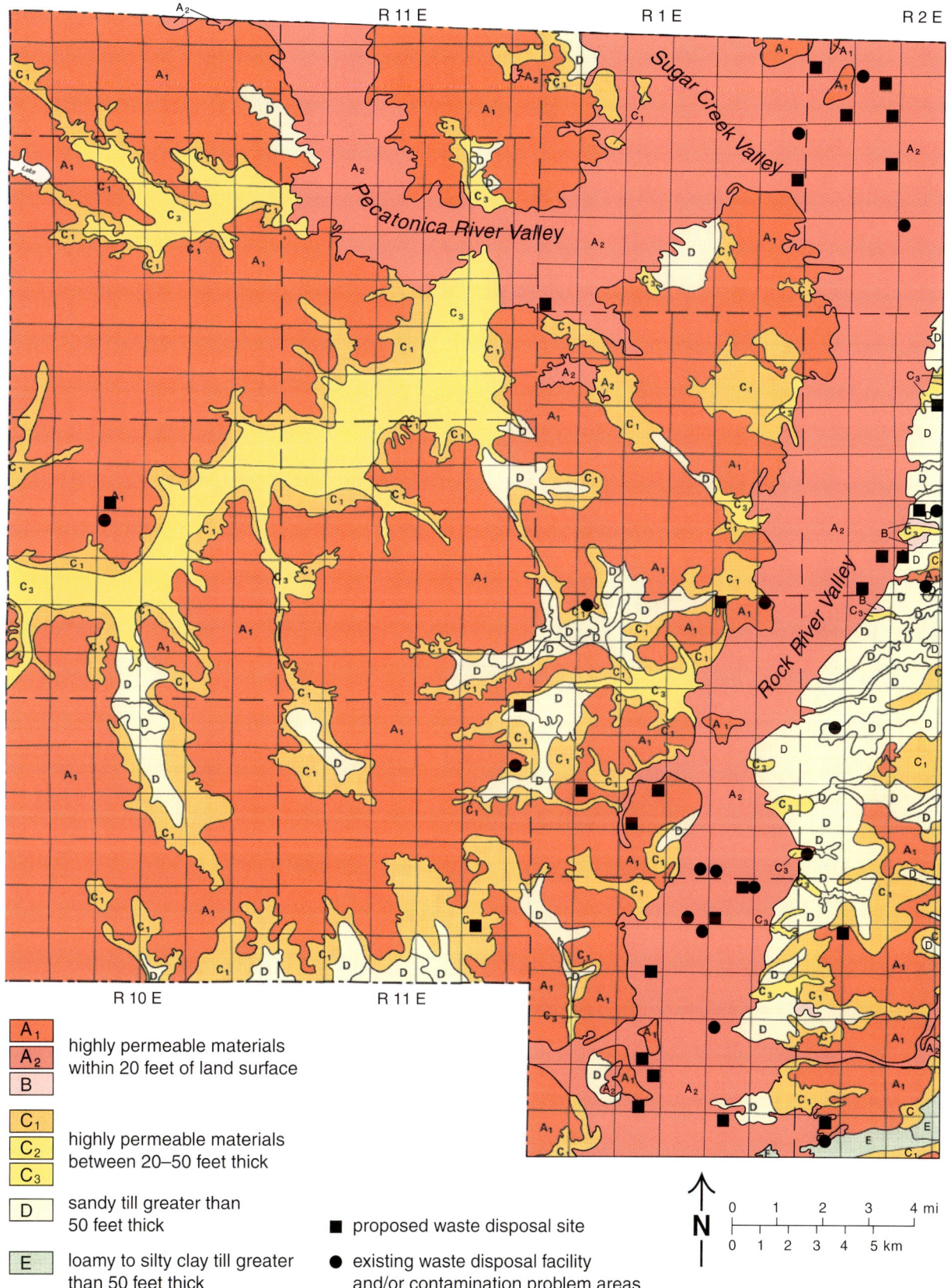

Figure 20-5 Aquifer sensitivity map for part of Winnebago County (Berg et al. 1984b). See Figure 20-6 for more detailed description of these map units.

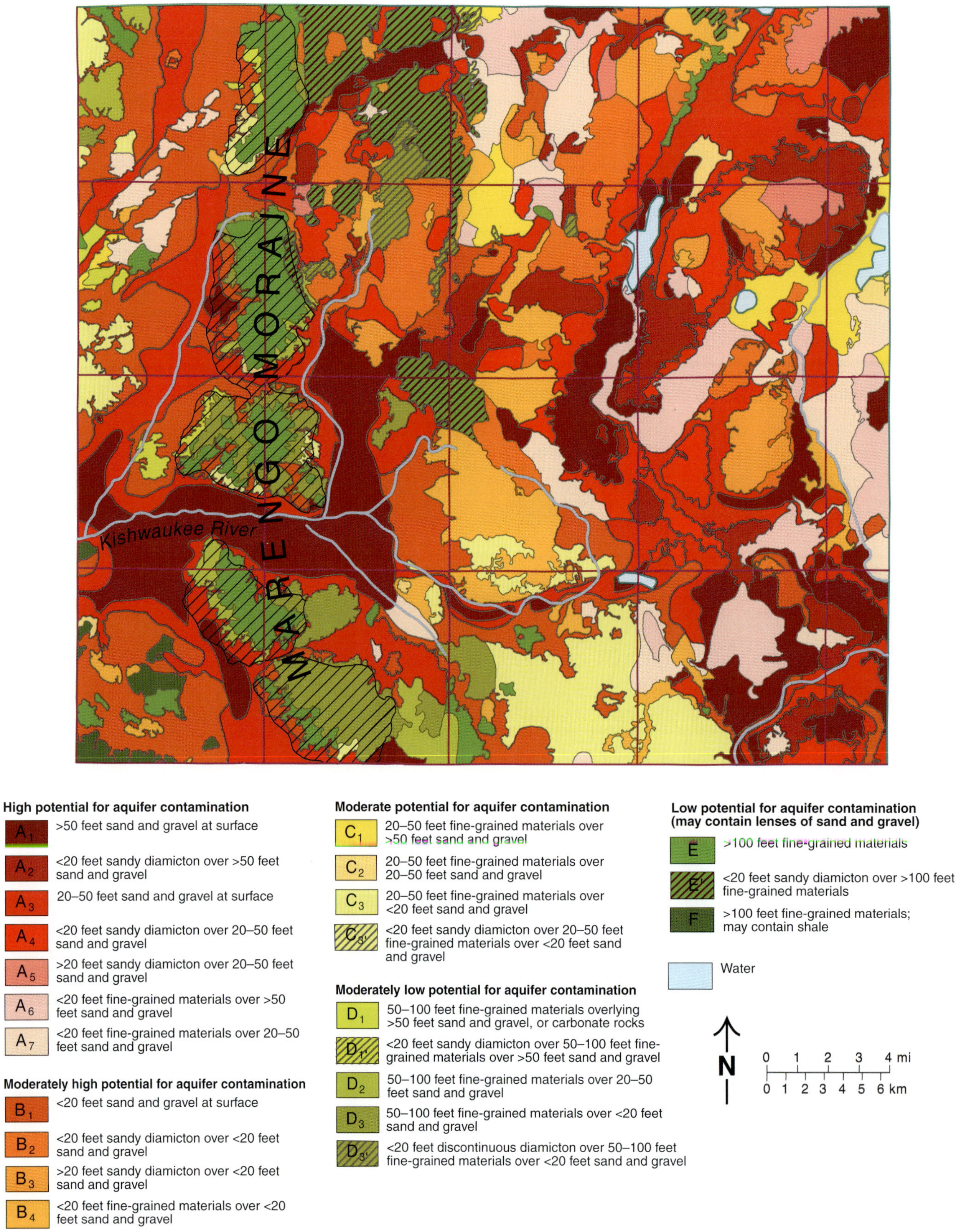

MARENGO MORAINE
Kishwaukee River
High potential for aquifer contamination
A1 >50 feet sand and gravel at surface
A2 <20 feet sandy diamicton over >50 feet sand and gravel
A3 20–50 feet sand and gravel at surface
A4 <20 feet sandy diamicton over 20–50 feet sand and gravel
A5 >20 feet sandy diamicton over 20–50 feet sand and gravel
A6 <20 feet fine-grained materials over >50 feet sand and gravel
A7 <20 feet fine-grained materials over 20–50 feet sand and gravel
Moderately high potential for aquifer contamination
B1 <20 feet sand and gravel at surface
B2 <20 feet sandy diamicton over <20 feet sand and gravel
B3 >20 feet sandy diamicton over <20 feet sand and gravel
B4 <20 feet fine-grained materials over <20 feet sand and gravel
Moderate potential for aquifer contamination
C1 20–50 feet fine-grained materials over >50 feet sand and gravel
C2 20–50 feet fine-grained materials over 20–50 feet sand and gravel
C3 20–50 feet fine-grained materials over <20 feet sand and gravel
C3' <20 feet sandy diamicton over 20–50 feet fine-grained materials over <20 feet sand and gravel
Moderately low potential for aquifer contamination
D1 50–100 feet fine-grained materials overlying >50 feet sand and gravel, or carbonate rocks
D1' <20 feet sandy diamicton over 50–100 feet fine-grained materials over >50 feet sand and gravel
D2 50–100 feet fine-grained materials over 20–50 feet sand and gravel
D3 50–100 feet fine-grained materials over <20 feet sand and gravel
D3' <20 feet discontinuous diamicton over 50–100 feet fine-grained materials over <20 feet sand and gravel
Low potential for aquifer contamination (may contain lenses of sand and gravel)
E >100 feet fine-grained materials
E' <20 feet sandy diamicton over >100 feet fine-grained materials
F >100 feet fine-grained materials; may contain shale
Water
N
0 1 2 3 4 mi
0 1 2 3 4 5 6 km

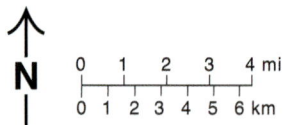

Figure 20-6 Aquifer sensitivity map of McHenry County (Curry et al. 1997).

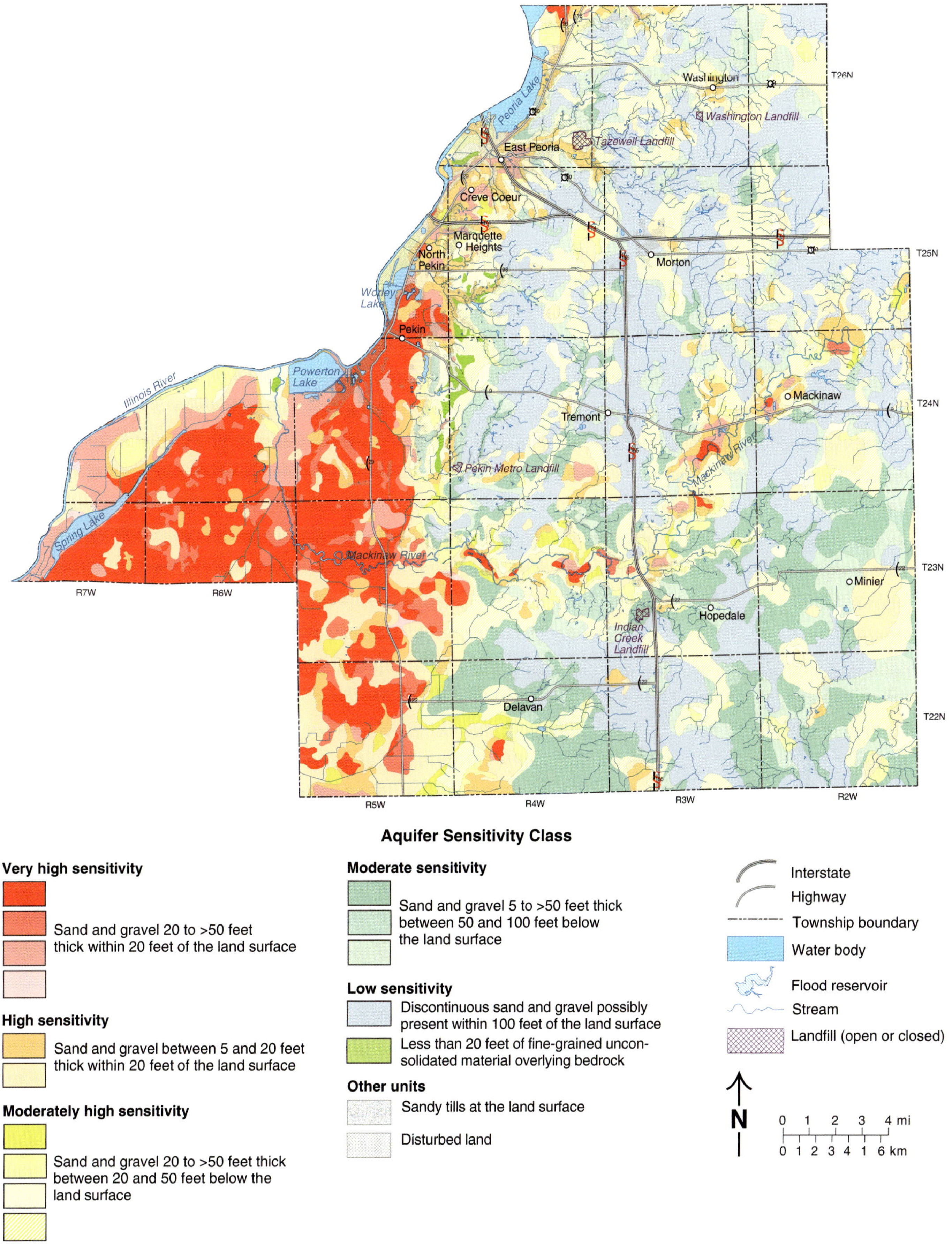

PROTECTING GROUNDWATER

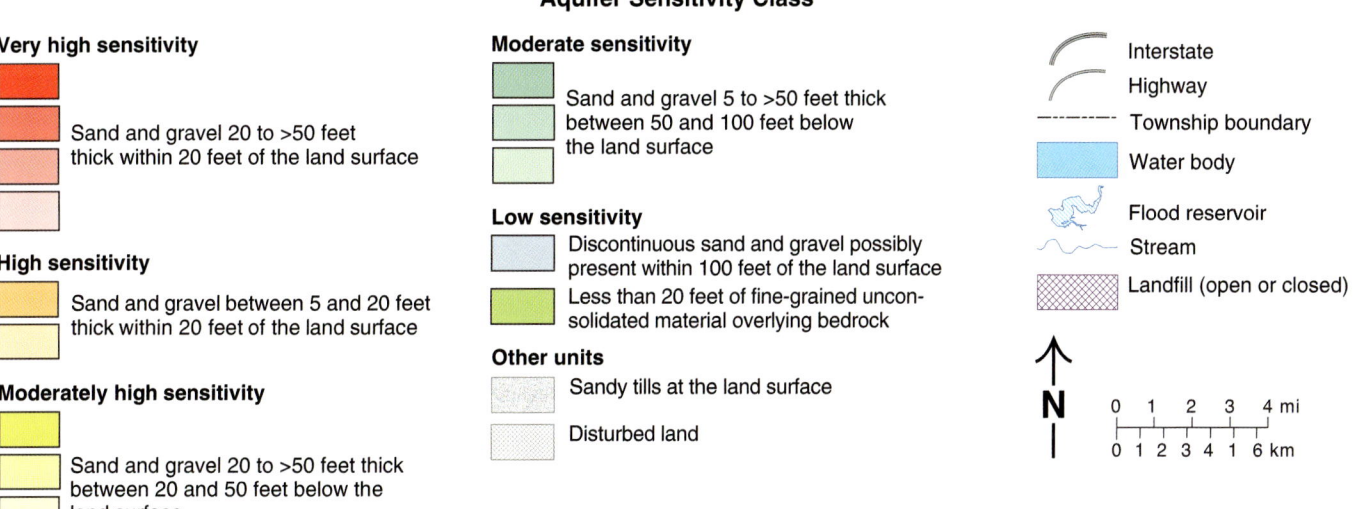

Figure 20-7 Aquifer sensitivity map of Tazewell County (Johnstone 2003).

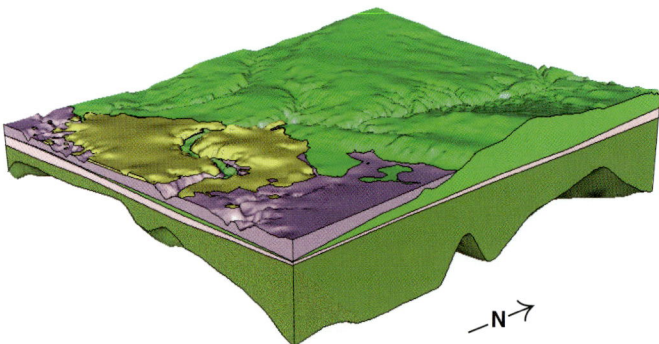

Figure 20-8 Block diagram of the Villa Grove Quadrangle showing sand and gravel aquifers in yellow (Abert et al. 2000).

groundwater resource protection strategies, because of their scale, the maps are less useful in helping county and municipal governments address specific local issues. Therefore, several aquifer sensitivity maps have been developed at the county and municipal scale to deal with problems associated with rapid population growth and accompanying environmental problems. Figures 20-5, 20-6, and 20-7 are examples of aquifer sensitivity maps that were prepared for Winnebago County (Berg 1984b), McHenry County (Curry et al. 1997), and Tazewell County (Johnstone 2003).

Winnebago County

Winnebago County is considered to have the most vulnerable groundwater resources in Illinois (Bhagwat and Berg 1991). Surface and near-surface aquifers are prevalent over most of the county, there is extensive industrial development, and serious waste disposal problems exist. Figure 20-5 shows aquifer sensitivity to a depth of 50 feet (15.2 m) for a portion of Winnebago County. Highly permeable bedrock or sand and gravel of the Rock River and Sugar Creek valleys lie within 20 feet (6.1 m) of land surface. Forty-two of the 46 contamination problem areas (shown on Figure 20-5 as black dots and squares) occur in these high-sensitivity areas. Regions of the Pecatonica River valley and tributary valleys to the Rock River also have highly permeable bedrock or sand and gravel; however, these materials lie between 20 and 50 feet (6.1 to 15.2 m) of land surface and are overlain by low-permeability fine-grained materials.

The map for Winnebago County is unique in that a follow-up investigation was conducted about ten years after mapping was completed to assess how the geological information had been used by health and land-use planners (Bhagwat and Berg 1991). The mapping provided a framework for understanding the environmental context for groundwater resource availability and protection issues by focusing attention so that education, inspection, and land-use planning could identify potential problems. The mapping also provided information about where to test for contaminants by identifying high-risk areas and providing further targeting for prevention initiatives. Positive contamination test results consistently occurred in areas where they were suspected based on the map and industrial development patterns.

McHenry County

McHenry County's population growth is one of the fastest in the United States, and the entire public water supply for its more than 300,000 residents comes from groundwater. An aquifer sensitivity map was created to help the county manage its water supply (Curry et al. 1997). Red areas on Figure 20-6 have thick sand and gravel near the land surface; orange and yellow areas have buried sand and gravel within 50 feet (15.2 m) of the surface (Kishwaukee River floodplain and glacial outwash deposits). Light green areas have sand and gravel buried between 50 feet (15.2 m) and 100 feet (30.5 m), and dark green areas lack aquifer materials within the upper 100 feet (30.5 m). These areas mostly occur in the Marengo Moraine, which is composed of thick tills and is found in the western portion of the county.

In addition to being a basis for the county's groundwater resource plan and recommendations, the McHenry County map has been used to determine whether land-use changes from agricultural to residential purposes detrimentally affected groundwater quality and whether municipal water wells and sewers would be more appropriate than private water wells and septic systems for subdivision development. The map was also used to assess the potential contamination from a proposed oil pipeline through the county; to evaluate the impact of garbage transfer stations; to evaluate water resource and contamination issues of electricity-generating peaker plants; to make general planning decisions on land use, transportation corridors, and industrial placements; and to evaluate newly proposed subdivisions.

Tazewell County

An aquifer sensitivity map was developed for Tazewell County, which was faced with several difficult land-use planning decisions (Johnstone 2003) (Figure 20-7). This map has a more comprehensive aquifer sensitivity classification system than is found on previous maps. The Tazewell County map focuses on the presence of sand and gravel or high-permeability bedrock aquifer materials, but the methodology for developing this map can accommodate those deposits lying within 100 feet (30.5 m) of land surface and also deeper deposits. This map currently is being used

PROTECTING GROUNDWATER

(Berg and Abert 1999). Residents of the Villa Grove Quadrangle have 100% reliance on groundwater as a drinking water source. Figure 20-8 shows a three-dimensional block diagram of the quadrangle geology; sand and gravel aquifers are shown as yellow. Figure 20-9 shows the derivative aquifer sensitivity map for the Villa Grove Quadrangle, which can be used to determine the potential for contamination of aquifers from municipal and hazardous waste disposal sites (particularly to assess existing sites and, more importantly, to screen for regions where the likelihood of finding suitable sites is highest), septic systems, and agricultural chemicals.

CONCLUSION

This chapter has discussed the sensitivity of the state's sand and gravel and bedrock aquifer material to become contaminated by various land-use practices. Specific examples have shown why geological information is needed to help planners and public health officials, engineering and geological consultants, developers, and the public make informed land-use decisions and address environmental problems. Aquifer sensitivity maps provide guidance to regulators seeking maximum protection of especially vulnerable groundwater resources while avoiding overprotection of resources in areas where natural environmental protection already exists. Additionally, the maps can and have been used to direct sensitive land-use practices (such as the siting of waste-disposal and industrial facilities) to areas with low aquifer contamination potential, thereby reducing future liabilities.

Establishing the three-dimensional geological framework from land surface to specific depths is a critical component in assessing the potential for various land-use activities to contaminate an aquifer (Thorleifson et al. 2010). Mapping allows for an analysis of the real distribution and variability of the water-holding and water-transmitting properties of aquifers and their degree of natural protection by overlying aquitards. Only after the geological framework is understood can the basic geological information be joined with hydrological data to assess the sensitivity of specific locations. Site-specific evaluations can determine where water wells will provide maximum yields and the extent to which aquifers are naturally protected by overlying fine-grained deposits. As the examples in this chapter have shown, basic three-dimensional maps provide (1) a practical basis for deriving information about aquifer sensitivities, and (2) a conceptual model to help evaluate potential contamination problems regionally and at specific sites.

As a final mapping step, geologists and hydrologists must work with regulators and land-use planners to help

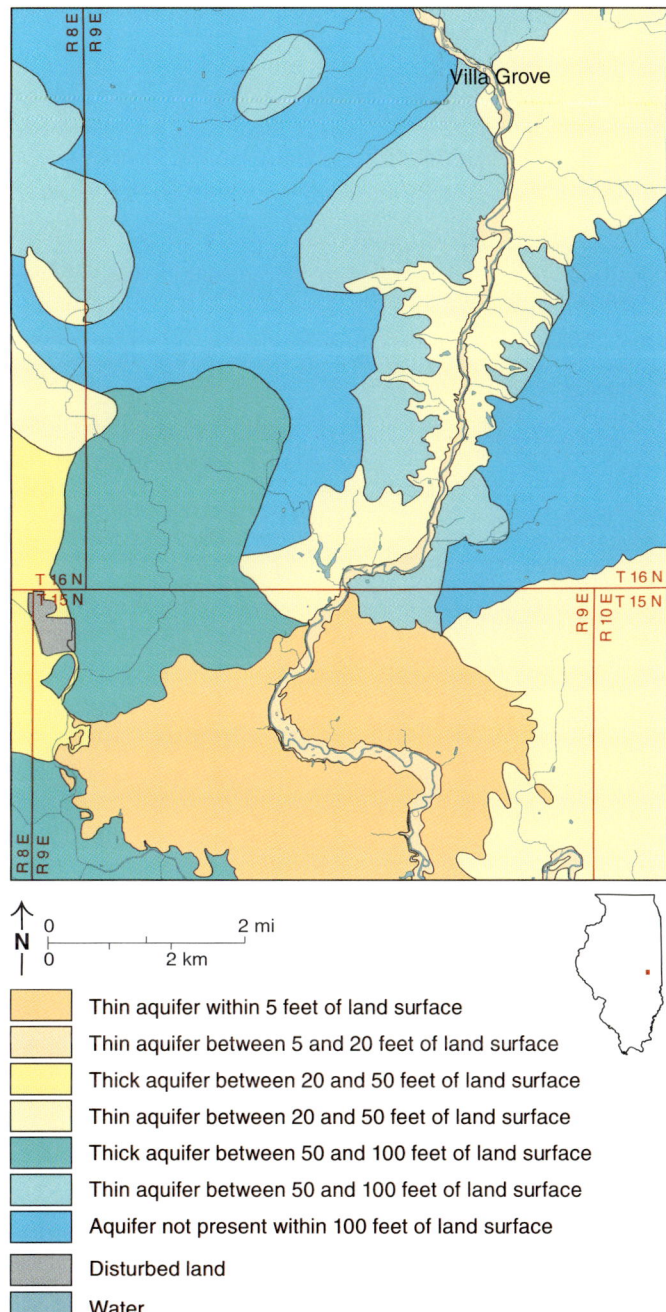

Figure 20-9 Aquifer sensitivity map of the Villa Grove Quadrangle (Berg and Abert 1999).

to help evaluate the vulnerability of waste disposal facilities in the county.

QUADRANGLE-SCALE MAP OF AQUIFER SENSITIVITY

In addition to statewide and county aquifer sensitivity maps, more detailed maps have been made for specific quadrangles at the 1:24,000-scale (1 inch [2.5 cm] on the map represents 2,000 feet [610 m] on the land surface), including the Villa Grove Quadrangle in Douglas County

them understand the three-dimensional geological framework and its implications for balancing water use with environmental protection. Such an understanding allows the development of groundwater flow models and the determination of the areas that contribute groundwater to wells (capture zones) over specific time periods. Well-informed decision makers who use the best and most-detailed scientific information as the basis for the most cost-effective land- and water-use decisions can maximize groundwater resource potential and minimize contamination. Recharge zones and areas immediately surrounding wells can be delineated and protected, and numerous state and federal regulations can be established for cleanup and water use based on quantities and ambient qualities.

References

Abert, C. C., C. P. Weibel, and R. C. Berg, 2000, Three-dimensional geologic mapping of the Villa Grove Quadrangle, Douglas County, Illinois, *in* D. R. Soller, ed., Digital Mapping Techniques '00—Workshop Proceedings, May 17–20, 2000, Lexington, KY: Reston, Virginia, U.S. Geological Survey, Open-File Report 00-325, p. 125–129.

Berg, R. C., 2001, Aquifer sensitivity classification for Illinois using depth to uppermost aquifer material and aquifer thickness: Illinois State Geological Survey, Circular 560, 14 p.

Berg, R. C., and C. C. Abert, 1999, General aquifer sensitivity map, Villa Grove Quadrangle, Douglas County, Illinois: Illinois State Geological Survey, Illinois Geologic Quadrangle Map, IGQ Villa Grove-AS, 1:24,000.

Berg, R. C., J. P. Kempton, and K. Cartwright, 1984a, Potential for contamination of shallow aquifers in Illinois: Illinois State Geological Survey, Circular 532, 30 p.

Berg, R. C., J. P. Kempton, and A. N. Stecyk, 1984b, Geology for planning in Boone and Winnebago Counties: Illinois State Geological Survey, Circular 531, 69 p.

Bhagwat, S. B., and R. C. Berg, 1991, Benefits and costs of geologic mapping programs in Illinois: Case study of Boone and Winnebago Counties and its statewide applicability: Illinois State Geological Survey, Circular 549, 40 p.

Curry, B. B., R. C. Berg, and R. A. Vaiden, 1997, Geologic mapping for environmental planning, McHenry County, Illinois: Illinois State Geological Survey, Circular 559, 79 p.

Illinois Groundwater Protection Act, 1987: Springfield, Illinois, Illinois General Assembly, P.A. 85-863, September 24, 1987, 48 p.

Johnstone, P. D., 2003, Aquifer sensitivity map of Tazewell County, Illinois: Illinois State Geological Survey, Open File Series, 2003-6d, 1:62,500.

Keefer, D. A., 1995, Potential for agricultural chemical contamination of aquifers in Illinois: 1995 revision: Illinois State Geological Survey, Environmental Geology 148, 28 p.

Keefer, D. A., R. C. Berg, and W. S. Dey, 1990, Potential for aquifer recharge in Illinois (appropriate recharge areas): Illinois State Geological Survey, 1:1,000,000.

Schock, S. C., E. Mehnert, M. E. Caughey, G. B. Dreher, W. S. Dey, S. Wilson, C. Ray, S. J. Chou, J. Valkenburg, J. M. Gosar, J. R. Karny, M. L. Barnhardt, W. F. Black, M. R. Brown, and V. J. Garcia, 1992, Pilot study: Agricultural chemicals in rural, private wells in Illinois: Illinois State Water Survey and Illinois State Geological Survey, Cooperative Groundwater Report 14, 80 p.

Soller, D. R., and R. C. Berg, 1992, A model for the assessment of aquifer contamination potential based on regional geologic framework: Environmental Geology and Water Sciences, v. 19, no. 3, p. 205–213.

Thorleifson, L. H., R. C. Berg, and H. A. J. Russell, 2010, Geological mapping goes 3-D in response to societal needs: GSA Today, v. 20, no. 8, p. 27–29.

U.S. Environmental Protection Agency, 1993, Ground water resource assessment: Washington, D.C., U.S. Environmental Protection Agency, Office of Water, EPA 813-R-93-003, 232 p.

Wetlands Geology 21

James J. Miner and Michael V. Miller

INTRODUCTION

Wetlands contribute greatly to the wealth of Illinois. The state's richest agricultural soils originally formed in wetlands that subsequently were drained for farming, and valuable resources such as coal and peat are mined from the deposits of ancient and modern wetlands. Existing wetlands perform economically and environmentally important functions, such as storing floodwaters, removing sediment and chemicals from surface water, recharging shallow aquifers, maintaining low flows in streams, and providing wildlife habitat and recreational opportunities (National Research Council 1995).

Wetlands are places that are continuously or recurrently flooded or saturated long enough to produce oxygen-deficient conditions in the soil. The anoxic environment creates hydric soils, which support vegetation uniquely suited to these conditions. Wetlands include swamps, marshes, bogs, and many other similar environments, but deep water bodies such as lakes and rivers generally are not considered wetlands. The definition of a wetland used here is adapted from the *Corps of Engineers Wetlands Delineation Manual* (Environmental Laboratory 1987), which is widely used in the United States to identify wetlands.

Due to gradients in temperature and moisture across the state, Illinois has a large variety of wetland types, from swamps in the south to bogs in the north, including forested and prairie-dominated wetlands (Figure 21-1). This rich wetland heritage includes the extensive cypress swamps of southern Illinois, which are one of only 29 "Wetlands of International Importance" in the United States, as currently designated by the international Ramsar Convention

Figure 21-1 Photographs of Illinois wetlands: **(a)** cypress swamp, Union County (photograph by Marshall Lake); **(b)** floodplain forest, Wabash County (photograph by Keith Carr); **(c)** sedge meadow, Lake County (photograph by James Miner); and **(d)** marsh, St. Clair County (photograph by Steven Benton).

on Wetlands of 1971. This chapter discusses the extent, origins, geological deposits, and functions of Illinois wetlands.

CURRENT AND FORMER EXTENT OF ILLINOIS WETLANDS

Prior to settlement by Europeans, wetlands occupied about 24%, or 8.7 million acres (3.5 million ha) (Taylor et al. 2009), of the total land area in Illinois (Figure 21-2). Wetlands were widespread, exceeding 20% of the land area in over 61 counties (Figure 21-3a) (Suloway and Hubbell 1994). High concentrations existed in northeastern, east-central, and southern Illinois, where wetlands exceeded 40% of land area in 18 counties. Only 11 counties in Illinois contained less than 10% wetlands by area, and those counties were located primarily in northwestern and west-central Illinois.

After settlement, according to data from the 1980s, wetlands had declined by about 90% to about 2.5% of the land area of the state (0.9 million acres [0.4 million ha]) (Figure 21-3b), a reduction of nearly 7.8 million acres (3.2 million ha) (McCorvie and Lant 1993, Suloway and Hubbell 1994). No counties afterward were composed of more than 15% wetlands by area, and nearly half of Illinois counties contained less than 3% wetlands by area. Remaining concentrations were located in northeastern Illinois, along the Illinois River, and in southern Illinois (Figure 21-4). Although more recent comprehensive data are not available, trends indicate losses continue in Illinois and nationwide, although at a decreased rate (Dahl 2000).

Nearly half of the counties in Illinois have lost more than 90% of their wetland area, and these counties mostly are located in rural settings (Figure 21-3c). In 10 counties in east-central Illinois, wetlands have been virtually eliminated where they originally occupied 40 to 61% of the land area in those counties (Suloway and Hubbell 1994). Settlers avoided the large wetland complexes in east-central Illinois (Figure 21-2) until the 1850s, when wealthy individuals and investment groups began to buy extensive areas and invest considerable funds to drain them (Prince 1997). Some of these individual drainage schemes were as large as 30,000 acres (12,000 ha) (Bogue 1959).

The main purpose of wetland drainage in Illinois was for agriculture. Widespread drainage activities began in Illinois in the mid to late 1800s, with more than 117,763 miles (189,521 km) of drainage tile laid between 1880 and 1895 alone (Prince 1997). By the 1930s, over 30% of all farmland in Illinois had been artificially drained (McCorvie and Lant 1993) whether wetland or not, and Illinois currently has the most artificially drained land of any state in the United States, almost 10 million acres (4 million ha) (Thompson 2003).

ORIGIN OF WETLANDS

Wetlands form where the hydrology, geology, and topography cause water inputs to be at least temporarily higher than water outputs, producing continuous or recurring saturation of the land surface. The interaction among these three conditions controls whether wetlands form. For example, clay-rich geological units generally impede infiltration and therefore facilitate wetland formation at the land surface, although only in places where the shape of the land

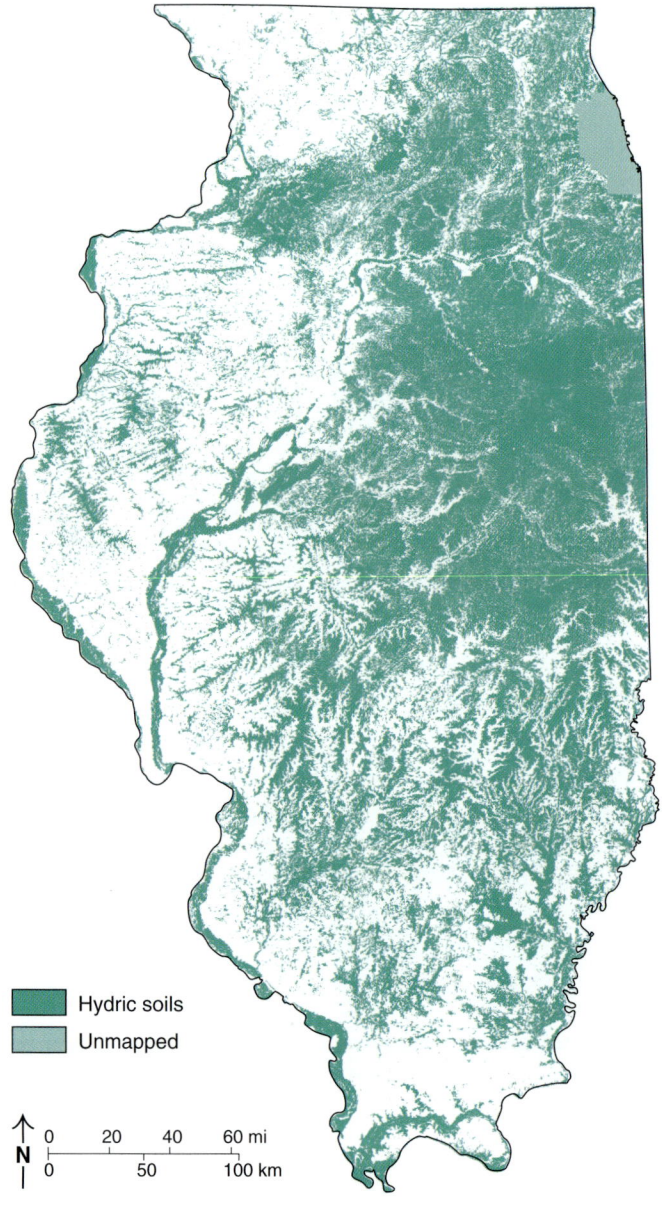

Figure 21-2 Approximate extent of presettlement wetlands, based on the extent of hydric soils, which may include some nonwetland areas (Natural Resources Conservation Service 2009). Portions of Cook County are not mapped. (Map by Adrianne Knight.)

WETLANDS GEOLOGY

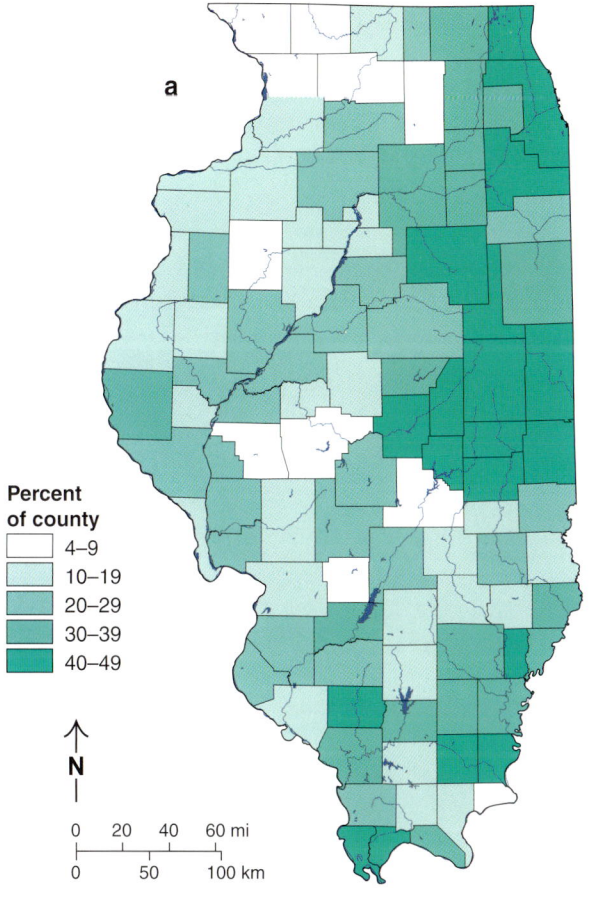

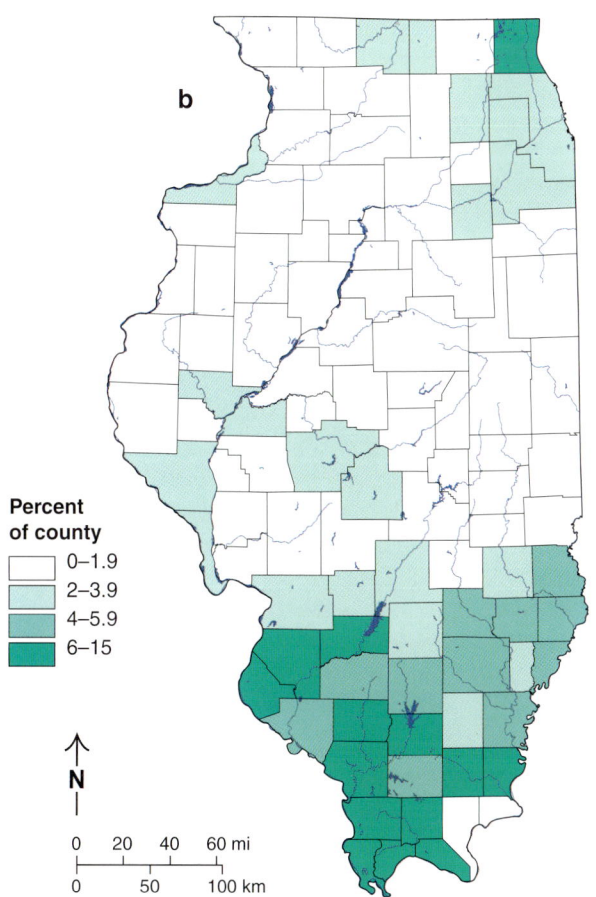

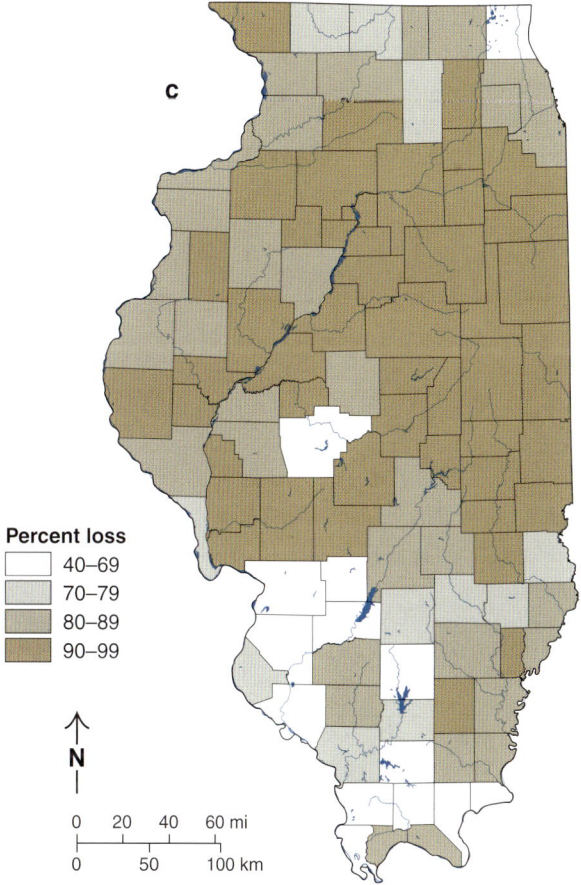

Figure 21-3 (**a**) Percentage of land area occupied by wetlands prior to settlement, by county; (**b**) percentage of land area occupied by wetlands in the 1980s, by county; and (**c**) percentage of original wetlands lost, by county (Suloway and Hubbell 1994).

allows water to collect. Similarly, coarse geological units such as sand can support wetlands by providing groundwater discharge, but only where the hydrology allows discharge; otherwise, sand deposits may provide drainage by facilitating infiltration.

Wetlands may form if continuous inundation or saturation occurs for as little as 5% of the growing season, which is as few as 9 days per year in Illinois, although recent changes suggest a minimum of 14 days (U.S. Corps of Army Engineers 2008). Many wetland types, however, require inundation or saturation for much longer in order to support certain plant species that require wetter conditions. Potential water inputs include precipitation, runoff, and groundwater discharge. Potential water outputs include infiltration, runoff, and evapotranspiration (evaporation plus transpiration by plants).

In Illinois, wetlands are most often saturated or inundated in the spring, when water inputs such as precipitation and runoff are high, and water losses by evapotranspiration are low (Hensel 1992). These seasonal inputs may saturate poorly drained areas and depressions or run off into streams and rivers, causing inundation in floodplains.

Although precipitation remains high during summer, increased evapotranspiration generally causes summertime drying in most wetlands, therefore limiting peat-forming wetlands to areas with year-round saturation most often associated with groundwater input. Lower evapotranspiration rates in fall and winter may allow saturation to return in some wetlands, despite the lower precipitation.

Hydrogeology also influences the character of a wetland. Sources and quality of water inputs greatly influence the duration of inundation and/or saturation as well as the types of vegetation that can inhabit a wetland. In Illinois, calcareous glacial deposits and carbonate bedrock are common, leading to surface water and groundwater that are generally alkaline or of near-neutral pH. Therefore, acidic wetlands or those with low ionic strength waters are not common. Further discussion is limited to associations with the following geologic settings and processes.

Although it is beyond the scope of this chapter to describe all combinations of conditions that can create wetlands, certain geological settings, events, and geomorphic processes have led to the development of Illinois' major wetland systems. The general geological setting and the processes that have formed the most prominent Illinois wetland systems are discussed next.

River Floodplains

The majority of existing wetlands in Illinois are floodplain forest wetlands (Figure 21-4). Major rivers, including the Mississippi, Illinois, Wabash, Ohio, Sangamon, Rock, and Kaskaskia, flowing within and adjacent to Illinois have produced large floodplain areas. In addition, many smaller rivers have floodplain wetlands. Although many floodplain wetlands have been cleared, leveed, ditched, and otherwise hydrologically altered, considerable forested floodplain wetland acreage remains.

Through time, the development of rivers and drainage networks can create wetlands through the formation of wide floodplains, natural levees, cut-off meanders, and deltas. As rivers downcut, they can expose coarse-grained sediment and bedrock in their valley walls, allowing groundwater discharge and the development of fens and other groundwater-supported wetlands. Rivers can change course, deposit fine-grained sediments in the floodplain, and have other impacts that increase inundation or saturation and form wetlands. Alternatively, rivers can also destroy wetlands as the river systems mature and increase drainage of the uplands.

As an example of how rivers can generate wetlands, Jo Daviess County in northwestern Illinois was not glaciated during the Quaternary Period, and millions of years of riverine erosion have created a well-developed drainage system. Due to efficient natural drainage, wetlands are almost absent in upland areas, but have formed in the river valleys as the floodplains have developed. Current and former wetlands in Jo Daviess County are shown in Figure 21-5.

During the Quaternary Period (Chapter 12), extensive wetlands were created in Illinois as major rivers, carrying high sediment loads from glacial meltwaters, deposited thick sediment successions in their floodplains (aggraded). Many tributary streams were dammed by this sediment, forming backwater lakes that in turn led to widespread inundation in tributary valleys and the deposition of fine-

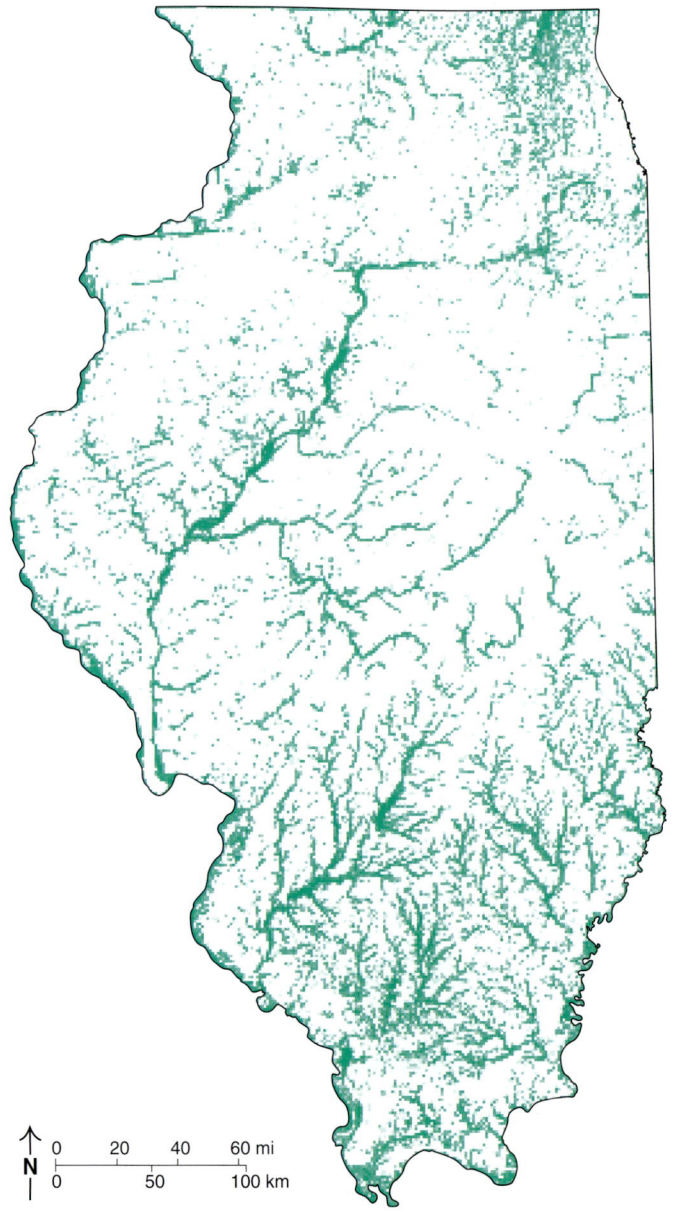

Figure 21-4 Mapped locations of modern wetlands (Suloway and Hubbell 1994). This map was produced using data collected in the 1980s by the National Wetlands Inventory program and shows individual wetlands. Some smaller wetlands are not visible at this scale.

grained sediments. The fine-grained sediment and the flat topography created extensive wetlands after these backwater lakes eventually drained, including floodplain forest wetlands, swamps, marshes, and other wetland types. Sediment deposited in these areas is generally mapped as Equality Formation silts and clays, and occurs in many places in Illinois, although this unit is especially common in southern Illinois (Figure 12-12).

Similarly, natural course migrations have caused the abandonment of river channels and floodplains, leading to the creation of major riverine wetland systems. The previously mentioned Ramsar-designated cypress swamps of the Cache River valley in southern Illinois were formed when the Ohio River shifted southward to its current position, abandoning its former channel in the Cache River valley (Frankie et al. 1997). The low gradient of the abandoned valley allowed the formation of vast wetland complexes, and this wetland complex is the largest remaining in Illinois (Frankie et al. 1997).

Human activities also have affected riverine wetland systems. For example, the Illinois River valley has been extensively altered by human activities. Lock-and-dam systems installed in the late 1800s and up to the mid-1900s raised the lowest river stages by several feet. The reversal of the Chicago River in 1900 contributed more discharge, raising the lowest river stages by as much as an additional 4 feet (1.2 m) (Thompson 2002) and raising flood heights even further. In response, farmers and communities along the river installed levees and pump stations to prevent flooding and to drain the floodplain for agriculture. A study from the early 1900s suggested that 80% of the floodplain's floodwater storage area had been removed by levees and that the floodway had been reduced to 25% of its original area (Thompson 2002). In addition, artificial drainage of land in the Illinois River watershed increased the rate and volume at which runoff was delivered to the river, further increasing flood heights on the unprotected land remaining in the floodplain. Many backwater lakes and unleveed areas have become permanently inundated by these changes.

Glacial Deposits

As discussed in Chapter 12, Quaternary Period, glaciers advanced into Illinois on numerous occasions during the Quaternary Period and left deposits that cover about 90% of Illinois (Willman et al. 1975). Glacial processes and their deposits create a landscape that promotes wetlands in a number of ways. Sediments deposited by glaciers generally have a poorly developed drainage network, thus allowing inundation over broad areas. Also, glacial till in Illinois is commonly rich in clay, which prolongs saturation by slowing infiltration. The glacial till surface can be hummocky, forming a number of isolated depressions that can trap water.

The most recent glaciation covered the northeastern quadrant of Illinois (Hansel and Johnson 1996) and formed a large number of presettlement wetlands (Figure 21-3a). Extensive wetlands also formed after older glaciations elsewhere in Illinois. Although many of those older wetlands in southern and western Illinois drained long ago as stream networks developed through time, their former locations can be identified by the geological deposits and hydric soil features that are preserved. Typical wetlands that form on glacial deposits include wet prairies, sedge meadows, and marshes. These wetlands generally contain shallow standing water during spring and dry up during summer. Deeper marshes may not dry completely every year.

Glacial Meltwater

In addition to the glacial deposits that favored wetland development in Illinois, glacial meltwater interacted with the landscape to form wetlands. Glacial moraines trapped meltwater, forming expansive glacial lakes. These lake basins had very low relief and poor drainage because of the silt and clay deposits that lined the lakebed, favoring the formation of large tracts of wetlands once they eventually drained.

One of the largest inland wetland complexes in the country was the Kankakee Marsh, located on over 3 million acres (1.2 million ha) in Indiana and extending into Illinois (Mitsch and Gosselink 2000). The marsh was located in the beds of a series of former lakes that formed

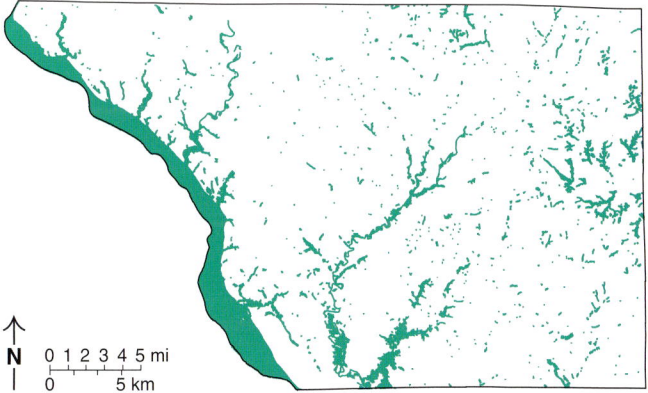

Figure 21-5 Current and former wetlands mapped in Jo Daviess County from the National Wetlands Inventory and soil survey data. Wetlands are dominantly located in floodplains, where they formed during floodplain development. Wetlands are rare in the uplands due to the formation of an extensive drainage network. Areas of open water are also included. (Figure prepared by Bonnie Robinson.)

when copious meltwaters and torrential floods from late Wisconsin glaciers (Willman and Frye 1970) inundated the landscape south of the Valparaiso Moraine. When the lakes receded, the poorly drained lake beds supported wetlands that persisted until the 1850s, when large-scale drainage efforts began. The vast majority of the Kankakee Marsh is now artificially drained and farmed, although attempts to restore and preserve small parts of the original marsh are under way. Other former lake basins and associated wetland complexes existed over parts of Cook, Grundy, Iroquois, Douglas, and neighboring counties. These former lakes are marked by silt and clay deposits of the Equality Formation (see Figure 12-12). Typically, shallow to deep marshes, wet prairie, and sedge meadows form in poorly drained areas such as former lake beds and undrained depressions.

Similarly, large blocks of glacial ice from retreating glaciers often remained behind and were left encased in sediments, especially in moraines. These blocks later melted to form kettle lakes or closed depressions that could be occupied by wetlands. These types of wetlands are concentrated in northeastern Illinois, although they formed elsewhere in Illinois after older glaciations. Many of these kettles eventually developed into bogs or deep marshes, and existing wetlands often have continuous standing water. Peat may form in these types of wetlands due to the sustained inundation.

Finally, glacial meltwaters eroded through sediments and bedrock, forming channels that may have become wholly or partly abandoned. These meltwater floods often facilitated wetland growth by exposing bedrock and coarse sediments, which may have allowed groundwater discharge, or by creating flat areas that drained poorly. Fens are found in many places along the bluffs of former glacial meltwater channels where groundwater discharge occurs, and peat can accumulate because of the continuous saturation by groundwater. Some major meltwater channels that formed in Illinois include the Green River Lowland, the Illinois River, Des Plaines River, Fox River, Cal-Sag (Calumet-Saganashkee) Channel, and numerous smaller channels that are now abandoned or reoccupied by small streams or ditches.

THE GEOLOGICAL RECORD OF WETLANDS IN ILLINOIS

Wetlands have left geological deposits in Illinois that range in age from ancient to modern. Descriptions of many of these geological deposits appear elsewhere in this book (especially Chapter 10, Pennsylvanian and Permian; Chapter 12, Quaternary Period; and Chapter 14, Coal), but they are summarized here to discuss their wetland-related origin.

Soils

Modern wetlands occur on about 900,000 acres (360,000 ha) in Illinois (2.5% of total land area). In these wetlands, deposited sediments such as silt, clay, and organic matter are altered by the anoxic conditions, creating what are known as hydric soils, which are characterized by mottling and gleyed colors produced by reduced iron and other elements. Due to the low decomposition rates in wetlands, hydric soils usually contain enough accumulated organic matter to be dark or black in color.

Wetlands that were drained after settlement occur at the land surface on another 7.8 million acres (3.2 million ha) of Illinois (21.5% of total land area). Many of these acres are located on deposits of the most recent (Late Wisconsin) glaciers, but many also occur on deposits of older glaciers. Fewer wetlands occur in the unglaciated parts of Illinois, except those wetlands formed by river processes, as discussed previously. Deposits in drained wetlands are similar to those in modern wetlands and have accumulations of silts and clays, organic material, peat, and other minerals. Even after drainage, hydric soil features may persist indefinitely, such as mottled or gleyed colors, nodules, and accumulations of organic matter.

The majority of drained wetlands have been turned into farmland, where the buildup of organic matter allows highly productive farming. Indeed, the most highly productive land area in Illinois and adjacent states comes from drained wetlands (Mitsch and Gosselink 2000).

Geosols

Wetlands that developed on the surface of older glacial deposits but have since been buried by more recent sediments are generally identified by buried soil horizons (geosols). Former wetlands are recognized by the dark or reduced color of silt and clay deposits, the presence of peat and other dark organic matter, mottling, and other hydric soil features. The Yarmouth, Sangamon, and Farmdale Geosols (see Chapter 12, Quaternary Period) are some of the prominent buried soils that contain wetland features. Some deposits of silt, clay, and organic matter of former wetlands are considerable enough in some places to be given individual stratigraphic names (e.g., Lierle Clay Member, Berry Clay Member). The extents of these deposits are poorly known in some instances because their exposures are limited.

Peat

The extent of peat deposits in Illinois is poorly mapped because they are generally relatively small, reflecting the size of the wetlands in which they were originally deposited. However, in certain areas of Illinois, peat deposits are large enough to be mapped and may extend for several square miles. Figure 21-6 shows Grayslake Peat deposits at the land surface in Illinois based on digital soil survey data (Natural Resources Conservation Service 2009). Peat occurs in every county covered by late Quaternary glaciers (Willman and Frye 1970) as well as some others. Continuous saturation is generally required for the accumulation of peat, which would otherwise oxidize and decay. In Illinois, most of these peat deposits are in locations where groundwater discharge occurs, which offsets the summertime drying that is typical of wetlands supported by runoff or rainfall.

Common wetland types that accumulate peat include fens, bogs, and deep marshes, which are generally saturated throughout the year. Many are related to channels or lakes formed by glacial meltwaters. Erosion of meltwater channels exposed aquifers, causing groundwater discharge that produced many peat-accumulating fens, especially in northeastern Illinois along the Fox and other rivers. Goose Lake Channel, a former glacial drainageway in Whiteside County between Fenton and East Clinton, contains several square miles of peat that is mined extensively. Peat also is common where sand is at land surface and groundwater discharges, such as along the Illinois River in Mason and Cass Counties and in the Green River Lowland in Bureau, Henry, and Whiteside Counties.

In places, peat deposited in former wetlands has been covered by more recent glacial deposits. Portions of geological deposits (such as the Robein Member of the Roxana Silt) contain peat, although no extensive deposits of peat have been mapped due to their limited exposures.

Coal

As one of the most valuable mined products in Illinois, coal has greatly influenced the state's development and economy. Details about coal, coal mining, and coal's economic importance are discussed extensively in Chapter 10, Pennsylvanian and Permian, and Chapter 14, Coal.

Most coal derives from organic matter that was deposited in wetlands that existed in Illinois during the Pennsylvanian Period from about 325 to 286 million years ago. During that time, the southern two-thirds of Illinois was part of a broad coastal plain, providing a landscape where extensive swampy wetlands formed (Wanless et al. 1969). Drainage from the east, north, and northwest came into the flat, poorly drained deltas, providing water to the swamps that formed there. In these wetlands, a wide array of plants and animals lived, such as tree ferns, lycopods, and insects. This highly productive environment allowed thick deposits of organic material to accumulate as peat. Through time, delta lobes shifted and ocean levels fluctuated, burying the peat for future transformation into coal.

Although no coal swamps exist today in Illinois, the cypress swamps in southern Illinois are perhaps the most physically similar type of wetland that remains. Although peat is forming in small wetlands today, no peat deposits as large as those of the Pennsylvanian Period have formed

Figure 21-6 Mapped deposits of Grayslake Peat, which include Adrian muck, Aurelius muck, Elvers silt loam, Houghton muck, Lena muck, Muskego muck, Muskego silty clay loam, Palms muck, Wallkill silt loam, Wallkill silty clay loam, and Organic Pits. This map is based on soil survey data alone (Natural Resources Conservation Service 2009) and does not utilize previous Illinois State Geological Survey mapping of Grayslake Peat. Thus, some differences occur in mapped extent. Portions of Cook County are not mapped. (Map by Adrianne Knight.)

in Illinois since that time. The Pennsylvanian age coal deposits are unique in Illinois' geological history and form a heritage that benefits Illinois citizens today.

FUNCTIONS OF WETLANDS

The functions of wetlands have been extensively studied nationwide, and many have been identified as having economic value to society. These include storing floodwaters, improving water quality, recharging shallow aquifers, maintaining base flows in rivers, providing recreation, providing wildlife habitat, and stabilizing river banks (National Research Council 1995). Although all of these wetland functions have been observed or are expected to occur in Illinois, only functions known from a study of Illinois wetlands are discussed in this section.

Wetlands moderate stream flows by decreasing flood peaks and discharge and by increasing low flows during summer. In one Illinois study, Demissie and Khan (1993) found that peak flood flows decreased by 3.7%, and total volume of flows decreased by 1.4%, with each 1% increase in wetland area in a watershed; in addition, each 1% increase in wetland area in a watershed increased summertime low flows by 7.9%. Because wetlands store floodwaters and reduce flood peaks, researchers have calculated that all of the floodwaters from the record Mississippi River flood of 1993 could have been stored if 13 million acres (5.3 million ha) of wetlands had been restored throughout the upper Mississippi River watershed (Mitsch and Gosselink 2000), thus greatly reducing or eliminating the flood. This amount of restoration would be less than twice the acreage of wetlands lost in Illinois alone.

Water quality is also improved by wetlands. Illinois wetlands have been shown to reduce suspended solids, nitrogen, and phosphorus (Hey et al. 1994), reducing impacts to the environment. Nitrogen, which is applied widely as fertilizer in Illinois and throughout the Midwest, has been implicated in creating a "dead zone" in the Gulf of Mexico (Mitsch and Gosselink 2000). It is hypothesized that the delivery of nitrogen to the Gulf via the Mississippi River causes algal blooms in the ocean that eventually die and decay, reducing oxygen over a large area and subsequently causing an observed decline in sea life. Wetland restoration is one potential solution for this problem because the anoxic conditions within wetlands can remove nitrogen from surface water by converting it to gaseous forms (i.e., denitrification). One Illinois experimental study found that total nitrogen in surface water was reduced by up to 74% after the water was routed through a wetland (Phipps and Crumpton 1994). Restoration of about 24 million acres (9.7 million ha) of wetlands and buffer strips throughout the Mississippi River watershed has been calculated as sufficient to eliminate this problem (Mitsch and Gosselink 2000).

Additionally, runoff from farm fields and other disturbed areas also damages native habitats by providing a high sediment load to rivers. Wetlands slow runoff, which causes deposition of entrained sediment (Fennessy et al. 1994), thus helping protect native plants and animals that are not adapted to high sediment levels in surface water. Heavy metals and other pollutants adsorbed onto the sediment particles are removed at the same time.

Wetlands also provide considerable habitat benefits to wildlife. Studies of Illinois flora and fauna suggest that many species utilize wetlands for at least some portion of their life cycle; 42% of native Illinois plant species are considered wetlands species, and 90% of amphibians inhabit wetlands (Levin et al. 2002). A considerable number of endangered plants and animals, including almost half of endangered reptiles, are found in wetlands. In one Illinois study, an experimental wetland creation project was associated with a dramatic increase in bird usage and nesting (Hickman 1994); the total number of birds observed increased by 4,000%, and the number of species of birds observed increased by 400%.

CONCLUSION

Wetlands greatly contribute to the economy and environment of Illinois. Wetlands have been widespread in the recent past and in geological history, and the deposits they have left are economically important, especially to the farming and mining sectors. Society now recognizes the services that wetlands perform, including flood storage, water-quality improvements, habitat, recreation, and many others. It is no longer federal policy to drain all wetlands, and assistance for those endeavors has largely been eliminated. The damage caused by wetland drainage has been recognized, leading to proposed large-scale federal efforts to restore rivers such as the Illinois River. Federal and state permitting systems now attempt to balance the economic benefits versus the damage caused by wetland drainage and to strive for no net loss of wetlands. Understanding the benefits of wetlands is necessary to allow for the wise use of Illinois' heritage of wetland resources without reducing the functions that wetlands perform for all of the state's citizens.

REFERENCES

Bogue, M. B., 1959, Patterns from the sod: Land use and tenure in the grand prairie, 1850–1900: Springfield, Illinois, Illinois State Historical Library, 327 p.

Dahl, T. E., 2000, Status and trends of wetlands in the conterminous United States, 1986–1997: Washington, D.C., Department of the Interior, U.S. Fish and Wildlife Service, 82 p.

Demissie, M., and A. Khan, 1993, Influence of wetlands on streamflow in Illinois: Champaign, Illinois, Illinois State Water Survey, Contract Report 561, 47 p.

Environmental Laboratory, 1987, Corps of Engineers wetlands delineation manual: Technical report Y-87-1 (on-line edition): Vicksburg, Mississippi, U.S. Army Corps of Engineer Waterways Experiment Station, 143 p. http://el.erdc.usace.army.mil/wetlands/pdfs/wlman87.pdf. Accessed June 27, 2006.

Fennessy, M. S., C. C. Brueske, and W. J. Mitsch, 1994, Sediment deposition patterns in restored freshwater wetlands using sediment traps: Ecological Engineering, v. 3, p. 409–428.

Frankie, W. T., R. J. Jacobson, J. M. Masters, N. L. Rorick, A. K. Admiraal, M. R. Jeffords, S. M. Post, M. A. Phillips, and E. Jones, 1997, Guide to the geology of the Mississippi Embayment Area, Johnson and Pulaski Counties, Illinois: Illinois State Geological Survey, Field Trip Guidebook 1997D, 72 p.

Hansel, A. K., and W. H. Johnson, 1996, Wedron and Mason Groups: Lithostratigraphic reclassification of deposits of the Wisconsin Episode, Lake Michigan Lobe area: Illinois State Geological Survey, Bulletin 104, 116 p.

Havera, S., 1999, Waterfowl of Illinois: Status and management: Champaign, Illinois, Illinois Natural History Survey, Special Publication 21, 628 p.

Hensel, B. R., 1992, Natural recharge of groundwater in Illinois: Illinois State Geological Survey, Environmental Geology 143, 33 p.

Hey, D. L., A. L. Kenimer, and K. R. Barrett, 1994, Water quality improvement by four experimental wetlands: Ecological Engineering, v. 3, p. 381–397.

Hickman, S., 1994, Improvement of habitat quality for nesting and migrating birds at the Des Plaines River Wetlands Demonstration Project: Ecological Engineering, v. 3, p. 485–494.

Lineback, J. A., 1979, Quaternary deposits of Illinois: Illinois State Geological Survey, 1:500,000.

Levin, G. A., L. Suloway, A. E. Plocher, F. R. Hutto, J. J. Miner, C. A. Phillips, J. Agarwal, and Y. Lin, 2002, Status and functions of isolated wetlands in Illinois: Champaign, Illinois, Illinois Natural History Survey, Special Publication 23, 16 p.

McCorvie, M. R., and C. L. Lant, 1993, Drainage district formation and the loss of midwestern wetlands, 1850–1930: Agricultural History, v. 67, no. 4, p. 13–39.

Mitsch, W. J., and J. G. Gosselink, 2000, Wetlands: New York, John Wiley and Sons, 920 p.

National Research Council, 1995, Wetlands: Characteristics and boundaries: Washington, D.C., National Academy Press, 307 p.

Natural Resources Conservation Service, 2009, Soil data mart. http://soildatamart.nrcs.usda.gov. Accessed July 20, 2010.

Phipps, R. G., and W. G. Crumpton, 1994, Factors affecting nitrogen loss in experimental wetlands with different hydrologic loads: Ecological Engineering, v. 3, p. 399–408.

Prince, H. C., 1997, Wetlands of the American Midwest: A historical geography of changing attitudes: Chicago, Illinois, University of Chicago Press, 395 p.

Suloway, L., and M. Hubbell, 1994, Wetland resources of Illinois: An analysis and atlas: Champaign, Illinois, Illinois Natural History Survey, Special Publication 15, 88 p.

Taylor, C. B., J. B. Taft, and C. E. Warwick, eds., 2009, Canaries in the catbird seat: The past, present, and future of biological resources in a changing environment: Champaign, Illinois, Illinois Natural History Survey, Special Publication 30, 306 p.

Thompson, J., 2002, Wetlands drainage, river modification, and sectoral conflict in the lower Illinois Valley, 1890–1930: Carbondale, Illinois, Southern Illinois University Press, 284 p.

Thompson, J., 2003, Pioneer dredging in the midwestern wet prairies: Bulletin of the Illinois Geographical Society, v. 45, no. 2, p. 3–32.

U.S. Army Corps of Engineers, 2008, Interim regional supplement to the Corps of Engineers wetland delineation manual: Midwest Region: Vicksburg, Mississippi, U.S. Army Corps of Engineers, Engineer Research and Development Center, Environmental Laboratory, Report no. ERDC/EL TR-08-27, 152 p.

Wanless, H. R., J. R. Baroffio, and P. C. Trescott, 1969, Conditions of deposition of Pennsylvanian coal beds, in E. C. Dapples and M. E. Hopkins, eds., Environments of coal deposition: Boulder, Colorado, Geological Society of America, Special Paper 114, p. 105–142.

Willman, H. B., E. Atherton, T. C. Buschbach, C. Collinson, J. C. Frye, M. E. Hopkins, J. A. Lineback, and J. A. Simon, 1975, Handbook of Illinois stratigraphy: Illinois State Geological Survey, Bulletin 95, 261 p.

Willman, H. B., and J. C. Frye, 1970, Pleistocene stratigraphy of Illinois: Illinois State Geological Survey, Bulletin 94, 204 p.

Geological Applications: Land Use and Environmental Hazards in Illinois

Beverly L. Herzog

The geological setting of a place is the fundamental framework on which life depends, furnishing the basic requirements of water, food, and shelter. Illinois is abundantly endowed with life-supporting geological materials, making the state one of the most fertile, resource-rich, and hospitable regions on Earth and providing a home for several million people.

The geological setting of the state, however, also holds the potential for serious natural hazards, such as floods, landslides, earthquakes, and radiation. Human activities, too, can affect geological materials in ways that pose hazards and that may result in diminished environmental quality. Destructive activities include contaminating soil and groundwater, accelerating erosion due to poor agricultural and construction practices, building structures and infrastructure on unstable terrain or floodplains, and degrading the environment during the extraction of mineral resources. In many cases, these activities have been so subtle and pervasive and have occurred over such long periods of time that they have nearly gone unnoticed. However, the long-term effect can be more harmful than sudden damage from more obvious natural disasters.

Fortunately, awareness of land use and environmental hazards by the citizens of Illinois has grown considerably over the past 30 years. The disciplines of environmental and engineering geology are providing valuable insights that will help in the long-term management of the state's geological hazards.

This part of the volume focuses on how geology impacts everyday activities and mitigates or exacerbates the effects of human actions on the environment.

Photograph by Joel M. Dexter.

Image on previous page: Aerial view of Grafton, Illinois, Jersey County, within the Mississippi River floodplain during the Great Flood of 1993.

Soils 22

Michael L. Barnhardt

INTRODUCTION

Illinois contains some of the most fertile and productive soils in the world. These high-quality soils are a legacy of the massive continental glaciers that repeatedly advanced and retreated across most of the state, altering the Illinois landscape through erosion and deposition. These geological processes of erosion and deposition, along with the torrents of water released from melting ice, created sediment mixtures that were spread across the landscape. In the wake of these glacial events, soil-forming processes began to alter the deposited sediments into the soils of today.

The state's glacially derived soils are a natural resource forming the foundation that supports the growth of all food production and of trees that supply wood and the raw material for paper products. Soils (unconsolidated sediments) help regulate water supply by storing it as groundwater and then slowly releasing it to streams, rivers, lakes, and oceans, thereby smoothing hydrologic discharge. Soils also act to filter water, improving its quality as the water percolates through it, leaving behind organic matter and chemicals that can enrich or contaminate the soil. Soils shape the development of ecosystems and their plant and animal populations. Human activities are influenced by soil properties that aid or limit the construction of buildings, the development of transportation networks, and the expansion of urbanized areas.

This chapter discusses the interacting factors and processes that control the development of Illinois soils, explains their occurrence on the landscape, and illustrates how information on soils can be used to plan for population growth and resource utilization. For more details on soil-related topics, consult chapters that deal more specifically with Illinois glacial history (Chapter 12, Quaternary Period), wetlands (Chapter 21, Wetlands Geology), construction limitations (Chapter 23, Siting and Design), and the influence of soil and sediment on water resources (Chapter 19, Groundwater Quality; Chapter 20, Protecting Groundwater; and Chapter 28, Karst Terrane).

LAND COVER AND LAND USE CLASSIFICATION

The modern cultural and biological landscape of Illinois is very different from that observed by the earliest European settlers. Emerging from the dense eastern forests, pioneers found the Illinois landscape covered by native prairie vegetation; forests were confined to better-drained, rolling upland sites and along bottomlands (Figure 22-1). Prairie vegetation then covered about 55% of the landscape, but the forests were already encroaching upon the prairies—as evidenced by the presence of moderately dark-colored soils under forests along the prairie-forest transition (Fehrenbacher et al. 1984).

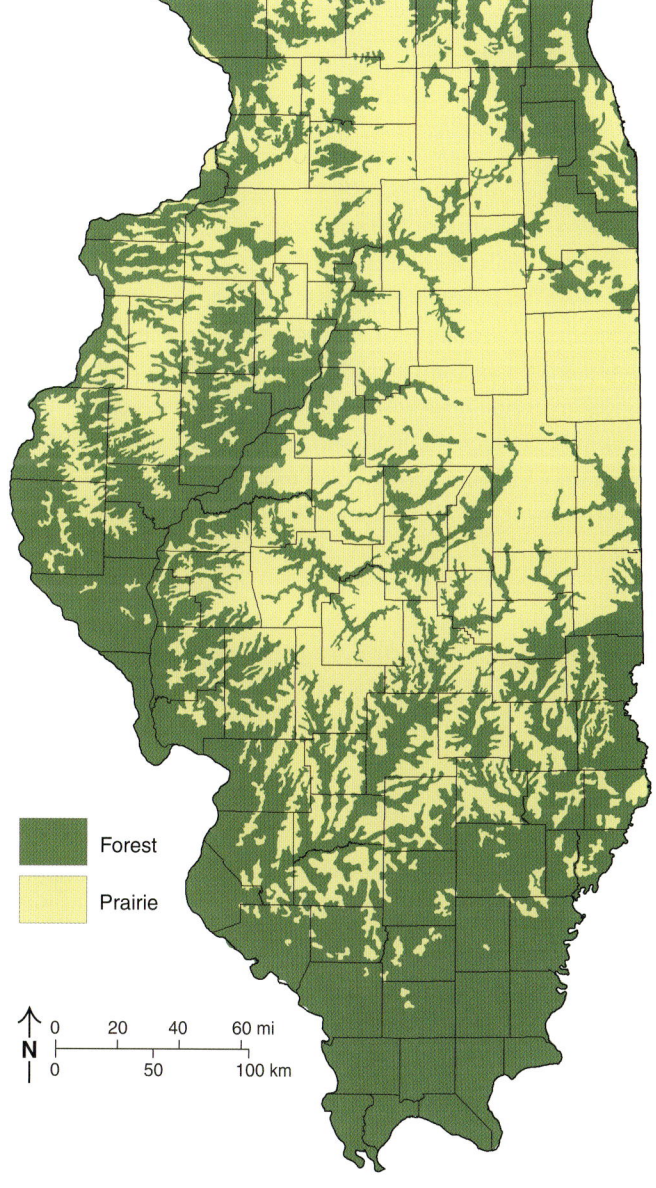

Figure 22-1 Native vegetation in Illinois (Fehrenbacher et al. 1984).

Today, virtually no native prairie remains in Illinois. Cropland covers more than 65% of the total surface area of about 36,065,963 acres (14,595,377 ha) and forest land and pasture cover about 23% (U.S. Department of Agriculture, National Agriculture Statistics Service 2002). Recent trends reflect the effects of population growth and subsequent urban expansion. About 58% of Illinois land is classified as prime farmland, of which about 89% is used as cropland. Since 1982, about 405,900 acres (164,268 ha) of prime farmland have been lost. About 33,000 acres (13,355 ha) of prime farmland are currently being lost each year. Continuing urban expansion suggests this trend will continue (U.S. Department of Agriculture, Natural Resources Conservation Service 1997).

Soil Classification

An area's land cover and land use are strongly related to the character of its soil, and the soil itself is a product of many environmental factors—physical, chemical, and biological processes. "Top soil," "black earth," "sediment," "loam," and "dirt" are some of the generic terms commonly used to describe or identify soil, but, depending on the individual and the use (e.g., agricultural, engineering), the term "soil" can imply a considerable range of information. Soils are not a material that is deposited; rather, they are developed by and are the result of the weathering of geological materials. Because soil-forming processes are most intense at or near the land surface, only the upper portion of a geological deposit will be altered by the various physical, chemical, and biological processes to such a degree that recognizable, distinctive layers with unique characteristics—known as soil horizons—are produced. The type, number, and character of soil horizons that develop at a specific location are influenced by climatic, topographic, geological, and biological factors and the amount of time over which they have interacted.

The Herrick silt loam is a widely occurring soil in southwestern Illinois and eastern Missouri that exhibits several distinctive soil horizons (Figure 22-2). A view of the soil profile for this soil shows that, at the land surface, organic matter from plants and organisms has accumulated to produce a dark-colored horizon called the A horizon. Underlying it is a whitish horizon called the E horizon (eluvial horizon), which is the result of organic material and clay particles being chemically and physically removed from this horizon by water percolating through the sediment. Next is the brownish B horizon, which has developed through the accumulation of organic and clay particles moved downward from the E horizon. Further differentiation of the B horizon occurs below and is reflected in color and texture differences. Below a depth of about 5 feet (1.5 m), modern soil-forming processes have had little influence, and the geological material (in this case, silt) is essentially unaltered and is called the C horizon. A virtually infinite number of variations can occur across the landscape, at many scales, and this continuum is reflected in the many different soils that can be found throughout Illinois and the world.

Different disciplines describe and classify soils (and sediments) differently depending on their objectives. Geologists may concentrate more on the texture, genesis, and location of the sediment, whereas soil scientists and engineers may focus more on the ability of the soil to support plant growth or man-made structures. Soils can be described in great detail to provide the greatest amount of use to the largest audience (Buol et al. 1997, Schoeneberger et al. 2002).

Soil Orders in Illinois

Soil patterns on the landscape have been classified. The resulting soil class hierarchy helps scientists better

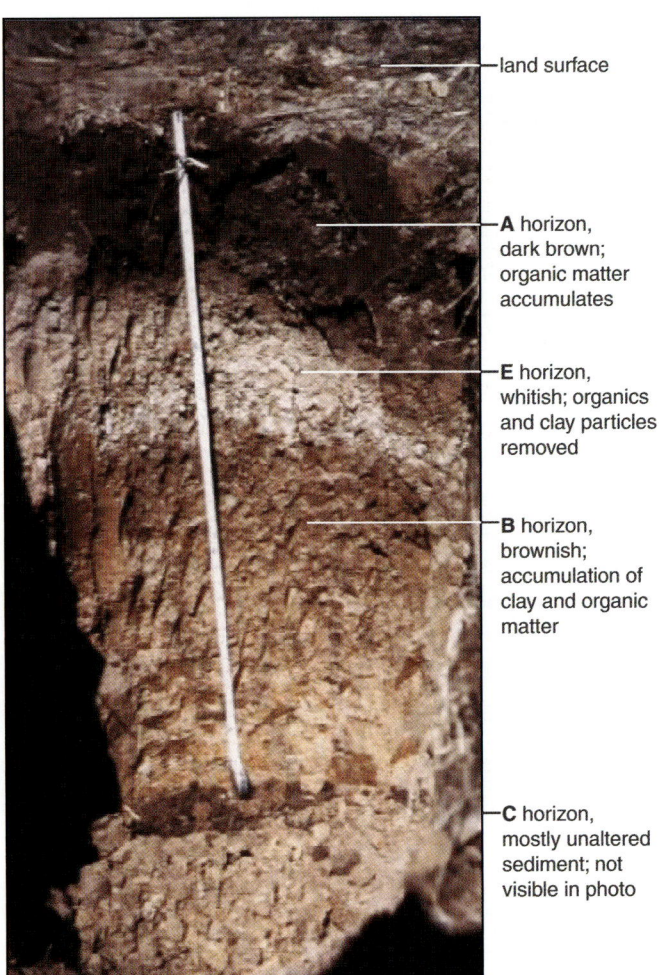

Figure 22-2 Soil horizons, Herrick silt loam. (Photograph courtesy of U.S. Department of Agriculture, Natural Resources Conservation Service.)

understand the relationships between different soils, their limitations, and their properties. The system used in the United States has six categories; of these, soil order is the highest rank (Struben and Lilly 1999). There are twelve orders worldwide, but only six occur in Illinois: Mollisols, which cover about 49% of the state, Alfisols (46%), Entisols (3.25%), Inceptisols (1.5%), Histosols (0.25%), and Ultisols (<0.1%) (Fehrenbacher et al. 1984).

Characteristics and Locations

Mollisols and Alfisols dominate the Illinois landscape (Figure 22-3). Generally speaking, the dark-colored Mollisols are found in central and northern Illinois, where native prairie grasses once covered the landscape, and along some bottomlands. In these areas, the vegetation's thick, deep root masses generated large amounts of organic matter that slowly decomposed and accumulated in the sediment, developing into highly productive, fertile, black soil (Figure 22-4). In contrast, the Alfisols are light-colored soils that most often develop beneath deciduous forest vegetation in the rolling topography that borders stream valleys (Figure 22-5). Alfisols are found primarily in the southern part of the state. Although generally less fertile than Mollisols, under best management practices, Alfisols can be very productive.

Entisols, Inceptisols, Histosols, and Ultisols cover only a small part of the Illinois landscape, but they create important and unique ecosystems and habitats. Entisols are most often found on floodplains where recently deposited sediments have not been subjected to soil-forming processes long enough to have modified the sediments into recogniz-

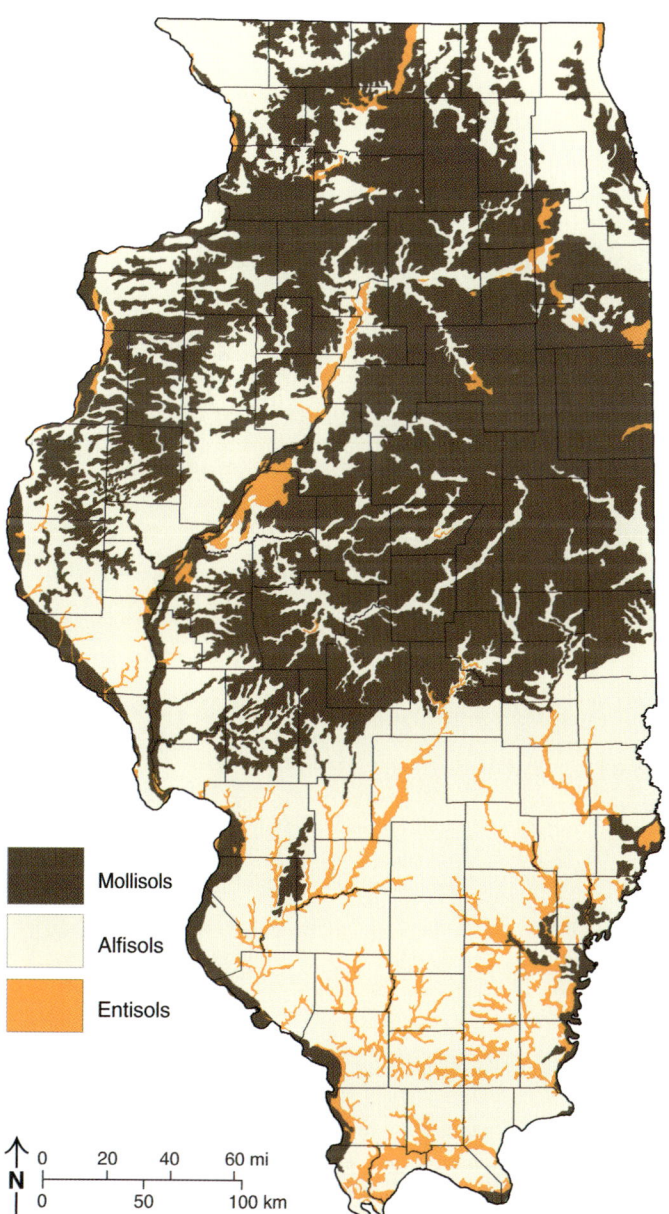

Figure 22-3   Major soil orders in Illinois (Fehrenbacher et al. 1984).

Figure 22-4   Mollisol, Catlin silt loam. Scale intervals are 10 cm (3.94 inches). (Photograph courtesy of U.S. Department of Agriculture, Natural Resources Conservation Service.)

able horizons. Entisols also form in very sandy areas along stream channels and on slopes where erosion has inhibited soil development. Inceptisols are generally restricted to steep slopes (Figure 22-6) and floodplains where minimal soil development has occurred, yet they have some soil horizon development that distinguishes them from Entisols. Histosols are soils that have developed in predominantly decomposed organic matter and include peats and mucks. Histosols are commonly wet and occupy low-lying areas on floodplains and upland depressions, mainly in northeastern Illinois. Although Histosols occupy less than 1% of the land surface in Illinois, they play an important role in local and regional ecosystems, interacting with surface and groundwater systems (see Chapter 21, Wetlands Geology). Ultisols resemble Alfisols, but Ultisols are more strongly weathered soils of generally lower fertility and are usually found on older landscapes. In Illinois, they are found only in a few locations in extreme southern and southwestern Illinois outside the southern limits of Pleistocene continental glaciers. Ultisols are not important soils in Illinois.

Soil Development Factors

Jenny (1941) stated that a given soil property is related to the interaction of climate, organisms, relief, parent material, and time. Each of these soil-forming factors involves many physical, chemical, and biological processes whose rate, intensity, and duration vary over time and space to create patterns of soils on scales ranging from global to local. A nearly infinite number of combinations are possible on a global scale, but, at smaller scales, one or two of the factors tend to dominate.

Climate

Climate is important in soil development because it influences the type and rate of physical and chemical weath-

Figure 22-5 Alfisol, Fayette silt loam. (Photograph courtesy of U.S. Department of Agriculture, Natural Resources Conservation Service.)

Figure 22-6 Inceptisol, Gosport silt loam. (Photograph courtesy of U.S. Department of Agriculture, Natural Resources Conservation Service.)

ering and the type of vegetation that grows in the area. The major components of climate are temperature and precipitation. Although Illinois generally has a humid, temperate climate, there are recognizable differences that influence soil development in different parts of the state. In Illinois, the annual mean temperature ranges from about 44°F (6.7°C) in the north to about 58°F (14.4°C) in the south. Normal annual precipitation ranges from about 33 inches (84 cm) in the north to more than 48 inches (122 cm) in the south (Illinois State Water Survey 2007). The amount of water in a given location affects how much water is available to move vertically into the surface sediment to promote the development of soil horizons. Amounts also affect how much water is available to move horizontally across the surface of the landscape, eroding sediment from one site and depositing it at another.

Local differences in climate can be found in some areas, such as where north- and south-facing slopes along a river valley exhibit different vegetation and soils due to the difference in the amount of solar radiation absorbed by sunny (south-facing) and shaded (north-facing) slopes. This difference in radiation affects the amount of water available for plant growth and soil development and, consequently, the amount of organic matter that is produced and its rate of decay. In some areas, these differences in soil thickness and type and in vegetation density can lead to the development of landslides, soil creep, and other forms of mass movement. Soils develop slowly, so most soil properties—such as texture, structure, soil horizons, and thickness—reflect the composite influences of climatic change that may span thousands of years.

Organisms

The factor organisms includes the combined effects of flora and fauna on and in the soil. Native vegetation cover in Illinois was mostly prairie and forest in the recent past (Figure 22-1), and the distribution of two soil orders, Mollisols and Alfisols (Figure 22-3), closely matches that of native prairie and forest, respectively. The fertile, black-to-dark brown Mollisols are the result of the deep, dense rootlet mass produced by the prairie grasses and the subsequent decomposition of the roots by organisms such as fungi and earthworms (Figure 22-4). Forests also produce large amounts of organic debris, such as leaves, but the majority of this material accumulates and decomposes on top of the soil rather than within it (Figure 22-5). The decomposed organic matter (humus) in Alfisols soils is more acidic than grassland humus; thus, leaching in forest soils is more aggressive.

Relief

Relief, or topography, takes into account how landscape factors affect soil development. Relief includes the degree of steepness—the slope gradient—and the total change in elevation at a location. Relief influences drainage, the amount of infiltration and runoff, and water table levels. At the local scale, drainage differences are usually the major reason for variations in soil development. Soils may be well drained at the top of the slope and poorly to very poorly drained at the base of the slope or in nearby flat-lying areas. Relief influences the degree of erosion by affecting the concentration, direction, and amount of water moving across the landscape. Figure 22-7 illustrates a landscape where drainage is important.

An examination of the *Surface Topography of Illinois* statewide map (Luman et al. 2003, Figure 12-2) reveals a landscape of considerable topography and character, which may surprise many individuals who perceive the state's surface to be flat and featureless. The state's physiographic variability interacts with climatic, biological, hydrological, and other processes to create myriad landscapes, which, upon more detailed inspection, reveal interesting micro-habitats and biological niches, such as sandy areas where cacti grow and where different soil moisture conditions on north-facing and south-facing slopes result in different vegetative covers.

Parent Material

Parent material is the physical material in which soil develops. The parent material is modified by the soil-forming processes. Although a few soils develop in decomposed

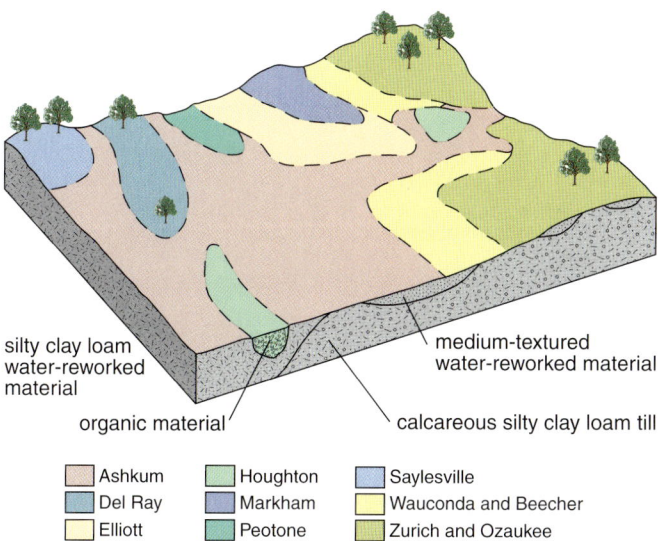

Figure 22-7 Landscape with drainage differences, Lake County, Illinois (Paschke and Alexander 1970).

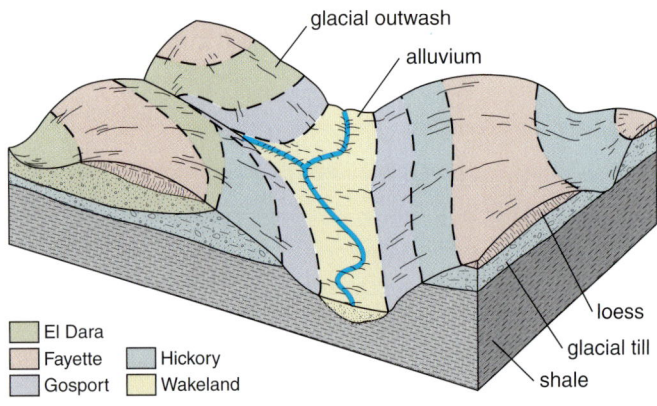

Figure 22-8 Relationship of soils and parent materials in Pike County, Illinois (Struben and Lilly 1999).

soils have developed completely or partially in loess. During and after the glacial episodes, wind lifted the silt and clay that had been deposited by glacial meltwater on large floodplains such as those along the Mississippi River and the Illinois River. This windblown silt was deposited like a blanket draped over the underlying landscape. Most of the state is covered by at least 5 feet (1.5 m) of silt. Loess commonly is thicker than 25 feet (7.6 m) in places along the eastern bluffs of major floodplains due to predominantly west-northwest winds (Fehrenbacher et al. 1986; Willman and Frye 1970). This ubiquitous material is a major reason for the high quality of Illinois soils: loess has high available moisture-storage capacity, ease of tilth, and high plant nutrient content (Fehrenbacher et al. 1984). The state soil,

organic material, most develop in mineral material derived from decomposed bedrock and sediments. Environmental elements begin to alter the parent material as soon as it is exposed. The rate and degree of subsequent alteration are related to the interacting soil-forming factors. Soil parent material has texture, particle size, and an origin or agent of deposition; soil is the result of all of the dynamic landscape processes.

After texture, color, structure, and other morphological attributes have been described, soil horizons can be interpreted. Determining the geological agent that deposited the genetic parent material and noting its position on the landscape help scientists understand soil relationships and make interpretations (Figure 22-8). Although bedrock is the ultimate source of sediment, many parent materials have been further altered by geological agents such as running water, glaciers, and wind. These and other processes have modified the parent material by abrading, crushing, and sorting it during erosion, transport, and deposition. Sediments produced by various geological processes can be differentiated based on their properties, grain size, distribution, arrangement or orientation of particles, and position on the landscape. For example, deposits of sand may be transported by wind (sand dunes), waves (beaches), running water (alluvium), glacial meltwater (outwash), or other geological agents.

With all other soil-forming factors held constant, different soils will develop in sediment with different textures (e.g., silt, sand, clay), regardless of the mechanism that deposited them. The soils of Illinois have developed in five mineral parent materials and organic matter parent materials (Figure 22-9) (Fehrenbacher et al. 1967, 1984).

Loess

A wind-deposited, predominantly silt-size sediment, loess is the principal parent material in Illinois; 63% of the

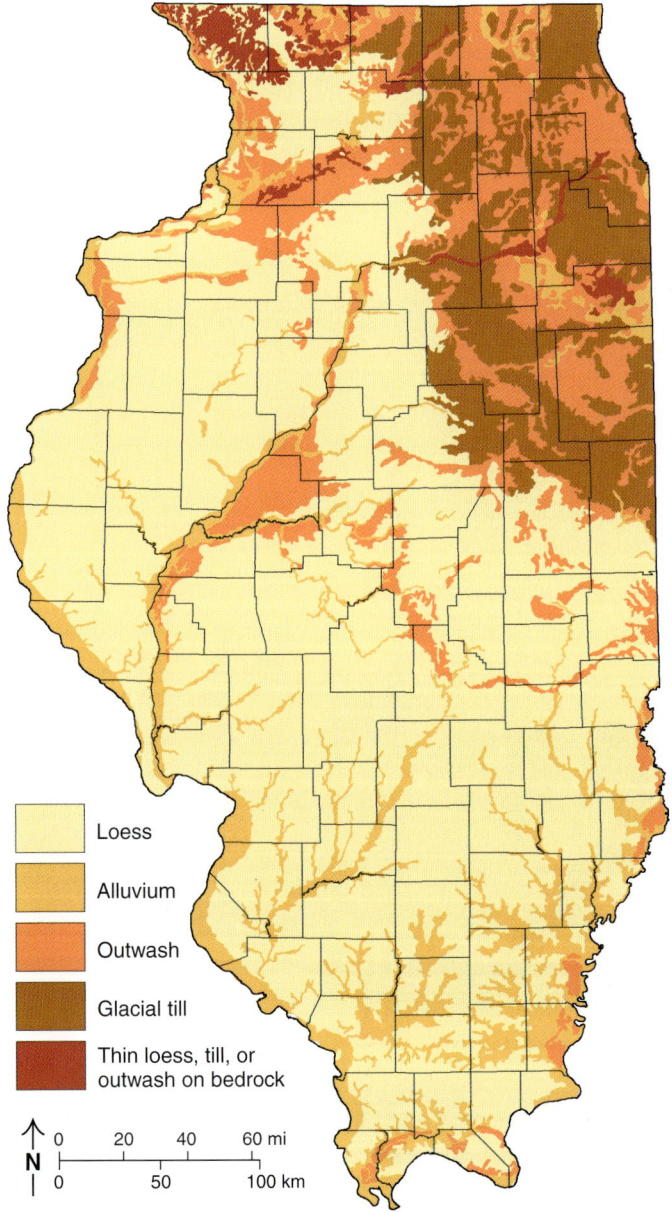

Figure 22-9 Extent of major soil parent materials in Illinois (Fehrenbacher et al. 1967).

the Drummer silty clay loam, was formed in 40 to 60 inches (100 to 150 cm) of loess or silty water-laid material and in the underlying stratified glacial outwash (Figure 22-10).

Till

Till, a mixture of sediment sizes deposited by or in contact with glaciers, ranges in size from clay to sand, with pebbles, cobbles, and boulders intermixed. Till accounts for 12% of the parent material in Illinois. Soils developed mostly or entirely within till are found mainly in northeastern Illinois where the texture of these deposits can range from clay, silty clay, silt loam, and sandy loam (Fehrenbacher et al. 1967, Wascher et al. 1960) (Figure 22-11).

Alluvium

Sediment deposited on floodplains by streams covers about 12% of the state. Known as alluvium, this generally medium- to fine-textured sediment is most common and thickest along the valleys of major rivers where its depositional history can be traced and interpreted from the stratified deposits of sand, silt, and clay intermixed with organic matter and, in some places, archaeological artifacts. If the sedimentation on the floodplains is episodic, previous soils may be buried by rapidly deposited sediments. Soil development processes would then be concentrated in the newer sediments, and the buried soil would be left as an artifact of prior soil development at the site. Sequences of buried soils can be observed on many floodplains.

Outwash

Outwash is the primary parent material for 8% of Illinois soils. The massive amounts of meltwater issuing from the marginal areas of the ice sheets that covered northeastern Illinois as recently as about 12,000 years ago deposited gravel, sand, silt, and clay in the floodplains of the major rivers and across the level uplands ahead of advancing glaciers. These outwash deposits may be tens of feet thick and are a major source of construction aggregate resources and groundwater, especially in the northeastern part of Illinois.

Bedrock

Bedrock accounts for only about 5% of the surface area in Illinois. Shale, limestone, and sandstone are the parent materials in areas where the glacial deposits are absent or where the loess cover is very thin. The texture of the bedrock type (e.g., sandstone vs. shale) strongly influences soil development and the initial characteristics of the soil.

Organic Material

Most commonly found as peat and muck in northeastern Illinois in depressions and low-lying wet areas such as marshes and wetlands, organic material accounts for 0.2% of the parent materials in the state. The cool climate and widespread occurrence of fine-textured sediments at the land surface and poorly drained depressions contribute to the development and maintenance of conditions that favor the accumulation of organic material.

Time

Time differs from other factors in soil development in that it exhibits no change in rate or intensity. Time is the control against which the climatic, biological, topographic, and geological processes can be measured. In northeastern Illinois, the glaciated landscape of the most recent glacial episode is considerably younger than that south and west of the Wisconsin Episode ice front, which was glaciated during the previous (Illinois) glacial episode (Chapter 12, Quaternary Period). Even though the overlying silt covering most of the state is similar in age, the underlying till is much older. The effect of time is often revealed in small, lo-

Figure 22-10 State soil of Illinois, Drummer silty clay loam. (Photograph courtesy of U.S. Department of Agriculture, Natural Resources Conservation Service.)

cal areas along floodplains where soil development has been minimal on the active floodplain because of recent and ongoing deposition of new sediment but more advanced on adjacent, higher stream terraces. These terraces were once part of a former floodplain that is now abandoned. Where a sequence of terraces is present, the degree of soil development tends to be greater as the age of the terrace increases.

THE SHIFTING LANDSCAPE— EROSION, TRANSPORTATION, AND DEPOSITION

Natural Processes

Erosion, transportation, and deposition can be a natural part of landscape development. A large number of geological processes may interact to produce a particular landscape, but, generally, only a few are responsible for the majority of the landscape's characteristics. In Illinois, most of the landscape is the result of continental ice sheets whose multiple advances and retreats reshaped the previously existing surface. Running water that emanated from the melting ice front eroded, moved, and redeposited some of these sediments on outwash plains and floodplains. During the glacial events, strong westerly winds picked up silt- and clay-size particles exposed on the flat, barren floodplains after large floods subsided. Today, wind erosion is largely attenuated by vegetation. Running water, however, continues to be a major agent in the cycle of erosion, transportation, and deposition in the state.

Cropland

The arrival of large numbers of European settlers in Illinois during the 1800s signaled the beginning of a major change in the state's land cover and an increase in the rate of soil erosion that exceeded the long-term geological rate. The widespread destruction of native vegetation, notably the cutting of forests for wood products and the conversion of forest and native prairie to cropland, accelerated erosion. Evidence of this increase can be seen in the layers of sediment deposited on many floodplains. In many places, a thick, dark brown or black soil horizon is found buried beneath many feet of recent silt and clay. This dark layer represents the former land surface and soil horizon at the time of settlement. Some floodplain sediment successions often exhibit multiple buried soil horizons, which suggests both periods of stability (soil horizon development) and instability (rapid deposition) at that location.

Cropland is particularly susceptible to sheet, rill, and gully erosion. The low plant density and frequent, widespread disturbance of the soil surface by cultivation in-

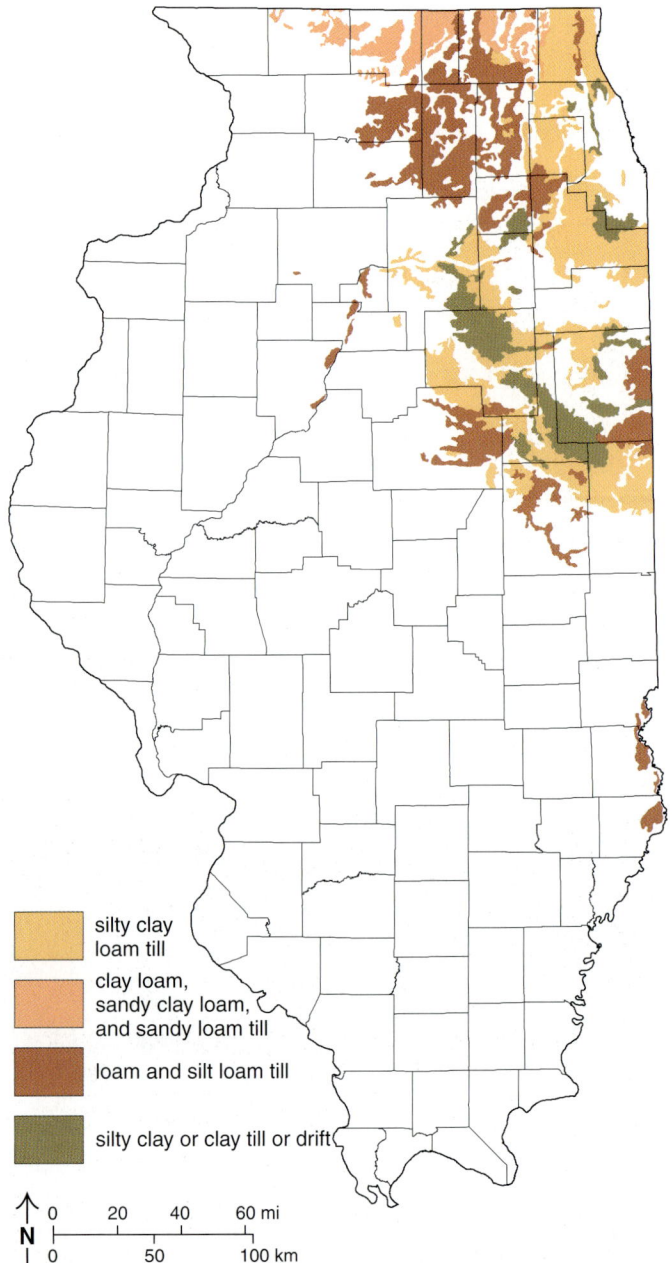

Figure 22-11 Textures of parent materials of glacial till soils in northeastern Illinois (Fehrenbacher et al. 1967).

crease the ability of raindrops to dislodge soil particles, resulting in a greater loss of sediment than that experienced with grass cover. However, this loss can be avoided through good land management. Between 1982 and 1997, average sheet and rill erosion rates decreased by 35% in Illinois (U.S. Department of Agriculture 2007), after the adoption of conservation tillage and the retirement of highly erodible cropland (Soil and Water Conservation Society 2001). Unfortunately, the sediment yielded by gully erosion is not included in the erosion total; therefore, in areas with active gully development, the amount of sediment being eroded is mostly likely being understated.

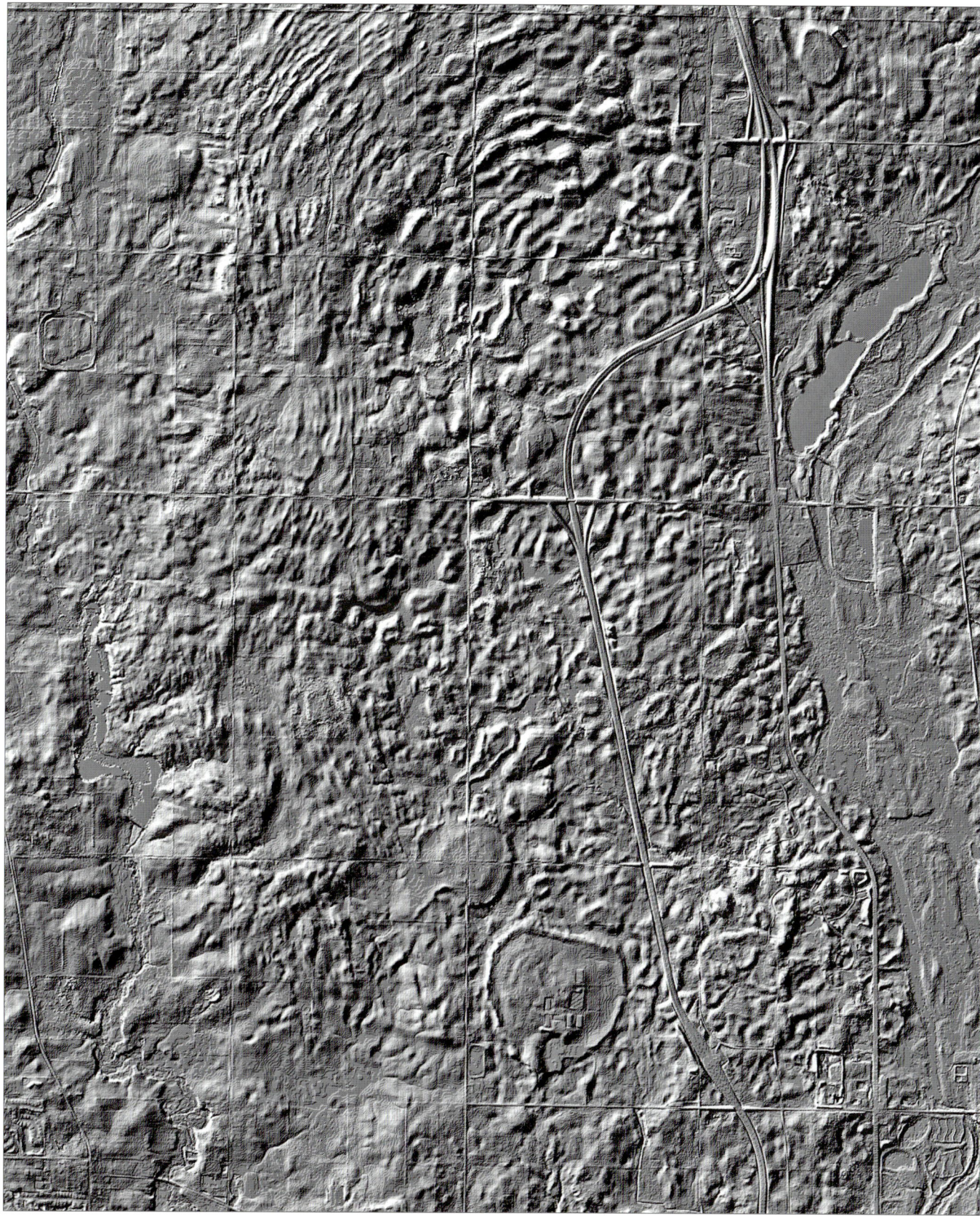

Figure 22-12 Landscape model of northeastern Lake County, Illinois. The Des Plaines River is at the far right, and the Wisconsin state border is at the top of the scene. Interstate 94 traverses north-south just west of the river. Note the pattern of linear to somewhat arcuate landforms (possibly ice-deformed sediments) in the upper center to the middle of the image along with other isolated circular features called ice-walled lakes. LIDAR data are courtesy of Lake County Department of Information and Technology.

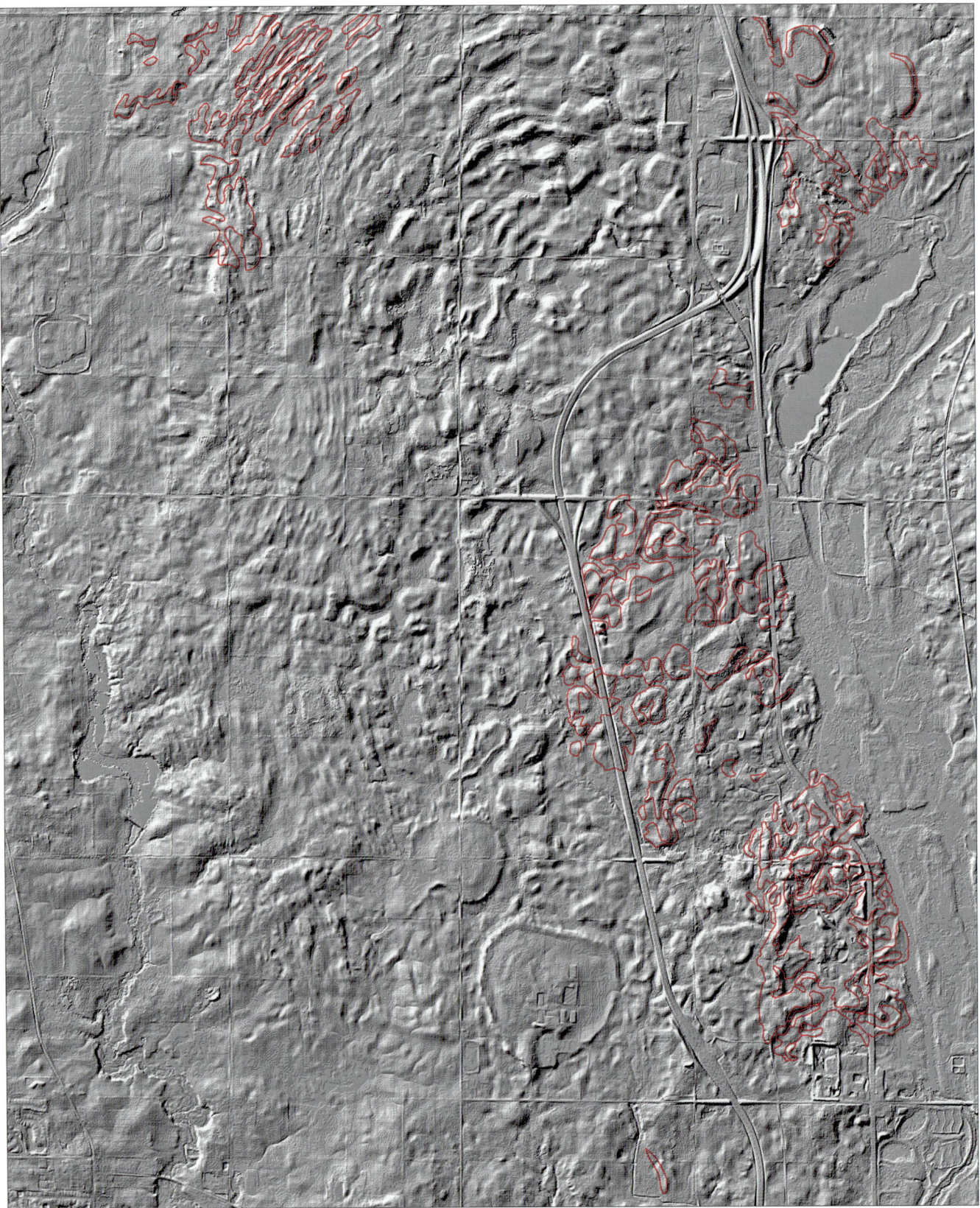

Figure 22-13 Overlay of the Ozaukee silt loam on the landscape model. Note how this soil is closely related to some of the curvilinear and circular features. Soil-landscape patterns often provide guidance for additional analysis and fieldwork to determine the underlying relationships. LIDAR data are courtesy of the Lake County Department of Information and Technology; soils data are from Soil Survey of Lake County, updated digital version of Paschke and Alexander 1970; the Digital Elevation Model (DEM) was developed by and soils overlay assembled by Donald E. Luman, Illinois State Geological Survey.

Gully Development and Floodplain Incision

Gully development in Illinois is most severe in areas with high relief, easily erodible sediment, and a disturbed land surface. These factors most often coincide along major river valleys where thick silt deposits occur and where deforestation or land conversion disturbs the soil. As more water runs off the surface and concentrates in small rills and gullies on steep hillslopes, the drainage network quickly deepens and expands, and eroded sediment is moved down the channels to be deposited on the neighboring floodplain. Many of these channels show a stair-step profile, indicating periodic disturbances, each of which causes a new cycle of channel downcutting. In many places, gullies often begin to tap into roadside drainage as they erode headward. In some areas, these headcuts can undermine roads and can be difficult to control. The increase in runoff and eroded sediment eventually leads to a buildup of sediment on the local floodplains and the burial of soils in those areas. Later, these same floodplains may experience increased drainage from surrounding uplands, which will start downcutting their channels, as has already occurred in the adjacent gullies. The stream then begins incising deeply into its floodplain. Changes in land use and climate are often the causes for renewed downcutting or erosion in gullies and floodplains. The sediment stored along gully channels and on adjacent floodplains is there only temporarily, awaiting the next phase of landscape changes to begin moving downstream.

INTEGRATING SOILS AND GEOLOGY FOR DECISION-MAKING

Changes in land use from urban growth, transportation network upgrades, groundwater utilization, and wetlands and open space preservation are but a few of the complex issues facing state, county, and municipal agencies. These agencies need a large variety of information concerning geology, soils, groundwater, and landscape to assist in their analyses and planning.

Figures 22-12 and 22-13 illustrate a basic approach to integrating topographic, cultural, and soils information by using a Digital Elevation Model (DEM), with a 2-foot vertical resolution, developed from LIght Detection And Ranging (LIDAR) data as a map base upon which a selected soil from the updated digital soil survey for Lake County, Illinois, has been overlain using Geographic Information Systems (GIS) technology. Individual soils occupy unique positions on the landscape, and their location and shape (geometry) provide information about the underlying sediments and landforms. By combining additional information, such as infiltration and runoff rates, erosion potential, construction limitations, and sediment textural data, a broad, yet detailed view of the landscape can be available for use by decision makers. Additional data layers can be combined to illustrate the spatial relationships between population density, transportation grids, open space, sensitive habitats, and land-use planning. Various hydrologic data can be displayed to illustrate conditions such as drainage and areas with flooding potential. With GIS, many different scenarios can be generated for analysis and planning. Increasingly, more in-depth analyses are needed to assist decision makers and to ensure the correct interpretation and application of geological information.

CONCLUSION

Soils will continue to play an important role in the economic, social, and environmental issues of Illinois. Even though soils have developed in only the uppermost part of the sediments that cover the land, they influence many issues related to land use and quality of life in Illinois. Wise decision making will be necessary to preserve this valuable resource, and good management practices will help ensure long-term fertility and ease of cultivation.

REFERENCES

Buol, S. W., F. D. Hole, and R. J. McCracken, 1997, Soil genesis and classification (4th ed.): Ames, Iowa, Iowa State University Press, 527 p.

Fehrenbacher, J. B., J. D. Alexander, I. J. Jansen, R. G. Darmody, R. A. Pope, M. A. Flock, E. E. Voss, J. W. Scott, W. F. Andrews, and L. J. Bushue, 1984, Soils of Illinois: University of Illinois at Urbana-Champaign, College of Agriculture, Agricultural Experiment Station and U.S. Department of Agriculture, Soil Conservation Service, Bulletin 778, 85 p.

Fehrenbacher, J. B., I. J. Jansen, and K. R. Olson, 1986, Loess thickness and its effect on soils in Illinois: University of Illinois at Urbana-Champaign, College of Agriculture, Agricultural Experiment Station and U.S. Department of Agriculture, Soil Conservation Service, Bulletin 782, 14 p.

Fehrenbacher, J. B., G. O. Walker, and H. L. Wascher, 1967, Soils of Illinois: University of Illinois at Urbana-Champaign, College of Agriculture, Agricultural Experiment Station and U.S. Department of Agriculture, Soil Conservation Service, Bulletin 725, 47 p.

Illinois State Water Survey, 2007, Climate maps (1971–2000 Normal) for Illinois. http://www.isws.illinois.edu/atmos/statecli/mapsv2/index.htm. Accessed November 5, 2009.

Jenny, H., 1941, Factors of soil formation: A system of qualitative pedology: New York, McGraw-Hill, 281 p.

Luman, D. E., L. R. Smith, and C. C. Goldsmith, 2003, Illinois surface topography: Illinois State Geological Survey, Illinois Map 11, 1:500,000.

Paschke, J. E., and J. D. Alexander, 1970, Soil survey of Lake County, Illinois: U.S. Department of Agriculture, Soil Conservation Service and University of Illinois at Urbana-Champaign, College of Agriculture, Agricultural Experiment Station, 82 p.

Schoenberger, P. J., D. A. Wysocki, E. C. Benham, and W. D. Broderson, 2002, Field book for describing and sampling soils, version

2.0: Lincoln, Nebraska, Natural Resources Conservation Service, National Soil Survey Center, 288 p.

Soil and Water Conservation Society, 2001, The state of the soil, *in* Soil fact packet: Ankeny, Iowa, Soil and Water Conservation Society. http://www.swcs.org/documents/filelibrary/pdfs/soil_fact_sheets/The_State_of_the_Soil_050505105719.pdf. Accessed September 28, 2010.

Struben, G. R., and M. E. Lilly, 1999, Soil survey of Pike County, Illinois: U.S. Department of Agriculture, Natural Resources Conservation Service and Illinois Agricultural Experiment Station, 305 p.

U. S. Department of Agriculture, Natural Resources Conservation Service, 1997, National Resources Inventory, prime farmland, revised December 2000. http://www.il.nrcs.usda.gov/technical/nri/prmfrmlnd.html. Accessed October 13, 2009.

U.S. Department of Agriculture, National Agricultural Statistics Service, 2002, Land cover of Illinois 1999–2000. http://www.isgs.illinois.edu/nsdihome/webdocs/landcover/index.html. Accessed November 5, 2009.

U.S. Department of Agriculture, National Agricultural Statistics Service, 2002, Land cover of Illinois 1999–2000. http://www.isgs.illinois.edu/nsdihome/webdocs/landcover/stats.html. Accessed October 13, 2009.

U.S. Department of Agriculture, National Agricultural Statistics Service, 2007, Cropland data layer (CDL), raster, georeferenced, categorized land cover data layer. http://www.isgs.illinois.edu/nsdihome/webdocs/landcover/index.html. Accessed November 5, 2009.

U.S. Department of Agriculture, Natural Resources Conservation Service, 1999, Soil taxonomy: A basic system of soil classification for making and interpreting soil surveys (2nd ed.), U.S. Department of Agriculture, Handbook 436: Washington, D.C., U.S. Government Printing Office, 869 p.

Wascher, H. L., J. D. Alexander, B. W. Ray, A. H. Beavers, and R. T. Odell, 1960, Characteristics of soils associated with glacial tills in northeastern Illinois: University of Illinois at Urbana-Champaign, College of Agriculture, Agricultural Experiment Station and U.S. Department of Agriculture, Soil Conservation Service, Bulletin 665, 155 p.

Willman, H. B., and J. C. Frye, 1970, Pleistocene stratigraphy of Illinois: Illinois State Geological Survey, Bulletin 94, 204 p.

Geological Factors in Siting and Design of Facilities and Infrastructure

23

Robert A. Bauer, Wen-June Su, and Nelson Kawamura

INTRODUCTION

Every man-made structure on Earth rests on or in geological materials. Many of these structures also require connections to land-based infrastructure. The stability of these structures and infrastructure depends on the soil or rock characteristics and the potential for geological hazards. The suitability of geological materials for any construction activity, including ease and stability of excavation, adequate bearing strengths, and drainage conditions, is a major concern. In Illinois, important soil and geological hazard characteristics to identify, avoid, or design for are compressible peat and muck deposits, poorly drained soils, shallow groundwater or flooding conditions, unstable slopes, high susceptibility to freeze-thaw cycles, high shrinking and swelling properties, soil corrosiveness, and soil amplification of ground shaking from earthquake activity. This chapter examines the characteristics and use of earth materials and geological hazards as they relate to homes and other structures and to everyday activities.

GENERAL SITING ISSUES

The long-term, trouble-free stability of a structure depends on the ability of its foundation to withstand its geological setting. To determine whether a location is suitable for construction, the earth materials of the site must first be characterized to determine

- the materials' bearing strength, that is, the ability to support the planned structure or excavation (e.g., cavern walls, tunnels, mined-out areas);
- the need for artificial support to contain weaker soil or rock;
- the depth of groundwater (water reduces strength of soil materials and enters excavations);
- the shrinking and swelling characteristics of the soil;
- the vulnerability of certain soils to erosion by moving water on or below the ground surface; and
- the direction and spacing of natural rock fractures that affect stability and groundwater flow and pressure.

Many of the very near-surface characteristics of Illinois soils, such as strength, shrinking and swelling, and moisture, can be found in county-based soil map publications of the U.S. Department of Agriculture, Natural Resource Conservation Service (formerly the Soil Conservation Service).

Shrinking and Swelling

The geological hazard that the largest segment of the population will experience is created by the actions of "reactive soils." The national annual cost of structural damage from subsidence or heaving—estimated at $6 billion to $8 billion—is likely to exceed the combined annual cost of damage caused by earthquakes, tornados, hurricanes, and floods (Holtz 1983). Approximately half of the homes built in the United States today are constructed on reactive soils. When exposed to certain physical or geological conditions, these soils undergo changes in shape that can cause foundation and structural damage to buildings. Drying of reactive soil, as during a prolonged dry spell or drought, causes the soil to shrink, leading to settlement, or subsidence, of the ground surface. Addition of water to soil causes it to swell, which leads to heaving of the ground surface and horizontal pressures on basement and foundation walls.

The magnitude of settlement and heave is related to the amount of change in water content of the soil. In Illinois, most soils are not highly expansive, but nearly all contain varying amounts of clays, which are reactive to the reduction in or addition of water. Two major contributors to the removal of water from clayey soils are drought and transpiration by plants, especially trees.

As an example, Perpich et al. (1965) reviewed damages in the Chicago area from excessive drying of soils by trees. Investigations of more than 200 structures and foundations indicated that damage began as the moisture content of the soil was reduced to less than about 13 to 21%. Differential settlements of 1 to 4 inches (2.5 to 10.2 cm) were measured across the structures. Damage to buildings was typically associated with large trees having a trunk diameter of over 10 inches (25.4 cm); small trees and shrubs were not a problem. In no cases were evergreen trees involved. The depth of desiccation varied between 6 and 15 feet (1.8 and 4.6 m) from the ground surface and between 1 and 11 feet (0.3 and 3.3 m) from the bottom of the foundation. Generally the trees were located 10 to 25 feet (3.1 to 7.6 m) away from the foundation.

The most efficient remedy for shrinkage caused by transpiration is to remove the offending trees. For example, Hammer and Thompson (1966) reported on foundation clay shrinkage caused by large trees in the officer housing area of Selfridge Air Force Base in Michigan. The houses in that area were constructed between 1933 and 1935; most were two-story brick structures with basements. The first subsidence damage to houses occurred during the 1952–1953 drought when sites with a concentration of large trees underwent ground subsidence in the form of a shallow, dish-shaped depression that severely damaged sidewalks, streets, and house foundations. Most damaged were houses situated between two large trees or where large trees grew between houses. During the early 1960s, a second period of drought occurred. Some large trees were removed from the area, which has experienced no further settlement to date.

An alternative approach to tree removal is to restore water to the soil and tree root system by irrigation through sand-filled holes and perimeter trenches. This solution was used for Chicago's St. Hedwig Church, which suffered the separation of one 150-foot-high Gothic tower from the main building. After the tower foundation area was irrigated, the tower returned to its original position, and the cracks closed (Holtz 1983).

Cycles of soil shrinking and swelling can cause another type of damage. During long dry periods, soils shrink away from the foundations of homes, leaving a gap between the soil and foundation. When this gap is open, activities near the foundation push soil particles down into the opening. When precipitation restores soil moisture, the ground swells and closes the space against the wall. The new material, though, takes up additional space, which increases pressure on the foundation wall. These pressures increase gradually over decades until concrete block basement walls are pushed inward, producing a horizontal crack in the wall, usually within 1 to 2 feet (0.3 to 0.6 m) of the top of the soil (Figure 23-1). This horizontal crack opens with moist soil conditions and closes as the soil pulls back from the foundation wall during prolonged dry periods.

Saturation and Piping

Although drying soil can lead to structural damage, too much water can be equally disastrous. As some soils become saturated with water, their strength and bearing capacity are reduced, and foundations may lose some support. Piping occurs when water percolates through a silty soil layer, removing these fine particles. Excessively wet soil and the subsurface movement of water due to saturation and piping can remove support for a foundation. The corner of the foundation may then drop downward slightly, rotating the bricks or concrete blocks outward and causing foundation cracks that start near the base of the corner and move upward and away from the corner, becoming wider as they move upward.

Both saturation and piping can be caused by downspouts that empty their water along foundation walls or corners or that drain into systems buried around the outside bases of foundation walls. It is best to direct downspout drainage as far from the structure's foundation as possible and to slope the ground so that water drains away from buildings.

Unlike shrinking and swelling, which are associated with clay-rich soils, piping occurs in soils that contain windblown silts (loess). Silt particles are very fine and erode easily. Buildings and other structures, such as sewer pipes

Figure 23-1 This concrete block foundation wall has been pushed inward into the basement by the pressure on the wall from decades of soil shrinking and swelling cycles. (Photograph by Robert A. Bauer.)

and old farm field drainage tiles, are susceptible to damage caused by piping. Water can move the fine silt particles into a small opening in a sewer line at a joint or crack. Old clay farm tiles can fill with soil, greatly reducing capacity. Over time, this erosion lowers the ground surface, creating a depression along the length of the pipe or a small sinkhole immediately above the opening in the pipe. Today many farm tile systems use a filter fabric, or sock, to prevent silts from entering the pipe.

MAJOR GEOLOGICAL HAZARDS

In addition to determining the suitability of earth materials, it is necessary to evaluate the potential for major geological hazards, which can affect a large area and damage many structures. Landslides and coal mine subsidence are discussed here. Geological hazards from earthquakes and karst terrane are discussed in detail in Chapter 27, Earthquakes, and Chapter 28, Karst Terrane.

Landslides

Landslides can occur throughout the state, primarily where steep slopes exist along rivers, creeks, and lakes. Many people build on or near the top of a slope because of the views (Figures 23-2 and 23-3), but, unfortunately, these landscapes can become unstable. Over time, rivers, creeks, and lakeshores shift and erode soils at the slope base, eventually (or suddenly!) causing slope failure. Humans contribute to the failure by cutting away the base of the slope for borrow materials or road cuts, by removing trees or plants that help to stabilize the slopes, or by increasing the amount of water draining to the slopes. Some slopes near the Mississippi and Illinois Rivers have thick layers of low-strength loess (windblown silt) that are very prone to surface or subsurface erosion from moving water and have a reduced strength when wet.

Loess and other low-strength soil materials on steep slopes need special attention in order to avoid slope failure and damage. Recommendations to mitigate slope stability problems include the following:

- Avoid construction near the top of steep slopes.
- Do not remove material from the base of slopes.
- Do not add weight to the top of the steep slopes by pushing soil material out over slopes to form more flat land to enlarge lot sizes.
- Do not construct septic fields near the top of slopes where drainage would saturate the area.

Figure 23-2 Homes along the Lake Michigan lakeshore showing unprotected shoreline and shoreline protected to reduce erosion and slope failure, Highland Park, Illinois. (Photograph by Michael J. Chrzastowski.)

Figure 23-3 Landslide effects on a home built at the top of a slope west of Peoria, Illinois. (Photograph by Robert A. Bauer.)

- Do not allow excessive watering, including downspout discharge, above or on steep slopes.
- Do not remove trees on steep slopes or near tops of slopes.
- Maintain plantings that have good root systems to hold soil together on the slopes.

Landslides and slope movements affect more than building structures. Transportation corridors—roads and railroads—can be affected (Figures 23-4 and 23-5). Slope movement can surprise excavators in some places (Figure 23-6). A statewide study of landslides by the Illinois State Geological Survey (Killey et al. 1985) showed that the average cost of repair was about $300,000 per landslide in 1982 dollars and about $625,500 in 2007 dollars.

Coal Mine Subsidence

Coal mine subsidence is the lowering of the ground surface as a result of a collapse of the mine opening or the supporting pillars in an underground coal mine. Depending on the depth and type of failure, subsidence at the

Figure 23-4 Failure of interstate embankment near the intersection of Interstate 57 and Interstate 70 after a month of excessive rainfall during May 1995. (Photograph by Robert A. Bauer.)

Figure 23-5 Railroad tracks south of Chester, Illinois, pushed out of alignment from landslide movements of the slope to the left of the tracks. (Photograph by Wen-June Su.)

ground surface can be either a large, dish-shaped sag or a pit (sinkhole) with nearly vertical walls.

Underground mines were developed in Illinois soon after settlers arrived in the area. Mining began between 1810 and 1840 in a few places along major waterways in the southern and north-central portion of the state and in the late 1860s in northeastern Illinois. Impacts of mine subsidence from underground coal mines was recorded in court cases as early as 1880 (Beck and Sigwerth 1980). The Illinois State Geological Survey has been investigating, documenting, and publishing information about coal mine subsidence since 1908. See Chapter 14, Coal, for additional information about coal history and subsidence.

Pit Subsidence

Pit subsidence usually occurs over mines that are less than 200 to 300 feet (61 to 91.4 m) deep and are overlain by soft bedrock only tens of feet thick. These pits generally develop when the mine roof collapses and the void works its way up through the overlying soft bedrock and glacial deposits to the ground surface. The ground moves mainly in one direction—downward (Figure 23-7). The pit on the ground surface may be deeper than the height of the mine void, however, because the soft glacial materials may be washed into the adjacent mine voids by groundwater or surface runoff.

Pits are commonly 6 to 8 feet (1.8 to 2.4 m) deep and 2 to 40 feet (0.6 to 12.2 m) in diameter, although most are less than 16 feet (4.9 m) across. Newly formed pits have steep sides with straight or bell-shaped walls (Figures 23-7 and 23-8). Because of their depth and the steepness of their walls, pits pose a special danger to people and animals. Anyone who falls into a pit may be injured and find it very difficult to get out because of the vertical walls.

The risk of damage to a structure from pit subsidence is low, however, because of the relatively small size of the pit. If a foundation bridges a pit, it might not be immediately affected but could eventually fail if not reinforced. A structure is more likely to be damaged if pit subsidence develops under a building corner, the support posts of a foundation, or other critical supports.

Pits are most likely to form at the surface after heavy rainfall or snowmelt. Water does not usually accumulate in the pits but drains down into the mine. A common treat-

Figure 23-6 Massive failure of slope in excavation in Chicago, Illinois. Note the landslide on the right with the scarp at top, formed when the soil slid downward into the excavation. More than half of the bottom of the excavation was covered by the slide, tearing up the wooden forms installed for pouring concrete. (Photograph by G. H. Otto.)

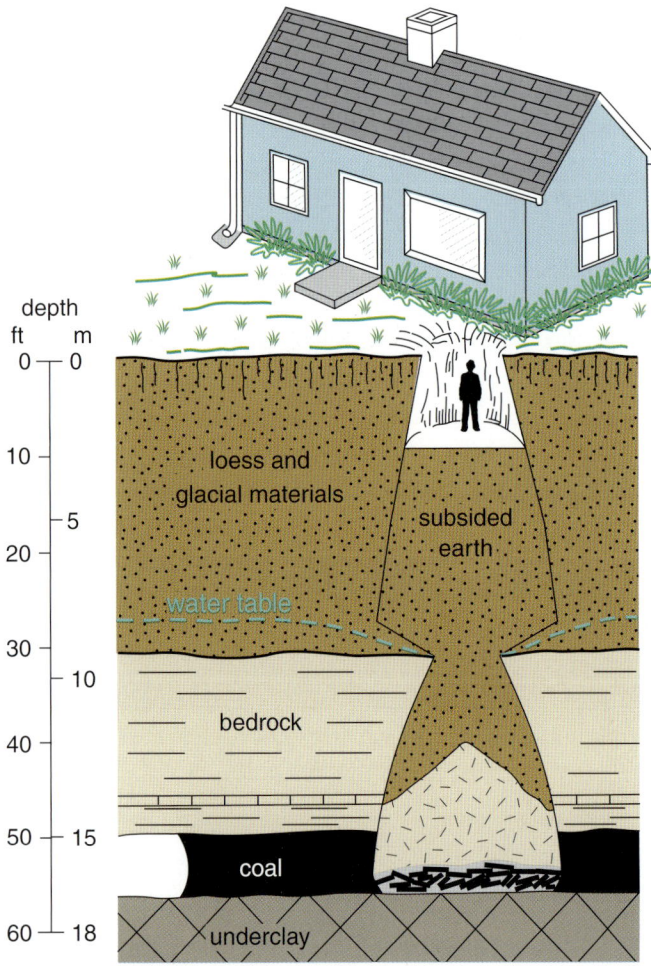

Figure 23-7 Characteristics of a typical pit subsidence event (Wildanger et al. 1980).

Figure 23-8 Pit subsidence in Belleville, Illinois. (Photograph by Philip J. DeMaris.)

ment is to fill the pit with sand or clay, cap the fill with a clayey soil in a slight mound to shed water, and compact the clay tightly so that its permeability is very low. Many pits have been filled in this way.

Sag Subsidence

Sag subsidence occurs as a gentle depression over a broad area (Figures 23-9 and 23-10) and can develop above mines at any depth. Sags are created suddenly (in a few hours or days) when coal pillars have deteriorated and suddenly collapse. Sags also can be created more gradually (over years) when pillars have slowly settled into the mine floor of soft underclay. Some sags may be as large as an entire mine panel—several hundred feet long and a few hundred feet wide. The maximum vertical settlement near the center of the sag is typically 2 to 4 feet (0.6 to 1.2 m). Sag subsidence also is caused by longwall mining, which allows the roof over the mined-out area to collapse predictably as the mining equipment advances.

Sag subsidence produces a fairly orderly pattern of tension cracks (ground pulled apart) around the outside edge of the sag (Figure 23-11b). Near the center of the sag, compression ridges form as the ground is squeezed by the upward bending of the land (Figure 23-11a). Ridges are observed less frequently than tension fractures because the area of compression is much smaller than the area of the sag. The compression area is where foundation walls are sometimes found being pushed inward into a basement or crawl space.

During sag subsidence, the ground moves in two directions, dropping vertically and moving horizontally toward the center of the sag. Consequently, the sag may be much broader at the ground surface than at the collapsed part of the mine. For example, a failure in a mine 160 feet (48.8 m) deep could cause minor surface subsidence more than 75 feet (22.9 m) beyond the edge of the undermined area. The deeper the mine, the larger is the affected ground surface in comparison with the area collapsed underground.

The type and extent of damage to surface structures from sag subsidence relate to their orientation and position within the sag. In the tension zone, large ground movements causing the development of cracks in the ground

SITING AND DESIGN

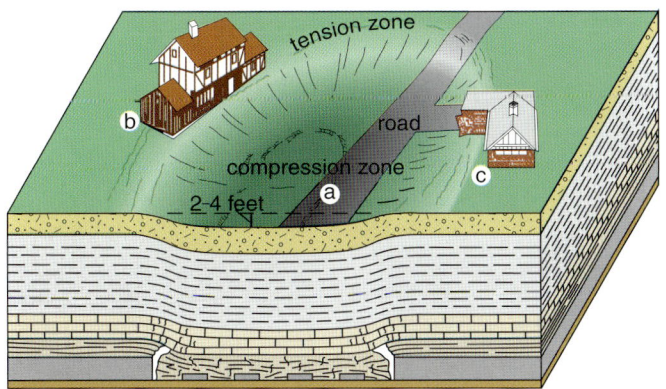

Figure 23-9 Characteristics of a sag subsidence event showing compression and tension areas with associated damage (Bauer 2006). The road is in the compression zone **(a)**, and asphalt has buckled. The wood frame house **(b)** is in the tension zone; the house's foundation has been pulled apart and dropped away from the superstructure in one corner. A brick house **(c)** in the tension zone would likely show cracks in walls, ceilings, and floors.

may damage buildings (Figure 23-11b, c) and roads as well as driveways, sidewalks, pipes, sewers, and utilities. Until subsidence has ceased and repairs can be made, structures may have to be temporarily supported, in part or totally, depending on their location in the sag.

Repair of Subsidence Damage

Each type of construction requires a different type of treatment for subsidence damage. However, because subsidence can continue to produce ground movements for months or even years after its onset, no permanent repairs should be made to any structure until the movements have slowed to the point where no further damage to the building will occur. Measurements made on and in the ground in the vicinity of the building can indicate when movements have slowed sufficiently.

Repair to structures generally involves releveling. The releveling technique is unique for each house because of differences in construction and access. Repair of most structures requires detaching the building from the slab or

Figure 23-11 Photographs of the sag subsidence features shown in Figure 23-9. **(a)** Road in compression zone with buckled asphalt in Hegeler, Illinois. **(b)** Wood frame house in tension zone in Energy, Illinois. **(c)** Brick house exterior in tension zone near Freeburg, Illinois. (Photograph a by Calhoun Smith; photographs b and c by Robert A. Bauer.)

Figure 23-10 Sag subsidence in Freeburg, Illinois, shows a dip in the road. The center of the sag lies near a water tank. The house in the center of the photograph has the corner nearest the center of sag dropped downward and moved toward the center of the sag. Note the misshapen garage doors produced by these movements. (Photograph by Robert A. Bauer.)

Figure 23-12 A 32-foot (9.8-m)-diameter tunnel cut out of dolomite 300 feet (91.4 m) under the ground surface for the combined sanitary and storm sewer overflow collection in Chicago. (Photograph by Robert A. Bauer.)

foundation to relieve stress to the frame (superstructure) and to allow releveling. This process prevents further damage to the frame (superstructure). While the structure is still moving, utilities such as sewer, water, gas, and electricity need to be examined and provisions made for each to accommodate continued movements without creating a break. The ground beneath the foundation should also be inspected for potential support problems or danger from moving water in the subsurface. When buildings have not been releveled, problems have been known to occur for the life of the subsidence movements, which may be months to years.

More detailed descriptions of repairs and damage can be found in *Mine Subsidence in Illinois: Facts for Homeowners* (Bauer 2006).

Determining Whether Property Is Undermined

Many citizens want to know if a home or property lies over a mine. The Coal Section of the Illinois State Geological Survey produces maps showing the locations of various types of underground coal mines. One set of maps is produced by county; these show mine outlines and street systems on the ground surface. Another set of maps covers various areas of the state at scales of the topographic quadrangle maps (1 map inch represents 2,000 feet [610 m] on the ground). These detailed maps have very accurate outlines of the mines superimposed over the topographic map features. All of these maps can be found on the Illinois State Geological Survey Web site (http://www.isgs.illinois.edu/).

USE OF UNDERGROUND SPACE

Every day, Illinois residents benefit from the use of underground space without giving it much thought. In urban

Figure 23-13 Old freight tunnels that lie 40 feet (12.2 m) under the streets of Chicago, Illinois. The tunnels were used to move coal, ash, mail, and merchandise to stores and buildings. This 1907 photograph was taken under the intersection of La Salle and Monroe Streets. (Photograph cropped from the *Chicago Daily News* negatives collection, Negative DN-0005301, courtesy of the Chicago Historical Society.)

areas, utilities—such as sanitary and stormwater sewers, water, gas, and in some cases power, telephone, and broadband cable—may be located underground. Knowledge of subsurface materials and their properties has allowed many successful uses of the subsurface in crowded urban areas. For example, in Chicago, the foundations of tall buildings must go down to stiff soils found at depths of 40 to 60 feet (12.2 to 18.3 m) or even to the more stable bedrock 100 to 150 feet (30.5 to 45.7 m) deep in order to withstand the building loads and wind forces that might otherwise topple the buildings.

Additionally, the subsurface has hundreds of miles of man-made tunnels that carry drinking water, including tunnels built in 1867 that run out under Lake Michigan to water intake structures. Other tunnels, including those in the Tunnel and Reservoir Plan (known as TARP; Figure 23-12), are up to 32 feet (9.75 m) in diameter and run for over 100 miles (161 km) at 300 feet (91.4 m) underground in bedrock to collect combined storm and sanitary waters during heavy rains to prevent them from polluting Lake Michigan.

Transportation is another use for underground space. The construction of the 62 miles (99.7 km) of old freight

SITING AND DESIGN

Figure 23-14 Excavation of clay below Chicago, Illinois, for the subway tunnel to house underground transit trains. (Photograph courtesy of Ralph Peck.)

tunnels 40 feet (12.2 m) under Chicago's streets began in 1901. Originally, these tunnels were intended for telephone and telegraph lines (Chicago Tunnel Terminal Corporation 1928) before being used for moving coal, ash, mail, and merchandise (Figure 23-13). They are now used as conduits for power and telecommunications lines. The tunnels were brought to prominence by the 1992 flood when the tunnel wall was breached under the Chicago River. The tunnels are closed to the public but can be seen in the movie *The Blues Brothers*. Chicago's subway train tunnels, which were dug by hand in layers of clay (Figure 23-14), are another subsurface use.

Throughout Illinois, underground space is used for storage. Underground mined-out limestone quarries (Figure 23-15) have been converted into storage facilities for

Figure 23-15 Underground area mined out for production of crushed rock near Prairie du Rocher, Illinois. Areas such as this are being turned into underground storage facilities in parts of the state. (Photograph by Robert A. Bauer.)

Figure 23-16 Chamber in shale for liquid petroleum gas storage 300 feet (91.4 m) underground near Kankakee, Illinois. The chamber is 200 feet (60.9 m) long, 60 feet (18.3 m) high, and about 30 feet (9.1 m) in diameter. (Photograph from the Illinois State Geological Survey collection.)

grain, refrigerated foods, and other products. In other parts of the country, underground rock mines have been turned into factories and offices. There are currently 14 mined-out caverns in nine counties in Illinois that are used for underground storage. These caverns are excavated in shale and limestone specifically for petrochemical storage (Figure 23-16) and are usually associated with a petrochemical manufacturing facility on the ground surface.

Natural gas is another everyday product stored underground in Illinois. A number of storage fields have been developed in naturally occurring porous rock layers where gas is accumulated for peak use during winter. The Illinois subsurface even hosts scientific projects. At Fermi National Accelerator Laboratory in Batavia, for example, tunnels and chambers in soil and rock contain equipment for accelerating subatomic particles (Figure 23-17).

CONCLUSION

The properties of earth materials are important factors in deciding where, what, and how to build structures and infrastructure in Illinois. It is important to understand how the cycles of shrinking and swelling and the move-

Figure 23-17 Main injection ring about 40 feet (12.2 m) underground at Fermi National Accelerator Laboratory near Batavia, Illinois. The large rectangular boxes running down the left side of the tunnel are the magnets that suspend the accelerating particles within the tube that is capped by the oval plug in the center of the magnets. (Photograph by Robert A. Bauer.)

ment of groundwater can affect the stability of surface and subsurface structures. The potential for and risk from geological hazards, such as landslides, subsidence, karst terrane and earthquakes, must be understood to ensure safer, more stable structures. As the ground surface becomes populated with even more structures, there is likely to be an ever increasing use of underground space, which will require knowledge and understanding of earth materials and their properties.

REFERENCES

Bauer, R. A., 2006, Mine subsidence in Illinois: Facts for homeowners: Illinois State Geological Survey, Circular 569, 20 p.

Beck, R. E., and S. Sigwerth, 1980, Illinois coal mine subsidence law: DePaul Law Review, v. 29, no. 2, p. 383–441.

Chicago Tunnel Terminal Corporation, 1928, What the freight tunnels mean to Chicago: Chicago Illinois, Chicago Tunnel Terminal Corporation, 32 p.

Hammer, M. J., and O. B. Thompson, 1966, Foundation clay shrinkage caused by large trees: Journal of the Soil Mechanics and Foundation Division, American Society of Civil Engineers (ASCE), v. 92, no. 6, p. 1–17.

Holtz, W. G., 1983, The influence of vegetation on the swelling and shrinking of clays in the United States of America: Geotechnique, v. XXXIII, no. 2, p. 159–163.

Killey, M. M., J. K. Hines, and P. B. DuMontelle, 1985, Landslide inventory of Illinois: Illinois State Geological Survey, Circular 534, 27 p.

Perpich, W. M., R. G. Lukas, and C. N. Baker, Jr., 1965, Desiccation of soil by trees related to foundation settlement: Canadian Geotechnical Journal, v. 1, no. 2, p. 23–39.

Wildanger, E. G., J. W. Mahar, and A. Nieto, 1980, Sinkhole type subsidence over abandoned coal mines in St. David, Illinois: Springfield, Illinois, Abandoned Mined Lands Reclamation Council, 88 p.

Geological Perspectives on Flooding   24

Michael J. Chrzastowski and Richard C. Berg

INTRODUCTION

Flooding is the most frequent of all geological hazards in Illinois. Floods occur when the water level in rivers, streams, lakes, and ponds rises high enough to overflow banks and inundate the adjoining land. Most flooding in Illinois occurs annually in recognized flood-prone areas, where planning and mitigation efforts can minimize the impacts of rising water levels. Even so, the average annual cost of damage from flooding in the state is nearly $1 million. When flooding is extreme, such costs can exceed tens of millions of dollars. In addition to damaging property, flooding disrupts commerce and transportation and causes loss of personal income. The most serious threat posed by flooding, though, is the potential for personal injury, including loss of life.

About 14% (roughly 7,400 square miles, or 19,000 km^2) of the total land area in Illinois is vulnerable to flooding by the state's extensive system of rivers, streams, and lakes (Illinois Department of Transportation, Division of Water Resources 1993) (Figure 24-1). Different settings are susceptible to several different types of floods:

- River floods, the most common type, typically occur in spring and result from excessive rainfall and snowmelt. The major rivers of the state—the Mississippi, Rock, Illinois, Kaskaskia, Wabash, and Ohio Rivers—can be expected to overflow some stretches of their banks every spring.
- Ice-jam floods appear in winter and early spring when floating ice piles up at river bends or against channel obstructions such as bridge supports. These ice jams act as dams, causing the water levels upstream to rise.
- Flash floods occur when thunderstorms bring rapid and extreme precipitation to part of a watershed. The resulting surges of water arrive with little or no warning to downstream localities. The volume and speed of such flows are extremely dangerous.
- Urban floods occur when a city's stormwater system cannot adequately handle excessive runoff. Basement flooding is a common threat posed by urban floods.
- Coastal floods are prevalent along some reaches of the Lake Michigan shore when onshore winds combine with a high lake level to submerge beaches and low-lying areas adjacent to the shore. During a winter storm in 1987, coastal flooding inundated a segment of Chicago's Lake Shore Drive (U.S. Highway 41), which required rerouting of both north- and southbound traffic (Figure 24-2).

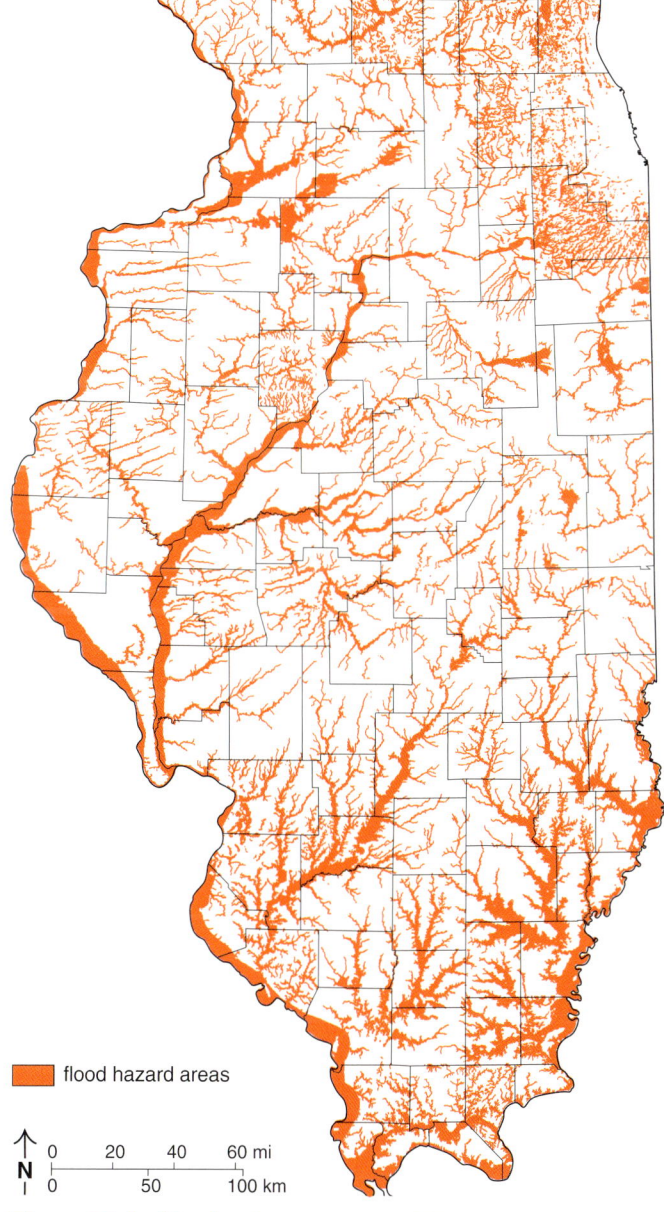

Figure 24-1 The flood-prone areas of Illinois have a distribution that mimics the state's drainage network (Illinois Department of Natural Resources Geospatial Data Clearinghouse at http://www.isgs.illinois.edu/nsdihome/).

Figure 24-2 Coastal flooding on Chicago's lakefront at Lake Shore Drive near North Avenue during high lake levels and strong onshore winds on February 8, 1987, led officials to redirect traffic away from the area. (Photograph from the Chicago Shoreline Protection Commission 1988.)

THE GREAT FLOOD OF 1993

The worst flooding in Illinois history to date occurred in 1993. The event was ranked as a 250-year flood (Visocky 1995); that is, the probability of a flood of similar magnitude occurring in any given year is 1 in 250. The Great Flood of 1993, as it is commonly called, was unprecedented in the United States. It was a regional flood that directly affected nine states (Illinois, Iowa, Kansas, Minnesota, Missouri, Nebraska, North Dakota, South Dakota, and Wisconsin) (*St. Louis Post-Dispatch* 1993, Bhowmik et al. 1994, Changnon 1996). The upper Mississippi and its tributaries bore the brunt of the deluge. In places in Illinois, water levels rose to nearly 23 feet (7 m) above flood stage, and areas were inundated that historically had never been flooded. Thirty-nine Illinois counties were declared both federal and state disaster areas, and five more counties were state disaster areas. The damage and clean-up costs in Illinois exceeded $1 billion, making the event the state's most costly natural disaster to date.

The main cause of the 1993 flooding was above-normal precipitation that began during the spring and continued through the summer. Compounding the impact, this rain fell on soils already saturated from heavy rains the previous fall. Moisture-laden air from the Gulf of Mexico drifted into the upper Mississippi watershed during the spring and summer. A persistent jet stream pattern pushed cool Canadian air against the front of the moist air coming from the south. As a result, thunderstorms formed along this convergence zone during much of the summer. In northwestern Illinois, rainfall was 150% of the 30-year average for January through July. Several locations in the state had 1993 annual rainfall totals that were the greatest since record-keeping began in the late 1800s (Wahl et al. 1993, Kunkel 1996, Rodenhuis 1996).

Several aspects of the Great Flood of 1993 were unusual and unprecedented. Flooding in Illinois is typically a spring event, but the deluge of 1993 continued throughout the summer. Rivers remained at flood stage for months rather than days or weeks. The entire length of the Illinois border along the Mississippi River was inundated. The high water on the Mississippi contributed to a backup along its tributaries, particularly the lower Illinois River. Multiple flood crests occurred at most locations, and these flood crests set new highs for the historical flood record (Figure 24-3). Because of extreme rainfall across northern Illinois, many rivers in that part of the state reached flood stage independently of the Mississippi River. Furthermore, the persistent downpours triggered urban flooding in parts of the Chicago metropolitan area.

GEOLOGICAL CONTROLS ON THE 1993 FLOOD

Precipitation patterns determined the distribution and amount of water that fell within the different watersheds impacted by the 1993 flood. However, geological factors, such as topography, slope, sediment, and stratigraphy, all played important roles in determining where flooding occurred and how floodwater moved as either surface water or groundwater.

Topography and Slope

In all river flooding, the relationship between water height and land elevation is critical in determining whether or not a specific location becomes inundated (Figure 24-4). Floodplains—the primary areas affected by river flooding—are prominent Illinois landscape features (Figure 24-1) (Luman et al. 2003). These broad, low-elevation areas adjacent to rivers have been shaped by glacial and modern river processes and successive flood events for thousands of years. Near East St. Louis, the Mississippi River floodplain is as much as 11 miles (17.7 km) wide, and much of the floodplain along the lower Illinois River is at least 3 miles (4.8 km) wide.

Most Illinois cities that are built on major floodplains, such as East St. Louis, are protected by flood walls, dikes, or tall and possibly reinforced earthen, urban levees. Most of the state's floodplains are drained wetlands now used for agriculture. These areas are protected by a system of agricultural levees that are lower and narrower than urban ones. The agricultural levees prevented the 1993 floodwaters from inundating many floodplain areas, although major emergency efforts were required to raise the crest elevation of the levees by adding sandbags or even straw bales blanketed with plastic sheeting. Some of these efforts were successful, but, in many other cases, levees eventually were overtopped. All that was needed to inundate hundreds of acres behind a levee was an area of slightly lower elevation along the levee crest where floodwater could begin to flow over and erode a levee breach. During the 1993 flood, the difference between floodplain elevation and flood height in places resulted in the floodplain being inundated to an average depth of 10 feet (3 m) or more.

In Illinois, most major rivers and their tributaries have gentle gradients consistent with the overall relief of the state. The 1993 flood demonstrated how these low gradients could contribute to flood conditions on a river and cause backflow up the valleys and floodplains of tributary rivers and streams. For example, much of Alton, in Madison County, is on high land that was not threatened by the flood. However, a major effort was needed to construct a sandbag dike to prevent Mississippi floodwater from extending upstream along a small tributary valley that crosses the Alton central business district (Figure 24-5). The most dramatic case of backflow flooding occurred along more than 80 miles (128.7 km) of the southern portion of the Illinois River, upstream from its junction with the Mississippi River. Here the Illinois Valley has a low slope gradient resulting from extensive and long-term deposition of slackwater sediments from both the Illinois and Mississippi Rivers. As high water on the Mississippi River caused backup along the Illinois River, the high water on the Illinois River in turn caused backflow into several of its tributaries. For a time, on the lower reach of the Illinois River, flow direction actually reversed as floodwater from the Mississippi River advanced upstream into the Illinois Valley.

Sediments and Stratigraphy

The sediments and sedimentary successions of the river channels and floodplains of Illinois are integral to flood dynamics, particularly in relation to groundwater flow. A key factor is the contrast between the permeability of clay, silt, and sand. Clay and silt deposits retard or limit groundwater flow, but sand layers can provide pathways for flow.

An important factor in the 1993 flood was the permeability of sand layers in relation to the integrity of levees.

Figure 24-3 The Great Flood of 1993 inundated the entire 6-mile (9.7-km)-wide floodplain near Lock and Dam No. 20 at Meyer, just north of Quincy. Few people in their lifetimes will ever see this entire floodplain submerged. (Photograph by Joel M. Dexter.)

Figure 24-4 Subtle differences in elevation can determine the limits of flooding, as is well demonstrated by this V-shaped high-water line caused by the slightly higher land elevation and the higher elevation along the crown of this roadway in the Illinois River valley in Greene County. (Photograph by Michael J. Chrzastowski.)

Figure 24-5 A sandbag dike across the low elevation of a stream valley prevented floodwater from inundating the central business district of Alton in 1993. (Photograph by Michael J. Chrzastowski.)

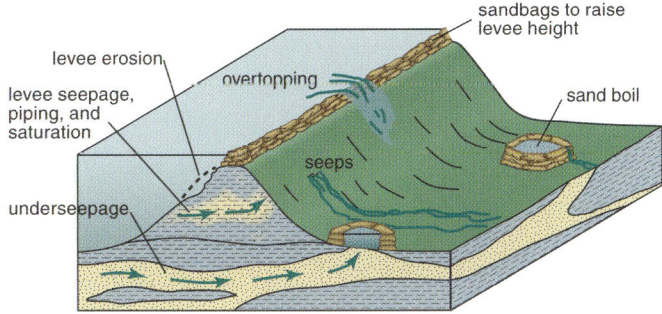

- ▬ floodplain sediments with low permeability (abundance of silt and clay)
- ▬ floodplain sediments with high permeability (abundance of sand)
- ▬ levee materials (varied combinations of sand, silt, and clay)

Figure 24-6 Floodwater has several means of moving under, through, and over levees (Chrzastowski et al. 1994).

Many of the agricultural levees along the Mississippi and Illinois River floodplains were built to rigorous engineering specifications; a clay core provided an internal, low-permeability barrier to groundwater flow. Other agricultural levees are simply mounded sediment dredged from the nearby floodplain or river channel; these levees typically contain internal lenses and layers of sand. For floods of short duration, water rising against these levees would not have sufficient time to infiltrate the sand layers. However, the prolonged high-water levels of the 1993 flood allowed water to percolate and pipe through or under some levees, resulting in seeps and sand boils on the floodplain side of the levee (Figure 24-6). Such groundwater flow through or under the levees in many places weakened their structure and contributed to subsidence, overtopping, and failure. During the 1993 flood, levee seeps and sand boils persisted for more than a week along the Sny Island Levee located south of Quincy, in Adams County (Figures 24-7 and 24-8).

Groundwater flooding is a special example of how sediments and stratigraphy can influence a flood. This process occurs when the weight of the higher water levels exerts enough downward pressure on the water table in a particular area to force it through layers of gravel and sand until it breaches the surface elsewhere. In groundwater flooding, the displaced groundwater actually wells upward through the ground. Prime examples of this process were seen in Mercer and Mason Counties in 1993.

The ground in western Mercer County, bounded by the Mississippi River, is characterized by a layer of sand and gravel that is more than 100 feet (30.5 m) thick (Piskin and Bergstrom 1975). During 1993, the Mississippi River was as much as 15 feet (4.6 m) higher than the ground behind some levees. The pressure from the massive amount of water pushed groundwater through the sand and gravel deposits under the levees. As a result, the already shallow water table rose appreciably. Although the local levees did not break during the flood, groundwater flooding occurred extensively throughout a leveed area of about 25,000 acres (101.2 km^2) within the Bay Island Drainage District.

Groundwater flooding also occurred behind levees of the Illinois and Sangamon River floodplains in Mason County where the surficial geology is characterized by a layer of sand and gravel up to 200 feet (61 m) thick. These sediments were deposited as glacial outwash in a broad lowland at the confluence of the ancient Mississippi River valley—now the Illinois River valley—and the ancient preglacial Mahomet-Teays Bedrock Valley (Melhorn and Kempton 1991). During 1993, areas away from the immediate influence of river flooding experienced serious groundwater flooding (Figure 24-9). From fall 1992 to fall 1993, excessive rainfall and lack of pumping by numerous irrigation wells caused the water table to rise more than 9 feet (2.7 m) in some areas (Clark 1994, Sanderson and Buck 1995).

Figure 24-7 Along the Sny Island Levee in 1993, straw was scattered to absorb water seeping through the levee, and straw bales were placed beneath the plastic sheeting to add levee height (Chrzastowski et al. 1994).

Figure 24-8 As sand boils began to form on the Sny Island Levee, they were ringed with sandbags to prevent the loss of sediment that could undermine the levee (Chrzastowski et al. 1994).

Figure 24-9 The 1993 groundwater flooding near Havana inundated areas where thick and well-drained sandy soils normally require irrigation. The flooding created this ironic situation of floodwater inundating an irrigation supply store. (Photograph by Michael J. Chrzastowski.)

GEOLOGICAL IMPACTS

A record flood, such as the Great Flood of 1993, would be expected to have widespread geological impacts, such as shifts in channel position, channel scouring and deposition, changes in the location and shape of river bars and islands, and meander cutoffs. Such geological impacts did occur, but they were not as widespread as the scale of the flood might have suggested. Much greater landscape change likely would have occurred if the river had been in a natural state free of levees, engineered navigation channels, and shoreline protection. Post-flood surveys of the Mississippi River channel revealed that local scouring and deposition changed its profile by several feet, and dredging was necessary in some locations to restore proper depths for barge navigation (Bhowmik 1996). A meander cutoff almost occurred near Miller City, in Alexander County, following a levee breach that sent floodwater flowing across the floodplain between River Miles 34 and 15 (Chrzastowski et al. 1994, Bhowmik 1996, Jacobson and Oberg 1997). However, the flow volume diverted through this breach was not sufficient to establish a new channel. If a meander cutoff had occurred, 19 miles (30.6 km) of existing river channel would have been isolated to become a backwater lake. The new channel segment would have been about 4 miles (6.4 km) long. The most common flood-induced landscape changes—from both erosion and deposition—occurred at and near levee breaches.

Erosion

As river water surged through the breach of a failing levee segment and inundated the exposed floodplain, levee materials dispersed in the extreme flow. The levee rupture

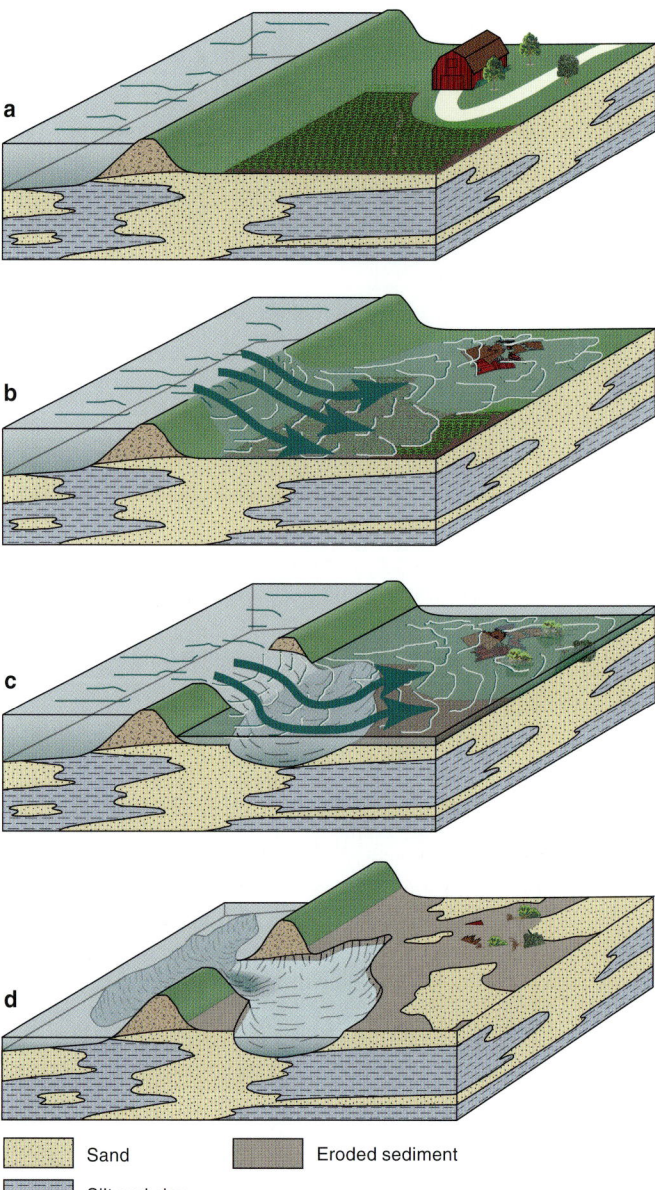

Figure 24-10 Stages leading to landscape change near a levee breach: **(a)** floodwater nears the top of the levee; **(b)** the levee is breached at a low point or zone of weakness; **(c)** water surges through the breach, erodes downward, and transports eroded sediment onto the floodplain; and **(d)** the flood aftermath reveals the deep scour hole, the nearby floodplain stripped by erosion, and the more distant floodplain buried by an extensive sand and silty sand sheet (Chrzastowski et al. 1994).

widened as the water continued to pour through, and the flow caused even greater erosion where it focused downward into the floodplain sediments, carving deep scour holes (Figure 24-10). A detailed survey of the scour hole at the levee breach at Kaskaskia Island measured as much as 50 feet (15.2 m) of erosion into the original floodplain. About 1 million cubic yards (about 800,000 m³) of floodplain sediment were eroded from this scour hole and transported onto the floodplain (Chrzastowski et al. 1994). The floodplain immediately surrounding these scour holes suffered substantial surface erosion. These erosional zones extended several hundred feet and included erosional scars as much as 2 feet (0.6 m) deep. Where cut into agricultural fields, the pattern and orientation of these erosional features suggested that plow furrows may have concentrated the floodwater flow and direction.

Deposition

A film of fine sand, silt, and clay is a typical flood signature left on buildings, vehicles, and roads when the water recedes. Because of the duration of the 1993 flooding, fine sediment as much as 6 inches (15.2 cm) thick was deposited in some places. Along the flood margin in flooded levee districts, in areas distant from the levee breaches, little if any sediment was deposited. However, many of these areas were covered by thick and extensive accumulations of plant debris washed from the flooded corn and soybean fields.

Beyond the erosional areas near levee breaches, wide expanses of deposited sand and silt were common (Figure 24-10). In some locations, these extensive deposits resembled a desert of sand. On the floodplain beyond the breach in the Columbia Levee in Monroe County, for instance, sand deposits nearly 1 foot (0.3 m) thick covered 760 acres (3.1 km²) and were spread up to nearly 1 mile (1.6 km) from the levee failure. Maximum sand thickness was 8 feet (2.4 m) (Chrzastowski et al. 1994). Although some of this deposited sand originated from the erosion of the levee or was already present in the river, the volume of sediment indicates that most of it likely came from the erosion of the scour holes at the levee breach. The abundance of sand suggests that the scour holes possibly developed in ancient, sand-filled channels.

Relocation of Valmeyer

The Great Flood of 1993 brought into focus the consequences of human modifications to the rivers and watersheds in Illinois. Channelization, urbanization, loss of wetlands, and construction of levee networks are a few of the changes Illinois waterways have undergone since people began to settle the state. Most importantly, the flood raised issues about settlement on the floodplains, flood risks, and costs in terms of life and property. In the post-flood recovery, the town of Valmeyer, in Monroe County, gained state and national attention for the decision by its residents—about 900 in all—to relocate the town from the floodplain to higher ground.

Prior to the flood, Valmeyer was located on the Mississippi River floodplain about 20 miles (32.2 km) downstream from East St. Louis. On August 1, 1993, the Harrisonville

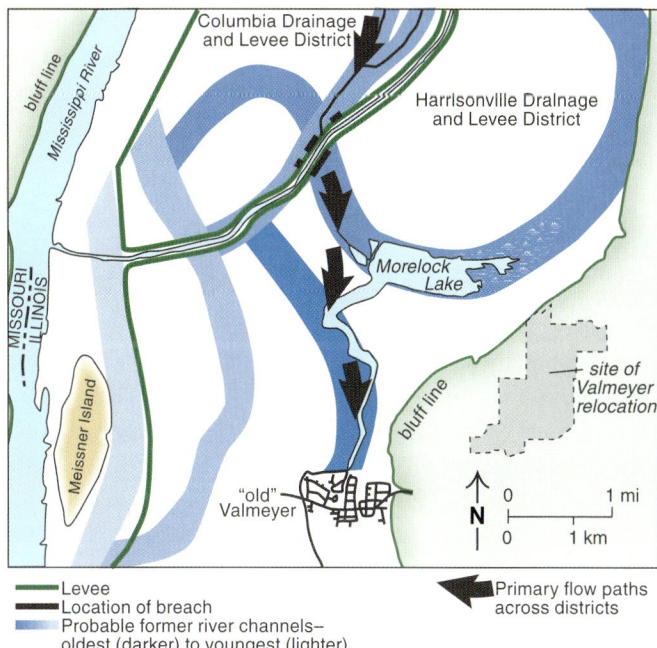

Figure 24-11 Levees north of Valmeyer failed in places over sand-filled, ancient channels. The topography along a relict channel also helped direct the floodwaters toward the town (Chrzastowski et al. 1994).

Levee broke about 3 miles (4.8 km) north of Valmeyer (Figure 24-11). A torrent of floodwater rapidly surged into the town and destroyed buildings, lifting and moving some off their foundations (Figure 24-12). Floodwater in the town eventually rose to a height of 20 feet (6.1 m). Rather than rebuilding on the floodplain and remaining vulnerable to future flood events, the citizens of Valmeyer voted to relocate the town. The new location was a 500-acre (2.02-km²) site about 1.5 miles (2.4 km) to the east and about 400 feet (120 m) higher atop the bluffs overlooking the Mississippi

Figure 24-12 The floodwaters raging through Valmeyer lifted this house off its foundation and hurled it into a tree. Corn and other plant debris in the tree branches (rectangular area in upper right corner) indicate the maximum floodwater height. (Photograph by Michael J. Chrzastowski.)

Valley (Erdmann and Bauer 1993, Watson 1996). Considerable media attention was devoted to the story of Valmeyer's relocation as this effort became a national model for hazard mitigation and post-flood response. By 2003, the 10-year anniversary of the 1993 flood, Valmeyer was a prospering community with a population nearly at its pre-flood level of 900.

GEOLOGICAL LESSONS

Of the many lessons learned from the 1993 flood, perhaps the most important are how the structural integrity of agricultural and urban levees influences flood dynamics, what the recovery time is for such an extreme event, and why wetlands are crucial to flood mitigation efforts.

Levees confined floodwaters to river channels and prevented their lateral dispersion across the floodplains. The hydraulic head created by this restriction produced torrential flows when levees did fail. The creation of such extreme floodwater heights is one example of how human modifications along the floodplains influenced flood dynamics. When a breach did occur, maximum flood height was reduced as floodwater spread across the floodplain, reducing the flood threat to locations farther downstream. Intentionally opening a levee or designing levee systems to be readily overtopped or breached at a critical flood stage can thus be an effective means of flood mitigation. During the 1993 flood, the opening of a levee upriver from the town of Prairie du Rocher, in Randolph County, diverted floodwaters and prevented the inundation of that town (Chrzastowski et al. 1994).

In the wake of the 1993 flood, insight was also gained concerning the recovery time for some geological systems following such an extreme event. For example, in Mason County where groundwater flooding was extensive, some fields were still flooded two years after the flood event. There was so much groundwater from 1993 that normal discharge of the water to nearby rivers and streams was very slow, even through the highly permeable sands and gravels.

The flood aftermath produced new and renewed efforts to improve overall floodplain management. The relocation of the town of Valmeyer gained substantial attention and became a model for moving individuals and an entire town out of flood-prone areas. Greater emphasis was placed on identifying suitable geological areas for wetlands restoration and construction to restore habitats and provide floodwater storage to mitigate the effects of future floods. Wetlands can provide temporary storage of floodwater and reduce floodwater height and flow rate. One of the major results of these efforts was the restoration of wetlands across a 2,600-acre (10.5-km^2) site along the floodplain of the Illinois River near Hennepin, in Putnam County. Like much of the floodplain of the Illinois River valley, for decades the site near Hennepin had been protected by levees and had been used for agriculture. Once pumping for land drainage was terminated, the floodplain soon reverted to a setting once common along the Illinois River valley that included wetlands, wet prairie, and backwater lakes.

CONCLUSION

The Great Flood of 1993 was one event in a long history of flooding that has shaped and reshaped the state's river valleys and floodplains. Much of this landscape evolution occurred in late glacial time when melting glacial ice discharged large volumes of water and contributed to torrential river flows that eroded adjacent uplands, widened channels, and deposited sand, silt, and gravel on floodplains and terraces. Although the magnitude of the 1993 flood does not compare with those of the floods of late glacial time that impacted the region from about 20,000 to 13,000 years ago, the 1993 flood was a catastrophic event that resulted in unprecedented damages and costs for the people of Illinois.

Climate controls the amount and distribution of precipitation an area receives, but geology is a major factor in the location and severity of flooding and in the kind of mitigation efforts that are undertaken to reduce the potential for loss of life and property. The 1993 flood clearly demonstrated the roles that topography and sediments can play in contributing to flooding, groundwater flow, seepage through levees, and levee stability. Because rivers have typically migrated across their floodplains over time, sand-filled channels are common floodplain features. These sand-filled, ancient river channels apparently contributed to the locations of several of the levee ruptures. It is common for segments of a levee that are built atop such channels to be potential weak points.

Flooding is a natural process in river dynamics and remains the most common geological hazard in Illinois. Although the 1993 event is currently the state's most severe flood in modern times, future flood events could rival or exceed it. The 1993 flood has established a reference point for improved planning and management of flood-prone areas in Illinois and has shown that land use across the floodplains and flood-prone areas in Illinois should be compatible with the inevitable occurrence of major flooding.

REFERENCES

Bhowmik, N. G., 1996, Physical effects—A changed landscape, *in* S. A. Changnon, ed., The great flood of 1993: Causes, impacts, and responses: Boulder, Colorado, Westview Press, p. 101–131.

Bhowmik, N. G., A. G. Buck, S. A. Changnon, R. H. Dalton, A. Durgunoglu, M. Demissie, A. R. Juhl, H. V. Knapp, K. E. Kunkel, et al., 1994, The 1993 flood on the Mississippi River in Illinois: Champaign, Illinois, Illinois State Water Survey, Miscellaneous Publication 151, 149 p.

Changnon, S. A., ed., 1996, The great flood of 1993: Causes, impacts, and responses: Boulder, Colorado, Westview Press, 321 p.

Chicago Shoreline Protection Commission, 1988, Recommendations for shoreline protection and recreational enhancement—Final report: City of Chicago, 52 p. plus appendices.

Chrzastowski, M. J., M. M. Killey, R. A. Bauer, P. B. DuMontelle, A. L. Erdmann, B. L. Herzog, J. M. Masters, and L. R. Smith, 1994, The Great Flood of 1993: Geologic perspectives on the flooding along the Mississippi River and its tributaries in Illinois: Illinois State Geological Survey, Special Report 2, 45 p.

Clark, G. R., 1994, Mouth of the Mahomet regional groundwater model, Imperial valley region of Mason, Tazewell, and Logan Counties, Illinois: Springfield, Illinois, Illinois Department of Transportation, Division of Water Resources Report, 70 p.

Erdmann, A. L., and R. A. Bauer, 1993, Geologic evaluation of the proposed new town site, Valmeyer, Illinois: Illinois State Geological Survey, Open File Series 1993-12, 42 p.

Illinois Department of Transportation, Division of Water Resources, 1993, Floodplain management, local floodplain administrator's manual: Illinois Department of Transportation, various pagings.

Jacobson, R. B., and K. A. Oberg, 1997, Geomorphic changes on the Mississippi River flood plain at Miller City, Illinois, as a result of the flood of 1993: Reston, Virginia, U. S. Geological Survey, Circular 1120-J, 22 p.

Kunkel, K. E., 1996, A hydroclimatological assessment of the rainfall, *in* S.A. Changnon, ed., The great flood of 1993: Causes, impacts, and responses: Boulder, Colorado, Westview Press, p. 52–67.

Luman, D. E., L. R. Smith, and C. C. Goldsmith, 2003, Illinois surface topography: Illinois State Geological Survey, Illinois Map 11, 1:500,000.

Melhorn, W. N., and J. P. Kempton, 1991, Geology and hydrogeology of the Teays-Mahomet Bedrock Valley systems: Boulder, Colorado, Geological Society of America, Special Paper 258, 128 p.

Piskin, K., and R. E. Bergstrom, 1975, Glacial drift in Illinois: Thickness and character: Illinois State Geological Survey, Circular 490, 35 p.

Rodenhuis, D. R., 1996, The weather that led to the flood, *in* S. A. Changnon, ed., The great flood of 1993: Causes, impacts, and responses: Boulder, Colorado, Westview Press, p. 29–51.

Sanderson, E. W., and A. G. Buck, 1995, Reconnaissance study of ground-water levels in the Havana lowlands area: Champaign, Illinois, Illinois State Water Survey, Contract Report 582, 63 p.

St. Louis Post-Dispatch, 1993, High and mighty: The great flood of 1993: Kansas City, Missouri, Andrews and McMeel (A Universal Press Syndicate Company), 95 p.

Visocky, A. P., 1995, Determination of 100-year ground-water flood danger zones for the Havana and Bath areas, Mason County, Illinois: Champaign, Illinois, Illinois State Water Survey, Contract Report 584 [16 p.].

Wahl, K. L., K. C. Vining, and G. J. Wiche, 1993, Precipitation in the Upper Mississippi River Basin, January 1 through July 31, 1993: [Washington, D.C.], U.S. Government Printing Office, Circular 1120-B, 13 p.

Watson, B., 1996, A town makes history by rising to new heights: Smithsonian, v. 27, no. 3, p. 110–120.

25 The Illinois Coast of Lake Michigan

Michael J. Chrzastowski

Introduction

The Illinois coast of Lake Michigan is one of the state's most dynamic geological settings. Coastal processes of waves, ice, and changing lake levels contribute to yearlong and seasonal erosion and deposition (accretion) along the beaches and across the nearshore lake bottom, although major change also can occur in days or even hours.

The social and economic importance of the Illinois coast cannot be overstated. This coast borders the most populous part of the state and includes some of the most valued real estate in Illinois. It is also the most densely populated coastal area in the entire Great Lakes region and has the region's highest degree of engineering and human modification. Notably, the historical development of Chicago, Waukegan, and the other municipalities along the shore was strongly influenced by the coastal geology.

Despite extensive urbanization, the Illinois coastal zone preserves an exceptional geological record of coastal evolution that spans about 14,000 years, beginning in late glacial time when an ancestral shoreline first formed. A series of ancient shoreline features document how the Illinois coast evolved through changing lake levels and shaping and reshaping by wave-induced erosion, transport, and deposition. The historical record of coastal change also provides numerous examples of how human modification has notably altered the shoreline configuration and has had an impact on local coastal processes. The continuing geological changes along this coast present ongoing engineering and management challenges that will face future generations.

Coastal Shape and Dynamics

The Illinois coastline extends 63 miles (101.4 km) along the southern reach of the western shore of Lake Michigan (Figure 25-1), 22 miles (35.4 km) of which is the City of Chicago shoreline. North of Chicago, the coast includes 14 near-lake municipalities, the U.S. Navy's Naval Training Center Great Lakes, and state-owned Illinois Beach State Park and North Point Marina. The municipalities from Lake Bluff south to Evanston are collectively called the North Shore. The state's jurisdiction extends across a portion of the lake, covering approximately 1,500 square miles (3,885 km²) of lake and lake bottom. The deepest water within the Illinois extent of Lake Michigan is about 490 feet (149.4 m).

Figure 25-1 The Illinois coast of Lake Michigan showing municipalities and harbors. The index map shows the configuration of the Illinois state line across Lake Michigan.

The Illinois coast contains three distinct shoreline reaches (Figure 25-1). From the Illinois-Wisconsin state line southward to Waukegan Harbor is a nearly linear, north-south reach that has been shaped exclusively by coastal deposition and erosion. From Waukegan Harbor southward to Wilmette Harbor, the shoreline follows a broad arc. This reach has been modified by coastal erosion, but the overall shape reflects the shape of the margin of the receding glacial ice that once occupied this part of the Lake Michigan basin. From Wilmette Harbor south to the Illinois-Indiana state line, the curving shore is part of a broad arc that continues along the Indiana shore. The overall shape along this reach has been strongly influenced by glacial processes and, more importantly, by the coastal erosion and deposition accompanying a history of lake-level change. Human modifications within this reach have transformed a once featureless shore to one with promontories, harbor embayments, and designed shorelines.

The prominent change in shoreline orientation that occurs near Wilmette Harbor relates to a natural headland on the coast centered about 1 mile (1.6 km) south of the harbor along the Evanston shore. This headland—historically known as Grosse Point (U.S. Lake Survey 1873)—is the site of the historic Grosse Point Lighthouse.

Geomorphic Divisions of the Illinois Coast

The Illinois coast consists of three geomorphic divisions: beach-ridge plain, bluff coast, and lake plain (Figure 25-2) (Chrzastowski et al. 1994).

Zion Beach-Ridge Plain

From the Wisconsin-Illinois state line southward to North Chicago is the Zion Beach-Ridge Plain. This expanse of sand and gravelly sand is up to 1 mile (1.6 km) wide at Zion. The surface of the plain has a ridge-and-swale topography of sand ridges and intervening marshy areas (Figure 25-3). Much of the plain is no more than 10 feet (3.1 m) above mean lake level. In east-west cross section, this sand body has a maximum thickness of 30 feet (9.1 m) near the shore and becomes thinner both landward and into the lake, resulting in an overall lenticular (convex lens) shape. The sand body overlies older lake-bottom deposits and till (Hester and Fraser 1973, Fraser and Hester 1974, Chrzastowski and Frankie 2000). Along its western margin, most of the plain is bordered by a 10- to 20-foot (3- to 6-m) low bluff that delineates the uplands from the plain. Much of the plain is preserved within Illinois Beach State Park where the distinctive surface features are the arcuate beach ridges and marshy swales that record successive shorelines formed as deposits of sand (accretion) advanced the plain southward and into the lake (Figure 25-3).

Bluff Coast

From North Chicago southward to Winnetka lies a bluff coast (Atwood and Goldthwait 1908). Here the shore primarily intercepts the Highland Park Moraine but includes a short segment along the Zion City Moraine and, in its northern extent, a high segment of the Chicago Lake Plain (Willman and Lineback 1970). The Highland Park Moraine is part of the Lake Border Morainic System, which is a series of parallel end moraines that formed just prior to the withdrawal of Wisconsin Episode glacial ice into the Lake Michigan basin. The bluff slopes range from nearly vertical to 25°. The maximum bluff height (about 90 feet or 27.4 m) occurs at Highland Park where coastal erosion has nearly intercepted the crest of the Highland Park Moraine. The bluffs are cut by a series of more than 50 V-shaped ravines that extend inland as much as 1 mile (1.6 km) from the bluff line and include intermittent streams. The bluff materials are predominantly a clayey till but include beds of glacial outwash and glacial lake sediments (Clark and Rudloff 1990). The average composition of the bluff materials is 10% sand, 42% silt, and 48% clay (Lineback 1974). Although shore protection now acts as armor for nearly all of the bluffs, bluff erosion was formerly the primary source of sand (and gravel) for the beaches. The eroded silt and clay, too fine to remain on the beaches, were further eroded, transported, and deposited into the deep basin of the lake (Colman and Foster 1994). Long-term bluff erosion rates from 1872 to 1987 averaged approximately 7.2 inches (18.3 cm) per year (Jibson et al. 1994). The low percentage of sand in the bluff materials resulted in the beaches along the Illinois coast being sediment starved, even when bluff erosion was extensive. A prominent change in bluff height of about 45 feet (13.7 m) occurs in Winnetka (south of Tower Road) at the southern limit of the Highland Park Moraine. From here southward to northern Evanston, the low bluff gradually diminishes in height, forming a transitional zone between the high bluffs of the North Shore and the lower elevation of the Chicago Lake Plain.

Chicago Lake Plain

Extending southward from the bluff coast to the Illinois-Indiana state line is the Chicago Lake Plain (Willman and Lineback 1970, Willman 1971). The plain continues eastward along the Indiana coast (Schneider and Keller 1970) where it is referred to as the Calumet Lake Plain (Chrzastowski and Thompson 1992, 1994). The plain is a low-slope, glacially formed surface that is an above-water

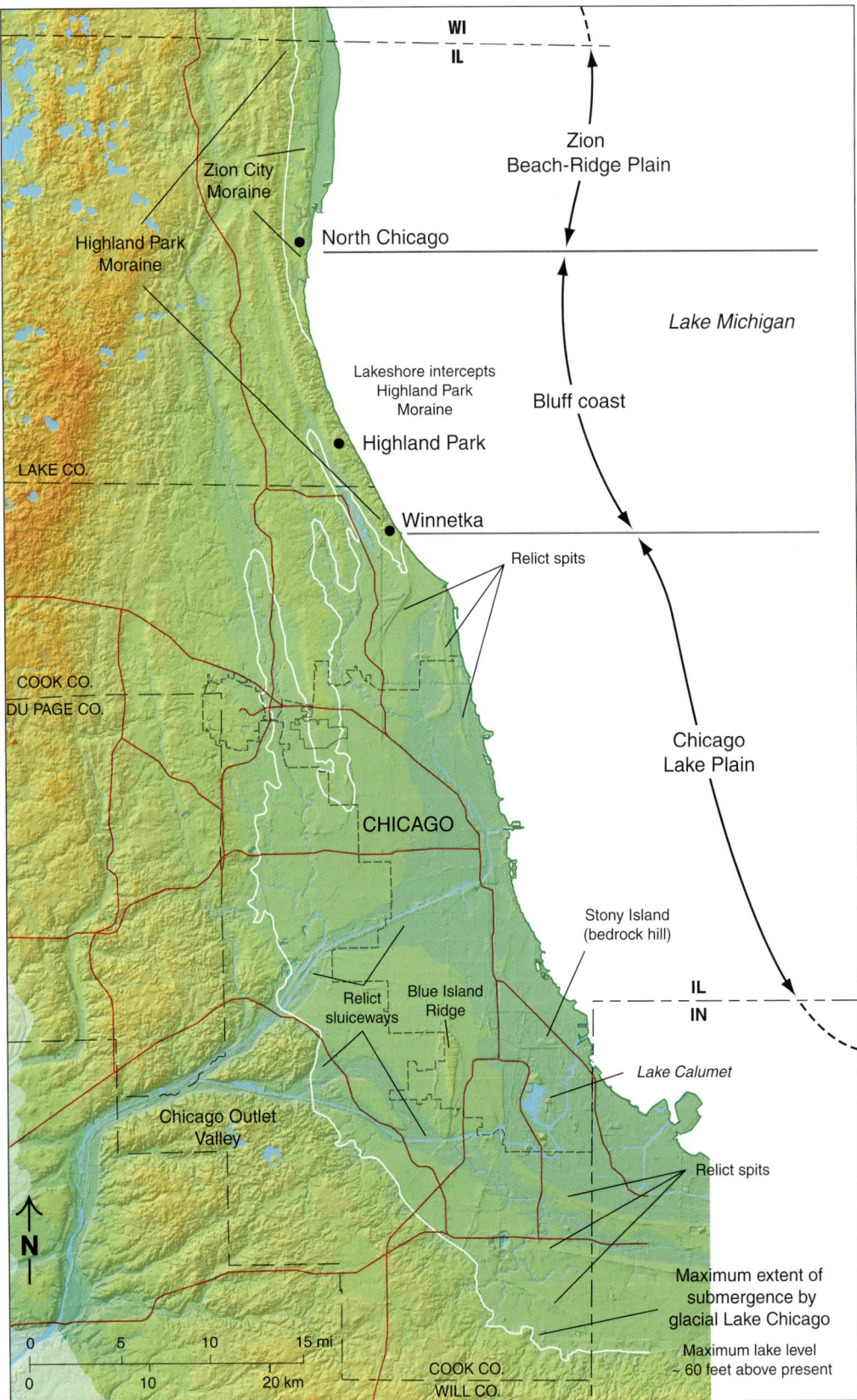

Figure 25-2 There are three geomorphic divisions of the Illinois coast: Zion Beach-Ridge Plain, bluff coast, and Chicago Lake Plain. Blue Island Ridge and Stony Island were islands in the ancestral Lake Michigan during different times of submergence of the Chicago Lake Plain. The Chicago Outlet Valley provided for the outflow of ancestral Lake Michigan during times of elevated lake levels. (Map modified from Chrzastowski 2005.)

continuation of the southern Lake Michigan lake floor. The plain was submerged to varying degrees during times of higher lake level, resulting in coastal deposits overlying the plain's till base. When lake water extended onto the Chicago Lake Plain, the Illinois coast was embayed (i.e., indented) in the Chicago area. Wave transport of beach and nearshore sand resulted in the formation of elongate sand peninsulas called "spits" that extended into this embayment. Within the limits of Chicago and several near-in suburbs, these ancient spits are preserved as sand ridges that rise as high as 20 feet (6 m) above the neighboring plain (Leverett 1897, Alden 1902, Bretz 1943, Willman and Lineback 1970, Willman 1971). The plain also includes Blue Island Ridge, an erosional remnant of one of the Lake Border Moraines, and Stony Island, a hill formed by protrusion of the regional bedrock of Silurian age Racine Dolomite (Willman 1971). A narrow band of Chicago lake plain extends along the uplands near the Illinois-Wisconsin state line and is related to the higher lake levels of late glacial time.

Coastal Processes

Wave action is the most important factor in the geological dynamics of the Illinois coast. Calm water conditions are common because of prevailing westerly (offshore) winds. When onshore winds generate waves, average wave heights along the nearshore are 1.5 to 2 feet (0.5 to 0.6 m), average maximum wave heights are 8 feet (2.4 m), and extreme waves rarely reach 10 to 12 feet (3.6 m) (U.S. Army Corps of Engineers 1953). The greatest fetch (i.e., the distance across water) is from the northeast quadrant. As a result, northeasterly waves have the greatest height and the greatest net influence on sediment movement. Waves from the southeast quadrant can move sediment northward along the Illinois coast, but northerly waves have the greater net effect. The result is that littoral transport, the wave-induced movement of sand, has a net southward transport along the Illinois beaches and nearshore lake bottom.

Annual and multiyear changes in lake level result from changes in the lake's water budget. Annual lake level fluctuation is about 1 foot (0.3 m); high water tends to occur in summer and low water in winter. The historical record of mean monthly lake level since 1918 shows a 6.3-foot (1.92-m) range between the historical low (March 1964) and the historical high (October 1986) (U.S. Army Corps of Engineers Detroit District 2003).

Each winter, ice influences the Illinois coast to varying degrees depending on temperature, wind, and wave action. A nearshore ice complex can protect the beaches from winter storm waves, but the lakeward edge of the ice can direct wave energy downward and cause lake-bottom erosion. Wave turbulence can bring eroded sand and gravel onto the ice where they can become frozen into the ice complex. Subsequent ice breakup can result in ice-rafting of the sediment along the Illinois shore or offshore (Barnes et al. 1994, Kempema 1998).

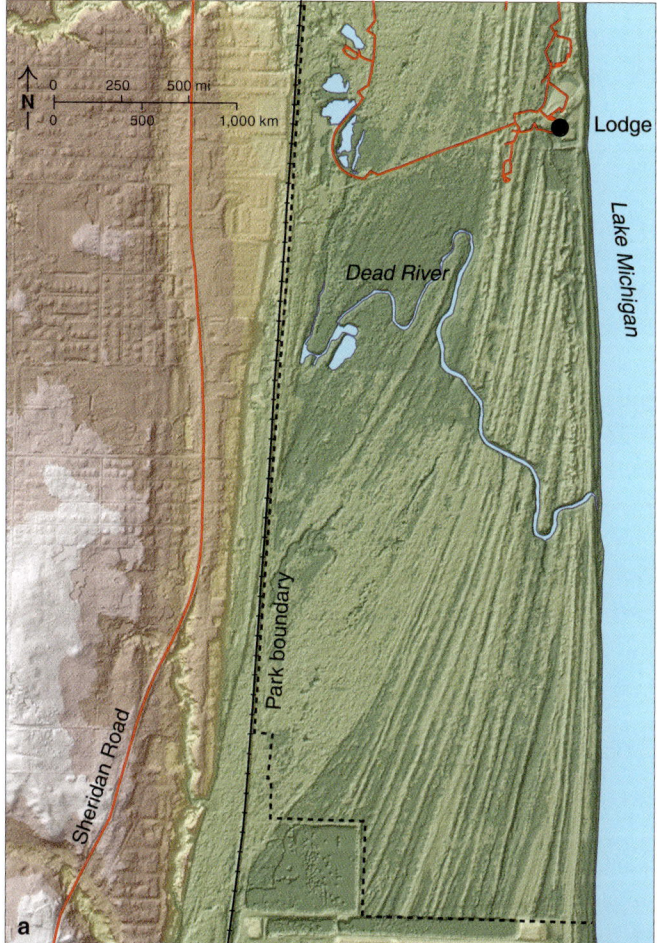

Figure 25-3 (a) Radar mapping (LIDAR) of the South Unit of Illinois Beach State Park provides a detailed view of the ridge-and-swale topography that records former shoreline positions. (LIDAR data courtesy of Lake County, Illinois.) (b) The mouth of the Dead River is seen in this May 2000 photo. (Photograph by M. J. Chrzastowski.)

A geological hazard along the Illinois coast is a fast-moving, tsunami-like wave called a "seiche." This basin-wide oscillation of lake water is caused by strong wind and/or atmospheric pressure changes associated with the rapid passage of fronts and thunderstorms. Seiches less than a foot (0.3 m) in height are common along the Illinois shore. An extreme seiche in June 1954 had a maximum height of 10 feet (3.1 m) at Chicago's North Avenue Beach. This seiche contributed to shore damage and the drowning of eight persons (Ewing et al. 1954; State of Illinois, Division of Waterways 1958; Recktenwald 1994).

COASTAL EVOLUTION

The geological evolution of the Illinois coast has had two phases. Initially, glacial processes established thick deposits of till and molded the overall topography of the moraines and lake plain. Then, as ice receded and an open-water area developed, the coastal processes of wave-induced erosion, transport, and deposition combined with changing lake level to shape and reshape the shore (Leverett 1897; Alden 1902; Goldthwait 1908; Bretz 1939, 1955; Willman 1971). Early in its evolution, the coast included spits, bays, and lagoons, and the shoreline in many places was landward of its present position. There were also times when the lake level was below its current level and much of the area now offshore was exposed land. The present Illinois shoreline was configured relatively recently during the past 1,000 to 2,000 years.

The Illinois coast began to form about 14,000 years before present (BP) as glacial ice receded from the end moraines that border the Chicago Lake Plain. The ice recession allowed development of a glacier-fed lake (glacial Lake Chicago) between the ice margin and the moraines. The moraines acted as a dam to hold lake levels at an elevation above the historical levels, submerging the Chicago Lake Plain. The Chicago Outlet Valley is a breach through the morainal uplands that permitted lakewater drainage (Figure 25-2). Wave erosion of the morainal uplands bordering the lake provided sand and gravelly sand for beach development. The oldest beaches and shoreline features along the landward margins of the plain (shorelines of the Glenwood phases) are about 60 feet (18.3 m) above the present lake level and up to 15 miles (24.1 km) inland from the present lakeshore (Figure 25-4a).

For the next 11,500 years, after the ancestral Illinois coast first formed, lake levels fluctuated widely, from tens of feet higher to hundreds of feet lower than historical lake levels (Figure 25-5). The lake level fluctuations resulted from ice recession that opened and closed different lake outlets at different elevations, fluvial erosion of outlet threshold elevations, major changes in the volume of water in the lake basin, and rising elevation of the lake outlets as the land rose (i.e., isostatic uplift) when it no longer was weighed down by glacial ice (Hansel and Mickelson 1988, Colman et al. 1994). High lake levels above the historical mean occurred in a series; peak elevations of each level were successively lower (Glenwood, Calumet, Nipissing, and Algoma phases). As a result, the ancient shorelines across the Chicago Lake Plain decrease in age and elevation toward the modern shore (Figure 25-4).

In its initial history as a lake embayment, the Chicago Lake Plain received wave-transported sediment from the north, resulting in spits on the northern part of the plain. Wave-transported sediments from the east resulted in spits on the southern part of the plain. By about 2,000 years BP, lower lake level and continued southward development of spits and barrier beaches along the Chicago shoreline resulted in littoral transport from the north extending to the Indiana shore. The last vestiges of submergence on the Chicago Lake Plain occurred in the Calumet area. Lake Calumet and the neighboring lakes (Wolf Lake, Lake George, and former Hyde and Bear Lakes) developed in low areas between the beach ridges built by littoral transport from the north (Figure 25-6). Ancient shorelines in the Calumet area are cut off (i.e., truncated) at sharp angles along the modern shoreline. These shorelines developed along a curving, embayed shore and were later eroded along their northern segments as the shoreline was reshaped to form the present coast.

The primary source for the large volume of sand and gravelly sand for the spits that built onto the northern part of the Chicago Lake Plain was erosion along the coastal bluffs to the north. Like the former shorelines in the Calumet area, former bluff positions are presently submerged in what is now the nearshore zone.

Coastal evolution along the far northern reach of the Illinois coast, from the Illinois-Wisconsin state line southward to North Chicago, has had a geological evolution unlike the rest of the Illinois shore. First, during the time of glacial Lake Chicago (Glenwood and Calumet phases), lagoons and barrier beaches developed on a narrow lake plain that is now a bench along the margin of this upland area. Second, a bluff coast developed when the lake level was below the threshold elevation of this upland bench (early Nipissing phases). Third, progressing from north to south, new land formed on the lakeward side of the bluffs as the Zion Beach-Ridge Plain advanced southward. Radiocarbon dating of swale marsh deposits indicates that the plain first advanced southward across the state line about 3,700 years BP (Larsen 1985). The oldest preserved part of

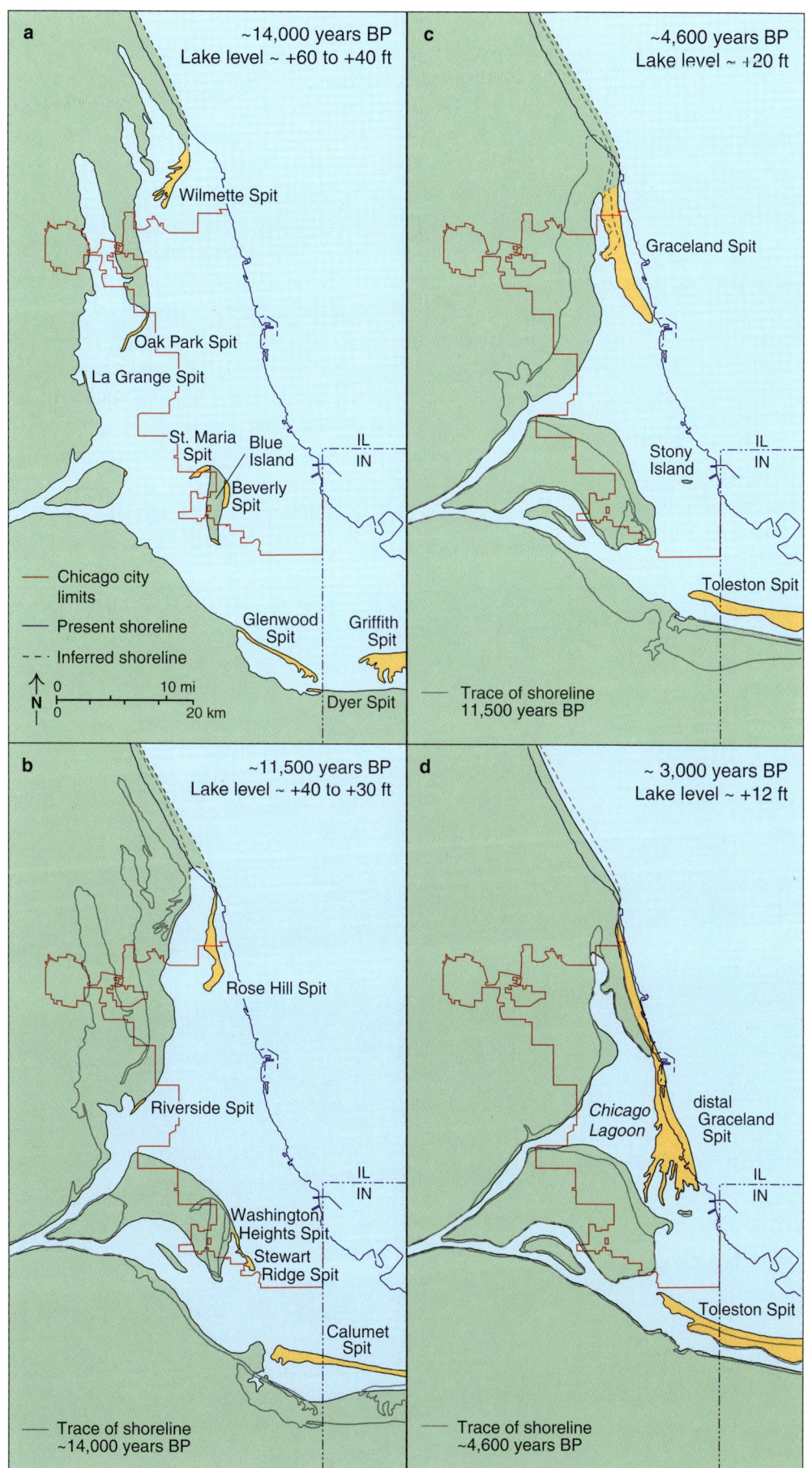

Figure 25-4 Successive shoreline configurations across the Chicago Lake Plain during the series of progressively lower lake levels and reduced submergence of the lake plain: (a) Glenwood I and II phases, (b) Calumet phase, (c) Nipissing I and Early Nipissing II phases, and (d) Late Nipissing II and Algoma phases (Chrzastowski and Thompson 1992, 1994).

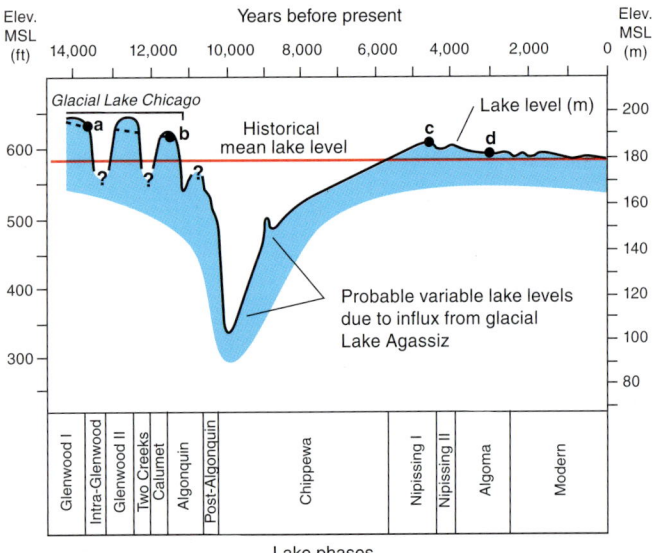

Figure 25-5 Generalized lake level curve for southern Lake Michigan. Times labeled **a, b, c,** and **d** correspond to times of shorelines shown in Figure 25-4. Elevation is mean sea level (Chrzastowski and Thompson 1992, 1994).

the plain is about 3 miles (4.8 km) north of the state line at Kenosha, Wisconsin. The plain has moved southward, experiencing erosion along its northern shore and deposition along its southern shore (Figure 25-7). The plain has also become narrower and more elongate with time. Over the past several hundred years, the northern part of the plain in Illinois has had considerable erosion. Here the plain includes old beach ridges that are tangential to and truncated by the present shore (Chrzastowski and Frankie 2000). These old ridges are aligned to former shorelines that were located as much as 1,000 feet (305 m) or more lakeward of the present shoreline.

When European settlement began along the Illinois coast during the early 1800s, a continuous, unobstructed beach of sand and gravelly sand extended along the entire coast. Even river mouths such as the mouth of the Chicago River were at times blocked by beach deposits as wave processes and littoral transport overwhelmed the ability of the river flow to maintain a channel across the beach. Geo-

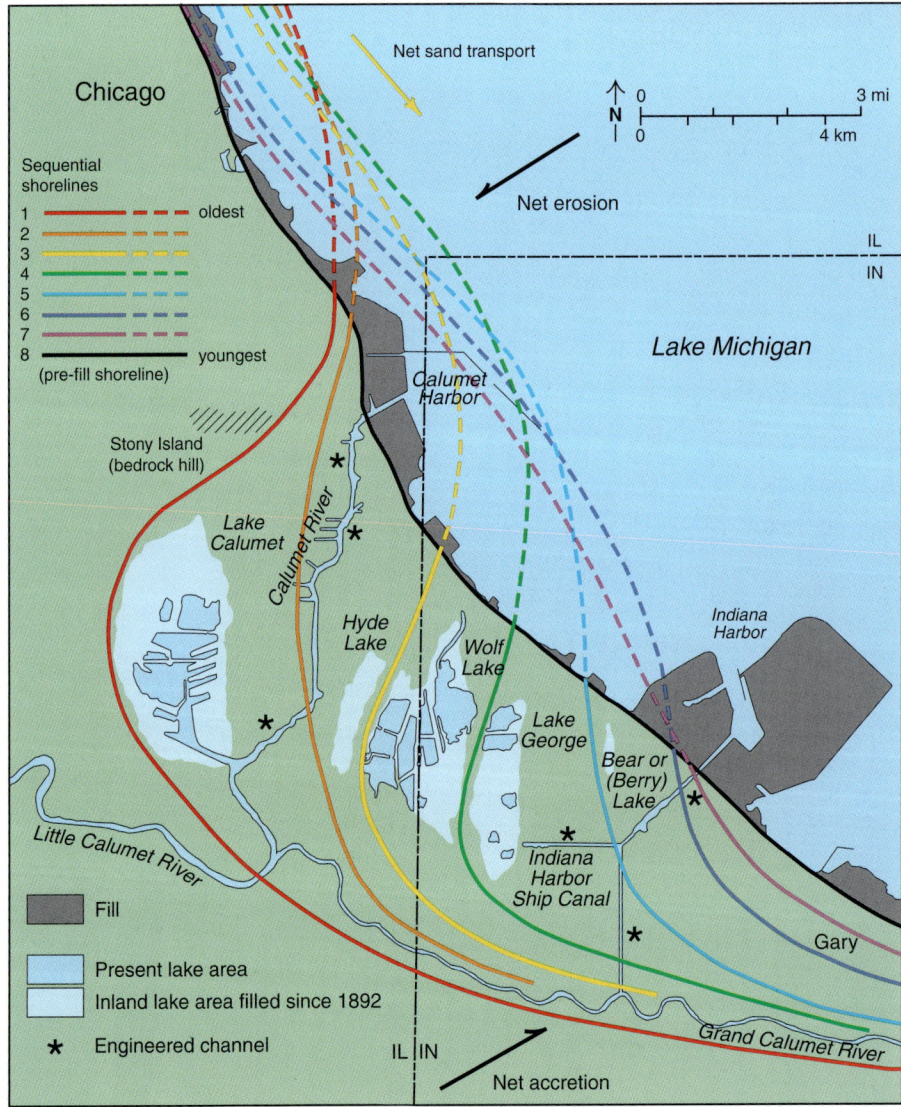

Figure 25-6 Changes to the shoreline configuration in the Calumet area near the Illinois-Indiana state line over the past 2,000 years.

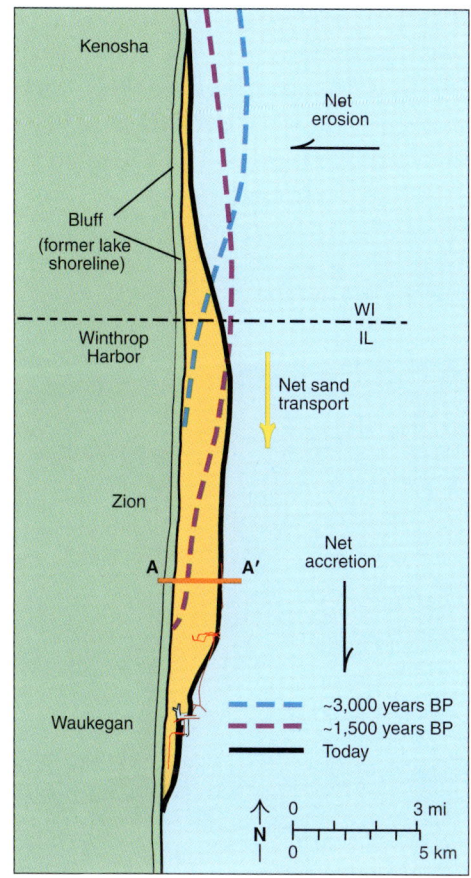

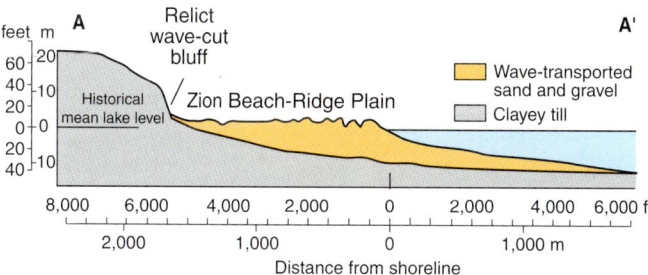

Figure 25-7 Illustration of the southward migration of the Zion Beach-Ridge Plain by erosion along its northern segment and accretion along its southern segment. In cross section, this sand body is thickest near the shore and thins both landward and lakeward. Abbreviation: BP, before present.

morphic indicators of littoral transport, such as deflected river mouths, indicate that the Illinois coast was part of a single littoral transport cell with net southerly transport originating north of Sheboygan, Wisconsin, and continuing southward to the central Indiana coast. Along this transport pathway, sediment sorting occurred, resulting in fine to medium sand reaching the Indiana shore. The Indiana Dunes were the terminus and the sediment sink for this net southerly littoral transport (Chrzastowski 1990a, Chrzastowski et al. 1994, Olyphant and Bennett 1994).

HUMAN MODIFICATIONS

The earliest documented human modifications along the Illinois coast occurred in the 1820s at the mouth of the Chicago River. When beach deposits blocked the river mouth, soldiers from nearby Fort Dearborn dug trenches across the beach to open river flow to the lake and alleviate high water and the threat of river flooding near the fort (Andreas 1884). The first coastal engineering along the Illinois shore also occurred at the mouth of the Chicago River. In 1833, U.S. Army Engineers began construction of a pair of timber and stone jetties (called the U.S. Government Piers or North Pier and South Pier) to straighten and stabilize the mouth of the river (Cram 1839, Larson 1979). Several phases of jetty lengthening and altered design were necessary in the continued attempt to prevent the river entrance from being blocked by littoral sediment. By 1869, nearly 70 acres of new land had accreted on the north (updrift) side of North Pier (Figure 25-8). Additional accretion occurred along the lake bottom, forming bars and shoals across the harbor entrance. The associated deficit of sand supply to the lakeshore south of the river required extensive shoreline protection for roads and railroads threatened by shore erosion (Rosenbaum 1981).

The construction of jetties for a harbor at Waukegan began in the 1880s and resulted in similar sand entrapment on the north (updrift) side of the harbor entrance (Figure 25-8). Sand entrapment also occurred on a lesser scale at the shore end of the breakwater built at Calumet Harbor (breakwater construction in 1904) and the harbor at Naval Training Center Great Lakes (outer breakwater construction in 1923).

As private property development increased along the North Shore during the late 1800s and early 1900s, groins were built to trap sand to form wider beaches and maintain beaches (Shabica et al. 2004). Continued residential development along the North Shore led to a greater amount of shore protection to prevent wave erosion along the toe of the bluff. By the 1980s, the continuous nature of the shore protection along the North Shore coastal bluffs had essentially eliminated bluff erosion as a source of new beach sediment.

The greatest degree of coastal modification has been along the Chicago shoreline. Extensive lakefill projects began in the late 1800s and peaked between 1920 and 1940. The purpose was to create new lakefront land for parks and open space (Wille 1972; Chrzastowski 1991, 2004). Three geological factors of the Chicago shore favored these lakefill projects: the gentle lake-bottom slope that facilitated nearshore filling, thick till deposits conducive to driving piles for fill-retention bulkheads, and abundant nearby sources of sand for clean fill. At least 5.5 square miles (14.2 km²) of artificial land has resulted, extending the land as much as 0.75 mile (1.21 km) into the lakeward side of the pre-fill shoreline (Figure 25-9). Marinas, promontories, lagoons, and even islands were designed and built. Much of this effort resulted from ambitious shoreline landscape designs first proposed in the 1909 publication of *Plan of Chicago* (Burnham and Bennett 1909, Chrzastowski 2008). Additionally, lakefill created 74 acres (30 ha) of new land for Northwestern University in Evanston during the 1960s.

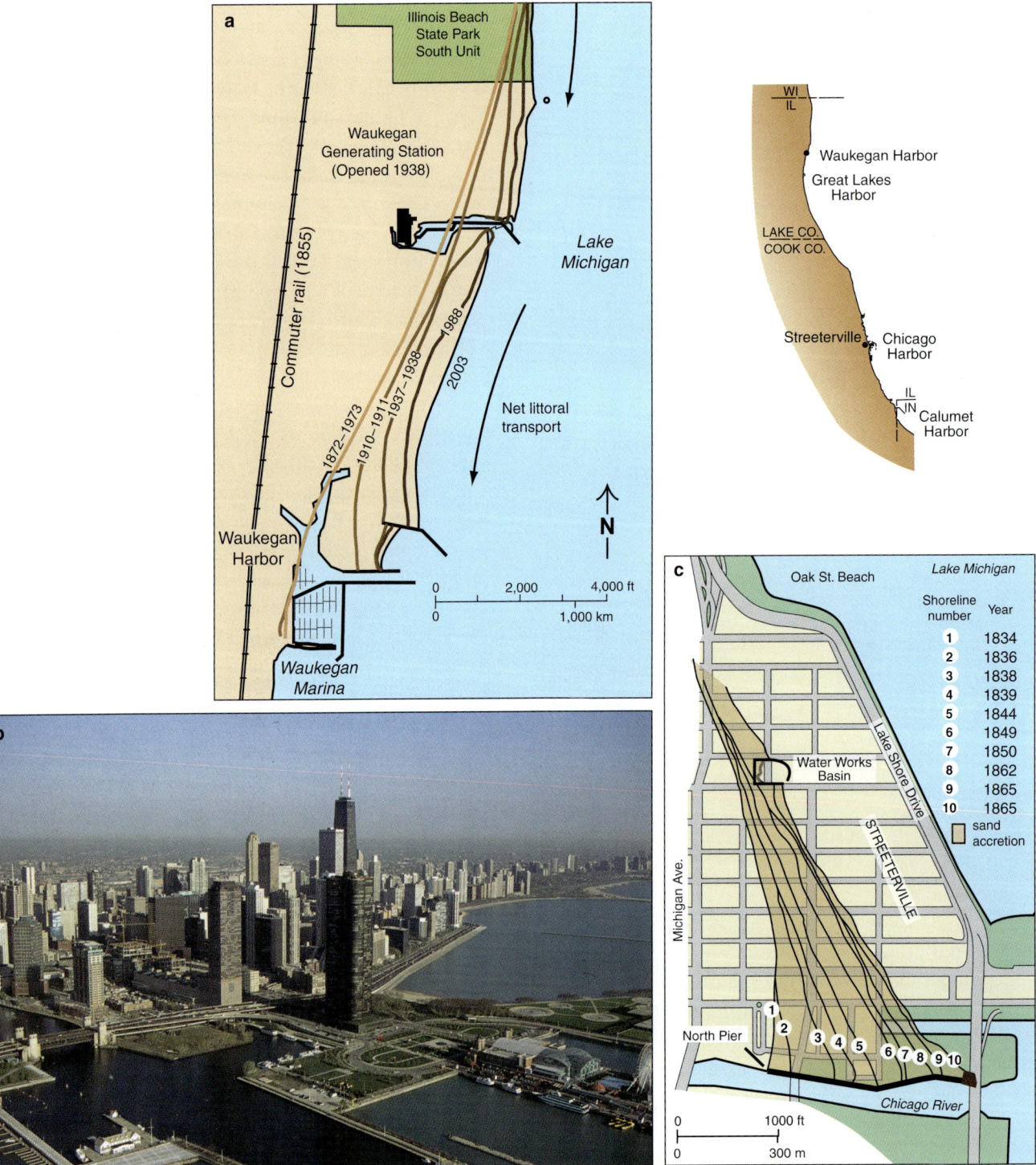

Figure 25-8 (a) Historical shoreline accretion on the north (updrift) side of jetties built to defend the entrance to Waukegan Harbor. (b) Chicago's Streeterville neighborhood, shown in May 2000, resulted from (c) sand accretion and lakefilling north (updrift) of the jetty built to defend the entrance to the Chicago River.

ILLINOIS COAST

These extensive engineering projects altered the dynamics of littoral transport along the Illinois coast. Before development, the entire Illinois coast was part of a continuous pathway for net southward transport. The subsequent construction of harbor jetties, groins, breakwaters, and lakeward filling for new land has introduced barriers and segmented transport along the Illinois shore (Chrzastowski et al. 1994). The historical impact of the loss of sand supply from bluff erosion has reduced the thickness and extent of sand along many beach and nearshore areas, particularly along the North Shore (Shabica and Pranschke 1994). The volume of littoral sediment in transport has been reduced along the entire Illinois coast, especially along the Chicago lakeshore. The 1834–1869 record of beach accretion against North Pier at the Chicago River (Figure 25-8) provides a means to calculate the minimum natural-state littoral transport along the central Chicago shore.

During the mid to late 1800s, approximately 98,000 cubic yards (74,927 m^3) per year was moving along the Chicago shore southward to the Indiana coast (Chrzastowski 1990b). Although several thousand cubic yards per year are still transported along Chicago's far north lakeshore, no appreciable transport occurs along the city's south lakeshore. There transport is limited to what can be eroded from the few artificial beaches along the lakeward edge of filled land. The complex of breakwaters and made land at Chicago Harbor prevents the Chicago south lakeshore from receiving littoral sediment from the north. These near-total barriers to littoral transport include the Chicago Harbor outer breakwaters constructed between 1889 and 1923 (Bottin 1988), the earth-filled Navy Pier (originally called Municipal Pier) constructed during 1915–1916, and the extensive made land filled between 1952 and 1954 for the James W. Jardine Water Purification Plant.

COASTAL MANAGEMENT CHALLENGES

Many of the erosion and accretion issues that have occurred along the Illinois coast during the past 150 years have been mitigated by engineering and coastal management. However, wave dynamics, sediment transport, nearshore ice, and fluctuations in lake level are all coastal processes that continue to impact the Illinois coast and will contribute to future issues and challenges. An unknown in future coastal management is how potential climatic changes will impact lake level.

Diminishing Sand Resources

A major management issue along the bluff coast is sediment starvation. Sand resources have diminished due to reduced supply from updrift sources and the elimination of a sand supply from bluff erosion. Where beaches are desired, engineered beaches that retain artificially supplied sand within semi-enclosed beach cells may be needed. Beach cell systems have been built along numerous private lakeshore properties along the North Shore (Figure 25-10).

Figure 25-9 Construction under way in September 1931 to build the lakeward-protruding land near Montrose Harbor on Chicago's north lakefront. (Photograph courtesy of Great Lakes Dredge and Dock Company, Oakbrook, Illinois.)

Larger-scale projects have been built along the municipal beach at Lake Forest in 1986 (Forest Park Beach) (Anglin et al. 1987, Chrzastowski and Trask 1995) and at Lake Bluff in 1991 (Sunrise Park Beach).

The greatest challenge for sand management along the Illinois coast is at Illinois Beach State Park. The southern 2 miles (3.2 km) of the park's South Unit preserve the last remaining shoreline in the state that is free of any armor or shore protection. Preventing net erosion along this shore is important to maintain this unique setting for the benefit of future generations and to protect the first designated nature preserve in Illinois. Erosion along the shore of the North Unit is the primary source for sand reaching the South Unit beach. Net erosion along this northern shore can be documented from the time of the earliest shoreline mapping along this reach (U.S. Lake Survey 1872). Shoreline recession on this shore is the greatest on the Illinois coast (U.S. Army Corps of Engineers 1953, State of Illinois Division of Waterways 1958), as much as 10 feet (3.1 m) per year (Jennings 1990). The net erosional trend relates to the long-term southward migration of the Zion Beach-Ridge Plain.

Preventing any net loss of park land requires a sustained program of annual beach nourishment to compensate for the volume of beach sand naturally removed by wave action. Since the late 1980s, sand stockpiles along the shore (called "feeder beaches") have been constructed at times at the north ends of the North and South Units of the state park. These stockpiles are allowed to erode and to nourish the downdrift beaches (Terpstra and Chrzastowski 1992, Chrzastowski and Frankie 2000). Long-term challenges for park management are identifying source areas for the nourishment sand, efficiently transporting it to the feeder beaches, and having the financial resources to sustain beach nourishment indefinitely. Studies of the coastal sand budget indicate that 80,000 cubic yards (61,000 m^3) per year is the minimum amount of sand required to supply the northern 2 miles (3.2 km) of erosional shore (Foyle et al. 1998).

Lakebed Downcutting

A growing problem is the erosion across the nearshore lake bottom, particularly along the North Shore, where the lake bottom is typically a sand lens up to several feet thick overlying till. Comparison of nearshore profiles and core data have documented a historical reduction in the thickness of the sand lens (Shabica and Pranschke 1994). In extreme cases, all or nearly all of the sand has been lost, resulting in localized exposure of till (Foster and Folger 1994). Wave erosion across the exposed till and abrasion of the till by shifting sand lowers the elevation of the till sur-

Figure 25-10 Example from May 2000 of engineered beaches built on the Illinois' North Shore of Lake Michigan that combine steel sheetpile groins and rubble-mound breakwaters to form beach cells that maintain artificially placed sand. (Photograph by Michael J. Chrzastowski.)

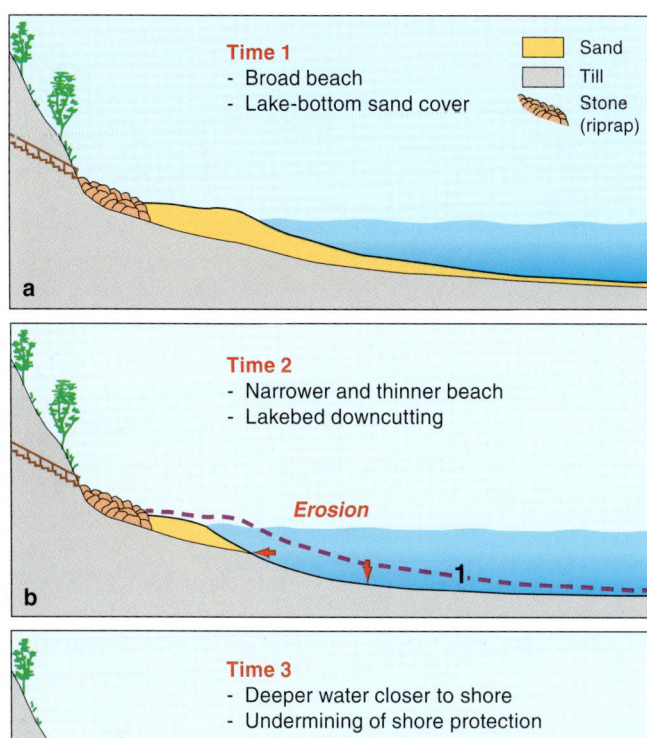

Figure 25-11 This schematic diagram shows the long-term process of lakebed downcutting. Erosion into the lakebed is a permanent loss of this glacially derived sediment.

face. The long-term result is deeper water in the nearshore area, making the beach and backshore susceptible to the impact of larger waves, which in turn contributes to more shore erosion and further exacerbates the lowering of the lake-bottom profile (Figure 25-11). This process of nearshore erosion is referred to as lakebed downcutting.

Although erosion of beach and nearshore sand can be reversed by a corresponding influx of sand, erosion of the lake bottom till is not reversible. Downcutting along the Illinois coast can undermine existing shore structures, create the need to build larger and more costly shore protection, and lead to the permanent loss of beach area. The average rate of lakebed downcutting at Lake Bluff has been 2.3 to 3.2 inches (5.8 to 8.2 cm) per year, the greatest along the Illinois shore (Chrzastowski and Trask 1995). A downcutting rate of 0.5 inch (1.22 cm) per year has been reported for the shore to the south between Lake Forest and Highland Park (Illinois Department of Transportation, Division of Water Resources 1980).

In the coming decades, mitigating and preventing additional lakebed downcutting will be an unprecedented coastal management challenge. A potential solution has been modeled: (1) dispersing cobbles across the shallow nearshore (i.e., average water depths less than 8 feet [2.4 m]) to protect the lakebed from additional erosion and (2) constructing a submerged breakwater at about the 9-foot (2.7-m) average water depth to trip waves and thereby reduce their energy across the shallow nearshore (U.S. Army Corps of Engineers Chicago District, personal communication; Baird and Associates 2000). The costs for such a protection and submerged breakwater project could be substantial. Any cost-benefit analysis needs to consider the costs of potential repairs to or replacement of existing groins, beaches, and shore protection if the lakebed downcutting continues unabated.

Conclusion

The Illinois coast is a dynamic and diverse setting that includes some of the most engineered shoreline in the Great Lakes region as well as shoreline in a near-natural state. Since the ancestral Illinois coast formed in late glacial time, the volume of sand along the beaches and nearshore areas has changed substantially. Extensive ancient spits and barrier reaches are evidence of the robust littoral transport volume once moved along this shore. Early efforts at coastal engineering, such as building harbors at the mouth of the Chicago River and at Waukegan, were challenged by the abundance of sand in littoral transport, leading to rapid accretion on the updrift side of the harbor jetties. Now the Illinois coast faces diminishing volumes of beach and nearshore sand as shore protection has eliminated sand replenishment from bluff and shore erosion. A sand deficit is increasing with time. The conservation of existing beach and nearshore sand resources is critical. In the coming decades, beach nourishment and construction of engineered beach systems will likely become increasingly important in maintaining beaches at select sites.

References

Alden, W. C., 1902, Description of the Chicago District, Illinois-Indiana Chicago Folio (Riverside, Chicago, Des Plaines, and Calumet Quadrangles): Washington, D.C., U.S. Geological Survey Atlas, Folio 81, 14 p.

Andreas, A. T., 1884, History of Chicago from the earliest period to the present time: Chicago, Illinois, A. T. Andreas Publisher, v. 1, 648 p.

Anglin, C. D., A. M. Macintosh, W. F. Baird, and D. J. Werren, 1987, Artificial beach design, Lake Forest, Illinois, *in* O. T. Magoon, ed., Coastal Zone '87, Proceedings of the Fifth Symposium on Coastal and Ocean Management: New York, American Society of Civil Engineers, p. 179–195.

Atwood, W. A., and J. W. Goldthwait, 1908, Physical geography of the Evanston-Waukegan region: Illinois State Geological Survey, Bulletin 7, 102 p.

Baird and Associates, 2000, Illinois shoreline, interim IV, lakebed paving armoring plan feasibility study final report (prepared for U.S. Army Corps of Engineers Chicago District): Madison, Wisconsin, Baird and Associates, 137 p.

Barnes, P. W., E. W. Kempema, E. Reimnitz, and M. McCormick, 1994, The influence of ice on southern Lake Michigan coastal erosion: Journal of Great Lakes Research, v. 20, no. 1, p. 179–195.

Bottin, R. R., Jr., 1988, Case histories of Corps breakwaters and jetty construction: Vicksburg, Mississippi, Department of the Army, Corps of Engineers, Waterways Experiment Station, Technical Report REMR-CO-3, 433 p.

Bretz, J H., 1939, Geology of the Chicago region, Part I, General: Illinois State Geological Survey, Bulletin 65, 118 p.

Bretz, J H., 1943, Chicago area geologic maps: Illinois State Geological Survey, Bulletin 65 maps, 24 sheets, 1:24,000.

Bretz, J H., 1955, Geology of the Chicago region, Part II, The Pleistocene: Illinois State Geological Survey, Bulletin 65, 132 p. plus maps.

Burnham, D. H., and E. H. Bennett, 1909, Plan of Chicago: Chicago, Illinois, The Commercial Club of Chicago, 164 p.

Chrzastowski, M. J., 1990a, Late Wisconsinan and Holocene littoral drift patterns in southern Lake Michigan, *in* P. W. Barnes, ed., Coastal sedimentary processes in southern Lake Michigan: Reston, Virginia, U.S. Geological Survey, Open-File Report 90-295, p. 13–18.

Chrzastowski, M. J., 1990b, Estimate of the natural-state littoral transport along the Chicago lakeshore, *in* P. W. Barnes, ed., Coastal sedimentary processes in southern Lake Michigan: Reston, Virginia, U.S. Geological Survey, Open-File Report 90-295, p. 19–26.

Chrzastowski, M. J., 1991, The building, deterioration and proposed rebuilding of the Chicago lakefront: Shore and Beach, April 1991, p. 2–10.

Chrzastowski, M. J., 2004, History of the uniquely designed groins along the Chicago lakeshore: Journal of Coastal Research, Special Issue 33, winter 2004, p. 19–38.

Chrzastowski, M. J., 2005, Chicagoland—Shaped by ice and water: Illinois State Geological Survey, poster.

Chrzastowski, M. J., 2008, "Make no little plans": Field trip guidebook for the American Shore & Beach Preservation Association 2008 National Conference: Illinois State Geological Survey, Guidebook 36, 42 p.

Chrzastowski, M. J., and W. T. Frankie, 2000, Guide to the geology of Illinois Beach State Park and the Zion Beach-Ridge Plain, Lake County, Illinois: Illinois State Geological Survey, Field Trip Guidebook 2000 C–D, 69 p. plus appendices.

Chrzastowski, M. J., and T. A. Thompson, 1992, Late Wisconsinan and Holocene coastal evolution of the southern shore of Lake Michigan, *in* C. H. Fletcher and J. F. Wehmiller, eds., Quaternary coasts of the United States: Tulsa, Oklahoma, Society for Sedimentary Geology (SEPM), Special Publication 48, p. 397–413.

Chrzastowski, M. J., and T. A. Thompson, 1994, Late Wisconsinan and Holocene geologic history of the Illinois-Indiana coast of Lake Michigan: Journal of Great Lakes Research, v. 20, no. 1, p. 9–26.

Chrzastowski, M. J., T. A. Thompson, and C. B. Trask, 1994, Coastal geomorphology and littoral cell divisions along the Illinois-Indiana coast of Lake Michigan: Journal of Great Lakes Research, v. 20, no. 1, p. 27–43.

Chrzastowski, M. J., and C. B. Trask, 1995, Nearshore geology and geologic processes along the Illinois shore of Lake Michigan from Waukegan Harbor to Wilmette Harbor: Illinois State Geological Survey, Open File Series 1995-10, 93 p.

Clark, P. U., and G. A. Rudloff, 1990, Sedimentology and stratigraphy of late Wisconsinan deposits, Lake Michigan bluffs, northern Illinois, *in* A. F. Schneider and G. S. Fraser, eds., Late Quaternary history of the Lake Michigan Basin: Boulder, Colorado, Geological Society of America, Special Paper 251, p. 29–41.

Colman, S. M., R. M. Forester, R. L. Reynolds, D. S. Sweetkind, J. W. King, P. Gangemi, G. A. Jones, L. D. Keigwin, and D. S. Foster, 1994, Lake-level history of Lake Michigan for the past 12,000 years—The record from deep lacustrine sediments: Journal of Great Lakes Research, v. 20, no. 1, p. 73–92.

Colman, S. M., and D. S. Foster, 1994, A sediment budget for southern Lake Michigan—Source and sink models for different time intervals: Journal of Great Lakes Research, v. 20, no. 1, p. 215–228.

Cram, T. J., 1839, Report on harbor improvements on Lake Michigan: Twenty-sixth Congress, First Session, Senate Document 140, v. 4, ser. 357, p. 16–22.

Ewing, M., F. Press, and W. L. Donn, 1954, An explanation of the Lake Michigan wave of 26 June 1954: Science, v. 120, no. 3122, p. 684–686.

Foster, D. S., and D. W. Folger, 1994, The geologic framework of southern Lake Michigan: Journal of Great Lakes Research, v. 20, no. 1, p. 44–60.

Foyle, A. M., M. J. Chrzastowski, and C. B. Trask, 1998, Erosion and accretion trends along the Lake Michigan shore at North Point Marina and Illinois Beach State Park: Illinois State Geological Survey, Open File Series 1998-3, 100 p.

Fraser, G. S., and N. C. Hester, 1974, Sediment distribution in a beach ridge complex and its application to artificial beach replenishment: Illinois State Geological Survey, Environmental Geology Note 67, 26 p.

Goldthwait, J. W., 1908, A reconstruction of water planes of the extinct glacial lakes in the Lake Michigan basin: Journal of Geology, v. 16, no. 5, p. 459–476.

Hansel, A. K., and D. M. Mickelson, 1988, A reevaluation of the timing and causes of high lake phases in the Lake Michigan basin: Quaternary Research, v. 29, no. 2, p. 113–128.

Hester, N. C., and G. S. Fraser, 1973, Sedimentology of a beach ridge complex and its significance in land-use planning: Illinois State Geological Survey, Environmental Geology Note 63, 24 p.

Illinois Department of Transportation, Division of Water Resources, 1980, Lake Michigan shore protection study, Illinois shoreline Wisconsin state line to Hollywood Boulevard, Chicago (Appendix 1, Assessment of conditions and processes): Springfield, Illinois, Illinois Department of Transportation, 266 p.

Jennings, J. R., 1990, 150 year erosion history at a beach ridge and dune plain on the Illinois Lake Michigan shore: Windsor, Ontario, Canada, International Association of Great Lakes Research, 33rd Conference, Program and Abstracts, p. 67.

Jibson, R. W., J. K. Odum, and J.-M. Staude, 1994, Rates and processes of bluff recession along the Lake Michigan shoreline in Illinois: Journal of Great Lakes Research, v. 20, no. 1, p. 135–152.

Kempema, E. W., 1998, Nearshore ice formation and sediment transport in southern Lake Michigan: Seattle, Washington, University of Washington, Ph. D. dissertation, 154 p.

Larsen, C. E., 1985, A stratigraphic study of beach features on the southwestern shore of Lake Michigan—New evidence of Holocene lake level fluctuations: Illinois State Geological Survey, Environmental Geology Note 112, 31 p.

Larson, J. W., 1979, Those army engineers, A history of the Chicago District U.S. Army Corps of Engineers: Chicago, Chicago District U.S. Army Corps of Engineers, 307 p.

Leverett, F., 1897, The Pleistocene features and deposits of the Chicago area: Chicago, Chicago Academy of Sciences, Geology and Natural History Bulletin 2, 86 p.

Lineback, J. A., 1974, Erosion of till bluffs—Wilmette to Waukegan, *in* C. Collinson, J. A. Lineback, P. B. DuMontelle, and D. C. Brown, eds., Coastal geology, sedimentology, and management: Illinois

State Geological Survey, Guidebook 12, p. 37–45.

Olyphant, G. A., and S. W. Bennett, 1994, Contemporary and historical rates of eolian sand transport in the Indiana Dunes area of southern Lake Michigan: Journal of Great Lakes Research, v. 20, no. 1, p. 153–162.

Recktenwald, W., 1994, The seiche of '54: Lakefront was caught off guard by a deadly inland tidal wave: Chicago Tribune, June 19, 1994, sec. 1, p. 15, 18.

Rosenbaum, J. G., 1981, Early problems with littoral drift at shoreline harbors on the Great Lakes: Transactions of the Wisconsin Academy of Sciences, Arts and Letters, v. 69, p. 121–134.

Schneider, A. F., and S. J. Keller, 1970, Geologic map of the 1° × 2° Chicago quadrangle, Indiana, Illinois, and Michigan, showing bedrock and unconsolidated deposits: Indiana Geological Survey, Regional Geologic Map 4, part B, 1:250,000.

Shabica, C., J. Meshberg, R. Keefe, and R. Georges, 2004, Evolution and performance of groins on a sediment starved coast—The Illinois shore of Lake Michigan north of Chicago, 1880–2000: Journal of Coastal Research, Special Issue 33, winter 2004, p. 39–56.

Shabica, C., and F. Pranschke, 1994, Survey of littoral drift sand deposits along the Illinois and Indiana shores of Lake Michigan: Journal of Great Lakes Research, v. 20, no. 1, p. 61–72.

State of Illinois, Division of Waterways, 1958, Interim report for erosion control— Illinois shore of Lake Michigan: Springfield, Illinois, 108 p.

Terpstra, P. D., and M. J. Chrzastowski, 1992, Geometric trends in the evolution of a small log-spiral embayment on the Illinois shore of Lake Michigan: Journal of Coastal Research, v. 8, no. 3, p. 603–617.

U.S. Army Corps of Engineers, 1953, Illinois shore of Lake Michigan, beach erosion control study: Eighty-third U. S. Congress, First Session, House document no. 28, 137 p. plus 21 sheets.

U.S. Army Corps of Engineers Detroit District, 2003, Subject: Long term average min-max water levels. http://www.lre.usace.army.mil/greatlakes/hh/greatlakeswaterlevels/historicdata/longtermaverage-min-maxwaterlevels. Accessed November 5, 2009.

U.S. Lake Survey, 1872, Survey of N. and N.W. Lakes, west shore of Lake Michigan: Rockville, Maryland, U.S. Lake Survey (NOAA National Ocean Service), field survey I-521, 1:20,000.

U.S. Lake Survey, 1873, Survey of N. and N.W. Lakes, west shore of Lake Michigan: Rockville, Maryland, U. S. Lake Survey (NOAA National Ocean Service), field survey I-551, 1:10,000.

Wille, L., 1972, Forever open, clear and free, The historic struggle for Chicago's lakefront: Chicago, Illinois, Henry Regnery Co., 175 p.

Willman, H. B., 1971, Summary of the geology of the Chicago area: Illinois State Geological Survey, Circular 460, 77 p.

Willman, H. B., and J. A. Lineback, 1970, Surficial geology of the Chicago region: Illinois State Geological Survey, Circular 460, plate 1, 1:250,000.

26 Natural Radiation

Richard A. Cahill

INTRODUCTION

Many people are unaware that naturally occurring radioactive elements are commonly found in geological materials, soils, and groundwater in Illinois. Natural occurrences of radon in homes and radium in some groundwater supplies in Illinois have been discussed in the press, in part because exposure to radioactivity is feared and poorly understood. The debate over siting of facilities to store radioactive wastes still has not been resolved. This chapter reviews basic information about radiation to clarify issues associated with radioactive elements and human exposure to them.

Radiation is a broad term that includes sunlight, heat, radio waves, and microwaves. It is also used to refer to the ionizing radiation given off by elements that contain unstable or radioactive atoms that change, or decay. With each decay of a radioactive element, energy is released, and, in most cases, a particle is ejected. Radioactive decay continues until the element has been transformed into a stable or nonradioactive element. Three types of energy released during radioactive decay are mentioned here: alpha, beta, and gamma. Each type has different intensities, strengths, and potential health effects. Slow-moving alpha particles are produced when a decaying element releases two protons and two neutrons. Alpha emitters are dangerous when they are ingested or inhaled. Beta particles are produced when electrons are released during radioactive decay. They move near the speed of light, have an electrical charge, and can be harmful when emitted inside the body. Gamma radiation is produced during radioactive decay as a way of releasing energy. Gamma radiation has no electrical charge but has great penetrating power that can damage tissue. The types of radioactive decay influence how these radioactive elements occur in the environment and their impact on public health.

RADIOACTIVE DECAY

Each radioactive element decays at a different rate, and the time it takes for an element to be reduced by one-half of its initial amount is called its half-life. An element's half-life can range from a fraction of a second to more than a billion years. For example, the half-life of cesium-137 (^{137}Cs), one of the elements produced in nuclear explosions, is 37 years.

Along with other radioactive fallout, ^{137}Cs was deposited over Illinois as a result of atmospheric testing of nuclear weapons between 1954 and 1969. The ^{137}Cs that was produced in 1967 is now half gone; by 2337, that ^{137}Cs will have gone through ten half-lives and could then be considered no longer present.

The concept of half-life is very important in predicting the effects that a particular radioactive material will have on the environment and on human health. In general, the shorter the half-life of an element is, the greater the risk, because there will be more decay events per unit of time. However, a highly radioactive element with a short half-life that occurs at very low concentrations does not pose as great a threat to health as a less radioactive, but more abundant element. Elements that have long half-lives, however, can pose a hazard for hundreds of years.

Radioactivity is measured in units of activity or number of decay events per unit of time. Radiation levels are commonly reported in Curies (Ci), where 1 Ci is defined as 37 billion disintegrations per second. Commonly encountered levels are in the range of picocuries (pCi); 1 pCi equals 0.037 disintegrations per second, or 2.22 disintegrations per minute. A newer unit now being used is the becquerel (Bq), and 1 Bq is equal to one disintegration per second; 1 Bq is equal to 27 pCi. Concentrations are then expressed as the activity per volume (picocuries per liter) or mass (picocuries per gram). The relative damage that radiation exposure causes to cells is measured in rem units. Regulations that are designed to limit the exposures of radiation commonly measure millirems (mrems) per year. The average annual exposure of an Illinois citizen from all sources is about 360 mrems. Of this total, natural radon and its decay products contribute 55% (Table 26-1).

HUMAN EXPOSURE TO RADIATION

The major (82%) source of exposure to humans is radiation from natural environmental sources (National Council on Radiation Protection and Measurements 1975). Exposure from terrestrial radiation arises from radioactive elements called radionuclides that are distributed in earth materials or have been transferred from the Earth to the atmosphere or hydrosphere. Water (as groundwater or soil moisture) plays an important role in the movement of radionuclides in geological materials. The significant natural

Table 26-1 Average annual radiation exposure per person in Illinois.

Source	Average dose (mrem)	Percent of total[1]
Natural	297	83
Radon	200	56
Internal	40	11
Terrestrial	30	8
Cosmic	27	8
Artificial	63	18
Diagnostic x-rays	39	11
Nuclear medicine	14	4
Consumer goods	9	3
All other	1	~0

[1]Values are rounded to the nearest whole number.

sources are potassium-40 (^{40}K), which can occur in many common rock-forming minerals, and the decay series of the two very long-lived radionuclides, thorium-232 (^{232}Th) and uranium-238 (^{238}U), which generally occur only in small concentrations in geological materials at the surface, such as soils.

In addition to these naturally occurring sources, about 18% of human exposure (National Council on Radiation Protection and Measurements 1975) is from human-made sources, including medical x-rays, radioactive materials from nuclear reactors and nuclear accelerators, and fallout from the atmospheric testing of nuclear weapons. Radioactive materials are used in smoke detectors, luminous watch dials, mantles in gasoline camping lanterns, and other applications.

Natural Radioactive Elements in Groundwater

Groundwater from aquifers in Illinois provides municipal supplies and well water for most of the state's residents (see Chapter 18, Aquifers). The most significant aquifers are (1) sand and gravel deposits in the glacial drift, (2) fractures and crevices in shallow Silurian or Ordovician dolomite bedrock (a hydraulic connection commonly exists between the glacial drift and the shallow bedrock directly beneath it), and (3) deep sandstones (St. Peter and Ironton-Galesville formations) of Ordovician and Cambrian bedrock. Domestic wells may be finished in confined dolomites in localities where the glacial drift sand and gravel aquifer or shallow bedrock does not provide an adequate domestic groundwater supply (Gilkeson et al. 1988).

Of public health concern are naturally occurring radioactive elements in groundwater that are members of the ^{238}U and ^{232}Th decay series. Radium (Ra) isotopes ^{226}Ra and ^{228}Ra are of primary concern. To meet U.S. Environmental Protection Agency (U.S. EPA) drinking water standards, the maximum contaminant level (MCL) for these two radium isotopes combined is 5.0 pCi/L. The MCL is 27 pCi/L (or 30 µg/L) for total uranium. Note that the standards for uranium concentrations in water are generally expressed as micrograms per liter because uranium is more of a chemical risk than a radiation risk due to its toxicity.

Research conducted on the distribution of radioactive isotopes in groundwater from the deep sandstones in northeastern Illinois (Gilkeson et al. 1983, 1984; Kay 1999) found high radium concentrations in groundwater from deep sandstone aquifers; concentrations were highest in groundwater from the Ironton-Galesville Sandstones, which constitute a major aquifer. Areas in Illinois where radium in excess of 5.0 pCi/L has been detected in public water supplies are shown in Figure 26-1.

The uranium and radon-222 (^{222}Rn) concentrations in groundwater of the lower Illinois River basin have been studied (Morrow 2001). Uranium concentrations ranged from less than 0.9 pCi/L to 15 pCi/L. Radon-222 is produced from the decay of ^{226}Ra. Concentration of ^{222}Rn ranged from less than 80 pCi/L to 1,300 pCi/L. The U.S. EPA has proposed two regulations for ^{222}Rn in water. The MCL is 300 pCi/L, and the alternative maximum contaminant level (AMCL) is 4,000 pCi/L (National Primary Drinking Water Regulations 1999, 2000). The ^{222}Rn concentrations exceeded the MCL in about half the samples tested but never exceeded the AMCL.

In December 2000, the Illinois Environmental Protection Agency indicated that 99 Illinois communities throughout the state were in violation of federal radium standards for drinking water. The communities were given three years to implement a plan to reduce radium levels. The U.S. EPA (2005) suggested several options to fix water supply problems, including connecting to noncontaminated water mains in neighboring towns and reverse osmosis.

Indoor Radon

Inquiries from the public and the press addressed to the Illinois State Geological Survey since 1986 indicate that much confusion exists about the hazards of radon and how to address them. This confusion results partly from the complex and highly technical nature of the subject. Also, the apparent disagreement among the scientific experts regarding the absolute and relative risks of radon exposure has led to some uncertainty and ambiguity concerning the merits of the risk values adopted by the U.S. EPA (Cohen 1988, National Research Council 1988, Nero 1989).

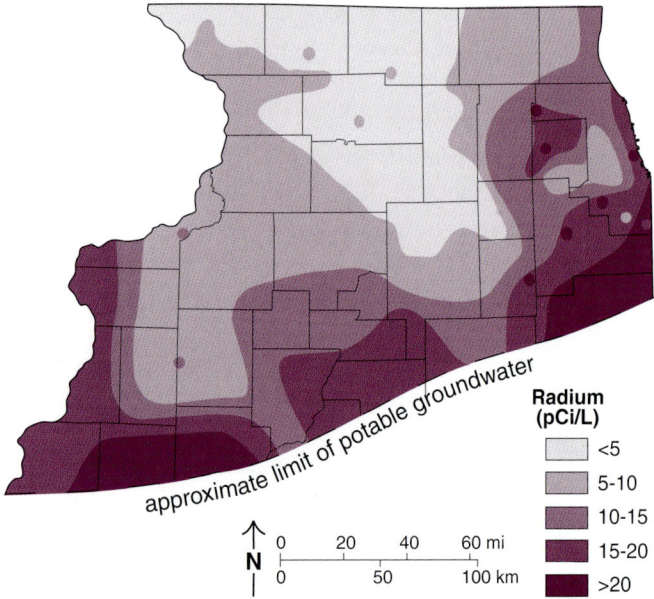

Figure 26-1 Areas in Illinois where radium concentrations in excess of the U.S. Environmental Protection Agency drinking water standard of 4 pCi/L have been detected in public water supplies from the Cambrian-Ordovician bedrock aquifers (Gilkeson et al. 1988).

For the indoor radon issue, the radionuclide of main concern, ^{222}Rn, is the radionuclide that is generally being referred to when the term "radon" is used. This discussion of radon refers to ^{222}Rn. Radon, a colorless, odorless, chemically inert gaseous element, is part of the ^{238}U decay series and has a half-life of 3.82 days. The parents, ^{226}Ra and ^{238}U, occur in trace amounts in most geological materials, including soil. Despite its relatively short half-life, radon gas can migrate out of the soil, resulting in outdoor air concentrations of radon ranging from 0.1 to 0.3 pCi/L (Nazaroff and Sextro 1989, Nazaroff and Teichman 1990). The decay products of radon are the short-lived radionuclides polonium-218 (^{218}Po), lead-214 (^{214}Pb), bismuth-214 (^{214}Bi), and polonium-214 (^{214}Po). The potential health risk from inhaling radon is from lung exposure to the ionizing radiation. Public concern, fueled by the publicity associated with the discovery in the northeastern United States of high concentrations of radon in some homes, prompted the U.S. EPA to establish a level of 4 pCi/L as the action level or guideline for U.S. homes and public buildings (U.S. EPA 1986). The Radon Pollution Control Act of 1988 states that "the national long-term goal of the United States with respect to radon levels in buildings is that the air within buildings should be as free of radon as the ambient air outside of buildings" (Congressional Record 1988). The U.S. EPA (2003a, 2003b) has continued to use 4 pCi/L as the guideline in part because affordable cost mitigation technologies cannot reduce radon below 2 pCi/L in most homes.

The dominant source of radon in air inside homes is soil gas (Figure 26-2) (Hopke 1987, Michel 1987, Nero 1989). The concentrations of radon in soil depend on the concentrations of uranium and/or radium and the fraction of radon that can escape or emanate from a soil grain into the pore spaces between the grains. The radon gas concentration in the soil gas is also controlled by the amount of water present in pore spaces, pore volume, and grain size of the soil particles (Hopke 1987, Nazaroff and Sextro 1989). For example, radon gas could dissolve in layers of water attached to soil particles or water filling pore spaces and could be elevated in soils with large pore volumes. Typical values for radon in soil air are 270 to 675 pCi/L, but values are in excess of 10,000 pCi/L in some places (Hopke 1987). The ease with which radon can infiltrate homes is controlled by the concentration of radon in the soil gas, the openings in the home, and the pressure differences between the structure and the surrounding soil (Hopke 1987, Michel 1987, Nero 1989). The atmospheric pressure inside a home relative to the exterior can decrease due to wind and temperature differences, mechanical ventilation systems such as forced-air heating and cooling, and ventilation rates. These factors can lead to a "stack effect" in which soil gas is drawn into the house through openings in the basement or crawl space. Michel (1987) has estimated that, if only 0.1% of the air in a building comes from soil gas containing an average concentration of radon (about 500 pCi/L), the resulting indoor radon level could be 1.3 pCi/L.

In response to reports of elevated concentrations of indoor radon in Illinois, the Illinois Department of Nuclear Safety (1986) published the report of the Governor's Radon Task Force. The report reviewed the available data and concluded that, although indoor radon is not as prevalent in Illinois as in some other states, about 25% of the tested Illinois residences had radon concentrations above the U.S. EPA recommended action level. Recent radon screening in all 102 counties of Illinois indicated that about 31% of all radon measurements exceed the 4 pCi/L guideline, but only about 1% exceeded 20 pCi/L (Illinois Department of Nuclear Safety 1988a, 1992). The U.S. EPA measured radon concentrations in Illinois homes using the same protocol as used for 45 other states and concluded that about 19% of the homes tested exceeded the 4 pCi/L guideline (Illinois Department of Nuclear Safety 1991). Recently, the Illinois Emergency Management Agency (2006) updated screening results for indoor radon based on measurements in 22,000 homes. The median concentration for indoor radon was 3.6 pCi/L (range was 0.4 to 179 pCi/L), and indoor radon in 46% of the homes tested exceeded 4 pCi/L

NATURAL RADIATION

Figure 26-2 Radon can enter a house in several ways (Eichholz 1987).

Several geological factors must be evaluated to accurately determine the potential of a geological material to supply radon to the indoor environment (Otton 1992, Schumann 1992). Both source and transport factors must be established prior to predictive mapping. The source data include the concentration and distribution of radium within the geological material. Transport factors include data on the material's permeability to air, water content, and grain size and knowledge of the presence of extensive joints or fractures in the surficial geological material (Gilkeson et al. 1988). Many of these factors are not known in sufficient detail to predict radon concentrations in a particular home. Several attempts have been made, however, to locate elevated radon levels based on geological factors (Hasenmueller 1988, Flood et al. 1990, Otton 1992, Schumann 1992).

A number of resources are available for homeowners to mitigate elevated radon levels (Illinois Emergency Management Agency 2006, U.S. EPA 2007a). Most systems used to reduce indoor radon are based on the collection of radon gas prior to entry in the building, the modification of air pressure differences between the lowest level of a building and surrounding soil, or the dilution of radon concentration by increased ventilation (Illinois Emergency Management Agency 2006). The Illinois Emergency Management Agency provides a list of licensed radon measurement professionals and names of licensed mitigation professionals trained to reduce radon.

OTHER SOURCES OF EXPOSURE

Naturally occurring radioactive materials generally contain radionuclides found in nature. Once these naturally occurring radioactive materials become concentrated through human activity, such as mineral extraction, it can become a radioactive waste (U.S. EPA 2008). There are two types of naturally occurring radioactive waste: discrete and diffuse. Discrete naturally occurring radioactive materials have a relatively high radioactivity concentration in a very small volume. An example would be a radium source used in medical procedures (U.S. EPA 2008). Diffuse naturally occurring radioactive materials have a much lower concentration of radioactivity but a higher volume of waste (U.S. EPA 2008). The high volume poses different disposal problems and may pose a health hazard because of multiple uses. As examples, oil and gas production may produce radioactive pipe scale (a residue left in pipes from drilling oil wells) and sludge that leave sites and equipment contaminated (Smith et al. 1996). Radiation-contaminated water treatment residue accumulates when radioactive material is filtered from drinking water during its purification process (U.S. EPA 2005). In Illinois, this waste may be used in agriculture as a soil conditioner or, in some cases, discharged into the state's waters (Illinois Environmental Protection Agency 2007).

Safe disposal of low-level radioactive waste requires compliance with stringent state and federal environmental regulations. Examples of this type of waste are medical waste, laboratory waste, wrist watches, contaminated soils, and smoke detectors. In 1983, the State of Illinois and the Commonwealth of Kentucky entered into the Central Midwest Interstate Low-Level Radioactive Waste Compact. This agreement was in response to a federal policy set out in the Low-Level Radioactive Waste Policy Act of 1980, which made each state responsible for ensuring that disposal capacity is available for certain categories of low-level radioactive waste generated within its borders (U.S. Congress 1980). Despite years of effort and the expenditure of hundreds of millions of dollars, states and compacts throughout the nation have been unsuccessful in developing new regional disposal facilities. Site selection has been the most contentious part of the process in each case. At this point, waste generators in most states are relying on the facilities in Utah and South Carolina. This appears to be the short-term posture in all states, including Illinois (Illinois Department of Nuclear Safety 2000).

Illinois currently has 11 operating nuclear power reactors at six nuclear power stations. These stations are monitored by the Illinois Emergency Management Agency, which is also responsible for protecting the public in the event of an accidental release of nuclear materials (Illinois Emergency Management Agency 2005). Some of the radioactive wastes generated by these plants are considered to be low-level radioactive waste. The nuclear fuel that is used to power the reactors is contained in metal tubes or rods that with time must be replaced. Spent fuel rods are currently being stored at eight sites in Illinois until a permanent storage or disposal facility is made operational by the federal government. Sites include six operational nuclear power plants, located at Braidwood, Byron, Clinton, Dresden, La Salle, and the Quad cities (Moline and Rock Island, Illinois; and Davenport and Bettendorf, Iowa). The Zion site is no longer operational. The eighth site, the General Electric facility at Morris, was intended as a reprocessing center but was never put in operation.

The Ottawa Radiation Areas site, located in La Salle County, Illinois, consists of 14 areas that have been polluted by radioactive materials. Those 14 areas, scattered throughout Ottawa and in places outside the city, were placed on the National Priorities List as a single site because the same wastes polluted each area. The pollution came from the Radium Dial Co. from 1918 to 1936 and Luminous Processes, Inc. from 1937 to 1978. These businesses made glow-in-the-dark dials for clocks and watches using radium-based paint. Building demolition materials and soils that were polluted with radioactive waste were used as fill material in the Ottawa area. Many of the 14 areas are residential and include buildings. To address this situation, the U.S. EPA (2003a) removed the polluted soil in the residential areas first and then began cleaning up the remainder.

Another example of industrial radioactive pollution is found in the Kerr-McGee and Reed-Keppler Park sites in West Chicago, Illinois. In 1931, Lindsay Light & Chemical Co. Rare Earths Facility in West Chicago was used to extract thorium and rare earth elements from monazite and other ores. Later, the property was used for the manufacture of gaslight mantles (which contain thorium). Ownership of the facility changed from Lindsay to American Potash & Chemical in 1958 and to Kerr-McGee Chemical Corp. in 1967. Operations at the property continued until 1973, when Kerr-McGee closed the facility. Radioactive materials were landfilled in what is now Reed-Keppler Park, an 11-acre (4.5-ha) site that apparently had been a gravel quarry. The U.S. Nuclear Regulatory Commission's contractor located contaminated areas within the landfill and around and under tennis courts adjacent to it. Contaminated materials around (not under) the tennis courts were excavated and moved onto an area of surface contamination, which was then fenced and posted (Illinois Department of Nuclear Safety 1988b). The groundwater quality of the site continues to be monitored for uranium concentration, and, based on a recent five-year review, there were no exceedances of uranium in drinking water. (U.S. EPA 2007b).

Agencies

The lead agency in Illinois for radiation issues is the Illinois Emergency Management Agency, Division of Nuclear Safety. Its Web site contains several useful fact sheets and lists contractors who are approved for radon measurement and remediation (http://www.state.il.us/idns). The U.S. Geological Survey is no longer engaged in active radon research projects; however, information is available on its Web site (http://energy.cr.usgs.gov/radon). The U.S. EPA provides detailed information on the various programs associated with radiation protection and other topics related to radiation (http://www.epa.gov/ebtpages/radiationandradioactivity.html). The U. S. Department of Energy has the responsibility for cleaning contaminated sites and disposing of radioactive waste left behind as a by-product of nuclear weapons production, nuclear-powered naval vessels, and commercial nuclear energy production. That agency's Web site (http://www.energy.gov/engine/content.do?BT_CODEEN_SS1) provides information on these issues.

Conclusion

This brief chapter is intended to provide basic information on issues relating to radiation in Illinois. Much of human exposure to radiation is from natural sources, and background information is needed to interpret radiation information.

References

Cohen, B. L., 1988, Correlation between mean radon levels and lung cancer rates in U.S. counties: A test of the linear-no threshold theory: Proceedings of the 1988 Symposium on Radon and Radon Reduction Technology: Denver, Colorado, U. S. Environmental Protection Agency.

Congressional Record, 1988, Radon Pollution Control Act of 1988, H.R. 2837: Congressional Record/House, October 5, 1988, H9634–H9645.

Eichholz, G. G., 1987, Human exposure, in C. R. Cothern, ed., Environmental radon: New York, Plenum Press, p. 131–172.

Flood, J. R., T. B. Thomas, N. H. Suneson, and K. V. Luza, 1990, Geologic assessment of radon-222 potential in Oklahoma: Radon potential map: Norman, Oklahoma, Oklahoma Geological Survey, MSP GM-32, 28 p.

Gilkeson, R. H., R. A. Cahill, and C. R. Gendron, 1988, Natural background radiation in the proposed Illinois SSC siting area: Illinois State Geological Survey, Environmental Geology Note 127, 47 p.

Gilkeson, R. H., K. Cartwright, J. B. Cowart, and R. B. Holtzman, 1983, Hydrologic and geochemical studies of selected natural radioisotopes and barium in groundwater in Illinois: Urbana-Champaign, Water Resources Research Center, University of Illinois, Report No. 83 0180, 3 p.

Gilkeson, R. H., E. C. Perry, J. B. Cowart, and R. B. Holtzman, 1984, Isotopic studies of the natural sources of radium in groundwater in Illinois: Urbana-Champaign, Water Resources Research Center, University of Illinois, Research Report No. 187, 50 p.

Hasenmueller, N. R. 1988, Preliminary geologic characterization of Indiana for indoor-radon survey: Bloomington, Indiana, Indiana Geological Survey, Report of Progress 32, 7 p.

Hopke, P. K., ed., 1987, The indoor radon problem explained for the layman, in Radon and its decay products: Washington, D.C., American Chemical Society, Symposium Series 331, p. 572–586.

Illinois Department of Nuclear Safety, 1986, Radon in Illinois: A report to Governor James R. Thompson from the Governor's Radon Task Force: Springfield, Illinois, Illinois Department of Nuclear Safety, 51 p.

Illinois Department of Nuclear Safety, 1988a, Radon in Illinois: A status report: Springfield, Illinois, Illinois Department of Nuclear Safety, 50 p.

Illinois Department of Nuclear Safety, 1988b, The continuing story of the Rare Earths Facility in West Chicago, radiological responsibilities: Springfield, Illinois, Illinois Department of Nuclear Safety, p. 4–6.

Illinois Department of Nuclear Safety, 1991, U.S. EPA radon screening results differ from findings of IDNS study: Springfield, Illinois, Illinois Department of Nuclear Safety, v. 2, 5 p.

Illinois Department of Nuclear Safety, 1992, Radon in Illinois: A status report, 1992 update: Springfield, Illinois, Illinois Department of Nuclear Safety, 62 p.

Illinois Department of Nuclear Safety, 2000, A report regarding the impacts and ramifications of low-level radioactive waste management issues in Illinois, executive summary: Springfield, Illinois, 4 p.

Illinois Emergency Management Agency, 2005, DNS Info: What is Spent Nuclear Fuel?: Springfield Illinois, 4 p. http://www.state.il.us/iema/publications/radioactive.asp. Accessed November 5, 2009.

Illinois Emergency Management Agency, 2006, Status report for radon in Illinois: Springfield, Illinois, 51 p.

Illinois Environmental Protection Agency, 2007, National Pollutant Discharge Permit of Discharge into Water of the State of Illinois: Springfield, Illinois, NPDES permit no. IL0074101, 4 p.

Kay, R. T., 1999, Radium in ground water from public-water supplies in northern Illinois: Reston, Virginia, U.S. Geological Survey, Fact Sheet 137-99, 4 p.

Michel, J. 1987, Sources, in C. R. Cothern, ed., Environmental radon: New York, Plenum Press, p. 81–130.

Morrow, W. S., 2001, Uranium and radon in ground water in the lower Illinois River basin: Urbana, Illinois, U.S. Geological Survey, Water-Resources Investigations Report 01-4056, 29 p.

National Primary Drinking Water Regulations, 1999, Radon-222, Proposed rule: Federal Register, v. 64, no. 211, p. 59295–59344.

National Primary Drinking Water Regulations, 2000, Radionuclides, Final rule: Federal Register, v. 65, no. 236, p.76707–76753.

National Council on Radiation Protection and Measurements, 1975, Natural background radiation in the United States: Bethesda, Maryland, Report No. 45, 163 p.

National Research Council, 1988, Health risks of radon and other internally deposited alpha-emitters—BEIR IV: Washington, D.C., National Academy Press, Committee on the Biological Effects of Ionizing Radiations, Board on Radiation Effects, Research Commission on Life Sciences, 624 p.

Nazaroff, W. W., and R. G. Sextro, 1989, Technique for measuring the indoor ^{222}Rn source potential of soil: Environmental Science and Technology, v. 23, no. 4, p. 451–458.

Nazaroff, W. W., and K. Teichman, 1990, Indoor radon—Exploring U.S. federal policy for controlling human exposures: Environmental Science and Technology, v. 24, no. 6, p. 774–782.

Nero, A. 1989, Earth, air, radon and homes: Physics Today, v. 42, no. 4, p. 32–39.

Otton, J. K., 1992, The geology of radon: Reston, Virginia, U. S. Geological Survey, General Interest Publication, 1992-0-326-248, 29 p.

Schumann, R. R., 1992, Geologic potential of the glaciated Upper Midwest, Proceedings of the 1992 Symposium on Radon and Radon Reduction Technology: Minneapolis, Minnesota, U. S. Environmental Protection Agency, VIII-3.

Smith, K. P., D. L. Blunt, G. P. Williams, and C. L. Tebes, 1996, Radiological dose assessment related to management of naturally occurring radioactivity materials generated by the petroleum industry: Environmental Assessment Division, Argonne National Laboratory, Argonne, IL, ANL/EAD-2, 69 p.

U.S. Congress, 1980, The Low-Level Radioactive Waste Policy Act, Title 42, Ch. 23, Sec. 2021–2021j, 35 p.

U.S. Environmental Protection Agency, 1986, A citizen's guide to radon: What it is and what to do about it, U.S. Environmental Protection Agency and U.S. Centers for Disease Control Pamphlet: Washington, D.C., U.S. Government Printing Office, OPA-86-004, 13 p.

U.S. Environmental Protection Agency, 2003a, NPL fact sheet for Ottawa, Illinois, radiation areas site: Chicago, Illinois, U. S. Environmental Protection Agency, Region 5, EPA ID3 ILD980606750, 2 p.

U.S. Environmental Protection Agency, 2003b, EPA assessment of risk from radon in homes: U.S. Environmental Protection Agency, Air and Radiation, EPA 402-R-03-003, 88 p.

U.S. Environmental Protection Agency, 2005, A regulators' guide to the management of radioactive residuals from drinking water treatment technologies: Office of Water, EPA 816-R-05-004, 81 p.

U.S. Environmental Protection Agency, 2007a, A citizen's guide to radon: The guide to protecting your family from radon: Washington, D.C., U.S. Environmental Protection Agency, U.S. EPA 402-K-07-009, 16 p.

U.S. Environmental Protection Agency, 2007b, Five-year review report for Kerr-McGee Reed-Keppler Park site, West Chicago, Du Page County, Illinois: Chicago, Illinois, U.S. Environmental Protection Agency, Region 5, 25 p.

U.S. Environmental Protection Agency, 2008, Environment, health and safety online: NARM and NORM radioactive waste: Office of Radiation and Indoor Air Radiation Protection Division: Washington, D.C. http://www.ehso.com/NuclearNORM.htm. Accessed November 5, 2009.

27 Earthquakes

Timothy H. Larson and Robert A. Bauer

INTRODUCTION

Most Illinois residents think that earthquakes do not pose a significant risk to the state. Even though nearly 500 earthquakes have been reported in Illinois, and 30 have caused damage, none has been larger than magnitude 5.5, and none has caused serious damage. Still, at least one earthquake is felt in the Chicago area every decade, and more frequent earthquakes are felt in the southern parts of the state. A 10 to 12% probability exists that a damaging earthquake might occur in the southwestern part of the state during the next half century (Frankel et al. 2002). It is therefore instructive to review the earthquake history of Illinois in order to understand the potential for possible future earthquake activity.

EARTHQUAKE MEASUREMENT AND OBSERVATION

Magnitude and *intensity* are terms used to describe the severity of an earthquake. Magnitude is a measure of the seismic energy released in the earthquake. It is calculated from measurements of the ground vibrations recorded by seismographs. Earthquake magnitudes are reported in logarithmic increments, which means that a magnitude 7 earthquake has about 30 times more energy than a magnitude 6 earthquake. An increase of 0.2 means that twice as much energy was released. The common term "Richter magnitude," named for the California seismologist who originated and popularized the measure, is technically only appropriately used for California earthquakes. Similar measures, with slight variations in the way they are calculated, have been developed for other areas or other purposes. Otto Nuttli (1973), at St. Louis University, developed a magnitude scale appropriate for the central United States. The Richter scale is one of several variations of magnitude that are reported. More recently, a universally appropriate scale, the moment magnitude, has become standard. In earthquake literature, the notation for the Richter magnitude is M_L, the Nuttli magnitude is m_{bLg} or M_N, and the moment magnitude is M_w (Bolt 1993).

Earthquakes in Illinois originate within the crust at depths generally up to 15 miles (25 km) (Heigold and Larson 1990). The vibrations move away from the point of origin (the hypocenter or focus) through the bedrock and then up though the unconsolidated earth materials on top of the bedrock. In the central United States, the bedrock is mostly flat-lying, old, intact, and rigid. Earthquake vibrations can travel very far through this material compared with travel distance through the young, broken, weak bedrock of the West Coast. Because of the bedrock difference, central United States earthquakes are felt and cause damage over an area 15 to 20 times larger than that area affected by California earthquakes with similar magnitudes.

Intensity is an evaluation of the effects brought about by an earthquake using observations of people in the affected area. Intensities are based on descriptive reports rather than instrumental readings. Several formal intensity scales have been proposed for use in different parts of the world. In the United States, earthquake intensities are reported using the 12-point Modified Mercalli Intensity scale (Table 27-1) (Wood and Neumann 1931).

Most of the 500 Illinois earthquakes have had relatively small magnitudes (2 to 4) that did not cause much damage but were felt over large areas. Larson (2001) documented the effects of one of these small earthquakes in north-central Illinois. Over 500 people responded to a questionnaire published in local newspapers. Their recollections of the 1999 earthquake (magnitude 3.5, centered near Amboy, southeast of Dixon) fit surprisingly well into the Modified Mercalli Intensity scale. Throughout the area, many people who felt the earthquake did not recognize it as an earthquake but attributed the sharp vibrations to a quarry blast or the passing of a particularly large vehicle (Intensity III). Closer to the epicenter, many people reported the sensation of something striking the house or falling onto it (Intensity IV). Others reported dishes and windows rattling but not breaking (Intensity IV).

The first modern seismograph network was established in the central United States during the 1960s (Heigold and Larson 1990). Prior to that time, most earthquake records were based on newspaper accounts and personal journals. Magnitudes of these older earthquakes can be estimated by comparing their intensities with the intensities from recent earthquakes that have known magnitudes. Magnitude estimates can be based on the maximum intensity of the earthquake (Nuttli and Herrmann 1978) or by the area over which the earthquake was felt (Sibol et al. 1987).

Damage to buildings, highways, power lines, pipelines, and other structures only partly depends on the amount of energy released during the earthquake. Certain kinds of earth materials resting on the bedrock amplify the earthquake ground motions. In Illinois, structures built on the thick, loose sediments of river floodplains are more likely to be damaged than structures on glacial till (stiff, pebbly clay) or bedrock. In fact, seismic intensity may increase one or more units on the Modified Mercalli Intensity scale if loose sediments are present. Also, loose sandy sediments with high moisture content can turn to liquid (liquefaction) when shaken enough.

Illinois Earthquake History

The map of Illinois earthquakes (Figure 27-1) was compiled from catalogs of historical earthquakes in the eastern, central, and mountain states (Stover et al. 1984, Armbruster and Seeber 1992, Stover and Coffman 1993). Because most of those earthquakes occurred before seismograph stations were established, their locations and magnitudes are only approximate. The compiled list was carefully reviewed to remove duplicate events. Finally, the list was adjusted to conform to the locations (± 0.02 degrees) and magnitudes (± 0.3 units) published by Stover and Coffman (1993) in their authoritative compendium of larger historical earthquakes throughout the United States. The list of more recent earthquakes was compiled from data searches of the U.S. Geological Survey (USGS) National Earthquake Information Center and regional seismological observatories, including the Tennessee Earthquake Information Center and the St. Louis University regional catalogs from 1974 to 1992, the St. Louis University and Central Mississippi Valley Earthquake Bulletin catalogs from 1993 to 1994, and the St. Louis University and University of Memphis Center for Earthquake Research and Information catalogs from

Table 27-1 Modified Mercalli Intensity scale.[1]

Intensity	Description	Percent of acceleration of gravity	Magnitude
I	Not felt except by a very few under especially favorable conditions.	<0.17	
II	Felt only by a few persons at rest, especially on upper floors of buildings. Delicately suspended objects swing.	0.17 to 1.4	
III	Felt quite noticeably by persons indoors, especially on upper floors of buildings. *Many people do not recognize it as an earthquake. Vibrations are similar to the passing of a truck.*[2]	1.4 to 3.9	
IV	Felt indoors by many, outdoors by few during the day. *At night, some awakened.* Dishes and windows rattled but not broken. Doors swing; walls make cracking sound. Sensation like a heavy truck striking a building.	3.9 to 9.2	~3.5
V	Felt by nearly everyone; many awakened. *Some dishes and windows broken; plaster cracked. Unstable objects overturned.*	3.9 to 9.2	
VI	Felt by all; many frightened. Some heavy furniture moved; a few instances of fallen plaster and cracked chimneys. Damage slight.	9.2 to 18	~4
VII	Damage negligible in buildings of good design and construction; slight to moderate in well-built ordinary structures. Considerable damage in poorly built or badly designed structures. Some chimneys broken.	18 to 34	
VIII	Damage slight in specially designed structures; considerable damage in ordinary substantial buildings with *partial collapse.* Damage great in poorly built structures. *Fall of chimneys, factory stacks, columns, monuments, and walls. Heavy furniture overturned.*	34 to 65	~6
IX	Damage considerable in specially designed structures; well-designed frame structures thrown out of plumb. Damage great in substantial buildings, with partial collapse. *Buildings shifted off foundations.*	65 to 124	
X	Some well-built wooden structures destroyed; most masonry and frame structures destroyed with their foundations. Rails bent.	>124	~8
XI	Few, if any, masonry structures remain standing. Bridges destroyed. Rails bent greatly.		
XII	Damage total. Lines of sight and level are distorted. Objects are thrown into the air.		

[1]Acceleration estimates are from Wald et al. (1999), and magnitude estimates are from Heigold and Larson (1990).
[2]Criteria that are particularly useful in the central United States are shown in italics.

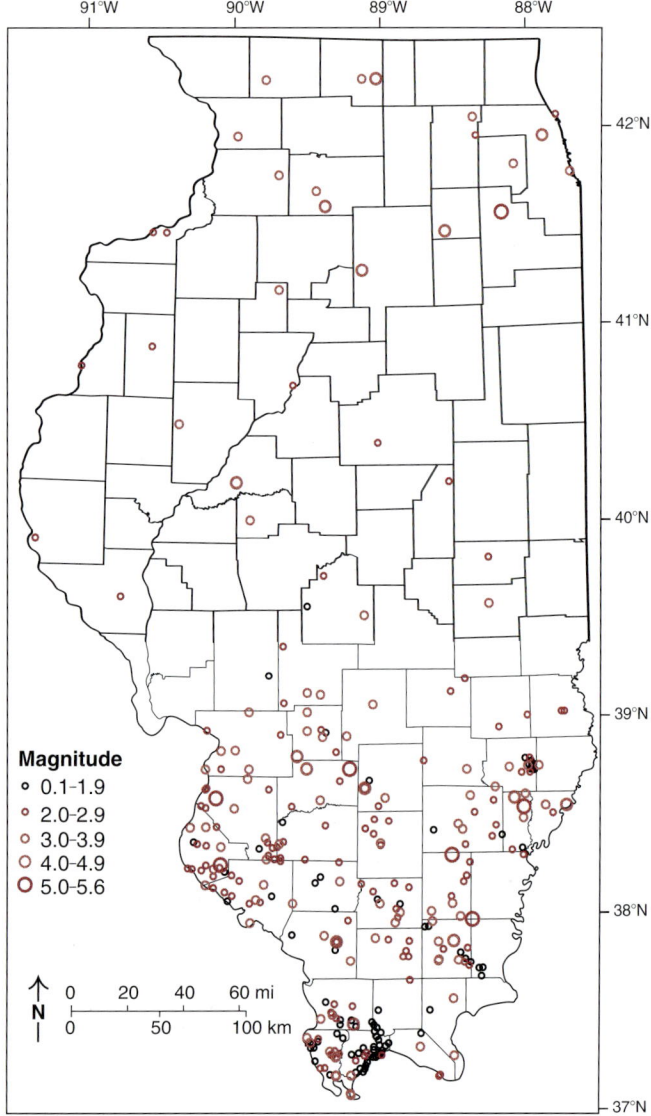

Figure 27-1 Earthquakes in Illinois from 1795 to December 2003. Earthquakes compiled from published catalogs (Stover et al. 1984, Armbruster and Seeber 1992, Stover and Coffman 1993) and unpublished databases available from the U. S. Geological Survey, University of Memphis, and St. Louis University.

1995 to the present. Figure 27-1 includes many small, instrumentally recorded earthquakes with magnitudes ranging from less than 1.0 to 2.5 that are not included in the national databases but are archived in the regional databases. Although the temporal and spatial coverage of these smaller earthquakes is not complete, they were included to provide a picture of seismicity in southern Illinois that would not be apparent from the larger earthquakes alone.

Discounting the smallest earthquakes, because the combined seismic energy released from them is minimal, the remaining 252 earthquakes shown on Figure 27-1 have a distinct, regional variation in their distribution. Eighty percent (202) occurred in the southern third of the state, south of latitude 39° N; 9.5% (24) occurred in the central part of the state between latitudes 39° N and 41° N; and another 10.3% (26) were located in the northern part of the state, north of latitude 41° N. Earthquakes in northern and central Illinois are widely scattered, seemingly random events. The effects from the larger earthquakes in Illinois have been briefly described by Stover and Coffman (1993), and the USGS maintains a complete catalog of U.S. earthquakes in the eastern, central, and mountain states from 1568 to 1986 (modified from Stover et al. 1984).

Northern Illinois Earthquakes: Fire at Aurora

One of the largest earthquakes in the state occurred in northern Illinois on May 26, 1909. The exact location of the magnitude 5.1 (estimated) earthquake is not known, but the largest intensities occurred in and near Aurora, where many chimneys fell, a stove overturned, gas lines broke, and a fire started. Although considerable excitement ensued, the Aurora fire was quickly extinguished and soon forgotten. Elsewhere, houses were jostled out of plumb in Beloit, Wisconsin, and brick walls cracked as far away as Bloomington, Illinois. The area encompassed by minor damage is shown in Figure 27-2. A somewhat smaller earthquake occurred nearby in 1912.

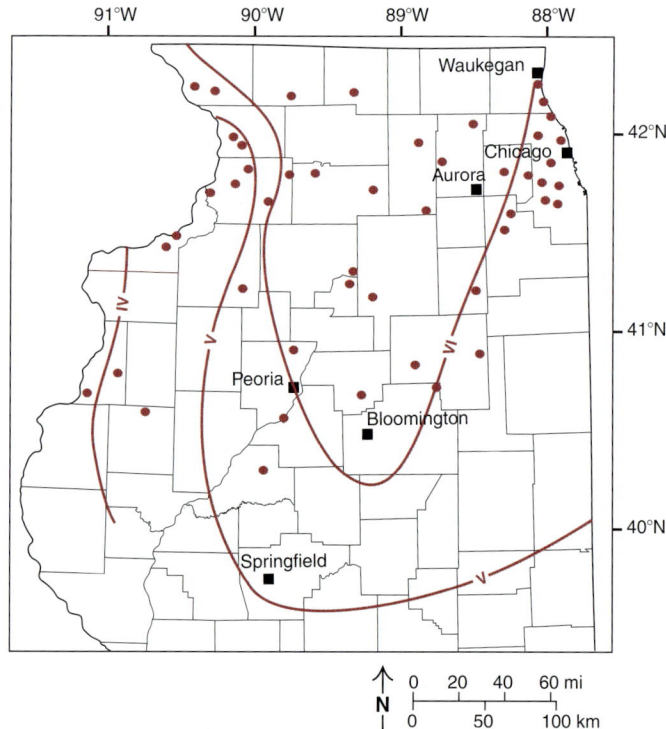

Figure 27-2 Intensity distribution from the 1909 earthquake in northern Illinois constructed from newspaper reports (Udden 1910, Heigold 1972).

Earthquakes

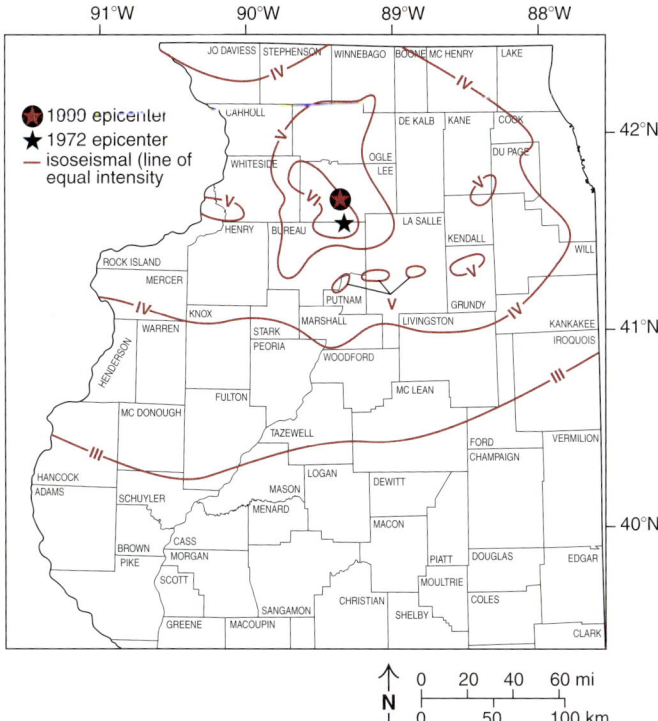

Figure 27-3 Intensity distribution from the 1972 earthquake in northern Illinois (Heigold 1972).

A magnitude 4.0 earthquake centered in north-central Illinois south of Rockford near the village of Amboy woke many Chicago area residents when it struck late at night on September 15, 1972 (Figure 27-3). Although felt over a very large area, the Intensity VI area was much smaller than that of the 1909 earthquake. Twenty-seven years later on September 2, 1999, a magnitude 3.5 earthquake occurred at nearly the same location as the 1972 earthquake. An isoseismal map (Figure 27-4) prepared from responses to questionnaires distributed to post offices and published in local papers indicated that the greatest intensities were several miles northwest of the epicenter, as located by regional seismographs (Larson 2001). Several small earthquakes occurred in the Rock Island area in the 1930s (Fryxell 1940, Heigold 1972); most were in Iowa, but two that were centered in Illinois are shown in Figure 27-1.

Central Illinois Earthquakes: Burglars in the Basement

For the most part, central Illinois is characterized by a lack of earthquakes. The largest known earthquake in that part of the state occurred at about 11:00 p.m. on July 19, 1909, between Petersburg and Havana. With an estimated magnitude of 4.8, this earthquake damaged chimneys and broke windows. Many residents were awakened and congregated outside after the unusual event. The *St. Louis Post-Dispatch* (1909) reported that "several who had gone to bed before the tremor arrived report that they experienced a vague Pullman berth sensation, upper or lower not specified. The sensation did not last long enough for the porter to come around for his tip."

Larger earthquakes centered to the south of the region are frequently felt in central Illinois. Residents typically have difficulty explaining the unusual sounds (Intensity III) or vibrations. The magnitude 6.2 (estimated) earthquake that occurred around 5:00 a.m. on October 31, 1895, near Cairo was widely felt in central Illinois. Residents in Champaign, Bloomington, Decatur, and Springfield were jostled out of their beds, some with the impression that burglars had broken into their basements or that their steam boilers had burst (*Champaign Daily Gazette* 1895; *Illinois State Journal* 1895).

Southern Illinois Earthquakes: Minor Damage, Scary Cemeteries

Nearly two dozen damaging earthquakes have been reported in southern Illinois, although seldom has the damage been more severe than fallen chimneys, broken windows, or cracked masonry walls. The largest earthquake in the central United States during the twentieth century occurred in southeastern Illinois on November 9, 1968. This magnitude 5.3 earthquake caused Intensity VII damage in the area of the epicenter northeast of Harrisburg (Stover and Coffman 1993). Chimneys were broken, foundations cracked, and bricks were thrown from masonry parapets.

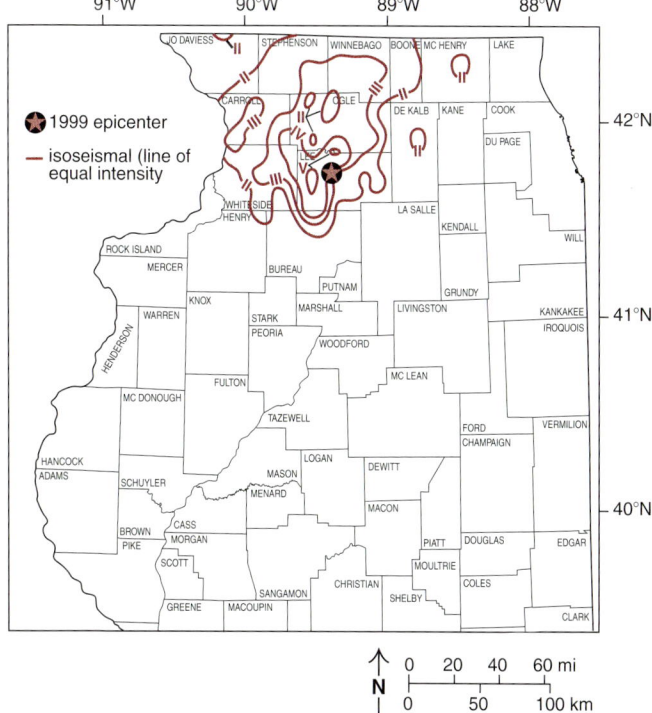

Figure 27-4 Intensity distribution from the 1999 earthquake in northern Illinois (Larson 2001).

The earthquake was felt by people in multistory buildings as far away as Boston, Massachusetts. Figure 27-5 shows the extent of the damaged area and the area where this earthquake was generally felt for most people (Gordon et al. 1970). One Galatia resident reported,

> At the time of the quake I was in the city cemetery 1 mile [1.6 km] east of town. The earth trembled and tombstones shook and I thought the dead were coming forth, and that this was it . . . . I came back into town and everyone was scared, chimneys were torn down, dishes fell out of cabinets and off tables and some refrigerator doors were even jarred open . . . (Gordon et al. 1970).

A subsequent survey of the area revealed that many tombstones and chimneys had been systematically rotated in response to the direction of the compressive waves (Gordon et al. 1970).

A magnitude 5 earthquake near Lawrenceville in 1987 is typical of the 18 other magnitude 4 or greater earthquakes in southern Illinois. The earthquake caused minor damage in the area and was felt throughout the region (Reagor and Brewer 1987, Langer and Bollinger 1991). Diagonal cracks in masonry walls (Figure 27-6) and loosened bricks (Figure 27-7) are common examples of the minor damage that can occur in earthquakes of these magnitudes.

Since the mid-1970s, hundreds of very small earthquakes have been recorded in southern Illinois. A particularly striking cluster of small earthquakes occurred in Pulaski County and neighboring Ballard County, Kentucky, between November 1983 and March 1984. During this period, more than 100 earthquakes were recorded in Pulaski County alone. Of these, 67 occurred in the 75 hours following a magnitude 3.3 earthquake on February 13, 1984. This cluster, or "swarm," near Olmsted is an indication of continuing seismicity in southern Illinois and of its links with the New Madrid Seismic Zone to the south (Figure 27-8).

Regional Earthquakes

During the winter of 1811–1812, a series of four very large earthquakes (estimated at greater than magnitude 7.0) and hundreds of smaller earthquakes rocked the central United States, including what is now Illinois. These earthquakes were centered in the New Madrid area of what is now Missouri, south and southwest of Illinois, where they caused severe damage. Frightened settlers reported various disturbances along the Ohio, Wabash, and Mississippi Rivers, including thousands of sand blows and massive landslides along the bluffs from what is now Memphis to about present-day Cairo (Street and Nuttli 1990). In addition, tem-

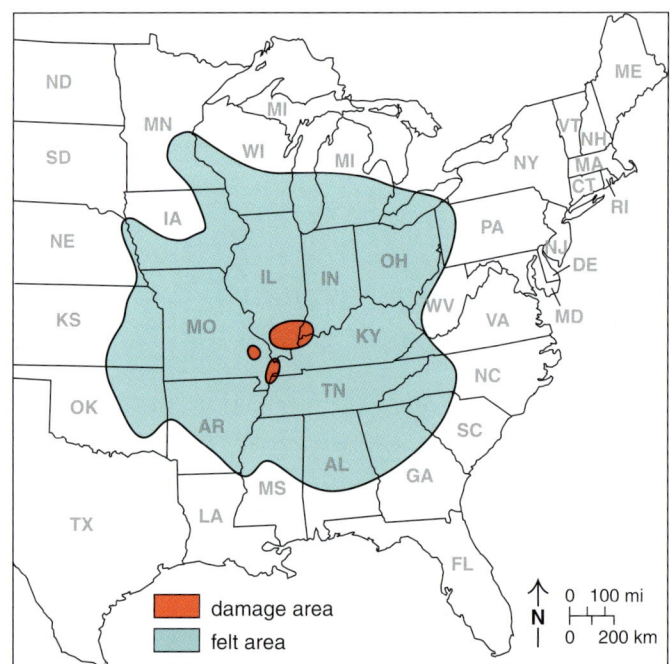

Figure 27-5 Damage area and felt area from the 1968 earthquake in southern Illinois (Gordon et al. 1970).

Figure 27-6 Diagonal masonry cracks in a church tower in Lawrenceville from a magnitude 5.0 earthquake (Intensity VI to VII) in 1987. (Photograph from the Illinois State Geological Survey collection.)

Figure 27-7 Bricks thrown from a chimney in Lawrenceville by a magnitude 5.0 earthquake (Intensity VI to VII) in 1987. (Photograph from the Illinois State Geological Survey collection.)

EARTHQUAKES

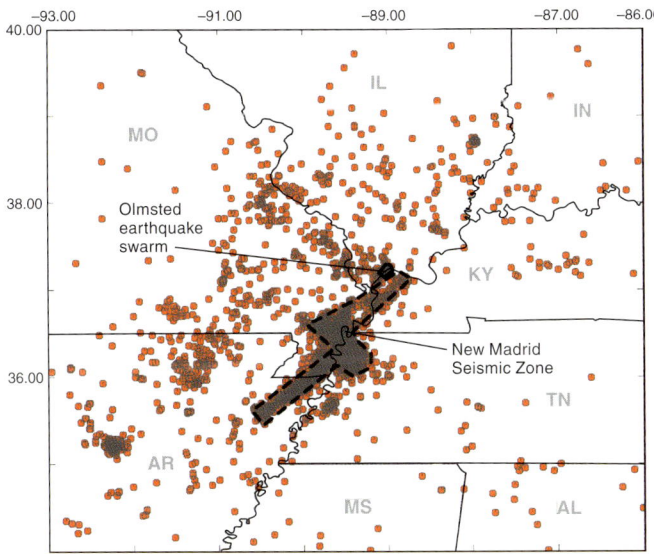

Figure 27-8 Central United States earthquakes from 1975 to 1995 showing the location of the Olmsted earthquake swarm in relation to the New Madrid Seismic Zone. Earthquake information was compiled from regional catalogs.

porary waterfalls formed in the Mississippi River and a large tract of land sank to form Reelfoot Lake in Tennessee.

Evidence of liquefaction from several prehistoric earthquakes from New Madrid north to Vincennes, Indiana, demonstrates that large earthquakes have occurred in the central United States over the past several thousand years (Obermeier et al. 1992, Tuttle and Schweig 1995). The frequency of recurrence and magnitude of these large earthquakes is the object of intense research (Johnston and Nava 1985, Newman et al. 1999). Estimates of recurrence range from 500 to 1,200 years and magnitude 7 to 7.5. Continuing small and moderate earthquakes have defined an area of seismicity known as the New Madrid Seismic Zone, which extends from Tennessee and Arkansas northward through Kentucky and Missouri to the southernmost counties of Illinois. The Olmsted earthquake swarm is an example of the ongoing, low-magnitude seismicity that characterizes the New Madrid Seismic Zone. Other moderately large earthquakes have occurred in surrounding states.

The most notable Illinois earthquake was one that struck near Cairo, Illinois, on Halloween morning October 31, 1895. This earthquake, with an estimated magnitude of 6.2, broke hundreds of chimneys and plate glass windows in Cairo (Hopper and Algermissen 1980). This was the same event that caused many central Illinois residents to check for burglars in their basements. Intensity in the Chicago area varied from IV to V; no severe damage was reported, and many people did not feel the earthquake. The *Chicago Tribune* (1895) reported the disruption to their state-of-the-art communications network:

> There are 12,000 telephones in Chicago, each one of which has a corresponding hinged "drop" in one of the stations. When the shock was felt in Chicago every one of these 12,000 drops fell at the same instant. It was a case of Mother Earth ringing up the telephones of Chicago, not one, but all of them at the same time, and demanding instant response. It was too big and vociferous a call for the attendants to answer, and all the startled girls could do was to shout back "Busy now" in a frightened but instinctive response.

The area damaged by the 1895 earthquake is shown in Figure 27-9. Based on what is known about the response of different sediments to earthquake vibrations, it is not surprising to see that the greatest damage from this earthquake was concentrated in the major river valleys that are filled with thick, loose sediments.

EARTHQUAKE HAZARDS

Although there have been nearly 500 earthquakes in Illinois during the last two centuries, only a few have caused damage (Modified Mercalli Intensity of VI or higher) or injuries. Larger earthquakes in the New Madrid region have caused more damage in Illinois than earthquakes originating in the state itself. The probability of future damaging earthquakes can be estimated based on the historical record of past earthquakes. Frankel (1995) and Frankel et al. (1996) have created maps of the largest probable ground shaking that might be exceeded over a 50-year period (Figure 27-

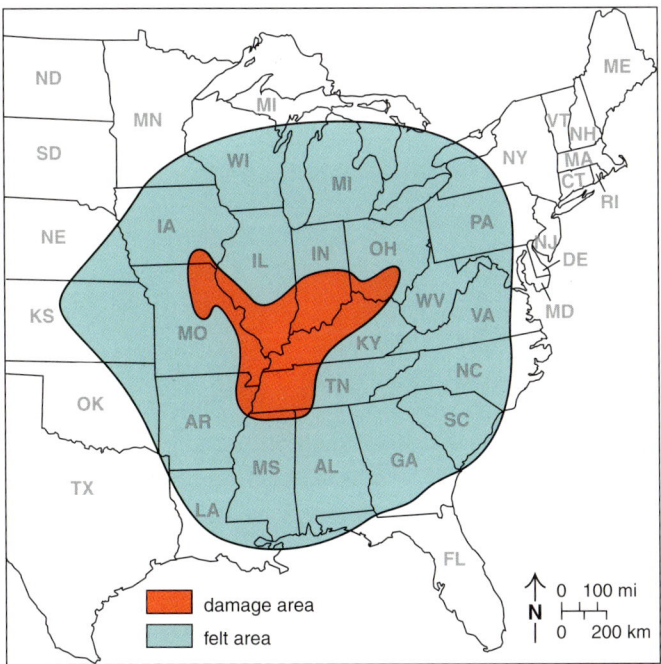

Figure 27-9 Damage area and felt area from the "Halloween earthquake" of 1895 (Hopper and Algermissen 1980).

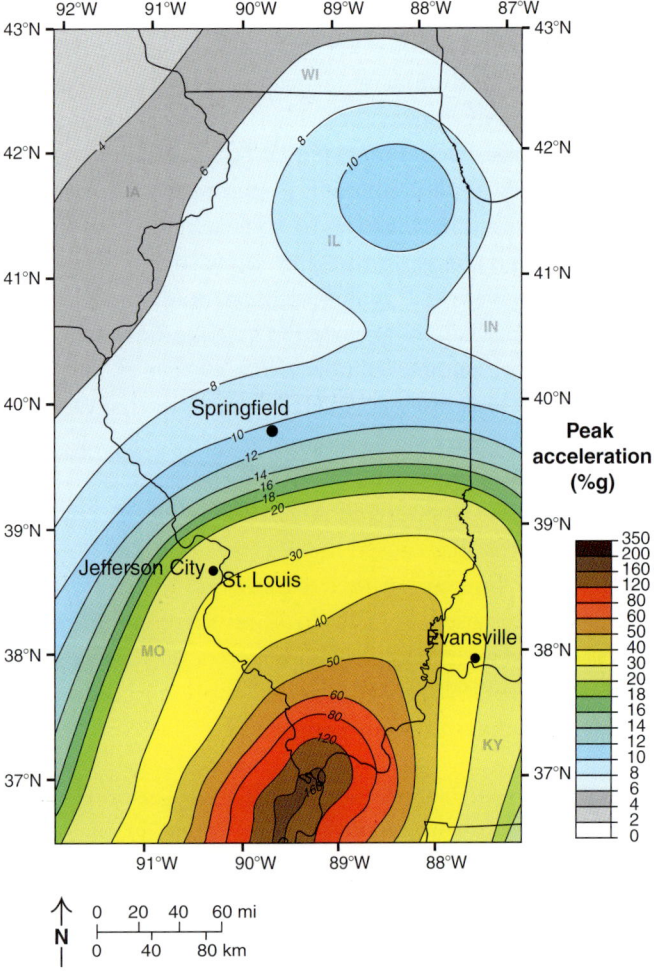

Figure 27-10 Earthquake hazard for Illinois given as maximum accelerations (percent of gravity) with a 2% probability of being exceeded within a 50-year window (Frankel et al. 1996).

10). The ground motions, plotted as accelerations, can be roughly converted to intensities using conversion values in Table 27-1.

For most of Illinois, the earthquake hazard is dominated by the possibility of large earthquakes recurring in the New Madrid Seismic Zone, south of Illinois. In this scenario, the maximum accelerations in the southernmost counties of Illinois would exceed 60% of gravity, or Modified Mercalli Intensity X. Although the earthquake hazard decreases to the north, there is a 2% probability during the next 50 years that damaging ground motions (accelerations greater than 10% of gravity, Modified Mercalli Intensity VII) could be experienced anywhere in the southern half of Illinois. Because of the record of minor to moderate earthquakes in northern Illinois, especially the 1909 Aurora earthquake, the earthquake hazard is a little greater in the western suburbs of Chicago.

However, the damage (the observed intensity) from future earthquakes can be mitigated by taking appropriate precautions, such as strengthening bridges and adding bracing in buildings, and by identifying hazards that exist in homes, schools, or places of business and then systematically removing or correcting each hazard. This is especially important in the southernmost counties of Illinois where earthquake hazards are greatest. Some common hazards include freestanding water heaters, stoves, and other gas or electric appliances that could move or fall during an earthquake; bookshelves or filing cabinets that are freestanding or bookshelves with objects stored above head level; water or gas pipes that are not horizontally braced; and large panes of glass that could shatter.

Other things that people can do to make their homes more earthquake-ready include keeping a few days' supply of food and water available; being sure the home has a fire extinguisher and smoke alarm; maintaining a properly equipped first aid kit, complete with enough prescription medication to last a few days to a few weeks; organizing and testing a family emergency plan that would help ensure each family member's survival; and knowing how to turn off the gas supply to the building.

If you are involved in an earthquake, remembering a few simple facts can greatly increase your chance of survival and can help to reduce the possibility of serious injury. If inside a building: stay inside and duck, cover, and hold onto any solid object such as a desk or table. Do not run out of a brick (masonry) building during the shaking—falling bricks and glass can kill or injure you. Protection inside a building is found next to or under heavy furniture. You should avoid being under ceiling fixtures, particularly hanging lights, or near large windows, which can shatter and send shards of glass flying inward. Large rooms with open-span ceilings or roofs are the most vulnerable to collapse and should be avoided. If outside during an earthquake, move far enough from buildings to avoid falling materials.

Conclusion

Except for far southern Illinois, earthquakes are no more than occasional curiosities in Illinois. Occasional small to moderate earthquakes originating in Illinois or nearby states are felt in most of the state but have caused little or no damage. However, very strong earthquakes have occurred and are likely to reoccur in the New Madrid Seismic Zone. These earthquakes could cause locally severe damage in southernmost Illinois and could be felt throughout the state.

REFERENCES

Armbruster, J., and L. Seeber, 1992, NCEER-91 earthquake catalog for the United States: Buffalo, New York, National Center for Earthquake Engineering Research, State University of New York (SUNY).

Bolt, B. A., 1993, Earthquakes: New York, New York, W. H. Freeman and Company, 331 p.

Champaign Daily Gazette, 1895, The earth shook: Champaign, Illinois, no. 3724, October 31, 1895, p. 1.

Chicago Tribune, 1895, Earth in a quiver: Chicago, Illinois, v. 54, no. 305, November 1, 1895, p. 4, 6.

Frankel, A., 1995, Mapping seismic hazard in the central and eastern United States: Seismological Research Letters, v. 66, no. 4, p. 8–21.

Frankel, A., C. Mueller, T. Barnhard, D. Perkins, E. V. Leyendecker, N. Dickman, S. Hanson, and M. Hopper, 1996, National seismic hazard maps, June 1996: Documentation: Reston, Virginia, U.S. Geological Survey, Open-File Report 96-532, 21 p.

Frankel, A. D., M. D. Petersen, L. S. Mueller, K. M. Haller, R. L. Wheeler, E. V. Legendecker, R. L. Wesson, C. S. Harmsen, C. H. Cramer, D. M. Perkins, and K. S. Rutstales, 2002, Documentation for the 2002 update of the national seismic hazard maps: Reston, Virginia, U.S. Geological Survey, Open-File Report 02-420, 33 p.

Fryxell, F. M., 1940, The earthquakes of 1934 and 1935 in northwestern Illinois and adjacent parts of Iowa: Bulletin of the Seismological Society of America, v. 30, no. 3, p. 213–218.

Gordon, D. W., T. J. Bennett, R. B. Herrmann, and A. M. Rogers, 1970, The south-central Illinois earthquake of November 9, 1968: Macroseismic studies: Bulletin of the Seismological Society of America, v. 60, no. 3, p. 953–971.

Heigold, P. C., 1972, Notes on the earthquake of September 15, 1972, in northern Illinois: Illinois State Geological Survey, Environmental Geology Notes 59, 15 p.

Heigold, P. C., and T. H. Larson, 1990, Seismicity of Illinois: Illinois State Geological Survey, Environmental Geology Notes 133, 20 p.

Hopper, M. G., and S. T. Algermissen, 1980, An evaluation of the effects of the October 31, 1895 Charleston, Missouri, earthquake: Reston, Virginia, U.S. Geological Survey, Open-File Report 80-778, 42 p.

Illinois State Journal, 1895: Trembled: Old Mother Earth had a seismic shiver: Springfield Illinois, November 1, 1895, p. 1–2.

Johnston, A. C., and S. J. Nava, 1985, Recurrence rates and probability estimates for the New Madrid Seismic Zone: Journal of Geophysical Research, v. 90, no. B8, p. 6737–6753.

Langer, C. J., and G. A. Bollinger, 1991, The southeastern Illinois earthquake of 10 June 1987: The later aftershocks: Bulletin of the Seismological Society of America, v. 81, no. 2, p. 423–445.

Larson, T. H., 2001, The earthquake of September 2, 1999, in northern Illinois: Big lessons from a small earthquake: Illinois State Geological Survey, Environmental Geology 153, 22 p.

Newman, A., S. Stein, J. Weber, J. Engeln, A. Mao, and T. Dixon, 1999, Slow deformation and lower seismic hazard at the New Madrid Seismic Zone: Science, v. 284, no. 5414, p. 619–621.

Nuttli, O. W., 1973, The Mississippi Valley earthquake of 1811 and 1812: Intensity, ground motion, and magnitudes: Bulletin of the Seismological Society of America, v. 63, no. 1, p. 227–248.

Nuttli, O. W., and R. B. Herrmann, 1978, State-of-the-art for assessing earthquake hazards in the United States: Credible earthquakes for the central United States: Vicksburg, Mississippi, U.S. Army Engineer Waterways Experiment Station, Miscellaneous Paper S-73-1, Report 12, 99 p.

Obermeier, S. F., P. J. Munson, C. A. Munson, J. R. Martin, A. D. Frankel, T. L. Youd, and E. C. Pond, 1992, Liquefaction evidence for strong Holocene earthquake(s) in the Wabash Valley of Indiana-Illinois: Seismological Research Letters, v. 63, no. 3, p. 321–335.

Reagor, G., and L. R. Brewer, 1987, Preliminary isoseismal map and intensity distribution for the southern Illinois earthquake of 10 June 1987: Reston, Virginia, U.S. Geological Survey, OFR 87-578, 3 p.

Sibol, M. S., G. A. Bollinger, and J. B. Birch, 1987, Estimation of magnitudes in central and eastern North America using intensity and felt area: Bulletin of the Seismological Society of America, v. 77, no. 5, p. 1635–1654.

St. Louis Post-Dispatch, 1909, St. Louis feels slight shock of an earthquake: St. Louis, Missouri, v. 61, no. 333, July 19, 1909, p. 1.

Stover, C. W., and J. L. Coffman, 1993, Seismicity of the United States 1568–1989 (rev.): Reston, Virginia, U.S. Geological Survey, Professional Paper 1527, 418 p.

Stover, C. W., B. G. Reagor, and S. T. Algermissen, 1984, United States earthquake data file: Reston, Virginia, U.S. Geological Survey, Open-File Report 84-225, 123 p.

Street, R. L., and O. W. Nuttli, 1990, The great Central Mississippi Valley earthquakes of 1811–1812: Lexington, Kentucky, Kentucky Geological Survey, Special Publication 14, Series XI, 15 p.

Tuttle, M. P., and E. S. Schweig, 1995, Archeological and pedological evidence for large prehistoric earthquakes in the New Madrid Seismic Zone, central United States: Geology, v. 23, no. 5, p. 253–256.

Udden, J. A., 1910, Observations on the earthquake in the Upper Mississippi Valley, May 26, 1909: Transactions of the Illinois State Academy of Science, v. 3, p. 132–143.

Wald, D. J., V. Quitoriano, T. H. Heaton, and H. Kanamori, 1999, Relationships between peak ground acceleration, peak ground velocity, and Modified Mercalli Intensity in California: Earthquake Spectra, v. 15, no. 3, p. 557–564.

Wood, H. O., and F. Neumann, 1931, Modified Mercalli Intensity scale of 1931: Bulletin of the Seismological Society of America, v. 21, p. 278–283.

28 Karst Terrane

Samuel V. Panno and C. Pius Weibel

INTRODUCTION

Illinois is known for its gentle landforms, relative flatness, and fertile croplands. Welcome exceptions to the topography of most of Illinois, away from towns with names like Flatville, are the karst regions of the state. Big Sink, Cave-In-Rock, Vishnu Springs, and Illinois Caverns—these diverse and unusually named natural features from southern and western Illinois are all related by their geological setting and their origins in karst terrane.

Karst refers to a geological environment with distinctive landforms (e.g., sinkholes, numerous and large springs, and sinking streams) and hydrology (e.g., subterranean drainage through crevices, conduits, and caves) arising from the solution of the host rock (Ford and Williams 1989). Most karst landscapes form when slightly acidic water from rain and snowmelt seeps through a thin soil cover, typically less than 50 feet (15 m), into relatively soluble limestone or dolomite bedrock. The term *soil* is used here and throughout the chapter as a catchall term that includes the surficial organic-rich layer (topsoil) and all of the unconsolidated material above bedrock. As the water seeps down through soil enriched in carbon dioxide (from the degradation of organic debris), the water becomes more acidic. This acidified water moves through fractures, joints, and bedding planes of bedrock, where it slowly (over thousands to tens of thousands of years or more) dissolves and enlarges these openings, creating sinkholes. In addition, conduits and caves begin to develop in the bedrock, forming a karst aquifer. This karstic bedrock aquifer, usually in carbonate rock (limestone or dolomite), contains fractures that have been widened by solution to form crevices and conduits through which groundwater flows (Figure 28-1). Eventually the groundwater discharges at springs that flow into surface streams.

Three components are essential for the formation of karst: (1) precipitation, (2) topographic relief, and (3) soluble bedrock at or near the land's surface. The average annual precipitation in Illinois ranges from about 35 to 48 inches (0.9 to 1.2 m), which is adequate for karst formation throughout the state. Most of the karst features occur near major rivers that have dissected the landscape sufficiently to provide the necessary topographic relief. For karst to form, the bedrock needs to be elevated above a stream so that rainwater and snowmelt can migrate downward through crevices and horizontally through fractures or along bedding planes toward stream valleys. As these pathways are enlarged by solution, karst features are formed in the bedrock. In Illinois, carbonate strata occur only in Paleozoic bedrock (from oldest to youngest): Ordovician limestone and dolomites, Silurian dolomites, Devonian limestones, Mississippian limestones, and a few Pennsylvanian limestone beds (see Chapters 7 to 10 for stratigraphic descriptions). Mississippian strata are the dominant host rock for karst, followed by Ordovician and Silurian strata. Karst features are less common in Devonian limestones due to their limited lateral extent and are rare in Pennsylvanian limestones because only a few of them are thick enough for meaningful karst development.

Because most of the state is covered with Quaternary glacial sediment (drift), ranging from almost none to 500 feet (152 m) thick (Piskin and Bergstrom 1975), there is very little exposed carbonate bedrock or exposed karst (without soil cover) in Illinois. Most of the karst consists of a combination of mantled (or covered) karst and fluvio-

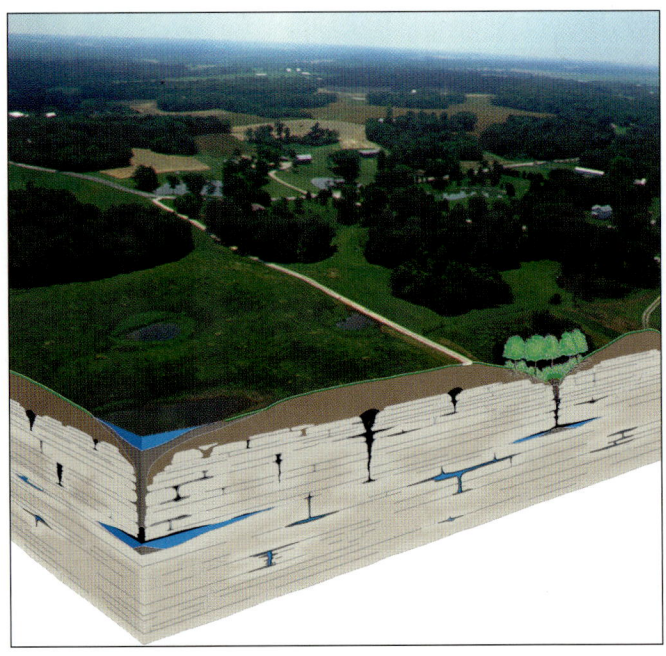

Figure 28-1 Block diagram of southern Illinois karst terrane showing sinkholes at the surface and crevices and bedding planes of the underlying karstic bedrock aquifer (Panno and Weibel 1998).

karst. *Mantled karst* is bedrock that contains karst features and is covered by a veneer or blanket deposit of more recent sediments that subdue the topography of the karst features. *Fluviokarst* is a karst landscape that includes both surface and subsurface drainage basins. Fluviokarst commonly occurs where the bedrock consists of alternating layers of carbonate and sandstone or shale. Surface streams develop on the insoluble rocks (e.g., sandstone and shale) and continue into or across areas where there are soluble carbonate rocks.

Karst Occurrence in Illinois

Karst features occur throughout Illinois, but most are clustered in five karst regions (Figure 28-2) constituting about 10% of the state. The karst regions contain the three most common karst features (caves, sinkholes, and springs) in areas where the bedrock is carbonate rock. Outside of these regions, karst features are not as abundant and are smaller in size. Not coincidentally, four of the five regions (all but the North-Central region) occur within or encompass areas of the state where glacial deposits are absent (Chapter 12, Quaternary Period) and the bedrock is exposed or covered by loess (windblown silt) as well as soil and colluvium. Loess deposits are very permeable, allowing surface waters to move rapidly downward, which tends to facilitate karst formation in the underlying bedrock. Conversely, thick glacial deposits tend to inhibit karst formation in the underlying bedrock because the clay-rich tills impede the downward movement of water.

In some places, glaciation inhibited karst development by infilling sinkholes, injecting conduits with till, or completely destroying the karst system by abrasion (Ford 1983). As glaciers receded, they stimulated karst development in other places by raising hydraulic gradients and infusing fractured carbonate bedrock with cold and very dilute water. The initiation of modern karst in Illinois has not been determined. However, evidence obtained from sequentially deposited cave sediments and stalactites and stalagmites in the Salem Plateau, dated using carbon-14 and uranium-thorium dating techniques, revealed a rate of cave development that Panno et al. (2008) extrapolated back to the time of initiation of core features. That evidence indicated that glacial activities during the Illinois Episode (between 190,000 and 130,000 years before present, coincident with glacial melting) initiated cave formation in the Salem Plateau (Panno et al. 2004) and could have reactivated karst formation elsewhere. Our data show that the caves were formed coincident with glacial melting between 190,000 and 130,000 years ago. That does not preclude the occurrence of karst features in the Salem Plateau prior to that event. Cores of cave features in the Kentucky area are 4.3 million years old (Mammoth Cave). Caves in Missouri are perhaps 1 million years old. The largest caves in southwestern Illinois appear to be about 150,000 years old.

Southern Illinois

The three karst regions in the southern half of the state—Shawnee Hills, Salem Plateau, and Lincoln Hills (Figure 28-2)—are similar in that the bedrock is mostly limestone-dominated Mississippian strata. These karst regions are bordered by large rivers (Mississippi, Ohio, and Wabash) and contain abundant, relatively large karst features. The Upper Mississippian bedrock typically consists of limestone alternating with either shale or sandstone, resulting in fluviokarst occurrence in the three regions, particularly the Shawnee Hills. Middle Mississippian strata comprise a succession of limestone deposits that attain a thickness of up to 300 feet (91.4 m) and are the host rock for Illinois' longest caves (Salem Plateau region) and its largest, most abundant sinkhole areas (Salem Plateau and Shawnee Hills regions). Small areas of exposed karst bedrock occur in both the Shawnee Hills and the Salem Plateau regions, particularly in southern Hardin County and western Monroe County (Figure 28-2 and Figure 2-12 in Chapter 2, Overview). Mantled karst is more common in the glaciated parts of the Salem Plateau and Lincoln Hills regions. Except for the northern boundary of the Lincoln Hills region, all of the boundaries of these karst regions correspond to changes in bedrock from limestone-dominated Mississippian strata to noncarbonate-dominated, mostly Pennsylvanian strata. Mississippian bedrock occurs near the surface north of the Lincoln Hills karst region, but only a few karst features have been identified there, mainly adjacent to the Mississippi River.

Northern Illinois

In the northern half of the state are the Driftless (named after the unglaciated Driftless Area) and North-Central karst regions (Figure 28-2), where the Ordovician and Silurian bedrock generally contains more dolomite than limestone, which makes it harder to dissolve. Although northern karst regions are bordered or transected by major rivers (the Mississippi and Rock Rivers), as they are in the southern part of the state, northern karst features are less common and smaller than those in the south. In the Driftless karst region, the bedrock strata consist of thick units of limestone alternating with thick units of shale or sandstone, resulting in the widespread development of fluviokarst. The glaciated North-Central region consists of mantled karst where the Rock River and its tributaries have eroded the drift, resulting in sinkholes and the exhumation of a buried karst terrane.

The most common karst feature in the Driftless karst area—caves—occurs primarily within Ordovician strata (Heyl et al. 1959). All of the mapped sinkholes occur in areas of Silurian bedrock and are small to moderate in size, less than 100 feet (30.5 m) in diameter. Smaller sinkholes have been reported to occur in the Ordovician strata (Heyl

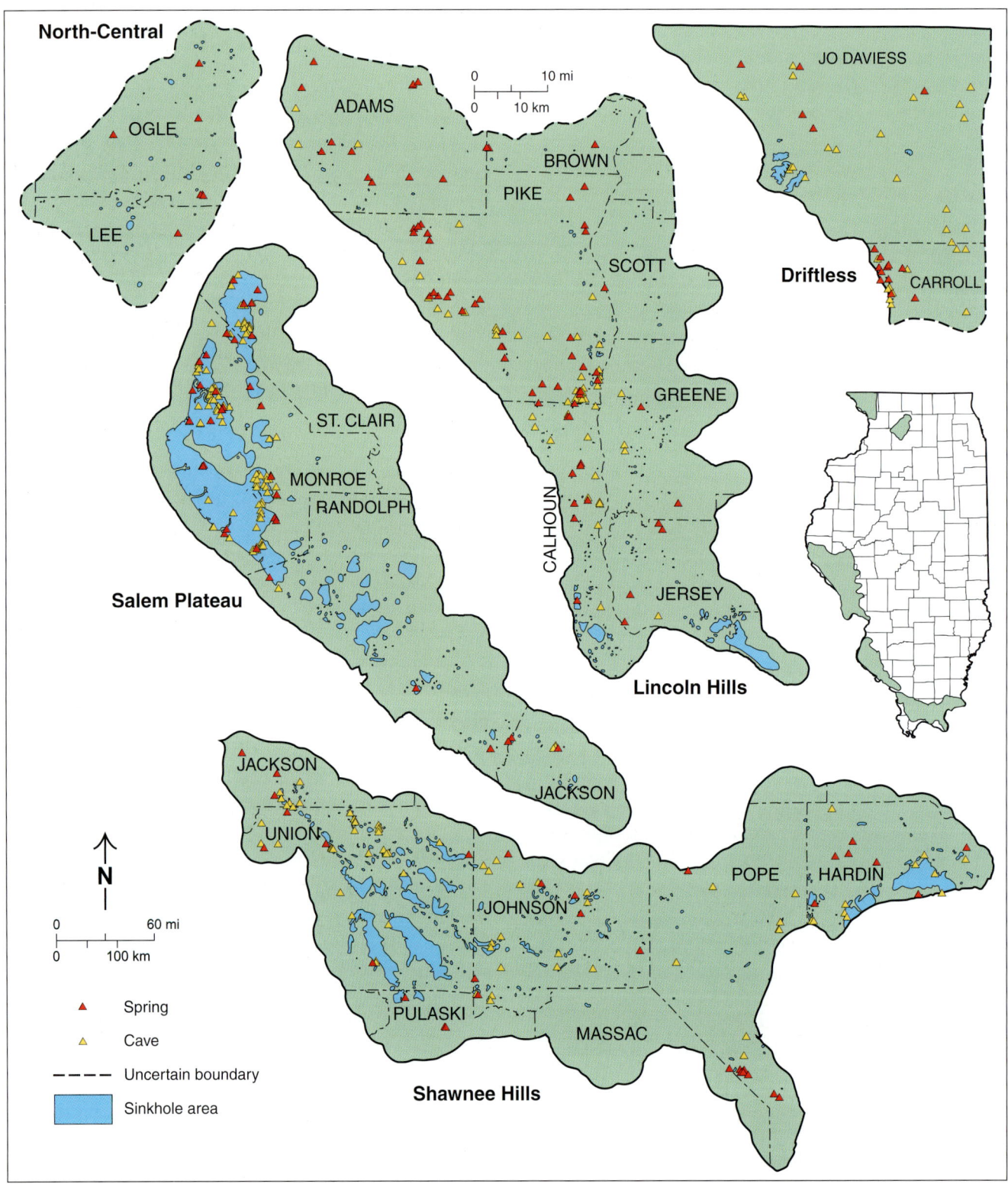

Figure 28-2 Locations of the five karst regions in Illinois as identified by the distribution of sinkholes, caves, and springs. Only the North-Central region occurs within an area of the state covered with glacial drift. Scale bar at lower left applies to North-Central, Driftless, Salem Plateau, and Shawnee Hills regions. The Lincoln Hills region has a different scale bar. (Modified from Panno et al. 1997, Weibel and Panno 1997.)

et al. 1959), and solution-enlarged, vertical joints and crevices are commonly found in road cut and quarry exposures of bedrock (Panno and Luman 2008). Springs are abundant (Trowbridge and Shaw 1916), although few have been mapped (Figure 28-2). Recently, Panno et al. (2009) showed that karst features (sinkholes and springs) are typically found along lineaments seen on LIDAR imagery in Jo Daviess County. These lineaments appear as aligned sinkholes in sediment overlying Silurian age dolomite and as aligned drainage patterns in sediment overlying Ordovician age dolomite. Because the orientation of lineaments, exposed crevices in road cuts and quarries, and major structural features of the state are nearly identical, it is likely that the lineaments are dominant crevices and part of the underlying karst aquifer. In the North-Central region, sinkholes are the dominant karst feature, and caves have been reported. Springs are common (Knappen 1926) but have not been adequately mapped. The boundaries of both karst regions are imprecise because much of northern Illinois is underlain by carbonate-dominated bedrock. Further mapping has identified additional features and increased the areas of these karst regions. For example, recent work in Will and Cook Counties has revealed abundant karst features consisting of cover-collapse sinkholes in thin glacial sediment overlying Silurian dolomite (Panno, unpublished letter report to Illinois Senator A. J. Wilhelmi, September 10, 2008).

PICTURESQUE SCENERY, ENVIRONMENTAL PROBLEMS, AND GEOLOGICAL HAZARDS

Karst terranes are beautiful and are among the most scenic landscapes in the state. They include rolling to rugged hills and sinkholes of many sizes (Figure 28-3), caves with exotic formations, and numerous and large springs.

Figure 28-3 An aerial view of karst terrane in southwestern Illinois, facing west toward Missouri, showing the high density of sinkholes (circular features generally filled with trees). (Photograph by Joel M. Dexter.)

Figure 28-4 (a) Surface-water drainage pattern typical of most of Illinois. (b) Surface drainage patterns in karst terrane are different in that the streams intersect sinkholes (swallow holes) and flow underground (sinking stream). Eventually, the water flowing through the bedrock conduits or caves flows to springs and discharges onto the surface in a low-lying area. (Photographs by Joel M. Dexter.)

Unfortunately, residents who live in karst regions also contend with the environmental problems and geological hazards inherent to these areas. In the east-central part of Illinois, the landscape is a generally flat, glaciated terrane where runoff from rainfall and snowmelt are carried away by streams and rivers to the Mississippi River (Figure 28-4a). The small percentage of precipitation that recharges both drift and bedrock aquifers must first, in most areas, migrate slowly through fine-grained glacial sediments. The slow movement of this water can provide sufficient time, filtration, and the chemical and biological mechanisms to remove surface-borne pollutants from the water. In karst regions, however, sinkholes and conduits in bedrock (Figure 28-4b) allow for runoff from rainfall and snowmelt to move rapidly into and through bedrock aquifers, allowing surface-derived contaminants in the aquifer. In this way, water resources and sensitive subterranean biological environments can become easily degraded. The rapid move-

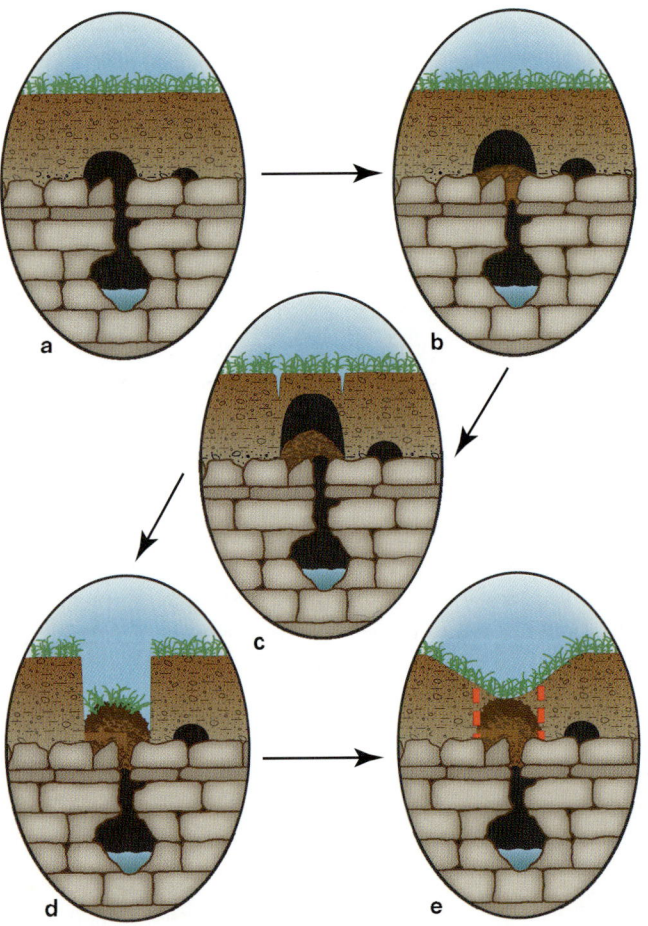

Figure 28-5 The formation of cover-collapse sinkholes begins when soil (unconsolidated material) collapses into a bedrock crevice or cavity **(a)**. The dome-shaped cavity formed at the base of the soil propagates to the surface **(b, c)**, ultimately resulting in a collapse of the surface material **(d)**. Subsequent erosion converts the cylinder-shaped hole into a bowl-shaped depression **(e)** (Panno et al. 2004).

Figure 28-6 This crevice in bedrock is exposed in a road cut along Interstate 39 in northern Illinois just south of Rockford. The crevice is about 6 inches (15 cm) wide and is partially filled with soil that fell in from above. (Photograph by Samuel V. Panno.)

ment of groundwater through the karst system also is responsible for several geological hazards, including sinkhole collapse and flash floods. The karst "plumbing" system has three main components: sinkholes and sinking streams, conduits and caves, and springs, although groundwater in the rock matrix also plays a role.

Sinkholes and Sinking Streams

A sinkhole is a naturally occurring depression in the land surface that, in Illinois, commonly forms as a result of the collapse of earth materials into a crevice or void in the bedrock (Figures 28-5 and 28-6). The area occupied by a typical sinkhole ranges from a few square feet to several hundred square feet. Sinkholes in southern Illinois are typically much larger than those found in northern Illinois. Initially, a sinkhole is a deep, steep-sided cylindrical hole that eventually erodes into a cone- or bowl-shaped depression in the land surface. The movement of groundwater within bedrock, which removes collapsing soil from the crevice and allows sinkhole development, is the most common mechanism for sinkhole formation. Other mechanisms include (1) collapse of a large void or cave in bedrock that creates the type of relatively large and deep sinkholes that occur in a few places in the state (Panno and Weibel 1998), and (2) solution of the bedrock surface along fractures or the intersection of fractures that slowly lowers the bedrock surface and results in slow subsidence of the land surface above it (Waltham et al. 2005). Sinkhole drains vary in size from the typical small conduits 3 to 6 inches (8 to 15 cm) wide to exceptionally large openings (up to tens of feet wide) that can drop precipitously into a relatively large conduit or cave. Sinkholes formed by the collapse of a cave ceiling are very steep-sided and are a danger to both humans and animals walking along the edge. Farmers in southern

Illinois have constructed fences around such sinkholes to protect their livestock.

Sinkholes can have single or multiple drains (e.g., compound sinkholes; when separate sinkholes coalesce). Disruption of the natural vegetation around a sinkhole during cultivation or construction can accelerate surficial erosion, increasing the size of an existing sinkhole and creating deep gullies along its flanks (Figure 28-7). The largest sinkhole in the state, and probably among the largest in North America, is the Big Sink in southeastern Hardin County (Figure 28-8). The sinkhole is located about 2 miles (3 km) north-northwest of the village of Cave-in-Rock. The Big Sink is shaped like a very irregular diamond with a north-south axis of about 2.5 miles (4 km) and an east-west axis of about 1.4 miles (2.3 km). The coalescence of many sinkholes formed this very large, shallow depression. The bottom of the sinkhole is filled with alluvium and is generally wet only seasonally. Historically, however, the sinkhole has been reported to have filled with water to form a large,

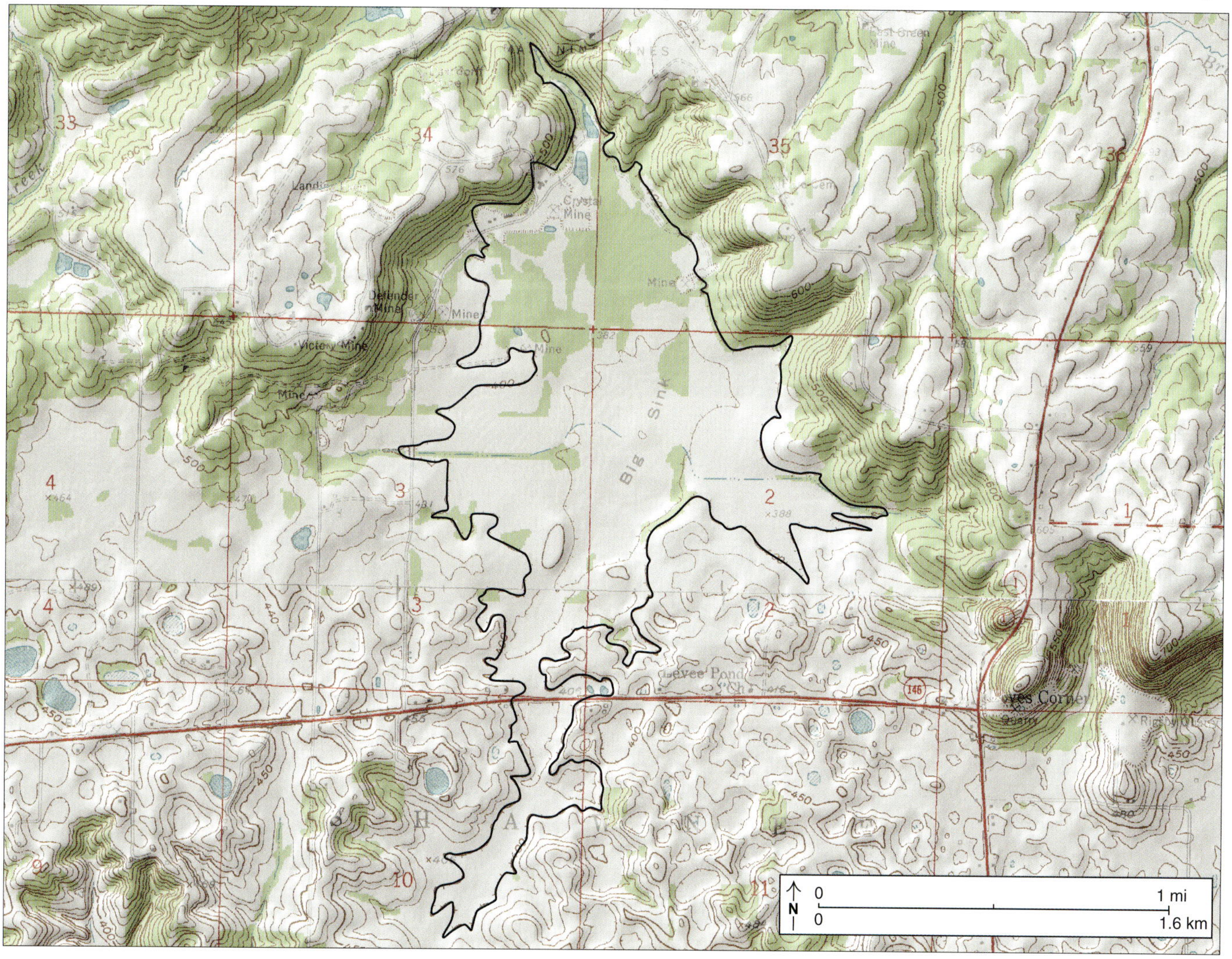

Figure 28-7 Serious erosion and soil loss, as shown beneath a road in the photograph, can occur in the soil adjacent to and within sinkholes. Sinkhole erosion can be accelerated by activities associated with land use (e.g., plowing of croplands, construction of houses and roads). (Photograph by Samuel V. Panno.)

Figure 28-8 Big Sink in Hardin County is the biggest sinkhole in Illinois and possibly in North America. The sinkhole is more than 1 mile (1.6 km) wide and more than 2 miles (3.2 km) long and is defined here as the closed 400-foot (123-m) contour on a 7.5-minute U.S. Geological Survey topographic map.

Figure 28-9 The fenced area in the foreground surrounds a water-filled sinkhole that, along with two other sinkholes, formed in the playground of the Dongola grade school during test pumping of a newly drilled municipal well across the street. After collapse, the sinkhole filled with water. (Photograph by Samuel V. Panno.)

Figure 28-10 Circular sinkhole ponds are picturesque features of karst regions of the state. (Photograph by Samuel V. Panno.)

Figure 28-11 A 15-foot (4.6-m)-diameter sinkhole formed overnight during a particularly heavy rainfall. The white, karstic limestone bedrock of the Ste. Genevieve Limestone is exposed. (Photograph by Samuel V. Panno.)

broad lake more than 1 mile (1.6 km) across that drained within a period of several months to several years (Currier 1944, Weller et al. 1952).

Sinkhole formation is typically a slow process that takes place underground and culminates in the sudden appearance of a circular, steep-sided hole at land surface. This process can be accelerated, however, when the water table is lowered by drought or excessive pumping of water wells, dewatering the sediment overlying a karst bedrock aquifer. Such sinkholes are probably best known in Florida, but recent and similar sinkhole formation in southern Illinois has been reported (Panno et al. 1994). In 1993, a new municipal well was installed in Dongola, in Union County, across the street from a public school. While the new well was being rapidly pumped for testing purposes, three sinkholes suddenly formed in the school's playground (Figure 28-9). Analysis of the local geology and pumping activities and a geophysical study of the site indicated that pumping the well had caused the water table in the unconsolidated sediment to be lowered from just below the surface to a depth of more than 50 feet (15 m). When the water table dropped below the karst bedrock surface, water began to be drained from the overlying sediments, initiating sediment collapse into pre-existing cavities within the limestone. As the collapse propagated upward, sinkholes suddenly appeared at the surface. Consequently, the well was abandoned. After pumping stopped, the water table returned to its previous depth near the surface, and no further sinkhole development occurred.

Surface water runoff readily flows into sinkholes where it either quickly drains to the bedrock aquifer or, if the drain of the sinkhole is plugged by sediment, forms a permanent or ephemeral pond. The risk of contamination from the surface is considerably greater in open sinkholes and much less if the sinkhole is totally or partially plugged. At the extreme, open sinkholes used for disposal of waste are miniature, uncapped, unlined, leaky landfills that introduce contaminants to the groundwater.

Ponded sinkholes (Figure 28-10) allow water to slowly seep through the unconsolidated sediment at the bottom of the pond where the water will undergo physical, chemical, and biological cleansing. Ponded sinkholes are common in the karst regions in the southern half of the state, and, on occasion, can drain catastrophically during or shortly after heavy rainfalls (Figure 28-11). Such an event can occur when the "plug" fails due to flooding of the pond (increased weight of water on the plug) or erosion of the plug from beneath by flowing groundwater. For example, a large sinkhole pond in St. Clair County drained catastrophically following a large rainstorm in the middle 1990s. Its water—along with its abundant and large fish—reportedly

Figure 28-12 Caves can be large and extensive beneath karst terrane. The stream passage shown is in Illinois Caverns in southwestern Illinois. Such cave streams can flood quickly and become dangerous during and following heavy rainfall events. (Photograph by Joel M. Dexter.)

Figure 28-13 The amphipod *Gammarus acherondytes* is a crustacean common to Illinois Caverns (Panno et al. 2004). This tiny (0.2 inch or 0.5 cm) organism lives in the stream and stream gravels and eats bacteria and organic debris brought from the surface. (Photograph by Frank M. Wilhelm.)

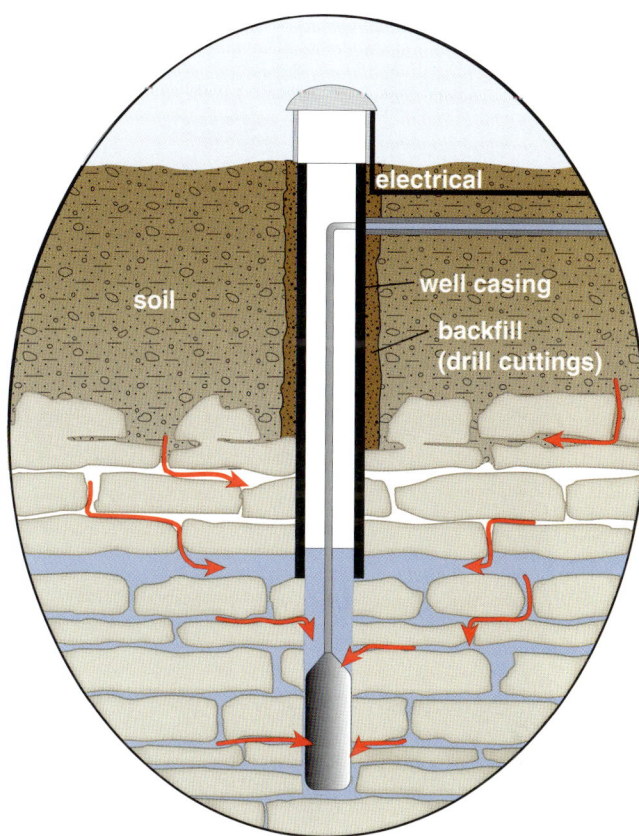

Figure 28-14 Wells constructed in creviced rock (limestone or dolomite) with 30 feet (9 m) or more of overburden (unconsolidated material) before the year 2000 were typically not cased much below the soil zone, thus providing an avenue for contaminated water (arrows) from the surface and the shallow part of the karst aquifer to enter the well and contaminate it.

discharged from a spring about 2 miles (3.2 km) from the pond site.

A worst-case scenario for groundwater contamination is a sinking stream (see Figure 28-4), which occurs when an entire stream flows into a sinkhole (also known as a swallow hole) and drains directly into an underground conduit system. Sinking streams can pose the greatest contamination potential to the karst aquifer because of the large amounts of water involved, particularly if the stream contains septic effluent or flows through a pasture used by livestock.

Fortunately, these streams are rare and are known only in a few areas of western Monroe County and southeastern Hardin County.

Caves and Conduits

After surface water infiltrates and flows into the karst bedrock and becomes groundwater, it begins to move laterally and downward through a system of conduits (including caves, a conduit big enough for an adult human to pass through), creating a karst aquifer. Unlike sinkholes and springs, most of the conduits that permeate karst bedrock

Figure 28-15 Falling Springs near Dupo (southwestern Illinois) is actually a small cave that discharges its water from a 50-foot (15-m)-high limestone bluff. (Photograph by Samuel V. Panno.)

Figure 28-16 This average-size spring, located at the base of a limestone bluff within Camp Vandeventer (owned by the Boy Scouts of America) in southwestern Illinois, is typical of karst springs in Illinois. (Photograph by Samuel V. Panno.)

cannot be seen. The conduit system can be directly studied in caves (Figure 28-12), at bedrock excavations, and at road cuts.

There are several hundred cave openings in Illinois, mostly in the southern half of the state, and some of the caves have miles of passages (Bretz and Harris 1961, Panno and Weibel 2000). Caves are habitats for unique animals, such as bats, salamanders, blind fish, and other unusual aquatic organisms (Figure 28-13). The cave environment is unusual because sunlight does not reach it, except near the entrance, and because the surrounding bedrock's insulating effects cause the temperature and humidity of the cave to be very stable. Many of the organisms that live in caves have special adaptations to the darkness and to the limited food supply. Unfortunately, degradation of the water quality in karst aquifers, of which the cave is part, can upset this delicate cave ecosystem, resulting in the loss of its inhabitants, which can include rare and endangered species.

Karst aquifers in Illinois consist of a network of conduits and, in many places, caves ranging in size from a 0.5-inch (1.3-cm)-wide crevice to a 25-square-foot (2.3-m^2) cave passage. Because of the rapid movement of surface water into the karst aquifer, area residents who draw groundwater from karst aquifers and karst springs for domestic use are at risk of ingesting contaminants—in sharp contrast to parts of the state where the water supply comes from more protected glacial sand and gravel deposits or deeper bedrock aquifers.

Special consideration should be given to water wells that are drilled into bedrock aquifers in karst terranes. Before 2000, water wells drilled in karst terrane were typically cased through the soil and to about 10 feet (3.05 m) into bedrock and then backfilled with drill cuttings (Figure 28-14). Contaminants in karst aquifers are typically stratified; the concentrations of contaminants are greatest closer to the surface and decrease with depth. Consequently, a well drilled before 2000 is generally open from the top of the bedrock to several hundred feet into the bedrock. The well can be fed by both shallow, contaminated groundwater and deeper, uncontaminated groundwater. Shallow groundwater contaminated with septic effluent, animal waste, fertilizer-tainted runoff, and other contaminants can seep into the well through shallow crevices and along bedding planes and contaminate deeper groundwater that would otherwise be isolated from such contaminants. The Illinois water well construction code was changed in August 2000. Today, in areas with creviced bedrock (karst areas) with less than 30 feet (9 m) of overburden, the well casing must extend a minimum of 40 feet (12 m) into firm rock. Further, the space between the drill hole and casing must be pressure grouted (Illinois Administrative Code 2009). In most cases, this technique should ensure that only the deeper, cleaner groundwater is pumped and that the tainted groundwater flowing through caves and conduits of the shallow karst aquifer is excluded.

Caves can contain spectacular speleothems (cave formations) that include stalagmites, stalactites, and flowstone. As a result, cave exploration (spelunking) is a popular recreational activity. Caves can be dangerous places, however, for explorers who are not experienced and properly equipped. Some caves can flood rapidly (potentially to the ceiling) during and following significant rainfall events. Only two Illinois caves are accessible to the public. Cave-in-Rock State Park contains a cave along the banks of the Ohio River. The opening is very large, and the cave is normally dry. Because it only extends for several hundred feet, it is easy to explore without any equipment other than a flashlight. The cave has a colorful history as a shelter for river pirates, counterfeiters, and other bandits during the late 1700s and early 1800s. The other accessible cave is Illinois Caverns, an Illinois Natural Area in Monroe County. With 6 miles (9.7 km) of explored passage and abundant cave formations, Illinois Caverns was once operated as a commercial cave known as "Mammoth Cave of Illinois." A permit and proper caving equipment are required to enter this cave (Panno et al. 2004).

Springs

Karst springs are points of natural discharge or resurgence (Figures 28-15 and 28-16) of the water that has entered and traversed through karst bedrock. Karst springs differ from non-karst springs in that the total output generally is very large, but the flow rate can be extremely variable. Flow rates typically increase quickly during large rainfall events and decrease, although more slowly, after the storms have ended. A spring in the Salem Plateau region, for example, normally flows at a rate of about 1,300 gallons (4,900 L) per minute but, during periods of high rainfall, can flow at more than 160,000 gallons (600,000 L) per minute.

The water quality of most karst springs is poor because the water moves quickly through the karst system and because natural cleansing mechanisms are absent. Despite the high potential for contamination, some karst springs have been used as retreats and resort spas. The Mineral Springs Hotel in Alton opened in 1914 over the site of a karst spring that was considered to have medicinal qualities (Taylor 1999). For the next decade, this ornate hotel was a very popular tourist site, and the water from its spring was bottled and shipped to several cities down the Mississippi River. The hotel closed in 1971, but the building was soon after restored and is currently being used as an antiques mall. A less successful, short-lived resort and village, Vishnu Springs, was built in a secluded valley in McDonough County in the late nineteenth century (Adams 1999). The largest building in the village, the Capital Hotel, was built next to a karst spring that was reported to have "curative powers." The village prospered for several years as the resort attracted crowds, some coming for health reasons and others for the social life. After about 20 years, the spa's popularity waned, and the resort eventually went out of business. Soon after, the village of Vishnu Springs was abandoned (Adams 1999).

Conclusion

Karst terrane results when near-surface bedrock is slowly dissolved by rainwater, snowmelt, and soil water. The resulting enlarged crevices and conduits can create an underground aquifer/drainage system, caves, large springs, and a surface pockmarked with sinkholes. Karst terrane in Illinois predominates in about 10% of the state, primarily in northwestern, western, and southern Illinois. With the beauty of the rolling hills, sinkholes, circular ponds, and picturesque springs comes potential water quality problems and geological hazards (sinkhole collapse and structural damage). People considering relocating into a karst region of Illinois should examine the area closely, keeping its inherent shortcomings in mind.

References

Adams, J., 1999, A lost village, the legacy of Vishnu Springs: Historic Illinois, v. 22, no. 2, p. 3–8.

Bretz, J H., and S. E. Harris, Jr., 1961, Caves of Illinois: Illinois State

Geological Survey, Report of Investigations 215, 87 p.

Currier, L. W., 1944, Geology of the Cave in Rock district, *in* Geological and geophysical survey of fluorspar areas in Hardin County, Illinois: Washington, D.C., United States Geological Survey, Bulletin 942, pt. 1, p. 1–72.

Ford, D. C., 1983, Effects of glaciations upon karst aquifers in Canada: Journal of Hydrology, v. 61, no. 1–3, p. 149–158.

Ford, D. C., and P. W. Williams, 1989, Karst geomorphology and hydrology: Boston, Massachusetts, Unwin Hyman Ltd., 610 p.

Illinois Administrative Code, 2009, Illinois water well construction code, Title 77: Public Health, chapter I, subchapter r, Part 920. http://www.ilga.gov/commission/jcar/admincode/077/07700920sections.html. Accessed October 21, 2009.

Heyl, A. V., A. F. Agnew, C. H. Behre, Jr., E. L. Lyons, and A. E. Flint, 1959, The geology of the upper Mississippi Valley zinc-lead district: Washington, D.C., United States Geological Survey, Professional Paper 309, 310 p.

Knappen, R. S., 1926, Geology and mineral resources of the Dixon Quadrangle: Illinois State Geological Survey, Bulletin 49, 141 p.

Panno, S. V., B. B. Curry, H. Wang, and K. C. Hackley, 2008, Glaciation-induced speleogenesis of Illinois' longest caves: Evidence based on cave sediments and speleothems: Quaternary Research, v. 61, no. 3, p. 301–313.

Panno, S. V., S. E. Greenberg, C. P. Weibel, and P. K. Gillespie, 2004, Guide to the Illinois Caverns State Natural Area: Illinois State Geological Survey, GeoScience Education Series 19, 108 p.

Panno, S. V., and D. E. Luman, 2008, Assessment of the geology and hydrogeology of two sites for a proposed large dairy facility in Jo Daviess County near Nora, IL: Illinois State Geological Survey, Open File Series 2008-2, 32 p.

Panno, S. V., D. E. Luman, T. H. Larson, and S. J. Taylor, 2009, Identification and characterization of karst terrane in Illinois' unglaciated region: Results of LIDAR imagery and ground penetrating radar in Jo Daviess County, northwestern Illinois. http://www.isgs.illinois.edu/research/karst-jd.shtml. Accessed October 21, 2009.

Panno, S. V., and C. P. Weibel, 1998, Karst landscapes of Illinois: Dissolving bedrock and collapsing soil: Illinois State Geological Survey, Geobit 7, 4 p.

Panno, S. V., and C. P. Weibel, 2000, Caves in Illinois: Our subterranean landscape: Illinois State Geological Survey, Geobit 11, 4 p.

Panno, S. V., C. P. Wiebel (sic), P. C. Heigold, and P. C. Reed, 1994, Formation of regolith-collapse sinkholes in southern Illinois: Interpretation and identification of associated buried cavities: Environmental Geology, v. 23, no. 3, p. 214–220.

Panno, S. V., C. P. Weibel, and W. B. Li, 1997, Karst regions of Illinois: Illinois State Geological Survey, Open File Series 1997-2, 42 p.

Piskin, K., and R. E. Bergstrom, 1975, Glacial drift in Illinois: Thickness and character: Illinois State Geological Survey, Circular 490, 35 p.

Taylor, T., 1999, Haunted Alton: History and hauntings of the riverbend region: Alton, Illinois, Whitechapel Press, 272 p.

Trowbridge, A. C., and E. W. Shaw, 1916, Geology and geography of the Galena and Elizabeth Quadrangles: Illinois State Geological Survey, Bulletin 26, p. 13–171.

Waltham, T., F. Bell, and M. Culshaw, 2005, Sinkholes and subsidence: Karst and cavernous rocks in engineering and construction: Chichester, United Kingdom, Praxis Publishing, 382 p.

Weibel, C. P., and S. V. Panno, 1997. Karst terrains and carbonate bedrock of Illinois: Illinois State Geological Survey, Illinois Map 8, 1:500,000.

Weller, J. M., R. M. Grogan, and F. E. Tippie, 1952, Geology of the fluorspar deposits of Illinois: Illinois State Geological Survey, Bulletin 76, 147 p.

Pollution of Groundwater and Surface Water

29

Samuel V. Panno, Richard C. Berg, and Walton R. Kelly

INTRODUCTION

Numerous substances may enter surface and groundwater and make the water unsuitable for drinking or cause ecological damage. Common contaminants include salts, suspended solids, chloride, nitrate, arsenic, heavy metals, pathogenic organisms, and a wide variety of organic chemicals including pesticides, petroleum compounds, and solvents (U.S. Environmental Protection Agency 2004a). Contamination sources can be natural (e.g., arsenic dissolved from aquifer materials) or human, in which case the contamination is referred to as pollution. Pollutants can enter the environment and natural waters via point sources, where a single point can be identified, or non-point sources, where the pollution occurs over a large area. This chapter discusses pollutants in Illinois arising from human sources, focusing on those that pollute groundwater resources. However, because all water resources are inextricably linked in the hydrologic cycle (see Chapter 18, Aquifers), surface water and atmospheric pollution are also discussed.

EXTENT OF WATER POLLUTION IN ILLINOIS

In Illinois, most contamination of aquifers comes from natural sources, such as elevated arsenic concentrations in sand and gravel aquifers in central Illinois, the presence of radium and barium in the deep bedrock aquifers of northern Illinois, and elevated concentrations of sodium and chloride in both shallow and deep aquifers from localized seeps or regional movement of basin brines into fresh groundwater. Shallow unconfined aquifers are the most susceptible to human sources of pollution. These shallow aquifers are generally sand and gravel but include shallow bedrock aquifers. The Illinois Environmental Protection Agency classifies the state's aquifers into four main categories: (1) sand and gravel, (2) shallow bedrock down to 500 feet (152.4 m), (3) deep bedrock, and (4) mixed sand and gravel and bedrock. The Illinois Environmental Protection Agency (2002a) reported that groundwater quality was unchanging or slightly improving, based on decreasing numbers of contaminant detections in three of the four major aquifer groups. Based on data collected from 1990 through 2000, only wells in shallow bedrock aquifers showed a slight upward trend in the number of contaminant detections.

The susceptibility of shallow groundwater to pollution depends on the geological and hydrologic conditions. The most well-protected shallow aquifers in Illinois are primarily in west-central and southern Illinois where fine-grained glacial deposits overlie shale. These deposits and rocks have low permeability, and liquids flow through them very slowly, essentially trapping them for many thousands to tens of thousands of years. During that time, toxic chemicals can degrade into less harmful compounds and become dispersed throughout these sediments and rocks. Disposal of hazardous waste should be avoided in areas where sand and gravel or bedrock aquifers are within 300 feet (91.4 m) of land surface and thus lack this long-term filter. Such shallow aquifers exist mainly in northern Illinois and above deep glacially filled bedrock valleys in central and east-central Illinois (Berg et al. 1989).

Table 29-1 lists sources of contamination that have impaired streams in Illinois. Most are point sources, such as hydromodification (channelization), municipal sources, and resource extraction. The most important source, agriculture, is primarily a non-point source.

Table 29-1 Sources of use impairment in Illinois streams based on data collected through 2000 (Illinois Environmental Protection Agency 2002a).

Source	Impaired miles
Agriculture	4,071
Hydromodification	2,013
Municipal point source	1,566
Resource extraction	1,079
Urban runoff and storm sewers	1,004
Habitat modification (other than hydromodification)	760
Combined sewer overflow	368
Industrial point source	348
Contaminated sediments	325
Construction	238
Natural sources	127
Highway maintenance and runoff	52
Land disposal	28
Collection system failure	26
Wildcat sewer	18
Atmospheric deposition	7
Recreation activities	7

POINT SOURCES OF POLLUTION

Point sources are specific, identifiable sites where contaminants are released to the environment. Common point sources include landfills, chemical spills, pipeline breaks, military bases, dry cleaners, and leaking underground storage tanks (Table 29-2). Pollutants from point sources commonly result in the most acute environmental problems, but the scope of the problem is usually fairly well defined. Once a point source is identified, a site assessment usually can be done to determine the extent of pollution and the direction of groundwater movement. Cleanup operations can then begin. However, cleanup efforts can be quite costly, depending on the type and quantity of the pollutant that must be removed and the geological and hydrologic conditions of the site (Killey and Berg 2004).

Hazardous Waste Sites

A hazardous waste generator is an industrial establishment that produces at least 2,200 pounds (about 1,000 kg) of hazardous waste or accumulates more than 220 pounds (100 kg) of contaminated cleanup material (U.S. Environmental Protection Agency 2001b). Regulated wastes (U.S. Environmental Protection Agency 2001a) include old chemicals, inorganic liquids (e.g, caustic solutions with metals and cyanides), organic liquids (e.g., waste oil), solids (e.g., ash and slag from incineration), organic solids (e.g., pesticides), sludges (e.g., degreasing sludge with metal filings), and gases.

Illinois is the third leading state in production of hazardous wastes (U.S. Environmental Protection Agency 2001b), generating about 3 million tons (2.7 million metric tonnes) annually, primarily from more than 1,000 large-quantity individual generators (fifth in the United States). Illinois companies manage more than 400,000 tons (363,000 tonnes) of hazardous waste (twelfth in the United States) at 86 facilities. Illinois also imports about 185,000 tons (168,000 tonnes) and exports about 228,000 tons (207,000 tonnes) of hazardous waste annually, ranking eighth in the nation in exports.

The U.S. Environmental Protection Agency has a Toxics Release Inventory program that collects data on releases of nearly 650 toxic chemicals and chemical categories in the air, surface water, and groundwater from manufacturing, metal and coal mining industries, electric utilities, and commercial hazardous waste treatment facilities (U.S. Environmental Protection Agency 2003b). Illinois ranks twelfth among all states for toxic releases. In 2001, almost 68,500 tons (62,000 tonnes) of toxic releases occurred from 1,376 facilities (Figure 29-1). Slightly more than half of these releases were air emissions, about 4,000 tons (3,600 tonnes) were discharged to surface water, about 1.5 tons (1.4 tonnes) were injected into underground wells, and about 22,500 tons (20,400 tonnes) were released to landfills managed and monitored according to the Resource Conservation and Recovery Act of 1976 and other on-site land areas.

Superfund Sites

Enactment of the Comprehensive Environmental Response, Compensation, and Liability Act (CERCLA) occurred late in 1980. Since that time, the U.S. Environmental Protection Agency has designated Superfund, or CERCLA, hazardous waste sites as high priority for clean-

Table 29-2 Major sources of groundwater pollution and key pollutants (Illinois Environmental Protection Agency 2002a).

Pollution source
Point source
Agricultural chemical facilities: *pesticides, nitrate*
Storage tanks, above ground: *petroleum compounds*
Storage tanks, below ground: *petroleum compounds*
Surface impoundments: *nitrate*
Waste piles: *nitrate*
Manufacturing and repair shops: *halogenated solvents, petroleum compounds*
Septic systems: *nitrate*
Animal feedlots
Drainage wells
Material stockpiles
Waste tailings
Injection wells
Landfills
Salt storage
Hazardous waste generators
Hazardous waste sites
Industrial facilities
Material transfer operations
Spills
Transportation of materials
Mining
Pipelines and sewer lines
Non-point source
Fertilizer applications: *pesticides, nitrate*
Pesticide applications: *pesticides, nitrate*
Septic systems: *nitrate*
Irrigation practices
Land application of waste
Mining and mine drainage
Road salting
Salt-water intrusion
Urban runoff

POLLUTION OF NATURAL WATERS

up and has imposed liability for their cleanup on responsible parties. The sites also are known as National Priority List (NPL) sites. As of 2004, 40 of these sites were located in Illinois, and 5 additional sites were being proposed as Superfund sites (Figure 29-2) (U.S. Environmental Protection Agency 2003a). Most are located in the northern quarter of Illinois. These sites contain a wide range of toxic substances including heavy metal sludges, oil, coal tar, and numerous organic compounds.

Radioactive Waste Sites

Radioactive wastes are by-products of nuclear power generation and other nuclear material uses. High-level radioactive waste results from the processing of spent nuclear fuel at nuclear reactors and contains fission products (i.e., radionuclides and "daughter" products produced by atom splitting), uranium, and plutonium. In Illinois, there are 11 operating commercial reactors at six sites. Three other reactors (Illinois Emergency Management Agency 2003) and one commercial low-level radioactive waste storage site (Sheffield, Bureau County) are closed, and one commercial high-level radioactive waste storage facility is full (General Electric site near the Dresden Nuclear Power Station, Grundy County).

Low-level radioactive wastes are generated from various activities at universities, medical facilities, and nuclear power plants. Such wastes include protective clothing, used rags, and luminous dials. During the early 1990s, Illinois produced more than 150,000 cubic feet (4,250 m^3) of low-level radioactive waste annually, about 88% of which was generated by industry or electric power utilities (League of Women Voters 1993). However, a four-year moving average of waste generation ending in 2002 shows that about 400 generators in Illinois now produce only about 43,000 cubic feet (1,200 m^3) of low-level radioactive waste (Illinois Emergency Management Agency 2002). The waste being shipped to commercial disposal facilities has also declined from about 135,000 cubic feet (3,800 m^3) in 1989 to about 65,000 cubic feet (1,800 m^3) in 1998 (Fuchs 1999).

During the late 1980s and early 1990s, the Illinois state government attempted to site a low-level radioactive waste disposal facility near Martinsville. The selected site was found to be unacceptable, mainly because sand and gravel aquifers were discovered in a narrow buried bedrock valley beneath the proposed site. Consequently, waste has continued to be stored at generator sites or has been shipped to disposal facilities in South Carolina and Utah. A centralized storage facility in Illinois is not under consideration at this time (Illinois Department of Nuclear Safety 2002).

Nonhazardous Municipal Solid Waste Landfills

Municipal solid waste is the garbage disposed of in landfills by households, commercial establishments, industry, and other institutions. The waste, generated at a rate of about 4.5 pounds (2 kg) per day per person, typically consists, by weight, of 36% paper, 12% yard waste, 11% food scraps, 11% plastic, 7.9% metals, 7.1% rubber, leather, and textiles, 5.5% glass, 5.7% wood, and 3.4% other (U.S. Environmental Protection Agency 2004b). Since the 1960s, the

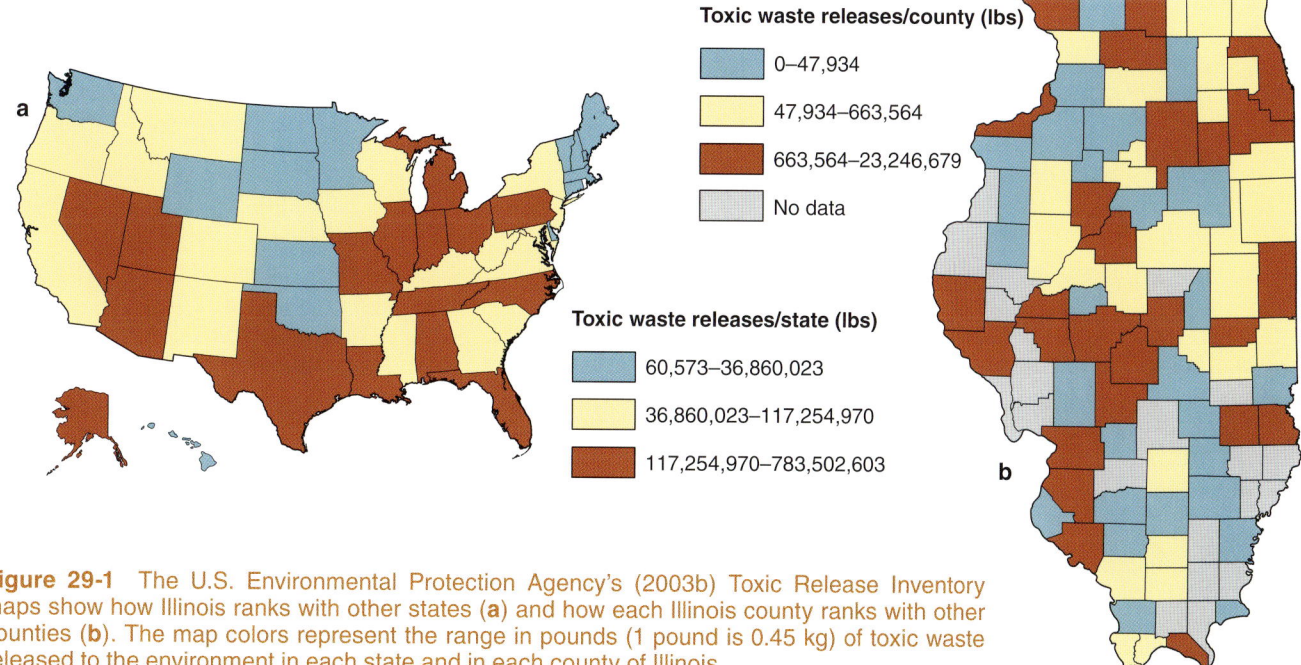

Figure 29-1 The U.S. Environmental Protection Agency's (2003b) Toxic Release Inventory maps show how Illinois ranks with other states (**a**) and how each Illinois county ranks with other counties (**b**). The map colors represent the range in pounds (1 pound is 0.45 kg) of toxic waste released to the environment in each state and in each county of Illinois.

siting of new municipal landfills and the pollution problems from existing landfill sites have been a statewide concern (Hughes 1967, Cartwright and Sherman 1969, Hughes et al. 1971, Hensel et al. 1991), particularly in areas where sand and gravel or bedrock aquifers are within 100 feet (30.5 m) of the land surface (Curry et al. 1997). Stringent regulations have been adopted for siting and managing municipal landfills, thereby reducing their pollution potential and the level of concern (Richard P. Cobb, Illinois Environmental Protection Agency, personal communication 2004).

As of 2005, 51 active landfills in Illinois received waste; their remaining capacity was estimated at about 14 years (Figure 29-3) (Illinois Environmental Protection Agency 2002b). However, in the Chicago metropolitan area and east-central Illinois, the existing landfills are estimated to reach capacity by about 2010. There are more than 3,400 abandoned landfills in Illinois (Figure 29-4). Cook County has the most, with more than 200. In 1999, the Illinois Environmental Protection Agency (2000) identified and designated 33 of the abandoned landfills for cleanup (Figure 29-5).

Because landfills may cause environmental degradation, statewide mapping (Berg and Kempton 1984; Chapter 18, Aquifers) was developed to help determine areas most vulnerable to pollution from municipal landfills. Susceptible areas are most prevalent in northern Illinois where sand and gravel or bedrock aquifers are within 50 feet (15 m)

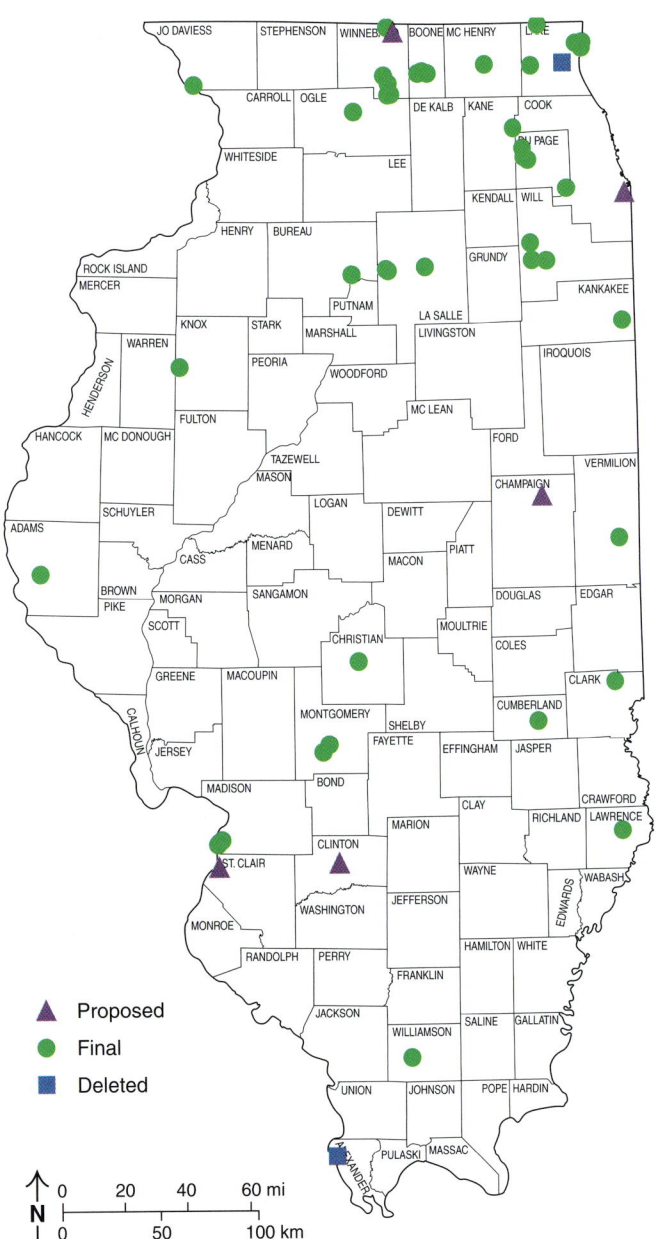

Figure 29-2 The location of Superfund (CERCLA) sites in Illinois (U.S. Environmental Protection Agency 2003a).

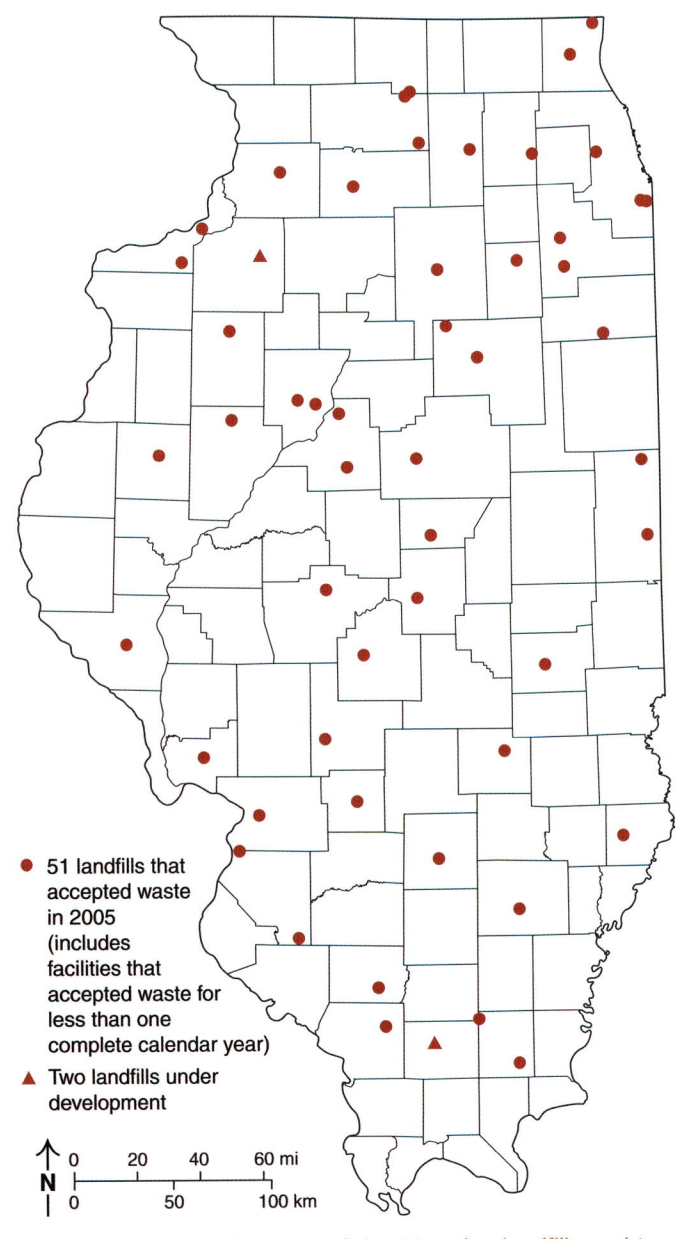

Figure 29-3 Location map of the 51 active landfills and two landfills actively being developed in Illinois (Illinois Environmental Protection Agency 2002b).

of the land surface and along Illinois' major rivers, where sand and gravel have been deposited in floodplains and on terraces (Figure 29-6).

Solid Waste Transfer Stations

Solid waste transfer stations collect wastes locally, consolidate it, and prepare it for shipment to landfills. Some transfer stations also separate materials for recycling. Transfer stations can be point sources of pollution if the waste is not handled properly and is allowed to spill, drain, or accumulate for long periods in regions of sensitive geology with aquifers at or near the land surface. The 80 active transfer stations in Illinois handle about 4.9 million tons (4.4 million tonnes) of waste; 54 of these facilities, handling about 4.7 million tons (4.26 million tonnes), are located in the Chicago metropolitan area.

Concentrated Animal Feeding Operations

Concentrated animal feeding operations (CAFOs) house dairy cattle, beef cattle, hogs, turkeys, or chickens in large numbers in a relatively small area. The U.S. Environmental Protection Agency (2004c) defines a CAFO for cattle at 700 mature dairy cows, 1,000 veal calves, 1,000 cattle other than the above two categories, 2,500 swine (≥ 55 pounds [≥ 25 kg]), 10,000 swine (<55 pounds [<25 kg]), 500 horses, 10,000 sheep or lambs, 55,000 turkeys,

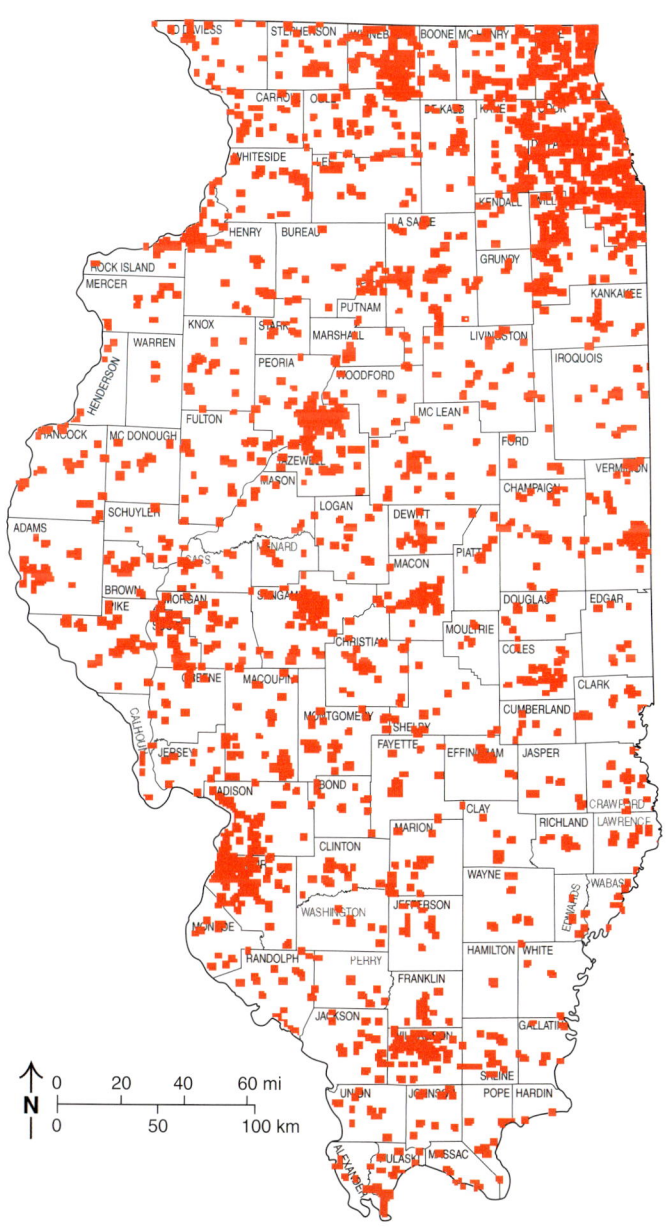

Figure 29-4 Location map of Illinois landfill sites including about 3,400 abandoned sites (Illinois Waste Management and Research Center 1997).

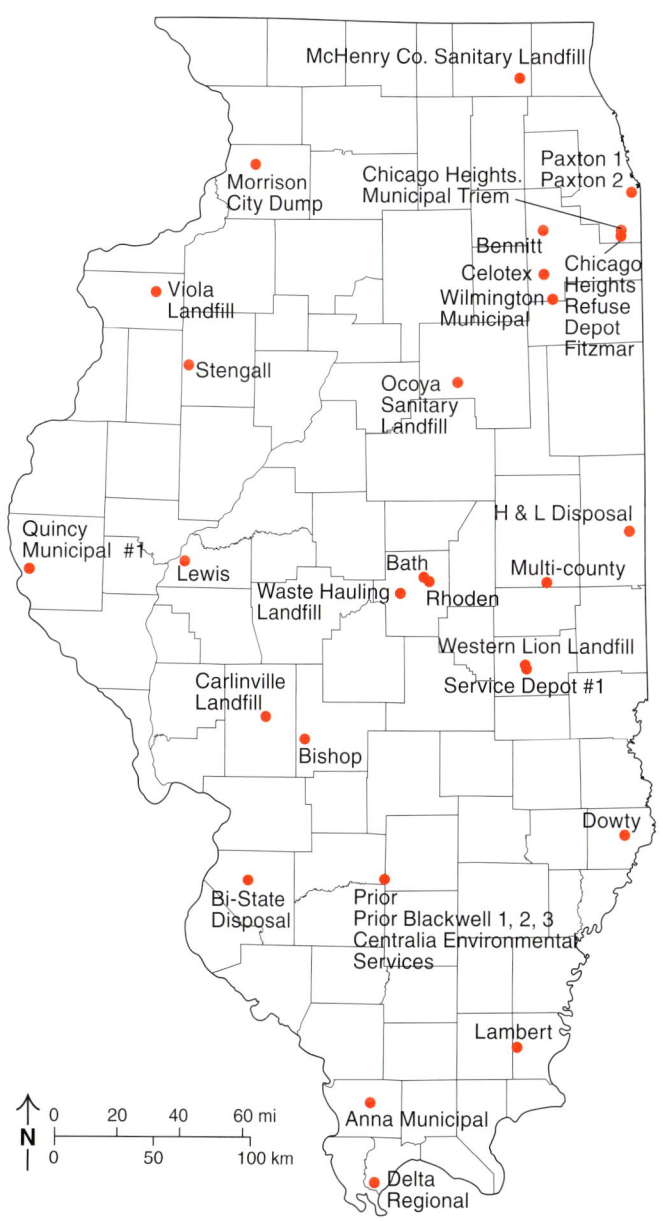

Figure 29-5 Location map of Illinois' 33 abandoned landfill sites that have been selected for cleanup (Illinois Environmental Protection Agency 2000).

30,000 to 125,000 chickens, and 5,000 to 30,000 ducks depending on breeds and manure handling system. Illinois currently ranks sixth in the nation for production of live animals and meat (U.S. Department of Agriculture 2004).

Animal waste contains very large concentrations of nutrients, including nitrogen, phosphorous, and potassium and is typically applied as fertilizer. Potential pollutants in the waste may include chloride, pathogenic organisms, pharmaceuticals, and heavy metals. Odor is an unavoidable by-product of raising livestock and is a concern for those living in the vicinity of a CAFO. Livestock waste is typically stored in lined lagoons, in deep pits below the animal facilities, or in surface slurry structures prior to being spread onto croplands.

Krapac et al. (2002) examined two hog finishing facilities in Illinois, one built on clay-rich glacial till and shale and the other on fractured sandstone. The facilities did not adversely affect groundwater underlying either facility with inorganic contaminants despite waste pit leakage. However, additional studies of facilities with waste lagoons and adjacent manure-fertilized croplands showed that fecal bacteria were present in shallow aquifers in the vicinity of the CAFOs either because of leakage from the facilities or application of waste as fertilizer on nearby fields (Krapac et al. 1998, 2002). Krapac et al. (1998, 2002) thought that the movement of fecal bacteria from these facilities could pose the greatest threat to human health, especially in areas where the geology did not protect underlying aquifers.

NON-POINT SOURCES OF POLLUTION

Non-point sources of pollution occur over wide geographic areas where pollutants enter the groundwater with relative uniformity and without specific, easily identified entry points. Non-point sources include chemicals and fertilizers applied to agricultural fields and residential lawns; sewage sludge, septage, and manure spread onto the land surface; and salt applications for road deicing. If multiple point sources cannot be identified easily (e.g., septic effluent emanating from a large residential area), then the pollution is considered non-point source. Because non-point pollution is typically widespread, it is difficult to remediate in groundwater or surface water. As with point sources of pollutants, the effects of non-point source pollutants depend on the underlying geology and hydrology of the area.

In Illinois, land use is dominated by agricultural activities. There are 76,000 farms covering about 28 million acres (113,312 km^2) of land, or about three quarters of the total land area of the state (Figure 29-7) (Illinois Department of Agriculture et al. 2002). The state ranks first in the production of soybeans and second in corn (Figure 29-7 and 29-8), ranks fourth in the production of hogs, and sixth in total agricultural production (Stuffaboutstates.com 2004). Not unexpectedly, agriculturally related environmental problems, including pollution of surface water and groundwater by nitrate and pesticides, are found throughout the state (McKenna et al. 1990).

Almost half of the state is in the Illinois River watershed; almost all of the entire lower Illinois River basin is in agricultural production, predominantly corn and soybeans (Warner 1998). The Illinois River discharges into the Mississippi River, which, in turn, empties into the Gulf of

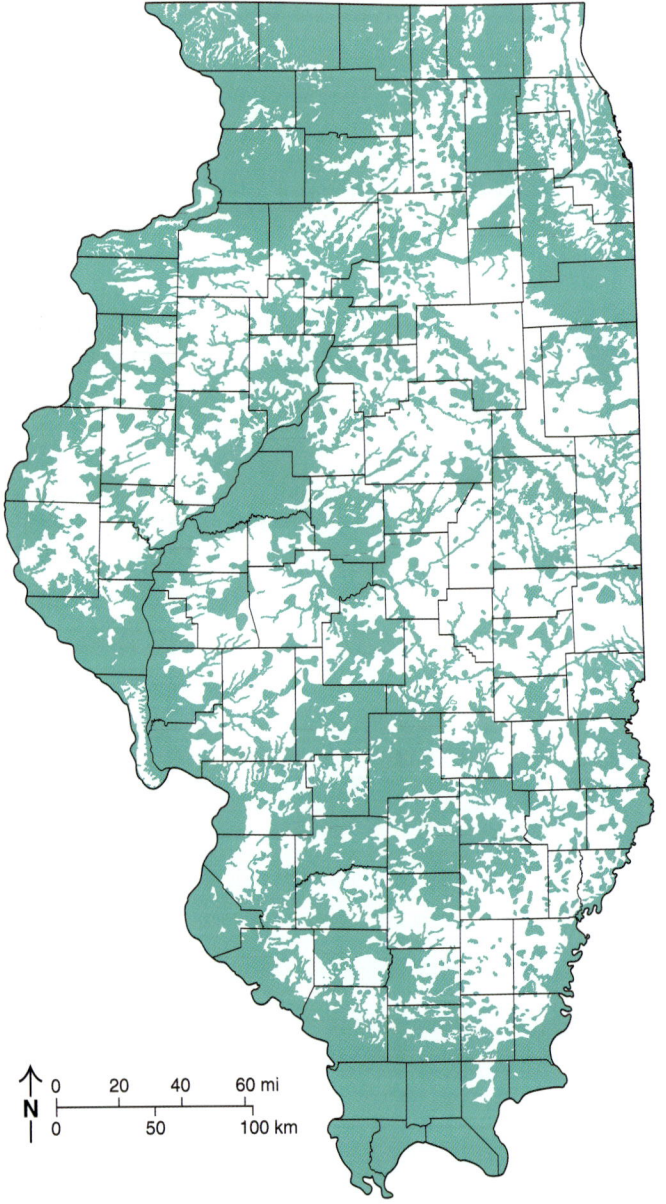

Figure 29-6 Aquifers in Illinois located within less than 50 feet (15 m) of land surface. These aquifers are particularly vulnerable to contamination from sources on the surface (Berg and Kempton 1984).

Mexico. Since the 1980s, a seasonal hypoxic (oxygen-poor) zone has been developing annually in the bottom waters of the Gulf of Mexico, threatening aquatic life and commercial fisheries. The generation of this hypoxic zone has been primarily attributed to the discharge of large amounts of nitrogen and other nutrients from the Mississippi River basin (Rabalais et al. 2002), much of it from Illinois and other midwestern states (David and Gentry 2000).

Despite environmental problems associated with agriculture, successful conservation measures and best management practices to reduce erosion and minimize groundwater contamination have been used extensively in Illinois' rural agricultural areas. For example, the Illinois Farm Bureau (Erickson 2000) and other agricultural institutions encourage the development of programs to help farmers implement best management practices to improve water quality, promote development of conservation buffers, develop educational campaigns regarding fertilizer nitrogen application, and advocate incentives for landowners to seal old wells.

These and other program efforts have reduced the amounts of agricultural chemicals in Illinois' lakes, streams, rivers, and groundwater. Consequently, Illinois farmers have saved millions of tons of soil that would have been lost to erosion using historic farming practices. Soil erosion of croplands has decreased by over one-third between 1982 and 1997 (U.S. Department of Agriculture 1997). Additionally, several thousand abandoned wells, mostly in rural areas, have been sealed during the past ten years (Figure 29-9), thereby eliminating conduits for pollution transport to underlying aquifers (Illinois Department of Natural Resources 2003).

Important Types of Pollutants in Illinois

Nitrogen

Nitrogen is a very common and important element and constitutes a significant portion of all living things. The most important natural source of nitrogen is natural organic matter, especially soil organic matter, and humans have greatly altered the Earth's nitrogen cycle, primarily through the production and application of synthetic fertilizers. Other important sources include animal and human waste and industrial discharge. In Illinois, nitrogen is predominantly a non-point pollutant found in agricultural regions.

More than 150 years of farming activities in Illinois have resulted in the loss of just less than half of the original nitrogen and organic matter of soils due to erosion and oxidation of organic matter (Czapar and Simmons 2002).

To replace what has been lost and to increase crop yields, nitrogen is applied to cropland annually in several forms: injected into the soil as anhydrous ammonia (the dominant mode of application in Illinois) and applied on the surface as dry ammonium nitrate pellets, dry urea, dry ammonium sulfate, a solution of urea and ammonium nitrate, and manure from livestock. Some of the applied ammonia is lost to the atmosphere by volatilization (Figure 29-10). The remaining ammonia adsorbs strongly to clay particles in the soil, where it is oxidized by bacteria to nitrate (a process called nitrification), which is taken up by plants as a nutrient. Unfortunately, nitrate is mobile in the environ-

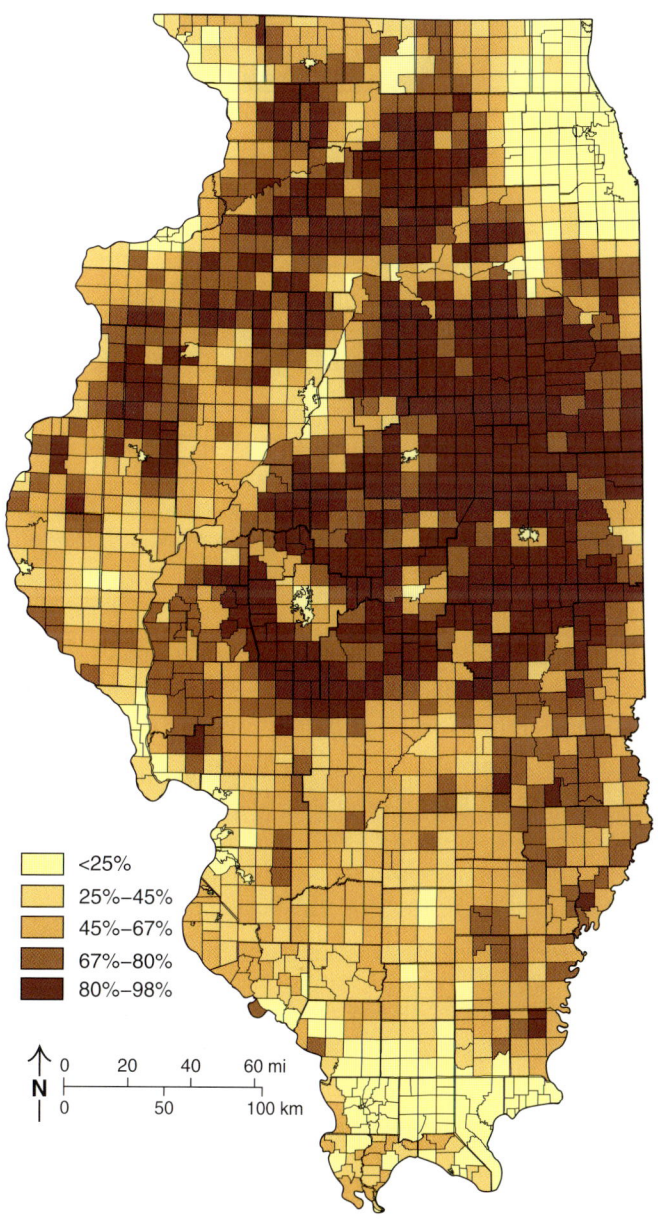

Figure 29-7 Townships of Illinois showing the percentage of land planted with corn and soybeans in 1977. Those areas with the largest percentages of corn and soybeans coincide with the potential for nitrate and pesticide pollution of shallow wells (McKenna et al. 1990).

Figure 29-8 Soybean fields and cornfields in east-central Illinois use pesticides and, in the case of corn, nitrogen fertilizer. Residual concentrations of these agrichemicals are often found in groundwater in underlying aquifers. (Photographs by Samuel V. Panno.)

ment, easily dissolving in water and being transported to groundwater and/or surface water bodies. In Illinois, nitrogen contributions to surface water bodies can be very high because a large percentage of farmland is artificially drained, which allows water with high concentrations of nitrate and other nutrients to migrate rapidly from the soil zone to surface water bodies without allowing for natural nitrate-removing reactions to occur (e.g., denitrification, by which bacteria convert nitrate to innocuous nitrogen gas). Several studies in Illinois have shown that as much as one third of the nitrogen applied to croplands can be lost annually to surface water (David et al. 1997, Keefer and Demissie 1999, Panno et al. 2003). Nitrate in surface water is primarily an ecological concern, promoting noxious algal blooms and suppressing dissolved oxygen levels.

Nitrate in groundwater may be more of a human health concern. Nitrate in drinking water in excess of the U.S. Environmental Protection Agency's drinking water standard (10 mg/L [parts per million] as nitrogen) may be toxic to babies, causing decreased oxygen concentrations in the blood. Elevated nitrate may also be responsible for increases in stomach cancer in adults (O'Riordan and Bentham 1993), increased risk of non-Hodgkin's lymphoma, and spontaneous abortions (Nolan 2001). Unfortunately, nitrate cannot be removed easily from water. Well owners whose water has nitrate concentrations exceeding the U.S. Environmental Protection Agency's recommended limit have few cost-effective options for nitrate removal. The main recommendation is that well owners not use the water for cooking or drinking, but purchase bottled water for consumption. If possible, connecting to a public water utility is usually the best option.

Leaching of agricultural field soils in Illinois' humid climate (Tisdale et al. 1993) is a major source of nitrate transport to groundwater and surface water (Figure 29-10). Concentrations of nitrate from naturally occurring sources are typically less than 2 mg/L (as nitrogen) in shallow groundwater (Panno et al. 2003). A study of nitrate contamination in private shallow wells throughout Illinois (Goetsch et al. 1992) showed that about one-third of the wells contained elevated concentrations of nitrate (above the background value of 2.0 mg/L). About 10% of wells had nitrate concentrations that exceeded the U.S. Environmental Protection Agency's limit for drinking water. Because most sources of nitrate are near the land surface, shallower wells were more likely to be contaminated with nitrate than the deeper wells.

Keefer (1995) combined depth-to-uppermost aquifer information with the rate at which water moves through the soil profile (upper 5 feet [1.5 m] of land surface) and soil drainage classes to derive a map showing potential for pollution of shallow aquifers by nitrate leaching (Figure 20-2). Those areas most sensitive to pollution by nitrate leaching occur mainly in northern Illinois from

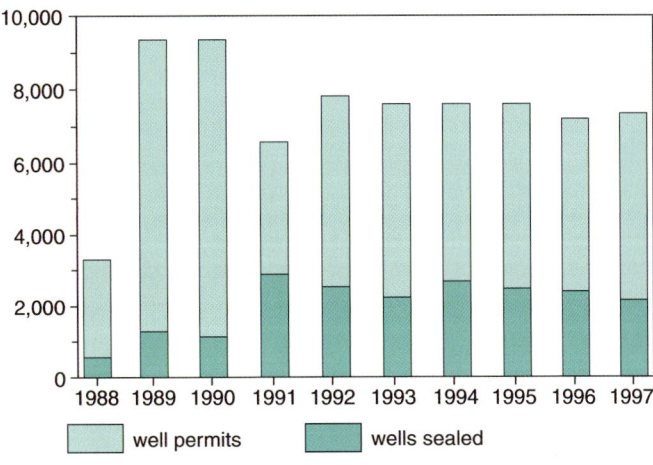

Figure 29-9 Number of new well permits and sealed abandoned wells in Illinois between 1988 and 1997 (Illinois Department of Natural Resources 2003).

POLLUTION OF NATURAL WATERS

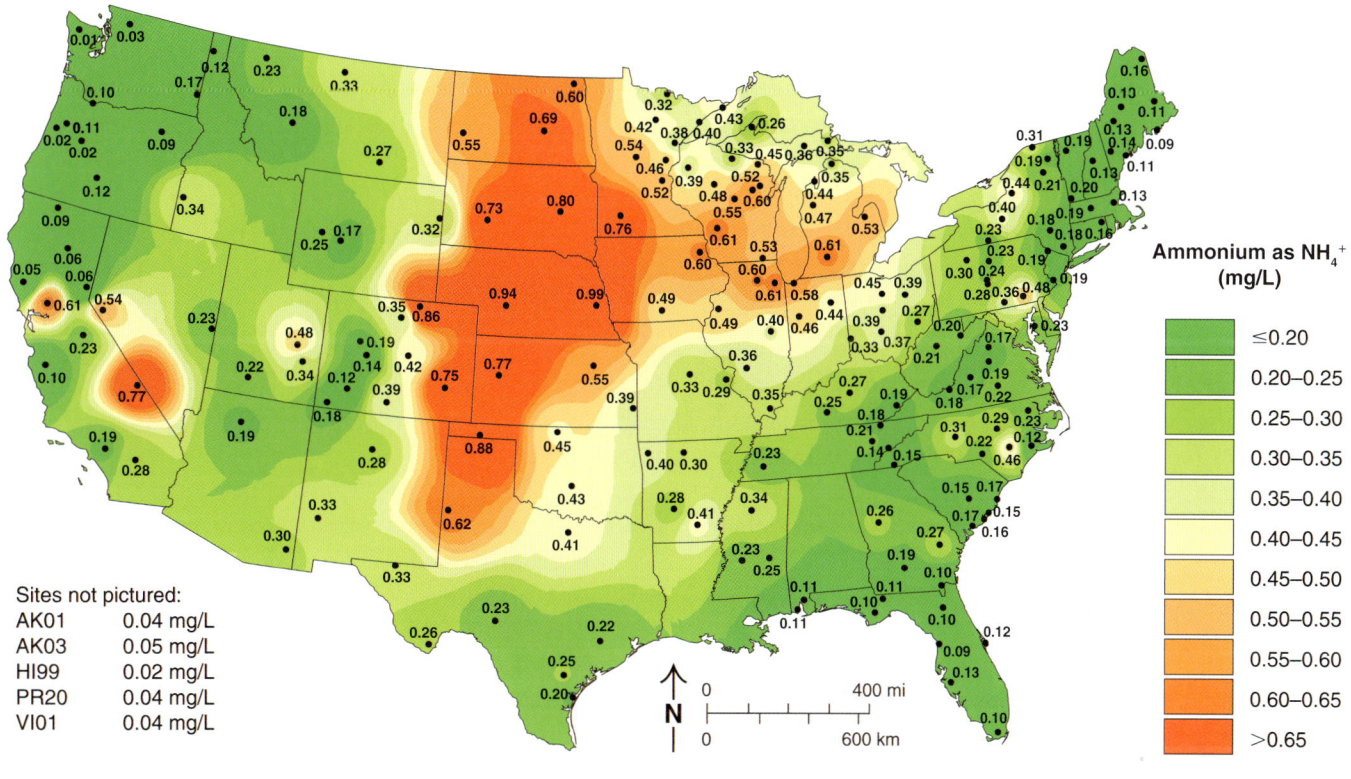

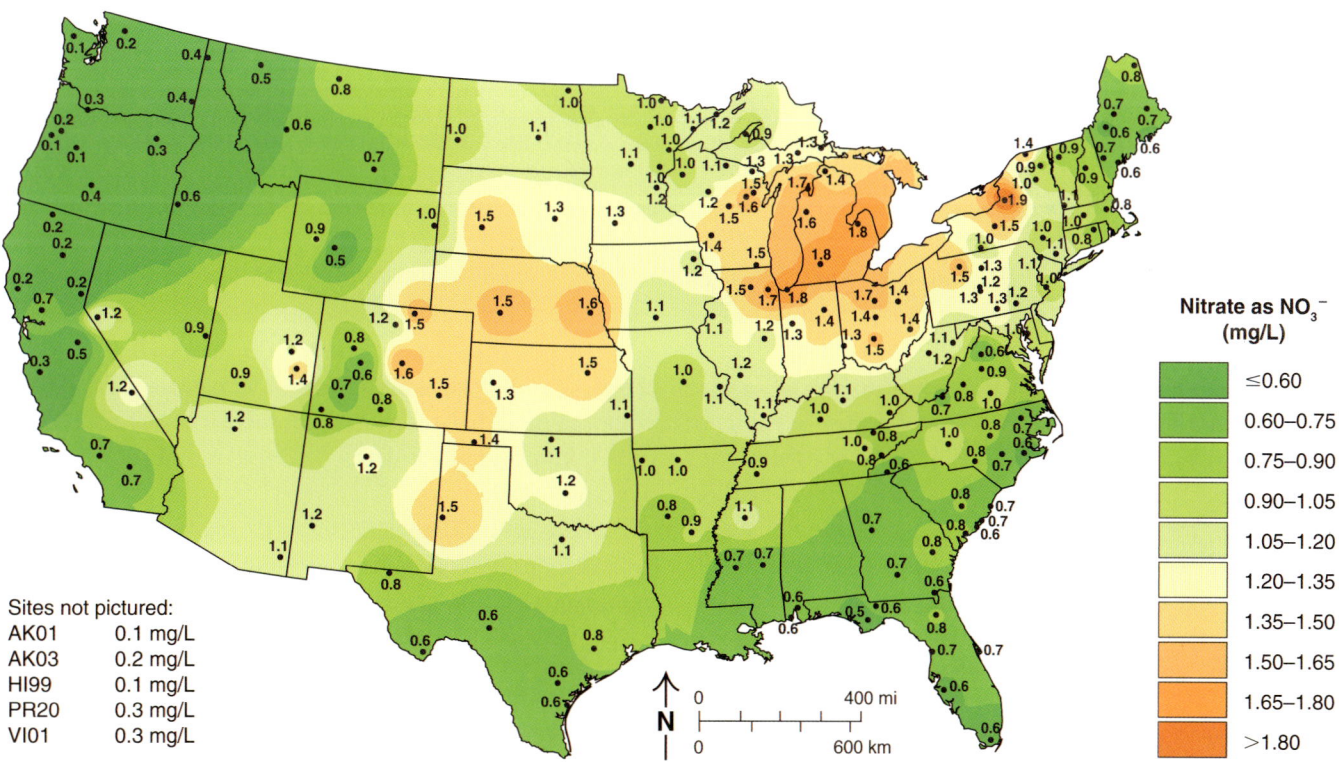

Figure 29-10 Inorganic nitrogen (as nitrate and ammonium) deposited across the United States, 2004; the greatest amount is deposited in the central United States, partially due to agricultural activities (data from National Atmospheric Deposition Program, National Trends Network, Illinois State Water Survey, Champaign).

Jo Daviess County to McHenry County, throughout the Chicago collar suburbs, and on the floodplains and terraces of Illinois rivers.

Other Nutrients

Phosphorous and potassium need to be added to agricultural soils in Illinois. Phosphorous is applied as a solid material, sometimes in combination with nitrogen; for example, processed phosphate fertilizers can contain phosphate, potassium, and nitrogen (as ammonium) in solid form. Phosphorous is readily taken up by soils, and organic matter and is generally found at very low concentrations in groundwater. However, phosphorus, even at low concentrations, can be an important contributor to eutrophication (depletion of dissolved oxygen) of lakes and streams (Hoeft and Peck 2002).

Pesticides

The term pesticide is used here to include both herbicides and insecticides. These weed- and insect-controlling chemical agents are used extensively on Illinois croplands, primarily on corn and soybeans (Figure 29-8). The U.S. Environmental Protection Agency has set drinking water standards for about 20 pesticides, several of which have been or are currently being used in Illinois: alachlor, atrazine, heptachlor, methoxychlor, and simazine. Atrazine was first used in 1958 and is currently the most commonly applied pesticide in the state (U.S. Environmental Protection Agency 2004d). In addition to being toxic to humans, these compounds can be harmful to aquatic biota (Nowell et al. 1999).

Keefer (1995) combined depth-to-uppermost aquifer information with the rate at which water moves through the soil profile (upper 5 feet of land surface), soil drainage classes, and soil organic matter classes to derive a map showing pollution potential of Illinois' shallow aquifers from pesticide leaching. Because soil organic matter attenuates pesticides, statewide aquifer sensitivity for pesticides is lower than for nitrate. However, the most vulnerable areas, as for nitrates, are in northern Illinois and along the floodplains and terraces of Illinois rivers. Chapter 20 (Protecting Groundwater) discusses in more detail the statewide vulnerability of shallow aquifers to both pesticides and nitrate.

Although the use of pesticides has increased crop yields significantly, their use also has created concerns over groundwater and surface water pollution. Recent studies in the midwestern United States and in Illinois have shown that shallow aquifers are most susceptible to pollution by pesticides. Goetsch et al. (1992) sampled hundreds of randomly selected wells throughout the state and found that about 10% of the rural, private wells in Illinois they sampled were polluted by pesticides and/or their degradation products. However, only a small percent of the samples exceeded the U.S. Environmental Protection Agency drinking water standards. Large-diameter bored and dug wells were most susceptible to pesticide pollution. More recently, Mehnert et al. (2003) seasonally sampled about 191 monitoring wells installed throughout Illinois. Those scientists found that shallow wells were more likely to be polluted with pesticides than deeper wells and that pesticide pollution was more likely to be present in groundwater samples from June through October following their application to fields.

A larger study of shallow wells throughout the midcontinental United States by the U.S. Geological Survey (Kolpin et al. 1994) showed that about one-fourth of the private wells were polluted with pesticides. The most frequently detected pesticide compounds were, in order of frequency, desmethylatrazine, atrazine, deisopropylatrazine, prometon, metolachlor, alachlor, simazine, metribuzin, and cyanazine. Notably, the first three pesticides are atrazine and its degradation products (metabolites). These metabolites are not regulated, and little is known about their toxicity to humans or aquatic biota.

Petroleum Compounds

Petroleum compounds include the hundreds of organic compounds found in petroleum products, such as crude oil, refined gasoline, and jet fuel. In general, the most important environmentally are those compounds that travel most rapidly in the environment. These include a set of aromatic compounds referred to as BTEX (benzene, toluene, ethylbenzene, and xylene) and MTBE (methyl tertiary butyl ether), a common gasoline octane enhancer. The BTEX compounds all have U.S. Environmental Protection Agency drinking water standards, and a standard for MTBE is pending.

There are a number of potential sources of petroleum pollutants to natural waters, including spills, leaky storage containers, and damaged pipelines. For example, in 1995, a dragline operator struck a gasoline pipeline in Marshall County that released 20,000 gallons (75,708 L), heavily polluting a wetland area. The most important source of petroleum pollutants, however, is leaking underground storage tanks. Underground storage tanks have a tendency to leak after they have been in the ground for extended periods, causing contamination of groundwater, surface water, and soils. Gasoline, diesel fuel, fuel oil, jet fuel, and waste oil constitute 96% of the substances that have leaked from these tanks (Illinois Environmental Protection Agency

2003). Over 22,000 incidents of pollution have been reported as of 2002 (over 6,000 from 1997 to 2002). However, there has been an active program to clean up these sites. In 2001 and 2002, more incidents have been remediated than reported, and over 17,000 acres (69 km^2) of land have been remediated since 1989 (about 1,400 acres [5.67 km^2] in 2002). Because leaking underground storage tank sites (known also as LUST sites) are most commonly associated with gasoline stations, the sites are distributed widely throughout the state, concentrated along highways and in higher population areas (Figure 29-11).

Because of its mobility, MTBE is of special concern (Illinois Environmental Protection Agency 2002c). Since 1994, MTBE has been found in 26 public water supply wells in 18 counties, including five wells in McHenry County and three in Tazewell County. Most of the wells are finished in sand and gravel and are located either within floodplains of major rivers or areas where sand and gravel is at or near the land surface.

Halogenated Solvents

Halogenated solvents include some very toxic pollutants, and about a dozen of them have U.S. Environmental Protection Agency drinking water standards, including carbon tetrachloride, dichloroethylene (DCE), and trichloroethylene (TCE). These compounds, which readily dissolve other organic compounds, are commonly used for degreasing, dissolving vegetable oils, decaffeinating coffee, and dry cleaning. Health effects linked to exposure to these chemicals include major organ (liver, kidney) problems and increased cancer risk.

Pollution of groundwater by halogenated solvents is almost always caused by a point source. Many industries and businesses use these compounds, so there are many potential sources. Perhaps the most famous case of groundwater pollution by solvents occurred in Woburn, Massachusetts, and was described in both the book and movie, *A Civil Action*.

Pharmaceuticals and Personal Care Products

The pharmaceuticals and personal care products used by humans and livestock for health or cosmetic reasons (PPCPs) include thousands of compounds, including antibiotics and other drugs and medications, fragrances, and cosmetics. The PPCPs are defined by the U.S. Environmental Protection Agency as emerging pollutants of interest. In recent years, these compounds have been turning up in many surface water bodies throughout the United States, generally at extremely low concentrations (parts per trillion or parts per billion). They generally enter the environment via discharge of treated sewage. Some PPCPs have been linked to negative ecological effects, such as feminization of fish. None of these compounds has been linked to human health effects at these low concentrations. The PPCPs are generally thought to not be a problem in groundwater because they tend to migrate slowly in the subsurface, but few studies have been conducted on this topic to date.

Pathogenic Microbes

Pathogenic microbes (bacteria, viruses, and protozoa) entering groundwater and surface water generally come from fecal sources as a result of poor sanitation. Lack of

Figure 29-11 Locations of Illinois' leaking underground storage tank sites (LUST sites). Notice that the sites follow the major roadways and outlines of towns (U.S. Environmental Protection Agency 2003a).

proper sanitation is the leading cause of water-borne disease and death in the world. Fortunately, the strict sanitation practices that were developed and implemented early in the twentieth century have made water-borne disease outbreaks very rare in Illinois and the United States. However, pathogenic microbes still enter the state's natural waters, primarily from private septic systems (Figures 29-12 and 29-13), livestock waste (Figure 29-14), and manure applied to croplands. Private septic systems and livestock waste are generally viewed as point sources. However, many animals spread out over several pastures, or numerous private septic systems in large subdivisions collectively act as non-point sources of pollution.

The Illinois Department of Public Health (1994) requires that well water contain no colonies of coliform bacteria in water samples. They also require that the water samples (about 0.1 of a quart, or 0.09 L, in volume) contain fewer than 100 colonies of aerobic bacteria to meet regulatory requirements for safe drinking water. Because most potential sources of pathogenic bacteria originate at or near the land surface, the susceptibility of underlying aquifers to pollution depends on the openness of the aquifers. In most of Illinois, aquifers are protected by overlying layers of materials of low permeability, such as fine-grained glacial till, that filter and impede the movement of fecal bacteria. However, a more open system, such as shallow karst, which is found in southwestern Illinois south of St. Louis, is typically poorly protected and can be recharged rapidly (Chapter 28, Karst Terrane). Springs and wells in the karst regions of Illinois generally have unacceptably high concentrations of fecal bacteria (Panno et al. 1996).

Many private septic systems are discharged onto the surface and either seep into the ground (and underlying aquifers) or run off to surface streams (Figures 29-12 and 29-13). The ability of private septic systems to minimize bacterial pollution of treated water is questionable. For example, over half the effluent from professionally maintained aeration-type septic systems sampled in southwestern Illinois in the mid-1990s by Panno et al. (1997) did not meet the minimum Illinois Department of Public Health (1994) requirement for fecal coliform for septic effluent. The dominant bacteria present were *Escherichia coli* and other enteric bacteria (those bacteria found in fecal matter) (Panno et al. 1997); their presence suggested that human viruses may also be present in the effluent (Geldreich 1996). Potentially, these systems could be a continual source of bacteria and viruses to underlying aquifers or surface water bodies (Panno et al. 1997).

Manure from small and large livestock operations (Figure 29-14) is applied to fields as fertilizer in order to use

Figure 29-12 The laterals of a leach field for a private septic system in southern Illinois are clearly defined by lush, green grass during an extended dry period. This situation occurs when the laterals are not draining properly and partially discharge at the surface, where rainwater can carry untreated sewage into nearby water bodies. (Photograph by Samuel V. Panno.)

Figure 29-13 Aerobic-type septic systems are the current choice of builders in Illinois. Discharge from these units is to the surface, where nutrient-enriched wastes can run off to surface water bodies and infiltrate into shallow aquifers. (Photograph by Samuel V. Panno.)

Figure 29-14 Livestock waste is either used as fertilizer for nearby fields or allowed to remain in the pastures where its nutrients, and bacteria, can seep into underlying aquifers. (Photograph by Samuel V. Panno.)

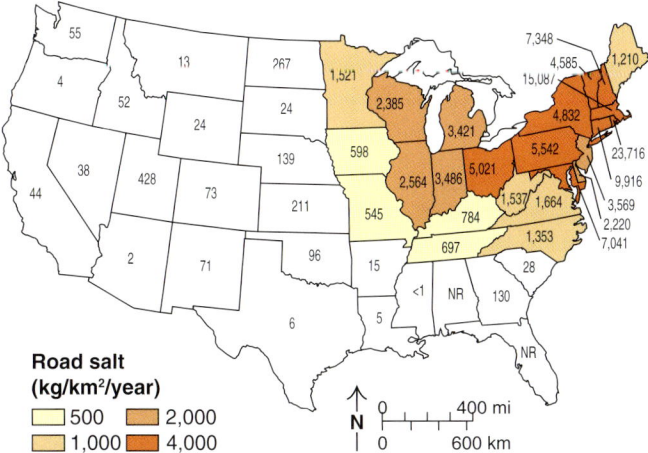

Figure 29-15 Map of the contiguous United States showing the average amount of road salt sold to each state annually (relative to the area of the state) and averaged over a three-winter period. This map shows that the amount of road salt used in Illinois averages over 2.76 tons/247 acres (2.5 tonnes/km²) per year. Most is used in the northern part of the state (Panno et al. 2004). Abbreviation: NR, not reported.

the waste. Manure is enriched in carbon, potassium, nitrogen, and phosphorous; each nutrient can be up to several percent of the total volume (University of Illinois Extension 2002). Manure is an ideal fertilizer for replacing nutrients in the soil. In addition to nutrients, however, manure contains abundant enteric bacteria that potentially could include *Salmonella, Escherichia coli* O157:H7, *Cryptosporidium* and *Leptospira interrogans* (from cattle), and *Balantidium coli* (from hogs) (Geldreich 1996).

If septic systems and manure-fertilized fields lie over a shallow aquifer, and the aquifer were to become contaminated with fecal-related bacteria, remediation efforts would be extremely difficult. When well water is contaminated with bacteria, residents generally are directed by the local health department to disinfect their wells by circulating chlorine bleach through the well and pipes. Although this procedure may be effective in disinfecting the well and plumbing over the short-term, it has little or no effect on a polluted aquifer, and pollution problems usually return (Panno et al. 1996).

Chloride

Chloride is not toxic to humans, but it can be toxic to some aquatic organisms, and elevated chloride concentrations can make drinking water unpalatable. A secondary drinking water standard of 250 mg/L has been established; water with higher concentrations has an unpleasant salty taste. There are a number of human sources of chloride that can enter natural waters, including road salt, water conditioning salt, sewage, and livestock manure.

In Illinois and other midwestern states, road salt is the predominant source of chloride in groundwater (Figure 29-15). In the Chicago region, over 270,000 tons (245,000 tonnes) of road salt are applied during an average winter (Keseley 2006). Rock salt, composed predominantly of halite (sodium chloride), was first used on roads after World War II, and the use of relatively large amounts began around 1960 (Salt Institute 2003). Rock salt is the primary road deicer used in northeastern Illinois; most is applied as brine solutions to pretreat roadways and prevent bridges from developing icy surfaces. Calcium chloride solutions are used when temperatures are very low (D. Johnson, personal communication 2003).

Chloride concentrations have been increasing in shallow groundwater in the Chicago region since around 1960, primarily due to road salt runoff (Kelly 2008, Panno et al. 2004). Although chloride concentrations in most wells are less than 250 mg/L, trends are discouraging. Large concentrations of chloride in surface water can adversely affect vegetation and biodiversity in sensitive areas (Panno et al. 1999), and concentrations have been increasing in the Illinois River since around 1960 (W. R. Kelly et al. 2010).

Conclusion

Almost all shallow groundwater and surface waters in Illinois contain pollutants that can result in ecologically impaired streams and in surface water and groundwater resources that require expensive treatment to achieve federal drinking water quality standards. Natural contaminants include arsenic, radium, barium, sodium, and chloride. Human sources of pollutants are described as having point or non-point sources.

Point sources of pollution in Illinois include Superfund sites, nuclear power stations, municipal landfills, and confined animal feeding operations. Non-point source pollutants, such as nitrate and pesticides from cropland, occur over very large areas and can uniformly enter surface water and groundwater. Other pollutants in Illinois include petroleum compounds from leaking pipelines and underground storage tanks, halogenated solvents from degreasing agents, pharmaceuticals and pathogenic microbes from human and livestock wastes, and sodium and chloride from road deicers (predominantly road salt). Many of these pollutants, once in drinking water, can pose health risks, and because non-point source pollutants are so widespread, they are much more difficult to control and remediate than those from point sources.

References

Berg, R. C., and J. P. Kempton, 1984, Potential for contamination of shallow aquifers from land burial of municipal wastes: Illinois State Geological Survey, Land Use Planning Map, 1:500,000.

Berg, R. C., H. A. Wehrmann, and J. M. Shafer, 1989, Geological and hydrological factors for siting hazardous or low-level radioactive waste disposal facilities: Illinois State Geological Survey, Circular 546, 65 p.

Cartwright, K., and F. B. Sherman, 1969, Evaluating sanitary landfill sites in Illinois: Illinois State Geological Survey, Environmental Geology Notes 27, 15 p.

Curry, B. B., R. C. Berg, and R. C. Vaiden, 1997, Geologic mapping for environmental planning, McHenry County, Illinois: Illinois State Geological Survey, Circular 559, 79 p.

Czapar, G. F., and F. W. Simmons, 2002, Water quality, in Illinois agronomy handbook (23rd ed.): Urbana-Champaign, University of Illinois Extension, p. 85–90.

David, M. B., and L. E. Gentry, 2000, Anthropogenic inputs of nitrogen and phosphorous and riverine export for Illinois, USA: Journal of Environmental Quality, v. 29, p. 494–508.

David, M. B., L. E. Gentry, D. A. Kovacic, and K. M. Smith, 1997, Nitrogen balance in and export from an agricultural watershed: Journal of Environmental Quality, v. 26, p. 1038–1048.

Erickson, N., 2000, Comments on the Mississippi River/Gulf of Mexico draft action plan for reducing, mitigating and controlling hypoxia in the northern Gulf of Mexico; Letter report to the U.S. Environmental Protection Agency (September 7, 2000): Springfield, Illinois, Illinois Farm Bureau, 5 p.

Fuchs, R. L., 1999, 1998 State-by-state assessment of low-level radioactive wastes received at commercial disposal sites: Idaho Falls, Idaho, U.S. Department of Energy, Idaho National Engineering and Environmental Laboratory, DOE/LLW-252, 132 p.

Geldreich, E. E., 1996, Microbial quality of water supply in distribution systems: New York, Lewis Publishers, 504 p.

Goetsch, W. D., D. P. McKenna, and T. J. Bicki, 1992, Statewide survey for agricultural chemicals in rural, private water-supply wells in Illinois: Springfield, Illinois, Illinois Department of Agriculture, 4 p.

Hensel, B. R., D. A. Keefer, R. A. Griffin, and R. C. Berg, 1991, Numerical assessment of a landfill compliance limit: Ground Water, v. 29, no. 2, p. 218–224.

Hoeft, R. G., and T. R. Peck, 2002, Soil testing and fertility, in Illinois agronomy handbook (23rd ed.): Urbana-Champaign, University of Illinois Extension p. 85–90.

Hughes, G. M., 1967, Selection of refuse disposal sites in northeastern Illinois: Illinois State Geological Survey, Environmental Geology Notes 17, 26 p.

Hughes, G. M., R. A. Landon, and R. N. Farvolden, 1971, Summary of findings on solid waste disposal sites in northeastern Illinois: Illinois State Geological Survey, Environmental Geology Notes 45, 25 p.

Illinois Department of Agriculture, U.S. Department of Agriculture and Illinois Agricultural Statistics Service, 2002, Illinois agriculture statistics: Springfield, Illinois, 158 p.

Illinois Department of Natural Resources, 2003, Water well permits issued and water wells sealed. http://dnr.state.il.us/orep/inrin/eq/well/permits.htm. Accessed July 19, 2006.

Illinois Department of Nuclear Safety, 2002, DNS 2001–2002 biennial report: Springfield, Illinois, Illinois Department of Nuclear Safety, 28 p.

Illinois Department of Public Health, 1994, Water well construction code 77: Springfield, Illinois, Illinois Administrative Code, Chapter I, v. 920, 42 p.

Illinois Emergency Management Agency, 2002, Annual survey report: Springfield, Illinois, Illinois Emergency Management Agency, Division of Nuclear Safety.

Illinois Emergency Management Agency, 2003, INFO: Department inspectors at nuclear power stations: Springfield, Illinois, Division of Nuclear Safety, IEMA-007-100-08/03, 4 p.

Illinois Environmental Protection Agency, 2000, Illinois FIRST abandoned landfill program: Springfield, Illinois, Illinois Environmental Protection Agency, Bureau of Land, IEPA/BOL/00-021, 32 p.

Illinois Environmental Protection Agency, 2002a, Illinois water quality report 2002: Springfield, Illinois, Bureau of Water, IEPA/BOW/02-006, 99 p.

Illinois Environmental Protection Agency, 2002b, Nonhazardous solid waste management and landfill capacity in Illinois: Springfield, Illinois, Bureau of Land, IEPA/BOL/02-020, 372 p.

Illinois Environmental Protection Agency, 2002c, Illinois groundwater protection program, biennial comprehensive status and self-assessment report: Springfield, Illinois, Bureau of Water, IEPA/BOW/02-001, 80 p.

Illinois Environmental Protection Agency, 2003, Leaking underground storage tank program 2002 annual report: Springfield, Illinois, Bureau of Land, IEPA/BOL/03-001, 15 p.

Illinois Waste Management and Research Center, 1997, Landfill sites of Illinois (GIS data), ed. 1.1: Champaign, Illinois.

Keefer, D. A., 1995, Potential for agricultural chemical contamination of aquifers in Illinois: 1995 revision: Illinois State Geological Survey, Environmental Geology 148, 28 p.

Keefer, L., and M. Demissie, 1999, Watershed monitoring for the Lake Decatur watershed: Champaign, Illinois, Illinois State Water Survey, Contract Report 637, 27 p.

Kelly, W. R., 2008, Long-term trends in chloride concentrations in shallow aquifers near Chicago: Ground Water, v. 46, no. 5, p. 777–781.

Kelly, W. R., S. V. Panno, K. C. Hackley, H.-H. Hwang, A. T. Martinsek, and M. Markus, 2010, Using chloride and other ions to trace sewage and road salt in the Illinois Waterway: Applied Geochemistry, v. 25, p. 661–673.

Keseley, S., 2006, Road salt is a slippery subject: Lake County Health Department and community Health Center Cattail Chronicles, v. 16, no. 1, p. 4–5.

Killey, M. M., and R. C. Berg, 2004, Land-use decisions and geology: Getting past "out of sight, out of mind": Illinois State Geological Survey, Geoscience Education Series 18, 68 p.

Kolpin, D. W., M. R. Burkart, and E. M. Thurman, 1994, Herbicides and nitrate in near-surface aquifers in the midcontinental United States, 1991: Reston, Virginia, U.S. Geological Survey, Water-Supply Paper 2413, 34 p.

Krapac, I. G., W. S. Dey, C. A. Smyth, and W. R. Roy, 1998, Impacts of bacteria, metals, and nutrients on groundwater at two hog confinement facilities: Proceedings of the National Ground Water Association Animal Feeding Operations and Ground Water: Issues, Impacts, and Solutions—A Conference for the Future: St. Louis, Missouri: Westerville, Ohio, National Ground Water Association, p. 29–50.

Krapac, I. G., W. S. Dey, W. R. Roy, C. A. Smyth, E. Storment, S. L. Sargent, and J. D. Steele, 2002, Impacts of swine manure pits on groundwater quality: Environmental Pollution, v. 120, p. 475–492.

League of Women Voters, 1993, The nuclear waste primer: Washington, D.C., League of Women Voters Education Fund, 170 p.

Luman, D., T. Tweddale, B. Bahnsen, and P. Willis, 2004, Illinois land cover: Illinois State Geological Survey, Illinois Map 12, 1:500,000.

McKenna, D. P., T. J. Bicki, W. S. Dey, D. A. Keefer, E. Mehnert, S. V. Panno, C. Ray, and S. D. Wilson, 1990, An initial evaluation of the impact of pesticides on groundwater in Illinois: Illinois State Geological Survey and Illinois State Water Survey, Cooperative Groundwater Report 12, 107 p.

Mehnert, E., W. S. Dey, D. A. Keefer, H. A. Wehrmann, S. D. Wilson, and C. Ray, 2003, Pesticide occurrence in shallow monitoring wells in Illinois (abs.): Kalamazoo, Michigan, Midwest Ground Water Conference Program and Abstracts: Westerville, Ohio, National

Water Association, October 1–3, p. 46.

Nolan, B. T., 2001, Relating nitrogen sources and aquifer susceptibility to nitrate in shallow ground waters of the United States: Ground Water, v. 39, no. 2, p. 290–299.

Nowell, L. H., P. D. Capel, and P. D. Deleanis, 1999, Pesticides in stream sediment and aquatic biota: Distribution, trends and governing factors, in R. J. Gilliom, ed., v. 4, Pesticides in the hydrologic system: New York, Lewis Publishers, 1001 p.

O'Riordan, T., and G. Bentham, 1993, The politics of nitrate in the UK, in T. P. Burt, A. L. Heathwaite, and S. T. Trudgill, eds., Nitrate—Processes, patterns and management: New York, John Wiley and Sons, p. 403–416.

Panno, S. V., K. C. Hackley, H. H. Hwang, S. Greenberg, I. G. Krapac, S. Landsberger, and D. J. O'Kelly, 2004, Characterization and identification of the sources of NaCl in groundwater and surface water: Illinois State Geological Survey, Open File Series 2005-1, 15 p.

Panno, S. V., W. R. Kelly, C. P. Weibel, I. G. Krapac, and S. L. Sargent, 2003, Water quality and agrichemical loading in two groundwater basins of Illinois' sinkhole plain: Illinois State Geological Survey, Environmental Geology 156, 36 p.

Panno, S. V., I. G. Krapac, C. P. Weibel, and J. D. Bade, 1996, Groundwater contamination in karst terrain of southwestern Illinois: Illinois State Geological Survey, Environmental Geology 151, 43 p.

Panno, S. V., V. A. Nuzzo, K. Cartwright, B. R. Hensel, and I. G. Krapac, 1999, Impact of urban development on the chemical composition of ground water in a fen-wetland complex: Wetlands, v. 19, no. 1, p. 236–245.

Panno, S. V., C. P. Weibel, I. G. Krapac, and E. C. Storment, 1997, Bacterial contamination of groundwater from private septic systems in Illinois' sinkhole plain: Regulatory considerations: Proceedings of the Sixth Multidisciplinary Conference on Sinkholes and the Engineering and Environmental Impacts of Karst, Springfield, Missouri: Oak Ridge Tennessee, P. E. Moreaux and Associates, Inc., p. 443–447.

Rabalais, N. N., R. E. Turner, and D. Scavia, 2002, Beyond science into policy: Gulf of Mexico hypoxia and the Mississippi River: Bioscience, v. 52, no. 2, p. 129–142.

Salt Institute, 2003, http://www.saltinstitute.org. Accessed July 19, 2006.

Stuffaboutstates.com, 2004, Total agricultural receipts ranked by state. http://stuffaboutstates.com/illinois/agriculture.htm. Accessed April 13, 2009.

Tisdale, S. L., W. L. Nelson, J. D. Beatonard, and J. L. Havlin, 1993, Soil fertility and fertilizers (5th ed.): New York, MacMillan Publishing Company, 634 p.

University of Illinois Extension, 2002, Illinois agronomy handbook (23rd ed.): Urbana-Champaign, University of Illinois, 321 p.

U.S. Department of Agriculture, 1997, National resource inventory. http://www.il.nrcs.usda.gov/technical/nri/find97.html. Accessed July 19, 2006.

U.S. Department of Agriculture, 2004, Illinois state fact sheet. http://www.ers.usda.gov/statefacts/il.htm. Accessed July 19, 2006.

U.S. Environmental Protection Agency, 2001a, National analysis: The national biennial RCRA hazardous waste report (based on 1999 data): Solid waste and emergency response (5305W), Washington, D.C., EPA530-S-01-001, PB2001-106313, 81 p.

U.S. Environmental Protection Agency, 2001b, Executive summary: The national biennial RCRA hazardous waste report (based on 1999 data): Solid waste and emergency response (5305W), Washington, D.C., EPA530-S-01-001, PB2001-106318, 10 p.

U.S. Environmental Protection Agency, 2003a, National priority list sites in Illinois. http://www.epa.gov/superfund/sites/npl/il.htm. Accessed July 19, 2006.

U.S. Environmental Protection Agency, 2003b, 2001 Toxics release inventory, state fact sheets: Washington, D.C., Office of Environmental Information (2810A), EPA 260-F-03-002, 4 p.

U.S. Environmental Protection Agency, 2004a, List of drinking water contaminants and MCLs. http://www.epa.gov/safewater/mcl.html. Accessed July 20, 2006.

U.S. Environmental Protection Agency, 2004b, Municipal solid waste: Basic facts. http://www.epa.gov/epaoswer/non-hw/muncpl/facts.htm. Accessed July 20, 2006.

U.S. Environmental Protection Agency, 2004c, Concentrated animal feeding operations (CAFO)—Final rule. http://cfpub.epa.gov/npdes/afo/cafofinalrule.cfm. Accessed July 20, 2006.

U.S. Environmental Protection Agency, 2004d, Pesticides: Topical and chemical fact sheets. http://www.epa.gov/pesticides/factsheets. Accessed July 20, 2006.

Warner, K. L., 1998, Water quality assessment of the lower Illinois River basin: Environmental setting: Reston, Virginia, U.S. Geological Survey, Water-Resources Investigation Report 97-4165, 50 p.

30 Surface Mine Reclamation

Timothy J. Kemmis, Robert A. Bauer, and Zakaria Lasemi

Introduction

The coal and aggregate mining industries are vital to the Illinois economy. The state's industrial growth has been fueled by Illinois coal for the past century and a half, and coal remains a crucial energy source for the state's electricity-generating plants. A growing economy also relies on the crushed stone and sand and gravel aggregate that are essential components of concrete and asphalt. Once these important mineral reserves have been mined, mine site reclamation is needed.

A century ago, land was cheap, and much of the state was rural. Coal mines, stone quarries, and sand and gravel pits were simply left abandoned when mining ceased. Today, mine reclamation is mandated by state and federal regulations, but the reasons for reclamation go beyond those regulations. The economics are obvious—land is valuable. Appearance also is important because mined land that is reclaimed and returned to use enhances its value for everyone in the community. Also, the necessity to protect and restore the environment is well understood. Depending on the specific mine type and site conditions, mine reclamation is engineered to control various environmental concerns, such as erosion and stormwater runoff, barren or eroding mine spoils and refuse piles, acid mine drainage, hazardous mine highwalls, and the safety of dilapidated mine buildings and equipment.

Mine reclamation methods in Illinois are varied and tailored to the type of mine—coal mine, sand and gravel pit, or stone quarry—and the type of mining method. The choice of final end use takes into consideration factors such as site location, mine characteristics, type of mining methods, land ownership, and community needs. Illinois has been a leader in mine reclamation, and several reclaimed sites have won national awards, such as those received by the Max McGraw Wildlife Foundation near Elgin (Figure 30-1) and Vulcan Materials' Casey Quarry, near the town of Casey in Clark County, east-central Illinois. Reclaimed Illinois mines have been converted to sites for housing developments and schools (Figure 30-2), recreation (including parks, athletic fields, biking and picnic areas, golf courses, and fishing and swimming lakes), shopping centers, industrial developments, row-crop agriculture, pasture, forests, and wildlife habitat.

Reclamation of Surface Coal Mines

The coal industry has long been an important component of the Illinois economy. Coal first fueled the rise of the state's railroad system and heated the nation's homes and businesses. Now coal is a major energy source for electricity-generating stations. More than 4,500 coal mines have operated since commercial mining began in Illinois about 1810, but fewer than 20 mines are currently active. Illinois coal is mined in either surface (formerly called strip mines) or underground mines (see Chapter 14, Coal).

Reclamation of land used for surface mining has two primary goals: (1) to return the land to productivity with a usable contour and (2) to protect the environment. Surface-mine reclamation considers several factors, including the effects of the surface-mining process on land-surface

Figure 30-1 This aerial photograph of the Max McGraw Wildlife Foundation site near Elgin in Kane County demonstrates how reclaimed sand and gravel pits can provide important lake and upland wildlife habitat in urban areas where natural wildlife areas have otherwise been largely lost to development. An active sand and gravel operation is shown in the background. (Photograph provided courtesy of Meyer Material Company, McHenry, Illinois.)

SURFACE MINE RECLAMATION

Figure 30-2 Modern reclamation means that former sand and gravel pits are not wastelands, but vital land that can be reclaimed for a variety of important purposes. These aerial photographs of a site near Carpentersville in Kane County show **(a)** the mined and initially reclaimed site and **(b)** the largely reclaimed and developed site with new housing developments, a middle school, and recreation lakes. (Photographs provided courtesy of Meyer Material Company, McHenry, Illinois.)

contours and overburden materials (the soils and rock above the coal), the amount and type of trace minerals in the overburden and coal, and state and federal reclamation regulations.

During the surface-mining process, the topsoil is removed and stockpiled for later replacement on the reclaimed ground. Next, a dragline or bucket wheel excavator moves the sediments above bedrock to the mined-out side of the pit (Figure 30-3). The bedrock above the coal seam is then broken up by drilling and blasting and removed by a dragline or shovel to the mined-out side of the pit. Next, the coal is excavated with a power shovel and loaded into haulage trucks (Figure 30-4). Before reclamation, the over-

Figure 30-3 Overburden is being excavated by a bucket wheel excavator and conveyed to the side of the pit already mined for coal. (Photograph from the Illinois State Geological Survey collection.)

burden is piled in ridges (Figure 30-5). In the past, these ridges were left in place, which significantly decreased land productivity and value, and trace minerals in the newly exposed overburden posed potential environmental dangers. At modern reclamation sites, after mining is completed, the ridges of overburden material are leveled, topsoil is replaced, and the land is reclaimed for its chosen use as farmland, parks, home sites, or other purpose (Figure 30-6).

Over the years, unreclaimed surface mines have been a source of water pollution. The overburden materials (gob) typically contain shale (clay-rich rocks) and poorer quality coal. At facilities that wash the coal to remove impurities, settling ponds concentrate impurities and fine coal removed by the washing process. Both the overburden rock

Figure 30-4 This truck is being loaded with coal by a power shovel while the dragline removes overburden. (Photograph from the Illinois State Geological Survey collection.)

Figure 30-5 Overburden was piled in ridges near this old surface mine. (Photograph by Timothy J. Kemmis.)

types and settling pond materials (slurry) typically contain sulfur-rich minerals, particularly pyrite and marcasite, which can react with rainfall and surface water to produce sulfuric acid. Eventually, the sulfuric acid may drain or percolate into surface water and groundwater resources. The resulting increase in water acidity can affect aquatic life and weaken concrete structures, such as bridge piers, retaining walls, utility pipes, and well casings (Nuhfer et al. 1993).

Because unreclaimed surface-mined land presents potential environmental harm, health and safety concerns, loss of land value and productivity, and lack of aesthetic appeal, state and federal regulations were enacted. Illinois' first reclamation law, the Open Cut Land Reclamation Act, was enacted in 1962, followed by the more stringent 1968 Surface-Mined Land Reclamation Act. In 1971, the Illinois Surface-Mined Land Conservation and Reclamation Act imposed some of the toughest restrictions in the nation. Before a permit could be approved, reclamation plans had to be filed with the county clerk, and a waiting period and public hearings had to follow. A 1975 amendment to the act included requirements for soil productivity. Many Illinois requirements were included in the 1977 Surface Mining Control and Reclamation Act. This federal law was incorporated into Illinois law by the Surface Coal Mining Land Conservation and Reclamation Act and went into effect in Illinois in 1983.

Today, Illinois coal mine reclamation is regulated by two different programs within the Illinois Department of Natural Resources, Office of Mines and Minerals. Sites currently being mined are regulated by the Land Reclamation Division, and areas mined before state regulations in 1962 are reclaimed by the Division of Abandoned Mined Land Reclamation.

To obtain a coal permit to surface mine coal in Illinois, companies must file a detailed report on the proposed mine area. The report must demonstrate that the surface-mining activity will have no environmental impact outside the mine area and that the land can and will be reclaimed after mining to a condition equal to or better than before mining. To support reclamation plans, companies must post a bond with the state for an amount considered sufficient to reclaim the land. As the land is reclaimed, the bond is returned. If the company fails to reclaim the land properly, the bond is forfeited, and the state uses the money for the reclamation.

All mines must report annually to the state on their mining and reclamation activities, and state inspectors visit the mines regularly to ensure that activities are in compliance with the permit. About 150,597 surface-mined acres (609.4 km^2) were reclaimed under these various reclamation acts during 1962 to 1991, the last date for which records were compiled (State of Illinois 1991).

Coal mines closed before 1962 were not regulated for mine reclamation. According to the Illinois Department of Natural Resources, Office of Mines and Minerals statistics, 103,181 acres (417.6 km^2) were mined in Illinois prior to the 1962 law (State of Illinois 1991). Because some of these abandoned mines posed significant health and safety problems, the Surface Mining Control and Reclamation Act of 1977 (Public Law 95-87) established the federal Abandoned Mine Land program to fund and address serious problems at coal mines abandoned prior to August 3, 1977. The program includes both underground and surface mines, and funds are derived from production fees on active coal mining. In Illinois, the program is administered by the Division of Abandoned Mined Land Reclamation of the Illinois Department of Natural Resources, Office of Mines and Minerals. Priorities are determined by the hazard or environmental problems the site presents. To date, the program has completed over 550 projects, addressed more than 775 mine sites, and reclaimed nearly 9,400 acres

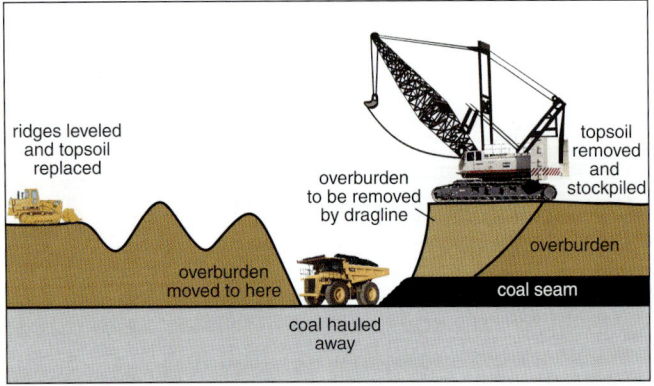

Figure 30-6 As part of the modern surface coal mining and reclamation process, overburden is removed by an excavator and placed on the mined-out side of the pit, where the ridges are contoured, and topsoil is replaced.

(38 km²) of land at a cost of approximately $146 million. The projects have included sealing 1,236 mine openings, removing 907 hazardous tipple structures, stabilizing 5,850 acres (23.7 km²) of gob and slurry and 1,900 acres (7.7 km²) of spoil, and restoring 1,400 acres (5.7 km²) of affected land and water.

Reclaimed surface mines in Illinois have been converted to farmland, parks, home sites, and other uses. Peabody Energy, for example, has converted a 6,000-acre (24-km²) former surface mine near Marissa in Randolph County into productive agricultural and forest land, and a wildlife habitat with a 300-acre (1.2-km²) wetland. The wetland was converted from the mine's refuse pond. It was reclaimed by dewatering the pond, treating the materials with a lime-neutralizing agent, and establishing the proper water levels for wetland plant development. This 300-acre (1.2-km²) wetland, surrounded by 40,000 trees, won the Illinois Department of Natural Resources 2003 Mine Land Reclamation Award and honorable mention from the Interstate Mining Compact Commission.

Sand and Gravel Pit Reclamation

Sand and gravel pits (and crushed stone aggregate mines discussed in the next section) supply aggregate, which is an essential component in the concrete and asphalt used to construct roads, bridges, homes, sidewalks, driveways, shopping malls, office buildings, factories, and other structures. Aggregate is also a primary material for sub-base and base course materials, backfill, and drainage and erosion control. The need for aggregate is high, and, because of the large capital equipment investment and local scarcity of deposits, most aggregate producers typically develop large sites capable of producing for at least 15 to 20 years.

The reclamation path to final end use takes several steps, involving consideration of the mining methods used, state regulations, community needs, site ownership, and planned use for the reclaimed property. Most open pit sand and gravel mine reclamation in Illinois is regulated under Illinois Administrative Code Title 62, Chapter I, Part 300, Surface Mined Land Conservation and Reclamation Act, which is administered by the Illinois Department of Natural Resources, Office of Mines and Minerals, Mine Safety and Training Division.

The sand and gravel mining method influences reclamation. Open pit sand and gravel is extracted in two principal settings: (1) above groundwater using wheel loaders and other equipment or (2) below groundwater by dredging, utilizing a suction apparatus or various mechanical devices such as a ladder or chain bucket, clamshell, dragline, or backhoe. Many sites use a combination of these techniques, initially developing all or parts of the site as an open pit above groundwater and then dredging suitable areas.

Two principal concerns are the surface topography and the steep working face. Reclamation consists of contouring and grading, local filling to connect differing surface elevations, and vegetating slopes to conform with state regulations, when applicable. Dredged sites usually have one or more lakes and ponds present after the sand and gravel have been removed from beneath the water table, and these are left for recreational use. The surrounding land is contoured and reclaimed within the pit boundaries.

Site reclamation is done systematically as the site is mined (Figure 30-7). Once mining of the first cell is completed, mining begins at the next cell. Topsoil and overburden from this new cell are then used to reclaim the previously mined cell. Initial site reclamation to state regulations then proceeds across the site as new cells are mined. Once mining and initial reclamation are completed by the sand and gravel producer, further reclamation may follow, using one of several options. In some cases, the producer is also the property owner who may either develop the property or sell it to commercial developers; other sites are leased by the producer, and it is the landowner's decision to either sell the site to developers or develop it.

The possibilities for using reclaimed sand and gravel pits are nearly limitless. At a site straddling the border between Carpentersville and Algonquin northwest of Chicago, Meyer Material Company leased over 200 acres (0.8 km²) and mined it for over 25 years. Then the company recontoured the land around a number of lakes formed by mining sand and gravel beneath the water table (Figure 30-2). The site was sold to developers. The land now has a

Figure 30-7 Reclamation at Meyer Material Company's active Algonquin sand and gravel operation in McHenry County. In the background is the reclaimed portion of the site with landscaping and a new home. A portion of the reclaimed but undeveloped part of the site is shown (right center) while mining continues in the foreground. (Photograph by Timothy J. Kemmis.)

centrally located middle school, a large tract of new single-family homes, and another tract of multi-family homes; development continues across the rest of the former site. Before long, evidence of this former sand and gravel pit will have essentially vanished, covered by new homes, a school, and lakes.

Another site on Illinois Route 31 north of Algonquin shows the benefits of reclamation. Although reclamation was planned, Meyer Material Company could not design the final surface contours until usable sand and gravel were found and mined. During the mid-1970s, when the mining began, the area was mostly surrounded by farm fields. Today, the site is ringed by several thousand homes. The company is filling in part of the pit next to the highway to prepare for commercial development, and the remainder will be contoured around the lakes remaining after dredging. After the company sells the land, new owners will develop it.

Many sand and gravel sites are successfully reclaimed for recreation and wildlife habitat. One example of this is the private Max McGraw Wildlife Foundation property located between Hoffman Estates, Elgin, and East Dundee (Figure 30-1). From the outset, this 164-acre (0.7-km^2) site was planned to be a wildlife reserve. As Meyer Material mined the site, it created lakes and reclaimed the slopes to the standards expressed by the Foundation. The Foundation then stocked the lakes and vegetated the site with plants that provide wildlife with food and habitat. The reclaimed site has won multiple state and national awards, including the 1986 Wildlife Conservation Award, Best Primary Wildlife Use Project, from the Wildlife Society and National Stone, Sand and Gravel Association and the 1987 State of Illinois Mined Land Reclamation Award–Aggregate Program. The site received honorable mention for both the Interstate Mining Compact Annual Non-Fuel National Mine Reclamation Award and the National Association of State Land Reclamationists Mined Land Reclamation Award. The site also provides wildlife habitat for an area where industrial and urban growth had caused the disappearance of nearly all of the natural wildlife areas.

Stone Quarry Reclamation

Stone is another resource that is vital to Illinois. Crushed stone is the main ingredient in concrete and the principal constituent of paving asphalt. Crushed stone also is used for water treatment plant filtration systems, shoreline erosion control, and agricultural lime application, among hundreds of other uses.

Crushed stone is generally quarried from open pit mines by blasting and crushing the rock. Because crushed stone is heavy and expensive to move, quarries should be located close to the communities they serve. Crushed stone quarries typically operate for 30, 50, 75, or even 100 years. Because of such long life spans, final quarry reclamation is not an everyday occurrence, but quarry reclamation, like

Figure 30-8 Aerial photographs show the operation and reclamation of Riverstone Group's Valley Quarry, opened in 1958, in southeastern Warren County near Abingdon: **(a)** initial development of the site (June 28, 1974); **(b)** Valley Quarry on June 27, 2003, showing the reclaimed lake, ongoing reclamation of the adjacent upland, the present quarry operation, and the processing area. (Aerial photography provided by Illinois Department of Transportation, Aerial Surveys Section.)

that for sand and gravel operations, is generally done in sequence. As a new part of the quarry is opened up, the previously mined area is reclaimed.

An example is the sequential development of the Riverstone Group's Valley Quarry in southeastern Warren County between Abingdon and Avon (Figure 30-8). The quarry first opened in 1958, and the initial part of the site was mined out. Production then shifted to another part of the site, a new processing area was developed, and reclamation began in the mined-out portion. Groundwater was allowed to flood parts of the former quarry floor, and the resulting lake was stocked with bass, bluegill, catfish, and crappie with the cooperation of the Illinois Department of Natural Resources. Trees have been planted on reclaimed slopes. The remainder of the lake was filled with soil and rock materials to make more natural contours and to establish a wetland habitat. Floating nesting sites were installed for Canadian geese, and wild turkeys were stocked on adjacent private properties. A log cabin (Figure 30-9) was built to hold meeting, lecture, and field trip participants.

Despite careful reclamation planning and construction, unexpected events can occur. At Valley Quarry, floodwaters from the adjacent Cedar Creek and Cedar Fork Creek spilled into the reclaimed lake, devastating the newly constructed wetland and introducing unwanted fish species to those that had been carefully stocked.

Reclaimed quarries typically provide important recreational resources for surrounding communities. Although the Valley Quarry is expected to operate long into the future, the reclaimed portion of the site is already being used for various activities, including Boy Scout activities that involve the study of wildlife and geology. The site also regularly hosts field trips, lectures, Junior Achievement gatherings, and Gem & Mineral Club meetings.

In short, quarries supply needed resources for the state's growing communities, and reclaimed sites provide wildlife and recreational resources that Illinois communities can enjoy for generations to come.

Underground Stone Quarry Reclamation

Underground mining of stone creates extensive open spaces with stable roof, floor, and wall conditions and relatively constant, moderate temperatures year-round. These spaces are desirable for storage and other industrial and commercial purposes.

A former underground limestone mine near Valmeyer in Monroe County has been converted into a business complex with nearly $3.5 million in grants from state and federal sources. Similar developments of underground mine sites at Quincy, Illinois, and Kansas City, Missouri, have been successful. A business complex is being developed at the former Columbia Quarry where more than 5 million square feet (0.5 km^2) in the former underground limestone mine are expected to be leased at a rate of $2 to $6 per square foot (0.09 m^2) (U.S. Geological Survey 2003). A 101,000-square foot (9,383.21 m^2), temperature-controlled warehouse facility is being constructed within the former mine. This facility may lead to a wide range of additional commercial activities including industrial, retail, and office space. Because reclamation of underground quarries has advantages for several kinds of businesses, this mine reclamation practice is expected to become more widespread in the years ahead.

Figure 30-9 A photograph of a log cabin and reclaimed lake at Riverstone Group's Valley Quarry. (Photograph by Timothy J. Kemmis.)

Conclusion

Today, former surface mines are not desolate, abandoned wastelands; they are reclaimed, revitalized sites that serve their local communities as parks, subdivisions, school sites, farm fields, golf courses, and wildlife refuges, among other uses. Mine reclamation is necessary for aesthetic appeal, environmental protection, and recovery of important property value.

Mine reclamation is specialized and must be engineered to the type of mine and the mining method. Although the reclamation method differs among coal mines, sand and gravel operations, and quarries, the results are the same—the mined land is appealingly restored and returned to beneficial use. Illinois' reclamation projects continue to be innovative and win national awards for their excellence. Today, mines and quarries supply the raw materials needed to sustain continued economic growth, and their reclamation provides new value to surrounding communities.

References

Nuhfer, E. B., R. J. Proctor, and P. H. Moser, 1993, The citizens' guide to geologic hazards: Arvada, Colorado, American Institute of Professional Geologists, 134 p.

State of Illinois, 1991, 1990, and 1991, Annual statistical report: Springfield, Illinois, Illinois Department of Natural Resources, Department of Mines and Minerals, 114 p.

U.S. Geological Survey, 2003, The mineral industry of Illinois: Reston, Virginia, 8 p. http//minerals.usgs.gov/mineral/pubs/state/2003/il-stmyb03.pdf. Accessed July 17, 2009.

Epilogue

This volume is a status report on the great body of knowledge that exists about Illinois geology. Assembling and condensing this knowledge into a single tome involved the synthesis of information dispersed in thousands of reports and maps published over the last hundred years. We salute those original authors whose research discovered and whose writing documented the geology of Illinois. We also salute the authors and editors of this volume, whose syntheses of this information have produced a magnificent result, a substantial work that celebrates and commemorates the 2005 centennial of the modern Illinois State Geological Survey (ISGS).

There is no end in sight for geological discovery or service in Illinois. Even after a century, much remains to be discovered, and the demand for geological insight and information is more urgent than ever for the economy of the state and the well-being of its citizens. The practical value to Illinois of the cumulative knowledge about the state's geology, represented in this volume, lies in the application of that knowledge to the major issues of economic development, environmental protection, resource conservation, sustainability, and hazards mitigation. The tasks of applying and continually adding to that geological understanding are fundamental to the mission and legislative mandate of the ISGS and the state's universities and are conducted cooperatively with the support of its industry, government, and citizens.

When this book project was conceived, the programmatic emphases of the Survey were environmental quality, groundwater and waste disposal issues, and expansion of detailed geological mapping of priority areas. Shortly thereafter, however, energy resource scientists recognized the need to help Illinois government and industries reduce the release of greenhouse gases from stationary sources, such as coal-fired power plants. A new line of research, with state and federal funding and industry cooperation, supported large-scale demonstration projects to lay the groundwork for possible industrial-scale sequestration of carbon dioxide in subsurface geological reservoirs in Illinois.

A century of geological research had provided the basis for and a headstart on sequestration and a wide range of future projects. Studies, especially those in support of the coal and petroleum industries on the composition, distribution, and characteristics of the rocks and fluids deep in the Illinois Basin, provide unique information to guide key strategies and decisions. These and other studies have added substantially to the state's large archive of geological samples; huge volumes of descriptive, analytical, and geophysical data; and hundreds of maps and reports on the detailed aspects of the geology. These form a strong technical basis for selection and characterization of sites for large-scale tests of geological sequestration of carbon dioxide or other future needs.

What will be the geological challenges of the next century? They will likely include

- the geochemical legacy of hundreds of years of human habitation, industrial production, and waste disposal in Illinois;
- the environmental impacts of a century of oil production and coal mining;
- growing demand for groundwater for an increasing population;
- expanding alternative energy and biofuel production;
- new transportation systems and corridors to carry people and freight;
- restoration of wetlands and floodplains along major rivers;
- sustained aggregate production in and near major metropolitan centers;
- agriculture adapting to the effects of changing climate; and
- the hazards of floods, landslides, and earthquake activity.

Whatever the future, geologists will continue to gather, map, and provide important, relevant, accessible, and usable geological knowledge and information to Illinois' citizens, industries, and governments apprising them of opportunities and challenges related to protecting their environment, ensuring their energy supplies, and minimizing impacts of geological hazards.

—E. Donald McKay III, Ph.D.
Director, Illinois State Geological Survey

Appendix I: Explanation of Lithologic Symbols

Symbol	Description
	Limestone
	Limestone reef
	Cherty limestone
	Oolitic limestone
	Bioclastic limestone
	Shaly or argillaceous limestone
	Sandy limestone
	Silty limestone
	Brecciated limestone
	Dolomitic limestone
	Dolomite (same variations as limestone)
	Sandstone
	Calcareous sandstone
	Dolomitic sandstone
	Gravel or conglomerate
	Shale or clay shale
	Silty shale
	Silt, siltstone
	Carbonaceous shale
	Coal
	Gypsum or anhydrite
	Underclay
	Glauconite
	Granite or rhyolite
	K-bentonite Bed
	Hardground omission surface
	Phosphatic grains and pebbles
	Soil
	Geodes or drusy quartz

Appendix II: Illinois Counties and Selected Cities

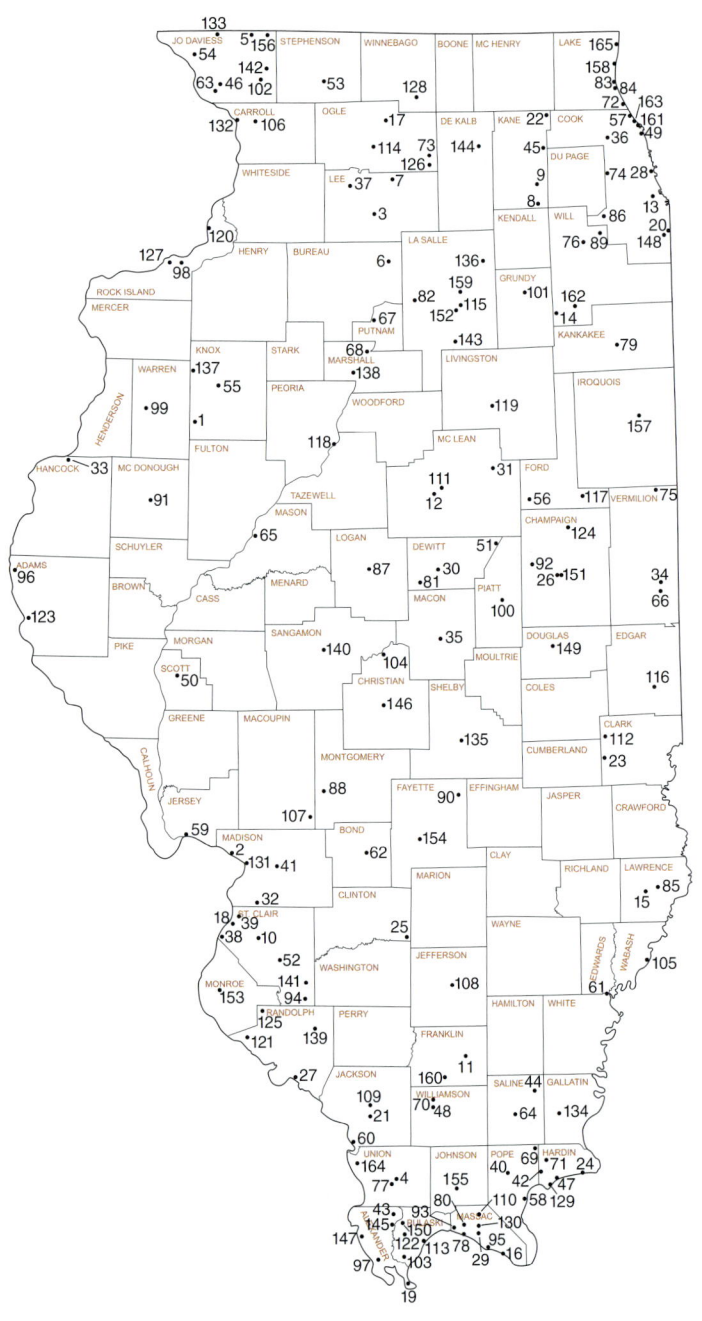

1	Abingdon	65	Havana	129	Rosiclare
2	Alton	66	Hegeler	130	Round Knob
3	Amboy	67	Hennepin	131	Roxana
4	Anna	68	Henry	132	Savannah
5	Apple River	69	Herod	133	Scales Mound
6	Arlington	70	Herrin	134	Shawneetown
7	Ashton	71	Hicks	135	Shelbyville
8	Aurora	72	Highland Park	136	Sheridan
9	Batavia	73	Hillcrest	137	Soperville
10	Belleville	74	Hillside	138	Sparland
11	Benton	75	Hoopeston	139	Sparta
12	Bloomington	76	Joliet	140	Springfield
13	Blue Island	77	Jonesboro	141	St. Clair
14	Braidwood	78	Joppa	142	Stockton
15	Bridgeport	79	Kankakee	143	Streator
16	Brookport	80	Kelley	144	Sycamore
17	Byron	81	Kenney	145	Tamms
18	Cahokia	82	La Salle	146	Taylorville
19	Cairo	83	Lake Bluff	147	Thebes
20	Calumet City	84	Lake Forest	148	Thornton
21	Carbondale	85	Lawrenceville	149	Tuscola
22	Carpentersville	86	Lemont	150	Ullin
23	Casey	87	Lincoln	151	Urbana
24	Cave in Rock	88	Litchfield	152	Utica
25	Centralia	89	Lockport	153	Valmeyer
26	Champaign	90	Loudon	154	Vandalia
27	Chester	91	Macomb	155	Vienna
28	Chicago	92	Mahomet	156	Warren
29	Choat	93	Maple Grove	157	Watseka
30	Clinton	94	Marissa	158	Waukegan
31	Colfax	95	Metropolis	159	Wedron
32	Collinsville	96	Meyer	160	West Frankfort
33	Dallas City	97	Miller City	161	Wilmette
34	Danville	98	Moline	162	Wilmington
35	Decatur	99	Monmouth	163	Winnetka
36	Des Plaines	100	Monticello	164	Wolf Lake
37	Dixon	101	Morris	165	Zion
38	Dupo	102	Morseville		
39	East St. Louis	103	Mounds		
40	Eddyville	104	Mt. Auburn		
41	Edwardsville	105	Mt. Carmel		
42	Eichorn	106	Mt. Carroll		
43	Elco	107	Mt. Olive		
44	Eldorado	108	Mt. Vernon		
45	Elgin	109	Murphysboro		
46	Elizabeth	110	New Columbia		
47	Elizabethtown	111	Normal		
48	Energy	112	Oilfield		
49	Evanston	113	Olmsted		
50	Exeter	114	Oregon		
51	Farmer City	115	Ottawa		
52	Freeburg	116	Paris		
53	Freeport	117	Paxton		
54	Galena	118	Peoria		
55	Galesburg	119	Pontiac		
56	Gibson City	120	Port Byron		
57	Glencoe	121	Prairie du Rocher		
58	Golconda	122	Pulaski		
59	Grafton	123	Quincy		
60	Grand Tower	124	Rantoul		
61	Grayville	125	Red Bud		
62	Greenville	126	Rochelle		
63	Hanover	127	Rock Island		
64	Harrisburg	128	Rockford		

List of Contributors

Michael L. Barnhardt, senior geologist, joined the Illinois State Geological Survey (ISGS) in 1990 after teaching for 13 years at Illinois State University, University of Wisconsin-Madison, and the University of Memphis. His major research interests are in geological mapping, glacial stratigraphy, and soil geomorphology. Since 2000, he has led a team of geologists in mapping the glacial geology of Lake County. He also is involved in the application of three-dimensional geological mapping to assist public agencies in decision-making for planning and other purposes. He received his M.S. degree in geomorphology and Quaternary environments from the University of Utah in 1973 and his Ph.D. in glacial and soil geomorphology from the University of Illinois in 1979.

Robert A. Bauer, engineering geologist and head of the Engineering and Coastal Geology Section, joined the Illinois State Geological Survey in 1976. He earned his M.S. degree in engineering geology from the University of Illinois in 1983. His career at the ISGS began with studies of rock mechanics, longwall mining, and land subsidence. His studies have expanded to include geotechnical assessments, environmental site assessments, and extensive involvement in seismic risk, vulnerability assessments, and earthquake preparedness.

Richard C. Berg is acting chief scientist at the Illinois State Geological Survey where he has worked as researcher and leader for more than 30 years. His major professional interests and publications include environmental and groundwater geology, glacial stratigraphy, and geological mapping. Berg has made important contributions in the application of three-dimensional geological mapping information to planning decisions, groundwater protection, land use, and public policy. Berg received his M.A. degree in geomorphology from Eastern Michigan University and his Ph.D. in soil geomorphology in 1979.

Subhash B. Bhagwat, mineral economist, received his B.S. degree in physics, mathematics, and geology from Fergusson College, Pune University, India, in 1962 and his M.S. degree in mining engineering from Technical University, West Berlin Germany, in 1968. He received his Ph.D. in mineral economics from Technical University, West Berlin, Germany, in 1974 and an M.B.A. in finance from the University of Illinois in 1986. He worked at the Illinois State Geological Survey from 1979 until his retirement in 2005. His more than 100 publications include studies related to coal market analysis, mining economics, economic geology, industrial minerals, water resource economics, policy and planning, mineral statistics, benefit-cost studies of geological mapping, and energy market analysis.

Richard A. Cahill retired from the Illinois State Geological Survey in 2007 after 33 years of service. His major research interests and publications have been in the areas of trace element distribution in recent sediments in rivers, lakes, and wetlands; determination of sediment rates and patterns using ^{137}Cs and ^{210}Pb; and evaluation and risk assessment of naturally occurring radioactivity. Cahill received his B.A. degree in chemistry in 1971 from Montclair State College, an M.S. degree in nuclear chemistry in 1974 from University of Maryland, and an M.S. degree in geochemistry from the University of Illinois in 1980.

Michael J. Chrzastowski, senior coastal geologist, joined the Illinois State Geological Survey in 1987. He received a B.S. degree in geology and a B.A. in oceanography from the University of Washington in 1974. After several years of work with the National Ocean Survey and the U. S. Geological Survey, he completed an M.S. degree in geology from Western Washington University in 1982 and a Ph.D. degree in coastal geology from the University of Delaware in 1986. His research, service, and outreach activities at the ISGS have focused on the coastal evolution and human modifications of the Illinois shoreline of Lake Michigan. He led the ISGS studies of geological impacts from the historic 1993 flooding along the Mississippi and Illinois Rivers.

B. Brandon Curry, Quaternary geologist at the Illinois State Geological Survey since 1984, has major research interests in glacial geology, environmental geology, and geological mapping as well as understanding Quaternary climates and lakes. Curry holds geology degrees from the University of California at Santa Barbara (B.A., 1980), Purdue University (M.S., 1984), and the University of Illinois at Urbana-Champaign (Ph.D., 1994).

Joseph A. Devera is a senior geologist at the Illinois State Geological Survey. He received his bachelor's degree from Northern Illinois University and master's degree from Southern Illinois University at Carbondale (SIU-C). His professional interests include geological mapping, invertebrate paleontology, and teaching at SIU-C.

Gary B. Dreher, analytical chemist, began working at the Illinois State Geological Survey in 1972, left for some

List of Contributors

years to work in the private sector, and returned to the ISGS until his retirement in 2005. Dreher's major research interests and publications are in the areas of trace element analyses and distributions and in the development and use of analytical methods. His analytical methods were applied to coal studies, lake sediments, limestones, Silurian reefs, shales, groundwater, and the statewide geochemical evaluation of soils.

Scott D. Elrick, geologist in the Coal Section, joined the Illinois State Geological Survey in 1999. He received his B.S. degree in geology from the University of Illinois in 1995 and his M.S. degree in geology from University of California–Riverside in 1998. His major research interests are in Pennsylvanian stratigraphy, coal depositional environments, and geological issues affecting coal mines.

Keith C. Hackley, isotope geochemist, is head of the Illinois State Geological Survey Isotope Geochemistry Section. Hackley joined the Illinois State Geological Survey in 1983. He received his B.S. degree in geology in 1978 from The Pennsylvania State University, followed by an M.S. degree in geology (1984) and a Ph.D. in geochemistry (2002) from the University of Illinois at Urbana-Champaign. He publishes in the areas of groundwater geochemistry and environmental science, especially using isotopic evidence to discern geochemical processes and determine the origin of environmental contaminants.

Ardith K. Hansel, Quaternary geologist, received her M.A. degree in geography at the University of Northern Iowa, followed by a Ph.D. degree in geography at the University of Illinois. She began work at the Illinois State Geological Survey in 1977 and retired from there 30 years later. During her productive career, Hansel made important contributions in Quaternary sedimentology, stratigraphy, and mapping; late Quaternary evolution of the Great Lakes Region; late glacial and postglacial history of the Lake Michigan lobe; classification of the deposits of the last glaciation; and lake level fluctuations in the Lake Michigan basin. She now resides in Port Angeles, Washington.

Beverly L. Herzog was assistant to the director for environmental initiatives at the Illinois State Geological Survey at the time of her retirement in 2010. In that role, she worked to increase the funding, visibility, and relevance of ISGS environmental programs. Her research interests and many publications focused on groundwater, including the monitoring and modeling of natural groundwater systems, characterization and mapping of Illinois' major aquifers, and remapping Illinois' bedrock surface. Herzog received her B.S. in geology from the University of Wisconsin-Oshkosh in 1976 and her M.S. degree in hydrology from Stanford University in 1978. She joined the ISGS in 1980.

Thomas G. Hildenbrand (deceased December 10, 2007) was a retired geophysicist from the U.S. Geological Survey. His work greatly increased the understanding of the crustal structure of North America. He championed the application of potential-field geophysics in order to identify relationships between crustal structure and geological hazards, ore genesis, and groundwater movement. He is especially known for his work in assessing the concealed geological framework of the New Madrid Seismic Zone and Yucca Mountain areas. His legacy also includes the many U.S. Geological Survey geophysicists mentored by him over the years.

Bryan G. Huff is a geologist in the Oil and Gas Section. He first joined the Illinois State Geological Survey in 1980. He left for graduate studies in 1981, earning B.S. and M.S. degrees in geology from the University of Illinois in 1981 and 1984, respectively. He worked as a visiting lecturer for the University of Illinois and in the private sector before rejoining the ISGS in 1986. His current duties include compiling petroleum statistics for the state, assisting industry and the public, and conducting research in petroleum reservoir characterization and modeling. He earned certification as a petroleum geologist by the American Association of Petroleum Geologists in 1995 and is a Licensed Professional Geologist in the State of Illinois.

Randall E. Hughes, geologist, retired from the Illinois State Geological Survey in 2003 after more than 25 years of service, first as a student. After a decade in industry, Hughes returned to the ISGS, working and publishing primarily in areas related to clay mineralogy and stratigraphy. He and his colleagues also made archaeologically important discoveries of carved stone materials used by Native Americans. He has remained active in both geology and archaeology since his retirement. Hughes received his B.S. in geology and a Ph.D. in clay mineralogy from the University of Illinois.

Russell J. Jacobson (aka "Dino Russ"), geologist, worked at the Illinois State Geological Survey for 34 years, from 1973 until his retirement in 2007. He has a B.A. degree in geology from the University of Northern Iowa and an M.S. degree in geology from the University of Illinois. His research interests include paleontology, mapping, outreach, coal geology, and, especially, dinosaurs and other vertebrate fossils. He created a popular dinosaur Web site and spent several weeks each summer teaching a field course in Utah, South Dakota, and Montana.

Nelson Kawamura worked as a geotechnical engineer at GeoConsult, Inc., San Juan, Puerto Rico, and is now with URS Corporation in Oakland, California. He received both his M.S. (1991) and Ph.D. (1998) in civil engineering (geotechnical) from the University of Illinois. As a graduate student, he worked at the Illinois State Geological Survey from 1993 to 1998, assisting projects related to longwall coal mines and landslides. He performed field investigations, rock mechanics testing, soil testing, and data analyses.

Karan S. Keith worked at the Illinois State Geological Survey from 2003 until 2006. First hired as an assistant clay mineralogist, Keith later became director of the ISGS Sedimentology Laboratory in 2004. Keith received a B.S. degree in geology and a B.S. degree in systematics and ecology, both summa cum laude from University of Kansas, 1983. In 1989, Keith received an M.S. degree in environmental science and a second M.S. degree in geology, both from Indiana University. She earned her Ph.D degree in environmental science in 2002 from Indiana University.

Walton R. Kelly, groundwater geochemist, has been at the Illinois State Water Survey since 1992. He received a B.S. degree in geology in 1981 from Duke University, an M.S. in geosciences from Case Western Reserve University in 1984, and a Ph.D. degree in environmental sciences (geochemistry emphasis) from the University of Virginia in 1993. He worked at the Nuclear Regulatory Commission in Silver Spring, Maryland, in the high-level and low-level nuclear waste programs between 1983 and 1987. He is an adjunct professor in geology at Illinois State University in Normal and the University of Fort Hare in Alice, South Africa. His research activities focus on groundwater and surface water pollutants, including nutrients, arsenic, chloride, and pathogens, and geochemical processes in aquifers.

Timothy J. Kemmis is a Quaternary geologist specializing in glacial and fluvial stratigraphy and sedimentology and in engineering geology. He is currently employed by AECOM, an international engineering firm, at its Sheboygan, Wisconsin, office. He has extensive field experience in glacial mapping, glacial and fluvial sedimentology, and the origin of jointing in glacial tills. He has been lead geologist and project manager for large-scale hydrogeological and foundation engineering investigations in the United States and abroad. He presents continuing education workshops for engineering geologists. Kemmis received his M.S. in geology from the University of Illinois and his Ph.D. from the University of Iowa.

Joanne Kluessendorf is director of the Weis Earth Science Museum at the University of Wisconsin-Fox Valley and adjunct professor at the University of Wisconsin Geology and Geography Department. Kluessendorf is an expert in carbonate sedimentology, stratigraphy, Silurian paleontology and geology, Paleozoic reefs, paleokarst, and ichnofossils. She received both her B.S. and Ph.D. degrees in geology from the University of Illinois.

Dennis R. Kolata has expertise and publications in invertebrate paleontology, physical and chemical stratigraphy, structural geology, tectonic history, crustal geology, and the origin and evolution of cratonic basins. His identification and correlation of Ordovician K-bentonite ash beds is regarded as an important contribution to the understanding of Lower Paleozoic geology in North America, Europe, and Argentina. Kolata received his B.S. degree in biology and chemistry (1968) and his M.S. degree in zoology (1970) from Northern Illinois University. He received his Ph.D. in geology from the University of Illinois in 1972. He retired from the Illinois State Geological Survey in 2004.

David R. Larson, hydrogeologist and head of the Hydrogeology Section, joined the Illinois State Geological Survey in 1991. His career includes 15 years as a hydrogeologist with the North Dakota State Water Commission in Bismarck and 2 years as a research hydrogeologist with the Nebraska Conservation and Survey Division, University of Nebraska–Lincoln. He spent 35 years in groundwater resources research. His principal professional interests focus on groundwater resources planning and management, conjunctive use of water resources, groundwater and surface water interactions, and hydrogeological mapping. He earned a B.A. degree in geology from the State University of New York College at Fredonia and an M.S. degree in geology from the University of Nebraska–Lincoln.

Timothy H. Larson, geophysicist, joined the Illinois State Geological Survey in 1980 after completing his B.S. and M.S. degrees in geology from Wheaton College (1978) and Northern Illinois University (1980), respectively. While working at the Survey, he completed his Ph.D. degree in geology from the University of Illinois. In addition to contributions to groundwater investigations and aquifer characterizations, Larson has used geophysics to improve characterization at waste-disposal facilities, to better understand the state's seismicity, and to assist the Survey's three-dimensional geological mapping program.

Zakaria Lasemi, senior geologist, is the head of the Illinois State Geological Survey Industrial Minerals and Resource Economics Section and interim director of the Earth Resources Center. He received his B.S. degree from the University of Shiraz, Iran; M.S. from the University of Illinois; and Ph.D. degree from Miami University. His major interests and publications have been in the area of carbonate diagenesis, especially microcrystalline limestones; the origin of recent dolomites; the depositional environment and sequence stratigraphy of Middle Mississippian carbonates in the Illinois Basin; the geology of aggregate resources; and the inventory and characterization of Illinois limestone for desulfurization.

List of Contributors

Morris W. Leighton led the Illinois State Geological Survey from 1983 until his retirement in 1994. During his time as chief, Leighton emphasized the Survey's efforts in working toward a healthy economy and a healthy environment. He also was senior editor of Memoir 51 of the American Association of Petroleum Geologists. Remaining very active after his retirement, Leighton was member and then chair of the Geological Society of America's Foundation Board and volunteered on behalf of the University of Illinois Department of Geology, the Association of American State Geologists, and the ISGS Centennial. Before coming to the Survey, Leighton worked as research and exploration geologist and manager for Exxon and its affiliates for 32 years beginning in 1951. In research, he specialized in the area of carbonate rocks and basin studies. He also worked for Esso from 1961 to 1983 overseeing international exploration and development in a number of countries. Leighton received his B.S. degree in geology at the University of Illinois in 1947 and his M.S. and Ph.D. degrees in geology at the University of Chicago in 1948 and 1951, respectively.

John H. McBride is a professor at Brigham Young University, where he teaches courses in geophysics. From 1995 until 2002, he worked in the Illinois State Geological Survey Sedimentary and Crustal Processes Section under the supervision of Dennis R. Kolata. McBride received B.S. and M.S. degrees from the University of Arkansas and a Ph.D. degree from Cornell University.

E. Donald McKay III, the director of the Illinois State Geological Survey and Illinois State Geologist, began his career at the ISGS in 1971 after receiving his B.S. degree in geology from Hanover College. He completed his M.S. (1975) and Ph.D. (1977) degrees in geology from the University of Illinois. His primary research interests are in glacial and Quaternary geology and environmental studies, supported by his knowledge in engineering properties of materials, hydrogeology, and Geographic Information Systems and digitization.

Donald G. Mikulic, geologist in the Industrial Minerals and Resource Economics Section, has been with the Illinois State Geological Survey since 1979. He has used his extensive knowledge of primarily Silurian, but also Devonian, dolomite to assist the aggregate industry, several museums, and educational efforts. His biostratigraphic and paleoecologic knowledge of those strata, both within Illinois and the region, is extensive. Mikulic received his B.S. degree from the University of Wisconsin-Milwaukee in 1975 and his Ph.D. degree from Oregon State University.

Michael V. Miller retired from the Illinois State Geological Survey in 2005 as the director of the Transportation and Environment Center, which comprised the Wetlands Geology and Environmental Site Assessment sections. Miller's research interests and publications included Lake Michigan bluff retreat rates, atmospheric and sediment chemistry and particle size analysis, and, most recently, the hydrogeology and geochemistry of natural and man-made wetlands and their mitigation and restoration. Miller received his undergraduate degree in 1967 and his M.S. degree in 1969 from the University of North Dakota at Grand Forks. He received his Ph.D. in physical geography from the University of Illinois in 1974.

James J. Miner received his B.S. degree in geology from the University of Illinois in 1987 and his M.S. degree in geology from Northern Illinois University in 1989. He first began working at the Illinois State Geological Survey in 1985, and currently he heads the Wetlands Geology Section. His professional interests include glacial geology, sedimentology, and geomorphology, and his research is focused on the hydrogeology of wetlands and natural areas as related to their occurrence, sustainability, and restoration.

W. John Nelson, although he officially retired from the Illinois State Geological Survey in 2007 after 33 years of service, continues working and publishing. Nelson received his B.S. degree in geology in 1971 from Williams College and his M.S. degree in geology from the University of Illinois in 1973. His work includes more than 100 publications and maps. His main research interests at the Survey are in coal geology, stratigraphy, structural geology, and regional tectonics.

Cheryl K. Nimz, managing writer and editor, has been publications coordinator at the Illinois State Geological Survey since 2000. She received her B.A. degree in English education in 1972 and her M.A. in English literature in 1973, both from University of Illinois. Prior to joining ISGS, she was a tenured secondary English teacher in Urbana, Illinois (1974–1977); editorial director and journal editor, American Dairy Science Association/Federation of Animal Science Societies (1977–1999); and production editor, college-level journals, National Council of Teachers of English (1999–2000). Professional interests include environmental sciences, publishing issues, Web standards, copyright, events management, and public education.

Rodney D. Norby received his B.S. degree from the University of North Dakota in 1967, his M.S. degree from Arizona State University in 1971, and his Ph.D. from the University of Illinois in 1976. His major research interest and extensive list of publications are in Paleozoic biostratigraphy, and he is very knowledgeable about Cambrian through Pennsylvanian faunas. After more than 32 years of service to the Illinois State Geological Survey, Norby continues, in his retirement, to curate the ISGS paleontological collection.

Samuel V. Panno, senior geochemist, received his B.S. degree in biology in 1972 from Eureka College, his B.S. in geology in 1976 from Oregon State University, and his M.S. degree in geology from Southern Illinois University. He had additional coursework in hydrogeology and groundwater chemistry at George Washington University in 1985. He researched the effects of radiation on sodium chloride at Brookhaven National Laboratory from 1979 through 1983 and was consultant to the U.S. Department of Energy on its High Level Nuclear Waste Program from 1984 to 1988. In 1988 he joined the Illinois State Geological Survey where his research interests are groundwater and surface water contamination, karst geology and hydrology, and earthquake activity of the New Madrid Seismic Zone.

William R. Roy, senior geochemist, began his career at the Illinois State Geological Survey in 1980 after receiving his B.S. (1971) and M.A. (1980) degrees in geology from Indiana University at Bloomington, Indiana. He received his doctorate in soil physical chemistry in 1985. Roy's research projects and published works have included studies of coal wastes; the chemical fate and movement of organic solvents, metals, and other wastes in soil and water; computer-assisted modeling of aqueous systems; groundwater contamination; soil cleanup objectives; radioactive wastes; and carbon sequestration.

Michael L. Sargent, geologist, received his B.S. degree in economics in 1965 from the University of Wisconsin-Madison and his M.S. degree in geology in 1970 from the University of Illinois. During his tenure at Illinois State Geological Survey, from 1971 to 2001, Sargent acquired in-depth knowledge of Cambrian, Ordovician, and Silurian stratigraphy in Illinois. His expertise and publications were valuable in describing drill core, providing insight to the structural and stratigraphic framework of the Illinois Basin, expanding the knowledge of meteorite impact structures, and curating the Illinois State Geological Survey Geological Samples Library and the Albert V. Carozzi paleontological collection.

Beverly Seyler was head of the Oil and Gas Section at the Illinois State Geological Survey at the time of her retirement in 2010. Her most recent work focused on underexplored Paleozoic reservoirs, improvement of databases and map products available to industry, and support of the oil reservoir component of the carbon sequestration project. Seyler received bachelor's (1973) and master's (1978) degrees from the State University of New York at Buffalo and an additional master's degree in geology from the State University of New York at Fredonia. After working in the private sector for seven years, Seyler joined the ISGS in 1980.

William W. Shilts is the executive director of the Institute of Natural Resource Sustainability at the University of Illinois. Shilts received his A.B. degree in geology from DePauw University, his M.S. degree in geology from Miami University, and his Ph.D. in geology from Syracuse University. From 1995 to 2008, during his tenure as chief of the Illinois State Geological Survey, he made detailed, three-dimensional geological mapping a priority, greatly strengthened the energy program, and emphasized the importance of earth science information as a component in responsible economic development and environmental security. Prior to 1995, Shilts worked for 30 years as a research scientist for the Geological Survey of Canada, leading studies in glacial geology, environmental geochemistry, mineral exploration, atmospheric contaminants, and the impacts of historic and prehistoric earthquakes. He has been or is presently an adjunct professor at Carleton University (Ottawa), University of Montreal, University of Illinois, Illinois State University, and the University of Québec at Montreal.

Wen-June Su, engineering geologist in the Engineering and Coastal Geology Section, joined the Illinois State Geological Survey in 1981. He earned his M.S. degree in geology from the National Taiwan University in 1976 and another M.S. degree in civil (geotechnical) engineering from the University of Illinois in 1986. His career began at Engineering Geology Consulting Firms in Taiwan in 1977 and at the ISGS in 1981. Su also taught a course of Engineering Geology at the Chinese Culture University in Taiwan in 1979. He has been involved in studies of soil and rock mechanics, land subsidence, geotechnical site assessment, seismic risk assessment, earthquake ground response analysis, and development of a Web-based engineering geology database.

Hong Wang, geologist and geoarcheologist, is the director of the Illinois State Geological Survey Radiocarbon and Optical Stimulated Luminescence Dating Laboratories. Wang joined the Illinois State Geological Survey in 1991. He received his B.S. (1981) and M.S (1984) degrees in geology from Northwestern University in the People's Republic of China, and he obtained the M.A. (1993) and Ph.D (1996) degrees in anthropology from the University of Illinois at Urbana-Champaign.

C. Pius Weibel is a senior geologist at the Illinois State Geological Survey. He received his B.S. degree in geology from the University of Wisconsin–Platteville in 1978 and his M.S. (1982) and Ph.D. (1988) degrees, also in geology, from the University of Illinois. He has been studying diverse aspects of the sedimentary geology of the Midwest for more than 25 years.

Index

A

A horizon, 234f[1], 253, 374f
abandoned landfill sites, 446, 447f
Abandoned Mine Land program, 460
Abandoned Mined Land Reclamation, Division of, 460
abandoned mines, 460
abandoned water wells, 449, 450f
Abbott Formation, 195f, 287f
Abingdon, 462f
abrasives, 316
Abrotocrinus coreyi, 178
Absaroka sequence, 64, 187–205
    chronostratigraphy, 65f
    cross sections, 61f, 66f, 128f, 141f
    geochronology, 67f
    Sloss sequence, 64, 187–205
    stratigraphic column, 188f
    structural geology, 62f
    tectonics, 62f, 79, 79f, 128f, 141f
absorbent materials, 211
Acadian Orogeny, 62f, 80f–81f, 84, 167
acceleration (seismic response), 425t, 430, 430f
accretion
    Lake Michigan shoreline, 410f, 411f, 412f
    North American craton, 77
    Precambrian terranes, 78f
acoustical waves, 285f
Acrocyathus species, 175f, 180
active faults, 105
Adams County
    groundwater, 332
    Mendon Moraine, 230
    petroleum reservoirs, 291
    pre-Illinois Episode deposits, 224
    Sonora formation, 168f, 178, 287f
Adeline esker, 231
Aden Consolidated Field, 293
aerial photographs
    glacial ridge, 232f
    Grafton, 371f–372f
    Great Flood of 1993, 371f–372f
    Illinois till plain, 232f
    karst terrane, 435f
    Lake Michigan shoreline, 387f, 407f, 412f, 413f, 414f
    Max McGraw Wildlife Foundation site, 458f
    Riverstone Group's Valley Quarry, 462f
    sand and gravel pit, 315f, 459f
    Thornton Quarry, 59f, 164f
Aeronian, 160f
Africa, paleogeography, 80f–81f
Aftonian. *See* pre-Illinois Episode
Agassizocrinus species, 182
age, 64f, 65
age determination
    Canadian Shield rocks, 78f
    caves, 433
    groundwater, 347
    Loveland Silt, 234
    paleoclimatology, 248–260
        Farm Creek, 250f
        fossils, 252f

    Keller Farm, 254f
    Quaternary materials, 219, 222, 227, 243
    speleothems, 255f, 256f
"Age of Fishes," 170
"Age of Reptiles," 206
aggregates, 15, 38–39, 309–322
    Galena Group, 151
    Illinois River valley, 237f
    outwash, 379
    Silurian, 164
agrichemicals, 33, 351, 353, 359, 444t, 449f, 450f, 451f
agricultural lime, 184, 312
agriculture
    best management practices, 449
    land use, 374, 449f, 450f
        drained wetlands, 396
        erosion, 380
        nitrogen, 449
        pollution, 340, 443, 443t, 448
    soils, 316, 361, 373, 421
air quality
    coal, 33, 280, 444
    geochemistry, 31
    radon, 420
alachlor, 452
Albertosaurus?, 209
Albion-Scipio Field (Mich.), 289
Alburnett Formation, 223f, 224, 227–228
Alexander County
    Exogyra costata, 211f
    geosols and silts, 228f
    Great Flood of 1993, 400
    Illinois low point, 70
    Kimmswick Limestone, 311
    McNairy Formation, 210, 210f
    Post Creek Formation, 209–210, 210f
    Thebes, 211, 227
    tripoli, 207, 316
Alexandrian
    on cross section (Ill.), 66f
    oil and gas fields, 291f
    reservoir rocks, 287f
    Silurian reef, relation to, 290f
    stratigraphic columns, 160f, 287f
    Tippecanoe subsequence II, 161
    western limit, 291f
Alfisols, 234, 375f, 375–376, 376f, 377
algae, 145, 150f, 151, 253, 302f
algal blooms, 368, 450
Algoma phase, 409f, 410f
Algonquin, 461, 461f, 462
Algonquin phase, 410f
Allard Limestone, 168f, 183f
Alleghany Orogeny, 62f, 167, 202
    Appalachians, 100, 177f, 182
    Mississippian, 84, 182
    Paleozoic, 80f–81f
    Pennsylvanian-Permian, 86, 188
alluvium
    Cahokia Formation, 229f, 242
    fan, 237f
    floodplains, 379

---

[1]Abbreviations: f, figure; t, table.

INDEX

Illinois River valley, 237f
Massac Creek graben, 109f
Mississippi River valley, 335
parent materials, 378f
Quaternary, 206f, 221f
Wabash Bedrock Valley aquifer, 335
alnoite, 100, 202
alpha particles, 418
Alto formation, 65f, 168f
Alto Pass, 75f–76f, 212
Alton
    Acrocyathus species, 180
    Chesterian, 181
    Colchester Coal Member, 268
    Great Flood of 1993, 397, 398f
    Mineral Springs Hotel, 441
aluminum oxide (alumina), 315
Amarillo-Wichita Uplift, 189f
Amboy earthquakes, 424, 427, 427f
American Bottoms, 317f
American Coal Company's Galatia Mine, 277f, 278
Ameura sauki, 191f
amino acids, racemization, 248
ammonia compounds, 338t
ammonium, 338t, 345, 451f
ammonium nitrate, 449
Ammonoidea, 209
amphibians, 190, 368
amphiboles, 132
Amphicoelia neglecta, 162f–163f
Amphigenia curta, 162f–163f, 172f–173f
Amphigenia species, 171
Amphipoda, 439f
Anadarko Basin, 170f
anaerobic bacteria, 33
anaerobic environment, 362, 366, 368
Ancell Group, 143f, 146–147, 149f, 330
    cross section, 148f
    lead-zinc deposits, 302f
    stratigraphic columns, 137f, 138f
ancestral Rocky Mountain Orogeny, 62f, 86, 93, 187, 193
ancient Illinois floodplain, 70, 71f
Ancona Anticline, 92f, 94, 94f
Andalusia sill, 239–240
andesitic magma, 77f
anhydrite
    Everton Formation, 145–146
    Jurassic, 207
    reservoir rocks, 287f
anhydrous ammonia. *See* nitrogen
animal waste
    concentrated animal feeding operations (CAFOs), 447–448
    feedlot pollutants, 444t, 447–448
    fertilizers, 454f, 454–455,
    nitrogen, 449
    pollution, 448, 449, 455
    sinkholes, 439
anions, 337
Anna, 181
Anna municipal landfill, 447f
Anna Shale, 188f, 272
Annelida, 211
Annularia species, 190f
anoxic conditions. *See* anaerobic environment
antacids, 312
Antarctica
    paleogeography, 125f
    Vostok Station ice cores, 251
Anthozoa (corals)
    Acrocyathus species, 175f, 180
    Cedar Valley Limestone, 173
    Devonian, 170
    Dutch Creek Sandstone, 171

Favosites species, 162f–163f, 171
in fluorspar deposits, 305f
Foerstephyllum species, 150f, 151
Galena Group, 151
Halysites species, 162f–163f
Hexagonaria profunda, 173
Hexagonaria species, 171
Microcyclus discus, 172f–173f, 173
Pennsylvanian, 189
Platteville Group, 149, 150f, 151
Pleurodictyum problematicum, 171, 172f–173f
Salem Limestone, 180
St. Laurent Formation, 173
Streptelasma species, 150f, 151
anthracite coal, 264
Anthracospirifer increbescens, 172f–173f, 182
antibiotic residues, 453
anticlines, 84, 93
    Cottage Grove Fault System, 97
    Fairfield Basin, 95
    Mattoon Anticline, 286f
    Pennsylvanian, 193
    petroleum reservoirs, 284, 284f, 287f, 292, 293
    Precambrian, 123
    structural traps, 288, 288f
    See also individual named anticlines
Antiquatonia species, 191f
Anvil Rock Sandstone Member, 195f, 200, 287f
apatite (calcium phosphate)
    in groundwater, 343
    in Permian rocks, 202
aphanitic texture, 125f, 132
Aphelaspis Zone (Iowa, Mo.), 142
Appalachian Basin, 60f, 91f
    Chesterian time, 170f
    paleogeography, 91f, 189f
    Paleozoic, 80f–81f
    Pennsylvanian, 189f
Appalachians, 60f
    Chesterian, 170f
    Cretaceous, 208f
    drainage, 195
    paleohydrology, 300f
    Pennsylvanian, 189f, 193f
    provenance, 182, 193, 211
    Triassic, 207f
    Valmeyeran, 177f
    See also Alleghany Orogeny
Apple River, 72t, 301f
Apple River Canyon State Park, 72t
aquifer vulnerability, 344f, 352f, 353–354, 354f, 356f, 358, 359, 446, 452
    karst, 435–436
    maps, 352f, 353f, 354f, 355f, 356f, 359f
    McHenry County, 356f, 358
    nitrates, 352f, 449f, 450, 450f, 451f
    open systems, 341f, 454
    permeability, 358
    pesticides, 353f, 449f, 450f
    salt-water intrusion, 344f
    shallow aquifers, 164, 331, 345f, 354f, 448, 448f
    Tazewell County, 357f
    total dissolved solids, 329f
    Villa Grove Quadrangle, 359f
    waste, 353, 355f
    Winnebago County, 355f, 358
aquifers, 325–334, 334f, 335–336, 337–339, 339f, 351
    bedrock valleys, 333–335
    chemical composition, 341f
    confined aquifers, 327, 327f, 334
    groundwater protection, 351–360
    groundwater quality, 329f, 330f, 331f, 338–339, 341f
    hydraulic conductivity, 325f, 325–326
    Mt. Simon Sandstone, 141

pre-Illinois Episode, 224
Quaternary, 243
radioactivity, 419
salt-water intrusion, 344f
sand and gravel, 332–333, 334f, 335–336
sensitivity. *See* aquifer vulnerability
"sole source" designation, 352
St. Peter Sandstone, 147, 331
unconfined aquifer, 327, 327f
Villa Grove Quadrangle, 338f, 359f
wetlands, 361, 368
yields, 329–336, 351
See also groundwater; water resources
aquitards, 325–327, 327f, 328–336, 339, 351
Eau Clair Formation, 330
groundwater quality, 338–339
hydraulic conductivity, 325f, 325–326, 327f
Maquoketa Group, 331
New Albany Shale Group, 332
Arab oil embargo, 26, 285
arches, 90, 93
See also individual named arches
Archimedes species, 179, 181f, 182
Archimedes wortheni, 181f
Arctinurus occidentalis, 162f–163f
areal extent
aquifers, 328–336
aquitards, 328–336
coal, 263, 264f
Cretaceous, 208f
Illinois coast, 404, 404f
Pennsylvanian, 189–201
Sangamon Geosol, 232
shallow aquifers, 353
Tippecanoe II subsequence, 158
See also individual named geological units
argillite, 224
Argyle Member, 232
Arkoma Basin, 60f
paleogeography, 91f
paleohydrology, 300f
Paleozoic, 82f
Permian, 300f
proto-Illinois Basin, 127f
arkose sandstone, 132, 140, 141
Arlington Moraine, 236f
arsenic
in coal, 266
in groundwater, 34, 338t, 344–345, 345f, 443
health effects, 344
Artemia species, 211
artesian aquifer. *See* confined aquifer (artesian aquifer)
artesian wells, 327
artifacts—flint clay (pipestone), 3, 6f, 317f
ash (combustion product)
bricks, 36, 37f
coal, 264–265
waste disposal, 280, 444
Ashkum series, 377f
Ashmore Tongue, 221f
Ashton
bedrock, 143f
Franconia Formation, 143
Oneota Dolomite, 145f
Potosi Dolomite, 144f
Ashton Anticline, 143f, 144
asphalt, 280
Asphaltinoides grandtowerensis, 171
Assumption Coal Member, 267, 267f
Assumption oil and gas fields, 291f
Asterotheca species, 190f
asthenosphere (upper mantle), 77, 77f
Athens Subepisode, 223f, 238–239

Atlantic Ocean
Cretaceous, 207, 208f
Jurassic, 208f
Permian to present, 80f–81f
Triassic, 206, 207f
Atlas Powder borehole, 99f
atmospheric precipitation
chemical composition, 338t
Great Flood of 1993, 396
hydrologic cycle, 328, 328f, 337
karst, 432
land subsidence, 388f, 389
radon infiltration, 420
soils, 377
wetlands, 363
atmospheric pressure, radon infiltration, 420
Atokan, 66f, 187–188, 189f
reservoir rocks, 287f
stratigraphic columns, 188f, 287f
Tradewater Formation, 197
atomic bomb, 21, 22f
atomic weights, 347
atrazine, 452
Aurora earthquake (1909), 426
Aurora Member, 223f
Australia, paleogeography, 125f
"Aux Vases lime," 168f, 287f
Aux Vases Sandstone, 181–182, 184
fluorspar deposits, 305f
groundwater, 338t, 343–344
Hicks Dome, 101f
petroleum reservoirs, 184, 287f, 294–296, 296f
stratigraphic columns, 168f, 287f
Avalonia, 80f–81f
Avon diatremes (Mo.), 84, 85f, 173

B

B horizon, 222f, 234f, 234–235, 374, 374f
Babylon Sandstone Member, 198
back-arc basin, 77f
Backbone Limestone, 163
Ste. Genevieve Fault System, 99f
stratigraphic columns, 65f, 160f, 287f
sub-Kaskaskia surface, 169f
backflow flooding, 397
bacteria, 453–454
Bailey Limestone, 65f, 163
reservoir rocks, 287f
Ste. Genevieve Fault System, 99f
stratigraphic columns, 160f, 169f, 287f
sub-Kaskaskia surface, 169f
Balantidium coli, 455
Ballard County (Ky.)
earthquake swarm, 428
seismic profile, 117f
Baltagnostus species, 140
Baltica, 80f–81f, 125f, 139f, 167
Bankston Fork Limestone, 188f, 200
Banner Formation, 223f, 225, 226, 227
barite, 100
barium, 443
"Barlow" limestone. *See* Beech Creek ("Barlow") Limestone
Barnes Creek, 106f
Barnes Creek Fault Zone, 106f
Barnes Creek faults, 106, 106f, 107t
barrier beaches, 408
barrier islands, 148, 148f
barriers
clay, 316
to littoral drift, 413
basalt (Precambrian), 125f
basaltic crust, 60

Index

base (cushion) gas, 288–289
basement rock. See Precambrian
basement walls, 385, 386, 386f, 388f, 421f
Bashkirian, 188f
basins, 90–91
    back-arc basin, 77f
    starved basin, 28, 169, 177f
    Triassic, 206
    See also individual named basins
Batavia
    building stone, 317
    Fermi National Accelerator Laboratory, 393, 394f
    quarry, 318f
Batestown Member, 221f, 223f, 236f, 238
Bath landfill, 447f
bats. See Chiroptera
batteries, 300
Battery Rock Formation, 195f
Battery Rock Sandstone, 188f
Bay Island Drainage District groundwater flooding, 399
Baylis Formation, 206f, 209, 209f
beach
    Illinois coast, 72t, 404f, 404–407, 407f, 411f, 417
    littoral erosion, 405, 413, 414
    nourishment, 414, 414f
    protection, 413, 414, 414f
    Quaternary deposits, 229f
    ridges, ancient, 72t
Beach Park, 404f
Bear (Berry) Lake, 408, 410f
bearing capacity (engineering geology), 385, 386
Beaver Bend Limestone Member, 296f
bed, 64, 64f
bedded replacement deposits, 303, 304f, 305
bedding planes. See planar bedding structures
bedrock, 60, 67–69
    aquifers, 299, 328–332, 338, 341f, 354
        karst terrane, 432f, 436f
        Mt. Simon Sandstone, 330f
        northern Illinois, 141, 142, 151, 329, 330, 354
        radioactivity, 348, 419, 420f
        sulfate, 346
        total dissolved solids, 329f, 331f
        yield, 329, 330f, 331f
    benches, 241f
    Cambrian-Ordovician, 143f, 299, 330–331
    Driftless Area, 217
    history of investigations, 20, 25f, 40
    ice movement, 216f, 216
    Illinois, 68f, 69f, 225f
    land subsidence, 389
    pedogenesis, 379
    on Quaternary cross section, 226f
    sinkholes, 436f
    speleothems, 255f
    topography, 225f
    valleys, 24, 225f, 333f, 333–335
    waste disposal, 33f
    wetlands, 366
Beech Creek ("Barlow") Limestone, 99f, 168f, 181, 183f, 296f
    fluorspar, 305f
    reservoir rocks, 287f, 295f, 296f
Beecher series, 377f
belemnites, 209
Bell Coal Member, 188f, 267f, 272
Bell Smith Springs, 72t
Bellair 500, 800, and 900, 287f
Belle Rive Coal Member, 201, 267f, 272
Belleville, 390f
Belvidere Member, 233–234
Benoist. See Yankeetown ("Benoist") Sandstone
Bennett landfill, 447f
benthic animals
    diatoms, 252f
    Foraminifera, 250f
Benton, 271
Benton-Fairfield sub-Pennsylvanian valley, 196f
Benton Hills (Mo.), 107f, 108–109
benzene, 452
Berry (Bear) Lake, 410f
Berry Clay Member, 234, 366
beryllium
    in coal, 266
    in groundwater, 347
beryllium-10, 248
best management practices, agriculture, 449
beta rays, 418
Bethel Sandstone, 180–184
    fluorspar deposits, 305f
    reservoir rocks, 287f, 297
    stratigraphic columns, 168f, 183f, 287f
Bettendorf (Iowa)
    nuclear facilities, 422
    Pennsylvanian, 189
    Rock Island Coal Member, 267
Beverly Field, 291f
Beverly Spit, 409f
bicarbonate, 337–338, 338t, 340–341
"Biehl," 188f, 287f
Big Bend, Illinois River, 239f
Big Clifty Sandstone, 168f, 182, 287f
"Big Paul" stripping shovel, 272–273
Big Sink, 432, 437, 437f
Bilyeu Member, 180f
biostratigraphy, 41
biostromes, 173
Birds Member, 160f
bird's-eye structures, 171
Bishop landfill, 447f
bismuth-214, 420
Bi-State Disposal landfill, 447f
bituminous coal, 16, 187, 188f, 264, 266
    See also coal; coal deposits; individual named coal members
Bivalvia
    Amphicoelia neglecta, 162f–163f
    Clayton Formation, 211
    Dutch Creek Sandstone, 170, 171
    Maquoketa Group, 154
    paleoclimatology, 253
    Porters Creek Formation casts, 211
    Racine Dolomite, 162f–163f
    Vanuxemia species, 150f, 151
"black earth." See soils
Black Powder Hollow, 213f
Black River Limestone. See Platteville Group (= Black River Limestone)
Black Warrior Basin, 60f, 91f
    paleohydrology, 300f
    Paleozoic, 82f
    proto-Illinois Basin, 127f
Blackland North Field, 291f
Blanding Dolomite, 160f, 161
blasting, 274
Blastoidea
    Metablastus species, 178–179
    Pentremites species, 175f, 182
blind faults, 115
Blocher Shale Member, 65f, 168f, 173, 174
    Hicks Dome, 101f
    New Albany Shale, 174f
Bloomington
    Danville Coal Member, 272
    earthquakes
        Aurora (1909), 426
        Cairo (1895), 427
    La Salle Anticlinorium, 93, 344f

Bloomington Moraine, 226f, 236f, 238
Bloomington Morainic System, 335
Bloomington Ridged Plain, 70, 71f
"Blue" dolomite, 302f
Blue Island, 409f
Blue Island Ridge, 406f, 407
Blue Ridge Province, 211
blue-green algae, 150f
bluffs
    Illinois coast, 405, 406f, 408, 411f
    Illinois River valley, 237f
    loess, 220f
    St. Peter Sandstone, 2
    state parks, 72t
body waves, 113f
Bogota Limestone, 188f
bogs, 235, 253, 361, 366, 367
    See also wetlands
boiler, 266
Bond County, 270
Bond Formation, 195f, 200–201
    coal seams, 267f
    reservoir rocks, 287f
    stratigraphic columns, 188f, 287f
Bonneterre Formation, 137f, 142
Boone County
    aquifers, 330, 331
    Illinois Episode, 231, 232
    urban planning, 33f
"Bootheel" region (Mo.), 113
Borden Delta Complex, 176–177, 177f
Borden (Osage), 287f
Borden Siltstone
    on cross section, 180f
    environment of deposition, 169, 176–177
    reservoir rocks, 293
    stratigraphic columns, 168f, 287f
    tectonics, 84
boron, 266
Bouguer anomalies, 129f
"boulder clay," 248
Bowling Green Dolomite, 160f, 161
bow-tie pattern (ultramafic composition), 100
Brachiopoda
    Amphigenia curta
        Clear Creek Formation, 162f–163f
        Grand Tower Formation, 172f–173f
    Amphigenia species, 171
    Anthracospirifer increbescens, 172f–173f, 182
    Antiquatonia species, 191f
    Campylorthis deflecta, 150f
    Campylorthis species, 151
    Carbondale Formation, 199
    Composita species, 191f
    Composita subquadrata, 182
    Dalmanella species, 150f, 151
    Derbyia species, 191f
    Diceromyonia species, 153f, 154
    Dictyoclostus species, 191f
    Eodevonaria arcuata, 162f–163f
    Eodevonaria species, 171, 172f–173f
    Glyptorthis species, 153f, 154
    Hebertella species, 153f, 154
    Hesperorthis concava, 150f
    Hesperorthis species, 151
    Hypsiptycha species, 153f, 154
    inarticulate Brachiopoda, 138, 140f
    Independatrypa independensis, 173
    Lepidocyclus species, 153f, 154
    Maquoketa Group, 154
    Marginifera species, 191f
    Marginirugus magnus, 175f, 179
    Megamyonia species, 154

    Megamyonia unicostata, 153f
    Mesolobus mesolobus, 191f
    Neochonetes granulifer, 191f
    Neochonetes species, 191f
    Neospirifer species, 191f
    Öpikina minnesotensis, 150f
    Öpikina species, 151, 153f, 154
    Orbiculoides newberryi, 191f
    Ordovician, 311f
    Pennsylvanian, 189, 191f
    Pentamerus species, 161, 162f–163f
    Pionodema species, 151
    Plaesiomys species, 153f, 154
    Platteville Group, 149
    Platymerella species, 160
    Platystrophia species, 153f, 154
    Pseudolingula species, 151
    Rafinesquina species, 151
    Rostricellula species, 151
    Salem Limestone, 180
    Schuchertera species, 162f–163f
    Sowerbyella punctostriata, 150f
    Sowerbyella species, 151
    Spirifer grimesi, 178
    Spirifer increbescens, 172f–173f
    Spirifer logani, 175f
    Spiriferida, 162f–163f, 173
    St. Louis Limestone, 180
    Stricklandia species, 161
    Strophomena plattinensis, 150f
    Strophomena species, 151, 153f, 154
    Thaerodonta species, 153f, 154
brackish water environment
    Clayton Formation, 197, 211
    coal swamps, 263
Braidwood
    Colchester Coal Member, 268
    mine safety, 278
    nuclear facility, 422
Brainard Shale, 138f, 154, 154f
Brandon Bridge Member, 160f, 161
brass, 300
breakwaters, 414f
breccias, 83, 100, 101, 108f, 171
Brereton Limestone Member, 188f, 195f, 198, 200, 272
bricks, 36, 37f, 280, 316
Bridgeport, 287f
Bridgeport (Ill.), 283f,
brine shrimp. *See Artemia* species
brines, 341f, 342t
Bristol Hill Coal Member, 201, 267f, 272
brittle star, 172f–173f
bronze, 300
Brookport, 106f
Brookville Dome, 96f
Broughton, 114–115
Brown County, 291
Browning, 287f
Brownsville sub-Pennsylvanian valley, 196f
Brunhes normal magnetic polarity, 223f
Brunhes-Matuyama reversal, 227
Brussels Quadrangle geological mapping, 105
Bryozoa
    Archimedes species, 179
    Archimedes wortheni, 181f
    Clayton Formation, 211
    Cystodictya species, 172f–173f, 180
    Devonian, 170
    encrusting Bryozoa, 171
    fenestrate Bryozoa, 178
    fluorspar deposits, 305f
    Galena Group, 151
    Maquoketa Group, 153f, 154

Index

Pennsylvanian, 189
Platteville Group, 149, 150f, 151
Prasopora species, 153f, 154
Prismopora serrulata, 175f, 182
Trepostomata species, 150f, 151, 153f, 154
Ullin Limestone, 178
Valmeyeran, 180
BTEX (benzene, toluene, ethylbenzene, xylene), 452
"Buchanan," 188f, 287f
Buckhorn Consolidated Field, 291, 291f, 292f
"Buff" dolomite, 302f
Buffalo Hart Moraine, 226f, 230
Buffalo Rock State Park, 72t, 147
building stone, 317–318, 318f, 319
    Chesterian, 184
    history, 20, 20f
    reclamation, 462f, 462–463, 463f
buildings
    construction, 373
    damage, 385–386
        earthquakes, 424–425, 425t, 428f, 429f
        foundations, 29, 386f, 388f
        land subsidence, 389–390, 391f, 391–392
Burdick Member, 228
Bureau County
    Green River Lowland, 72t, 333
    Illinois River valley, 237f, 333
    mine safety, 278
    radioactive waste, 445
    Shelburn Formation, 200
    valley train deposits, 238
    wetlands, 367
buried features
    hills, 101–102
    impact structures, 100, 101f
    reefs, 101–102
    stream valleys, 25f
    See also aquifers; bedrock: valleys
Burlington Limestone, 176, 180f
    commodity, 312
    Dizygocrinus species, 178
    Melonechinus multiporus, 178
    New Albany Shale, 174f
    Spirifer grimesi, 177–178
    stratigraphic column, 168f
    Valmeyeran paleogeography, 177f
Burlington Shelf, 177f
Burlington-Keokuk Limestone, 168f, 175f, 177f, 177–178
burrows
    Clayton Formation, 211
    McNairy Formation, 210f, 211
    Pinicon Ridge Member, 171
Burtschi, 287f
button-shaped coral, 172f–173f, 173
Byron
    nuclear facility, 422
    near Plum River Fault Zone, 96

C

C horizon, 234f, 374, 374f
Cache Bedrock Valley, 333, 333f, 335
Cache River State Natural Area, 72t
Cache River valley cypress swamps, 365
cadmium, 266
CAFOs (concentrated animal feeding operations), 447–448
Cahokia Creek, 109f
Cahokia Formation, 223f, 229f, 242
Cahokia mound builders, 316
Cairo
    earthquakes, 427, 428–429
    Eocene stratigraphic units, 212
Calamites species, 190f
calcareous algae
    Asphaltinoides grandtowerensis, 171
    Devonian, 170
Calceocrinus species, 151
calcium carbonate (calcite)
    calcium fluoride, 305
    Chesterian concretions, 183
    in coal ash, 265
    groundwater quality, 340, 346f
    limestone, 309, 311f
    pre-Illinois deposits, 225
    speleothems, 255f, 256, 256f
    water hardness, 340
calcium
    in clay, 315
    groundwater quality, 337, 338, 338t, 340–341, 341f, 342f, 343
    supplements, 312
calcium chloride, 455
calcium fluoride, 305, 343
calcium-magnesium bicarbonate, 338f, 340–341, 341f
calderas, 42, 81, 133, 137
Calhoun Coal Member, 201, 267f, 272
Calhoun County
    carbonate rocks, 168
    Joachim Dolomite, 149
    Pancake Hollow Section, 228
caliche, 183
California earthquakes, 424
Cal-Sag Channel, 242, 366
Calumet area, 408, 410f
Calumet City, 209
Calumet Harbor (Ind.)
    Illinois coast, 404f, 410f, 412f
    sand entrapment, 411
Calumet Lake Plain (Ind.), 405
Calumet phase, 408, 409f, 410f
Calumet River, 410f
Calumet Spit, 409f
calymenid trilobite, 153f
Cambrian, 136–147, 147f, 148–157
    aquifers, 299, 330–331
        radioactivity, 348, 419, 420f
        sulfate, 346
    bedrock, 68f
    chronostratigraphy, 65f, 66f, 67f
    cratonic embayment, 82f
    cross sections
        Illinois, 66f
        Illinois and eastern Indiana, 61f
        Illinois and western Kentucky, 61f
    Des Plaines buried impact structure, 62, 63f, 92f, 100–101
    faults, 97, 100, 113
    Fluorspar Area Fault Complex, 100
    fossils, 138–139, 140f, 142
    gas storage, 288
    groundwater quality, 339f
    Illinois Basin, 143
    Mt. Simon Sandstone, 124f, 128f, 139
    Ogle and Lee Counties, 143f
    Pascola Arch, 86f
    proto-Illinois Basin, 79–81, 127f
    Rough Creek-Shawneetown Fault System, 100
    Sauk III subsequence, 142–143
    sea level, 137
    stratigraphic columns, 137f, 138f, 287f
    sub-Tippecanoe surface, 147f
    tectonics, 62f, 80f–81f
    total dissolved solids, 339t
Camel Rock, 69f
Camp Vandeventer spring, 440f
Campanian Stage, 206f
Campbell Hill, 13f
Campylorthis deflecta, 150f, 151

Canadian (Series), 66f, 287f
Canadian Shield, 170, 182, 193
    Cambrian, 155
    Chesterian, 170f
    Galena Group, 151
    incised valleys, 195
    Labrador ice-dispersal center, 216f, 216, 238
    Ordovician, 139f, 155
    paleogeography, 91f, 139f
    Pennsylvanian, 193f
    Precambrian, 77–78, 78f
    Quaternary, 216f, 216
Canton Shale, 188f, 194f
Cap au Grès Faulted Flexure, 62, 63f, 92f, 95
    neotectonics, 107f
    Quaternary faults, 105
    tectonics, 62f, 84, 85f
Cape Girardeau (Mo.), 304f
Cape La Croix Shale, 137f, 154–155
Cape Limestone, 137f
Capital Hotel, 441
caprock, 289
Captain Mine, 278
"captive" mines, 277
carbon dioxide
    groundwater, 341, 341f, 346
    speleothems, 253, 255f
carbon isotopes, 253, 256, 258
carbon tetrachloride, 453
carbon-14 dating
    cave sediment, 433
    Farmdale Geosol, 239
    groundwater, 347
    history of investigations, 27, 31f
    New Madrid earthquakes (1811–1812), 109–110
    paleoclimatogy, 244f, 248, 249t, 252f, 254f, 255f
    Quaternary sediment, 236f
    speleothems, 433
    water, 34f
carbonate rocks
    aquifers, 338
        vulnerability, 351–360
    argillaceous texture
        Brandon Bridge Dolomite, 161
        Decorah Formation, 151–152
        Glenwood Formation, 148–149
        Offerman Member, 161
        Racine Dolomite, 161
        St. Laurent Formation, 173
        Sugar Run Dolomite, 161
    belt, 139f
    deposition, 84, 167
    fossils, 177f
    karst, 432, 433
    minerals, 341
    northern Illinois, 311
    Ordovician, 160
    Paleozoic, 216f
    Salem Limestone, 294
    Silurian, 289, 290
    turbidites, 163
    Valmeyeran, 177f
    wetlands, 364
carbonatites, 100, 202
Carbondale, 198
Carbondale Formation, 195f, 198–199, 199f
    coal seams, 267f
    Hicks Dome, 101f
    Murphysboro Coal Member, 267, 267f
    petroleum reservoirs, 287f, 297
    Rough Creek-Shawneetown Fault System, 99f
    stratigraphic columns, 188f, 287f
carbonic acid, 255f

Carboniferous, 8f, 65f, 66f, 168f, 188f
Cardiomorpha species, 191f
Carlinville landfill, 447f
Carlinville Limestone, 188f
Carlyle, 287f
Carlyle Lake, Illinois Episode, 230f
Carpentersville, 459f, 461
"Carper sand," 168f, 180f, 287f, 293
Carrier Mills Shale, 188f
Carroll County
    lead-zinc deposits, 301
    Mississippi Palisades State Park, 72t, 165f
    Plum River Fault Zone, 96f
Carter Oil Company, 17f
Carthage Bedrock Valley, 333, 333f
Carthage (Shoal Creek) Limestone Member, 188f, 195f, 200, 201
Cary substage, 22
Caryocrinites species, 162f–163f
"Casey," 188f, 287f
Casey oil well, 11f
Casey Quarry reclamation award, 458
Caseyville Formation, 187, 189, 195f, 195–197, 197f
    coal seams 267f
    Garden of the Gods State Park, 121f, 197f
    Hicks Dome, 101f
    petroleum reservoirs, 287f, 297
    Rough Creek-Shawneetown Fault System, 99f
    stratigraphic columns, 188f, 287f
Caseyville Group, 195f
Cass County wetlands, 367
Castle Rock State Park, 72t, 148
catalysts (clay minerals), 19
cations, 337
Catlin silt loam, 375f
Catskill Delta, 169f
Cave Hill Shale Member, 168f, 183f
Cave-in-Rock
    area minerals, 304, 304f, 305
    Chesterian, 181
    fluorspar, 307
    karst, 432
Cave-in-Rock District, 304f
Cave-in-Rock State Park, 72t, 441
caves
    age determination (Mo.), 433
    Cave-in-Rock State Park, 72t
    Illinois Caverns, 33f, 72t, 432, 439, 441
    Illinois Caverns State Natural Area, 72t
    karst, 432, 433, 434f, 439–441
    paleoclimatology, 253, 255f
Cayugan, 66f, 160f, 287f
Cedar Valley Limestone, 168f, 173, 174f, 290f
Cedaria species (Iowa), 142
Celotex landfill, 447f
cement materials, 280, 311f, 312, 313, 316, 319, 319f
Cenozoic, 61f, 65, 65f, 66f, 67f, 79f, 115
Center for Research on Sulfur in Coal, 32
Central Fault System (Ky.), 63f
Central Lowlands Province, 70, 71f
Central Midwest Interstate Low-Level Radioactive Waste Compact, 421
Central Mississippi Valley Earthquake Bulletin regional earthquake data, 425–426
Central United States Earthquake Consortium (CUSEC), 36
central United States earthquakes, 424–431
    1810 to 1995, 113f
    1975 to 1995, 429f
Centralia
    mine safety, 278
    Springfield Coal Member, 270
Centralia Environmental Services, 447f
Centralia Field
    Geneva Dolomite, 293
    Kimmswick ("Trenton") Limestone, 289

Index

Centralia sequence
    Eastern Granite-Rhyolite Province, 127
    history of investigations, 41
    Illinois Basin, 128f, 133, 141f
    Illinois cross section, 66f
    Mt. Simon Sandstone, 81
    Precambrian, 79f, 80f–81f, 81, 125f, 128f, 141f
    proto-Illinois Basin, 79, 81, 127f
    seismic profiles, 81, 82f, 126f, 127, 129f
Cephalopoda
    Carbondale Formation, 199
    Dawsonoceras annulatum, 162f–163f
    Endoceras species, 151
    Endolobus species, 172f–173f
    Galena Group, 150f, 151
    Maquoketa Group, 154
    Muensteroceras species, 175f
    Nautiloidea, 170
    Pennsylvanian, 189
    Richardsondoceras species, 150f, 151
    Shakopee Dolomite, 145
ceramics
    clay mineralogy, 19
    claystone, 315
    fluorite, 303
    limestone, 313
    magnesia, 313
    tripoli, 316
CERCLA (Comprehensive Environmental Response, Compensation, and Liability Act; 1980) legislation, 29t
CERCLA sites, 33f, 444–445, 446f
Chaenomya species, 191f
Champaign
    Cairo earthquake (1895), 427
    La Salle Anticlinorium, 93
    Mahomet aquifer, 344f
    Sangamon Arch, 93
Champaign County
    Champaign Moraine, 236f
    groundwater, 21f, 338t, 342, 344f
    Mammut americanum, 242f
Champlanian, 287f
channels
    coal deposits, 270, 270f, 271f, 271
    drainage, 383
    human activity, 401
    to Lake Michigan, 410f
    petroleum reservoirs, 296f, 297
    point source pollution, 443
    Quaternary sediment, 220f
    See also meltwater outflow channels
Chapel Coal Member, 188f, 200
Charles Mound, 70, 71f, 165
Charleston District, 271f
Charleston Quarry, 220f
Chatfieldian Stage
    Millbrig K-bentonite Bed, 152
    stratigraphic columns, 137f, 138f
Chatsworth Bedrock Valley, 333
Chatsworth Moraine, 236f
Chattanooga Shale, 169f
chemical analysis, 40–41
chemical pollutants
    leaking storage tanks, 351
    spills, 351, 444, 448
    uranium, 419
chemical precipitation
    oxygen isotopes, 253, 256
    speleothems, 255f
Chenopodiaceae, 252f
Cherry Mine disaster, 278
chert
    Baylis Formation, 209
    Clayton Formation, 211
    Devonian, 163
    Eocene, 212f
    Grover Gravel, 214
    on Illinois cross section, 66f
    Mounds Gravel, 213
    Post Creek Formation, 210f
    reservoir rocks, 287f
    Shakopee Dolomite, 144–145
    Silurian aquifers, 331
    St. Laurent Formation, 173
    Wilcox Formation, 212
Chester, 180, 181, 388f
Chesterian, 180–184, 184f
    cross sections
        Illinois, 66f
        Ste. Genevieve Fault System, 99f
    fossils, 175f
    history of investigations, 14
    outcrops, 178f, 180, 181
    petroleum reservoirs, 287f, 296
    Rough Creek-Shawneetown Fault System, 99f
    sediments, 170f, 182
    stratigraphic columns, 168f, 183f, 188f, 287f
    sub-Pennsylvanian valleys, 196f
Chicago
    aquifer vulnerability, 452
    Des Plaines buried impact structure, 62, 63f, 92f, 100–101
    engineering geology
        slope stability, 389f
        soil dessication, 385–386
        St. Hedwig Church, 386
        underground installations, 29, 101, 392, 392f, 393f
    floods, 35f, 35–36, 396f
    history of investigations, 7, 15, 35–36
    Illinois coast, 404f, 405–406, 406f, 407–410, 410f, 411–412, 412f, 413f, 413–417
    industrial minerals, 320
        Devonian, 158
        dolomite quarries, 38
        lime, 164
        quarries, 313, 320f
        Silurian stone, 158, 164
        Thornton Quarry, 38f
    Metropolitan Water Reclamation District of Greater Chicago, 164
    road salt, 455
    Tunnel and Reservoir Plan (TARP), 29, 164
    Wadsworth Formation, 238
    waste management, 392f, 446, 447, 447f
Chicago Harbor, 404f, 412f, 413
Chicago Heights landfills, 447f
Chicago Lagoon, 409f
Chicago Lake Plain, 71f, 405, 406f, 407, 408, 409f
Chicago Outlet, 241f, 242
Chicago Outlet Valley, 406f, 408
Chicago River, 410, 411, 412f
    floods, 393
    wetlands, 365
Chicago, Rock Island, and Pacific Railroad, 147f, 149f
Chicago-O'Hare International Airport, 404f
Chicago-O'Hare Reservoir, 220f
Chicago Water Tower, 165
Chippewa phase, 410f
Chironomidae, 252f, 253
Chiroptera, 440
chloride, 337, 342–343, 448, 455
chlorine
    in clay, 315
    in coal, 266
    in groundwater, 338t, 347
    in water, 338t
chlorofluorocarbons, 303, 346
Choat, 106f, 107t

Cholaster peculiaris, 172f–173f
Chouteau Limestone, 168, 174f, 176, 180f
   stratigraphic columns, 168f, 287f
Christian County
   aquifer vulnerability, 353
   drumlins, 231
   Herrin Coal Member, 271
   roof control, 275f
chromium, 266
chronostratigraphy, 60f, 62f, 63f, 64f, 64–65, 65f, 66f, 66–67, 67f, 78f
Cincinnati Arch, 169, 170f
Cincinnatian, 60, 152–155
   fossils, 182
   generalized geological column, 287f
   Illinois cross section, 66f
   Kaskaskia sequence, 171f
   paleogeography, 91f
   stratigraphic columns, 137f, 138f, 287f
   structural geology, 60f, 63f, 91f
   tectonics, 62f
Claiborne Formation, 212
Claiborne/Wilcox Formation, 206f
clams. *See* Bivalvia
Clark County
   Danville Coal Member, 272
   petroleum, 11, 15, 283
   reclamation, 458
classification
   aquifer vulnerability, 352f, 353f, 354f, 355f, 356f, 357f
   Caseyville Formation, 197
   earthquakes, 424–425, 425t
   Glasford Formation, 232
   groundwater protection, 342t, 352–353
   Pennsylvanian, 193–194, 195f
   physiographic divisions, 22
   Quaternary, 10, 13, 220–221, 221f, 222, 223f, 232
   Tradewater Formation, 198
clastic belts, 183
clastic wedges
   Carbondale Formation, 199, 199f
   Decorah Formation, 152
clasts, 338
Clathrospira species, 150f, 151
clay
   Alto Pass, 212
   aquitards, 327, 351
   Aux Vases Sandstone, 296
   Baylis Formation, 209
   bedrock valleys, 334, 335
   Equality Formation, 229f
   floods, 400f, 401
   in groundwater, 340, 343
   hydraulic conductivity, 326t, 327
   on Illinois cross section, 66f
   Illinois Episode, 229f
   Kaskaskia Bedrock Valley, 335
   liners, 32, 316
   McNairy Formation, 210
   meltwater, 333
   moisture, 30f, 315
   parent materials, 380f
   Pennsylvanian, 187
   Porters Creek Formation, 211, 211f
   Post Creek Formation, 210
   Quaternary sediments, 226f, 236f
   research program, 19f, 19–20
   soil moisture, 315–316, 385–386
   in soils, 374f, 377
   speleothems, 256f
   Tradewater Formation, 198
   underground installations, 393f
   wetlands, 363, 365, 366
   Wilcox Formation, 212

"clay bed," 302f
Clay City Anticline, 63f, 92f, 196f
   oil fields, 286f
   during Pennsylvanian, 193
   structural traps, 95
Clay City Field, 17, 19f, 284–285
clay deposits, 315–316, 317f
   in coal ash, 265
   contaminant filter, 354
   economic value, 319f
   flint clay (pipestone), 317
   history, 7, 19, 317f
   lead-zinc deposits, 303f
   mining, 311f
   pet litter, 211
Claypool, 287f
claystone
   dikes ("horsebacks"), 270, 272
   Pennsylvanian, 315
   Tradewater Formation, 198
Clayton Formation, 206f, 210f, 211
   Lamna cuspidata, 211f
   Turritella species, 211f
Clean Air Act (1963; amended 1970, 1990), 29t
clean coal technologies, 279, 280
Clean Water Act (amended 1977), 29t, 361
Clear Creek Formation, 65f, 159, 163, 170, 240
   fossils, 162f–163f
   reservoir rocks, 287f
   Ste. Genevieve Fault System, 99f
   stratigraphic columns, 160f, 287f
   sub-Kaskaskia surface, 169f
   tripoli, 316
cleats, 279
climate, 243, 248–250, 250f
   See also paleoclimatology
climatic controls
   erosion, 383
   flooding, 402
   soils, 376–377
Clinton
   Chesterian rocks, 181
   Mahomet aquifer, 344f
   nuclear facility, 422
   Weldon Springs State Park spring, 342f
Clinton Member, 232
Clore Formation, 168f, 183f, 287f, 297
coal, 263–282
   aquifers, 332
   aquitards, 332
   areal extent, 263–264, 264f
   Bond Formation, 201
   Carbondale Formation, 198, 199f
   clastic wedge, 199f
   coal balls, 41, 190, 192f
   coal seams, 267f, 267–272
      as aquitards, 326
      numbers, 266
      thickness, 264
   cyclothems, 194f
   Des Plaines buried impact structure, 100–101
   environment of deposition, 188, 189, 263–264
   history, 15, 20, 31f, 36, 263
   hydraulic conductivity, 326
   in Illinois cross section, 66f
   Mattoon Formation, 201
   moisture, 264
   stratigraphic columns, 188f, 267f, 287f
   trace elements, 266
   Tradewater Formation, 198
   wetlands, 361, 367
coal deposits, 263–282
   by county, 264f

coal bed methane, 278–279, 280
coal rank, 201, 265, 265f
coal seams, 267f, 267–272
coke coal, 278
Colchester Coal Member, 268, 269f
degasification, 279
economic value, 263, 266, 277, 280, 319f
for electricity, 263, 277
exploration, 266, 269
fuel, 263, 264
Herrin Coal Member, 265f, 271f, 271–272
history of investigations, 11, 15, 16, 17f, 21f, 31f, 36
preparation, 32, 277, 277f
relation to reservoir rocks, 287f
Springfield Coal Member, 270f, 270
synthetic fuel, 279
transportation, 273f, 277
See also individual named coal members
coal mines. *See* mining, coal; individual named coal mines
"Coal Measures," 187
coal reserves. *See* coal deposits
coal resources. *See* coal deposits
coal swamps, 41, 192
coal tar, 279, 445
coal waste, 31, 270, 277, 278, 279–280
ash, 264–265
bricks, 36, 37f
gob, 279, 459, 461
pollutants, 32f, 444, 444t
coal-fired power plants, 313
coastal environment, 404–417
Chicago River mouth, 411
erosion, 30, 404–408
sedimentation, 405
shoreline configuration, 408–411, 413–415
Coastal Plain Province, 70, 71f
coastal plains, 80f–81f, 189, 367
cobalt, 266
COCORP (Consortium for Continental Reflection Profiling), 125f, 126f
COGEOMAP program, 40
coke, 22f, 267, 277, 278, 279
Colchester Coal Member, 195f, 199, 265, 268, 269f
on electric logs, 296f
stratigraphic columns, 188f, 267f
"Cole," 168f, 287f
Coles County
Charleston Quarry, 220f
Fox Ridge State Park, 72t
Colfax, 272
coliform bacteria, 454
collections, 28f, 43–44
Collinsville, 228
colluvium, 221f, 242
Colmar-Plymouth Field, 291f
Columbia Drainage and Levee District, 401f
Columbia Levee failure, 401
Columbia Quarry, 463
combustion processes, 280
Commerce geophysical lineament, 107f, 108–109, 113f
Composita species, 191f
Composita subquadrata, 182
Clore Formation, 172f–173f
Comprehensive Environmental Response, Compensation, and Liability Act (CERCLA), 29t, 33f, 444–445, 446f
concentrated animal feeding operations (CAFOs), 447–448
concrete, 184
condensation, 328, 328f
confined aquifer (artesian aquifer), 325f, 327, 327f, 334
confining beds. *See* aquitards
Conodonta
Carbondale Formation, 199
Polygnathus cristata, 173
research programs, 28

Tippecanoe II subsequence, 161
Valmeyeran, 176
Conostichus ichnospecies, 175f
conservation tillage, 380
Consortium for Continental Reflection Profiling (COCORP) imaging, 125f, 126f
contaminants. *See* pollutants
continuous mining machines (continuous miners), 274, 274f
conularids, 171
Cook County
landfills, 446
Lemont till, 220f
Mt. Simon Sandstone aquifer, 141
Ordovician-Silurian boundary, 154f
sinkholes, 435
Thornton Quarry, 59, 59f, 72t, 164f
wetlands, 366
See also Chicago
Cooks Mills Consolidated, 294
cooperative groundwater reports, 43
copper, 266
Cora Limestone, 168f, 183f
corals. *See* Anthozoa (corals)
corn, 449f, 450f
corrosion, 346
cosmetics, 300, 453
cosmic rays, 419t
Cottage Grove Fault System, 97–98, 98f
Fairfield Basin, 91
Illinois Basin, 62
Illinois-Kentucky Fluorspar District, 304f
Pennsylvanian-Permian, 85f, 86, 202
Permian, 201f, 202
petroleum exploration, 286f, 289
structural geology, 62f, 63f, 92f, 196f
tectonics, 62f
Williamson County, 98f
county maps, 354–358, 392
cratonic embayment, 82f
cratonic margin, 82, 85f, 127f
cratons, 85f, 127f, 158
Crawford County
coal seams, 200
petroleum reservoirs, 15, 284, 296f
Crepicephalus (Iowa, Mo.), 142
Cretaceous
aquifers, 332
bedrock, 68f
chronostratigraphy, 65, 65f, 66f, 67f
cross section, 66f
fossils, 209, 209f, 210, 211, 211f
glacial extent, 216f, 216
Mallard Creek, 108f
Mesozoic Era, 207–211
paleogeography, 208f
sediments, 208f, 209f
seismic profiles, 115
shorelines, 208f
stratigraphic column, 206f
structural geology, 62f
tectonics, 62f, 80f–81f, 209
Mississippi Embayment, 87, 93
New Madrid Rift System, 80f–81f, 81
Ozark Dome, 91
Pascola Arch, 93
Crevice Cave (Mo.), 250f, 258
crevice deposits
dolomite, 301
galena, 303f
karst, 432, 432f
lead deposits, 301
mine roof, 111f
zinc, 301

crevice exposure, 436f
Crinoidea
    Abrotocrinus coreyi, 178
    Agassizocrinus species, 182
    Cupulocrinus angustatus, 153f
    Cupulocrinus gracilis, 150f
    debris, 178
    Devonian, 170
    Dizygocrinus gorbyi, 178
    Endelocrinus fayettensis, 191f
    Eratocrinus coxanus, 178
    fragments, 178, 311f
    Maquoketa Group, 153f
    Mississippian, 172f–173f, 175f, 177–180, 182
    Onychocrinus species, 172f–173f, 182
    Ordovician, 150f, 153f
    Pennsylvanian, 191f
    Phanocrinus species, 172f–173f, 182
    Platteville Group, 150f
    Platycrinites penicillus, 175f, 180
    Platycrinites species, 182
    Pterotocrinus species, 172f–173f, 182
    stems, 180
    stick-like, 170
    Synbathocrinus species, 178
    Talarocrinus species, 175f, 182
    Taxocrinus species, 182
    Tricoelocrinus woodmani, 178
    Zeacrinites species, 182
    Zeacrinites wortheni, 175f
Crocodylidae 209, 211
cropland. *See* agriculture: land use
cross-bedding, 196f, 197, 210f, 213f
Croweburg Coal, 268
Crown Fault, 111f
Crown Number 1 Mine, 275f
Crown Point Phase, 236f, 240, 242
crushed stone. *See* stone, crushed
crust, 77, 77f, 125f, 137f
    See also plate tectonics; tectonics
Crustacea, 190
Cryptosporidium species, 455
Cryptozoic, 65, 65f, 67f
crystal fractionation, 101
"crystalline crust," 123, 341f
Cuba, 287f
Cumberland County
    petroleum exploration, 15
    Webster Branch, 189
Cumberland-Tennessee River, ancient, 224f
Cupulocrinus angustatus, 153f
Cupulocrinus gracilis, 150f
Cupulocrinus species, 151, 154
Curlew Limestone, 188f
cushion (base) gas, 288–289
cyanazine, 452
cyanide, 444
cyclothems, 23, 41, 193–194, 194f, 195f, 197
Cypress Sandstone, 182, 183f, 295f
    fluorspar deposits, 305f
    oil and gas, 184
    petroleum reservoirs, 294–295, 297
    reservoir rocks, 287f
    stratigraphic columns, 168f, 183f, 287f
cypress swamp ecology, 361f, 365, 367
Cystodictya species, 172f–173f, 180
Cystoidea, 162f–163f

D

Dagley, 287f
Dallas City, 318f
Dalmanella species, 150f, 151
damage
    earthquakes, 424–425, 427, 428f, 429f
    land subsidence, 388–392, 391f
    soil moisture, 385–386
Danville
    Danville Coal Member, 195f, 272
    Hegeler Member, 227
    Mahomet aquifer, 344f
Danville Bedrock Valley, 333
Danville Coal Member, 195f, 272
    Shelburn Formation, 200
    stratigraphic columns, 188f, 267f
Darien Moraine (Wisc.), 240
Darriwilian Stage, 137f, 138f
data
    Precambrian, 123–134
    well records, 43–44, 326
Davenport (Iowa)
    nuclear facilities, 422
    Pennsylvanian, 189
    Rock Island Coal Member, 267
Davis Coal Member, 195f, 198, 268
    Jader Coal's No. 4 Mine, 273f
    stratigraphy, 188f, 267f
Davis Formation, 137f, 195f
Davis Shale, 137f, 138f, 142
Dawsonoceras annulatum, 162f–163f
DCE (dichloroethylene), 453
De Kalb County, 330, 331
"dead" ice, 217
Dead River, 407f
"dead zone" (Gulf of Mexico), 368
Decatur, 291, 344f, 427
Decatur Field and Decatur North Field, 291, 291f
Decatur sublobe, 236f
deciduous forest, 249t, 257f, 258
décollements, 86
Decorah Formation, 151–152
    lead-zinc deposits, 302f
    stratigraphic columns, 137f, 138f
deep aquifers, 328–332, 354
    groundwater movement, 328f
    groundwater quality, 330f, 345
    groundwater vulnerability, 351, 354f
deep marsh. *See* wetlands
deep seismic sounding, 113f, 125–128
deep tunneling project (Chicago), 101
Deerpath Trail, Glacial Park, 233f
Degonia Sandstone, 181, 182
    fossils, 182
    reservoir rocks, 287f, 297
    stratigraphic columns, 168f, 183f, 287f
deicing chemicals, 455
Deicke K-bentonite Bed, 137f, 139, 151
deisopropylatrazine, 452
Dekoven Coal Member, 268
    Jader Coal's No. 4. Mine, 273f
    stratigraphic columns, 188f, 267f
Del Ray series, 377f
Delavan Member, 221f, 223f, 238
Delaware Basin, 170f
delta oxygen-18 notation, 251
Delta regional landfill, 447f
deltaic sedimentation
    Carbondale Formation, 199f
    Chesterian, 183f
    environment of deposition, 189
    Kaskaskia II subsequence, 169
    Mississippi Embayment, 93
    Pennsylvanian, 187–188
deltas (wetlands), 364
Delwood Coal Member, 188f
Delwood Formation, 195f

Index

denitrification, 345, 346, 368
Dennis Hollow, 176
depauperate zone, 154
depocenter, 82f, 216f, 216
depth
    aquifer vulnerability, 341f, 351–360, 452
    deep aquifers, 328–332
    Herrin Coal Member, 271
    Precambrian, 123, 126
Derby-Doerun Formation, 137f, 142–143
Derbyia species, 191f
Des Plaines buried impact structure, 62, 63f, 92f, 100–101
Des Plaines River landscape model, 381f, 382f
Des Plaines River valley
    glacial environment, 240, 242, 366
    Silurian, 165
desmethylatrazine, 452
Desmoinesian, 66f, 187–188, 188f
    paleoclimatology, 192
    reservoir rocks, 287f
    Tradewater Formation, 197
detrital deposits
    limestones, 290, 290f
    sedimentation, 137, 139f, 143
    Taconic Highlands, 154
Detroit River Formation, 168f
Devonian, 68f, 158–166, 167–186
    aquifers, 332
    chronostratigraphy, 65f, 66f, 67f
    cross sections
        Illinois, 66f
        Illinois and eastern Indiana, 61f
        Illinois and western Kentucky, 61f
        Kaskaskia sequence, 168f
        Ste. Genevieve Fault System, 99f
    dolomite, 312f
    fossils, 162f–163f, 170, 171, 172f–173f, 173
    Hicks Dome, 101f
    Kaskaskia sequence, 167–186
    limestone, 312f, 432
    New Albany Shale, 99f, 286
    oil and gas fields, 291f
    Pascola Arch, 86f
    petroleum traps, 292–293, 293f
    Plum River Fault Zone, 96
    reservoir rocks, 287f
    Rough Creek-Shawneetown Fault System, 98–99, 99f, 100
    Sangamon Arch, 291
    Shawnee National Forest, 72t
    Silurian reefs, 289, 290f
    southern Illinois, 163
    Ste. Genevieve Fault System, 98, 99f
    stratigraphic columns, 160f, 168f, 287f
    stratigraphic units, 14
    structural geology, 62f, 84, 85f
    sub-Kaskaskia surface, 169f
    tectonics, 62f, 84, 85f
    Tippecanoe II subsequence, 158–166
    total dissolved solids, 339f
    tripoli, 164–165, 311f, 316, 320
diachronism, 64, 64f, 66f, 66–67, 67f
    Dutchtown Limestone, 149
    Joachim Dolomite, 149
    pre-Illinois Episode, 226
    Quaternary, 222, 226, 226f, 236f
    sediment, 219
    St. Peter Sandstone, 146
diamicton, 219, 232, 327
    aquifer vulnerability, 351–356, 356f, 357–360
    hydraulic conductivity, 327
    Illinois Episode, 229–232
    Illinois River valley, 237f
    Lake Michigan lobe, 225, 238

Lemont till, 220f
    pre-Illinois Episode, 224–226
    Quaternary, 223f, 226f, 229f, 236f, 237f, 333
    sand and gravel pits, 220f
    Wedron Quarry, 231f
    Wisconsin Episode, 235–238
    See also till
diatoms, 252f, 253
diatremes, 100, 201–202
Diceromyonia species, 153f, 154
dichloroethylene (DCE), 453
Dickerson borehole, 99f
Dictyoclostus species, 191f
diesel fuels, 452
Digital Elevation Model (DEM), Lake County, 39, 382f, 383
Dikelocephalus species, 143
dikes, 109–110
    Cottage Grove Fault System, 98
    Hicks Dome, 100, 101f
    southeastern Illinois, 201–202
dikes (levees), 396, 398f
dinosaurs, 206, 209
dirt. *See* soils
discharge
    groundwater, 328, 328f
    radioactive waste, 421
    to wetlands, 361
dissected landscape, 225f, 230f
Dissected Till Plains Section, 70, 71f
dissolved oxygen, 344, 450
Ditomopyge species, 191f
Division of Abandoned Mined Land Reclamation, Illinois Department of Natural Resources, 460
Dixon
    La Salle Anticlinorium, 93
    Platteville Group, 151
    St. Peter Sandstone, 313
Dixon Springs Graben, 304f
Dizygocrinus gorbyi, 178
dolomite (dolostone), 20, 145–146, 309–313
    aggregates, 164
    aquifers, 326, 331, 332, 338, 351
    aquitards, 326, 332
    Bonneterre Formation, 142
    carbonate rocks, argillaceous texture, 148–149
    Cedar Valley Limestone, 173
    chert, 144, 145
    Chicago tunnels, 392f
    clasts, 338
    Derby-Doerun Formation facies, 142–143
    Devonian, 312f
    Eau Claire Formation, 142, 330
    Galena Group, 151, 152f
    Geneva Dolomite, 171
    hydraulic conductivity, 326, 326t
    Illinois cross section, 66f
    karst, 433–434
    lead-zinc deposits, 301, 302f, 303f
    mining, 311f
    New Richmond (Roubidoux) Sandstone, 144
    Ordovician, 311f
    Pinicon Ridge Member, 171
    pre-Illinois Episode sediment, 225
    quarries, 311f
    reservoir rocks, 287f
    Sauk sequence, 137
    Silurian, 312f
    speleothems, 255f
    St. Laurent Formation, 173
    stone, crushed, 312f
    stratigraphic column, 287f
    Vulcan Materials quarry, 152f
    See also individual named geological units

domes, 90, 101–102, 123f, 288
Dongola grade school, 438f
Douglas County
    Dover Moraine, 236f
    Herrin Coal Member, 271
    Quaternary sediment, 219, 222f
    Tuscola Quarry, 94, 222f
    Villa Grove Quadrangle, 358f, 359, 359f
    wetlands, 366
Downeys Bluff Limestone (lower Paint Creek Formation), 181–182, 183f
    electrical logging, 296f
    fluorspar deposits, 305f
    stratigraphic column, 287f
Dowty landfill, 447f
"Drab" dolomite, 302f
drainage
    Chicago Outlet Valley, 408
    floods, 395f
    foundations, 386
    karst, 435f
    landscapes, 231f, 377, 377f
    Quaternary history, 224f
    reef location, 292
    wells, 444t
    wetlands, 362, 364, 365, 365f, 366, 396
Drakesboro sub-Pennsylvanian valley, 196f
drape folds, 93, 101–102
    structural traps, 288, 292
dredging
    Montrose Harbor, 413f
    navigation, 400
Dresbachian Stage, 137f, 138f, 140–143
Dresden Nuclear Power Station, 422, 445
drift. *See* Quaternary
drift gas, 283, 346
Driftless Area, 69, 69f, 217, 219, 225f, 229f
    glacial extent, 216f
    Illinois Episode, 233
    karst, 433–434, 434f
    pre-Illinois Episode, 224
    research programs, 9, 10f
    Silurian, 159
drill cores
    analyses, 133f
    Des Plaines buried impact structure, 101
    hardground omission surface, 151f
    logs, 15
    Permian, 201f
    Precambrian, 123, 125f, 132
    Quaternary, 236f, 243
drumlins, 230, 231, 231f
Drummer (Illinois state soil), 378–379, 379f
Drummond Member, 160, 160f
dry cleaning industry, 444
Du Quoin Monocline, 62f, 63f, 92f, 95, 171f, 304f
    environment of deposition, 198
    Fairfield Basin, 91
    Illinois and eastern Indiana cross section, 61f
    Illinois Basin, 62
    Kaskaskia sequence, 171f
    oil fields, 286f
    seismic profiles, 114
    Springfield Coal Member, 270f, 271
    sub-Pennsylvanian valleys, 196f
    tectonics
        faults, 93
        Illinois-Kentucky Fluorspar District, 304f
        Mississippian, 84, 85f
        Pennsylvanian, 84, 85f, 189f, 193f
        Permian, 202
Dubuque Formation, 138f, 152, 302f
Dudley, 287f
Dunbarella species, 191f

Dunderbergia Zone (Mo.), 142
dunes
    aquifer vulnerability, 353
    Calumet Harbor, 411
    Chesterian, 183
    Illinois River valley, 239f
    Jurassic, 207
    Kankakee County, 353
    Quaternary, 72t, 217, 229f, 238, 239f
Dunleith Formation, 138f, 302f
DuPage River valley, 165
Dupo, 346f, 440f
Dupo Field, 289
Dutch Creek Sandstone Member, 159, 170–171
    petroleum reservoirs, 287f, 292, 293
    stratigraphic columns, 168f, 287f
Dutchtown Limestone, 146, 149
    in cross sections, 137f, 148f
    reservoir rocks, 287f
Dyer Spit (Ind.), 409f
Dykersburg Shale
    clastic wedges, 199
    cyclothems, 194f
    Galatia channel, 194, 270, 270f
    Galatia Mine, 112f
Dykstra, 287f
dysaerobic environment, 84
Dysart paleosol (Iowa), 227–228

E

E horizon, 234f, 374, 374f
Eads Bridge, 165
Eagle Creek State Park, 72t
Eagle Valley-Moorman Syncline, 304f
earthen dams, 396
earthquakes, 105–116, 424–430, 430f, 431
    near Aurora (1909), 426f
    California, 424
    central United States, 424–431
    effects, 109–110, 424–425, 426, 428f, 429, 429f
    epicenters
        along Commerce geophysical lineament, 107f, 109
        in New Madrid Seismic Zone, 113, 113f
        recorded (1974–1994), 87
    faults, 112–116
    focus, 115f, 424
    geological hazards, 429–430, 430f
    great earthquakes, 105, 109–110, 114–115, 428–429
    history in Illinois, 113f, 425–426, 426f, 427–428
    intensity, 424, 425t, 426f
    Lawrenceville, 428f
    magnitude, 424, 425t, 428f
    Modified Mercalli Intensity scale, 425t
    New Madrid earthquakes (1811–1812), 109–110
    northern Illinois (1972, 1999), 427f
    Nuttli magnitude scale, 424
    Quaternary sediments, 243
    research programs, 30, 36
    seismic profiles, 113–114, 114f, 115f, 115–116, 116f
    southern Illinois, 428f
    swarm (Olmsted), 107f, 113f, 115–116, 428, 429, 429f
    Wabash River valley, 110
    Wabash Valley Fault System, 97
    Wabash Valley Fault Zone, 115f
    See also New Madrid Rift System; New Madrid Seismic Zone
earthworms, 377
East St. Louis area
    carbon-14 dating, 253, 254f, 255f
    floods, 396, 401–402
    Keller Farm section, 253, 254f, 255f
    Mississippi River floodplain, 396
East-Central Iowa Basin, 160–161

Index

Eastern Granite-Rhyolite Province, 77–78, 78f, 81
    paleogeography, 125f
    Precambrian, 123, 124, 133
Eastern Interior Basin. *See* Illinois Basin
"Eastern Kansan till," 227
eastern provenance, 225
Eastern Shelf, 91f, 200
Eau Claire Formation, 142, 330
    Brachiopoda, 140f
    stratigraphic columns, 137f, 138f, 287f
Echinodermata
    Calceocrinus species, 151
    Cupulocrinus species, 151, 154
    Platteville Group, 149
Echinoidea
    in drift, 209
    Melonechinus multiporus, 178, 179f
    Salem Limestone, 180
economic value
    Chesterian rocks, 184
    Devonian-Mississippian rocks, 167
    geological mapping, 39
    geological structures, 90
    Illinois coast, 404
    industrial minerals, 16–23, 309, 316–319, 319f, 320
    Precambrian deposits, 124
    resource planning, 358, 359
    Silurian and Lower Devonian rocks, 164–165
    sulfur in coal, 265
    water resources, 39
    See also coal deposits; individual named resources
ecosystems
    caves, 440
    soils, 373, 374, 376
Ectomaria species, 150f, 151
Eddyville, 304f
Edenian Stage, 137f, 138f, 152
Edgewood Creek reservoir rocks, 287f
Ediacaran, Metazoa, 138
Edinburg fields, 291f
Edwards County, 268
Edwardsville, 220f
Eichorn, 304f
Eifelian, 168f, 170
El Dara series, 378f
El Niño–La Niña cycles, 248, 255f
El Niño–Southern Oscillation (ENSO) cycles, 253
Elco
    silica, 209f
    tripoli, 207
Eldorado, 270
Eldredgeops (Phacops)
    Dutch Creek Sandstone, 171
    St. Laurent Formation, 173
Eldredgeops (Phacops) cristata Hall, 172f–173f, 173
electric conductivity, 347
electric power production, 263, 277
electrical earth resistivity, 20
electrical logging, 266, 296f
Elgin, 458f
Elizabeth, 301f
Elizabethtown
    fluorspar, 304f, 307
    Hicks Dome, 101f
Elkhart, 270
Elliott series, 377f
Elmhurst Sandstone, 138f, 330
Elvinia Zone, 142
Elwood Dolomite 160, 160f
Embarras Bedrock Valley, 333f
embayments
    Chicago Lake Plain history, 407, 408
    Pennsylvanian, 189
    proto-Illinois Basin, 127f
    See also Mississippi Embayment
Eminence Dolomite, 143, 147f
    aquifers, 330
    stratigraphic columns, 137f, 138f, 287f
emissions (coal combustion), 32, 280
Empire District, 304f
Emsian, 160f
en echelon faults, 96, 96f
Encrinurus species, 150f, 151
encrustations, 341
end moraines, 229f
    Illinois Episode, 233
    Quaternary, 217, 219, 231f, 236f
    Wisconsin Episode, 230, 235–238
    See also moraines
endangered species, 368
Endelocrinus fayettensis, 191f
Endoceras species, 151
Endolobus species, 172f–173f
Energy (Ill.), 391f
Energy Reorganization Act 1974 (Nuclear Regulatory Commission established), 20, 44
Energy Resources of Indiana electric log, 296f
Energy Shale
    clastic wedges, 199
    above Herrin Coal Member, 192f
    roof control, 271, 272
    stratigraphic column, 188f
Enfield sub-Pennsylvanian valley, 196f
engineering geology, 373, 385–394
    building stone, 164
    clay, 316
    Lake Michigan shoreline, 410f, 411–412, 412f, 413, 413f, 414f
    landslides, 34f, 387–388, 388f
    littoral drift, 413
    navigation channels, 400
    Quaternary sediment, 243
    research programs, 15, 23, 30f, 34–35
    site assessment
        aquifer vulnerability, 326, 353
        engineering properties, 385
        facilities, 44, 385–394
        pollution, 444
        radioactive waste, 421
        skyscrapers, 392
    slope stability, 387
    soil moisture, 373, 374, 377
ENSO (El Niño–Southern Oscillation) cycles, 253
Enterobacteriaceae, 454
Enterprise subsequence, 79f, 81, 127, 128f, 140, 141f
Entisols, 375f, 375–376
Entrada Sandstone (Utah), 207
environment of deposition
    Cambrian and Ordovician, 139
    Carbondale Formation, 199
    Caseyville Formation, 195–196
    Centralia sequence, 82f
    Clayton Formation, 211
    Galena Group, 151
    Joachim Dolomite, 149
    Jurassic, 207
    limestone, 309
    Maquoketa Group, 154–155
    McNairy Formation, 211
    Mounds Gravel, 213–214
    Mt. Simon Sandstone, 82f
    Pennsylvanian, 187–189, 192–194, 263–264, 316
    Pleistocene, 313
    Quaternary, 221f, 237f
    Salem Limestone, 294
    Ste. Genevieve Limestone, 294
    Tippecanoe II subsequence, 158

Triassic, 206–207, 207f
Valmeyeran, 176–177
See also marine environment; individual named geological units
Eocene
Illinois cross section, 66f
Massac Creek graben, 109f
stratigraphic column, 206f
Wilcox Formation, 212, 212f
Eodevonaria arcuata, 162f–163f
Eodevonaria species, 171, 172f–173f
Eoleperditia fabulites, 150f
eolian sediments, 22, 219, 221f
eon, 64f, 65f
eonothem, 64f, 65f
epidote, 225
episode, 64f
epoch, 64f, 65f
Equality Fault Zone, 342
Equality Formation
Mammut americanum, 242f
Quaternary, 221f, 223f, 229f
wetlands, 365, 366
Wisconsin Episode, 238
Equality Springs, 343f
era, 64f, 65f
erathem, 64f, 65f
Eratocrinus coxanus, 178
Erie Phase, 240
erosion
benches, 241f
best management practices, 449
floodplain incision, 383
Great Flood of 1993, 400–401
groundwater quality, 340f
Hudson Episode, 242
Illinois Episode, 230
Lake Michigan shoreline, 387f, 410f
landscape effects, 380
levees, 399f, 400f, 400–401
Mississippian, 86
Paleozoic Era, 86
Pascola Arch, 87, 208
pre-Sauk sequence, 139
scour, 401
shorelines, 410f
sinkholes, 437f
soil, 380, 383
topography, 377
Triassic, 206, 207
Valmeyeran, 178f
Zion Beach-Ridge Plain, 411f
Escherichia coli, 454, 455
eskers
Deerpath Trail, Glacial Park, 233f
fans, 217, 230, 233, 233f
Illinois Episode, 230f, 233, 233f
Kaskaskia River watershed, 230, 230f, 231
Quaternary, 217, 219
Wisconsin Episode, 233f
Esmond Member, 232, 233–234
estuaries
Caseyville Formation, 197
Chesterian, 184f, 184
Pennsylvanian, 189
ethylbenzene (BTEX component), 452
eustacy. *See* sea level
eutrophication, 452
Evanston, 404f, 405, 412
Evansville (Ind.), 196f, 304f
evaporation. *See* hydrologic cycle
Everton Formation, 145–146
Ancell Group facies, 148f
stratigraphic columns, 137f, 287f

Excello Shale, 188f, 194f
Exeter, 233f
Exline Limestone, 188f, 200
Exogyra costata, 211f
extinction
Devonian, 173
Mississippian, 84
Pennsylvanian, 192
Wisconsin Episode, 242

F

Fabaeformiscandona rawsoni, 252f
Fairfield Basin, 62, 63f, 91, 92f, 93, 189, 196f
Bond Formation, 200–201
Carbondale Formation, 198
Herrin Coal Member, 271, 271f, 272
Mattoon Formation, 201
mineral resources, 299–300, 300f
paleohydrology, 300f
paleogeography, 91f
Patoka Formation, 200
Shelburn Formation, 200
Ullin Limestone, 293–294
Falling Springs, 346f, 440
false-color infrared satellite imagery, 237f
Famennian, 168f, 173
far-field effects, 86
Farina sub-Pennsylvanian valley, 196f
Farm Creek section, 13, 22, 250f
Farm Ridge Moraine, 236f
Farmdale Geosol, 239
Charleston Quarry, 220f
geochronology, 223f
Thebes, 228f
wetlands, 366
Farmer City, 344f
Farmersville Crown No. 1 Mine, 275f
Farmington Shale, 188f
fault block structures
Sauk I subsequence, 140
Wabash Valley Fault System, 97
western Kentucky cross section, 66f
fault reactivation
at Broughton, 114–115
history of investigations, 30
New Madrid Seismic Zone, 81
Permian, 202
southern Illinois, 115
See also neotectonics
faults, 60, 92f, 94f, 95–96, 113
active faults, 105, 207f
Cottage Grove Fault System, 97, 98f
displacing Mounds Gravel, 107t
en echelon, 96, 96f
Fairfield Basin, 93f
folds, 84, 95
Hicks Dome, 100
Illinois-Kentucky Fluorspar District, 303, 304f
Mounds Gravel, 107t
normal, 95, 97, 212f
Ozark Dome area, 91
Pennsylvanian, 187, 193
petroleum reservoirs, 288, 288f
Porters Creek Formation, 212f
"positive flower structures," 97
post-Pennsylvanian, 93
Precambrian, 42
reverse faults, 114
Rough Creek Graven, 128f
salt-water intrusion, 342, 343
Sandwich Fault Zone, 97f
seismic profiles, 126f

Ste. Genevieve Fault System, 75, 99f
strike-slip faults, 95, 97, 98f, 111
sub-Pennsylvanian valleys, 196f
tectonics, 80f–81f
thrust faults, 86, 96, 96f, 111, 111f, 124f
Triassic, 207f
Wisconsinan, 106
See also individual named fault systems; individual named fault zones
Favosites species, 162f–163f, 171
Fayette County
    aquifer vulnerability, 353
    sand and gravel deposits, 233f
Fayette series, 378f
Fayette silt loam, 376f
Fayville, pit exposure, 210f
Federal Insecticide, Fungicide, Rodenticide Act (FIFRA) (1972), 29t
federal regulations. *See* laws and legislation; regulations; individual named acts
feldspar, 198
felsic composition, 123
fenestrate bryozoans, 178
fens, 364, 366, 367
    See also wetlands
Fermi National Accelerator Laboratory, 393, 394f
Fern Glen Formation, 168f, 177, 180f
fertilizers
    animal waste, 454, 454f
    aquifer vulnerability, 326, 352f, 354
    pollutants, 444t, 448, 449
    wetlands, 368
field studies (historical)
    1839, 7
    1898, 10f
    1900s, 13f
    1910s, 13f
    1920s, 13f, 14f
    1930s, 17f, 20f
    1940s, 21f, 22f
    1960s, 25f
    1990s, 42f
fills, 410f, 411–412, 413
Findlay Arch, 60f, 91f
fine-grained materials
    aquifer protection, 351, 353
    Galesville Sandstone, 142
    petroleum reservoirs, 296
    till, 341f
first commercial natural gas discovery, 95
first Illinois geological survey, 7–9
fish. *See* Pisces
Fishhook Anticline, 63f, 92f
Fishhook Field, 291f
fission products, 445
Flannigan Coal Member, 188f, 201
"Flat Gap Limestone," 65f, 160f, 163
"flats," 303f
Flexicalymene species, 154
flint clay (pipestone), 3, 6f, 316, 317f
floodplains, 396
    alluvium, 379
    aquifer vulnerability, 353
    earth movements, 425
    Entisols, 375–376
    forests, 361f
    Histosols, 376
    horizons, 380
    Illinois River valley, 237f
    Inceptisols, 376
    incision, 383
    land management, 402
    loess, 378
    Meyer, 397f
    nitrates, 452

    wetlands, 364–365, 365f
    See also Great Flood of 1993
floods, 395–403
    caves, 441
    Chicago area, 35f, 35–36
    geological hazards, 395f
    Illinois Caverns, 439f
    Lake Michigan (1987), 35f
    Meyer, 397f
    Ordovician, 41
    paleoclimatology, 72t, 250–251
    tunnels, 35–36
    wetlands, 361, 363, 365, 368
    Wisconsin Episode, 237f
    Woodstock Phase, 240
    See also geological hazards; Great Flood of 1993
Florence Quadrangle, 233f
flowstone, 253, 441
flue gases, desulfurization of, 27
fluid inclusions, 300
fluid migration, 299–300
fluorides, 343–344
fluorine
    in bedded replacement deposits, 305
    in coal, 266
    in groundwater, 338t, 343
    history of investigations, 22f, 26
fluorite, 301, 303, 303f
    Fluorspar Area Fault Complex, 100
    stratigraphic units, 305f
Fluorspar Area Fault Complex, 62, 62f, 63f, 92f, 100
    faults, 105–107, 107f, 108
    fluorspar deposits, 100
    Mesozoic, 86f, 87
    Permian, 86f
    tectonics, 62f, 100, 107f
    Triassic, 206
fluorspar deposits, 100, 299, 300, 301–305, 305f
    Hardin County, 303f
    Illinois-Kentucky Fluorspar District, 300f, 303, 304f
    mining, 6, 7, 11, 21, 304f, 306f, 306–307
Fluorspar District faults
    flute, 232f
    New Madrid Seismic Zone, 116
    Permian, 201f, 202
fluvial processes
    Chicago Outlet, 408
    incision, 213–214
    sedimentation
        Caseyville Formation, 197
        Chesterian, 184f
        Illinois Episode, 234f
        Jurassic, 207
        Lake Michigan lobe, 240
        sandstone, 183f
        Wedron Quarry, 221f
        Woodstock Phase, 240
    Triassic, 207f
fluviokarst, 432–434
flux
    clay, 316
    fluorspar, 11–12, 303
    limestone, 313
    magnesia, 313
fly ash. *See* ash (combustion product)
focus (earthquakes), 424
Foerstephyllum species, 150f, 151
Fogelpole Cave, 250f, 256, 256f
folds, 90–95
    Devonian strata, 292
    drape folds, 93, 101–102, 288, 292
    Illinois Basin overview, 60
    in Illinois, 92f

Pennsylvanian, 187, 193
salt-water intrusion, 343
structural traps, 97, 286, 288–297
See also individual named structures
Foraminifera
Clayton Formation, 211
in drift, 209
Globoendothyra baileyi, 180
paleoclimatology, 251
Porters Creek Formation, 211
Tippecanoe II subsequence, 161
Valmeyeran, 176
Ford Station Limestone, 168f, 183f
Forest City Basin, 60f, 91f
Forest Park Beach, 414
forests
on floodplains, 361f
paleoclimatology, 249t, 252f, 253, 257f, 258
soils, 373, 373f, 375–376, 377
formation (geological unit), 64, 64f, 220
See also individual named geological formations
Forreston Dome, 96f
Fort Atkinson Limestone, 138f, 154
Fort Dearborn, 411
Fort Dodge Formation, 208f
Fort Dodge (Iowa), 207
Fort Payne Formation, 168f, 169, 176–177, 177f, 178
reservoir rocks, 287f, 293
Rough Creek-Shawneetown Fault System, 99f
stratigraphic columns, 168f, 177, 180f, 287f
Fortville Fault (Ind.), 63f
fossils
Cambrian, 138–139, 140f, 142
Decorah Formation, 151–152
Devonian, 162f–163f, 170, 171, 172f–173f, 173
in drift, 209
fossil ice wedges, 240
Maquoketa Group, 153f, 154
Mazon Creek, 190
Mississippian, 172f–173f, 175f, 176, 177, 178, 179, 179f, 180, 181f, 182, 311f
Ordovician, 139, 143, 145, 146, 148, 150f, 151, 153f, 154
packstone, 173
paleoclimatology, 251, 252f, 253
Pennsylvanian, 189–190, 190f, 191, 191f, 192f, 197, 199, 201
Platteville Group, 151
pollen, 14, 23f, 210, 211, 212, 253
Quaternary, 235, 240, 242, 242f, 251, 252f, 253, 258
relative age, 64
Silurian, 160, 161, 162f–163f
spores, 14, 23, 23f
Tippecanoe I subsequence, 139
See also individual named fossils
foundations (engineering geology), 29, 385, 386, 386f, 388f
Fox Ridge State Park, 72t
Fox River, 144, 146f
Fox River torrents, 240
Fox River valley
Silurian, 165
valley train deposits, 313
Woodstock Phase, 240
Fox River wetlands, 366
fractures
aquifers, 331, 351
coal seams, 111f, 111–112
engineering geology, 385
karst, 432
paleosols, 183
petroleum reservoirs, 289,
salt-water intrusion, 342
Fraileys Shale (Golconda shale), 182, 183f
fluorspar deposits, 305f
reservoir rocks, 287f, 297

stratigraphic column, 168f
Francis Creek Shale Member, 199
Colchester Coal Member, 268
Mazon Creek fossils, 190
stratigraphic columns, 168f, 188f
Franconia Formation, 142–143, 143f, 147f
aquifers, 330
Sandwich Fault Zone, 97f
stratigraphic columns, 137f, 138f, 287f
Franconian, 137f, 138f, 142–143
Franklin County, 227, 268
Franklin Creek, 144
Franklin Grove, 143f
Franklin paleosol (Iowa), 227–228
Frasnian Stage, 168f, 173
Fredonia Limestone Member
Chesterian, 183f
fluorspar deposits, 305, 305f
"McClosky" oolites, 168f, 294
reservoir rocks, 287f
stratigraphic columns, 168f, 183f, 287f
Freeburg, 391f
Freeman Coal Mining Company, Crown No. 1 Mine, 275f
Freeport, 301, 301f
Freon (hexafluoroethane), 303
freshwater algae, 253
Friendsville Coal Member, 201, 267f, 272
Friendsville Mine, 273f
frost heaving, 385
froth flotation, 32
fuel—synthetic, 279, 280
fuller's earth, 319–320
Fulton County, 270
fungi, 377
fungicides, 26
fusulinids, 189
future
Illinois coal, 280
industrial minerals mining, 320
research programs, 44–45

G

gabbro, 125f
Gabriceraurus species, 150f, 151
Galatia channel, 194, 194f, 270, 270f
Galatia Mine, 277f
coal balls, 192f
economic value, 277f, 278
roof fall, 112f
Galena, 299f, 301, 301f
Galena Group, 143f, 151–154, 300, 301
dolomite, 152f, 311
fossils, 150f, 151
hardground omission surface, 151f
under La Salle Anticlinorium, 94f
lead-zinc deposits, 299f, 301, 301f, 302f, 303f
limestone, 311
mining, 6
petroleum reservoirs, 287f, 289
Sandwich Fault Zone, 97f
stratigraphic columns, 137f, 138f, 287f
sub-Kaskaskia surface, 169f
Wise Lake Formation, 152f
See also individual named geological units
Galena Lead-Zinc District, 6–7
Galena-Platteville Groups, 330–331
La Salle Anticlinorium, 94f
lead-zinc deposits, 6–7, 301, 301f, 302f, 303f
Galesburg, 267
Galesburg Plain, 70, 71f
Galesville Sandstone, 142
aquifers, 330, 419

radium, 419
stratigraphic columns, 137f, 138f
Gallatin County
    Davis Coal Member, 268, 273f
    Dekoven Coal Member, 268, 273f
    Garden of the Gods Recreation Area, 120f–121f, 197f
    Jader Coal's No. 4. Mine, 273f
    Pennsylvanian, 189
    Rough Creek Fault System, 98, 99f
    zinc ore, 300
Gamachian Stage, 137f
gamma particles, 418
gamma-ray logs, 266
Gammarus acherondytes, 439f
Garden of the Gods Recreation Area
    Camel Rock, 69f
    Caseyville Formation, 197f
    Shawnee National Forest, 72t
Gary (Ind.), 410f
garnet, 202, 225
gas. *See* natural gas
Gasconade Dolomite. *See* Oneota (Gasconade) Dolomite
gases
    in groundwater, 346–347
    waste disposal, 444
gasoline, 300, 452
Gastropoda
    Clathrospira species, 150f, 151
    Clayton Formation, 211
    climatic proxy, 252f, 253
    Devonian, 170
    Ectomaria species, 150f, 151
    Hormotoma major, 150f
    Hormotoma species, 151
    Lophospira species, 150f, 151
    Maclurites species, 150f, 151
    Maquoketa Group, 154
    Natacopsis species, 191f
    Oneota (Gasconade) Dolomite, 144
    paleoclimatology, 252f, 253
    Pennsylvanian, 189, 191f
    Phragmolites species, 150f, 151
    Pupilla muscorum, 252f
    Shakopee Dolomite, 145
    shells, 248
    Tetranota species, 150f, 151
    Trepospira species, 191f
    Turritella species, 211f
    Vertigo modesta, 252f
    Worthenia species, 191f
Gedinnian Stage, 159
gems, economic value, 319f
general circulation models, 250, 251, 253
Geneva Dolomite Member, 170–171
    Miletus Field, 292
    reservoir rocks, 37, 287f, 292
    stratigraphic columns, 168f, 287f
Gentry Landing, 195
Geochelone crassiscutata, 235, 258
geochemical profiles
    microfossils, 251
    neutron activation analysis, 24f
    Precambrian rocks, 124
geochronology, 59–60, 64f, 64–65, 65f, 66f, 66–67, 67f, 78–88
    Cambrian, 136–157
    Devonian, 158–166, 167–186
    Illinois coast, 404, 408–411
    Kaskaskia sequence, 167–186
    Mesozoic, 206–211
    Mississippian, 167–186
    Ordovician, 136–157
    overview, 59–76
    Pennsylvanian, 187–205
    Permian, 187–205
    Precambrian, 123–135
    Quaternary, 216–223, 223f, 224f, 224–247
    Silurian, 158–166
    Tertiary, 211–215
    Tippecanoe II subsequence, 158–166
    wetlands, 366
Geographic Information Systems (GIS), 44, 383
geological attractions in Illinois, 72, 72t
geological hazards
    aquifer vulnerability, 351–359
    earthquakes, 429–430, 430f
    engineering geology, 387–392
    floods, 395, 395f, 396, 401–402
    hazardous waste, 313, 353, 444, 444t
    history of investigations, 34–35
    Illinois overview, 70–71
    karst, 435–439
    landslides, 388
    radiation, 418–419
    seiches, 408
    slope stability, 389
    underground mining, 274
    See also earthquakes; floods; land subsidence; landslides
geological mapping. *See* mapping
Geological Records Unit, Illinois State Geological Survey, 44f
geomorphology, Illinois coast, 405–407
geophones, 284, 285f
geophysical methods
    Aux Vases Sandstone, 295–296
    coal deposits, 266
    research programs, 17f, 20, 21, 41f
geopotential-field data, 128f, 128–129, 129f, 130f, 131f, 131–132
geosols, 222, 222f, 223f, 228f, 234f, 366
    See also Farmdale Geosol; Sangamon Geosol; Westburg Geosol; Yarmouth Geosol
germanium, 280
giant beaver, 242
Giant City State Park, 197f
giant tortoise, 235, 258
Gibson City, 344f
Gimlet Sandstone, 188f
Girardeau Limestone, 137f, 155
GISP and GISP2 records, 251, 255f
Givetian Stage, 168f, 170
glacial environment
    Illinois Episode, 224, 229, 232–234
    pre-Illinois Episode, 226–229
    Wisconsin Episode, 238–242
glacial extent
    glacial sequences, 219–220
    landscape development, 380
    paleoclimatology, 248
    Quaternary, 10f, 216f, 216–217, 224f, 229f
    Salem Plateau region, 433
    Shelby Phase, 239
    Tippecanoe I sequence, end of, 158
    wetlands, 367
glacial features
    drumlins, 230, 231
    dunes, 217, 229f, 238, 239f
    eskers, 217, 219, 230, 230f, 231, 233, 233f
    glacial lake plains, 217, 229f, 231f
    glacial lakes, 224f, 381f, 382f
    at Glacial Park, 72t, 233f
    gravel hills, 230
    kettles, 217, 233f, 366
    moraines, 10f, 72t, 219, 229f, 235–236, 237f, 238, 365, 408
glacial geology, 9–11, 69, 216–260
    geochronology, 223f
glacial ice volume, 227
glacial Lake Agassiz, 410f
glacial Lake Chicago, 242, 406f, 408, 409f, 410f

glacial Lake Milan, 224f, 239
Glacial Park, 72t, 233f
glacial phases, 236f
glacial processes, 216–247
  history of investigations, 39
  pedogenesis, 373
glacial sediment, 220f
  aquifers, 334f, 351, 353, 357f, 358
  bluff coast, 405
  Charleston Quarry, 220f
  Chicago Lake Plain, 405, 407
  history of investigations, 10f, 40
  Illinois cross section, 66f
  karst, 433
  lake plain, 408
  mapped extent, 250f
  meltwater
    clay, 333
    silt, 333
    soil parent materials, 379
    wetlands, 365
  paleoclimatology, 249t, 250f
  paleohydrogeology, 328
  pedogenesis, 249t
  Quaternary, 216–247
  ridges, 230, 232f
  sand and gravel, 220f, 221f, 233f, 237f
  shorelines, 405
  sinkholes, 435
  stratigraphic column, 206f
  wetlands, 365
  See also outwash; till; individual named moraines
glaciated terrain, 218f
Glasford buried impact structure, 63f, 92f, 100, 101
Glasford Formation, 221f, 223f, 229f, 231
glass materials
  dolomite, 312
  limestone, 312
  St. Peter Sandstone, 147, 314
  tripoli, 316
"glass rock," 302f
glauconite
  Clayton Formation, 210f, 211
  Franconia Formation, 142–143
  Owl Creek Formation, 211
  Porters Creek Formation, 211
  St. Laurent Formation, 173
Glen Dean Limestone, 181–182
  Hicks Dome, 101f
  Mississippian fossils, 175f
  Prismopora serrulata, 175f, 182
  reservoir rocks, 287f
  Ste. Genevieve Fault System, 99f
  stratigraphic columns, 168f, 183f, 287f
Glen O. Jones Lake, 99f
Glen Park limestone. *See* Horton Creek (= "Glen Park") Formation
Glenarm Field, 291f
Glencoe, 404f
Glenwood Formation, 96f, 146, 148–149
  Ancell Group, 330
  cross section, 148f
  lead-zinc deposits, 302f
  stratigraphic column, 138f
Glenwood phases, 408, 409f, 410f
Glenwood Spit, 409f
Glenwood-St. Peter Sandstone unit, 330–331
Globoendothyra baileyi, 180
Glyptambon gassi, 162f–163f
Glyptorthis species, 153f, 154
gneiss, 125f
Golconda
  Chesterian, 181
  fluorspar deposits, 304f, 305f

Golconda Formation, 168f, 183f, 296f, 305f
Golconda lime. *See* Haney Limestone (Golconda lime)
Golconda shale. *See* Fraileys Shale (Golconda shale)
Goldengate Consolidated Field, 293
Gondwana, 80f–81f, 139f
Goose Creek Vein, 304f
Goose Lake Channel, 367
goosefoot, 258
Goreville Limestone Member, 168f, 183f, 188f
Gorham sub-Pennsylvanian valley, 196f
Gorstian Stage, 160f
Gosport series, 378f
Gosport silt loam, 376f
Governor's Radon Task Force, 420
grabens, 83, 106, 107t
  Eocene strata, 212
  Fluorspar Area Fault Complex, 100
  Massac Creek, Massac County, 107t, 109f
Graceland Spit, 409f
Grafton
  building stone, 164, 317
  Gravicalymene celebra, 162f–163f, 163
  Great Flood of 1993, 371f–372f
grainstone, 151, 151f
  fossils, 171, 293–294
Grand Calumet River, 410f
Grand Detour Formation
  lead-zinc deposits, 302f
  stratigraphic columns, 137f, 138f
Grand Tower, 181
Grand Tower Limestone, 159, 170–171
  Geneva Dolomite, 292
  Illinois Basin, 171f
  petroleum reservoirs, 287, 293
  Ste. Genevieve Fault System, 98
  stratigraphic columns, 168f, 287f
granites
  cross sections, 61f, 66f
  glacial transport, 224
  intrusions, 125f
  Precambrian rocks, 123, 124f, 125f
  stratigraphic columns, 137f, 138f, 287f
  tectonics, 60
granodiorites, 123
granophyre, 125f, 132, 133f
granulometry (grain size analyses), 232
Grape Creek, 272
Graptolithina, 153f, 154
grasses, 253, 258
Grassy Creek Shale, 174
  cross section, 174f
  Hicks Dome, 101f
  stratigraphic columns, 65f, 168f
Grassy Knob Chert, 163
  reservoir rocks, 287f
  Ste. Genevieve Fault System, 99f
  stratigraphic columns, 65f, 160f, 287f
  sub-Kaskaskia surface, 169f
gravel
  aquifers
    hydraulic conductivity, 326, 326t
    Ohio River Bedrock Valley, 335
    vulnerability, 351–360
    Wabash Bedrock Valley, 335
    water supply, 332
  Illinois cross section, 66f
  Illinois River valley, 237f
  Mounds Gravel, 213
  Post Creek Formation, 209–210, 210f
  tills, 238
  Wedron Quarry, 221f
  See also sand and gravel
Gravicalymene celebra, 162f–163f, 163

Index

Gravicalymene species, 153f, 154
gravity field, 108–109, 124–125, 129, 129f, 207
gravity surveys, 24
"Gray" dolomite, 302f
Grayslake Peat, 367
Grayville, 270
Grayville-Lawrenceville Anticline, 196f
Great Depression, history of investigations, 16, 16f
Great Flood of 1993, 35, 35f, 396–402
    Alton, 398f
    floodplain, 397f
    Grafton, 371f–372f
    Havana, 400f
    Illinois River valley, 398f
    levees, 398f, 399f, 400f, 401f
    Lock and Dam No. 20, 397f
    Meyer, 397f
    road, 398f
    Valmeyer, 401f
    wetlands, 368
Great Lakes drainage, 241f
Great Lakes Harbor, 404f, 412f
Great Lakes Section, 70, 71f
Green Bay ice lobe, 240
Green River Lowland, 72t
    aquifers, 333, 334, 334f
    physiographic regions, 70, 71f
    wetlands, 366, 367
Greene County, 398f
Greenland GISP, GISP2, and GRIP records, 251, 255f
greenstone, 224
Greenup Limestone, 188f
Greenville, 196f, 270
Grenville Front Tectonic Zone, 125f
Grenville Province, 78f, 238
Griffith Spit (Ind.), 409f
Griggsville Plain, 70, 71f
Grindstaff Formation, 195f
GRIP and GRIP2 records, 251
groins, 414f
Grosse Point, 404f, 405
Grosse Point Lighthouse, 405
ground-penetrating radar, 276
groundwater, 325–336, 351–360
    aquifers, 330f, 331f, 334f, 338–339
    aquitards, 338–339
    deep aquifers, 330f
    depth categories, 329
    discharge, 342f
    groundwater type, 337
    history of investigations, 15, 24, 24f, 25f, 44–45
    Kelly municipal well, 21f
    McHenry County, 358
    outwash, 379
    Quaternary, 219, 243
    radioactive isotopes, 419
    reservoir rocks, 29, 288, 288f
    resource planning, 44–45, 358
    salinity, 338, 342t, 342–343, 344f, 351
    sand and gravel, 237f
    soils, 373
    springs, 342f
    stratigraphic traps, 288f
    Villa Grove Quadrangle, 359
groundwater movement, 327–328, 328f, 341f
    contaminant plumes, 32
    floods, 397, 399, 399f, 400f
    hydraulic conductivity, 338t, 347
    hydraulic gradient, 328
    hydrologic cycle, 328, 328f
    isotopes, 347
    karst, 432, 435–436, 438f, 439f, 439–441
    models, 43, 325f, 341f, 359
    radionuclide movement, 418
    recharge, 347
    salt-water intrusion, 344f, 444t
    soils, 373, 385
    wetlands, 363
groundwater pollution, 33f, 34f, 337, 352, 439, 443–457
    arsenic, 34, 344–345, 345f, 443
    chloride, 342–343
    deicer, 455, 455f
    nitrates, 352f, 449f, 450f
    pesticides, 353f, 449f, 450f, 452
    sulfuric acid, 460
    waste disposal, 355f, 356f, 357f, 454, 454f
groundwater protection, 29t, 351–360, 385
groundwater quality, 337–350, 351–360
    best management practices, 449
    bicarbonate, 340–341
    calcium, 340–341
    chemical composition, 337, 338t, 339–340, 341f
    chloride, 342–343
    depth, 354f
    fluoride, 343–344
    gases, 346–347
    isotopes, 347–348
    magnesium, 340–341
    radium, 420f
    redox (reduction-oxidation), 341f, 344, 345, 347f
    salt-water intrusion, 344f, 444t
    sodium, 342–343
    soils, 373
    suspended solids, 340, 340f
    total dissolved solids, 329f, 331f, 338t, 339f, 341f, 342t, 347
    type, 338t
    water hardness, 340–341
    See also karst
group, 64, 64f, 220
Grove Church Shale, 194–195
    stratigraphic columns, 168f, 183f, 188f, 287f
Grover Gravel, 213f, 214
    stratigraphic column, 206f
    tectonics, 105
    See also Mounds Gravel
Grundy County
    aquifers, 330, 331
    Dresden Nuclear Power Station, 445
    fossils, 140f, 190f
    Houchin Creek Coal Member, 270
    Hyolithes, 139, 140f
    Mazon Creek fossils, 190
    mines, 190
    wetland, 366
Guarantee Trust borehole, 98f
Gulf Coastal Plain, 60f, 80f–81f, 207
Gulf of Mexico
    geochronology
        Cretaceous, 80f–81f, 208f
        Jurassic, 208f
        Permian to present, 80f–81f
        Triassic, 206, 207f
    hypoxia, 368, 448–449
    wetlands, 368
Gulfian, 206f
gullies, 380, 383
Gunter Sandstone, 137f, 138f, 144, 147f
Guttenberg Formation, 302f
gypsum, 207

H

H & L Disposal, 447f
habitat
    cave, 440
    Illinois streams, 443t

Index

reclamation, 320, 458f, 459f, 461, 462–463
soils, 377
wetlands, 361, 368
Hadrosauridae (Mo.), 209
Haeger Member, 223f, 236f, 238, 240, 242
Hagarstown Member, 229f
Hagler borehole, 99f
half-grabens, 79f
halite, 455
Hall Limestone, 188f
halogenated hydrocarbons, 444, 453
Halysites species, 162f–163f
Hamburg oolite bed, 174f
Hamilton County
    Broughton earthquake (1968), 114–115
    Davis Coal Member, 268
    Dekoven Coal Member, 268
    Ullin Limestone, 178
Hamp Mines, 304f
Hancock County
    Archimedes wortheni, 181f
    building stone, 318f
    paleoclimatology, 227
    pre-Illinois Episode deposits, 224
    Sonora formation, 168f, 178, 287f
Haney Limestone (Golconda lime)
    fluorspar deposits, 305f
    Prismopora serrulata, 182
    reservoir rocks, 287f
    stratigraphic columns, 168f, 183f, 287f
Hannibal Shale, 168f, 174f, 174–176
Hanover, 301f
Hanover Limestone, 188f, 194f
harbors, 404f
hard water. *See* water quality: water hardness
hardground surfaces, 41, 151f, 152
"Hardin," 168f, 173–174, 287f
Hardin County
    Caseyville Formation type section, 195
    Cave-in-Rock State Park, 72t
    fluorite, 303f, 307, 343
    Fluorspar Area Fault Complex, 100
    galena, 301f
    groundwater, 338t
    Hicks Dome, 100, 101f, 202
    karst, 433
        Big Sink, 437, 437f
        groundwater pollution, 439
    sphalerite, 301f
    Valmeyeran, 177, 338t
Hardinsburg Sandstone, 182, 183
    plant fossils, 182
    reservoir rocks, 287f, 297
    stratigraphic columns, 168f, 183f, 287f
    Wabash Valley clastic belt, 183
"hardpan" paleosols, 183
Harkness Silt Member, 223f
Harmattan Member, 223f
Harrisburg
    fluorspar mining, 304f
    Houchin Creek Coal Member, 270
    Omaha Dome, 202
    Rough Creek-Shawneetown Fault System, 75
    Springfield Coal Member, 270
Harrisonville Drainage and Levee District, 401f
Harrisonville Levee, 401–402
Harristown Field, 291f
Hartco sub-Pennsylvanian valley, 196f
Harvard sublobe, 236f
Havana
    dunes, 239f
    groundwater floods, 400f
    Illinois River valley, 226f

Mahomet aquifer, 333
Quaternary aquifers, 333
sand and gravel aquifers, 334f
Hazardous and Solid Waste Amendments (1984), 29t
hazards. *See* geological hazards
health effects
    abandoned mines, 460
    arsenic, 34, 338t, 344–345, 345f, 443
    earthquake preparations, 430
    floods, 395
    fluoride, 343
    groundwater pollution, 337, 443–457
    groundwater supply, 351, 453–454
    halogenated solvents, 453
    karst aquifer, 440–441
    radioactivity, 418–419, 419t
    sinkholes, 436
heavy metals
    CERCLA sites, 445
    concentrated animal feeding operations (CAFOs), 448
Hebertella species, 153f, 154
Hegeler, land subsidence, 391f
Hegeler Member, 223f, 227
helium, 347
Henderson County, 177
Hennepin, 237f, 402
Henry, 237f
Henry County
    Green River Lowland, 72t
    valley train deposits, 238
    wetlands, 367
Henry Formation, 221f, 223f, 229f, 238
Henryville bed, 174f
heptachlor, 452
herbicide. *See* pesticides
Herod, 304f
Herrick silt loam, 374, 374f
Herrin, 271
Herrin Coal Member, 198, 199, 265, 265f, 271f, 271
    coal balls, 192f
    coke coal, 278
    Cottage Grove Fault System, 98f
    roof control, 201f
    stratigraphic columns, 188f, 267f
Hesperorthis concava, 150f
Hesperorthis species, 151
Heterocypris punctata, 235, 258
hexafluoroethane (Freon), 303
Hexagonaria profunda, 173
Hexagonaria species, 171
"Hibbard," 168f, 287f
Hickory Hills Member, 223f
Hickory soil series, 378f
Hicks, 304f
Hicks Dome, 63f, 92f, 100, 101f, 201f, 201–202
    fluorspar mining, 304f
    Illinois and western Kentucky cross section, 61f
    Illinois-Kentucky Fluorspar District, 304f
    mineralization, 300
    tectonics, 62, 62f, 85f, 86
"high coal," 264
high point (Ill.), 70, 71f
Highland Park, 387f, 404f, 405, 406f
Highland Park Moraine, 405, 406f
highstands 197
Highwood, 404f
Hillcrest, 143f
Hillery Member, 223f
Hillside, 154f
Hirnantian, 137f
history
    clay, 316
    coal, 263

INDEX

distinguished scientists, 4f, 5f
earthquakes, 113f, 425–428
fluorite mining, 306–307
geological investigations, 3–58
industrial minerals, 316–319
lead-zinc mining, 305–306
oil and gas, 283–298
paleoclimatology, 249t
Quaternary research programs, 217–219, 242–243
soils, 373
stone transportation, 318–319
Histosol, 375–376
Hobbs Creek Fault Zone, 106f
"Hoing" sandstone, 173, 168f, 173, 287f
Holocene
    faults, 108
    gravel, 108f
    Mallard Creek, 108f
    Massac Creek graben, 109f
    stratigraphic column, 206f
    See also Hudson Episode
Homerian Stage, 160f
Hoopeston, 344f
Hopwood Farm, 258
horizons, 222f, 234f, 253, 366, 374, 374f, 380
horizontal drilling, 37, 292–293
Hormotoma major, 150f
Hormotoma species, 151
horn coral, 150f
hornblende, 343
Hornsby District, 271, 271f
"horsebacks." *See* claystone: dikes ("horsebacks")
Horseshoe Quarry, 99f
"Horseshoe upheaval," 98–100, 99f
horsts, 100
Horton Creek (= "Glen Park") Formation, 168f, 174f, 174–176
Houchin Creek Coal Member, 270
    cyclothems, 194f
    stratigraphic columns, 188f, 267f
Houghton, 377f
Hudson Bay, 216f
Hudson Episode, 223f, 229f, 238, 242
    See also Holocene
Hulick Member, 223f, 231, 232, 233
human activity
    floodplains, 401
    human waste, 449
    Illinois coast, 405, 411–413
    landslides, 387
    radioactivity, 419
    soils, 373
    water pollution, 337, 443–457
    wetlands, 365
human health. *See* health effects
hummocks, 233, 240
humus, 377
Hunton Limestone Group, 66f, 330
Huron-Erie glacial lobe, 216f, 216, 225, 233
    Trafalgar Formation, 238
    Wisconsin Episode, 235, 236f, 239, 240
Hyde Lake, 408, 410f
hydrates, 315
hydraulic conductivity, 325, 326t, 326–327
    groundwater movement, 327–328, 328f
    hydraulic head, 325f
    hydrologic cycle, 328f
    model, 325f
    soils, 354
    See also aquifers; permeability
hydraulic fracturing
    Carper Sandstone, 293
    coal bed methane, 279
    Glenwood-St. Peter Sandstone, 331

    reservoir rocks, 285, 288
    St. Peter Sandstone, 314
hydric soils, 328, 361, 362f, 366
hydrocarbons, 37, 313
hydrofluoric acid, 303
hydrogen isotopes, 346, 347
hydrogen sulfide, 341f, 346
hydrogeology, 326–327
    aquifer vulnerability, 351–360
    floodplains, 364
    hydrologic cycle, 328, 328f, 376
    karst, 432
    landfills, 30
    resource planning, 336
    wetlands, 364
hydrologic cycle, 328, 328f, 376
hydrothermal conditions, 201
Hymera (VI) Coal (Ind.). *See* Jamestown Coal Member
Hyolithes, 138, 139, 140f, 142
hypocenter (earthquakes), 424
hypoxic zone, 448–449
Hypsiptycha species, 153f, 154

I

Iapetus Ocean, 80f–81f, 139f, 155
Ibexian, 137f, 138f, 143–145
ice contact
    drift, 240
    drift ridges, 230f
    Quaternary landforms, 217
    sand and gravel deposits, 229f
ice core age, 255f
ice jam flooding, 395
ice margin
    bedrock valleys, 224
    depositional environment, 219, 221f, 229f
    glacial extent, 216f, 216–217, 224f, 229f
    glacial Lake Chicago, 408
    glacial sequences, 219, 220f, 221f, 222f
    Illinois Episode, 229f
    sediments, 220f
    source areas, 216f
    till lithology changes, 240
    Wisconsin Episode, 229f, 236f, 242
ice rafting, 407
ice-deformed sediments, 381f, 382f
ice-marginal sediments, 220f
ice-walled lakes, 381f
ichnofossils, 175f, 197
igneous activity. *See* magmatism
igneous rocks
    aphanitic texture, 125f, 132
    Canadian Shield, 78f
    Cretaceous, 208f
    dikes, 86, 87
    domes, 92f, 304f
    intrusions, 125, 132
    Permian, 201–202
    Precambrian rocks, 132
    pre-Illinois Episode diamictons, 224
Illaenurus species, 143
Illaenus species, 151
Illiana Member, 236f
Illinoian
    faults, 107t
    till plain, 229
Illinois and Michigan Canal and locks, 165
Illinois and Michigan Canal Heritage Corridor, 165
Illinois and Michigan Canal State Trail, 148
Illinois Basin, 37, 60, 60f, 62–64, 78, 78f, 91
    brines, 348
    Chesterian, 169, 170f, 183

coal deposits, 263
Devonian, 167
environment of deposition, 145
Iowa Basin, 171
Kaskaskia sequence, 169, 171f
mineral resources, 299–300, 300f
Mississippian, 84
Mt. Simon Sandstone, 137–138
New Albany Shale, 169f, 174f
paleohydrology, 300f
Paleozoic Era, 80f–81f, 81, 83, 128f, 133, 141f
Pascola Arch, 87, 208
Patoka Formation, 200
petroleum, 283–298
    history of investigations, 16–17, 19f, 38
    Kimmswick ("Trenton") Limestone, 289
    oil fields, 286f
Precambrian, 78f, 125f
Sauk sequence, 137
Ste. Genevieve Fault System, 98
structural geology, 91f
Tippecanoe II subsequence, 158
Valmeyeran, 176
Wabash Valley Fault Zone, 115f
See also individual named topics
Illinois Beach State Park, 72t, 404f, 412f
    environmental management, 414
    LIDAR methods, 407f
    Zion Beach-Ridge Plain, 405
Illinois Bedrock Valley, 333f, 334–335
Illinois Caverns, 33f, 432, 439, 441
Illinois Caverns State Natural Area, 72t
Illinois Clean Coal Institute, 32
Illinois coast, 404f, 404–406, 406f, 407–417
Illinois Department of Health, waste disposal sites, 30
Illinois Department of Natural Resources
    coal mine reclamation, 460
    Division of Abandoned Mined Land Reclamation, 460
    Office of Mines and Minerals, Mine Safety and Training Division, 461
Illinois Department of Nuclear Safety, 420
Illinois Department of Public Health, 454
Illinois Department of Transportation, 36
Illinois Emergency Management Agency
    nuclear facilities, 421–422
    radiation, 422
    radon, 420, 421
Illinois Environmental Protection Agency
    groundwater protection
        groundwater quality, 443
        legislation, 353
        planning, 354, 358
        radium, 419
        surface water, 354
Illinois Episode, 226f, 229f, 229–234, 234f
    drainage, 224f
    geochronology, 223f
    geosols, 234f, 234–235
    interglacial environment, 234
    linear stream valleys, 230
    neotectonics, 107t
    paleoclimatology, 250f, 257f
    topography, 233f
    Tuscola Quarry, 222f
Illinois Groundwater Protection Act (1987), 29t, 351, 353
Illinois Pesticide Act (1990; amended 1997), 29t
Illinois' Pesticide Management Plan, 353–354
Illinois River, 224
    Buffalo Rock State Park, 72t
    chloride, 455
    floodplains, 364, 399, 452
    floods, 395
    Great Flood of 1993, 35, 396, 397
    groundwater, 333, 419
    Kress Member, 148f
    loess, 378
    Matthiessen State Park, 72t
    Pennsylvanian, 189
    Shakopee Dolomite, 148f
    Shelburn Formation, 200
    Silurian, 159, 164
    slope stability, 387
    Starved Rock State Park, 72t
    valley train deposits, 313
    wetlands, 364, 366, 367, 402
Illinois River valley, 226f, 229f, 333f
    aquifers, 333, 353
    backwater lake, 237f
    bedrock valley, 225f
    Big Bend, 239f
    dunes, 239f
    false-color infrared satellite imagery, 237f
    floods, 398f, 399
    glacial Lake Chicago, 240, 241f, 242
    human activity, 365
    loess, 240
    Sangamon Geosol, 234f
    sedimentation, 240
Illinois River watershed, 448
Illinois Route 3, 228f
Illinois satellite image map, 39f
Illinois State Capitol building, 165, 318, 318f
Illinois State Geological Survey
    first Survey (1851–1875), 7–9
    history of investigations, 3–58
    modern Survey (1905–present), 12–45
    Survey gap (1875–1905), 9
Illinois State Water Survey, 25, 354, 358
Illinois Surface Mined Land Conservation and Reclamation Act (1971, 1975 amendment), 460, 461
Illinois till plain, 229f, 232f
Illinois water well construction code (2000), 441
Illinois-Kentucky Fluorspar District, 299, 300, 300f, 303, 304f, 306–307
illite, 19, 315
    discovery and characterization, 19
    pre-Illinois deposits, 225
    Wisconsin Episode tills, 238
impact features, 62, 63f, 92f, 100–101, 101f, 202
Inarticulata
    Carbondale Formation, 199
    Lombard Dolomite Member, 140f
    northern Illinois, 142
Inceptisols, 375–376, 376f
incised valleys
    Carbondale Formation, 184, 199f
    Caseyville Formation, 197
    Chesterian, 183, 183f
    cyclothems, 194f
    Galatia channel, 194f
    Pennsylvanian, 193–194
    Quaternary, 224, 225f, 227
    sub-Pennsylvanian surface, 195
Independatrypa independensis, 173
independent oil and natural gas producers, 26
Indiana Dunes (Ind.), 404f, 411
Indiana Harbor, 410f
Indiana Harbor Ship Canal, 410f
indicators, 64
    Independatrypa independensis, 173
    Microcyclus discus, 172f–173f, 173
    paleoclimatology, 258
industrial facilities, 358, 443, 443t, 444t, 449
industrial minerals, 309–321
    economic value, 27, 38f, 319f, 319–320
    fluorite, 305–307
    history of investigations, 3, 6, 11–12, 15, 38
    Illinois Basin, 60

INDEX

lead-zinc deposits, 305–307
mines, 311f, 312f
radioactive by-products, 421
sand, 311f
transportation, 318–319
infill drilling, 37
infiltration
   radon, 420
   relief, 377
   wetlands, 362
infrared stimulated luminescence, 234
infrastructure
   history of investigations, 10f, 13, 13f, 17f
   industrial minerals, 309, 310f
   site exploration, 385–394
injection ring, 394f
injection wells, 444t
Inland No. 2 Mine roof fall, 112
Inman sub-Pennsylvanian valley, 196f
inner shelf, 177f
inorganic compounds, 444
inorganic nitrogen, 451f
Insecta
   fossils, 190
   wetlands, 367
insecticide. *See* pesticides
insolation, 248, 250–251
intensity, 424, 425t
interglacial environment
   climate, 235
   Hudson Episode, 242
   pre-Illinois Episode, 226–229
   research programs, 10
   Sangamon Episode, 234–235, 257f, 258
   timetable, 223f
Interior Low Plateaus Province, 70, 71f
Interstate 39, 152f
Interstate 57, 388f
Interstate 70, 388f
Interstate 80, 59f, 163–164
Interstate 94, 381f, 382f
Interstate Group, 304f
intra-Glenwood phase, 410f
intrusions, 61f, 62f, 63f, 125f
ion exchange
   clay minerals, 315
   groundwater, 341f
   radioactivity, 418
ions, 337
Iowa Basin, 158, 171, 171f
Iowa River, ancient, 224f, 225f
Iowa River valley, ancient, 333, 333f
Iowa substage, 22
iron
   clay minerals, 315
   in groundwater, 338t, 340, 345, 346f
   suspended solids, 340
   well screens, 345, 346f
iron ores, 129
Ironton Sandstone, 137f, 138f, 142
   aquifer, 330
   radium, 419
   See also Galesville Sandstone
Iroquois County
   artesian wells, 327
   dunes, 238
   wetlands, 366
Iroquois Moraine, 235, 236f, 238
irrigation, 444t
Isabel, 287f
Ischadites species, 151
islands, 412
isoseismic maps, 426f, 427f

isostatic rebound, 242, 408
Isotelus iowensis, 153f
Isotelus species, 151, 154
isotope analysis
   atomic weight, 347
   groundwater, 347–348
   history of investigations, 31, 34
   methane, 346
   paleoclimatology, 251
   Quaternary research programs, 242–243
   radioactive decay, 347
isotopes, 347
   beryllium-10 (^{10}Be), 234
   oxygen, 250f, 251, 253, 256, 347
   radioactive, 418, 445
isotravel times, 82f
Iuka sub-Pennsylvanian valley, 196f

J

Jackson County
   Giant City State Park, 197f
   Murphysboro Coal Member, 265, 267
   Pine Hills area, Shawnee National Forest, 72t
   Ste. Genevieve Fault System, 99f
Jackson Formation (Ky.), 212
"Jackson sand"
   electrical logging, 296f
   reservoir rocks, 287f, 296f, 297
   stratigraphic columns, 168f, 287f
Jader Coal's No. 4 Mine, 273f
Jake Creek, 287f
James W. Jardine Water Purification Plant, 413
Jamestown Coal Member, 200, 267f, 272, 287f
Jefferson's ground sloth, 242
Jeffersonville Limestone (Ind., Ky.), 170
jellyfishes, 190
Jersey County
   carbonate rocks, 168
   Great Flood of 1993, 371f–372f
   New Richmond (Roubidoux) Sandstone, 144
   Pere Marquette State Park, 72t
Jersey Landing, historical engraving, 8f
jetties, 411, 412f
Jo Daviess County
   Apple River Canyon, 72t
   aquifer vulnerability, 450
   Charles Mound 70, 71f
   Driftless Area, 224, 229f
   Galena Group, 151
   high point (Ill.), 70, 71f
   lead-zinc deposits, 299f, 301, 301f, 305–306
   LIDAR methods, 435
   pre-Illinois deposits, 224
   Silurian, 160f
   stratigraphic column, 160f
   wetlands, 364, 365f
Joachim Dolomite, 146, 148f, 149
   stratigraphic columns, 137f, 138f, 287f
Johnson County
   Cache River State Natural Area, 72t
   Pennsylvanian, 194–195
joints, 111, 111f, 432
Joliet
   building stone, 164, 317, 318f
   Illinois State Capitol building, 318f
Joliet Dolomite, 160f, 161, 162f–163f
Joliet Penitentiary, 165
Joliet sublobe, 236f
Jonesboro, 178, 212f
Joppa, 105, 106f
Joppa Limestone Member, 305, 305f
"Jordan," 106f, 188f, 287f

Jordan Sandstone, 138f, 143
Jurassic, 62f, 65f, 67f, 91, 202, 207, 208f
juvenile crust, 125f

K

kames, 217, 219, 230f, 238
Kane County, 458f, 459f
Kankakee
    La Salle Anticlinorium, 93
    liquid petroleum gas storage, 393f
    Silurian, 165
Kankakee Arch, 60, 60f, 63f, 78f, 91, 91f
    Devonian, 170
    Kaskaskia sequence, 171f
    paleohydrology, 300f
    tectonics, 62f, 84, 85f
Kankakee County
    aquifer vulnerability, 353
    dunes, 238, 353
    Kankakee River State Park, 72t, 165
Kankakee dolomite, 160f, 160, 161
Kankakee Formation, 287f, 289, 291, 292f
Kankakee Marsh, 365–366
Kankakee Plain, 70, 71f
Kankakee River State Park, 72t, 165
Kankakee River valley, 165, 240
Kankakee torrents, 240
Kansan. *See* pre-Illinois Episode
Kansas City (Mo.), 463
kaolinite, 315
Karnak Limestone Member, 183f, 294–295
    fluorspar deposits, 305f
    reservoir rocks, 287f
    stratigraphic columns, 168f, 183f, 287f
karst, 432–434, 434f, 435–436, 436f, 437–438, 438f, 439–442
    aerial photograph, 435f
    deformation, 107–108
    drainage, 435f
    groundwater pollution, 454
    groundwater quality, 340, 340f
    habitat, 435–436
    history of investigations, 33, 34f
    hydrogeology, 432f, 438f, 439–440
    Jo Daviess County, 435
    springs, 441
    stratigraphic traps, 293f
    See also sinkholes
Kasimovian, 188f
Kaskaskia Bedrock Valley, 333, 333f
Kaskaskia Bedrock Valley aquifer, 335
Kaskaskia I subsequence, 65f, 168f, 170–176
Kaskaskia II subsequence, 168f, 176–184
Kaskaskia Island levee, 400–401
Kaskaskia River
    eskers, 231, 233f
    floods, 395
    watershed, 230, 230f
    wetlands, 364
Kaskaskia River valley
    eskers, 233f
    outwash, 231
Kaskaskia sequence, 64, 128f, 167–168, 168f, 169–171, 171f, 172–186
    chronostratigraphy, 65f
    cross sections, 66f, 128f, 141f
    deposition, 79, 79f
    geochronology, 67f
    Illinois Basin, 171f
    stratigraphic column, 168f
    tectonics, 62f
K-bentonites, 40–41, 137f, 138f, 139, 151, 152
Keewatin center, 216, 216f, 224
Keller Farm section, 253, 254f, 255f

Kellerville Field, 291, 291f
Kellerville Member, 223f, 231, 232, 234
Kelley site, 106, 106f, 107, 107t, 108f
Kempton Bedrock Valley, 333
Kendall County
    Galena-Platteville aquifer, 331
    Illinois coast, 404f
    Shakopee Dolomite, 145
Kenilworth, 404f
Kenney, 344f
Kenosha (Wisc.), 411f
Kentland (Ind.) impact structure, 63f, 92f, 100
Kentucky
    low-level radioactive waste, 421
    Permian, 201
Keokuk Limestone, 177–178, 180f
Kerr-McGee Company
    Galatia Mine, 112f,
    pollution, 422
kettles, 217, 233f, 366
Kewanee Group, 195f
Kickapoo, 287f
Kimmswick ("Trenton") Limestone
    hardground omission surface, 151f
    lead-zinc deposits, 302f
    limestone, 34, 311
    petroleum, 95, 151, 287f, 289
    stratigraphic columns, 137f, 138f, 287f
    structural geology, 60, 60f
    underground mining, 311
Kincaid Field, 291f
Kinderhookian, 84, 168
    Illinois cross section, 66f
    stratigraphic columns, 168f, 287f
kink folds, 112, 112f
Kinkaid Limestone, 181, 182
    Hicks Dome, 101f
    stratigraphic columns, 168f, 183f, 188f, 287f
Kirkwood, 168f, 287f
Kishwaukee River
    floodplain, 358
    Wise Lake Formation, 152f
knobs, 101–102
Knobs Group, 65f, 66f, 168f
Knox County
    drumlins, 231
    Houchin Creek Coal Member, 270
Knox Group, 142–143, 143f, 144, 144f, 145, 145f, 146, 146f, 147f
    cross sections, 66f, 148f
    Pascola Arch, 86f
    seismic profile, 114f, 126f
    stratigraphic column, 137f
Kress Member, 138f, 147, 148f

L

La Grange Spit, 409f
La Salle
    Colchester Coal Member, 268
    Danville Coal Member, 272
    Herrin Coal Member, 272
    Illinois River valley, 237f, 239f
    Matthiessen State Park, 195
    nuclear facility, 422
    Pennsylvanian, 189
    Shakopee Dolomite, 148f
    sphalerite, 306
    St. Peter Sandstone, 148f, 313–314
La Salle Anticlinal Belt, 62f
La Salle Anticlinorium, 61f, 63f, 92f, 93–94, 94f, 95, 196f
    Ancona Anticline, 94f
    Atokan, 189f
    Carbondale Formation, 98

Index

coal rank, 264
Fairfield Basin, 91
faults, 93
folds, 94f
groundwater, 343, 344, 344f
Illinois Basin, 62
Mahomet aquifer, 342, 344f
Mattoon Formation, 201
paleohydrology, 300, 300f
petroleum reservoirs, 284, 286f, 289
seismic profiles, 114
tectonics
    Mississippian, 84, 85f
    Ordovician, 83
    Pennsylvanian, 85f, 189f, 193
    Permian, 202
La Salle County
    Ancona Anticline, 94, 94f
    aquifers, 330, 331
    Buffalo Rock State Park, 72t
    Charleston Quarry, 220f
    Gunter Sandstone, 144
    Illinois River valley, 237f
    La Salle Anticlinorium, 94, 94f
    Matthiessen State Park, 72t, 195, 323f–324f
    New Richmond (Roubidoux) Sandstone, 144, 146f
    Old Salt Well, 342
    radioactivity, 422
    Shakopee Dolomite, 145, 147f, 148f
    Shelburn Formation, 200
    St. Peter Sandstone, 2, 147, 148f, 149f, 241f, 313–314
    Starved Rock State Park, 2, 72t, 241f
    Wedron Quarry, 220f, 221f
La Salle Limestone, 188f, 201
Labrador center
    diamicton, 225
    Illinois Episode, 232–233
    Quaternary, 216f, 216
    Wisconsin Episode, 238
lacustrine environment
    Illinois River valley, 237f
    Lake Michigan lobe, 240
    paleoclimatology, 252f
    Quaternary, 221f, 229f
    Triassic, 207f
"Lafayette" Gravel. See Mounds Gravel
lagoons, 189, 408, 412
Lake Agassiz, 410f
Lake Bluff, 404f
Lake Border Morainic System
    Blue Island Ridge, 407
    bluffs, 405
    Crown Point Phase, 242
    Quaternary, 236f
    Wisconsin Episode, 405
Lake Calumet, 406f, 408, 410f
Lake Chicago, 242, 406f, 408, 409f, 410f
Lake County
    aquifer vulnerability, 353
    Digital Elevation Model, 383
    drainage, 377f
    Illinois Beach State Park, 72t, 404f, 405, 407f, 412f, 414
    Illinois coast, 404f, 404–406, 406f, 407–410, 410f, 411f, 411–417
    model, 381f, 382f
    wetlands, 361f
Lake Forest, 404f
    Forest Park Beach, 414
Lake George, 408, 410f
Lake Michigan
    bottom, 411–412, 415f
    fills, 411–412, 412f, 413f
    flooding, 395
    history of investigations, 30, 35f
    Illinois Beach State Park, 72t, 404f, 405, 407f, 412f, 414
    Montrose Harbor, 413f
    shoreline (Ill.), 404f, 404–406, 406f, 407–410, 410f, 411f, 411–417
    shoreline stabilization, 35f, 387f
    water supply, 392
    Wisconsin Episode, 236f
Lake Michigan geochronology
    ancestral drainage
        Chicago Outlet Valley, 406f, 408
        glacial Lake Chicago, 242, 406f, 408, 409f, 410f
        Hudson Episode, 241f, 242
    Chicago Lake Plain, 407
    geomorphology, 406f
    history of investigations, 7, 31
    lake level changes, 242, 407, 409f, 410f, 411f
    shoreline, 408, 410f, 412f
    shoreline stabilization, 414–415, 415f
Lake Michigan lobe, 216–217, 216f, 225, 236f, 239
    end moraines, 235, 236f
    Illinois Episode, 229, 233
    paleoclimatology, 249t
    pre-Illinois Episode, 227
    Wisconsin Episode, 229f
    Woodstock Phase, 240
Lake Milan, 224f, 239
Lake Muddy, 224f
lake plains, 217, 408
    Illinois Episode, 229f, 230f, 231
    Wisconsin Episode, 229f, 231f, 235
Lake Saline-Embarras, 224f
Lake Shore Drive flooding, 395–396, 396f
lakes
    agrichemicals, 354
    Equality Formation, 238
    groundwater movement, 328
    mine reclamation, 462f, 462–463, 463f
    paleoclimatology, 257f
    sediment, 235, 253
    shoreline, 410f
    wetlands, 365, 366
    See also individual named lakes
Lambert landfill, 447f
laminations
    clay, 236
    dolostones, 171
    silt, 236f
Lamna cuspidata, 211f
Lamotte Sandstone (Mo.), 141
lamprophyre, 202
land cover, 363f, 373–374, 380
land management
    erosion, 380
    floods, 395
    government agencies, 383
    groundwater protection, 351, 358
    history of investigations, 44
    McHenry County, 358
    mine reclamation, 273, 458
    planning, 383
    soils, 373
    wetlands, 362, 363f, 366
    Winnebago County, 358
    See also engineering geology; geological hazards; land subsidence; land use
Land Reclamation Division, Office of Mines and Minerals, Illinois Department of Natural Resources, 460
land subsidence
    Belleville, 390f
    compression, 390, 391f
    levees, 399
    mining, 275, 276f, 387, 388–389, 390, 391f, 391–392
    mitigation, 385
    road, 391f

site assessment, 385–391
slope stability, 387f, 387–388, 388f
land use, 70–72
   agriculture, 449f, 450f
   aquifers, 326, 351, 352, 358
   environmental geology, 29
   erosion, 383
   soils, 373–374
   wetlands, 362, 363f, 364f, 366
landfills, 445–446, 446f, 447, 447f
   clay liners, 316
   leachate
      aquifer pollution, 351
      geochemical profiles, 348
      history of investigations, 30, 31
   pollutants, 444, 444t
   radioactive waste, 422
   site exploration, 29, 446
   Tazewell County, 357f
   waste management, 29, 446
   Winnebago County, 355f, 358
landforms
   Illinois River valley, 237f
   Quaternary, 216–247, 381f, 382f
landscapes
   ancient, 351
   drainage, 377, 377f
   floods, 400
   glaciated terrain, 218f
   habitat, 377
   Holocene, 69–70
   Illinois Episode, 229–230, 230f, 231
   Illinois River valley, 237f
   model, 380, 381f, 382f, 383
   pre-Illinois Episode, 224, 225f
   Quaternary, 216–217
   soils, 373
   Wisconsin Episode, 235–238
landslides, 107–108, 387–388, 388f
   New Madrid earthquakes (1811–1812), 428–429
   Porters Creek Formation, 212f
   soils, 377
   transportation corridors, 388
   See also engineering geology; geological hazards
Laramide Orogeny, 93
   See also ancestral Rocky Mountain Orogeny
Laurentia
   Cambrian, 139f, 155
   Ordovician, 155
   paleogeography, 80f–81f, 81, 125f, 139f
   Phanerozoic, 79
   Precambrian, 79, 136
   sub-Kaskaskia unconformity, 167
   Tippecanoe II subsequence, 158
Laurentide ice sheet
   paleoclimatology, 251, 253, 258
   uranium-series dating, 248, 256
Lawrence County
   Danville Coal Member, 272
   oil and gas, 15, 284
Lawrenceville
   earthquake (1987), 428, 428f
   La Salle Anticlinorium, 93
   Survant Coal Member, 269
laws and legislation
   Central Midwest Interstate Low-Level Radioactive Waste Compact, 421
   Clean Air Act (1963; amended 1970, 1977, 1990), 29t
   Clean Water Act (amended 1970, 1977), 29t, 297
      discharge to wetlands, 361
   Comprehensive Environmental Response, Compensation, and Liability Act (CERCLA [Superfund]; 1980), 29t, 33f, 444–445, 446f
   Energy Reorganization Act (1974)
   environmental issues, 29
   Federal Insecticide, Fungicide, Rodenticide Act (FIFRA), 29t
   federal regulations
      radium in drinking water, 419
      reclamation, 459
      water well disinfection, 352
      wetland restoration, 368
   groundwater protection, 351, 353
   hazardous and solid waste amendments (1984), 29t
   Illinois Groundwater Protection Act (1987), 29t, 351, 353
   Illinois Pesticide Act (amended 1997)
   Illinois Surface-Mined Land Conservation and Reclamation Act (1971, 1975 amendment), 460, 461
   Low-Level Radioactive Waste Policy Act (1980 amendments), 29t, 421
   mining, coal, 279, 280
   National Cooperative Geologic Mapping Act (1992), 40
   National Environmental Policy Act (1969 amendment), 29t
   National Primary Drinking Water Regulations (NPDWR), U.S. Environmental Protection Agency, 339–340
   National Priority List sites, 444–445
   National Secondary Drinking Water Regulations (NSDWR), U.S. Environmental Protection Agency, 340
   Nuclear Waste Policy Act (1982 amendment), 29t
   Radon Pollution Control Act (1988), 420
   Resource Conservation and Recovery Act (1976), 29t
   Safe Drinking Water Act (1974), 29t
   Solid Waste Disposal Act (1965; amendments 1980), 29t
   Surface Coal Mining Land Conservation and Reclamation Act (1977), 460
   Surface-Mined Land Conservation and Reclamation Act (1971; amended 1975), 460, 461
   Surface-Mined Land Reclamation Act (1968), 460
   Surface Mining Control and Reclamation Act (1977), 29t, 460
   Toxic Substances Control Act (1976), 29t
   Water Pollution Control Act (1965), 29t
   See also regulations
lead, 300–301
   in coal, 266
   Fluorspar Area Fault Complex, 100
   health effects, 300
   mining, 11–12, 152, 299–308
   pollution, 313
   stratigraphy, 302f
lead-214, 420
lead-zinc deposits, 299–301
   mining, 299f, 305–306
   stratigraphy, 302f
Lead-Zinc District, 299–300, 301f
leaf imprints, 211
Leaf River Anticline, 96f
Leaf River esker, 231
leaking underground storage tanks (LUST), 351, 444t, 452–453, 453f
Lee Center Member, 232–234
Lee County
   aquifers, 330, 331
   bedrock, 143f
   Franconia Formation, 143
   Green River Lowland, 72t
   Gunter Sandstone, 144
   New Richmond (Roubidoux) Sandstone, 144
   Oneota Dolomite, 143f, 144, 145f
   Platteville Group, 151
   Potosi Dolomite, 143, 143f, 144f
   Shakopee Dolomite, 145
   valley train deposits, 238
Lee Vein fluorspar mining, 304f
Leemon Formation, 161, 137f
Lehigh Stone Company quarry, 20f
Lemont, 164, 317
Lemont Formation, 221f, 223f, 229f, 236f, 238
Lemont till, 220f, 236f, 240
lenses
   Herrin Coal Member, 271
   Springfield Coal Member, 270

501

Index

Lepidocyclus species, 153f, 154
Lepidodendron species, 190f
Leptospira interrogans, 455
levees, 364, 365, 399, 400
    floods, 399, 399f, 400f, 402
    Great Flood of 1993, 397
    human activity, 401
    infiltration, 399, 399f
    sandbags, 398f
    Valmeyer, 401f
Levias Limestone, 168f, 183f, 305f
Lewis landfill, 447f
Lexington Coal (Mo.), 271
LIDAR methods
    Illinois Beach State Park–South Unit, 407f
    Jo Daviess County, 435
    Lake County, 381f, 382f
Lierle Clay Member, 366
lignite, 210–211, 264
lime, 312, 313
    economic value, 319, 319f
    industrial facilities, 311f
    Mississippian, 312
    Silurian, 164
lime kiln, 318f
lime mudstones
    Galena Group, 151
    Geneva Dolomite, 171
    Platteville Group, 149, 151f
limestone, 309–312
    aquifers, 326, 332, 338
        vulnerability, 351
    aquitards, 326
    bedded replacement deposits, 305
    Camp Vandeventer, 440f
    Carbondale Formation, 199, 199f
    clasts, 338
    Devonian, 332
    Everton Formation, 145–146
    historic engraving, 8
    hydraulic conductivity, 326
    Illinois cross section, 66f
    Jurassic, 207
    Mattoon Formation, 201, 320f
    Mississippian, 332, 433
    Moccasin Springs Formation, 162, 163
    Pennsylvanian, 187, 188, 189, 332
    Pinicon Ridge Member, 171
    pre-Illinois Episode, 224
    Sauk sequence, 137
    Silurian reef, 290f
    soils, 379
    speleothems, 255f
    Valmeyeran, 177f
    See also individual named geological units
limestone deposits, 309–313
    bedded replacement deposits, 305
    coal balls near, 192f
    creviced mine roof, 111f
    Devonian, 312f
    fluorspar deposits, 304f, 305f
    history of investigations, 15, 20, 38
    lead-zinc deposits, 302f
    Mississippian, 311f, 312f
    Ordovician, 312f
    Pennsylvanian, 312f
    petroleum traps, 288f
    quarries, 311f
    reservoir rocks, 287f
    Salem Limestone, 179–180
    total dissolved solids, 339t
    underground storage, 393
Limnocythere friabilis, 258
Limnocythere varia, 252f
Lincoln, 344f
Lincoln Anticline, 63f, 84, 85f, 92f, 95, 107f
Lincoln Hills region, 433, 434f
Lincoln Hills Section, 70, 71f
Lindley, 287f
Lindsay Light & Chemical Co. Rare Earths Facility, 422
lineaments, 435
Lingle formation, 65f, 168f, 287f, 291f
liquefaction. *See* soils: liquefaction
liquid petroleum gas storage, 393f
Litchfield, 11, 283
Litchfield Coal Member, 267, 267f
lithium, 266
lithology. *See* individual named geological units
lithosphere, 60, 77f
lithostratigraphic units, 64, 64f, 65f, 66f, 64–67
    Quaternary, 27, 220–221, 221f, 222, 223f
Lithostrotionidae, 305f
litter (absorbent), 211
Little Calumet River, 410f
Little Wabash Bedrock Valley, 333f
littoral drift, 407, 408, 410f, 411f, 413
livestock
    pollution, 454, 454f
    sinkhole hazard, 436–437
Livingston County, 94, 94f
Livingston Limestone. *See* Millersville (Livingston) Limestone Member
Livingston Phase, 236f, 239, 240
Llandovery Stage, 160f
loam, 238, 374, 377f, 378–379, 379f, 380f
    See also soils
lobsters, 211
Lochkovian Stage, 159, 160f
Lock and Dam No. 20, 397f
lock-and-dam systems impacts, 365
Lockport, 318f
loess, 327, 378, 378f
    Baylis Formation, 209
    bluffs, 220f
    drumlins, 231f
    hydraulic conductivity, 326t, 327
    karst, 433
    Keller Farm section, 254f, 255f
    Loveland Silt, 232
    outcrops, 22f
    paleoclimatology, 253
    Pancake Hollow Section, 228
    Peoria and Roxanna Silts, 229f
    Pere Marquette State Park, 72t
    Quaternary, 219, 226f
    snail shells, 248
    soil parent materials, 378, 378f
    soil piping, 386–387
    stratigraphic column, 206f
    Thebes, 227, 228f
    Wedron Quarry, 221f
    Wisconsin Episode, 229f, 234f, 239, 240
Lombard Dolomite Member, 138f, 140f
longwall mining, 268, 275–276, 276f, 390
Lonsdale Limestone, 188f
Lophospira species, 150f, 151
Louann Salt, 207, 208f
Louden Anticline, 62f, 63f, 92f, 95, 196f
    neotectonics, 107f, 114, 114f
    oil fields, 286f
    Shelburn Formation, 200
    tectonics, 62f
        Mississippian, 84, 85f
        Pennsylvanian, 84, 85f, 193
        Permian, 202
Louden (Loudon) Field, 18, 284–285, 293

Index

Loudon, 115
Louisiana Limestone, 65f, 101f, 168, 168f, 174, 174f
Louisiana (Mo.), 93
Louisville Limestone, 162f–163f
Loveland Silt
    geochronology, 223f
    Illinois Episode, 232, 234
    Kelley site, 108f
    neotectonics, 107, 107t
    Thebes, 228f
low point (Ill.), 70
Lowden State Park, 72t
Lowell Coal. *See* Survant Coal Member
"Lower Buff," 302f
Lower Kittanning (Ohio, Pa.), 268
"Lower *Receptaculites* Zone," 302f
"Lower Renault," 168f
lowlands
    Chesterian, 170f
    Pennsylvanian, 189f
    Quaternary, 219
Low-Level Radioactive Waste Policy Act (as amended 1980), 29t, 421
low-level waste, 34, 445
low-sulfur coal
    Colchester Coal Member, 268
    Danville Coal Member, 272
    history of investigations, 15
    Murphysboro Coal Member, 267
    relation to gray shale, 26, 265
    Springfield Coal Member, 270f
Ludfordian Stage, 160f
Ludlow Stage, 160f
Luminous Processes Inc., 422
luminous watches, 445
Lusk Creek Fault System, 62, 87
Lusk Creek Fault Zone
    fluorspar mining, 304f
    structural geology, 63f, 92f
    tectonics
        Fluorspar Area Fault Complex, 100
        Hicks Dome, 101f
        Illinois-Kentucky Fluorspar District, 304f
        Pennsylvanian-Permian, 85f, 86
        Quaternary sediment, 106f
        Rough Creek-Shawneetown Fault System, 98
Lusk Formation, 195f
LUST (leaking underground storage tanks), 452–453, 453f
Lycopodiaceae
    tree bark, 190f
    in wetlands, 367

M

Maastrichian Stage, 206f, 211
Macedonia Formation, 195f
Mackinaw Bedrock Valley
    ancient Mississippi River valley, 225f, 333f, 333–334
    aquifers, 224, 333, 338t
    Illinois Episode, 226f
    pre-Illinois Episode, 226f
Mackinaw River valley
    cross section, 226f
    groundwater, 338t, 344–345
Maclurites species, 150f, 151
Macomb, 268
Macoupin County
    Crown Fault, 111f
    Herrin Coal, 271
    Illinois Episode, 231
    mine roof fractures, 111, 111f
Macoupin Limestone, 188f
made land. *See* fills
Madison County
    Acrocyathus species, 175f, 180
    drill core analysis, 133f
    Great Flood of 1993, 397
    liquefaction, 109f
    loess, 220f
    Marine Field, 289
    Marine reef, 163
    Precambrian drill core, 124f
    pre-Illinois Episode, 228
Madisonville sub-Pennsylvanian valley, 196f
mafic composition, 131, 132
magmas, 77, 77f
magmatism
    coal rank, 201
    Cretaceous, 207
    intrusions, 61f, 62f, 90
    Mississippi Embayment, 93
    Permian, 300
    Precambrian, 123–124
magnesium
    clay minerals, 315
    in groundwater, 337, 338t, 340–341, 341f
    water hardness, 340
magnesium oxides, 313
magnetic anomalies
    Commerce geophysical lineament, 108–109
    maps, 128–129, 129f, 130f, 131, 131f, 132
magnetic fields
    Brunhes-Matuyama reversal, 223f, 227
    diamicton, 225
    geopotential maps, 129
    paleomagnetism, 124–125, 129, 130f, 131f, 207
magnetic lineation, 60
magnetite, 202
magnitude, 424, 425t
Mahomet, 344f
Mahomet aquifer, 333–334, 338t, 354
    methane, 347f
    Onarga Valley, 346
    salt-water intrusion, 342, 344f
    sulfate, 346
Mahomet Bedrock Valley, 225f, 227, 333, 333f
Mahomet Sand, 223f, 226
Mahomet-Teays Bedrock Valley, 224, 225f, 399
Main oil field, 15
Mallard Creek faults, 106, 106f, 107t, 108f
Mammoth Cave Group, 61f, 66f, 168f
"Mammoth Cave of Illinois," 441
Mammoth Cave (Ky.), 433
Mammut americanum, 242f
manganese
    in coal, 266
    in groundwater, 338t, 345
manganese oxides, 340, 346f
Manhattan Project, 21
Mankato substage, 22
"Mansfield," 188f, 287f
mantle (Earth), 60
manures. *See* animal waste
Maple Grove faults, 106f, 107t
mapping
    aquifer vulnerability, 351–360
    coal deposits, 266
    faults, 113
    field work, 243
    history of investigations, 12–14, 25, 40, 40f
    Precambrian, 129
    Quaternary, 243
    radon, 421
Maquoketa Formation, 137f, 287f
Maquoketa Group, 83, 152f, 154–155
    Brainard Formation, 154f
    Buckhorn Consolidated Field, 292f

Index

Des Plaines buried impact structure, 101
fossils, 153f
lead-zinc deposits, 302f
oil and gas fields, 291f
shale, 311
Silurian reef, 290f
Ste. Genevieve Fault System, 99f
stratigraphic columns, 138f, 168f, 287f
sub-Kaskaskia surface, 169f
Marathon-Ouachita Orogenic belt, 60f
Marble Hill (Mo.), 208–209, 209f
marcasite, 460
Marcus Dolomite, 160f, 161, 162f–163f
Marengo Moraine, 249t
groundwater quality, 356f
Henry Formation, 238
Marengo Phase, 236f, 239
McHenry County, 356f
Villa Grove Quadrangle, 358
Marengo Phase, 236f, 239
Marginifera species, 191f
Marginirugus magnus, 175f, 179
Margraf Member, 161
marinas, 412
marine environment
Caseyville Formation, 197
Chesterian, 170f, 184f
Clayton Formation, 211
Devonian, 163, 169f
fossils
drift, 209, 209f
Pennsylvanian, 189–191, 199
Galena Group, 151
Jurassic, 207, 208f
limestone
Chesterian, 183f
cyclothems, 194f
Jurassic, 207
marine transgression surface, 41
Mississippi Embayment, 93
New Albany Shale, 169f
Ordovician, 139
Paleozoic, 83
Pennsylvanian, 187–189, 189f, 190–193, 193f, 194f, 194–201, 263–264, 265
Porters Creek Formation, 211
Ste. Genevieve Limestone, 294
Upper Devonian, 173
Marine Oxygen Isotope Stages, 223f, 227, 229, 234, 235
Marine Field, Madison County, 19, 289
Marine reef, 163, 290f
Marion County
drill core analysis, 133f
Herrin Coal Member, 271
Miletus Field, 292
"Quality Circle," 271
Salem Field, 261f–262f
Springfield Coal Member, 270
Marion sub-Pennsylvanian valley, 196f
Marissa, 461
marker beds, 64, 289–290
geosols, 222f, 234f
Markgraf Member, 160f
Markham series, 377f
marl, 207
Marseilles Morainic System, 231f, 236f
Marshall County
Illinois Episode tills, 234
Illinois River valley, 237f
pollution, 452
Marshall-Sidell Syncline, 63f, 92f
marshes, 361, 361f, 365, 366
gas, 283

swales, 405
See also wetlands
"Martinsville," 168f, 287f
Martinsville disposal site, 445
Mason County
aquifer vulnerability, 353
Great Flood of 1993, 402
Illinois River wetlands, 367
valley train deposits, 238
Mason Group, 229f
mass spectrometry, 256
Massac County
faults, 106f
Fluorspar Area Fault Complex, 100, 106
Kelley site, 108
Massac Creek graben, 109f
neotectonics, 105
Massac Creek, 106f, 107, 107t, 109f
mastodon molar, 242f
Material Service Corporation. *See* Thornton Quarry
Matthiessen State Park, 72t, 323f–324f
La Salle Anticlinorium, 94f
Pennsylvanian, 195
St. Peter Sandstone, 147, 313–314
Mattoon anticlinal field, 294
Mattoon Anticline, 286f
Mattoon Formation, 188f, 195f, 200, 201
coal seams, 267f
stratigraphic columns, 188f, 287f
Mattoon sub-Pennsylvanian valley, 196f
Matuyama reversed magnetic polarity, 223f, 227
Max McGraw Wildlife Foundation reclamation efforts, 458, 462
maximum contaminant level, 419
See also radium, in groundwater
Maysvillian, 137f, 138f
Mazon Creek fauna
fossils, 190
Francis Creek Shale Member, 268
Tullimonstrum gregarium (Illinois state fossil), 192f
"McClosky"
oolites, 294
reservoir rocks, 287f
stratigraphic columns, 168f, 287f
McCormick Anticline, 62f, 201f
Mesozoic, 86f, 87
Pennsylvanian-Permian, 85f, 86
Permian, 201f
structural geology, 62f, 86f
tectonics, 62f, 86f, 87
McCormick Group, 195f
McCraney Limestone, 168f
McDonough County
Illinois Episode landforms, 231, 231f, 232f
Vishnu Springs, 441
McHenry County
aquifer vulnerability, 353, 356f, 358, 450
Glacial Park, 72t, 233f
reclamation, 461f, 461–462
sanitary landfills, 447f
McLean County
Gunter Sandstone, 144
Mahomet aquifer, 338t
Moraine View State Recreation Area, 72t
McLeansboro Formation. *See* McLeansboro Group
McLeansboro Group, 195f, 200–201
Springfield Coal Member, 270
stratigraphic column, 188f
sub-Pennsylvanian valleys, 196f
McNairy Formation, 210, 211
faulting, 106
Kelley site, 108f
Mallard Creek, 108f
Massac Creek graben, 109f

near Olmsted, 208
Fayville pit, 210f
stratigraphic column, 206f
meanders, 364, 400
Mecca Quarry Shale Member, 188f, 268
Media Anticline, 84, 85f, 92f, 95
medical wastes, 313
medicines, 453
Megamyonia species, 154
Megamyonia unicostata, 153f
Meissner Island, 401f
Melonechinus multiporus
    Burlington Limestone, 178
    St. Louis Limestone, 178, 179f
meltwater outflow channels
    Chicago Outlet, 241f, 242, 406f, 408
    Illinois Episode, 233f
    Kankakee River State Park, 72t
    Quaternary, 217, 235, 237f, 239
    Starved Rock State Park, 241f
    wetlands, 366, 367
    Woodstock Phase, 240
member, 64, 64f, 220
Memphis (Tenn.), 428
Menard Limestone, 181
    Ste. Genevieve Fault System, 99f
    stratigraphic columns, 168f, 183f, 287f
Mendon Moraine, 226f, 230
Meppen Limestone, 168f, 177–178
Meramecian. *See* Valmeyeran
Mercer County, 399
mercury, 266, 280, 313
Mesolobus mesolobus, 191f
Mesozoic, 61f, 65, 65f, 66f, 67f, 79f, 206f, 206–211
    Ozark Dome, 91
    Pascola Arch, 141f
    stratigraphic column, 206f
    structural geology, 85f, 86f
    tectonics, 85f, 86f
Metablastus species, 178–179
metallurgical coke, 21, 278
metallurgy, 313
metamorphic rocks, 78f, 216f
Metazoa, 138
methane
    coal mining, 276, 278–279
    groundwater quality, 341f, 346, 347f, 348
"Method of Modern Analogs," 248
methoxychlor, 452
methyl tertiary butyl ether (MTBE), 452, 453
metolachlor, 452
metribuzin, 452
Metropolis
    Massac Creek graben, 109f
    Quaternary faults, 105, 106, 106f, 107f
Metropolis Formation, 106, 107, 108f
Metropolitan Water Reclamation District of Greater Chicago, 164
Mexico, paleogeography, 125f
Meyer, 397f
Meyer Material Company reclamation site, 461, 461f, 462
mica
    in groundwater, 343
    McNairy Formation, 211
    muscovite, 210
    Permian, 202
    Tradewater Formation, 198
Michigan Basin, 60f, 91
    paleohydrology, 300f
    Paleozoic, 80f–81f, 81
    Pennsylvanian, 189f
    Tippecanoe II subsequence, 158, 160, 161
Michigan Subepisode, 223f, 236f
Microcyclus discus, 172f–173f, 173

microfossils, 180, 251, 252f, 253
Microsauria, 182
Middle Fork watershed seeps, 34f
Middle Illinois Bedrock Valley, 333, 333f
"Middle *Receptaculites* Zone," 302f
midges, 252f, 253
Midway Group, 206f
Mifflin Formation, 137f, 138f, 302f
Milankovitch theory, 248, 250
Miletus Field, 292
military bases, 444
Mill Shoals Field, 293
Millbrig K-bentonite Bed, 137f, 138f, 139, 152
Miller City, 400
Millersville (Livingston) Limestone Member, 188f, 195f, 200
Mine Safety and Training Division, Office of Mines and Minerals, Illinois Department of Natural Resources, 461
mineral industries. *See* industrial minerals
Mineral Springs Hotel, 441
minerals
    aquifers, 329, 330–332, 341f
    pedogenesis, 378
    Permian, 300
    Upper Mississippi Valley Lead-Zinc District, 301f
    wetlands, 366
    See also individual named mineral resources
mines. *See* mines, coal; mining, coal; mining, fluorspar; mining, industrial minerals; mining, lead-zinc
mines, coal
    locations, 264f, 392
    orientation, 111–112
mining, coal
    auger mining, 276
    bucket wheel excavator, 459f
    coal seams, 265f, 267f, 267–269, 269f, 270f, 270–271, 271f, 272
    continuous miner, 274f
    dragline, 459f
    drift mines, 273, 276f, 389, 390
    Herrin Coal Member, 111f, 271, 271f, 272
    highwall mining, 276
    history, 7, 15, 35, 36f, 389
    land subsidence, 275, 388–390
    laws and legislation, 277–278, 278f, 279–280, 460–461
    licensing, 460
    maps, 392
    power shovel, 459f
    reclamation, 273, 458–464
    safety, 274, 276, 278
    shearer, 276f
    spoil, 461
    stack scrubbers, 280
    surface mining
        coal seams, 267–268, 271
        equipment, 26f, 272–273, 273f, 459f
        Grape Creek, 272
        Mazon Creek fossils, 190
        reclamation, 458, 458f, 459f, 459–460, 460f, 461–464
    tipple, 277, 461
    transportation, 277
    underground mining, 264f, 273–276
        faults, 111f
        longwall mining, 268, 275–276, 276f, 390
        roof control, 105, 110–111, 111f, 112, 112f, 263–264, 270, 274, 275f, 275–276
        room-and-pillar mining, 274f, 274–275, 275f, 320f
    Wabash Valley Fault System fault blocks, 97
mining, dredge, 233f
mining, fluorspar, 306f, 306–307
    deposits, 303–304, 304f, 305
    genesis, 299
    specimen, 303f
    stratigraphic units, 305f
mining, industrial minerals

INDEX

building stone, 318f
clay, 315, 317f
economic value, 309, 310f, 311–312, 316–319, 319f, 322
extraction sites, 311f
lime kiln, 318f
peat, 316
reclamation, 320f, 458f, 459f, 461–463
sand and gravel deposits, 313–314, 314f, 315, 315f, 320f
shale, 315–316
stone, crushed, 312f, 318f
  tripoli, 316
  urban geology, 320f, 320–321
mining, lead-zinc, 152, 299, 299f, 305–306
  cross sections, 303f
  deposits, 301, 303f
  galena specimen, 301f
  genesis, 299–300
  mining, 305–306
  paleohydrology, 300f
  stratigraphic units, 302f
  Upper Mississippi Valley Lead-Zinc District, 301f
Minooka Moraine, 236f
Miocene, 206f
Mississippi Embayment, 62f, 79, 79f, 80f–81f, 81, 93, 209
  Cretaceous, 79, 79f, 209f
  Illinois-Kentucky Fluorspar District, 304f
  McNairy Formation sediment source area, 211
  Mesozoic, 141f
  neotectonics, 107f, 113f, 115
  paleohydrology, 300f
  Paleozoic, 128f
  Precambrian, 128f, 141f
  Quaternary deformation, 105
  reservoir rocks, 286f
  stratigraphy, 141f
  structural geology, 60, 60f, 62f, 63f, 91f, 92f, 286f
  sub-Pennsylvanian valleys, 196f
  subsidence, 79, 79f, 87f, 87–88
  tectonics, 62f, 141f
Mississippi Palisades State Park, 72t, 165, 165f
Mississippi River
  ancient, 224f, 225f, 234
  bank slopes, 387
  channels, 224, 239
  Chesterian, 181
  Devonian, 159
  drainage, 249t
  floodplains, 364, 371f–372f, 401–402
  floods, 395
  Galena-Platteville aquifer, 331
  Great Flood of 1993, 35, 35f, 371f–372f, 396–402
  historic engraving, 8f
  Illinois low point, 70
  karst, 433
  Kaskaskia sequence, 167
  loess parent materials, 378
  Melonechinus multiporus Norwood and Owen, 179f
  Mississippi Palisades State Park, 72t, 165, 165f
  Pennsylvanian, 189
  Pere Marquette State Park, 72t
  Pine Hills area, Shawnee National Forest, 72t
  pollution, 448
  Quaternary sediment, 226f, 228f
  Silurian, 72t, 159
  transportation, 164
  valley train deposits, 313
  Valmeyeran, 177
  watersheds, 368
Mississippi River Arch, 60, 60f, 63f, 78f, 92
  paleogeography, 91f
  paleohydrology, 300, 300f
  tectonics, 62f
Mississippi River valley, 226f

bedrock valleys, 225f, 333f, 333–334
Grover Gravel, 213f, 214
loess, 22f, 232, 240
Mounds Gravel, 213f, 214
sand and gravel aquifers, 226, 333–336
sedimentation, 22f, 232, 234, 240
valley train deposits, 238
Mississippian, 167–186
  Acadian Orogeny, 80f–81f
  aquifers, 332
  aquitards, 332
  bedrock geology, 68f
  chronostratigraphy, 65f, 66f, 67f
  cross sections
    Illinois, 66f
    Illinois and eastern Indiana, 61f
    Illinois and western Kentucky, 61f
  Des Plaines buried impact structure, 101
  Fort Payne Formation, 99f
  fluorite mineralization, 303
  fossils, 172f–173f, 175f, 176, 177, 178, 179, 179f, 180, 181f, 182, 311f
  Hicks Dome, 101f
  limestone, 311f
    caves, 72t
    karst, 432, 433
    mining, 311–312
    state parks, 72t
  petroleum reservoirs, 37, 287f, 289, 293–296, 296f, 297
  Pine Hills area exposures, 72t
  Ste. Genevieve Fault System, 98
  stratigraphic columns, 168f, 287f
  structural geology, 62f, 85f
  sub-Kaskaskia surface, 169f
  tectonics, 62f, 85f
  near Thebes, 228f
  total dissolved solids, 339f
Missouri, Plattin (= Platteville Group), 151
Missouri Arch, 60f
Missouri cave, age-dating, 433
Missouri River valley, 232
Missourian (Series), 66f, 187, 188f, 193f, 194, 287f
mitigation
  aquifers, 351–352
  radon, 420, 421
  wetlands, 36
Moccasin Springs Formation, 160f, 161, 163, 287f
models, digital
  Geographic Information Systems (GIS), 49
  hydrogeology, 43, 325f, 341f, 359
  landscape, 381f, 382f
  paleoclimatology, 251
  Quaternary research programs, 243
  three-dimensional visualization, 40, 44, 358f
modern analogs, 257f
modern landscapes, 69–70
modern phase, 410f
modern soil
  Hudson Episode, 242
  Quaternary, 222f, 223f, 228f
Modesto Formation, 195f, 200, 287f
Modified Mercalli Intensity scale, 424, 425t
Mohawkian, 145–152
  Illinois cross section, 66f
  stratigraphic columns, 137f, 138f
moisture
  clay, 315
  coal, 264
  loess, 378
  soils, 377, 385–386
Moline
  nuclear facility, 422
  Pennsylvanian exposure, 189
  Rock Island Coal resources, 267

506

Mollisols
    Catlin silt loam, 375f
    Illinois Episode sediment, 234
    prairie grass, 375–376
    Sangamon Episode sediment, 234
Mollusca
    Galena Group, 151
    Maquoketa Group, 154
    Platteville Group, 149, 151
    St. Louis Limestone, 180
molybdenum, 266
moment magnitude, 424
monazite, 422
Monmouth, 267
monoclines, 84, 92f, 93, 95, 193, 304f
    See also individual named monoclines
Monroe County
    Fogelpole Cave, 250f, 256f, 256
    groundwater
        chemical composition, 338t
        pollution, 33f, 439
    Illinois Caverns State Natural Area, 72t
    karst, 433
    Mississippian rocks, 176, 177
    oil fields, 342
    saline springs, 342
    Valmeyer
        Great Flood of 1993, 401f, 401–402
        Kimmswick Limestone, 311
        underground mining, 320f, 463
Montgomery County
    aquifer vulnerability, 353
    drumlins, 231
    Freeman Coal Mining Company, Crown No. 1 Mine, 275f
    Raymond Basin, 258
Monticello, 344f
montmorillonite, 315
Montrose Harbor, 413f
Montrose/Wilson Avenue Beach, 413
Moorman Syncline (Moorman Trough) (Ky.), 63f, 91, 91f, 92f, 196f
    Illinois-Kentucky Fluorspar District, 304f
    relation to oil fields, 286f
Moraine View State Recreation Area, 72t
moraines, 10f, 219, 229f, 235–236, 236f, 237–238, 408
    Chicago Outlet Valley, 408
    Illinois River valley, 237f
    in state parks, 72t, 233f
    wetlands, 365
Morelock Lake, 401
Morrill Act (1862), 9
Morris, nuclear facility, 422
Morrison area, 161
Morrison city dump, 447f
Morrison Formation (Utah), 207
Morrowan, 187–188
    Illinois cross section, 66f
    reservoir rocks, 287f
    stratigraphic columns, 188f, 287f
    tectonics, 192
Morseville, 301f
Mosalem Formation, 160f, 161
Moscovian, 188f
Moses gravel pit, 212f
Mounds, 212f
Mounds Gravel, 213f, 213–214
    Massac County cross sections, 108f, 109f
    neotectonics, 106, 107, 107t
    stratigraphic column, 206f
    See also Grover Gravel
Mount Hill Country, 70, 71f
Mount Morris, 143f
Mountain Glen, 178
Mt. Auburn Field, 291, 291f
Mt. Carroll, 301, 301f
Mt. Carmel
    Friendsville Coal Member, 201
    Friendsville Mine, 273f
    lead-zinc deposits, 301
    Springfield Coal Member, 270
Mt. Carmel Fault System, 63f, 286f, 288
Mt. Carmel Sandstone, 188f
Mt. Olive, 269
Mt. Rorah Coal Member, 188f, 267f, 272
Mt. Simon Sandstone, 79f, 82f, 124, 128f, 139, 141
    Ancona Anticline, 94f
    aquifers, 299, 330, 330f
    Cambrian, 140
    Centralia sequence, 81, 82f, 128, 128f
    environment of deposition, 139
    gas storage, 289
    reservoir rocks, 287f
    seismic profiles, 126f
    stratigraphic columns, 137f, 138f, 287f
    sub-Sauk sequence, 139
Mt. Vernon
    Herrin Coal Member, 271
    Opdyke Coal Member, 201
Mt. Vernon-Ste. Marie sub-Pennsylvanian valley, 196f
MTBE (methyl tertiary butyl ether), 452, 453
Muensteroceras species, 175f
multi-county landfill, 447f
multivariate analyses, 253
Municipal Pier (now Navy Pier), 413
municipal waste, 443t, 445–447
    See also waste disposal
Murphysboro, 267, 267f
Murphysboro Coal Member, 188f, 265, 267f, 267–268
Murray Bluff Sandstone Member, 195f
Muscatatuck Group, 65f, 168f
muscovite, 210
Mystic Coal (Iowa), 271

N

Nachusa Formation, 137f, 138f
Namurian, 168f, 188f
Nashville Dome
    paleogeography, 91f, 189f
    structural geology, 60f, 91f, 171f
    tectonics, 62f, 78f, 189f, 192–193
Natacopsis species, 191f
National Cooperative Geologic Mapping Act (1992), 40
National Environmental Policy Act (amended 1969), 29t
National Primary Drinking Water Regulations (NPDWR), U.S. Environmental Protection Agency, 339–340
National Priority List sites, 444–445
National Secondary Drinking Water Regulations (NSDWR), U.S. Environmental Protection Agency, 340
Native Americans
    flint clay (pipestone), 3, 6f, 316, 317f
    lead-zinc deposits, 3, 305
natural gas, 283–298
    Ancona Anticline, 95
    coal bed methane, 279
    gas storage, 288, 393
    history of investigations, 11, 29
    New Albany Shale, 167–168, 173–176
    oil and gas fields, 291f
    reservoir rocks, 184, 187, 288f, 288–289
Nautiloidea, 170
Navajo Sandstone (Utah), 207
Naval Training Center Great Lakes, 404f, 411
Navy Pier (formerly Municipal Pier), 413
Nebraskan. *See* pre-Illinois Episode
Neda Formation, 138f
Negli Creek Limestone, 168f, 183f

INDEX

Nemaha Anticline, 193f
Neochonetes granulifer, 191f
Neochonetes species, 191f
Neospirifer species, 191f
neotectonics, 105–119
    Commerce geophysical lineament, 108–109
    earthquakes, 80f–81f, 109–110, 112–116
    features, 107f
    mapping, 42
    Quaternary faults, 105–109
    See also earthquakes
Neuropteris species, 190f
neutron activation analysis, 24
New Albany Shale, 169f, 173–176
    aquitard, 332
    cross sections
        Rough Creek-Shawneetown Fault System, 99f
        Ste. Genevieve Fault System, 99f
        stratigraphic units, 174f
    Devonian, 84, 167–169
    Hicks Dome, 101f
    history of investigations, 13, 26, 37–38
    Marine reef, 290f
    marker horizon, 289–290, 291, 293f
    Mississippian, 84, 167–169
    paleogeography, 169f
    reservoir rocks, 167, 173–176, 286, 287f, 292
    seismic profiles, 126f
    stratigraphic columns, 65f, 168f, 287f
    structural geology, 293f
New Burnside Anticline, 62f, 85f, 86, 87, 201f
New City Field, 291f
New Columbia, 106f, 107t
New Harmony (Ind.), 19f
New Harmony Field, 18
New Haven Coal Member, 200
New Madrid area
    earthquakes (1811–1812), 428–429
    neotectonics, 108–110
New Madrid Rift System, 60, 80f–81f, 81–82, 82f, 83, 137–138
    cross sections, 66f, 79f
    Everton Formation, 145–146
    Illinois Basin subsidence, 62
    neotectonics, 80f–81f, 81, 107f
    Sauk I subsequence, 140
    sedimentation, 79f, 80f–81f, 81
    stratigraphy, 128f
    structural geology, 60, 60f, 62f, 141f
    tectonics, 62f, 79, 80f–81f, 87f, 141f
    Tippecanoe I subsequence, 145
New Madrid Seismic Zone
    Commerce geophysical lineament, 109
    earthquakes, 80f–81f, 81, 88, 113, 113f
    epicenters (1974–1994), 87f
        New Madrid earthquakes (1811–1812), 108
        Olmsted earthquake swarm, 429f
        southern Illinois, 428–429
    Fluorspar Area Fault Complex, 100
    Fluorspar District, 116
    history of investigations, 30
    neotectonics, 88, 107f
    Reelfoot Rift, 105
New Orient Mine, 278
New Richmond (Roubidoux) Sandstone, 137f, 138f, 144, 146f, 147f
New York
    Acadian Orogeny, 84
    Ordovician delta, 83
    Niagaran, 161
        Illinois cross section, 66f
        oil and gas fields, 291f
        Silurian reef, 290f
        stratigraphic columns, 160f, 287f
    nickel, 266

Nimtz Member, 232
Nipissing phases, 408, 409f, 410f
nitrates
    aquifer vulnerability, 351, 352f, 354, 450
    in groundwater, 337, 338t, 341f, 345
    pollutants, 348, 444t, 449, 449f, 450, 450f
    in United States (2004), 451f
    in wells, 450
nitrogen
    in groundwater, 341f, 345, 346, 347
    pollutants, 448–452
    wetlands, 368
No. 1 coal. *See* Rock Island Coal Member
No. 2 coal. *See* Colchester Coal Member
No. 2A coal. *See* Survant Coal Member
No. 4 coal. *See* Houchin Creek Coal Member
No. 5 coal. *See* Springfield Coal Member
No. 6 Block Coal (West Virginia), 268
No. 6 coal, 195f
No. 7 coal. *See* Danville Coal Member
No. 11 Coal (Kentucky), 271
Noix Oolite, 137f, 155, 161
nonfuel minerals. *See* industrial minerals
non-point sources, 33, 449–449
Normal, 344f
North American Plate, 60, 77, 80f–81f, 88
    neotectonics, 110
    Tippecanoe I subsequence, 139
North American Pollen Database, 253
North American stratigraphic sequences, 65f, 67f
North Chicago, 404f, 406f
North Pier, 411, 412
North Point Marina, 404f
North Shore, Lake Michigan, 411, 413, 414f
North-Central karst region, 433–434, 434f, 435
Northeast Missouri Arch, 60f, 62f, 84, 85f
northeastern depocenter, 225
northern Centralia subsequence, 126f
northern sub-Centralia sequence, 128f
Northwestern University, 412
nuclear accelerators, 419
nuclear energy, 419, 421–422
nuclear facilities, 30, 422, 445
nuclear medicine, 419t
Nuclear Regulatory Commission, 20, 44
Nuclear Waste Policy Act (as amended 1982), 29t
nuclear weapons, 418, 419, 422
Nuttli magnitude, 424

O

Oak Grove Limestone, 188f
Oak Park Spit, 409f
Oak Street Beach (Chicago), 412f
Oakley Field, 291f
oceanic crust, 77, 77f, 80f–81f
oceanic records, 227, 250f, 251
Oconee Coal Member, 201
Ocoya sanitary landfill, 447f
Odontocephalus species, 171
Offerman Member, 160f, 161
Office of Mines and Minerals, 460, 461
Ogle County
    bedrock, 143f
    bedrock aquifer, 330
    Castle Rock State Park, 72t, 148
    Franconia Formation, 143
    Gunter Sandstone, 144
    Leaf River, 231
    Leaf River Anticline, 96f
    Leaf River esker, 231
    Lee County, 144
    Lowden State Park, 72t

Oneota (Gasconade) Dolomite, 144
Plum River Fault Zone, 96, 96f
Potosi Dolomite, 143
Rock River Hill Country, 72t
Sandwich Fault Zone, 70, 71f, 96, 97f
Shakopee Dolomite, 145
St. Peter Sandstone, 147, 313
Ogle Member, 223f, 231, 232, 233
"Ohara," See Karnak Limestone Member
Ohara limestone. See Karnak Limestone Member
Ohio Oil Company well log, 296f
Ohio River
   cypress swamp ecology, 365
   faults near, 106f, 117f
   floodplains, 364
   floods, 395
   karst, 433
   low point (Ill.), 70
   McNairy Formation, 211
   Pennsylvanian, 189
   Porters Creek Formation, 211–212
   Valmeyeran, 177
Ohio River Bedrock Valley aquifer, 335
Ohio River bluffs
   Caseyville Formation, 195
   St. Louis Limestone, 72t
Ohio River valley, 72t, 232
oil. See petroleum engineering; petroleum exploration; petroleum products
oil and gas. See petroleum engineering; petroleum exploration; petroleum products
oil and gas fields
   near Bridgeport, 283f
   in Cypress Sandstone, 295f
   in Illinois, 286f
   See also individual named fields
"oil rock," 302f
Oilfield, 283
OIS. See Marine Oxygen Isotope Stages
Old Faithful (Ill.), 342f
Old Timers Lead Mine, 12f
oldest rock exposed in Illinois, 143
Oligocene strata (none), 212
olivine, 202
Olmsted
   earthquake swarm (1984), 107f, 113f, 115–116, 428, 429, 429f
   faults, 208, 209f
   Clayton Formation, 211
   Porters Creek Formation, 211–212, 212f
   seismic profiles (Kentucky), 117f
Omaha Dome, 63f, 86, 92f, 201f
   Illinois-Kentucky Fluorspar District, 304f
   relation to oil fields, 202, 286f
   tectonics, 62, 62f, 85
Omphghent Member, 228
Onarga Bedrock Valley, 333, 346
Oneota (Gasconade) Dolomite, 143f, 144, 145f, 147f
   stratigraphic columns, 137f, 138f, 287f
Onychocrinus species, 172f–173f, 182
ooids, 311f
oolite
   in fluorspar deposits, 305f
   Mississippian limestone, 311f
   Platteville Group, 149
   reservoir rocks, 287f
   tidal bars, 181, 294
oolitic texture
   chert, 144
   grainstone lenses, 294
   limestone, 142, 305f
   packstone, 173
Opdyke Coal Member, 201, 267f, 272
Open Cut Land Reclamation Act (1962), 460
open pit mining
   fluorite, 307
   land subsidence, 388–392, 390f
   Mounds Gravel, 213
   reclamation, 462f, 462–463, 463f
   Wilcox Formation, 212
Ophiomorpha species, 211
Öpikina minnesotensis, 150f
Öpikina species, 151, 153f, 154
optically stimulated luminescence dating, 248
Orbiculoides newberryi, 191f
Orchard Creek Shale, 137f, 154–155
Ordovician, 62f, 136–157
   aquifers, 330–331
      radium, 348, 420f
      sulfate, 346
   areal extent, 147f
   bedrock, 68f, 143f, 433–434
   boundary, 154f
   carbonate traps, 289
   Castle Rock State Park, 72t
   chronostratigraphy, 65f, 66f, 67f
   cross sections, 61f, 66f, 99f, 339f
   dolomite
      Apple River Canyon State Park, 72t
      aquifer, 419
      karst, 432
   dolomite deposits, 311, 311f, 312f
      La Salle Anticlinorium, 94f
      lead-zinc deposits, 301
   fossils, 139, 143, 145, 146, 148, 150f, 151, 153f, 154
   Illinois Basin, 60, 60f, 83, 133
   impact structures, 101
   Kankakee Arch, 91
   K-bentonites, 40–41, 137f, 138f, 139f, 151, 152
   under La Salle Anticlinorium, 94f
   lead-zinc deposits, 300–301, 301f, 302f
   Lee County, 143f
   limestone, 432
   limestone deposits, 311, 312f
   Lowden State Park, 72t
   Maquoketa Shale Group, 291
   marine environment, 139
   Ogle County, 143f
   Pascola Arch, 86f
   petroleum traps, 289, 293f
   Platteville Group, 311
   reservoir rocks, 287f
   Rock River Hill Country, 72t
   Silurian reefs, 290f
   St. Peter Sandstone, 101, 145, 313
   Ste. Genevieve Fault System, 99f
   stratigraphic columns, 137f, 138f, 287f
   stratigraphy, 14, 195
   structural geology, 85f
   sub-Kaskaskia surface, 169f
   sub-Tippecanoe I surface, 147f
   Taconic Orogeny, 80f–81f
   tectonics, 62f, 83, 85f
   Tippecanoe I subsequence, 145–155
   total dissolved solids, 339f
   See also individual named geological units
ore deposits. See individual named mineral resources
Oregon
   Franconia Formation, 143
   Sandwich Fault Zone, 97f
   St. Peter Sandstone, 147
organic carbon, 255f
organic compounds, 366
   CERCLA sites, 445
   groundwater quality, 341f
   Histosols, 375–376
   landscape drainage, 377f
   Mollisols, 375

Index

nitrogen, 345
paleoclimatology, 248
pedogenesis, 377, 378
pesticides, 354
in soils, 374f, 377f, 379
sulfur, 265
suspended solids, 337, 340
waste disposal, 444
orogeny, 90, 250–251
    See also Acadian Orogeny; Alleghany Orogeny; ancestral Rocky Mountain Orogeny; Laramide Orogeny; Ouachita Orogeny; Taconic Orogeny
Osage. See Borden Siltstone
Osagean, 168f
Ostracoda
    Eoleperditia fabulites, 150f
    Fabaeformiscandona rawsoni, 252f
    Heterocypris punctata, 235
    Limnocythere friabilis, 258
    Limnocythere varia, 252f
    paleoclimatology, 252f, 253, 257f, 258
    Pennsylvanian, 189, 193f
    Pinicon Ridge Member, 171
    Platteville Group, 149
Ostreidae
    Exogyra costata, 211f
    Ophiomorpha species, 211
Otis Member, 171
Ottawa
    Illinois River valley, 237f
    St. Peter Sandstone, 147
Ottawa Megagroup, 66f
Ottawa Radiation Areas, 422
Ottawa Silica Sand Company, 27f
Ouachita Fold Belt, 300f
Ouachita Mountains
    Cretaceous, 208f
    Jurassic, 208f
    Pennsylvanian, 192
    structural geology, 60f, 91f
    Triassic, 207f
Ouachita Orogeny
    Mississippian, 86
    Pennsylvanian, 192
    Tradewater Formation, 197
Ouachita Trough
    Atokan, 189f
    Chesterian, 169, 170f
    incised valleys, 195
    Pennsylvanian, 189f
    Valmeyeran, 177f
outwash, 326, 378f
    aggregates, 379
    aquifer vulnerability, 358
    bedrock aquifers, 334, 335
    bluff coast, 405
    fans, 229f, 230f
    Illinois River valley, 177f
    Kaskaskia River valley, 231
    Quaternary aquifers, 333
    shallow aquifers, 335–336
    soil parent materials, 378f
    terraces, 217, 237f
    valley train deposits, 335
outwash plains
    Green River Lowland, 333
    sand and gravel, 229f, 313, 314f
overburden, 26f, 459f, 459–460, 460f, 461
Owl Creek Formation, 206f, 211
oxidation, 341f, 344, 346, 347
oxygen isotope ratio ($^{18}O:^{16}O$), 250f, 251, 253, 255f, 256
Ozark Dome (Mo.), 60, 60f, 62f, 63f, 78f, 91, 91f
    Chesterian, 170f

    Cretaceous, 208, 208f
    Illinois Basin, 98, 171f, 208
    paleohydrology, 300f
    Pennsylvanian, 189, 189f
    Ste. Genevieve Fault System, 98
    Valmeyeran time, 177f
    tectonics, 62f, 78f, 84, 85f, 192–193
Ozark Plateaus Province, 70, 71f
Ozark Uplift, 182, 193f, 208f
Ozaukee series, 377f
Ozaukee silt loam, 382f

P

Pacific Ocean (paleogeography), 207f, 208f
packstone, 151–152
Paducah (Ky.), 304f
paint
    lead-zinc deposits, 300
    St. Peter Sandstone, 314
    tripoli, 316
Paint Creek Formation (Downeys Bluff), 168f, 181, 183f, 287f, 297
Paint Creek Sand ("Paint Creek sand"), 168f, 287f
Paleocene, 66f, 206f, 211f
paleochannels
    electrical logging, 296f
    structural traps, 293f
    Wedron Quarry, 220f
paleoclimatology, 248, 249t, 250–253, 254, 260
    fossils, 251, 252f
    geochronology, 248, 249t, 250, 257f
    GISP2 record, 255f
    global change, 251
    insolation, 250–251
    Jurassic, 207
    loess, 253, 254f
    oxygen isotope ratio, 250f
    paleosols, 253, 254f
    pedogenesis, 376–377
    Pennsylvanian, 192–194
    pollen, 23f, 28f, 107, 249t, 252f, 257f, 258
    Quaternary, 243, 248–256, 257f, 258–260
    Sangamon Episode, 234
    Sauk sequence, 136
    speleothems, 253, 255f, 256, 256f, 258
paleogeography
    Cambrian, 136, 139f
    Chesterian, 169, 170f
    Devonian, 167
    Laurentia, 125f, 138, 139f
    New Albany Shale, 169f
    Ordovician, 136, 139f
    Pennsylvanian, 192
    Precambrian, 139f
    Sauk sequence, 138
    Tippecanoe I subsequence, 139
    Valmeyeran, 177f
paleohydrology, 196f, 216f, 300f
paleoindian, 3
paleontology
    history of investigations, 7, 11, 14, 28f, 41
    See also individual named fossils; individual named periods
paleosols, 219
    beryllium-10 (^{10}Be), 248
    Chesterian, 183, 183f
    Galatia Mine, 192f
    "hardpan," 183
    Illinois Episode, 228, 231–232
    Keller Farm section, 253, 254f
    loess, 253, 254f
    northern Illinois, 39
    paleosol A, 228
    Pancake Hollow Section, 228

Pike Soil, 233, 234
pre-Illinois Episode, 224–226
Quaternary studies, 206f, 219, 223f, 226f, 236f
Sangamon Episode, 234f, 234–235
Tuscola Quarry, 222f
Wisconsin Episode, 236f
paleovalleys, 196, 224, 225f, 291, 292f
Paleozoic, 65, 66f, 67f
    Alleghany Orogeny, 80f–81f
    bedrock, 68f, 115
    fossils, 11, 23
    glacial extent, 216f
    Massac Creek graben, 109f
    paleohydrology, 216f
    petroleum reservoirs, 287
    Precambrian, 123f, 128f
    proto-Illinois Basin, 127f
    sedimentary basin, 123f
    seismic profiles, 113–114, 115f
    tectonics, 62–64, 80f–81f, 83, 141f
    See also individual named geological time periods; individual named geological units
Palestine Sandstone, 168f, 183f
    fossils, 182
    reservoir rocks, 287f, 297
Palzo Sandstone Member, 195f
Pancake Hollow Section, 228
Pangea, 62f, 80f–81f, 81, 84, 87, 192, 206
Paoli Limestone, 168f, 183f
Paradise (No. 12) Coal (Ky.), 272
parent materials, 377–378, 378f, 379, 380
Paris, 181, 270
Partlow, 287f
Pascola Arch, 60, 60f, 62f, 63f, 91f, 87, 93
    Cambrian to Permian, 127f
    Cretaceous, 79, 79f, 81, 87, 208, 208f
    cross sections, 66f, 128f, 141f
    Illinois Basin, 87, 91, 93
    Paleozoic strata, 87
    Pennsylvanian, 79, 79f
    Permian, 87
    Precambrian, 125f, 128f
    proto-Illinois Basin, 127f
    stratigraphic units, 141f
    tectonics, 62f, 80f–81f, 141f
"patch reefs," 170, 171
pathogenic microbes, 448, 453–455
Patoka Formation, 195f, 200, 267f
    See also Shelburn-Patoka Formation
patterned ground, 240
Paw Paw Bedrock Valley, 333f, 334
Paxton, 344f
Paxton landfills, 447f
Peabody Coal Company
    No. 10 coal mine, 275f, 278
    River King Mine, 273, 273f
Peabody Energy surface mine reclamation, 461
peaker plant, 358
Pearl Formation, 223f, 229f, 231
Pearlette 'O' (Lava Creek B) volcanic ash bed, 227
peat
    Chesterian, 183–184
    coal, 263, 264
    cyclothems, 194f
    Histosols, 375–376
    Pennsylvanian swamps, 187–188, 194f
    Quaternary sediment, 226
    wetlands, 361, 366–367
peat deposits, 311f, 316, 319f, 320
Pecatonica Formation, 83, 149–151
    lead-zinc deposits, 302f
    stratigraphic columns, 137f, 138f
Pecatonica River valley aquifer, 358

Peddicord Tongue, 221f
pedogenesis. *See* soils
pedostratigraphic units, 222, 223f
pelecypods
    Cardiomorpha species, 191f
    Chaenomya species, 191f
    Dunbarella species, 191f
    paleoclimatology, 253
    Pennsylvanian, 189
    Porters Creek Formation, 211
Pennsylvanian, 187–205, 195f
    aquifers, 332
    Atokan, 189f
    bedrock, 68f
    Carbondale Formation, 198–199, 199f, 200
    chronostratigraphy, 65f, 66f, 67f
    claystones, 315
    climate, 192
    Cottage Grove Fault System, 97
    cross sections
        Illinois, 66f
        Illinois and eastern Indiana, 61f
        Illinois and western Kentucky, 61f
        Ste. Genevieve Fault System, 99f
    cyclothems, 193–194, 194f
    Des Plaines buried impact structure, 101
    Du Quoin Monocline formation, 95
    fossils, 189f, 189–190, 190f, 191, 191f, 192f, 197, 199, 201
    freshwater lakes, 189
    Hicks Dome, 101f
    Pascola Arch, 93
    petroleum reservoir, 296f
    Pine Hills, 72t
    Plum River Fault Zone, 96
    sandstone, 149f
    sea level changes, 193–194
    Ste. Genevieve Fault System, 98, 99f
    stratigraphic columns, 188f, 287f
    stratigraphic units, 14, 194–201
    structural geology, 62f, 85f
    tectonics, 62f, 85f
    total dissolved solids, 339f
    wetlands, 367
    See also coal; individual named geological units
Pennsylvanian deposits, 299–300
    limestone, 312, 312f, 432
    petroleum reservoirs, 287f, 294–295, 296f, 297
    shale bricks, 316
    See also coal deposits
Pennyrile Fault System (Ky.), 63f, 189f
    deformation, 84, 85f, 86f,
    Illinois-Kentucky Fluorspar District, 304f
Pennyrile Fault Zone, 62f
Pentamerus species, 161, 162f–163f
Pentremites species, 175f, 182
Peoria
    Dover Moraine, 236f
    Illinois River valley, 226f
    landslides, 388f
    Rock Island Coal Member, 267
    Springfield Coal Member, 271
Peoria Silt (loess)
    history of investigations, 10f
    Keller Farm section, 254f, 255f
    Kelley site, 108f
    meltwater, 240
    neotectonics, 108
    paleosols, 254f, 255f
    Quaternary, 221f, 223f, 228f, 229f
    Wedron Quarry, 221f
Peoria sublobe, 236f
Peotone series, 377f
Pere Marquette State Park, 72t, 165

Index

peridotites, 100, 202
periglacial environment, 221f
period, 64f, 65
permafrost, 240, 249t
permeability, 287, 325, 351
  barrier, 288
  clay, 316
  levees, 397, 399
  reservoir rocks, 286
  suspended solids, 340
  See also hydraulic conductivity
Permian, 187, 188, 201f, 201–202
  bedrock, 68f
  chronostragraphy, 65f, 66f, 67f
  Illinois cross section, 66f
  ores, 299–300
Permian tectonics, 62f, 80f–81f, 84
  Fluorspar Area Fault Complex, 100
  Hicks Dome, 100, 101f
  Illinois Basin closure, 93
  Pascola Arch uplift, 87, 93
  Rough Creek-Shawneetown Fault System, 100
  structural geology, 85f, 86f
  ultramafic dikes, 101f
Perry County, 267, 278
personal care products, 453
Peru, 115, 237f, 239f
Peru Monocline, 63f, 83, 84, 85f, 92f
  neotectonics, 107f
  oil fields, 286f
  seismic profiles, 114
  relation to St. Peter Sandstone, 149f
Pesotum Bedrock Valley, 333
pesticides
  aquifer vulnerability, 326, 351, 353f, 354
  pollutants, 444, 444t, 449f, 450f, 452
pet litter, 315
Petersburg, 427
petrified wood, 211
Petro, 287f
petrographic textures, 132
petroleum engineering
  development techniques, 284, 285f
  history, 11f, 15, 18f, 18–19, 19f, 37, 283f, 283–285
  hydraulic fracturing, 285
  Precambrian rocks, 123, 124
  primary recovery, 288
  secondary recovery, 285, 288
  seismic profiles, 113–114, 125–126, 284–285
  waterflooding, 285, 285f
  well casings, 284, 284f
  well logging, 296f
petroleum exploration, 293–298
  anticlines, 95
  Cypress Sandstone, 295f
  Fairfield Basin, 91
  history, 11, 11f, 15, 17f, 17–18, 18f, 19f, 26, 37f, 37–38, 283f, 284f
  Kimmswick Limestone, 151
  Marine reef, 163
  oil and gas fields, 291f
  structural geology, 286f, 291f, 292f
  New Albany Shale, 167–179, 286, 290f, 292, 293f
  reef structures, 102, 289–290, 290f, 291
  reservoir rocks, 287f, 287–292, 292f, 293–297
  Ste. Genevieve Limestone, 181
  stratigraphic-structural traps, 94, 288f
    Chesterian, 184
    Devonian, 292–293
    Mississippian, 181, 288f, 293–295, 295f, 296f, 296–297
    Ordovician, 289
    Pennsylvanian, 187, 297
    Permian, 202
    Precambrian, 132

  on Sangamon Arch, 291, 291f
  Silurian, 164, 289, 290f, 290–291, 292f, 293f
  See also individual named oil and gas fields
petroleum products
  economic value, 18–19, 283–284, 284f, 285, 285f, 287f, 319f
  pollutants, 351, 358, 421, 444, 445, 452–453
  underground storage, 288–289, 393, 452
Petroleum Technology Transfer Council, 37
Peyton Formation colluvium, 242
pH, of groundwater, 338t, 364
Phanerozoic, 65, 65f, 66f, 67f, 79
Phanocrinus species, 172f–173f, 182
pharmaceuticals, 300, 448, 453
phase, 64f
phosphorous
  coal, 266
  eutrophication, 452
  wetlands, 368
Phragmolites species, 150f, 151
physiographic divisions, 22
physiographic regions, 70, 71f
Piasa Limestone, 98f, 188f, 200
Piatt County, 342
Piatt Member, 221f, 223f, 238
Piedmont Province, 211
Pike County
  carbonates, 168
  groundwater, 332
  parent materials, 378f
  pre-Illinois Episode, 224
Pike Soil (paleosol), 233, 234
Pine Hills, 72t
Pinicon Ridge Member, 171
pinnacle reefs, 289
Pionodema species, 151
pipelines
  earthquake effects, 424–425
  pollutants, 444, 444t
pipestone (flint clay), 3, 6f, 316, 317f
piping, 386–387, 399f
Pisces, 440
  Carbondale Formation, 199
  Devonian, 170
  Mazon Creek fauna, 190
  Pinicon Ridge Member, 171
  Porters Creek Formation, 211
"pitches," 303f
Pittsfield, 189, 209, 291f
Pittsfield Anticline, 63f, 92f, 95
Placodermi, 170
Plaesiomys species, 153f, 154
plagioclase, 133f
Plaines Member, 160f
planar bedding structures, 432, 432f
planktonic taxa, 252f
plants. *See* vegetation
plastic materials, 300
plate tectonics, 60, 62f, 77f, 77–79, 80, 80f–81f, 81–90, 110, 128, 139
  See also neotectonics; tectonics
Platteville Group (= Black River Limestone), 94f, 148f, 149–151
  dolomite, 311
  fossils, 150f, 151
  hardground omission surface, 151f
  La Salle Anticlinorium, 94f
  lead-zinc deposits, 302f
  Lee County, 143f
  limestone, 311
  Ogle County, 143f
  Sandwich Fault Zone, 97f
  stratigraphic columns, 137f, 138f, 287f
Platteville Group (= Plattin, in Mo.), 137f, 138f, 151
Plattin Limestone, 137f
Platycrinites penicillus, 175f, 180

Platycrinites species, 182
Platymerella species, 160
Platystrophia species, 153f, 154
Pleasantview, 287f
Pleistocene
    Fluorspar Area Fault Complex, 100
    history of investigations, 27
    Holocene boundary, 249t
    Illinois cross section, 66f
    Massac Creek graben, 109f
    sand and gravel, 313
    stratigraphic columns, 66f, 206f
    tectonics, 105
    See also Quaternary
Pleurodictyum problematicum, 171, 172f–173f
Pliocene, 66f, 206f, 250f
Plum River Fault Zone, 62, 63f, 84, 85f, 92f, 95, 96, 96f
plutonium, 445
point bars, 296
point sources, 33, 443, 444t, 444–448
polar regions, 251
polish
    St. Peter Sandstone, 314
    tripoli, 316
pollen
    paleoclimatology, 23f, 28f, 107, 249t, 252f, 257f, 258
    Sangamon Geosol, 235
pollutants, 444t, 449–455
    groundwater, 325–336, 342f, 344f, 345f, 351–360
    Illinois Caverns, 33f
    karst, 435–436
    radioactive isotopes, 347, 351
    research programs, 30, 32–33
    streams, 443t
    waste disposal, 33f
pollution, 444t
    hydrogeology, 443
    industrial facilities, 358, 443, 444t, 449
    sanitation systems, 309, 351, 452, 453–454
    water, 33, 337, 364, 368, 441, 443–457
    See also pesticides; radioactive waste
polonium, 420
Polygnathus cristata, 173
Pomona Fault, 99f
Pond Creek Coal Member, 98f
Pontiac, 272
Pope County
    Bell Smith Springs, 72t
    fluorite production, 307
    Fluorspar Area Fault Complex, 100
    Fort Payne Formation, 178
Pope Group, 66f, 101f, 168f, 188f
pore water, 30f
porosity, 286–289, 340
porphyritic texture, 132
Port Byron stone industry, 164
Port Huron Phase, 242
Porters Creek Formation, 206f, 211–212, 212f, 315
portland cement. *See* cement materials
positive flower structures, 97, 98f, 99f
Post Creek, 106f
Post Creek Formation, 209–210, 210f
    faults, 106f, 107t
    stratigraphic column, 206f
post-Algonquin phase, 410f
potability, 329, 331f
potassium, 337, 338t, 452
potassium feldspar, 132, 133f
potassium-40, 419
potentiometric surface, 327, 327t
Potosi Dolomite, 143, 143f, 144f, 147f
    aquifers, 330
    stratigraphic columns, 137f, 138f, 287f

"Pottsville," 188f, 287f
Pottsville Formation, 195f
Pounds Formation, 195f
Pounds Sandstone, 188f, 195–197
power plants, 280
power shovel, 459f
Pragian Stage, 160f
prairie
    grasses, 375–376, 377
    soils, 373f
    vegetation, 252f, 257f, 258, 373
Prairie du Chien Group, 143f, 143–145, 148f
    aquifers, 330
    stratigraphic column, 138f
Prairie du Rocher
    *Acrocyathu*s species, 180
    Chesterian, 181
    Great Flood of 1993, 402
    underground storage, 393f
Prasopora species, 153f, 154
Precambrian, 41, 123–124, 124f, 125f, 125–135
    basement rock, 123
    chronostratigraphy, 65f, 66f, 67f
    core drilling, 125f, 132
    cross sections
        Illinois, 66f
        Illinois and eastern Indiana, 61f
        Illinois and western Kentucky, 61f
    data management, 132–133
    deep seismic sounding, 79–81, 113–114, 114f, 125–128, 127f
    drill core analysis, 133f
    groundwater quality, 341f
    Illinois Basin, 63, 78f, 79–81, 127f, 128f, 133
    magnetic anomalies, 128, 129f, 130f, 131f, 131–132
    Metazoa, 138
    Mt. Simon Sandstone, 124f
    paleogeography, 139f
    Sauk sequence, 137
    sedimentary basin, 123f
    stratigraphic columns, 137f, 138f, 287f
    stratigraphy, 141f
    structural geology, 62f
    tectonics, 62f, 77–78, 78f, 79, 79f, 80f–81f, 141f
    total dissolved solids, 339f
    Wabash Valley faults, 97
precipitation. *See* atmospheric precipitation; chemical precipitation
preglacial events, 222f, 222–224
pre-Illinois Episode, 222, 223f, 224–226, 226f, 227–229
    drainage, 224f
    landscapes, 225f
    Mallard Creek, 108f
    paleoclimatology, 250f
    Quaternary sediment, 229f
    sand and gravel aquifers, 224
    Tuscola Quarry, 222f
pre-Sauk sequence, 65f, 67f
pressure transducer, 30f
Pridoli, 160f
primary recovery, 288
Princess No. 6 coal seam (Ky.), 268
Princeton Bedrock Valley, 224, 225f, 333, 333f
Princeton sublobe, 236f
principal compressive stress axis, 110f, 110–111, 111f
Prior landfill, 447f
Prior Blackwell landfill, 447f
Prismopora serrulata, 175f, 182
prodeltaic deposits, 173, 178
Productida, 178
production
    coal deposits, 263, 266, 277, 278, 278f, 280
    fluorite deposits, 307
    hazardous waste, 444
    industrial minerals, 27, 309, 320

Index

natural gas, 283–284, 284f
petroleum, 283–284, 284f
sand and gravel deposits, 313, 319f
sphalerite, 306
stone, crushed, 319f
productive geological units, 288–297
proglacial sediment, 219, 221f, 238, 240
prometon, 452
promontories, 412
property damage
    earthquakes, 415t, 427–428, 428f
    floods, 396, 396f, 397f, 398f, 400f, 401f
    land subsidence, 388–390, 390f, 391f, 391–392
    landslides, 387f, 387–388, 388f, 389f
    moisture, soils, 385–386
    radon, 420, 421f
    slope stability, 387f
    See also geological hazards
Prospect Hill Siltstone, 168f
Proterozoic Eon/Eonothem, 65, 65f, 66f, 67f, 137f, 138f
Protista, 453–454
proto-Illinois Basin, 79–80, 80f–81f, 81, 127f, 139
Proviso Siltstone, 138f
proxy records. *See* paleoclimatology
Pseudolingula species, 151
Pterotocrinus species, 172f–173f, 182
public field trip (1920), 14f
Pulaski, 211
Pulaski County
    Cache River State Natural Area, 72t
    earthquake swarm, 115–116, 428
    faults, 100, 106f
    Lamna cuspidata, 211f
    lime kiln, 318f
    Turritella species, 211f
    Wilcox Formation, 212, 212f
pump stations, 365
Pupilla muscorum, 252f
Putnam County
    Illinois Episode tills, 234
    Illinois River valley, 237f, 239f
    Sangamon Geosol, 234f
    wetlands, 402
Putnam Phase, 236f, 239, 240
pyrite (iron sulfide)
    Carbondale Formation, 198–199
    groundwater, 341f
    history of investigations, 31
    mining, 265, 277, 277f, 279
    pollution, 460
pyroxene group, 202

Q

Quad cities (Bettendorf and Davenport, Iowa; Moline and Rock Island, Ill.)
    building stone, 164
    earthquakes, 427
    nuclear facility, 422
    Pennsylvanian, 189
    Rock Island Coal Member, 188f, 267, 267f
quadrangle maps
    aquifer vulnerability, 359
    coal mines, 392
"Quality Circle," 271, 271f
quarries
    Batavia stone quarry, 318f
    Casey Quarry, Vulcan Materials, 458
    Charleston Quarry, 220f, 320f
    Chicago area, 320, 320f
    Columbia Quarry, 463
    dolomite, 311f, 312f
    near Hillside, 154f
    Horseshoe Quarry, 99f
    Illinois, 311f
    Lehigh Stone Company quarry, 20f
    limestone, 311f, 312f
    Mounds Gravel, 213f
    Potosi Dolomite, 143
    reclamation, 29, 462f, 462–463, 463f
    Riverstone Group Valley Quarry, 462f, 462–463, 463f
    stone, crushed, 310f, 312f, 318f
    Sycamore Quarry, 152f
    Thornton Quarry, 38f, 59f, 72t, 163–164, 164f, 165
    Tuscola Quarry, 94, 222f, 219, 240
    underground storage, 393, 462–463
    urban environment, 38, 320f, 462f, 462–463, 463f
    Vulcan Materials quarry, 152f,
    Wedron Quarry, 220f, 221f, 240
quartz
    coal ash, 265
    galena in, 301f
    igneous rocks, 132
    Post Creek Formation, 210
    in silica sand, 313
    sub-Mt. Simon Sandstone, 133f
    Wilcox Formation, 212
quartz arenite
    Caseyville Formation, 196–197
    Dutch Creek Sandstone, 171
    Grover Gravel, 214
    lower Tradewater Formation, 198
quartz sand
    Baylis Formation, 209
    environment of deposition, 187
    Everton Formation, 145–146
    Gunter Sandstone, 144
    "Hoing" sandstone, 173
    Ironton-Galesville Sandstones, 142
    Mt. Simon Sandstone, 141
    New Richmond (Roubidoux) Sandstone, 144
    Post Creek Formation, 210
    St. Laurent Formation, 173
    St. Peter Sandstone, 146–148, 149f
    Tradewater Formation, 198
    unnamed facies of Grand Tower Limestone, 171
quartzites
    Post Creek Formation, 210
    Precambrian rocks, 125f
    pre-Illinois deposits, 225
Quaternary, 108, 216–223, 223f, 224–229, 229f, 230–232, 233–247
    aquifers, 219, 332–335
    bedrock topography, 222–224, 225f
    chronostratigraphy, 65, 65f, 66f, 67f
    cross sections
        Illinois, 66f
        Illinois and eastern Indiana, 61f
        Illinois and pre-Illinois Episode, 226f
        Illinois and western Kentucky, 61f
        Shelbyville to Lake Michigan, 236f
    drainage, 224f
    faults, 105–106, 106f, 107f, 107t, 107–108, 108f
    Fluorspar Area Fault Complex, 106
    fossils, 209, 209f, 235, 239, 240, 242, 242f, 251, 252f, 253, 258
    glacial ridge, axis, 232f
    groundwater floods, 399
    history of investigations, 9, 10f, 13, 39–40, 40f, 217–219, 242–243
    Hudson Episode, 242
    Illinois Episode, 229–234
    Illinois River valley, 237f
    isotope ratio, 250f
    Jo Daviess County wetlands, 364
    landforms, 217, 236f, 381f, 382f
    landscapes, 225f
    Massac County, 107, 107t
    neotectonics, 107, 107t

overview, 69
paleoclimatology, 248–250, 250f, 251–260
paleosols
    Illinois Episode, 231–232
    pre-Illinois Episode, 222f, 224–226
    Sangamon Episode, 222f, 234–235
    Wisconsin Episode, 238
Pulaski County, 106f
preglacial events, 222–224
pre-Illinois Episode, 224–225, 225f, 226, 226f, 227–229
sand dunes, 72t
Sangamon Episode, 234–235
Sangamon Geosol, 234f
sediment, 219–220, 220f, 224–226, 229f, 238
soils, 223f, 228f, 380f
stratigraphic columns, 206f, 221f, 223f
surface topography, 105–109, 218f, 233f
time-distance diagram, 236f
tongues, 40, 220, 238
Tuscola Quarry, 222f
Wedron Quarry, 221f
wetlands, 365, 367
Wisconsin Episode, 231f, 235–242
See also individual named geological units; individual named landforms; loess; paleoclimatology
Québec (Canada), paleogeology, 125f
Quimbys Mill Formation, 137f, 138f, 302f
Quincy
    Baylis Formation, 209
    on cross section, 226f
    Great Flood of 1993, 397f
    municipal landfill, 447f
    underground mine reclamation, 463

R

Raccoon Creek Group, 188f, 195f
racemization, 248
Racine Dolomite, 161, 163
    fossils, 162f–163f
    Stony Island, 407, 409f
    stratigraphic column, 160f
radiation, 418–423
    history of investigations, 33
    public health, 419t
    regulations, 418
    shields, 300
radioactive decay
    absolute age, 65
    geological materials, 418
    groundwater, 347, 419
    rates, 347
radioactive fallout, 419
radioactive isotopes, 418, 445
radioactive pollution, 421, 422
radioactive waste
    military, 422
    sites, 418, 445
    sludge, 421
radiocarbon dating. *See* carbon-14 dating
Radiolaria, 209, 211
radium, in groundwater, 348, 418, 419, 420f, 422, 443
radium-based paint pollution, 422
Radnor Member, 223f, 231, 232, 233–234
radon
    health effects, 419t
    in homes, 33, 418, 419, 420, 421f
    parent elements, 420
    radon-222, 419
Rafinesquina species, 151
railroads
    coal transportation, 277
    landslides, 388
    shoreline protection, 411
    stone transportation, 164
rainfall, *See* atmospheric precipitation
Ramsar Convention on Wetlands (1971), 361–362, 365
Randolph County
    Acrocyathus species, 180
    Chesterian, 181
    Great Flood of 1993, 402
    reclamation, 461
    Sangamon Soil, 13f
Rantoul, 344f
rare earths, 124
Raum Fault Zone, 106f
Raymond Basin, 249t, 250f, 258
Receptaculites oweni, 150f
Receptaculites species, 151, 302f
recharge, 328, 328f, 352, 359
reclamation
    awards, 458, 462
    mines, 279, 320, 320f, 460, 462
    polluted aquifer, 352
    quarries, 458, 461, 461f, 462f, 462–463, 463f
    regulations, 459
    sand and gravel pits, 320f, 458f, 459f, 461–463
    surface mines, 273, 320, 320f, 458–460, 460f, 461–464
    underground limestone mines, 463
recreation
    reclaimed quarries, 462f, 462–463f, 463f
    speleology (caving), 441
    streams, 443t
    wetlands, 361, 368
Red Bud, 13f
reduction-oxidation (redox), 339, 341f, 344, 347f
Reed-Keppler Park, 422
reefs, 66f, 158, 161, 288f, 289–290, 290f, 291
Reelfoot Lake (Tenn.), 429
Reelfoot Rift
    Cambrian, 113, 127f, 137–138
    Commerce geophysical lineament, 109
    Cretaceous, 87f
    earthquakes, 87, 87f
    Holocene, 87f
    Illinois Basin, 133
    Illinois-Kentucky Fluorspar district, 304f
    Mississippian, 182–183
    neotectonics, 107f
    New Madrid Rift System, 82f
    paleohydrology, 300f
    Pascola Arch, 86f
    Permian, 127f
    Permian-Mesozoic Era, 86f
    Precambrian, 83, 123, 125f, 128f, 137–138
    proto-Illinois Basin, 127f
    Sauk I subsequence, 140
    stratigraphic column, 137f
    structural geology, 60, 60f, 62f, 63f, 93
    tectonics, 62f, 79f, 87f, 105, 141f
regional planning
    earthquakes, 428–429
    floods, 395–403
    groundwater quality, 24, 42–43, 325–336, 337–350, 351–360
regulations
    bacteria, 453–454
    Illinois Pesticide Management Plan, 353–354
    landfills, 446
    radiation exposure, 418
    reclamation, 459
    U.S. Environmental Protection Agency standards
        aquifer protection, 353–354
        arsenic, 344
        CERCLA sites, 444–445
        drinking water, 339–340, 343, 344, 345, 347, 348, 419, 450, 452
        fluoride, 343

iron, 345
manganese, 345
National Primary Drinking Water regulations, 339–340
National Priority List sites, 445, 446, 446f
National Secondary Drinking Water regulations, 340
nitrogen compounds, 345, 450
pesticides, 452
radiation, 422
radium, 348, 419
radon, 419, 422
solid waste, 446
sulfate, 346
total dissolved solids, 347
Toxic Release Inventory program, 444
water wells
abandoned wells, 450f
construction code (2004), 441
disinfection, 352
permits, 450f
wetlands, 361, 368
See also laws and legislation
relative time, 64, 65, 65f
relief, 123
remediation
aquifers, 352–353
lakebed downcutting, 415
radioactivity, 422
water supply, 419, 448
wells, 455
wetlands, 36, 368
remote sensing, 39, 123, 237f
Renault Limestone, 182, 296f, 297
fluorite, 303
fluorspar deposits, 305f
reservoir rocks, 287f
Reptilia, 209, 211, 235, 258, 368
reservoir rocks, 287f
See also individual named geological units; petroleum exploration
resistivity, 20f, 25f
Resource Conservation and Recovery Act (RCRA) (1976), 29t, 444
reverse faults. *See* faults: reverse faults
Reynoldsburg Coal Member, 188f, 267f, 272
Rhizocorallium ichnospecies, 175f
rhizoliths, 253
Rhoden landfill, 447f
Rhuddanian Stage, 160f
rhyolites
cross sections, 61f, 66f
McNairy Formation, 210
porphyry, 101, 225
Precambrian, 123, 124f, 125f
stratigraphic columns, 137f, 138f, 287f
rhythmites, 197
Richardsondoceras species, 150f, 151
Richmondian Stage, 137f, 138f, 160f
Richter magnitude, 424
Ridenhower Formation
Chesterian, 182, 183f
fluorspar deposits, 305f
reservoir rocks, 287f
ridges, 232f, 233f
sandstone reservoirs, 297
stratigraphic columns, 168f, 183f, 287f
ridge-and-swale topography, 405, 407f
"ridged drift"
Illinois Episode, 229, 229f, 230
linear aquifers, 335–336
"Ridgley," 188f, 287f
rift basins, 82f, 127f, 207f
rift system, 130f
rift zones, 77, 127, 207f
See also tectonics
rills, 380, 383

riprap, 179f, 415f
rivers
banks, 368
base flows, 368
channel floods, 397, 400, 402
Illinois topography, 218f
nitrates, 452
See also individual named rivers
River King Mine, 272–273, 273f
river valleys
agrichemicals, 354
alluvium, 379
ancient Iowa River valley, 333, 333f
ancient Ohio River valley, 72t
aquifer vulnerability, 353, 354
Cache River valley, 365
Des Plaines River valley, 165, 240, 242, 366
DuPage River valley, 165
erosion, 217, 387
floods, 395, 401
Fox River valley, 165, 240, 313
ground motion, 429
groundwater discharge, 328
Kankakee River valley, 165, 240
Kaskaskia River valley, 231, 233f
loess, 240
Mackinaw River valley, 226f, 338, 344–345
meltwater, 217, 240
Missouri River valley, 232
Ohio River valley, 232
Pecatonica River valley, 358
pollution, 452
terraces
aquifer vulnerability, 353
Illinois River valley, 237f
Mounds Gravel, 214
Woodstock Phase, 240
wetlands, 364, 365
See also Illinois River valley; Mississippi River: ancient; Mississippi River valley; Wabash River valley
Riverside Spit, 409f
Riverstone Group's Valley Quarry, 462f, 462–463, 463f
roads
deicer, 351, 444t, 448, 455, 455f
drainage, 383
earthquake effects, 424–425
erosion, 383
gravel, 184
land subsidence, 390, 391f
landslides, 388
maintenance, 443t
runoff, 443t
shoreline protection, 411
surfacing materials, 213, 280, 312
Robein Member, 221f, 223f, 367
Robinson strata, 287f, 296f
Roby Field and Roby West Field, 291f
Rochelle, 143f
Rock Bedrock Valley, 225f, 333, 333f, 334, 358
Rock Creek Graben, 106f, 304f
Rock Island
building stone, 164
earthquakes, 427
nuclear facility, 422
Pennsylvanian, 189
Rock Island Coal Member, 188f, 267, 267f
Rock Island Arsenal stone, 165
Rock Island Coal Member, 188f, 267, 267f
rock record. *See* stratigraphy
Rock River, 224
floods, 395
karst, 433–434
Sandwich Fault Zone, 97f

valley train deposits, 313
wetlands, 364
Rock River Hill Country, 70, 71, 72t
Rockdale Moraine, 236f
Rockford, 152f, 436
Rockford Limestone (Ind.), 175f
Rodinia, 81, 80f–81f, 82f, 136
"rolls"
    Herrin Coal Member, 271
    Springfield Coal Member, 270
Rome Trough, 80f–81f, 81, 125f
Romeo Member, 160f, 161
roof control
    bolting, 111f, 274, 275f
    Kerr-McGee Company Galatia Mine, 112f
    longwall mining, 275–276, 276f, 390
    Peabody Coal Company No. 10 mine, 275f
    Springfield Coal Member, 270
    underground mines, 105, 110, 112
room-and-pillar mining, 274f, 274–275, 320
rootlets, 253
roots
    in paleosols, 183
    in Post Creek Formation, 210f
    respiration, 255f, 341f, 346
    in underclay, 193–194, 194f
Rose Hill Spit, 409f
"Rosiclare." *See* Spar Mountain (Rosiclare) Sandstone Member
Rosiclare, 7, 101f, 304f, 306, 307, 338t, 343
Rosiclare District, 303–304, 304f
Rostricellula species, 151
Roubidoux Sandstone. *See* New Richmond (Roubidoux) Sandstone
Rough Creek Fault System, 201, 201f, 202, 304f
Rough Creek Graben, 60, 60f, 79f
    Cambrian, 127f, 137–138
    Centralia sequence, 82f
    Cretaceous, 87f
    earthquake epicenters (1974–1994), 87
    Holocene, 87f
    Mesozoic, 83, 86f
    mineral resources, genesis, 299–300
    New Madrid Rift System, 82f
    paleohydrology, 300f
    Pascola Arch, 86f
    Permian, 83, 86f, 127f
    Precambrian, 123, 125f, 133, 137–138
    proto-Illinois Basin, 127f
    Sauk I subsequence, 140
    stratigraphic column, 137f
    stratigraphic units, 128f
    structural geology, 86f
    subsidence, 133
    tectonics, 86f, 87f, 141f
Rough Creek-Shawneetown Fault System, 98–100
    Fairfield Basin, 91
    Fluorspar Area Fault Complex, 100
    Illinois Basin, 62
    oil fields, 286f
    petroleum traps, 288f
    structural geology, 62f, 63f, 92f
    sub-Pennsylvanian valleys, 196f
    tectonics, 62f
        Devonian, 85f
        Devonian-Mississippian, 84
        Mesozoic, 86f, 87
        Pennsylvanian-Permian, 85f, 86
        Permian, 86f
        Triassic, 206
Round Knob, 106, 106f, 107t
Roxana, 109f
Roxana Silt, 239
    Kelley site, 108f
    Quaternary, 221f, 223f, 228f, 229f, 234f

wetlands, 367
Royal Center Fault (Ind.), 63f
Royalton coal mine disaster, 278
runoff, 328, 328f, 377, 443t
    sinkholes, 438
    wetlands, 363, 368

S

Safe Drinking Water Act (1974), 29t
sag subsidence. *See* land subsidence
Saginaw lobe, 240
Salem Anticline, 93, 95
    oil fields, 286f
    seismic profiles, 114
    structural geology, 63f, 92f, 196f
    tectonics, 62f, 193, 202
Salem Field, 18, 261f–262f, 284–285, 289, 293
Salem Limestone, 38, 169, 179–180, 312
    aggregates, 179–180
    cross section, 180f
    fossils, 179–180
    reservoir rocks, 287f
    stratigraphic columns, 168f, 287f
    structural traps, 293, 294
    Valmeyeran, 177, 180f
Salem Oil Field (1940), 18f
Salem Plateau, 433, 434f
Salem Plateau Section, 71f
Salem-Dale, 19f
Saline County
    coal balls, 192f
    Davis Coal Member, 268
    Dekoven Coal Member, 268
    Equality Springs, 343f
    faults, 111f
    Garden of the Gods, Shawnee National Forest, 72t
    Rough Creek-Shawneetown Fault System, 99f
    Saline River seep, 343
Saline County Coal Company, 98f
Saline River
    Caseyville Formation, 195
    Rough Creek-Shawneetown Fault System, 99f
    Saline County seep, 343
Salmonella species, 455
salt
    deicer, 351, 448
    pollution, 351, 444t, 455
    supply, 343, 343f
    salt-water intrusion, 341f, 342, 343, 344f, 444t
Salt Fork of the Vermilion River, 343
"salt works" (1800s), 343
Sample Sandstone, 168f, 287f, 296f
sand
    aquifers, 332, 335
    Clayton Formation, 210f
    floods, 397, 400f, 401
    hydraulic conductivity, 326, 326t
    Illinois cross section, 66f
    Illinois River valley, 237f
    lakebed downcutting, 415f
    littoral drift, 407, 410f, 411f, 412f
    McNairy Formation, 210f, 210–211
    Quaternary, 229f
    river channels, 402
    shorelines, 405, 410f, 411, 411f, 412f, 413–414
    Silurian reef, 290f
    Wedron Quarry, 221f
    Zion Beach-Ridge Plain, 405, 411f
sand deposits, 311f, 313–315
    economic value, 319f
    foundry sand, 9, 314
sand and gravel, 226f, 229f, 233, 236f, 237f, 411f

Index

sand and gravel aquifers, 325, 332–334, 334f, 335–336, 351, 354
   aquifer vulnerability, 334, 358
   block diagram, 358f
   groundwater processes, 341f, 342f
   groundwater quality, 338
   Illinois River valley, 237f
   Mississippi River valley, 335
sand and gravel deposits, 15, 313, 314f, 315, 315f, 319f
   Baylis Formation, 209
   dredge mining, 233f
   economic value, 319, 319f
   environment of deposition, 238, 313, 314f
   floods, 397–400
   Henry Formation, 229f, 238
   Illinois Episode, 233, 233f
   Illinois River valley, 237f
   outwash, 238, 335
   Quaternary, 40f, 220f, 229f
   shaded relief map, 233f
   on valley trains, 314f
   Wisconsin Episode, 236f
sand and gravel pits, 220f, 310f, 311f, 315f, 320f, 320–321
   reclamation, 320f, 458f, 459f, 461f, 461–462
sand blows, 109f, 109–110, 428–429
sand boils, 399f
sand ridges, 407
Sandoval reef, 290f
sandstone
   aquifers, 329, 330f, 332, 338, 351, 419
   aquitards, 332
   Carbondale Formation, 199f
   Caseyville Formation, 189, 196–197, 197f, 199f
   Chesterian, 180–182, 183f
   cyclothem, 194f
   Eau Claire Formation, 142
   economic value, 319f
   Everton Formation, 145–146
   Franconia Formation, 142
   Glenwood Formation, 148–149
   above Herrin Coal Member, 272
   hydraulic conductivity, 326, 326t
   Illinois cross section, 66f
   isotopes, 419
   Mattoon Formation, 201
   Mt. Simon Sandstone, 141, 330f
   Patoka Formation, 200
   pedogenesis, 379
   petroleum traps, 288f
   Post Creek Formation, 210
   reservoir rocks, 287f
   Sauk sequence, 137
   St. Peter Sandstone, 146–148
   total dissolved solids, 339t
   Tradewater Formation, 189, 197f, 198
   See also individual named geological units
Sandwich Fault Zone, 63f, 92f, 95, 96, 97f
   Cambrian-Ordovician, 143f
   Illinois and western Kentucky cross section, 61f
   Mississippian, 84, 85f
   Pennsylvanian, 84, 85f
Sangamon and Peoria weathered zones, 10f
Sangamon Arch, 63f, 91f, 93
   Devonian, 170, 171
   Illinois Basin, 62, 171f
   Kankakee Formation, 289
   Kaskaskia sequence, 171f
   oil and gas fields, 291f
   petroleum reservoirs, 19, 289, 291
   tectonics, 62f, 84, 85f
Sangamon County, 231
Sangamon Episode, 234–235
   geochronology, 223f
   paleoclimatology, 249t, 257f, 258
   Quaternary drainage, 224f
   Thebes, 228f
Sangamon Geosol, 223f, 232, 234–235
   faults, 106–107, 107t, 108f
   Quaternary, 228f, 234f
   Tuscola Quarry, 222f
   wetlands, 366
   Winnebago Formation, 234–235
Sangamon River floodplains, 364, 399
Sangamon Soil, 13f
Sanitary and Ship Canal, 164
Sankoty aquifer, 333–334, 335
Sankoty Sand Member, 223f, 226
satellite imagery, 39, 39f, 237f
saturated zone. *See* water table
saturation
   soils, 386–387, 396
   wetlands, 361, 363, 367
Sauk I subsequence, 128f, 137f, 140
Sauk II subsequence, 128f, 137f, 138f, 140–142
Sauk III subsequence, 137f, 138f, 142–145, 148f
Sauk sequence, 64, 65f, 66f, 67f, 136–157
   environment of deposition, 79, 79f, 139f
   gas storage, 288
   Precambrian, 128f
   stratigraphic columns, 137f, 138f
   stratigraphic units, 139–141, 141f, 142–145
   structural geology, 62f
   tectonics, 62f, 141f
Saukiella species, 143
Savannah, 165f, 301f
savannas, 257f
Saverton Shale, 174
   cross section, 174f
   Hicks Dome, 101f
   stratigraphic column, 168f
Saylesville series, 377f
Scales Mound, 301f
Scales Shale, 138f, 154
schists, 125f
Schuchertera species, 162f–163f
Schuyler County, 291
Schweizer Member, 160f
scintillation methods, 31f
Scott County, 233f
Scottsburg Limestone, 168f, 183f
Scottville Limestone, 200
scour, 400f, 400–401
sea level
   Cambrian-Ordovician transition, 143
   Chesterian, 170, 183–184
   Devonian, 170
   Kaskaskia sequence, 167–170
   Mesozoic, 207f, 208f
   Mississippian, 84–86
   Pennsylvanian, 187, 189f, 193–194
   sea-floor spreading, 60, 77, 77f
seals
   abandoned mines, 461
   gas storage, 289
   structural traps, 288, 288f
   wells, 449
seawater, 300, 338t
secondary porosity, 291
secondary recovery, 19, 37, 285, 288
sedge meadow, 361f, 365, 366
   See also wetlands
sedimentary basin, 123f
sedimentary rocks
   Cambrian, 82f
   Chesterian time, 169, 170f, 182
   Cretaceous, 208f, 209f
   delta, 84

Illinois Basin overview, 60
Jurassic, 208f
Pennsylvanian, 189, 193
Permian (none), 201
Precambrian, 124f, 125, 125f, 132, 216f
sedimentation
  Kaskaskia sequence, 167–170
  New Madrid Rift System, 80f–81f, 81
  Ozark Dome burial, 91
  succession, 64, 77, 137f
  Triassic, 207f
sediments
  floods, 397–400, 400f, 401
  Hudson Episode, 242
  Illinois Episode, 226f, 228f, 231–232
  littoral drift, 405, 411, 413
  paleoclimatology, 248, 250f, 255f
  pedogenesis, 378
  pre-Illinois Episode, 224–226, 226f
  Quaternary, 216f, 219–220, 220f, 221f, 228f, 229f, 230, 238
  Sangamon Geosol, 234–235
  wetlands, 364–365, 368
  Wisconsin Episode, 238
  See also loess; soils
seed ferns
  Neuropteris species, 190f
  Sphenopteris species, 190f
seeds, 253
Seelyville Coal Member, 98, 118f, 267f, 268
seiches, 408
seismic profiles, 24, 116f
  Ballard County (Ky.), 117f
  Centralia sequence, 81, 82f, 127
  deep-seated structures, 113f, 113–115
  Hicks Dome, 100
  Illinois Basin, 41–42
  La Salle Anticlinorium, 94
  Louden Anticline, 114f
  New Madrid Rift System, 82f, 83
  Precambrian rocks, 123, 124–125, 125f, 126, 126f, 127f, 132
  proto-Illinois Basin, 81
  Quaternary sediment, 243
  Sandoval reef, 290f
  seismic methods, 284f, 284–285, 285f
  structural traps, 284, 290f
seismic risk, 116f, 424
  neotectonics, 42
  New Madrid Seismic Zone, 113f
  See also earthquakes
seismic stratigraphy, 40
seismographs, 110–111, 424
selenium, 266
Selfridge Air Force Base, 386
Sellersburg Limestone, 174f
Selmier Shale, 65f, 101f, 168f, 174, 174f
semi-anthracite coal, 264
septage, 351, 448
septic systems, 351, 353, 359, 444, 454, 454t
sequence, 64, 65f
series, 64f, 65
Serpukhovian, 188f
Service Depot #1 landfill, 447f
setback zone, 352
Seventy-Six Shale Member, 160f, 161
sewage, 313, 387, 443t, 455
sewage sludge, 351, 448
sewer pipes, 386–387
Sexton Creek Limestone, 160f, 161, 287f
shaded relief map, 39
shaft mines, 274
Shakopee Dolomite, 144–145, 147f, 148f, 149f
  Sandwich Fault Zone, 97f
  stratigraphic columns, 137f, 138f, 287f

shale
  aquifers, 331–332
  aquitards, 326, 351, 356f
  Blocher Shale Member, 101f, 168f, 173, 174, 174f
  Bonneterre Formation, 142
  bricks, 316
  Carbondale Formation, 198–199, 199f, 200
  Caseyville Formation, 197
  above coal seams, 111
  cross section, 66f
  Davis Shale, 142
  Decorah Formation, 151–152
  Des Plaines buried impact structure, 101
  in drift, 209
  Eau Claire Formation, 142, 330
  Everton Formation, 145–146
  fluorspar deposits, 304f
  gas storage, 393
  Glenwood Formation, 148–149
  gob, 459
  above Herrin Coal Member, 111f
  hydraulic conductivity, 326, 326t
  Illinois cross section, 66f
  industrial minerals, 315–316
  lead-zinc deposits, 302f
  liquefied petroleum gas storage, 393f
  Maquoketa Group, 83, 152f, 154, 331
  Mattoon Formation, 201
  paleogeography, 177f
  parent materials, 378f, 379
  Patoka Formation, 200
  Pennsylvanian, 187, 188, 189, 316, 332
  Permian, 188
  petroleum traps, 288f, 292–293, 293f
  Platteville Group, 151
  radium, 348
  reservoir rocks, 287f
  Rough Creek-Shawneetown Fault System, 98–100
  Sauk sequence, 137
  Silurian reef, 290f
  St. Laurent Formation, 173
  Tradewater Formation, 198
  underground storage, 393
  Valmeyeran, 177f
  See also individual named geological units
shale—black fissile
  Carbondale Formation, 198
  cyclothems, 194f
  Excello Shale, 194f
  McLeansboro Group, 200
  Pennsylvanian, 189
  roof control, 111f, 111–112, 112f, 272, 275f
  Rough Creek-Shawneetown Fault System, 98–100
shale—gray
  Carbondale Formation, 198–199, 199f, 200
  cyclothems, 194f
  above Danville Coal Member, 272
  research programs, 26
  sulfur, 200, 265
shallow depth aquifers
  aquifer vulnerability, 32–33, 352f, 353, 353f, 355f, 356f, 357f, 358–360, 452
  bedrock aquifers, 331f
  dolomite, 331
  hydraulic conductivity, 354
  hydrogeology, 328f
  iron, 345
  manganese, 345
  nitrogen, 345
  sand and gravel aquifers, 335–336
  Silurian, 164
shallow water environment
  Chesterian, 170f

Index

Pennsylvanian, 189f
shark
    fossils, 211, 211f
    teeth, 180, 209, 211, 211f
shatter cones, 101
Shawnee Hills, 181, 189, 434
Shawnee Hills karst region, 433, 434f
Shawnee Hills Section, 71f
Shawnee National Forest, 69f, 72t, 75f–76f
Shawneetown
    Rough Creek-Shawneetown Fault System, 98
    Survant Coal Member, 268
Shawneetown Fault, 304f
Shawneetown Fault Zone, 101f
Sheboygan (Wisc.), 404f, 411
sheet erosion, 380
Sheffield radioactive waste storage site, 445
Sheinwoodian Stage, 160f
Shelburn Formation, 99f, 195f, 199, 200
Shelburn-Patoka Formation
    coal seams, 267f
    Hicks Dome, 101f
    stratigraphic column, 188f
Shelby County, 72t
Shelby Phase, 236f, 239
Shelbyville, 236f
Shelbyville Coal Member, 201
Shelbyville Moraine, 226f, 236f, 238, 249t
shelf-edge banks, 158, 162, 163
shelves (structural terraces), 90
Sheridan, 146f
Shetlerville Limestone Member, 168f, 183f, 305f
Shoal Creek Limestone. *See* Carthage (Shoal Creek) Limestone Member
shoals, 411
shore protection, 405
shorelines
    bars, 296
    Chicago Lake Plain, 409f
    Glenwood phases, 408
    groins, 411
    history of investigations, 30, 35, 35f
    Illinois Beach State Park, 407f
    Lake Michigan, 405, 408f, 409f, 410f, 412f
    paleogeography, 136–137, 207f, 208f
    Zion Beach-Ridge Plain, 405
Shumway Limestone, 188f
siderite, 190, 199
Siggins, 188f, 287f
silcrete, 210f
silica (quartz)
    clay, 315
    coal ash, 265
    groundwater, 337, 338t
    kaolinite, 315
    sand, 27f, 313–315
siliciclastics, 18, 142–143, 154, 177f, 290
silicon dioxide, in groundwater, 338t
Siloam Field, 291, 291f
silt
    aquitards, 351
    Batestown Member, 238
    bedrock valleys, 334, 335
    cross section, 66f
    environment of deposition, 333, 401
    glacial lake sediments, 365
    floods, 400f
    hydraulic conductivity, 326t, 327
    Illinois River valley, 237f
    McNairy Formation, 210
    Peoria Silt, 229f
    Post Creek Formation, 210
    Quaternary, 226f, 229f, 236f
    Roxana Silt, 229f
    in speleothems, 256, 256f
    suspended solids, 340
    wetlands, 365, 366
    Wilcox Formation, 212
    See also loess
siltstone
    aquifers, 331, 332
    aquitards, 332
    Carbondale Formation, 199f
    Carper Sandstone, 293
    Caseyville Formation, 197
    cyclothems, 194f
    Eau Claire Formation, 330
    above low-sulfur coal, 265
    Maquoketa Group, 83
    Pennsylvanian, 189
    Sauk sequence, 137
    St. Laurent Formation, 173
Silurian, 158–166
    Acadian Orogeny, 80f–81f
    bedrock, 68f
    bedrock aquifers, 330f, 331–332, 419
    chronostratigraphy, 65f, 66f, 67f
    cross sections
        Illinois and eastern Indiana, 61f
        Illinois and western Kentucky, 61f
        Illinois, 66f
        Ste. Genevieve Fault System, 99f
    Des Plaines buried impact structure, 101
    fossils, 160, 161, 162f–163f
    Illinois Basin, 195
    karst, 433–434
    Mississippi Palisades State Park, 165f
    northern Illinois, 311
    Ordovician boundary, 154f
    Pascola Arch, 86f
    Racine Formation, 407
    research programs, 14
    Ste. Genevieve Fault System, 99f
    stratigraphic columns, 160f, 287f
    stratigraphy, 159–163
    structural geology, 62f
    sub-Kaskaskia surface, 169f
    tectonics, 62f
    total dissolved solids, 339f
    See also individual named geological units
Silurian deposits
    dolomite, 311, 312f
    lead-zinc deposits, 302f
    limestone, 293f, 312f
    oil and gas fields, 291, 291f
    reservoir rocks, 287f
    stratigraphic traps, 292f, 293f
    Thornton Quarry, 38f, 59f, 72t, 163–164, 164f, 165
Silurian dolomite
    aquifers, 419
    deposits, 311, 312f
    karst, 432
    Mississippi Palisades State Park, 72t
    stratigraphic traps, 292f, 293f
Silurian reefs, 289–290, 290f, 291, 291f
    limestone deposits, 309–311
    Marine reef, 290f
    Niagaran, 290f
    paleontology, 163
    Sandoval reef, 290f
    schematic diagram, 290f
    seismic profile, 290f
    stratigraphic traps, 19, 289
    structural geology, 102, 288
    Thornton Quarry, 59, 59f, 72t, 164f
silver deposits, 100

simazine, 452
sinkholes, 432f, 432–433, 434f, 436f, 436–439
    Big Sink, 437f
    Dongola grade school, 438f
    drainage, 438f, 438–439
    Eocene, 212
    erosion, 437f
    groundwater quality, 328, 340, 340f
sinking streams, 432, 436–439
site assessment. *See* engineering geology: site assessment
slag, 280, 444
"slates," 198
slickensides, 96
slope stability, 387f, 387–388, 388f, 389f
slopes
    floods, 396–397
    Inceptisols, 375–376
    pedogenesis, 377
Sloss sequences, 28, 136, 159
sludge, 444
sluiceways—relict, 406f
slurries, 279
smectite, 315
smelting, 278
Smithboro Member, 223f, 231, 232
snails. *See* Gastropoda
Sny Island Levee, 399f
sodium
    clay component, 315
    dietary concerns, 342
    in groundwater, 337, 338t, 342–343
sodium bicarbonate type groundwater, 338t, 338–339
sodium chloride
    deep aquifers, 342
    groundwater pollution, 443
sodium chloride type groundwater, 337
soils, 373–384, 385–386
    Alfisols, 234, 375, 375f, 376f, 377
    clay, 385–386
    drainage, 377f, 452
    Entisols, 375f, 375–376
    forest, 373f
    geosols, 222, 222f, 223f, 228f, 234f, 366
    groups, 374–375, 375f
    Histosols, 375,
    horizons, 222f, 234f, 253, 366, 374, 374f, 380
    Hudson Episode, 242
    Illinois Surface-Mined Land Conservation and Reclamation Act (1975 amendment), 460
    Inceptisols, 375, 376, 376f
    karst, 432
    Keller Farm section, 255f
    land cover, 373
    land use, 373
    landscapes, 377f, 382f
    leaching, 377
    liquefaction, 105, 107f, 109f, 425, 429
    moisture, 377, 385–386
    Mollisols, 234, 375, 375f, 377
    movement, 377
    organic matter, 248, 452
    Ozaukee silt loam, 382f
    paleoclimatology, 252f, 255f
    paleosols, 189, 210, 210f
    parent materials, 377–378, 378f, 379, 380
    peat, 316
    pedogenesis, 255f, 373, 374, 376–377
    pedostratigraphy, 379
    piping, 386–387
    prairie, 373f
    pressure damage, 385–386, 386f
    profiles, 234–235
    quality, 373, 378
    radon, 420
    reclamation, 460f
    Sangamon Geosol, 222f, 234f, 235
    saturation, 386–387
    shrinking-swelling, 385–386, 386f
    sinkholes, 437f
    soil gas, 420
    speleothems, 255f
    stratigraphic units, 27
    suspended solids, 340
    texture, 374, 378, 380f
    thorium-232, 419
    topographic relief, 377
    Ultisols, 234, 375, 375f
    vegetation, 373f
    wetlands, 362f, 366
    Yarmouth Geosol, 222f
    zone, 255f, 341f
solder, 300
"sole source" designation, 352
Solid Waste Disposal Act (1965, 1980 amendments), 29t
solid waste, 280, 447
    See also waste disposal
sonar methods, 35f
Sonora formation, 168f, 178, 287f
Soperville, 270
sorbents, 32
sound waves. *See* acoustical waves
source rocks, 286
South America, 80f–81f
South Pier jetty construction, 411
southern central plains orogen, 125f
Southern Shelf, 196f
Sowerbyella punctostriata, 150f
Sowerbyella species, 151
soybeans, 449f, 450f
span, 64f
Spar Mountain (Rosiclare) Sandstone Member, 183f, 294–295
    fluorspar deposits, 305f
    reservoir rocks, 287f
    sediments, 182
    stratigraphic columns, 168f, 183f, 287f
Sparland, 272
Sparta, 11
Sparta Shelf, 62f, 63f, 91f, 93
    Devonian, 170, 171
    Illinois Basin, 62, 171f
    tectonics, 62f
Spechts Ferry Shale, 138f, 302f
speleology (caving), 441
speleothems, 441
    carbonic acid, 255f
    paleoclimatology, 253, 255f, 256f, 256–258,
    uranium-series dating, 248, 256, 256f
sphalerite, 300–301, 301f, 303f
Sphenopteris species, 190f
spherulitic texture, 132
spiders, 190
Spirifer grimesi, 178
Spirifer increbescens, 172f–173f
Spirifer logani, 175f
Spiriferida
    Clear Creek Formation, 162f–163f
    Spirifer grimesi, 178
    Spirifer increbescens, 172f–173f
    Spirifer logani, 175f
    St. Laurent Formation, 173
spits, 406f, 407, 408, 409f
Split Rock, 147–148, 149f
"splits." *See* lenses
Spoon Formation, 195f, 287f
spores, 23f, 28f, 41
Spring Valley

Index

Colchester Coal Member, 268
Herrin Coal Member, 272
springs
   Camp Vandeventer, 440f
   Equality Spring, 342f
   Falling Springs, 346f, 440f
   groundwater discharge, 328, 342f, 346f
   karst feature, 432–433, 434f, 440f
   Old Faithful spring (Ill.), 342t
   salinity, 6, 342, 343, 343f
   Weldon Springs State Park, 342f
Springfield
   Chesterian, 181
   coal deposits, 278
   earthquakes, 110, 427
   petroleum reservoirs, 291
   Sangamon Arch, 93
Springfield Coal Member, 270f, 270–271
   Carbondale Formation, 198
   clastic wedges, 199
   coke, 278
   Cottage Grove Fault System, 98f
   cyclothems, 194f
   stratigraphic columns, 188f, 267f
   sulfur, 265, 270f
Springfield East Field, 291f
Springfield Plain, 70, 71f
Springville Shale, 168f, 169, 176, 178, 180f
spruce, 249t, 252f, 253
St. Clair, 438–439
   Falling Springs groundwater, 346f
   marsh, 361f
   Peabody Coal Company's River King surface mine, 273f
St. Clair County, 273f, 346f
St. Clair Limestone, 160f, 161, 287f
St. Croixan, 137f, 138f, 287f
St. David Limestone, 188f, 194f
St. Francois Mountains (Mo.), 81, 91f, 123, 125f, 139
St. Hedwig Church (Chicago), 386
St. Jacob Field, 289
St. Jacob Oil Pool, 124f
St. James Field, 293
St. Laurent (Alto and Lingle) Formation, 65f, 168f, 171f, 173, 174f
St. Louis Limestone, 180
   fluorspar deposits, 303, 304, 305f
   fossils, 175f, 178, 179f
   Hicks Dome, 101f
   Melonechinus multiporus Norwood and Owen, 178, 179f
   reservoir rocks, 287f, 293
   stratigraphic columns, 168f, 287f
   Valmeyeran, 177
St. Louis Metro East, 320
St. Louis (Mo.) area, 38, 249t
St. Louis University regional earthquake data, 425–426
St. Maria Spit, 409f
St. Peter Sandstone, 2, 7, 72t, 145, 146–148, 149f
   Ancell Group, 148f
   bedrock aquifer, 299, 330, 419
   crystal fractionation, 101
   Sandwich Fault Zone, 97f
   Shakopee Dolomite, 148f
   Starved Rock, Starved Rock State Park, 241f
   stratigraphic columns, 137f, 138f, 287f
St. Peter Sandstone deposits, 7, 27f, 313–314
   gas storage, 289
   lead-zinc deposits, 302f
   Ottawa Silica Sand Company, 27f
   petroleum reservoirs, 287f
   quartz sand, 149f
   stratigraphic column, 287f
   Wedron Quarry, 221f
stable isotopes, 227, 347
"stack effect," 420
stage, 64f, 65
stages 2, 3, 5, and 6, 137f, 138f
stagnated glacier, 217, 233
stalactites, 253, 441
stalagmites, 253, 256, 256f, 441
Starrs Cave Limestone, 168f
starved basin, 177f
Starved Rock Lock and Dam, 148f
Starved Rock Sandstone Member, 138f, 147, 148f
Starved Rock State Park, 2, 14f, 72t
   erosion, 241f
   Old Salt Well, 342
   St. Peter Sandstone, 147, 241f, 313–314
State Historical Library and Natural History Museum, 9
State of Illinois
   Central Midwest Interstate Low-Level Radioactive Waste Compact, 421
   regulations
      Pesticide Management Plan, 353–354
      reclamation, 459
      water wells, 352, 441, 450f
   state capitol, 318f
   state fossil, 190, 192f
   state mineral, 303
   state parks, 72t, 165
   state soil, 378–379, 379f
STATEMAP program (1992–2000s), 40
statewide aquifer vulnerability, 353–354, 358
Ste. Genevieve Fault System, 84, 85f, 92f, 98, 99f
Ste. Genevieve Fault Zone
   Illinois Basin, 62, 171f
   Illinois-Kentucky Fluorspar District, 304f
   structural geology, 62f, 63f
   tectonics, 62f, 171
Ste. Genevieve Limestone, 168f, 169, 181, 312
   electrical logging, 296f
   fluorspar deposits, 303, 304, 305f
   groundwater, 338t
   Hicks Dome, 101f
   petroleum resources, 184, 293, 294
   reservoir rocks, 287f
   Ste. Genevieve Fault System, 99f
   stratigraphic columns, 168f, 183f, 287f
Stengall landfill, 447f
Stephanian, 188f
Stephen A. Forbes State Park, 37
Stephenson County
   bedrock aquifer, 330
   lead-zinc deposits, 301
   Silurian, 160f
Sterling Member diamicton, 231, 232, 233–234
Stewart Ridge Spit, 409f
Stewart Vein System, 304f
stillstands, 237f
Stockton, 301f
stone, crushed
   economic value, 319f, 319, 320
   mine reclamation, 461
   quarries, 310f, 312f
   Silurian, 164
stone, riprap, 415f
stone industry. *See* building stone
Stonefort Limestone, 188f, 195f,
Stony Island, 406f, 407, 409f, 410f
storage
   coal bed methane, 279
   jet fuel leakage, 451
   leaking underground storage tanks, 351, 444t, 452–453, 453f
   natural gas, 288
   radioactive waste, 418
   underground mines, 393
stormwater, 351, 392, 443t
stratigraphic code, 27
stratigraphic traps, 286, 288, 288f, 291, 293f

stratigraphy, 61f, 64–65, 65f, 66f, 66–67, 67f , 123–260
    Absaroka sequence, 187, 188f, 189–205
    Cambrian, 139–157
    Carbondale Formation, 199f
    Chesterian, 180–182
    Devonian, 158–160, 160f, 161–166, 168f, 170–176
    Handbook of Illinois Stratigraphy (1975), 27–28
    history of investigations, 14, 40
    Kaskaskia sequence, 168f
    Mississippian, 167, 168f, 169–186
    New Albany Shale, 174f
    Ordovician, 137f, 138f, 139–157
    Pennsylvanian, 188f, 194–201
    Permian, 201–202
    petroleum reservoirs, 287f
    Precambrian, 123–135
    Quaternary, 216–247
    Sauk sequence, 137f, 138f, 145
    Silurian, 160f, 158–160, 160f, 161–166
    Tejas I and II subsequences, 65f, 67f
    Tejas sequence, 206f
    Tippecanoe I subsequence, 137f, 138f, 145–155
    Tippecanoe II subsequence, 159, 160f, 160–163
    Valmeyeran, 177
stream valleys, drainage, 231f
streams
    agrichemicals, 354
    erosion, 217
    groundwater movement, 328
    in karst, 432, 436–439
    pollution, 443t
    sand, 376
    soil, 375
    wetlands, 368
Streator, 272
Streeterville neighborhood (Chicago), 412f
Streptelasma species, 150f, 151
Stricklandia species, 161
strike-slip faults, 95, 97, 98f, 111
strip mining. *See* surface mining
stromatolites, 138, 171
Stromatoporoidea, 170, 171, 173
strontium, 347
Strophomena plattinensis, 150f
Strophomena species, 151, 153f, 154
structural geology, 60f, 60–63, 63f, 64, 75, 90–91, 91f, 92, 92f, 93–104
    anticlines, 95
    arches, 91–93
    bedrock surface, 91–93
    Cap au Grès Faulted Flexure, 92f, 95
    Cottage Grove Fault System, 92f, 97–98
    domes, 90–91
    Du Quoin Monocline, 92f, 95
    Fairfield Basin, 91, 91f, 92f
    faults, 92f, 95–96
    Fluorspar Area Fault Complex, 92f, 100
    folds, 92f, 93–95
    Hicks Dome, 92f, 100, 101
    history of investigations, 23, 41
    Illinois Basin, 91, 91f
    impact structures, 92f, 100–101
    Kankakee Arch, 91, 91f
    La Salle Anticlinorium, 92f, 93–95
    Marine reef, 290f
    Media Anticline, 92f, 95
    midcontinental United States, 60f
    Mississippi Embayment, 91f, 92f, 93
    Mississippi River Arch, 91f, 92
    Moorman Syncline (Trough), 91, 92f
    oil fields, 286f
    Ozark Dome, 91, 91f
    Pascola Arch, 91f, 93
    Permian, 201f
    Pittsfield Anticline, 95
    Plum River Fault Zone, 96
    reefs, 101 102, 288, 289–290, 290f
    Rough Creek-Shawneetown Fault System, 92f, 98–100
    Sandwich Fault Zone, 96
    Sangamon Arch, 91f, 93
    Sparta Shelf, 93
    Ste. Genevieve Fault System, 98
    synclines, 304f
    tectonics, 62f, 79f, 90–104
    Valmeyer Anticline, 92f, 95
    Wabash Valley Fault System, 96–97
    Waterloo-Dupo Anticline, 95
    Western Shelf, 92f, 93
    Wisconsin Arch, 91f, 91–92
    See also individual named structures
structural traps, 97, 286, 288–297
sub-Absaroka unconformity, 28, 66f
subbituminous coal rank, 264
sub-Brandon Bridge unconformity, 159
sub-Centralia sequences, 79f, 128f
sub-Cretaceous geological map, Pascola Arch, 86f
subduction, 60, 77f, 86
subglacial environment
    drainage, 219
    erosion, 240
    glacial extent, 216f
    sediments, 219, 221f
    sub-Kaskaskia surface, 169f
sub-Kaskaskia unconformity, 28, 66f, 159, 170
sublobes, 236f
submarine fans, 169
sub-Mt. Simon Sandstone, 131f, 133f
sub-Offerman unconformity, 159
sub-Pennsylvanian surface, 195, 196f
sub-Sauk surface, 66f, 139
    See also Precambrian
subsidence, 80f–81f
    Cambrian, 79, 81
    Centralia sequence, 81
    Illinois Basin
        Cambrian through Pennsylvanian, 91
        Ordovician, 139
        Paleozoic, 82f, 83
        Tippecanoe II subsequence, 158
    Mt. Simon Sandstone, 81
    New Madrid Rift System, 83, 429
    Pascola Arch, 208
    Permian, 84
    Precambrian, 79, 81, 128
    Valmeyeran time, 176
    See also land subsidence
sub-Silurian unconformity, 289
sub-Tejas unconformity, 66f
sub-Tippecanoe I unconformity, 66f
sub-Tippecanoe II unconformity, 66f, 159
sub-Tippecanoe sequence unconformities, 28
sub-Tippecanoe surface, 66f, 145, 147f
subways, 393, 393f
sub-Zuni unconformity, 66f
Sugar Creek valley, 358
Sugar Run Dolomite, 160f, 161
sulfates
    in coal, 265
    water quality, 337, 338t, 341, 345
sulfur
    in coal, 31, 265, 270f, 271f, 313
    regulations, 265
    in shale, 200
    water quality, 347, 460
sulfuric acid, 279, 460
Sullivan Mach., 98f
supercontinent, breakup, 62f

INDEX

Superfund. *See* CERCLA legislation; CERCLA sites
supergroup, 64, 64f
Surface Coal Mining Control and Reclamation Act (1977), 460
Surface Coal Mining Land Conservation and Reclamation Act (1983), 460
Surface-Mined Land Conservation and Reclamation Act (1971), 460
Surface-Mined Land Reclamation Act (1968), 460
surface mining, 272–273, 273f, 459, 459f, 460f
    Colchester Coal Member, 268
    Davis Coal Member, 268
    Dekoven Coal Member, 268
    equipment, 26f, 272–273, 273f, 459f
    Grape Creek, 272
    Mazon Creek fossils, 190
    Murphysboro Coal Member, 267
    production values, 272
    reclamation, 458–459, 459f, 460f, 460–461, 461f, 462–464
    sand and gravel deposits, 315f, 458f, 459f
    Springfield Coal Member, 271
    stone, crushed, 312f
Surface Mining Control and Reclamation Act (1977) (Public Law 95-87), 29t, 460
surface water
    drainage, 435f
    hydrologic cycle, 328, 328f
    pollution, 444, 450, 452, 460
    wetlands, 361
Surface-Mined Land Reclamation Act (1968), 460
surficial geology. *See* Quaternary
Survant Coal Member, 188f, 267f, 268–269
suspended solids, 340, 340f, 368
swales, 405, 408
swamps, 41, 192, 263, 361
    See also Pennsylvanian; wetlands
Sweeney Dolomite, 160f, 161
Sweetland Creek Shale, 65f, 101f, 168f, 174, 174f
Sycamore, 152f
Sylamore (= Hardin in Mo.) Sandstone, 65f, 101f, 168f, 173, 174f
Synbathocrinus species, 178
synclines, 304f
synthetic fuel, 279, 280
system, 64f, 65

T

Tabb Fault System, 304f
Table Grove Moraine, 226f, 230
tabulate coral, 171, 172f–173f, 173
Taconic Highlands, 154
Taconic Orogeny, 62f, 80f–81f, 81, 83, 151
Talarocrinus species, 175f, 182
Tamms, 210f
Tamms Group, 65f, 160f
Tampico aquifer, 334–335
Tar Springs Sandstone, 182
    Hicks Dome, 101f
    reservoir rocks, 287f
    sandstone reservoirs, 297
    stratigraphic columns, 168f, 183f, 287f
TARP (Tunnel and Reservoir Plan), 29, 164, 392
Taxocrinus species, 182
Taylorville sub-Pennsylvanian valley, 196f
Tazewell County
    aquifer vulnerability, 357f, 358–359
    Farm Creek section, 250f
    Mackinaw Valley aquifer, 338t
    methyl tertiary butyl ether, 453
Tazewell substage, 22
TCE (trichloroethylene), 453
Teays-Mahomet River, 224f
tectonics, 60, 62f, 75f, 77f, 77–79, 80f–81f, 80–89, 96
    Applachian uplift, 299
    Cambrian, 79, 80, 80f–81f, 81, 155
    Carbondale Formation, 198
    Cottage Grove Fault System, 98f
    cratons, 85f, 127f, 158
    Cretaceous, 80f–81f, 87–88, 207
    Devonian, 83–94, 292
    history of investigations, 28, 30, 41
    Illinois Basin, 23, 41, 63, 83–87
    Illinois-Kentucky Fluorspar District, 304f
    Jurassic, 207
    measurement, 110–111
    Mesozoic, 86f, 87
    Mississippi Embayment, 87–88, 128f, 141f
    Mississippian, 84–87, 182–183
    neotectonics, 105–119
    New Madrid Rift System, 81–83, 128f, 141f
    Ordovician, 83–85, 85f, 94, 155
    orogenies, 86
    Ouachita uplift, 299
    Pascola Arch, 7, 87, 128f, 141f
    Pennsylvanian, 86, 188, 192–193
    Permian, 84–87, 85f, 86f, 88, 202
    Phanerozoic, 79
    Porters Creek Formation, 212f
    Plum River Fault Zone, 96
    Precambrian, 79, 80, 80f–81f, 81, 128f, 133
    rifting, 77f, 80f–81f, 90, 98–100, 207
    Sauk sequence, 137
    Ste. Genevieve Limestone, 181
    stratigraphic units, 141f
    structural geology, 62f, 78–79, 79f, 90–104
    structural traps, 293
    Tertiary, 87–88
    Tippecanoe I subsequence, 139
    Tradewater Formation, 197
    Triassic, 206–207, 207f
    See also faults; individual named structures
tectonic plates. *See* plate tectonics
Tejas I subsequence, 65f, 67f
Tejas III subsequence, 65f, 67f
Tejas sequence
    cross sections, 66f, 128f, 141f
    Sloss sequence, 64
    stratigraphic column, 206f
    structural geology, 62f, 79f
    tectonics, 62f, 64, 79f, 141f
Telychian Stage, 160f
temperature
    groundwater, 24
    ocean floor, 227
    soils, 377
Teneriffe Silt, 223f, 229f
Tennessee Earthquake Information Center data, 425–426
Tentaculites species, 153f, 154
terraces
    aquifer vulnerability, 353
    Illinois River valley, 237f
    Mounds Gravel, 214
    Woodstock Phase, 240
Tertiary, 211–215
    bedrock, 68f, 332
    chronostratigraphy, 65, 65f, 66f, 67f
    Fluorspar Area Fault Complex, 100
    fossils, 211, 211f, 212
    Illinois cross section, 66f
    Massac Creek faults, 107, 107t
    Mississippi Embayment, 87, 93
    stratigraphic column, 206f
    structural geology, 62f
    tectonics, 62f
Tete des Morts Dolomite (East Central Iowa Basin), 160f, 161
Tetranota species, 150f, 151
Texaco Silverman, 293
Thaerodonta species, 153f, 154
Thaleops ovata, 150f

Thaleops species, 151
Thebes
  loess, 227
  McNairy Formation, 210f, 211
  Mounds Gravel, 213f
  Quaternary, 228f
Thebes Sandstone, 137f, 154–155
thermal analysis, 19, 256
thorium
  in coal, 266
  in monazite, 422
thorium-232, 419
Thornton Quarry, 38f, 59f, 72t, 163–164, 164f, 165
three-dimensional models
  hydrogeology, 359
  Illinois River valley, 237f
  mapping, 40, 352
  porosity, 289
  Quaternary, 243
  seismic methods, 37, 292
  Villa Grove Quadrangle, 359
thrust faults
  Pennsylvanian-Permian, 86
  Plum River Fault Zone, 96
  southern Illinois, 111
  underground coal mine, 111f
tidal bar deposition, 294–295
tidal ridges, 296
till, 219, 238, 240, 377f, 380f
  aquitards, 351
  Chicago Lake Plain, 407
  end moraines, 235, 236f
  glacial deposits, 250f, 365
  hydraulic conductivity, 326t
  hydrogeology, 341f
  Illinois Episode, 234f
  lakebed downcutting, 415f
  pre-Illinois Episode, 229f
  Putnam Phase, 240
  Lake Michigan shoreline, 405, 411f, 411–412
  Lemont till, 220f
  Livingston Phases, 240
  soil parent materials, 377, 378f, 379, 380f
  Tuscola Quarry, 222f
  Wedron Group, 229f
  Wisconsin Episode, 236f, 238
  Yorkville till, 240
  Zion Beach-Ridge Plain, 411f
  See also diamicton
till plain, 229f, 230, 230f, 231f, 232f, 235, 237f
Till Plains Section, 70, 71f
Tilton Member, 223f
time factor, 64f
  hydrogeology, 328f
  pedogenesis, 377, 379
  Quaternary, 219, 222, 235, 236f
  stratigraphy, 27, 64
tin, 266
Tinley Moraine, 236f, 242
Tioga K-bentonite Bed, 168f, 170
Tippecanoe I subsequence, 65f, 67f, 136–157
  cross section, 148f
  stratigraphic columns, 137f, 138f
  stratigraphy, 145–155
Tippecanoe II subsequence, 65f, 67f, 158, 159, 160–166
  stratigraphic column, 160f
Tippecanoe sequence, 62f, 66f, 79, 79f, 128f
  eroded extent, 147f
  Illinois cross section, 66f
  Sloss sequence, 64
  structural geology, 62
  tectonics, 62f
Tiskilwa Formation

  cross section, 236f
  Quaternary deposits, 223f, 229f, 236f, 238
  Wedron Quarry, 220f, 221f
Toledo Anticline, 196f
Toleston Spit, 409f
Tolu Arch (Kentucky), 85f, 86
toluene, 452
tongues, proglacial sediment, 220, 221f, 238
tonnage (coal), 277–278, 278f
Tonti Sandstone Member, 138f, 147, 148f
topographic relief
  flooding relationship, 396–397
  Illinois, 70, 71f
  karst formation, 432
  Precambrian surface, 123
  soil development, 377
  wetlands, 365
topography
  bedrock, 222–224, 225f
  Illinois, 39, 70f, 218f
  Illinois Episode landforms, 233f
  Illinois River valley, 237f
  mapping, 12
  reclamation site, 461
  shoreline, 407f
  unglaciated areas, 217
topsoil. *See* soils
tortoise. *See* Reptilia
total dissolved solids, 329, 329f, 337, 338t, 339f, 341f, 342t, 347
Tournasian, 168f
Toxic Release Inventory, 444, 445f
toxic releases in Illinois, ranking, 444
Toxic Substances Control Act (TSCA) (1976), 29t
trace fossils, 175f
Tradewater Formation, 195f, 197–198
  coal seams, 267f
  Giant City State Park, 197f
  Hicks Dome, 101f
  Pennsylvanian, 189, 297
  reservoir rocks, 287f
  Rough Creek-Shawneetown Fault System, 99f
  stratigraphic columns, 188f, 287f
Tradewater Group, 195f, 198
Trafalgar Formation, 229f, 238
Transcontinental Arch
  Chesterian time, 169, 170f
  paleogeography, 139f
  Pennsylvanian, 189f, 193
  structural geology, 60f, 91f
transition beds, 305f
transpiration
  clay soils, 385–386
  hydrologic cycle, 328, 328f
transportation
  floods, 395
  industrial minerals, 309
  pollutants, 444t
  soil engineering properties, 373
  stone, 318–319
travel time, 126, 128f, 129f
tree ferns
  Asterotheca species, 190f
  wetlands, 367
trees, 385–388
  paleoclimatology, 248, 249t, 252f
Tremadocian Stage, 137f, 138f
Trempealeauan Stage, 137f, 138f, 142–143
Tremper Mound (Ohio), 317f
"Trenton" limestone. *See* Kimmswick ("Trenton") Limestone
Trepospira species, 191f
Triassic, 62f, 65f, 67f, 202, 206–207, 207f
trichloroethylene (TCE), 453
Tricoelocrinus woodmani, 178

INDEX

Trilobita
   Ameura sauki, 191f
   calymenid, 153f
   Cambrian, 138–139
   Dikelocephalus species, 143
   Ditomopyge species, 191f
   Eldredgeops (Phacops), 171, 172f–173f, 173
   Encrinurus species, 150f, 151
   Flexicalymene, 154
   Franconia Formation, 143
   Gabriceraurus species, 150f, 151
   Gravicalymene celebra, 162f–163f, 163
   Gravicalymene species, 153f, 154
   Illaenus species, 151
   Isotelus iowensis, 153f
   Isotelus species, 151, 154
   Devonian, 170
   Odontocephalus species, 171
   Pennsylvanian, 189
   Platteville Group, 149
   (pygidium) *Arctinurus occidentalis*, 162f–163f
   (pygidium) *Glyptambon gassi*, 162f–163f
   Saukiella species, 143
   St. Louis Limestone, 180
   Thaleops ovata, 150f
   Thaleops species, 151
tripoli deposits, 164–165, 207, 311f, 316, 319f, 320
tritium, 347
Trivoli Sandstone, 188f, 200, 287f
troughs, 213f
Troutman Member, 160f
Trowbridge Coal Member, 201
Troy Bedrock Valley, 224, 225f, 333, 333f, 334
Troy District, 271, 271f
Tullimonstrum gregarium (Illinois state fossil), 190, 192f
tundra, 249t
Tunnel and Reservoir Plan (TARP), 29, 164, 392
turbidites, 28
Turinian Stage, 137f, 138f
Turner Mine Shale, 188f, 194f
Turritella species, 211f
turtle. *See* Reptilia
Tuscaloosa Formation. *See* Post Creek Formation
Tuscola, 236f, 344f
Tuscola Quarry, 94, 219, 222f, 240
Two Creeks phase, 410f
Two Rivers Phase, 242
two-dimensional seismic profile, 290f
Tygett Sandstone, 168f, 182, 183f
type sections
   Caseyville Formation, 195
   Mississippian, 167

U

U.S. Army Engineers jetty construction, 411
U.S. Environmental Protection Agency standards
   aquifer protection, 353–354
   arsenic in groundwater, 344
   CERCLA sites, 444–445
   drinking water, 339–340, 343, 419
   groundwater protection, 351
   iron, 345
   manganese, 345
   municipal solid waste, 445
   National Priority List sites, 445, 445f
   nitrogen compounds for health, 345
   radium in groundwater, 348
   radiation resources, 422
   total dissolved solids in drinking water, 347
   Toxic Release Inventory program, 444
U.S. Geological Survey
   history of investigations, 9
   National Earthquake Information Center data, 425–426
   radiation resources, 422
U.S. Government Piers, 411
U.S. Highway 41 (Chicago), 395–396, 396f
Ullin lime kiln, 318f
Ullin Platform, 177f
Ullin ("Warsaw") Limestone, 168f, 169, 177, 178, 180f
Ullin ("Warsaw") Limestone deposits
   industrial minerals, 312
   reservoir rocks, 287f, 293–294
Ultisols, 234, 375–376
ultramafic composition, 100, 101f, 202
unconfined aquifers, 327, 327f
unconformities, 64
   Chesterian, 183
   Matthiessen State Park, 195
   Precambrian, 126f
   reservoir rocks, 287f
   stratigraphic traps, 288, 289, 291, 293f
   sub-Absaroka, 28, 66f
   sub-Brandon Bridge, 159
   sub-Kaskaskia, 28, 66f, 159, 170
   sub-Kress Member, 148f
   sub-Offerman, 159
   sub-Sauk surface, 139
   sub-Tippecanoe I, 66f
   sub-Tippecanoe II, 66f, 159
   sub-Tippecanoe surface, 28, 145, 147
   sub-Zuni, 66f
   tectonics
      Chesterian, 182
      Pennsylvanian, 193
underclay
   Carbondale Formation, 199, 199f
   cyclothems, 194f
   Galatia Mines, 192f
   land subsidence, 390
   Pennsylvanian, 189, 194f
   stratigraphic column, 287f
underground installations
   Chicago skyscraper foundations, 392
   Fermi National Accelerator Laboratory, 393, 394f
   freight tunnels (Chicago), 392, 392f
   mines, 393f
   sewers, 392, 392f
   transit trains, 393, 393f
underground mining
   fluorite, 307
   limestone, 38, 311f, 320f, 462–463
      history of investigations, 38
      Valmeyer, 320f
   liquid petroleum gas storage, 393f
   roof control, 105
   sequential use, 462–463
   urban areas, 320, 320f
   See also mining, coal
underground space, 45, 392–393
underground storage
   limestone mine, 393, 393f, 463
   natural gas, 141, 288, 393
   Prairie du Rocher, 393f
undermining. *See* land subsidence
Union County
   Alto Pass, Shawnee National Forest, 75f–76f
   cypress swamps, 361f
   Dongola sinkhole, 438
   Eocene gravel, 212f
   McNairy Formation, 210f, 211
   Pennsylvanian, 189
   sand and gravel, 212f
   Ste. Genevieve Fault System, 99f
   tripoli deposits, 207, 316
University of Chicago, 9

University of Illinois, 9
University of Memphis Center for Earthquake Research and Information (Tenn.), 425–426
unnamed bioclastic facies, 171
unnamed diamictons, 227
unnamed red beds, 208f
uplands
    Chesterian, 170f
    Illinois River valley, 237f
    Pennsylvanian, 189f
    Quaternary, 219
uplifts. *See* tectonics; structural geology
"Upper Carboniferous." *See* Pennsylvanian
Upper Mississippi Valley District, 299–300, 300f
Upper Mississippi Valley Lead-Zinc District, 301f
upper Pope Group, 101f
"Upper *Receptaculites* Zone," 302f
"upper Renault," 168f
Uptons Cave Syncline, 96f
uranium
    in coal, 266
    in groundwater, 34, 348, 419
    pollution, 422, 445
    public health, 419
uranium-series dating, 248, 256, 256f
uranium-thorium dating, 433
urban geology
    floods (Chicago), 395, 396
    land use, 374
    mining, 310f, 315f, 320f
    quarries, 38, 310f
    rivers, 401
    runoff, 443t, 444t
    soils, 373
    urban planning, 29, 309, 320f, 320–321
    watersheds, 401
Urbana, 242f, 344f
urea nitrogen, 449
Utica
    Peru Monocline, 83
    Shakopee Dolomite, 147f
    St. Peter Sandstone, 149f
utilities
    drill cores, 123
    land subsidence, 390
    radioactive waste, 445
    underground space, 392

V

Valley Quarry, 462f, 463, 463f
valley train sediments
    ancient Mississippi Valley, 238
    Illinois Episode, 231
    Illinois River valley, 238
    Quaternary aquifers, 333
    sand and gravel deposits, 229f, 238, 313, 314f
    Wabash River valley, 238
Valmeyer, 176
    relocation, 401f, 401–402
    underground limestone mine, 320f, 463
Valmeyer Anticline, 63f, 92f, 95, 311
Valmeyeran, 66f, 176–177, 178f
    Chesterian, 178f
    groundwater, 338t
    limestones, 338t
    paleogeography, 177f, 178f
    reservoir rocks, 287f, 293
    stratigraphic columns, 168f, 287f
    tectonics, 84, 99f, 169
Valparaiso Moraine, 236f, 240, 242
vanadium, 266
Vandalia, 233f, 279

Vandalia Arch, 171f
Vandalia Member, 223f, 231, 232, 233, 234
Vandalia sub-Pennsylvania valley, 196f
Vanuxemia species, 150f, 151
Varna Moraine, 236f
vegetation
    Chesterian, 182
    Devonian, 170
    at European settlement, 380
    Illinois River valley, 237f
    loess, 378
    Mollisols, 375
    paleoclimatology, 235, 252f, 253, 256, 257f
    peat, 316
    Pennsylvanian, 189–190, 190f
    slope stability, 387–388
    soils, 373, 373f
    water quality, 364
    wetlands, 368
ventilation
    mines, 276
    radon, 420
vermiculite, 315
Vermilion Bedrock Valley, 225f
Vermilion County
    artesian wells, 327
    Danville Coal Member, 272
    Herrin Coal Member, 271
    Mahomet aquifer, 338t
    Middle Fork watershed, 34f
    saline seep, 343
Vermilionville Sandstone, 188f
Vernon Fork Member (Ind.), 171
Vertigo modesta, 252f
Vienna, 180
Vienna Limestone, 168f, 181, 183f, 287f
Vigo Coal Company Friendsville Mine, 273f
Villa Grove Quadrangle, 222f, 358f, 359, 359f
Viola landfill, 447f
Vincennes (Ind.), large earthquake evidence, 110
Virgilian, 66f, 187–188, 188f, 189, 287f
viruses, 453–454
Visean, 168f
Vishnu Springs, 432, 441
volcanic ash, 139, 151
volcanic features, 77f, 81
volcaniclastics, 124f, 132
volcanism, 83, 207
Vostok Station, ice cores, 251
Vulcan Materials quarries, 152f, 458

W

Wabash Bedrock Valley aquifer, 335
Wabash County, 268, 361f
Wabash River
    faults, 115f
    floods, 395
    karst, 433
    sand and gravel, 238, 313
    wetlands, 364
Wabash River valley
    earthquake studies, 110
    Illinois Episode loess, 232, 240
    valley train deposits, 238, 313
Wabash Valley Fault System, 96–97
    relation to Illinois Basin, 62
    relation to oil fields, 286f
    stratigraphic traps, 288
    structural geology, 63f
    tectonics, 86f, 87, 182–183
Wabash Valley Fault Zone
    earthquake (1968), 115f

Index

Illinois-Kentucky Fluorspar District, 304f
Triassic, 206
Wabash Valley Seismic Zone, 113f, 115
wackestone, 149, 151–152
Wadsworth Formation, 223f, 229f, 236f, 238
Wakeland series, 378f
Walche Limestone, 168f, 183f
Walpole sub-Pennsylvanian valley, 196f
Walshville Channel, 271, 271f
Waltersburg Formation, 182
    fossils, 182
    reservoir rocks, 287f
    sandstone reservoirs, 297
    stratigraphic columns, 168f, 183f, 287f
Wapella East Field, 291f
Wapsipinicon Limestone, 168f, 171, 290f
Warren County
    drumlins, 231, 231f
    stone quarry reclamation, 462f, 462–463, 463f
    Valmeyeran, 177
"Warsaw." See Ullin ("Warsaw") Limestone
Warsaw Formation, 178–179, 180f
    fossils, 175f, 178, 181f
    Mississippian, 312
    paleogeography, 177f
    stratigraphic column, 168f
"Warsaw lime," 168f
Washington Heights Spit, 409f
waste disposal, 443–457
    aquifer vulnerability, 326, 352f, 353, 355f, 356f, 357f, 358, 359, 448f
    clay liners, 316
    concentrated animal feeding operations, 448, 454f, 454–455
    environmental geology, 29, 33f
    pollution source, 443t, 444t
    septic systems, 351, 353, 359, 444, 444t, 454, 454f
    solid waste, 280, 445–447
    storage tank leakage, 452–453, 453f
    See also coal waste; landfills
Waste Hauling landfill, 447f
wastewater treatment, 309, 313, 453–454
water exploration, 20, 25f
water pollution, 443–457
    deicer, 455, 455f
    extent, 443
    karst springs, 441
    salt, 351, 444t, 455
    salt-water intrusion, 342, 343, 344f, 444t
    surface mines, 459
    wetlands, 364, 368
Water Pollution Control Act (1965), 29t
water pressure damage, 386, 386f
water quality, 337–350
    arsenic, 34, 344–345, 345f, 443
    best management practices, 449
    bicarbonate, 340–341
    calcium, 340–341
    chloride, 342–343
    fluoride, 343–344
    gases, 346–347
    geochemical methods, 337, 338t, 339–340
    isotopes, 347–348
    magnesium, 340–341
    pollution, 443–457
    radium, 420f
    redox (reduction-oxidation), 344, 345
    sodium, 342–343
    soils, 373
    sulfuric acid, 460
    suspended solids, 340, 340f
    total dissolved solids, 329f, 331f, 338f, 339f, 341f, 347
    waste disposal, 355f, 357f, 454, 454f
    water hardness, 313, 340–341, 345
    See also groundwater quality; surface water

water resources
    economics, 39
    McHenry County, 358
    urban planning, 383
    Winnebago County, 358
water supply. See aquifers; groundwater; surface water
water table (saturated zone), 273
    aquifer, 327, 327f
    groundwater flooding, 399, 400
    hydrologic cycle, 328, 328f
    speleothems, 255f
    topographic relief, 377
    unconfined aquifer, 327, 327f
Water Tower (Chicago), 165
water treatment
    hydrogen sulfide, 346
    radioactivity, 421
    wastewater, 313
water wells
    abandoned wells, 450f
    aquifers, 327
    construction code (2000), 441
    history of investigations, 20, 24, 44f
    karst, 439f, 440–441
    large-diameter wells, 452
    nitrates, 449f, 450, 450f
    pesticides, 449f, 450f, 452
    radioactive elements, 419
    regulations, 352, 450f
    toxic releases, 444
    well screens, 345
Water Works Basin (Chicago), 412f
water-borne disease, 453–454
waterfall, 323f–324f
waterflooding
    history of investigations, 19, 19f, 37
    petroleum, 284f, 285f, 288
    reservoir rocks, 285
Waterloo Field, 289
Waterloo-Dupo Anticline, 62f, 63f, 84, 85f, 92f, 95
watersheds
    human activity, 401
    "sole source" designation, 352
    wetlands, 368
Watseka, 344f
Wauconda series, 377f
Waukegan, 404f, 411f
Waukegan Generating Station, 412f
Waukegan Harbor, 404f, 411, 412f
Waukegan Marina, 412f
waves, 30, 407, 408, 411f
Waveland Golf Course, 413f
Waverly Field, 291f
Wayne County
    groundwater, 338t
    petroleum, 293
Wayside Sandstone, 188f
weathered materials
    sediment, 248
    Ultisols, 375–376
weathering
    Farm Creek section, 22, 341f
    groundwater quality, 337
    pedogenesis, 374
    Sangamon Geosol, 234, 234f
    Thebes, 228f
    Tuscola Quarry, 222f
Webster Branch, 189
Wedron, 236f
Wedron Group, 223f, 229f, 236f, 238
Wedron Quarry, 220f, 221f, 240
"Weiler," 168f, 287f
Weldon Springs State Park, 342f

Wenlock, 161f
Wenlock Stage, 160f
"West Baden clastic belt," 182
West Chicago, 422
West Frankfort
   Herrin Coal Member, 271
   mine safety, 278
   New Orient Mine, 278
West Franklin Limestone Member, 188f, 195f, 200
West Lebanon till (Ind.), 227
Westburg Geosol (Iowa), 223f, 227–228
Western Interior Basin, 189f
Western Lion landfill, 447f
Western Shelf, 93
   Bond Formation, 201
   Carbondale Formation, 198
   Herrin Coal deposits, 271, 271f, 272
   Illinois Basin, 62
   Kimmswick ("Trenton") Limestone, 289
   Shelburn Formation, 200
   structural geology, 63f, 91f, 92f
   sub-Pennsylvanian valleys, 196f
"Westfield," 168f, 287f
Westfield Field, 289
Westphalian, 188f
wet prairies, 365, 366
wetlands, 361f, 361–364, 364f, 365f, 365–369
   complexes, 365–366
   drainage, 361, 362, 365f, 366, 396, 401
   evapotranspiration, 363–364
   extent, 362, 362f, 363f, 364f
   floodplains, 364–365, 365f
   floods, 402
   glacial sediment, 365
   human activity, 373, 380, 410
   hydrogeology, 36, 328, 362–364
   mitigation, 36, 402
   organic compounds, 379
   peat, 366–367, 367f
   Pennsylvanian, 367
   presettlement, 362, 362f, 363f, 365
   soils, 362f, 366
   surface mine reclamation, 461
Wheaton Morainal Country, 70, 71f
White County
   Davis Coal Member, 268
   Dekoven Coal Member, 268
   Kimmswick Limestone, 151f
"white top," Herrin Coal Member, 272
Whitebreast Coal (Iowa), 268
Whiterockian, 66f, 137f, 138f, 140, 145–152
Whiteside County
   ancient Mississippi Valley, 238
   Goose Lake Channel, 367
   Green River Lowland, 72t
   Silurian, 160f
   wetlands, 367
Wilcox Formation, 206f, 212, 212f
wildcat drilling (1865), 11
Wilhelmi Formation, 160, 160f
Will County
   sinkholes, 435
   surface mines, 190
Williamson County
   Cottage Grove Fault System, 98f
   Davis Coal Member, 268
   Dekoven Coal Member, 268
   Murphysboro Coal Member, 267
Willis Coal Member, 267f, 272
Williston Basin, 170f, 189f
Wilmette, 404f
Wilmette Harbor, 404f, 405
Wilmette Spit, 409f
Wilmington
   Colchester Coal Member, 268
   Houchin Creek Coal Member, 270
Wilmington municipal landfill, 447f
Wilson, 287f
Wilsonville, 33
wind
   erosion, 380
   proglacial sediment, 219
   radon, 420
   transport, 378
   Triassic, 207f
   waves, 407
windblown silt. *See* loess
Windrow Formation (Iowa), 209
Winfield Quadrangle, 105
Winnebago County
   aquifer vulnerability, 330, 355f, 358
   drumlins, 231
   urban planning, 33f
   waste disposal, 355f
Winnebago Formation
   Illinois Episode, 231, 232, 239
   Quaternary, 223f, 229f
   Sangamon Geosol, 234
Winnetka, 404f, 405, 406f
Winslow Member, 227, 231
Winthrop Harbor, 30f, 404f, 411f
Winthrop Member, 223f
Wisconsin Arch, 60, 60f, 63f, 91f, 91–93
   Forest City Basin, 91–92
   Michigan Basin, 91–92
   paleogeography, 189, 189f
   paleohydrology, 300f
   Pennsylvanian, 189f
   structural geology, 62f, 91f, 91–93
   tectonics, 62f, 78f, 84, 85f
Wisconsin Driftless Section, 70, 71f, 225f, 229f
Wisconsin Episode, 22, 235–242
   drainage, 224f, 231f
   eskers, 233f
   geochronology, 223f, 236f
   glacial extent, 236f, 250f
   gravel, 108f
   Illinois River valley, 237f
   Keller Farm section, 254f, 255f
   Lake Border Morainic System, 405
   loess, 234f
   neotectonics, 107, 107t
   paleoclimatology, 250f, 254f, 257f, 258
   Quaternary, 223f, 229f
   Roxana Silt, 234f
   sand and gravel deposits, 314f
   till plain, 226f, 229f
   Tuscola Quarry, 222f
   Wedron Quarry, 221f
   western Illinois, 231f
   wetlands, 366
Wise Lake Formation, 152, 152f
   lead-zinc deposits, 302f
   stratigraphic column, 138f
Wise Ridge Coal Member, 267f, 272
Woburn (Mass.), 453
Wolf Creek Formation, 223f, 224, 229f
   paleoclimatology, 227
   paleosols, 227–228
Wolf Lake
   shoreline, 408, 410f
   silica deposits, 209f
   Ste. Genevieve Fault System, 99f
   tripoli deposits, 207
Wolfcampian, 66f
Womac Coal Member, 200

Index

Woodbury Limestone, 188f, 189
Woodstock Moraine, 240
Woodstock Phase, 236f, 240, 242
woolly mammoth, 242
World War II impacts, 20–21, 21f
worms, 189, 190
Worthenia species, 191f

X

x-ray diffraction analysis, 19, 27f, 31f
x-ray exposure, 419, 419t
xylene pollution, 452

Y

Yankeetown ("Benoist") Sandstone, 181–182
    electrical logging, 296f
    fluorspar deposits, 305f
    reservoir rocks, 287f, 297
    stratigraphic columns, 168f, 183f, 287f
Yarmouth Episode, 222f, 223f, 224f, 226, 228, 228f
Yarmouth Geosol, 223f, 226, 228
    Thebes road cut, 228f
    Tuscola Quarry, 222f
    wetlands, 366
Yarmouthian. *See* pre-Illinois Episode

yield, of aquifers, 328–332
Yorkville Member, 221f, 223f, 236f, 238, 240

Z

Zeacrinites species, 182
Zeacrinites wortheni, 175f
Zeigler coal mine disaster, 278
zinc deposits, 300–301
    in coal, 266
    Fluorspar Area Fault Complex, 100
    history of investigations, 11–12
    mining, 152, 299–308
zinc oxide, 300
zinc sulfide, 301
zinc-based alloys, 300
Zion, 404f, 422
Zion Beach-Ridge Plain, 405, 406f, 408–410, 411f
Zion City Moraine, 405, 406f
Zion Nuclear Power Station, 30f
Zuni sequence, 65f, 66f, 67f, 128f, 206f
    Mississippi Embayment, 79f
    Sloss sequence, 64
    structural geology, 62f
    tectonics, 62f, 128f, 141f
Zurich series, 377f